COMPREHENSIVE INSECT PHYSIOLOGY BIOCHEMISTRY AND PHARMACOLOGY

Volume 6

NERVOUS SYSTEM: SENSORY

COMPREHENSIVE INSECT PHYSIOLOGY BIOCHEMISTRY AND PHARMACOLOGY

Volume 6

NERVOUS SYSTEM: SENSORY

Executive Editors

G. A. KERKUT
Department of Neurophysiology,
University of Southampton, UK

L. I. GILBERT
Department of Biology,
University of North Carolina, USA

PERGAMON PRESS

OXFORD · NEW YORK · TORONTO · SYDNEY · PARIS · FRANKFURT

UK	Pergamon Press Ltd., Headington Hill Hall, Oxford OX3 0BW, England
USA	Pergamon Press Inc., Maxwell House, Fairview Park, Elmsford, New York 10523, USA
CANADA	Pergamon Press Canada Ltd., Suite 104, 150 Consumers Road, Willowdale, Ontario M2J 1P9, Canada
AUSTRALIA	Pergamon Press (Aust.) Pty. Ltd., P.O. Box 544, Potts Point, N.S.W. 2011, Australia
FRANCE	Pergamon Press SARL, 24 rue des Ecoles, 75240 Paris, Cedex 05, France
FEDERAL REPUBLIC OF GERMANY	Pergamon Press GmbH, Hammerweg 6, D-6242 Kronberg-Taunus, Federal Republic of Germany

First edition 1985

Library of Congress Cataloging in Publication Data

Main entry under title:
Comprehensive insect physiology, biochemistry, and pharmacology.

Contents: v. 1. Embryogenesis and reproduction.
1. Insects—Physiology—Collected works. I. Kerkut, G. A.
II. Gilbert, Lawrence I. (Lawrence Irwin), 1929–

QL495.C64   1984   595.7′01   83–25743

British Library Cataloguing in Publication Data

Comprehensive insect physiology, biochemistry and pharmacology.
1. Insects
I. Kerkut, G. A.   II. Gilbert, Lawrence I.
595.7   QL463

ISBN 0–08–030807–4 (volume 6)
ISBN 0–08–026850–1 (set)

Filmset by Filmtype Services Ltd., Scarborough
Printed in Great Britain by A. Wheaton & Co. Ltd., Exeter

Contents

Contents

Foreword

Aristotle was enchanted by the phenomenon of insect metamorphosis and the early microscopists such as Robert Hook, Marcello Malpighi, Anton van Leeuwenhoek, René de Réaumur and Pieter Lyonet were fascinated by the structure and function of the different parts of insects and made some of the first important contributions to our knowledge of insect physiology. More detailed functional studies were made by Borelli in his book "De Motu Animalium", published in 1680, and his interpretation of insect walking patterns remained in our textbooks until 1955.

In general, the 18th and 19th century research workers were more concerned with the morphology and classification of insects, though physiologists such as Claude Bernard and naturalists such as John Lubbock and Henri Fabre were always interested in the functional analysis of insects.

One of the milestones in the study of insect physiology was the publication by Wigglesworth of his small book on insect physiology in 1934. This was stimulated by an appreciation of the way in which studies on the basic physiology of insects were necessary before one could understand and ultimately control the activity of insect pests of man and crops.

Wigglesworth initially studied medicine and then carried out research at the London School of Hygiene and Tropical Medicine. His innate gift for planning simple but fundamental experiments on *Rhodnius* led rapidly to an increase in our knowledge about moulting, the control of larval and adult stages, and provided the foundation for insect endocrinology. Furthermore he inspired a group of co-workers who later played a key role in the application of modern techniques to solve the problems of insect physiology and biochemistry.

Wigglesworth's "Insect Physiology" was followed by a more detailed and full-sized textbook, "Principles of Insect Physiology", which was published in 1947 and is now in its 7th edition (1972).

The three-volume edition of "Physiology of Insecta", a multi-authored work edited by Morris Rockstein, was published in 1964 and a new edition in six volumes followed in 1973.

The study of insect biochemistry developed more slowly, partly because there was no special distinction between physiology and biochemistry; the investigator just used the methods available for his studies. David Keilin started his studies working on insects: "From 1919 onwards I had been actively engaged in the study of the anatomy of the respiratory system, respiratory adaptation and respiration of dipterous larvae and pupae. Among the vast amount of material I was investigating, special attention was given to the larvae of *Gasterophilus intestinalis*". For these studies Keilin developed a method for the spectroscopic analysis of respiratory pigments of insect pupae under the microscope, which ultimately led to the discovery of the cytochromes.

The pteridines were discovered in insect pigments, and the one gene–one enzyme hypothesis of Beadle and Tatum, which was the cornerstone of molecular biology, was a result of biochemical and genetic analysis of *Drosophila*.

The rapid expansion of biochemistry after 1945 led to many more workers studying insect biochemistry and the first textbook on the subject by Darcy Gilmore was published in 1961. This was followed by the multi-authored "Biochemistry of Insects", edited by Morris Rockstein, in 1978.

The first evidence that a steroid hormone acts at the level of the gene came from the studies of Clever and Karlson in the 1960s on the puffing by the polytene chromosomes of *Chironomus*.

Though insect physiologists and biochemists initially published their papers in journals such as *Biological Bulletin, Journal of Biological Chemistry, Biochemical Journal, Journal of Physiology, Journal of Experimental Zoology, Journal of Experimental Biology, Roux' Archiv für Entwicklungsmechanik*, and *Zeitschrift für vergleichende Physiologie*, the great expansion of insect physiology and biochemistry from 1945 onwards led to the establishment of journals and other periodicals specialising in insects, such as the *Journal of Insect Physiology, Insect Biochemistry, Annual Review of Entomology*, and *Advances in Insect Physiology*.

It is also fitting to mention the work of other pioneers in the study of insect physiology and biochemistry, such as Autrum, Bounhiol, Bodenstein, Butenandt, Chadwick, Dethier, Fraenkel, Fukuda, Joly, Lees, Karlson, Kopec, Piepho, Richards, Roeder, Berta Scharrer, Snodgrass and Williams; these and many others laid the foundations of the subject and all following research workers have stood on the shoulders of these giants.

In July 1980 a meeting was held at Pergamon Press in Oxford to discuss the possibility of publishing a series of volumes on insect physiology, biochemistry and pharmacology. The idea was to produce 12 volumes that would provide an up-to-date summary and orientation on the physiology, biochemistry, pharmacology, behaviour and control of insects that would be of value to research workers, teachers and students. The volumes should provide the reader with the classical background to the literature and include all the important basic material. In addition, special attention would be given to the literature from 1950 to the present day. Emphasis would be given to illustrations, graphs, EM pictures and tabular summaries of data.

We were asked to act as Executive Editors and by December 1980 we had produced a 27-page booklet giving details of the aims and objectives of the project, details of the proposed volumes and chapters, suggested plans within the chapters, abbreviations, preparation of diagrams and tables, and journal citations to ensure uniformity of presentation as far as possible. This booklet was sent to authors of the chapters and their comments invited. By the middle of 1981 most of the chapters had been assigned to authors and the project was under way. The details of the volumes and the chapters they contain are given on the following pages so that the reader can see the contents of each of the other volumes.

In addition, there is a final volume, Volume 13, which is the Index Volume. Although each volume will contain its own subject index, species index and author index, Volume 13 will contain the combined subject, species and author indexes for all 12 text volumes so that any material in these volumes can be rapidly located.

All references in the volumes are given with full titles of papers, journal, volume, and first and last pages. The references to the authors in the text are given with their initials so that it is clear that the text refers to D. Smith and not, say, to A. Smith. There are more than 50,000 references to the literature, more than 10,000 species of insect referred to, and all should be readily found in the 12 different volumes.

There are 240 authors of the 200 chapters in the volumes and they have produced a series of very readable, up-to-date, and critical summaries of the literature. In addition, they have considered the problems associated with their subject, indicated the present state of the subject and suggested its developmental pathway over the next decade.

We are very grateful to our colleagues for the efficient way that they have met the challenge and the deadlines in spite of their many other commitments.

This series of volumes will be very useful to libraries, but an important case can be made that the books should be considered as research instruments. A set of volumes should also be available in the laboratory for constant reference. They will provide the research worker with an account of the literature and will always be instantly available for consultation. For this reason they should be considered as research equipment equally important as microscopes, oscilloscopes or spectrophotometers.

The volumes should save research workers many weeks of time each year in that not only will they provide an awareness of the literature and the background, and so save valuable research time, but the full index to authors, subject and species, and the full literature references, should also make it much easier to write reports and papers on their own new research work.

It is hoped that these volumes will do much to strengthen the case for insects as a source of research material, not only because insects are important medical and agricultural pests (over 200 million people at present have malaria: insects eat or destroy about 20% of planted food crops), but also because in many cases insects are the ideal unique research material for studying and solving fundamental biological problems.

G. A. KERKUT
Southampton

L. I. GILBERT
Chapel Hill

Preface to Volume 6

This volume (Volume 6) describes the structure and function of the sensory systems in insects. These are the antennae and sensilla; mechanoreceptors; chemoreceptors; the visual system; hearing and sound; gravity; and the time sense (clocks and rhythms).

The last two chapters are integrative and deal with (a) multimodal sensory convergence — the way that information from the different sensory systems, *i.e.* vision, sound, and tactile receptors, interact to produce a final motor response; and (b) the visual guidance of flies during flight, *i.e.* how the fly uses its visual system in flying and reacts with objects and other flies in the environment.

The general anatomy of the insect central nervous systems has been described in the previous volume (Volume 5) which includes accounts of insect nerve action potential and synapse physiology; the development of the nervous system; the structure of neurones and glial cells; the blood–brain barrier; nerve tissue culture; the structure of ganglia and brain; tactile receptors in the anal cerci and their relation to sensory neurones and interneurones in the ganglia; giant fibre afferents; extra ocular photoreceptors; the functional aspects of walking, running, swimming, and flying; aerodynamics; and an analysis of flight in *Calliphora*.

Other volumes also contain material of interest to the sensory physiologist. In Volume 3 there is an account of the physiology of the respiratory system. In Volume 4 there is a chapter on thermoregulation, and also a chapter on the regulation of feeding behaviour.

Volume 9 on Behaviour has chapters on the neurobiology of pheromone perception, sex pheromones, alarm pheromones, trail pheromones, and aggregation pheromones, with details of the chemicals involved and the sensory thresholds to these pheromones. There are also chapters on feeding behaviour, colour change, courtship and mating behaviour, and the chemical control of behaviour.

These chapters show the sensory system in action often working at maximum sensitivity, detecting very small quantities of chemicals and bringing about striking behaviour activity of the whole animal.

Insects have a wide range of size so that the smallest insects are smaller than the largest protozoa, and the largest insects are bigger than the smallest mammal. Students of squid giant axon physiology are humbled to hear that the squid giant axon is large enough to allow three of the smallest insects to march side by side along the axon.

The relatively small size of insects has some advantages. For example, the compound eye, with its high resolution without the limitation of a single foveal region, is often at its maximum effective size in insects. If a large animal such as Man had to have a coumpound eye with high absolute angular resolution, then the compound eye would be of

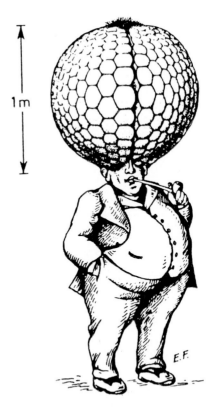

1 m

Fig. 1. What a man would look like if he had a compound eye of the same overall resolution as the simple lens eye. To simplify the drawing only 100 facets per eye are shown: there should have been one million per eye. (From Kirschfield, K. (1976). The resolution of lens and compound eyes. In *Neural Principles of Vision*. Edited by F. Zettler and R. Weiler. Pages 354–370. Springer-Verlag, Berlin.)

a size equivalent to a sphere of 1 metre diameter (Figure 1).

The study of insect sensory physiology offers several advantages over the study of sensory physiology of other animals. It is often easier to set up an isolated sensory preparation from an insect and obtain clear electrical recordings of the activity of single sensory units. It is also easier to study the interactions between these sensory units in the periphery in relation to their sensory neurones and interneurones in the ganglia in the central nervous system. Furthermore, as studies on the auditory system have shown, the different afferent inputs are tone coded and may end in different parts of the arborizations of the interneurone in the neuropile. Study of the interactive functions of such systems will provide better understanding of the way in which the neuropile brings about selective transfer of information to the higher centres.

Many of the interneurones are multimodal; they can transfer information coming from different sensory modalities. The nature of the sensory information transmitted depends on the previous inputs to the interneurone as well as the pattern of simultaneous sensory inputs.

Though large multimodal neurones and interneurones are known in other animal groups, the precise manner in which they function is not clear and the insect multimodal neurones offer the best chance of understanding how such neurones can act as a central processing unit (CPU) modifying reflex activity in the CNS and bringing about complex variable behaviour to what at first looks like simple repeatable sensory inputs.

G. A. KERKUT
Southampton

Contributors to Volume 6

DeVoe, R. D.
Department of Visual Sciences, School of
Optometry, Indiana University,
Bloomington, IN 47405, USA

Horn, E.
Zoologisches Institut II, Universität
(T. H.) Karlsruhe, Kaiserstrasse 12,
D-7500 Karlsruhe 1, Federal Republic of
Germany

Järvilehto, M.
Department of Zoology, University of
Oulu, Linnanmaa, 90570 Oulu 57,
Finland

Land, M. F.
Biology Building, School of Biological
Sciences, University of Sussex, Falmer,
Brighton, Sussex BN1 9QG, UK

Larsen, O. N.
Institute of Biology, Odense University,
Campusvej 55, DK-5230 Odense M,
Denmark

McIver, S. B.
Ramsay Wright Zoological Laboratories,
Department of Zoology, University of
Toronto, 25 Harbord Street, Toronto,
Ontario M5S 1A1, Canada

Michelsen, A.
Institute of Biology, Odense University,
Campusvej 55, DK-5230 Odense M,
Denmark

Morita, H.
Department of Biology, Kyushu
University, Faculty of Science 33,
Fukuoka 812, Japan

Page, T. L.
Department of General Biology,
Vanderbilt University, Nashville,
TN 37235, USA

Shiraishi, A.
Department of Biology, Kyushu
University, Faculty of Science 33,
Fukuoka 812, Japan

Trujillo-Cenóz, O.
Instituto de Investigaciones, Biológicas
Clements Estable, Avde Italia No 3318,
Montevideo, Uruguay

Wehrhahn, C.
Max-Planck-Institut für Biologische
Kybernetik, Spemannstrasse 38, D-7400
Tübingen, Federal Republic of Germany

White, R. H.
Department of Biology, University of
Massachusetts at Boston, Harbor
Campus, Boston, MA 02125, USA

Zacharuk, R. Y.
Department of Biology, University of
Regina, Regina, Saskatchewan S4S 0A2,
Canada

Contents of All Volumes

Contents of All Volumes

1 Antennae and Sensilla

R. Y. ZACHARUK

University of Regina, Regina, Saskatchewan, Canada

1 INTRODUCTION

Insects are a very large group of widely dispersed, highly mobile, heterotrophic animals with diverse habits and habitats that often change with each developmental stage. Some are predators, but most are preyed upon. With this demanding and hazardous lifestyle they have especial need to develop and maintain mechanisms for adequately and specifically monitoring their external environment and their body parts for a co-ordinated function of the whole. Of most general import are exteroceptors for vision or electromagnetic waves in the infrared to ultraviolet range, for touch and sound and related pressures and vibrations, for various chemicals in air or solution, and for geo-, thermo-, and hygroreception, and proprioceptors for monitoring movement and function of body parts. Given the diversity within the group, many differences can be expected between individual insect forms and developmental stages in number and distribution of various structural and functional types of sensilla. Some may be peculiar to an insect form; others may be related to its development, habits or habitat.

The body plan poses special problems peculiar to insects and related animal forms in the development and maintenance of exterosensilla. The tough, often thick and hard, and not readily permeable integumental exoskeleton precludes an adequate access of stimuli to sensing cells without some transcuticular conduction or transfer mechanism specific for each type of stimulus. This is subject to surface wear and tear, may be a point of water loss of particular consequence to terrestrial forms in dry habitats, and is shed with the exuvium at each molt. Thus, along with the structural and functional mechanisms needed for accepting stimuli, generating a nerve impulse message and conducting this message to an appropriate receiving cell, insect exterosensilla must also have mechanisms to maintain the cuticular stimulus transfer parts functional and impermeable to significant water loss between molts, and to replace these parts with minimal interruption of sensory function during a molt. The small body size, a consequence of the exoskeletal body plan in highly mobile terrestrial animal forms, limits the surface and internal space available for a sensing and integrating system. The insect lifestyle demands a considerable amount of sensory input and integration, but this must be done by a cell complement limited in number and size.

However, the body plan also provides some significant structural and functional advantages for the insect sensory system. In a small body, and with a limited number of cells involved, transmission fibres are short and interconnections few. Message delivery throughout the body can be very rapid. This gives insects a capacity for short reaction times and high activity rates with appropriate co-ordination. Segmentation, with flexible membranous joints between segmental exoskeletal plates, provides sites admirably suited to simple but effective proprioceptors for monitoring position and movement of body parts and action of associated muscles. These may sense through movement of surface cuticular projections, tension on attachments to the inner wall, or shear stresses in the cuticle in the area of the joint. The same or similar joint sensilla might also monitor the pull of gravity on body parts above or lateral to the joint, or detect pressures, vibrations or compression waves that may be transferred to them by body parts peripheral to the joint. Although their bodies may be small, the various moveable, segmented appendages of insects provide much additional space for sensilla. These can extend the sensillar field by their length away from the body, enhance sense of direction by bilateral probing, and may be lengthened, enlarged or architecturally modified for specific sensory needs.

In perceiving stimuli a single sensillum with but one or two neurons and appropriate stimulus transfer and acceptor devices may be enabled to monitor and discriminate between several different stimuli; that is, be multimodal. By virtue of its peculiar structure and placement of sensilla a body part or appendage, in transferring a particular stimulus to several specific sensilla on it, may itself act as a sense organ that is only responsive to that stimulus *in toto*. It may also function multimodally by involving different combinations of sensilla for different stimuli. Such devices may contribute to peripheral integration as well as discrimination of sensory information, in individual sensilla and sensory fields. The need for an elaborate and bulky central nerve system is reduced as a consequence.

Much information was amassed by 1935 on structure, morphological types, distribution and innervation of sensilla in the various sensillar fields of insects. This was based on, but limited by, the resolution of the light microscope. Functions were deduced primarily from structure, location, and behavioral response with and without ablation or other incapacitation of sensillar fields. Snodgrass, R. (1926, 1935) provides reviews of this earlier literature, and McIndoo, N. (1931) typifies the morphological/behavioral approach that was most often used in studying sensilla of individual insect forms. With the increasing advances and application of electrophysiological monitoring techniques first, and electron microscopy somewhat later in the next 30 years, physiologists contributed a more precise and definitive knowledge of function in individual cells, sensilla and sensillar fields, and morphologists clarified many aspects of structure at the organelle and cellular levels. However, these two novel technologies were not integrated in many studies. Function was often elaborated in sensilla that were poorly characterized structurally, and the converse was true at least to the same extent. This has been increasingly rectified in the past 15 years. Integrated structural and functional information is now available for a number of types of sensilla in several insect forms. The following are but a few of a number of general reviews which summarize the progression of advancements in our understanding of structure and function in insect sensilla in the past 45 years: Dethier, V. (1953, 1963), Bullock, T. and Horridge, G. (1965), Wigglesworth, V. (1972),

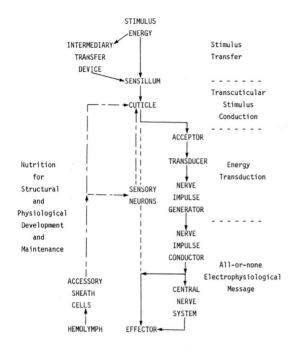

FIG. 1. Summary schematic pathways of function and the mechanisms involved in insect cuticular sensilla.

and in Rockstein, M. (1974). Vinnikov, Ya. (1974) interrelates sensory reception primarily in vertebrates and insects. Currently, several research groups are applying computer analysis in physiological or behavioral studies on specific sensilla or sensillar fields.

In this chapter the functional morphology of the various fields and types of sensilla that have been reported in insects is reviewed in a comparative way, excluding eyes. The latter are amply covered in chapters 4 to 8. Definitive and, to some extent, putative chemo-, thermo-, and hygroreceptive sensilla are emphasized, but specific aspects of the physiology of chemoreception are deferred to chapter 3. Mechanosensilla, including those that monitor sound and gravity, are dealt with specifically in chapters 2, 9 and 10. They are included here primarily in the considerations of the generalized typology and homology of parts of sensilla, and their involvements in sensillar fields. The basic structural mechanisms and functional pathways involved are summarized cursorily in Fig. 1.

2 SENSORY FIELDS AND INNERVATION

Insect exteroceptors are most numerous on the appendages of the head (antennae and mouthparts), thorax (legs and wings) and anal segments (cerci and genitalia). These appendages usually also bear some proprioceptors with external or internal stimulus transfer devices. Extero- and proprioceptors also occur in specific patterns but generally lower densities on sclerites of the head, thorax and abdomen. These parts have additional internal proprioceptors in specific locations intersegmentally and associated with organs such as muscle, epidermis and alimentary tract. Their gross patterns of innervation and segmental connections to the central nerve system are basically similar among the insect forms.

2.1 Antennae

Insect antennae are a pair of bilateral preoral appendages of the head, which are basically different in segmentation and musculature from the appendages of the postoral body segments. They occur in all insects except Protura, but are reduced to a tubercle or absent in larvae of the higher Hymenoptera and Diptera. The basic three-part structure is uniform among the insect groups. The basal part is a single segment termed the scape, which inserts in a membranous antennal socket on the cranium. In many but not all insects it is partially supported by, and articulates on, a pivot termed the antennifer in a manner resembling a ball joint. It is moved by one or two pairs of opposing muscles that originate on the tentorium, or on the cranial wall in some larval forms, and insert within. The second part also consists of a single segment termed the pedicel. It is attached to the scape by membranous cuticle, and is usually moveable in a manner resembling a hinge joint by one pair of opposing muscles that originate in the scape. The movements of these two joints together usually enable omnidirectional movements of the antennae. The remaining distal part is termed the flagellum. Of the three parts it is the most variable. The greatest variations are in length, from a single short segment to a long shaft made up of many subsegments, and in form of subsegments along the shaft of individual antennae, between antennae of the two sexes of individual insect forms,

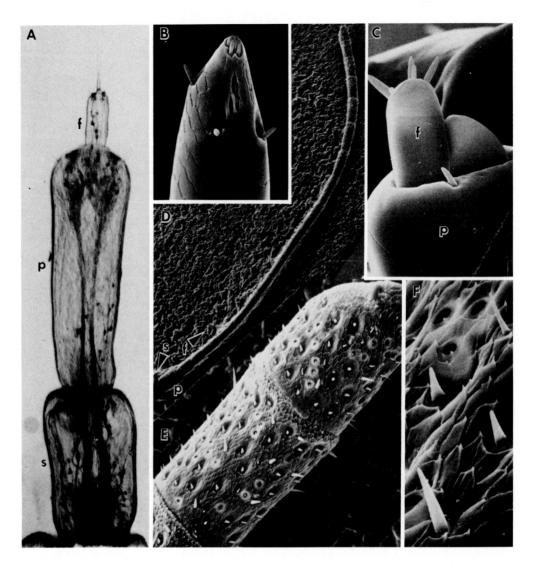

FIG. 2. Antennae. **A**: LM, *Tenebrio molitor* larva, methylene blue wholemount (× 150). **B**: SEM, flagellar tip, dytiscid larva, *Laccophilus biguttatus* (× 1700). **C**: SEM, pedicel tip and flagellum, elaterid larva, *Ctenicera destructor* (× 1700). **D–F**: SEM, grasshopper, *Melanoplus bivittatus*, whole antenna (× 17), terminal two segments (× 170), and a portion of the penultimate segment (× 1700), respectively. f, flagellum; p, pedicel; s, scape. (**A** is from Bloom, J. *et al.*, 1982a, courtesy of the authors and The National Research Council of Canada.)

and among antennae of the different insect groups. Various types of antennae are identified on this basis, from the simpler thread-like (setaceous and filiform) to the more intricate club-like (clavate and capitate), lamellate, pectinate and plumose forms. Membranous cuticle attaches the flagellar subsegments to one another, some of the lateral elaborations such as pectinae to the subsegments, and

the flagellum to the pedicel, enabling some articulation at these joints.

Imms, A. (1939) identified two main categories of insect antennae based, in part, on flagellar musculature. In Collembola and Diplura each flagellar subsegment is inserted by two groups of muscles that originate in the preceding subsegment, or in the pedicel for the basal subsegment. Insects with such antennae are capable of elaborate flagellar movement, and they lack the special mechanoreceptive Organ of Johnston. In the remaining insect forms there are no intrinsic muscles or muscle insertions in the flagellum, but the subsegments and their branches are capable of some movement. This can only be passive under an external pressure, or under changing internal pressures by the hemolymph. Such antennae have Johnston's organ developed to varying degrees in the pedicel.

For more details on general antennal structure see Snodgrass, R. (1935), Schneider, D. (1964) and, for forms of antennae, a work on descriptive entomology such as Richards, O. and Davies, R. (1977). Some simple antennae and variations in their sensory fields are illustrated in Fig. 2.

The antennal integument consists of annular segmental sclerites jointed by membranes of cuticle and all underlaid by epidermal cells. The cuticle of sensilla is sclerotized either in continuity with an antennal sclerite, or joined to it by a ring or socket of membranous cuticle. The epidermal layer lines the antennal cuticle continuously and usually thinly throughout, including the elaborated projections of the flagellar subsegments, except for perforations for each cuticular sensillum. These holes are usually under the sensillar cuticle and its socket, and match their area in size. Most of the epidermis is a simple layer of squamous cells, but where space is restricted and sensilla are numerous it may be a pseudostratified layer of columnar cells, as in the pedicel and reduced flagellum of a larval mosquito antenna (Zacharuk, R. et al., 1971). In such restricted regions only the distal cytoplasmic extensions might line the cuticle; the cell bodies would underlay other epidermal cells or their extensions in a more spacious basal segment. The cells associated with each sensillum are in discrete bundles either in the epidermal layer, partly or entirely suspended in the antennal lumen under the epidermis, or united with neighboring sensillar cell bundles into a large multi-sensillar cell mass in the lumen.

The antennal lumen is continuous from the cranial hemocoel at the base to the terminal cell masses at the ends of the flagellum and of its processes if present. This comprises the epidermis-bound antennal hemocoel. The tracheae, muscles, nerves and sensillar cell bundles extend through or into, abut, or are in close proximity to, the hemolymph within it. All these tissues are separated from the blood by a continuous connective tissue membrane, which is made up of the basement membranes of the epidermis and tracheocytes and the lamellae of the nerves and muscles blended into one another (Schneider, D. 1964). This forms a blood–cell barrier throughout, and is the initial blood barrier for most of the sensory neurons. The blood is pumped into the antenna through a blood vessel that extends to the end of the flagellum, usually without branching, by a pulsating ampulla in the head. It leaves the antennal vessel through slits in the wall and returns to the body through the hemocoels. In the lamellate antenna of *Melolontha melolontha* the antennal blood vessel extends a branch into each flagellar lamella, and each branch opens at the end of the lamellar hemocoel (Pass, G. 1980). The antennal heart is dilated by an elastic connective tissue band, drawing blood from the cranial hemocoel through a valvular ostium, and is contracted by an adjacent compressor muscle that is innervated from the antennal nerve through a small peripheral ganglion. The antennal hearts of most of the other insects studied are also directly innervated, but the source of the nerve and the dilator–compressor mechanism are variable and most of them differ from the above. For details on these see chapter 6, volume 3; Jones, J. (1977); or the discussion and references by Pass, G. (1980).

In *M. melolontha* the antennal lamellae apparently are closed or spread to expose more sensillae by changes in blood flow through the antennal blood vessels and in blood pressures in the hemocoels. Movement of flagellar subsegments and their branches in other insect antennae that do not have intrinsic flagellar muscles is presumed to be effected hydraulically in a similar way. Vibratory antennal movements are common, and these tend to increase in frequency in "excited" animals. The frequency of antennal vibration often is matched to the frequency of wing beat (Callahan, P. 1975). Both the antennal ampullae and the basal antennal muscle mechanisms are directly innervated, and either or both could be involved in these vibratory movements.

The whorls of long hairs on the plumose antennae of male *Anopheles stephensi*, which are folded against the antennal shaft much of the time, are erected outwardly by a more unique mechanism (Nijhout, H. and Sheffield, H. 1979). Each whorl of hairs is attached to the edge of a deeply slitted, ring-like annulus, which is composed of a cuticle different from the rest of the antennal cuticle, is probably rich in basophilic protein, and is attached to the antennal shaft by a chitinous flange. The annulus is subtended by large, irregular, richly microvillate cells. The operative mechanism suggested is that the underlying cells increase the pH of the medulla of the annulus, leading to its hydration by fluids also provided by these cells. The swelling of the hydrated medulla spreads or opens the cortical slit, and this pushes the attached whorl of hairs outward into a fully erected position. This effector mechanism is under direct neural control (Nijhout, H. 1977), and is significant for the perception by sound of females by males during periods of mating.

Insect antennae are well tracheated. Usually one or two tubes enter each antenna from the cephalic trunks and extend the length of the lumen. In *Blabera craniifer* the two tubes that enter the antenna divide in the scape and the four tubes span the antennal length (Urvoy, J. 1963). The main tracheae extend individual branches into the lateral extensions of the flagellar subsegments, and form many small branches along their length that ramify along the walls and over the muscles and nerves.

The gross innervation of insect antennae is fairly standard (Figs 2A and 3A). One nerve trunk for each antenna arises in the deutocerebrum. It gives off a small motor branch to the antennal muscles in the anterior part of the head. Beyond this the nerve is primarily or wholly sensory. It enters the scape where it divides into two main branches approximately equal in size. Both usually continue to the antennal tip, but in some larval forms with a reduced flagellum one nerve terminates in the flagellum and the other in the pedicel. In antennae with flagellar lamellae or pectinae a small branch from one of the main branches innervates each subsegmental extension. The antennal trunk nerve may give off other small nerves in the head, such as the motor nerve to the antennal heart of *Melolontha melolontha* noted above. In elaterid larvae a small sensory branch innervates cranial sensilla around the antennal socket, and another small branch connects the antennal trunk nerve with a small

peripheral ganglion on the labral nerve (Zacharuk, R. 1962a).

The fine details of innervation vary with the form of antennae and the number and pattern of sensilla on them. These are elaborated for insects such as *Bombyx mori* adults (Schneider, D. and Kaissling, K. 1957), *Speophyes lucidulus* larvae (Corbière-Tichané, G. 1973), and the cockroaches *Blaberus craniifer* (Urvoy, J. 1963) and *Periplaneta americana* (Petryszak, A. 1975b). In general, axons from individual or several adjacent sensilla form small nerve branches that join a main branch soon after leaving the sensillar cell bundles, and this occurs along the length of the antenna. In larvae of *Tenebrio molitor*, Bloom, J. *et al.* (1982a) noted that dendrites and cell bodies of some sensilla extend into the main antennal nerve branch along with axons from other sensilla.

Wigglesworth, V. (1959) estimated that there were about 15 times as many sensory neurons in the terminal antennal segment of the bug *Rhodnius prolixus* as there were axons in the nerve leaving the segment. Dethier, V. *et al.* (1963) reported that there are several hundred-fold more sensory cells in a blowfly antenna than there are axons in the antennal nerve. Both concluded that fusion of sensory axons is extensive in insect antennae. In more recent counts from ultrastructural preparations of antennae of *Scolytus multistriatus* beetles by Borg, T. and Norris, D. (1971), *Aedes aegypti* mosquitoes by McIver, S. (1978), and *Tribolium* larvae by Behan, M. and Ryan, M. (1978), all concluded that axonal fusion was highly unlikely. The last lists four other reports with the same conclusion. No examples of fusion were seen in the extensive literature surveyed. Until shown ultrastructurally, we may assume that fusion of sensory axons does not occur peripherally.

The central pathways and connections of antennal sensory nerve fibers were examined in a few insects, and are exemplified by the following. In *Locusta migratoria* and *Periplaneta americana* (Ernst, K. *et al.*, 1977), most of the sensory axons from the flagellum terminate in the ipsilateral deutocerebral glomerulus, where both convergent and divergent connections are made with branched processes from deutocerebral neurons. A small number of the sensory axons terminate in the dorsal lobe. Boeckh, J. *et al.* (1975) estimated that in *Periplaneta* more than 120,000 sensory axons

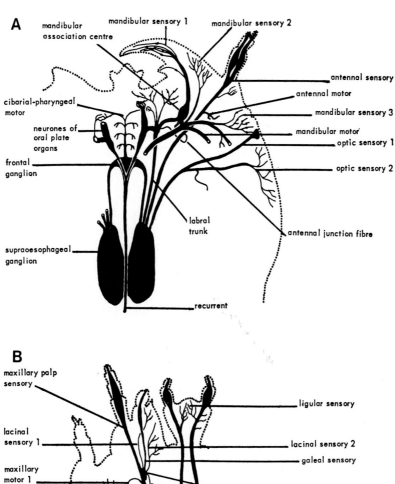

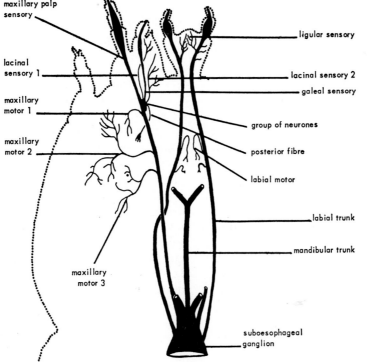

FIG. 3. Innervation of the cephalic sensory fields in an elaterid larva. **A**: dorsal; **B**: ventral. (From Zacharuk, R. 1962a, by permission of the Wistar Institute Press.)

from antennal sensilla terminate in the deuto-cerebral glomerulus, where they interact with widely branched processes of 490 deutocerebral neurons of two functional types; "B-cells" that respond only by excitation and only to stimulation of the ipsilateral antenna by odors, and "W-cells" that respond by excitation, inhibition, on, off, or on–off modes to odor, mechanical or thermal stimuli. Only 260 of these interneurons extend their axons into the protocerebrum. Masson, C. and Strambi, C. (1977) identified two specific areas, more or less over-lapping and with some differences in metabolic activity, in the deutocerebrum of the wasp *Polistes gallicus* and the ant *Camponotus vagus*, one a chemi-cal and the other a mechanical sensory center. These centers and the above neuronal types may be related. In *Locusta* the axons from mechanosensilla in the pedicel project into the protocerebrum and the subesophageal ganglion (Gewecke, M. 1979). The central pathways and terminations of antennal sensory fibers in worker honeybees, *Apis mellifera*, are similar to the above (Pareto, A. 1972; Suzuki, H. 1975). Pareto, A. noted that the central primary terminations were all ipsilateral except in the subesophageal ganglion where they were contralateral. Also, antennal sensory fibers ter-minated directly on motor fibers in the dorsal lobe, seemingly forming synaptic contacts not involving interneurons. In the antennal lobes of the moth *Manduca sexta*, Matsumoto, S. and Hildebrand, J. (1981) detail two classes of central neuron; local sensory processing interneurons, and output neurons that relay this sensory information to higher centers in the brain, with a distinct sexual dimorphism. Two ultrastructural types of synaptic junctions are described by Schurmann, F. and Wechsler, W. (1970) in the glomerular neuropile of the *Locusta migratoria* deutocerebrum, but their relationship to cell type or function is uncertain.

The primary function of antennae is to bear ex-terosensilla, and to extend these ahead of the body and bilaterally in approximately equal proportions. This seems to be the sole function in the majority of insects. In a few forms they are modified for other purposes, such as to form channels for refilling the air reservoirs in some water beetles, to capture and hold prey in certain water beetle larvae, and for a physical role during mating in fleas and Collembola (Schneider, D. 1964).

Researchers who have studied the distribution of sensilla on insect antennae generally agree that it is not random. There is little doubt that their patterns of distribution result from morphogenetic field ef-fects, but there has been little work specific to this. The following generalized distribution pattern has emerged from studies to date. The scape and pedicel bear primarily mechanosensilla of several types, which function as extero- or proprioceptors. They monitor the active movements of these segments and the passive deflections, vibrations or other movements of the flagellum at its base. Where the flagellum is much reduced, as in many larval forms, there may be chemosensilla and thermo- or hygroreceptors at the tip of the pedicel. The greatest number of the antennal sensilla and of different sensillar types occur on the flagellum, especially in adult insects. Their density tends to be significantly higher on the more distal than on the proximal seg-ments, and is often higher on the lateral extensions than on the bodies of the flagellar subsegments. A great diversity of types and subtypes of sensilla, both morphological and functional, have been described from the antennae of diverse insects. The differentiation of many of these is based on the form and size of their cuticular parts. Most of these can be placed into one of the ten broad morphological categories listed by Schneider, D. (1964). Their oc-currence, number and distribution differs between insect species and, to a variable extent, between stages and sexes within a species. The distribution pattern of each type of sensillum is fairly specific within an insect form, and its field may overlap the fields of some but not of other different sensillar types. However, information on form, exposed area, distribution and function of antennal sensilla in relation to antennal morphometry is available for but a few insects and, even in these, is not always complete, and the functional significance of such morphometric parameters is not yet clear. Some of the more recent literature which presents specific information and discusses other literature relevant to the above generalizations is given in Table 1. Among the studies on ultrastructural morphometry of antennae and their sensilla published since this Table was prepared are those of Walther, J. (1981) comparing male, female and worker *Formica rufa* and a few other ant species, Bland, R. (1982) on the grasshopper *Hypochlora alba*, Craig, D. and

Baty, H. (1982) on simuliid larvae, Hallberg, E. (1982) on the beetle *Ips typographus*, and Schaller, L. (1982) on the cockroach *Leucophaea maderae*.

Despite the limited comparative morphometric data available for the different antennal forms of different insects, as evident in this table, a general direct correlation is apparent between the size of an antenna and the number of sensilla on it. This is to be expected given a need for a certain minimum antennal surface area related to the size of the cuticular part of each sensillum, and for a specific volume within the antenna related to the size of the cellular bundle associated with each sensillum. Access by these cells to nutrients in the hemolymph and tracheae is also a consideration. However, the correlation is neither universal nor proportional.

The capacity of an antenna for sensilla is increased in a variety of ways in different insects. These include lengthening the flagellum, as in *Monochamus* and *Periplaneta,* widening some or all of the flagellar subsegments, as in *Necrophorus* beetles (Boeckh, J. 1962) or in the males of *Apis,* and lateral branching of the flagellar subsegments, as in the saturniid moths *Bombyx* and *Telea*. In *Necrophorus*, sensilla with small surface cuticular parts are as dense as 1 per 25 μm^2 of surface area on the terminal segment. Based on data from Schneider, D. and Kaissling, K. (1957), Schneider, D. (1964) notes that the actual surface area of the *Bombyx* branched flagellum that they examined was 29 mm^2, that of the shaft was only 4.8 mm^2 and, without branches, would have to be six times as long to match the branched flagellum in surface area. The mean density of sensilla on the flagellum of this moth was 1 per 1400 μm^2, but the sensilla on parts of the branches were up to 1 per 200 μm^2 in density. He concluded that, given the surface area and volume, the antennal flagellum of *Bombyx* perhaps could accommodate twice as many sensilla as it does, but not seven times as many if it were to match the sensillar density on the flagellar club of *Necrophorus*. In the noctuid *Trichoplusia ni* the mean density of all sensilla on its filiform flagellum is 1 per 400 μm^2, but on the more distal segments (Mayer, M. *et al.,* 1981) the density of only the hairlike sensilla is 1 per 120 μm^2, a density significantly greater than that on the branches of *Bombyx* flagella. Thus, it seems that while surface area and volume of antennae play some role in controlling the number and distribution patterns of flagellar sensilla, other factors are also involved. The latter may play a greater role in considerations of antennal form as well as some role related to its size.

Antennae are often considered to be the nostrils of insects. For an especially sensitive olfactory function, Kaissling, K. (1971) stresses the important of an extended outline area and an effective subdivision of form in insect antennae. The sieve-like bipectinate antennae of the moths *Telea polyphemus* and *Bombyx mori* are well suited to trap odor molecules with a high "filter coefficient". This is particularly true of the males, which have antennal outline areas of 85 and 6 mm^2, respectively. The corresponding outline areas of the females is 18 and 5.5 mm^2. The larger odor filters of the males, with their patterns of specific odor receptors, is related to their need to find their female mates at a distance. While the basic surface area of the male *Bombyx* moths examined by Steinbrecht, R. (1970) was 24 mm^2, he estimated that the actual surface area of insensitive cuticle with surface sculpturing taken into account is at least 3 times greater, or about 75 mm^2. The surface area of the exposed sensory cuticle of the olfactory pegs and hairs was about 12 mm^2, or about 14% of the total surface area of sensitive and insensitive cuticles. The hairs that detect the female sex pheromone cover the free spaces between the flagellar branches. They are believed to adsorb most of the odor molecules carried by the airstream that is filtered through the antennae. In these moths, therefore, the form of their antennae combined with a suitable placement of sensilla in fields on it and an adequate surface area of sensitive cuticle seem to be most important for their efficiency or sensitivity of odor perception. In comparison with the above moths, the outline area of an antenna of *Necrophorus vespilloides* is 0.7, of *Apis mellifera* drones 1.12 and of queens 0.56, of *Sarcophaga* 0.5, and of *Drosophila melanogaster* only 0.01, in mm^2 (Kaissling, K. 1971). The efficiency of their odor perception is presumably much reduced accordingly.

Insect antennae are often also referred to as feelers. They may literally function as such in many insects, especially in those with very long flagella such as the cerambicid beetle *Monochamus* and in the cockroaches. The concentration of several types

Table 1: *Morphometry of antennal flagella and their sensilla in selected insects.*

Insect, stage	adult sex	Flagella characteristics			Number of sensilla		Total sense cells	Reference
		Subsegments	Length (mm)	Diameter (mm)	Types	Total		
Coleoptera								
Aphaenops cryticola	M	9 + 2*	3.5		8	1,798		Juberthie, C. and
	F	9 + 2*	3.5		8	1,833		Massoud, Z. (1977)
Monochamus	M	9	69.0		6	25,000	39,000	Dyer, L. and Seabrook,
notatus	F	9	29.2		7	19,000	30,000	W. (1975)
Nebria brevicollis	M,F	9	5.4		7	5,785	8,000	Daly, P. and Ryan, M. (1979)
Tenebrio molitor	M,F	9	3.0		7	4,360	9,025	Harbach, R. and Larsen, J. (1977)
Last larval instar		1 (much reduced)			6	9	38	Bloom, J. *et al.* (1982a)
Diptera								
Culicoides furens	F	13	0.58		5	235	900	Chu-Wang, I. *et al.* (1975)
Simulium rugglesi	M	9	0.35		6	813		Mercer, K. and McIver,
	F	9			6	1,463		S. (1973)
Stomoxys calcitrans	M,F	1	0.54,0.6		4	4,840	15,000	Lewis, C. (1971)
Wyeomyia smithii	M	13	1.12		6	394		McIver, S. and Hudson,
	F	13	1.13		6	430		A. (1972)
Hemiptera								
Cimex lectularius	M,F	2 + 2*	1.8		9	250	370	Steinbrecht, R. and Müller, B. (1976)
Homoptera								
Acyrthosiphon pisum								
1st nymph		3 + 2*	1.3		4	48		Shambaugh, G. *et al.*
3rd nymph		4 + 2*	2.5		4	123		(1978)
apterous		4 + 2*	4.8		4	113		
alate		4 + 2*	4.8		4	132		
Hymenoptera								
Apis mellifera								
drone worker		11	4.0	0.21	10	19,721	338,859	Esslen, J. and Kaissling,
		10	2.8	0.29	10	6.364	64,889	K. (1976)
Lepidoptera								
Bombyx mori	M	35–41		$(24\,\text{mm}^2)^s$	3†	24,500		Steinbrecht, R. (1970)
	F	35–41		$(19\,\text{mm}^2)^s$	3†	16,000		
Choristoneura	M	46	4.9	0.07	4	2,777		Albert, P. and Seabrook,
fumiferana	F	45	5.1	0.07	4	1,754		W. (1973)
Cydia nigricans	M	55.5	3.87	0.09	5	17,200		Wall, C. (1978)
	F	53.8	3.55	0.08	5	11,700		
Heliothis zea	M	82	13.0	0.13	4	12,850		Jefferson, R. *et al.* (1970)
	F	82	13.0	0.13	4	12,565		
Telea polyphemus	M	33	18.0	8.0^w	6	67,000	150,000	Boeckh, J. *et al.* (1960)
	F	30	13.0	2.5^w	5	13,300	35,000	
Trichoplusia ni	M	75	10.1	$(3.6\,\text{mm}^2)^s$	7	9,017		Mayer, M. *et al.* (1981)
	F	75	10.1	$(3.6\,\text{mm}^2)^s$	7	9,191		

Table 1: *Morphometry of antennal flagella and their sensilla in selected insects—Continued.*

Insect, stage	adult sex	Flagella characteristics			Number of sensilla		Total sense cells	Reference
		Subsegments	Length (mm)	Diameter (mm)	Types	Total		
Orthoptera								
Blaberus craniifer	M,F	100 +			3†	50,000	137,000	Lambin, M. (1973)
Melanoplus	M	23			3†	1,876		Riegert, P. (1960)
bivittatus	F	23			4	1,951		
5th instar		22			4	1,517		
3rd instar		18			4	589		
1st instar		11			4	423		
Melanoplus	M			$(6\,mm^2)^s$	3†	3,926	96,183	Slifer, E. *et al.* (1959)
differentialis	F			$(8\,mm^2)^s$	3†	3,956		

* Includes scape and pedicel.
† Olfactory sensilla only.
s Surface area not considering sculptures.
w Greatest width including pectinae.

of sensilla noted at the flagellar tips of many insects and the reports of several of the authors listed in Table 1, that more sensilla of certain types are distributed on the ventral than on the other aspects of the flagellar subsegments, are in keeping with this function. In the hymenopteran *Itoplectis conquisitor*, Borden, J. *et al.* (1973) noted a terminal cluster of unique sensilla, cylindrical pegs with truncate tips, which they propose may perceive stimuli in the act of antennal tapping. In addition to tactile stimuli, insects may perceive chemo-, hygro- and thermo-stimuli by antennal feeling.

The sensitive cuticle of some types of sensilla is typically located in shallow or deep pits in the antennal cuticle, as will be detailed in section 4. This principle is extended to a unique placement of fields of sensilla in internal vesicles within specific flagellar subsegments of the beetle *Ptomaphagus* (Peck, S. 1977). The vesicles have slit openings to the exterior. The novel fluted pegs within these vesicles are presumed to be olfactory. If they are, their specificity and mode of access to odor molecules in the air around the antenna would be of interest.

Complete morphometric data on an antenna and its sensilla are of paramount importance to an understanding of the specificity, thresholds and transduction processes of an antennal sensory system. Studies towards this end are well advanced on the pheromone receptor system of *Bombyx mori*, and are initiated on the antennal sensilla of *Trichoplusia ni* and, to varying extents, of a few

other insects. It is evident from the preceding that much more information of a more comparative nature is needed on the various types of sensilla in a variety of antennal and insect forms towards an understanding of the generalized structural and functional mechanisms and principles involved.

The developmental processes in insect antennae were examined in a number of insects. Imms, A. (1940) reviewed the earlier literature and concluded that the segmented antennae with flagellar musculature of the Apterygota, excepting Thysanura, lengthen at a molt by division of the apical segment. The antennae of Thysanura and of the Pterygota, which have annulated flagella without musculature, lengthen by division of the basal flagellar segment. In certain Orthoptera and Odonata some of the adjacent basal or all of the flagellar subsegments also divide.

Most of the more recent developmental studies on insect antennae are on those of various cockroaches. In *Blaberus craniifer* (Urvoy, J., 1963), *Blattella germanica* (Campbell, F. and Priestley, J., 1970), *Leucophaea maderae* (Schafer, R., 1973) and *Periplaneta americana* (Haas, H., 1955; Schafer, R. and Sanchez, T., 1973), new subsegments are added to the flagellum at each molt. The basal flagellar segment, the meriston, divides into 5 to 30 subsegments at each molt. These meristal subsegments produce more subsegments by binary division at the next molt. The growth potential is offset somewhat by the mechanical loss of terminal subsegments during the molting process and perhaps from other causes, so that the net gain is less than production by division. In *Leucophaea*, for example, there are about 47 flagellar subsegments at hatching, and about 119 in the adult. With the divisions noted and if no losses had occurred, Schafer, R., estimated that the adult antenna would have about 190 subsegments. The above studies of Haas, H. Schafer, R., and Urvoy, J. also show that nymphal cockroaches have a strong potential for antennal regeneration, including varying complements of sensilla. Even if an antenna is removed at the base of the scape, the antennifer or head capsule can regenerate a new one with the original types of sensilla.

The postembryonic development of the antennae and their sensilla in selected stages of *Leucophaea* and *Periplaneta* is given in Table 2. The density of the sensilla on the antennal surface changes little, or even decreases slightly, through the nymphal instars, but increases significantly during the molt to the adult stage. This is due primarily to an increase in the number of olfactory pegs and hairs. In *Leucophaea* there is no antennal sexual dimorphism, and the sensillar density increases similarly in both sexes in the molt to the adult stage. In *Periplaneta* the antennae may be slightly larger in the adult males than in the females, but the males acquire a disproportionately much higher number of presumably pheromone-sensitive olfactory sensilla. The density of these sensilla remains about the same in females but approximately doubles in males in the molt to the adult stage (Schafer, R. and Sanchez, T., 1976a).

Wigglesworth, V. (1940) originally proposed that the density of cuticular sensilla is controlled by a zone of inhibition exerted by each sensillum around itself. By analogy with other embryological field systems, Lawrence, P. (1970) suggested that groups of epidermal cells determine the development of new sensilla. In the model presented by Richelle, J. and Ghysen, A. (1979) for *Drosophila*, when an inducer secreted by imaginal cells, chaetogen, reaches threshold levels and triggers bristle formation in an epidermal cell, this cell prevents neighboring cells from being induced. All the above depend on density effects. Schafer's data on antennae of *Leucophaea* indicate that the rapid addition of olfactory sensilla is inhibited until released at the terminal molt. He suggested that the inhibiting factor may be

Table 2: Development of the antennal flagellar segments and sensilla in Leucophaea maderae *(Schafer, R., 1973) and* Periplaneta americana *(Schafer, R. and Sanchez, T., 1973).*

| Instar | Number of segments | | Surface area (mm²) | | Number of sensilla | | | |
| | | | | | Total | | Olfactory | |
	L.m.	P.a.	L.m.	P.a.	L.m.	P.a.	L.m.	P.a.
1	47.5	46	5.2	2.3	3,756	3,900	2,410	2,700
3	61	60	7.7	4.0	4,589	4,600	2,910	3,200
5	83	67	16.6	7.9	8,825	5,600	6,100	3,900
6	97	75	23.4	8.9	14,010	7,000	10,460	4,900
7	109.5	100	42.7	15.9	20,585	9,600	15,890	6,500
9		116		22.3		16,100		11,800
10		121		30.3		19,200		14,500
11 M		138		34.2		24,000		16,300
F		138		34.2		24,000		16,300
Adult								
M	120	135	43.8	41.0	33,734	46,200	27,800	39,000
F	119	139	43.8	37.1	33,734	29,300	27,800	22,100

hormone. In a subsequent study with *Periplaneta*, Schafer, R. and Sanchez, T. (1976b) confirmed the inhibitory effects of this hormone on the formation of many new olfactory sensilla in males at the terminal molt.

The development of the antennae and their sensilla in Thysanoptera is detailed by Heming, B. (1975) for *Frankliniella fusca* and *Haplothrips verbasci*. In the former the antenna undergoes metamorphosis externally within the antennal exuvium of the preceding stage in reorganizing from the larval to the adult form. In the latter, the antennal cell mass is withdrawn into the head capsule during the larval–propupal molt and remains there until everted at ecdysis. Further reorganization occurs externally in the propupal and pupal stages to the adult form. The type, number and distribution of sensilla are very similar in the larvae but are very different between the adults of the two species, even though some of the larval sensilla are retained in the adult antenna. The flagellum has five segments in the larval stages and six segments in the adults of both species.

In the three-segmented larval antennae of many Lepidoptera and Coleoptera, the flagellum is represented by the much-reduced terminal segment. It remains so from hatching to the last larval instar, with little change in number and types of sensilla on it. The flagellum of the male of the moth *Manduca sexta* (Sanes, J. and Hildebrand, J., 1976a) is about 20 mm long and consists of about 80 subsegments. It bears about 10^5 sensilla and contains about

2.5 × 10⁵ sense cells. The pupal antennae form from imaginal discs that are at the base of the larval antenna. During metamorphosis of the antennae in the pupa, the flagellar subsegments develop synchronously and sensilla arise and differentiate in recognizable regions on each subsegment. In most Coleoptera and other less derived endopterygote insects the development of the adult antenna seems to follow a somewhat similar pattern (Snodgrass, R., 1954). In holometabolous insects with apodous larvae, the adult antennae develop from antennal imaginal discs that are invaginated into the larval head at about the time of hatching and are everted during the larval–pupal molt. In *Calliphora*, for example, the antennal sites may be represented by a pair of minute tubercles, each bearing a single, large "antennensinnesorgane" through the larval stage (Richter, S., 1962). The three-segmented adult antennae and their numerous sensilla differentiate from the antennal imaginal discs as noted above and detailed by Schoeller, J. (1964).

2.2 Mouthparts

The mouthparts are the cephalic parts and appendages that are involved in feeding and food ingestion. These are the labrum, mandibles, maxillae, labium and hypopharynx. They are structurally modified to various extents among insects primarily in relation to their mode of feeding. From a sensory standpoint, while the antennae are most often viewed as the insect's feelers and organs of smell, some of the mouthparts are considered as its organs of taste. However, there are also other types of sensilla in their sensory fields, including various types of mechanosensilla to monitor external forces and the positions and movements of the parts themselves.

The typical labrum is a simple plate-like structure attached along the posterior margin to the clypeus by a membrane or suture. Its inner surface, often referred to as the epipharynx, forms the roof of the preoral cavity. The cranial hemocoel extends into the space between. Through it extend the labral retractor muscles, nerves and tracheae. Thus, the tissues, including the epidermis and sensillar cells, have access to the required nutrients through the intervening basement membranes. The labral sensilla are typically distributed in bilaterally symmetrical fields on each side of the midline (Fig. 4).

The labral sensillar fields are quite simple in the larval forms studied. In the lepidopterans *Choristoneura fumiferana* (Albert, P., 1980) and *Pieris brassicae* (Ma, W., 1972), the outer surface has 12 mechanosensitive hairs and one or two campaniform organs. The inner surface has six short mechanosensitive hairs, two campaniform organs and two chemosensitive papillae. Larvae of Elateridae (Zacharuk, R., 1962a) and of *Speophyes lucidulus* (Corbière-Tichané, G., 1973) have fewer hairs and more papillae on the epipharynx, and many campaniform organs and short hairs in addition to 12 long hairs externally. The sensilla are much more diverse and numerous on the labrum of grasshopper adults (Cook, G., 1972; Chapman, R. and Thomas, J., 1978; Bland, R. 1982), the cricket *Acheta domesticus* (Rohr, W., 1982), the cockroaches *Periplaneta americana* (Petryszak, A. 1975a) and *Blaberus craniifer* (Moulins, M., 1968, 1971a, b), the odonatans *Aeshna interrupta lineata* (Pritchard, G., 1965) and *Libellula depressa* and *Libellula quadrimaculata* (Petryszak, A., 1977), and the termites *Calotermes flavicollis* (Richard, G., 1951) and *Schedorhinotermes putovius* (Quennedey, A., 1975). The outer surface has numerous varisized hairs and campaniform organs. All are typical mechanosensilla, but some of the hairs in grasshoppers are putative, and those of *Schedorhinotermes* are proven, contact chemoreceptors also. On the epipharyngeal surface there are several paired fields, some extensive, of various types of hairs, pegs and papillae. Some of the hairs are typical mechanosensilla. The other sensilla have presumed or proven chemosensitivity. In mouthparts that are modified considerably for piercing and sucking, the labrum still forms the dorsal wall of the food channel. In the mosquito *Aedes aegypti* (Lee, R., 1974), there are four contact chemosensory pegs in females and two campaniforms in both sexes at or near the labral tip. The epipharynx in the cibarium has six chemosensory papillae, two campaniforms, and four to seven mechanosensory hairs in both sexes. This sensillar field is similar in other mosquito species (Lee, R. and Davies, D., 1978; Uchida, K., 1979) and in *Calliphora erythrocephala* (Rice, M., 1973), but in the latter there are many more tactile hairs. In females of the blackfly

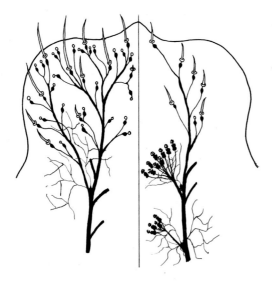

FIG. 4. Sensillar fields and innervation of an insect labrum;
left, dorsal; right, ventral or epipharynx.

Simulium venustum the types and pattern of labral and cibarial sensilla more closely resemble those of mosquitoes (Sutcliffe, J. and McIver, S., 1982). The epipharynx has 14 papillae in the aphid *Brevicoryne brassicae* (Wensler, R. and Filshie, B., 1969), 20 papillae in the aphid *Macrosteles fascifrons* (Baskus, E. and McLean, D., 1982), and about 180 short hairs in worker *Apis mellifica* (Galic, M., 1971). Both types are presumed chemosensilla.

Two trunk nerves, one from each side of the tritocerebrum, serve the labrum. Each innervates the ipsilateral half. Individual branches from it usually innervate each plaque or field of epipharyngeal sensilla. The main nerve arborizes into many fine branches, which go to sensilla on the external surface and to individual sensilla on the epipharynx. In elaterid larvae there is a junctional complex on each labral trunk nerve. It contains a large neuron that may be an interneuron, and forms connections with the ipsilateral antennal and mandibular nerves and with the recurrent nerve just behind the frontal ganglion (Fig. 3A).

The paired mandibles of most insects with biting mouthparts are more or less triangular structures, very heavily sclerotized, and with muscles inserted in the base. They are usually flattened dorsoventrally and have a convex outer margin. The inner margin is variably toothed to the apex. The mandibular sensory fields and their innervation are basically quite similar among the insect forms. There are two or more hairs on the outer surface and, in adult Isoptera (Richard, G., 1951), Orthoptera (Petryszak, A., 1975a; Chapman, R. and Thomas, J., 1978), Odonata (Petryszak, A., 1977) and a few others, additional fields of hairs on the dorsal and ventral surfaces. Several to many small campaniforms are on the outer surface and one or more very large ones are on the dorsal side. Pore canal organs terminate in the thick cuticle of the teeth, perpendicular to the surface. Zacharuk, R. (1962a) initially thought they were chemosensitive, but they were later proven to be scolopophoroid (Zacharuk, R. and Albert, P. 1978). Hamon, M. (1961) noted small pegs on the outer surface and near the apex of larval and adult mandibles of the water beetle *Dytiscus*. Possible mandibular chemosensitivity is noted in a few other reports, but none were seen that provide proof of it. The mandibles of *Aedes* adults, which are modified into stylets and present only in females, have no reported sensilla putative mechanosensory, but two "genal hairs" were noted by Sutcliffe, J. and McIver, S. (1982) on the articular processes of the mandible in female *Simulium venustum*.

Each mandible is innervated by a trunk nerve that arises either directly from the subesophageal ganglion or, as in elaterid larvae (Fig. 3), branches from a single median nerve near that ganglion. Each nerve gives off one or more motor branches to the mandibular muscles and some sensory branches to sensilla on the cranial integument. The terminal, solely sensory branch bifurcates at or below the base of the mandible. The medial branch serves primarily the scolopophorous organs of the teeth. The outer branch arborizes to the other mandibular sensilla. In elaterid larvae there is a small ganglion at the base of the medial nerve, from which a connection is made to the labral nerve (Fig. 3A). The cephalic hemocoel extends into the lumen of each mandible, which is lined by an epidermis that is especially thick and pseudostratified in the region of the teeth. The teeth are subject to considerable wear, and this may affect the tips of the scolopophorous organs. A high level of continual secretory activity is necessary for maintenance and repair, more so than usual for other parts of the integument and other sensilla (Zacharuk, R., 1979).

The maxillae of insects are a pair of head appendages that are positioned laterally or ventro-laterally to the mouth. Their basic limb structure is preserved to varying extents in insects with biting mouthparts, but is greatly reduced or modified in other insect forms. Typically, there are two basal articulating segments, the proximal cardo and the distal stipes. Two lobes are attached to the stipes distally, the mesal lacinia and the outer galea. The most conspicuous appendage of the maxilla is the palp. It is attached to the side of the stipes either directly or through a palpiger, and consists of one to seven segments that are usually directed forward. The cephalic hemocoel extends through the basal segments into the lumens of the three appendages. There are no known reports of maxillary accessory circulatory organs; circulation is probably aided by active movements of the parts. All are inserted by muscles, including the segments of the palps. Each maxilla is served by a trunk nerve from the subeso-phageal ganglion (Figs 3B and 5). This nerve branches at or below the base of the maxilla, and a main branch extends into each appendage where it arborizes to the fields of sensilla on different parts. The branch to the palp usually bifurcates at the base into an inner and outer branch, each of which usually extends to the tip of the palp. In elaterid larvae there is a small ganglion at the point of origin of the sensory branches to the maxillary appendages. The maxillary motor nerve branches arise from the trunk nerve proximal to this ganglion.

The cardo and stipes bear only trichoid hairs and campaniform organs over their surfaces. Scolopo-phorous organs were also noted in them in some insects. All these sensilla are presumed to be mechanosensitive. Petryszak, A. (1975a) identified small pegs on the mesal posterior surface of the stipes near its junction with the lacinia in *Periplaneta americana* whose function, not specified, may be chemosensory.

The lacinia is very variable in structure and function among insect forms. It is not evident in the larva of *Choristoneura fumiferana* (Albert, P., 1980). In elaterid larvae (Zacharuk, R., 1962a) it is a well-developed lobe, which bears several innervated chaetal hairs and many non-innervated hairs and spines along the medial surface. The innervated chaetae are believed to be mechanosensory. In

Fig. 5. Sensillar fields and innervation of an insect maxilla, dorsal, with the palp, galea and lacinia left to right.

grasshoppers (Chapman, R. and Thomas, J. 1978), *Periplaneta americana* (Petryszak, A., 1975a) and *Libellula depressa* (Petryszak, A., 1977) the lacinia has apical and medial cuticular teeth and, particularly in the latter two insects, an extensive field of innervated spines and hairs along its medial surface. Scattered campaniform organs also occur over the surface. In the latter two insects scolopophoroid pore canal organs were also noted in the lacinial teeth, similar to those described from their mandibles. One of the main functions for which lacinia where developed is to aid the mandibles in holding and masticating the food and, as noted in the earwig *Forficula auricularia* (Popham, E. 1979), also to carry it into the cibarial cavity. The mechanosensilla reported are in keeping with the needs for specific tactile stimuli and proprioception for these physical functions. There are no known reports that clearly demonstrate a lacinial chemosensory function. However, it is a suitable location for "taste" sensilla in some insects, and this should be monitored in future studies.

The galea is a single lobe in some insects and two-segmented in others. It is separated from the lacinia in grasshoppers and the cockroach *P. americana*, but is fused with it along some or all of its length in other insects. There is a great variation in number and some in the general types of galeal sensilla among insects. A common pattern is along the lines of that described for the beetle *Entomoscelis americana* by Sutcliffe, J. and Mitchell, B. (1980). Usually there are campaniform organs or mechano-sensitive hairs or both on the outer surface, and hairs or pegs with dual mechano- and contact chemosensitivity on a membranous tip. A few other sensillar types with uncertain sensitivities are also reported for some insects. In the Orthoptera noted above the number of galeal sensilla is much greater than in beetles, but the distribution pattern is quite similar. The mechanosensilla are predominantly on the upper, lower and lateral surfaces, and the putative chemosensilla are concentrated at the tip. In contrast, in *Libellula* the galea is entirely fused with the lacinia and all the sensilla on the part presumed to be galeal are of the type usually found on the lacinia. In *Entomoscelis* the larval galea has many fewer sensilla than that of the adult, but the types are remarkably similar. Sutcliffe, J. and Mitchell, B. (1980) believe that some if not all the larval sensilla are carried through to the adult stage. The function of the galea seemingly varies among insects, and is probably related to a large extent to the food and feeding habits. In some Odonata it seems to be wholly physical in grasping and ingesting food. In the majority of insects it seems to be involved more in a sensory role, in sampling the texture and chemical content of the food. In the beetle studied by Sutcliffe, J. and Mitchell, B., they noted organs that secreted some substance through pores in the cuticle onto the surface of the galea. They suggest that the secretion may prevent oral juices and food particles from sticking to the galea and fouling its sensilla.

The palp of an insect maxilla is considered to be primarily a sensory appendage. In its simplest form it consists of a single segment that arises directly from the outer distal edge of the stipes, as exemplified in *Libellula*. In the larva of the lepidopteran *Choristoneura* it is similarly simple but two-segmented. In the majority of insect forms it has three to five segments and a more elaborate pattern of sensilla. The mechanosensilla are usually distributed along the shaft, with surface hair plates and internal scolopidia in the intersegmental joint areas, and campaniform organs and vari-sized tactile hairs scattered sparsely to densely over the surface of the segments. The chemosensilla are primarily or wholly concentrated at or near the tip of the terminal segment (Figs 5 and 6). Some of these also have mechanosensitivity. Sensilla with other presumed but uncertain sensitivities are also reported to be among the terminal cluster of chemosensilla. There are few sensilla in the terminal cluster in larvae of Coleoptera and Lepidoptera, and a very large number of small pegs and hairs in one or more apical and subapical clusters in Orthoptera. The mechanosensilla along the shaft of the latter are also much more numerous than in most other insect forms. The descriptions by Bland, R. (1982), Blaney, W. and Chapman, R. (1969), Klein, U. (1982), Petryszak, A. (1975a) and Malz, D. and Hintze-Podufal, C. (1979) typify the sensillar distribution on the maxillary palp of this insect group. The pattern in Coleoptera is much simpler, as reported for the beetle *Dendroctonus ponderosae* by Whitehead, A. (1981) and *Ips typographus* by Hallberg, E. (1982). Another type of sensillum with a finger-like cuticular projection that is sunken along its length in a groove of the cuticular wall of the palp regularly occurs individually or in fields in the sidewall of the terminal segment of the palp of larval and adult Coleoptera. Their varied forms and fields in adults are presented by Honomichl, K. (1980). A sensillum of this type was also noted in the larva of *Choristoneura fumiferana* by Albert, P. (1980). While shown to be mechanosensitive by Zacharuk, R. *et al.* (1977), Honomichl, K. and Guse, G. (1981) believe that this sensillum is sensitive to stimuli other than chemical or mechanical. While the other cuticular sensilla of the maxillary palp are on sclerotized cuticle and occasionally on intersegmental membrane, the terminal sensillar clusters are invariably on patches of flexible cuticle as described by Altner, H. (1975) for *Periplaneta americana*. This allows individual sensilla to yield inwardly under pressure and for the entire cluster to be depressed into the tip (cf. Figs 6A and 6B). The most numerous chemosensilla in these clusters are of the type considered to be contact or taste receptors usually also with a mechanosensitive unit, but

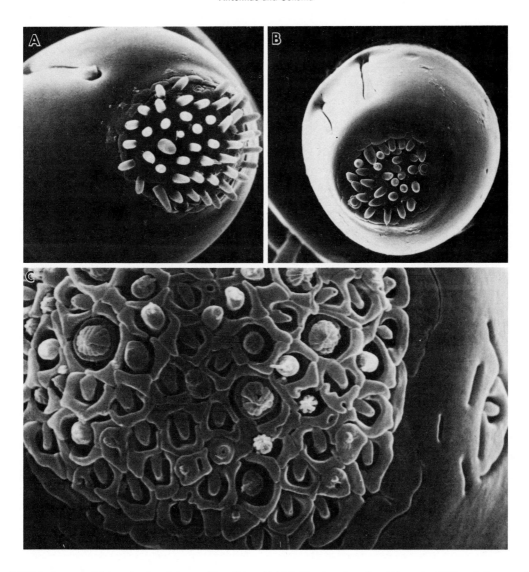

FIG. 6. SEM of sensory fields at the tips of the maxillary (**A**) and labial (**B**) palps of an elaterid larva ($\times 1700$) and about one-half of the tip of the labial palp (**C**) of a dytiscid larva ($\times 1750$).

chemosensilla of an olfactory type were also noted (Blaney, W., 1977; Klein, U., 1982).

The maxillae tend to be considerably reduced or modified in insects with piercing and sucking mouthparts. In adult Diptera the palps are retained in a fairly typical form. They are club-shaped appendages in mosquitoes (McIver, S. and Charlton, C. 1970) and *Calliphora vicina* (van der Starre, H. and Tempelaar, M., 1976), and are five-segmented in the former. In both forms there are tactile bristles and short pegs or capitate hairs with porous cuticles. These are sensitive to CO_2 in the former and chemosensitive to certain odors in the latter. In *Culicoides* (Rowley, W. and Cornford, M., 1972) the short capitate porous sensilla are arranged in clusters in pits on the palp, and are fewer in number in males than in females. The maxillae of simuliid larvae (Craig, D. and Borkent, A., 1980) retain more of the basic structure, with a basal segment, a one-segmented palp and a fused galeal–lacinial lobe.

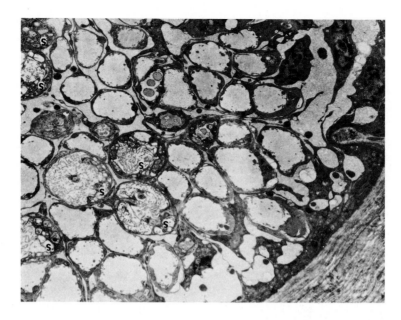

Fig. 7. TEM of about one-quarter of the terminal segment of the labial palp in an elaterid larva with transport hemolymph sinus cells in t.s. between the sensillar cell bundles (s) (× 3500). (From Bellamy, F. and Zacharuk, R., 1976, by courtesy of the authors and The National Research Council of Canada.)

There are typical mechanosensilla on all parts and presumed chemosensilla on the galeal part of the lobe and in a typical terminal cluster on the palp. There are a variety of peg types in this cluster, as evident from the surface scans presented. The fine structure of nine types of maxillary sensilla is reported by McIver, S. and Siemicki, R. (1982) for larvae of the mosquito *Toxorhynchites brevipalpis*.

The labium consists of a pair of cephalic appendages that are serially homologous with the maxillae, but which are fused together medially to various degrees in the different insect forms. It consists of a basal mentum, usually divided by a suture into a proximal post- and a distal prementum, two pairs of anterior lobes that arise from the distal end of the prementum, the outer paraglossae and the inner glossae, and a pair of lateral appendages, the palps, that arise from the anterolateral edges of the prementum and are directed forward. The palps consist of one to four segments in addition to a basal palpiger when this is present. They are primarily sensory in function. The glossae alone may be fused into a medial glossal lobe, or all four lobes may be fused into a single medial lobe that is often referred to as the ligula. All the lobes are usually inserted

with muscles and can be moved actively. The cephalic hemocoel extends into the lumens of the mentum and its appendages, and the circulation of hemolymph seems to be similar to that noted above for the maxillae. In the labial palps of elaterid larvae (Bellamy, F. and Zacharuk, R., 1976), much of their lumens are filled distally with sensillar and epidermal cells. This separates the terminal cellular elements and lymph sinuses from the basal hemocoelic fluid by a considerable cellular mass. Numerous elongate goblet-like cells traverse this cell mass. Their bases abut the hemocoel and their elongate, microvillate sinuses open distally into a terminal lymph cavity. Their apparent function is to transport materials from the hemocoel to the tip of the palps for nourishment of the sensilla (Fig. 7). The labium is innervated by two trunk nerves from the subesophageal ganglion, one to each half. In the more basic, divided forms, each trunk nerve branches in the mentum, and a main branch extends into each of the three appendages (Fig. 8). The palpal branch divides at the base into two branches that extend to the tip. In a highly fused labium, as in elaterid larvae (Fig. 3B), the trunk nerve extends into the palp, giving off only one small branch to the sensilla on the ipsilateral half of the ligula.

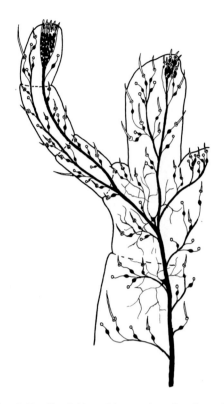

FIG. 8. Sensillar fields and innervation of an insect labium, dorsal, left half, with palp, paraglossa and fused glossa from left to right.

The number, types and bilateral distribution patterns of sensilla on the labium resemble to a certain extent those on the equivalent parts of the maxillae in the same insect form, but there is considerable variation among insects. This is related in part to the degree of labial fusion. In Orthoptera the sensory fields on the shaft of the three-segmented labial palps resemble those of the maxillary palps, but the concentration of terminal chemosensilla is limited to a single concentric field at the tip (Petryszak, A. 1975a; Chapman, R. and Thomas, J., 1978; Malz, D. and Hintze-Podufal, C., 1979). Their paraglossae and glossae have fewer of the heavy bristles noted on their maxillary counterparts, the galea and lacinia, and have more pegs with presumed chemosensitivity, particularly along the apical borders. They also have no scolopidial pore canals. This is in keeping with their primary sensory function and little physical testing or manipulation of food. In the larva of *Choristoneura fumiferana* (Albert, P. 1980), there are only two sensilla on the labial palp,

both mechanosensitive and at the tip. There is no mention of sensilla on its ligula. The structure of the labium and the labial sensilla of elaterid larvae (Bellamy, F. 1973) (Fig. 6b) are not too different from those of the adult *Dendroctonus ponderosae* beetle (Whitehead, A., 1981). The ligula of the former bears only tactile hairs and specialized surface campaniform organs (Zacharuk, R., 1962a). In these Coleoptera the sensilla in the terminal cluster of the labial palp seem identical to those of the maxillary palp but are fewer in number, and the lateral digitiform organs occur on both appendages. In the Odonatan *Libellula depressa* and *Libellula quadrimaculata* (Petryszak, A., 1977) the paraglossae are enlarged and plate-like, the glossae are fused into a small median plate, and all are covered with vari-sized hairs, campaniform organs and papillae. There are no palps. A unique type of large, cup-like sensillum is distributed among hairs and pegs on the terminus of the labial palp in *Lepisma* and *Thermobia* (Larink, O., 1978), but the details of its structure and its function are only partly known (Larink, O., 1982). The terminus of the labial palp of *Forficula* is covered with cuticular plaques, each centered by a papilla. These are presumed to sense the surface texture and hardness of food (Popham, E., 1979), but the details of their internal structure are unknown and their function has not been tested. It would be of interest to determine the structural details and functions of these novel sensilla in both above insect groups.

In insects with mouthparts modified into a sucking proboscis the labium is considerably and variably modified. It forms the terminal labellum in a number of insect forms, with transverse pseudotracheal grooves on an enlarged, flattened tip in some of these. In flies it bears a field of contact chemo- and mechanosensitive hairs and chemosensory interpseudotracheal papillae. The hairs are undoubtedly the most intensively studied chemosensilla of insects (Dethier, V., 1976). They are characterized and mapped in *Calliphora erythrocephala* (Peters, W., 1963; Maes, F. and Vedder, C., 1978), and *Phormia regina* (Wilczek, M., 1967), which have about 130 in marginal rows. Peters noted about 120 interpseudotracheal papillae in the former, and details the innervation of the labellar sensilla from the two labial nerves. In the stable fly *Stomoxys calcitrans* (Adams, J. and Forgash, A.,

1966) there are only about 60 labellar contact chemosensilla. Similar hairs were noted on the labellum of the mosquito *Culiseta inornata* (Pappas, L. and Larsen, J., 1976) and of other mosquito species. Whitehead, A. and Larsen, J. (1976) described chemosensory hairs on the labial palps and glossae and pegs on the former in the honeybee *Apis mellifera*. The tips of the labium of the plant bugs *Lygus lineolaris* (Hatfield, L. and Frazier, J. 1980) and *Dysdercus intermedius* (Gaffal, K., 1981) bear about 12 chemosensory pegs on each lobe, but the pegs on the labial tips of aphids are only mechanosensory (Wensler, R., 1977).

The hypopharynx in most insects is a tongue-like median lobe that is situated behind the mouth and forms the floor of the cibarium. It is innervated bilaterally by a pair of hypopharyngeal nerves from the subesophageal ganglion in cockroaches, by a single median nerve from this ganglion in the termite *Calotermes*, and by a branch from each fronto-labral nerve in worker honeybees, *Apis mellifica*. If the sensory fields on the ventral wall of the cibarium in other insects with a proboscis are homologous with those on the hypopharynx of the honeybee, they are similarly innervated from the trito-cerebrum through the frontal or labral nerves as tabulated by Eaton, J. (1979). The hypopharyngeal sensory fields are well detailed for cockroaches, particularly those of *Blaberus craniifer* (Moulins, M., 1968, 1971a, b). They have two or three pairs of plates or fields of chemosensitive short pegs or papillae medially on the posterior half of the pharyngeal surface, and one or two pairs of lateral elongate fields of short hairs and campaniforms. The surface of the anterior part is devoid of cuticular sensilla, but has a subepidermal network of branched dendrites from multipolar neurons that are at the ends of each hypopharyngeal nerve and its anteriormost branches. The surface sensillar fields are innervated by more proximal branches of each trunk nerve. The hypopharyngeal sensory fields are basically similar in grasshoppers. In the worker honeybee (Galic, M. 1971), there are only two bilateral fields of very short putative chemosensory hairs; each field has about 46 sensilla. In mosquitoes there are usually four papillae on the ventral wall of the cibarium (Uchida, K., 1979) and in the moth *Trichoplusia ni* (Eaton, J., 1979) there are two pairs of sensory fields that contain a total of about 24

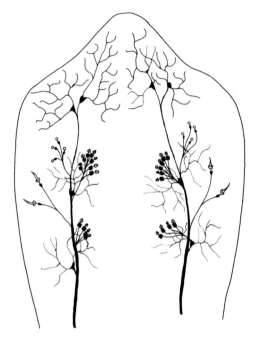

Fig. 9. Sensillar fields and innervation of an insect hypopharynx, dorsal.

papillae of two types on the floor of the cibario-pharyngeal pump. The basic sensory fields and their innervation in a typical insect hypopharynx are shown in Fig. 9.

2.3 Legs and Wings

The legs of insects are appendages of the thorax adapted primarily for walking and running. In specific groups they are modified for other functions also, such as grasping prey, swimming, burrowing or alighting. To monitor these functions they have extensive fields of diverse mechanosensilla on the surface and within for extero- and proprioception. Special organs on parts of the legs in certain insects are sensitive to sound. Discussion of all these mechanosensilla is deferred to Chapters 2, 8 and 9. Coverage here is limited to sensilla that respond to stimuli other than mechanical.

The legs were shown to be organs of taste in a few insects. This modality is concentrated primarily on the tarsi. Hairs or pegs with a structure and inner-vation typical of contact chemosensilla were noted in certain Diptera and Lepidoptera, in cockroaches, and in the honeybee. They are undoubtedly more universal in occurrence in these and other insect groups. Those of certain flies are the most intensively studied (Dethier, V., 1976). Several types of contact chemosensilla were characterized on the basis of structure, and mapped in *Stomoxys* (Adams, J. and Forgash, A., 1966), *Phormia terraenovae* (Hansen, K. and Heumann, H., 1971), and *Phormia terraenovae*, *Calliphora vicina* and *Musca domestica* (van der Wolk, F., 1978), among others. In these and other insects the chemosensilla are usually most numerous on the forelegs. Van der Wolk counted 304, 188 and 373, respectively, on the tarsi of each foreleg in the above three flies examined, and Adams, J. and Forgash, A. (1966) counted about 175, 150, and 120 per leg on the fore, mid, and hind legs, respectively, of *Stomoxys calcitrans*. The latter counts include about 15 chemosensilla on the tibia and one or two on the femur. Males of this species have more tarsal chemosensilla than do females; the reverse is true but to a lesser extent in *Phormia regina* (Wilczek, M., 1967). The hairs on the sides of the tarsi are longer than the medial bottom hairs, so that all tend to contact a flat substrate at the same time. All are also mechanosensory. The tarsi of mosquitoes (Pappas, L. and Larsen, J., 1976; McIver, S. and Siemicki, R., 1978) and blackflies (Sutcliffe, J. and McIver, S., 1976; McIver, S. *et al.*, 1980) also bear considerable numbers of contact chemosensilla of several types. Some also have a basal mechanosensory dendrite and others do not. One or two of these sensillar types may also be adapted for olfaction. The forelegs of butterflies are usually much reduced and not adapted for walking or alighting. They are used for tapping plant leaf surfaces, and may be primarily sensory. A few types of contact chemosensilla were identified on the legs of several butterfly species (Ma, W. and Schoonhoven, L., 1973; Calvert, H., 1974; Städler, E., 1978). Hairs that were examined internally also have a mechanosensory dendrite. The chemosensilla are implicated in tasting host plants primarily for their suitability as oviposition sites. Whitehead, A. and Larsen, J. (1976) noted several types of sensilla on the tarsi of the worker honeybee *Apis mellifera*. At least one of these has the internal structure of a typical gustatory hair with a dual contact chemo- and mechanosensory modality. Among the many sensilla that are on the legs of the orthopteran *Gryllus domesticus*, Fudalewicz-Niemczyk, W. *et al.* (1980) noted one type of small hair that is innervated by four neurons and could be chemosensitive. However, it occurred on all the leg segments except the tarsus, and its detailed structure and specific function need to be further clarified. In the proturan *Acerentomon majus*, where the first pair of legs are believed to carry on the function of the missing antennae, putative olfactory sensilla occur on the foretarsis (Dallai, R. and Nosek, J., 1981).

The sites of temperature reception on the tarsi of *Periplaneta americana* were found to be on the arolium and pulvilli by Kerkut, G. and Taylor, B. (1957), but specific sensilla were not identified. Reinouts van Haga, H. and Mitchell, B. (1975) localized temperature reception in two hairs on the tarsi of the forelegs of *Glossina morsitans*. Special infrared receptors are clustered in pits near the middle coxal cavities of the beetle *Melanophila acuminata* (Evans, W., 1966, 1975). The sensilla are interspersed by wax glands, which extrude wax fibres that form a mat over the pit. Thus, temperature is monitored by the legs in some insects; this may be more general, but the sensilla or receptor sites are for the most part poorly known.

Each leg is innervated by two nerves from the segmental ganglion. Both extend to the end of the tarsus, for the most part separately except for a short distance of fusion where they pass through the trochanter. Usually there are accessory ampullae at the bases of the legs that ensure adequate circulation of hemolymph into the legs for the nutrition of the sensilla and other tissues.

The wings of insects are sac-like extensions of the thoracic integument that are flattened into thin pads when fully formed. Primary longitudinal and smaller cross veins provide circular hemocoelic channels though which tracheae and nerves extend to the different parts of each wing. Accessory ampullae are reported to occur at the base of the wings in some insects. These are undoubtedly necessary for adequate circulation of nutrients in insects that have extensive sensillar fields on their wings. Most of the studies on number, types and distribution of wing sensilla are on Orthoptera. Albert, P. *et al.* (1976) provide some of this information for a grasshopper, and give references to some of the other studies. These are concentrated primarily on cockroaches. All the sensilla seem to be typical

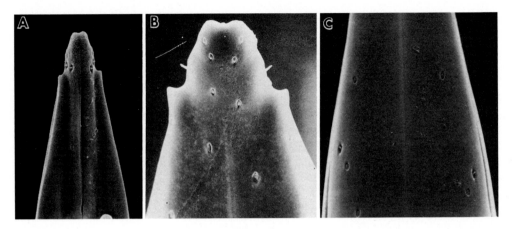

FIG. 10. SEM of sensory fields on the ovipositor tip of a dipterous plant parasite, *Urophora affinis*. **A**: ventral view (× 530); **B**: dorsal view (× 1350); **C**: field just below **B** (× 750).

mechanosensory organs that are involved in monitoring the mechanics of flight by extero- or proprioception. However, contact chemosensilla were identified on the wings of *Phormia regina* and shown to respond to NaCl and sucrose (Angioy, A. *et al.*, 1981). The dendrites of the cuticular sensilla are unusually short, and the hairs, pegs and campaniforms are arranged primarily along the veins. This is presumably for access to nutrients as well as for a rigid base. No studies are available on the internal fine structure of wing sensilla. These would be of great interest, especially with respect to their cellular mechanisms for nutrition and maintenance of function.

2.4 Genitalia and anal cerci

The genitalia in insects are composed of one or more modified terminal abdominal segments. Sensilla on male genitalia have received little attention, but those on the terminalia of the female ovipositor were studied in a few insects. All have mechanosensory hairs, pegs or campaniforms, and some of those examined are, or seem to be, organs of taste or olfaction also. The ovipositors of *Phormia regina* (Wallis, D. 1962) and *Musca autumnalis* (Hooper, R. *et al.*, 1972) bear bilateral sensory fields with several types of sensilla. Some of these were shown to be olfactory in the former and structurally resemble olfactory sensilla in the latter. On each of the paired lateral leaflets of the ovipositor of *Lucilia*

cuprina there are 14 tactile hairs in three size groups and one campaniform, two olfactory pegs and five contact chemosensitive hairs. Thus, the ovipositor in this insect functions as an organ of both taste and smell (Rice, M., 1976). In comparing the sensory fields on the ovipositor of *Psila rosae* and *Delia brassicae*, Behan, M. and Ryan, M. (1977) counted 110 presumably mechanosensory hairs in the former, and 245 such hairs in the latter along with 20 styloconic and four basiconic pegs believed to be chemosensory. The terminalia of both male and female *Aedes aegypti* bear several types of sensilla, all of which are mechanosensory (Rossignol, P. and McIver, S. 1977). The ovipositor valves of *Locusta migratoria* have six types of sensilla, at least one of which is a contact chemosensillum (Rice, M. and McRae, T. 1976). Some parasitic wasps can distinguish host species by probing and sensing with the ovipositor. Its sensillar field was examined in *Orgilus lepidus* (Hawke, S. *et al.*, 1973) and *Biosteres longicaudatus* (Greany, P. *et al.*, 1977). The tips of both ovipositors have mechanosensitive intracuticular campaniforms, and dome-shaped surface sensilla with a mechanosensory and chemosensory dendrites. There were no sensory cuticular projections seen above the surface of the ovipositor tips (i.e., Fig. 10). The sting and ovipositor of other Hymenoptera have greater numbers of more types of sensilla, including vari-sized hairs and pegs (Hermann, H. and Douglas, M. 1976). Some of those on the sting are presumed to sample the host chemically during probing.

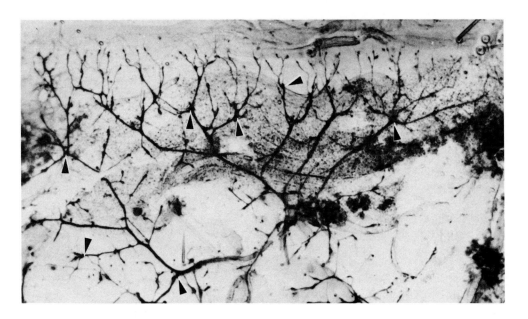

FIG. 11. LM, inside view, methylene blue whole-mount of the left half of an elaterid larval tergum with its arborized innervation of the sensillar fields; Type II nerve cell bodies are at arrowheads (× 90).

The typical cerci are a pair of true appendages of the last abdominal segment that have undergone considerable modification. Usually they are long, filamentous, and many-segmented, and are primarily sensory in function, but are modified for clasping or for other purposes in some insects. Most of the work on cercal sensory fields and sensilla has been with representatives of Orthoptera, especially crickets and cockroaches. In these insects there are many vari-sized hairs and campaniform organs distributed along the shaft. Their density is considerably greater on the ventral than on the dorsal side. One type of hair is long and filiform and, among the various sensory fields of insects, most common to cerci. The longer hairs, each with one or more campaniforms near the base, are mechanosensilla. In *Gryllus bimaculatus* (Schmidt, K. and Gnatzy, W. 1972) the cerci also bear many thick- and thin-walled short hairs, both of which have the structure of typical contact chemosensilla. Thus, cerci also serve as appendages for taste in this insect and probably in others also, i.e., in the praying mantid (Ball, E. and Stone, R. 1982). Kränzler, L. and Larink, O. (1980) found a variety of sensillar types in the extensive sensory field of the cerci of

Thermobia domestica. Some of these show a terminal pore in surface view, and are probably taste hairs also.

The cerci of *Periplaneta americana* are innervated by paired nerves from the last abdominal ganglion (Petryszak, A. 1975c). Each nerve forms four branches before it enters the cercus, and all extend to the tip. The two dorsal branches serve the dorsal sensilla, and the two much thicker ventral branches serve the more extensive sensory field along the ventral surface. There are no known reports of pulsatile ampullae for circulating hemolymph into the cerci in insects. In their study of the postembryonic development of cerci in *Acheta domesticus*, Edwards, J. and Chen, S. (1979) noted "only minor tidal movements" associated with abdominal movements in the cercal lumen, and wondered how the developing tissues within were nourished adequately. The nutrition of sensillar development during a molt and for maintenance of sensory function is similarly problematical in the postembryonic stages.

2.5 Other sensory fields

In addition to the varied and extensive sensory

fields that insects are endowed with on their body appendages, the integument of all parts of the body also bears various types of sensilla. Almost all of these are hairs of various sizes and shapes, and intracuticular campaniform organs. As far as is known, they are all mechanosensory. Their distribution is in fairly regular and predictable bilateral patterns and numbers in any one insect form. Each bilateral field is innervated by one of the bilaterally paired nerves from the corresponding segmental ganglion, which arborizes to all the sensilla in the field as exemplified in Fig. 11. The sensilla are most concentrated in the areas of joints and near the various intersegmental membranes, often as so-called hair plates or campaniform clusters, as detailed by Markl, H. (1962) for bees and other Hymenoptera. Within, there is an extensive network of multipolar nerve cells and their processes associated with the basement membranes of the epidermis. These cells and their processes are regularly associated with the nerve branches serving the cuticular sensilla. Some are implicated in mechano- or thermosensing, and others in neurosecretion (Finlayson, L. 1968). Scolopophorous organs or various types of stretch receptors are associated with joints, muscles and other moveable parts internally. Detailed discussion of these is outside the scope of this chapter.

A few body sensilla were reported that are not, or are believed not to be, solely mechanosensitive. The postantennal organ of Collembola is located in pits on the head near the antennae (Altner, H. and Thies, G. 1976). It resembles an olfactory sensillum in structure, but thermo- and hygro-sensitivity were other modalities suggested. The larva of the housefly, *Musca domestica*, has dorsal, terminal and ventral organs on cephalic lobes, which contain typical chemosensory and mechanosensory sensilla (Chu, I. and Axtell, R. 1971; Chu-Wang, I. and Axtell, R. 1972a,b). Similar organs are present in the fly *Erioischia brassicae* (Ryan, M. and Behan, M. 1973). These organs may be vestiges of the sensillar fields of the antennae and some of the mouthparts rather than cranial sensilla. Nymphs of Plecoptera have bulbous structures on the abdominal gills that may be involved in chemo- or osmosensing (Kapoor, N. and Zachariah, K. 1978). A more unique sensor was identified by Bitsch, J. and

Palévody, C. (1974) in the vesicular gland of the thysanuran *Machilis*. They speculated that it may be an osmo- or chemo-sensitive interoceptor. There are no other known reports of interoceptors with such modalities in insects. The setae noted by Lum, P. and Arbogast, R. (1980) in the spermathecal gland of female *Plodia interpunctella* have the fine structure of typical mechanosensory hairs.

3 SENSORY ORGANS

Two types of sensory nerve cells were identified by Zawarzin, A. (1912). Their fine structure and relationships were clarified since, and his classification still seems to be the most suitable today in the light of these (Fig. 12). His Type I neurons are bipolar and with a dendrite that has a ciliary structure and is for the most part unbranched. The extreme tip of its cilium is simple, branched or lamellate in different types of sense organs, almost all of which are associated in one way or another with the cuticle. His Type II neurons are bipolar or multipolar, with dendrites that arborize into many fine branches along their lengths and that typically do not have a ciliary structure. These neurons are generally associated with the epidermis, the epithelium of the alimentary tract or its muscles, other muscles and organs, and the peripheral nerves. Both types are ensheathed by glial cells to varying extents, but the Type I neurons typically have at least two other cells associated with them peripherally. They and their associated cells are of epidermal origin. The specific origin of the Type II neurons is still uncertain. The sense organs that we normally term sensilla in insects typically contain Type I neurons.

An insect sensillum is suitably defined as a sense organ that has one or more bipolar Type I neurons associated with cuticular parts or vestiges of these that extend above the surface of the cuticle or are within or beneath it, and the dendrite(s) of which are enveloped by at least two associated cells that form the cuticular parts or have other functions. Most workers agree that all insect sensilla evolved from an integumental hair or seta. Thus, the basic components of all sensilla should be homologous. The interpretation of homologies among sensilla with supra- or intracuticular parts is usually easy.

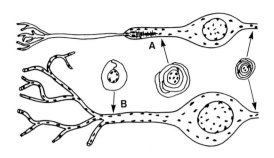

FIG. 12. **A**: Type I neuron with a terminally branched dendritic ciliary segment. **B**: bipolar Type II neuron with a small part of its terminal dendritic branches. Between are typical cross-section views of their dendritic and axonal cell sheaths.

The homologies are more difficult to see in sensilla that evolved to subcuticular positions, and in which the cuticular parts are lost or greatly modified and the relationships of the enveloping cells have changed significantly. However, most of the latter sensilla still maintain an association with the cuticle through an accessory cell if not directly. In accordance with the above criteria, the term sensillum should not be used for sense organs with Type II neurons. Most of the types of insect sensilla reported have either direct or indirect association with cuticle, and are considered here as cuticular. Reports of sensilla with no cuticular association are rare.

4 CUTICULAR SENSILLA

Historically, the various types of insect sensilla were identified on the bases of the form of their cuticular parts and their position on, within or under the cuticle. The classification for nine basic types was brought together by Snodgrass, R. (1926, 1935). Schneider, D. (1964), in his review of antennal sensilla, added another type, and two or three others have appeared in the literature since then. Except for the minor addition of types and some division of types into subtypes, the basic scheme of Snodgrass is very much in use today. It has a purely morphological basis, primarily as interpreted within the limits of resolution of the light microscope. Through the years workers continually felt the need to ascribe functions to the various types of sensilla. Initially this was by inference from structure,

position on specific parts or appendages of the body, or behavior of the insect after some form of incapacitation of specific sensory fields. More recently, functions of specific types of sensilla were identified electrophysiologically in some insects, and were inferred for other sensilla in these and other insects by a structural or distributional association with the known functional types. Inference of function from structure and position of sensilla was later corroborated electrophysiologically for some types of sensilla, but not for others. Inference by ultrastructural association with known functional types can be similarly erroneous, especially when the morphological knowledge is superficial.

The following is a descriptive summary of the nine basic types of sensilla elaborated by Snodgrass, and of some of the types that were added later. Their primary distinguishing features are outlined in Fig. 13. Included are some of their functions as known currently. The following reviews give additional details for most of these types and interpretations of subtypes: Slifer, E. (1961, 1970), Ivanóv, V. (1969), Sinoir, Y. (1969), McIver, S. (1975), Altner, H. (1977b), and Zacharuk, R. (1980a).

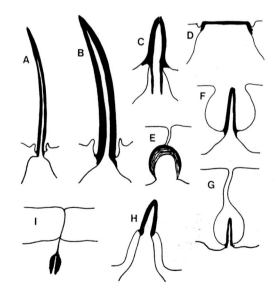

FIG. 13. Primary cuticular features of the various types of sensilla. **A**: trichodea; **B**: chaetica; **C**: basiconica; **D**: placodea; **E**: campaniformia; **F**: coeloconica; **G**: ampullacea; **H**: styloconica; **I**: scolopophora.

Sensilla trichodea are setiform hairs usually freely moveable on a basal membrane but with a variable basal insertion on the cuticle. They vary greatly in length, and their diameter is generally directly related to length. The thickness of their sidewalls and porosity varies with function. They are solely mechanosensitive, dually mechano- and contact chemosensitive, olfactory, or thermosensitive, with one to several neurons.

Sensilla chaetica are bristles or spines generally set in a socket, and are much like the trichoid hair type but are much heavier and have thicker walls. Most are generally regarded as tactile, but chemosensitive sensilla are identified with this type as well. They are also innervated by one or more neurons.

Sensilla basiconica are basically trichoid hairs that are much reduced in length and changed in form to peg- or cone-like, and some are identified as papillae. Their range of demonstrated sensitivities include hygrosensitivity in addition to those listed under trichoid hairs above. They are innervated by one to several neurons.

Sensilla coeloconica are basiconic pegs or cones that are set on the floor of relatively shallow depressions or pits in the cuticle. Although some have a neuron with a dendritic tubular body typical of mechanosensilla, they are most often reported to be chemo-, thermo-, or hygrosensitive.

Sensilla ampullacea are much like the coeloconic type but the pit in the cuticle is much deeper and often has a narrower opening to the surface; that is, it is more flask-shaped. Functions similar to those of the coeloconic sensilla are assumed for these sensilla also, but few are proven. These and the coeloconic sensilla generally have two to several neurons.

Sensilla squamiformia are innervated scales or, in other words, trichoid hairs that resemble scales in external form. Because of the gradations in form that occur between hair-like and scale-like, it is often difficult to identify these with one type or the other. Some are stated to be mechano- and others chemosensitive, and have one or more neurons.

Sensilla campaniformia are variously described as vesicles, papillae, domes, cupolae, sense pores, or umbrella or bell organs. The most common form of the innervated cuticular part is like a dome or cupola positioned at, or at various depths below, the surface of the cuticle and usually in association with the cuticular layers or lamellae. All are mechanosensors.

Sensilla placodea have a plate-like sensory cuticle that is usually level with the surface of the surrounding cuticle, or slightly raised or depressed, and attached to it by a ring of membranous cuticle. Those tested are chemosensory in function. They are usually innervated by several to many neurons.

Sensilla scolopophora are typically subcuticular sensilla with no supra- or intracuticular parts, but they usually maintain an attachment to the cuticle either through a vestige of cuticle or through an accessory attachment cell. They occur individually in different parts of the body, most often in the region of intersegmental joints, or in groups of two to many, forming composite organs such as the Organ of Johnston in the antennal pedicel. Each unit sensillum is typically innervated by one to three neurons. These are termed *sensilla scolopalia* by Schneider, D. (1964) and, when attached at both ends to monitor vibrations or sound, are commonly referred to as chordotonal organs.

Sensilla styloconica are pegs, cones or squat hairs inserted at the tip of a conical or cylindrical projection of insensitive cuticle. This type was included with the above list by Schneider, D. (1964), and is a common sensillum in Lepidoptera and occurs also in other forms. The terminal cuticular projections are innervated by one or a few neurons, and are assumed or proven to be mechano- or chemosensitive.

Descriptions of sensilla with novel or unique projecting cuticular parts continue to be added to the literature. Many of these are difficult to identify with any one of the above 10 types. The tendency is to name them as new types in an effort to classify them. Ernst, K. (1972a) proposed the type name *sensillum coelosphaericum* for an organ with a unique, ball-like sensory cuticle set in a coeloconic-like pit on the antennae of a beetle. Callahan, P. (1975) likened a sensillar cuticular projection on a moth antenna to a shoehorn, and Den Otter, C. *et al.* (1978) compared the form of a somewhat similar projection on the antenna of another moth to a rabbit's ear. The former typed it as *sensillum auricillicum* and the latter, *sensillum auricularium*, in their respective reports. Another sensory projection on the antennae of a cave beetle, which does not seem to differ greatly from the above two types in basic form and appearance, was identified as a *sensillum basiconicum inflatum* by Juberthie, C. and Massoud, Z. (1977). As is true for the earlier 10 types, the new additions are based on external form, and give little if any insight into their functional morphology or modality. A further proliferation of types based on the original criteria could encumber the classification scheme for insect sensilla and be of little significance.

There is no doubt that whether the sensory cuticle of a sensillum projects far out, little, or is flat on the surface, is in a shallow or deep pit or embedded among the cuticular lamellae, or is large, small or irregular in cross-sectional or surface area, all are significant to that sensillum in determining to some extent what stimuli it will be exposed to, or trap and conduct to its transducer mechanism. To this extent a generalized typing into one of the first 10 categories is informative. However, whether a sensory projection is a chaetal or trichoid hair, is long, medium or short or is a peg, cone or cupola, and whether the pit it is in is shallow or deep, are all very much subject to individual interpretation. It is imperative that comparative morphometric data are provided for the cuticular parts of sensilla under study to eliminate such individual biases, as proposed by Mayer, M. *et al.* (1981). With the recent advances in our knowledge of function and its correlation with structure in specific sensilla, there is now also a need and an information base for a general and simple classification scheme for insect sensilla that is related to function. This scheme should allow for future additions or subdivisions as warranted by advances in knowledge of functional morphology without major changes to the initial generalized framework. It should also allow for a continued use of the existing classification system to incorporate its descriptive values and to enable comparisons with the existing literature.

A simple typology that seems to meet these new criteria was proposed by Altner, H. (1977a) and was incorporated by Zacharuk, R. (1980a) in a review of chemosensilla. It is based on the premises,

which are supported by recent findings, that absence of conduction pores in the sensory cuticle precludes chemosensitivity, a single pore limits chemosensitivity primarily to taste, and many pores enable olfaction. It requires ultrastructural knowledge, which for some sensilla may be derived by scanning (SEM) but for most needs to be corroborated by transmission (TEM) electron microscopy. To be of value it also requires a knowledge of modality or specificity of response derived electrophysiologically that can be correlated directly with the ultrastructure of any given sensillum. Direct correlation will be assured if function is monitored first in a responsive sensillum, and the same sensillum is then characterized ultrastructurally. Until an adequate information base is amassed through such direct correlations, inferences by association could detract from any typology that is based on functional morphology. For the remainder of this chapter the typology of sensilla is based, in principle, on Altner's proposal, under the general categories *aporous*, *uniporous*, and *multiporous*. Elements of the existing typology are incorporated as appropriate.

4.1 Cuticular parts

The parts of an insect sensillum that are considered to be cuticular are those that are shed with the exuvium at a molt in a developmental stage, are homologous to these in the adult stage, or are vestiges that are interpreted as homologous even though they are not shed with the exuvium, as in some subepidermal sensilla. These include the external or intracuticular sensory cuticle, its socket or insertion in the surrounding insensitive cuticle, and the sleeves or sheaths that extend inward around the distal parts of the dendrite.

Most workers generally agree on the cuticular nature of the sensory cuticle and its insertion, whether it be a peg, hair, plate or dome. The layers of the sensory cuticle are continuous with those of the surrounding insensitive cuticle. In some sensilla the wall is solely epicuticular, with a thin, outer cuticulin coated with lipoidal secretions, and a thicker, amorphous inner protein epicuticle. In other sensilla, particularly those with thicker walls in the cuticular covering, there is an inner layer of well-sclerotized procuticle. The lipoidal pore canals are general in the body cuticle of most insects, and are usually evident in these insects in the walls of the sensillar cuticle, but are modified to varying extents. These canals usually contain pore filaments or tubules with a lipoidal wall and a non-lipoidal core (Zacharuk, R. 1971). In some sensilla the procuticle is demarcated from the protein epicuticle by swirled packets of these tubules or of the pore canals they are in. In most sensilla these pore canals open to the lumen within the sensory cuticle (Fig. 17). In some they extend downwards the length of the wall and open to the sinus under the sensory cuticle (Fig. 18).

The interpretation and terminology of the inward-projecting sleeve has been more problematic, due in part to an insufficient resolution initially of its relations to the terminal cuticular parts, the neuronal dendrites, and the accessory enveloping cells, within the limits of the light microscope. Terms most often applied to it are sense rod, scolops, scolopale, cuticular sheath, dendritic sheath, and "Stift" or "Stiftkörperchen" by German authors. It must be of a cuticular nature because it is shed at a molt, but its specific composition is not known. It is lengthened during a molt, but is not affected noticeably by the molting fluid (Zacharuk, R. 1962b). Certain bacteria rapidly degrade all the layers of procuticle but do not affect the sheath or the epicuticle (R.Y. Zacharuk, unpublished observations). In extensive examinations by TEM it resembles the superficial cuticulin layer in composition and density, and seems to be continuous with it in some uniporous sensilla and at molting scars. These observations suggest that it is an invagination of the surface layer of epicuticle — that is, its inner surface represents the outer surface of the sensory cuticle, and its composition should therefore resemble cuticulin. In sensilla that have a less dense outer sleeve around the terminus of the dense sheath, this would be an extension of the protein epicuticle or procuticle. The dense inner sheath, the equivalent of the scolopale, is referred to here as the dendritic sheath in keeping with its current use by some authors. However, it is associated only with the ciliary portion of the dendrites, and ciliary sheath may be more appropriate. The less dense but usually thicker outer sheath below the base of the sensory cuticle is termed the cuticular sleeve. It is obvious that the dendritic sheath, and possibly also the cuticular sleeve in some sensilla,

form a significant part of the stimulus transfer mechanism between the outer sensory cuticle and the enclosed dendrite in mechanosensilla. It may have some physical role in sensilla with other modalities, but this is not as obvious.

4.1.1 APOROUS

This category includes all sensilla that do not have a permeable pore in their sensory cuticle, but many have the typical cuticular pore canal system with pore filaments that extend through the protein epicuticle to minute pore openings at the surface. Representatives of all the 10 existing types of sensilla have this characteristic, although the placoid type probably are flattened cupolae and would be most subject to individual interpretation in distinguishing them from the campaniform type. The majority of aporous sensilla are mechanoreceptive. Some are known to be hygro-, xero-, or thermosensitive. Only one type is chemosensory (Altner, H. et al., 1977), but all may be involved in the so-called general chemical sense, that is, respond to a very strong concentration of a chemical in their environs, through the lipoidal pore canal system. Some of the variations in their cuticular parts and their homologies are shown in outline in Fig. 14, and in surface view in Fig. 15. Probably the only instance of a problematic homology is the dendritic cap (scolopale cap or "Stift") of the subepidermal scolopophorous sensilla. There is now little doubt that it is homologous with the dendritic sheath of other sensilla. Where it is embedded in the cuticle (amphinematic) it is shed at a molt, as in the mandible of an elaterid larva. Where its direct attachment with the cuticle is lost, and the connection is through an attachment cell (mononematic), it is not molted (Schmidt, K. 1974). In a few instances the dendritic cap has no attachment to the cuticle, direct or indirect, but is attached instead, through an attachment cell, to the epidermis (Chu-Wang, I. and Axtell, R. 1972a), to a basement membrane complex beneath the epidermis (Bloom, J. et al., 1981), or to the sensillar ganglion beneath the tip in the labium of an elaterid larva (Bellamy, F. 1973). In the last, it has lost all association with the integument, and is truly subepidermal. In all other structural aspects it resembles scolopophora that are attached to the cuticle. Given the above homology,

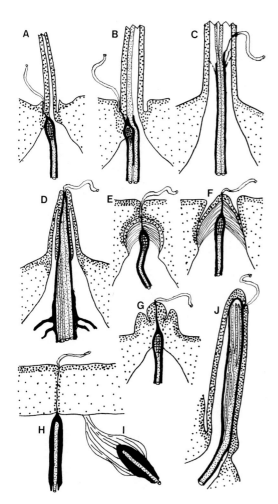

FIG. 14. Terminal cuticular parts and dendritic associations in various aporous sensilla, with exuvial dendritic sheath attached to the sealed molting pore or scar. **A**: mechanosensory hair; **B**: mechano- and putative chemosensory hair; **C**: putative chemosensory hair; **D**: thermo- and hygrosensory peg; **E–G**: cupola, papillate cone, and "ball-on-tee" campaniform organs; **H**: amphinematic and **I**: mononematic scolopophorous organs; **J**: palpal digitiform sensilla of Coleoptera.

pulled out or broken off when the exuvium is cast off. This results in a molting scar on the surface of the new sensory cuticle. In a surface view by SEM it may appear as a distinct pore when void of any vestige of the old sheath, or as an extrusion from a pore when a vestige of the old sheath remains (Fig. 15). In both cases this could easily be interpreted as evidence of a permeable pore that is open or extruding a liquor, and the sensillum can easily be mistaken for a uniporous type without further corroborative study. Also, new sensilla are added at each molt and, in holometabolous insects, most of the sensilla may be reorganized as "new" sensilla through the pupal stage. These sensilla would have no cuticular parts to shed, and should therefore have no molting pore or scar. Thus, one insect specimen may well have a number of sensilla of an identical structural and functional type, and some of these will have a molting pore and others will not. Thus, SEM evidence alone is of little value in distinguishing sensilla to functional–morphological types, particularly aporous from uniporous.

The majority of the aporous sensilla are mechanosensitive. For further discussion of their cuticular parts see McIver (this volume). Three types are identified that do not have the typical mechanosensory morphology: trichoid hairs with one or two dendrites in the shaft, short pegs or hairs with dendrites from several neurons that terminate within the shaft or beneath the base, and the digitiform hairs that are most common to Coleoptera. These are detailed further here.

Two slightly different examples of trichoid hairs that are each innervated by two neurons are outlined in Fig. 14b and c. The first was described from the antennal tip in mosquito larvae, and is rather long for the small size of the antenna. In *Aedes aegypti* (Fig. 16) one dendrite terminates with the tip of the dendritic sheath at the base of the hair in a tubular body, and the other extends naked from the base where it exits from the sheath to the tip of the hair appressed to one side of the lumen (Zacharuk, R. and Blue, S. 1971a). It has no tubular body. In *Toxorhynchites brevipalpis* the innervation is similar, but the partly exposed dendrite has a basal tubular body (Jez, D. and McIver, S. 1980). In the former a dual modality was presumed, mechanical by the basal dendrite and chemical by the dendrite in the shaft through the lipoidal pore filament

either the term scolopale should be used in place of dendritic sheath in all sensilla, including the various kinds of chemosensilla, or the term dendritic sheath (or cap), or ciliary sheath (or cap) should be applied universally.

During a molt from one stage to another, the dendritic sheath that is attached to the exuvial cuticle is pulled through the forming new cuticle of the sensillum at the point of attachment of the replacement sheath. The last vestige of the old sheath is

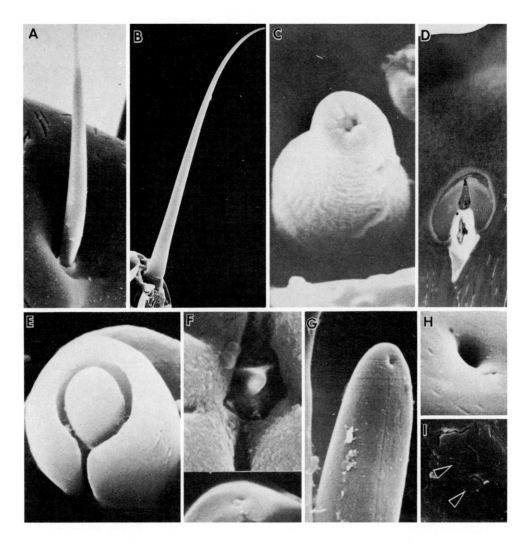

Fig. 15. SEM (except **D**) of some of the sensilla shown in Fig. 14. **A**: hair in 14A ($\times$ 590); **B**: hair in 14C ($\times$ 750); **C**: peg in 14D ($\times$ 15,000); **D**: campaniform in 14E (TEM, $\times$ 1500); **E**: campaniform in 14G ($\times$ 10,100); **F**: campaniform in 14F ($\times$ 9500), with apical molting scar in lower inset ($\times$ 18,500); **G**: digitiform organ in 14J ($\times$ 12,000); **H**: surface pore of a campaniform organ as in 14E and in **D** above ($\times$ 11,400); **I**: surface cuticular indications of points of insertion of two amphinematic scolopophorous sensilla as in 14H, at arrowheads ($\times$ 2100). (**B** and **C** are from Bloom, J. *et al.,* 1982 a and b, **G** from Zacharuk, R. *et al.,* 1977; and **I** from Zacharuk, R. and Albert, P. 1978, respectively; by courtesy of the authors and The National Research Council of Canada.)

system in the epicuticular wall. In the latter, both dendrites were presumed to be mechanosensitive. Both assumptions should be withheld and await functional proof. The molting pore was not identified in both larvae. In the larva of *Tenebrio molitor* (Fig. 17) both dendrites extend into the hair shaft within the sheath and exit it about one-third of the way up to extend naked for a short distance in the shaft. The molting scar is near the tip of the sheath

in the sidewall (Bloom, J. *et al.,* 1982a). The hair is smooth-walled but the sidewalls are traversed by a typical lipoidal pore filament system. The function is not specified, but its wall resembles that of the purported aporous hair on the antennae of *Periplaneta americana*, with two branched dendrites, that responds to alcohols (Altner, H. *et al.,* 1977). They suggest that the alcohol conduction mechanism through the hair wall to the dendrites is

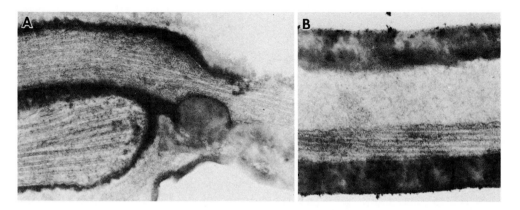

FIG. 16. TEM, l.s., of a hair as in Fig. 14B. **A**: base of hair where one dendrite terminates in a tubular body and the other exits the sheath into the hair shaft (× 40,100); **B**: the naked dendrite appressed to the wall in the hair shaft (× 30,300). (From Zacharuk, R. and Blue, S. 1971a, courtesy of the authors and by permission of The Wistar Institute Press.)

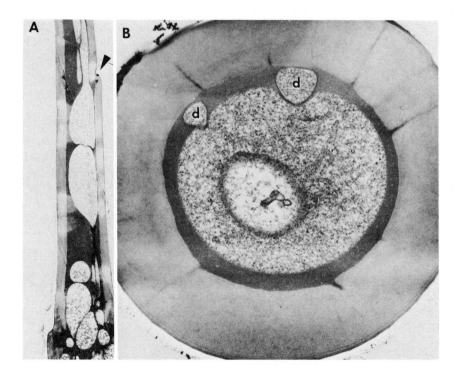

FIG. 17. TEM of a hair as in Figs 14C and 15B. **A**: l.s. of basal two-thirds with ballooned dendrites (a fixation artifact) in the hair shaft naked beyond the sealed molting pore at arrowhead (× 3250). **B**: t.s. beyond the molting pore with the two naked dendrites (d), and lipoidal pore filaments in the wall of the hair (× 29,000). (From Bloom, J. *et al.,* 1982a, by courtesy of the authors and The National Research Council of Canada.)

the cuticular pore tubule or pore filament system, which is typical of the sidewalls of most purely tactile hairs also. The difference is that hairs that are solely mechanosensitive typically have a dendritic tubular body, and the dendrite is always completely enclosed and separated from the lymph in the surrounding sensillar sinus. In the type of hairs under discussion, one or both dendrites are open to the lymph in the lumen of the hair shaft. In the aporous hair of *Periplaneta*, however, the exposed dendrites are branched, which is typical of olfactory sensilla. From the descriptions given, all of these hairs are aporous by SEM, all have one or both dendrites exposed outside the dendritic sheath by TEM, and one report gives evidence for chemosensitivity. As suggested by Bloom and co-workers, there may be a need to create a distinctive subtype within the aporous category for this sensillar form. Before this is considered, the function of the larval mosquito and beetle hairs will have to be established, the surface of the *Periplaneta* hair should be re-examined for porosity in thinly coated specimens by SEM, and similar complete information should be obtained for this type of hair in other insect forms.

Aporous and primarily coeloconic pegs were determined to be hygro- and thermosensitive in *Periplaneta americana* (Yokohari, F. 1981), *Apis mellifera* (Yokohari, F. *et al.*, 1982), *Carausius morosus* (Altner, H. *et al.*, 1978), *Aedes aegypti* (Davis, E. and Sokolove, P. 1975; McIver, S. and Siemicki, R. 1979), and *Locusta migratoria* (Altner, H. *et al.*, 1981). Similar sensilla were noted in the bedbug, *Cimex lectularius* (Steinbrecht, R. and Müller, B. 1976), the hymenopteran *Neodiprion sertifer* (Hallberg, E. 1979), in aphids (Bromley, A. *et al.*, 1979), and in the larva of *Tenebrio molitor* (Bloom, J. *et al.*, 1982b). These were presumed to have similar modalities by association. In the last-named species, Roth, L. and Willis, E. (1951) determined that the larvae are hygrosensitive and that the sensors are on the antennae, and it is most likely that the blunt peg described by Bloom and co-workers is the one involved. Its structure is basically similar to that of the characterized sensilla above. The micrographs of it, which are at hand, are used here to illustrate the cuticular parts and dendritic associations in this type of sensillum (Figs 15c and 18). Some mosquitoes have similar pegs in ampullae as well as in coeloconic pits.

In this type of sensillum in all the above insects the dendritic sheath lines the lumen to the apex where it is closed. The molting pore is at the apex and, where seen in section, contains a plug similar in nature to the dendritic sheath. This pore is large and distinct in some of the insects, but is hidden by terminal flutes or finger-like cuticular projections in others. If the surface of the sidewalls has grooves,

these are very shallow. They are innervated by two to four neurons. In some insects the ciliary dendritic segments of all four dendrites extend to the apex of the peg, but in most of the above, only one or two do so. The other ciliary segments terminate at or below the peg base. The dendrites that extend into the peg are typically packed with longitudinal microtubules and completely fill the lumen. In most of the above insects, one of the sub-basal dendrites is lamellate. In *Tenebrio* and *Carausius* there is in addition an even shorter ciliary segment from one of the four neurons that maintains the $9 \times 2 + 0$ microtubule configuration along the entire length, resembling the structure of the equivalent segments in mononematic scolopophorous sensilla. All four ciliary segments in *Periplaneta*, which extend to the tip of the peg, somewhat resemble the equivalent segments in amphinematic scolopophorous sensilla. There is usually some elaboration of the cuticle around the base of the peg, with slitted cuticle and a fibrous matrix beneath the insertion of the peg. In *Tenebrio* there is also an elaborate pore canal and pore tubule system oriented longitudinally in the wall of the peg. The canals open to, and the tubules extend into, the sensillar sinus beneath the base. There are also elaborate projections from the dendritic sheath into this sinus near this region. The skeletal rod complex (scolopale rods) in the inner sheath cell is heavy, also resembling that of scolopophorous sensilla.

Where it was tested electrophysiologically, this type of sensillum responded to one or both of the following changes in both temperature and humidity: increasing temperature (hot), decreasing temperature (cold), increasing humidity (moist), and decreasing humidity (dry). The mechanism of response to changes in temperature is uncertain, but there is some evidence to support an assumption that the neuron with the lamellate dendrite is involved. This will be elaborated further in the discussion of sensory neurons. The response to humidity is believed to be by a mechanosensory mechanism. Altner, H. *et al.* (1981) suggest that the cuticle in the peg and at its insertion, or a substance associated with this cuticle, is hygroscopic, and undergoes physical changes with changes in humidity. These physical changes in the cuticle lead to a mechanical deformation of the dendrites associated with the peg to initiate the response. Yokohari, F. (1981)

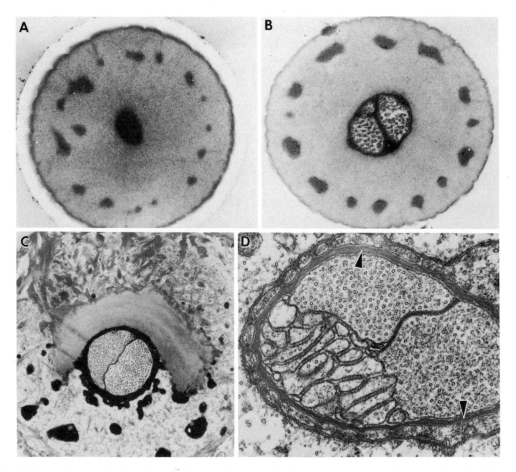

FIG. 18. TEM, t.s. of a peg as in Figs 14D and 15C, near the tip through the plug in the central apical pore in **A** (× 43,500); through the dendritic tips in the chamber below the plug in **B** (× 34,500), with the peripheral longitudinal pore canal system of the wall seen in both; just below the base of the peg in **C** (× 15,400); and at the proximal end of the dendritic sheath (arrowheads), where a third lamellate dendrite terminates, in **D** (× 38,800). (From Bloom, J. *et al.*, 1982b, courtesy of the authors and The National Research Council of Canada.)

proposes that the mechanosensing mechanism may be basically similar to that operative in scolopophorous sensilla, given the several structural similarities. How the truncate and little modified basal cilium is involved, where present, is uncertain. Bloom, J. *et al.* (1982b) raise the possibility that the pore tubule system of the peg's cuticular pore canals may be a conduction system for moisture to the hygroscopic cuticle or the hygroscopic substances associated with it.

The digitiform sensilla that were characterized by Zacharuk, R. *et al.* (1977) in an elaterid larva and by Guse, G. and Honomichl, K. (1980) and Honomichl, K. and Guse, G. (1981) in several

species of adult Coleoptera, by SEM and TEM, are more problematic with respect to function. They normally lie in longitudinal grooves in the outer walls of the maxillary and labial palps, in the distal segment, but their tips are everted by hemocoelic pressure in elaterid larvae. In some adult beetles they are so enclosed that they physically cannot be everted. A single neuron innervates the sensillum. Its ciliary dendritic segment extends to the tip of the hair encased in a dendritic sheath that is closed at the apex of the hair lumen. There is an obvious molting pore near the tip in all the larval sensilla and in some of those in the adult (Fig. 15G), but it is plugged with a dense material similar to that of the dendritic sheath. The dendrite is branched within the hair and has connections to the dendritic sheath in places. A lumen extends alongside the ensheathed dendrite the length of the hair in the elaterid larvae and some adult beetles, but not in other adult species. In electrophysiological monitoring, Zacharuk and co-workers noted a burst of impulses when the hair was moved, and a reduced but regular discharge to vibration. The latter could be maintained for several hours. However, without an adequate anti-vibration base, the specific modality for the sustained response was not determined with certainty. Honomichl and Guse suggest thermo- or hygrosensitivity or response to CO_2 as possible modalities for this sensillum in the adult beetles. It is the only sensillum with a terminally branched dendrite that is known to be mechanosensitive, but its response to other modalities should be tested further in preparations suitably isolated from vibrations. The sensillum in elaterid larvae did not respond to a variety of chemicals applied to it.

One other type of aporous peg occurs at the tip of the antennae in the larva of the mosquito *Toxorhynchites brevipalpis* (Jez, D. and McIver, S. 1980); it is innervated by two neurons. One dendrite extends to the tip of the peg where it forms a few folds, and the other terminates in a lamellate form below the base. Functions related to it by association are thermo- or proprioceptive chemosensitivity. On the antennae of larval *Aedes aegypti* a peg, seemingly homologous to the above, is innervated by only one neuron and the peg is purported to have a terminal pore. The dendrite terminates at the apex in a few folds, but the cilium is multibranched and swirled in an exceptionally large, fluid-filled ciliary sinus just above its insertion into the proximal dendritic segment (Zacharuk, R. and Blue, S. 1971b). The function suggested for it is osmosensory. This sensillum should be examined further in more larval mosquito species for clarification of structure and function.

4.1.2 UNIPOROUS

Sensilla with uniporous sensory cuticles have an external appearance of hairs that are usually short to medium-long, pegs, papillae, small plates, or simply pores in a cuticular depression. All have in common one permeable pore that is at or near the tip of extended sensory cuticles, or central in plates. This

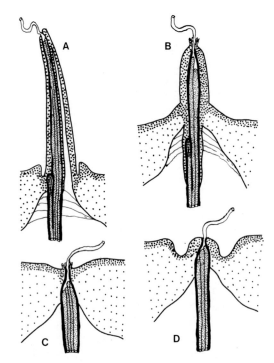

FIG. 19. The cuticular parts, dendritic associations and the open molting pore for the exuvial dendritic sheath of four types of uniporous sensilla. **A, B**: mechano- and chemosensory hair and pegs; **C,D**: chemosensory plate or pore and cone or papilla.

opening also serves as the molting pore, which is not closed by an impermeable plug after the exuvial dendritic sheath is shed (Fig. 19). Some hairs and pegs have an apical sculpturing around the pore. This varies from simple grooves towards the pore to elaborate finger-like cuticular projections that surround the opening and hide it from surface view. In others the pore is a simple hole with no obvious surrounding sculptures (Fig. 20). The latter condition is most usual in papillae and plates, but there are instances where obvious cuticular projections surround the central pore in these also. Some examples of uniporous sensory cuticles are shown in surface view in Fig. 21. In such surface scans the uniporous short hairs and pegs can be easily confused with the aporous temperature–humidity pegs, as can the uniporous sunken domes and cones be confused with similarly-shaped campaniform sensilla if the molting scars are pore-like in the latter sensillar types.

The shape of the pore varies from round to simple or stellate slits. Their width varies from 0.01 to 0.2

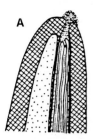 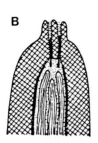

FIG. 20. Dendritic terminations and pore structure in two types of uniporous sensilla. **A**: a two-chambered hair with a simple pore opening; **B**: a one-chambered peg with finger-like sculptures around a slitted pore opening. (From Zacharuk, R. 1980, by permission of Annual Reviews Inc.)

μm when measured in TEM sections, and this depends on their form to some extent. The pore, whether simple or surrounded by shallow or elaborate cuticular sculptures, is usually apical or just subapical. The above variations are generally most pronounced between diverse insect forms, but they are also present among similar sensilla in a specific field in one insect. For example, in the flies *Phormia terraenovae*, *Calliphora vicina*, and *Musca domestica*, van der Wolk, F. (1978) identified three morphological types of uniporous hairs in the tarsal fields, the so-called A, B and D hairs. All three were present in *Phormia*, but only the B and D hairs were found in the latter two species. The pore in the A hair is rhombic and subapical. In the B hair it is oval and apical. In the D hair the pore is rectangular and near the apex of a terminal micropapilla beneath an undulated cap. The diagonal width in types A and B is 0.2–0.25 μm and in type D it is 0.1–0.15 μm. At least as many functional types were identified electrophysiologically in blowflies and mosquitoes by other workers, but how this relates to the type of pore is uncertain. A more unusual pore location in a uniporous sensillum was reported by McIver, S. *et al.* (1980) in a bifurcate hair on the tarsi of female *Simulium venustum*. It is on the flared base of the two flattened lobes, with a short groove leading to it; that is, it is on the side of the hair and considerably below the tips.

The fine structure of the pore is often difficult to resolve and interpret in sclerotized cuticles because of problems in obtaining thin sections through it. In suitable preparations the dendritic sheath is fused to the inner wall of the sensory cuticle at the base of the pore. In simple pores it is obviously continuous with the surface cuticulin layer through the pore. Some or all the dendrites from the innervating neurons extend within the dendritic sheath towards the tip, and at least one or two of them terminate at or just below the base of the pore. In addition to this dendritic chamber, uniporous hairs usually contain a second chamber that extends the length of the hair to near the tip. The dendritic chamber is often appressed to the wall along one side of the hair along much of its length, especially distally, and is separated from the other chamber by a thin cuticular septum as well as by the continuous dendritic sheath. Such hairs are referred to as two-chambered. The second chamber is continuous with the sensillar sinus basally, and is sometimes referred to as the sensillar chamber or channel. From it pore tubules or filaments extend through the wall to the surface of the hair. These are presumably the cuticular lipoidal pore canal system, which sequesters substances from the lymph in the sensillar chamber and deposits them on the surface. Extensive extrusions of a viscous substance occur through the lateral walls in the labellar and tarsal hairs of *Phormia regina* (Angioy, A. *et al.*, 1980), probably through this pore tubule system. Such pore tubule channels are not distinct between the two chambers in the hair. The wall between them is presumed to be impervious to larger molecules, but may be permeable to certain small ions (Broyles, J. *et al.*, 1976; Matsumoto, D. and Farley, R. 1978). In other uniporous sensilla the dendritic chamber extends to the tip through the middle and is surrounded by the sensillar chamber along much of its length.

Reports of extrusion of a viscous substance through the pore at the tip of the hair are common (i.e., Moulins, M. 1971b; Hansen, K. and Heumann, H. 1971; Dethier, V. 1972). It is believed to be a mucopolysaccharide from the dendritic chamber. This substance can be easily extruded by gentle pressure and viewed under a light microscope (Slifer, E. *et al.*, 1957). Certain ions freely leach out of the dendritic chamber through the pore and its contents (Broyles, J. *et al.*, 1976). In addition to this substance several workers report the presence of a more structured matrix in the pore. In the labellar hair of *Calliphora erythrocephala* it is sieve-like (Stürckow, B. 1971), in the styloconic peg on the maxillary palp of *Pieris brassicae* larvae it is

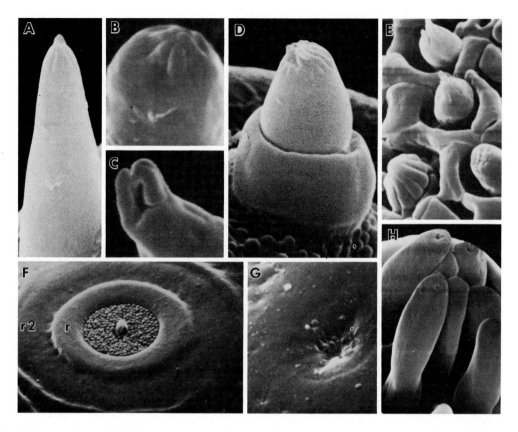

Fig. 21. SEM, cuticles of several types of uniporous sensilla. **A**: pointed-tipped peg with a grooved apical pore (× 9200); **B**: blunt-tipped peg with a slitted apical pore with finger-like projections closed around it (× 32,500); **C**: peg as in **A** with a simple apical pore (× 29,000); **D**: styloconic peg with apical pore as in **B** (× 9800); **E**: coeloconic pegs with variable apical sculptures as in **B** (× 3400); **F**: plate with a slitted central pore surrounded by a tuft of cuticular fingers (× 7000); **G**: pore as in **F** in a simple cuticular depression (× 4200); **H**: peg cluster in a pit with simple apical pores, as in Fig. 19A (× 3400). **A**, **B**, **D** and **E** are as in Fig. 19B; **F** and **G**, are as in Fig. 19C. (**A**, **C** and **F** are from Bloom, J. *et al.*, 1982a,b, by courtesy of the authors and The National Research Council of Canada.)

fenestrated (Ma, W. 1972), and in a similar peg on the galea of *Mamestra brassicae* larvae the pore contains many small filaments that extend inward towards the dendrites apparently from the dendritic sheath lining the pore (Gaffal, K. 1979). In the more sculptured tip of a peg on the labial palp of an elaterid larva, where the outer part of the pore consists of stellate slits, typical pore tubules extend inward from the base of the slits into the dendritic chamber (Bellamy, F. 1973). The viscous extrusion presumably fills the pore continually and serves as the conduction medium for chemical stimulants from the surface to the dendrites in the chamber beneath. The other structures that occur in the pores of some sensilla but not others may aid in the conduction of certain chemicals and not others,

thus conferring a selectivity to the conduction mechanism and a specificity of response to the sensillum. It could be the basis, in part, for the observed differences in specificity of similar uniporous sensilla in corresponding fields between sexes, as in the nectar-feeding males and blood-feeding females of mosquitoes (Lewis, C. 1972) and in the larvae of various species of the lepidopteran *Yponomeuta* (van Drongelen, W. 1979). However, this is purely an assumption at present. Whether the conduction mechanism is involved in the specificity of uniporous sensilla, and to what extent, must await specific studies on function and response correlated with adequate information on fine structure for an answer.

Uniporous sensilla are usually innervated by four

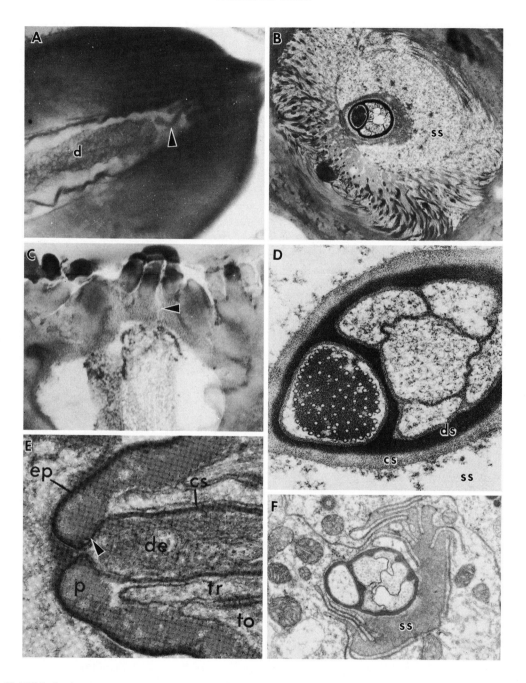

Fig. 22. TEM of uniporous sensilla. **A**: l.s. of tip of peg as in Fig. 21B with pore tubules (arrowhead) extended from a pore slit towards the dendrite (d) beneath (× 46,000); **B**: ensheathed dendrites, t.s., in the sensillar sinus just below the base of a peg as in Figs 19A and 21A (× 12,000); **C**: dendrites below a slitted pore (arrowhead) in a sensillum as in Figs 19C and 21F,G, l.s. (× 17,800); **D**: dendrites as in **B** enlarged, with the mechanosensitive tubular body separated from five chemosensory dendrites within the dendritic sheath (ds) which is enclosed in a cuticular sleeve (cs) (× 63,700) **E**: l.s. of a papillate sensillum as in Fig. 19D, with the dendritic sheath (cs) around the dendrite (de) continuous with the cuticulin layer of epicuticle (ep) of the papilla (p) (arrowhead) and enclosed by an inner (tr) and outer (to) sheath cell (× 60,000); **F**: t.s of a sensillum as in **B** near the base of the sensillar sinus below the end of the cuticular sleeve (× 12,000); ss, sensillar sinus. (**A** is by courtesy of F.W. Bellamy and the University of Regina; **B, D** and **F** are from Bloom, J. *et al.,* 1982a, by courtesy of the authors and The National Research Council of Canada; **E** is by permission of the authors Wensler, R. and Filshie, B., 1969 and The Wistar Institute Press.)

to six neurons, but this is variable within the wider range of one to ten between different types of sensilla in different insects and between similar sensilla in the same field in one insect. The hairs and many of the pegs have a dual chemo- and mechanosensitivity. In these the dendrite from one of the neurons becomes separated from the others by a septum within the dendritic sheath below the base, is closely appressed to the sheath along its terminal length, and terminates in a typical mechanosensory tubular body at the base of the peg or hair. It inserts through its dendritic sheath into the base of the sensory cuticle. The dendrites from the other neurons extend in the separate larger dendritic chamber toward the apical pore of the sensillar cuticle, and are presumed to be chemosensory. In the cockroach *Blaberus craniifer*, Lambin, M. (1973) noted uniporous hairs on the antennae with only one or two neurons. In the former the single dendrite always extended into the dendritic chamber toward the apical pore, and these hairs were not mechanosensitive. In the latter, one dendrite extended toward the apical pore and the other terminated in a typical tubular body at the base, and these hairs had the dual modality. The chemosensory dendrites are usually not compartmented from one another in their dendritic chamber as the mechanosensory dendrite is from them, although incomplete septa between them are common. However, in *Tenebrio molitor* one or more of the chemosensory dendrites are compartmented by the dendritic sheath at the base of the sensory cuticle (Harbach, R. and Larsen, J. 1977), and in the termite *Schedorhinotermes putovius*, all the dendrites are compartmented from one another at the base but not in the hair shaft (Quennedey, A. 1975). In the larva of the mosquito *Aedes aegypti* one of the four dendrites becomes apposed to the dendritic sheath and partly enclosed and separated from the others, but extends with the others into the shaft of the peg to terminate about midway along its length still open to the dendritic chamber, but in a groove in the sheath and tightly apposed to it (Zacharuk, R. and Blue, S. 1971a). At the base of the hair it atypically resembles a tubular body, in this way differing also in structure from the other dendrites. At the tips of the palps where the sensilla are on a flexible cuticle, some are mounted on top of a cylinder of hard cuticle through a flexible surface membrane and inner suspensory fibers.

This special insertion mechanism seems to be related to the conduction of the mechanical stimulus to the dendritic tubular body in the base in an otherwise freely flexible surrounding membrane (Klein, U. and Müller, B. 1978). The uniporous plates and papillae, and some of the pegs, generally do not have a neuron with a dendritic tubular body and are only chemosensory.

Some of the features of uniporous sensilla described above, as viewed by TEM in a few sensilla, are illustrated in Fig. 22. Some of the reports and sensilla on which the preceding is based, with a summary of a few of their characteristics, are listed in Table 3. Hansen, K. and Heumann, H. (1971) similarly summarize some of the earlier literature in tabular form.

The chemosensitivity of the uniporous sensilla is generally regarded to be through contact with chemicals in solution. Thus, they are most often termed contact, gustatory or taste sensilla. However, they were shown to respond also to vapors or olfactory stimuli in *Phormia* (Dethier, V. 1972) and the larva of *Manduca sexta* (Städler, E. and Hanson, F. 1975). In the bifurcate hair of the blackfly *Simulium venustum*, McIver, S. *et al.* (1980) noted that the exudate from the lateral pore spread out over the flattened surface from the groove. They suggest that this increases the adsorptive surface area for trapping stimulant molecules in a uniporous sensillum, and that this may be an instance of this type of sensillum being modified for an olfactory rather than primarily a gustatory function.

4.1.3 MULTIPOROUS

Sensilla that are considered here as multiporous are those that have many obvious pores in their sensory cuticle when examined in surface view by SEM and corroborated in sections by TEM. They are usually hairs, pegs or various forms of plates. The latter are often modified into a variety of forms in different insect groups. Altner, H. (1977a) referred to these as wall pore sensilla, and differentiated two basic types: single-walled sensilla with pore tubules, and double-walled sensilla with spoke canals. In surface view the former is pitted and the latter is grooved longitudinally, and most of the multiporous sensilla (MP) can be easily identified as one type or the other

Table 3: Some characteristics of uniporous sensilla in various insects

Insect	Location	Type	Pore type, size (μm)	Neurons	Reference
Diptera					
Aedes aegypti	tarsi	hair C1		5C	McIver, S. and Siemicki, R.
		C2		4C, 1M	(1978)
		C3		4C	
Aedes aegypti larva	antenna	peg	slits	3C, 1?	Zacharuk, R. and Blue, S.
					(1971a)
Culiseta inornata	tarsi	hair		4C	Pappas, L. and Larsen, J. (1976)
	labellum	hair T1		4C, 1M	
		T2		2C, 1M	
		T3		2–5?	
Phormia regina	labellum	hair	simple	4C, 1M	Felt, B. and Vande Berg, J.
					(1976)
Phormia terraenovae	tarsi	hair	grooves, 0.1	4C, 1M	Hansen, K. and Heumann, H.
					(1971)
Stomoxys calcitrans	labellum and tarsi	hair	sculptured, 0.05–0.2	4C, 1M	Adams, J. *et al.* (1965)
Musca domestica larva	terminal	papilla	simple	2C, 1M	Chu-Wang, I. and Axtell, R.
					(1972a)
	ventral	plate	simple	2C	Chu-Wang, I. and Axtell, R.
					(1972b)
Lepidoptera					
Danaus gillipus	antenna	long hair		4C, 1M	Myers, J. (1968)
Mamestra brassicae larva	galea	peg	simple, 60 nm^2	4C, 1M	Gaffal, K. (1979)
Pieris brassicae	tarsi	hair	simple	4C, 1M	Ma, W. and Schoonhoven, L.
					(1973)
Orthoptera					
Blaberus craniifer	cibarium	papilla	simple, 0.1–0.2	2–5C	Moulins, M. (1968, 1971b)
	antenna	long hair	simple	2–5C, 1M	Lambin, M. (1973)
Gryllus bimaculatus	cerci	thin hair	simple, 0.35	4–5C, 1M	Schmidt, K. and Gnatzy, W.
		thick hair	0.2	1–2C, 1M	(1972)
Grasshoppers, 3 spp.	antenna	long peg	simple	4–6C	Slifer, E. *et al.* (1957)
Locusta migratoria	Max. palp	peg	slits on crest	6–10C	Blaney, W. *et al.* (1971)
	cibarium	papilla	simple	5C	Cook, G. (1972)
Other insect groups					
Brevicoryne brassicae	cibarium	papilla	simple	3–5C	Wensler, R. and Filshie, B.
					(1969)
Cimex lectularius	antenna	hair A1		4C, 1M	Steinbrecht, R. and Müller, B.
					(1976)
Dysdercus intermedius					
adult	labium	chaetica	grooved, 50 nm^2	5C, 1M	Gaffal, K. (1979)
		basiconic	fluted 15 nm^2	2–3C, 1M	
larva	antenna	chaetica	simple, 4–8 nm^2	4C, 1M	
Lygus lineolaris	labium	peg	simple	3 or 5C	Hatfield, L. and Frazier, J.
					(1980)
Leptinotarsa decemlineata	palps and	peg	sculptured	C	Mitchell, B. and Schoonhoven,
larva	galea	peg	simple	C	L. (1974)
Tenebrio molitor	antenna	chaetica	simple, subapical	5C, 1M	Harbach, R. and Larsen, J.
					(1977)
Orgilus lepidus	ovipositor	dome	simple, 0.08	4–5C, 1M	Hawke, S. *et al.* (1973)
Schedorhinotermes putovius	labrum	long hair		2–5C, 1M	Quennedey, A. (1975)
soldier	labrum	short hair		4C, 1M	
Campodea sensillifera	labium	hair	simple	6–9C, 1M,	Bareth, C. and Juberthie-Jupeau, L. (1977)

nm^2 under pore size is cross-sectional area;
C neurons have dendrites in the channel open to the pore; M neurons have a tubular body inserted at the base.

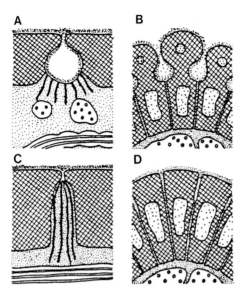

FIG. 23. Detail of the pore canal systems in the sensory cuticles of thin-walled (**A**) and thick-walled (**C**) MPP, and distinctly (**B**) and indistinctly (**D**) grooved MPG sensilla. (From Zacharuk, R., 1980, by permission of Annual Reviews Inc.)

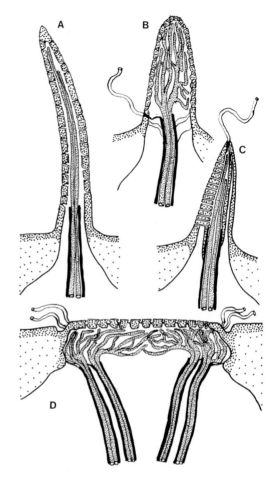

FIG. 24. Cuticular parts, dendritic associations and the location of usually sealed molting pores for the exuvial dendritic sheaths. **A:** thick-walled MPP; **B:** thin-walled MPP; **C:** MPG with left wall on a groove with pore channels and right wall between two grooves though a sensillar channel; **D:** a placoid, composite MPP with a thin sensory cuticle.

on this basis. For this reason they will be referred to here as multiporous pitted (MPP) and multiporous grooved (MPG) sensilla after Zacharuk, R. (1980a). The basic structural differences between the two types of sensory cuticles and the one most common variation within each type are outlined in Fig. 23. A few forms of sensory cuticles are shown in surface view in Fig. 25. The interrelationships of their cuticular parts are outlined in Fig. 24.

All the MPP sensory cuticles have in common many round pores or slits at the surface. In side view they are somewhat funnel-shaped, flaring outward from a narrowed aperture just below the surface. The two basic variations occur in the structure of the wall and of the conduction mechanism beneath the effective pore, but there are gradations between them in one feature or another. The most general type among the insect groups and stages has a relatively thin-walled sensory cuticle, within an approximate range in thickness of 0.1–0.3 μm. This is sometimes exceeded, especially when the area of sensory cuticle is large. The effective diameter of the pores, at the narrowest point, often exceeds 20–25 nm, and their density is high, often exceeding 15 μm^{-2}. Each pore usually opens into a small circular chamber referred to as a pore-kettle by most

workers, the bottom of which generally abuts the sensillar sinus beneath the cuticle. From the base of these kettles, dense pore tubules that average about 15 nm in diameter extend downward into the sinus toward the dendrites in a typical sensillum of this type, but there are many reports of no pore tubules in sensilla similar to this type in other respects. The number of pore tubules per pore or pore-kettle generally exceeds 15 and may be as high as 50 or 60. These sensilla are sometimes innervated by many neurons, exceeding 10, but only two or three neurons are common. The dendrites are typically branched terminally, and the branches ramify in close proximity to the sensory cuticle. In one form of thin-walled MPP sensillum one of the dendrites is lamellate within the sensory cuticle. The second variant is generally referred to as a thick-walled MPP sensillum. The wall of the sensory cuticle often tapers in thickness from about $0.8–1.0\,\mu m$ at the base to about $0.25–0.4\,\mu m$ near the tip. Their pore size is smaller, usually less than 20 nm and often in the range 6–15 nm. The pores are less numerous, with a density less than 15 and often in the range two to eight per μm^2. The pore-kettles are much reduced, and there are often cylindrical pockets that extend from the underlying sinus toward the pore. Pore tubules are usually present in this variant, but there are few per pore, usually less than 15 and often in the range of three to ten. They extend from the base of the small surface pore funnel into the underlying sinus through the cuticle of the wall and the inner pocket where present. This variant can sometimes be mistaken for a smooth-walled or aporous sensillum because of the small funnels and pore size. The density of the pores is low, usually less than 15 and often in the range three to ten per μm^2. These sensilla are not innervated by many neurons, usually two to five, and the dendrites are not branched within the sensory cuticle chamber. The cuticle is usually in the form of a hair, and this variant is typically present only in adults, often more numerous in males than in females.

The MPP sensilla of *Bombyx mori* are among the most intensively studied (Steinbrecht, R., 1973), and well exemplify the above. A long hair has a wall $0.32\,\mu m$ thick, an 8.5 nm pore size, four to six pore tubules per pore, and a pore density of two to seven per μm^2 in males and two to five per μm^2 in females. It is innervated by two unbranched dendrites and there is a sexual dimorphism in their terminations within the hair. A large peg has a wall $0.12\,\mu m$ thick, a 20 nm pore size, 12–23 pore tubules per pore, and a pore density of 20 per μm^2. The latter exemplifies well the features of a thin-walled, and the former of a thick-walled, MPP sensillum. Steinbrecht noted two thick-walled and three thin-walled variants in *Bombyx*, and sexual dimorphism in at least one of the former. The wide range in morphometric variations in these sensilla among insects is evident in Table 4. This stresses the need for careful correlative structural–functional studies on each individual sensillum if we are to understand the significance of

such variations and the specific mechanisms that are involved.

The MPP sensilla are considered to be olfactory. Their stimulus trapping and conduction mechanism has received much attention, especially the pore tubule system (i.e., Steinbrecht, R. and Müller, B., 1971; Zacharuk, R., 1971; Slifer, E., 1972; Steinbrecht, R., 1969, 1973; Schneider, D., 1971; Altner, H., 1977b; Keil, T., 1982). The pore tubules were initially enigmatic structures. They were distinct and often continuous between the pore and the dendritic membrane in some specimens, with some suggestion that they were evaginations of the dendritic membrane. In other preparations the tubules were less distinct or showed no membrane contacts or, in a significant number of sensilla and insects, were absent. Preparation techniques were and still are problematical. However, there is little doubt now that they are common to many MPP sensilla but are not developed in some forms, are an integral part of the pore system in the sensory cuticle, and are shed with the exuvium at a molt. The general consensus now is that they are a component part of the sensory cuticle, but the extent to which they form contacts with the dendritic membranes in a functioning sensillum is still uncertain. Where present and distinct they are inserted in the surface layer of the cuticle and open to the surface, whether the cuticular surface is invaginated into slits or pore-kettles or not. Where they are not distinct they may have been disrupted into a granular or dense homogeneous matrix during preparation for examination or replaced by such a matrix (cf. Fig. 26A and B), or they may be undeveloped in specific sensilla. In elaterid larvae, Zacharuk noted that the component of the pore tubule wall is lipoidal and is removed similarly from the sensory pore tubules and from the pore canal and pore filament system of insensitive cuticle if not stabilized prior to processing for TEM. He concluded that the pore tubule system of sensory cuticle is a modified cuticular pore canal system and is homologous with that of insensitive cuticle. In both systems there is a central core about 2.5 nm thick that is not lipoidal, which was always more distinct in the sensory pore tubules. Given these structural similarities and homologies, one should expect similarities in the mechanisms for their formation and maintenance by the epidermal cells and by the sensillar cells that

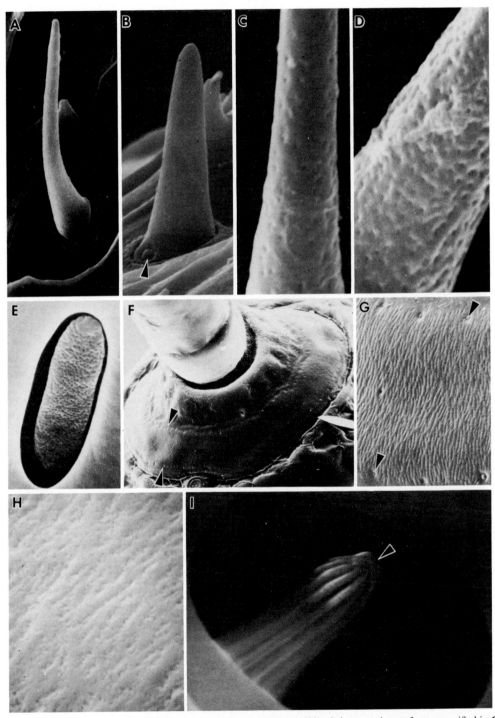

FIG. 25. SEM of MP sensilla. **A:** thick-walled, and **B:** thin-walled MPP (× 4500), their respective surfaces magnified in **C** and **D** (× 18000); **E:** a coeloconic MPP (× 14,000); **F:** a composite placoid sensillum (× 640), the surface of which is enlarged in **G,** with slitted pores (× 3200); **H:** surface of another composite MPP with pores at the base of short, shallow grooves (× 40,000); **I:** a coeloconic MPG peg as in Figs 23B and 24C (× 17,800). **A** and **C** are as in Figs 23C and 24A; **B, D** and **E** as in 23A and 24B; **F–H** as in 24D; arrowheads denote sealed molting pores of the dendritic sheaths. (**F** and **G** are from a manuscript in preparation by J. W. Bloom, A. E. Holodniuk and R. Y. Zacharuk, by courtesy of the authors.)

are derived from them. The sensory pore tubules are initially formed by the cell that forms the sensory cuticle but, when this retracts, are further elaborated and maintained either through the sensillar sinus or the dendritic terminations, through granules, vesicles or directly connected secretory strands. The latter, when connected to dendrites, may serve for inward conduction of stimulant molecules (Keil, T., 1982) as well as outward secretion (Fig. 26C–E).

The prime role of the cuticular pore canal system is considered to be transport of lipoidal substances from the epidermis to the surface to maintain an adequate superficial layer that will minimize moisture loss. This protective layer is particularly important over the very thin cuticles of MPP sensilla and is undoubtedly maintained by the pore tubules. This is evidenced by the surface secretions noted on and around the pores in uncleaned specimens viewed by SEM and TEM. In addition the substances that they transport to the surface, and any that diffuse through the thin cuticle, are presumed to form a continuous thin layer there. The primary components of this layer may be lipoidal (Zacharuk, R., 1971) or proteinaceous (Seabrook, W., 1977) or both. Steinbrecht, R. (1973) suggests that this layer traps specific stimulating molecules, which then diffuse two-dimensionally to the nearest pore and one-dimensionally along the pore tubules directly to a connected dendrite or into the sensillar fluid. In the latter they would diffuse three-dimensionally to the nearest dendritic membrane.

A few MPP sensilla have less typical sensory cuticles or pore conduction mechanisms than those described above. In *Cimex lectularius* (Steinbrecht, R. and Müller, B., 1976) and *Oncopeltus fasciatus* (Harbach, R. and Larsen, J., 1976) there is one type of hair that has a thick wall without pores on the side facing the antenna and a thin wall with a low density of pores on the outer side. The other features resemble those of a thick-walled MPP sensillum. In the hymenopteran *Camponotus vagus* (Masson, C. *et al.*, 1972), one type of peg has the bottom half of its wall much thickened inwardly and without pores, and the thinner top half has but a few pores subapically. This abrupt inner change in wall thickness is similar in the MPP sensillum on the ovipositor of the fly *Musca autumnalis* (Hooper, R. *et al.*, 1972), but this sensillum is also unusual in that

there are no pore tubules and a continuous thin layer of cuticle separates the relatively large and superficially open pore-kettles from the underlying sinus and dendrites. The stimulating molecules must diffuse through this cuticle and the sensillar liquor to the dendrites. In the latter two sensilla the dendritic sheath is fused to the inside of the thick-walled base and the dendrites emerge from the open end and branch in the enlarged outer chamber. Thus, this chamber is not open to the sensillar sinus beneath the peg. This is in marked contrast to the more typical termination of the dendritic sheath in the more usual types of MPP sensilla. In these the sheath terminates at or near the base of the sensory cuticle, and the dendrites emerge from the open end into the sensory chamber. The distal end of the sheath is usually suspended from the base of the sensory cuticle by attachment fibres, or one side of its attaches at or near the base, at which point the molting scar may be visible externally (Fig. 25B). In *Folsomia* the molting scar is about one-third of the way up the shaft of a MPP hair (Slifer, E. and Sekhon, S., 1978). Regardless of the mode and point of attachment, the sensillar chamber where the dendrites terminate is continuous with the sensillar sinus beneath, around the dendritic sheath and through its suspensory fibres. (Fig. 24B).

In the postantennal organ of Collembola the pore-kettles have wide external openings much like those in the ovipositor sensillum of *Musca*, but the base of each kettle is perforated by about 25 pores that are 5–10 nm in diameter. These provide channels from the kettle to the underlying sensillar chamber. A dense secretion often underlies the sensory cuticle and fills the kettles. This apparently serves as the outward secreting and inward stimulus conducting mechanism (Altner, H. and Thies, G., 1972, 1976). Each organ is innervated by one neuron with two ciliary dendritic segments, each of which branches under the sensory cuticle. There is no dendritic sheath, and there is a tendency for the nerve cell body to be incorporated into the protocerebrum in some species. The cupuliform organ of *Campodea* has four sensilla with spheroid sensory cuticles located in a pit at the tip of the antenna. The pore mechanism and its relationships with the underlying dendrites somewhat resemble those of the collembolan postantennal organ, but each sensillum in *Campodea* has three neurons with two ciliary

Table 4: *Some characteristics of multiporous sensilla in various insects*

Insect	Location	Type, wall thickness (μm)	Pore Size (nm)	Pore Density (μm^2)	PT per pore	Neurons	Reference
Necrophorus spp.	antenna	hair 1, 0.2	15	100	5–6	1B	Ernst, K. (1969)
		hair 2, 0.17	10	100	8	2B	
Tenebrio molitor	antenna	peg 0.2		450*	yes	2B	Harbach, R. and Larsen, J. (1977)
		peg 0.12		1090*	7	2B	
		peg G, 0.35		214*	5U		
Aedes aegypti	antenna	hair, thick		few		2U	McIver, S. (1978)
		hair, thick		few		1U	
		hair, thin		many		2B	
	palp	peg, 0.25	18	12	yes	2B, 1L	McIver, S. (1972)
Anopheles stephensi	antenna	peg G, coeloconic				4–5B	Boo, K. and McIver, S. (1976)
		peg G, A, surface				2U	
		peg G, B, surface				3–4U	
Culex pipiens	palp	peg, 0.25	15	23	yes	3	McIver, S. (1972)
Culicoides furens	antenna	hair, 0.13	15	15	no	12B	Chu-Wang, I. *et al.* (1975)
		hair, 0.2	10	3	no	1–3U	
		peg G	10			3–4U	
		peg G	10			5U	
	palp	pit-peg	20	30	no	1B, 1L	
Calliphora vacina	palp	peg, 0.2				3B	van der Starre, H. and Tempelaar, M. (1976)
Haematobia irritans	antenna	peg, 0.4	17	3	yes	3–5U	White, S. and Bay, D. (1980)
		hair, 0.15	20	20–35	10–20	2–4B	
		peg G, 0.4	30		1–3U		
Musca autumnalis	antenna	hair, 1.0–0.3	17	20	yes	2–4U	Bay, D. and Pitts, C. (1976)
		peg 1, 0.2	37	35	15–20	1B	
		peg 2, 0.2	37	60	15–20	1B	
		peg G, 0.1	30			2U	
		pit-peg, 0.1	37	50		1B	
	ovipositor	peg, 1.0–0.3	170			1B	Hooper, R. *et al.* (1972)
Stomoxys calcitrans	antenna	peg, 0.1–0.2	30	950*	25–40	2–3B	Lewis, C. (1971)
		peg, 0.1	40	1500*	40	1L	
		peg G	5			2U	
		hair, 0.3		500*	5–8	6U	
Ptilocerembia spp.	antenna	hair, thick	10		yes	3U	Slifer, E. and Sekhon, S. (1973)
		hair, thin	10		yes	4–15B	
Aphidius smithi	antenna	plate, 0.2	8		no	37B	Borden, J. *et al.* (1978)
Camponotus vagus	antenna	peg, 0.65–0.3	30		3–5	30B	Masson, C. *et al.* (1972)
Chalcidoidea, 17 sp.	antenna	plate, thin		10	yes	50B	Barlin, M. and Vinson, S. (1981)
		thick		4	yes	50B	
Neodiprion sertifer	antenna	hair, 0.3–0.6 (male)			yes	8–12B	Hallberg, E. (1979)
		(female)			yes	5–6B	
		hair, 0.3		many	yes	10–14B	
		peg G				3–4U	
Cimex lectularius	antenna	hair E1, thick	5–6	2†	yes	1–3U	Steinbrecht, R. and Müller, B. (1976)
		E2, thick		1†		2U	
		D, thin		10†	yes	6–19B	
		peg G, C	20	3†		4–5U	
Oncopeltus fasciatus	antenna	hair S, thick		1,000*	4	3–8	Harbach, R. and Larsen, J. (1976)
		A, thick/thin		500*	4	2–3	
		hair, thick		670*	1	2–3	
		peg, thin		6350*	5–6	40B	
		peg G		240*		3–5	
Danaus gillippus	antenna	peg, 0.8–0.3	40	few	yes	1–3U	Myers, J. (1968)
		peg, 0.17	40	many	yes	3–5B	
		peg G	60			2U	
		peg G	30–40			4–5U	

Table 4: *Some characteristics of multiporous sensilla in various insects*—Continued

Insect	Location	Type, wall thickness (μm)	Pore Size (nm)	Pore Density (μm²)	Pore PT per pore	Neurons	Reference
Arenivaga spp.	antenna	peg, 0.25	10	many	yes	7B	Hawke, S. and Farley, R. (1971a)
		peg, 0.32	10	more	4–12	6B	
		peg G, 0.3	20			3U	
		peg G, 1.0	20		2U		
Blaberus craniifer	antenna	peg 0.35–0.1	20		yes	1–2B	Lambin, M. (1973)
		peg G	30			4U	
Locusta migratoria	max. palp	peg, 0.24–0.35	51	many	yes	15B	Blaney, W. (1977)
Periplaneta americana	antenna	hair		few	yes	1–3U	Toh, Y. (1977)
		peg, 0.2–0.3	8–15	many	yes	4B	
		peg G				1–5U	
Frenesia missa	antenna	plate 0.2	18	uneven		2B	Slifer, E. and Sekhon, S. (1971b)

* Number of pores per hair or peg.
† Number of pores per section.
G type sensilla have a grooved wall; B, branched and U, unbranched neurons.
PT, pore tubules.

segments each. There is a dendritic sheath in only one of the four sensilla (Juberthie-Jupeau, L. and Bareth, C., 1980). The sensory cuticles of the two types of sensilla coelosphaerica reported by Ernst, K. (1972a) from antennae of *Necrophorus* have a pore conducting mechanism that also resembles that of the postantennal organs above. One has a porous sieve-like base in the kettles as in Collembola, and the other has an extended vesicle filled with a fenestrated material beneath the kettles. This sensillum is innervated by two neurons with branched terminations in the chamber within the sensory cuticle, but the dendritic branches are completely encased in a dendritic sheath that, in one type, is attached apically at a molting scar. Presumably there are pores in the dome of the sheath through which the odorous molecules can penetrate to the dendrites. A secretory-conduction mechanism that also seems to be similar in principle to the above occurs in the "olfactory vesicles" or Hamann's organ on the antennae of the cave-dwelling Bathysciinae (Corbière-Tichané, G., 1974; Accordi, F. and Sbordoni, V., 1977), but the pore pits are much deeper in the highly involuted thin sensory covering and the pores line the walls as well as the bases of the elongated pits. In the pseudoculus of the proturan *Eosentomon* the sensory cuticular dome is clefted in a regular pattern and pore tubules seem to extend from the cleft bases toward the terminations of the dendritic ciliary segments of the two innervating neurons or a cytoplasmic extension

of the single cell that envelops each neuron and also terminates under the sensory cuticle (Haupt, J., 1972). There seems to be no dendritic sheath. These pore conduction systems probably do not conserve moisture well. They seem to be limited to some but not all sensilla of the more primitive insects in moist habitats, or to sensory cuticles in deep cuticular pits with small openings in other insects. They also show that pore tubules are not a required component of a transcuticular conduction system for chemical stimuli.

Two or more sensilla are sometimes associated together in a specialized area of cuticle. An example already given is the location of a campaniform organ at the base of a type of mechanosensory hair on cockroach cerci. Altner, H. and Thies, G. (1978) describe a multifunctional complex of sensilla on the antennal tip of a collembolan. More often a number of sensilla of the same type are clustered in a pit or depressed area of cuticle. The olfactory pits on the antennae of *Musca autumnalis* are an example, some of which contain up to 200 similar thin-walled MPP pegs (Bay, D. and Pitts, C. 1976). Each retains an individual sensory cuticle identity externally, but the internal cellular arrangment is uncertain. Some large sensilla are believed to have evolved from such a cluster of similar sensilla whose sensory cuticles were combined into one continuous covering. The number of constituent sensilla is identifiable from the number of individually wrapped neuronal groups beneath.

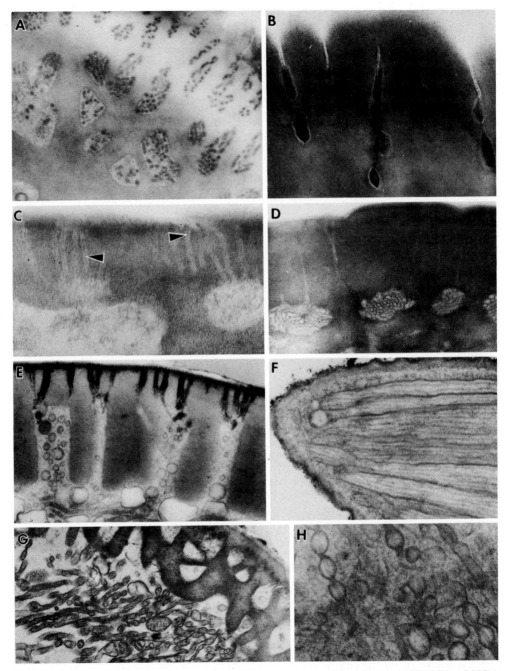

FIG. 26. TEM of MPP and related sensilla. **A, B:** tangential sections of sensory cuticle surfaces as in Fig. 25H and 25G ($\times 53,000$ and $\times 42,000$ respectively), with pore tubules to slit pores in the former and a dense secretion to surface slits in the latter; **C, D:** comparison of the tubule pores of the same type of sensillum as in **A** with the pore canal and filament system of insensitive cuticle, with lipoidal substances removed in both ($\times 108,000$ and $\times 105,000$, respectively) — arrowheads denote 2.5 nm non-lipoidal cores of the pore tubules in the former; **E:** secretion into the pore tubules in a newly-molted sensillum as in **A** ($\times 24,000$); **F:** sensory cuticle of a composite sensillum on the antenna of *Aedes* larvae homologous to that in **A** and **B** above from terrestrial beetle larvae, without cuticular pores or pore tubules ($\times 27,000$); **G, H:** oblique sections of sensory cuticle as in **A** with associated dendritic branches showing beading ($\times 15,000$ and 39,000, respectively). (**B,** courtesy of the authors, Bloom, J. *et al.*, manuscript in preparation; **C-E, G, H,** are from Zacharuk, R. 1971, courtesy of The National Research Council of Canada; **F,** courtesy of the authors, Zacharuk, R. *et al.*, 1971, and by permission of The Wistar Institute Press.)

In elaterid larvae the large cone of MPP sensory cuticle on the pedicel alongside the flagellar segment (Fig. 2C) is innervated by 36 neurons that are in 12 groups of about three neurons each. Each group is separated from the others by an individual dendritic sheath and inner sheath cell wrap. These units are interspersed by 12 accessory cells, and the complex is enclosed by an outer layer of 12 more accessory cells. The latter 24 cells subtend a common sensillar sinus beneath the cone. The dendrites exit the open ends of their respective sheaths near the base of the cone, branch repeatedly, and form a common mass of dendritic branches under the sensory cuticle. The dendritic sheaths are suspended by fibers from the non-porous cuticular base, and form 12 molting scars around the circumference at a molt (Scott, D. and Zacharuk, R. 1971a). The multiple molting scars are often distinct in a SEM view (Fig. 25F and G). These relationships are outlined in Fig. 24D. The following lists the insect, form of sensory cuticle, neurons per group × the number of groups, and the reference for some other composite sensilla of this type:

Aphid spp., sunken plates (rhinaria), 2–3 × 1–6, Bromley, A. et al. (1979);

Tribolium spp., placoid, 10–11 × 9–17, Behan, M. and Ryan, M. (1978);

Speophyes lucidulus, cone, 2 × 6, Corbière, G. (1969);

Musca domestica larva, 3 × 7, Chu-Wang, I. and Axtell, R. (1971);

Pyrops candelaria, folded placoid, 6–10 × 25–35, Lewis, C. and Marshall, A. (1970).

The circumfilia of the midge Contarinia sorghicola are long MPP filaments coiled around the surface of an antennal segment (Slifer, E. and Sekhon, S., 1971a). It branches from several hair-like sockets and may be a more unique composite sensillum in which the hair shafts are fused but retain their individual sockets. The cellular arrangements beneath each socket are unknown. In the placoid types the pore tubules generally insert at the base of short surface slits (Fig. 26A), but there are no pore tubules in Speophyes and a dense amorphous secretion instead in Tenebrio molitor (Fig. 26B).

The rhinaria of aphids are often surrounded by a variously elaborated "picket fence" of microtrichia which may be protective and perhaps a filtering

device (Shambaugh, G. et al., 1978). Placoid sensilla or so-called "pore-plates" of this type are very common on the antennae of lamellicorn beetles also. Here their cuticular parts are in a great variety of forms, externally and internally. These variations are useful in the systematics of the group, as they are in aphids, but it is not known whether they also have a functional significance (Meinecke, C. 1975). The antennal cupuliform organ of Stylops (Brandenburg, J. and Matuschka, F. 1976) also seems to be composite, the dome of which is innervated by four individually ensheathed neurons. The antennal sensory cones in larvae of the mosquitoes Aedes aegypti (Zacharuk, R. et al., 1971) and Toxorhynchites (Jez, D. and McIver, S. 1980) are innervated by 11 or 12 neurons that are ensheathed in six groups of one to three neurons each. The thin sensory cuticle covering the dendritic branches in these composite sensilla is unusual in that it has no pore system. If chemosensory, the stimulating molecules would have to diffuse from its aquatic habitat through the rather homogeneous cuticle to the dendritic branches that are packed within (Fig. 26F).

The cupuliform sensilla that are on the abdominal gills of nymphs of the plecopteran Thaumatoperia alpina are each innervated by two neurons. Their ciliary segments are branched within the dome of sensory cuticle. This cuticle has a fenestrated structure throughout its thickness and regularly spaced microtubule-like channels on the surface, with no other obvious pore system. These sensilla are believed to have an osmo- rather than a chemosensory function. Probably the most unusual type of sensory structure is reported by Slifer, E. and Sekhon, S. (1980) from the tip of the antennal flagellum in the human louse, Pediculus humanus. The cuticle around the bases of other terminal sensilla is porous and is underlaid by a thin layer of dendritic branches. The pores seem to be permeable, and they believe these are chemosensory areas of membranous cuticle that augment the functions of the few other chemosensilla that were noted without exposing extra surfaces of porous sensory cuticle to moisture loss.

The grooved multiporous sensilla (MPG) are characterized by longitudinal grooves in the wall. Two basic variants occur. In one the grooves are well developed in surface view and the narrowed bottom channel opens into a longitudinally cylindrical chamber that appears much like a pore-kettle in cross-section (Fig. 23C). The ridges between the grooves are elongated to a point in some forms, giving it a stellate appearance in section. In the second variant the grooves are much narrower and not as distinct in surface view, and the slits do not widen appreciably inward (Fig. 23D). In both variants pores, often referred to as spoke-canals, extend inward to the dendritic chamber. Most MPG appear double-walled because of longitudinal channels that extend from the sensillar sinus beneath towards the tip, in the wall between all the grooves and their spoke canals. These sensilla do not have pore tubules. A dense substance from the dendritic chamber fills the spoke-canals and apparently flows out over the grooved surfaces. This is presumed to be the trapping and conduction mechanism for chemical stimulants. The spoke-canals are 20–30 nm in diameter (or width if slit-like) in most sensilla but canals as thin as 5 nm and as wide as 60 nm are reported. In two instances the number of pores per sensory cuticle was estimated at just over 200. Most are innervated by two or five neurons or by a number within this range.

The MPG sensilla are not numerous in most insects and are located primarily on the antennae. Although hairs are reported, most are coeloconic pegs, in shallow or deep pits. The grooved porous surface extends the length of the cuticular process in some (Fig. 25I) and only the terminal half in others. The wall thickness of double-walled pegs and hairs is about 0.3–0.4 μm, but a single-walled variant is reported by Bay, D. and Pitts, C. (1976) with a wall thickness of only 0.1 μm. In some the wall appears double basally and single and thin distally. The dendrites extend into the shaft of the sensory cuticle and terminate within the porous portion near the tip. The open distal end of the dendritic sheath usually ends at the base of the porous grooves where it fuses radially with the sensory cuticle. It is molted through an apical pore (Fig. 24C). Some workers report that this pore is plugged after a molt and others indicate that it is open and permeable. Detailed descriptions of MPG sensilla are much fewer than of MPP sensilla. Hawke, S. and Farley, R. (1971a, b) and Steinbrecht, R. and Müller, B. (1976) give good descriptions and illustrations of both variants. Some features of others are given in Table 4, and in the general reviews by Altner, H. (1977a, b).

A significant number of MPP and a few MPG sensilla were shown to be chemosensory and primarily olfactory. Of the former type, the thick-walled variants seem to be the more selective and the ones that are stimulated by pheromones. Some are referred to as "specialist" sensilla on this basis. The thin-walled variants generally have a wider range of chemosensitivity, and are sometimes referred to as "generalists". However, more correlated studies on structure and function are needed before any such generalizations can be accepted. Some of the thin-walled MPP pegs with a lamellate dendrite in the shaft were shown to respond to CO_2 in the biting flies. In *Periplaneta americana* some of the grooved and double-walled sensilla are thermo- and hygrosensory and others are thermo- and chemosensory (Altner, H. *et al.*, 1977).

4.2 Sensory neurons

The sensory nerve cells that innervate all the types of sensilla reviewed above are basically similar in form, structure, and their general relationships to the proximal length of the dendritic sheath and the innermost enveloping cell. All are typical bipolar cells with the cyton located peripherally within or just under the epidermis near the sensory cuticle it innervates. The axon extends to the nearest nerve branch from the tapered and usually little differentiated proximal pole of the cyton, and the dendrite extends toward the dendritic sheath and the sensory cuticle from the opposite pole that is tapered to varying degrees in different sensilla (Fig. 27). The dendrite is differentiated into two distinct segments, with the point of demarcation usually near the middle of its length. The proximal segment resembles the cell body in cytoplasmic contents, and the distal segment has the appearance of a modified cilium and is often referred to as the ciliary segment or cilium. With the exception of some unique sensilla in the more primitive insects which have two ciliary segments arising from the proximal segment, as noted in the preceding section, all other sensilla have only one ciliary process per neuron. This differs markedly from the ciliated neurons of a number of other animal phyla (Barber, V., 1974).

The dendritic ciliary segment has a $9 \times 2 + 0$ arrangement of microtubules near its point of origin, and this is maintained to the tip in most scolopophorous sensilla and in one truncate dendrite in some aporous temperature–humidity sensilla discussed in that section (Figs 28B and 31H). In most of the other sensilla the segment enlarges toward its point of entry into the dendritic sheath, some or all of the doublet microtubules separate, and there is a variable increase in their number. They also become evenly dispersed throughout the cross-sectional area (Fig. 31C–E, F, G). This is shown best in well-fixed specimens (Steinbrecht, R., 1980). There is no apparent decrease in number of tubules to the tips of dendrites that terminate in a tubular body. In other unbranched dendrites there is a gradual decrease in their number to the dendritic tip from the widest point of the segment, usually at some point within the dendritic sheath beneath the base of the sensory cuticle. Where terminal branches are formed, one or more tubules extend into, and to the tip of, each branch, the number apparently directly related to the size of the branch. In one MPP peg of *Musca autumnalis*, the single large dendrite produces one branch at a time, and each branch receives just one tubule (Bay, D. and Pitts, C. 1976). This is one of the more unique modes of branching. In dendrites with lamellate terminations, some are truncate and terminate below or alongside the dendritic sheath. In other sensilla they terminate within the sheath below the base of the sensory cuticle, or in the sensillar chamber in certain MPP sensilla. The last is implicated in sensing CO_2. In the former two types of terminations the lamellate dendrites are believed to be temperature sensors, perhaps sensing radiation in the infrared range (Corbière-Tichané, G. and Bermond, N. 1972; Corbière-Tichané, G., 1977), or sudden changes in temperature (Loftus, R. and Corbière-Tichané, G., 1981). The tubules extend into the lamellae in some sensilla but not in others. There is a considerable variation in the central and distal diameters of the ciliary segments among the dendrites of a multineuronal sensillum and between those of different types of sensilla, but the significance of this is not known.

Most workers report microtubules as the sole organelles of the distal dendritic segments. However, there are several reports of vesicular inclusions occurring also in uniporous and multiporous sensilla (Moulins, M., 1968; Lewis, C. and Marshall, A., 1970; Chu-Wang, I. and Axtell, R., 1971; Scott, D.

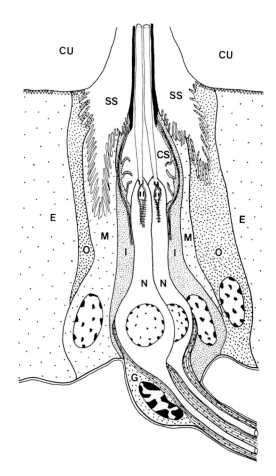

Fig. 27. Outline of the cellular components and sinuses in a typical insect sensillum; CS, ciliary sinus; CU, cuticle; E, epidermis; G, basal glial sheath cell; I, inner or dendritic sheath cell; M, intermediate sheath cell; N, nerve cells; O, outer sheath cell; SS, sensillar sinus.

and Zacharuk, R., 1971b; Marshall, A., 1973). They are most abundant as multivesicular bodies (MVB) or clusters of small free vesicles near the origin of the segment (Fig. 28D–F), but were noted also along the length of the segment to its termination beneath the sensory cuticle. Marshall noted similar vesicles to be produced by Golgi complexes in the cyton, and MVB are common near the tip of the proximal dendritic segment in many sensilla (Fig. 28G). They seem to be picked up here and transported to the sensillar chamber. Some are electron-lucent and others very dense. Their contents may be the materials needed to maintain the pore tubules and the secreted surface layer of the sensory cuticle (Fig. 26E).

The origin of the microtubules in excess of the 9×2 configuration is uncertain. Slifer, E. and Sekhon, S. (1969b) suggest that they form by division of some of the original 18 after they separate, but this has not been confirmed. In the composite antennal cone sensillum of elaterid larvae the number of doublets at the base of the segment exceeds the basic 9×2 by several multiples of 9. Ciliary configurations of 36×2, 45×2, and 54×2 were usual (Scott, D. and Zacharuk, R., 1971b). These originate from a basal body, and fewer tubules need to be formed distally. Dendritic blebbing or beading is common, particularly in the dendritic branches of thin-walled MPP sensilla (Slifer, E. and Sekhon, S., 1969a; Steinbrecht, R., 1980), but the microtubules are continuous through the constricted parts (Fig. 26G and H). This does not seem to be an artifact, and may be an indication of an activity such as was noted by Ochs, S. (1963) in mammalian nerve fibers when they were subjected to a small stretch.

Most workers report a marked change in the diameter of the dendrite from the thick proximal segment with a truncate tip to the thin ciliary segment that inserts centrally (Fig. 28A and C). Steinbrecht, R. (1980) believes that this is an artifact of fixation because in freeze-fixed section the taper of the proximal to the distal segment is smooth and gradual, and the contours of all cell membranes are much more even throughout. Directly beneath the cilium, and in line with it, are two centriole-like basal bodies, the distal at the base and the proximal below it. The nine tubule doublets originate from the distal body and extend symmetrically and peripherally into the cilium. A thin dense "collar" lines the circle of tubules just above the distal body (level 1 in Fig. 28B). This ciliary collar is very well developed in some sensilla (Fig. 28D–F). In some sensilla also the distal body is incorporated into the base of the cilium and collar (Fig. 28D), and in a number of reported sensilla it is not evident. In the larva of *Musca domestica* some of the ciliary tubules originate in the distal and some in the basal body in sensilla in the ventral and terminal organs (Chu-Wang, I. and Axtell, R., 1972a, b). From within the distal body, ciliary rootlets extend proximally around the proximal body to which they attach and into the cytoplasm of the dendrite for varying distances. They are not continuous with the tubule triplets of the distal body in mosquitoes (Boo, K., 1981) and also seem to originate from the inner wall of this body in other sensilla. These rootlets are almost universally cross-striated in a regular periodic pattern, but Quennedey, A. (1975) reports no rootlet cross-striations in the labral uniporous sensilla of the termite *Schedorhinotermes*. The rootlets are especially well developed in the scolopophorous sensilla; in chordotonal types they are usually fused into a solid striated rod that extends close to or into the cyton, but in the mandibular pore canal organs they form junctions with the walls of the proximal dendritic segment. Gaffal, K. and Bassemir, U. (1974) excellently illustrate and discuss the variations in structure of the basal ciliary apparatus in several types of insect sensilla, and the freeze-fracture characteristics of some of the ciliary components are detailed by Menco, B. and van der Wolk, F. (1982).

The cytoplasmic inclusions in the proximal dendritic segment differ little from those in the cyton. There is usually a small area of finely granular plasm around the basal ciliary apparatus around which vesicles and MVB are often reported. These are much more numerous in newly molted than in intermolt specimens, which suggests that much of their contents are used in the development and maintenance of the sensillar cuticular parts. Some may contribute materials to the transduction mechanism but this is less apparent (Scott, D. and Zacharuk, R., 1971b; Zacharuk, R. and Albert, P., 1978). From here proximally the cytoplasm of the dendrite and cyton has much rough and smooth endoplasmic reticulum, longitudinal microtubules, some of which are associated with the ciliary rootlets, and it is especially rich in mitochondria. Golgi complexes are abundant in the cell body, and some workers report so-called onion bodies here as well (Cook, G. 1972). These consist of several membranous lamellae usually within an enveloping membrane. Daly, P. and Ryan, M. (1981) noted an association between these bodies, rough endoplasmic reticulum and Golgi complexes, and suggest that they are part of a protein-synthesizing system. The nuclei of the sensory neurons are typically large, circular, central and with finely dispersed chromatin granules, an indication of a metabolically highly active cell (Fig. 29B).

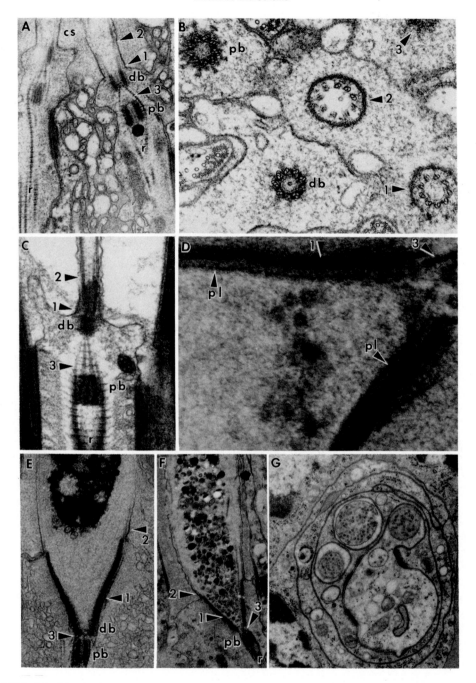

FIG. 28. TEM of dendrites in the region of the ciliary sinus. **A:** l.s. of two dendrites of a uniporous sensillum as in Fig. 19A, in the ciliary sinus (cs) with distal (db) and proximal (pb) basal bodies, ciliary rootlets (r) and arrowheads denoting levels 1 at the ciliary collar, 2 above the collar where additional tubules begin to add to the $9 \times 2 + 0$ configuration, and 3 between the two basal bodies ($\times 24,800$); **B:** t.s. of the same sensillum as **A**, with five dendrites cut at levels indicated in **A** ($\times 29,000$); **C:** same as **A** in a scolopophorous sensillum ($\times 31,000$); **D:** l.s. of the ciliary collar (pl) lining the ciliary tubules in a dendrite of a composite MPP, with regions as in **A**; arrowheads denote apparent pores in the lining ($\times 48,000$); **E,F:** larger area of dendrites similar to that in **D**, with secretory vesicles in the ciliary segments above the collar; levels as in **A** ($\times 17,700$ and $\times 15,000$); **G:** t.s. of a proximal dendritic segment below the level of pb, with MVB in dendritic evaginations, from a newly molted campaniform sensillum ($\times 24,000$). (**A-C**, from Bloom, J. *et al.*, 1981, 1982a; **D-F**, from Scott, D. and Zacharuk, R. 1971b, by courtesy of the authors and The National Research Council of Canada.)

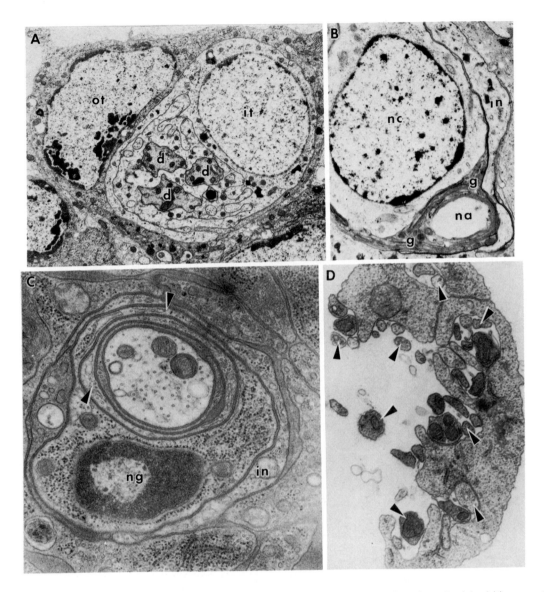

FIG. 29. TEM, t.s. through the basal nuclei of the intermediate (it) and outer (ot) sheath cells and proximal dendritic segments (d) wrapped separately by the inner sheath cell (**A**, × 4600); through the cyton (nc) of one and an axon (na) of another neuron separately wrapped by a glial (g) and surrounded by the inner (in) sheath cell (**B**, × 7000); through an axon from a campaniform organ in the region of a glial cell nucleus (ng) partly enclosed by a remnant of the inner sheath cell (in), with arrowheads at two unlapped glial wraps (**C**, × 17,200); and of the naked axons from sensilla at the tip of the antenna of an *Aedes aegypti* larva (arrowheads at some), near the antennal base where they are beginning to be wrapped by a glial cell (**D**, × 27,500). (**D**, from Zacharuk, R. *et al.*, 1971, by permission of The Wistar Institute Press.)

The axon typically has only mitochondria and evenly distributed longitudinal microtubules in a light axoplasmic ground substance, but occasionally vesicles are also reported (Fig. 29C). They are ensheathed from the cyton into and within the nerve by several wraps of glial cells, but in the larva of *Aedes aegypti* (Zacharuk, R. *et al.*, 1971) the bases of the cytons and a distal length of their axons of all the sensilla are naked in the antennal hemocoel and begin to be wrapped near the antennal base (Fig. 29D).

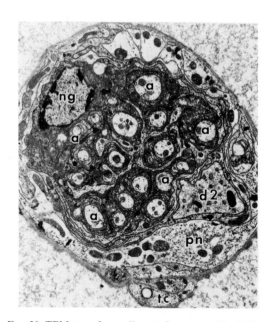

FIG. 30. TEM, t.s. of a small nerve from a sensillar field; a, axon; d2, dendrite of a Type II neuron; ng, glial cell nucleus; pn, perineurial cell; tc, tracheocyte (× 6800).

In *Cimex* the axons from some of the sensilla are similarly unwrapped distally (Steinbrecht, R. and Müller, B., 1976). The physiological significance of this is not known but should be investigated. The axons and their cell sheaths are invested by a basement membrane in the hemocoel, which is continuous with the bounding basement membranes of surrounding tissues. Collections of axons in larger nerves are enclosed by a perineurial cell layer under the basement membrane (Fig. 30).

4.3 Sheath cells and sinuses

The initial belief based on light microscopy was that there were three accessory cells associated with each sensillum: a trichogen cell that provided the inner wrap of the neuron(s) and formed the sensory cuticle, a tormogen cell that wrapped the trichogen cell and formed the socket of the sensory cuticle, and a basal neurilemma or glial cell that ensheathed the axons distally (Snodgrass, R., 1926, 1935). The ultrastructural evidence now available indicates a considerable variability among sensilla in the number of accessory cells, and some significant features that all sensilla have in common. The cells

and their relationships most commonly reported in some aporous and most uniporous and multiporous sensilla are outlined in Fig. 27. Three cells — an inner, an intermediate, and an outer — successively envelop the dendritic and distal cytonal parts of the neurons. A basal glial cell ensheathes the basal part of the cytons and the axons from them. Some sensilla, such as a MPP peg on the antennae of *Necrophorus* (Ernst, K., 1972b) has four sheath cells around the dendritic region, and in a large campaniform sensillum on the mandible of elaterid larvae there are six (R. Y. Zacharuk, unpublished observations). Many aporous sensilla and a few of the other types have only two accessory cells in this region, as proposed initially. The basal glial cell seems to be common to all sensilla except in the rare instances where the axons are naked distally. This cell is absent in the antennal sensilla of *Aedes* larvae, where the axons are naked, as noted in the preceding section.

The innermost distal sheath cell is variously termed in the literature as the trichogen cell, neurilemma cell, inner sheath cell, dendritic sheath cell, or thecogen cell. The homologous cell in scolopophorous sensilla is usually termed the scolopale cell. Typically, it laps once around the proximal end of the dendritic sheath and the neuron(s) from here to some part of the neuronal cyton. In some sensilla, as in all the larval *Tenebrio molitor* sensilla on the antenna (Bloom, J. *et al.*, 1982a), it does not lap, but is in the form of a cylinder. Apparently the dendrites grow through the cell during development in this instance (Fig. 31G). This cell secretes the dendritic sheath (Ernst, K., 1972b) and, where there are only two distal sheath cells, may also be involved in forming the sensory cuticle (Wensler, R. and Filshie, B. 1969). Because of the possible variation in the formative functions of the sheath cells, the terms inner or dendritic sheath are the least confusing for this cell. Distally the cytoplasmic wrap of this cell is closely apposed to the outer wall of the dendritic sheath, first as projecting cytoplasmic fingers and, before the sheath ends, as a continuous sleeve (Fig. 31C–E, F–H). Longitudinal microfibrils and microtubules are associated with the apposition with the dendritic sheath. From here to just beyond the tips of the proximal dendritic segments this cell encloses a small sinus variously termed the inner, basal,

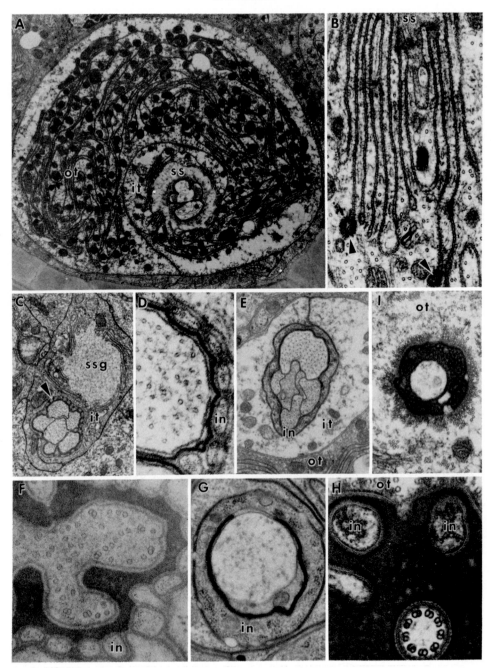

Fig. 31. TEM, t.s. **A:** sheath cells at the base of the dendritic sheath (×7000); **B:** coated vesicles at arrowheads associated with the lamellate or microvillate wall of the outer sheath cell bordering the sensillar sinus (×41,200); **C-H:** through the proximal end of the dendritic sheath — **C:** fingerlike distal projections of the inner sheath cell enclose the sheath (arrowhead) in a composite MPP sensillum, (×8000); **D:** finger-like projections enlarged showing longitudinal filaments within (×60,800); **E:** proximal to **C** in a uniporous hair, where inner sheath cell wrap is continuous around the sheath (×14,000); **F,** as in **D** in a campaniform organ (×60,800); **G:** proximal to **F** in a similar sensillum, with a continuous, unlapped inner sheath cell wrap (×28,000); **H:** base of the dendritic cap in a scolopophorous organ (×85,500); **I:** at the tip of the dendritic cap where microtubules in the attachment cell are associated with the cap (×40,000): in, inner sheath cell; it, intermediate sheath cell; ot, outer sheath cell; ss, sensillar sinus; ssg, goblet-like sinus in the intermediate cell, which opens into the sensillar sinus. (**A, B, D, E, H, I,** from Bloom, J. *et al.*, 1981, 1982, courtesy of the authors and The National Research Council of Canada.)

trichogen cell, or ciliary sinus. The liquor in this sinus, usually finely granular and fairly lucent, bathes the tips of the proximal segments and the ciliary segments of the dendrites from here into the dendritic sheath to their tips. Thus, in sensilla such as the uniporous and MPG, where the open distal end of the sheath is radially fused with the inner wall of the sensory cuticle, the liquor in the terminal dendritic or sensory chamber is continuous with that in the ciliary sinus. In some instances, such as that reported by Bareth, C. and Juberthie-Jupeau, L. (1977) for a sensillum in *Campodea sensillifera*, the ciliary sinus liquor is very dense and homogeneous. This was also noted atypically in one out of about 25 scolopophorous sensilla in the mandible of elaterid larvae (Zacharuk, R. and Albert, P., 1978), but its significance is uncertain. The inner sheath cell forms a number of well-developed longitudinal junctions with the side walls of the proximal dendritic segments near their tips. Clusters of microtubules and microfibrils in the inner sheath cell are associated with each of these junctions (Fig. 32). From this point inward this sheath cell wraps each dendrite and cyton separately with inner cytoplasmic extensions, and terminates to one side of the neuronal cyton in an expansion that contains its nucleus. Other prominent organelles are mitochondria, MVB and scattered vesicles, rough endoplasmic reticulum and scattered microtubules. It usually extends some microvilli into the ciliary sinus and seems to secrete into it.

The discrete complexes of microfibrils and microtubules, which are most developed at the junctions with the proximal dendritic segments, are best developed in the chordotonal sensilla, less so in the mandibular scolopophorous sensilla, and in order of decreasing complexity, in the aporous temperature–humidity receptors, various chemosensilla, and large and small mechanosensory hairs and campaniform sensilla (Fig. 32). In the scolopophorous sensilla they are usually referred to as scolopale rods. Some of their component tubules and fibrils seem to be continuous from their association with the dendritic sheath to their associations with the dendritic junctions proximally, but their number increases particularly in the proximal half of their length. They are generally assumed to be skeletal structures, forming a protective and supportive "cage" around the ciliary sinus and the den-

dritic segments within. Coupled with the dendritic sheath, this skeletal cage extends from the ciliary rootlets to the distal end of the dendritic sheath, primarily around the ciliary dendritic segment. The microfibrils are about 6 nm in diameter (Fig. 32E and F), resembling actin fibers in this respect (Bloom, J. *et al.*, 1982b). If they are similarly contractile in the scolopale rods, they may also play some role in the initiation of nerve impulses. They would also be expected to contract during chemical fixation and give the telescoped, truncate appearance to the tip of the proximal dendritic segment that is not evident when the freeze-substitution method is used (Steinbrecht, R. 1980).

The intermediate and outer sheath cells seem to be basically similar in structure and function. The former secretes the sensory cuticle, that is, it is the functional trichogen cell, and the latter forms the base or socket as the functional tormogen cell. Where an additional cell is present, the three cells are involved in the formative function. In sensilla with only the inner and one other sheath cell, the inner sheath cell apparently takes on some of the functions of the trichogen cell. Thus, use of the terms related to the role in forming a cuticular part could lead to confusion in the absence of developmental information and given the variability among sensilla. Regardless of the number present, the first or innermost makes at least one complete lap around the inner or dendritic sheath cell, and the others lap it in succession in a similar manner (Figs 27, 29A and 31). They subtend a sinus beneath the base of the sensory cuticle and around the dendritic sheath, which is sealed peripherally by the apical cytoplasm of the outermost cell against the surrounding cuticle. The outer wall of the terminal part of the dendritic sheath cell is usually left exposed to this sinus also. The walls of the cells that abut this sinus are sparsely to profusely lamellate or microvillate, depending on the type of sensillum and the insect. The size of the sinus is also very variable. It is usually larger than the ciliary sinus, is very large in some, and very reduced or not obvious in a few. It is referred to as the upper, the receptor lymph, or the sensillar sinus. The last term is used here. These cells successively draw to one side near the level of the neuronal cytons and terminate in an expansion that contains their nucleus (Fig. 29A). Their proximal ends abut, or are not far removed from,

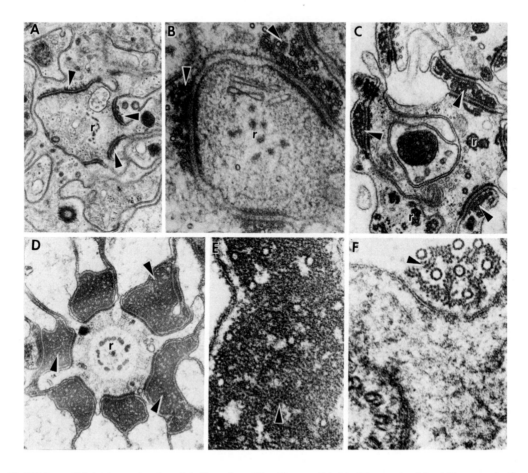

FIG. 32. TEM, t.s. of skeletal or scolopale rods in the region of the ciliary rootlets (r) of the proximal dendritic segments; the rods are in the inner sheath cell and are best developed where this cell junctions with the dendrite, at arrowheads in **A-D**. **A:** in a small campaniform organ ($\times 28,000$); **B:** in a uniporous peg ($\times 62,000$); **C:** in an aporous putative thermo- and hygrosensory peg ($\times 35,000$); **D:** in a mononematic scolopophorous organ ($\times 23,500$); **E,F:** enlargement of these rods in a scolopophorous and a uniporous sensillum to show their longitudinal microtubular and fibrillar composition (fibrils at arrowheads) ($\times 75,200$ and $\times 131,000$, respectively. (**B, D-F**, from Bloom, J. *et al.*, 1981, 1982, courtesy of the authors and The National Research Council of Canada.)

the underlying hemocoel, from which they are separated by a basement membrane.

In addition to their role in forming the cuticular parts, the accessory cells outside the dendritic sheath cell are implicated in a major physiological role. There is mounting evidence that they sequester selected ions and other substances from the underlying hemolymph and secrete into the sensillar sinus (Phillips, C. and Vande Berg, J., 1976a,b). These secretions include mucopolysaccharides (Gnatzy, W., 1978) and selected ions that maintain a transepithelial potential between the sensillar sinus and the base of the sensillum at the hemocoel,

presumably potassium in particular (Küppers, J. and Thurm, U., 1979; Thurm, U. and Küppers, J., 1980; Kaissling, K. and Thorson, J., 1980). The sensillar sinus is continuous with the space alongside or around the dendritic sheath in the sensory cuticles of uniporous sensilla, the lumen in the outer cuticle of many aporous sensilla, with the intergroove sinuses in the double-walled MPG, and with the dendritic chamber in or under the sensory cuticles of most MPP sensilla. The general consensus is, therefore, that these outer sheath cells maintain a chemical environment appropriate to the functioning of the stimulus conduction and transduction processes in the ciliary dendritic segments, their terminations, and the covering sensory cuticle. Whether the dendritic sheath, which forms a barrier between the dendrites and the sensillar sinus in many types of sensilla, is permeable or not is uncertain. Broyles, J. *et al.* (1976) believe that it is selectively permeable to certain ions in the contact chemosensory hair of *Phormia regina*, and that the sensillar sinus, through its extension into the lumen of the hair, acts as an ion reservoir for the adjacent dendritic chamber. There are also many ultrastructural indications that material from this sinus diffuses into the cuticle or through the pore canal or pore stimulus conduction system to maintain the sensory cuticle and its surface coating. In considering the nature of the sensillar liquor, therefore, both the physiological activities of the dendrites relating to their sensing function and the developmental and maintenance needs of the sensory cuticle and its associated parts should be kept in mind. The secretory cells involved are rich in mitochondria, various vesicles and MVB, and coated vesicles and other particles associated with their apical lamellae and microvillae that border the sensillar sinus (Fig. 31A and B), indicative of a major secretory role. They probably also transport materials to the enclosed dendritic sheath cell and, through it, to the ciliary sinus.

The attachment or cap cell of scolopophorous sensilla is believed to be homologous to the outer sheath cell of other types of sensilla that also have only two distal accessory cells. It has few orgenelles that are usually indicative of high secretory or other metabolic activity, but it contains numerous longitudinal microtubules and fibrils. Some of these attach to the dendritic cap proximally and extend with others to its apical point of attachment. Its function here seems to be primarily physical (Fig. 31I; Schmidt, K., 1974; Bloom, J. *et al.*, 1981).

The basal glial cell is near or below the basal region of the neuronal cytons. Its cell body is small, with a small irregular nucleus in which the chromatin is condensed into very large granules. It contains few mitochondria but fairly abundant rough endoplasmic reticulum. The cytoplasm is flattened and attenuated distally and proximally, and wraps around the basal parts of the cyton, always inserting under the wrap of the inner sheath cell around the circumference so that all parts of

every neuron proximal to the base of its cilium are separated from one another by a cytoplasmic layer of either the one sheath cell or the other. Proximally it separates all the axons by several individual cytoplasmic layers around each (Fig. 29B and 30). In sensilla that have an unlapped dendritic sheath cell, there is a tendency for the initial glial wraps around the axon also to be unlapped (Fig. 29C). As in the former, the axon presumably grows through the glial cell in development.

5 SUBCUTICULAR SENSORY SYSTEMS

Some chordotonal sensilla may be classed as subcuticular. Those associated with the sensillar ganglia in the antenna of *Tenebrio molitor* larvae (Bloom, J. *et al.*, 1981) and in the labial palp of elaterid larvae (Bellamy, F., 1973) are attached to basement membranes lining the hemocoel or bounding the sensillar cells and their nerves. However, their structure resembles that of the scolopophorous sensilla that still maintain attachments to the cuticle or in the epidermal layer, and they are presumed to have a related function and a similar developmental origin of parts.

The dendrite associated with the vesicular gland of *Machilis* has a distal ciliary segment and a basal ciliary apparatus in the thicker proximal segment resembling that of neurons of cuticular sensilla, but there are no cuticular parts or accessory cells associated with it other than the gland cell it innervates. The origin and homologies of this sensory system are enigmatic. It seems to be a purely interoceptive Type I bipolar neuron, and is presumed to be osmo- or chemosensing (Bitsch, J. and Palévody, C. 1974).

All other internal sensory systems reported in insects seem to be innervated by bi- or multipolar Type II neurons or derivatives of them. Their fields and associations were discussed earlier. They are common to all insects and have a general distribution throughout the body (i.e., Whitten, J., 1963; Finlayson, L., 1968; Fifield, S. and Finlayson, L., 1978). The dendrites are typically multiterminal (Fig. 33A and B; Osborne, M., 1963; Moulins, M., 1974), and commonly extend along nerves with the axons from other neurons and may be confused with them (Fig. 33C). They have similar inclusions

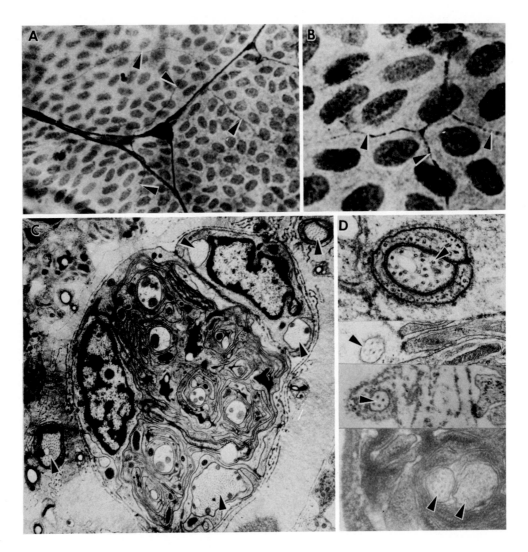

FIG. 33. **A:** LM of a tripolar Type II neuron associated with a small subepidermal sensory nerve, its fine dendritic branches at arrowheads, silver-stained whole mount (×600); **B:** LM of the Type II dendritic branches in a methylene blue whole mount (×2100); **C:** TEM, t.s. of a small subepidermal sensory nerve with Type II dendritic branches in and alongside it at arrowheads (×7200); **D:** terminations of Type II dendritic branches at arrows, from top to bottom, in the hemocoel still in a glial cell wrap (×22,100), and unwrapped in the hemocoel, in basement membrane, and within a cell in the epidermis (to the right in all four instances) (×10,100, ×12,200 and ×17,200, respectively).

and resemble one another in appearance in section. However, the Type II neuronal dendrites generally contain more vesicles, and these and the mitochondria tend to be distributed peripherally near the plasma membrane. A glial sheath cell is associated with the dendrites and their branches, but it tends to form only one lapped wrap around them rather than several typical of the axonal sheath. The fine terminations are usually naked (Fig. 33D). Some of

the Type II neurons associated with lateral nerves have abundant granules and are considered neurosecretory. Those associated with other tissues, such as epidermis and muscle, are considered to be mechanosensory. Whether this is their primary or only role in their association with epidermis is uncertain. They were indirectly implicated in temperature perception in conjunction with thin patches of fenestrated cuticle in grasshoppers, but

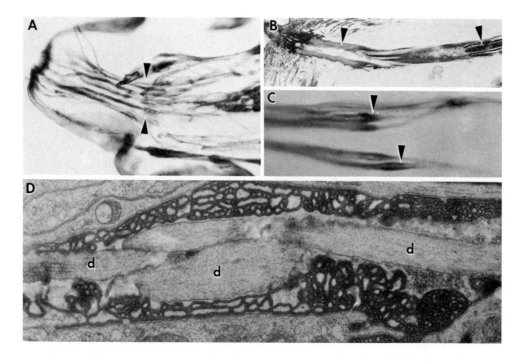

FIG. 34. Moulting sensilla. **A**: LM, l.s. of a maxillary palp, with the exuvial pegs connected to the new pegs forming at the level of the arrowheads by dendrites in their elongated sheaths (× 550); **B**: TEM, l.s. of a uniporous peg, with the cuticular sleeve at the left and the tubular body in the dendritic sheath at the right arrowhead (× 3000); **C**: exuvial dendritic sheaths after ecdysis, with the bases of the original dendritic sheaths at arrowheads; lengths of thinner sheaths extending to the right were formed during the molt, LM of a methylene blue whole-mount (× 1700); **D**: thin and very long but much-folded new dendritic sheath, TEM, l.s., forming beneath the base of the original sheath (to the left), enclosing the ciliary dendritic segments (d) and surrounded by the inner sheath cell that is forming it (× 21,500). **B** and **D**, courtesy of D. A. Scott and the University of Regina; **C**, from Zacharuk, R., 1962b, by permission of The Wistar Institute Press.)

this was found to be non-specific (Makings, P., 1964).

Associations of Type II neurons with the nerve branches from sensory fields were fairly frequently noted in light microscope studies. In TEM studies their dendrites are more rarely reported to be associated with the cells of cuticular sensilla (Moulins, M. 1971b; Cook, G., 1972). The naked dendritic terminations of one such neuron innervate the circumference of the sheath cells of the scolopophorous sensilla in the mandible of elaterid larvae (Zacharuk, R., 1980b). A neuroendocrine role is proposed for these terminations, perhaps in co-ordinating the secretory activities of these cells. The nature of the associations of the Type II neurons and their role, particularly in the epidermis and the cuticular sensilla, deserve much further attention.

6 DEVELOPMENT, MAINTENANCE AND AGING

A detailed discussion of the development of insect sensilla is beyond the scope of this chapter. Only a brief summary and an introduction to some of the literature is presented. In the developing cerci of *Acheta domesticus* embryos, "pioneer axons" and an associated glial cell pathway appear first (Edwards, J. and Chen, S., 1979). The sensillar neurons and associated cells and cuticular parts are differentiated from the epidermal cells during the last embryonic molt to the first instar, and the axons follow the pathway of the pioneer axons and glial cells to the last abdominal ganglion. In the moth *Manduca sexta* virtually the entire population of antennal sensilla in the adult is differentiated during the pupal stage. The sensillar neurons differentiate from epidermal cells by division and quickly develop the axonal and ciliated dendritic processes with a glial cell sheath. The axons extend along a thin "pioneer" pupal nerve toward the supraesophageal ganglion, and the dendrites extend through the epidermis to the cuticular parts formed by the accessory cells that also ensheath them. These cells also differentiate from epidermal cells. All the neurons and sensilla develop synchronously (Sanes, J. and Hildebrand, J. 1976b). In the absence of their axonal synaptic targets the sensilla and their sensory neurons still develop and respond to stimulation normally (Sanes, J. et al., 1976). Both chemo- and mechanosensilla begin to respond to stimulation several days before eclosion. The number responding increases in magnitude to the time of emergence and changes little for 3 days after that, but responses to different stimuli do not develop synchronously (Schweitzer, E. et al., 1976).

The processes of molting and replacement of parts lost by ecdysis was studied in a variety of sensilla and insects (i.e., Wensler, R. and Filshie, B., 1969; Blaney, W. et al., 1971; Gnatzy, W. and Schmidt, K., 1972; Ernst, K., 1972b; Gnatzy, W. and Romer, F., 1980). The accessory cells involved in the secretion of new cuticular parts lost at ecdysis were discussed in section 4.3. In the larva of Barathra brassicae mechanosensory hairs on the exuvial cuticle continued to respond until 10–30 min before ecdysis began. After ecdysis the new hairs attained maximum sensitivity as soon as they were erect (Gnatzy, W. and Tautz, J., 1977). How sensilla with other modalities respond during a molt is not known. In general, the structural integrity of the ciliary segments of the dendrites seems to be maintained within the exuvial dendritic sheath until it is shed. At this point, the length within the sheath is lost with it and a new functional dendritic termination is formed in the new sheath. To maintain the structural and functional integrity of the dendrites in the widening ecdysial space, a considerable length of dendritic sheath is elaborated during the molt, which is also shed with the exuvium (Fig. 34; Zacharuk, R. 1962b). In newly molted insects there are usually numerous dense inclusions in the cytoplasm of the sheath cells. These are most probably lysosomes that are autolyzing organelles no longer needed after the cuticular parts are formed.

The physical maintenance of the cuticular parts, as well as the functional maintenance of the sensory response, must continue after ecdysis. As already discussed, this seems to occur from the outer accessory cells through the sensillar sinus and the cuticular pore canal system, and by the dendritic sheath cell and the distal dendritic segments through the ciliary sinus. Researchers should be cognizant of the dual function of all the sensillar cells, including the neurons, in the physical maintenance of the cuticular parts as well as the physiological maintenance of sensory function.

In electrophysiological studies there has been a tendency to discard preparations of sensilla that do not perform. Why do they not respond? Some may not because of a closure mechanism operated by hydraulic pressure from the sensillar sinus as proposed by Blaney, W. et al. (1971) for the pore of a uniporous peg in Locusta migratoria. Other contact chemosensilla with cuticular fingers around the pore may have a similar closure mechanism, opening and closing the pore to chemical stimuli by spreading or clustering tightly, as illustrated by Mitchell, B. and Schoonhoven, L. (1974). There is mounting evidence now that sensilla undergo a developmental and an aging process that has a significant bearing on their response (Saxena, K., 1967). It can be altered by the foods or chemicals present in their environment during development (Dethier, V. and Goldrich, N. 1971). The previous nutritional history of the insect may be a significant factor, particularly in behavioral studies (Davies, D. and Lall, S. 1970). The age of an adult insect after eclosion is significant with respect to the number of sensilla that respond (O'Ceallachain, D. and Ryan, M., 1977; Stoffolano, J. et al., 1978; Seabrook, W. et al., 1979; Ross, R. et al., 1979). Generally there is an increase in the number of chemosensilla that respond after eclosion to a maximum, followed by a decline, but this varies between sexes and between sensilla in different locations in a particular field. In the uniporous sensilla on the labella and tarsi of Phormia regina, increased losses of a viscous substance through the lateral walls, presumably with damage and age, decreases the extrusion of substances through the apical pore (Angioy, A. et al., 1980). It is suggested that this may be the cause for loss of response with age and an increase in the electrical resistance of sensilla that still respond. There are no other known changes in structure in the stimulus conduction, energy transduction, or impulse generating systems that can be correlated with the development or loss of response in insect sensilla with age.

7 TRANSDUCTION MECHANISMS

In mechanosensilla, to summarize current theories and models, the stimulus is physically conducted or transferred through appropriate cuticular linkages or other attachments to the dendrite enclosed in its sheath or cap, causing a physical deformation in its form and membrane. This results in a change in membrane potential, which generates the nerve impulses if adequate. The details are discussed in the next chapter. A similar principle is suggested for hygrosensilla, with changes in the dimensions or form of some hygroscopic cuticular component linked to the dendritic sheath providing the force for the deformation of the dendrite. Temperature changes affect the response characteristics and activities of nerve tissue generally, and may be perceived by all sensory neurons of both types. In addition there are specific sensilla that are sensitive to hot or cold, and these typically have a neuron with a lamellate distal dendritic segment. A biochemical transduction mechanism resembling that in organs of vision is one proposal, particularly in response to radiant heat, as discussed by Corbière-Tichané, G. (1977). There is still much open to question in the functional morphology of such receptors.

In both gustatory and olfactory chemosensilla the general assumption is that the stimulant molecules are trapped by a solvent from the dendritic chamber that coats the pores and surrounding surfaces of sensory cuticle, diffuse in this substance into the dendritic chamber and onto the membranes of the dendritic terminations. Specific stimulants combine with a protein acceptor specific to it, and this initiates a chain of events that leads to the initiation of nerve impulses. (Steinbrecht, R., 1973; Altner, H. 1977b; Gaffal, K., 1979). The conducting solvent may even be modal specific in some sensilla. Moulins, M. and Noirot, Ch. (1972) suggest that it may act as an ion exchanger in the pore of uniporous sensilla, which implicates it also in the energy transduction process. The response is reversibly blocked in both uniporous and multiporous

sensilla by organic mercurials, an indication that membrane acceptor protein or some other moiety in the transduction process contains –SH (Shimada, I. et al., 1972; Rozental, J. and Norris, D. 1973; Frazier, J. and Heitz, J. 1975). Zacharuk, R. (1962c) noted that, although there are –SS– or –SH compounds in the dendritic regions of cuticular sensilla, they are far more concentrated in the axons from these sensilla, but not in ocellar or efferent axons. Perhaps the mercurials block nerve impulse conduction as well as, or in addition to, their initiation.

The microtubules in the distal dendritic segment are also implicated directly in the transduction process by transmitting an event initiated by the stimulus at the dendritic tip to the proximal dendrite where a number of workers believe the nerve impulse is initiated (Atema, J., 1973, 1975; Moran, D. et al., 1977). Vinblastine and colchicine disrupt the distal dendritic microtubules in the taste hairs of blowflies, and the response of these sensilla to stimulation is altered or blocked by such treatment (Matsumoto, D. and Farley, R., 1978). Microtubule function is also disrupted by organic mercurials (Dustin, P., 1978). Adenylate cyclase activity is present near the dendritic tips, suggesting the possibility of a cAMP-mediated microtubule involvement in the transduction process, as discussed by Olsen, R. (1975) and Gaffal, K. (1979). Villet, R. (1978) presents a model involving adenyl cyclase in olfactory transduction, and Ferkovich, S. et al. (1982) implicate esterases on and beneath sensory cuticles in preventing surface accumulation, and clearing olfactory receptor sites, of stimulating molecules. The recent literature on receptor membrane function in olfaction and gustation is surveyed by Ma, W.-C. (1981). The freeze-fixation techniques for TEM sections (Steinbrecht, R., 1980), with and without prior stimulation, selected treatment, and electrophysiological monitoring, may provide an insight into the role of microtubules in the transduction process of all types of sensilla. Chemical fixatives stimulate chemosensilla into a burst of activity before the cells die (P. J. Albert and R. Y. Zacharuk, unpublished observations), and may mask comparisons between stimulated and nonstimulated sensilla.

Septate and gap junctions are common between proximal dendritic segments in multineuronal sensilla as well as in the laps of, and among, the accessory sheath cells (Moulins, M. and Noirot, Ch., 1972). Gap and tight junctions between the distal dendritic segments are also reported, and one instance of a gap junction between two axons from one sensillum was noted in *Tenebrio molitor* larvae (Bloom, J. *et al*., 1982b). These should allow for some peripheral integration of the sensillar response. Moulins and Noirot also suggest that the Type II neuronal terminations in cuticular sensilla may be involved in a centrifugal control, inhibiting or modulating their afferent activity.

8 LOCALIZATION AND INTEGRATION OF FUNCTION

Various specific aspects of the location and integration of sensillar function were covered in preceding sections. Others are dealt with in other chapters. Some aspects are briefly introduced or summarized here.

Mention was made earlier that an entire appendage may function as a sensillum for a particular response. Antennae are an excellent example in their function as a whole unit in monitoring air currents (Gewecke, M. and Heinzel, H. 1980), in gravity reception (Horn, E. and Kessler, W. 1975), and in monitoring sound in male mosquitoes, as discussed earlier. Other body parts are similarly integrated to function as a whole unit in mechanosensing as extero- or proprioceptors.

In chemosensing the antennae have the greatest concentration of MPP and MPG sensilla of any sensory fields. The MPG and specialist MPP are almost exclusively on antennae. Thus, they are the primary olfactory organs in long- and short-range sensing of mates, hosts, oviposition sites, and the like. There are usually also some uniporous sensilla for near-field tasting by contact. Some MPP also are on the tips of the maxillary and labial palps and on the ovipositor of some insects. These are probably second-order olfactory fields used in close-range probing. All the chemosensilla on the tarsi, most of those on the tips of the labial palps and the maxillary palps and galea, and all of those in the cibarium (on the epi- and hypopharynx) are uniporous types that taste by contact. There are also some on the ovipositor and cerci. Those in the cibarium and on the ovipositor are third-order, and on the other appendages, second-order, organs of taste. For example the butterfly *Pieris brassicae* (Behan, M. and Schoonhoven, L., 1978) and the mosquito *Aedes aegypti* (Davis, E., 1977) monitor the attractiveness of oviposition sites by antennal olfaction, and the butterfly *Delia brassicae* tastes the host plant surface with tarsal hairs before ovipositing (Städler, E., 1978). The cibarial papillae are second check points for completion of ingestion after it is initiated through taste sensilla on tarsi, maxillae and labium in phytophagous (De Boer, G. *et al*., 1977) and hematophagous (Salama, H., 1966) insects. The number and distribution of chemosensilla on the antennae and palps in various species of *Culicoides* is related to their host preference (Braverman, Y. and Hulley, P. 1979), and the cibarial sensilla control the swallowing of the bloodmeal in all. The MPP sensilla that respond to CO_2 are usually on the antenna of adults, but in elaterid and other beetle larvae they are on the maxillary and labial palps (Doane, J. and Klinger, J., 1978).

Specific thermo- and hygrosensitive sensilla are primarily on the antennae of insects. In three species of Lepidoptera larvae there is at least one additional thermosensitive neuron in the maxillary palp (Schoonhoven, L., 1967) and, as already indicated, neurons sensitive to temperature changes also occur in tarsi and other parts of the body. In *Drosophila melanogaster*, while the primary hygrosensilla are on the antennae, there are some also on the palps and proboscis (Syrjämäki, J., 1962). This author discusses extensively the humidity responses of these and other insects.

For the past several years there has been an increasing trend toward applying the techniques used in the mass of information derived from the more recent studies on individual sensilla to studies on the mechanisms of integration of all the afferent information by an insect, and its total response to it. In considering the various chemosensory processes involved in feeding, Schoonhoven, L. (1977) indicates that the sensory fields, by smell and taste, monitor the phagostimulants, deterrents and the nutritive quality of the food. They transmit this information in a complex pattern of nerve activities from the antennae to the brain, from the outer mouthparts to the subesophageal ganglion, and from the cibarium to the brain and frontal ganglion, where it is integrated for appropriate response. Each insect has a physiologically characteristic receptive system that is never completely identical with that of any other species, as demonstrated by van der Pers, J. (1982) in moths of *Yponomelita* spp. Dethier, V. (1977) suggests that there is also considerable peripheral integration, in the number and kinds of sensilla, the band width of each, character of each tuning curve, absolute threshold and rate of adaptation, synergism and inhibition. These will be detailed further in chapter 3. However, the possibility of interneurons in small peripheral ganglia should not be discounted.

REFERENCES

ACCORDI, F. and SBORDONI, V. (1977). The fine structure of Hamann's organ in *Leptodirus hohenwarti*, a highly specialized cave Bathysciinae (Coleoptera, Catopidae). *Int. J. Speleol. 9*, 153–165.

ADAMS, J. R. and FORGASH, A. J. (1966). The location of the contact chemoreceptors of the stable fly, *Stomoxys calcitrans* (Diptera: Muscidae). *Ann. Ent. Soc. Amer. 59*, 133–141.

ADAMS, J. R., HOLBERT, P. E. and FORGASH, J. (1965). Electron microscopy of the contact chemoreceptors of the stable fly, *Stomoxys calcitrans* (Diptera: Muscidae). *Ann. Ent. Soc. Amer. 58*, 909–917.

ALBERT, P. J. (1980). Morphology and innervation of mouthpart sensilla in larvae of the spruce budworm, *Choristoneura fumiferana* (Clem.) (Lepidoptera: Tortricidae). *Canad. J. Zool. 58*, 842–851.

ALBERT, P. J. and SEABROOK, W. D. (1973). Morphology and histology of the antenna of the male eastern spruce budworm, *Choristoneura fumiferana* (Clem.) (Lepidoptera: Tortricidae). *Canad. J. Zool. 51*, 443–448.

ALBERT, P. J., ZACHARUK, R. Y., and WONG, L. (1976). Structure, innervation, and distribution of sensilla on the wings of a grasshopper. *Canad. J. Zool. 54*, 1542–1553.

ALTNER, H. (1975). The microfiber texture in a specialized plastic cuticle area within a sensillum field on the cockroach maxillary palp as revealed by freeze fracturing. *Cell. Tiss. Res. 165*, 79–88.

ALTNER, H. (1977a). Insect sensillum specificity and structure: an approach to a new typology. In *Olfaction and Taste VI* (Paris). Edited by J. Magnen and P. MacLeod. Pages 295–303. Information Retrieval, London.

ALTNER, H. (1977b). Insektensensillen: Bau- und Funktionsprinzipien. *Verh. Dtsch. Zool. Ges. 70*, 139–153.

ALTNER, H. and THIES, G. (1972). Reizleitende Strukturen und Ablauf der Häutung an Sensillen einer euedaphischen Collembolenart. *Z. Zellforsch. Mikr. Anat. 129*, 196–216.

ALTNER, H. and THIES, G. (1976). The postantennal organ: A specialized unicellular sensory input to the protocerebrum in apterygotan insects (Collembola). *Cell Tiss. Res. 167*, 97–110.

ALTNER, H. and THIES, G. (1978). The multifunctional sensory complex in the antenna of *Allacma fusca* (Insecta). *Zoomorphol. 91*, 119–131.

ALTNER, H., ROUTIL, CH. and LOFTUS, R. (1981). The structure of bimodal chemo-, thermo-, and hygroreceptive sensilla on the antenna of *Locusta migratoria*. *Cell. Tiss. Res. 215*, 289–308.

ALTNER, H., SASS, H. and ALTNER, I. (1977). Relationship between structure and function of antennal chemo-, hygro-, and thermoreceptive sensilla in *Periplaneta americana*. *Cell Tiss. Res. 176*, 389–405.

ALTNER, H., TICHY, H. and ALTNER, I. (1978). Lamellated outer dendritic segments of a sensory cell within a poreless thermo- and hygroreceptive sensillum of the insect *Carausius morosus*. *Cell. Tiss. Res. 191*, 287–304.

ANGIOY, A. B., LISCIA, A. and PIETRA, P. (1981). Electrophysiological responses of wing chemosensilla in *Phormia regina* (Meig.) to NaCl and sucrose. *Boll. Soc. It. Biol. Sper. 17*, 588–594.

ANGIOY, A. M., YIN, L. R-S., LISCIA, A. and PIETRA, P. (1980). Lateral extrusions of viscous substance in the labellar and tarsal chemosensilla of *Phormia regina* (Meig.). *Boll. Soc. It. Biol. Sper. 16*, 1851–1856.

ATEMA, J. (1973). Microtubule theory of sensory transduction. *J. Theor. Biol. 38*, 181–190.

ATEMA, J. (1975). Stimulus transmission along microtubules in sensory cells: an hypothesis. In *Microtubules and Microtubule Inhibitors*. Edited by M. Borgers and M. de Brabander. Pages 247–257. North-Holland, Amsterdam.

BACKUS, E. A. and MCLEAN, D. L. (1982). The sensory systems and feeding behavior of leafhoppers. I. The aster leafhopper *Macrosteles fascifrons* Stål (Homoptera, Cicadellidae). *J. Morphol. 172*, 361–379.

BALL, E. E. and STONE, R. C. (1982). The cercal receptor system of the praying mantid, *Archimantis brunneriana* Sauss. I. Cercal morphology and receptor types. *Cell Tiss. Res. 224*, 55–70.

BARBER, V. C. (1974). Cilia in sense organs. In *Cilia and Flagella*. Edited by M. A. Sleigh. Pages 403–433. Academic Press, London.

BARETH, C. and JUBERTHIE-JUPEAU, L. (1977). Ultrastructure des soies sensorielles des palpes labiaux de *Campodea sensillifera* (Conde et Mathieu) (Insecta: Diplura). *Int. J. Insect Morphol. Embryol. 6*, 191–200.

BARLIN, M. R. and VINSON, S. B. (1981). Multiporous plate sensilla in antennae of the Chalcidoidea (Hymenoptera). *Int. J. Insect Morphol. Embryol. 10*, 29–42.

BAY, D. E. and PITTS, C. W. (1976). Antennal olfactory sensilla of the face fly, *Musca autumnalis* DeGreer (Diptera: Muscidae). *Int. J. Insect Morphol. Embryol. 5*, 1–16.

BEHAN, M. and RYAN, M. F. (1977). Sensory receptors on the ovipositor of the carrot fly (*Psila rosae* (F.)) (Diptera: Psilidae) and the cabbage root fly (*Delia brassicae*) (Wiedemann) (Diptera: Anthomyiidae). *Bull. Ent. Res. 67*, 383–389.

BEHAN, M. and RYAN, M. F. (1978). Ultrastructure of antennal sensory receptors of *Tribolium* larvae (Coleoptera: Tenebrionidae). *Int. J. Insect Morphol. Embryol. 7*, 221–236.

BEHAN, M. and SCHOONHOVEN, L. M. (1978). Chemoreception of an oviposition deterrent associated with eggs in *Pieris brassicae*. *Ent. Exp. Appl. 24*, 163–179.

BELLAMY, F. W. (1973). Ultrastructure of the labial palp and its associated sensilla of the prairie grain wireworm *Ctenicera destructor* (Brown) (Elateridae: Coleoptera). Ph.D. thesis, University of Regina, Regina, Saskatchewan.

BELLAMY, F. W. and ZACHARUK, R. Y. (1976). Structure of the labial palp of a larval elaterid (Coleoptera) and of sinus cells associated with its sensilla. *Canad. J. Zool. 54*, 2118–2128.

BITSCH, J. and PALÉVODY, C. (1974). Mise en évidence de récepteurs sensoriels dans les glandes vésiculaires des Machilides (Insecta, Thysanura). Etude ultrastructurale. *C. R. Acad. Sci. Paris, Série D, 278*, 2643–2646.

BLAND, R. G. (1982). Morphology and distribution of sensilla on the antennae and mouthparts of *Hypochlora alba* (Orthoptera: Acrididae). *Ann. Ent. Soc. Amer. 75*, 272–283.

BLANEY, W. M. (1977). The ultrastructure of an olfactory sensillum on the maxillary palps of *Locusta migratoria* (L.). *Cell Tiss. Res. 184*, 397–409.

BLANEY, W. M. and CHAPMAN, R. F. (1969). The anatomy and histology of the maxillary palp of *Schistocerca gregaria* (Orthoptera, Acrididae). *J. Zool. 157*, 509–535.

BLANEY, W. M., CHAPMAN, R. F. and COOK, A. G. (1971). The structure of the terminal sensilla on the maxillary palps of *Locusta migratoria* (L.), and changes associated with moulting. *Z. Zellforsch. Mikr. Anat. 121*, 48–68.

BLOOM, J. W., ZACHARUK, R. Y. and HOLODNIUK, A. E. (1981). Ultrastructure of a terminal chordotonal sensillum in larval antennae of the yellow mealworm, *Tenebrio molitor* L. *Canad. J. Zool. 59*, 515–524.

BLOOM, J. W., ZACHARUK, R. Y. and HOLODNIUK, A. E. (1982a). Ultrastructure of the larval antenna of *Tenebrio molitor* L. (Coleoptera: Tenebrionidae): structure of the trichoid and uniporous peg sensilla. *Canad. J. Zool. 60*, 1528–1544.

BLOOM, J. W., ZACHARUK, R. Y. and HOLODNIUK, A. E. (1982b). Ultrastructure of the larval antenna of *Tenebrio molitor* L. (Coleoptera: Tenebrionidae): structure of the blunt tipped peg and papillate sensilla. *Canad. J. Zool. 60*, 1545–1556.

BOECKH, J. (1962). Elektrophysiologische Untersuchungen an einzelnen Geruchsrezeptoren auf den Antennen des Totengrabers (*Necrophorus*, Coleoptera). *Z. Vergl. Physiol.* 46, 212–248.

BOECKH, J., ERNST, K. D., SASS, H. and WALDOW, U. (1975). Coding of odor quality in the insect olfactory pathway. In *Olfaction and Taste V*. Edited by D. A. Denton and J. P. Coghlan. Pages 239–245. Academic Press, New York.

BOECKH, J., KAISSLING, K.-E. and SCHNEIDER, D. (1960). Sensillum und Bau der Antennengeissel von *Telea polyphemus* (Vergleiche mit weiteren Saturniden: *Antheraea*, *Platysamia* und *Philosamia*). *Zool. Jb. Anat.* 78, 559–584.

BOO, K. S. (1981). Discontinuity between ciliary root processes and triple microtubules of the distal basal body in mosquito sensory cilia. *Korean J. Ent.* 11, 5–18.

BOO, K. S. and McIVER, S. B. (1976). Fine structure of surface and sunken grooved pegs on the antenna of female *Anopheles stephensi* (Diptera: Culicidae). *Canad. J. Zool.* 54, 235–244.

BORDEN, J. H., MILLER, G. E. and RICHERSON, J. V. (1973). A possible new sensillum on the antennae of *Itoplectis conquisitor* (Hymenoptera: Ichneumonidae). *Canad. Ent.* 105, 1363–1367.

BORDEN, J. H., ROSE, A. and CHORNEY, R. J. (1978). Morphology of the elongate sensillum placodeum on the antennae of *Aphidius smithi* (Hymenoptera: Aphidiidae). *Canad. J. Zool.* 56, 519–525.

BORG, T. K. and NORRIS, D. M. (1971). Ultrastructure of sensory receptors on the antennae of *Scolytus multistriatus* (Marsh.). *Z. Zellforsch. Mikr. Anat.* 113, 13–28.

BRANDENBURG, J. and MATUSCHKA, F. (1976). Feinstruktur der kuppelformigen Chemorezeptoren auf den Fuhlern von Facherflugler-Mannchen (Strepsiptera: Stylopidae: Stylops). *Ent. Germ.* 2, 341–349.

BRAVERMAN, Y. and HULLEY, P. E. (1979). The relationship between the numbers and distribution of some antennal and palpal sense organs and host preference in some *Culicoides* (Diptera: Ceratopogonidae) from southern Africa. *J. Med. Ent.* 15, 419–424.

BROMLEY, A. K., DUNN, J. A. and Anderson, M. (1979). Ultrastructure of the antennal sensilla of aphids. I. Coeloconic and placoid sensilla. *Cell Tiss. Res.* 203, 427–442.

BROYLES, J. L., HANSON, F. E. and SHAPIRO, A. M. (1976). Ion dependence of the tarsal sugar receptor of the blowfly *Phormia regina*. *J. Insect Physiol.* 22, 1587–1600.

BULLOCK, T. H. and HORRIDGE, G. A. (1965). *Structure and Function in the Nervous Systems of Invertebrates*. 1st edn. W. H. Freeman, London.

CALLAHAN, P. S. (1975). Insect antennae with special reference to the mechanism of scent detection and the evolution of the sensilla. *Int. J. Insect Morphol. Embryol.* 4, 381–430.

CALVERT, H. (1974). The external morphology of foretarsal receptors involved with host discrimination by the nymphalid butterfly, *Chlosyne lacinia*. *Ann. Ent. Soc. Amer.* 67, 853–856.

CAMPBELL, F. L. and PRIESTLEY, J. D. (1970). Flagellar annuli of *Blatella germanica* (Dictyoptera: Blattellidae). Changes in their numbers and dimensions during postembryonic development. *Ann. Ent. Soc. Amer.* 63, 81–88.

CHAPMAN, R. F. and THOMAS, J. G. (1978). The numbers and distribution of sensilla on the mouthparts of Acridoidea. *Acrida* 7, 115–148.

CHU, I-WU and AXTELL, R. C. (1971). Fine structure of the dorsal organ of the house fly larva, *Musca domestica* L. *Z. Zellforsch. Mikr. Anat.* 117, 17–34.

CHU-WANG, I-WU and AXTELL, R. C. (1972a). Fine structure of the terminal organ of the house fly larva, *Musca domestica* L. *Z. Zellforsch. Mikr. Anat.* 127, 287–305.

CHU-WANG, I-WU and AXTELL, R. C. (1972b). Fine structure of the ventral organ of the house fly larva, *Musca domestica* L. *Z. Zellforsch. Mikr. Anat.* 130, 489–495.

CHU-WANG, I-WU, AXTELL, R. C. and KLINE, D. L. (1975). Antennal and palpal sensilla of the sand fly *Culicoides furens* (Poey) (Diptera: Ceratopogonidae). *Int. J. Insect Morphol. Embryol.* 4, 131–149.

COOK, A. G. (1972). The ultrastructure of the A1 sensilla on the posterior surface of the clypeo-labrum of *Locusta migratoria migratorioides* (R and F). *Z. Zellforsch. Mikr. Anat.* 134, 539–554.

CORBIÈRE, G. (1969). Ultrastructure et électrophysiologie du lobe membraneux de l'antenne chez la larve du *Speophyes lucidulus* (Coléoptère). *J. Insect Physiol.* 15, 1759–1765.

CORBIÈRE-TICHANÉ, G. (1973). Sur les structures sensorielles et leurs fonctions chez la larve de *Speophyes lucidulus*. *Ann. Spéléol.* 28, 247–265.

CORBIÈRE-TICHANÉ, G. (1974). Fine structure of an antennal sensory organ ("vesicule olfactive") of *Speophyes lucidulus* Delar. (cave Coleoptera of the *Bathysciinae* subfamily). *Tissue Cell* 6, 535–550.

CORBIÈRE-TICHANÉ, G. (1977). Données et hypotheses sur la fonction du recepteur sensoriel en lamelles chez les Coleopteres cavernicoles. *Bull. Soc. Zool. France* 102, 31–38.

CORBIÈRE-TICHANÉ, G. and BERMOND, N. (1972). Comparative study on the lamellated nervous structures in the antenna of certain Coleoptera. *Z. Zellforsch. Mikr. Anat.* 127, 9–33.

CRAIG, D. A. and BORKENT, A. (1980). Intra- and inter-familial homologies of maxillary palpal sensilla of larval Simuliidae (Diptera: Culicomorpha). *Canad. J. Zool.* 58, 2264–2279.

CRAIG, D. A. and BATZ, H. (1982). Innervation and fine structure of antennal sensilla of Simuliidae larvae (Diptera:Culicomorpha). *Canad. J. Zool.* 60, 696–711.

DALLAI, R. and NOSEK, J. (1981). Ultrastructure of sensillum t₁ on the foretarsus of *Acerentomon majus* Berlese (Protura: Acerentomidae). *Int. J. Insect Morphol. Embryol.* 10, 321–330.

DALY, P. J. and RYAN, M. F. (1979). Ultrastructure of antennal sensilla of *Nebria brevicollis* (Fab.) (Coleoptera: Carabidae). *Int. J. Insect Morphol. Embryol.* 8, 169–181.

DALY, P. J. and RYAN, M. F. (1981). Ultrastructural relationships of the lamellar or onion body in an insect sensillum. *Int. J. Insect Morphol. Embryol.* 10, 83–87.

DAVIES, D. M. and LALL, S. B. (1970). Differences in gustatory responses of newly emerged and field-caught females of *Hybomitra lasiophthalma* (Diptera: Tabanidae) to sucrose solutions. *Ann. Ent. Soc. Amer.* 63, 1192–1193.

DAVIS, E. E. (1977). Response of the antennal receptors of the male *Aedes aegypti* mosquito. *J. Insect Physiol.* 23, 613–617.

DAVIS, E. E. and Sokolove, P. G. (1975). Temperature responses of antennal receptors of the mosquito, *Aedes aegypti*. *J. Comp. Physiol.* 96, 223–236.

DE BOER, G., DETHIER, V. G. and SCHOONHOVEN, L. M. (1977). Chemoreceptors in the preoral cavity of the tobacco hornworm, *Manduca sexta*, and their possible function in feeding behaviour. *Ent. Exp. Appl.* 21, 287–298.

DEN OTTER, C. J., SCHUIL, H. A. and VAN OOSTEN, A. S. (1978). Reception of host-plant odours and female sex pheromone in *Adoxophyes orana* (Lepidoptera: Tortricidae): electrophysiology and morphology. *Ent. Exp. Appl.* 24, 570–578.

DETHIER, V. G. (1953). Vision. Mechanoreception. Chemoreception. In *Insect Physiology*. Edited by K. D. Roeder. Pages 488–576. John Wiley, New York.

DETHIER, V. G. (1963). *The Physiology of Insect Senses*. Methuen, London; John Wiley, New York.

DETHIER, V. G. (1972). Sensitivity of the contact chemoreceptors of the blowfly to vapors. *Proc. Natl. Acad. Sci. U.S.A.* 69, 2189–2192.

DETHIER, V. G. (1976). *The Hungry Fly*. Harvard University Press, Massachusetts.

DETHIER, V. G. (1977). Gustatory sensing of complex mixed stimuli by insects. In *Olfaction and Taste VI*. Paris. Edited by J. LeMagnen and P. MacLeod. Pages 323–331. Information Retrieval, London.

DETHIER, V. G. and GOLDRICH, N. (1971). Blowflies: alteration of adult taste responses by chemicals present during development. *Science* 173, 242–244.

DETHIER, V. G., LARSON, J. R. and ADAMS, J. R. (1963). The fine structure of the olfactory receptors of the blowfly. In *Olfaction and Taste I*. Edited by Y. Zotterman. Pages 105–110. Pergamon Press, Oxford.

DOANE, J. F. and KLINGER, J. (1978). Location of CO_2-receptive sensilla on larvae of the wireworms *Agriotes lineatus-obscurus* and *Limonius californicus*. *Ann. Ent. Soc. Amer.* 71, 357–363.

VAN DRONGELEN, W. (1979). Contact chemoreception of host plant specific chemicals in larvae of various *Yponomeuta* species (Lepidoptera). *J. Comp. Physiol.* 134, 265–279.

DUSTIN, P. (1978). *Microtubules*. Springer-Verlag, Berlin and New York.

DYER, L. J. and SEABROOK, W. D. (1975). Sensilla on the antennal flagellum of the sawyer beetles *Monochamus notatus* (Drury) and *Monochamus scutellatus* (Say) (Coleoptera: Cerambycidae). *J. Morphol.* 146, 513–532.

EATON, J. L. (1979). Chemoreceptors in the cibario-pharyngeal pump of the cabbage looper moth, *Trichoplusia ni* (Lepidoptera: Noctuidae). *J. Morphol. 160*, 7–16.

EDWARDS, J. S. and CHEN, S.-W. (1979). Embryonic development of an insect sensory system, the abdominal cerci of *Acheta domesticus*. *Wilhelm Roux's Archiv. 186*, 151–178.

ERNST, K.-D. (1969). Die Feinstruktur von Riechsensillen auf der Antenne des Aaskäfers *Necrophorus* (Coleoptera). *Z. Zellforsch. Mikr. Anat. 94*, 72–102.

ERNST, K.-D. (1972a). *Sensillum coelosphaericum*, die Feinstruktur eines neuen olfaktorischen Sensilientypes. *Z. Zellforsch. Mikr. Anat. 132*, 95–106.

ERNST, K.-D. (1972b). Die Ontogenie der basiconischen Riechsensillen auf der Antenne von *Necrophorus* (Coleoptera). *Z. Zellforsch. Mikr. Anat. 129*, 217–236.

ERNST, K.-D., BOECKH, J. and BOECKH, V. (1977). A neuroanatomical study on the organization of the central antennal pathways in insects. II. Deutocerebral connections in *Locusta migratoria* and *Periplaneta americana*. *Cell Tiss. Res. 176*, 285–308.

ESSLEN, J. and KAISSLING, K.-E. (1976). Zahl und Verteilung antennaler Sensillen bei der Honigbiene (*Apis mellifera* L.). *Zoomorphol. 83*, 227–251.

EVANS, W. G. (1966). Morphology of the infrared sense organs of *Melanophila acuminata* (Buprestidae: Coleoptera). *Ann. Ent. Soc. Amer. 59*, 873–877.

EVANS, W. G. (1975). Wax secretion in the infrared sensory pit of *Melanophila acuminata* (Coleoptera: Buprestidae). *Quaest. Ent. 11*, 587–589.

FELT, B. T. and VANDE BERG, J. S. (1976). Ultrastructure of the blowfly chemoreceptor sensillum (*Phormia regina*). *J. Morphol. 150*, 763–784.

FERKOVICH, S. M., OLIVER, J. E. and DILLARD, C. (1982). Pheromone hydrolysis by cuticular and interior esterases of the antennae, legs, and wings of the cabbage looper moth *Trichoplusia ni* (Hübner). *J. Chem. Ecol. 8*, 859–866.

FIFIELD, S. M. and FINLAYSON, L. H. (1978). Peripheral neurons and peripheral neurosecretion in the stick insect, *Carausius morosus*. *Proc. R. Soc. Lond. B 200*, 63–85.

FINLAYSON, L. H. (1968). Proprioceptors in invertebrates. In *Invertebrate Receptors*. Edited by J. D. Carthy and G. E. Newell. Pages 217–249. Academic Press, London.

FRAZIER, J. L. and HEITZ, J. R. (1975). Inhibition of olfaction in the moth *Heliothis virescens* by the sulfhydryl reagent fluorescein mercuric acetate. *Chem. Senses and Flavor. 1*, 271–281.

FUDALEWICZ-NIEMCZYK, W., OLEKSY, M. and ROŚCISZEWSKA, M. (1980). The peripheral nervous system of the larva of *Gryllus domesticus* L. (Orthoptera) Part III. Legs. *Acta Biol. Cracov. Series: Zool. 22*, 51–63.

GAFFAL, K. P. (1979). An ultrastructural study of the tips of four classical bimodal sensilla with one mechanosensitive and several chemosensitive receptor cells. *Zoomorphol. 92*, 273–291.

GAFFAL, K. P. (1981). Terminal sensilla on the labium of *Dysdercus intermedius* Distant (Heteroptera: Pyrrhocoridae). *Int. J. Insect Morphol. Embryol. 10*, 1–6.

GAFFAL, K. P. and BASSEMIR, U. (1974). Vergleichende Untersuchung modifizierter Cilienstrukturen in den Dendriten mechano- und chemosensitiver Rezeptorzellen der Baumwollwanze *Dysdercus* und der Libelle *Agrion*. *Protoplasma 82*, 177–202.

GALIC, M. (1971). Die Sinnesorgane an der Glossa, dem Epipharynx und dem Hypopharynx der Arbeiterin von *Apis mellifica* L. (Insecta, Hymenoptera). *Z. Morphol. Tiere 70*, 201–228.

GEWECKE, M. (1979). Central projection of antennal afferents for the flight motor in *Locusta migratoria* (Orthoptera: Acrididae). *Ent. Gen. 5*, 317–320.

GEWECKE, M. and HEINZEL, H.-G. (1980). Aerodynamic and mechanical properties of the antennae as air-current sense organs in *Locusta migratoria* I. Static characteristics. *J. Comp. Physiol. 139*, 357–366.

GNATZY, W. (1978). Tormogen cell and receptor-lymph space in insect olfactory sensilla. *Cell Tiss. Res. 189*, 549–554.

GNATZY, W. and ROMER, F. (1980). Morphogenesis of mechano-receptor and epidermal cells of crickets during the last instar, and its relation to molting-hormone level. *Cell Tiss. Res. 213*, 369–391.

GNATZY, W. and SCHMIDT, K. (1972). The fine structure of the sensory hairs on the cerci of *Gryllus bimaculatus* Deg. (Saltatoria, Gryllidae). IV. Ecdysis of short bristles. *Z. Zellforsch. Mikr. Anat. 126*, 223–239.

GNATZY, W. and TAUTZ, J. (1977). Sensitivity of an insect mechanoreceptor during moulting. *Physiol. Ent. 2*, 279–288.

GREANY, P. D., HAWKE, S. D., CARLYSLE, T. C. and ANTHONY, D. W. (1977). Sense organs in the ovipositor of *Biosteres (Opius) longicaudatus*, a parasite of the Caribbean fruit fly *Anastrepha suspensa*. *Ann. Ent. Soc. Amer. 70*, 319–321.

GUSE, G.-W. and HONOMICHL, K. (1980). Die digitiformen Sensillen auf dem Maxillarpalpus von Coleoptera. II. Feinstruktur bei *Agabus bipustulatus* (L.) und *Hydrobius fuscipes* (L.). *Protoplasma 103*, 55–68.

HAAS, H. (1955). Untersuchungen zur Segmentbildung an der Antenne von *Periplaneta americana* L. *Roux' Arch. Entwickl. Mech. 147*, 434–473.

HALLBERG, E. (1979). The fine structure of the antennal sensilla of the pine saw fly *Neodiprion sertifer* (Insecta: Hymenoptera). *Protoplasma 101*, 111–126.

HALLBERG, E. (1982). Sensory organs in *Ips typographus* (Insecta: Coleoptera). Fine Structure of the sensilla of the maxillary and labial palps. *Acta Zool. (Stockh.) 63*, 191–198.

HALLBERG, E. (1982). Sensory organs in *Ips typographus* (Insecta: Coleoptera) – Fine structure of antennal sensilla. *Protoplasma 111*, 206–214.

HAMON, M. (1961). Contribution a l'étude de la morphogenese sensorinerveuse des Dytiscidae. (Insectes Coleopteres). *Ann. Sci. Naturelles. 12ᵉ Serie. 3*, 153–171.

HANSEN, H. and HEUMANN, H. (1971). Die Feinstruktur der tarsalen Schmeckhaare der Fliege *Phormia terraenovae* Rob.-Desv. *Z. Zellforsch. Mikr. Anat. 117*, 419–442.

HARBACH, R. E. and LARSEN, J. R. (1976). Ultrastructure of sensilla on the distal antennal segment of adult *Oncopeltus fasciatus* (Dallas) (Hemiptera: Lygaeidae). *Int. J. Insect Morphol. Embryol. 5*, 23–33.

HARBACH, R. E. and LARSEN, J. R. (1977). Fine structure of antennal sensilla of the adult mealworm beetle, *Tenebrio molitor* L. (Coleoptera: Tenebrionidae). *Int. J. Insect Morphol. Embryol. 6*, 41–60.

HATFIELD, L. D. and FRAZIER, J. L. (1980). Ultrastructure of the labial tip sensilla of the tarnished plant bug, *Lygus lineolaris* (P. de Beauvois) (Hemiptera: Miridae). *Int. J. Insect Morphol. Embryol. 9*, 59–66.

HAUPT, J. (1972). Ultrastruktur des Pseudoculus von *Eosentomon* (Protura, Insecta). *Z. Zellforsch. Mikr. Anat. 135*, 539–551.

HAWKE, S. D. and FARLEY, R. D. (1971a). Antennal chemoreceptors of the desert burrowing cockroach, *Arenivaga* sp. *Tissue Cell 3*, 649–664.

HAWKE, S. D. and FARLEY, R. D. (1971b). The role of pore structures in the selective permeability of antennal sensilla of the desert burrowing cockroach, *Arenivaga* sp. *Tissue Cell 3*, 665–674.

HAWKE, S. D., FARLEY, R. D. and GREANY, P.D. (1973). The fine structure of sense organs in the ovipositor of the parasitic wasp, *Orgilus lepidus* Muesebeck. *Tissue Cell 5*, 171–184.

HEMING, B. S. (1975). Antennal structure and metamorphosis in *Frankliniella fusca* (Hinds) (Thripidae) and *Haplothrips verbasci* (Osborn) (Phlaeothripidae) (Thysanoptera). *Quaest. Ent. 11*, 25–68.

HERMANN, H. R. and DOUGLAS, M. E. (1976). Comparative survey of the sensory structures on the sting and ovipositor of hymenopterous insects. *J. Georgia Ent. Soc. 11*, 223–239.

HONOMICHL, K. (1980). Die digitiformen Sensillen auf dem Maxillarpalpus von Coleoptera. I. Vergleichend-topographische Untersuchung des kutikularen Apparates. *Zool. Anz. 204*, 1–12.

HONOMICHL, K. and GUSE, G.-W. (1981). Digitiform sensilla on the maxillar palp of Coleoptera III. Fine structure in *Tenebrio molitor* L. and *Dermestes maculatus* De Geer. *Acta Zool. 62*, 17–25.

HOOPER, R. L., PITTS, C. W. and WESTFALL, J. A. (1972). Sense organs on the ovipositor of the face fly, *Musca autumnalis*. *Ann. Ent. Soc. Amer. 65*, 577–586.

HORN, E. and KESSLER, W. (1975). The control of antennae lift movements and its importance on the gravity reception in the walking blowfly, *Calliphora erythrocephala*. *J. Comp. Physiol. 97*, 189–203.

IMMS, A. D. (1939). On the antennal musculature in insects and other arthropods. *Quart. J. Mic. Sci. 81*, 273–320.

IMMS, A. D. (1940). On growth processes in the antennae of insects. *Quart. J. Mic. Sci. 81*, 585–593.

IVANOV, V. P. (1969). The ultrastructure of chemoreceptors in insects. *Trudy Vses. Ent. Obsh. 53*, 301–333.

JEFFERSON, R. N., RUBIN, R. E., McFARLAND, S. U. and SHOREY, H. H. (1970). Sex pheromones of noctuid moths. XXII. The external morphology of the antennae of *Trichoplusia ni, Heliothis zea, Prodenia ornithogalli*, and *Spodoptera exigua*. *Ann. Ent. Soc. Amer. 63*, 1227–1238.

JEZ, D. H. and McIVER, S. B. (1980). Fine structure of antennal sensilla of larval *Toxorhynchites brevipalpis* Theobald (Diptera: Culicidae). *Int. J. Insect Morphol. Embryol. 9*, 147–159.

JONES, J. C. (1977). *The Circulatory System of Insects*. Thomas. Springfield.

JUBERTHIE, C. and MASSOUD, Z. (1977). L'equipement sensoriel de l'antenne d'un Coleoptere troglobie, *Aphaenops cryticola* Linder (Coleoptera: Trechinae). *Int. J. Insect Morphol. Embryol. 6*, 147–160.

JUBERTHIE-JUPEAU, L. and BARETH, C. (1980). Ultrastructure des sensilles de l'organe cupuliforme de l'antenne des Campodes (Insecta: Diplura). *Int. J. Insect Morphol. Embryol. 9*, 255–268.

KAISSLING, K.-E. (1971). Insect olfaction. In *Handbook of Sensory Physiology*. Vol. 4, pt 2, Pages 351–431. Springer, Berlin and New York.

KAISSLING, K. E. and THORSON, J. (1980). Insect olfactory sensilla: structural, chemical and electrical aspects of the functional organization. In *Receptors for Neurotransmitters, Hormones and Pheromones in Insects*. Edited by D. B. Satelle and L. M. Hall. Pages 261–282. Elsevier/North-Holland Biomedical Press, Amsterdam.

KAPOOR, N. N. and ZACHARIAH, K. (1978). The internal morphology of sensilla on the abdominal gills of *Thaumatoperla alpina* (Plecoptera: Eustheniidae; subfamily: Thaumatoperlinae). *Canad. J. Zool. 56*, 2194–2197.

KEIL, T. A. (1982). Contacts of pore tubules and sensory dendrites in antennal chemosenilla of a silkmoth: demonstration of a possible pathway for olfactory molecules. *Tissue Cell 14*, 451–462.

KERKUT, G. A. and TAYLOR, B. J. R. (1957). A temperature receptor in the tarsus of the cockroach, *Periplaneta americana*. *J. Exp. Biol. 34*, 486–493.

KLEIN, U. (1981). Sensilla of the cricket palp. Fine structure and spatial organization. *Cell Tiss. Res. 219*, 229–252.

KLEIN, U. and MÜLLER, B. (1978). Functional morphology of palp sensilla related to food recognition in *Gryllus bimaculatus* (Saltatoria, Gryllidae). *Ent. Exp. Appl. 24*, 491–495.

KRÄNZLER, L. and LARINK, O. (1980). Postembryonale Veränderungen und Sensillenmuster der abdominalen Anhänge von *Thermobia domestica* (Packard) (Insecta: Zygentoma). *Braunschw. Naturk. Schr. 1*, 27–49.

KÜPPERS, J. and THURM, U. (1979). Active ion transport by a sensory epithelium. I. Transepithelial short circuit current, potential difference, and their dependence on metabolism. *J. Comp. Physiol. 134*, 131–136.

LAMBIN, M. (1973). Les sensilles de l'antenne chez quelques blattes et en particulier chez *Blaberus craniifer* (Burm.). *Z. Zellforsch. Mikr. Anat. 143*, 183–206.

LARINK, O. (1978). Sensillenmuster auf dem Labium von Lepismatiden (Insecta: Zygentoma). *Zool. Anz. Jena. 201*, 341–352.

LARINK, O. (1982). Das Sensillen-Inventar der Lepismatiden (Insecta:Zygentoma). *Braunschw. Naturk. Schr. 3*, 493–512.

LAWRENCE, P. A. (1970). Polarity and patterns in the postembryonic development of insects. *Adv. Insect Physiol. 7*, 197–266.

LEE, R. (1974). Structure and function of the fascicular stylets, and the labral and cibarial sense organs of male and female *Aedes aegypti* (L.) (Diptera, Culicidae). *Quaest. Entomol. 10*, 187–215.

LEE, R. M. K. W. and DAVIES, D. M. (1978). Cibarial sensilla of *Toxorhynchites* mosquitoes (Diptera: Culicidae). *Int. J. Insect Morphol. Embryol. 7*, 189–194.

LEWIS, C. T. (1971). Superficial sense organs of the antennae of the fly, *Stomoxys calcitrans*. *J. Insect Physiol. 17*, 449–461.

LEWIS, C. T. (1972). Chemoreceptors in haematophagous insects. In *Behavioural Aspects of Parasite Transmission*. Edited by E. U. Canning and C. A. Wright. Pages 201–213. Academic Press, London.

LEWIS, C. T. and MARSHALL, A. T. (1970). The ultrastructure of the sensory plaque organs of the antennae of the Chinese lantern fly, *Pyrops candelaria* L. (Homoptera: Fulgoridae). *Tissue Cell 2*, 375–385.

LOFTUS, R. and CORBIÈRE-TICHANÉ, G. (1981). Antennal warm and cold receptors of the cave beetle, *Speophyes lucidulus* Delar., in sensilla with a lamellated dendrite. I. Response to sudden temperature change. *J. Comp. Physiol. 143*, 443–452.

LUM, P. T. M. and ARBOGAST, R. T. (1980). Ultrastructure of setae in the spermathecal gland of *Plodia interpunctella* (Hübner) (Lepidoptera:Pyralidae). *Int. J. Insect Morphol. Embryol. 9*, 251–253.

MA, W.-C. (1972). Dynamics of feeding responses in *Pieris brassicae* Linn. as a function of chemosensory input: a behavioural, ultrastructural and electrophysiological study. *Meded. Landbouwhogesch. Wageningen, 72–11*.

MA, W.-C. (1981). Receptor membrane function in olfaction and gustation: implications from modification by reagents and drugs. In *Perception of Behavioral Chemicals*. Edited by D. M. Norris. Pages 267–287. Elsevier/North-Holland Biomedical Press, New York.

MA, W.-C. and SCHOONHOVEN, L. M. (1973). Tarsal contact chemosensory hairs of the large white butterfly *Pieris brassicae* and their possible role in oviposition behaviour. *Ent. Exp. Appl. 16*, 343–357.

MAES, F. W. and VEDDER, C. G. (1978). A morphological and electrophysiological inventory of labellar taste hairs of the blowfly *Calliphora vicina*. *J. Insect Physiol. 24*, 667–672.

MAKINGS, P. (1964). "Sliffer's Patches" and the thermal sense in Acrididae (Orthoptera). *J. Exp. Biol. 41*, 473–497.

MALZ, D. and HINTZE-PODUFAL, C. (1979). Bau und Verteilung von Sinnesorganen auf dem Labium und der Maxille der Schabe *Gromphadorhina brunneri* Butler (Blaberoidea, Oxyhaloidae). *Zool. Beitr. 25*, 81–100.

MARKL, H. (1962). Borstenfelder an den Gelenken als Schweresinnesorgane bei Ameisen und Anderen Hymenopteren. *Z. Vergl. Physiol. 45*, 475–569.

MARSHALL, A. T. (1973). Vesicular structures in the dendrites of an insect olfactory receptor. *Tissue Cell 5*, 233–241.

MASSON, C. and STRAMBI, C. (1977). Sensory antennal organization in an ant and a wasp. *J. Neurobiol. 8*, 537–548.

MASSON, C., GABOURIAUT, D. and FRIGGI, A. (1972). Ultrastructure d'un nouveau type de récepteur olfactif de l'antenne d'insecte trouve chez la fourmi *Camponotus vagus* Scop. (Hymenoptera, Formicinae). *Z. Morphol. Tiere. 72*, 349–360.

MATSUMOTO, D. E. and FARLEY, R. D. (1978). Alterations of ultrastructure and physiology of chemoreceptor dendrites in blowfly taste hairs treated with vinblastine and colchicine. *J. Insect Physiol. 24*, 765–776.

MATSUMOTO, S. G. and HILDEBRAND, J. G. (1981). Olfactory mechanisms in the moth *Manduca sexta*: response characteristics and morphology of central neurons in the antennal lobes. *Proc. R. Soc. Lond. B 213*, 249–277.

MAYER, M. S., MANKIN, R. W. and CARLYSLE, T. C. (1981). External antennal morphometry of *Trichoplusia ni* (Hübner) (Lepidoptera: Noctuidae). *Int. J. Insect Morphol. Embryol. 10*, 185–201.

McINDOO, N. E. (1931). Tropisms and sense organs of Coleoptera. *Smithsonian Misc. Coll. 82*, 1–70.

McIVER, S. B. (1972). Fine structure of pegs on the palps of female culicine mosquitoes. *Canad. J. Zool. 59*, 571–576.

McIVER, S. B. (1975). Structure of cuticular mechanoreceptors of arthropods. *Ann. Rev. Ent. 20*, 381–397.

McIVER, S. (1978). Structure of sensilla trichodea of female *Aedes aegypti* with comments on innervation of antennal sensilla. *J. Insect Physiol. 24*, 383–390.

McIVER, S. and SIEMICKI, R. (1982). Fine structure of maxillary sensilla of larval *Toxorhynchites brevipalpis* (Diptera:Culicidae) with comments on the role of sensilla in behavior. *J. Morphol. 171*, 293–303.

McIVER, S. and CHARLTON, C. (1970). Studies on the sense organs on the palps of selected culicine mosquitoes. *Canad. J. Zool. 48*, 293–295.

McIVER, S. and HUDSON, A. (1972). Sensilla on the antennae and palps of selected *Wyeomyia* mosquitoes. *J. Med. Ent. 9*, 337–345.

McIVER, S. and SIEMICKI, R. (1978). Fine structure of tarsal sensilla of *Aedes aegypti* (L.) (Diptera: Culicidae). *J. Morphol. 155*, 137–156.

McIVER, S. and SIEMICKI, R. (1979). Fine structure of antennal sensilla of male *Aedes aegypti* (L.). *J. Insect Physiol. 25*, 21–28.

McIVER, S., SIEMICKI, R. and SUTCLIFFE, J. (1980). Bifurcate sensilla on the tarsi of female blackflies, *Simulium venustum* (Diptera: Simuliidae): contact chemosensilla adapted for olfaction? *J. Morphol. 165*, 1–11.

MEINECKE, C.-C. (1975). Riechsensillen und Systematick der Lamellicornia (Insecta, Coleoptera). *Zoomorphol. 82*, 1–42.

MENCO, B. Ph.M. and VAN DER WOLK, F. M. (1982). Freeze-fracture characteristics of insect gustatory and olfactory sensilla. I. A comparison with vertebrate olfactory receptor cells with special reference to ciliary components. *Cell Tiss. Res. 223*, 1–27.

MERCER, K. L. and McIVER, S. B. (1973). Studies on the antennal sensilla of selected blackflies (Diptera: Simuliidae). *Canad. J. Zool. 51*, 729–734.

MITCHELL, B. K. and SCHOONHOVEN, L. M. (1974). Taste receptors in Colorado beetle larvae. *J. Insect Physiol. 20*, 1787–1793.

MORAN, D. T., VARELA, F. J. and ROWLEY, J. C. (1977). Evidence for active role of cilia in sensory transduction. *Proc. Natl. Acad. Sci. USA 74*, 793–797.

MOULINS, M. (1968). Les sensilles l'organe hypopharyngien de *Blabera craniifer* Burm. (Insecta, Dictyoptèra). *J. Ultrastruct. Res. 21*, 474–513.

MOULINS, M. (1971a). La cavité préorale de *Blabera craniifer* Burm. (Insecte, Dictyoptère) et son innervation: etude anatomo-histologique de l'epipharynx et l'hypopharynx. *Zool. Jb. Anat. 88*, 527–586.

MOULINS, M. (1971b). Ultrastructure et physiologie des organes épipharyngiens et hypopharyngiens (Chimiorécepteurs cibariaux) de *Blabera craniifer* Burm. (Insecte, Dictyoptère). *Z. Vergl. Physiol. 73*, 139–166.

MOULINS, M. (1974). Recepteurs de tension de la region de la bouche chez *Blaberus craniifer* Burmeister (Dictyoptera: Blaberidae). *Int. J. Insect Morphol. Embryol. 3*, 171–192.

MOULINS, M. and NOIROT, CH. (1972). Morphological features bearing on transduction and peripheral integration in insect gustatory organs. In *Olfaction and Taste IV*. Edited by D. Schneider. Pages 49–55. Wissenschaftliche Verlagsgesellschaft MBH, Stuttgart.

MYERS, J. (1968). The structure of the antennae of the Florida queen butterfly, *Danaus gilippus berenice* (Cramer). *J. Morphol. 125*, 315–328.

NIJHOUT, H. F. (1977). Control of antennal hair erection in male mosquitoes. *Biol. Bull. 153*, 591–603.

NIJHOUT, H. F. and SHEFFIELD, H. G. (1979). Antennal hair erection in male mosquitoes: a new mechanical effector in insects. *Science 206*, 595–596.

O'CEALLACHAIN, D. P. and RYAN, M. F. (1977). Production and perception of pheromones by the beetle *Tribolium confusum*. *J. Insect Physiol. 23*, 1303–1309.

OCHS, S. (1963). Beading phenomena of mammalian myelinated nerve fibers. *Science 139*, 599–600.

OLSEN, R. W. (1975). Filamentous protein model for cyclic AMP-mediated cell regulatory mechanisms. *J. Theor. Biol. 49*, 263–287.

OSBORNE, M. P. (1963). An electron microscope study of an abdominal stretch receptor of the cockroach. *J. Insect. Physiol. 9*, 237–245.

PAPPAS, L. G. and LARSEN, J. R. (1976). Gustatory hairs on the mosquito, *Culiseta inornata*. *J. Exp. Zool. 196*, 351–360.

PARETO, A. (1972). The spatial distribution of sensory antennal fibres in the central nervous system of worker bees. *Z. Zellforsch. Mikr. Anat. 131*, 109–140.

PASS, G. (1980). The anatomy and ultrastructure of the antennal circulatory organ in the cockchafer beetle *Melolontha melolontha* L. (Coleoptera, Scarabaeidae). *Zoomorphol. 96*, 77–89.

PECK, S. B. (1977). An unusual sense receptor in internal antennal vesicles of *Ptomaphagus* (Coleoptera: Leiodidae). *Canad. Ent. 109*, 81–86.

PETERS, W. (1963). Die Sinnesorgane an den Labellen von *Calliphora erythrocephala* M. (Diptera). *Z. Morph. Okol. Tiere. 55*, 259–320.

PETRYSZAK, A. (1975a). The sensory peripheric nervous system of the *Periplaneta americana* (L.) (Blattoidea). I. Mouth parts. *Zesz. Nauk. UJ. Prace Zool. 20*, 41–84.

PETRYSZAK, A. (1975b). The sensory peripheric nervous system of *Periplaneta americana* (L.) (Blattoidea). Part II. Antenna. *Acta Biol. Cracov. Zool. 18*, 257–263.

PETRYSZAK, A. (1975c). The sensory peripheric nervous system of *Periplaneta americana* (L.) (Blattoidea). Part III. The legs, cerci, and styli. *Acta Biol. Cracov. Zool. 18*, 265–276.

PETRYSZAK, A. (1977). The sense organs of the mouth parts in *Libellula depressa* L. and *L. quadrimaculata* L. (Odonata). *Acta Biol. Cracov. Zool. 20*, 87–100.

PHILLIPS, C. E. and VANDE BERG, J. S. (1976a). Directional flow of sensillum liquor in blowfly (*Phormia regina*) labellar chemoreceptors. *J. Insect Physiol. 22*, 425–429.

PHILLIPS, C. E. and VANDE BERG, J. S. (1976b). Mechanism for sensillum fluid flow in trichogen and tormogen cells of *Phormia regina* (Meigen) (Diptera: Calliphoridae). *Int. J. Insect Morphol. Embryol. 5*, 423–431.

POPHAM, E. J. (1979). The micro-structure of the maxillary and labial papillae of *Forficula auricularia* (Dermaptera). *J. Zool. 188*, 353–355.

PRITCHARD, G. (1965). Sense organs in the labrum of *Aeshna interrupta lineata* Walker (Odonata: Anisoptera). *Canad. J. Zool. 43*, 333–336.

QUENNEDEY, A. (1975). The labrum of *Schedorhinotermes* minor soldier (Isoptera, Rhinotermitidae) morphology, innervation and fine-structure. *Cell Tiss. Res. 160*, 81–98.

REINOUTS VAN HAGA, H. A. and MITCHELL, B. K. (1975). Temperature receptors on tarsi of the tsetse fly *Glossina morsitans* West. *Nature 255*, 225–226.

RICE, M. J. (1973). Cibarial sense organs of the blowfly, *Calliphora erythrocephala* (Meigen) (Diptera: Calliphoridae). *Int. J. Insect Morphol. Embryol. 2*, 109–116.

RICE, M. J. (1976). Contact chemoreceptors on the ovipositor of *Lucilia cuprina* (Wied.), the Australian sheep blowfly. *Aust. J. Zool. 24*, 353–360.

RICE, M. J. and McRAE, T. M. (1976). Contact chemoreceptors on the ovipositor of *Locusta migratoria* L. *J. Aust. Ent. Soc. 15*, 364.

RICHARD, G. (1951). L'innervation et les organes sensoriels des pièces buccales du termite a cou jaune (*Calotermes flavicollis* Fab.). *Ann. Sci. Nat. Paris, Zool.^{e} Série 13*, 397–412.

RICHARDS, O. W. and DAVIES, R. G. (1977). *Imms General Textbook of Entomology*. 10th edn. Vol. 1. Chapman & Hall, London.

RICHELLE, J. and GHYSEN, A. (1979). Determination of sensory bristles and pattern formation in *Drosophila* I. A model. *Devel. Biol. 70*, 418–437.

RICHTER, S. (1962). Unmittelbarer Kontakt der Sinneszellen cuticularer Sinnesorgane mit der Aussenwelt. Eine Licht- und Elektronenmikroskopische Untersuchung der Chemorezeptorischen Antennensinnesorgane der *Calliphora*-Larven. *Z. Morphol. Okol. Tiere. 52*, 171–196.

RIEGERT, P. W. (1960). The humidity reactions of *Melanoplus bivittatus* (Say) (Orthoptera, Acrididae): antennal sensilla and hygro-reception. *Canad. Ent. 92*, 561–570.

ROCKSTEIN, M. (editor) (1974). *The Physiology of Insecta*. 2nd edn. Vol. 2. Academic Press, New York.

ROHR, W. (1982). Bau und Verteilung der Sensillen auf der Innenseite des Clypeolabrum von *Acheta domesticus* L. (Insecta:Ensifera) während der postembryonalen Entwicklung. *Braunschw. Naturk. Schr. 3*, 513–531.

ROSS, R. J., PALANISWAMY, P. and SEABROOK, W. D. (1979). Electroantennograms from spruce budworm moths (*Choristoneura fumiferana*) (Lepidoptera: Tortricidae) of different ages and for various pheromone concentrations. *Canad. Ent. 111*, 807–816.

ROSSIGNOL, P. A. and McIVER, S. B. (1977). Fine structure and role in behavior of sensilla on the terminalia of *Aedes aegypti* (L.) (Diptera: Culicidae). *J. Morphol. 151*, 419–438.

ROTH, L. M. and WILLIS, E. R. (1951). Hygroreceptors in Coleoptera. *J. Exp. Zool. 117*, 451–487.

ROWLEY, W. A. and CORNFORD, M. (1972). Scanning electron microscopy of the pit of the maxillary palp of selected species of *Culicoides*. *Canad. J. Zool. 50*, 1207–1210.

ROZENTAL, J. M. and NORRIS, D. M. (1973). Chemosensory mechanism in American cockroach olfaction and gustation. *Nature 244*, 370–371.

RYAN, M. F. and BEHAN, M. (1973). Cephalic sensory receptors of the cabbage-root fly larva, *Erioischia brassicae* (B.) (Diptera: Anthomyiidae). *Int. J. Insect Morphol. Embryol. 2*, 83–86.

SALAMA, H. S. (1966). The function of mosquito taste receptors. *J. Insect Physiol. 12*, 1051–1060.

SANES, J. R. and HILDEBRAND, J. G. (1976a). Structure and development of antennae in a moth, *Manduca sexta*. *Devel. Biol. 51*, 282–299.

SANES, J. R. and HILDEBRAND, J. G. (1976b). Origin and morphogenesis of sensory neurons in an insect antenna. *Devel. Biol. 51*, 300–319.

SANES, J. R., HILDEBRAND, J. G. and PRESCOTT, D. J. (1976). Differentiation of insect sensory neurons in the absence of their normal synaptic targets. *Devel. Biol. 52*, 121–127.

SAXENA, K. N. (1967). Some factors governing olfactory and gustatory responses of insects. In *Olfaction and Taste II*. Edited by T. Hayashi. Pages 799–819. Pergamon Press, Oxford.

SCHAFER, R. (1973). Postembryonic development in the antenna of the cockroach, *Leucophaea maderae*: growth, regeneration and the development of the adult pattern of sense organs. *J. Exp. Zool. 183*, 353–364.

SCHAFER, R. and SANCHEZ, T. V. (1973). Antennal sensory system of the cockroach, *Periplaneta americana*: postembryonic development and morphology of the sense organs. *J. Comp. Neurol. 149*, 335–354.

SCHAFER, R. and SANCHEZ, T. V. (1976a). The nature and development of sex attractant specificity in cockroaches of the genus *Periplaneta* I. Sexual dimorphism in the distribution of antennal sense organs in five species. *J. Morphol. 149*, 139–158.

SCHAFER, R. and SANCHEZ, T. V. (1976b). The nature and development of sex attractant specificity in cockroaches of the genus *Periplaneta*. II. Juvenile hormone regulates sexual dimorphism in the distribution of antennal olfactory receptors. *J. Exp. Zool. 198*, 323–336.

SCHALLER, L. (1982). Structural and functional classification of antennal sensilla of the cockroach, *Leucophaea maderae*. *Cell Tiss. Res. 225*, 129–142.

SCHMIDT, K. (1974). Die mechanorezeptoren im Pedicellus der Eintagsfliegen (Insecta, Ephemeroptera). *Z. Morphol. 78*, 193–220.

SCHMIDT, K. and GNATZY, W. (1972). Die Feinstruktur der Sinneshaare auf den Cerci von *Gryllus bimaculatus* Deg. (Saltatoria, Gryllidae). III. Die kurzen Borstenhaare. *Z. Zellforsch. Mikr. Anat. 126*, 206–222.

SCHNEIDER, D. (1964). Insect Antennae. *Ann. Rev. Ent. 9*, 103–122.

SCHNEIDER, D. (1971). Specialized odor receptors of insects. In *Gustation and Olfaction*. Edited by G. Ohloff and A. F. Thomas. Pages 45–60. Academic Press, New York.

SCHNEIDER, D. and KAISSLING, K.-E. (1957). Der Bau der Antenne des Seidenspinners *Bombyx mori* L. II. Sensillen, cuticulare Bildungen und innerer Bau. *Zool. Jahrb. Abt, Anat. Ontog. Tiere 76*, 223–250.

SCHOELLER, J. (1964). Recherches descriptives et experimentales sur la céphalogenese de *Calliphora erythrocephala* (Meigen), au cours des developpements embryonnaire et postembryonnaire. *Arch. Zool. Exptl. Gen. 103*, 1–216.

SCHOONHOVEN, L. M. (1967). Some cold receptors in larvae of three Lepidoptera species. *J. Insect Physiol. 13*, 821–826.

SCHOONHOVEN, L. M. (1977). On the individuality of insect feeding behaviour. *Proc. K. Ned. Akad. Wet. (C) 80*, 341–350.

SCHURMANN, F. W. and WECHSLER, W. (1970). Synapses in the antennal lobes of *Locusta migratoria*. *Z. Zellforsch. Mikr. Anat. 108*, 563–581.

SCHWEITZER, E. S., SANES, J. R. and HILDEBRAND, J. G. (1976). Ontogeny of electroantennogram responses in the moth, *Manduca sexta*. *J. Insect Physiol. 22*, 955–960.

SCOTT, D. A. and ZACHARUK, R. Y. (1971a). Fine structure of the antennal sensory appendix in the larva of *Ctenicera destructor* (Brown) (Elateridae: Coleoptera). *Canad. J. Zool. 49*, 199–210.

SCOTT, D. A. and ZACHARUK, R. Y. (1971b). Fine structure of the dendritic junction body region of the antennal sensory cone in a larval elaterid (Coleoptera). *Canad. J. Zool. 49*, 817–821.

SEABROOK, W. D. (1977). Insect chemosensory responses to other insects. In *Chemical Control of Insect Behavior: Theory and Application*. Edited by H. H. Shorey and J. J. McKelvey, Pages. 15–43. Wiley, New York.

SEABROOK, W. D., HIRAI, K., SHOREY, H. H. and GASTON, L. K. (1979). Maturation and senescence of an insect chemosensory response. *J. Chem. Ecol. 5*, 587–594.

SHAMBAUGH, G. F., FRAZIER, J. L., CASTELL, A. E. M. and COONS, L. B. (1978). Antennal sensilla of seventeen aphid species (Homoptera: Aphidinae). *Int. J. Insect Morphol. Embryol. 7*, 389–404.

SHIMADA, I., SHIRAISHI, A., KIJIMA, H. and MORITA, H. (1972). Effects of sulphhydryl reagents on the labellar sugar receptor of the fleshfly. *J. Insect Physiol. 18*, 1845–1855.

SINOIR, Y. (1969). L'ultrastructure des organes sensoriels des insectes. *Ann. Zool. Ecol. Anim. 1*, 339–356.

SLIFER, E. H. (1961). The fine structure of insect sense organs. *Int. Rev. Cytol. 11*, 125–159.

SLIFER, E. H. (1970). The structure of arthropod chemoreceptors. *Ann. Rev. Ent. 15*, 121–142.

SLIFER, E. H. (1972). Pores in the thin-walled chemoreceptors of the grasshopper. *Acrida 1*, 1–5.

SLIFER, E. H. and SEKHON, S. S. (1969a). Nodes on insect sensory dendrites. *27th Ann. Proc. EMSA*. Pages 242–243.

SLIFER, E. H. and SEKHON, S. S. (1969b). Some evidence for the continuity of ciliary fibrils and microtubules in the insect sensory dendrite. *J. Cell Sci. 4*, 527–540.

SLIFER, E. H. and SEKHON, S. S. (1971a). Circumfila and other sense organs on the antennae of the sorghum midge (Diptera, Cecidomyiidae). *J. Morphol. 133*, 281–302.

SLIFER, E. H. and SEKHON, S. S. (1971b). Structures on the antennal flagellum of a caddisfly, *Frenesia missa* (Trichoptera, Limnephilidae). *J. Morphol. 135*, 373–388.

SLIFER, E. H. and SEKHON, S. S. (1973). Sense organs on the antennal flagellum of two species of Embioptera (Insecta). *J. Morphol. 139*, 211–226.

SLIFER, E. H. and SEKHON, S. S. (1978). Sense organs on the antennae of two species of Collembola (Insecta). *J. Morphol. 157*, 1–20.

SLIFER, E. H. and SEKHON, S. S. (1980). Sense organs on the antennal flagellum of the human louse, *Pediculus humanus* (Anoplura). *J. Morphol. 164*, 161–166.

SLIFER, E. H., PRESTAGE, J. J. and BEAMS, W. (1957). The fine structure of the long basiconic sensory pegs of the grasshopper (Orthoptera, Acrididae) with special reference to those on the antenna. *J. Morphol. 101*, 359–398.

SLIFER, E. H., PRESTAGE, J. J. and BEAMS, H. (1959). The chemoreceptors and other sense organs on the antennal flagellum of the grasshopper (Orthoptera: Acrididae). *J. Morphol. 105*, 145–192.

SNODGRASS, R. E. (1926). The morphology of insect sense organs and the sensory nervous system. *Smithsonian Misc. Coll. 77*, 1–80.

SNODGRASS, R. E. (1935). *Principles of Insect Morphology*. McGraw-Hill, New York.

SNODGRASS, R. E. (1954). Insect metamorphosis. *Smithsonian Misc. Coll. 122*, 1–124.

STÄDLER, E. (1978). Chemoreception of host plant chemicals by ovipositing females of *Delia (Hylemya) brassicae*. *Ent. Exp. Appl. 24*, 511–520.

STÄDLER, E. and HANSON, F. E. (1975). Olfactory capabilities of the "gustatory" chemoreceptors of the tobacco hornworm larvae. *J. Comp. Physiol. 104*, 97–102.

STEINBRECHT, R. A. (1969). Comparative morphology of olfactory receptors. In *Olfaction and Taste III*. Edited by C. Pfaffmann. Pages 3–21. Rockefeller University Press.

STEINBRECHT, R. A. (1970). Zur Morphometrie der Antenne des Seidenspinners, *Bombyx mori* L.: Zahl und Verteilung der Riechsensillen (Insecta, Lepidoptera). *Z. Morphol. Tiere 68*, 93–126.

STEINBRECHT, R. A. (1973). Der Feinbau olfaktorischer Sensillen des Seidenspinners (Insecta, Lepidoptera) Rezeptorfortsatze und reizleitender Apparat. *Z. Zellforsch. Mikr. Anat. 139*, 533–565.

STEINBRECHT, R. A. (1980). Cryofixation without cryoprotectants. Freeze substitution and freeze etching of an insect olfactory receptor. *Tissue Cell 12*, 73–100.

STEINBRECHT, R. A. and MÜLLER, B. (1971). On the stimulus conducting structures in insect olfactory receptors. *Z. Zellforsch. Mikr. Anat. 117*, 570–575.

STEINBRECHT, R. A. and MÜLLER, B. (1976). Fine structure of the antennal receptors of the bed bug, *Cimex lectularius* L. *Tissue Cell 8*, 615–636.

STOFFOLANO, J. G., DAMON, R. A. and DESCH, C. E. (1978). The effect of age, sex and anatomical position on peripheral responses of taste receptors in blowflies, genus *Phormia* and *Protophormia*. *Exp. Gerontol. 13*, 115–124.

STÜRCKOW, B. (1971). Electrical impedance of the labellar taste hair of the blowfly, *Calliphora erythrocephala* MG. *Z. Vergl. Physiol. 72*, 131–143.

SUTCLIFFE, J. F. and McIVER, S. B. (1976). External morphology of sensilla on the legs of selected black fly species (Diptera: Simuliidae). *Canad. J. Zool. 54*, 1779–1787.

SUTCLIFFE, J. F. and McIVER, S. B. (1982). Innervation and structure of mouth part sensilla in females of the black fly *Simulium venustum* (Diptera:Simuliidae). *J. Morphol. 171*, 245–258.

SUTCLIFFE, J. F. and MITCHELL, B. K. (1980). Structure of galeal sensory complex in adults of the red turnip beetle, *Entomoscelis americana* Brown (Coleoptera, Chrysomelidae). *Zoomorphol. 96*, 63–76.

SUZUKI, H. (1975). Antennal movements induced by odour and central projection of the antennal neurones in the honey-bee. *J. Insect Physiol. 21*, 831–847.

SYRJÄMÄKI, J. (1962). Humidity perception in *Drosophila melanogaster*. *Ann. Zool. Soc. "Vanamo". 23*, 1–72.

THURM, U. and KÜPPERS, J. (1980). Epithelial physiology of insect sensilla. In *Insect Biology in the Future*. Edited by M. Locke and D. S. Smith. Pages 735–763. Academic Press, New York.

Toh, Y. (1977). Fine structure of antennal sense organs of the male cockroach, *Periplaneta americana. J. Ultrastruct. Res. 60,* 373–394.

Uchida, K. (1979). Cibarial sensilla and pharyngeal valves in *Aedes albopictus* (Skuse) and *Culex pipiens pallens* Coquillett (Diptera: Culicidae). *Int. J. Insect Morphol. Embryol. 8,* 159–167.

Urvoy, J. (1963). Étude anatomo-fonctionnelle de la patte et de l'antenne de la blatte *Blabera craniifer* Burmeister. *Ann. Sci. Natur. Zool. Paris, 12ᵉ Serie, 5,* 287–414.

van der Pers, J. N. C. (1982). Comparison of single cell responses of antennal sensilla trichodea in the nine european small ermine moths (*Yponomeuta* spp.). *Ent. Exp. Appl. 31,* 255–264.

van der Starre, H. and Tempelaar, M. J. (1976). Structural and functional aspects of sense organs on the maxillary palps of *Calliphora vicina. J. Insect Physiol. 22,* 855–863.

van der Wolk, F. M. (1978). The typology and topography of the tarsal chemoreceptors of the blow flies *Calliphora vicina* Robineau-Desvoidy and *Phormia terranovae* Robineau-Desvoidy and the housefly *Musca domestica* L. *J. Morphol. 157,* 201–210.

van Drongelen, W. (1979). Contact chemoreception of host plant specific chemicals in larvae of various *Yponomeuta* species (Lepidoptera). *J. Comp. Physiol. 134,* 265–279.

Villet, R. H. (1982). Mechanism of insect sex-pheromone sensory transduction: role of adenyl cyclase. *Comp. Biochem. Physiol. 61,* 389–394.

Vinnikov, Ya. A. (1974). *Sensory Reception, Cytology, Molecular Mechanisms and Evolution.* Molecular Biology, Biochemistry and Biophysics, 17. Springer-Verlag, Berlin.

Wall, C. (1978). Morphology and histology of the antenna of *Cydia nigricana* (F.) (Lepidoptera: Tortricidae). *Int. J. Insect Morphol. Embryol. 7,* 237–250.

Wallis, D. I. (1962). Olfactory stimuli and oviposition in the blowfly, *Phormia regina* Meigen. *J. Exp. Biol. 39,* 603–615.

Walther, J. R. (1981). Die Morphologie und Feinstructur der Sinnesorgane auf den Antennengeisseln der Männchen, Weibchen und Arbeiterinnen der Roten Waldameise *Formica rufa* Linné 1758 mit einem Vergleich der Antennalen Sensillenmuster weiterer Formicoidea (Hymenoptera). Ph.D. dissertation, Free University of Berlin.

Wensler, R. J. (1977). The fine structure of distal receptors on the labium of the aphid, *Brevicoryne brassicae* L. (Homoptera). *Cell Tiss. Res. 181,* 409–422.

Wensler, R. J. and Filshie, B. K. (1969). Gustatory sense organs in the food canal of aphids. *J. Morphol. 129,* 473–491.

White, S. L. and Bay, D. E. (1980). Antennal olfactory sensilla of the horn fly, *Haematobia irritans irritans* (L.) (Diptera: Muscidae). *J. Kansas Ent. Soc. 53,* 641–652.

Whitehead, A. T. (1981). Ultrastructure of sensilla of the female mountain pine beetle, *Dendroctonus ponderosae* Hopkins (Coleoptera: Scolytidae). *Int. J. Insect Morphol. Embryol. 10,* 19–28.

Whitehead, A. T. and Larsen, J. R. (1976). Ultrastructure of the contact chemoreceptors of *Apis mellifera* L. (Hymenoptera: Apidae). *Int. J. Insect Morphol. Embryol. 5,* 301–315.

Whitten, J. M. (1963). Observations on the cyclorrhaphan larval peripheral nervous system: muscle and tracheal receptor organs and independent peripheral Type II neurons associated with the lateral segmental nerves. *Ann. Ent. Soc. Amer. 56,* 755–763.

Wigglesworth, V. B. (1940). Local and general factors in the development of "pattern" in *Rhodnius prolixus* (Hemiptera). *J. Exp. Biol. 17,* 180–200.

Wigglesworth, V. B. (1959). The histology of the nervous system of an insect *Rhodnius prolixus.* I. The peripheral nervous system. *Quart. J. Mic. Sci. 100,* 285–298.

Wigglesworth, V. B. (1972). *The Principles of Insect Physiology.* 7th edn. Chapman and Hall, London.

Wilczek, M. (1967). The distribution and neuroanatomy of the labellar sense organs of the blowfly *Phormia regina* Meigen. *J. Morphol. 122,* 175–201.

Yokohari, F. (1981). The sensillum capitulum, an antennal hygro- and thermoreceptive sensillum of the cockroach, *Periplaneta americana. Cell Tiss. Res. 216,* 525–543.

Yokohari, F., Tominaga, Y. and Tateda, H. (1982). Antennal hygroreceptors of the honey bee, *Apia mellifera* L. *Cell Tiss. Res. 226,* 63–73.

Zacharuk, R. Y. (1962a). Sense organs of the head of larvae of some Elateridae (Coleoptera): their distribution, structure and innervation. *J. Morphol. 111,* 1–34.

Zacharuk, R. Y. (1962b). Exuvial sheaths of sensory neurones in the larva of *Ctenicera destructor* (Brown) (Coleoptera, Elateridae). *J. Morphol. 111,* 35–47.

Zacharuk, R. Y. (1962c). Some histochemical characteristics of tissues in larvae of *Ctenicera destructor* (Brown) (Coleoptera, Elateridae), with special reference to cutaneous sensilla. *Canad. J. Zool. 40,* 733–746.

Zacharuk, R. Y. (1971). Fine structure of peripheral terminations in the porous sensillar cone of larvae of *Ctenicera destructor* (Brown) (Coleoptera, Elateridae), and probable fixation artifacts. *Canad. J. Zool. 49,* 789–799.

Zacharuk, R. Y. (1979). Some ultrastructural characteristics of mandibular cuticle in an elaterid larva (Coleoptera). *Canad. J. Zool. 57,* 1682–1692.

Zacharuk, R. Y. (1980a). Ultrastructure and function of insect chemosensilla. *Ann. Rev. Ent. 25,* 27–47.

Zacharuk, R. Y. (1980b). Innervation of sheath cells of an insect sensillum by a bipolar type II neuron. *Canad. J. Zool. 58,* 1264–1276.

Zacharuk, R. Y. and Albert, P. J. (1978). Ultrastructure and function of scolopophorous sensilla in the mandible of an elaterid larva (Coleoptera). *Canad. J. Zool. 56,* 246–259.

Zacharuk, R. Y., Albert, P. J. and Bellamy, F. W. (1977). Ultrastructure and function of digitiform sensilla on the labial palp of a larval elaterid (Coleoptera). *Canad. J. Zool. 55,* 569–578.

Zacharuk, R. Y. and Blue, S. G. (1971a). Ultrastructure of the peg and hair sensilla on the antenna of larval *Aedes aegypti* (L.). *J. Morphol. 135,* 433–456.

Zacharuk, R. Y. and Blue, S. G. (1971b). Ultrastructure of a chordotonal and a sinusoidal peg organ in the antenna of larval *Aedes aegypti* (L.). *Canad. J. Zool. 49,* 1223–1229.

Zacharuk, R. Y., Yin, L. R.-S. and Blue, S. G. (1971). Fine structure of the antenna and its sensory cone in larvae of *Aedes aegypti* (L.). *J. Morphol. 135,* 273–298.

Zawarzin, A. (1912). Histologische Studien über Insekten. III. Über das sensible Nervensystem der Larven von *Melolontha vulgaris. Zeit. Wiss. Zool. 100,* 447–458.

2 Mechanoreception

SUSAN B. McIVER

University of Toronto, Toronto, Ontario, Canada

1 INTRODUCTION

Mechanoreception is the perception of distortion of the body caused by mechanical energy which may originate either externally or internally. External mechanical energy varies from continuous pressure to rapid oscillations. Internal mechanical energy is due to forces generated by activities of the muscles. Perception of the various forms of mechanical energy results in the senses of gravity, pressure, touch, position, current, vibration, and hearing. These mechanical senses are involved in more behavioral activities, for example, locomotion, posture, feeding, orientation, mating and oviposition, than the chemical or visual senses. The senses of gravity and hearing have been particularly well studied in insects and are considered in separate chapters in this series.

The rigid exoskeleton of insects is well suited for detection of mechanical stimuli and a number of specialized structures have been developed for this purpose. These structures occur both inside and outside the body and singly, in small groups, or in large groups that function as specialized organs.

The mechanosensilla may be classified according to structure, function or the types of behavior they mediate. One structural type may have more than one function and a specific function may be involved in several types of behavior.

Perception of mechanical stimuli, as other sensory stimuli, involves (1) *coupling* of the stimulus to the sensitive neuron, (2) *transduction* or conversion of the mechanical energy of the stimulus to the electrical energy of the impulse and (3) *coding* or development of the specific temporal characteristics of the electrical impulse. There is a decrease in specificity of these steps from coupling to coding; that is, the principles that govern coding are more similar in the various kinds of mechanosensilla and indeed for sensilla of other modalities than are those involved in coupling. The most specific step, coupling, is determined primarily by the morphology of the sensillum. Consequently, this chapter is organized around the various structural types of mechanosensilla with concomitant discussion of their physiological features and roles in behavior.

Mechanosensilla occur as two general types (Table 1). Type I are those sensilla with bipolar neurons that have regions assuming the structure of modified cilia and dendrites associated with the cuticle or its invaginations. Type I may be subdivided into (a) those sensilla with an external cuticular portion, the cuticular mechanosensilla, and (b) those associated with only the inner aspect of the cuticle and lacking an external cuticular component, the scolopidial sensilla, commonly grouped into chordotonal organs. Type II are sensilla with multiterminal neurons which lack ciliary regions and which are associated with connective tissues, the inner surface of the body wall, muscles, and the walls of the alimentary canal, but not the cuticle. Type I sensory neurons differentiate *de novo* within the epidermis; the axon grows inward and the cell body remains peripheral (Bullock, T. and Horridge, G., 1965). It was tacitly assumed that Type II sensory neurons originate in the same manner until Rice, M. (1975) proposed alternatively that the Type II neurons may have originated within the central nervous system and then migrated outwards to innervate mesodermal and epidermal tissues. The possibility of different origins should be given serious consideration because of the large anatomical and physiological differences between the two types of cells (Table 1).

Table 1: Comparison of Types I and II sensilla

Type I (uniterminal)	Type II (multiterminal)
Bipolar neuron	multipolar neuron: dendrites contain mitochondria and neurotubules; frequent occurrence of axon collaterals
Neuron with ciliary structure	neuron without ciliary structure
Dendrites associated in some manner with cuticle	dendrites associated primarily with tissues of mesodermal origin
Possess specialized sheath cells; axons covered by glial cells	lack sheath cells; axons covered by glial cells
Receptor regions of dendrites separated from haemolymph and bathed in a distinct fluid, the composition and constancy of which are probably controlled by the sheath cells	receptor regions of dendrites exposed to haemolymph
Mainly phasic–tonic*	phasic–tonic, more tonic, slower adaptation rate
Differentiated *de novo* within epithelium	perhaps derived from neurosecretory cells with soma in peripheral region

* Trichoid hair plates are an exception in that they supply tonic information for proprioceptive purposes

A common feature of insect sensilla is that the cell bodies are located in the periphery, near to the place of stimulus reception. In the migratory locust, *Locusta migratoria*, and the cockroach, *Periplaneta americana*, are multiterminal sensilla that have their cell bodies located inside the third thoracic ganglion (Bräunig, P. and Hustert, R., 1980; Collin, S., 1981). This finding indicates that at least a small number of cell bodies in insect ganglia may belong to sensory neurons, rather than to motoneurons, interneurons or neurosecretory cells. Elucidation of the ontogeny of these sensilla could determine whether they develop like Type I uniterminal sensilla from late ectodermal cells during embryogenesis with cell bodies migrating into the CNS, or they are derived from neuroblasts like motoneurons and interneurons (Bräunig, P. and Hustert, R., 1980).

R. Snodgrass's 1926 paper, with a reference list of 74 papers, was the first review and synthesis of the literature on insect sense organs, including the mechanosensilla. The earliest paper he cited to deal primarily with insect sense organs is that of O. vom Rath in 1838. From the time of vom Rath until shortly after World War II the light microscope was the only possible tool and the only readily available one until the 1960s for investigating the structure of insect sense organs. The increasingly widespread accessibility of both scanning and transmission electron microscopes during the past two decades has permitted an ever more detailed understanding of the structure of mechanosensilla. The parallel development and refinement of electrophysiological recording techniques which were pioneered by Pringle, J. (1938a,b), have resulted in a basic understanding of the function of mechanosensilla. Quantitative aspects of transduction are being revealed by mathematical analyses and computer simulations. Results of studies using recently developed techniques, that permit tracing of axons and interneurons and recording of their electrical activity within the central nervous system, are providing the basis for understanding integration of sensory information.

2 TYPE I, UNITERMINAL SENSILLA

A uniterminal sensillum is derived from a mother epidermal cell which divides to form the sheath and nerve cells (Clever, U., 1958; Jägers-Röhr, E., 1968; Wigglesworth, V., 1953). Three basic forms of Type I sensilla occur; namely, chordotonal organs, campaniform sensilla, and hair sensilla or clear derivative of a hair such as a scale, filament or peg. The hair and campaniform sensilla are considered to be cuticular mechanosensilla (McIver, S., 1975). A fully differentiated hair or campaniform sensillum consists of (a) external cuticular components, (b) sensory neuron(s), and (c) sheath cells. Chordotonal organs lack an external cuticular component consisting entirely of neuron(s) and sheath cells. Berlese, A. (1909) and Demoll, R. (1917) suggested that hair and campaniform sensilla and chordotonal organs are homologous structures. Schmidt, K. (1973) supported this concept with evidence from studies on fine structure, ontogeny and molting. Apparently the hair sensillum was the original type. Reduction of the hair-shaft to a small plate would produce the campaniform condition and further movement along the same axis would result in the chordotonal situation (Fig. 1). Intermediate forms between hairs and campaniform sen-

silla have been observed on the wings of fruit-flies (Lees, A., 1942).

2.1 Chordotonal organs

Chordotonal organs are sensory structures composed of special sensilla, the scolopidia, that are associated with a support. A scolopidium consists of one or more sensory cell(s) and two accessory cells, the scolopale and attachment cells (Moulins, M., 1976). The distinguishing feature of scolopidia is the *scolopale* which is an intracellular structure composed of electron-dense material deposited around longitudinally oriented microtubules. Graber, V. (1882) was the first to note a distinction between two types of scolopidia. An *amphinematic* scolopidium is drawn into a distal thread, a *monematic* one lacks such a thread. Monematic scolopidia are subintegumental — that is, without any direct relationship to the cuticle — whereas amphinematic scolopidia may be either integumental or subintegumental. As proposed by Howse, P. (1968), those organs in which the scolopidia are contained in connective tissue strands that span or

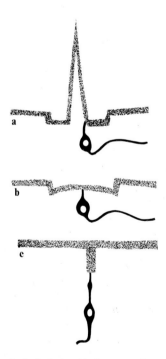

FIG. 1. Hypothetical relationship between subtypes of uniterminal sensilla. **a**: Hair, presumably most primitive; **b**: campaniform; **c**: scolopidium. (Redrawn from Rice, M., 1975.)

connect two structures are commonly referred to as "connective chordotonal organs".

Chordotonal organs are known from a large number of insect orders and probably occur in all insects. Chordotonal organs are found in almost all parts of the insect body (Finlayson, L., 1976; Moulins, M., 1976; Wales, W., 1976; Wright, B., 1976) and vary in complexity from simple organs in which the sensory element consists of only one neuron or a few neurons, such as those in the abdominal segments of insects (Finlayson, L., 1976) to highly developed tympanal and Johnston's organs — for example, the Johnston's organ of male mosquitoes, each of which is composed of about 7000–7500 scolopidia (Boo, K. and Richards, A., 1975a). Chordotonal organs function as proprioceptors or vibration receptors, commonly being the "ears" of insects. Hearing in insects is covered later in this volume, so emphasis here is on those chordotonal organs with a proprioceptive function.

2.1.1 FINE STRUCTURE OF A TYPICAL SCOLOPIDIUM

Moulins, M. (1976) provided a comprehensive review of the literature on the ultrastructure of chordotonal organs. The following description is adapted from his work.

2.1.1.1 *General organization of a scolopidium.* Proximally in a scolopidium glial cells are wrapped around the cell bodies and axons of the neurons (Fig. 2). Just distally the scolopale cell envelops the inner segment and ciliary region of the dendrite(s). The attachment all basally encloses the distal region of the scolopale cell and apically wraps around the distal region of the ciliary region (Fig. 2). Mesaxons are formed by the accessory as well as the glial cells. An electron-dense, extracellular, apical structure that takes the form of either a cap or a tube surrounds the most distal part of the outer dendritic segment.

2.1.1.2 *Sensory neurons.* The sensory neuron is composed of an axon, cell body, and a dendrite that is divided into inner and outer segments (Figs 2 and 3). Moulins, M. (1976) used the terms dendrite and centriolar derivative for the inner and outer segments respectively, and subdivided the latter into ciliary and distal segments (Figs 2 and 3). The outer segment contains a region assuming the structure of a modified cilium with a $9 \times 2 + 0$ arrangement of

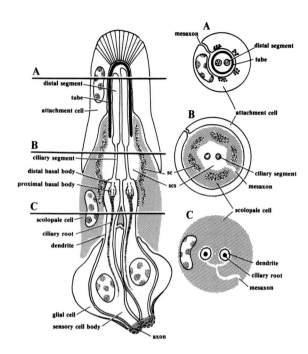

FIG. 2. Diagram of longitudinal section through a scolopidium (left) with transverse sections at various levels (right) as indicated by arrows. sc = Scolopale; scs = scolopale space. (Redrawn from Moulins, M., 1976.)

microtubules (ciliary segment of Moulins, M., 1976). An unusual neuron associated with the type B scolopidium in the Johnston's organ of mosquitoes do have the usual ciliary structure, having become transformed into a packet of many microtubules (Boo, K. and Richards, A., 1975a). The microtubules originate from those in the distal basal body and in each doublet one microtubule possesses a dense core and bears a pair of "arms". In the femoral chordotonal organ of a grasshopper Moran, D. *et al.* (1977) reported the occurrence of a ciliary necklace (Fig. 4), a small area between the ciliary region and the distal basal body which is seen in cross-section as champagne-glass-shaped structures linking the microtubular doublets to the dendritic membrane (Gilula, N. and Satir, P., 1972). Ciliary necklaces may be widespread, perhaps even universal, in scolopidia. Their small length, and the quality of preparation required to demonstrate their existence, may have precluded frequent mention of them in the literature. Likely ciliary necklaces are shown in Fig. 5, Boo, K. (1981); Fig. 9, Boo, K. and Richards, A. (1975a) and Fig. 6f, Ball, E.

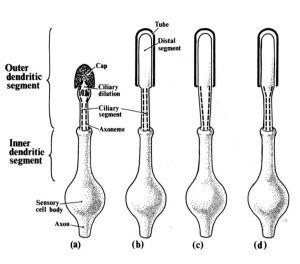

FIG. 3. Various types of scolopidial neurons. Type I (**a**) lacks a distal prolongation (distal segment) of the dendrite and has a ciliary dilation which contains microtubules and electron dense material. Type II (**b–d**) has a distal prolongation and lacks a ciliary dilation. Subtypes (**c**) and (**d**) are the paraciliary and ciliary cells of Whitear, M., 1962. (Redrawn after Moulins, M., 1976.)

(1981). Young, D. (1973) used the term "ciliary base" to include the ciliary necklace region.

Moulins, M. (1976) recognized two types of scolopidial sensory neurons on the basis of the presence or absence of a prolongation (distal segment) of the dendrite past the ciliary region. Type I (Fig. 3) lacks a distal prolongation and a dilation occurs in the subterminal area of the ciliary region. At the dilation the diameter of the dendrite increases with a concomitant increase in the diameter of the ring of internal microtubules. Electron-dense material occurs between the doublets of microtubules. The dilation is perhaps comparable to the apical tubular body of hair and campaniform sensilla. Distal to the dilation the microtubules lack dense cores and arms, and quite near the tip the internal structure of the dendrite becomes disorganized. Type I scolopidia occur in many organs — for example, tympanal organs, chordotonal organs of legs and mouthparts and Johnston's organs. Except for the last organ, the Type I scolopidia are always associated with an apical cap.

Type II scolopidia lack any dilation containing electron-dense material and there is a definite distal prolongation of the dendrite beyond the ciliary region (Fig. 3). In the prolonged region the diameter of the dendrite is larger than in the ciliary region and there are many microtubules. The tip of the dendrite is always ensheathed by a tube. The Type II scolopidia are known from certain chordotonal organs of legs (Moulins, M., 1976).

Distally in the inner dendritic segment there are two basal bodies with the typical $9 \times 3 + 0$ configuration of microtubules. The two innermost microtubules of each triplet continue as the doublets found in the ciliary region. A periodically banded ciliary root, the axial filament of light microscopists, extends proximally from the basal bodies. The ciliary roots of scolopidia are usually more prominent that those in other sensilla and may run proximally the length of the dendrite and through the cell body into the axon (Moulins, M., 1976). The ciliary rootlet is sometimes divided proximally at the level of the cell body and always distally in association with the basal bodies. Moulins, M. (1976) stated that the distal divisions give rise to a ring of nine processes. Each process may possibly be associated with one triplet of the basal body. The ciliary root processes and the triplet microtubules of the distal basal body in the scolopidia of the Johnston's organ of mosquitoes are discontinuous (Boo, K., 1981). This is in contrast to the situation in motile or mammalian cilia. The nine concentric processes of the root converge distally at an eccentric point just inside the most proximal part of the distal basal body. This association results in a bilateral symmetry in which the medial axis connects the tip of the ciliary root and the geometrical centre of the basal body (Boo, K., 1981). In the auditory scolopidia of the weta, *Hemideina crassidens*, an orthopteran, Ball, E. (1981) observed that the nine ciliary root processes pass around the outside of the proximal basal body and then rejoin at the level of the distal basal body.

2.1.1.3 *Scolopale cell.* The scolopale cell extends from the base of the inner dendritic segment to just distal to the ciliary region of the neuron. At the level of the inner segment the scolopale cell is closely apposed to the dendrite, being connected distally by zonula adherans (belt desmosomes) (Moulins, M., 1976) (Fig. 2). At the ciliary region there is an extracellular space, the scolopale space, between the scolopale cell and the dendrite. In the distal region the scolopale cell is closely associated laterally with the attachment cell and medially with the apical extracellular structure — that is, the cap or tube. The

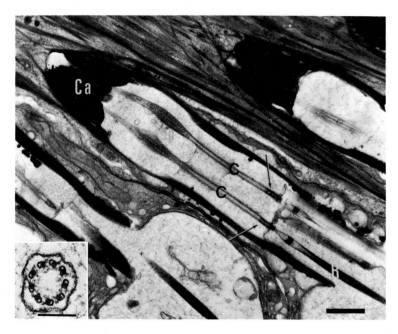

FIG. 4. High-voltage electron micrograph of a longitudinal section through the paired dendrites and cilia of a chordotonal sensillum from a minimally stimulated femoral chordotonal organ of *Melanoplus bivittatus*. Note that the ciliary bases are straight. R = Rootlet in distal dendrite; C = cilium; Ca = cap; arrow indicates ciliary necklace. Inset shows cross-section through cilium; note (dynein?) arms attached to A-subfibers. (Scale = 2 μm; inset scale = 20 μ.) (Reproduced from Moran, M. *et al.*, 1977 with permission.)

scolopale cell may be highly vacuolated and has a labyrinth of extracellular cavities on the side of the scolopale space. These features are indicative of a secretory function and this cell, like its equivalent the trichogen cell, in hair and campaniform sensilla, probably regulates the ionic composition of the fluid in the adjacent extracellular space.

The scolopale is an intracellular structure which cannot be compared with the dendritic sheath in chemosensilla or other mechanosensilla. The latter sheath is always extracellular and is comparable to the apical extracellular structure of scolopidia. The term scolopale has unfortunately been used for various kinds of extracellular structures. The substance of the scolopale is composed of electron-dense material deposited between longitudinally oriented microtubules. The entire scolopale is a cylinder that extends around the dendrite(s) from the level of the distal part of the inner segment to just distal to the ciliary region. The scolopale may be irregular in outline and is frequently fenestrated,

especially at the level of the scolopale space. These fenestrations give rise to the scolopale rods described by various authors. In these cases the space between the rods, that is, the fenestrations, is continuous with the scolopale space.

2.1.1.4 *Attachment cell.* The attachment cell encloses the apical extracellular structure and provides the apical anchorage of the scolopidium to the support. In chordotonal organs where the scolopidia lack lateral connections between each other, the most common type in insects, the attachment cell takes three forms:

(1) the attachment cell is an epidermal cell and the extracellular apical structure of the scolopidium is inserted in the cuticle;

(2) the attachment cell is an epidermal cell, but the apical structure is not inserted in the cuticle;

(3) the attachment cell becomes a subepidermal cell and is connected at its apex to the base of an epidermal cell.

2.1.2 Distribution

2.1.2.1 Head

(a) *Mouthparts and larval antenna.* Connective chordotonal organs have been reported from the mouthparts of several species (Wales, W., 1976) and probably occur in the mouthparts of all insects. Recent fine structural descriptions are given by Zacharuk, R. and Albert, P. (1978) for organs in the mandibles of elaterid beetle larvae; by Wensler, R. (1974) for ones in the mandibles of adult aphids; by Pappas, L. and Larsen, J. (1976) for the labellar chordotonal organs of adult mosquitoes; and by Lee, J.-K. *et al.* (1980) for the labial palp of adult butterflies. In all these insects the chordotonal organs are believed to monitor bending of the particular structure in which they are located. One of the two neurons in the mandibular organ of the elaterid larva produced impulses with a higher amplitude and at a higher rate than the other (Zacharuk, R. and Albert, P., 1978).

Eggers, F. (1928) and Howse, P. (1968) reviewed the occurrence of connective chordotonal organs in the antennal segments of insects. In two species of termites there is a connective chordotonal organ in the scape and one in the pedicel and in the beetle, *Hydropsyche longipennis*, an organ traverses the pedicel to insert on a cuticular projection at the base of the third segment (Howse, P., 1968). The arista of the dipterans *Micropeza* and *Diopsis* are probably sensitive to touch owing to scolopidia which terminate mesially at the articular membrane of the funicule and the arista (Dudel, H., 1974).

The fine structure of chordotonal organs in the larval antenna of a mosquito (Zacharuk, R. and Blue, S., 1971) and a beetle (Bloom, J. *et al.*, 1981) has been described. The organ in the beetle is unusually attached in that there is no contact or connection between the fully developed, microtubule-filled attachment cell and the cuticle. Rather the attachment cell anchors the sensillum in a basement membrane-like extracellular matrix. This organ probably monitors the position or movement of the antennal third segment with respect to the second segment, and may do so by detecting motion transmitted either directly or through changes in hemocoelic pressures on the basement membrane-like extracellular matrix in which the attachment cell is inserted (Bloom, J. *et al.*, 1981).

(b) *Johnston's organ.* In 1855 Johnston described an organ in the swollen pedicel of *Culex* mosquitoes which bears his name. Snodgrass, R. (1935) defines the Johnston's organ as "an organ of scolopophorous type located in the second segment, or pedicel, of the antennae of nearly all insects". As discussed by Boo, K. and Richards, A. (1975a), this broad definition is commonly followed although some authors have restricted the term to the peripheral scolopidia and use another term ("central organ" or "connective chordotonal organ") for centrally located single scolopidia. It seems desirable to retain Snodgrass's definition of Johnston's organ and to devise new and functionally descriptive terms for the various types of scolopidia which may occur.

The structure of Johnston's organ of immatures as well as adults has been described from light and electron microscopic studies on a number of species (Table 2). The degree of development varies considerably, reaching its greatest complexity in male mosquitoes and chironomid midges which use Johnston's organ to detect the flight tone of the female. In both sexes of most insects the pedicel is not particularly swollen and contains a few to several hundred scolopidia. The pedicel of male mosquitoes and chironomids, however, is not only greatly enlarged to accommodate the thousands of scolopidia, but also an extensive arrangement of apodemes (prongs) for their internal attachment (Fig. 5). The degree of sexual dimorphism of Johnston's organ has been used as an indicator of the relative importance of sound detection and airflow and vibrations in the lives of blackflies (Boo, K. and Davies, D., 1980) and sawflies (Hallberg, E., 1981), respectively.

In most pterygote species as well as the thysanuran, *Lepisma*, the scolopidia are each innervated by three neurons. Some dipterans also have scolopidia with two neurons; for example, mosquitoes (Boo, K. and Richards, A., 1975a,b; Risler, H. and Schmidt, K., 1967), blackflies (Boo, K. and Davies, D., 1980) and *Drosophila* (Uga, S. and Kuwabara, M., 1965). The apterygotes (Order Archaeognatha) *Machilis* and *Dilta* possess scolopidia with two neurons whose outer dendritic segments (sensory cilia) are less well developed than those in the more advanced species (Schmidt, K., 1975).

Table 2: Structural studies on Johnston's organ

Order and Genus	Stage	Technique	Comments	References
Archaeognatha				
Machilis	adult	TEM	detailed study	Schmidt, K. (1975)
Dilta	adult	TEM	detailed study	Schmidt, K. (1975)
Petrobius brevistylis	adult	TEM	general observations	Kinzelbach-Schmidt, B. (1968)
Thysanura				
Lepisma saccharina	adult	TEM	detailed study	Schmidt, K. (1975)
Lepisma saccharina	adult	TEM	general observations	Kinzelbach-Schmidt, B. (1968)
Ctenolepisma lineata pilifera	adult	TEM	general observations	Kinzelbach-Schmidt, B. (1968)
Thermobia domestica	adult	TEM	general observations	Kinzelbach-Schmidt, B. (1968)
Ephemeroptera				
Cleon dipterum	adult	TEM	detailed study	Schmidt, K. (1974)
Cleon dipterum	subadult	TEM	detailed study	Schmidt, K. (1974)
Baetis	adult	TEM	detailed study	Schmidt, K. (1974)
Epeorus	adult	TEM	detailed study	Schmidt, K. (1974)
Ephemera	adult	TEM	detailed study	Schmidt, K. (1974)
Dictyoptera				
Periplaneta americana	adult	TEM	detailed study	Toh, Y. (1981)
Isoptera				
Zootermopsis angusticollis	larva	LM	general description of scolopidia	Howse, P. (1965)
Calotermes flavicollis	larva	LM	ontogeny	Richard, G. (1957)
Phasmatodea				
Carausius morosus	larva	LM	ontogeny	Richard, G. (1957)
Sipyloidea sipylus	larva	LM	ontogeny	Richard, G. (1957)
Hemiptera				
Oncopsis flavicollis	adult	TEM	detailed study	Howse, P. and Claridge, M. (1970)
Aphis pomi	adult	TEM	detailed study	Bromley, A. *et al.* (1980)
Neuroptera				
Chrysopa carnea	adult	TEM	detailed study	Schmidt, K. (1969a)
Coleoptera				
Tenebrio	adult	TEM	details of ciliary structure	Schmidt, K. (1970)
Speophyes lucidulus	adult	TEM	detailed study	Corbiére-Tichané, G. (1975)
Speophyes lucidulus	larva	TEM	detailed study	Corbiére-Tichané, G. (1971)
Hydropsyche longipennis	adult	LM	thorough description	Debauche, H. (1935)
Dytiscus marginalis	adult	LM	thorough description	Lehr, R. (1914)
Mecoptera				
Panorpa	adult	TEM	details of ciliary structure	Schmidt, K. (1970)
Panorpa communis	adult ♂ ♀	TEM	detailed study	Leonowitch, S. and Ivanov, V. (1978)
Lepidoptera				
Pieris	adult	TEM	details of ciliary structure	Schmidt, K. (1970)
Manduca sexta	adult	TEM	detailed study	Vande Berg, J. (1971)
Bombyx mori	adult	LM	general description	Schneider, D. and Kaissling, K. (1957)
Diptera				
Chironomus	adult	LM	general description	Child, C. (1894)
Chironomus anthracinus, Chironomus riparius	adult ♂ ♀	TEM	effect of nematode parasite on intersexual features; detailed study	Schmidt, K. and Wülker, W. (1980)
Simulium vittatum	adult ♂ ♀	TEM	detailed study	Boo, K. and Davies, D. (1980)
Anopheles stephensi	adult ♂	TEM	detailed study	Boo, K. (1980)
	adult ♂	LM	thorough description	Risler, H. (1953, 1955)
	adult ♂ ♀	TEM	details of ciliary structure and basal bodies	Boo, K. (1981)

Table 2: Structural studies on Johnston's organ—Continued

Order and Genus	Stage	Technique	Comments	References
Culex pipiens	adult ♂	LM	thorough description	Risler, H. (1955)
	adult ♂ ♀	LM	thorough description	Johnston, C. (1855)
	adult	SEM	general observations	Risler, H. (1977)
Aedes aegypti	adult ♂	LM	thorough description	Risler, H. (1955)
	adult ♂	TEM	structure of scolopidia	Risler, H. and Schmidt, K. (1967)
	adult ♂	TEM	excellent, detailed study	Boo, K. and Richards, A. (1975a)
	adult ♂	TEM	general observations	Risler, H. (1977)
	adult ♀	TEM	excellent, detailed study	Boo, K. and Richards, A. (1975b)
	pupa	TEM	development	Schmidt, K. (1967)
Bibio	adult	TEM	details of ciliary structure	Schmidt, K. (1970)
Drosophila melanogaster	adult	TEM	detailed study	Uga, S. and Kuwabara, M. (1965)
Hymenoptera				
Neodiprion sertifer	adult ♂ ♀	SEM, TEM	detailed study	Hallberg, E. (1981)
Athalia	adult	TEM	details of ciliary structure	Schmidt, K. (1970)
Camponotus vagus	adult	TEM	detailed study	Masson, C. and Gabouriaut, D. (1973)
Mellinus	adult	TEM	details of ciliary structure	Schmidt, K. (1970)
Variety of species in a number of orders	adult, larva	LM	general observations	Eggers, F. (1924, 1928)
Variety of species in a number of orders	adult	LM	comparative cytology	Debauche, H. (1936)

The most common type of scolopidium in ectognath insects is amphinematic. In Ephemeroptera all the scolopidia composing Johnston's organ are mononematic while in *Aedes aegypti* (Boo, K. and Richards, A., 1975a,b) and *Simulium vittatum* (Boo, K. and Davies, D., 1980) only a few are mononematic, the majority being amphinematic. Mononematic scolopidia are known also from the Johnston's organ of a termite (Howse, P., 1968), *Chrysopa* sp. (Schmidt, K., 1969a), and *Lepisma* sp. (Kinzelbach-Schmitt, B., 1968). Although exceptions are known, the amphinematic scolopidium possessing three neurons is the most widespread type of scolopidium in Johnston's organ.

Several different functions have been attributed to Johnston's organ:

(1) a proprioceptor relating antennal position and movements to the rest of the body in numerous species;
(2) a flight speed indicator in bees and mosquitoes;
(3) a gravity detector in *Dytiscus* and mosquitoes;
(4) a prey detector in *Notonecta*;
(5) an air current detector in mosquitoes; and
(6) a sound receptor in mosquitoes and chironomid midges (Boo, K. and Richards, A., 1975b).

Schwartzkopff, J. (1974) reviewed the relevant physiological literature.

As originally suggested by Johnston, C. (1855), many male mosquitoes respond to the flight tone of females, but not vice-versa, and Johnston's organ is

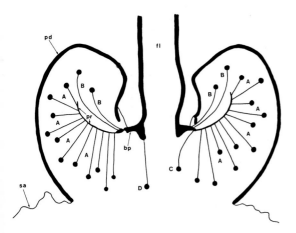

FIG. 5. Diagram of longitudinal section of Johnston's organ of male *Aedes aegypti*. A, B, C, D = Types A, B, C and D scolopidia, bp = basal plate, fl = flagellum, pd = pedicel, pr = prong, sa = scape. (Reprinted by permission of the Editor, from the *J. Med. Ent. 19*, 508, 1982.)

responsible for this behavior of the male as demonstrated by several behavioral and electrophysiological studies (Keppler, E., 1958a,b; Roth, L., 1948; Tischner, H. and Schief, A., 1954; Wishart, G. and Riordan, D., 1959; Wishart, G. *et al.*, 1962). These studies have been analyzed and discussed by Clements, A. (1963), Schwartzkopff, J. (1974) and Belton, P. (1974). Subsequent work has focused on the role of Johnston's organ in localization of sound by male mosquitoes, and on the function of specific types of scolopidia within these organs.

Since the Johnston's organs are exposed and close together, mosquitoes must use a mechanism for the localization of sound different from that of vertebrates, whose widely separated ears detect differences in intensity and phase as the sound waves strike them. Clements, A. (1963) reviewed the several existing theories on sound localization by mosquitoes. More recently Belton, P. (1974) reported that there are probably at least two phenomena involved in determining the harmonic content of the electrical response of Johnston's organ. These are (a) the mechanical resonance of the flagellum, determined largely by the properties of Johnston's organ, and (b) summation of electrical events occurring in functionally distinct groups of cells. In addition, Belton, P. (1974) presented electrophysiological evidence supporting the idea of Wishart, G. *et al.* (1962) that information obtained from the phase of the electrical activity of Johnston's organ could be used by the mosquito to obtain information on the direction of the sound source.

Mosquitoes apparently locate sound in a way analogous to triangulation or radio direction-finding. Belton, P. (1974) outlined the way in which triangulation could occur as follows. With the two Johnston's organs pointing slightly above the line of flight and with the antennal flagella at an angle to each other of about 65°, male *Aedes aegypti* can rapidly assess the position of the flying female (Wishart, G. *et al.*, 1962). If a point source of sound were at an angle greater than 40° the flagella would be deflected in phase. With the wavelength of sound greater than the distance between the flagella, the antennae would be pushed in the same direction almost simultaneously. If the source of sound were directly in front of the antennae, the flagella would be deflected out of phase, because neurons in identical positions in the two Johnston's organs in the

plane of the direction of propagation of the sound would be pulled and pushed by their respective flagella. The phase of the signals produced by the neurons would indicate whether the sound was within 30° of the line of flight or greater than 40° from it, resulting in almost immediate awareness of the position of a source of sound by the mosquito. The plane of maximal vibration of the flagella would indicate the azimuth of the sound source, and the slight divergence of the planes of vibration of the flagella would permit the mosquito to extrapolate these planes and the sound would be pinpointed in a manner similar to triangulation.

The results of K. Boo and co-workers' detailed studies on both sexes of two species of mosquitoes and one species of blackflies have aided in understanding the probable function of each type of scolopidium found within the Johnston's organs. About 97% of the scolopidia in the pedicel of male *A. aegypti* are Type A, which are amphinematic with two neurons and attach to the underside of the prongs (Fig. 5). Type B account for about 3% of the total scolopidia, are amphinematic with three neurons and attach to the upper side of the prongs (Fig. 5). In the central portion of pedicel there are three additional scolopidia, the "single scolopidia", that are completely independent of the great mass of scolopidia of types A and B. Two of the "single scolopidia" are termed type C and one type D. Both types are mononematic with two neurons and attach to the epidermis under the basal plate at the points indicated in Fig. 5. The Johnston's organ of female *A. aegypti* has three types of scolopidia, types A, B, and C, but lacks the type D scolopidium observed in the male's pedicel (Boo, K. and Richards, A., 1975b). The basic structure and location of each type in the female is similar to its counterpart in the male. Although the same structural components, except for the single type D scolopidium, occur in the Johnston's organ of the female, their general organization and development are relatively poorer than those in the organ of the male.

The general organization, types of scolopidia, and fine structure of Johnston's organ of male and female *Anopheles stephensi* are generally similar to that of the corresponding sex of *A. aegypti* (Boo, K., 1980). However, in *A. stephensi* there is considerable doubt as to whether type D scolopidia, which are observed in the female as well as the male, should be considered distinct from the type C scolopidia. Male and female *Simulium vittatum* have three types of scolopidia in their Johnston's organ: types A, B, and C (Boo, K. and Davies, D., 1980). The basic structure and location of each type are similar to, although less well organized than, these types in adult mosquitoes.

Boo, K. and Richards, A. (1975b) suggested that types A and B scolopidia function in detection and monitoring of antennal movements and in perception of air current during flight. These kinds of amphinematic scolopidia are widespread in insects and probably have the same function in all species. Owing to their intimate association with the flagellular blood vessel, the type C scolopidia may be involved in a proprioceptive function connected with circulation. The type D scolopidia of male *A. aegypti* may function in hearing (Boo, K. and Richards, A., 1975b), although verification awaits appropriate behavioral and electro-physiological experiments. Boo, K. and Davies, D. (1980) inferred a common function for type B scolopidia in all insects; namely, detecting air currents indirectly by deflection of the antenna. In the blackfly, type C scolopidia could be joint receptors to perceive changes in the position of the flagellum against the pedicel. The simuliid type C scolopidia, unlike those in culicids are not associated with the antennal blood vessel (Boo, K. and Davies, D., 1980).

The antennae control flight in locusts (Gewecke, M., 1970); aphids, (Johnson, B., 1956); bees (Heran, H., 1959); mosquitoes (Bässler, U., 1958); flies (Burkhardt, D. and Schneider, D., 1957; Gewecke, M., 1967; Hollick, F., 1940; Schneider, P., 1965); cockroaches (Yagodin, S., 1980) and some moths (Gewecke, M. and Niehaus, M., 1981; Niehaus, M., 1981) (Table 3). As indicated by Gewecke, M. (1974), the antennae of these insects are very different in form but similar in function. In all species it is the flagellum that receives the stimulus — that is, air currents generated during flight — and conducts the aerodynamic forces to the pedicel. The static characteristics of these aerodynamic properties of the locust antenna have been investigated in detail (Gewecke, M. and Heinzel, H., 1980). Associated with the pedicel are various mechanoreceptors, Johnston's organ, other chordotonal organs and campaniform

Table 3: *Species in which Johnston's organ is involved in flight control*

Insect	References
Dictyoptera	
Periplaneta americana	Yagodin, S. (1980)
Orthoptera	
Locusta migratoria	Gewecke, M. (1970, 1972b,c)
Hemiptera	
Lepidoptera	
Aglais urticae	Gewecke, M. and Niehaus, M. (1981)
	Niehaus, M. (1981)
Diptera	
Aedes aegypti	Bässler, U. (1958)
Calliphora erythrocephala	Gewecke, M. (1967b)
Calliphora vicina	Schneider, P. (1965)
Muscina stabulans	Höllick, F. (1940)
Hymenoptera	
Apis mellifica	Heran, H. (1959)

organs, that perceive the movements of the flagellum with respect to the pedicel (Gewecke, M., 1974).

In the locust, air-current-sensitive hair patches on the head and the antennae control, at least in part, wing-beat frequency, wing-stroke angles, lift, and flight speed (Gewecke, M., 1975). In free-flying locusts the hairs and antennae appear to affect the flight speed in opposite ways. The hair patches stimulate flight speed and the antennae reduce it, thereby being the sensory units of a negative feedback mechanism (Gewecke, M., 1975). The antennae are apparently not important in the control of flight in the horizontal plane by locusts (Gewecke, M. and Philippen, J., 1978).

In the small tortoiseshell moth, *Aglais urticae*, the flight speed in relation to the air is controlled by mechanoreceptors in the proximal part of the antennae, including Johnston's organ (Niehaus, M. and Gewecke, M., 1978). The antennal angle during flight is about 43° and independent of air speed up to $2.0\,\mathrm{ms}^{-1}$ (Fig. 6) (Gewecke, M. and Niehaus, M., 1981). Under normal flight conditions the passive antennal deflection is below 0.2° (Fig. 6). A constantly held, specific torque in the pedicel–flagellum join apparently permits the antennae to work as sensory units of a negative feedback mechanism by which flight speed is adjusted (Niehaus, M., 1981).

2.1.2.2 *Abdomen and thorax.* The abdomen and thorax of some insects contain tympanal organs. In some

Hemiptera these organs are found in the meso- and metathorax and in many Lepidoptera in the meta-thorax. The first or second abdominal segments of short-horned grasshoppers, cicadas, and the lepidopteran families Pyralidae and Geometridae and the seventh abdominal segment of axiid moths bear tympanal organs (Howse, P., 1968). In the meso- and metathorax of a number of aquatic heteropterans are "scolopophorous organs" which may function in hearing and/or detecting changes in water pressure (Anderson, B., 1980).

Simple connective chordotonal organs consisting of a single or a very few scolopidia have been described in every abdominal segment of insects (Finlayson, L., 1976). The organs occupy constant positions, have always been found in mid-ventral, ventrolateral, lateral or dorsal position, or in all these positions, and apparently occur in substantial numbers — for example, 45 pairs of simple chor-dotonal organs occur in the larva of *Drosophila* (Finlayson, L., 1976).

The longitudinal, ventrolateral chordotonal organs of the stick insect, *Carausius morosus*, each possess two neurons and are attached to the cuticle by a single strand (Orchard, I., 1975). These organs respond in a highly phasic manner to both stretch-ing and subsequent relaxation of the attachment strand. Although the chordotonal organs are sen-sitive to substrate vibrations, they are activated by ventilatory movements. It seems likely that the ventrolateral chordotonal organs of the stick insect provide information about either the longitudinal or vertical muscles during the expiratory phase of ventilation; inspiration may be controlled by lateral chordotonal organs (Orchard, I., 1975).

2.1.2.3 *Wings.* In the wings connective chordotonal organs are known to be associated with the articu-lation, a tracheal bladder or tympanum, and in dip-terans, the halteres (Howse, P., 1968). The physi-ology and role in behavior of these organs are poorly understood. It seems reasonable that they are important in flight, but some may have other func-tions (Wright, B., 1976). For example, Huber, F. (1974) found that the single chordotonal organ at the wing-base of a cricket responds specifically to stridulatory movements, and Möss, D. (1967) sug-gested that the same organ is a vibration receptor.

2.1.2.4 *Legs.* The legs of insects are well supplied with chordotonal organs of varying structural

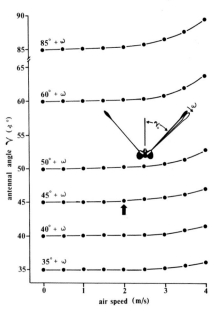

FIG. 6. Antennal angle (γ) as a function of air speed in three inactive *Aglais urticae*. After fixing the proximal, actively movable joints of one antenna in different positions (when the air current equals 0 m/s $\gamma = \gamma_0$; $35° \leqslant \gamma_0 \leqslant 85°$), the ante-nnal angle is increased with increasing air speed by the angle of passive antennal deflection (ω):$\gamma = \gamma_0 + \omega$. *Arrow* marks the value of γ in fast flying normal animals. (Redrawn from Gewecke, M. and Niehaus, M., 1981.)

complexity (Wright, B., 1976). Their distribution in the leg has been mapped for several orders (Debai-seaux, P., 1938). The types of chordotonal organs known from the legs are: (a) fairly simple connective chordotonal organs that occur in all parts of the leg; (b) the femoral chordotonal organ, a connective chordotonal organ that is particularly well developed in the Orthoptera; and (c) the subgenual and complex tibial organs that are located on the tibiae of some insects.

(a) *Simple connective chordotonal organs.* The connective chordotonal organ at the tibio-tarsal joint of *Periplaneta* divides into a main branch and two side branches, and consists of three types of scolopidia that give rise to a total of 26 neurons (Young, D., 1970). One Type 1 scolopidium is in each of the side branches and two occur proximally in the main branch. Type 2 scolopidia are dis-tributed along the main branch and Type 3 are located distally in the main branch. Both Types 1 and 2 have two neurons per scolopidia. The Type 1 neurons are 15–20 μm in diameter and have heavily sheathed dendrites, while the Type 2 neurons are 8–15 μm in diameter and are less well sheathed. Type 3 scolopidia have a single neuron, about 10 μm in diameter, with a short, weakly sheathed dendrite (Young, D., 1970).

The cockroach tibio-tarsal organs respond to downward and backward deflection of the tarsus (Young, D., 1970). Type 1 scolopidia show a unidirectional phasic and tonic response to extreme deflection of the tarsus. The smaller Type 2 scolopidia show a undirectional tonic response to the full range of deflection of the tarsus. Young, D. (1970) suggested that the differences in adaptation between these two types of scolopidia are probably not related to differences in mechanical attachment. The adequate stimulus of the scolopidia is probably an increase in their longitudinal tension.

The tarsal scolopidial organs (connective chordotonal organs) of the backswimmer, *Notonecta*, each possesses eight neurons and is divided into distal and proximal scoloparia (Weise, K. and Schmidt, K., 1974). Three phasic–tonic neurons are in the proximal and five purely phasic cells occur in the distal scoloparium. The distal scoloparium is anchored on a movable part of the claw-hinge mechanism, a feature that enlarges the effective range of the receptor part and reverses the direction of the stimulus-effective movements (Wiese, K. and Schmidt, K., 1974).

The tarsal scolopidial organs of Notonectidae (Wiese, K., 1974) as well as Gerridae (pond-striders) are used to detect surface waves for localizing their quarry, which has dropped into the water (Schwartz-kopff, J., 1974). In both cases prey-catching behavior is optimally activated and oriented by surface waves with the frequency range of 20–200 Hz. The adequate stimulus appears to be wave amplitude coming to 0.5 μm at threshold. Adult *Notonecta* react more sensitively than the nymphal stages (Schwartzkopff, J., 1974). The organs in all three leg pairs are equally sensitive (Wiese, K., 1972).

(*b*) *Femoral chordotonal organ.* In Orthoptera the femoral organ has two distinct sensory regions, commonly called distal and proximal scoloparia, that are structurally different (Fig. 7). In the locust's proximal scoloparium there are over 200 small neurons and in the grasshopper's organ 300 to 400 neurons with diameters ranging from 10 to 12 μm.

The neurons in the proximal scoloparium are arranged in a regular cone-shaped array. In contrast the distal scoloparium of both the locust and grasshopper contain many fewer cells, 50 and 150, respectively, that are much larger in diameter (12–20 μm) and are scattered throughout the scoloparium. In the grasshopper each scolopidium contains two neurons (Moran, D. *et al.*, 1975; Slifer, E. and Sekhon, S., 1975).

Differences in structure of the femoral organ are known that probably reflect adaptation of the metathoracic leg for jumping (Wright, B., 1976). In the locust the femoral organ of the metathoracic leg is a single scoloparium containing only 24 cells, and is located in the distal part of the femur (Usher-wood, P. *et al.*, 1968). In pro- and mesothoracic legs the femoral organ is located proximally. The metathoracic scoloparium may be the homolog of the distal scoloparium of the pro- and mesothoracic legs (Wright, B., 1976).

The pro- and mesothoracic femoral chordotonal organs of the grasshopper are purely tonic mechanoreceptors (Moran, D. *et al.*, 1977). The U-shaped response curve recorded from the organs indicates that progressive flexion or extension from the resting joint angle of 90° increases the response frequency of individual neurons and also recruits additional units. The response curve for the femoral chordotonal organs of the locust is also U-shaped (Burns, M., 1974), but both phasic and tonic responses are common (Burns, M., 1974; Usherwood, P. *et al.*, 1968). The frequency of the tonic discharge varies with the femur–tibia angle while that of the phasic response is related to the velocity of angular

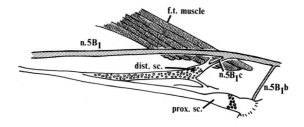

Fig. 7. Diagram of a typical prothoracic or mesothoracic femoral chordotonal organ drawn with its apodeme to the left. dist. sc. = Distal scoloparium attached to the cuticle and the flexor tibiae muscle (f.t. muscle); prox. sc. = proximal scoloparium attached only to the cuticle; n. 5B_1b, n. 5B_1c$_1$ = branches of nerve 5B_1 innervating the chordotonal organ. (Redrawn after Burns, M., 1974.)

tibial movements greater than 3°. The locust femoral chordotonal organs may also serve as receptors for muscle tension, and apparently mediate phasic resistance reflexes in all three extensor tibiae motoneurons and tonic reflexes in the extensor "slow" neuron (Burns, M., 1974).

In the walking stick insect the femoral chordotonal organs are involved in feedback control of the femur–tibia joint (Cruse, H., 1981; Cruse, H. and Pflüger, H., 1981). Bässler, U. and Storrer, J. (1980) postulated that the neural basis of the femur–tibia control system in *Carausius morosus* involves two separate channels, one channel for fast and one for slow inputs, from the chordotonal organs to each muscle. The slow and fast channels may correspond to slow and fast motoneurons that supply the muscles.

In the orthopteran insect, *Hemideina femorata*, the femoral chordotonal organ mediates intersegmental reflexes in five muscles of the mesothoracic leg tibia–tarsus joint and the coxa–trochanter joint (Field, L. and Rind, F., 1981). This is in addition to the classical resistance reflexes in the two muscles of the femur–tibia joint. One of the intersegmental reflexes — that is, the trochanteral levator — is apparently a resistance reflex. The other four intersegmental reflexes appear to be locomotory and function to augment hind leg retraction as well as to help lift the tarsal claws off the substrate during protraction of the hind leg.

Bässler, U. (1979) studied the effects of crossing the receptor apodeme of the femoral chordotonal organs on various behaviors of male locusts and grasshoppers, and concluded that the type of motor program in use determined the response to a particular afference; that is a program-dependent reaction. When the apodeme was moved from its natural origin on the tibia, dorsal to the axis of rotation of the joint, and fixed to the apodeme of the flexor tibiae muscle, the femoral chordotonal organ signalled the opposite of the real movement of the tibia. During walking meso- and metathoracic legs with crossed receptor apodemes were often held up with extended, immobile femur–tibiae joints while the other legs took many steps. Insects with operated metathoracic legs could jump well, but rarely did so. In operated insects the number and duration of songs were smaller, although only a few, small differences were noted on the way the legs

were used during singing.

(c) *Subgenual organ.* The subgenual organ is located in the proximal tibial region of each leg. The shape varies. It is fan-shaped in Orthoptera and Dictyoptera, club-shaped in certain Hymenoptera, club-shaped or cup-shaped in termites, and diffuse and apparently unattached distally in Lepidoptera (Howse, P., 1968). The subgenual organ is apparently lacking in Coleoptera, Hemiptera, and Diptera (Howse, P., 1968).

The subgenual organ of the cockroach, *Blaberus discoidalis*, is a thin, fan-shaped flap of tissue slung across the dorsal blood space of the tibia at right angles to the leg's long axis beneath the knee in each walking leg (Fig. 8) (Moran, D. and Rowley, J., 1975b). Approximately 50 scolopidia, each with one neuron, are associated with each organ. Most of the mass of the subgenual organ is derived from flattened, disc-shaped accessory cells. In the subgenual organ of the termite, *Zootermopsis angusticollis*, numerous tendril-like processes bind the accessory cells to each other. Some of the processes are wound together to form a stalk that attaches the organ to the wall of the tibia (Howse, P., 1965).

Subgenual organs detect vibrations within the substrate upon which the animal stands. In *Periplaneta americana* the maximum sensitivity of the subgenual organ is between 1000 and 5000 Hz and the threshold amplitude lies between 10^{-7} and about 10^{-10} cm, depending upon the position of the leg (Schnorbus, H., 1971). The adequate stimulus for the organ is acceleration and indeed Schnorbus, H. (1971) likened the organ to an accelerometer; that is, to a membrane with a weighted center whose motion is dampened by the surrounding haemolymph.

(d) *Complex tibial organs.* On the tibia of crickets and long-horned grasshoppers (Grylloidea and Tettigonoidea) are "complex tibial organs". These organs on the prothoracic legs function in hearing and are the tibial tympanal organs of Howse, P. (1968) and others. They usually consist of a subgenual organ, an intermediate organ and a crista acoustica (Ball, E., 1981; Ball, E. and Field, L., 1981; Houtermans, B. and Schumacher, R., 1974; Howse, P., 1968). Sometimes the latter is called the tympanal organ (Friedman, M., 1972). Ball, E. and Field, L. (1981) compared the terminology and number of scolopidia in the complex tibial organs of

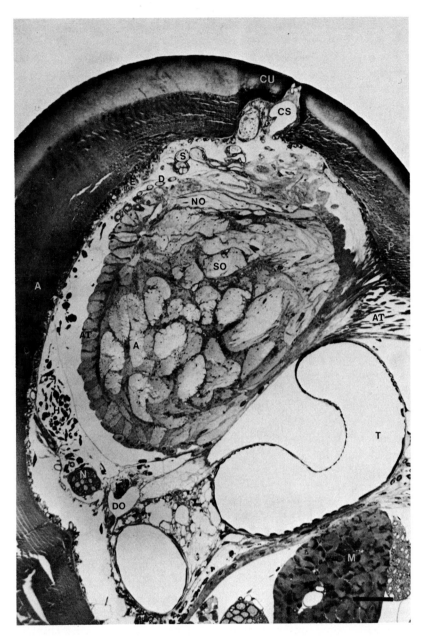

FIG. 8. Light micrograph of an epoxy cross-section through the tibia of *Blaberus discoidalis* illustrating the position of the subgenual organ within the leg. White A = Anterior, black A = accessory cell, AT = attachment cell, CS = campaniform sensillum, CU = cuticle, D = dendrite, DO = distalorgan, E = epidermis, M = muscle, N = nerve, NO = nebenorgan, P = posterior, S = soma, SO = subgenual organ, T = trachea. (Scale = 100 μm.) (Reproduced from Moran, D. and Rowley, J., 1975b with permission.)

gryllids, tettigoniids and stenopelmatids. In *Tettigonia viridissima* there are no signs of beginning tympanisation in the meso- and metathoracic legs and the subgenual organs of all three leg pairs are almost equally developed, an indication for the independent function of tympanisation (Houtermans, B. and Schumacher, R., 1974). There is also a progressive decrease in the number of scolopidia of the intermediate organs and cristae acusticae from the pro- to the metathoracic legs.

Kalmring, K. *et al.* (1978) simultaneously stimulated the complex tibial organs in the prothoracic legs of the tettigoniid, *Decticus verrucivorus*, with airborne sound and vibration and used glass micropipettes to record the activity of single units in the tympanal and subgenual nerves at the point where they enter the prothoracic ganglia. The receptor units are characterized by their spontaneous activity, threshold dB, characteristic frequency, and slope of the response/intensity curve. Kalmring, K. and co-workers could divide the receptors into three main physiological groups; namely, pure sound receptors, mixed sound and vibration receptors, and pure vibration units, but had difficulty in ascribing any of these to a particular part of the complex tibial organ.

2.1.3 TRANSDUCTION

An evident feature of the neurons of almost all scolopidia is some form of ciliary dilation in the dendrite. The material within the dilation varies considerably (Young, D., 1973). Howse, P. (1968) suggested that for most scolopidia the adequate stimulus may be stretch of the membrane of this dilation by flexion at the scolopale–cap junction. From an ultrastructural study on the proximal femoral chordotonal organ of the grasshopper Moran, D. *et al.* (1975) postulated that slight bending of the dilation of the cilium can cause an *active stroke* in the cilium which travels to its base where the movement distorts the cell membrane near the basal body. Anderson, B. (1980) discussed the fact that stretch of the dilation membrane may be brought about by *longitudinal stretch*, such as would occur for scolopidia located in joints, or *compression*, as might be the case for scolopidia situated in organs with concentrically folded sensory membranes — for example the scolopophorous organs of bugs.

Young, D. (1970) considered the adequate stimulus for scolopidia in the tibio-tarsal connective organ of the cockroach to be stretch of the ciliary apparatus. In this study, as well as one on the sensory cilium of an insect auditory receptor, Young, D. (1970, 1973) stressed the probable importance of the ciliary base/basal body region, especially the ciliary necklace, in transduction. Indeed the ciliary

necklace is emerging as a region of considerable functional significance in sensilla as well as motile cilia. As discussed by Wright, K. (1983) the ciliary necklace may be important in stimulus induction in nematode and vertebrate as well as arthropod sensilla of various modalities. In recent studies on motile cilia:

(1) the ciliary necklace has been demonstrated to be capable of anionic binding (Anderson, R. and Hein, C., 1977), strengthening N. Gilula and P. Satir's (1972) suggestion that the subunits of the ciliary necklace might be sites of selective ion permeability;

(2) the possibility that the subunits of the ciliary necklace might be Ca^{2+} pump proteins has been raised (Fischer, G. *et al.*, 1976);

(3) the concept that the ciliary necklace may form a boundary between two different states of cellular membrane has been introduced (Boisvieux-Ulrich, E. *et al.*, 1977); and

(4) the importance of the ciliary necklace in transduction of mechanosensitive motile cilia has been considered in detail (Wiederhold, M., 1976).

In the femoral chordotonal organ of grasshoppers, a purely tonic mechanosensillum, Moran, D. *et al.* (1977) found that the most conspicuous change at maximum stimulation is the production of a pronounced bend at the base of the sensory cilia. The sequence of events during sensory transduction in this organ appears to be:

(1) When the ligament is pulled, the cilium tip is laterally displaced.

(2) Mechanical displacement of the cilium tip induces active sliding between adjacent doublets of the axoneme.

(3) Active sliding produced ciliary bending.

(4) The bend is propagated to the movable distal basal body at the base of the cilium.

(5) Bending of the ciliary base distorts the cell membrane in the region of the ciliary necklace, causing local changes in ionic permeabilities leading to the production of the generator current.

The possibility that the structures responsible for

motility in motile cilia may be involved in mechanosensory transduction, including in scolopidia, has been considered for some time. The main objection to this hypothesis has been based on the ultrastructural differences between the ciliary region in mechanosensilla and motile cilia. In sensilla the ciliary region lacks (a) spokes on the dynein arms, (b) a central sheath, and (c) a central pair of tubules, consistent features of motile cilia. These differences caused Moran and colleagues to change their suggestion that sensory cilia respond to mechanical stimulation with an active stroke (Moran, D. *et al.*, 1975) — that is, a co-ordinated bend-propagation employed by motile cilia in which dynein arms provide the force for sliding, and radial spokes that interconnect the doublets with the central sheath transduce sliding into bending, to the concept that the sensory cilia are incapable of an active stroke, but are capable of active sliding (Moran, D. *et al.*, 1977). Interestingly, Crouau, Y. (1980) observed bridges with the A subfibers of the doublets and similarity of the shape and dimension between the pairs of membrane protuberances and the spoke heads in antennal chordotonal organs of the crustacean, *Antromysis juberthiei* (Mysidacea). If these or similar features are found in insect scolopidia, then the original suggestion of Moran and colleagues that an active stroke occurs may be the correct one. In contrast, Boo, K. (1981) presented evidence that the cilia of scolopidia are passively involved in sensory transduction.

2.2 Cuticular mechanosensilla

2.2.1 STRUCTURE

2.2.1.1 *Cuticular parts.* The cuticular part of most of the cuticular mechanosensilla occur as two basic shapes, namely, *hair-shaped* and *campaniform*, whereas this part of some sensilla has an *unusual structure*. Hair-shaped sensilla commonly are called sensilla chaetica, Bohm's bristles or sensilla trichodea. Trichobothria belong to this group also, but because of unusual features are herein discussed separately. Hair sensilla are the most abundant, widespread and extensively investigated type of mechanosensillum and may function solely as mechanosensilla or as chemosensilla as well. Cam-

paniform sensilla generally appear as a small cuticular cap surrounded by a ring of raised cuticle, and function as proprioceptors responding to strains in the exoskeleton.

(a) *Mechanosensilla*. Hair sensilla which are solely mechanosensitive typically bear no pores or openings in the hair wall and are innervated by one neuron. The hair may vary considerably in length, is usually drawn to a sharp tip and exteriorly may bear cuticular sculpturings such as grooves or spicules. The hair usually arises from a socket, the structure of which may vary and upon which much of the coupling process is dependent. In many hairs, for example, large sensilla chaetica on the thorax of blowflies (Keil, T., 1978), the hair is attached to the socket by an articulating or connecting membrane, which may consist of the rubber-like protein resilin (Thurm, U., 1964a), and on occasion have layers of differing electron-densities (McIver, S., 1972). The walls of the socket are cuticular and may bear inward-projecting ribs or diaphragms. The sockets of hairs on hard-bodied adults are usually large, sturdy structures whereas those on softer-bodied immatures frequently are not as well developed; for example, antennal hair sensilla of mosquito *Toxorhynchites brevipalpis* larvae (Jez, D. and McIver, S., 1980). Interestingly, the sockets of hair sensilla on the eyes of adult praying mantis (Zack, S. and Bacon, J., 1981); honey bees (*Apis mellifera*) (Herbstsommer, D. and Schneider, L., 1979); tiger beetles (*Cicindela tranquebaria*) (Kuster, J., 1980); and houseflies (*Musca domestica*) (Chi, C. and Carlson, S., 1976) are less developed than those usually observed on adults. The cuticular projections as well as the height and diameter of the socket clearly restrict the movement of the hair (McIver, S., 1975).

Hair sensilla of adults of a number of species were examined in serial sections by Gaffal, K. *et al.* (1975) to determine the structural polarities and their influence on stimulus transmission — that is, coupling. They found the hairs to be characterized by three cuticular elements; namely, joint membrane, socket septum, and usually suspension fibers (Fig. 9) and in all cases studied at least one of these elements forms the basis for a structural bilateral symmetry along whose plane of symmetry the direction line of the maximum receptor sensitivity lies. The same authors provide valuable tables of terms used to describe socket structures in mechanosensilla with and without hair shafts.

From a study on the mechanics and electrophysiology of hair sensilla of *Calliphora* Theiss, J. (1979) reported that when the hair is displaced to progressively greater angles in the direction of greatest restoring force, the bristle joint behaves like a non-linear spring until the shaft meets the edge of the socket. When released the shaft returns to the resting position with a non-oscillatory, strongly damped movement. The pivot point lies about in the middle of the ball joint at the base of the hair shaft. Consequently, the suspension fibers are stretched parallel to the course of their component fibrils. In the case of bending of the hair shaft from 5° to 6°, the range of Young's modulus is of the order of 10^{10} dyn cm^{-2} = 1 GN m^{-2}, a value two or three orders of magnitude higher than that of resilin (Theiss, J., 1979).

The features of the socket in some hair sensilla play a smaller, if any, role in coupling than in those sensilla just discussed. In some the tubular body terminates well into the hair shaft; for example, the scalpal hairs of adult mosquitoes (McIver, S. and Siemicki, R., 1981). In such sensilla coupling must occur by bending of the hair shaft at the level of the

tubular body. The sockets of these hairs, as well as the antennal hairs of larval mosquitoes, are not well developed. In the larval hairs the dendritic tip containing the tubular body is flared and attaches directly to the hair base in a manner reminiscent of a ball and socket (Jez, D. and McIver, S., 1980).

Trichobothria (Fadenhaare, thread-hairs) usually have long, filamentous hairs and a more complex socket than common hair sensilla. The socket has the shape of a deep cup or flask with cuticular projections on the inner surface. A large socket orifice permits the hair a wide angle of deflection. The hair attaches to a membrane which forms the base of the socket. Single innervation of trichobothria appears to be the rule in insects and scorpions, although multiple innervation is known from other arthropods (McIver, S., 1975), including Pauropoda which have eight neurons per trichobothrium (Haupt, J., 1978). In insects the dendrite is known to attach to the membrane that forms the floor of the cup-shaped socket, or may end in an ecdysial canal (McIver, S., 1975).

As with other hair sensilla, coupling in trichobothria is initiated by leverage from the hair shaft. The elastic forces which return the lever to its starting point have to act perpendicularly to the lever and at some distance away from the center of gyration. All three components of the articulating apparatus — that is, joint membrane, suspension fibers and socket septum — have structural features which possibly influence directional dependence of the opposing forces (Gaffal, K. and Theiss, J., 1978). In abdominal trichobothria of bugs Gaffal, K. (1976) considered that the structural polarities of the socket septum largely determine the directional sensitivity of one functional type, and that the degree of stability of the socket septum may influence the response patterns, that is, phasic–tonic and phasic, as well as the "dynamical decay of stimulus". In crickets the socket septum limits the extent to which tibial trichobothria may be displaced (Gaffal, K. and Theiss, J., 1978).

Using cercal trichobothria of *Gryllus bimaculatus* Gnatzy, W. and Tautz, J. (1980) demonstrated that the direction of best mobility of the hair is the direction in which the neuron is depolarized, and that this direction can be determined entirely by morphological criteria. In particular, the hair can be deflected farthest from the resting position in the

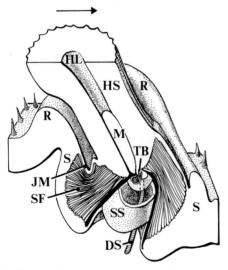

FIG. 9. Diagram of bristle on the head of *Calliphora* bisected except for the socket septum. Arrow indicates direction of maximum sensitivity with cut out sections. DS = Dendritic sheath, HL = hair lumen, HS = hair shaft, JM = joint membrane, M = cuticular material inside HL, R = rim of socket, S = socket, SF = suspension fibers, SS = socket septum, TB = tubular body. (Redrawn after Gaffal, K. *et al.*, 1975.)

direction of a cuticular peg at the hair base, which projects toward the lumen of the hair and marks the flat side of the tubular body. Deflection of the hair shaft in the opposite direction is limited by a fibrous cushion which exerts a counter-pressure. In a cockroach the oscillation direction of each trichobothrium is apparently determined by the shape of the hair base (Gnatzy, W., 1976).

Trichobothria are sensitive to small variations in air currents (Nicklaus, R., 1965). Tautz, J. (1979) proposed that such air currents are a form of particle oscillation in a medium; in particular, a sound field. Cercal trichobothria of *Gryllus* show regular or irregular oscillations depending on intensity of vibrations of the medium (Gnatzy, W. and Tautz, J., 1980). At higher stimulus intensities, that is, about $100\,\mu m$ at $100\,Hz$, the hairs flutter irregularly in various direction, at slightly lower intensities preferentially in the plane of best mobility, and at even lower intensities in the plane of stimulus vector. The best mobility is in the range of $100–200\,Hz$ and in the plane of best mobility the maximal angle of deflection from the resting position is $5.3 \pm 1.4°$ (Gnatzy, W. and Tautz, J., 1980). Cercal trichobothria of the cockroach, *Periplaneta americana*, are sensitive to gentle wind puffs that elicit escape behavior from a toad, a natural predator (Camhi, J. and Tom, W., 1978a,b). Thoracical trichobothria of a caterpillar detect medium vibration in the near-field of predatory wasps. A flying wasp can stimulate the trichobothrial neurons from a distance of maximally $70\,cm$ (Tautz, J., 1978; Tautz, J. and Markl, H., 1978).

Fletcher, N. (1978) produced a first-order linear model for the behavior of hairs in a sound field which accounts for many of the measurements of Tautz, J. (1977) on the caterpillar, *Barathra brassicae*. Fletcher's analysis helps to explain K. Draslar's (1973) report that the length of the hair is related to the frequencies perceived. The hairs should have a length roughly in inverse proportion to the square root of the frequencies they are designed to detect. The radius of the hair is not important, it merely needs to enable the hair to remain stiff. The structure of the socket region should insure an elastic restoring force sufficient to give a resonance frequency for the hair and its associated air load that is close to the frequency to be detected.

Henson, B. and Wilkens, L. (1979) derived a differential equation for the motion of mechanosensory hairs of animals in a fluid environment where the hair–fluid coupling is due to viscous forces. Although the equation model is specifically applied to crayfish sensilla in an aqueous medium, the assumptions of the model are also valid in air for trichobothria on the cerci of crickets (Dumpert, K. and Gnatzy, W., 1977) and on the thorax of a caterpillar (Tautz, J., 1977). The only approximation Henson, B. and Wilkens, L. (1979) used in the solution of the mathematical model was the assumption of small angles. The final solutions indicate that this small angle approximation is indeed valid, and the results are consistent with qualitative and quantitative evidence from the crayfish as well as the insects.

Scales or sensilla squamiformia are widespread in insects and usually are not innervated (McIver, S., 1975). In Lepidoptera, however, certain scales on the wings, abdomen (Bullock, T. and Horridge, G., 1965) and antennae (Schneider, D. and Kaissling, K., 1957) are innervated, as are the numerous body scales on several species of Machilidae (Archaeognatha) and the thysanuran, *Lepisma saccharina* (Larink, O., 1975, 1976).

(b) *Mechano- and chemosensilla.* In insects thick-walled hair sensilla, which function as contact chemosensilla, frequently contain a mechanosensitive neuron. These sensilla may occur widely on the body surface, but are most common on the tarsi, mouthparts and antennae (Slifer, E., 1970). Typically the hair has a blunt tip perforated by a single pore. The chemosensitive dendrites extend the length of the hair to the pore, while the mechanosensitive dendrite attaches to one side of the hair base (McIver, S., 1975). Commonly there are four chemosensitive neurons and one mechanosensitive neuron. In the antennal sensilla chaetica of a cockroach the response of the four chemosensitive neurons is phasic–tonic with a very pronounced phasic component and that of the one mechanosensitive neuron is exclusively phasic (Ruth, E., 1976). Exceptions to the "four and one" rule include a contact chemosensillum on the antennae of an ant that has two mechanosensitive neurons (Dumpert, K., 1972) and bristles on the cerci of a cockroach with up to five chemosensitive neurons (Füller, H. and Ernst, A., 1977). An

Table 4: *Structural features of some campaniform sensilla*

Order and species	Location	Number	Shape	Cap layers	Sheath cells	References
Ephemeroptera						
Cleon dipterum	pedicel		oval	4 in diagram	3	Schmidt, K. (1974)
Dictyoptera						
Blaberus giganteus, Blaberus discoidalis	tibia	groups of 6	oval			Chapman, K. *et al.* (1975) Chapman, K. and Duckrow, R. (1975)
Blaberus discoidalis	tibia	groups of 6	oval			Mann, D. and Chapman, K. (1975) Pringle, J. (1938a,b) Spinola, S. and Chapman, K. (1975)
Blattella germanica, Blaberus discoidalis	tibia	groups of 6	oval	inner and outer cuticular layers with a space in between	2	Moran, D. and Rowley, J. (1975a) Moran, D. *et al.* (1971, 1976)
Periplaneta americana	tibia	1 group of 3–5 and another of 6–10 sensilla	oval			Zill, S. and Moran, D. (1980)
Periplaneta americana	femur, in cuticular wall of a spine	1/spine	round 10–15 μm diameter	cuticular terminal plug with a ring of flexible cuticle		Chapman, K. (1965) French, A. and Sanders, E. (1981)
Isoptera						
Zootermopsis sp.	assoc. with sternal gland	200	round	at least 2	2	Stuart, A. and Satir, P. (1968)
Orthoptera						
Gryllus bimaculatus	cercus, in proximity to sockets of filiform hairs	groups of 3	1 round, 5 μm diam. 2 elliptical, 1.5 μm × 2.5 μm	outer cuticular middle spongy inner cuticular	3 enveloping 1 glial	Gnatzy, W. and Schmidt, K. (1971) Dumpert, K. and Gnatzy, W. (1977)
Gryllus campestris	eye, clefts betwen corneal facets	random distribution	oval	outer dense, homogeneous inner fibrous	2	Muller, M. *et al.* (1978)
Locusta migratoria	pedicel	70 comprising Hick's organ	elliptical			Heinzel, H. and Gewecke, M. (1979)
Locusta migratoria	almost all parts of body	many	round or oval			Knyazeva, N. (1974)
Hemiptera						
Aphis pomi	pedicel	1	round	outer cuticular inner fibrous		Bromley, A. *et al.* (1980)
Neuroptera						
Chrysopa	apex of pedicel	5		2	3	Schmidt, K. (1969b)

Table 4: *Structural features of some campaniform sensilla* – Continued

Order and species	Location	Number	Shape	Cap layers	Sheath cells	References
Coleoptera						
Cetonia aurata, Geotrupes sylvaticus	hind wings	many, arranged in fields	oval	3	3	Pfau, H. and Honomichl, K. (1979)
Hymenoptera						
Apis mellifera Orgilus lepidus	head ovipositor	type A	elliptical round 1 μm diameter	2 dendrite inserts directly into cuticle		Thurm, U. (1964a) Hawke, S. *et al.* (1973)
	ovipositor	type B		dentrite inserts into fibrous band beneath cuticle		Hawke, S. *et al.* (1973)
Diptera						
Toxorhynchites brevipalpis (larvae)	antenna	2	round	dendrite inserts directly into cuticle		Jez, D. and McIver, S. (1980)
Anopheles stephensi	maxillary palp	1/palp	round	outer 0.26 μm thick, similar to exocuticle middle 0.44 μm thick, spongy inner 0.34 μm thick, fibrous		McIver, S. and Siemicki, R. (1975)
Drosophila melanogaster	halteres	scabellum = 45 pedicel = 80	oval	2 (c2, c4)	2	Chevalier, R. (1969)
Boettcherisca peregrina	halteres	many		2	2	Uga, S. and Kuwabara, M. (1967)
Calliphora erythrocephala	halteres	numerous in basal and scalpal regions	oval	outer dense cuticle middle narrow, filamentous inner spongy, some sensilla with 'hinged' caps	2	Smith, D. (1969) Thurm, U. *et al.* (1975)

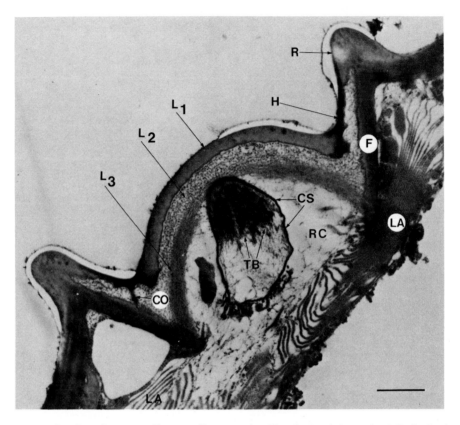

FIG. 10. Transverse section through a campaniform sensillum on palp of female *Anopheles stephensi*. Projecting inwardly from the ring (R) of raised cuticle is a flange (F). A hinge (H) connects ring (R) with outer layer (L_1) of cap. Between the hinge and flange are connections (CO) which originate from former. Middle spongy layer (L_2) and inner fibrous layer (L_3) are visible as are tubular body (TB) and cuticular sheath (CS). Beneath cap is receptor-lymph cavity (RC) and lamellae (LA) of tormogen cell. (Scale = 1 μm.) (Reproduced from McIver, S. and Siemicki, R., 1975 with permission.)

unusual situation occurs in the short bristles on the cerci of crickets in which the tubular body, a characteristic feature of cuticular mechanosensilla, is located at the hair base within one of the chemosensitive dendrites (Schmidt, K. and Gnatzy, W., 1972). In the terminal sensilla on the maxillary palps of the grasshopper, Blaney, W. and Chapman, R. (1969) observed a tubular body-like structure in one of the dendrites which has no specialized attachment at the hair base. Another type of unusual situation was reported from the antennae of aphids in which the dendrite of one of the three to five neurons innervating a sensillum extend into the hair lumen while at the hair base one ends in a classical tubular body and the others terminate unspecialized.

Although some differences have been noted, coupling occurs in the same general way as in solely mechanosensitive hair sensilla. In labellar taste hairs of blowflies, Matsumoto, D. and Farley, R. (1978) reported that when the hair is bent toward the distal end of the proboscis the hair base over the tubular body is elevated and the distal end of the tubular body is bent dorsally. The socket septum, as well as the suspension fibers, apparently serves to direct the force of the mechanical stimulus to specific regions of the tubular body or prevent the application of excessive force to the dendrite. These structures and the joint membrane may also help bring the hair shaft back to the resting position. In contrast to the hypothesis of Gaffal, K. *et al.* (1975) and Gaffal, K. (1976) the socket septum does not compress the dendritic ending regardless of the direction in which the hair is bent (Matsumoto, D. and Farley, R., 1978).

(c) *Campaniform sensilla*. Berlese, A. (1909) first

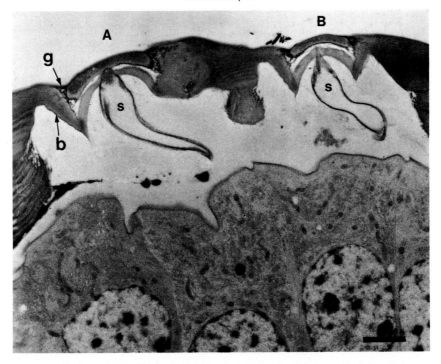

FIG. 11. The caps of two campaniform sensilla are shown in a 1/2 μm thick cross-section through the tibia of *Blaberus* species. The tip of the sensory process (S) is shaped like a flat paddle; its major axis is parallel to the plane of section of cap A, and perpendicular to the plane of section of cap B. Note that the buttress (b) is much thicker than the gasket (g) and that there are inner and outer layers of cap cuticle. (Scale = 1 μm.) (Reproduced from Moran, D. and Rowley, J., 1975a with permission.)

used the term campaniform to describe sensilla previously known as dome, bell and cupola-shaped structures. These sensilla occur widely on the body of adult insects, frequently being concentrated near joints or on structures subject to cuticular distortion, such as legs and halteres (Table 4). Immatures of holometabolous as well as hemimetabolous insects may have campaniform sensilla.

Externally a campaniform sensillum generally appears as a small cuticular cap surrounded by a ring of raised cuticle. Each sensillum is characteristically innervated by one neuron (Fig. 10 and 11). Usually the dendrite attaches to the center of the cap, but eccentric placement is known (Müller, M. *et al.*, 1978; Schneider, D. and Kaissling, K., 1957). Variation in size and structure of the cuticular portions is common, and determines the mechanics of coupling, relating to the specific characteristics of the stimulus perceived at the particular location of the sensillum. The cap may be circular or oval, raised or sunken, and flat, arched, or formed into a small peg (McIver, S., 1975). The degree of elevation of the cap is related to the type of reaction of a given stimulus (Gettrup, E., 1973). Cuticular hinges may attach the cap to the body cuticle (McIver, S. and Siemicki, R., 1975; Smith, D., 1969). Hemimetabolous insects may have a pore, the ecdysial canal, in the cap. In holometabolous insects at least one instance is known of a pore occurring in the cap (Moeck, H., 1968). The ring of raised cuticle surrounding the cap may be incomplete, as occurs in sensilla on the palps of a locust (Blaney, W. and Chapman, R., 1969) or absent, as is known from a mosquito larva (Jez, D. and McIver, S., 1980). This reduction is most likely correlated with the flexibility of the cuticle.

Considerable differences in the structure of the cap were originally revealed by light microscopy (Sihler, H., 1924; Snodgrass, R., 1926). Sometimes the dendrite inserts directly into the body cuticle (Fig. 10), but usually there are two or three distinct layers (Table 4) (Fig. 11). In the caps with two layers the outer one is modified exocuticle and the inner one is spongy or fibrous, which may be the rubber-like protein, resilin (Thurm, U., 1964a). In the three-layer caps the outer two are similar to the corresponding ones in two-layer caps while the innermost layer may be cuticular, fibrous or spongy (Table 4).

Campaniform sensilla with round caps respond to strains in the cuticle from all directions, whereas oval sensilla are directionally selective (Pringle, J., 1961). The level of the excitation induced in the campaniform sensilla on the halteres of flies and the pedicel of locusts is dependent upon the *direction* as well as the *intensity* of the stimulus (Gewecke, M., 1972a; Heinzel, H. and Gewecke, M., 1979). Thurm, U. *et al.* (1975) demonstrated electrophysiologically that the fan-shaped dentritic tip of campaniform sensilla on the halteres of *Calliphora* are stimulated by a monoaxial compression directed perpendicular to their elongated transverse axis. In contrast, displacements parallel to that plane are ineffective. Using replicas made from stimulated and unstimulated states of living sensilla, Thurm and co-workers found that in the stimulated state there is a $0.2 \mu m$ maximum decrease in the width of the cuticular clefts in which the dendritic tips terminate.

In *Periplaneta americana* the tibial campaniform sensilla occur as two subgroups with mutually perpendicular cap orientation (Zill, S. and Moran, D., 1980). The long axis of the caps of the proximal subgroup are oriented perpendicular to the tibial long axis, whereas the caps of the distal subgroups are parallel to the tibia. The proximal sensilla (Fig. 12) respond only to dorsal bending and fire upon axial compression, while the distal ones are sensitive only to ventral bending and fire upon axial tension. Both subgroups respond simultaneously but weakly to imposed torques (Zill, S. and Moran, D., 1980). Contraction of flexor and extensor muscles stimulates the proximal and distal sensilla, respectively. The same campaniform sensilla trigger reflexes that constitute a negative feedback system (Zill, S. *et al.*, 1981). The reflex function of both subgroups of these sensilla in rapid walking is limited by phase shifts relative to slow extensor firing (Zill, S. and Moran, D., 1981).

The actual process of coupling has been studied most thoroughly in the tibial campaniform sensilla of cockroaches. Two structural components of the cap appear to be especially important (Moran, D. and Rowley, J., 1975a).

(1) The "cuticular collar", a reinforcing ring, is an integral part of the substructure of the cap (Fig. 11) and confers rigidity by being firmly attached to the inner cuticular layer (L_2) into which the dendrite inserts. As a result the stabilized cap may move as a solid unit in response to local cuticular deformations.

(2) Each end of the oval dome of the cap is firmly attached by "buttresses" to the surrounding exoskeleton. This feature underlies the directional sensitivity of oval campaniform sensilla by physically conveying cuticular strains to the cap and the attached dendrite. Low-amplitude cuticular deformations passing through the cap's major axis would be conducted to the dome by the buttresses; those propagated parallel to the cap's minor axis would be filtered out by the thinner, more flexible cuticle.

Transmissal of the energy in the cuticular strains produced physiologically during proprioception to the dendrite is by indentation rather than bulging of the domed cap (Chapman, K. *et al.*, 1973). Punctate stimulation during a proprioceptive discharge increases the response rather than decreases it, indicating that proprioceptive stimuli do indeed indent the cap (Fig. 13) (Spinola, S. and Chapman, K., 1975). Based on compliance and sensitivity measurements, Spinola, S. and Chapman, K. (1975) estimated that the magnitude of the indentation associated with strong proprioceptive discharge to be in the range of 10–50 nm. In a more detailed study of compliance and sensitivity of the sensilla, Chapman, K. and Duckrow, R. (1975) reported that the magnitude of cap indentation associated with moderately brisk discharge frequencies of the order of 100 Hz, well within the range of linear encoding, is of the order of a few tens of nanometers.

Chapman, K. *et al.* (1973), and Spinola, S. and Chapman, K. (1975), argued that punctate or proprioceptive stimulation in campaniform sensilla does not excite by stretching the dendrite as Pringle, J. (1938b) originally suggested, or by bending it in the manner of hair and other mechanosensilla (Rice, M. *et al.*, 1973; Thurm, U., 1965), because cap indentation would likely cause axial compression of the dendrite. However, Chapman, K. and Duckrow, R. (1975) pointed out that the intricate bilayer cuticular structure of the cap as described by Moran, D. *et al.* (1971) and Moran, D.

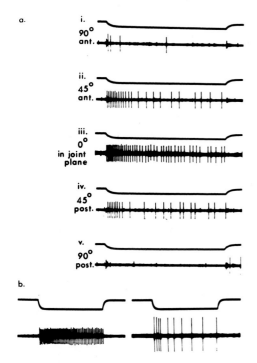

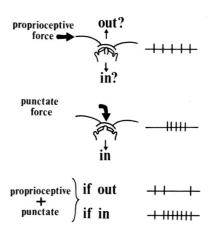

FIG. 13. Principle of the critical experiment to demonstrate that a proprioceptive stimulus indents the cap of a campaniform sensillum. Middle column represents a cross section through the cap of a sensillum; right hand column, the afferent discharges expected in response to proprioceptive force (top), punctate force (middle) and combination of the two stimuli (bottom). Punctate stimulation is known to indent the cap (in) while proprioceptive force may cause the cap to bulge (out) or indent (in). Combination of the two stimuli should lead to a decrease or gap in the sensillum's discharge if the cap bulges during proprioception (if out), but should lead to an increase if the cap normally indents (if in). (Redrawn from Spinola, S. and Chapman, K., 1975.)

FIG. 12. Response of a proximal campaniform sensillum on the tibia of *Periplaneta americana* to bending forces in different planes. **a**: The femoro-tibial joint was held fully extended and bending forces were applied to the distal end of the tibia (i) 90° anterior to the plane of joint movement, (ii) 45° anterior, (iii) 0° (in the joint plane), (iv) 45° posterior, (v) 90° posterior. The response is maximal in the plane of the joint and declines in other planes. **b**: Punctate stimulation of the cuticular caps of Group 6. Two sensilla were recordable in nerve 5r8: one, a proximal sensillum matches the spike height recorded in **a**; the second, a distal sensillum, did not respond to any dorsal bending force. (Reproduced from Zill, S. and Moran, D., 1980 with permission.)

and Rowley, J. (1975a) may serve as a mechanical linkage which pinches the flattened tubular body, in an analogous manner to the mechanism proposed by Thurm, U. (1964, 1965) for the bending of hair plate sensilla. Resolution of the exact mechanism of coupling cuticular strain to the dendrite awaits more explicit knowledge of the mechanical properties of the terminal "buttresses" and the "cuticular collar" as well as the two cuticular layers in the cap.

Campaniform and hair sensilla may be closely associated in various ways to form functional units. In tactile spines on cockroach legs a single campaniform sensillum is located in the wall of the hair at the junction with the articular membrane. The tactile spine itself is not innervated, but movements

of the hair stimulate the campaniform sensillum which functions as the sensory element (Chapman, K., 1965). Based on detailed fine structural observations French, A. and Sanders, E. (1981) suggested that the term "socket sensillum" be used for the campaniform sensillum. On the antenna of a mosquito larva two hairs and a campaniform sensillum may function in concert to monitor range of water pressure (Jez, D. and McIver, S., 1980). Campaniform sensilla are in close proximity to the sockets of trichobothria in species of *Gryllus bimaculatus* (Gnatzy, W. and Schmidt, K., 1971; Sihler, H., 1924) and *Gryllotalpa, Schistocerca* and *Periplaneta* (Dumpert, K. and Gnatzy, W., 1977), in *Acheta domesticus* (Edwards, J. and Palka, J., 1974) and *Teleogryllus oceanicus* (Bentley, D., 1975), but not in the notopteran, *Grylloblatta* (Edwards, J. and Mann, D., 1981). In crickets, Gnatzy, W. and Schmidt, K. (1971) found the number of campaniform sensilla to be directly proportional to the diameters of the trichobothria; the smallest have none and the largest have five, distributed asymmetrically around the socket.

In *Gryllus bimaculatus*, Dumpert, K. and Gnatzy, W. (1977) demonstrated electrophysiologically that the campaniform sensilla on the cercus trigger the kicking response, and that through the functional coupling of trichobothria and campaniform sensilla the working range of these combined organs is considerably extended. Spike potentials could be recorded from the neurons of the trichobothria as long as these were deflected in a weak air current, but not during permanent deflection in strong air streams when they touch the inner wall of their sockets. The neurons of the campaniform sensilla respond phasically to deflection of the sockets in either the proximal or distal direction. Only the trichobothria respond to speeds of air currents up to 1.9 m s^{-1}. Stronger currents up to 3.6 m s^{-1} deflect the trichobothrial sockets and elicit a response of the campaniform sensilla.

(d) *Unusual structures*. As knowledge of cuticular mechanoreceptors expands, structures are being found which do not conveniently fit into presently conceived categories. Examples of these unusual structures follow. On the mouthparts of beetle larvae typical mechanosensitive neurons innervate sensilla tigella and structures with conical bases distally divided into cones and crescents (Corbière-Tichané, G., 1971b). A broad spectrum of campaniform-like sensilla exists. Several of these, including pedunculate forms (Finlayson, L., 1972; Zacharuk, R., 1962), occur in soft-bodied larvae. In the antennae of adult aphids are campaniform-like "joint receptors" (Bromley, A. *et al.*, 1980). Variation on the contact chemosensilla theme is provided by spots, pits, papillae and unclassified sensilla or housefly larvae (Chu, I.-Wu. and Axtell, R., 1971, Chu-Wang, I.-Wu. and Axtell, R., 1971) and domes on wasp ovipositors (Hawke, S. *et al.*, 1973). All of these have a pore to the exterior and are innervated by neurons with features characteristic of chemo- and mechanoreceptors.

In the hemipteran *Nepa cinerea* there are three pairs of organs that sense pressure under water. Each organ contains approximately 100 modified hair sensilla and each sensillum bears a large flattened parasole in place of the hair (Bonke, D., 1975). Single neurons are phasic–tonic and the intensity response curve is linear within the physiological range of stimuli. As determined behaviorally, the receptor threshold for a pressure difference is 180 μmH$_2$O (Bonke, D., 1975). Other modified hairs, club-shaped ones, that occur on the cerci of certain cockroaches (Hartman, H. *et al.*, 1979; Roth, L. and Slifer, E., 1973) and crickets (Bischof, H., 1975) are sensitive to gravity. In the cricket the single neuron in each sensillum has phasic–tonic properties and deflection of the hair along the preferential plane produces the largest changes of spike frequencies. The height of the phasic response is correlated to the rising time of the stimulus. Sinusoidal stimuli up to 300 Hz causes stimulus-synchronized spike trains (Bischof, H., 1975).

Digitiform sensilla occur on the palps of larval elaterid beetles (Doane, J. and Klinger, J., 1978; Zacharuk, R. *et al.*, 1977) and of adult mountain pine beetles, *Dendroctonus ponderosae* (Whitehead, A., 1981). In *Ctenicera destructor* each of the six elongate pegs is positioned in a longitudinal groove (Fig. 14), has a plugged subapical pore, and is innervated by one neuron. These pegs respond electrophysiologically to contact and vibratory stimuli (Fig. 15) which would aid the larva in monitoring activity within the subterranean tunnels in which they live and prey capture or predator evasion, respectively (Zacharuk, R. *et al.*, 1977).

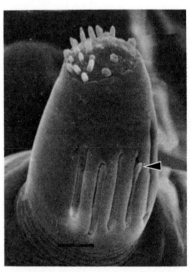

F$_{\text{IG}}$. 14. Lateral aspect of the distal segment of a labial palp of larval *Ctenicera destructor* showing four digitiform pegs each within a longitudinal groove in the cuticular wall, the tip of one of which is extended outward (arrowhead). (Scale = 10 μm.) (Reproduced from Zacharuk, R. *et al.*, 1977 with permission.)

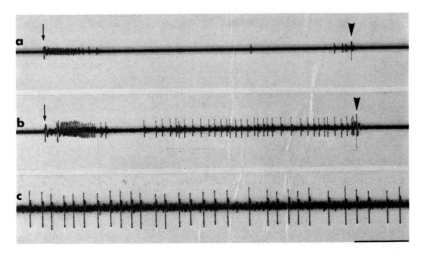

FIG. 15. Electrophysiological recordings from a digitiform peg on palp of larval *Ctenicera destructor*. **a**: Sidewall recording with a glass micropipette containing 0.01 M D-(+)-glucose in 0.08 M NaCl, applied at arrow and removed at arrowhead, during low nighttime building activity of subterranean tunnels. **b**: As in (**a**) but micropipette contained 0.08 M NaCl only. **c**: A short sample of a recording from a tungsten wire electrode taken from a sequence several seconds after the electrode was applied to the base of a peg during high daytime building activity. Scale bar (lower right corner) = 0.5 s. (Reproduced from Zacharuk, R. *et al.*, 1977 with permission.)

2.2.1.2 Sensory neuron. The sensory neuron is composed of an axon, cell body and dendrite that is divided into inner and outer segments by the ciliary region. The cell body has a large spherical nucleus, relatively pale cytoplasm (Gnatzy, W., 1978), probably due to a high water content (Moran, D. *et al.*, 1971), and the usual complement of cell organelles. In trichobothria the main mass of ribosomes is found in the form of polyribosomes rather than attached to the few cisternae in the endoplasmic reticulum (Gnatzy, W., 1978). The numerous, well-developed Golgi complexes display inner lamellae and Golgi vesicles (Gnatzy, W., 1978; Bernays, E. *et al.*, 1976) and have acid phosphatase (Gnatzy, W., 1978). The variously shaped mitochondria are found predominantly near the nucleus. Sometimes typical microtubules extend from the dendrites into the cell bodies for about 3–4 μm (Bernays, E. *et al.*, 1976).

Proximally, the cell body gives rise to the axon which extends uninterrupted to the central nervous system. The axonal cytoplasm has less density of structure than that of the cell body. Microtubules and mitochondria are the most frequently reported organelles in the cytoplasm. Cytolysosomes (Gnatzy, W., 1978) as well as microfibers and multivesicular bodies (Hansen, K. and Heuman, H., 1971) may also occur. Numerous recent investigations have determined the pathway of the axons in the central nervous system, their subsequent connection with interneurons and organization of the latter; for example, studies on wind-sensitive head hairs of locusts (Bacon, J. and Tyrer, M., 1979; Tyrer, M. *et al.*, 1979), various leg sensilla of cockroaches and grasshoppers (Zill, S. *et al.*, 1980), campaniform sensilla in wild-type and mutant *Drosophila* (Burt, R. and Palka, J., 1982), hairs, hair plates and campaniform sensilla as well as chordotonal organs and multiterminal sensilla of locusts (Bräunig, P. *et al.*, 1981; Hustert, R. *et al.*, 1981; Pfluger, H. *et al.*, 1981).

Distally from the cell body the dendrite extends to the cuticular portion of the sensillum. Within the inner segment the usual assortment of organelles, as well as glycogen, occurs; whereas the outer segment contains only microtubules, filaments, vesicles and vacuoles (McIver, S., 1975). There is usually a concentration of mitochondria in the distal portion of the inner segment. The dendrite typically narrows at the ciliary region and distally expands to a diameter not greater than that of the inner segment. In this region, however, a bulbous dilation occurs in some campaniform sensilla and an expanded area with lobe-like extensions has been found in a few hair sensilla (McIver, S., 1975). The outer segment is encased in a dendritic sheath which extends down to the ciliary region and basally may appear as a porous "felt-work"; for example, hair sensilla on the terminalia of adult mosquitoes (Rossignol, P. and McIver, S., 1977).

The ciliary region is composed of nine sets of peripherally located doublet tubules, but lacks the central pair characteristic of motile cilia, with the resulting ciliary formula of $9 \times 2 + 0$ (McIver, S., 1975). Central tubules are occasionally observed (Chevalier, R., 1969; Stuart, A. and Satir, P., 1968). The outer member of each peripheral doublet lacks the arms found in motile cilia. Distally the ciliary microtubules extend into the outer dendritic segment while basally they connect with a centriole-like structure, the basal body, which has nine sets of peripheral triplet tubules. Arising from the basal body are periodically banded rootlets which extend basally into the inner dendritic segment. In most papers two basal bodies have been reported, although several workers have found only one. When two basal bodies are present, rootlets have been observed extending from each. Sometimes ciliary tubules as well as rootlets originate from both basal bodies (Chu-Wang, I.-Wu. and Axtell, R. 1972a).

In the transition region between inner and outer segments, unusual modifications of the dendrite have been reported. One is an aggregate of cords termed connecting body in a campaniform sensillum (Uga, S. and Kuwabara, M., 1967) and another is a granular body in a bristle (Keil, T., 1978). Fibrillar bodies which consist of an accumulation of fibers through which the microtubules pass are known from housefly larvae (Chu-Wang, I.-Wu. and Axtell, R., 1972a,b) and adult blowflies (Smith, D., 1969).

2.2.1.3 *Tubular body.* The characteristic feature of cuticular mechanosensilla is an accumulation of microtubules usually in the distal region of the outer segment. The importance of the microtubules was emphasized when Thurm, U. (1964a,b, 1965) reported from work on hair and campaniform sensilla of honeybees that a structure he called the tubular body most likely was the site of sensory transduction. The tubular bodies studied by Thurm occur at the distal end of the outer segment and consist of 50–100 tubules lying parallel to one another in an electron-dense material. Subsequently, consider-

able variation in complexity and structure of tubular bodies has been demonstrated in numerous insect species. McIver, S. (1975) recommended that the concept of the tubular body encompass the observed spectrum of the amount of electron-dense material as well as the number and configuration of microtubules. Although the tubular body almost always has been reported occurring in the distal region, its location within the outer dendritic segment may vary. An antennal hair sensillum on mosquito larvae has been described in which there is a large number of microtubules in an electron-dense material at the hair base, but the dendrite continues into the hair shaft where distally it has fewer microtubules and no electron-dense material (Jez, D. and McIver, S., 1980).

The reported diameters of tubular bodies range from about $0.4 \mu m$ to $2.5 \mu m$ (Table 5). The number of microtubules varies from a few dozen to about 1000; most of the tubular bodies studied have a few hundred (Table 5). Arrangement of the microtubules occurs in all degrees of complexity from the simple, consisting of a single row of peripheral microtubules (Rice, M. *et al.*, 1973) to the very complicated (e.g. Gaffal, K. and Hansen, K., 1972; Gnatzy, W. and Tautz, J., 1980; Keil, T., 1978; Moran, R. *et al.*, 1971; Smith, D., 1969). Frequently, the microtubules have a definite ordered arrangement; for example, in primarily hexagonal arrays (Chi, C. and Carlson, S., 1976), but some apparently do not show a regular pattern or dense packing (Müller, M. *et al.*, 1978). Variation in arrangement of microtubules within the same tubular body is known; for example, in the antennal hair sensilla of the red cotton-bug the peripheral tubules are 100–200 Å apart and occur in pallisades, whereas the central ones are 300 Å apart and are hexagonally arranged (Gaffal, K. and Hansen, K., 1972). Another example comes from the labellar hair sensilla of blowflies where the microtubules at the periphery of the tubular body are closely applied to each other and conform with the rounded end of the tubular body, whereas the ones in the center are straight and are arranged longitudinally (Matsumoto, D. and Farley, R., 1978). In another instance the microtubules occur in two distinct planes (Jez, D. and McIver, S., 1980) (Fig. 16). French, A. and Sanders, E. (1979, 1981) reported that the distal ends of the microtubules often terminate in line with

Table 5: Selected characteristics of some tubular bodies*

Species	Type of sensillum	Location	Diameter of tubular body	No. of micro-tubules	Neuro-filaments	References
Dictyoptera						
Periplaneta americana	trichobothria	cercus	2.5 μm	*ca.* 1000	—	Nicklaus, R. *et al.* (1968)
Blaberus discoidalis	camp. sensillum	tibia	—	350–1000	—	Moran, D. *et al.* (1971)
Orthoptera						
Gryllus bimaculatus	trichobothria	cercus	trich. with 2 camp. sensilla *ca.* 1.6 μm; trich. with 3 camp. sensilla, *ca.* 2.2 μm	*ca.* 500 *ca.* 700		Gnatzy, W. and Tautz, J. (1980)
Acheta domesticus	trichobothria	tibia	1.5–2.4 μm	—	—	Gaffal, K. and Theiss, J. (1978)
Schistocerca gregaria	club-shaped hair	pronotum, head	—	200–300	—	Bernays, E. *et al.* (1976)
Hemiptera						
Dysdercus intermedius	tactile hair	antenna	0.7–0.8 μm	100–150	—	Gaffal, K. and Hansen, K. (1972)
Dysdercus intermedius	taste hair	antenna	0.5–0.7 μm	150–200	—	Gaffal, K. and Hansen, K. (1972)
Brevicoryne brassicae	distal receptor	labium	*ca.* 0.4 μm	*ca.* 80	—	Wensler, R. (1977)
Coleoptera						
Cicindela tranquebarica	tactile hair	eye	*ca.* 0.6 μm	—	—	Kuster, J. (1980)
Diptera						
Aedes aegypti	tactile hair	terminalia	*ca.* 0.5 μm	*ca.* 100	—	Rossignol, P. and McIver, S. (1977)
Musca domestica	tactile hair	eye	1.5 μm	*ca.* 400	+	Chi, C. and Carlson, S. (1976)
Calliphora vicina	bristle	thorax	1.5 μm	400	+	Keil, T. (1978)
	camp. sensillum	haltere	—	*ca.* 230	+	Völker, W. (1978)
Glossina	LR7 receptor	mouth parts	—	*ca.* 30	—	Rice, M. *et al.* (1973)

* Adapted from Gnatzy, W. and Tautz, J. (1980) and updated.

each other, forming a flat end to the tubular body before the overlying cell membrane, and that there are often apparent dislocations between such bundles of microtubules with a dendrite (Figs 17 and 18). This latter observation suggests that the microtubules may slide relative to one another (French, A. and Sanders, E., 1979). Jez, D. and McIver, S. (1980) observed in campaniform sensilla that the microtubules flare, are bent, separate into groups which terminate at two distinct sites (Fig. 19), and have associated electron-dense "beads", which are probably proteins (Fig. 16).

In some tubular bodies, filaments as well as microtubules occur. Interestingly, reports of filaments in insect tubular bodies are apparently restricted to studies on dipteran species, namely adult blowflies (Hansen, K. and Heuman, H., 1971; Keil, T., 1978; Matsumoto, D. and Farley, R., 1978; Smith, D., 1969; Völker, W., 1978) and adult (Chi, C. and Carlson, S., 1976) and larval (Chu-Wang, I. -Wu. and Axtell, R., 1972a,b, 1973) houseflies. Filaments are known to occur in tubular bodies of arachnid species (Foelix, R. and Chu-Wang, I. -Wu., 1972).

The amount of electron-dense material varies from virtually nothing (Rice, M. *et al.*, 1973) to being present in a copious, compact form, for example, trichobothria of crickets (Gnatzy, W. and

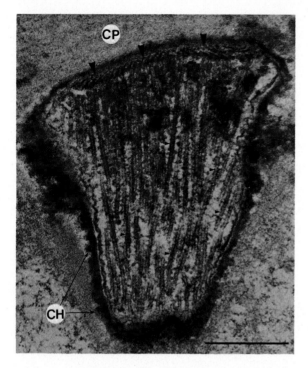

FIG. 16. Longitudinal section through tubular body and cuticular sheath (CH) of campaniform sensillum in midregion of antenna of fourth-instar larval *Toxorhynchites brevipalpis*. Note microtubules oriented in two planes; obliquely (large arrowheads) under cap (CP) and longitudinally (small arrowheads). "Beads" associated with microtubules are probably proteins. (Scale = 0.5 μm.) (Reproduced from Jez, D. and McIver, S., 1980 with permission.)

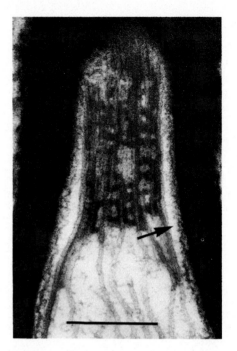

FIG. 17. A section through the short axis of the elliptical tubular body of a sensillum in the trochanteral hair plate of *Periplaneta americana*. Note that the dense matrix appears to form cross bridges between the microtubules. There is an electron-lucent zone (arrow) between the microtubules and the cell membrane which underlies the dendritic sheath. (Scale = 0.5 μm.) (Reproduced from French, A. and Sanders, E., 1979 with permission.)

Tautz, J., 1980). In many instances the electron-dense material is present in intermediate amounts and has a striated appearance due to its arrangement in layers perpendicular to the axis of the microtubules (Figs 16, 17 and 18). As a result, the microtubules are connected with one another by "bridges" of dense material. In the tactile spines of cockroaches the most peripheral microtubules appear linked together by electron-dense material and form a boundary for the tubular body inside the dendritic membrane (French, A. and Sanders, E., 1981). Particularly evident bands of dense material occur at the periphery of the tubular body of cockroach trichobothria (Gnatzy, W., 1976). In trichobothria of *Acheta domesticus* the materials interspersed between the microtubules show differences in staining; the most heavily stained component appears in the form of longitudinally oriented cords (diameter 30–50 nm) whereas the less

heavily stained component seems to be dispersed irregularly (Gaffal, K. and Theiss, J., 1978). The combination of the electron-dense material and microtubules evidently results in a relatively stable cytoskeleton (Gnatzy, W. and Tautz, J., 1980). In spite of the obvious importance of the electron-dense material, especially with regard to the viscoelastic properties of the tubular body, its chemical composition is unknown.

The relationship between the outermost microtubules in the tubular body and the dendritic membrane, and in turn the membrane and the surrounding dendritic sheath, must be of considerable significance in the final stages of coupling and in transduction itself. The peripheral microtubules may be connected in some way to the dendritic membrane, for example by dense patches (Chevalier, R., 1969), by filaments (Gaffal, K. and Hansen, K., 1972), by bridges composed of less electron-dense material than those between the inner microtubules (Matsumoto, D. and Farley, R.,

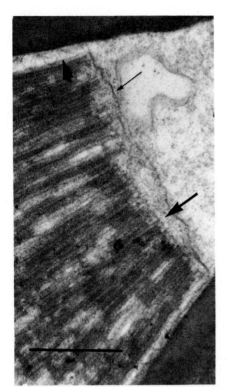

FIG. 18. The tip of a tubular body from trochanteral hair plate of *Periplaneta americana* showing the alignment of a bundle of microtubule endings (large arrow) while other microtubule endings approach the enveloping membrane very closely (small arrow). There is a distinct dislocation between these bundles. Note also the lucent area between the microtubules and the dendritic sheath (arrowhead). (Scale = 0.5 μm.) (Reproduced from French, A. and Sanders, E., 1979 with permission.)

1978) and by strongly stained granules which are 30–40 nm in diameter and are the site of origin of microtubules (Gaffal, K. and Theiss, J., 1978). In campaniform sensilla of crickets the microtubules were observed to emerge from the tubular body and to adhere to the dendritic membrane (Gnatzy, W. and Tautz, J., 1980). At a different level there is also direct coupling between the peripherally located microtubules and the dendritic membrane in the form of weakly osmiophilic bridges which are arranged at regular intervals of approximately 20 nm (Gnatzy, W. and Tautz, J., 1980).

In other sensilla the peripheral microtubules are not connected to the dendritic membrane. In hair plates and tactile spines of cockroaches there is normally an electron-lucent zone between the perimeter of the tubular body and the dendritic membrane (Figs 17 and 18) (French, A. and San-

ders, E., 1979, 1981) and in campaniform sensilla located *at the base* of the antenna of mosquito larvae there is a considerable intracellular space between the sides of the tubular body and the dendritic membrane (Jez, D. and McIver, S., 1980) (Fig. 19). Interestingly, in the mosquito larva a campaniform sensillum which is located *in the mid region* of the antenna has a tubular body in which the peripheral microtubules are closely apposed in all areas to the dendritic membrane (Fig. 16). All of these differences clearly indicate that the relationship between the tubular body and the dendritic membrane varies, probably depending on the location of the sensillum and rigidity of the cuticle.

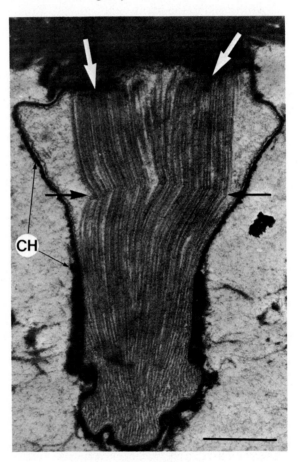

FIG. 19. Longitudinal section through tubular body of campaniform sensillum at base of antenna of larval *Toxorhynchites brevipalpis*. Note that microtubules are bent (black arrowheads) and attach to cap at two sites (white arrowheads) and that distally the cuticular sheath (CH) is flared away from microtubules. (Scale = 1 μm.) (Reproduced from Jez, D. and McIver, S., 1980 with permission.)

The dendritic membrane and dendritic sheath have been reported to be connected by trivial desmosomal connections (hemidesmosomes?) (Uga, S. and Kuwabara, M., 1967) and large granules (Hawke, S. *et al.*, 1973). In trichobothria of crickets the distance between the dendritic membrane and dendritic sheath is 3–5 nm and both structures are joined by tiny bridges (*ca.* 3 nm in diameter) which are densely packed at intervals of 3–4 nm (Gaffal, K. and Theiss, J., 1978).

Just proximal to the tubular body the dendritic sheath is often highly convoluted or narrows considerably. This structural arrangement probably serves to hold the tubular body in place, thereby imparting more rigidity to the system. Recent reports include the interfacetal hair sensilla of houseflies (Chi, C. and Carlson, S., 1976), hair plates (French, A. and Sanders, E., 1979) and trichobothria (Gnatzy, W., 1978) of cockroaches and campaniform sensilla of crickets (Müller, M. *et al.*, 1978). In the latter sensillum at the level of the infolding of the dendritic sheath, the dendritic membrane is bordered by granules which are *ca.* 20 nm in diameter, consist of an electron-dense, amorphous material and are regularly spaced. In some cases microtubules appear in close association with the granules (Müller, M. *et al.*, 1978). French, A. and Sanders, E. (1979) observed that convolutions in the dendritic sheath often produce cavities which are partially or completely separated from the interior or exterior spaces, and suggested that these are actually pores in the dendritic sheath and would allow a free exchange of electrolytes to the dendrite while maintaining the structural strength of the sheath.

2.2.1.4 *Sheath cells.* Surrounding the neuron are two or more cells which are known collectively as sheath cells. In sensilla with two sheath cells the terms trichogen and tormogen are commonly used. When three cells occur they are numbered or referred to, inner to outer, as neurilemma or dendritic sheath cell, trichogen cell, and tormogen cell. Glial cells may be associated with the neuronal cell body and/or axon; for example, two glial cells envelop the cell body of a trichobothrial neuron (Gnatzy, W., 1976). Differences in number of sheath cells in a fully differentiated sensillum may be attributable to:

(1) degeneration of a cell during development as reported for the trichogen cell in a hair sensillum (Keil, T., 1978);
(2) thoroughness of investigation; and
(3) association of epidermal cells which may even participate in formation of particular parts of the sensillum but which have not arisen by differential cell divisions of the "mother stem cell" (Gnatzy, W., 1978).

Schmidt, K. (1973) presented evidence from electron microscopic studies that the sheath cells of hairs and campaniform sensilla as well as scolopidia are homologous.

In development the innermost cell, be it the trichogen or the dendritic sheath cell, produces the dendritic sheath (Gnatzy, W., 1976, 1978; Slifer, E., 1970). During the entire larval stage of crickets the dendritic sheath cell has massively developed rough endoplasmic reticulum, an abundance of secretory vesicles and dense cytoplasm. These features are probably all causally related to the formation of the dendritic sheath (Gnatzy, W., 1978). The trichogen cell forms the hair (Bernays, E. *et al.*, 1976; Gnatzy, W., 1976, 1978; Keil, T., 1978) or equivalent structure (Moran, D. *et al.*, 1976). During morphogenesis of the new hair, microtubules serve as a cytoskeleton and probably control the flow of vesicles, which contain phenol oxidase, and microfibrils are involved in the surface sculpturing (Gnatzy, W., 1978). The trichogen cell also forms other structural elements such as the "cup" and "strut" which are important in the spatial orientation of the dendrite (Gnatzy, W., 1978). In hair sensilla of adult blowflies the trichogen cell contains a polytene nucleus and is completely reduced around the time of emergence (Keil, T., 1978). In hair sensilla of locusts the cytoplasm of the trichogen cell is basally packed with rough endoplasmic reticulum and scattered ribosomes, and it is generally more electron-dense than that of the surrounding cells (Bernays, E. *et al.*, 1976). The tormogen cell builds the socket region. An oval nucleus and numerous ribosomes occur in the basal region of the tormogen cell, whereas apically there are many mitochondria and microtubules (Bernays, E. *et al.*, 1976).

Upon completion of development the sheath cells withdraw leaving a fluid-filled extracellular space, the receptor–lymph cavity (Nicklaus, R. *et al.*,

1967), beneath the cuticle. The receptor–lymph cavity may be divided into chambers (Müller, M. *et al.*, 1978) and frequently projects deeply into the tormogen cell (e.g. Gnatzy, W., 1976). In locusts the fluid in the receptor–lymph cavity of a hair sensillum is most electron-dense at the time of hatching and appears to become less so with time (Bernays, E. *et al.*, 1976). Surfaces of the sheath cells, particularly the tormogen cell, contacting the extracellular space bear numerous lamellae or microvilli.

2.2.1.5 *Cell connections.* Various types of linkages, including regular, septate and macula adherens desmosomes and septate junctions, connect the sheath cells to each other, to the dendrite and to the surrounding epidermal cells (McIver, S., 1975). A regular arrangement of three types of cell linkages occurs in several, if not all, cuticular mechanosensilla (Thurm, U. and Küppers, J., 1980). Outermost are macula adherens desmosomes which connect the epithelial cells to the sheath cells and the latter to the dendrite (Fig. 20). In the middle are septate junctions which link the same three cell types, and innermost are gap junctions which link all the epithelial and sheath cells, but not the inner sheath cell to the neuron (Fig. 20).

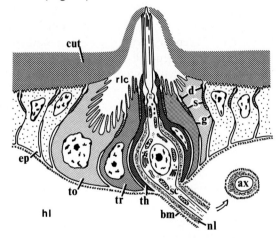

FIG. 20. Diagram of cells and intercellular junctions of an insect sensillum; the outer segment characterized as that of a mechanosensillum. ax = Axon, bm = basement membrane, cut = cuticle, d = desmosome, ep = epidermis, g = gap junction, hl = haemolymph, nl = neurilemma cell, rlc = receptor–lymph cavity, s = septate junction, sc = sensory cell, th = thecogen cell, to = tormogen cell, tr = trichogen cell. (Redrawn from Thurm, U. and Küppers, J., 1980.)

2.2.2 MOLTING

During morphogenesis and molting the fate of various portions of the sensillum, especially the dendrite with possible loss of function, is of particular interest. Shedding of the dendritic sheath has been reported for several types of mechanosensilla (McIver, S., 1975). There is clear evidence that the tips of the dendrites of hair (Gnatzy, W., 1978; Gnatzy, W. and Schmidt, K., 1972a,b; Schmidt, K. and Gnatzy, W., 1971) and campaniform sensilla (Gnatzy, W. and Romer, F., 1980) of crickets, campaniform sensilla of cockroaches (Moran, D., 1971), scales of a thysanuran (Larink, O., 1976), and hair sensilla of a caterpillar (Gnatzy, W. and Tautz, J., 1977) are lost at ecdysis. Light microscopy revealed that the distal portion of the dendrite remains connected to the old cuticle during apolysis, presumably permitting functional continuity until an advanced state in molting (Wigglesworth, V., 1953). Electron microscopic studies have shown that at least in the campaniform and hair sensilla of crickets (Gnatzy, W., 1978; Gnatzy, W. and Romer, F., 1980; Gnatzy, W. and Schmidt, K., 1972a,b; Schmidt, K. and Gnatzy, W., 1971) and hair sensilla of a caterpillar (Gnatzy, W. and Tautz, J., 1977) the tubular body stays in contact with the old cuticle during apolysis by an extension of the dendrite, and a new tubular body is formed below the old one in the same dendrite (Fig. 21). The dendritic extension containing the old tubular body usually leaves the new cuticle via the ecdysial canal (Fig. 21). An ecdysial canal was not found associated with scales of molting *Lepisma saccharina* (Larink, O., 1976). In hair sensilla the ecdysial canal is located above or near the base of the hair and in most campaniform sensilla in the middle of the cap. In campaniform sensilla on the eye of a cricket the ecdysial canal is eccentrically located (Müller, M. *et al.*, 1978). When the tubular body is located within one of the chemosensitive dendrites in a contact chemosensillum, it is shed through the apical pore along with the distal portion of that dendrite (Schmidt, K. and Gnatzy, W., 1972).

The sensitivity of thoracic filiform hairs of *Barathra brassicae* caterpillars to a 300 Hz tone was tested in the standing wave of a Kundt's tube (Gnatzy, W. and Tautz, J., 1977). Throughout most of the larval instar the threshold was $2.0 \pm 0.3 \, \mu m$

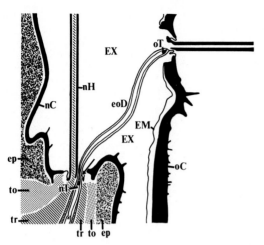

FIG. 21. Diagram of longitudinally sectioned old and new trichobothria on a *Barathra brassicae* caterpillar. Dendritic sheath cell (short, closely spaced lines). eoD = Elongated outer dendritic segment, oC = old cuticle, nC = new cuticle, EM = ecdysial membrane, EX = exuvial space, nH = new hair, oT = old tubular body, nT = new tubular body, ep = epidermis, to = tormogen cell, tr = trichogen cell; arrow indicates point of break off of the elongated dendritic segment when the caterpillar strips off the exuvium. (Redrawn from Gnatzy, W. and Tautz, J., 1977.)

particle displacement amplitude until 1–2 h before ecdysis when it rose to $6.8 \pm 1.3\,\mu m$ and at 10–30 min before the beginning of ecdysis no reaction to sound could be detected (Fig. 22). Once the old cuticle was shed maximum sensitivity returned as soon as the replacement hairs were erect. Consequently, the caterpillar sensilla are not functional for only 30–60 min during actual ecdysis. In contrast the numerous body scales of *Lepisma saccharina* are insensitive for several days (Larink, O., 1976). This difference in length of nonfunctional periods probably reflects the roles the sensilla play in behavior. The filiform hairs of *Barathra* warn the caterpillar of approaching danger by detecting natural sounds generated by predators such as wasps, whereas in *Lepisma* the scales are general tactile organs. Interestingly, removal of the scales in *Lepisma* apparently induces molting (Larink, O., 1976).

The morphogenesis of cercal filiform hairs and campaniform sensilla of crickets has been studied in considerable detail (Gnatzy, W., 1978; Gnatzy, W. and Romer, F., 1980). The hairs pass through six developmental stages during the last larval stage (Gnatzy, W., 1978). The cytoplasm of the neuron in

each hair sensillum contains numerous polysomes found free in the cytosol, indicating a high level of protein synthesis for cellular metabolism. Possibly cytoplasmic material is exchanged between the neuron and glial cells through evaginations of the latter. Ultrahistochemical tests demonstrated the presence in the neurons of acid phosphatase in the lysosomes as well as in various components of the Golgi apparatus. During molting the appearance of the neuronal cytoplasm was constant, reflecting a maintained functional state (Gnatzy, W., 1978). In studying the relation between morphogenesis of campaniform sensilla and molting-hormone level, Gnatzy, W. and Romer, F. (1980) found that at the beginning of the instar the level of β-ecdysone falls and prior to apolysis the concentration of α-ecdysone rises, reaching an intermediate peak after apolysis is complete. The maximum hormone concentration of *ca.* $2000\,ng\,g^{-1}$ is mainly due to an increase in β-ecdysone and occurs after the cuticulin layer is deposited. While the proecdysial cuticle is forming the hormone titer is reduced; β-ecdysone is still the chief component (Gnatzy, W. and Romer, F., 1980).

2.2.3 TRANSDUCTION

2.2.3.1 *Possible mechanisms and characteristics of transduction.* U. Thurm's (1964a,b) proposal that the tubular body is the site of sensory transduction in cuticular mechanosensilla is generally accepted. Direct proof is provided by such studies as D. Moran and colleagues' (1976) on campaniform sensilla. In

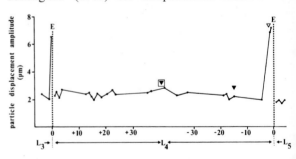

FIG. 22. Threshold of behavioral reaction relative to the molting cycle of one *Barathra brassicae* caterpillar. Abscissa: time after (+ hours) and before (− hours) ecdysis (E). Ordinate, air particle displacement amplitude (peak-value). $L_{3,4,5}$ = larval stage 3, 4 and 5. ▼. Apolysis begins; ▼, replacement hair is formed and cuticulum layer is laid down on the epidermis cells; ▽, new hair is ready. (Redrawn after Gnatzy, W. and Tautz, J., 1977.)

considering the various suggested mechanisms for transduction one must keep in mind the four fundamental structural features of the dendritic tip, namely: (1) dendritic membrane, (2) microtubules, (3) electron-dense material, and (4) physical connections, if any, between the microtubules themselves, the peripheral microtubules and the dendritic membrane, and the dendritic membrane and surrounding dendritic sheath. Chemical disassembly studies have demonstrated that the microtubules within the tubular body are probably essential for transduction (Moran, D. and Varela, F., 1971; Matsumoto, D. and Farley, R., 1978; Schafer, R. and Reagan, P., 1981) and sophisticated electron microscopic observations have revealed that the distances between the microtubules change when pressure is applied to the tubular body (Matsumoto, D. and Farley, R., 1978). In the taste hairs of blowflies, colchicine and vinblastine treatment for an hour or more increased response latency, decreased spike amplitude and decreased the number of spikes produced by each stimulus and caused loss of microtubules and some membrane disruption (Matsumoto, D. and Farley, R., 1978). In tibial spines of cockroaches colchicine, a microtubule-active agent, reversibly inhibits electrical activity, whereas lumicholchicine, a colchicine analog which does not bind to microtubule protein, does not inhibit responses (Schafer, R. and Reagan, P., 1981).

Thurm, U. (1964a,b, 1965) proposed that transduction was initiated by *compression* of the microtubules within the tubular body. Smith, D. (1969) suggested that subsequent to compression the microtubules function in an analogous, but reverse, way compared to myofilaments in a muscle cell. Deformation of microtubules would result in transient ionic changes responsible for transduction. Other suggestions for the mechanism of transduction based on the compression concept are "strain-activated mechanochemical engine in reverse" (Moran, D. *et al.*, 1971) and that the microtubules act as a piezoelectric crystal spanning the bending hair to the dendrite, thus inducing current flow from the mechanical perturbations of the hair (Strickler and Bal, 1973, discussed by Matsumoto, D. and Farley, R., 1978).

Another line of thought is that the microtubules form a cytoskeleton across which the dendritic membrane is stretched, causing change in shape of pores in the membrane with resulting increase in ionic conductance and depolarization (Rice, M., 1975; Rice, M. *et al.*, 1973). Important corollaries of the *membrane stretch theory* are:

(1) the exact amount of stretching exacted by unit movement will depend on the physical and spatial characteristics of membrane and cytoskeleton;
(2) compliance of the microtubular cytoskeleton is a key factor in this model, regulating not only the sensitivity of the dendritic terminal but also the adaptation characteristic of the receptor; and
(3) the most sensitive receptors have a short, highly organized and rigid microtubular cytoskeleton within the tubular body.

French, A. and Sanders, E. (1979) suggested that this concept of a relationship between the complexity of the tubular body and its sensitivity should be extended to include the *time-varying* viscoelastic properties of the tubular body and the spaces surrounding it.

Recent studies by French and colleagues on the trochanteral hair plates and tactile spines of cockroaches have given considerable insight into the mechanism of transduction. The individual sensilla within the hair plate may be divided into two physiological types (French, A. and Wong, R., 1976). Type I are larger in size, produce larger recorded action potentials, give a strongly dynamic (phasic) response to deflection, and exhibit a definite threshold frequency for sinusoidal stimulation below which no response can be elicited. Above this threshold frequency the linear frequency response function amplitude increases linearly with frequency while the phase of the response led the stimulus by 90° over the frequency range investigated. Type II hairs are smaller in size, produce smaller recorded action potentials, give both dynamic and static (tonic) responses, and have no threshold frequency and a smaller phase lead over the stimulus. Both types of sensilla display strongly non-linear behavior in several ways:

(1) The sensitivity is greatest if the sensillum is deflected in one direction, while deflecting it in the opposite direction usually elicits no response. This unidirectionality is often called

rectification in a function analogy to electrical rectifiers.

(2) Sinusoidal stimuli cause entrainment of the action potentials to fixed times within each cycle. This is usually called phase locking.

(3) Random stimulation elicits a low-frequency response which cannot be produced by sinusoidal stimuli, so that there must be non-linear interaction between different components in the input signal (French, A. and Wong, R., 1976).

The analysis with random (white noise) displacement to examine the non-linear behavior of type I sensilla was extended by French, A. and Wong, R. (1977). From the recorded afferent action potentials, the first- and second-order frequency response functions between the stimulus and the response were computed, together with their inverse Fourier transforms, the time domain Wiener kernels. The Wiener functional expansion is a general method for the analysis of non-linear systems with memory, where "memory" indicates that the output of such a system depends not only upon its current input but on some portion of the past history of the input. The behavior of type I sensilla could be minimally accounted for by a cascade of two functional elements, where the first is a linear element affected by the past history of the input signal (memory) and the second is a non-linear element with no memory. The linear element is quite similar to a time differentiator or velocity detector, while the non-linear element is like a rectifier which transmits the velocity signal only during flexion of the leg. Differentiation occurs before rectification. The overall behavior of the femoral tactile spine is similar to that of the type I sensilla in the trochanteral hair plate, although the linear element of the spine behaves rather differently (French, A., 1980).

In both the type I hairs and the tactile spine the linear frequency response apparently arises from the properties of the viscoelastic tubular body and rectification occurs when the deformed dendritic membrane transduces mechanical movement into electrical changes (French, A. and Wong, R., 1977; French A., 1980). A possible mechanism for rectification is a selective sliding of the microtubules relative to one another; that is, if the links between the microtubules act as ratchets, that prevent sliding

when the tubular body is rotated away from the joint membrane but allow it in the opposite direction, then there would be a different force produced upon the tubular body in the two situations (French, A. and Sanders, E., 1979). These hair sensilla lack cuticular features which structurally determine unidirectionality (rectification) as occur in trichobothria and some campaniform sensilla. Presumably then two mechanisms exist for determining unidirectionality in cuticular mechanosensilla. One occurs during coupling and involves cuticular structures that restrict movement to a particular plane, and the other happens during transduction and requires the participation of some element in the dendritic membrane or selective sliding of the microtubules.

Relatively small frequencies have been used in most studies on insect mechanosensilla that have reported frequency response functions which demonstrate fractional exponents of frequency (Chapman, K. and Smith, R., 1963; Chapman, K. et al., 1979; French, A., 1980; French, A. and Wong, R., 1976). In their study of transduction in campaniform sensilla Chapman, K. et al. (1979) used a frequency range of 1 mHz to 100 Hz with sinusoidal stimulation and found a frequency response function of the form given in equation 1 (Fig. 23) with an exponent of approximately 0.5 over the range of nearly 5 decades. Using the cockroach femoral tactile spine (Fig. 24) French, A. and Kuster, J. (1981) extended the test frequency range nearly 4 decades and found that the same relationship is applicable with an exponent of approximately 0.5 gain (Fig. 25). The variation in the exponent of frequency with stimulus amplitude is not large, but is quite distant and may be related to the cytoplasm-filled spaces which occur between the microtubules of the tubular body and between the tubular body and dendritic membrane.

$$H_1(j\omega) = g(j\omega)^k$$

FIG. 23. $H_1(j\omega)$ is the linear frequency response function at natural frequency ω, g is the gain at $\omega = 1$ radian/s, k is an exponent of frequency and $j = \sqrt{-1}$

To determine the effects of temperature on transduction in the tactile spine French, A. and Kuster, J. (1982) measured the frequency response function at temperatures in the range of 10–40°. At temperatures a few degrees Celsius outside this range

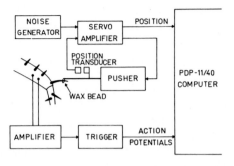

FIG. 24. The experimental arrangement for stimulating and recording from a femoral tactile spine of *Periplaneta americana*. The pusher consisted of a small audio loudspeaker and the position transducer was made from an infrared light-emitting diode and an infrared phototransistor. Action potentials were recorded by a pair of insect pins pushed through the femur close to the afferent nerve. Positive displacement was in the direction from the pusher towards the leg. (Reproduced from French, A. and Kuster, J., 1981 with permission.)

transduction ceases. Apparently the effect of temperature on transduction is to multiply the entire response by a constant factor, independent of frequency, at each temperature. The multiplication factor increases with warming up to about 35° and then decreases rapidly. The activation energy of transduction is about 18.6 kcal mol^{-1} (French, A. and Kuster, J., 1982) which is in good agreement with the estimate of 16–20 kcal mol^{-1} for transduction in the Pacinian corpuscle (Ishiko, N. and Lowenstein, W., 1961). The exponent of frequency remains at approximately 0.5 regardless of temperature, indicating a system which performs semi-differentiation.

Two models or possible types of mechanism have been suggested to account for fractional differentiation, that is, provided an adequate strength of stimulus in μm displacement, the sensitivity or gain in impulses per second increases with the frequency of stimulation (French, A., 1980). One mechanism involves some form of *travelling wave phenomenon*, such as is found in coaxial electrical cables (Moore, R., 1960) where relationships involving square-root functions of frequency often arise. The second mechanism involves a system in which *multiple parallel paths* lie between the input and output, and where there is a suitable distribution of system parameters among the different paths (Thorson, J. and Biederman-Thorson, M., 1974). One feature of the multiple parallel paths model is that it can

produce any value of the fractional exponent k. However, the work in which wide frequency ranges have been used indicates a fairly narrow range of exponents for the responses of mechanoreceptors to displacement, usually close to 0.5 (French, A. and Kuster, J., 1981, 1982). There are physical processes which behave as semi-differentiators — for example, diffusion, heat, conduction, acoustic waves and electrical transmission lines — and all of these processes are characterized by the presence of *travelling waves* (French, A. and Kuster, J., 1982). Mechanical analogs can be generated which behave similarly, and which generate a pressure which is the semi-differential of displacement. French, A. and Kuster, J. (1982) state that the tubular body of the tactile spine or the campaniform sensillum can be modelled in this way if it is assumed that the microtubules resist deformation elastically and are coupled to each other by viscous elements. A significant feature of this model is that changes in elasticity or viscosity do not change the exponent k,

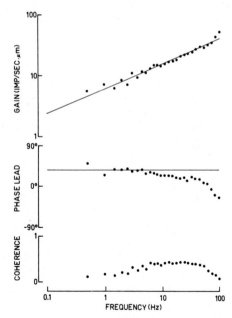

FIG. 25. The frequency response and coherence functions for sensory transduction by the femoral tactile spine of *Periplaneta americana* during stimulation by a randomly varying movement with a bandwidth from 0 to 125 Hz. The continuous lines are a fitted relationship of the form $H_1(j\omega) = 2.90(j\omega)^{0.41}$. This relationship was fitted by minimizing the mean square error between the experimental and predicted gain data. The phase prediction is based only upon the gain data. (Reproduced from French, A. and Kuster, J., 1981 with permission.)

which is always 0.5 (French, A. and Kuster, J., 1982).

Based on existing morphological and physiological evidence transduction by direct mechanical distortion of the dendritic membrane is the most plausible process. The temperature sensitivity of transduction strongly supports the concept that the mechanism is in the dendritic membrane because the electron-dense material and microtubules of the tubular body probably would be quite insensitive to temperature (French, A. and Kuster, J., 1982). These findings fit the travelling wave model which is a simple extension of the membrane stretch theory to include the time-varying mechanical impedance of the tubular body so that the force exerted on the membrane by displacement follows the fractional differentiation relationship observed in the electrical output of the receptor (French, A. and Kuster, J., 1982). In this model temperature would not be expected to affect significantly the viscoelastic elements, that is, the electron-dense material and microtubules, and would certainly not change the exponent k. In contrast temperature would be expected to change the pressure-sensitive conductance of the dendritic membrane and thus the overall sensitivity of the receptor (French, A. and Kuster, J., 1982). As discussed by French, A. and Sanders, E. (1981), the possible existence of other, more complex mechanisms for transduction, such as internal chemical transmitters and active microtubular sliding, cannot be completely discounted, but at present they seem unlikely.

The various physical connections between the dendritic sheath and the dendritic membrane are probably involved in the last stage of coupling of the mechanical stimulus to the site of transduction. The connections between the dendritic membrane and peripheral microtubules of the tubular body would serve as anchor points to the underlying microtubular cytoskeleton and aid in the entire dendritic sheath–dendritic membrane–tubular body system acting as a functional unit. The variation in closeness of association of the peripheral microtubules and the dendritic membrane in tubular bodies — for example, the tubular bodies in the two antennal campaniform sensilla of mosquito larvae (Jez, D. and McIver, S., 1980) — would be an important morphological feature underlying the spectrum of strength and dynamic behavior of the

transmembrane pressure during deformation of the tip of the dendrite.

The suggestion of Lewis, C. (1970) that the viscoelastic properties of tubular bodies in cuticular mechanosensilla may be largely responsible for determining the dynamic behavior of these sensilla has been substantiated by the various studies of French and colleagues. A specific example, as discussed by French, A. and Kuster, J. (1981), is the possibility that in the tactile spine the fluid-filled spaces between microtubules of the tubular body and the tubular body and the dendritic membrane may cause the behavior of the system to move closer to that of a purely elastic system and thus reduce the observed exponent of frequency. The total number, density, and pattern of the microtubules, amount and arrangement between the microtubules of the electron-dense material and presence or absence of microfilaments are the structural features underlying the observed variation in the dynamic behavior. As originally proposed (Rice, M. et al., 1973), the membrane stretch theory implied that the microtubular cytoskeleton is rigid. However, recent evidence indicates that at least some flexibility may occur; for example, the changes in spatial relationships of the microtubules during stimulation (Matsumoto, D. and Farley, R., 1978) and the possibility of sliding of bundles of microtubules upon application of pressure (French, A. and Sanders, J., 1979).

Transducer operations in the directionally sensitive cercal trichobothria (thread-like hair sensilla) and bristles of the cockroach, *Periplaneta americana*, have been simulated in a digital computer (Buno, W. et al., 1981a,b) (Fig. 26). The trichobothria were spontaneously active, purely phasic, did not respond to sustained displacements and with small sinusoidal displacements behaved as a linear, second-order lead system sensitive to velocity. With larger amplitudes the trichobothria showed prominent non-linear features with minimal consequences of displacements at the extremes. Second-order response components were found in the responses of the trichobothria to other waveforms (Buno, W. et al., 1981b). The bristles were also purely phasic but were not spontaneously active (Buno, W. et al., 1981a). Sinusoidal analysis of the bristles suggested the behavior of a first-order lead system with corner frequencies distributed between 8 and 20 Hz. This is a first approximation

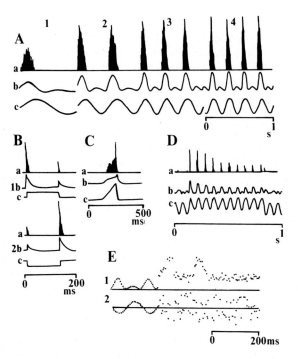

FIG. 26. Computer simulation of dynamic properties of cercal bristle sensilla of *Periplaneta americana*. **A**: Sine waves of increasing frequency (from 1 to 4). Peri-stimulus times histograms, generator potential averages, and input waveforms, (a), (b) and (c), respectively (as in (**B**), (**C**) and (**D**). **B**: Responses evoked by long-duration pulses of opposite directions (1 and 2, respectively). **C**: Effects of constant velocities. **D**: Mixed pulse-sine wave displacements. **E**: Effects of sine wave–white-noise displacements; the generator potential average (1) and a single-stimulus wave form (2) are shown in each record. The control response evoked by the sine wave, the effects of adding white noise, and the response evoked by the white noise alone are shown at left, middle and right, respectively. Note the direct current bias in the generator potential added by the white noise. (Redrawn after Buno, W. *et al.*, 1981a.)

because of the important non-linearities. A roughly similar behavior was observed in response to ramp-like displacements. Non-linearities occurred both at the level of the generator potential and of spike generation. For the cercal bristles Buno, W. *et al.* (1981a) suggested a mechanical origin for the phasic and non-linear behaviors at the generator potential level and that other non-linearities may be accounted for by the lack of spontaneous activity and the threshold nature of the spike generator. Conceptually, the functioning of the bristles can be considered as a linear element followed by non-linear elements (Buno, W. *et al.*, 1981a). For both the trichobothria and bristles the computer simulations

were not only descriptive in showing a close fit to the biological responses, including transduction, but could also be used in a predictive way to plan future experiments (Buno, W. *et al.*, 1981a,b).

2.2.4 CODING

2.2.4.1 *Receptor potential.* Transduction results in production of a receptor potential across the dendritic membrane. The extensive investigations of Thurm and associates (Erler, G. and Thurm, U., 1977, 1978, 1981; Keil, T., 1979; Keil, T. and Thurm, U., 1979; Küppers, J., 1974; Küppers, J. and Thurm, U., 1975, 1979; Thurm, U., 1970, 1974a,b; Thurm, U. and Küppers, J., 1980; Thurm, U. and Wessel, G., 1979) on the organization and physiology of the epithelia of insect sensilla, especially the campaniform sensilla of fly halteres, have elucidated some of the possible events in the origin of the receptor potential. Basic to their theory is the existence of a potential difference across the epithelium, the *transepithelial potential* (TEP), at the receptor site.

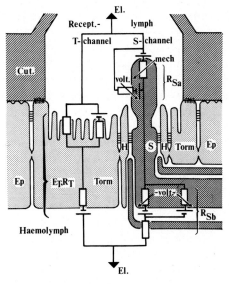

FIG. 27. Equivalent circuit of a mechanosensitive sensillum; cable properties of epithelium and axon neglected. Cut = Cuticle, El = electrodes, Ep = unspecialized epidermal cell, H = thecogen cell, S = sensory cell, Torm = tormogen and trichogen cell, mech = mechanically controlled resistance, E_T, R_T = effective electromotive force and resistance, respectively, of the T-channel, R_{Sa}, R_{Sb} = effective apical and basal resistances, respectively, of the S-channel. (Redrawn from Thurm, U. and Küppers, J., 1980.)

The arrangement of cells and intercellular junctions considered by Thurm and associates to be important in the production of TEPs are illustrated in Fig. 20 (see section 2.2.1.5 for details). The septate junctions probably prevent communication between the receptor–lymph cavity and the hemolymph, and the gap junctions are known to establish low-resistance intercellular pathways. As a result of the types and arrangements of cell junctions, the neuron is surrounded by an epithelial compartment of electrically interconnected cells (see Fig. 1, Küppers, J. and Thurm, U., 1979). The equivalent circuit is shown in Fig. 27.

The TEP is usually between 20 and 80 mV with the cuticular side positive, and it decreases with adequate receptor stimulation. TEPs were found in all groups of hemi- and holometabolous insects tested, but not in spiders (Thurm, U. and Wessel, G., 1979). TEPs occur in chemo- as well as mechanosensilla. The TEP is dependent on oxidative metabolism, as readily demonstrated by anoxia. The mechanism is thought to be an electrogenic cation pump rather than simple diffusion because the TEP recovers very quickly upon resupply of oxygen. In several insect species Küppers, J. (1974) found the K^+ concentration in the receptor–lymph cavity to be quite high ($100–200 \, \text{m mol} \, l^{-1}$); in fact, in a campaniform sensillum it is thirteen times that found in the haemolymph. The site of the electrogenic K^+ pump is probably the microvilli on the surfaces of the tormogen cell that contact the receptor–lymph cavity. The microvilli show indications of secretory activity — namely, occurrence of numerous mitochondria, presence near the plasma membrane of small particles which may have a role in transmembrane ion transport and the existence of vesicles and vacuoles (McIver, S., 1975). Possibly some Na^+ as well as K^+ is pumped from the cytoplasm into the receptor–lymph cavity (Küppers, J., 1974; Thurm, U. and Küppers, J., 1980). The production of the receptor current is closely linked to the potential difference generated by the tormogen cell. The change in TEP observed upon stimulation of the sensillum may indicate an increase of current through the sensory cell, which in turn would produce impulses by depolarization of more proximal regions of the dendritic membrane.

From studies on the tibial hair sensilla of crickets Erler, G. and Thurm, U. (1981) concluded that a transepithelial current traverses the sensory cell mainly via the membrane of the outer dendritic segment and leaves the cell at the basal side of the epithelium via the membrane of the basal part of the inner dendritic segment. Entry of the current into the cell is by conductances which are increased by adequate mechanical stimuli and are quite high in the mechanically unstimulated state. Possibly some apical conductance is controlled by the membrane voltage as is a rectifying K^+ conductance. Erler, G. and Thurm, U. (1981) also concluded that the regular site of impulse initiation is in the inner dendritic segment, although areas in membranes of both inner and outer dendritic segments are able to initiate impulses repetitively.

Using a hair sensillum of *Periplaneta americana* which was mechanically stimulated with repetitive step displacements of the spine and irrigated with standard or modified salines for several hours, Bernard, J. *et al.* (1980) found evidence for a barrier between hemolymph (blood) and the sensory terminal. The junctional complexes of the sensillar and epidermal cells may constitute the anatomical support of the barrier. Stretch disrupts the barrier and Bernard, J. *et al.* (1980) postulated that the disrupted zone is in the distal region of the axon.

The receptor potential is non-propagated, negative-going, varies directly with the magnitude of the stimulus and shows no overshoot when returning to baseline (Wolbarsht, M., 1960). It is "slow" in comparison with the subsequent "fast" action potential (Wolbarsht, M., 1960) and spreads electrotonically only (Schwartzkopff, J., 1974). Bernard, J. and Pinet, J.-M. (1973) reported that, for electrotonic transmission of a receptor potential to occur, the resistance of the dendrite alone, or of the dendrite and dendritic sheath, should be very high. The impulse rate of both the receptor and action potential is directly proportional to temperature (Smola, U., 1970). The size of the receptor potential can vary continuously, indicating that apparently a virtual threshold does not exist (Schwartzkopff, J., 1974). The receptor potential most likely originates in the vicinity of the tubular body. Sometimes it can be demonstrated that the cell membrane may develop a second sensitive area at the axon hillock (Schwartzkopff, J., 1974; Thurm, U., 1963, 1970; Wolbarsht, M., 1960). If this area is depolarized by the spreading receptor potential, a propagated nerve impulse results which is termed the "generator potential".

From an analysis of electrophysiological responses Spencer, H. (1974) developed a compartmental model for the trochanteral hair sensilla of *P. americana*. Compartment I is the area from the hair tip to the socket, compartment II extends from the tubular body that inserts on one side of the hair shaft to the basal area of the soma, and compartment III is from the latter region through the distal part of the axon. The functional events that occur in each compartment are displacement and initiation of generator and action potentials in compartments I, II, and III, respectively. Spencer, H. (1974) did not distinguish between receptor and generator potentials. In the power function that exists for the sensilla the values of the exponential term and of the constant appear to reflect the degree of coupling, respectively, between the stimulus and the generator region and between the generator potential and spike-initiating region (Spencer, H., 1974).

Guillet, J. and Boistel, J. (1968) reported that an electrical current acting alone can induce propagated spikes, indicating that it acts directly at the level of the initiation site of the impulses by modifying the membrane potential. In a subsequent study on a phasic–tonic hair sensillum Bernard, J. and Guillet, J. (1972) found that the amplitude of the receptor potential increases when the receptors are polarized by a current passing in the direction of dendrite to axon, except for the second phase. With the current flowing in the opposite direction the amplitude is reduced and even inverted.

To establish the transfer function Guillet, J. (1975) applied only rectangular steps to phasic mechanosensilla on locust legs. The dynamic component of the receptor potential increases with the logarithm of the intensity of the stimulation to a maximum which is not due to a mechanical feature. The receptor potential then decreases under the maintained stimulus as a power function of time. During the dynamic phase Guillet, J. (1975) was unable to establish a linear relationship between the amplitude of the receptor potential and the instantaneous frequency of the spikes. Even for small stimulations the frequency is limited by the refractory period. The relationship between the two phenomena becomes linear during the adaptation period which starts more rapidly for the instantaneous frequency than for the receptor potential. This decrease in frequency is also a power function (Guillet, J., 1975).

2.2.4.2 *Action potentials*. Action potentials are propagated along the axons at a rate of 0.5 to 3 m s^{-1} and are all-or-none in response. After repolarization, the membrane requires about 1 s for regeneration (Schwartzkopff, J., 1974). The upper limit of transmitted pulse rate which is about 1000 s^{-1} and the repetition rate of the action potential are functions of time and/or the amplitude of the receptor (generator) potential (Schwartzkopff, J., 1974). Wolbarsht, M. (1960) reported that changes in the impulse size are due to changes in the membrane resistance of the receptor site, and similarly Guillet, J. and Bernard, J. (1972) found that the first positive component of the nerve spike increases with the receptor potential and was attributable primarily to a diminution of the dendritic membrane resistance. With antidromic stimulation the amplitude of the negative phase of the spike appears to be correlated with the polarization of the dendritic membrane, although when bursts of action potentials are applied, the relation is more complex, including a depressive influence of a given spike on the following spike (Guillet, J. *et al.*, 1980). As indicated by Schwartzkopff, J. (1974), the information originally contained in the mechanical stimulus is recorded in the action potential and transmitted eventually to CNS by the temporal pattern of identical pulses.

Two functional types of mechanosensilla can be distinguished on the basis of the temporal behavior of the neurons. The *phasic* or *velocity*-sensitive type yields impulses only while the stimulus is changing, responding in proportion to the rate by which the stimulus changes. During stimulation phasic receptors may fire repetitively at a very high rate (Wolbarsht, M., 1960). The *tonic* or *pressure*-sensitive neurons show a repetitive discharge during a static deformation in addition to responding during a changing stimulus. Ideally the frequencies of the nerve impulses are always in proportion to the degree of mechanical deformation. The most common receptor is a combination of the two basic types — that is, phasic–tonic — and can detect changes in the sensitivity of the mechanical forces as well as continuously determining their size.

A variety of hair sensilla that give purely phasic responses (Neumann, H., 1975; Specht, U., 1977; Spencer, H., 1974; Tautz, J., 1978) show the following certain fundamental characteristics (Theiss, J., 1979).

(1) The sensilla are not spontaneously active. Interestingly, some purely phasic trichobothria have been shown subsequently to have a resting activity that appears to be due to neuronal factors since it was not abolished by preventing hair movements (Buno, W. et al., 1981a).
(2) The frequency maximum rises with increasing rate of displacement as it does also for the phasic component of hair sensilla with phasic–tonic responses.
(3) The maximum is reached within a displacement section of 3° above the threshold angle.
(4) The same number of impulses are always elicited by a given angular displacement, a feature that may be important in understanding the nature of stimulus transformation.
(5) Impulse number increases with angular displacement.

Theiss, J. (1979) further noted that the various purely phasic hair sensilla differ distinctly in their morphologies with resulting differences in restoring forces, threshold angle and directional characteristics. In addition there is considerable variation in discharge dynamics. The time course of discharge rate during ramp stimuli may be described by a simple exponential function in some sensilla (Theiss, J., 1979) and by a simple power function in others (Tautz, J., 1978), whereas still other sensilla respond with constant impulse frequency during the movement phase of the stimulus (Neumann, H., 1975; Specht, U., 1977).

Adaptation is the change during stimulation of the relationship by which properties of the mechanical force are translated into corresponding properties of the receptor potential and subsequently into action potential patterns (Schwartzkopff, J., 1974) and is of fundamental importance in the functioning of a mechanosensillum. Using campaniform sensilla Chapman, K. et al. (1979) reported that the large size of the power coefficient for the sensory discharge indicates a major component of sensory adaptation must be somewhere between the cap and the axon. The adaptation rate of hair sensilla of a cockroach varies with the size of hair (Pumphrey, R., 1936). In trichobothria of a caterpillar, Tautz, J. (1978) reported that at sine stimulation the neuron adapts only slowly. At a stimulus frequency of 100 Hz the response begins to decrease after 5 s and completely disappears after 20 s.

Mann, D. and Chapman, K. (1975) used sinusoidal mechanical stimuli applied to campaniform sensilla to explore the adaptation of both the transcuticular receptor potential and the action potential relative to the indenting force, and expressed these as linear transfer functions for the combined time-dependence of coupling and transduction together and for coding. Subsequently, Chapman, K. et al.. (1979) determined the linear transfer function for cap compliance explicitly, as a measure of time-dependent coupling, and to assess its contribution to adaptation of the action potential. With punctate, sinusoidal mechanical stimuli at forcing frequencies ranging from 0.003 Hz to 100 Hz most sensilla stiffened viscoelastically as a power of frequency with median power coefficient −0.058 for compliance, and with indentation lagging force by a constant phase angle, median −8.4°. These results indicate broadly distributed viscoelastic rate constants in individual sensilla. Interestingly, the extent of viscoelasticity also varied among sensilla with about a third being purely elastic with constant compliance and a zero phase angle. Chapman, K. et al. (1979) concluded that cap compliance exerts a mild filtering action throughout the entire time-course of adaptation in viscoelastic sensilla, but no role in the adaptation of purely elastic ones.

Among the several mechanisms for receptor adaptation discussed by Buno, W. et al. (1981b) are mechanical filters represented by structures surrounding the dendrite such as the inner and outer receptor–lymph cavities (Gnatzy, W., 1976; Nicklaus, R. et al., 1967) and elastic cellular structures. Another is that an axonal encoder determines the impulse frequency (Mann, D. and Chapman, K., 1975). Still another line of thought is that the spike generator contributes to the rate sensitivity because in certain phasic receptors the neuron adapts during depolarizations with externally applied currents (Guillet, J. and Bernard, J., 1972).

The Hodgkin–Huxley equations predict a maximum temperature for action potential propagation at about 35° (Huxley, A., 1959). Chapman, K. and Pankhurst, J. (1967) reported the sensory axons of cockroaches to cease conduction in the range of 35–42°. In the axon of the tactile spine of cockroaches French, A. and Kuster, J. (1982) could not record action potentials a few degrees outside the range of 10–40°. Sensitivity of the tactile spine axon changes with temperature, and French, A. and Kuster, J. (1982) suggested that the action potential encoder and the sensory axon could contribute to these changes by alteration of the threshold for action potential generation or by failure of action potential propagation.

3   TYPE II, MULTITERMINAL SENSILLA

Multiterminal sensilla may be divided into two general groups. Some are oriented in that they are attached to accessory structures such as a special receptor muscle, a strand or tube of tissue, or a combination of muscle and tissue, and others are unoriented being without accessory structures. Finlayson, L. (1976) considered it likely that the oriented stretch receptors have evolved from multiterminal neurons on the body wall or attached to various tissues. It is appropriate, therefore, to consider those multiterminal sensilla without accessory structures and then to discuss the sensilla which have accessory components.

3.1   Multiterminal neurons without accessory structures

3.1.1   SUBEPIDERMAL NEURONS

Viallanos, H. (1882) was the first to describe a system of subepidermal multiterminal neurons in insects, and M. Osborne's (1963a) work on the larva of the blowfly *Phormia* is the most comprehensive description. References to partial description of the system in other insects are given by Alexandrowicz, J. (1957) and Osborne, M. (1963b, 1964). The subepidermal multiterminal neuron system has been called a subepidermal plexus, but there is no fusion of the processes from subepidermal neurons and no evidence of synaptic associations (Finlayson, L., 1976).

The topography of the sensilla of *Phormia* can be

seen in Fig. 28. In the blowfly larva the number of neurons in each segment is surprisingly constant: 24 in the prothorax, 28 in the mesothorax and the metathorax, and 30 in each abdominal segment. Osborne, M. (1963b) reported that the whole body wall of the larva is covered by a fine meshwork of dendritic processes from these multiterminal sensilla. In contrast, in nymphs of the house cricket, *Gryllus domesticus*, the subepidermal neurons do not, as a rule, form an extensive meshwork, even in the soft abdomen, but lie singly or in small groups consisting of a few cells (Knyazeva, N. *et al.*, 1980). In the blowfly larva most of the dendrites ramify under the epidermis, but some end on muscles (Osborne, M., 1963b). Subsequently Osborne, M. (1964) observed in electron micrographs of the terminations of the neurons that they send fine processes through the basement membrane and into invaginations of the epidermal cells. Neurilemma cells cover the cell body and its processes, except for the dendritic terminations, which are therefore in intimate contact with the epidermal cells.

Finlayson and Osborne (in Finlayson, L., 1968) found that the two most dorsal subepidermal neurons of an abdominal segment fire continuously, and can be stimulated to increase their discharge frequency by pulling on the cell body. The main

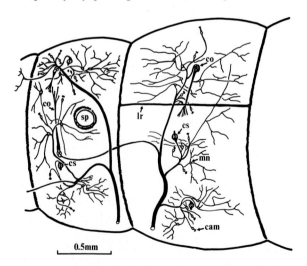

0.5mm

FIG. 28. Lateral view (from inside) of the right half of the pro- and mesothorax of a *Phormia regina* larva, to show the topography of the sensory neurons and sensilla. cam = Campaniform sensillum, cs = cuticular sensillum (possibly a chemosensillum), co = chordotonal organ, lr = longitudinal stretch receptor, mn = multipolar neuron, sp = spiracle. (Redrawn after Osborne, M., 1963b.)

function of the neurons is likely to respond to deformations of the body wall caused by external forces, thereby being exteroreceptors, although they must also respond to changes in body shape owing to the activity of the muscular system.

In each segment of the abdomen of adult male and female *Glossina morsitans* are three ventral body wall neurons (Anderson, M. and Finlayson, L., 1978). These neurons are placed on that part of the abdominal wall which undergoes the greatest extension at feeding. Using suction electrodes Anderson, M. and Finlayson, L. (1978) found that these ventral wall neurons respond to mechanical stimulation and presumably play a primary role in monitoring abdominal distension. Interestingly, evidence indicates that molting in the bug *Oncopeltus fasciatus* is triggered by stimulation of abdominal multiterminal sensilla (Nijhout, H., 1979) as in bloodsucking Reduviidae (Anwyl, R., 1972). These molt-triggering neurons may well be quite similar to the ventral body wall neurons of *G. morsitans*.

3.1.2   NEURONS LOCATED MORE DEEPLY IN THE BODY

Some multiterminal peripheral neurons which are distinct from connective tissue strand organs and muscle receptor organs are situated more deeply in the insect body than the subepidermal neurons. They are associated with a variety of tissues (Finlayson, L., 1976). A group of neurons of this type has been described in the lateral region of each abdominal segment of the blowfly larva (Osborne, M., 1963b) and the tsetse fly larva (Finlayson, L., 1968). There are usually three multiterminal neurons in each segment. The topography of these neurons in the blowfly larvae is shown in Fig. 29.

In the blowfly the most anterior nerve is situated in a connective tissue capsule attached to the posterior edge of the transverse intersegmental muscle (Osborne, M., 1963b). The second nerve lies near the motor nerve into which the sensory process from the anterior cell runs. The third nerve is located within the nerve that innervates the lateral transverse muscles. Sometimes a fourth cell lies adjacent to the third one. From each of these cells one process, presumably the axon, always runs in the nerve toward the central nervous system (Osborne, M., 1963b).

A pair of multiterminal neurons that lack accessory structures occurs in the bursa copulatrix of the

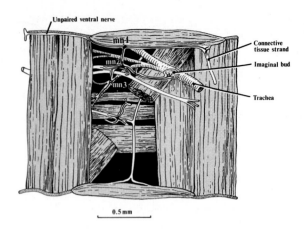

FIG. 29. Drawing of the musculater of the lateral region of the second abdominal segment of a *Phormia regina* larva to show the topography of neurons mn1, mn2 and mn3 and the distribution of their processes. mn1, mn2 and mn3 = Multipolar neurons. (Redrawn after Osborne, M., 1963b.)

female cabbage white butterfly, *Pieris rapae crucivora* (Figs 30 and 31). Signals induced by stretching of the bursa copulatrix upon transfer of the spermatophore mediate the behavioral change from acceptance to refusal of mating by the female (Sugawara, T., 1979). The frequency of spontaneous afferent impulses increases ten times when the bursa is stretched by a spermatophore. For experimental analysis Sugawara, T. (1979) filled the bursa artificially by injection with silicone oil. A phasic response occurred during injection and a tonic response at sustained expansion (Figs 32 and 33). When the volume of oil injected into the bursa exceeded 3 μl, the impulse frequency increased proportionally to the logarithm of the volume (Sugawara, T., 1979). Although the volume of a normal spermatophore is about 6 μl, the bursa could be injected with as much as 16 μl of silicone oil (Fig. 33). Females displayed the mate-refusal posture only if the spermatophore substance ejaculated had been more than half a volume of full spermatophore or if the bursa was expanded by the injection of more than 4 μl silicone oil. Apparently the tonic response of the stretch receptor plays the main role in the switchover of the behavior.

The cell body of each multiterminal neuron is about 10 μm in diameter and lies on the edge of the muscular region in the anterolateral wall of the bursa (Fig. 30). Several dendrites extend radially

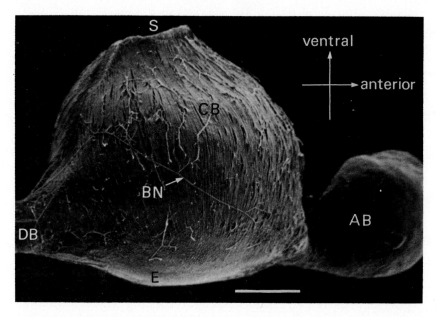

FIG. 30. Scanning electron micrograph of the bursa copulatrix of a mated female *Pieris rapae crucivora* (ventral side up). The main body of the bursa, or corpus bursae (CB), is spherically inflated by a spermatophore. On the ventral region lies a signum (S not clearly shown) from which muscle fibers extend radially to the dorsal epithelial (E). Dorsal epithelial cells are exposed. A bursal nerve (BN) runs spirally from the posterior to the anterior portion. AB = Appendix bursae; DB = ductus bursae. (Scale = 500 μm.) (Reproduced from Sugawara, T., 1981 with permission.)

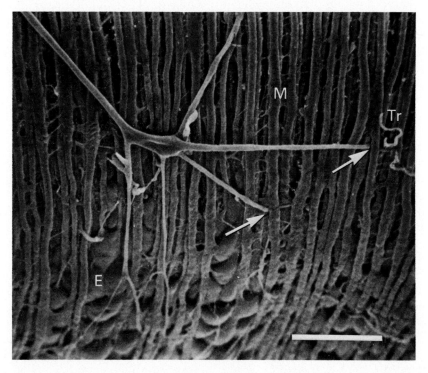

FIG. 31. Scanning electron micrograph of the typical bursal nerve ending. Some processes (arrows) pass beneath the muscle fibers (M) others terminate on the epithelium (E). Tr = tracheole. (Scale = 50 μm.) (Reproduced from Sugawara, T., 1981 with permission.)

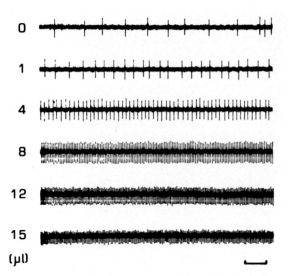

FIG. 32. Examples of tonic response in a single nerve–bursa preparation. Volume of injected silicone oil, given by the numerals at the head of each line, was increased step by step from upper to lower line. (Time scale = 1 s.) (Reproduced from Sugawara, T., 1979 with permission.)

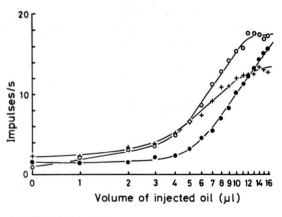

FIG. 33. Relationship between volume of silicone oil injected into the corpus bursae and frequency of tonic afferent impulses of the bursal nerve. Three examples are plotted. The data of open circles from the same recordings as in Fig. 32. Note that abscissa is graduated logarithmically. (Reproduced from Sugawara, T., 1979 with permission.)

into the muscle layer (Fig. 31) and are ensheathed except for their naked distal tips that terminate in the basement membrane of the epithelial cell. At repeated points on the epithelial cells on the muscle fibers the dendrites attach to them in such a manner that their basement membranes are fused. The 0.1–0.2 μm in diameter dendrite often forms 1–2 μm varicosities and consequently appears varicose or beaded. The varicosities, which are more prevalent at anchor points, contain a number of mitochondria. Only microtubules occur in the interconnecting parts of the dendrite. A thin layer of sheath cells that often appears as only the double membrane of the cells covers the varicosities, and becomes progressively thinner in those varicosities near the tip of the dendrite.

3.1.3 MULTITERMINAL NEURONS ON OR IN NERVES

Multiterminal neurons have been found on or inside of (a) nerves associated with the spiracles of tsetse flies (Finlayson, L., 1966); (b) the median abdominal nerve of the blowfly, *Phormia* (Gelperin, A., 1971); (c) the labellar nerve of the blowfly, *Calliphora* (Peters, W., 1962; Stürckow, B. *et al.*, 1967); and (d) a variety of nerves in the stick insect (Finlayson, L. and Orchard, I., 1977; Orchard, I.,

1976; Orchard, I. and Finlayson, L., 1976, 1977; Orchard, I. and Osborne, M., 1977). Many of these multiterminal neurons are also neurosecretory — for example, those studied by Gelperin, A. (1971); Orchard, I. (1976); Orchard, I. and Finlayson, L. (1977); and Orchard, I. and Osborne, M. (1977). The cell bodies of neurosecretory neurons situated on the link nerve in the stick insect are electrically excitable and produce overshooting action potentials, whereas the cell bodies of non-neurosecretory neurons located on peripheral nerves are electrically inexcitable (Orchard, I. and Finlayson, L., 1977).

In blowflies the nerve-associated multiterminal neurons that have been studied are apparently involved in various aspects of feeding. In the labellar nerve of *Calliphora* there are about 30 multiterminal neurons (Peters, W., 1962) from which Stürckow, B. *et al.* (1967) recorded responses to mechanical stimulation caused by movement of the proboscis, bending of hairs, or changing the rate of flow of solutions used to stimulate chemosensilla. In the abdomen of *Phormia regina* are neurons in nerves that Gelperin, A. (1971) showed by recording with suction electrodes to respond to stretch. The frequency of discharge of these neurons alters as the crop contracts and expands. Gelperin, A. (1971) postulated that extension of the crop when the blowfly feeds is detected by mechanosensitive neurons, not neurosecretory ones, within individual nerves in the extensively cross-connected nerve network of the abdomen.

The ionic requirements for the action potentials of several nerve-associated neurons have been investigated in the stick insect. The link nerve neurosecretory neurons (LNNs) have their cell bodies and processes laying superficially on major peripheral nerves and are in direct contact with the hemolymph which has a low sodium concentration. The anomaly of how these neurosecretory cells function in such a low sodium environment was reconciled by Orchard, I. (1976) who used intracellular microelectrodes to provide convincing evidence that the inward current is caused by calcium ions. Subsequently, Orchard, I. and Osborne, M. (1977) concluded from studies using extracellular electrodes that calcium is the major charge carrier in the inward current in neurosecretory axons associated with neurohemal tissue on the lateral branch of the median nerve (Fig. 34). This neurosecretory system differs from the LNNs in having the cell bodies inside the CNS and the axons terminating in the neurohemal tissue on the transverse nerves. These neurosecretory axons, like the LNNs, are in ionic contact with the hemolymph (see Fig. 5, Orchard, and Loughton, vol. 7 in this series). Orchard, I. and Osborne, M. (1977) also found that small amounts of sodium and magnesium are necessary to maintain electrical activity, and that magnesium is a competitive inhibitor of the calcium current.

In contrast the action currents in the axons of the dorsal longitudinal stretch receptors which have a blood/brain barrier are carried by sodium ions (Finlayson, L. and Orchard, I., 1977), as occurs in most other insect axons. Therefore, the stick insect has two types of peripheral neurons. The neurosecretory neurons are exposed directly to the hemolymph, except for a tenuous, discontinuous Schwann cell, and are dependent upon calcium for their action potentials. The non-neurosecretory neurons are separated from hemolymph by glial cells and the perineurium cell layer, and have sodium-dependent action potentials.

Using suction electrodes Orchard, I. and Finlayson, L. (1976) described the electrical activity of neurosecretory and mechanosensitive neurons in the stick insect. The neurosecretory cells which lie on the "link" nerve (Fig. 35) are spontaneously active in completely isolated preparations and fire with a regular low frequency of less than 1 imp s^{-1} or in small bursts of 12 imp s^{-1} (Fig. 36). The action potentials of the neurosecretory cells are characteristically of long duration (2 to 10 ms) and have a slow conduction velocity (0.15–0.25 m s^{-1}). In contrast the duration of the action potentials of motor or sensory fibers are of shorter duration (0.6–0.8 m s) and have a slower conduction velocity (0.54–0.7 ms) (Fig. 36). In the neurosecretory cells the action potentials are propagated from the region of the cell body towards the terminals and pass along all the major nerves in the periphery (Orchard, I. and Finlayson, L., 1976).

Non-neurosecretory cells which lie on peripheral nerves respond to stretching of the nerves upon which they lie, or of nerves which branch in the immediate vicinity (Orchard, I. and Finlayson, L., 1976) (Fig. 35). The action potentials of these mechanosensitive cells are propagated away from the cell body towards the central nervous system (Fig. 36). Orchard, I. and Finlayson, L. (1976) termed these neurons the "peripheral nerve stretch receptors".

3.1.4 Multiterminal Neurons Associated with Sensilla

Multiterminal neurons are associated with the axons from scolopophorous sensilla in the mandibles of *Locusta migratoria* (LeBerre, J. and Louveaux, A., 1969) and the beetle, *Ctenicera* (Zacharuk, R., 1980), and the cibarial chemosensilla of *Blaberus* (Moulins, M., 1971) and *Locusta* (Cook, A., 1972, as interpreted by Zacharuk, R., 1980). In *Blaberus craniifer* other type II neurons occur with connective tissue strands in the wall of

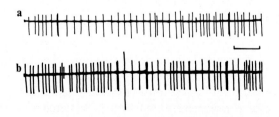

FIG. 34. The effects of calcium ions on the electric activity of neurosecretory nerves in *Carausius*. **a**: Spontaneous activity in normal saline; **b**: same preparation after 30 s in a high calcium saline (75 mM). Note the increase in amplitude of the action potentials. (Scale bar = 0.5 s, 100 μV.) (Redrawn after Orchard, I. and Osborne, M., 1977.)

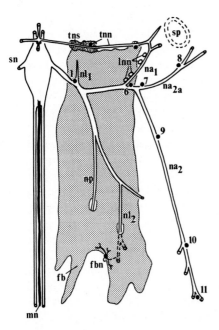

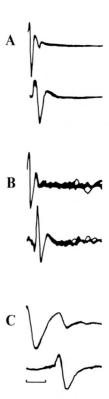

FIG. 35. Diagram of innervation of one side of an abdominal segment of *Carausius morosus* showing the position of neurons associated with major nerves. fb = Fat body, fbn = fat body neuron, lnn = link nerve neurosecretory neurons, mn = median nerve, na, na₂, na₂ₐ, nl₁, nl₂ and np = nerves labeled according to Marquhardt's terminology, sn = segmental nerve, sp = spiracle, tnn = transverse nerve neuron, tns = transverse nerve neurohaemal swelling: 1, 6, 7, 8, 9, 10, 11 labeled number of peripheral non-neurosecretory neurons. (Redrawn from Orchard, I. and Finlayson, L., 1976.)

FIG. 36. Simultaneous recordings from two sites on nerve na₂ of *Carausius* showing spike from: **A**: motor neuron; **B**: dorsal longitudinal stretch receptor neuron; **C**: link nerve neurosecretory neuron. The lower trace in each of **A**, **B**, and **C** is triggered by the action potential in the upper trace in **A**, **B** and **C**. The electrodes were 0.6 mm apart. All traces filmed at the same sweep speed. Note the slow conduction velocity of the neurosecretory action potential indicated in **C**; note also the apparent long duration of this action potential. (Time mark = 2 ms.) (Redrawn from Orchard, I. and Finlayson, L., 1976.)

the cibarium (Moulins, M., 1971, 1974), a structure reminiscent of strand stretch receptors.

In each mandible of larval *Ctenicera* is a centrally situated multiterminal bipolar type II neuron (Zacharuk, R., 1980). It is wrapped by a glial cell to the base of the two terminal scolopophorous sensilla in the terminal mandibular tooth. The terminal branches of the neuron are naked and extend along the outer surfaces of the inner and outer sheath cells and of the adjacent surfaces of the epidermal cells around both sensilla. Peripheral mitochondria, longitudinal microtubules and clear and various dense vesicles occur in the dendrite and its branches. Newly molted larvae have more dense vesicles than intermolt larvae. In the sheath and epidermal cells adjacent to the naked dendritic branches are unique plates of endoplasmic reticulum and vesiculating bodies. Possibly this neuron controls the secretory activities of the sensillar sheath cells and adjacent

epidermal cells through release of appropriate chemical mediators (Zacharuk, R., 1980).

3.2 Multiterminal neurons attached to accessory structures

The accessory component of a multiterminal neuron may be a strand or tube of connective tissue, or a special receptor muscle, or a combination of connective tissue and muscle (Finlayson, L., 1968). Types of neurons are frequently designated on the basis of the kind of accessory structure, e.g. "strand stretch receptor" or "muscle receptor". The simplest type is a strand of connective tissue to which the neuron is attached, as occurs in Dictyoptera, Plecoptera, and Coleoptera (Finlayson, L., 1976).

In the exopterygote orders of Phasmida, Orthoptera and Hemiptera, multiterminal neurons and their connective tissue accessory components are attached to ordinary muscles of the body. In the Orthoptera (Slifer, E. and Finlayson, L., 1956) the muscle fiber to which the neuron is attached has less conspicuous striations than the normal fibers of the rest of the muscle band. In the endopterygote orders of Neuroptera, Trichoptera and Lepidoptera a special receptor muscle occurs that is independent of the ordinary muscles of the body.

3.2.1 MOUTHPARTS (SCLEROTIZED) AND VISCERA

Among the sclerotized mouthparts stretch receptors are known from the mandibles of two species of beetles (Honomichl, K., 1976, 1978a,b) and the labium and labrum of a termite (Richards, G., 1951). The motion of the mandible of *Dermestes maculatus* is probably controlled by about 10 multiterminal neurons which send dendritic processes into the interior of the third mandibular muscle where they periodically ramify in the levels of the Z bands (Honomichl, K., 1976) (Fig. 37). Possibly the dendritic terminations detect muscular activity by detecting the ionic changes associated with contraction.

In *Oryzaephilus surinamensis* the mandibular movement is presumably controlled by three distinct multiterminal receptors. The ventral muscle

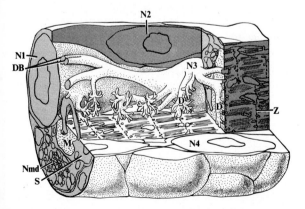

FIG. 37. Diagram of stretch receptor associated with mandible of *Dermestes maculatus*. D = Dendrite; DB = dendritic bundle; M = tentoriomandibularis muscle; N 1, 2, 3, 4 = neurons 1, 2, 3, 4; Nmd = mandibular nerve; S = Schwann cell; Z = Z band surface. (Redrawn from Honomichl, K., 1976.)

receptor consists of a muscle fiber and eight multiterminal neurons. The muscle fiber goes from the anterior tentorium arm to the ventral basis of the mandible. One group of dendrites extends into the interior of the muscle while another group forms a compact cord along the ventral side of the muscle. The mandibular tendon receptor has three neurons, whose dendrites are stretched between the anterior tentorium arm and the tendon of the mandibular adductor muscle. The third receptor is composed of two muscle fibers between the anterior arm of the tentorium and the dorsal base of the mandible, and a neuron, which sends a dendrite to the muscle insertion. The ramifications of the dendrite lie between the touching surfaces of muscle fibers and epidermal cells.

3.2.2 PREORAL CAVITY

Stretch receptors have been described associated with various parts of the preoral cavity of a number of species including a termite, *Calotermes flavicollis* (Richards, G., 1951), a cockroach, *Blaberus craniifer* (Moulins, M., 1966, 1974), a tsetse fly, *Glossina austeni* (Rice, M., 1970), and the blowflies *Calliphora erythrocephala* (Rice, M., 1970; Rice, M. and Finlayson, L., 1972) and *Phormia regina* (Gelperin, A., 1967). In the preoral cavity (cibarium) of *B. craniifer* are four complex stretch receptors, one in the epipharynx and three in the hypopharynx (Rice, M., 1970). These receptors are associated with transverse tissue strands that are attached at each extremity to the integument by a conjunctivo-epidermo-cuticular junction identical to the myo-epidermo-cuticular junction of muscle attachments. Each receptor may have up to fourteen multiterminal cells with a variety of morphological properties. Some of these cells occur in anatomical units that are not ganglia. Moulins, M. (1974) suggested that these complex receptors are probably able to insure some peripheral analysis of the mechanical events of feeding.

Three multiterminal neurons are on each side of the anterior wall of the cibarial pump of the tsetse fly and the blowfly (Rice, M., 1970). Simulated cibarial pumping evokes bursts of action potentials. The discharge frequency of each neuron is proportional to the extent of indentation of the anterior wall and to the point of stimulation. The activity patterns of the neurons differ with the region of the anterior wall stimulated (Rice, M., 1970). In the blowfly the slowly adapting cibarial pump neurons monitor sinusoidal stimuli up to a frequency of 23 Hz (Rice, M. and Finlayson, L., 1972). They monitor the amplitude and direction of each stimulus, being insensitive to the velocity of stretch. However, rebound is related to the velocity of relaxation, a likely important factor in the neural coordination of cibarial pumping (Rice, M. and Finlayson, L., 1972). In both the blowfly and the tsetse fly the cibarial pump multiterminal neurons probably provide a means of accurately measuring the volume of food ingested and their sensory input may consequently influence the production of digestive enzyme and satiation behavior (Rice, M., 1970).

3.2.3 GUT

The foregut of insects is innervated by an abundance of sensory nerves including numerous multiterminal ones (Wales, W., 1976). In *Acheta domesticus* the foregut proprioceptors respond to gut movements, being responsive to cyclical movements but not to maintained tension (Möhl, B., 1972). Stretching the foregut of *Acheta* causes a decrease in motor activity in the oesophageal nerves. At higher rates of stretch a short increase may precede this decrease. A qualitative change does not occur upon removal of the ventricular and hypocerebral ganglia of the stomodaeal nervous system, the nervous system that innervates the foregut of insects.

In the hindgut of the larval beetle, *Oryctes nasicornis*, are multiterminal neurons that monitor the volume of the rectal ampulla and the flow of faeces through the anus (Nagy, F., 1974, as cited by Wales, W., 1976). Other multiterminal neurons occur in association with the extrinsic muscle of the rectal ampulla. These are a muscle receptor organ with a single multiterminal neuron and a stretch receptor with two multiterminal neurons, a primary cell located proximally and a smaller, secondary cell located distally. The muscle receptor organ and stretch receptor attach between the lateral muscles of the rectal ampulla and the wall of the ampulla, and reflexly excite the muscles.

3.2.4 ABDOMEN AND THORAX

Abdominal stretch receptors in insects were first seen by Rogosina, M. (1928) and Hertweck, H. (1931) in larval dragonflies and fruit-flies, respectively, but not until their discovery by Finlayson, L. and Lowenstein, O. (1955) in a moth were they recognized as such. Stretch receptors in the abdomen have been extensively studied: those in the thorax less so. The literature on thoracic and abdominal receptors has been reviewed by Finlayson, L. (1968, 1976) and Osborne, M. (1970).

Complexity of stretch receptors of the body segments ranges from simple innervated connective tissue strands to the complicated muscle receptor organ of Lepidoptera (Finlayson, L., 1968) which resembles the vertebrate muscle spindle (Finlayson, L. and Lowenstein, O., 1958). Usually stretch receptors consist of a strand of connective tissue in which the dendrites of a single neuron ramify (Finlayson, L., 1968). Most often the cell body is located in or on the connective tissue strand but it may be situated on the neighboring epidermis or in the nerve itself (Osborne, M. and Finlayson, L., 1962).

Three types of abdominal stretch receptors have been found; *viz.*, dorsal and ventral longitudinal and vertical. A pair of dorsal longitudinal stretch receptors has been found in each full-sized abdominal segment in species from 14 orders of insects (Anwyl, R., 1972; Finlayson, L. and Lowenstein, O., 1958; Osborne, M., 1963b; Osborne, M. and Finlayson, L., 1962; Slifer, E. and Finlayson, L., 1956). These receptors consist of connective tissue plus neuron, and some have a muscular component as in Orthoptera, Neuroptera, Trichoptera, and Lepidoptera. In some insects the receptor occurs between the intersegmental folds, and in others the anterior attachment is to the tergal epidermis and the posterior attachment to the intersegmental fold. The receptors with a muscle component extend between the intersegmental folds.

Dorsal longitudinal receptors occur in the thorax of adult stick insects (Osborne, M. and Finlayson, L., 1962); stonefly larvae (Osborne, M. and Finlayson, L., 1962); moth larvae (Finlayson, L. and Lowenstein, O., 1960); and blowfly larvae (Osborne, M., 1963b). Similar receptors have been observed in the thorax of the desert locust that may be homologous with the dorsal abdominal series

(Gettrup, E., 1962, 1963; Wilson, D. and Gettrup, E., 1963). The widespread occurrence of the dorsal longitudinal series of stretch receptors probably indicates that they were derived from structures in the ancestors of insects or originated at an early stage in the evolution of insects (Finlayson, L., 1968).

Vertical receptors are situated approximately at right angles to the dorsal longitudinal receptors. With the exception of *Rhodnius prolixus* (Anwyl, R., 1972), a vertical receptor occurs in all the exopterygote insects which have been examined and in the endopterygote order Coleoptera (Osborne, M. and Finlayson, L., 1962). These receptors never have a muscular component consisting only of a neuron and connective tissue. Interestingly in those insects in which the dorsal longitudinal receptor has a muscular component the vertical receptor is absent except in the larva of the alder-fly *Sialis* (Neuroptera) in which the vertical receptor consists of the neuron alone (Finlayson, L., 1976).

A series of paired ventral longitudinal receptors occurs in the blowfly larva (Osborne, M., 1963b). Corresponding receptors are lacking in the tsetse fly (Finlayson, L., 1972). Each ventral longitudinal receptor consists of a tubular connective tissue strand and a neuron. Both the dorsal and ventral longitudinal receptors span the entire segment but differ in their mode of attachment. The ventral receptor is anchored posteriorly to the outer edge of an oblique muscle and anteriorly by a three-point attachment. Two points are on the epidermis of its own segment and one point is in the adjacent segment on the inner edge of an oblique muscle. A single muscle fiber enters the strand of connective tissue both anteriorly and posteriorly to the oblique muscle.

3.2.4.1 *Fine structure of abdominal stretch receptors.* Finlayson, L. (1968, 1976) discusses in detail the structure, topography and physiology of abdominal stretch receptors in a variety of insects. Attention here is limited to the dorsal longitudinal stretch receptors of the cockroach (Osborne, M., 1963a) and the moth (Osborne, M. and Finlayson, L., 1965) since these are the only two receptor organs which have been investigated with the electron microscope. Both consist of a single multiterminal neuron with its dendrite associated with connective tissue. In addition, the moth receptor has a muscular component so that it is indeed a "muscle receptor

organ". Consequently, the cockroach and moth receptors illustrate the two morphological types of abdominal stretch receptors. Figures 38 and 39 illustrate the general anatomical features of the two receptors. The muscle receptor organs of larval and pupal *Antheraea pernyi* each consist of a multipolar neuron whose dendrites are wrapped by Schwann cells and are associated with a tube of connective tissue, the fiber-tract (Osborne, M. and Finlayson, L., 1965). In turn, the fiber tract is attached to the edge of a single muscle fiber. In the lumen of the fiber tract are the tract cell and the dendrites. In the central region are two giant nuclei; one is the nucleus of the muscle cell and the other is the nucleus of the tract cell. Bundles of dense connective tissue fibrils, the fiber bundles, occur at the junction between the fiber tract and the muscle cell. The fiber bundles terminate towards the end of the fiber tracts. The dendrites run alongside the fiber bundles and throughout their length branch repeatedly, resulting in smaller dendrites, the terminal regions of which lack Schwann cells. These "naked" dendritic terminals are closely apposed to the peripheral surface of the fiber bundles or penetrate the connective tissue of the fiber bundles.

The receptor of the cockroach, *Blaberus craniifer*, consists of a single multipolar neuron associated

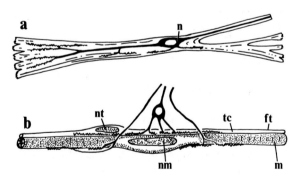

FIG. 38. Diagram to show the general structure of the cockroach (**a**) and moth (**b**) longitudinal stretch receptors as they appear after staining in methylene blue. The cockroach, *Blaberus*, receptor consists of a single multiterminal neuron (n) with its dendrites associated with a strand of connective tissue. The moth, *Antheraea*, receptor consists of a muscle fiber (m) along one side of which runs a tube of connective tissue, the fiber tract (ft). The lumen of the tube is occupied by the tract cell (tc). Both the giant nucleus (nm) of the muscle fiber, and smaller fiber tract cell nucleus (nt) are shown. The multiterminal neuron sends its dendrites into the fiber tract. Motor innervation of the receptor muscle is also shown. (Redrawn after Osborne, M., 1970.)

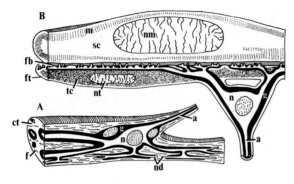

Fig. 39. Diagrams to show the main ultrastructural features of the cockroach (**A**) and moth (**B**) stretch receptors. Only the regions in the vicinity of the sensory neuron (n) are shown. In the cockroach the neuron and glial cell sheath (g) are embedded in the connective tissue strand (ct). The tips of the dendrites are without the glial cell sheath, and are buried in the matrix of the connective tissue. No connections are observed between the naked dendritic tips (nd) and the connective tissue fibrils (f). In the moth the dendritic terminals are also naked and are associated with dense accumulations of connective tissue fibers, the fiber bundles (fb). In the central region the muscle fiber is filled with clear sarcoplasm (sc), the sarcoplasmic core, and the giant nucleus (nm). The fiber tract (ft) contains the main dendrites and the tract cell (tc), whose nucleus (nt) is shown. (a) = axon of sensory cell. (Redrawn after Osborne, M., 1970.)

with a strand of connective tissue (Fig. 39) (Osborne, M., 1963a). The axon, cell body, and dendrites of the neuron are ensheathed by Schwann cells. The tips of the dendrites lack Schwann cell investment, are embedded in the connective tissue matrix and contain prominent mitochondria. Unlike the naked dendritic endings of the moth's muscle receptor organs, those of the cockroach receptor are not connected to the connective tissue fibrils.

3.2.4.2 *Physiology of muscle receptor organs.* The muscle receptor organs of the pupal moth *Antheraea pernyi*, as well as those of the larva of the dragonfly, *Aeschna juncea*, and of the cockroach, *Blaberus craniifer*, supply information both on displacement and on the rate of change of length, that is, velocity of stretch (Lowenstein, O. and Finlayson, L., 1960). The same authors found that there apparently is a basic minimum resting discharge which is evoked when an isolated caterpillar muscle receptor organ is put under slight tension. In high Na^+/low Mg^{2+} this basic discharge frequency is around 20–30 imp s^{-1}. In a saline which more closely resembles the moth hemolymph (Weevers, R., 1966a), however,

the resting discharge is lower in frequency.

The muscle receptor organ is a tonic receptor that maintains a resting discharge frequency for a considerable time (Finlayson, L. and Lowenstein, O., 1958; Weevers, R., 1966b). Weevers, R. (1966b) found that the complex response of the caterpillar muscle receptor organ has not only "position" but also "movement" and "acceleration" components. The position response is very linearly related to the length of the receptor and the movement response is related to stretching. These changes likely protect the sense organ during rapid stretching and would also "take up the slack" when it was released (Weevers, R., 1966c).

Weevers, R. (1966c) found that stimulation of the muscle via its nerve resulted in an increased sensory discharge frequency for the muscle receptor organ, and that the extent of this excitation was similar in hemolymph and in pupal saline. If connections with the CNS were intact, stimulation of a muscle receptor organ resulted in a transient reflex inhibition of the tonic discharge of its receptor muscle (Weevers, R., 1966c). When sensory stimulation ceased, tone of the receptor muscle became transiently elevated.

A major function of the caterpillar muscle receptor organ is to activate a negative feedback stretch reflex in the muscles in the body wall (Weevers, R., 1965, 1966d). Stretching of a single muscle receptor organ results in clear reflex changes in activity of at least 32 motor units. The latency of the reflex differs only slightly from one muscle to another. The response has both tonic and phasic components which correlate well with the magnitudes of the same components in the sensory discharge. The reflex pathway is via synaptic connections in the ganglion of the segment anterior to the stimulated receptor and responding muscles (Weevers, R., 1966d).

3.2.5 WINGS AND LEGS

Multiterminal receptors occur at leg joints and at the wing base of a number of insect groups, and probably are universal throughout the Class. These receptors usually consist of one or two neurons whose dendrites are commonly associated with a connective tissue strand, and although the cells may differ in detail, they are generally similar to the multiterminal stretch receptors of abdominal segments (Wright, B., 1976).

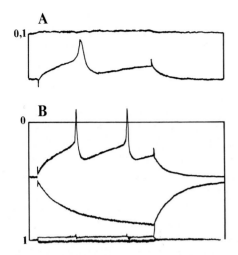

FIG. 40. Electrical excitability in the developing stretch in an embryo of *Schistocerca* species. **A**: Action potential elicited by a current pulse in a 60% embryo. The resting potential of this cell is *ca*. 50 mV. Zero potential (0) and the current monitor (1) are in the upper trace. **B**: Overshooting action potentials in a 70% embryo. Both hyperpolarizing and depolarizing pulses of the same magnitude are applied to the cell. The resting potential of this cell is ca. 55 mV and the input resistance as measured from this record is 700 MΩ. Voltage calibration: 40 mV. (Current calibration = 1 nA; time calibration = 40 ms.) (Redrawn from Heathcote, R., 1981.)

Differentiation of the grasshopper wing hinge stretch receptor has been followed throughout embryogenesis (Heathcote, R., 1981). The axon grows from the peripheral cell body to the CNS where it undergoes extensive arborization. The earliest time an action potential could be elicited by current injection was at 60% embryogenesis (Fig. 40). These early potentials have a very low threshold (less than 0.1 nA) as do ones recorded at 65%, 70% and 75% embryogenesis. The latter potentials usually overshoot the zero potential (Fig. 40) indicating that, unlike many other insect neurons, this sensory cell has an active cell membrane (Heathcote, R., 1981).

At the base of each wing in locusts is a single-celled stretch receptor neuron that monitors wing elevation and contributes to the control of the flight motor output (Altman, J. and Tyrer, N., 1978a). The hindwing stretch receptor projects to the second and third thoracic ganglia and the forewing stretch receptor to all three thoracic ganglia. Both stretch receptors send fine axons into the abdominal connectives. Within the ganglia Altman, J. and Tyrer, N. (1978a) have mapped for each stretch

receptor its arborization which is entirely ipsilateral and mainly in the dorsal neuropile being divided into medial, mediolateral and lateral branches. Within the neuropile there may be labelled sites which the growing tips of the stretch receptor neurons seek out (Altman, J. and Tyrer, N., 1978b). The central arborizations are very consistent from one individual to another. Only three "mistakes" were observed in a number of stretch receptor projections (Altman, J. and Tyrer, N., 1978b).

The stretch receptor neurons of the wing hinge are apparently multifunctional with differential information transfer resulting from both the spatial distribution of synaptic connections with the motor neurons and from filtering caused by low safety factors at branch junctions. Information in the lateral branching may be used for general excitation and control of firing frequency of the motor neurons, and in the medial branch for wing control and coordination (Altman, J. and Tyrer, N., 1978a).

Gettrup, E. (1962, 1963) and Pabst, H. (1965) found that in locusts the hinge stretch receptor in the region of the subalar sclerite responds to elevation of the wing, and during flight produces a few strokes near the end of each upstroke. The initial phasic response adapts to a tonic level, if the wing is elevated, and the frequency depends upon the extent of elevation. Consequently, the stretch receptor provides information concerning the amplitude and velocity of each wing upstroke.

The various multiterminal receptors in the legs of locusts have been particularly well studied (Bräunig, P., 1982; Bräunig, P. *et al.*, 1981; Coillot, J., 1974, 1975; Coillot, J. and Boistel, J., 1968, 1969; Williamson, R. and Burns, M., 1978). In *Schistocerca gregaria* three stretch receptors are situated at the level of the tibiofemoral joint of the jumping leg; two of the receptors have two multipolar cells and the third only one (Coillot, J. and Boistel, J., 1968). The activity of the receptors is separable into phasic–tonic and tonic components. The former is related to the amplitude and velocity of the movements and the latter to the stretch level of the receptors (stretch increasing with femoro-tibial angle) (Coillot, J. and Boistel, J., 1969). These receptors probably function in the control of the locust jump (Coillot, J. 1974, 1975).

Similar receptors exist in the tibio-femoral joint of the non-jumping mesothoracic legs of the same locust species (Williamson, R. and Burns, M., 1978). Three of the five multiterminal neurons respond to tibial movement and position in a phasic–tonic manner and mediate a weak reflex upon two of the extensor tibia motoneurons (Williamson, R. and Burns, M., 1978). Theophilidis, G. and Burns, M. (1979) showed that a single multipolar neuron in the cuticular end of the distal flexor tibiae muscle of the pro- and mesothoracic legs of *Schistocerca* is a tonic receptor for active and passive tension in the muscle fibers to which it is attached. This muscle tension receptor generally causes reflex excitation of flexor motoneurons and inhibition of the slow extensor neuron, although the sign of the reflexes can be reversed (Theophilidis, G. and Burns, M., 1979). The central projections of strand receptors as well as multiterminal sensilla in the proximal leg joints of both *Locusta* and *Schistocerca* have been determined (Bräunig, P. *et al.*, 1981).

Using both sinusoidal and pseudo-random mechanical stimuli for linear systems analysis Kuster, J. and French, A. (1983) studied the dynamic behavior of a multipolar mechanosensillum in the femoro-tibial joint of *Locusta migratoria*. The sensillum is a tonic receptor and dynamic movements of the joint position produces modulation of the firing rate around the mean level which allows faithful reproduction of the input signal in the probability of action potential occurrence. At frequencies of about 1 Hz and above, the response increasingly phase-locks to a repetitive signal. The dynamic behavior of the sensillum is well characterized as fractional differentiation for sinusoidal stimuli up to 1 Hz and for random stimuli up to 10 Hz which supports the original findings of Coillot, J. (1974). Consequently, there is a frequency response function which can be represented by the gain at a frequency of 1 radian per second and a fractional exponent of frequency.

In orthopterans from five different families Bräunig, P. (1982) found one strand receptor associated with the trochantin while two others are situated in the coxa. In addition to these sense organs, the coxa contains a multiterminal stretch receptor which spans the coxotrochanteral joint. Similar neurons occur in the coxae of termites (Denis, C., 1958). In *Periplaneta americana* a large multiterminal stretch receptor near the condyle of the femoro-tibial articulation responds with low-frequency, slow-adapting trains of impulses to tibial levation (Guthrie, D., 1967). Fast-adapting receptors may also occur in the same region (Guthrie, D., 1967).

3.3 Transduction

The absence of cilia and accessory structures in multiterminal sensilla indicates that the mechanical stimulus probably acts directly on the dendritic membrane. Rice, M. *et al.* (1973) suggested that the effective stimulus is a very slight deformation of the naked dendritic tip so as to increase membrane capacitance of conductance. Sugawara, T. (1981) raised the point that it is difficult to understand how impulse generation in multiterminal sensilla, in which the dendrites are quite long in proportion to the cell body, can be due only to the deformation of the small naked dendritic tips. For the butterfly bursal stretch receptor Sugawara, T. (1981) postulated that the chain of varicosities, which are often covered only by the double membrane of a Schwann cell, in the dendrite transduce the stimulus of stretch by deformation of their membranes. Mitochondria in the varicosities may control the ionic level as well as supply the necessary metabolic energy. If the stretch applied to the dendrite causes the deformation of all varicosities, then the membrane may depolarize as proposed for the frog muscle spindle by Katz, B. (1961).

Kuster, J. and French, A. (1983) found that fractional differentiation occurs in multiterminal sensilla as it does in Type I sensilla (Chapman, K. and Smith, R., 1963; French, A., 1980; French, A. and Kuster, J., 1981; French, A. and Wong, R., 1977). This suggests that transduction may be quite similar in the two types of sensilla. The gain at any frequency is always about five times less in Type II than in Type I sensilla, indicating that the dendritic membrane may be stretched less efficiently in the multiterminal sensilla to allow for an increase in ions for very small displacement. This would presumably be related to the lack of accessory structures to give rigidity to the Type II sensilla.

4 SUGGESTIONS FOR FUTURE WORK

The efforts of a number of investigators working

primarily in the past three decades have elucidated the fundamentals of the structure and function of the various types of mechanosensilla, and in certain instances have provided quite sophisticated insights into their dynamic properties. Fruitful areas for future endeavors include determining the embryonic derivation of the multiterminal sensilla as well as the adaptive significance of the cuticular portions of the unusual cuticular mechanosensilla. Relatively few electrophysiological studies have been conducted on chordotonal organs. It would be most interesting to know if fractional differentiation occurs during transduction in chordotonal organs. If so, is the gain at any frequency closer to that of cuticular mechanosensilla than multiterminal sensilla, as would be expected on the basis of degree of relatedness of the sensillar types? The ciliary necklace region of the dendrite in both scolopidia and cuticular mechanosensilla warrants more attention. Does it have the same structure and degree of prominence in the two types of sensilla? Does the ciliary necklace participate in transduction, in the production of a generator potential, or any other functional step? Considerable insight into the dynamics of the tubular body would result from knowing the chemical composition of the electron-dense material, and the details of the interaction between the microtubules themselves and between the peripheral microtubules and the dendritic membrane during deformation. Are there rachets on the microtubules that account for directionality in some hair sensilla? Computer simulation of the dynamic properties of chordotonal organs, as well as the multiterminal sensilla, would enhance understanding of these sensilla. What is the significance of particular numbers or groups of sensilla? Would it be possible and of any value to do computer simulations of the total mechanosense of an insect?

ACKNOWLEDGEMENTS

The author thanks Mrs. Nina Murray for secretarial assistance, and the Medical Research Council (Canada), grant No. MT 2909, for financial support.

REFERENCES

ALEXANDROWICZ, J. S. (1957). Notes on the nervous system in the Stomatopoda. V. The various types of sensory nerve cells. *J. Mar. Biol. Assoc. U.K. 36*, 603–628.

ALTMAN, J. S. and TYRER, N. M. (1978a). The locust wing hinge stretch receptors I. Primary sensory neurones with enormous central arborizations. *J. Comp. Neurol. 172*, 409–430.

ALTMAN, J. S. and TYRER, N. M. (1978b). The locust wing hinge stretch receptors II. Variation, alternative pathways and "mistakes" in the central arborizations. *J. Comp. Neurol. 172*, 431–440.

ANDERSON, B. R. (1980). A scanning electron microscope study of the mesothoracic and metathoracic scolopophorous organs of *Lethocerus* (Belostomatidae–Heteroptera). *J. Morph. 163*, 27–35.

ANDERSON, M. and FINLAYSON, L. H. (1978). Topography and electrical activity of peripheral neurons in the abdomen of the tsetse fly (*Glossina*) in relation to abdominal distension. *Physiol. Ent. 3*, 157–167.

ANDERSON, R. G. W. and HEIN, C. E. (1977). Distribution of anionic sites on the oviduct ciliary membrane. *J. Cell Biol. 72*, 482–492.

ANWYL, R. (1972). The structure and properties of an abdominal stretch receptor in *Rhodnius prolixus*. *J. Insect Physiol. 18*, 2143–2154.

BACON, J. and TYRER, M. (1979). The innervation of the wind-sensitive head hairs of the locust, *Schistocerca gregaria*. *Physiol. Ent. 4*, 301–309.

BALL, E. E. (1981). Structure of the auditory system of the weta *Hemideina crassidens* (Blanchard 1851) (Orthoptera, Ensifera, Gryllacridoidea, Stenopelmatidae). 2. Ultrastructure of the auditory sensilla. *Cell Tiss. Res. 217*, 345–359.

BALL, E. E. and FIELD, L. H. (1981). Structure of the auditory system of the weta *Hemideina crassidens* (Blanchard, 1851) (Orthoptera, Ensifera, Gryllacridoidea, Stenopelmatidae). 1. Morphology and histology. *Cell Tiss. Res. 217*, 321–343.

BÄSSLER, U. (1958). Versuche zur Orientierung der Stechmücken: Die Schwarmbildung und die Bedeutung des Johnstonschen Organs. *Z. Vergl. Physiol. 41*, 300–330.

BÄSSLER, U. (1979). Effects of crossing the receptor apodeme of the femoral chordotonal organ on walking, jumping and singing in locusts and grasshoppers. *J. Comp. Physiol. 134*, 173–176.

BÄSSLER, U. and STORRER, J. (1980). The neural basis of the femur-tibia-control-system in the stick insect *Carausius morosus*. I. Motoneurons of the extensor tibiae muscle. *Biol. Cybern. 38*, 107–114.

BELTON, P. (1974). An analysis of direction finding by male mosquitoes. In *Experimental Analysis of Insect Behavior*. Edited by L. B. Brown. Pages 139–148. Springer, Berlin and New York.

BENTLEY, D. (1975). Single gene cricket mutations: effects on behavior, sensilla, sensory neurons, and identified interneurons. *Science 187*, 760–764.

BERLESE, A. (1909). *Gli insetti. I. Embriologia e morfologia*. Societa Editrice Libraria, Milan.

BERNARD, J. and GUILLET, J. C. (1972). Changes in the receptor potentials under polarizing current in two insect receptors. *J. Insect Physiol. 18*, 2173–2187.

BERNARD, J. and PINET, J.-M. (1973). Origine des potentiels propagés dans un mécanorécepteur á dendrite longue. *J. Comp. Physiol. 87*, 361–377.

BERNARD, J., GUILLET, J. C. and COILLOT, J. P. (1980). Evidence for a barrier between blood and sensory terminal in an insect mechanoreceptor. *Comp. Biochem. Physiol. 67A*, 573–579.

BERNAYS, E. A., COOK, A. G. and PADGHAM, D. E. (1976). A club-shaped hair found on the first-instar nymphs of *Schistocerca gregaria*. *Physiol. Ent. 1*, 3–13.

BISCHOF, H. J. (1975). Club-shaped hairs in cerci of cricket *Gryllus bimaculatus* acting as gravity receptors. *J. Comp. Physiol. 98*, 277–288.

BLANEY, W. M. and CHAPMAN, R. F. (1969). The anatomy and histology of the maxillary palp of *Schistocerca gregaria*. *J. Zool. 157*, 509–535.

BLOOM, J. W., ZACHARUK, R. Y. and HOLODNIUK, A. E. (1981). Ultrastructure of a terminal chordotonal sensillum in larval antennae of the yellow mealworm, *Tenebrio molitor* L. *Canad. J. Zool. 59*, 515–524.

BOISVIEUX-ULRICH, E., SANDOZ, D. and CHAILLEY, B. (1977). A freeze-fracture and thin section study of the ciliary necklace in quail oviduct. *Biol. Cellulaire 30*, 245–252.

BONKE, D. (1975). Der Bau und die Antwortcharakteristik des Schirmrezeptors auf dem Statoorgansystem von *Nepa cinerea* L. (Hemiptera, Rhynchota). *Verh. Dtsch. Zool. Ges. 1974*, 42–45.

BOO, K. S. (1980). Antennal sensory receptors of the male mosquito, *Anopheles stephensi. Z. Parasitenkd. 61*, 249–264.

BOO, K. S. (1981). Discontinuity between ciliary root processes and triple microtubules of distal basal body in mosquito sensory cilia. *Korean J. Ent. 11*, 5–18.

BOO, K. S. and DAVIES, D. M. (1980). Johnston's organ of the black fly *Simulium vittatum* Zett. *Canad. J. Zool. 58*, 1969–1979.

BOO, K. S. and RICHARDS, A. G. (1975a). Fine structure of the scolopidia in the Johnston's organ of male *Aedes aegypti* (L.). (Diptera: Culidiae). *Int. J. Insect Morph. Embryol. 4*, 549–566.

BOO, K. S. and RICHARDS, A. G. (1975b). Fine structure of scolopidia in Johnston's organ of female *Aedes aegypti* compared with that of the male. *J. Insect Physiol. 21*, 1129–1139.

BRÄUNIG, P. (1982). Strand receptors with central cell bodies in the proximal leg joints of orthopterous insects. *Cell Tiss. Res. 222*, 647–654.

BRÄUNIG, P. and HUSTERT, R. (1980). Proprioceptors with central cell bodies in insects. *Nature 283*, 768–770.

BRÄUNIG, P., HUSTERT, R. and PFLÜGER, H. J. (1981). Distribution and specific central projections of mechanoreceptors in the thorax and proximal leg joints of locusts. I. Morphology, location and innervation of internal proprioceptors of pro- and metathorax and their central projections. *Cell Tiss. Res. 216*, 57–77.

BROMLEY, A. K., DUNN, J. A. and ANDERSON, M. (1980). Ultrastructure of the antennal sensilla of aphids. II. Trichoid, chordotonal and campaniform sensilla. *Cell Tiss. Res. 205*, 493–511.

BULLOCK, T. H. and HORRIDGE, G. A. (1965). *Structure and Function in the Nervous System of Invertebrates.* Freeman, San Francisco.

BUNO, W., JR., MONTI-BLOCH, L. and CRISPINO, L. (1981a). Dynamic properties of cockroach cercal "bristlelike" hair sensilla. *J. Neurobiol. 12*, 101–121.

BUNO, W., JR., MONTI-BLOCH, L., MATEOS, A. and HANDLER, P. (1981b). Dynamic properties of cockroach cercal "threadlike" hair sensilla. *J. Neurobiol. 12*, 123–141.

BURKHARDT, D. and GEWECKE, M. (1965). Mechanoreception in Arthropoda: The chain from stimulus to behavioral pattern. *Cold Spr. Harb. Symp. Quant. Biol. 30*, 601–614.

BURKHARDT, D. and SCHNEIDER, G. (1957). Die Antennen von *Calliphora* als Anzeiger der Fluggeschwindigkeit. *Z. Naturf. (B) 12*, 139–143.

BURNS, M. D. (1974). Structure and physiology of the locust chordotonal organ, *J. Insect Physiol. 20*, 1319–1339.

BURT, R. and PALKA, J. (1982). The central projections of mesothoracic sensory neurons in wildtype *Drosophila* and *bithorax* mutants. *Devel. Biol. 90*, 99–109.

CAMHI, J. M. and TOM, W. (1978a). Escape behavior of the cockroach *Periplaneta americana*. 1. Turning response to wind puffs. *J. Comp. Physiol. 128*, 193–202.

CAMHI, J. M., TOM, W. and VOLMAN, S. (1978b). Escape behavior of the cockroach *Periplaneta americana*. 2. Detection of natural predators by air displacements. *J. Comp. Physiol. 128*, 203–212.

CHAPMAN, K. M. (1965). Campaniform sensilla on the tactile spines of the legs of the cockroach. *J. Exp. Biol. 42*, 191–203.

CHAPMAN, K. M. and DUCKROW, R. B. (1975). Compliance and sensitivity of a mechanoreceptor of the insect exoskeleton. *J. Comp. Physiol. 100*, 251–268.

CHAPMAN, K. M. and PANKHURST, J. H. (1967). Conduction velocities and their temperature coefficients in sensory nerve fibres of cockroach legs. *J. Exp. Biol. 46*, 63–84.

CHAPMAN, K. M. and SMITH, R. S. (1963). A linear transfer function underlying impulse frequency modulation in a cockroach mechanoreceptor. *Nature 1977*, 699–700.

CHAPMAN, K. M., DUCKROW, R. B. and MORAN, D. T. (1973). Form and role of deformation in excitation of an insect mechanoreceptor. *Nature 244*, 453–454.

CHAPMAN, K. M., MOSINGER, J. L. and DUCKROW, R. B. (1979). The role of distributed viscoelastic coupling in sensory adaptation in an insect mechanoreceptor. *J. Comp. Physiol. 131*, 1–12.

CHEVALIER, R. L. (1969). The fine structure of campaniform sensilla on the halteres of *Drosophila melanogaster. J. Morph. 128*, 443–464.

CHI, C. and CARLSON, S. D. (1976). The housefly interfacetal hair. Ultrastructure of a presumed mechanoreceptor. *Cell Tiss. Res. 166*, 353–363.

CHILD, C. M. (1894). Ein bisher wenig beachtetes antennales Sinnesorgan der Insekten mit besonderer Berucksichtigung der Culiciden und Chironomiden. *Z. Wiss. Zool. 58*, 475–528.

CHU-WANG, I.-WU. and AXTELL, R. C. (1971). Fine structure of the dorsal organ of the housefly larva, *Musca domestica* L. *Z. Zell. Mikr. Ant. 117*, 17–34.

CHU-WANG, I.-WU. and AXTELL, R. C. (1972a). Fine structure of the terminal organ of the housefly larva, *Musca domestica* L. *Z. Zell. Mikr. Ant. 127*, 287–305.

CHU-WANG, I.-WU. and AXTELL, R. C. (1972b). Fine structure of the ventral organ of the housefly larva, *Musca domestica* L. *Z. Zell. Mikr. Ant. 130*, 489–495.

CHU-WANG, I.-WU. and AXTELL, R. C. (1973). Comparative fine structure of the claw sensillum of a soft tick, *Arga (Persicargas) arboreus* Kaiser, Hoogstraal and Kohls, and a hard tick, *Amblyomma americanum* (L.). *J. Parasitol. 59*, 545–555.

CLEMENTS, A. N. (1963). The *Physiology of Mosquitoes*. Pergamon, London.

CLEVER, U. (1958). Untersuchungen zur Zelldifferenzierung und Musterbildung der Sinnesorgane und des Nervensystems in Wachsmottenflügel. *Z. Morph. Tiere 47*, 201–248.

COILLOT, J. P. (1974). Analyse du codage d'un mouvement periodique, par des recepteurs a l'etirement d'un insecte. *J. Insect Physiol. 20*, 1101–1116.

COILLOT, J. P. (1975). La conversion analogique numerique des recepteurs a l'etirement de la patte métathoracique du criquet *Schistocerca gregaria. J. Insect Physiol. 21*, 423–434.

COILLOT, J. P. and BOISTEL, J. (1968). Localisation and description de recepteurs a l'etirement au niveau de l'articulation tibio-femorale de la patte sauteuse du criquet, *Schistocerca gregaria. J. Insect Physiol. 14*, 1661–1667.

COILLOT, J. P. and BOISTEL, J. (1969). Etude de l'activite electrique propagee de recepteurs a l'etirement de la patte metathoracique du criquet, *Schistocerca gregariaa. J. Insect Physiol. 15*, 1449–1470.

COLLIN, S. P. (1981). A proprioceptor with central cell bodies in the cockroach, *Periplaneta americana* (Insecta). *Zoomorphology 98*, 227–231.

COOK, A. G. (1972). The ultrastructure of the A1 sensilla on the posterior surface of the clypeo-labrum of *Locusta migratoria migratorioides* (R and F). *Z. Zell. Mikr. Ant. 134*, 539–554.

CORBIÉRE-TICHANÉ, G. (1971a). Ultrastructure des organes chordotonaux des pieces cephaliques chez la larve du *Speophyes lucidulus* Delar. (Coleoptere Cavernicole de la sous-famille des Bathysciinae). *Z. Zell. Mikr. Ant. 117*, 275–302.

CORBIÉRE-TICHANÉ, G. (1971b). Ultrastructure du systeme sensoriel de la maxille chez la larve du Coleoptere cavernicole *Speophyes lucidulus* Delar. *J. Ultrastruct. Res. 36*, 318–341.

CORBIÉRE-TICHANÉ, G. (1975). L'organe de Johnston chez l'imago de *Speophyes lucidulus* Delar. (Coleoptere cavernicole de la sous-famille des Bathysciinae). Ultrastructure. *J. Mic. 22*, 55–68.

CROUAU, Y. (1980). Comparison of a new structure associated with the membrane of 9 + 0 cilia of chordotonal sensilla with the central structure of motile cilia and flagella. *Biol. Cellulaire 39*, 349–352.

CRUSE, H. (1981). Is the position of the femur-tibia joint under feedback control in the walking stick insect? I. Force measurements. *J. Exp. Biol. 92*, 87–95.

CRUSE, H. and PFLÜGER, H. J. (1981). Is the position of the femur-tibia joint under feedback control in the walking stick insect? II. Electrophysiological recordings. *J. Exp. Biol. 92*, 97–107.

DEBAISEAUX, P. (1938). Organes scolopidiaux des pattes d'insectes. II. *La Cellule 47*, 77–202.

DEBAUCHE, H. (1935). Les organes sensoriels antennaires de *Hydropsyche longipennis. Cellule 44*, 45–83.

DEBAUCHE, H. (1936). Etude cytologique et comparee de l'organe de Johnston des insectes. II. *Cellule 45*, 75–148.

DEMOLL, R. (1917). *Die Sinesorgane der Arthropoden ihr Bau und ihre Funktion.* Vieweg, Braunschweig.

DENIS, C. (1958). Contribution a l'etude de l'ontogenese sensori-nerveuse du termite *Calotermes flavicollis* Fab. *Insectes Sociaux 5*, 171–188.

DOANE, J. F. and KLINGER, J. (1978). Location of CO_2-receptive sensilla on larvae of the wireworms *Agriotes lineaus-obscurus* and *Limonius californicus. Ann. Ent. Soc. Amer. 71*, 357–363.

DRASLAR, K. (1973). Functional properties of trichobothria in the bug *Pyrrhocoris apterus* (L.). *J. Comp. Physiol. 84*, 175–184.

DUDEL, H. (1974). Scolopidien im Funiculus von Dipteren. *Zool. Jb. Ant. 92*, 188–196.

DUMPERT, K. (1972). Bau und Verteilung der Sensillin auf der Antennengeissel von *Lasius fulginosus* (Latr.). *Z. Morph. Tiere 73*, 95–116.

DUMPERT, K. and GNATZY, W. (1977). Cricket combined mechanoreceptors and kicking response. *J. Comp. Physiol. 122*, 9–25.

EBNER, I. and BÄSSLER, U. (1978). Zur Regelung der Stellung des Femur-Tibia-Gelenkes im Mesothorax der Wanderheuschrecke *Schistocera gregaria* (Forskal). *Biol. Cybern. 29*, 83–96.

EDWARDS, J. S. and MANN, D. (1981). The structure of a cercal sensory system and ventral nerve cord of *Grylloblatta*. A comparative study. *Cell Tiss. Res. 217*, 177–188.

EDWARDS, J. S. and PALKA, J. (1974). The cerci and abdominal giant fibres of the house cricket *Acheta domesticus*: I. Anatomy and physiology of normal adults. *Proc. Roy. Soc. London B. 185*, 83–121.

EGGERS, F. (1924). Zur Kenntnis der antennalen stiftführenden Sinnesorgane der Insekten. *Z. Morph. Okol. Tiere 2*, 259–349.

EGGERS, F. (1928). Die stiftführenden Sinnesorgane. *Zool. Bausteine 2*, 1–353.

ERLER, G. and THURM, U. (1977). Eine vereinfachte Methode zur Ableitung von Rezeptorpotentialen und Nervenimpulsen epidermaler Mechanorezeptoren von Insekten. *Z. Naturforsch. 32c*, 1029–1030.

ERLER, G. and THURM, U. (1978). Die Impulsantwort epithelialer Rezeptoren in Abhängigkeit von der transepithelialen Potentialdifferenz. *Verh. Dtsch. Zool. Ges. 71*, 279.

ERLER, G. and THURM, U. (1981). Dendritic impulse initiation in an epithelial sensory neuron. *J. Comp. Physiol. 142*, 237–249.

FIELD, L. H. and RIND, F. C. (1981). A single insect chordotonal organ mediates inter- and intra-segmental leg reflexes. *Comp. Biochem. Physiol. 68A*, 99–102.

FINLAYSON, L. H. (1966). Sensory innervation of the spiracular muscle in the tsetse fly (*Glossina morsitans*) and the larva of the wax moth (*Galleria mellonella*). *J. Insect Physiol. 12*, 1451–1454.

FINLAYSON, L. H. (1968). Proprioceptors in the invertebrates. *Symp. Zool. Soc. London 23*, 217–249.

FINLAYSON, L. H. (1972). Chemoreceptors, cuticular mechanoreceptors, and peripheral multiterminal neurones in the larva of the tsetse fly (*Glossina*). *J. Insect Physiol. 18*, 2265–2276.

FINLAYSON, L. H. (1976). Abdominal and thoracic receptors in insects, centipedes and scorpions. In *Structure and Function of Proprioceptors in the Invertebrates*. Edited by P. J. Mill. Pages 153–211. Chapman & Hall, London.

FINLAYSON, L. H. and LOWENSTEIN, O. (1955). A proprioceptor in the body musculature of Lepidoptera. *Nature 176*, 1031.

FINLAYSON, L. H. and LOWENSTEIN, O. (1958). The structure and function of abdominal stretch receptors in insects. *Proc. Roy. Soc. London B. 148*, 433–449.

FINLAYSON, L. H. and ORCHARD, I. (1977). The ionic regulation of action potentials in the axon of a stretch receptor neuron of the stick insect (*Carausius morosus*). *J. Comp. Physiol. 122*, 45–52.

FISCHER, G., KANESHIRE, E. S. and PETERS, P. D. (1976). Divalent cation affinity sites in *Paramecium aurelia. J. Cell Biol. 69*, 429–442.

FLETCHER, N. H. (1978). Acoustical response of hair receptors in insects. *J. Comp. Physiol 127*, 185–189.

FEOLIX, R. F. and CHU-WANG, I.-WU. (1972). Fine structural analysis of palpal receptors in the tick *Amblyomma americanum* (L.). *Z. Zell. Mikr. Ant. 129*, 548–560.

FRENCH, A. S. (1980). Sensory transduction in an insect mechanoreceptor: linear and nonlinear properties. *Biol. Cybern. 38*, 115–123.

FRENCH, A. S. and KUSTER, J. E. (1981). Sensory transduction in an insect mechanoreceptor: extended bandwidth measurements and sensitivity to stimulus strength. *Biol. Cybern. 42*, 87–94.

FRENCH, A. S. and KUSTER, J. E. (1982). The effects of temperature on mechanotransduction in the cockroach tactile spine. *J. Comp. Physiol. 147*: 251–258.

FRENCH, A. S. and SANDERS, E. J. (1979). The mechanism of sensory transduction in the sensilla of the trochanteral hair plate of the cockroach, *Periplaneta americana. Cell Tiss. Res. 198*, 159–174.

FRENCH, A. S. and SANDERS, E. J. (1981). The mechanosensory apparatus of the femoral tactile spine of the cockroach, *Periplaneta americana. Cell Tiss. Res. 219*, 53–68.

FRENCH, A. S. and WONG, R. K. S. (1976). The responses of trochanteral hair plate sensilla in the cockroach to periodic and random displacements. *Biol. Cybern. 22*, 33–38.

FRENCH, A. S. and WONG, R. K. S. (1977). Nonlinear analysis of sensory transduction in an insect mechanoreceptor. *Biol. Cybern. 26*, 231–240.

FRIEDMAN, M. H. (1972). A light and electron microscopic study of sensory organs and associated structures in the foreleg tibia of the cricket, *Gryllus assimilis. J. Morph. 138*, 263–328.

FÜLLER, H. and ERNST, A. (1977). Die Ultrastruktur der cercalen Cuticularsensillen von *Periplaneta americana* (L.). *Zool. Jb. Anat. 98*, 544–571.

GAFFAL, K. P. (1976). The stimulus transmitting apparatus in the trichobothria of the bugs *Pyrrhocoris apterus* L. and *Dysdercus intermedius* (Dist.) and its influence on the dynamic of excitation in these sensilla. *Experientia 32*, 166–168.

GAFFAL, K. P. and HANSEN, K. (1972). Mechanoreceptive Strukturen der antennalen Haarsensillen der Baumwollwanze *Dysdercus intermedius* Dist. *Z. Zell. Mikr. Ant. 132*, 79–94.

GAFFAL, K. P. and THEISS, J. (1978). The tibial thread-hairs of *Acheta domesticus* L. (Saltatoria, Gryllidae). The dependence of stimulus transmission and mechanical properties on the anatomical characteristics of the socket apparatus. *Zoomorphologie 90*, 41–51.

GAFFAL, K. P., TICHY, H., THEISS, J. and SEELINGER, G. (1975). Structural polarities in mechanosensitive sensilla and their influence on stimulus transmission (Arthropoda). *Zoomorphologie 82*, 79–103.

GELPERIN, A. (1967). Stretch receptors in the foregut of the blowfly. *Science 157*, 208–210.

GELPERIN, A. (1971). Abdominal sensory neurons providing negative feedback to the feeding behaviour of the blowfly. *Z. Vergl. Physiol. 72*, 17–31.

GETTRUP, E. (1962). Thoracic proprioceptors in the flight system of locusts. *Nature 193*, 498–499.

GETTRUP, E. (1963). Phasic stimulation of a thoracic stretch receptor in locusts. *J. Exp. Biol. 40*, 323–333.

GETTRUP, E. (1973). Stimulus transmission in cuticular mechanoreceptors. *Naturwissenschaften 60*, 52–53.

GEWECKE, M. (1967). Die Wirkung von Luftströmung auf die Antennen und das Flugverhalten der Blauen Schmeissfliege (*Calliphora erythrocephala*). *Z. Vergl. Physiol. 54*, 121–164.

GEWECKE, M. (1970). Antennae: another wind-sensitive receptor in locusts. *Nature 225*, 1263–1264.

GEWECKE, M. (1972a). Antennen und Stirn-Scheitelhaare von *Locusta migratoria* L. als Luftströmungs-Sinnesorgane bei der Flugsteuerung. *J. Comp. Physiol. 80*, 57–94.

GEWECKE, M. (1972b). Die Regelung der Fluggeschwindigkeit bei Heuschrecken und ihre Bedeutung für die Wanderflüge. *Verh. Dtsch. Zool. Ges. 65*, 247–250.

GEWECKE, M. (1974). The antennae of insects as air-current sense organs and their relationship to the control of flight. In *Experimental Analysis of Insect Behavior*. Edited by L. B. Brown. Pages 100–113. Springer, Berlin and New York.

GEWECKE, M. (1975). The influence of the air-current sense organs on the flight behaviour of *Locusta migratoria. J. Comp. Physiol. 103*, 79–95.

GEWECKE, M. and HEINZEL, H. G. (1980). Aerodynamic and mechanical properties of the antennae as air-current sense organs in *Locusta migratoria. J. Comp. Physiol. 139*, 357–366.

GEWECKE, M. and NIEHAUS, M. (1981). Flight and flight control by the antennae in the small tortoiseshell (*Aglais urticae* L., Lepidoptera). *J. Comp. Physiol. 145*, 249–256.

GEWECKE, M. and PHILIPPEN, J. (1978). Control of the horizontal flight-course by air-current sense organs in *Locusta migratoria. Physiol. Ent. 3*, 43–52.

GILULA, N. B. and SATIR, P. (1972). The ciliary necklace. A ciliary membrane specialization. *J. Cell Biol. 53*, 494–509.

GNATZY, W. (1976). The ultrastructure of the thread-hairs on the cerci of the cockroach *Periplaneta americana* L.: The intermoult phase. *J. Ultrastruct. Res. 54*, 124–134.

GNATZY, W. (1978). Development of the filiform hairs on the cerci of *Gryllus bimaculatus* Deg. (Saltatoria, Gryllidae). *Cell Tiss. Res. 187*, 1–24.

GNATZY, W. and ROMER, F. (1980). Morphogenesis of mechanoreceptor and epidermal cells of crickets during the last instar, and its relation to molting-hormone level. *Cell Tiss. Res. 213*, 369–391.

GNATZY, W. and SCHMIDT, K. (1971). Die Feinstruktur de Sinneshaare auf den Cerci von *Gryllus bimaculatus* Deg. (Saltatoria, Gryllidae) I. Faden — und Keulenhaare. *Z. Zell. Mikr. Ant. 122*, 190–209.

GNATZY, W. and SCHMIDT, K. (1972a). Die Feinstruktur de Sinneshaare auf den Cerci von *Gryllus bimaculatus* Deg. IV. Die Häutung der kurzen Borstenhaare. *Z. Zell. Mikr. Ant. 126*, 223–239.

GNATZY, W. and SCHMIDT, K. (1972b). Die Feinstruktur der Sinneshaare auf den Cerci von *Gryllus bimaculatus* Deg. V. Die Häutung der langen Borstenhaare an der Cercusbasis. *J. Mic. 14*, 75–84.

GNATZY, W. and TAUTZ, J. (1977). Sensitivity of an insect mechanoreceptor during molting. *Physiol. Ent. 2*, 279–288.

GNATZY, W. and TAUTZ, J. (1980). Ultrastructure and mechanical properties of an insect mechanoreceptor: Stimulus-transmitting structures and sensory apparatus of the cercal filiform hairs of *Gryllus*. *Cell Tiss. Res. 213*, 441–463.

GRABER, V. (1882). Die chordotonalen Sinnesorganes und das Gehör der Insekten. *Arch. Mikr. Ant. Entwicklungsmech. 20*, 506–640.

GUILLET, J. C. (1975). Relations stimulus-réponse dans le cas d'un méchanorécepteur d'insecte á adaptation totale. *J. Insect Physiol. 21*, 1355–1364.

GUILLET, J. C. and BERNARD, J. (1972). Shape and amplitude of the spikes induced by natural or electrical stimulation in insect receptors. *J. Insect Physiol. 18*, 2155–2171.

GUILLET, J. C. and BOISTEL, J. (1968). Effets comparatifs de stimulations naturelles et électriques sur l'activité de méchanorécepteurs d'insectes. *C. r. Séanc. Soc. Biol. 162*, 227–233.

GUILLET, J. C., BERNARD, J., COILLOT, J. P. and CALLEC, J. J. (1980). Electrical properties of the dendrite in an insect mechanoreceptor: effects of antidromic or direct electrical stimulation *J. Insect Physiol. 26*, 755–762.

GUTHRIE, D. M. (1967). Multipolar stretch receptors and the insect leg reflex. *J. Insect Physiol. 13*, 1637–1644.

HALLBERG, E. (1981). Johnston's organ in *Neodiprion sertifer* (Insecta: Hymentopera). *J. Morph. 167*, 305–312.

HANSEN, K. and HEUMAN, H. G. (1971). Die Feinstruktur der tarsalen Schmeckhaare der Fliege *Phormia terraenovae* Rob.-Desv. *Z. Zell. Mikr. Ant. 117*, 419–422.

HARTMAN, H. B., WALTHALL, W. W., BENNETT, L. P. and STEWART, R. R. (1979). Giant interneurons mediating equilibrium reception in an insect. *Science 205*, 503–505.

HAUPT, J. (1978). Ultrastruktur der Trichobothrien von *Allopauropus (Decepauropus)* (Pauropoda). *Abh. Verh. Naturwiss. Ver. Hamburg 21–22*, 271–277.

HAWKE, S. D., FARLEY, R. D. and GREANY, P. D. (1973). The fine structure of sense organs in the ovipositor of the parasitic wasp, *Orgilus lepidus* Musebeck. *Tissue Cell 5*, 171–184.

HEATHCOTE, R. D. (1981). Differentiation of an identified sensory neuron (SR) and associated structures (CTO) in grasshopper embryos. *J. Comp. Neurol. 202*, 1–18.

HEINZEL, H. G. and GEWECKE, M. (1979). Directional sensitivity of the antennal campaniform sensilla in locusts. *Naturwissenschaften 66*, 212–213.

HENSON, B. L. and WILKENS, L. A. (1979). A mathematical model for the motion of mechanoreceptor hairs in fluid environments. *Biophys. J. 27*, 277–286.

HERAN, H. (1959). Wahrnehmung und Regelung der Fluggeschwindigkeit bei *Apis mellifica*. *Z. Vergl. Physiol. 42*, 103–163.

HERBSTSOMMER, D. and SCHNEIDER, L. (1979). Untersuchungen an den Augenborsten der Honigbiene *Apis mellifera* L. (Insecta, Hymenoptera) I. Feinbau der grossen Sinnesborsten. *Zool. Jb. Physiol. 83*, 106–125.

HERTWECK, H. (1931). Anatomi und Variabilitat des Nervensystems und der Sinnesorgane von *Drosophila melanogaster* (Meigen). *Z. Wiss. Zool. 139*, 559–663.

HOLLICK, F. S. J. (1940). The flight of dipterous fly *Muscina stabilans*. *Phil. Trans. Roy. Soc. B 230*, 357–390.

HONOMICHL, K. (1976). Feinstruktur enies Muskelrezeptors im Kopf von *Dermestes maculatus* DeGeer (Insecta, Coleoptera). *Zoomorphologie 85*, 59–71.

HONOMICHL, K. (1978a). Feinstruktur eines dritten, nicht-ciliären Propriozeptors an der Mandibel von *Oryzaephilus surinamensis* (L.) (Insecta, Coleoptera). *Protoplasma 96*, 149–156.

HONOMICHL, K. (1978b). Feinstruktur zweier Propriozeptoren im Kopf von *Oryzaephilus surinamensis* (L.) (Insecta, Coleoptera). *Zoomorphologie 90*, 213–226.

HOUTERMANS, B. and SCHUMACHER, R. (1974). Zur Morphologie der atympanalen tibialen Scolopalorgane von *Tettigonia viridissima* L. (Orthoptera, Tettigoniidae). *Z. Morph. Tiere 78*, 281–297.

HOWSE, P. E. (1965). The structure of the subgenual organ and certain other mechanoreceptors of the termite *Zootermopsis angusticollis*. *Proc. Roy. Ent. Soc. A 40*, 137–146.

HOWSE, P. E. (1968). The fine structure and functional organization of chordotonal organs. *Symp. Zool. Soc. Lond. 23*, 167–198.

HOWSE, P. E. and CLARIDGE, M. F. (1970). The fine structure of Johnston's organ of the leaf-hopper, *Oncopsis flavicollis*. *J. Insect Physiol. 16*, 1665–1675.

HUBER, F. (1974). Neural integration (central nervous system). In *The Physiology of Insecta*. Edited by M. Rockstein. Vol. 4, pages 3–100. Academic Press, New York.

HUSTERT, R., PFLÜGER, H. J. and BRÄUNIG, P. (1981). Distribution and specific central projections of mechanoreceptors in the thorax and proximal leg joints of locusts. III. The external mechanoreceptors: The campaniform sensilla. *Cell Tiss. Res. 216*, 97–111.

HUXLEY, A. F. (1959). Ion movements during nerve activity. *Ann. N.Y. Acad. Sci. 81*, 221–246.

ISHIKO, N. and LOWENSTEIN, W. R. (1961). Effects of temperature on the generator and action potentials of a sense organ. *J. Gen. Physiol. 45*, 105–124.

JÄGERS-RÖHR, E. (1968). Untersuchungen zur Morphologie und Entwicklung der scolopidial Organe bei der Stabheuschrecke *Carausius morosus* Br. *Biol. Zentralblatt 87*, 393–409.

JEZ, D. H. and MCIVER, S. B. (1980). Fine structure of antennal sensilla of larval *Toxorhynchites brevipalpis* Theobald (Dipter: Culicidae). *Int. J. Insect Morph. Embryol. 9*, 147–159.

JOHNSON, B. (1956). Function of the antennae of aphids during flight. *Aust. J. Sci. 18*, 199–200.

JOHNSTON, C. (1855). Auditory apparatus of the *Culex* mosquito. *Quartz. J. Mic. Sci. 3*, 97–102.

KALMRING, K., LEWIS, B. and EICHENDORF, A. (1978). The physiological characteristics of the primary sensory neurons of the complex tibial organ of *Decticus verrucivorus* L. (Orthoptera, Tettigonioidae). *J. Comp. Physiol. 127*, 109–121.

KATZ, B. (1961). The terminations of the afferent nerve fibre in the muscle spindle of the frog. *Phil. Trans. Roy. Soc. B 243*, 221–240.

KEIL, T. (1978). Die Makrochaeten auf dem Thorax von *Calliphora vicina* Robineau-Desvoidy (Calliphoridae, Diptera). *Zoomorphologie 90*, 151–180.

KEIL, T. (1979). Rutheniumrot-Färbung sensorischer Einheiten der Insekten-Epidermis. *Eur. J. Cell Biol. 19*, 78–82.

KEIL, T. and THURM, U. (1979). Die Verteilung von Membrankontakten und Diffusionsbarrieren in epidermalen Sinnesorganen von Insekten. *Verh. Dtsch. Zool. Ges. 72*, 285.

KEPPLER, E. (1958a). Über das Richtungshören von Stechmücken. *Z. Naturforsch. B 13*, 280–284.

KEPPLER, E. (1958b). Zum Hören von Stechmücken. *Z. Naturforsch. B 13*, 285–286.

KINZELBACH-SCHMITT, B. (1968). Zue Kenntnis der antennal chordotonalorgane der Thysanuren (Thysanura, Insecta). *Z. Naturforsch. 23b*, 289–291.

KNYAZEVA, N. I. (1974). Campaniform sensilla of *Locusta migratoria* L. (Orthoptera, Acrididae). *Ent. Rev. 53*, 62–66.

KNYAZEVA, N. I., FUDALEWICZ-NIEMCZYK, W. and ROSCISZEWSKA, M. (1980). Proprioceptors in the nymph of the house cricket *Gryllus domesticus* (Orthoptera, Gryllidae). *Ent. Rev. 58*, 16–19.

KÜHNE, R. (1982). Neurophysiology of the vibration sense in locusts and bushcrickets: response characteristics of single receptor units. *J. Insect Physiol. 28*, 155–163.

KÜPPERS, J. (1974). Measurements on the ionic milieu of the receptor terminal in mechanoreceptive sensilla of insects. In *Mechanoreception*. Edited by J. Schwartzkopff. Pages 387–394. *Abh. Rhein. Westf. Akad. Wiss. Opladen*.

KÜPPERS, J. and THURM, U. (1975). Humorale Steuerung eines Ionentransportes an epithelialen Rezeptoren von Insekten. *Verh. Dtsch. Zool. Ges. 67*, 46–50.

KÜPPERS, J. and THURM, U. (1979). Active ion transport by a sensory epithelium I. Transepithelial short circuit current, potential difference, and their dependence on metabolism. *J. Comp. Physiol. 134*, 131–136.

KUSTER, J. E. (1980). Fine structure of the compound eyes and interfacetal mechanoreceptors of *Cicindela tranquebarica* Herbst (Coleoptera: Cicindelidae). *Cell Tiss. Res. 206*, 123–138.

KUSTER, J. E. and FRENCH, A. S. (1983). Sensory transduction in a locust multipolar joint receptor: The dynamic behaviour under a variety of stimulus conditions. *J. Comp. Physiol. 150*, 207–215.

LARINK, O. (1975). Zur Schuppenbildung von *Lepisma saccharina* L. (Insecta, Zygentoma). *Verh. Dtsch. Zool. Ges. 197*, 205–208.

LARINK, O. (1976). Entwicklung und Feinstruktur der Schuppen bei Lepismatiden und Machiliden (Insecta, Zygentoma und Archaeognatha). *Zool. Jb. Ant. 95*, 252–293.

LEBERRE, J. R. and LOUVEAUX, A. (1969). Equipment sensoriel des manidbules de la larve du premier stade de *Locusta migratoria* L. *C.R. Acad. Sci. 268*, 2907–2910.

LEE, J.-K., KIM, W.-K. and KIM, C.-W. (1980). Fine structure of the chordotonal organ and development of the olfactory nerves in the tip of labial palp of *Pieris rapae* L. *Korean J. Ent. 10*, 21–31.

LEES, A. D. (1942). Homology of the campaniform organs on the wings of *Drosophila melanogaster*. *Nature 150*, 375.

LEHR, R. (1914). Die Sinnesorgane der beiden Flügelpaare von *Dytiscus marginalis*. *Z. Wiss. Zool. 110*, 87–150.

LEONOWITCH, S. A. and IVANOV, V. P. (1978). The fine structure of Johnston's organ in the scorpionfly *Panorpa communis* (Mecoptera, Panorpidae). *Zool. Zh. 57*, 214–221.

LEWIS, C. T. (1970). Structure and function in some external receptors. *Symp. Roy. Ent. Soc. 5*, 59–76.

LOWENSTEIN, O. and FINLAYSON, L. H. (1960). The response of the abdominal stretch receptor of an insect to phasic stimulation. *Comp. Biochem. Physiol. 1*, 56–61.

MANN, D. W. and CHAPMAN, K. M. (1975). Component mechanisms of sensitivity and adaptation in an insect mechanoreceptor. *Brain Res. 97*, 331–336.

MARQUHARDT, F. (1930). Beiträge zur Anatomie der Muskulatur und der peripheren Nerven von *Carausius (Dixippus) morosus* Br. *Zool. Jb. Anat. 66*, 63–128.

MASSON, C. and GABOURIAUT, D. (1973). Ultrastructure de l'organe de Johnston de la Fourmi *Camponotus vagus* Scop. (Hymenoptera, Formicidae). *Z. Zell. Mikr. Ant. 140*, 39–75.

MATSUMOTO, D. E. and FARLEY, R. D. (1978). Comparison of the ultrastructure of stimulated and unstimulated mechanoreceptors in the taste hairs of the blowfly *Phaenicia serricata*. *Tissue Cell 10*, 63–76.

MCIVER, S. B. (1972). Fine structure of the sensilla chaetica on the antennae of *Aedes aegypti*. *Ann. Ent. Soc. Amer. 65*, 1390–1397.

MCIVER, S. B. (1975). Structure of cuticular mechanoreceptors of arthropods. *Ann. Rev. Ent. 20*, 381–397.

MCIVER, S. and SIEMICKI, R. (1975). Campaniform sensilla on the palps of *Anopheles stephensi* Liston (Diptera: Culicidae). *Int. J. Insect Morph. Embryol. 4*, 127–130.

MCIVER, S. and SIEMICKI, R. (1981). Structure of selected body setae of immature and adult *Aedes aegypti* (Diptera: Culicidae). *Mosquito News 41*, 552–557.

MOECK, H. A. (1968). Electron microscopic studies of antennal sensilla in the ambrosia beetle *Trypodendron lineatum* (Olivier) (Scolytidae). *Can. J. Zool. 46*, 521–556.

MÖHL, B. (1972). The control of foregut movements by the stomatogastric nervous system in the European house cricket *Acheta domesticus* L. *J. Comp. Physiol. 80*, 1–28.

MOORE, R. K. (1960). *Travelling-Wave Engineering*. McGraw-Hill, New York.

MORAN, D. T. (1971). Loss of the sensory process of an insect receptor at ecdysis. *Nature 234*, 476–477.

MORAN, D. T., CHAPMAN, K. M. and ELLIS, R. A. (1971). The fine structure of cockroach campaniform sensilla. *J. Cell Biol. 48*, 155–173.

MORAN, D. T. and ROWLEY, J. C., III (1975a). High voltage and scanning electron microscopy of the site of stimulus reception of an insect mechanoreceptor. *J. Ultrastruct. Res. 50*, 38–46.

MORAN, D. T. and ROWLEY, J. C., III (1975b). The fine structure of the cockroach subgenual organ. *Tissue Cell 7*, 91–106.

MORAN, D. T. and VARELA, F. G. (1971). Microtubules and sensory transduction. *Proc. Nat. Acad. Sci. 68*, 757–760.

MORAN, D. T., ROWLEY, J. C., III and VARELA, F. G. (1975). Ultrastructure of the grasshopper proximal femoral chordotonal organ. *Cell Tiss. Res. 161*, 445–457.

MORAN, D. T., ROWLEY, J. C., III., ZILL, S. N. and VARELA, F. G. (1976). The mechanism of sensory transduction in a mechanoreceptor. Functional stages in campaniform sensilla during the molting cycle. *J. Cell Biol. 71*, 832–847.

MORAN, D. T., VARELA, F. J. and ROWLEY, J. C. III (1977). Evidence for active role of cilia in Sensory transduction. *Proc. Nat. Acad. Sci. 74*, 793–797.

MÖSS, D. (1967). Proprioceptoren im Thorax und Abdomen von Grillen (Orthoptera, Gryllidae). *Verh. Dtsch. Ges. Zool. Anz. Suppl. 31*, 742–749.

MOULINS, M. (1966). Presence d'un recepteur de tension dans l'hypopharynx de *Blabera craniifer* Burm. (Insecta: Dictyoptera). *C.R. Acad. Sci. Paris 262*, 2476–2479.

MOULINS, M. (1971). Ultrastructure et physiologie des organes epipharyngiens et hypopharyngiens (Chimirecepteurs cibariaux) de *Blabera craniifer* Burm. (Insecte, Dictyoptere). *Z. Vergl. Physiol. 73*, 139–166.

MOULINS, M. (1974). Recepteurs de tension de la region de la bouche chez *Blaberus craniifer* Burmeister (Dictyoptera: Blaberidae). *Int. J. Insect Morph. Embryol. 3*, 171–192.

MOULINS, M. (1976). Ultrastructure of chordotonal organs. In *Structure and Function of Proprioceptors in the Invertebrates*. Edited by P. J. Mill. Pages 387–426. Chapman & Hall, London.

MÜLLER, M. L., HONEGGER, H. W., NICKEL, E. and WESTPHAL, C. (1978). The ultrastructure of campaniform sensilla on the eye of the cricket, *Gryllus campestris*. *Cell Tiss. Res. 195*, 349–357.

NAGY, F. (1974). Le systeme neuromusculaire et sensoriel de l'intestin posterieur et de la region proctodeal, chez le larvae d'*Oryctes nasicornis* L. (Col. scarabeidae). Thèse, Universite de Dijon, France.

NEUMANN, H. (1975). Untersuchungen zur Struktur und Elektrophysiologie mechanorezeptiver Sensillen auf den Vordertibien von *Gryllus bimaculatus*. Dissertation. Mathematisichnaturwissenschaftliche Fakultät der Universität Köln.

NICKLAUS, R. (1965). Die Erregung einzelner Fadenhaare von *Periplaneta americana* in Abhängigkeit con der Grösse und Richtung der Auslenkung. *Z. Vergl. Physiol. 50*, 331–362.

NICKLAUS, R., LUNDQUIST, P. G. and WERSÄLL, J. (1967). Elektronmikroskopie am sensorischen Apparat der Fadenhaare auf den Cerci der Schabe *Periplaneta americana*. *Z. Vergl. Physiol. 56*, 412–415.

NICKLAUS, R., LUNDQUIST, P. G. and WERSÄLL, J. (1968). Die Ubertragung des Reizes auf den Fortsatz der Sinneszelle bei den Fadenhaaren von *Periplaneta americana*. *Verh. Dtsch. Ges. Zool. Anz. Suppl. 31*, 578–584.

NIEHAUS, M. (1981). Flight and flight control by the antennae in the small tortoiseshell (*Aglais urticae* L., Lepidoptera) III. Flight mill and free flight experiments. *J. Comp. Physiol. 145*, 257–264.

NIEHAUS, M. and GEWECKE, M. (1978). The antennal movement apparatus in the small tortoiseshell (*Aglais urticae* L., Insecta, Lepidoptera). *Zoomorphologie 91*, 19–36.

NIJHOUT, H. F. (1979). Stretch-induced moulting in *Oncopeltus fasciatus*. *J. Insect Physiol. 25*, 277–281.

ORCHARD, I. (1975). Structure and properties of an abdominal chordotonal organ in the stick insect (*Carausius morosus*) and the cockroach (*Blaberus discoidalis*). *J. Insect Physiol. 21*, 1491–1499.

ORCHARD, I. (1976). Calcium dependent action potentials in a peripheral neurosecretory cell of the stick insect. *J. Comp. Physiol. 112*, 95–102.

ORCHARD, I. and FINLAYSON, L. H. (1976). The electrical activity of mechanoreceptive and neurosecretory neurons in the stick insect *Carausius morosus*. *J. Comp. Physiol. 107*, 327–338.

ORCHARD, I. and FINLAYSON, L. H. (1977). Electrically excitable neurosecretory cell bodies in the periphery of the stick insect, *Carausius morosus*. *Experientia 33*, 226–228.

ORCHARD, I. and OSBORNE, M. P. (1977). The effects of cations upon the action potentials recorded from neurohaemal tissue of the stick insect, *J. Comp. Physiol. 118*, 1–12.

OSBORNE, M. P. (1963a). An electron microscope study of an abdominal stretch receptor of the cockroach. *J. Insect Physiol. 9*, 237–245.

OSBORNE, M. P. (1963b). The sensory neurons and sensilla in the abdomen and thorax of the blowfly larva. *Quart. J. Mic. Sci. 104*, 227–241.

OSBORNE, M. P. (1964). Sensory nerve terminations in the epidermis of the blowfly larva. *Nature 201*, 526.

OSBORNE, M. P. (1970). Structure and function of neuromuscular junctions and stretch receptors. *Symp. Roy. Ent. Soc. 5*, 77–100.

OSBORNE, M. P. and FINLAYSON, L. H. (1962). The structure and topography of stretch receptors in representatives of seven orders of insects. *Quart. J. Mic. Sci. 103*, 227–242.

OSBORNE, M. P. and FINLAYSON, L. H. (1965). An electron microscopic study of the stretch receptor of *Antheraea pernyi* (Lepidoptera, Saturniidae). *J. Insect Physiol. 11*, 703–710.

PABST, H. (1965). Elektrophysiologische Untersuchung des Streckrezeptors am Flügelgelenk der Wanderheuschrecke *Locusta migratoria. Z. Vergl. Physiol. 50*, 498–541.

PAPPAS, L. G. and LARSEN, J. R. (1976). Labellar chordotonal organs of the mosquito *Culiseta inornata* (Williston) (Diptera: Culicidae). *Int. J. Insect Morph. Embryol. 5*, 145–150.

PETERS, W. (1962). Die propriorezeptiven organe am Prosternium und an den labellan von *Calliphora erythrocephala. Z. Morph. Okol. Tiere 51*, 211–226.

PFAU, H. K. and HONOMICHL, K. (1979). Die campaniformen Sensillen des Flügels von *Cetonia aurata* L. und *Geotrupes silvaticus* Panz. (Insecta, Coleoptera) in ihrer Beziehung zur Flügelmechanik und Flugfunktion. *Zool. Jb. Ant. 102*, 583–613.

PFLÜGER, H. J., BRÄUNIG, P. and HUSTERT, R. (1981). Distribution and specific central projections of mechanoreceptors in the thorax and proximal leg joints of locusts. I. The external mechanoreceptors: Hair plates and tactile hairs. *Cell Tiss. Res. 216*, 79–96.

PRINGLE, J. W. S. (1938a). Proprioception in insects I. A new type of mechanical receptor from the palps of the cockroach. *J. Exp. Biol. 15*, 101–113.

PRINGLE, J. W. S. (1938b). Proprioception in insects II. The action of the campaniform sensilla on the legs. *J. Exp. Biol. 15*, 114–131.

PRINGLE, J. W. S. (1961). Proprioception in arthropods. In *The Cell and the Organism*. Edited by J. A. Ramsay and V. B. Wigglesworth. Pages 256–282. Cambridge University Press.

PUMPHREY, R. J. (1936). Slow adaptation of a tactile receptor in the leg of the common cockroach. *J. Physiol. 87*, 6–7.

RATH, O. VOM (1838). Uber die Hautsinnesorgane der Insekten. *Z. Wiss. Zool. 46*, 413–454.

RICE, M. J. (1970). Cibarial stretch receptors in the tsetse fly (*Glossina austeni*) and the blowfly (*Calliphora erythrocephala*). *J. Insect Physiol. 16*, 277–289.

RICE, M. J. (1975). Insect mechanoreceptor mechanisms. In *Sensory Physiology and Behavior*. Edited by R. Galun, P. Hillman, I. Parnas and R. Werman. Pages 135–165. Plenum, New York.

RICE, M. J. and FINLAYSON, L. H. (1972). Response of blowfly cibarial pump receptors to sinusoidal stimulation. *J. Insect Physiol. 18*, 841–846.

RICE, M. J., GALUN, R. and FINLAYSON, L. H. (1973). Mechanotransduction in insect neurons. *Nature (New Biol.) 241*, 286–288.

RICHARD, G. (1951). L'innervation et les organes sensoriel des pieces buccales du termite a cou jaune (*Calotermes flavicollis* Fab.). *Ann. Sci. Nat. (Zool.) 11ᵉ Serie 13*, 397–412.

RICHARD, G. (1957). L'ontogenese des organes chordotonaux antennaires de *Calotermes flavicollis* (Fab.) *Insectes sociaux 4*, 106–111.

RISLER, H. (1953). Das Gehörorgan der Männchen von *Anopheles stephensi* Liston (Culicidae). *Zool. Jb. Abt. Ant. Ont. Tiere 73*, 165–186.

RISLER, H. (1955). Das Gehörorgan der Männchen von *Culex pipiens, Aedes aegypti* und *Anopheles styliensi* eine vergleichend morphologische Untersuchung. *Zool. Jb. Abt. Ant. Ont. Tiere. 74*, 478–490.

RISLER, H. (1977). The construction of the auditory organ in male mosquitoes. *Fortsch. Zool. 24*, 143–147.

RISLER, H. and SCHMIDT, K. (1967). Der Feinbau der Scolopidien in Johnstonschen Organ von *Aedes aegypti* L. *Z. Naturforsch. 22b*, 759–762.

ROGOSINA, M. (1928). Über das periphere Nervensystem der *Aeschna*-Larve. *Z. Zell. Mikr. Ant. 6*, 732–758.

ROSSIGNOL, P. A. and MCIVER, S. B. (1977). Fine structure and role in behavior of sensilla on the terminalia of *Aedes aegypti* L. (Diptera: Culicidae). *J. Morph. 151*, 419–438.

ROTH, L. M. (1948). A study of mosquito behavior. An experimental study of the sexual behavior of *Aedes aegypti*. *Amer. Midl. Nat. 40*, 451–487.

ROTH, L. M. and SLIFER, E. H. (1973). Spheroid sense organs on the cerci of polyphagid cockroaches. *Int. J. Insect Morph. Embryol. 2*, 13–24.

RUTH, E. (1976). Elektrophysiologie der Sensilla Chaetica auf den Antennen von *Periplaneta americana. J. Comp. Physiol. 105*, 55–64.

SCHAFER, R. and REAGAN, P. D. (1981). Colchicine reversibly inhibits electrical activity in arthropod mechanoreceptors. *J. Neurobiol. 12*, 155–166.

SCHMIDT, K. (1967). Die Entwicklung der Scolopidien im Johnstonschen Organ von *Aedes aegypti* während der Puppenphase. *Verh. Dtsch. Zool. Ges. 48*, 750–762.

SCHMIDT, K. (1969a). Der Feinbau der stiftführenden Sinnesorgane im Pedicellus der Florfliege *Chrysopa* Leach (Chrysopidae, Planipennia). *Z. Zell. Mikr. Ant. 99*, 357–388.

SCHMIDT, K. (1969b). Die campaniformen Sensillen im Pedicellus der Florfliege (*Chrysopa*, Planipennia). *Z. Zell. Mikr. Ant. 96*, 478–489.

SCHMIDT, K. (1970). Vergleichend morphologische Untersuchungen über den Feinbau der Ciliarstrukturen in den Scolopidien des Johnstonschen Organs holometabaler Insekten. *Verh. Dtsch. Zool. Ges. 64*, 88–92.

SCHMIDT, K. (1973). Vergleichende morphologische Untersuchungen an Mechanoreceptoren der Insekten. *Verh. Dtsch. Zool. Ges. 66*, 15–25.

SCHMIDT, K. (1974). Die Mechanoreceptoren im Pedicellus der Eintagsfliegen (Insecta, Ephemeroptera). *Z. Morph. Tiere 78*, 193–220.

SCHMIDT, K. (1975). Das Johnstonsche Organ der primär flügellosen Ectognatha (*Lepisma*, Zygentoma; *Machilis*, Archaeognatha). *Cytobiologie 11*, 153–171.

SCHMIDT, K. and GNATZY, W. (1971). Die Feinstruktur der Sinneshaare auf den Cerci von *Gryllus bimaculatus* Deg. II. Die Häutung der Faden-und Keulenhaare. *Z. Zell. Mikr. Ant. 122*, 210–226.

SCHMIDT, K. and GNATZY, W. (1972). Die Feinstruktur der Sinneshaare auf den Cerci von *Gryllus bimaculatus* Deg. III. Die kurzen Borstenhaare. *Z. Zell. Mikr. Ant. 126*, 206–222.

SCHMIDT, K. and WÜLKER, W. (1980). Parasitäre Intersexualität des Johnstonschen Organes in den Antennen von *Chironomus* (Dipt.) *Z. Parasitenkd. 64*, 1–15.

SCHNEIDER, D. and KAISSLING, K. E. (1957). Der Bau der Antenne des Seiden-spinners *Bombyx mori* L. II. Sensillen, cuticulare Bildungen und innerer Bau. *Zool. Jb. Abt. Ant. Ont. Tiere 76*, 223–250.

SCHNEIDER, P. (1965). Vergleichende Untersuchungen zur Steuerung der Fluggeschwindigkeit bei *Calliphora vicina* Rob.-Desvoidy (Diptera). *Z. Wiss. Zool. 173*, 114–173.

SCHNORBUS, H. (1971). Die Subgenualen Sinnesorgane von *Periplaneta americana*: Histologie und Vibrationsschwellen. *Z. Vergl. Physiol. 71*, 14–48.

SCHWARTZKOPFF, J. (1974). Mechanoreception. In *The Physiology of Insecta*. Edited by M. Rockstein. Vol. 2, pages 273–352. Academic Press, New York.

SIHLER, H. (1924). Die Sinnesorgane an der Cerci der Insekten. *Zool. Jb. Ant. 45*, 519–580.

SLIFER, E. H. (1970). The structure of arthropod chemoreceptors. *Ann. Rev. Ent. 15*, 121–142.

SLIFER, E. H. and FINLAYSON, L. H. (1956). Muscle receptor organs in grasshoppers and locusts (Orthoptera, Acrididae). *Quart. J. Mic. Sci. 97*, 617–620.

SLIFER, E. H. and SEKHON, S. S. (1975). The femoral chordotonal organs of a grasshopper, Orthoptera, Acrididae. *J. Neurocytol. 4*, 419–438.

SMITH, D. S. (1969). The fine structure of haltere sensilla in the blowfly, *Calliphora erythrocephala* (Meig.), with scanning electron microscopic observations on the haltere surface. *Tissue Cell 1*, 443–484.

SMOLA, U. (1970). Rezeptor-und Aktionspotentiale der Sinneshaare auf dem Kopf der Wanderheuschrecke *Locusta migratoria. Z. Vergl. Physiol. 70*, 335–348.

SNODGRASS, R. E. (1926). The morphology of insect sense organs and the sensory nervous system. *Smithson. Misc. Coll. 77*, 1–77.

SNODGRASS, R. E. (1935). *Principles of Insect Morphology*. McGraw-Hill, New York.

SPECHT, U. (1977). Funktionsmorphologie und Elektrophysiologie der Sinnesborsten auf den Cerci der Schabe *Periplaneta americana*. Dissertation. Naturwissenschaftliche Fakultät der Technischen Universität Carolo-Wilhelmina zu Braunschweig.

SPENCER, H. J. (1974). Analysis of the electrophysiological responses of the trochanteral hair receptors of the cockroach. *J. Exp. Biol. 60*, 223–240.

SPINOLA, S. M. and CHAPMAN, K. M. (1975). Proprioceptive indentation of the campaniform sensilla of cockroach legs. *J. Comp. Physiol. 96*, 257–272.

STRICKLER, J. R. and BAL, A. K. (1973). Setae of the first antennae of the copepod *Cyclops scutifer* (Sars): Their structure and importance. *Proc. Nat. Acad. Sci. 70*, 2656–2659.

STUART, A. M. and SATIR, P. (1968). Morphological and functional aspects of an insect epidermal gland. *J. Cell Biol. 36*, 527–549.

STÜRCKOW, B., ADAMS, J. R. and WILCOX, T. A. (1967). The neurones in the labellar nerve of the blowfly. *Z. Vergl. Physiol. 54*, 268–289.

SUGAWARA, T. (1979). Stretch reception in the bursa copulatrix of the butterfly, *Pieris rapae crucivora*, and its role in behaviour. *J. Comp. Physiol. 130*, 191–199.

SUGAWARA, T. (1981). Fine structure of the stretch receptor in the bursa copulatrix of the butterfly, *Pieris rapae crucivora*. *Cell Tiss. Res. 217*, 23–36.

TAUTZ, J. (1977). Reception of medium vibration by thoracal hairs of caterpillars of *Barathra brassicae* L. (Lepidoptera, Noctuidae) I. Mechanical properties of receptor hairs. *J. Comp. Physiol. 118*, 13–31.

TAUTZ, J. (1978). Reception of medium vibration by thoracal hairs of caterpillars of *Barathra brassicae* L. (Lepidoptera, Noctuidae) II. Response characteristics of the sensory cell. *J. Comp. Physiol. 125*, 67–77.

TAUTZ, J. (1979). Reception of particle oscillation in a medium — an unorthodox sensory capacity. *Naturwissenschaften 66*, 452–461.

TAUTZ, J. and MARKL, H. (1978). Caterpillars detect flying wasps by hairs sensitive to airborne vibrations. *Behav. Ecol. Sociobiol. 4*, 101–110.

THEISS, J. (1979). Mechanosensitive bristles on the head of the blowfly mechanics and electrophysiology of the macrochaetae. *J. Comp. Physiol. 132*, 55–68.

THEOPHILIDIS, G. and BURNS, M. D. (1979). A muscle tension receptor in the locust leg. *J. Comp. Physiol. 131*, 247–254.

THORSON, J. and BIEDERMAN-THORSON, M. (1974). Distributed relaxation processes in sensory transduction. *Science 183*, 161–172.

THURM, U. (1963). Die Beziehungen zwischen mechanischen Reizgrössen und stationären Erregungszuständen bei Borstenfeld Sensillen von Bienen. *Z. Vergl. Physiol. 46*, 351–382.

THURM, U. (1964a). Mechanoreceptors in the cuticle of the honey bee: Fine structure and stimulus mechanism. *Science 145*, 1063–1065.

THURM, U. (1964b). Das Rezeptorpotential einzelner mechanorezeptorischer Zellen von Biene. *Z. Vergl. Physiol. 48*, 131–156.

THURM, U. (1965). An insect mechanoreceptor. Part I: Fine structure and adequate stimulus. *Cold Spr. Harb. Symp. Quant. Biol. 30*, 75–82.

THURM, U. (1970). Untersuchungen zur functionellen Organisation sensorischer Zellgruppen. *Verh. Dtsch. Zool. Ges. 64*, 79–88.

THURM, U. (1974a). Basics of the Generation of Receptor Potentials in Epidermal Mechanoreceptors of Insects. In *Mechanoreception*. Edited by J. Schwartzkopff. Pages 355–385. *Abh. Rhein. Westf. Akad. Wiss. Opladen*.

THURM, U. (1974b). Mechanisms of electrical membrane responses in sensory receptors, illustrated by mechanoreceptors. 25. *Mosbacher Colloquium Ges. Biol. Chemie*.

THURM, U. and KÜPPERS, J. (1980). Epithelial Physiology of Insect Sensilla. In *Insect Biology in the Future*. Edited by M. Locke and D. S. Smith. Pages 735–763. Academic Press, New York and London.

THURM, U., STEDTLER, A. and FOELIX, R. (1975). Reizwirksame Verformungen der Terminalstrukturen eines Mechanorezeptors. *Verh. Dtsch. Zool. Ges. 1974*, 37–41.

THURM, U. and WESSEL, G. (1979). Metabolic-dependent transepithelial potential differences at epidermal receptors of arthropods. I. Comparative data. *J. Comp. Physiol. 134*, 119–130.

TISCHNER, H. (1953). Über den Gehörsinn von Steckmücken. *Acustica 3*, 335–343.

TISCHNER, H. and SCHIEF, A. (1954). Fluggerrausch und Schallwahrnehmung bei *Aedes aegypti* L. (Culicidae). *Verh. Dtsch. Zool. Ges. 51*, 453–460.

TOH, Y. (1981). Fine structure of sense organs on the antennal pedicel and scape of the male cockroach, *Periplaneta americana*. *J. Ultrastruct. Res. 77*, 119–132.

TYRER, N. M., BACON, J. P. and DAVIES, C. A. (1979). Sensory projections from the wind-sensitive head hairs of the locust *Schistocerca gregaria*. Distribution in the central nervous system. *Cell Tiss. Res. 203*, 79–92.

UGA, S. and KUWABARA, M. (1965). On the fine structure of the chordotonal sensillum in antenna of *Drosophila melanogaster*. *J. Elect. Mic. 14*, 173–181.

UGA, S. and KUWABARA, M. (1967). The fine structure of the campaniform sensillum on the haltere of the fleshfly, *Boettscherisca peregrina*. *J. Elect. Mic. 16*, 304–312.

USHERWOOD, P. N. R., RUNION, H. I. and CAMPBELL, J. I. (1968). Structure and physiology of a chordotonal organ in the locust leg. *J. Exp. Biol. 48*, 305–323.

VANDE BERG, J. (1971). Fine structural studies of Johnston's organ in the tobacco hornworm moth, *Manduca sexta* (Johannson). *J. Morph, 133*, 439–456.

VIALLANOS, H. (1882). Recherches sur l'histologie des insectes. *Ann. Sci. Nat. 14*, 1–348.

VÖLKER, W. (1978). Elektronenmikroskopische Untersuchungen über den Reizübertragungsmechanismus eines Mechanorezeptors. Diplomarbeit. Universität Munster.

WALES, W. (1976). Receptors of the Mouthparts and Gut of Arthropods. In *Structure and Function of Proprioceptors in the Invertebrates*. Edited by P. J. Mill. Pages 213–241. Chapman & Hall, London.

WEEVERS, R. DE G. (1965). Proprioceptive reflexes and the co-ordination of locomotion in the caterpillar of *Antheraea pernyi* (Lepidoptera). In *The Physiology of the Insect Central Nervous System*. Edited by J. E. Treherne and J. W. L. Beament. Pages 113–124. Academic Press, London.

WEEVERS, R. DE G. (1966a). A lepidopteran saline: effects of inorganic cation concentration on sensory, reflex and motor responses in a herbivorous insect. *J. Exp. Biol. 44*, 163–175.

WEEVERS, R. DE G. (1966b). The physiology of a lepidopteran muscle receptor. I. The sensory response to stretch. *J. Exp. Biol. 44*, 177–194.

WEEVERS, R. DE G. (1966c). The physiology of a lepidopteran muscle receptor. II. The function of the receptor muscle. *J. Exp. Biol. 44*, 195–208.

WEEVERS, R. DE G. (1966d). The physiology of a lepidopteran muscle receptor. III. The stretch reflex. *J. Exp. Biol. 45*, 229–249.

WENSLER, R. J. D. (1974). Sensory innervation monitoring movement and position in the mandibular stylets of the aphid, *Brevicoryne brassicae*. *J. Morph. 143*, 349–363.

WENSLER, R. J. (1977). The fine structure of distal receptors on the labium of the aphid, *Brevicoryne brassicae* L. (Homoptera). *Cell Tiss. Res. 181*, 409–422.

WHITEAR, M. (1962). The fine structure of crustacean proprioceptors. I. The chordotonal organs in the legs of the shore crab, *Carcinus maenas*. *Phil. Trans. Roy. Soc. B. 245*, 291–325.

WHITEHEAD, A. T. (1981). Ultrastructure of sensilla of the female mountain pine beetle, *Dendroctonus ponderosae* Hopkins (Coleoptera: Scolytidae). *Int. J. Morph. Embryol. 10*, 19–28.

WIEDERHOLD, M. L. (1976). Mechanosensory transduction in sensory and motile cilia. *Ann. Rev. Biophys. Bioeng. 5*, 39–62.

WIESE, K. (1972). Das mechanorezeptorische Beuteortungssystem von *Notonecta*. I. Die Funktion des tarsalen Scolopidialorgans. *J. Comp. Physiol. 78*, 83–102.

WIESE, K. (1974). The mechanoreceptive system of prey localization in *Notonecta*. II. The principle of prey localization. *J. Comp. Physiol. 92*, 317–325.

WIESE, K. and SCHMIDT, K. (1974). Mechanorezeptoren im Insektentarsus. Die Konstruktion des Scolopidialorgans bei *Notonecta* (Hemiptera, Heteroptera). *Z. Morph. Tiere 79*, 47–64.

WIGGLESWORTH, V. B. (1953). The origin of sensory neurones in an insect, *Rhodnius prolixus*. *Quart. J. Mic. Sci. 94*, 93–112.

WILLIAMSON, R. and BURNS, M. D. (1978). Multiterminal receptors in the locust mesothoracic leg. *J. Insect Physiol. 24*, 661–666.

WILSON, D. M. and GETTRUP, E. (1963). A stretch reflex controlling wingbeat frequency in grasshoppers. *J. Exp. Biol. 40*, 171–185.

WISHART, G. and RIORDAN, D. F. (1959). Flight responses to various sounds by adult males of *Aedes aegypti* (L.) (Diptera: Culicidae). *Canad. Ent. 91*, 181–191.

WISHART, G., VAN SICKLE, G. R. and RIORDAN, D. F. (1962). Orientation of the males of *Aedes aegypti* (L.) (Diptera: Culicidae) to sound. *Canad. Ent. 94*, 613–626.

WOLBARSHT, M. L. (1960). Electrical characteristics of insect mechanoreceptors. *J. Gen. Physiol. 44*, 105–122.

WRIGHT, B. R. (1976). Limb and wing receptors in insects, chelicerates and myriapods. In *Structure and Function of Proprioceptors in the Invertebrates.* Edited by P. J. Mill. Pages 323–386. Chapman & Hall, London.

WRIGHT, K. A. (1983). Nematode chemosensilla: form and function. *J. Nemat. 15*, 151–158.

YAGODIN, S. V. (1980). Role of antennae in flight maintenance of the cockroach *Periplaneta americana. J. Evol. Biochem. Physiol. 16*, 30–36.

YOUNG, D. (1970). The structure and function of a connective chordotonal organ in the cockroach leg. *Phil. Trans. Roy. Soc. B. 256*, 401–426.

YOUNG, D. (1973). Fine structure of the sensory cilium of an insect auditory receptor. *J. Neurocytol. 2*, 47–58.

ZACHARUK, R. Y. (1962). Sense organs of the head of larvae of some Elateridae: Their distribution, structure and innervation. *J. Morph. 111*, 1–34.

ZACHARUK, R. Y. (1980). Innervation of sheath cells on an insect sensillum by a bipolar type II neuron. *Canad. J. Zool. 58*, 1264–1276.

ZACHARUK, R. Y. and ALBERT, P. J. (1978). Ultrastructure and function of scolopophorous sensilla in the mandible of an elaterid larva (Coleoptera). *Canad. J. Zool. 56*, 246–259.

ZACHARUK, R. Y. and BLUE, S. G. (1971). Ultrastructure of a chordotonal and a sinusoidal peg organ in the antenna of larval *Aedes aegypti* (L.) *Canad. J. Zool. 49*, 1223–1229.

ZACHARUK, R. Y., ALBERT, P. J. and BELLAMY, F. W. (1977). Ultrastructure and function of digitiform sensilla on the labial palp of a larval elaterid (Coleoptera). *Canad. J. Zool. 55*, 569–578.

ZACK, S. and BACON, J. (1981). Interommatidial sensilla of the praying mantis: Their central neural projections and role in head-cleaning behavior. *J. Neurobiol. 12*, 55–65.

ZILL, S. N. and MORAN, D. T. (1980). The exoskeleton and insect proprioception I. Responses to tibial campaniform sensilla to external and muscle-generated forces in the American cockroach, *Periplaneta americana. J. Exp. Biol. 91*, 1–24.

ZILL, S. N. and MORAN, D. T. (1981). The exoskeleton and insect proprioception III. Activity of tibial campaniform sensilla during walking in the American cockroach, *Periplaneta americana. J. Exp. Biol. 94*, 57–75.

ZILL, S. N., MORAN, D. T. and VARELA, F. G. (1981). The exoskeleton and insect proprioception II. Reflex effects of tibial campaniform sensilla in the American cockroach, *Periplaneta americana. J. Exp. Biol. 94*, 43–55.

ZILL, S. N., UNDERWOOD, M. A., ROWLEY, J. C., III and MORAN, D. T. (1980). A somatotopic organization of groups of afferents in insect peripheral nerves. *Brain Res. 198*, 253–269.

3 Chemoreception Physiology

HIROMICHI MORITA and AKIO SHIRAISHI

Kyushu University, Fukuoka, Japan

INTRODUCTION

Chemoreception is generally classified into contact and distance chemoreceptions. Contact chemoreception is when the source of chemicals to be detected is in contact with the chemoreceptors, whereas distance chemoreception is when the source is distant from the receptors. Taste (gustation) and smell (olfaction) precisely correspond to contact and distance chemoreceptions in humans, respectively, and are also referred to in insects. In analogy with terrestrial vertebrates, olfactory receptors are highly sensitive to volatile (therefore airborne) materials and gustatory receptors are moderately sensitive to substances in aqueous solutions. Once olfaction and taste are introduced into terrestrial insects in this way, these senses should be defined in aquatic insects according to the sense organs or sensilla homologous to those in terrestial ones. Thus, the sense mediated by antennal sensilla is olfactory, even when chemicals are detected in aqueous solutions. This is true only if the antenna is the olfactory organ, but it is well known that the antenna carries contact as well as distance chemoreceptors. If we then restrict ourselves to define the modality of sense after the homologous sensilla, we could call it olfaction when the sense is mediated by thin-walled multipored cone type of sensilla. However, distance chemoreceptors of aquatic insects are not necessarily contained in completely the same type of sensilla as those of terrestrial insects: a cone-shaped sensillum on the antenna of a mosquito larva is covered by a 50 nm thick cuticle without pores (Zacharuk, R. *et al.*, 1971). It is not definitely certain that this type of sensilla of mosquito larva is homologous to the thin-walled multipored cone sensilla of terrestrial insects. Furthermore, a single sensillum does not necessarily contain receptors of the same modality. For example, the sensillum on the labellum of the fly, which is usually called the chemosensillum or the chemosensory hair, contains a mechanoreceptor as well as chemoreceptors. This fact implies that it is difficult to classify the modality after the type of sensillum. Thus, distinction between olfaction and taste is not always clear in insects. However, such terms are customarily used, and will be used here when there is no likelihood of confusion.

Another problem is whether the water and hygroreceptors are chemoreceptors. As shown later, both the receptors might be mechanoreceptors in transduction mechanisms. However, these receptors are not described in Chapter 2 (Mechanoreception) this volume. Therefore, the water receptor is considered here as a contact chemoreceptor, and the hygroreceptor as a distance chemoreceptor.

1   RECEPTOR POTENTIAL AND INITIATION OF IMPULSES

Although many works have been reported on insect olfaction, (cf. McIndoo, N., 1914), von Frisch, K. (1921) is the first investigator to ingeniously demonstrate in the honeybee that the antenna bears olfactory receptors. He trained honeybees to find the sugar solution associated with an odor, and showed that the trained honeybees lost the ability to discriminate between the odor and indifferent ones, after both antennae were amputated. Such a work

was followed by studies of electroantennogram (EAG) found by Schneider, D. (1957). The electroantennogram is a slow potential change recorded between the base and tip of an antenna during an odorous stimulus, and is considered to be composed of the receptor potentials elicited simultaneously in many olfactory receptors. The next step of investigation was the recording of responses of a single olfactory sense cell in a single chemosensillum (Morita, H. and Yamashita, S., 1961; Schneider, D. and Boeckh, J., 1962).

Physiology of the contact chemoreception in insects is considered to have begun after the discovery of Minnich, D. (1921) that the stimulation of tarsi by sugar solutions evoked the unmistakable response of proboscis extension in nymphalid butterflies. Minnich, D. also found the chemical sensitivity in the tarsi of certain muscid flies (1926a) and of the blowfly (1929). Such works were followed by extensive behavioral studies of Dethier, V. and his coworkers (reviewed by Dethier, V., 1955) on the physiology of the contact chemoreceptors of the blowfly. Minnich, D. (1926b, 1931) further described the proboscis extension response to sugar stimulations of the tip of a sensillum located on the aboral surface of the labellum of the flies *Phormia* and *Calliphora*. From the tip of this labellar chemosensory hair, Hodgson, E. *et al.* (1955) recorded impulses responding to chemical stimulations, which was the first electrophysiological work of chemosensory cells in insects. Because of its large size, the labellar chemosensillum hair has been most extensively and precisely studied as to receptor potential and impulse initiation.

1.1   Morphology

The structure of the contact chemosensillum on the labellum of the blowfly, *Phormia regina*, is shown in Fig. 1. For simplicity, only one chemosensory cell is shown, although the single sensillum contains four chemosensory cells and one mechanosensory cell. Each chemosensory cell (*N*) sends a dendritic outer segment (*D*, or sensory cilium) to the tip of the sensillum through the inner lumen which is continuous with the ciliary sinus (*cs*) (or sheath cell lumen), and their axons to the subesophageal ganglion. The sensory cilium of the mechanosensory cell (not shown) terminates at the base of the sensillar shaft as the tubular body, and its axon follows the same route to the ganglion as the chemosensory cells. These sensory cells are wrapped by glial cells (*G*) proximally from the level of the base of the dendrite, by the inner sheath cell (*I*) at the inner segment of dendrite (dendritic inner segment), and surrounded by fluid of the ciliary sinus distally up to the distal ends of the dendritic outer segment (up to the tip pore of the sensillum in the case of chemosensory cells). The trichogen (intermediate sheath cell, *M*) and tormogen (outer sheath cell, *O*) cells concentrically envelop the inner sheath cell on the one hand, and form on the other hand the sensillar sinus (*ss*) confluent to the outer lumen of the sensillar shaft. Thus, the sensillum is partitioned by the cuticular wall into two lumina: the inner lumen filled with the dendritic outer segment of the chemosensory cells and the outer lumen filled with nothing but fluid. This fluid is probably secreted by trichogen and tormogen cells. The tarsal contact chemosensillum has the same structure as described above. Such a sensillum appears as a longitudinally two-toned (dark and light) shaft under the microscope through transmitted light, and is called the two-toned hair. The two-toned hairs are the contact chemosensilla in all the species of flies ever examined. Tarsi of other insects, for example, the butterfly, bear single-toned contact chemosensilla, where the cuticular sheath surrounding the inner lumen is completely separated from the outermost wall of cuticle, forming concentric inner and outer lumina. Commonly for both types of sensilla, the outer lumen constitutes the pathway for an electric current to efficiently flow from the receptor membrane to the impulse initiation site on chemical stimulation.

In contrast to contact chemosensilla, distance (olfactory) chemosensilla of insects have porous cuticles with thin (0.1–0.3 μm) or thick (0.2–1.0 μm) walls, and the distal part of the sensory cilium highly branches in some cases. Except for such adaptations for efficient capture of stimulant molecules, the basic relation between the sensory cell and supporting structure is very similar to the case of contact chemosensory cells. The dendritic outer segment of several sensory cells are bundled by the cuticular sheath, which disappears just distal to the root of the cilium. Thus, the outer segments are surrounded by the ciliary

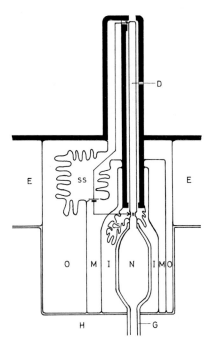

FIG. 1. Scheme of structure of the contact chemosensillum of the blowfly. Simplified as though only one chemoreceptor cell sends the dendritic outer segment to the tip of the sensillum. *cs*, ciliary sinus; *D*, dendric outer segment; *E*, epidermal cell; *G*, glia cell; *H*, hemolymph cavity; *I*, inner sheath cell; *M*, intermediate sheath cell; *N*, receptor cell; *O*, outer sheath cell; *ss*, sensillar sinus.

sinus distally up to the base of the chemosensory area which is exposed to the fluid of the outer lumen continuous with the sensillar sinus (see Fig. 24 A, B in chapter 1, this volume). Such situations are the same as in the contact chemoreceptor, as far as the generator current for impulses is concerned.

The sensilla, which are electrophysiologically ascertained to contain hygroreceptors, possess no pores. These sensilla are composed of a triad of two hygroreceptors (dry and moist receptors) and one thermoreceptor. Electron microscopy has revealed that the dendritic outer segments of two of the three receptor cells extend into the peg and terminate just beneath the tip, forming the tubular body-like structure. Within the peg, the outer segments are tightly packed by cuticular sheath, which is wrapped in turn by cuticular wall. There is no space between the sheath and the wall, and the outer segments do not seem to be directly exposed anywhere to the fluid of the sensillar sinus (or outer lumen).

In all the types of sensilla mentioned, the cilium part (connecting cilium) which connects the inner

and outer segments of the dendrite of the receptor
cell, is directly exposed to the ciliary sinus fluid. The
outer segment is partially (in olfactory receptor) or
totally (in gustatory or humidity receptor) shielded
by the cuticular sheath. The distal part of the outer
segment is thus exposed directly or faced indirectly
to the sensillar sinus fluid.

There is strong evidence that the sensillar sinus (*ss*
in Fig. 1) is 50 to 100 mV positive with reference to the
hemolymph cavity (*H*) (Thurm, U., 1972;
Wolbarsht, M., 1958; Wieczorek, H., 1982). The
membranes of tricogen and tormogen cells facing the
sensillar sinus are considered to be responsible for
this potential, acting as an electrogenic pump.
Provided that the other membranes do not act as such
a special pump, the sensillar sinus should be 50 to
100 mV positive with reference to the ciliary sinus.
Thus, the part of the connecting cilium of the receptor
cell is to be under strong catelectrotonus with
reference to the distal portion, if the sheath wrapping
the distal portion is electrically conductive at least locally (see below). This may surely assist the initiation of
impulses somewhere near the connecting cilium.

1.2 Chemosensory Impulses

When Morita, H. *et al.* (1957) first recorded impulses from the tip of the tarsal chemosensory hair
of the butterfly, *Vanessa indica*, they were puzzled
by the polarity of the recorded impulses. The recorded impulses showed an increase in positivity at the
sensillar tip with reference to the base. It seemed
reasonable to assume that chemical stimulations
were established at the tip, near to which impulses
were initiated and conducted proximally to the base
along the outer segment of the sense cell. It was
because the outer segment was very long (100–150
μm) compared with its diameter (perhaps less than
1 μm). At that time, there was no work on the
ultrastructure of this type of sensillum, but Eltringham, H. (1933) demonstrated by light microscopy
that some dye diffused from the tip into the sensillum on the tarsus of Lepidoptera.

The mystery of this polarity of impulses is resolved by assuming that the impulses are initiated at
the proximal part of the dendrite (Dethier, V., 1956;
Hodgson, E., 1958; Wolbarsht, M., 1958). Tateda,
H. and Morita, H. (1959) cut the long labellar
chemosensory hair of the blowfly, *Lucilia*, and

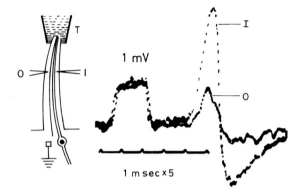

Fig. 2. Records of the same impulse obtained from the inner
(*I*) and outer (*O*) lumina of the labellar chemosensillum of the
fleshfly, *Boettcherisca peregrina*. *T* (left) is the capillary containing stimulus solution (Hori and Morita, unpublished).

recorded impulses from the cut end of the hair
remaining on the labellum, but never between the
tip and the cut end of the hair isolated from the
labellum however long the isolated portion might
be. They also recorded impulses through the side
wall of the hair, a part of which was cracked by a
glass needle. The amplitude of impulses recorded
from the sensillar tip, the indifferent (earthed)
electrode being inserted within the labellum, was
extremely reduced when the cracked part was
earthed. To the contrary, the amplitude of impulses
recorded from the cracked part was not so small
compared with those from the hair tip (inside of
labellum being earthed in both cases). All these
results are consistent with the assumption that the
impulse is initiated at a region proximal to the
cracked part. This assumption is justified below by
more precise experiments.

Figure 2 shows the records of the same impulse
simultaneously picked up by two glass capillary
microelectrodes, one (*O*) being impaled into the
sensillum from the convex (light) side and the other
(*I*) from the concave (dark) side at the same level
from the base. An indifferent electrode was inserted
into the proboscis and was earthed through the low
output resistance of a calibration pulse generator.
The shapes of records of a rectangular calibration
pulse of 1 mV (positive with reference to the
ground) and 2 ms represent the fidelity of the amplifier used. The spike potential (labeled with *I*)
recorded with the electrode *I* was about 3 mV (at the
positive peak) while the one (labeled with *O*)

recorded with the electrode O was less than 1 mV. This fact clearly indicates that there are two electrically separated rooms in the sensillum, one corresponding to the inner lumen and the other to the outer lumen. That is, the partition wall between the inner and outer lumina has an electrically high resistance.

When two electrodes were inserted into the outer lumen (Fig. 3), the spike potential recorded with the distal electrode (a) was always larger than that recorded with the proximal one (b), but both the spikes never exceeded, in amplitude, the spike recorded from the tip of the sensillum. This fact indicates that an electric current flows from the tip to the base of the sensillum within the outer lumen during the positive phase of the impulse. On the other hand, when two electrodes were inserted from the concave side (the tips of the electrodes were located within the inner lumen), the spike potential recorded from the distal level was not necessarily larger than that from the proximal one. In the case of Fig. 4, the distal potential (a) was smaller than the proximal one (b) at the peak, i.e., net current flowed distally from b to a during the rising phase of the impulses. In general, the spike potential recorded from the tip of the sensillum were smaller than that recorded from the inner lumen. These results indicate that electric currents flowed distally from the recording position to the sensillar tip as well as proximally to the base within the inner lumen during the rising phase of the impulse (cf. Fig. 8A). It should be noticed that the electric currents mentioned above were all extracellular, as the tips of the microelectrodes can never be assumed to be impaled into the dendritic outer segment. The locus generating action potential works as a sink of these extracellular currents, intensities of which are proportional to the conductances of their pathways. The membrane current flowing out at point a in Fig. 4 followed two distinct pathways, one being distally and the other proximally directed within the inner lumen. Clearly in this case; the distally directed current was larger than the proximally directed one, because the net current flowed distally between a and b. The conductance of the distally directed pathway was, therefore, higher than that of the proximally directed one at a. This implies that the inner lumen is continuous with the outer lumen somewhere near the tip as well as near the base of the sensillum.

The continuity between the inner and outer lumina at the sensillar tip is demonstrated by a

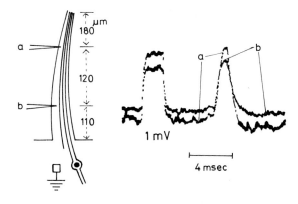

FIG. 3. Records of the same impulse obtained from different positions in the outer lumen (Hori and Morita, unpublished).

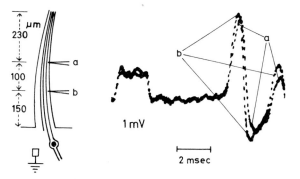

FIG. 4. Records of the same impulse obtained from different positions in the inner lumen (Hori and Morita, unpublished).

completely different experiment. The tip of the sensillum was brought into contact with $AgNO_3$ solution under a microscope. Precipitation of AgCl was observed to grow toward the sensillar base as $AgNO_3$ diffused in. In some cases, the precipitation was so dense in the outer lumen that it made a plug, beyond which no diffusion occurred, while growth of precipitation was observed in the inner lumen. Diffusion of Ag^+ ion across the partition wall was never observed at any position proximal to the outer plug (Fig. 5). Thus, it is certain that the outer lumen is connected with the inner lumen near the sensillar tip, although electron microscopy has not found any perforation on the partition wall near the tip. Altner, H. and Prillinger, L. (1980) have stated that the cuticular sheath does not impede diffusion, citing the experiments with lanthanun (Keil, T. and Thurm, U. 1979), where the tracer was not retained by the sheath. Furthermore, in one type of sen-

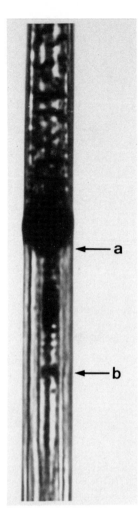

FIG. 5. Penetration of AgNO$_3$ from the sensillum tip into the inner and outer lumina. Plug resulting from dense precipitation of AgCl in the outer lumen and the front of Ag$^+$ diffusion in the inner lumen are indicated by arrows a and b, respectively.

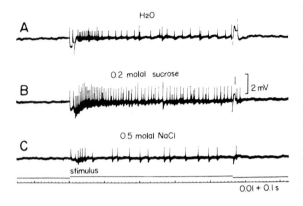

FIG. 6. Responses of a single labellar chemosensillum of the fleshfly. The stimulus solution given is indicated on each record (Morita, H. *et al.*, 1966).

sillum, in which the olfactory function is well established (Ernst, K., 1972; Waldow, U., 1973), the cuticular sheath completely encloses the dendritic outer segment. These results might indicate that electron microscopy cannot detect the difference in the structure between diffusible and non-diffusible cuticular sheaths.

There are many reasons to believe that the impulses initiated near the sensillar base are conducted distally toward the sensillar tip as well as proximally toward the central nervous system. Morita, H. (1959) cooled the sensillum locally at a point between its tip and base, and the resultant spike potential recorded from the tip showed a reduction selectively or exclusively in its falling rate. Since the cooling should reduce both the rising and the falling rates of the impulse arising at a *fixed* area of the sensory cell, he concluded that the falling phase of the spike potential recorded from the sensillar tip represented the impulse conducted distally along the outer segment within the sensillum. Such an antidromic invasion of impulses was further demonstrated by the simultaneous recordings of impulses from the tip and side wall of the sensillum (Morita, H. and Yamashita, S., 1959). The difference between the spikes recorded from the tip and side wall could be satisfactorily explained by the antidromic invasion of impulses. This conclusion was supported by Wolbarsht, M. and Hanson, F. (1965). They recorded the spike potentials from the sensillar tip which was kept in contact with a salt solution containing one of the local anesthetics, xylocaine, cocaine, etc., or tetrodotoxin, and observed that the diphasic (positive to slightly negative-going) spike potential changed to the monophasic (positive-going only) one as the local anesthetic diffused into the sensillum.

Quite recently, de Kramer, J. *et al.* (1983) studied biphasic spikes in an olfactory (pheromone) sensillum (s. trichodea) on the antenna of *Antheraea polyphemus*. They observed that an anodal current above 50 pA at the sensillar tip increased the resistance between the sensillar sinus and the hemolymph cavity, and that this increase in resistance was accompanied by a decrease in the recovery rate of the negative phase of the spike. They adapted an electrically equivalent circuit and demonstrated that a monophasic spike generated in the sensory neurone could be transformed into a biphasic spike recorded between the sensillar sinus and the hemolymph cavity. They cut the sensillar tip to record spike potentials or apply electric currents, and the properties of the dendritic outer segment might be altered, so that the antidromic conduction of impulses, which would have normally functioned, might be abolished. It seems, however, reasonable to speculate that the dendritic outer segment of the olfactory cell is incapable of impulse conduction because of its full differentiation into receptor membrane.

Impulses recorded from the outer lumen of the labellar chemosensillum of the fleshfly, *Boettcherisca peregrina*, are shown in Fig. 6, as a typical set of responses to chemical stimuli applied to the sensillar tip. We can see a train of small spikes of high frequency and that of median-sized spikes of low frequency during stimulation with H_2O (record *A*). Stimulation with 0.2 molal sucrose evoked trains of large and small spikes, both being of fairly high frequency (record *B*). When stimulated by 0.5 molal NaCl, only one of the chemoreceptor cells in the sensillum discharged a train of spikes of median size (record *C*). It is concluded that the small spikes originated in the water receptor cell, the median-sized ones in the salt receptor, and the large ones in the sugar receptor. This is the general rule in the spike height relation in the fleshfly. But generally in the blowfly, *Phormia regina*, the salt receptor discharges the large, the sugar receptor the median, and the water receptor the small spikes.

It is quite understandable that the water receptor responds to aqueous sucrose solution, but it may be somewhat surprising that the salt receptor responds to pure water (see section 1.5.2). On the contrary, the discharge of impulses in the water receptor is completely inhibited by NaCl at concentrations higher than 0.5 M.

1.3 Receptor potential

Figure 7 shows the responses of the sugar and the water receptors to glucose stimulations given to a single labellar chemosensillum of the fleshfly, *Boettcherisca peregrina*. Two microelectrodes were inserted into the outer lumen, and the potential changes between the electrodes (the proximal one

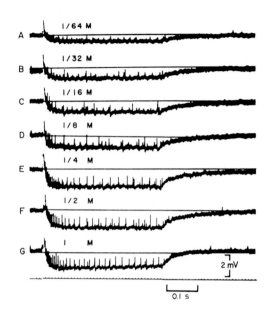

FIG. 7. D.C. records of responses of a single labellar chemosensillum of the fleshfly. Stimulus was glucose solution of the concentration indicated (Hori and Morita, unpublished).

was earthed) were recorded through a D.C. amplifier. It is possible to obtain D.C. records with a single microelectrode impaled into the outer lumen, the indifferent electrode being located within the proboscis. Compared with this single microelectrode method, the double microelectrodes method greatly reduces irregular fluctuations in potential, which might be related to secretory activities of tormogen and trichogen cells (Thurm, U., 1972; and see discussion below). In addition to impulses previously described, a potential change sustained during the stimulus was recorded. This potential change represents an increase in negativity at the distal electrode with respect to the proximal one, and becomes larger with higher intensities of stimulus. Therefore, this slow potential is considered to be the receptor potential(s) of the stimulated chemosensory cell(s). It is important to notice that an electric current flows toward the tip within the outer lumen during the receptor potential, which acts as a generator of the chemosensory impulses at the base of the sensillum.

Though several authors have discussed the generation of impulses by the receptor potential (recently discussed by Broyles, J. *et al.*, 1976; Maes, F., 1977), it seems that none of them gives a satisfactory solution to this problem. One of the causes of the confusion in understanding the electrophysiology of the chemosensillum is in the view-point of authors who do not consistently treat the recorded electrical activities as the extracellular ones. Morita, H. (1969, 1972a) described the generator current after the cable theory, but not fully quantitatively. Some of his equations were incomplete, and therefore correct expressions of the currents associated with the receptor potential will be given here, since this is the only way to understand the recorded potentials.

The contact chemosensillum of the fly is represented by A in Fig. 8. The depolarization, $V_m(0)$, occurs at the tip of the outer segment ($x=0$) during a chemical stimulation. Owing to this depolarization, the membrane currents (i_m, current intensity per unit length) flow, and are divided into distally and proximally directed external currents (i_{0_1} and i_{0_2}, respectively). As discussed before (section 1.2.), the inner and outer lumina are assumed to communicate with each other at the sensillum tip. The partition wall ends at $x=a$, where the outer segment is somewhat tightly wrapped by the sheath continuous with the partition wall. Just proximal to this point, the outer dendritic segment is constricted and connected to the inner dendritic segment of the sensory cell. For simplicity, let us assume that the partition wall and sheath are perfect insulators for electric currents. Then, the longitudinal current, I_o, constant independently of x, flows within the outer lumen. For $x < a$, the membrane potential change at x, $V_m(x)$, is expressed as:

$$V_m(x) = V_1 \exp(-x/\lambda) + V_2 \exp(x/\lambda), \quad (1)$$

and the extracellular longitudinal current within the inner lumen at x, $i_o(x)$, satisfies the relation:

$$dV_m(x)/dx = -i_o(r_i + r_o) - I_o r_i, \quad (2)$$

where λ is the length constant, r_i and r_o longitudinal intra- and extracellular resistances, respectively, per unit length within the inner lumen. V_1 and V_2 are determined by the boundary conditions as:

$$-i_o(a) - I_o + \int_a^\infty i_m(x)dx = 0, \quad (3)$$

$$V_o(a) = \int_0^a r_o i_o(x)dx = I_o R_o, \quad (4)$$

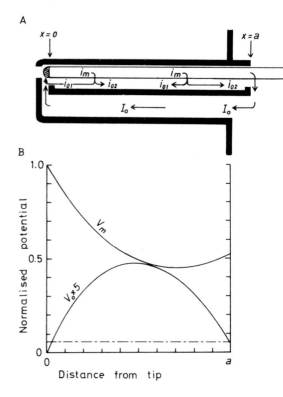

FIG. 8. Illustration of currents associated with depolarization at the tip of the chemoreceptor cell. x, distance from the receptor tip; i_m, membrane current; i_{o_1} and i_{o_2}, longitudinal current within the inner lumen toward the tip and base, respectively; I_o, longitudinal current within the outer lumen; V_o, potential at x within the inner lumen; V_m, the membrane potential displacement at x.

where $V_o(x)$ denotes the extracellular potential at x within the inner lumen, R_o the longitudinal resistance of the outer lumen in total, and the directions of currents are chosen as positive as arrows indicate (the direction of $i_o(x)$ is the same as i_{o_1}).

An example of the values of $V_m(x)$ and $V_o(x)$ is shown in Fig. 8B. The parameters are chosen as $R_o/r_o = 0.1$, $r_i/r_o = 2$, $a/\lambda = 2$, $\lambda'/\lambda = 0.5$, and $r_i/r_{o,a} = 0.1$, where λ' represents the length constant for $x > a$, and $r_{o,a}$ the value of r_o at $x = a$. The value of $\lambda/\lambda' = 0.5$ corresponds to constriction at the connecting cilium just proximal to the end of the partition wall, and $r_i/r_{o,a} = 0.1$ corresponds to tight wrapping of the sheath at $x = a$. The other values were selected by examining the dimensions of structures in electron micrographs of the sensillum. The indifferent electrode is equipotential with $V_o(a)$, and therefore the results in Fig. 9B indicate that the tip of the sensillum is $-V_o(a)$ with reference to the base. The receptor potential ordinarily recorded at x within the outer lumen should be negative by $V_0(a)\, x/a$, if the outer lumen is completely uniform throughout its length. On the other hand, the receptor potential recorded from the inner lumen should be positive almost everywhere except in the tip region. All of these predictions of this figure are ascertained by actual records, and the increase of V_m at $x = a$ clearly explains how the depolarization at the tip of the outer segment is efficiently transmitted to the basal region, where the impulses are initiated.

Thus, the two-lumina system of the contact chemosensillum is a very effective adaptation for the faraway transmission of membrane potential displacement at the tip to the basal region. The best way for transmission is to make the membrane resistance infinitely high in the dendritic outer segment, so that there is no leak current through the membrane of the outer segment. This might be impossible without myelination as occurs in vertebrate axons. The olfactory sensilla are adapted so that chemoreception may proceed everywhere in the outer segment. Here also, the best way to converge the generator current into the spike initiation site is the same as in the contact chemosensillum, and was actually assumed by Kaissling, K. (1971).

As mentioned in section 1.1, the tip of the sensillum is positive (not larger than 100 mV) with reference to the inside of the labellum in the resting state. Therefore, this potential may assist the intracellular longitudinal current increasing in intensity, but the results illustrated in Fig. 8B still hold.

Because of its large size, the labellar chemosensillum is the only sensillum which has been studied by the precise electrophysiological technique. As pointed out in section 1.1, the single-toned chemosensilla, such as those on the tarsi of butterflies, also have inner and outer lumina, the latter of which may serve as the pathway for generator current to flow efficiently to initiate impulses in the basal region. In this case also, the receptor potential is recorded by the electrode inserted into the sensillum as a potential change negative with reference to the base, accompanied by impulses during the stimulation (Morita, unpublished).

1.4 Impulse frequency vs. receptor potential

The sustained potential during the stimulation in Fig. 7 was described as the receptor potential. However,

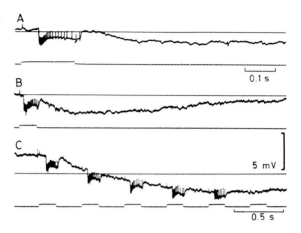

FIG. 9. Very slow potential other than the receptor potential elicited by 0.25 M NaCl. 21°C, *Lucilia* (Morita, H. and Yamashita, S., 1966).

it is not self-evident that the sustained potential accompanied by impulses is the receptor potential, because there are many sources of potential changes especially in the contact chemosensillum. First of all, we have to examine the possibility that the sustained potential is the diffusion potential. In fact, a very slow potential change is recorded when the tip of the sensillum is stimulated by high concentrations of NaCl. As can be seen in Fig. 9, this potential which is negative with reference to the base, grows till after the end of the stimulus of 0.25 M NaCl and attains an amplitude far larger than the potential sustained during the stimulus. Close examinations reveal that the same potential changes of opposite polarity are induced by water or non-electrolyte solutions, although their amplitudes are so small as to be just detectable. Such a property suggests that this very slow potential change is related to the diffusion of salt, but its polarity is opposite to that expected in an ordinary aqueous NaCl solution. Some special characteristics of the sensillar tip pore (for example, negatively charged wall) may have to be assumed to account for this opposite polarity.

To ascertain that the potential sustained during the stimulus is the receptor potential, the relation between the amplitude of the sustained potential and the impulse frequency should be examined. Generally, there are several receptor cells in a single sensillum, and a single stimulus might evoke responses in more than one receptor cell. This difficulty is resolved as follows in the contact chemosensillum. As can be seen in Fig. 7, both the impulse frequency and the

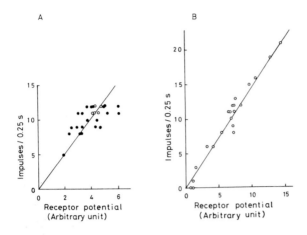

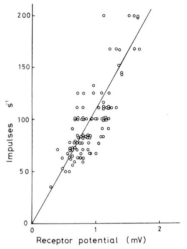

FIG. 10. Relation between impulse frequency and receptor potential in the steady state, estimated as shown in the text (water and sugar receptors, in **A** and **B**, respectively) (Hori and Morita, unpublished).

FIG. 11. Correlation between impulse frequency and receptor potential in the sensillum, where the water receptor was inactive (Morita, H. and Yamashita, S., 1966).

amplitude of the sustained potential are in a steady state during the period from 0.15 to 0.35 s after the beginning of stimulus. In record *A* of Fig. 7, only water receptor impulses are seen, and the sustained potential can be assumed to be the water receptor potential. Assuming further that the amplitude of the water receptor potential is proportional to the frequency of the water receptor impulses in the steady state, we can estimate the water receptor potential by counting the number of water receptor impulses occurring together with the sugar receptor impulses at high concentrations of glucose. Subtracting the estimated water receptor potential from the sustained potential, we can obtain the sugar receptor potential. In Fig. 10B, the impulse frequency is plotted against the amplitude of receptor potential thus estimated for the sugar receptor. The results are expressed by the straight line crossing the origin of ordinates. Based on this proportionality, we can estimate reversely the amplitude of water receptor potential; this is plotted as closed circles in Fig. 10A. Thus, all points plotted can be regarded as being on the straight line. Such a linear relation was also ascertained in the salt receptor cell.

The results mentioned here strongly suggest that a linear relation exists between the impulse frequency and the receptor potential, but do not prove it. However, Morita, H. and Yamashita, S. (1966) demonstrated the linear relation in the sugar receptor of the green bottlefly, *Lucilia*, where only the sugar receptor happened to function in the sensillum (Fig. 11).

Therefore, it is most probable that a linear relation exists between the impulse frequency and the receptor potential in every contact chemosensory cell.

Generally, the response magnitude is measured by the impulse frequency, and therefore its linear relation to the receptor potential is important especially for discussions of primary processes in the receptor membrane.

In the honeybee, Kaissling, K. (1971) shows a relation between the impulse frequency and receptor potential in the receptor of queen substance (Fig. 35 in Kaissling, K., 1971). The relation is not linear. This is not surprising, because the slow potential change evoked by odor stimulation is most probably contaminated with EAG. In contrast to contact chemoreceptor, it is very difficult to stimulate selectively only the sensillum under experimentation.

1.5 Response magnitude *vs.* stimulus intensity

Figure 12 shows the typical concentration–response relations for sucrose, glucose and fructose in a single labellar sugar receptor of the fleshfly, *Boettcherisca peregrina*. Sucrose evokes the highest response at every concentration. The response to fructose is higher than that to glucose below 0.3 M, but lower above 0.4 M. The maximum response differs with different sugars: those to glucose and to fructose are on average about 70 and 45% of that to sucrose (Morita, H. and Shiraishi, A., 1968).

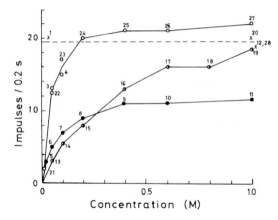

FIG. 12. Intensity–response curve in a single receptor for sucrose ($\bigcirc$), glucose ($\pmb{\bigcirc}$) and fructose ($\bullet$). 0.2 M sucrose ($\times$) was given as control at times. Number attached to each symbol indicates the order of stimulation (Morita, H. and Shiraishi, A., 1968).

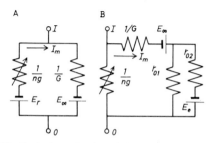

FIG. 13. Equivalent circuits of the receptor potential, **A** proposed by Morita, H. (1969) and **B** for Thurm's proposal. I_m, membrane current across the receptor membrane; I and O, inside and outside of the cell across the receptor membrane, respectively; ng, receptor membrane conductance dependent on stimulus; G, membrane conductance independent of stimulus; E_∞, resting receptor membrane potential; E_r, reversal potential; E_e, potential different between sensillar and ciliary sinuses; r_{O1} and r_{O2}, shunting and series resistances for E_e, respectively.

Beidler, L. (1954) published a theory describing the relation between response magnitude and stimulus intensity in a group of chemoreceptors of the tongue of the rat. This theory assumes a proportionality between the response magnitude and the number of receptor sites occupied by molecules or ions of stimulus. When each receptor site is occupied by one molecule or ion of stimulus, the response magnitude, r, is expressed as:

$$r = r_\infty/(1 + K/a), \qquad (5)$$

where r_∞ is the maximum response, a the concentration of stimulus, K the dissociation constant of the stimulus–receptor complex. Since then, many workers have examined the response in insect

chemoreceptor after this theory (Evans, D. and Mellon, D., 1962b; Morita, H. and Shiraishi, A., 1968; cf. Gillary, H., 1966).

Equation (5) is completely the same in form as the Michaelis–Menten equation for the enzyme kinetics. In the contact chemosensillum, the potential drop due to the current, I_o in Fig. 8A, was referred to as the receptor potential. Every chemosensory cell in insects makes such currents flow when chemically stimulated. Therefore, we have to examine whether the receptor potential is proportional to the number of the stimulus–receptor site complexes.

Figure 13A is an equivalent circuit, where I and O represent the inside and outside, respectively, of the sensory cell at the receptor locus. On the left and right sides, respectively, the membrane conductances (ng and G) dependent on and independent of stimuli are shown. The potential at O with reference to I is defined as E_o when the membrane current I_m flows for n, the number of the channels opened by stimulus. For simplicity, it is assumed that $I_m = 0$ and $E_o = E_\infty$, when $n = 0$ is at rest. Defining $V = E_o - E_\infty$ (depolarization for $V < 0$, hyperpolarization for $V > 0$) and $V_m = E_r - E_\infty$, we obtain:

$$V = V_m/(1 + G/ng), \qquad (6)$$

and

$$I_m = -GV. \qquad (7)$$

Any extracellular current is proportional to I_m, and consequently is proportional to V. This result indicates that all the potential changes, intra- and extracellularly recorded, are proportional to each other, and can be called the receptor potential.

If we assume that the channel is opened by one stimulus molecule occupying the receptor site of the channel, we obtain the equation for intensity–response relation as:

$$V = V_\infty/(1 + K_b/a), \qquad (8)$$

$$K_b = K/(1 + sg/G), \qquad (9)$$

where s is the total number of the channels, and K is the dissociation constant of the stimulus–receptor site complex. V_∞ is the value of V at $a = \infty$, i.e., the maximum response at the receptor membrane. Since the impulse frequency, r, is proportional to V as shown in Figs 10 and 11, equation (8) can be rewritten as:

$$r = r_\infty/(1 + K_b/a). \qquad (8')$$

Obviously, this is the simplest case that can be supposed when channels are opened by the stimulus, having their own electromotive force or reversal potential, E_r.

Figure 13B represents the situation adopted by Kaissling, K. (1971) according to Thurm's theory. It is assumed that the channel opened by the stimulus does not have its own electromotive force ($E_r = 0$). E_e represents the potential difference between the sensillar and ciliary sinuses; r_{O_1} is the shunting resistance and r_{O_2} the series resistance. As mentioned in Fig. 1, E_e promotes the membrane current I_m across the receptor membrane. Let us consider an ideal case, where $r_{O_1} = \infty$ and $r_{O_2} = 0$. Then, $V = E_o - (E_\infty + E_e) = -I_m/G$ and $V_m = -(E_\infty + E_e)$, so that $V = V_m/(1 + G/ng)$. These are completely the same relations obtained in Fig. 13A, except for the value of V_m. Taking E_e into account in Fig. 13A, i.e., inserting E_e between E_r and E_∞ in the circuit of A, V_m also coincides for both circuits when E_r is 0. Thus, the relation of equation (9) clearly holds for the ideal case of Thurm's theory.

As shown by Fig. 12, the same receptor gives rise to different maximum responses to different sugars. For different sugars, different values can be assumed for E_r, n_∞ and/or g in Fig. 13, where n_∞ is the number of channels activated by an infinitely high concentration of stimulus. It is obvious that the maximum response is different if any of the above three constants is different for different sugars. Here, E_r is examined with the results of responses affected by electric currents. Equation (6) shows that the response is proportional to V_m ($= E_r - E_\infty$), which can be evaluated with the depolarizing current counterbalancing the receptor potential.

As described in detail by Morita, H. (1972a), the receptor membrane can be de- or hyperpolarized by electric currents which are made to flow between the sensillum tip and the inside of the labellum. The sensillum tip was continuously exposed to a flow of 0.1 M NaCl solution, and the sugar receptor was stimulated only when a jet of 0.4 M sucrose or fructose dissolved in 0.1 M NaCl was blown from each container (capillary) on to the sensillum tip. Under the current-clamp condition, the intensity of current could be kept constant (within a few per cent change) before, during and after the sugar stimulation. As shown by Fig. 12, sucrose and fructose of the above-mentioned concentration evoked almost

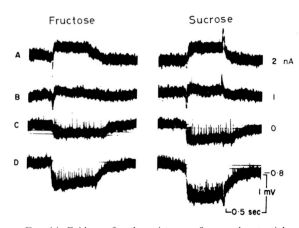

FIG. 14. Evidence for the existence of reversal potential. Records under the influence of electric current (Morita, Amakawa and Uehara, unpublished). Anodal (< 0) and cathodal (> 0) currents were applied at the sensillar tip.

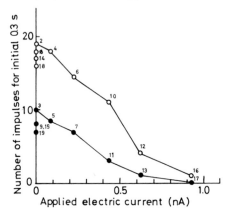

FIG. 15. Response of sugar receptor as a function of intensity of electric current (Morita and Hori, unpublished). A single sugar receptor was stimulated by 0.4 M sucrose ($\bigcirc$) and by 0.4 M fructose ($\bullet$). Number attached to each symbol indicates stimulation order.

the maximum responses, and the response to sucrose was about twice as large as that to fructose. Therefore, the intensity of depolarizing current counterbalancing the response to sucrose should be about twice as large as that to fructose, if the E_r value only is responsible for the response difference.

Figure 14 shows the receptor potential recorded under the above-mentioned condition. It can be seen that the depolarizing current reversed both the receptor potentials at 1 nA. The impulses responding to sucrose and to fructose were also abolished at about 1 nA as shown in Fig. 15. It cannot be determined whether E_r is the same for sucrose and fructose because of the poor accuracy in experiments,

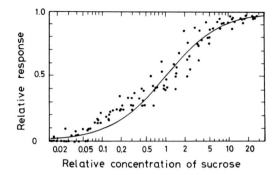

FIG. 16. Intensity–response curve of sugar receptor stimulated by sucrose (Morita, H. and Shiraishi, A., 1968). Normalized so that the maximum response (r_∞) and the half maximum concentration (K_b) are unity for each receptor. The continuous line is the theoretical curve obtained from equation (8').

but it is certain from the slope in Fig. 15 that V_m for sucrose is not twice as large as that for fructose. Quite different results were obtained for the salt receptor (see section 1.5.2).

As shown above, the relation of intensity–response is obtained by assuming the most simple or an ideal case. Therefore, it is quite understandable that the actual intensity–response relation deviates significantly from the prediction of equation (8'). Here, one example obtained in the fleshfly, *Boettcherisca peregrina*, is shown in Fig. 16 (Morita, H. and Shiraishi, A., 1968), where this equation describes well the response of the labellar sugar receptor to sucrose. However, the same sugar receptor does not respond to glucose like this. Such specific intensity–response curves are mentioned below.

1.5.1 SUGAR RECEPTOR

The response of the labellar sugar receptor to glucose is not described by equation (8'). Figure 17 shows the results obtained in the fleshfly, *Boettcherisca peregrina* (Morita, H. and Shiraishi, A., 1968). Let us assume that the channel is opened by its receptor site occupied by two molecules of glucose. Then, the binding reaction proceeds in two steps as,

$$A + S \overset{K_1}{\rightleftharpoons} AS, \quad A + AS \overset{K_2}{\rightleftharpoons} A_2 S,$$

where K_1 and K_2 are dissociation constants in the first and second steps, respectively. Defining $\alpha = K_2/K_1$ and $c = [A]/K_1$, we calculate the number (n) of $A_2 S$. Introducing the calculated n into equation (6), we obtain:

$$V/V_\infty = r/r_\infty = 1/(1 + \alpha'/c + \alpha'/c^2), \quad (10)$$

where $\alpha' = \alpha/(1 + sg/G)$. The continuous line in Fig. 17 is calculated from equation (10), adopting $\alpha' = 1.0$. The results of fructose stimulation are also expressed by equation (10), where the value of α' is 4.0. Thus, it can be supposed that the receptor site of the channel has two subunits, each being occupied by one monosaccharide molecule. Disaccharides such as sucrose and maltose would occupy these subunits by one molecule, and satisfy equation (8); monosaccharides such as glucose and fructose would occupy the subunits by two molecules, and satisfy equation (10). Mannose is known as an unique monosaccharide. In spite of a very weak stimulating effect, it is a strong competitive inhibitor for fructose according to the behavioral study on the blowfly, *Phormia regina*, by Dethier, V. *et al.* (1956). Their conclusion is verified in the labellar sugar receptor of the fleshfly: mannose definitely has an inhibitory effect on the response to fructose at concentrations, where mannose does not have any stimulating effect by itself (Morita, H. and Shiraishi, A., 1968). Such a result suggests that only one of the subunits occupied by the mannose molecule purely exhibits inhibition.

The model expressed by equation (10) should be called 1:2 complex model, while that by equation (8) should be called 1:1 complex model. These models well explain the responses to di- and monosaccharides, respectively, but are not proved to be actual processes. In fact, these models do not explain the reason why the maximum responses to sucrose, glucose and fructose are different (see Fig. 12). Morita, H. and Shiraishi, A. (1968) explained this fact by adopting a model for allosteric transitions (Monod, J. *et al.*, 1965). In this model, the maximum response, r_∞ or V_∞, is modified by the affinity of the receptor site for the stimulus in the inactive state.

Other factors may operate in determination of the maximum response, and will be discussed later.

1.5.2 SALT RECEPTOR

Evans, D. and Mellon, D. (1962a,b) were the first to demonstrate in the labellar chemosensillum of the blowfly that quantitatively reproducible responses were obtained if the duration of stimulus was limited to less than 1 s.

They (1962b) examined the response of the salt receptor according to Beidler's theory (equation 5).

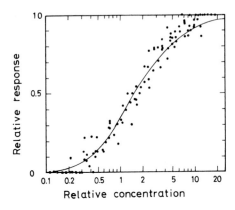

FIG. 17. Intensity–response curve for glucose stimulation (Morita, H. and Shiraishi, A., 1968). The continuous line is drawn after equation (10). See the text as to definition of the relative concentration (*c*).

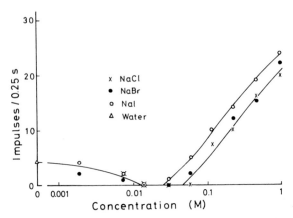

FIG. 18. Intensity–response curve in the salt receptor (Hori, unpublished).

Equation (5) was re-arranged as $a/r_\infty = a/r_\infty + K/r$ (Beidler's taste equation), where a was the concentration of salt used for the stimulus. They plotted a/r against a, and obtained a straight line crossing the X-axis at $-K$ with the slope of $1/r_\infty$. As already mentioned in Beidler's theory, K is the dissociation constant between the salt and the receptor site, but they found that the graphically obtained K values varied with receptors. They could not explain the reason for this variation.

Gillary, H. (1966a) re-examined the intensity–response relation of the same salt receptor in NaCl stimulation, and showed the linear relation between the response and the logarithm of the salt concentration over a range of 0.4 to 5.0 M.

Rees, C. (1968) studied the same problem, and explained the results assuming that the membrane potential is established across the receptor membrane according to the concentrations and permeability coefficients of the related ion species. The equation for the membrane potential was given by Hodgkin, A. and Katz, B. (1949). Assuming that the intracellular concentration (activity) of Cl^- was 0.05, the ratio of the permeability coefficients $P_{Na} : P_K : P_{Cl} : P_{Br} : P_I = 1.40 : 1.00 : 0.80 : 0.67 : 0.19$ was shown to give the best fit to the actual response.

In all the above-mentioned works, the concentrations of salts used for the stimulus were relatively high, i.e., above 0.1 M. This is because the salt receptor discharges no impulses at about 0.1 M. The side-wall recording enables us to record responses to very dilute salt solution up to pure water. Figure 18 shows that the salt receptor discharges a few impulses at extremely low concentrations of salts. The receptor potential was positive (hyperpolarization of the receptor membrane) when the receptor was stimulated by high concentrations of choline chloride or tris-(hydroxymethyl) amino-methane chloride (Hori, unpublished).

These results suggest that inorganic cations and anions de- and hyperpolarize the receptor membrane, respectively, as suggested by Beidler, L. (1967) in mammals. Under the condition of the side-wall recording (the recording capillary electrode contained 0.1 M NaCl), the salt receptor spontaneously discharges impulses at a low frequency. The hyperpolarization by anions mentioned above may result from the closing of channels which open during the unstimulated period, i.e., a decrease in n in Fig. 13. Cations may increase the number, n, of open channels, and the reversal potential, E_r, may also vary with concentrations of salts used for stimulus (Rees, C., 1968).

Maes, F. (1977) quantitatively studied the electric effects on the response of the labellar salt receptor of *Calliphora vicina* to various concentrations of NaCl or KCl. The electric current anodal at the sensillar tip increased the response to salt stimuli, while cathodal currents decreased the response. In contrast to the sugar receptor (section 1.5.1), the cathodal current necessary to just abolish the response was larger for higher stimulus concentrations. This fact suggests that the channel is opened by cations, but the reversal potential is different because of the different concentration of salts on the outside of the receptor membrane (the value of E_r in Fig. 13A is more negative at higher cation concentrations), even if the relative permeability coefficients for ions are constant.

1.5.3 PHEROMONE RECEPTOR

Pheromone receptors are known as chemoreceptors of extremely high and specific sensitivity. Among them, the bombykol receptor of the male silkworm moth, *Bombyx mori*, is the most extensively studied receptor, partly because bombykol is the first pheromone to be purified and then determined in chemical constitution (Butenandt, A. *et al.*, 1959).

Kaissling (unpublished; referred to in Kaissling, K., 1971) measured adsorption of bombykol by *Bombyx* antenna using tritiated bombykol synthetized by Kasang, G. (1968). The results showed that as much as 27% of the bombykol was filtered out of the airstream when it passed through the antenna. Of the total adsorbed bombykol, 75% was adsorbed on the surface of sensilla trichodea. This implies that 20% of bombykol contained in the airstream passing through the antenna was caught by the chemosensilla. Kaissling, K. (1971) calculated the number of bombykol molecules adsorbed on the sensillum, knowing the concentration of bombykol in the airstream at the threshold determined by Kaissling, K. and Preisner, E. (1970). It was revealed that 0.68 of molecules was sufficient for a single receptor cell to discharge 0.34 impulses above the noise level, assuming that each of the chemosensilla contained two bombykol receptor cells.

Figure 19 shows a kind of intensity–response curve for bombykol, where the percentage of receptor cells which responded with one or more impulses is plotted against the stimulus intensity (open circles). Here, the intensity is given as $\log_{10}$ of the amount of bombykol loaded on a piece of filter paper as the stimulus source, to which the bombykol concentration in the stimulus airstream was ascertained to be proportional. The broken lines are theoretical curves obtained from the Poisson equation as:

$$P_n = e^{-m} \sum_{k=n}^{\infty} (m^k/k!)$$

$$= 1 - e^{-m} \sum_{k=0}^{n-1} (m^k/k!), \tag{11}$$

where P_n is the probability of occurrence of n or more molecules hitting the single receptor cell, and m is the average number of molecules hitting the

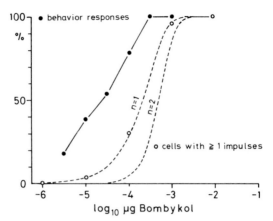

Bombyx mori

FIG. 19. Responses of the male silkworm moth to the sex attractant, bombykol (Redrawn from Kaissling, K., 1971). Broken lines were calculated after equation (11).

cell. The results in Fig. 19 clearly indicate that the best fit was obtained in the case of $n = 1$ (notice that the slope is steeper with larger n). Thus, the percentage of responding receptor cells coincided with the probability that one or more molecules of bombykol hit the single receptor cell. This result is consistent with the assumption that one molecule of bombykol is sufficient for excitation of the receptor cell.

The solid circles in Fig. 19 represent the percentage of the male individuals which responded to the given stimulus with fluttering of wings. The curves show that 50% of behavioral response was obtained at the stimulus intensity where less than 8% of the receptors were excited, and that the behavioral response was maximal at about 50% of the receptor response. Thus, the relation between behavioral and receptor responses is clearly non-linear, and no simple relation has been elucidated.

2 SPECIFICITIES OF RECEPTORS

2.1 Sugar receptor

Many insects distinguish sugar solutions from water, and ingest them. Since the classical work of von Frisch, K. (1935) with the honeybee (*Apis mellifera*), it has been assumed that a single kind of receptor is involved in the ingestion of sugars. This receptor is called a sugar receptor, although it is responsive to substances other than sugars. von Frisch, K. (1935)

Table 1: *Stimulating effectiveness (relative to sucrose)*

Classification	Substance	Effectiveness	
		honeybee[a]	blowfly[b]
Alcohol			
Linear	erythritol	0	0
	mannitol	0	0
	sorbitol	0	0
	dulcitol	0	0
Cyclic	quercitol	0	
	inositol	0.06–0.1	0.05
Monosaccharide			
Pentose	arabinose*	approx. 0.03	0.02
	arabinose	approx. 0.03	0.07
	xylose	0	0.02
Methyl pentose	fucose	0.125	0.1
	rhamnose*	0	0
Hexose	glucose	0.25–0.5	0.08
	galactose	0.03–0.06	0.02
	mannose	approx. 0.02	0.002
	fructose	0.25–0.5	2
	sorbose	0	0.07*
Disaccharide	sucrose	1	1
	trehalose	0.25–0.5	0.07, 0.05–0.1[c]
	maltose	0.5–1	2, 1[c]
	cellobiose	0	0.002, 0.02–0.1[c]
	gentiobiose	0	0.01–0.003[c]
	lactose	0.02–0.03	0[c]
	melibiose	0.04	0
Trisaccharide	raffinose	approx. 0.5	0.5, 0.05–0.1[c]
	melezitose		0.15
Polysaccharide	glycogen	0	

* L-isomer, all others D-isomer
[a] From von Frisch, K. (1934)   [b] From Dethier, V. (1955)   [c] From Pflumm, W. (1972).

defined the sweetness s of substance A so that s M sucrose was equivalent to 1 M substance A in the ingestion response of the honeybee. The value s was less than unity for all substances tested (Table 1, $s = 1$ for sucrose by definition).

The results in Table 1 show:

1. Sugar receptor is more specific in the honeybee than in humans. The honeybee does not respond to the artificial sweeteners of man such as saccharin, dulcin etc.
2. The size of molecule is important. Attractive sugars are trisaccharides at largest, while they are pentoses at smallest.
3. Sugars with α-glycosidic linkages are more effective than the corresponding β-anomers.
4. The pyran ring oxygen is not essential, since inositol, a cyclohexan derivative, is attractive.

All these conclusions were confirmed by Dethier, V. (1955) also in the blowfly, *Phormia regina* (Table

1). However, they could not find configurations that were common to the effective sugars.

2.1.1   GLUCOSE SITE AND FRUCTOSE SITE

Evans, D. (1963) re-evaluated the earlier data for *Phormia* (Hassett, C. *et al.*, 1950; Dethier, V., 1955), assuming that there were at least two types of combining sites, glucose and fructose sites, in the sugar receptor of *Phormia*. This assumption was based on the synergism between glucose and fructose, on competition between mannose and fructose (Dethier, V., 1955; Dethier, V. *et al.*, 1956), and on selective depression of the sensitivity of the adult flies either to fructose or to glucose by rearing them in the presence of one or the other sugar at the larval stage (Evans, D., 1961). Evans, D. (1963) proceeded to analyze in detail the structural requirement for the glucose site, introducing a view

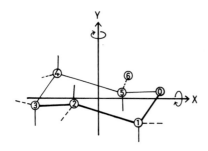

FIG. 20. Cl conformation of pyranose ring. Solid and broken lines drawn from ring carbons (not between ring carbons) represent axial and equatorial directions, respectively. Another conformation 1C is in rotation symmetry with C1 (180° rotation about axis X). Y is another rotation axis perpendicular to X.

of the conformation of pyranose ring (Fig. 20). He postulated that equatorial hydroxyl groups on C_3 and C_4 alone were essential for the glucose site.

As for the fructose site, Kijima, H. (1970) and Morita, H. (1972b) suggested that β-D-fructopyranose would be effective, but Hanamori, T. et al. (1974) studied the response of single labellar sugar receptors of the fleshfly, Boettcherisca peregrina, during mutarotation after D-fructose was dissolved in water, and found that β-D-fructofuranose is effective in D-fructose solution. They tried also to find structures of monosaccharides and their derivatives required for effectiveness in pyranose form. There are generally two stable conformations, C1 and 1C, in pyranoses; 1C can be superimposed over C1 by 180° rotation about one axis (X in Fig. 20) or by 60° rotation and then by every 120° rotation about another axis (Y in Fig. 20). As seen in the latter case, the ring oxygen is regarded as equivalent to carbon. This is justified because myo-inositol (cyclohexanol) is effective. Operating such rotations, we can find a pattern of residues distributed over ring carbons of pyranose common to effective sugars as shown in Table 2. The stimulating effectiveness should be determined by the concentration–response curve in single receptor cells, which is generally expressed by an empirical equation as:

$$r = r_\infty/(1 + K_b^n/a^n),$$

where r is the response, r_∞ the maximum response, K_b the concentration at which a half maximum response is obtained, a the concentration, and n the Hill coefficient. Among the three parameters, r_∞

and K_∞ are in particular closely related to the stimulating effectiveness. We have such complete sets of these parameters only for a few sugars; therefore, 50% threshold concentration (E_{50}) for ingestion response instead of K_b and r_∞ are listed in Table 2.

Sugars and their derivatives in group A that have equatorial OH at C-2, C-3 and C-4 are very effective. All of them are considered to be effective to the glucose site in this configuration. Only one exception is D-idose, which should be re-examined. An axial aglycon on C-1 increases effectiveness (methyl α-D-glucoside), while the same equatorial one decreases it (methyl β-D-glucoside). Hanamori, T. et al. (1972) showed in single labellar sugar receptors of Boettcherisca peregrina:

1. The hydrophilic aglycons in α-configuration such as ethylene glycol increased the maximum response (sucrose produces the highest maximum response as fructofuranosyl α-D-glucopyranoside), while hydrophobic aglycons such as ethyl and p-nitrophenyl groups decreased it.

2. The affinity for the sugar receptor site (measured with the reciprocal of K_b value) was increased by both hydrophilic and hydrophobic aglycons in α-configuration. The value of K_b was 0.0019 ± 0.0007 (S.D.) M for p-nitrophenyl α-D-glucopyranoside vs. 0.15 ± 0.03 M for α-D-glucopyranose. 2-O-Methyl-D-glucose and 2-deoxy-D-glucose displace the equatorial OH group on C-2 by hydrogen and by OCH_3 group, respectively. Such displacements slightly reduce both the affinity and the maximum response. myo-Inositol and D-chiro-inositol are listed in this group. Inositols as well as glycosides listed in the Table are definitely in the form of cyclohexan, while sugars have other possible forms such as furanose and open chain. Therefore, inositols are useful for studies of effective structure in the form of pyranose. Thus, the ring oxygen can be replaced by carbon with equatorial or axial hydroxyl group.

Sugars in group B are all fairly effective, and have equatorial hydroxyl groups at C-2, C-3 and C-4 as in group A, but have oxygen at C-5 and carbon at 0 position after operation of rotations. As mentioned above, ring oxygen can be replaced by carbon with hydrogen and hydroxyl group. Therefore, α-L-arabinose might be effective to the glucose site in this configuration, but the others

Table 2: Classification and stimulating effectiveness of pyranoses in flies

Group	Compound	Chair[a] form	C-1[b] Eq., Ax.	C-2 Eq., Ax.	C-3 Eq., Ax.	C-4 Eq., Ax.	C-5 Eq., Ax.	O Eq., Ax.	50% threshold in M	Maximum response (sucrose = 1)
	Methyl α-D-glucoside	C1	H, OCH_3	OH, H	OH, H	OH, H	CH_2OH, H	O	0.069[c]	0.79[f] 0.71[g]
	α-D-Glucose	C1	H, OH	OH, H	OH, H	OH, H	CH_2OH, H	O	0.132[d]	0.74[g]
	β-D-Glucose	C1	OH, H	OH, H	OH, H	OH, H	CH_2OH, H	O	0.132[d]	0.66[g]
	D-Glucose	C1	(H, OH)	OH, H	OH, H	OH, H	CH_2OH, H	O	0.132[c]	0.72[g]
	L-Glucose	1C	CH_2OH, H	OH, H	OH, H	OH, H	(H, OH)	O	0.140[c]	0.50[f]
	L-Sorbose	1C	H, H	OH, H	OH, H	OH, H	(OH, CH_2OH)	O	0.132[d]	0.54[f]
	Poligalitol	1C	H, H	OH, H	OH, H	OH, H	CH_2OH, H	O	0.337[c]	0.65[f,g]
A	L-Xylose	1C	H, H	OH, H	OH, H	OH, H	(H, OH)	O	0.440[c]	
	D-Xylose	C1	(H, OH)	OH, H	OH, H	OH, H	H, H	O	0.440[c]	
	Methyl β-D-glucoside	C1	OCH_3, H	OH, H	OH, H	OH, H	CH_2OH, H	O	—[c]	0.22[f]
	D-Idose	1C	H, CH_2OH / (H, OH)	OH, H	H, OH	H, OH	(H, OH)	O	—[c]	0.27[g]
	2-Deoxy-D-glucose	C1	(H, OH)	H, H	OH, H	OH, H	CH_2OH, H	O	0.165[d]	0.17[f]
	Myo-inositol		H, OH	OH, H	OH, H	OH, H	CH_2OH, H	OH, H	0.194[c]	0.36[f]
	D-Chiro-inositol		H, OH	OH, H	OH, H	OH, H	OH, H	H, OH	—[c]	0.12[f]
	2-0-Methyl-D-glucose	C1	(H, OH)	OCH_3, H	OH, H	OH, H	CH_2OH, H	O	0.264[d]	0.38[f]
	D-Galactose	C1	H, OH	OH, H	OH, H	(OH, H)	O	CH_2OH, H	0.50[c]	0.53[h]
B	β-D-Fucose	C1	H, OH	OH, H	OH, H	OH, H	O	CH_3, H	0.264[c]	
	α-L-Arabinose	C1	H, OH	OH, H	OH, H	OH, H	O	H, H	0.536[c]	
	D-Fructose	1C	H, H	H, OH	OH, H	OH, H	(CH_2OH, OH)	O	0.0058[e]	0.65
C_1	D-Arabinose	1C	H, H	H, OH	OH, H	OH, H	(H, OH)	O	0.144[e]	0.66[i]
	L-Fucose	1C	CH_3, H	H, OH	OH, H	OH, H	(H, OH)	O	0.087[e]	0.63[f]
	D-Mannose	C1	(H, OH)	H, OH	OH, H	OH, H	CH_2OH, H	O	7.59[c]	
	D-Tagatose	C1	(OH, CH_2OH)	H, OH	OH, H	OH, H	H, H	O	—[e]	
	D-Altrose	1C	H, CH_2OH / (H, OH)	H, OH	H, OH	OH, H	(H, OH)	O	—[c]	
C_2	D-Lyxose	C1	H, H	H, OH	H, OH	OH, H	CH_2OH, H	O	42.27[e]	
	Methyl α-D-Mannoside	1C	H, OCH_3	H, OH	OH, H	OH, H	(H, OH)	O	—[e]	[f,i]
	L-Chiro-inositol	C1	H, OH	H, OH	OH, H	OH, H	CH_2OH, H	OH, H		[f]
	Methyl α-D-fructoside	1C	H, H	H, OH	OH, H	OH, H	OCH_3, CH_2OH	O		[f]
	Methyl β-D-fructoside	1C	H, H	H, OH	OH, H	OH, H	CH_2OH, OCH_3	O		[f]

Table 2: — Continued

	Chair[a] form	C-1[b] Eq., Ax.	C-2 Eq., Ax.	C-3 Eq., Ax.	C-4 Eq., Ax.	C-5 Eq., Ax.	O Eq., Ax.	50% threshold in M	Maximum response (sucrose = 1)
D-Allose	C1	(H, OH)	OH, H	H, OH	OH, H	CH$_2$OH, H	O		0.24[f]
D-Ribose	C1	(H, OH)	OH, H	H, OH	OH, H	H, H	O	8.99[c]	
D-Gulose	1C	H, CH$_2$OH	OH, H	OH, H	H, OH	(H, OH)	O	—[e]	
D 3-Deoxy-D-glucose	C1	(H, OH)	OH, H	H, H	H, OH	CH$_2$OH, H	O		—[f]
3-0-Methyl-D-glucose	C1	(H, OH)	OH, H	OCH$_3$, H	OH, H	CH$_2$OH, H	O	—[d]	
Methyl α-D-alloside	C1	H, OCH$_3$	OH, H	H, OH	OH, H	CH$_2$OH, H	O		0.12[f]
Epi-inositol	C1	H, OH	OH, H	H, OH	OH, H	OH, H	OH, H		—[f]
α-D-galactose	C1	H, OH	OH, H	OH, H	H, OH	CH$_2$OH, H	O		—
L-Rhamnose	1C	CH$_3$, H	OH, H	OH, H	H, OH	(H, OH)	O	—[c]	—[f]
E 4-0-Methyl-D-glucose	C1	(H, OH)	OH, H	OH, H	OCH$_3$, H	CH$_2$OH, H	O		—[f]
Methyl α-D-galactoside	C1	H, OCH$_3$	OH, H	OH, H	H, OH	CH$_2$OH, H	O		—[f]
Allo-inositol	C1	H, OH	OH, H	H, OH	H, OH	OH, H	H, OH		—[f]

Rearranged from Hanamori, T. *et al.* (1974)

[a] Stable chair form (C1 or 1C); when the degree of stability is similar both forms are indicated. (Reeves, R., 1950).

[b] The carbon numbers of D-glucose when sugars were superimposed over C1 conformation of α-D-glucopyranose.

[c] Cited in Dethier, V., (1955).

[d] Calculated from the results obtained by Evans, D. (1963).

[e] Approximated from the results of Pflumm, W. (1971, 1972).

[f] Jakinovich, W. *et al.* (1971).

[g] Hanamori, T. *et al.* (1972).

[h] Shimada *et al.* (unpublished).

[i] Hanamori (unpublished).

C$_1$ This group may be effective in the furanose forms.

— indicates non-effective. Eq. and Ax., equatorial and axitial positions, respectively.

() indicates both α- and β-anomer exist.

have fairly large residues on carbon at the ring oxygen position. Replacement of a ring carbon by oxygen is rather a drastic change in structure, and these sugars (including α-L-arabinose) are not considered to be effective strictly to the glucose site (see below).

Sugars classified into C_1 and C_2 have axitial hydroxyl group at C-2. This change introduced into D-glucopyranose results in quite contradictory effects between C_1 and C_2 groups. It acts as if the affinity for the glucose site were greatly improved in D-fructose of group C_1, but as if the affinity is greatly reduced in group C_2. The contradiction is solved by the results of Hanamori, T. et al. (1974) showing that fructose does not stimulate the sugar receptor in pyranose form, but does in furanose form. It follows that the axitial hydroxyl group at C-2 in pyranose form, greatly or even completely reduced the affinity for the glucose site.

By treatment of the chemosensillar tip with 0.5 mM p-chloromercuribenzoate (PCMB) for 3 min, Shimada, I. et al. (1974) could abolish the response to D-glucose leaving the response to D-fructose intact. This demonstrates directly the existence of glucose and fructose sites; glucose binds to the glucose site, and fructose binds to the fructose site. They examined several sugars after the PCMB treatment, and found that responses to D-fructose and D-fucose were unchanged, those to L-arabinose, L-sorbose, D-xylose, L- and D-glucose were completely abolished, and responses were left more intact in the order of D-galactose (61%), D-arabinose (38%) and L-fucose (23%). Sugars that did not stimulate after the treatment are considered to be effective to the glucose site in the form of pyranose, while those which still stimulated are considered to be effective to the fructose site in the form of furanose. Sucrose (the response was depressed to 16%) and maltose (depressed completely) are to be considered effective mainly and completely to the glucose site, respectively.

Now, it is certain that all sugars classified into groups B and C can possibly work as furanose. D-mannose has been well known to strongly compete against fructose, but not so strongly against glucose and sucrose (Dethier, V., 1955; Dethier, V. et al., 1956; Morita, H. and Shiraishi, A., 1968; Omand, E. and Dethier, V., 1969). Considering these facts, the structure required for stimulation of the glucose

site is more strict than expected: every equatorial hydroxyl group at C-2, C-3 and C-4 is almost essential. However, some sugars and their derivatives are effective, still but only slightly, in spite of lack of an equatorial hydroxyl group in one of the three positions. In group A, 2-deoxy-D-glucose and 2-O-methyl-D-glucose displace it by H and OCH_3, respectively, at C-2. In group D, the equatorial hydroxyl group is replaced by the axitial one at C-3 (D-allose, D-ribose, and methyl α-D-alloside). These facts suggest that the glucose site may not recognize adequate molecules by forming firm bonds with the equatorial hydroxyl groups, but may recognize them by their profiles as a whole in a strict sense of lock and key.

The fructose site is now regarded as the furanose site, and has a specificity higher than the glucose (pyranose) site. Methyl α- and β-D-fructo-furanosides are non-effective for the furanose site, (Jakinovich, W. et al., 1971), while the corresponding glucopyranosides are effective for the pyranose site. However, some discrepancies exist between the results of Hanamori, T. et al. (1974) and of Shimada, I. et al. (1974), and structural requirements for the fructose site are left open for further studies.

2.1.2 OTHER SITES

Shiraishi, A. and Kuwabara, M. (1970) demonstrated that certain L-amino acids (valine, leucine, isoleucine, methionine, phenylalanine, tryptophane) stimulate the labellar sugar receptor of the fleshfly. This result was confirmed by Goldrich, N. (1973) in the blowfly. Shimada, I. (1975) observed that the response of the labellar sugar receptor of the fleshfly to phenylalanine and valine was left unchanged after the PCMB treatment mentioned above. He thought that amino acids stimulated the sugar receptor by combining to the fructose site. Later, Shimada, I. and Isono, K. (1978) found that treatment with pronase (10 mg ml^{-1}) for 2 min differentially abolished the responses to valine, leucine and isoleucine, leaving those to phenylalanine, tryptophan and fructose intact. They concluded that there was a third site for aliphatic amino acids and carboxylates, in addition to the fructose site for amino acids with the cyclic group as well as fructose. Structural requirements for the third site (T-site) and for the fructose site (F-site) were studied with many derivatives and dipeptides (Shimada, I., 1978; Shimada, I. and Tanimura, T., 1981), and adequate structures for T- and F-sites were proposed. The proposed structure of the F-site, however, was quite different from that of the fructose site originally assigned for furac-tofuranose. Therefore, the F-site and the fructose site would be different entities.

The trehalose site was detected by genetic dimorphism in *Drosophila melanogaster* (Tanimura, T. *et al.*, 1982). No difference in the sensitivity to glucose, fructose and sucrose was found between the trehalose high-sensitivity ($T-1$) and the low-sensitivity (*Oregon-R*) strains. Genetic analysis showed that the *Tre* gene is responsible for the difference in the trehalose sensitivity and is closely linked to *cx* (13.6) on the *X* chromosome.

2.2   Water receptor

The water receptor is defined as the receptor which responds best to water and the response of which is depressed by the presence of solutes in water. The existence of the water receptor in the blowfly has been demonstrated for a long time by behavioral studies, but first recorded by Wolbarsht, M. (1957). Evans, D. and Mellon, D. (1962a) tried to find the factor controlling its response. In sucrose solution, the steady state response was inhibited in direct proportion to the log of the osmotic pressure over a range of 0.2 to 200 atm. Other non-electrolytes examined, glycerol and mannose, inhibited the water receptor response, but the effect could not be simply correlated with the osmotic pressure of solutions. Such results suggest that the water receptor works as a kind of osmometer, where the receptor membrane is impermeable to sucrose but is partially permeable to glycerol or mannose. However, electrolytes inhibited the water receptor response more sharply and at lower concentrations. Therefore, this receptor is not a simple osmometer that detects the change of the mechanical pressure within the cell.

Rees, C. (1970) tested the hypothesis that electrokinetic streaming potentials, E_s, are responsible for the water receptor response:

$$E_s = \zeta PD/4\pi\eta\gamma$$

where ζ is the zeta potential across the charged surface layer lining the pore, P is the osmotic pressure difference across the membrane, D is the dielectric constant of the material filling the pore, η is the viscosity of the material within the pore, γ is the specific conductance of the pore contents. Stimulation with electrolytes showed that the response was almost independent of the concentration, species and valency of the anion in solution, but strongly dependent upon cationic valency and concentration. It was suggested that cations governed the streaming potential by modifying the pore-surface ζ potential and the pore conductance. The effect of pH was to inhibit response below pH 5.0 and above pH 11.0. Negatively charged groups lining the pore with an isoelectric point at about pH 3.7 could be responsible for these effects. Approximately isosmotic concentration (5.5 Osmol kg^{-1}) of non-electrolytes (urea, glucose, glycerol, sucrose) much greater than those of electrolytes were needed to inhibit the response fully. Non-electrolytes may thus modify the streaming potential by reducing P in the above equation, and inhibit the response.

Wieczorek, H. and Köppl, R. (1978) have found a quite unexpected fact that some type of sugars re-activate the water receptor which is completely inhibited by salts. In Fig. 21 the water receptor responses re-activated by sugars are illustrated together with the sugar receptor responses, where the sugars were dissolved in 0.05 M sodium citrate solution. This concentration of sodium citrate completely inhibited the water receptor response, and the maximum response to each sugar was normalized so that the response to 1 M glucose was unity. The bar represents the range of the mean value of response at the confidence level above 95%. It is clear from this figure that the physico-chemical properties of the solution are not responsible for this reactivation, because D- and L-isomers reactivated differently. Highest reactivations were obtained in D-fructose, D-fucose and D-galactose, and fructofuranose of D-fructose was ascertained to be effective in the reactivation (Wieczorek, H., 1980). These facts strongly suggest that the water receptor possesses the furanose site which has been postulated to exist in the sugar receptor (Hanamori, T. *et al.*, 1974; Shimada, I. *et al.*, 1974).

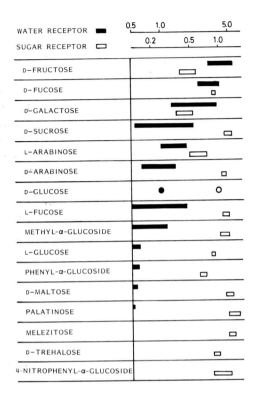

FIG. 21. Reactivation of the water receptor by sugars (Redrawn from Wieczorek, H. and Köppl, R., 1978). Responses of sugar receptor are shown for reference. The values are compared in the maximum response, relative to the value for D-glucose (1 M). Length of bar represents 95% confidence limit of the mean.

2.3 Salt receptor

It is generally accepted that monovalent inorganic cations take a principal part in stimulation of the salt receptor (Evans, D. and Mellon, D., 1962a), though anions are not indifferent (Steinhardt, R., 1965; Gillary, H., 1966b; Rees, C., 1968). These cations are considered to be adsorbed to a strongly acidic group, since the response to salts is inhibited only at pH below 3 (Evans, D. and Mellon, D., 1962a; Gillary, H., 1966a). Such an acidic group could be regarded as the receptor site in the salt receptor.

Gillary, H. (1966b) has shown in the labellar salt receptor of Phormia that the stimulating effectiveness is greatest for K^+ and decreases with either increase or decrease in atomic number, i.e., $K^+ > Na^+ > Rb^+ > Cs^+ > L^+$. den Otter, C. (1972a) obtained a similar result, $K^+ > Rb^+ > Cs^+ >$

$Na^+ > L^+$, in Calliphora vicina. Morita, H. and Yamashita, S. (1966) showed that the salt receptor of Calliphora vomitoria and Lucilia caeser was hyperpolarized by inhibitors such as quinine, $CaCl_2$ and $MgCl_2$. The hyperpolarization turned into depolarization accompanied by impulses (off-discharge) after the end of application of divalent cations, while the hyperpolarization lasted for a while after the end of quinine application. Rees, C. and Hori, N. (1968) have shown that $CaCl_2$ hyperpolarizes the salt receptor membrane above 1 mM but slightly depolarizes it below 1 mM. They showed the inhibitory effectiveness in the order, $Be^{2+} > Ca^{2+} > Sr^{2+} > Mg^{2+} > Ba^{2+}$. Taking these results into consideration, den Otter, C. (1972a,b,c) proposed highly polarizable negative groups as a candidate for the receptor site in the salt receptor. It should be noticed, however, that some of the ions carry the generator current through the receptor membrane, and that "stimulating effectiveness" involves both affinity for the receptor site and permeability to the activated receptor membrane.

Maes, F. and Bijpost, S. (1979) revealed discrimination of "salt taste" quality with the classical conditioning in Calliphora vicina. Blowflies were trained to discriminate between various concentrations of LiCl, KCl and NH_4Cl (reinforced with 0.5 M sucrose) from 1 M NaCl (not reinforced) given to the labellum. At all concentrations tested, blowflies responded to these three salts with proboscis extension significantly better than to 1 M NaCl. den Otter, C. (1972a) found that effective order of monovalent inorganic cations differed somehow among the salt receptors of the blowfly. Some differentiation might exist among them.

"Anion receptor" has been mentioned by Steinhardt, R. (1965) in relation to the fifth cell in the chemosensillum of Phormia. He postulated that this receptor is a real neuron which sends rejection signals into the central nervous system. In fact, the spikes supposed to occur in the fifth cell were recorded in stimulations by foreign monovalent cations such as Rb^+, Cs^+ and L^+ (Fig. 1 in Gillary, H., 1966b), and this receptor might send signals in order to reject such foreign substances.

2.4 Olfactory receptors

The specificity for odor substances in insects has

Table 3: Olfactory receptors studied in single cells

Species	Behaviour	Substance	Receptor (sensillum)	Author
Mosquito *Aedes aegypti*	Oviposition	Ethyl and isopropyl acetate Methyl and ethyl butyrate Ethyl propionate	One type of cell (blunt tipped type A2-11)	Davis, E. (1976)
	Host perception	Lactic acid (repellent, *N*, *N*-diethyl-*m*-toluamide, DEET)	Two cells respond to lactic acids; one cell, inhibited by lactic acid (grooved-pegs, type A3). All three cells, inhibited by DEET.	Davis, E. and Sokolove, P. (1976)
Silk moth *Bombyx mori*	Mating	*trans*-10, *cis*-12-Hexadecadien-1-ol (bombykol)	One type of cell (sensilla trichodea)	Kaissling, K. and Priesner, E. (1970) Schneider, D. *et al.* (1967)
Wild silk moth *Antheraea polyphemus*	Mating, Maximum at 9(A) + 1(B)	*trans*-6, *cis*-11-Hexadecadienyl acetate (A) *trans*-6, *cis*-11-Hexadecadien-1-al (B)	A and B cells (sensilla trichodea) A cell for (A), B cell for (B)	Kochansky, J. *et al.* (1975) Kaissling, K. (1979)
Gypsy moth *Lymantria dispar*	Mating, reduced at 10(A) + 1(B)	*cis*-7, 8-Epoxy-2-methyl octadecene (A) 2-Methyl-*cis*-7-octadecene (B) etc.	Similar A and B cells (sensilla trichodea); both (A) and (B) excite A cell	Bierl, B. *et al.* (1970); Carde, R. *et al.* (1973, 1975) Schneider, D. *et al.* (1977)
Summer fruit tortricid *Adoxophyes orana*	Mating, maximum at 9(A) + 1(B)	*cis*-11-Tetradeceyl acetate (A) *cis*-9-Tetradeceyl acetate (B)	A and B cells (sensilla trichodea): A cell responds to (A) rather than to (B); B cell to (B) only	Meyer, G. *et al.* (1972) Minks, A. *et al.* (1973) Persons, C. and Ritter, F. (1975) Voerman, S. *et al* (1976) den Otter, C. (1977)
Honey bee worker *Apis mellifera*	Inhibition of queen production	9-Oxo-*trans*-2-decenoic acid produced in queen	One cell type (sensilla placodea)	Kaissling, K. (1969) Kaissling, K. and Renner, M. (1968)
	Daily life	At least 7 groups of odors	At least 7 groups of cells (sensilla placodea)	Lacher, V. (1964) Vareschi, E. (1971)
Locust *Locusta migratoria*	Feeding	Hexenoic acids > Hexenals > Hexenols	One cell type of green-odor receptor (sensilla coeloconica)	Boeckh, J. (1967) Kafka, W. (1970)
Redbanded leafroller *Argyrotaenia veluntinana*	Mating: flying upwind at 92(C) + 8(T); maintenance of flight in (T); landing, wing fanning etc. in (D)	*cis*-11-Tetradecenyl acetate (C) *trans*-11-Tetradecenyl acetate (T) Dodecyl acetate (D) etc.	A and B cells (sensilla trichodea); both respond to both (C) and (T), but A cell rather to (C) B cell rather to (T). (D) is a synergist.	Roelofs, W. and Comeau, A. (1968, 1971) Bartell, B. and Roelofs, W. (1973) Roelofs, W. *et al.* (1975) Baker, T. *et al.* (1976) O'Connell, R. (1972, 1975)

been studied electrophysiologically and behaviorally. Electrophysiological works are divided into two classes: one is based on the individual cell activity and the other is on the electroantennogram (EAG). There are still not many types of olfactory cells that have been individually studied, but they are classified into "specialists" and "generalists" (Schneider, D. *et al.*, 1964). According to the original definition, specialists have an extremely high specificity for odorants, while generalists have a broad specificity, and different generalists show different but partially overlapping odor spectra.

Listed in Table 3 are only the olfactory receptors that were studied in single cells, since pheromones were studied behaviorally and are fully described in volume 9, this series. Among the receptors listed, only the bombykol receptor is strictly a specialist. It is highly specialized to bombykol, so that stereoisomers CT, CC and TT at positions 10 and 12 (C = *cis*, T = *trans*), respectively, are $10–10^2$

times less effective than bombykol in EAG test (cf. 10^4–10^5 times less effective in the behavior test) (Schneider, D. et al., 1967). In general, the sex pheromone in one species consists of several components, and each of several types of receptor cell responds more or less to every component. For instance, the sex pheromone of *Adoxophyes orana* consists of at least two components: cis-9- and cis-11-tetradecenyl acetate (cis-9- and cis-11-TDA). The optimum ratio is 9 : 1 of cis-9- and cis-11-TDA for the field trap. There are two types of receptor cell in sensilla trichodea; A cells respond to both cis-11 and cis-9-TDA but rather better to cis-11-TDA. B cells respond only to cis-9-TDA. No synergism exists between the two isomers for A nor B cells (den Otter, C., 1977).

The situation is more complicated in *Argyrotaenia veluntinana*. The sex pheromone consists of at least three components, c-11-TDA, t-11-TDA and DDA (c = *cis*, t = *trans*, TDA = tetradecenyl acetate, DDA = dodecyl acetate). Any one of the components alone is ineffective. A mixture of 92% c-11-TDA and 8% t-11-TDA is effective in initiating upwind flight toward the odor source, t-11-TDA is important in maintaining flight, and DDA increases occurrence frequency of landing, fluttering, etc. in the final stage of mating. There are two olfactory cells A and B in a sensillum trichodeum, being classified with spike height. Cell A responds to c-11-TDA better than cell B, and cell B to t-11-TDA better than cell A. However, this is only a result of statistics. Cell A responds to c-11-TDA better than cell B in one sensillum, while both equally respond in another. Responses of cell A to c-11-TDA are synergistically augmented by DDA, and inhibited by t-11-TDA (O'Connell, R., 1972). Individual A and B cells in different sensilla show sensitivities to different spectra of odors, but data are insufficient to conclude how many kinds of receptor sites work on the receptor membrane of individual cells (O'Connell, R., 1975). The coding system working in the mating behavior is open to future studies.

Besides sex attractant receptors, the green-odor receptor of the locust is one of the most extensively studied receptors. The odor spectra were studied with the intensity–response curves for many analogs or derivatives of hexenal (Boeckh, J., 1967; Kafka, W., 1970). This receptor is most sensitive to hexenoic acids (*trans*-2, *cis*-2, *trans*-3), and its odor spectrum is fairly broad. The back-bone of the hexane chain and the terminal functional group seems important: effective in the order -COOH > -CHO > -CH$_2$OH. It is also open to further studies whether multiple receptor sites exist on the receptor membrane.

The receptor of the queen substance was studied by Kaissling, K. (1969) and Kaissling, K. and Renner, M. (1968). This receptor reacts also to caproic acid, hexenal and hexenol, but not to hexanal; these substances are about 10^4 times less effective than the queen substance.

von Frisch, K. (1919) showed that the honeybee can discriminate odors as well as man can. The olfactory receptors have broad odor spectra, varying from cell to cell (Lacher, V., 1964). It seemed at first that individual receptors differed from one another, and grouping or classification of receptors of definite types was impossible. Vareschi, E. (1971), however, could classify the receptors into at least 7 groups, each of which reacted to a certain spectrum of odors. The spectra of the groups show little or no overlapping. Such groups of receptors give the basis of odor discrimination in the honeybee. A CO$_2$-receptor was also found by Lacher, V. (1964) in sensilla ampullacea.

2.5 Hygroreceptor

Lacher, V. (1964) recorded impulses responding to humidity changes from sensilla coeloconica or ampullacea on the honeybee antenna. Since then, hygroreceptor activity has been reported in median sensilla coeloconica of larvae of *Manduca sexta* (Dethier, V. and Schoonhoven, L., 1968), sensilla coeloconica of *Locusta migratoria* (Waldow, U., 1970), and sensilla basiconica of *Aedes aegypti* (Kellogg, F., 1970) and of *Periplaneta americana* (Altner, H. et al., 1973).

Yokohari, F. et al. (1975) recorded impulses of the moist and dry receptor from a poreless basiconic-like sensillum. Marking this sensillum with the hole of electrode penetration, they examined it with scanning electron microscopy, and found that the sensillum is capped with a cuticular structure at its tip. This type of sensillum was named sensillum capitulum, which contains the moist, dry and cold receptors as a triad. In the same way, such a triad was also found in a poreless peg in *Carausius* (Altner, H. *et al.*, 1978; Tichy, H., 1979), and in a capped poreless coeloconic sensillum (sensillum coeloconicum) in *Apis* (Yokohari, F. *et al.*, 1982).

Thus, hygroreceptors occur in poreless sensilla, and water vapor can not possibly reach the receptor membrane in contrast to the olfactory receptor. The hygroreceptors of *Periplaneta* were extensively investigated in particular, and it was found that they respond to the relative humidity like a hair hygrometer (Yokohari, F. and Tateda, H., 1976). The mechanical deformation of this sensillum could cause impulse discharge in the hygroreceptors (Yokohari, F., 1978), and it is therefore safely assumed that these hygroreceptors work as mechanoreceptors. In these receptors, indeed, the cilia differentiate into structures similar to tubular bodies, which are characteristic for mechanoreceptor cells. Absorption of water vapor into the cuticular wall should result in some deformation of the sensillum. Such a deformation should result in turn in a certain deformation of the hygroreceptor cells, the outer segments of which are tightly wrapped by cuticular wall with electron-dense material. Therefore, these cells should respond to humidity changes if they are mechanosensitive.

3 BEHAVIORS RELATED TO CHEMORECEPTION

Many insects prefer special types of foods, which commonly produce characteristic odors from numerous chemical compounds (Frankel, G., 1969). These odors attract the insects from a distance and contribute to the final recognition and elicit feeding, oviposition and other specific behavioral responses. Taste sense plays a role in ingestion of foods. Behavior varies greatly at different stages of the life cycle in insects. They eat principally during larval stage. Movements of many adult insects are often precise, rapid, complex and predictable. Their behaviors are sometimes stereotyped and sometimes flexible. The silkworm moth shows only mating responses and does not eat anything during adult life. The fly shows behavioral responses of feeding, drinking and mating. The memory trace plays important roles in the establishment of behavioral response of the honeybee (von Frisch, K., 1967).

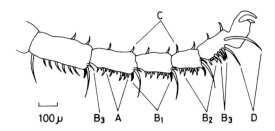

FIG. 22. A side view of the tarsus of a prothoracic leg of the blowfly. For clarity, tactile spines are omitted.

The adult blowfly, *Phormia regina*, requires only water and carbohydrate for its maintenance. During the reproductive period, the female fly ingests proteins. All other necessary materials are bequeathed from the larval stage. Therefore, blowfly feeding is reduced to its lowest common denominator and, on the face of it, offers a simple system for analysis of essential aspects of feeding (Gelperin, A., 1971; Dethier, V., 1976). Therefore, special attention has been paid to the feeding behavior in the fly.

3.1 Distribution of chemosensory cells

The chemosensory hairs are located on the ventral and ventrolateral surface of the tarsi of the fly. A few also occur on the dorsal surface of the tibia. Grabowski, C. and Dethier, V. (1954) have morphologically classified tarsal chemosensory hairs into four types in the blowfly, *Phormia regina*. Shiraishi, A. and Tanabe, Y. (1974) proposed to classify on the functional basis. The average number of hairs on the legs of the blowfly is 308, 208 and 147 on each of the first, second and third pair of legs, respectively. They range in length from 28 to 264 μm. Figure 22 shows the side view of the tarsus of the right prothoracic leg of the blowfly.

A-Type hairs are located on the ventral part of the tarsus and are the shortest among all tarsal chemosensory hairs. This type of hair contains the sugar and salt receptor cells. Discharges from these receptor cells are highly sensitive to change in ambient temperature. The sugar or salt receptor cells discharge no impulses when stimulated either by 1 M sucrose or 1 M NaCl at ambient temperatures of 21° and below. When the ambient temperature is raised above 21°, both receptor cells respond to stimulation either by 1 M sucrose or 1 M NaCl solution. B_2-Type hairs are located on the ventral side of the tarsus and respond only to sugar stimulation. A pair of B_3-type hairs are distal lateral projections of each of the proximal four tarsomeres and extra pairs of B_3-type hair are forward on the ventral surface of the 5th tarsomere. B_3-Type hairs respond to sugar and salt stimulations. The length of these hairs is intermediate among all tarsal chemosensory hairs. D-Type hairs are the largest in size among tarsal chemosensory hairs and are located in two pairs on the ventral side of the first and fifth tarsomer and in one pair on the other tarsomeres. This type of hair contains sugar and salt receptor cells. Wolbarsht, M. and Dethier, V. (1958) have shown that a row of chemosensilla is located on both the ventral and dorsal side of the wings of the blowfly. Angioy, A. *et al.* (1981) have confirmed that the wing chemosensilla can respond to sucrose or NaCl stimulation.

On the outer surface of the labellum of the blowfly there are on average 245 chemosensory hairs in the male and 257 in the female. They range in length from 41 to 429 μm. Their distribution and neuroanatomical connections have been described in detail by Wilczek, M. (1967). The labellar chemosensory hairs can be divided into five different groups. Every chemosensory hair, irrespective of the external appearance, contains three kinds of chemosensory cells, i.e., sugar, salt and water receptor cell. Labellar and tarsal chemosensory hairs are not the only contact chemoreceptors associated with feeding in the blowfly, but there are 132 interpseudotracheal papillae on the inner surface of the labellum (Wilczek, M., 1967). When all labellar chemosensory hairs except the papillae are covered with paraffin, the blowfly is still capable of discrimination between water, sugar and salt at first contact before the solution is ingested (Dethier, V., 1955). The papillae are homologous with the labellar chemosensory hairs and generally similar in basic structure (Dethier, V. and Hanson, F., 1965).

There may be still more chemoreceptor cells further back in the pharyngeal region. Arab, Y. (1957) found that the fly was still capable of some discrimination even after all labellar chemoreceptor cells had been removed. A search using the blowfly has still failed to reveal any likely candidates for the chemoreceptors. Stocker, R. and Schorderet, M. (1981) have revealed more than three different kinds

of sense cells on the internal mouthparts of *Drosophila melanogaster*. They are dorsal cibarial sense cells, ventral cibarial sense cells, and labial sense cells.

These results are summarized in Table 4. Studies with an electronmicroscope have now provided a fairly complete picture of the fine structure of the labellar, tarsal and interpseudotracheal chemosensory hairs or papillae. Each of the labellar and tarsal chemosensory hairs contain five bipolar neurons, one of which terminates at the proximal portion of a sensillum (mechanoreceptor), but dendrites from the other four neurons extend to the distal portion of a sensillum (Larsen, J., 1962a; Hansen, K. and Heumann, H., 1971). All labellar chemosensory hairs contain chemosensory cells which respond to chemicals of three different modalities. Tarsal chemosensory hairs respond in a different manner to chemicals. The interpseudotracheal papillae contain four bipolar neurons, one of which terminates at the proximal portion of a sensillum mechanoreceptor, but the dendrites from the other three neurons extend to the distal portion of the sensillum (Larsen, J., 1962b).

The olfactory receptors play a role in the feeding behavior of the fly. The principal site of the olfactory chemosensory hairs in the fly is the antennae. In the absence or impairment of the antennae, olfactory acuity is drastically reduced. Boeckh, J. *et al.* (1965) found pegs (sensilla basiconica) on the surface of antennae of male and female blowfly, *Calliphora erythrocephala*. They responded with a low threshold to meat, carrion and cheese and to some alcohols, aldehydes and mercaptans. Kaib, M. (1974) found olfactory sensilla in pits on the antennae of the blowfly, *Calliphora vicina*. Receptor cells that responded to meaty odors are insensitive to flowery odors, and vice versa. On the basis of their reaction spectra, the receptor cells sensitive to meaty odors may be subdivided into six different types and the receptor cells sensitive to flowery odors into three different types.

Wallis, D. (1962) investigated the sense organs on the ovipositor of the blowfly, *Phormia regina*, M. The morphological and electrophysiological evidence suggests that the chemosensory pegs contain the ovipositor olfactory receptor cells mediating oviposition. These chemosensory cells are also strongly stimulated by solutions of NaCl, Na_2CO_3

Table 4: Classification of contact chemosensory hairs and their characteristics

Hair type	Blowfly, *Phormia regina*													*Drosophila melanogaster* DSC & VCS*
	Tarsal Hair						Wing hair	Labellar hair					ITP	
	A	B₁	B₂	B₃	C	D		LL	L	I	M	d		
Stimulating substance — Sugar	○		○	○	○	○	○	○	○	○	○	○	○	
Water		○						○	○	○	○	○		
Salt	○		○	○	○		○	○	○	○	○	○	○	
Number of sensilla per leg or labellum	120	8	106	16	42	14		20	22	20	55	65	132	
Connection to CNS	Thoracic ganglion (*via* Leg nerve)						Thoracic ganglion (*via* Wing nerve)	Subesophageal ganglion (*via* Labial nerve)					Subesophageal ganglion (*via* Labial nerve)	Tritocerebram (*via* Labrafrontal nerve)
Behavioral function	Proboscis extension							Proboscis extension and Labellar lobe opening					Sucking	Swallowing

○: Presence of response.   ITP: Interpseudotracheal papillae.
DCS: Dorsal cibarial sense cell.   VCS: Ventral cibarial sense cell.
* No electrophysiological studies.

and $(NH_4)_2CO_3$. Dethier, V. (1972) found that contact chemoreceptor cells on the legs and labellum of the blowfly, *Phormia regina* that normally respond to aqueous solution of sapid substances also respond to compounds in the gaseous state. Effective vapors include organic and inorganic acids and various unrelated nonpolar compounds. There are many suggested theories that some chemosensory hairs located on the antennae respond to solutions of sucrose or NaCl (for example, antenna of cockroach, Yokohari, F., personal communication). Biological meanings of these phenomena including Dethier's results must be studied in the future.

3.2 Nerve and muscle of the head and proboscis in the blowfly

Dethier, V. (1961) investigated the nerve and muscle of the proboscis in the blowfly. Yano, T. (in preparation) re-examined the results from functional aspects. The proboscis consists of three sections; the proximal section or rostrum, the middle section or haustellum, and the distal oral disc or labellum. The nerves leading from the brain are the paired antennal nerve, the median oscellar nerve and the paired labrofrontal nerve. Another large pair of labial nerves connect with the sub-esophageal ganglion. The proboscis is innervated by the labrofrontal and the labial nerves.

Labrofrontal nerves extend from the brain and esophagus and each branches quickly. Part of the branch, the frontal ganglion connective, curves anteriorly and dorsally meeting the corresponding branch of the opposite side and uniting in a single nerve. After the two frontal ganglion connectives fuse, there is a swelling in the region of the first loop. It is probably the frontal ganglion, the recurrent nerve, which proceeds along the dorsal surface of the esophagus and joins the hypocerebral ganglion. Small nerves branch to the corpus allatum and corpus cardiacum. The large nerves from the hypocerebral ganglion extend to the crop and to the proventriculus. Another branch of the labrofrontal nerve except the frontal ganglion connective is called the labral nerve. The branches of two medial nerves unite on the anterior surface of the esophagus and the fused nerve extends to the cibarial pump muscles. Another lateral branch passes ventrally on either side of the esophagus, each giving a branch to the protractor of rostrum (van der Starre, H., 1977), passing lateral of these muscles, entering the flexor of labrum.

The large paired labial nerves extend from the subesophageal ganglion. Each immediately divides into branches. The main nerve trunk innervates the retractors of the furca, the retractors of the paraphyses, and the transverse muscle of the haustellum, and is also the principal sensory trunk from the labellum. A small branch of the labial nerve innervates the retractors of the rostrum. The principal branch of the labial nerve, the lateral branch, immediately subdivides. The dorsal branch sends small branches to the flexors of the haustellum and to the accessory retractors of the rostrum, then goes up into the cranial cavity to the trachea and fat body. The remaining branch passes to the flexors of the haustellum and the accessory retractors of the rostrum. Thereafter, the branch subdivides to send sensory fibers to the maxillary palpi and motor fibers to the extensors of the haustellum and the adductors of apodemes. The labial nerve sends at the distal portion small branches to the retractor of furca, to the retractor of paraphysis and to the transverse muscle of haustellum. Figure 23 shows the nerve and muscle system of the head and the proboscis in the feeding response of the blowfly.

3.3   Structure of the central nervous system with reference to chemoreception

The insect brain, supra-esophageal ganglion, has three main parts; protocerebrum, deutocerebrum and tritocerebrum. The protocerebrum possesses many separate neuropile masses (Lane, N., 1974). These include the optic ganglia, ocellar centers, central body, protocerebral bridge, corpora pedunculata (mushroom bodies) and pars intercerebralis. The optic ganglia are concerned with the inward flow of excitation from the compound eyes. The central body is suggested to have an association function, representing the source of pre-motor outflow from the brain to ventral nerve cord. The protocerebral bridge is a medial neuropilar mass lying across the front of the brain and posterior to the pars intercerebralis. The mushroom bodies are paired neuropilar areas associated with unipolar nerve cells called Kenyon cells. The Kenyon cells send fibers to the calyx and to the alpha and the beta lobes. The calyces and alpha lobes receive tracts from the sensory centers of the brain while the beta and gamma lobes send tracts to the motor regions. The mushroom bodies are thought to be the site of the most complex sensory input in the brain as well as the site of its translation into behavioral patterns (Mancini, G. and Frontali, N., 1967). The pars intercerebralis is in the dorsal median region. The component nerve cell bodies are neurosecretory and produce a hormonal product in the form of

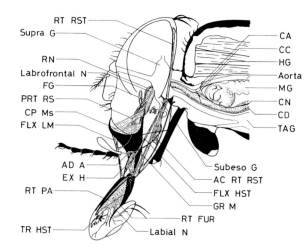

FIG. 23. Diagram of head, proboscis and thorax of the blowfly showing alimentary canal, endocrine complex and the relationship between nervous system and muscular system. AC RT RST is accessory retractor of rostrum, AD A adductors of apodeme, CA corpus allatum, CC corpora cardiaca, CD crop duct, CN crop nerve, cp Ms cibarial pump muscles, EX H extensors of haustellum, FG frontal ganglion, FLX HST flexor of haustellum, FLX LM flexor of labrum, GR M gracilis muscle, HG hypocerebral ganglion, Labial N labial nerve, Labrofrontal N labrofrontal nerve, MG midgut, PRT RS protractor of rostrum, RN recurrent nerve, RT FUR retractor of furca, RT PA retractor of paraphysis, RT RST retractor of rostrum, Subeso G subesophageal ganglion, Supra G supraesophageal ganglion, TAG thoracicoabdominal ganglion, TR HST transverse muscle of haustellum.

neurosecretory granules. The corpora cardiaca are mainly responsible for the storage and release of hormonal substances from the pars intercerebralis.

The deutocerebrum consists of antennal glomeruli formed from the endings of the antennal sensory nerves and the antennal motor centers. From these antennal lobes an olfactory tract of fibers runs to the mushroom body. The tritocerebrum forms the ventral part of the brain, sending nerves to the labrum and to the digestive tract, forming the stomatogastric or sympathetic system.

Strausfeld, N. (1976) has performed histological studies on the nervous system of the housefly brain. But the detailed nerve network to chemoreception has still not been elucidated. Under the existing circumstances, it is valid to consider a scheme representing information flow from sensory cells to muscles. The scheme shown in Fig. 24 was drawn with reference to results reported by Strausfeld and many other workers (cf. Dethier, V., 1976). The cibarial sense cells and the labial sense cells proposed by Stocker, R. and Schorderet, M. (1981) on the internal mouthparts of *Drosophila* are not found among other large fly groups. Neurons from these sense cells project into the tritocerebrum through the labral and the labrofrontal nerve. The interpseudotracheal papillae chemoreceptor cells and the labellar chemoreceptor cells project into the subesophageal ganglion through the labial nerve. The tarsal and wing chemoreceptor cells form a projection pattern in the thoracico-abdominal ganglion. They go through the ventral nerve cord and the subesophageal ganglion and reach the tritocerebrum. Information from chemoreceptor cells is gathered and transmitted to the calyx of the mushroom body through the antenno-glomerular tract. The antennal chemoreceptors show a projection pattern in the antennal lobe and the neurons go through the long olfactory fibers and the antenno-glomerular tract and reach the calyx of the mushroom body.

Information sent from the mushroom body goes through the protocerebrum for integration and reaches the subesophageal ganglion and the ventral nerve cord as the motor output. As described previously, the protocerebrum possesses many separate neuropile masses but any relationship between the neuropile masses and chemical information flows is not definitely demonstrated. The hypocerebral ganglion regulates the movement of a digestive organ and hormonal secretion. Information is transmitted to or from the brain through the frontal ganglion and the recurrent nerve.

3.4 Feeding response in the blowfly

Feeding in the blowfly can be divided into many component responses. As previously described, the blowfly is thought to have at least five different contact chemoreceptor cells. In fact, these different chemosensory inputs trigger different behavioral component responses. The initial step of the proboscis extension response is considered to be triggered mainly by impulses from D-type and B_2-type sugar receptor cells on the tarsi (Kawabata, K. and Shiraishi, A., 1977; Smith, D. *et al.*, 1983). Angioy, A. *et al.* (1981) have observed that the proboscis extension occurs

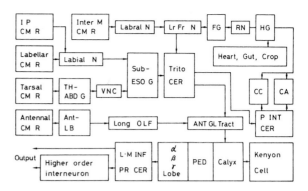

FIG. 24. Schematic representation of the nervous system and the information flow in the blowfly feeding response. Antennal CMR is antennal chemoreceptor, ANT-LB antennal lobe, ANT GL tract antenno glomerular tract, CA corpus allatum, CC corpora cardiaca, FG frontal ganglion, HG hypocerebral ganglion, Inter M CM R internal mouthparts chemoreceptor, I P CM R interpseudotracheal papillae chemoreceptor, Labellar CM R labellar chemoreceptor, Labial N labial nerve, Labral N labral nerve, L M INF PR CER lateral and medial inferior protocerebrum, Long OLF long olfactory fibers, Lr Fr N labrofrontal nerve, PED Pedunculus, P INT CER pars intercerebralis, PN recurrent nerve, Sub-ESO G subesophageal ganglion, Tarsal CM R tarsal chemoreceptor, TH-ABD G thoracico-abdominal ganglion. Trito CER tritocerebrum, VNC ventral nerve cord.

following stimulation of wing hairs with sucrose. The rostrum protractor contracts with the tarsal sugar receptor impulses (van der Starre, H., 1977). Therefore, the response after application of sucrose solution to the tarsi or the wing is thought to be initiated mainly by the contraction of this muscle system.

Proboscis extension results in stimulation of chemoreceptor cells on the labellum, and further extension of the proboscis and opening of the labellar lobes occur. The haustellum is extended by the contraction of its extensors and the adductors of apodemes. The oral disc is extended by the contraction of the retractors of the paraphysis and the transverse muscle of the haustellum. These responses are the further extension of the proboscis mentioned above. Labellar lobes are opened by the contraction of the furca retractors (Pollack, G., 1977). The proboscis also extends on direct application of sugar solution only to whole labellum of the blowfly (Minnich, D., 1931). Dethier, V. (1955) reported that the proboscis extension response occurs on direct stimulation of a single labellar hair by sucrose solution. When a single sugar receptor cell of LL-type or L-type hair is stimulated by sucrose solution, the fly certainly extends the proboscis and then opens the labellar lobes. In many cases, the fly only moves the labellar lobes on stimulation of a sugar receptor cell in the labellar hairs other than LL- and L-type hairs. The behavioral response to stimulation of the labellar sugar receptor cell mainly consists of extension of the haustellum and opening of the labellar lobe (Getting, P., 1971).

Sucking is accomplished by the action of the dilators of the cibarial pump (Rice, M., 1970; Rice, M. and Finlayson, L., 1972). The cibarial pump muscles work with impulses from sugar receptor cells in the interpseudotracheal papillae located at the groove surface of the labellum.

Chemoreceptors located at the internal mouthpart are supposed to trigger food searching behavior or the fly's dance. Under natural conditions, a fly is usually brought into contact with food sources by olfactory orientation. There is another search pattern triggered by the intake of a drop of sucrose. When a flies tarsi suddenly encounters a drop of sucrose on a horizontal surface, the fly immediately halts and turns toward the point source of stimulation so that the mouth parts are brought over the spot. The fly extends the proboscis and ingests a drop of sucrose. The fly remains in this position as long as the mouthparts are adequately stimulated. Thus, the locomotory response to continuous stimulation is a complete cessation of movement. However, the amount of sucrose present in a $0.1 \mu l$ droplet is insufficient to provide either satiation or continuous stimulation. It does result in a complex behavioral response, which may last for as long as 90 s after stimulation has ended. During this time the fly walks in a series of loops and spirals rather than straight lines. Dethier reported this behavior, and called it a "dance", describing its general characteristics in the blowfly, *Phormia regina*, M. (Dethier, V., 1957). Mourier, H. (1964) expanded Dethier's findings to the housefly, *Musca domestica*. The action is completely stereotyped, rather than purposeful. This is demonstrated by the fact that a fly which is held in the hand and stimulated with sugar, immediately on being released on a horizontal surface,

begins the searching action on the spot with no relation to the spatial location of the former stimulus.

Figure 25 shows a typical dance pattern after intake of a $1 \mu l$ droplet of 0.5 M sucrose in the 5-day old housefly deprived of sucrose for 24 h. The rate of turning is the greatest as the dance begins, and decreases slowly thereafter. Most of the dances occur within a circle with a radius of 6 cm. This dance can also be elicited by water or crude protein extract. Namely, a given feeding stimulant is capable of eliciting dancing only when the fly has been deprived of that substance. The intensity of dance as the distance walked in a spiral path is a function of the fly's state of deprivation with respect to the eliciting stimulus, and of the concentration of stimulating solution (Nelson, C., 1977). The flies that have experienced long periods of deprivation dance greater distances compared to less-deprived flies when they encounter a sugar droplet. When the period of deprivation is held constant and only the concentration of the stimulus is varied, higher concentrations of sucrose always elicit longer dances than do the lower concentrations. The degree of satiation needed to inhibit dancing is less than that needed to inhibit feeding. Thus, it is evident that dancing is correlated with relatively high levels of deprivation. Although naturalistic observations on the fly are lacking, the dance can function in the laboratory as an effective search pattern with several different patterns of food distribution.

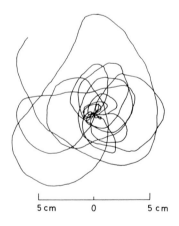

Fig. 25. The pattern of poststimulatory locomotion after ingestion of $1 \mu l$ of 0.5 M sucrose in the housefly, *Musca domestica* (by courtesy of Fukushi, T.).

3.5 Distribution of the behavioral threshold

Much of the feeding behavior of the blowfly is explained in terms of taste threshold. The measurement of taste threshold which we have employed is usually proboscis extension response. The usual method of threshold measurement consists of determining the concentration of the test substance to which 50% of a population of flies exhibit a feeding response and proboscis extension response (Evans, D. and Barton-Brown, L., 1960). Usually, 'ascending' thresholds are obtained for individual flies, beginning at a concentration low enough that none of the flies will respond, and median acceptance threshold of a population of flies are estimated by the procedure described by Bliss, C. (1938).

Figure 26 shows representative regression lines for the blowflies after starvation or raising on different concentrations of sucrose for five or six days. The starved flies were raised on 0.1 M sucrose for five days and then starved for 24 h before experimentation. In one series of experiments about 50 individuals were used. Stimulating sucrose solution was applied to a pair of prothoracic legs and extension of the proboscis was observed for each fly. In Fig. 26 the straight line is shifted to the right with higher raising concentration of sucrose. The tarsal threshold does not elevate more than about 0.2 M sucrose even if the flies are raised on sucrose of more than 2.0 M. Then, the 24 h starved flies were given 1 M sucrose for 30 min, and after 30 min rest period, the behavioral tests were performed to determine the tarsal thresholds. None of the flies extended the proboscis after the application of more than 2.0 M sucrose to the prothoracic legs. Another group of flies starved for 24 h were given 0.1 M sucrose for 30 min. After a 30 min rest period the behavioral tests were conducted. In this case, about 80% of the 50 experimental flies responded to sucrose (Fig. 26). The median acceptance threshold for sucrose is about 0.8 M. It is confirmed with the experiment described above and the other useful information (cf. Dethier, V., 1955) that the tarsal threshold is elevated from 0.001 M to 1.0 M for sucrose depending on the feeding state of the experimental flies.

Distribution of the labellar threshold on application of sucrose solution to the whole labellum in the blowfly after starvation or raising on different

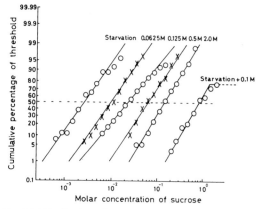

FIG. 26. Distribution of the acceptance thresholds as a function of the sucrose concentration. Flies used for experiments were starved for 24 h or were raised on different concentrations of sucrose for 6 days.

concentrations of sucrose shows low values compared with those of the corresponding tarsal threshold in Fig. 26 (Shiraishi, A. and Yano, T., 1984). Results obtained by a LL-type hair stimulation show almost the same distribution as those of tarsal threshold for sucrose. The labellar threshold does not elevate more than about 0.1 M for sucrose when behavioral tests are conducted with 2.0 M sucrose fed flies after 24 h starvation. As described before, the blowfly extends the rostrum and a part of the haustellum by tarsal stimulation. The blowfly also extends the haustellum and opens the labellar lobes by labellar stimulation. All these different responses are grouped under the same expression, the proboscis extension response. Neuronal mechanisms for behavioral threshold regulation are thought to be different between the tarsal and the labellar system.

Nakashima et al. (in preparation) investigated the sucking duration and amount of sucrose on stimulation of sugar receptor cells located at interpseudotracheal papillae in the blowfly after starvation or feeding on 0.1 M and 2.0 M sucrose, respectively. The blowfly was held by the wings with a small clothes-pin. Sucrose solution was applied to legs, and sucking duration and amount of sucrose ingested in one trial was measured. The average minimum sucrose concentration, at which 50 flies initiated sucking, were 0.005 M for starved flies, 0.06 M for flies raised on 0.1 M sucrose and 0.5 M for flies raised on 2.0 M sucrose. These concentrations were clearly higher than those for the initiation

of proboscis extension of the tarsal and labellar systems. The starved fly ingested about 19 μl of 1.0 M
sucrose during 200 s. The fly raised on 0.1 M sucrose
ingested about 13 μl during 55 s. The fly raised on
2.0 M sucrose ingested about 5 μl during 5 s. Sucking duration and amount of sucrose ingested in one
trial showed linear relationship for log molar concentration of sucrose.

3.6 Regulation of feeding responses in the blowfly

Gelperin, A. (1971) proposed a model for regulation of the feeding response in the blowfly. There are
two sets of internal receptors which provide
negative feedback to the feeding response. One set
has been identified as stretch receptors located in a
branch of the recurrent nerve (Gelperin, A., 1967).
The second set is located in the abdomen (Gelperin,
A., 1970). The cells responsible for this effect appear
to be nerve cord stretch receptors responsive to tension in the abdominal nerve which is suspended over
and stretched by the crop. In the brain, excitatory
input from the sugar receptors is balanced against
inhibitory input from the foregut and the abdominal stretch receptors, and the outcome determines
whether or not feeding occurs. The idea which constitutes the basis of the model is partially correct but
the actual regulation is more complicated.
Ultimately we are lacking in the detailed wiring
diagram in the central nervous system for interactions among the peripheral sensory receptors which
trigger response, the internal receptors which
regulate response, and the motor neurons which
determine the form of response. Neurophysiological information is not as yet available but several
types of behavioral experiments indicate which kind
of regulation system for responses must be present.

On the sensory side of the mechanism in the blowfly,
the role of the sugar receptor impulses in the feeding
component response must be explained initially.
Figure 27 shows average concentration–response
curves of the sugar receptor cells in tarsal D- and
B_2-type and labellar LL-type hairs and interpseudotracheal papillae for sucrose. The number of
impulses is counted for the initial 0.1 s after the
beginning of stimulus. The data shown in Fig. 27 are
averages of between 10 and 20 experiments. Sensitivity of sugar receptor cells in D-type hair is the
highest among four receptor types. It has already

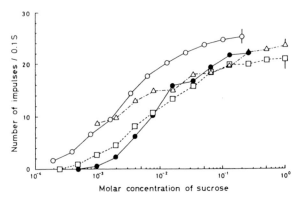

FIG. 27. Responses of tarsal D-($\bigcirc$) and B_2-type ($\bullet$), labellar
LL-type ($\square$) and intrerpseudotracheal papillae sugar receptor cells ($\triangle$) of blowflies raised on 0.1 M sucrose solution.
These data represent the averages between 10 and 20 receptors for each test solution.

been described that impulses of these different sugar
receptors trigger different components of response.

Shiraishi, A. *et al.* (in preparation) investigated
relationships between acceptance threshold, latency
and tarsal sugar receptor impulses in the proboscis
extension response of the blowfly after starvation or
feeding on different concentrations of sucrose. The
latency was defined as the time elapsing until initiation of sucking after the beginning of application of
a behavioral threshold concentration of sucrose to
a pair of prothoracic legs or to a single LL-type
labellar hair. The latency for tarsal stimulation was
measured as follows. A 6 cm watchglass was filled
with sucrose dissolved into 0.01 M NaCl. A small
piece of platinum plate connected to an input of
preamplifier with a fine silver wire was immersed in
sucrose solution. On the other hand, the fly was held
by the wings with a small aluminum clothes-pin
connected to an indifferent terminal of the
preamplifier with copper wire. When a pair of
prothoracic legs were forced to contact with sucrose
solution in the watchglass, the fly initiated extension of proboscis. After a chain of behavioral
responses, initiation of sucking occurs. A time
course of these behavioral responses could be
traced by means of electrical signals. A tough
artifact was seen on application of sucrose to legs
and motor action potentials on sucking could be
easily distinguished. The latency is prolonged when
tarsal threshold is elevated. For example, the value
of latency for tarsal threshold of 0.002 M sucrose
in the starved flies was about 0.76 s. The value of

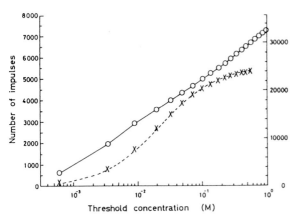

FIG. 28. Total number of impulses for the median threshold concentration of sucrose. Impulses form D-type (○) and B₂-type (x) sugar receptor cells during the latency were calculated and plotted for the threshold concentration of sucrose. The scales on the left and right Y-axes are for D-type and for B₂-type hairs, respectively.

latency for tarsal threshold of 0.28 M sucrose in the flies raised on 2.0 M sucrose was about 1.4 s.

The number of impulses discharged by the 244 B_2-type and the 28 D-type sugar receptor cells of a pair of prothoracic legs can be obtained on the basis of the concentration response curves of sugar receptor cells of B_2- and D-type hairs, the latencies, and the tarsal thresholds after starvation or raising on different concentrations of sucrose. These results are summarized in Fig. 28. From many experimental evidences, mainly sugar receptor impulses of D-type hairs are considered to trigger the proboscis extension response, but sugar receptor impulses of B_2-type hairs are only to assist D-type hairs. The latency for a single LL-type labellar hair stimulation does not change under different raising conditions, the value being about 0.35 s. The duration of sucking is proportional to impulses of interpseudotracheal papillae sugar receptor cell.

Figure 29 shows the simultaneous recording of sugar receptor impulses from a single LL-type hair located on the left side of the labellum (tip recording method) and of impulses recorded from the cut end of the labial nerve located on the contralateral side of that labellar hair (suction electrode method). The range of sucrose concentration within the two dotted lines shows the distribution of thresholds in which about 80% of 50 flies extended their proboses. Impulses from the labial nerve are divided into five different types according to their

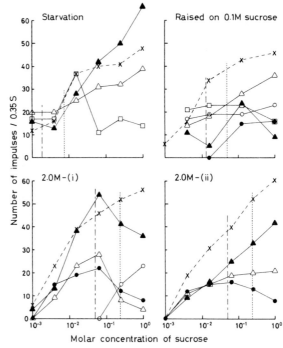

FIG. 29. Impulses of a single LL-type sugar receptor cell and of motor neurons located in the labial nerve. Each result was obtained with a single preparation of the blowfly. Motor impulses were divided arbitrarily into five types according to their height. (x) sugar receptor impulses; (□) type 1 impulses; (△) type II; (▲) type III; (○) type IV; (●) type V. The range of sucrose concentration within the two dotted lines shows the distribution of thresholds in which about 80% of 50 flies extended their proboscis.

spike heights. When the labellar hair of water satiated flies is stimulated by NaCl or distilled water, no impulse from the labial nerve can be observed. Firing frequency and pattern of motor impulses recorded from the labial nerve varies under different raising conditions. The result is thought to indicate that muscle system used for proboscis extension response might be different under different raising conditions.

We re-examined effects of cutting of either the recurrent nerve or the ventral nerve cord on the feeding response in the blowfly (Shiraishi, A. *et al.*, 1979; Shiraishi, A. and Yano, T., 1984). The threshold for tarsal stimulation is rather elevated in the experimental flies whose recurrent nerves are cut before the experiment, compared with those in the control flies. The median acceptance thresholds for sucrose after starvation, raising on 0.1 M and 2.0 M sucrose, respectively, are 0.0018 M, 0.019 M and

0.115 M for the control flies and 0.025 M, 0.048 M and 0.15 M for the experimental flies. In behavior, cutting of the recurrent nerve induces hyperphagia (Dethier, V. and Gelperin, A., 1967). These results indicate that information transmitted to the central nervous system through the recurrent nerve have dual effects for the feeding response. One concerns the positive feedback mechanism for the proboscis extension response to stimulation of tarsal sugar receptor cells, and the other concerns the negative feedback mechanism for ingestion of sucrose. The cutting of the ventral nerve cord influences ingestion of sucrose. The fly continuously ingests sucrose and does not stop ingestion during sucrose stimulation of sugar receptor cells in interpseudotracheal papillae. The abdominal stretch receptor is also thought to be involved in the stop mechanism of the ingestion of sucrose. The threshold for the proboscis extension response to stimulation of the labellar sugar receptor cells is not affected by cutting either the recurrent nerve or the ventral nerve cord (Getting, P. and Steinhardt, R., 1972; Yano, in preparation).

The concentration of blood trehalose is maintained by a condition of balance between the intake of carbohydrate and synthesis of trehalose on the one hand and metabolic utilization and other loss of this sugar on the other (Friedman, S., 1978). The medial neurosecretory cells located at the pars intercerebralis have an important role in the regulation of carbohydrate metabolism. Hormones produced at the medial neurosecretory cells are transported to the corpus cardiacum through the cardiac-recurrent nerve which joins the recurrent nerve. Evidence has been presented that the corpus cardiacum of the fly is the reservoir of neurohormones. If the region where the medial neurosecretory cells are located is stimulated with electric pulse (5–10 V, 0.05–0.1 mA) then in less than 10 min after stimulation the blowfly shows a significant rise in the hemolymph trehalose level *in vivo* (Normann, T. and Duve, H., 1969). This result suggests that the corpus cardiacum may be activated via the brain to release the hypertrehalosemic hormone, the glucagon-like peptide, provided that the nerve connection between the brain and the corpus cardiacum, the cardiac-recurrent nerve, is intact (Kramer, K. *et al.*, 1980). When the medial neurosecretory cells are removed or the cardiac

recurrent nerve is cut, blood trehalose level increases (Chen, A. and Friedman, S., 1977). Hypertrehalosemia also occurs after corpus cardiacectomy of the adult blowfly, *Phormia regina*, with no attendant change in fat body glycogen. The pars intercerebralis is shown to contain a hypotrehalosemic hormone, insulin-like peptide, the release of which depends on the integrity of the stomatogastric nervous system (Duve, H. and Thorpe, A., 1979; Duve, H. *et al.*, 1979). Cardiacectomy causes the fly to lower the turnover rate of hemolymph trehalose for 10 min, and seems to decrease the capability for synthesizing trehalose from hemolymph glucose. Therefore, the hypertrehalosemic condition in the cardiacectomized flies is thought to be a result of the absence of the hormone in the blood.

When the medial neurosecretory cells are removed or prevented from secretion, the fly decreases and slows locomotor activity (Green, G., 1964a,b). These perhaps result from impeded uptake of trehalose by muscle. The hypertrehalosemia and the hyperphagia seem to occur consistently after the cutting of the recurrent nerve or removal of the corpus cardiacum. Since the hypertrehalosemia cannot be the cause of the hyperphagia, these symptoms must result from different functions of the medial neurosecretory cell hormones. Thus, the medial neurosecretory cell hormones have a diversity of functions, and it should be determined which effects are secondary to which, when results of the medial neurosecretory cells extirpation are considered. After all, the control of feeding behavior in adult blowfly appears to be even more complex than is currently believed (cf. Gelperin, A., 1971; Dethier, V., 1976).

REFERENCES

ALTNER, H. and PRILLINGER, L. (1980). Ultrastructure of invertebrate chemo-, thermo-, and hygroreceptors and its functional significance. *Int. Rev. Cytol.* 67, 69–139.

ALTNER, H., ERNST, K. D., KOLNBERGER, I. and LOFTUS, R. (1973). Feinstruktur und adaquatr Reiz bei Insektensensillen mit Wandporen. *Werh. dtsch. Zool. Ges.* 66, 48–53.

ALTNER, H., TICHY, H. and ALTNER, I. (1978). Lamellated outer dendritic segments of a sensory cell within a poreless thermo and hygroreceptive sensillum of the insect *Carausius morosus*. *Cell Tissue Res.* 191, 284–304.

ANGIOY, A. M., LISCIA, A. and PIETRA, P. (1981). Some functional aspects of the wing chemosensilla in *Phormia regina* (Meig.) (*Diptera Calliphoridae*). *Monitore. zool. ital.* (N. S.) 15, 221–228.

ARAB, Y. M. (1957). Ph. D. Thesis Johns Hopkins University Baltimore, Maryland.

BAKER, T. C., CARDÉ, R. T. and ROELOFS, W. L. (1976). Behavioral responses of male *Argyrotaenia velutinana* (*Lepidoptera, Torticidae*) to components of its sex pheromone. *J. Chem. Ecol. 2*, 333–353.

BARTELL, R. J. and ROELOFS, W. L. (1973). Inhibition of sexual response in males of the moth *Argyrotaenia velutinana* by brief exposures to synthetic pheromone or its geometrical isomer. *J. Insect Physiol. 19*, 655–661.

BEIDLER, L. M. (1954). A theory of taste stimulation. *J. Gen. Physiol. 38*, 133–139.

BEIDLER, L. M. (1967). Anions influences on taste receptor response. *Olfaction and Taste*, II. Edited by T. Hayashi. Pages 133–148. Pergamon Press, Oxford.

BIERL, B. A., BEROZA, M. and COLLIER, C. W. (1970). Potent sex attractant of the gypsy moth, its isolation, identification and synthesis. *Science 170*, 87–89.

BLISS, C. I. (1938). The determination of the dosage–mortality curve from small numbers. *Quart. J. Pharm. 11*, 192–216.

BOECKH, J. (1967). Reaktionsschwelle, Arbeitsbereich und Spezifitat eines Gruchrezeptors auf der Heuschreckenantenne. *Z. Vergl. Physiol. 55*, 378–406.

BOECKH, J., KAISSLING, K.-E. and SCHNEIDER, D. (1965). Insect olfactory receptors. *Cold Spring Harbor Symp. Quant. Biol. 30*, 263–280.

BROYLES, J. L., HANSON, F. E. and SHAPIRO, A. M. (1976). Ion dependence of the tarsal sugar receptor of the blowfly. *J. Insect Physiol. 22*, 1587–1600.

BUTENANDT, A., BECKMANN, R., STAMM, D. and HECKER, E. (1959). Über den Sexual-Lockstoff des Seidenspinners *Bombyx mori*. Reindarstellung und Konstitution. *Z. Naturforsch. 14b*, 283–284.

CARDÉ, R. T., DOANE, C. C., GRANETT, J. and ROELOFS, W. L. (1975). Disruption of pheromone communication in the gypsy moth: Some behavioral effects of disparlure and an attractant modifier. *Environ. Entomol. 4*, 793–796.

CARDÉ, R. T., ROELOFS, W. L. and DOANE, C. C. (1973). Natural inhibition of the gypsy moth sex attractant. *Nature 241*, 474–475.

CHEN, A. C. and FRIEDMAN, S. (1977). Hormonal regulation of trehalose metabolism in the blowfly, *Phormia regina*: Interaction between hypertrehalosemic and hypotrehalosemic hormones. *J. Insect Physiol. 23*, 1223–1232.

DAVIS, E. E. (1976). A receptor sensitive to oviposition site attractants on the antennae of the mosquito, *Aedes aegypti*. *J. Insect Physiol. 22*, 1371–1376.

DAVIS, E. E. and SOKOLOVE, P. G. (1976). Lactic acid-sensitive receptors on the antennae of the mosquito, *Aedes aegypti*. *J. Insect Physiol. 22*, 1371–1376.

DE KRAMER, J. J., KAISSLING, K. E. and KEIL, T. (1983). Passive electrical properties of insect olfactory sensilla may produce the biphasic shape of spike. *ISOT VIII*.

DEN OTTER, C. J. (1972a). Differential sensitivity of insect chemoreceptors to alkali cations. *J. Insect Physiol. 18*, 109–131.

DEN OTTER, C. J. (1972b). Interaction between ions and receptor membrane in insect taste cells. *J. Insect Physiol. 18*, 389–402.

DEN OTTER, C. J. (1972c). Mechanism of stimulation of insect taste cells by organic substances. *J. Insect Physiol. 18*, 615–625.

DEN OTTER, C. J. (1977). Single sensillum responses in the male moth *Adoxophyes orana* (F.v.R.) to female sex pheromone components and their geometrical isomers. *J. Comp. Physiol. 121*, 205–222.

DETHIER, V. G. (1955). The physiology and histology of the contact chemoreceptor of the blowfly. *Quart. Rev. Biol. 30*, 348–371.

DETHIER, V. G. (1956). Chemoreceptor mechanism. Molecular structure and functional activity of nerve cells. Publ. No. 1, Amer. Inst. Biol. Sci., Washington, D.C.

DETHIER, V. G. (1957). Communication by insects, physiology of dancing. *Science 125*, 331–336.

DETHIER, V. G. (1961). The nerves and muscles of the proboscis of the blowfly, *Phormia regina* Meigen in relation to feeding responses. *Smithsonian Inst. Misc. Collections 137*, 157–174.

DETHIER, V. G. (1972). Sensitivity of the contact chemoreceptors of the blowfly to vapors. *Proc. Natl. Acad. Sci. U.S.A. 69*, 2189–2192.

DETHIER, V. G. (1976). *The Hungry Fly*. Harvard University Press, Cambridge, Massachusetts.

DETHIER, V. G. and GELPERIN, A. (1967). Hyperphagia in the blowfly. *J. Exp. Biol. 47*, 191–200.

DETHIER, V. G. and HANSON, F. E. (1965). Taste papillae of the blowfly. *J. Cell. Comp. Physiol. 65*, 93–100.

DETHIER, V. G. and SCHOONHOVEN, L. M. (1968). Evaluation of evaporation by cold and humidity receptors in caterpillars. *J. Insect Physiol. 14*, 1049–1054.

DETHIER, V. G., EVANS, D. R. and RHOADES, M. V. (1956). Some factors controlling the ingestion of carbohydrates by the blowfly. *Biol. Bull. 111*, 204–222.

DUVE, H. and THORPE, A. (1979). Immunofluorescent localization of insulin-like material in the median neurosecretory cells of the blowfly. *Cell Tissue Res. 200*, 187–191.

DUVE, H., THORPE, A. and LAZARUS, N. (1979). Isolation of material displaying insulin-like immunological and biological activity from the brain of the blowfly, *Calliphora vomitoria*. *Biochem. J. 184*, 221–227.

ELTRINGHAM, H. (1933). On the tarsal sense organs of *Lepidoptera*. *Trans. Roy. Ent. Soc.*, Lond. *81*, 33–36.

ERNST, K.-D. (1972). Sensillum coelosphaericum, die Feinstruckture eines neuen olfactorischen Sensillentyps. *Z. Zellforsch. 132*, 95–106.

EVANS, D. R. (1961). Depression of taste sensitivity to specific sugars by their presence during development. *Science 133*, 327–328.

EVANS, D. R. (1963). Chemical structure and stimulation by carbohydrates. *Olfaction and Taste I*. Edited by I. Y. Zotter. Pages 165–192. Pergamon Press, Oxford.

EVANS, D. R. and BARTON-BROWN, L. (1960). The physiology of hunger in the blowfly. *Am. Midland Naturalist 64*, 282–300.

EVANS, D. R. and MELLON, DeF. (1962a). Electrophysiological studies of a water receptor associated with the taste sensilla of the blowfly. *J. Gen. Physiol. 45*, 487–500.

EVANS, D. R. and MELLON, DeF. (1962b). Stimulation of a primary taste receptor by salt. *J. Gen. Physiol. 45*, 651–661.

FRAENKEL, G. (1969). Evaluation of our thoughts on secondary plant substances. *Proc. 2 Int. Symp. Insect and Hostplant*. Pages 473–486. Edited by J. de Wilde and L. M. Schoonhoven North-Holland, Amsterdam-London.

FRIEDMAN, S. (1978). Trehalose regulation, one aspect of metabolic homeostasis. *Ann. Rev. Entomol. 23*, 389–407.

GELPERIN, A. (1967). Stretch receptors in the foregut of the blowfly. *Science 157*, 208–210.

GELPERIN, A. (1970). Abdominal sensory neurons providing negative feedback to the feeding behavior of the blowfly. *Z. vergl. Physiol. 72*, 17–31.

GELPERIN, A. (1971). Regulation of feeding. *Ann. Rev. Entomol. 61*, 367–378.

GETTING, P. A. (1971). The sensory control of motor output in fly proboscis extension. *Z. vergl. Physiol. 74*, 103–120.

GETTING, P. A. and STEINHARDT, R. A. (1972). The interaction of external and internal receptors on the feeding behavior of the blowfly, *Phormia regina*. *J. Insect Physiol. 18*, 1673–1681.

GILLARY, H. L. (1966a). Stimulation of the salt receptor of the blowfly. I. NaCl. *J. Gen. Physiol. 50*, 337–350.

GILLARY, H. L. (1966b). Stimulation of the salt receptor of the blowfly. III. The alkali halides. *J. Gen. Physiol. 50*, 359–368.

GOLDRICH, N. R. (1973). Behavioral responses of *Phormia regina* (Meigen) to labellar stimulation with amino acids. *J. Gen. Physiol. 61*, 74–88.

GRABOWSKI, C. J. and DETHIER, V. G. (1954). The structure of the tarsal chemoreceptors of the blowfly, *Phormia regina* Meigen. *J. Morph. 94*, 1–20.

GREEN, G. W. (1964a). The control of spontaneous locomotor activity in *Phormia regina* I. Locomotor activity patterns of intact flies. *J. Insect Physiol. 10*, 711–726.

GREEN, G. W. (1964b). The control of spontaneous locomotor activity in *Phormia regina*. II. Experiments to determine the mechanism involved. *J. Insect Physiol. 10*, 727–752.

HANAMORI, T., SHIRAISHI, A., KIJIMA, H. and MORITA, H. (1972). Stimulation of labellar sugar receptor of the fleshfly by glucosides. *Z. vergl. Physiol. 76*, 115–124.

HANAMORI, T., SHIRAISHI, A., KIJIMA, H. and MORITA, H. (1974). Structure of monosaccharides effective in stimulation of the sugar receptor of the fly. *Chemical Senses and Flavour. 1*, 147–166.

HANSEN, K. and HEUMANN, H. G. (1971). Die Feinstruktur der tarsalen Schmeckhaare der Fliege, *Phormia terraenovae* Rob.-Desv. *Z. Zellforsch. 117*, 419–442.

HASSETT, C. C., DETHIER, V. G. and GANS, J. (1950). A comparison of nutritive values and taste thresholds of carbohydrates for the blowfly. *Biol. Bull. 99*, 446–453.

HODGKIN, A. L. and KATZ, B. (1949). The effects of sodium ions on the electrical activity of the giant axon of the squid. *J. Physiol. 108*, 37–77.

HODGSON, E. S. (1958). Electrophysiological studies of arthropod chemoreception. III. Chemoreceptors of terrestrial and freshwater arthropods. *Biol. Bull. 115*, 114–125.

HODGSON, E. S., LETTVIN, J. Y. and ROEDER, K. D. (1955). Physiology of a primary chemoreceptor unit. *Science 122*, 417–418.

JAKINOVICH, W. JR., GOLDSTEIN, I. J., VON BAUMGARTEN, R. J. and AGRANOFF, B. W. (1971). Sugar receptor specificity in the fleshfly, *Sarcophaga bullata*. *Brain Res. 35*, 367–378.

KAFKA, W. A. (1970). Analyse der molekularen Wechselwirkung bei der Erregung einzelner Riechzellen. (Elktrophysiologie einzelner Rezeptor zellen auf der antenne von *Locusta migratoria*). *Z. vergl. Physiol. 70*, 105–143.

KAIB, M. (1974). Die Fleisch- und Blumenduftrezeptoren auf der Antenne der Schmeissfliege *Calliphora vicina*. *J. Comp. Physiol. 95*, 105–121.

KAISSLING, K.-E. (1969). Kinetics of olfactory receptor potentials. In *Olfaction and Taste*, III. Edited by C. Pfaffman. Pages 52–70. Rockefeller University Press, New York.

KAISSLING, K.-E. (1971). Insect olfaction. *Handbook of Sensory Physiology*. Vol. IV. Chemical Senses. Edited by L. M. Beidler. Pages 351–431. Springer-Verlag, Berlin, Heidelberg, New York.

KAISSLING, K.-E. (1979). Recognition of pheromones by moths, especially in saturniids and *Bombyx mori*. In *Chemical Ecology: Odour Communication in Animals*. Edited by F. J. Ritter. Pages 43–56. Elsevier/North-Holland Biomedical Press, Amsterdam.

KAISSLING, K.-E. and PRIESNER, E. (1970). Die Riechschwelle des Seidenspinners. *Naturwissenschaften 57*, 23–28.

KAISSLING, K.-E. and RENNER, M. (1968). Anntennale Rezeptoren für Queen-Substance und Sterzelduft bei der Honigbiene. *Z. vergl. Physiol. 59*, 357–361.

KASANG, G. (1968). Tritium-Markierung des Sexuallockstoffes Bombykol. *Z. Naturforsch. 23b*, 1331–1335.

KAWABATA, K. and SHIRAISHI, A. (1977). Variation of acceptance thresholds in the blowfly by increasing sugar concentrations in the food. *J. Comp. Physiol. 118*, 33–49.

KEIL, T. A. and THURM, U. (1979). Die Verteilung von Membrankontakten und Diffusionsbarrieren in epidermalen Sinnesorganen von Insekten. *Verhandlungen dt. zool. Ges. 72*, 285.

KELLOGG, F. E. (1970). Water vapour and carbon dioxide receptor in *Aedea aegypti*. *J. Insect Physiol. 16*, 99–108.

KIJIMA, H. (1970). Taste in insects. (In Japanese). *Kagaku 40*, 523–530.

KOCHANSKY, J., TETTE, J., TASCHBERG, E. F., CARDÉ, R. T., KAISSLING, K. E. and ROELOFS, W. L. (1975). Sex pheromone of the moth, *Antheraea polyphemus*. *J. Insect Physiol. 21*, 1977–1983.

KRAMER, K. J., TAGER, H. S. and CHILDS, C. N. (1980). Insulin-like and glucagon-like peptides in insect hemolymph. *Insect Biochem. 10*, 179–182.

LACHER, V. (1964). Elektrophysiologische Untersuchungen an einzelnen Rezeptoren für Geruch, Kohlendioxyd, Luftfeuchtigkeit und Temperatur auf den Antennen der Arbeitsbiene und der Drohne (*Apis mellifica*). *Z. vergl. Physiol. 48*, 587–623.

LANE, N. J. (1974). The organization of insect nervous systems. *Insect Neurobiology*. Edited by J. E. Trehanene. Chapter 1, pp. 1–61. North Holland Publishing Company, Amsterdam and Oxford.

LARSEN, J. R. (1962a). The fine structure of the labellar chemosensory hairs of the blowfly, *Phormia regina* Meigen. *J. Insect Physiol. 8*, 683–691.

LARSEN, J. R. (1962b). Fine structure of the interpseudotracheal papillae of the blowfly. *Science 139*, 347.

MAES, F. W. (1977). Simultaneous chemical and electrical stimulation of labellar taste hairs of the blowfly, *Calliphora vicina*. *J. Insect Physiol. 23*, 453–460.

MAES, F. W. and BIJPOST, S. C. A. (1979). Classical conditioning reveals discrimination of salt taste quality. *J. Comp. Physiol. 133*, 53–62.

MANCINI, G. and FRONTALI, N. (1967). Fine structure of the mushroom body neuropile of the brain of the roach, *Periplaneta americana*. *Z. Zellforsch. 83*, 334–343.

MCINDOO, N. E. (1914). The olfactory sense of the honeybee. *J. Exp. Zool. 16*, 265–346.

MEYER, G. M., RITTER, F. J., PERSONS, C. J., MINKS, A. K. and VOERMAN, S. (1972). Sex pheromones of summer fruit tortrix moth *Adoxophyes orana*. Two synergistic isomers. *Science 175*, 1469–1470.

MINKS, A. K., ROELOFS, W. L., RITTER, F. J. and PERSONS, C. J. (1973). Reproductive isolation of two tortricid moth species by different ratios of a two-component sex attractant. *Science 180*, 1073–1074.

MINNICH, D. E. (1921). An experimental study of the tarsal chemoreceptors of two nymphalid butterflies. *J. Exp. Zool. 33*, 173–203.

MINNICH, D. E. (1926a). The chemical sensitivity of the tarsi of certain muscid flies. *Biol. Bull. 51*, 166–178.

MINNICH, D. E. (1926b). The organs of taste on the proboscis of the blowfly, *Phormia regina* Meigen. *Anat. Rec. 34*, 126.

MINNICH, D. E. (1929). The chemical sensitivity of the legs of the blowfly, *Calliphora vomitoria* Linn., to various sugars. *Z. vergl. Physiol. 11*, 1–55.

MINNICH, D. E. (1931). The sensitivity of the oral lobes of the proboscis of the blowfly, *Calliphora vomitoria* Linn., to various sugars. *J. Exp. Zool. 60*, 121–139.

MONOD, J., WYMAN, J. and CHANGEUX, J.-P. (1965). On the nature of allosteric transition, a plausible model. *J. Mol. Biol. 12*, 88–118.

MORITA, H. (1959). Initiation of spike potentials in contact chemosensory hairs of insects. III. D.C. stimulation and generator potential of labellar chemoreceptor of *Calliphora*. *J. Cell. Comp. Physiol. 54*, 189–204.

MORITA, H. (1969). Electrical signs of taste receptor activity. *Olfaction and Taste, III*. Edited by C. Pfaffman. Pages 370–381. Rockefeller University Press, New York.

MORITA, H. (1972a). Primary processes of insect chemoreception. *Adv. Biophys. 3*, 161–198.

MORITA, H. (1972b). Properties of the sugar receptor site of the blowfly. *Olfaction and Taste, IV*. Edited by D. Schneider. Pages 357–363. Wissenschaftliche Verlagsgesellschaft mbH, Stuttgart.

MORITA, H. and SHIRAISHI, A. (1968). Stimulation of the labellar sugar receptor of the fleshfly by mono- and disaccharides. *J. Gen. Physiol. 52*, 559–583.

MORITA, H. and YAMASHITA, S. (1959). The back-firing of impulses in a labellar chemosensory hair of the fly. *Mem. Fac. Sci., Kyushu Univ. Ser. E (Biol.) 3*, 81–87.

MORITA, H. and YAMASHITA, S. (1961). Receptor potentials recorded from sensilla basiconica on the antenna of the silkworm larvae, *Bombyx mori*. *J. Exp. Biol. 38*, 851–861.

MORITA, H. and YAMASHITA, S. (1966). Further studies on the receptor potential of chemoreceptor of the blowfly. *Mem. Fac. Sci., Kyushu Univ. Ser. E. (Biol.) 4*, 83–93.

MORITA, H., DOIRA, S., TAKEDA, K. and KUWABARA, M. (1957). Electrical response of contact chemoreceptor on tarsus of the butterfly, *Vanessa indica*. *Mem. Fac. Sci., Kyushu Univ., Ser. E. (Biol.), 2*, 119–139.

MORITA, H., HIDAKA, T. and SHIRAISHI, A. (1966). Excitatory and inhibitory effects of salts on the sugar receptor of the fleshfly. *Mem. Fac. Sci., Kyushu Univ. Ser. E. (Biol.). 4*, 123–135.

MOURIER, H. (1964). Circling food-searching behavior of the house fly (*Musca domestica* L.). *Vidensk. Medd. dansk. nat. Foren. 127*, 281–294.

NELSON, C. M. (1977). The blowfly's dance. Role in the regulation of food intake. *J. Insect Physiol. 23*, 603–611.

NORMANN, T. C. and DUVE, H. (1969). Experimentally induced release of a neurohormone influencing hemolymph trehalose level in *Calliphora erythrocephala (Diptera)*. *Gen. Comp. Endocrinol. 12*, 449–459.

O'CONNELL, R. J. (1972). Responses of olfactory receptors to the sex attractant, its synergist and inhibitor in the red-banded leaf roller, *Argyotaenia veluntinana*. In *Olfaction and Taste IV*. Edited by D. Schneider. Pages 180–186. Wissenshaftliche Verlagsgesellschaft mbH, Stuttgart.

O'CONNELL, R. J. (1975). Olfactory receptor responses to sex pheromone components in the redbanded leafroller moth. *J. Gen. Physiol. 65*, 179–205.

OMAND, E. and DETHIER, V. G. (1969). An electrophysiological analysis of the action of carbohydrates on the sugar receptor of the blowfly. *Proc. Natl. Acad. Sci. USA. 62*, 136–143.

PERSONS, C. J. and RITTER, F. J. (1975). Binary sex pheromone mixtures in Tortricidae. Role of positional and geometrical isomers. *Z. Aug. Ent. 77*, 342–346.

PFLUMM, W. (1971). 'Zur Reizwirksamheit von Monosacchariden bei der Fliege *Phormia terraenovae*'. *Z. vergl. Physiol. 74*, 411–426.

PFLUMM, W. (1972). Molecular structure and stimulating effectiveness of oligosaccharides and glycosides. In *Olfaction and Taste, IV*. Edited by D. Schneider. Pages 364–370. Wissenschaftliche Verlagsgesellschaft mbH, Stuttgart.

POLLACK, G. S. (1977). Labellar lobe spreading in the blowfly. Regulation by taste and satiety. *J. Comp. Physiol. 121*, 115–134.

REES, C. J. C. (1968). The effect of aqueous solutions of some 1:1 electrolytes on the electrical response of the type 1 ("salt") chemoreceptor cell in the labella of *Phormia. J. Insect Physiol. 14*, 1331–1364.

REES, C. J. C. (1970). The primary process of reception in the type 3 ("water") receptor cell of the fly, *Phormia terraenovae. Proc. Roy. Soc. Lond. B. 174*, 469–490.

REES, C. J. C. and HORI, N. (1968). The effect of electrolytes of the general formula XCl_2 on the response of the type 1 labellar chemoreceptor of the blowfly, *Phormia. J. Insect Physiol. 14*, 1499–1513.

REEVES, R. E. (1950). The shape of pyranose rings. *J. Am. Chem. Soc. 72*, 1499–1506.

RICE, M. J. (1970). Cibarial stretch receptors in the tsetse fly (*Glossina austini*) and the blowfly (*Calliphora erythrocephala*). *J. Insect Physiol. 16*, 277–289.

RICE, M. J. and FINLAYSON, L. H. (1972). Response of blowfly cibarial pump receptors to sinusoidal stimulation. *J. Insect Physiol. 18*, 841–846.

ROELOFS, W., HILL, A. and CARDÉ, R. T. (1975). Sex pheromone components of the red-banded leafroller, *Argyrotaenia velutinana* (Lepidoptera, Tortricidae). *J. Chem. Ecol. 1*, 83–89.

ROELOFS, W. L and COMEAU, A. (1968). Sex pheromone perception. *Nature 220*, 600–601.

ROELOFS, W. L. and COMEAU, A. (1971). Sex pheromone perception, synergists and inhibitors for the red-banded leaf roller attractant. *J. Insect Physiol. 17*, 435–448.

SCHNEIDER, D. (1957). Elektrophysiologische Untersuchungen von Chemo- und Mechanorezeptoren der Antenne des Seidenspinners, *Bombyx mori* L. *Z. vergl. Physiol. 40*, 8–41.

SCHNEIDER, D. and BOECKH, J. (1962). Rezeptorpotential und Nervenimpulse einzelner olfactorischer Sensillen der Insektenantenne. *Z. vergl. Physiol. 45*, 405–412.

SCHNEIDER, D., BLOCK, B. C., BOECKH, J. and PRIESNER, E. (1967). Die Reaktion der mannlichen Seidenspinner auf Bombycol und seine Isomeren, Elektroantennogramm und Verhalten. *Z. vergl. Physiol. 54*, 192–209.

SCHNEIDER, D., KAFKA, W. A., BEROZA, M. and BIERL, B. A. (1977). Odor receptor responses of male gypsy and nun moths (*Lepidoptera, Lymantriidae*) to disparlure and its analogues. *J. Comp. Physiol. 113*, 1–15.

SCHNEIDER, D., LACHER, V. and KAISSLING, K.-E. (1964). Die Reaktionsweise und das Reaktions-spektrum von Riechzellen bei *Antheraea pernyi* (Lepidoptera, Saturniidae). *Z. vergl. Physiol. 48*, 632–662.

SHIMADA, I. (1975). Two receptor sites and their relation to amino acid stimulation in the labellar sugar receptor of the fleshfly. *J. Insect Physiol. 21*, 1675–1680.

SHIMADA, I. (1978). The stimulating effect of fatty acids and amino acid derivatives on the labellar sugar receptor of the fleshfly. *J. Gen. Physiol. 71*, 19–36.

SHIMADA, I. and ISONO, K. (1978). The specific receptor site for aliphatic carboxylate anion in the labellar sugar receptor of the fleshfly. *J. Insect Physiol. 24*, 807–811.

SHIMADA, I. and TANIMURA, T. (1981). Stereospecificity of multiple receptor sites in a labellar sugar receptor of the fleshfly for amino acids and small peptides. *J. Gen. Physiol. 77*, 23–29.

SHIMADA, I., SHIRAISHI, A., KIJIMA, H. and MORITA, H. (1974). Separation of two receptor sites in a single labellar sugar receptor of the flesh-fly by treatment with *p*-chloromercuribenzoate. *J. Insect Physiol. 20*, 605–621.

SHIRAISHI, A. and KUWABARA, M. A. (1970). The effects of amino acids on the labellar hair chemosensory cells of the fly. *J. Gen. Physiol. 56*, 768–782.

SHIRAISHI, A., YANO, T. and NAKASHIMA, M. (1979). Effects of interaction of the recurrent nerve on feeding behavior of the blowfly, *Phormia regina*, M. *Functions of the Brain*. Edited by M. Ito. Volume 2, pp.341–342. Kodansha Co. LTD. Tokyo.

SHIRAISHI, A. and TANABE, Y. (1974). The proboscis extension response and tarsal and labellar chemosensory hairs in the blowfly. *J. Comp. Physiol. 92*, 161–179.

SHIRAISHI, A. and YANO, T. (1984). Neuronal control of the feeding behavior in the blowfly. *Animal Behavior*. Edited by K. Aoki, S. Ishi and H. Morita. Pages 83–93, Japan Scientific Societies Press, Tokyo, Springer-Verlag., Berlin, Heidelberg, New York, Tokyo.

SMITH, D. V., BOWDAN, E. and DETHIER, V. G. (1983). Information transmission in tarsal sugar receptors of the blowfly. *Chemical Senses. 8*, 81–101.

STEINHARDT, R. A. (1965). Cation and anion stimulation of electrolyte receptors of the blowfly, *Phormia regina. Amer. Zool. 5*, 651.

STOCKER, R. F. and SCHORDERET, M. (1981). Cobalt filling of sensory projections from internal and external mouthparts in *Drosophila. Cell Tissue Res. 216*, 513–523.

STRAUSFELD, N. J. (1976). *Atlas of an Insect Brain*. Springer-Verlag, Berlin, Heidelberg, New York.

TANIMURA, T., ISONO, K., TAKAMURA, T. and SHIMADA, I. (1982). Genetic dimorphism in the taste sensitivity to trehalose in *Drosophila melanogaster. J. Comp. Physiol. 147*, 433–437.

TATEDA, H. and MORITA, H. (1959). Initiation of spike potential in contact chemosensory hairs of insects. I. The generation site of the recorded spike potentials. *J. Cell. Comp. Physiol. 54*, 171–176.

THURM, U. (1972). The generation of receptor potentials in epithelial receptors. *Olfaction and Taste IV*. Edited by D. Schneider. Wissenshaftliche Verlagsgesellschaft mbH, Stuttgart.

TICHY, H. (1979). Hygro- and thermoreceptive triad in antennal sensillum of the stick insect, *Carausius morosus. J. Comp. Physiol. 132*, 149–152.

VAN DER STARRE, H. (1977). On the mechanism of proboscis extension in the blowfly *Calliphora vicina. Neth. J. Zool. 27*, 292–298.

VARESCHI, E. (1971). Duftuntersheidung be der Honigbiene – Einzelzellableitungen und Verhaltensreaktionen. *Z. vergl. Physiol. 75*, 143–173.

VOERMAN, S., MINKS, A. K. and GOEWIE, E. A. (1976). Specificity of the pheromone system of *Adoxophyes orana* and *Clepsis spectrana. J. Chem. Ecol. 1*, 423–429.

VON FRISCH, K. (1919). Über den Geruchsinn der Biene und seine blütenbiologische Bedeutung. *Zool Jahrb. 37*, 1–238.

VON FRISCH, K. (1921). Über den Sitz des Geruchsinnes bei Insecten. *Zool. Jahrb. Abt. f. allg. Zool. u. Physiol. 38*, 449–516.

VON FRISCH, K. (1935). Über den Geschmacksinn der Biene. *Z. vergl. Physiol. 21*, 1–156.

VON FRISCH, K. (1967). *The Dance Language and Orientation of Bees*. The Belknap Press of Harvard University Press, Cambridge, Massachusetts.

WALDOW, U. (1970). Electrophysiologische Untersuchungen an Feuchte-, Trocken- und Kalterezeptoren auf der Antenne der Wanderheuschrecke *Locusta. Z. vergl. Physiol. 69*, 249–283.

WALDOW, U. (1973). Electrophysiologie eines neuen Aasgeruchrezeptors und seine Bedeutung für das Verhalten des Totengrabers (*Necrophorus*). *J. Comp. Physiol. 83*, 415–424.

WALLIS, D. I. (1962). The sense organs on the ovipositor of the blowfly, *Phormia regina* Meigen. *J. Insect Physiol. 8*, 453–467.

WIECZOREK, H. (1980). Sugar reception by an insect water receptor. *J. Comp. Physiol. 138*, 167–172.

WIECZOREK, H. (1982). A biochemical approach to the electroenic potassium pump of insect sensilla: Potassium sensitive ATPases in the labellum of the fly. *J. Comp. Physiol. 148*, 303–311.

WIECZOREK, H. and KÖPPL, R. (1978). Effect of sugars on the labellar water receptor of the fly. *J. Comp. Physiol. 126*, 131–136.

WILCZEK, M. (1967). The distribution and neuroanatomy of the labellar sense organs of the blowfly *Phormia regina* Meigen. *J. Morph. 122*, 175–202.

WOLBARSHT, M. L. (1957). Water taste in *Phormia*. *Science 125*, 1248.

WOLBARSHT, M. L. (1958). Electrical activity in the chemoreceptors of the blowfly. II. Responses to electrical stimulation. *J. Gen. Physiol. 42*, 413–428.

WOLBARSHT, M. L. and DETHIER, V. G. (1958). Electrical activity in the chemoreceptors of the blowfly. I. Response to chemical and mechanical stimulation. *J. Gen. Physiol. 42*, 393–412.

WOLBARSHT, M. L. and HANSON, F. E. (1965). Electrical activity in chemoreceptors of the blowfly. III. Dendritic action potentials. *J. Gen. Physiol. 48*, 673–683.

YOKOHARI, F. (1978). Hygroreceptor mechanism in the antenna of the cockroach *Periplaneta americana*. *J. Comp. Physiol. 124*, 53–60.

YOKOHARI, F. and TATEDA, H. (1976). Moist and dry hygroreceptors for relative humidity of the cockroach, *Periplaneta americana* L. *J. Comp. Physiol. 106*, 137–152.

YOKOHARI, F., TOMINAGA, Y. and TATEDA, H. (1982). Antennal hygroreceptors of the honey bee, *Apis mellifera* L. *Cell Tissue Res. 226*, 63–73.

YOKOHARI, F., TOMINAGA, Y., ANDO, M. and TATEDA, H. (1975). An antennal hygroreceptive sensillum of the cockroach. *J. Electron Microscop. 24*, 291–293.

ZACHARUK, R. Y., YIN, L. R. and BLUE, S. G. (1971). Fine structure of the antenna and its sensory cone in larvae of *Aedes aegypti* (L.). *J. Morphol. 135*, 273–298.

4 The Eye: Development, Structure and Neural Connections

OMAR TRUJILLO-CENÓZ

Instituto de Investigaciones Biológicas, Montevideo, Uruguay

1 INTRODUCTION

Many arthropods use the information conveyed by light to cope with the challenge of a rapidly changing environment. Prey capture, predator evasion and sexual-partner recognition are very frequently eye-mediated behavioral activities. In the particular case of insects only the so-called faceted or compound eyes are involved in movement perception and form recognition.

Before dealing with the histology and development of the insect visual system it is worth stressing that the term "eye", when referring to compound eyes, always conveys a certain degree of anatomical uncertainty. There is no doubt concerning the superficial limits of these eyes, but the situation is much less clear when one tries to determine their boundaries in depth. Undisputable landmarks indicating the separation of the "eye proper" from the brain tissues are lacking and, to a certain extent, they are a matter of convention.

For the neuroanatomist mainly interested in the organization of the visual neural circuits, the term "eye" implies not only a mosaic of photoreceptors (and the associated dioptric systems) but also neuronal layers and synaptic neuropiles. Such a meaning for the word "eye" is implicit in S. R. Cajal's work (1909, 1910, 1917; Cajal, S. and Sánchez, D., 1915) on the visual system of invertebrates. (The same conception can be found in the publications by Radl, E. (1912) and Zawarzin, A. (1913, 1925)).

The Spanish author was impressed by the analogies encountered in the visual circuits of species pertaining to dissimilar zoological groups. Consequently, he disregarded discrepancies at gross anatomy level and envisaged the insect compound eye as composed of the same neural units he found in other eyes. In addition to the dioptric layer, the compound eye has, according to Cajal, a complex retina divided into three portions named "peripheral, intermediate and deep retinas". These last two regions exhibit the histological organization of typical nervous centers. Recent investigations have revealed that these neural zones also contain conspicuous concentrations of neurosecretory cells (Prasad, O., 1981).

On the other hand, for most entomologists, embryologists and neurophysiologists, the visual neuropiles do not form part of the compound eyes but integrate with the brain.

Unfortunately, developmental data do little in the selection of the most coherent terminology. In the best known species of holometabolous insects the peripheral and the deeper portions of the visual system have a different and independent origin. The photoreceptor layer arises, like the body appendages, from the imaginal discs while the visual neuropiles arise from the larval brain. In hemimetabolous species there is histological evidence indicating a common origin for both the photoreceptor layer and the associated neuropiles. They arise by delamination of the embryonic procephalic lobe.

Since in this chapter special emphasis will be given to the neural organization of the compound eye, it seems convenient to adhere to Cajal's conception. However, the widely accepted nomenclature introduced by Bullock, T. and Horridge, G. (1965) will be used for practical purposes.

In order to facilitate an accurate correlation between classical terms and those employed in the present account, a comparative table (Table 1) has been included. It is complemented with Fig. 1 which shows, schematically, the main anatomical regions of the insect visual system.

It is possible now, with the aid of well-defined semantics, to begin the study of the structural organization of the insect visual system.

The distal region of each compound eye is composed of a collection of sensilla known as "the ommatidia". Each ommatidium consists of a dioptric system and a group of photoreceptors enveloped by pigment cells. The number of ommatidia in each compound eye varies from a few in worker ants to several thousands in dragonflies. Their size also varies among insects and sometimes within the same eye. In species which rely on vision for hunting it is common to find, in the frontal region of the eye,

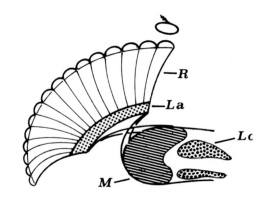

FIG. 1. The visual system of insects consists of a retina (R) and three visual ganglia. In dipterans the first ganglion or lamina (La) lies just beneath the retina basement membrane. The second and the third visual ganglia (medulla (M) and lobula complex (Lo)) form part of the brain mass.

Table 1: Terms employed by classical authors to name the different portions of the insect visual system

Bullock, T. and Horridge, G. (1965)	Hickson, S. (1885)	Viallanes, H. (1887)	Cuccati, J. (1888)	Kenyon, F. (1897)	Zawarzin, A. (1913)	Cajal, S. and Sánchez, D. (1915)
Retina	eye	retine		retina	auge	retina periférica
Lamina	periopticon	lame ganglionaire		outer fibrillar body	erstes ganglion	retina intermedia
First chiasma		chiasme externe	optische nerve	first or outer chiasma	auberes chiasma	quiasma externo
Medulla	epiopticon	masse médullaire externe	auberegeschichteter Korper	middle or second fibrillar body	zweites ganglion opticum	retina profunda
Second chiasma		chiasme interne	innere Kreuzung der Augenauschewellungen	inner chiasma	inneres chiasma	quiasma interno
Lobula	opticon	masse médullaire externe	innerergeschichteter Korper	inner or third fibrillar mass	drittes ganglion opticum	lóbulo óptico

The nomenclature introduced by Bullock, T. and Horridge, G., used in this chapter, appears as a guide in the first column

a population of ommatidia larger than those over the rest of the organ. In parallel, these larger ommatidia have their optical axes gathered in such a way that the region of space directly facing the insect is sampled by an increased number of sensory units. It will be shown in this chapter that this foveal region has its neural counterpart in the visual ganglia.

There are also species in which the differences among the ommatidia are so marked as to justify the use of terms like "double" or "subdivided" compound eyes. Typical examples are found in male specimens of mayflies (Ephemeroptera) and blackflies (Diptera, Simuliidae). On other occasions, such as in whirligig beetles (Coleoptera, Gyrinidae), each compound eye is divided by a chitinous ridge into a dorsal and a ventral portion.

Finally, it should be mentioned that in many species the eye surface bears innervated bristles and campaniform sensilla. These non-photic receptors seem to be involved in the estimation of flight velocity (bees), measurement of the sun angle (bees) and in triggering of the head-cleaning reflexes (mantids).

It is a reasonable assumption that acquaintance with the adult structures facilitates understanding of the developmental process. Consequently, it was considered convenient to close each one of the chapter sections with the respective developmental data. This is a feasible task when dealing with the retina or the lamina but a difficult one when considering the medulla or the lobula. Information concerning the neural maturation of these visual neuropiles is scarce, and still remains practically reduced to the data reported by Sánchez y Sánchez, D. (1916, 1918, 1919a,b).

2   THE RETINA

As already stated, the retina is composed of a variable number of ommatidia. These are arranged in space to form the two hemispheric bodies commonly known as the compound eyes. In many insects the shapes of these organs do not form a regular geometrical figure. Usually, the radius at the frontal region is greater than elsewhere. Each ommatidium has a dioptric system conveying light towards the photoreceptive structures. It has, in addition, various kinds of pigment cells ensheathing both the optical and the neural ommatidial components. In some species light capture is further improved by a "tapetum lucidum" which reflects back to the photoreceptive structures the fraction of non-absorbed photons.

Even though the major goal of this chapter is to cover the neural organization of the compound eye, it seems necessary, for the sake of completeness, to refer briefly to its non-neural components. A more detailed description of these latter can be found in the authoritative reviews by Bullock, T. and Horridge, G. (1965); Goldsmith, T. and Bernard, G. (1974); Autrum, H. (1975); and Carlson, S. and Chi, C. (1979).

2.1 Dioptric and catadioptric structures, pigment cells

The corneal lenses represent specialized transparent portions of the chitinous exoskeleton. In different insect orders some of the nocturnal species bear, on the outer surfaces of their corneal lenses, small conical projections termed "corneal nipples" (Bernhard, C. *et al.*, 1965). It has been suggested that the air–chitin layer resulting from the presence of these structures diminishes glares and improves light transmission.

Optical as well as electron microscope studies have demonstrated that the corneal lenses are not homogeneous. In *Apis*, for instance, each lens consists of three layers whose refractive indexes range from 1.490 (outer layer) to 1.435 (inner layer).

In other species (like tabanids and dolicopodid flies) the occurrence of multiple layers of dissimilar refractive indexes give origin to bright structural colors. It has been postulated that in these circumstances the corneal lenses function as interference color filters.

Since the publication of S. Exner's work in 1891 it is generally accepted that compound eyes are divisible, on morphological and physiological grounds, into two main types:

(1) the apposition eyes, in which there is practically no separation between the corneal layer and the photoreceptive structures; and
(2) the superposition eyes, in which a clear space is interposed between these two ommatidial components.

In apposition eyes, optical coupling between the lens and the photoreceptors is, in most species, mediated by a solid crystalline cone (eucone eyes). There are also insects, such as higher dipterans, in which such a coupling is obtained by means of a gelatinous substance contained within a two-cell, cup-like container (pseudocone eyes). Finally, there are species in which solid cones or gelatinous pseudocones are lacking. In these cases four flat transparent cells occupy the place of the crystalline cone (acone eyes).

The crystalline cone, present in the so-called eucone eyes, consists of four specialized cells containing a highly refractive transparent substance.

As already mentioned, in superposition or clear-zone eyes there is a long distance between the corneal lens and the photoreceptive structures. This space is traversed by thin refractile structures: the crystalline tracts. It has been proposed that the crystalline tracts may act as wave guides directing light to the subjacent rhabdoms. At this point it is appropriate to devote a few lines to deal with the dioptric system of the firefly, *Lampyris*. In this beetle, carefully studied by Exner, S. (1891), the corneal lens projects inwards forming a sort of solid chitinous crystalline cone (processus corneae). This peculiar feature allows the retina to be cleared from its cellular elements and still preserve most of the optical properties of the light-gathering system. Exner, taking advantage of this anatomical characteristic, was able to prove experimentally that parallel light entering through many corneal lenses gives origin to an erect image. The chitinous cones of lampyrid beetles are optically inhomogenous structures. They are composed of concentric lamellae whose refractive indexes decrease from the center (1.520) to the periphery (1.375) in an approximately parabolic manner.

The refractive index gradient determines that a non-axial ray, reaching the distal portion of the cone normally, will be deviated towards its axis. Taking into consideration the optical and geometrical characteristics of the dioptric structures of *Lampyris*, Exner inferred correctly that the optical system behaves as a double-length lens cylinder. Exner's findings have been confirmed in various species of coleopterans and lepidopterans.

At present there is general agreement that the clear-zone eyes function in the dark as Exner proposed. In these circumstances the crystalline tracts do not interfere in the formation of a focused superposition image. However, in an illuminated environment pigment migration interferes with the pathways of light rays outside the crystalline tracts. Then the optic system behaves in a way similar to apposition eyes.

Besides the usual dioptric components, some insects have a reflecting layer in their eyes: the tracheal tapetum. Electron microscope investigations have furnished detailed information concerning the tapetum microstructure. In lepidopterans, at the base of each rhabdom there are several cytoplasmic plates (modified taenidiae) which alternate with air spaces. It has been proposed that these multilayered structures function as quarter wave-length interference reflector filters. Apparently, each rhabdom (or groups of rhabdoms) may have a filter with different optical characteristics from those of neighboring rhabdoms. This may be the origin of the different colors observed in different eye regions. Such a point of view has been challenged by more recent studies. Ribi, W. (1979a), working on pierid butterflies, found that the dissected tapetum "reflects turquoise to yellow-green all over the eye, even in the ventral and medial eye regions which show (in the intact animal) a deep red glow colour". According to the same author, red glow does not result from the interference-filter characteristics of the tapetum but by the presence of red pigment granules occurring in some photoreceptors.

A summary of the relevant information concerning dioptric and catadioptric eye components can be found in Table 2.

The inner portions of the dioptric apparatus, as well as the cell bodies of the photoreceptors forming an ommatidium, are enveloped by pigment cells. Following the classical description of Johnas, W. (1911) it has become customary to recognize the existence of three kinds of pigment cells commonly named: primary, accessory and basal pigment cells. The use of terms like "primary" or "accessory" connotes functional hierarchies of a doubtful biological value. It therefore seems advisable to employ purely descriptive terms, which advantageously convey information about actual characteristics of the cells.

In higher Diptera there are three kinds of cells containing pigment granules (besides the photoreceptors). These are:

(1) the two cells forming the pseudocone wall;
(2) the large pigment cells extending from the inner surface of the corneal surface to the basement membrane; and
(3) the small pigment cells whose nuclei lie below the basement membrane (Fig. 2).

The large pigment cells of *Phaenicia* contain purple granules while the small cells contain a yellow pigment. According to Chi, C. and Carlson, S. (1976a) the large pigment cells play, in addition to their screening function, the important role of suspensory elements of the ommatidium. Electron micrographs show that the cytoplasm of the large pigment cells contains small dense particles with the

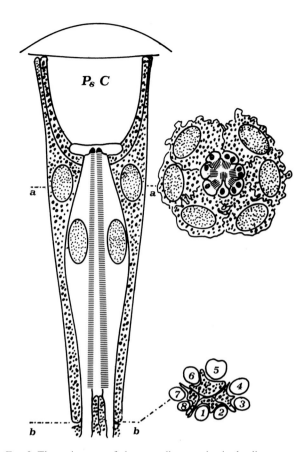

FIG. 2. The main types of pigment cells occurring in the dipteran ommatidium. Six large pigment cells run from the corneal lens layer to the retina basement membrane. A cross section at the level of *a–a* shows the nuclei of the large pigment cells. Close to the basement membrane (*b–b* level) four thick processes arising from the small basal cells invade the ommatidial cavity. The wall of pseudocone cavity (Ps-C) is composed of two cells which contain pigment granules. 1–8, Photoreceptor axons.

appearance of glycogen. These particles are absent both in the pseudocone and in the small pigment cells.

It is a well-known feature that, in superposition eyes, light induces dramatic morphological changes in the large pigment cells which modify the optical and functional characteristics of the retina.

2.2 Photoreceptors

The first steps leading to the generation of light-induced signals take place in the photoreceptors or retinula cells. Each photoreceptor consists of three well-differentiated portions:

Table 2: Dioptric and catadioptric structures

	Genera	Findings	References
CORNEAL LENS			
	several, covering different orders; selected examples: *Prodenia*, *Sphinx*, (Lepidoptera); *Myrmelon*, (Neuroptera); *Phryganea*, (Trichoptera); *Aedes*, (Diptera)	corneal nipples functioning as an antireflection coat	Bernhard, C. *et al.* (1965)
	Apis (worker) (Hymenoptera)	optical constants; the corneal lens consists of three layers of dissimilar refractive index	Varela, F. and Wiitanen, W. (1970)
	several, selected examples: *Hybromitra*, *Dolichopus* (Diptera)	interference filters consisting of systems of alternating dense and rare layers.	Bernhard, G. and Miller, W. (1968)
	Calliphora (Diptera)	several layers of different optical densities	Seitz, G. (1968)
CRYSTALLINE CONE			
	several, covering different orders; selected examples: *Periplaneta* (Dictyoptera), *Dytiscuss* (Coleoptera), *Tabanus* (Diptera), *Tipula* (Diptera), *Forficula* (Dermaptera)	eucone, pseudocone and acone eyes.	Grenacher, G. (1879)
	several, covering different orders; selected examples: *Lampyris* (Coleoptera), *Sphinx* (Lepidoptera), *Vespa* (Hymenoptera)	apposition and superposition eyes	Exner, S. (1891)
	Ephestia (Lepidoptera)	evidence that visual information is processed by a superposition dioptric apparatus	Kunze, P. (1970)
	Ephestia (Lepidoptera)	optical constants; the refractive index decreases from the axis to the periphery	Hausen, K. (1973)
	Ephestia (Lepidoptera)	optical constants; superposition of images at the level of the rhabdoms layer	Cleary, P. *et al.* (1977)
	Apis (worker) (Hymenoptera)	optically homogeneous	Varela, F. and Wiitanen, W. (1970)
Processus corneae	*Lampyris* (Coleoptera)	optical constants; double-length lens cylinder	Exner, S. (1891)
Processus corneae	*Phausis* (Coleoptera)	layers with different optical density; highest refractive index in the centre	Seitz, G. (1969a)
Pseudocone	*Calliphora* (Diptera)	optically homogeneous, isotropic	Seitz, G. (1969b)
CRYSTALLINE TRACT			
	Photuris (Coleoptera)	fine structure of the crystalline tracts; these may act as wave guides transmitting light to the photoreceptors (in light and dark conditions)	Horridge, G. (1968)
	several, lepidopterans; as example: *Hyalophora*.	light is largely confined to the tracts that function as wave guides	Döving, K. and Miller, W.(1969)

Table 2: Dioptric and catadioptric structures—Continued

	Genera	Findings	References
TAPETUM			
	moths?	eye-glow originates in the tracheal tapetum	Leydig, F. (1864)
	Euptychia, Anartia, Danaus (Lepidoptera)	fine structure of specialized tracheoles functioning as interference filters	Miller, W. and Bernard, G. (1968)
	Colias, Gonopteryx, Pieris (Lepidoptera)	fine structure of the tapetum; coloured retinula cell pigments determine the eye-glow hue	Ribi, W. (1979a)

(1) the soma bearing both the transducer structures and the metabolic machinery of the cell;

(2) the axonic segment invading the distal portion of the lamina; and

(3) the terminal synaptic portion establishing connections with different kinds of visual neurons.

The retina basement membrane usually represents a distinct landmark indicating the separation of the soma from the axonic and synaptic portions.

This section only concerns the photoreceptor somata since the other portions of the cell will be described when dealing with the lamina.

Since the pioneer investigations by Grenacher, G. (1879), it is known that the retinula cells bear long, longitudinally oriented refractive structures. These refractive rods are the "rhabdomeres". They usually integrate a single unit, then named "rhabdom". The rhabdomeres are the light-conducting, light-absorbing organelles of the insect photoreceptors.

With improvements in electron microscope techniques the fine structure of the rhabdomeres was revealed. The first studies already showed that each rhabdomere consists of thousands of closely packed tubules aligned at right angles to the long axis of the photoreceptor. Each tubule is about 400–500 Å in diameter and 1–0.5 μm long. The tubules actually represent finger-like evaginations of the photoreceptor plasma membrane which in cross-section exhibit a characteristic honeycomb aspect (Fig. 3).

At the rhabdomere level the plasma membrane seems to consist of globular subunits similar to those described by Nilsson, S. (1965) in the outer segments of frog rods. The significance of these membrane subunits is a matter of discussion. For some authors they merely represent radiation

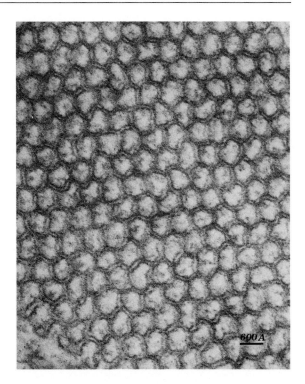

FIG. 3. Each rhabdomere consists of thousands of closely packed microvilli. When cross-sectioned they exhibit a typical honeycomb appearance. *Phaenicia.*

artifacts produced by the electron beam. However, using freeze-fracture techniques, membrane particles have been identified in the rhabdomeres of hymenopterans and dipterans. Rows of elongated particles (80 Å wide, 250 Å long) occur on the endoplasmic face (E face) of the rhabdomeric microvilli. In addition, spherical particles (85 Å diameter) were found in the protoplasmic face (P face). It has been proposed that these particles may represent photopigment molecules or some of their intermediate products.

It is pertinent to mention that there is compelling microspectrophotometric evidence indicating that in arthropods the visual pigments are mainly located in the rhabdomeric microvilli. Moreover, theoretical considerations support the hypothesis that the small diameter of each microvillus prevents free rotational movements of the rhodopsin molecules (assuming their asymmetrical shape). This feature favors the longitudinal alignment of the visual pigments with respect to the microvillus axis. It is commonly accepted that this preferential array of photopigments constitutes the molecular basis of polarized light sensitivity.

Photopigments with different spectral properties occur in different photoreceptors. These receptors are not distributed in different ommatidia but usually coexist in the same ommatidium. Electrophysiological records followed by cell marking, indicated that in *Periplaneta* "ultraviolet" and "green" units form part of a single ommatidium (Mote, M. and Goldsmith, T., 1971). It has been proposed that it is possible to identify different color receptors combining chromatic adaptation and conventional electron microscopy. Selective swelling, distortion and hyperdensity of microvilli should occur in the color-adapted cells. Obviously, this method requires a very accurate standardization of the electron microscope technical procedure (particularly manipulation and fixation of the retinal tissues).

Photostable sensitizing or "antenna" pigments also occur in the rhabdomeres. Both spectrophotometric and electrophysiological data indicate that the six "short" photoreceptors of the fly ommatidium achieve their high ultraviolet sensitivity by means of a sensitizing pigment. (A photostable pigment has also been identified in one of the two "long" photoreceptors.)

Measurements of the refractive indexes of the rhabdomeres in flies and bees have yielded values around 1.4, while the surrounding cytoplasm has a lower refractive index ($1,339 \pm 0.002$). These optical conditions allow the rhabdomeres to propagate light along their major longitudinal axis (Exner, S., 1891). Light is transmitted in specific field patterns called "modes" (as it occurs in the outer segments of vertebrate photoreceptors).

The perirhabdomeric cytoplasm seems to be a very active cellular region. It undergoes striking changes when the insect passes from a light to a dark environment (and vice-versa). Orthopterans and dyctiopterans have been extensively studied in this regard by Horridge, G. and Barnard, P. (1965) and Butler, R. and Horridge, G. (1973). When locusts or cockroaches are kept in the light and the eye tissues are likewise fixed under light, the rhabdomeres appear surrounded by a non-vacuolated cytoplasm rich in mitochondria. If the insect is maintained in the dark for 10–15 min before fixing, and the retina is fixed in dim red light, a palisade of vacuoles is found around the rhabdom. It has been speculated that this sleeve of clear vacuoles may assist in maintenance of a sharp difference between the refractive index of the rhabdom and that of the nearby cytoplasm. The morphological changes in the perirhabdomeric cytoplasm control both the angular sensitivity and absolute sensitivity of the retinular cells (Snyder, W. and Horridge, G., 1972). However, this kind of cytoplasmic photoreaction merits further study, since there are reports describing unexpected, well-developed palisades of clear vacuoles in fully light-adapted locusts.

Another point of interest is the action of light on the rhabdomere itself. Besides the rhabdomere movements triggered by light in water hemiptera, structural changes have been described in a variety of species. It is known for example, that the morphology of the mosquito rhabdomeres follows a diurnal rhythm.

More recently, in a series of experiments performed in grasshoppers and mantids (genus *Valanga* and *Orthodera*, respectively), Horridge, G. *et al.* (1981) found that the rhabdomeres of these insects appear larger during night than during day. When these night-state photoreceptors are illuminated, their rhabdomeres become disorganized over the course of about 1 h. If the insects are maintained in the dark, breakdown of the rhabdomeres is not avoided. This suggests that microvilli turnover is also controlled by an intrinsic, light-independent rhythm. The morphological modifications of the rhabdomeres have been correlated with changes of the photoreceptor field size and also with its responses to light. The results tend to indicate that in these species the compound eyes become more sensitive during the night.

Valanga and *Orthodera* represent species in which light causes remarkable structural changes. In other species such as tipulids or blowflies light-induced modifications are less dramatic. The reader interested in daily rhythms and the turnover of photoreceptor membranes in arthropods is referred to a recent review by Waterman, T. (1982).

Light flux in the rhabdomeres is controlled by the movement of small pigment granules lying in the cytoplasm of the photoreceptors. In the dark-adapted state the granules appear randomly scattered in the cytoplasm. Strong illumination elicits their migration towards the rhabdomeres. Experiments performed in dipterans have demonstrated that concentration of the granules close to the rhabdomeres is followed by a decrement of the transmitted light. In some way, the "pipe light" represented by the illuminated rhabdomere is optically punctured by the aggregated granules. This mechanism allows the automatic control of the light flux along the photoreceptive structure.

Since the rhabdomeres are considered the sites of initiation of the bioelectrical events (Lasansky, A. and Fuortes, M., 1969) it is particularly important to obtain information concerning the size of the potential pool of ions available at this level. By means of the use of electron-dense tracers it has been determined that in the locust the main rhabdomal extracellular space reservoir is represented by minute sacs lying at the rhabdom periphery. Measurements indicate that this rhabdomal reservoir accounts for approximately 1% of the total retina extracellular space. The major extracellular space compartment (70%) lies in lacunae between ommatidia. Obviously these data are valid for insects with a fused-type ommatidia in which there is not a central ommatidial cavity. Based on images derived from the use of lanthanum as an electron-dense tracer, Chi, C. and Carlson, S. (1981) have concluded that the ommatidial cavity in *Musca* is part of the extracellular space.

When considering the spatial distribution of the rhabdomeres two main types of ommatidial organization become apparent. In most insects the rhabdomeres pertaining to different photoreceptors form a single central photoreceptive structure known as a fused rhabdom (Fig. 4). In two insect orders (Diptera and Hemiptera) another type of ommatidia has evolved. In Brachycera and in aquatic Hemiptera the photoreceptors bear separated rhabdomeres which behave as optically

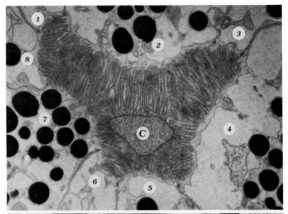
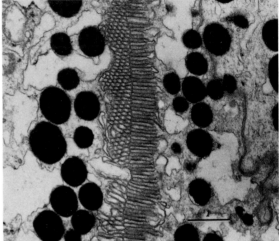

Fig. 4. Cross- and longitudinal sections through a fused-rhabdom ommatidium. Note that in this species (*Periplaneta*) the crystalline cone (C) penetrates the distal portion of the rhabdom (upper electron micrograph). As seen in the lower one, the rhabdomeric microvilli of neighboring photoreceptors lie in close proximity. 1–8, Photoreceptor cells. (Scale bar in μm.)

independent structures. In dipteran species there is a large space occupying the center of each ommatidium (Fig. 5).

In the fused-type rhabdom the possibilities of optical as well as electrical coupling are increased. According to A. Snyder's theoretical studies (1973), each photoreceptor can have a high absolute sensitivity and still preserve a narrow spectral sensitivity curve due to optical coupling.

The first investigators interested in the histological organization of compound eyes noticed that not all photoreceptors composing an ommatidia are morphologically alike. This is particularly evident in the ommatidia of higher dipterans.

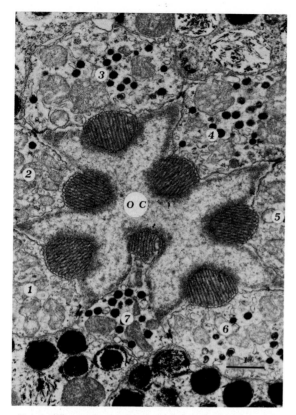

FIG. 5. Cross-section passing through the apical portion of an ommatidium of the fly *Phaenicia* sp. Note that the seven rhabdomeres are separated from each other leaving a large ommatidial cavity (OC). 1–7, Photoreceptor cells. (Scale bar in μm.)

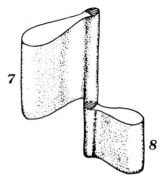

FIG. 6. Schematic representation of the superior (7) and inferior central cells (8). Note that the rhabdomere microvilli are orthogonally arranged.

In some species such as *Musca*, *Phaenicia* and *Drosophila*, each ommatidium has two central photoreceptors (the superior and inferior central cells, 7–8) whose rhabdomeres are aligned in tandem, thus sharing the same optical axis (Melamed, J. and Trujillo-Cenóz, O., 1968). In these two cells the rhabdomeric microvilli are orthogonally arranged as schematically shown in Fig. 6.

A variation of this basic dipteran type of organization was found in flies of the family Dolicopodidae (Trujillo-Cenóz, O. and Bernard, G., 1972). Two types of central rhabdoms occur in the dolicopodid *Sympicnus*. Half of the ommatidia exhibit the usual orthogonal pattern described in the central rhabdom of most dipterans. In the other half, the microvilli forming the superior rhabdomere (cell 7) lie in parallel to those composing the inferior one (cell 8). A similar situation has been described in the water-strider *Gerris lacustris*.

(Schneider, L. and Langer, H., 1969). The specializations of the central rhabdom, both in dolicopodids and aquatic hemipterans, seem to be useful to diminish glare and to improve vision through the water. It should also be mentioned that electron microscopical investigations on staphylinid beetles of the species *Creophilinus erythrocephalus* have revealed that in these insects the fused rhabdom consists of alternating layers of mutually perpendicular microvilli (Meyer-Rochow, V., 1972).

Another characteristic related to the precise array of the photoreceptors is the occurrence in some retinas of a dorsoventral symmetry. Dietrich, W. (1909) found that in dipterans a zig-zag line separates two populations of enantiomorphic ommatidia.

In recent years several papers have been published indicating that in hymenopterans the photoreceptors are twisted along their long axes. The same phenomenon has been reported to occur in the blowfly *Calliphora*. However, these findings are not unanimously accepted. It is argued that photoreceptor (and rhabdomere) twisting may result from osmotic or mechanical damage to the retinal cells. Since theoretical studies indicate that rhabdomere twist would reduce polarization sensitivity of the photoreceptors, this matter becomes a critical point in any model concerning polarized light detection.

It is not an easy task to give a wholly satisfactory classification of the various types of photoreceptor arrays found in the ommatidia of insects.

The somatic portions of the different photoreceptors (from 8 to 12) grouped in a single ommatidium, show a variable morphology. They can be short or long, wide or narrow, possessing well-developed rhabdomeres or only a few poorly organized microvilli. In *Dytiscus*, for example, there is a single photoreceptor cell body bearing a large rhabdomere. There are, in addition, six photoreceptors which contribute to the formation of a cross-shaped rhabdom lying in the proximal portion of the ommatidium. Close to the retina basement membrane occurs a small 8th basal cell whose rhabdomere occupies, at this level, the ommatidium center. Horridge, G. *et al.* (1970) introduced the term "tiered retina" to name this kind of stratified arrangement of the photoreceptor cells. One obvious functional consequence of the stratification of the photoreceptors is that the deeper cells receive light that has been filtered by the more superficial elements. This feature may affect the spectral characteristics of the deeper cells. As already mentioned, the spatial arrangement of the photoreceptors (and of their respective rhabdomeres) can also influence the polarization sensitivities. This matter has been the subject of theoretical investigations by Snyder, A. (1973).

In order to complete the information about the photoreceptors it is necessary to devote a few lines to the fine structure of the intercellular contacts at ommatidial level. This becomes a point of interest when considered in the context of physiological reports indicating electrical coupling between photoreceptors.

In the drone bee such coupling is strong (similar to that occurring in *Limulus*) whereas in the locust electrical interactions are weak. It has been postulated that this kind of phenomenon may be subserved by tip-to-tip or lateral contacts between rhabdomeric microvilli. However, conclusive evidence is lacking and the whole matter remains an open field for research. In open-rhabdom ommatidia such as those found in dipterans, the photoreceptors seem to be electrically insulated. Electron microscope investigations by Chi, C. and Carlson, S. (1981) have revealed that in *Musca domestica* the photoreceptors are connected by means of desmosomes and focal tight junctions. Table 3 offers a synthesized view of the information concerning insect photoreceptors.

3 DEVELOPMENT OF THE RETINA

3.1 General

One of the most characteristic features of insect life is the fact that in numerous species the recently hatched individual is morphologically different from the adult. Important changes of form, included under the general term "metamorphosis", usually occur at the end of the larval stage.

It is well known, however, that the magnitude of these changes is quite different in the various taxonomic groups within the class. Some insects emerge from the egg showing the general body-form of the imago. They are often regarded as having no metamorphosis (Ametabola). On the other hand, most of the exopterygotes pass through an incomplete metamorphosis (Hemimetabola). In this group the life-styles of the juvenile forms and the adults are very similar. This justifies the existence in the young of the same complete repertory of sense organs found in the fully developed specimens (including well-developed compound eyes).

Finally, there is a third group of insects in which the life cycle encompasses a series of larval stages differing markedly from the adult (Holometabola).

At the end of the active period of the larval life there is usually a quiescent instar interposed between the last larval molt and the adult. This instar is the pupa. Within its cuticle the most prominent features of the metamorphosis occur. Unlike the juvenile forms of the preceding groups, the larvae of holometabolous insects are devoid of typical compound eyes.

In Ametabola and Hemimetabola the compound eyes complete their basic structural organization within the egg. A dissimilar situation is found in Holometabola, in which the compound eyes only attain structural maturity after a sequence of developmental changes in step with the larval growth. As a rule, however, the most dramatic morphogenetic events occur within a relatively short period of insect life coincident with the early pupal period. The fact that important developmental events can occur within a large, soft-bodied larva instead of taking place in the interior of a small impermeable egg, has determined that most studies concerning the development of compound eyes

Table 3: *Photoreceptors*

	Findings	Genera	References
RHABDOMERES			
fine structure	closely packed microvilli	*Musca, Apis, Erebus*	Fernández-Morán, H. (1956)
		Sarcophaga, Anax	Goldsmith, T. and Philpott, D. (1957)
		Drosophila	Wolken, J. *et al.* (1957)
		Apis	Goldsmith, T. (1962)
		Sarcophaga, Lucilia	Trujillo-Cenóz, O. and Melamed, J. (1966a)
		Musca	Boschek, C. (1971)
	twisting	*Apis*	Grundler, O. (1974)
		Myrmecia	Menzel, R. and Blakers, M. (1976)
		Calliphora	Smola, U. and Tscharntke, H. (1979)
		Ptilogyna	Williams, C. (1981)
	no twist	*Apis, Calliphora*	Ribi, W. (1979b)
	ordered membrane particles	*Myrmecia*	Nickel, E. and Menzel, R. (1976)
		Drosophila	Harris, W. *et al.* (1976)
		Drosophila	Schinz, R. *et al.* (1978)
		Musca	Chi, C. and Carlson, S. (1979)
	extracellular space 1% of the total	*Locusta, Valanga*	Shaw, S. (1978)
	the ommatidial cavity is part of the extra-cellular space	*Musca*	Chi, C. and Carlson, S. (1981)
optical properties	refractive index		
	1.347	*Apis* (worker)	Varela, F. and Wiitanen, W. (1970)
	1.349	*Calliphora*	Seitz, G. (1968)
	1.365 ± 0.006	*Calliphora*	Stavenga, D. (1974)
	1.370–1.40	*Musca* (white m.)	Kirschfeld, K. and Snyder, W. (1975)
	—		De Vries, H. (1956)
	wave guides	*Apis*	Varela, F. and Wiitanen, W. (1970)
		Drosophila	Franceschini, N. and Kirschfeld, K. (1971)
		Musca	Kirschfeld, K. and Snyder, W. (1975)
	dichroism	—	Laughlin, S. *et al.* (1975)
		Musca	Kirschfeld, K. and Snyder, W. (1975)
	birefringence	*Apis, Volucella*	Menzer, G. and Stockhammer, K. (1951)
		Calliphora	Seitz, G. (1969b)
		Musca (white m.)	Kirschfeld, K. and Snyder, W. (1975)
photochemical characteristics	photolabile pigments	*Calliphora* (white m.)	Langer, H. and Thorell, B. (1966)
	photostable sensitizing pigments	*Musca* (white m.)	Kirschfeld, K. *et al.* (1977)
photodynamic phenomena	expansion and contraction	*Notonecta*	Ludtke, H. (1953)
		Dytiscus, Lethocerus	Walcott, B. (1969)
	volume changes	*Aedes*	Brammer, J. and Claren, B. (1976)
	structural changes	*Ptilogyna*	Williams, D. (1980)
		Valanga, Orthodera	Horridge, G. *et al.* (1981)
CYTOPLASM			
fine structure	in addition to the common cell organelles:		
	0.15–0.20 μg granules	*Musca*	Kirschfeld, K. and Franceschini, N. (1969)
		Musca	Boschek, C. (1971)
	Multivesicular bodies	*Sarcophaga, Lucilia*	Trujillo-Cenóz, O. and Melamed, J. (1966a)
		Apis (worker)	Varela, F. and Porter, K. (1969)
	red pigmented granules	*Pieris, Colias, Gonopteryx*	Ribi, W. (1979a)
optical properties	refractive index		
	perirhabdomeric 1.339 outer cytoplasm 1.343	*Apis*	Varela, F. and Wiitanen, W. (1970)
	— 1.341	*Calliphora*	Seitz, G. (1968)
	(assumed) 1.34	*Musca* (white m.)	Kirschfeld, K. and Snyder, W. (1975)

Table 3: Photoreceptors—Continued

	Findings	Genera	References
photodynamic phenomena	under light the small cytoplasmic granules move toward the rhabdomeres	*Musca* *Apis* *Drosophila, Musca*	Kirschfeld, K. and Franceschini, N. (1969) Kolb, G. and Autrum, H. (1972) Franceschini, N. and Kirschfeld, K. (1976)
	formation of a palisade of clear perirhabdomeric vacuole in the dark	*Locusta* *Periplaneta*	Horridge, G. and Barnard, P. (1965) Butler, R. and Horridge, G. (1973)
INTERCELLULAR JUNCTIONS fine structure	desmosomes desmosomes, focal tight junctions	*Musca* *Musca*	Boschek, C. (1971) Chi, C. and Carlson, S. (1981)

have been made on holometabolous species (particularly dipterans and lepidopterans). In these insects the eye primordia are amenable not only to most of the histological techniques but also to experimental procedures.

Furthermore, when looking for a developmental system in which a multidisciplinary approach (embryological, genetical and biochemical) could be applied, higher Diptera appear as one of the best-suited materials. In such a context the origin and further maturation of the dipteran eye has been the subject of detailed investigations.

Therefore, despite the complex anatomical changes occurring in the cephalic portions of the dipteran larvae (Snodgrass, R., 1953, 1954; Schoeller-Raccaud, J., 1977) these insects are considered good models for illustrating the development of the compound eye. When considered pertinent, the general picture emerging from the studies on dipterans will be complemented with data obtained from other insect groups.

3.2 The retina anlage

In higher Diptera, at a time more or less coincident with the formation of the cellular blastoderm, a peculiar segregation of two cell populations occurs. One of these cell populations will give origin to the larval tissues, the other will form the primordia of the adult organs.

These primordia appear during larval life as well delimited epithelial sacs (the imaginal discs) or are represented by less conspicuous cell aggregations (the imaginal rings or the nests of abdominal histoblasts).

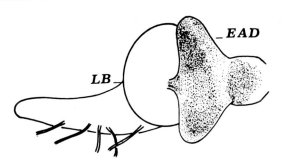

FIG. 7. This drawing shows that in the third-instar larva the eye-antenna discs lie closely apposed to the cerebral ganglia. LB, Larval brain; EAD, eye-antenna disc.

Owing largely to the pioneer investigations of Weismann, A. (1864, 1866) and Kunckel d'Herculais, J. (1882), it is now a well-established fact that the retina of muscoid dipterans derive from a pair of imaginal discs lying close to the cerebroid ganglia of the larva (Fig. 7) (each disc giving rise to a single retina and the homolateral antenna). In addition modern studies, mainly based on experiments with disc fragments, have proved that the eye-antennal discs also contain the primordia for the ocelli, the ptilinum and for other regions of the head capsule (Ouweneel, W., 1970).

At the end of larval life the eye-antennal discs can be easily recognized as whitish translucent organs attached to the cerebroid ganglia by means of short optic stalks. A mature disc consists of around 30,000 cells (data from *Drosophila*). However, it is generally admitted that each disc derives from a few primitive cells (approximately 20) forming some sort of "closed system" which grows by intrinsic mitotic activity and not by the external supply of cells.

The eye-antenna primordia become histologically distinct in the embryo as "a pair of small

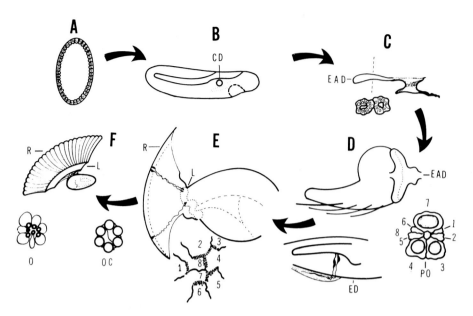

Fig. 8. This sequence shows the developmental story of the compound eye. **A**: At blastoderm stage, approximately 20 cells represent the initial eye-disc cell population. **B**: During embryonic development two cephalic discs (CD) arise from the frontal sacs. **C**: In the first-instar larva the eye-antennal discs (EAD) appear as two elongated sacs attached to the cephalic armature. **D**: At the end of the larval life the EAD lie closely apposed to the larval brain. At this stage some of the imaginal cells are already neurally differentiated (as shown in ED). They appear forming cell clusters (primitive ommatidia). **E**: After puparium formation the rhabdomeres start differentiation while the lamina-anlage initiates its migration towards the retina. **F**: In the fully developed eye, the retina (R) consists of closely packed ommatidia (O) and the lamina (L) of closely packed optical cartridges (OC).

outgrowths at the postero-lateral corners of the frontal sac" (Anderson, D., 1963).

At this developmental stage they are usually known as the "cephalic discs". It is worth noting that the terms employed in the first embryological descriptions by Pratt, H. (1900) and Snodgrass, R. (1924) do not correspond exactly to those employed nowadays.

When following the developmental history of the retina primordium during larval life, one sees that in recently hatched larvae the eye-antennal discs appear as two minute slender epithelial pouches lying dorsally with respect to the pharyngeal armature (Fig. 8c.) These tiny organs are very difficult to recognize under a dissecting microscope but they can be identified with certainty in properly stained histological preparations. The exploration of series of cross-sections passing through the cephalic extremity of a first-instar larva, shows a non-boundary transition between the epithelium investing the pharyngeal sclerites and the eye-antenna primordia. A transverse section passing through one of these small imaginal discs shows a single

layer of cells resting upon a thin basement membrane. Under the electron microscope the cells exhibit the usual appearance described for the embryonic, undifferentiated ectodermal elements. Two kinds of membrane specializations (zonula adherens and septated desmosomes) have been consistently found at the cell boundaries (Melamed, J. and Trujillo-Cenóz, O., 1975).

In the second-instar discs the same cytological organization is maintained, but the morphogenetic changes leading to a visible separation between the eye and the antennal anlage have now commenced.

The retina primordium exhibits a relatively simpler organization in insects other than brachiceran dipterans. In Nematocera, for example, the retina develops from a prospective "eye region" lying anterior to the larval stemmata. There is at this level a peculiar aggregation of epidermal cells which constitutes the "optic placode" (White, R., 1961). Unlike higher dipterans these primodial cells do not form closed sacs but lie "in series" with the larval epidermis.

Finally, it should be mentioned that in Hemimetabola both the retina and its associated visual centers originate from the first procephalic lobe. At this region the embryonic ectoderm becomes thicker and soon splits into two cellular sheets. The external sheet represents the retina-anlage or optic placode while the inner one represents the future visual neuropiles (lamina and medulla). As with the imaginal discs of dipterans, the optic placode gives rise to the different cellular types found in the adult ommatidium.

During post-embryonic development the retina grows by the addition of new ommatidia to the anterior edge of the expanding organ. The origin of the new ommatidia is a matter of discussion. In *Oncopeltus*, for example, they arise by recruitment of undetermined epidermal cells. A similar mechanism was suggested by Hyde, C. (1972) to explain the post-embryonic development of the retina in *Periplaneta americana*. However, more recent studies support the existence of a budding zone as originally proposed by Bodenstein, D. (1953). This zone may represent a remnant of the embryonic optic placode.

3.3   Differentiation of the dioptric apparatus and pigment cells

At the moment at which the first cuticular layer covering the prospective retina is secreted, there are no clear morphological differences between the putative corneagenous cells and the other ommatidial precursors. Electron microscope images suggest that all cells whose apical portions reach the surface of the retina-anlage contribute to the secretion of the cuticuline layer. Small electron-dense patches appearing at the tips of the apical microvilli constitute the first signs of secretory activity. During subsequent developmental stages various layers of fine fibrillar material are added.

In the particular case of nocturnal lepidopterans bearing "corneal nipples", the cuticuline layer at early developmental stages shows irregularly spaced evaginations. These evaginations are occupied by the distal segments of the underlying microvilli. According to Gemne, G. (1966), "the formation of the nipple anlage seems to be completed before the deposition of the rest of the cornea".

Investigations by Perry, M. (1968) on *Drosophila melanogaster* have revealed that in the last stages of the lens formation the Semper, or cone cells, are important elements. They seem to derive from a group of non-neural elements covering the tip of each proto-ommatidium when it moves inwards. At this moment the cone cells are in direct contact with the cuticuline layer. As development proceeds the cone cells move away from the corneal layer leaving (in dipterans) a cup-shaped cavity. In the phasmid *Carausius morosus* the cells which will give rise to the crystalline cone send out proximal processes which run between adjacent immature photoreceptors (Such, J., 1969a,b). These processes probably represent the "roots of the cone" present in the fully developed ommatidium.

The origin of the pigment cells has been investigated employing [^{3}H]thymidine and recessive mutations. These studies indicated that the photoreceptors and the large pigment cells arise from different precursors. The developing pigment cells are not difficult to recognize in the third-instar eye discs of *Phaenicia* or *Drosophila*. Occasionally Golgi-stained, they appear as elongated elements running from the basement membrane to the disc surface. More information has been provided by the electron microscope. Cross-sections of the imaginal epithelium show complex systems of cytoplasmic lamellae enveloping each one of the proto-ommatidia. These lamellae originate from the immature pigment cells which, at this developmental stage, still lack characteristic pigment granules. When compared with other cell components of the eye disc the future pigment cells contain few microtubules (Perry, M., 1968). The pigment granules will appear during later pupal stages.

3.4   Differentiation of the photoreceptor cells

As it has been advanced in earlier pages, in cyclorraphous dipterans larval growth is accompanied by important morphogenetic changes occurring at the level of the imaginal discs. These events are particularly noticeable in the eye-antennal discs which begin their development as minute epithelial sacs and terminate exhibiting a complex leaf-like appearance. In addition to the aforementioned morphogenetic changes, important structural modifications occur at a cytological level. These

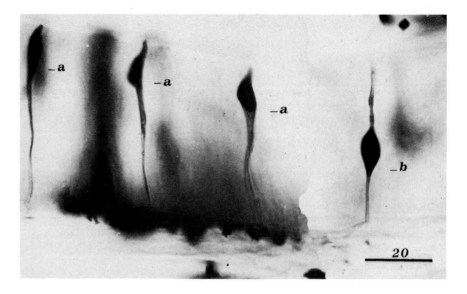

FIG. 9. Four neurally differentiated imaginal cells. Three of them (a) exhibit short apical dendrites. There is a fourth cell (b) with a long apical dendrite. Photographic collage, Golgi procedure. Third-instar eye-disc of *Phaenicia*. (Scale bar in μm.)

represent the morphological expression of the activation of sets of genes which have remained "silent" and now take command of the metabolism of the imaginal cells. There is compelling evidence that in insects the "activation signals" are of a hormonal nature.

The first morphological changes announcing the differentiation of the ommatidial units can be observed in third-instar blowfly larvae, 50–60 h after hatching. The temporospatial course of the differentiation process can be followed with the aid of a modification of one of Gomori's techniques routinely employed for detecting alkaline phosphatase. Ommatidial differentiation starts in a small region of the imaginal epithelium, very close to the insertion point of the optic stalk. The existence of a single differentiation center has also been described in other species of insects such as *Aedes, Ephestia, Notocneta* and *Carausius*.

Turning to cyclorraphous dipterans, studies on *Phaenicia* and *Drosophila* have indicated that a maturation wave spreads out from the differentiation center and advances toward the anterior and lateral portions of the disc. The "wave front" is represented by a transverse groove which behaves as a clear and visible landmark of the maturative phenomenon.

It is important to note that Gomori's technique provides a general panoramic view of the spread of the differentiation wave. However, it lacks the resolution to show fine cytological details. These can be revealed by the complementary use of the Golgi procedure and serial-section electron microscopy. By means of these more powerful histological tools a "neuronal stage" was found in the course of photoreceptor differentiation in insects. Sánchez y Sánchez, D. (1916, 1918) was the first to publish conclusive images of such a stage in the developing retina of *Pieris brassicae*. Golgi images similar to those found by the Spanish author in lepidopterans can be observed in the eye-antennal discs of dipterans. As shown in Fig. 9 the future photoreceptors appear as typical bipolar neurons with the dendrites ending at the epithelium surface and the axons running toward the optic stalks. A more detailed analysis permits recognition of two main populations of protophotoreceptors differing in the relative positions of the cell bodies with respect to the epithelium surface. Most of them have the cell bodies close to the epithelium surface and therefore the apical dendrites are short. There are, in addition, protophotoreceptors lying deeply in the epithelium and connected to the surface by means of relatively long dendrites. Owing to the peculiar staining characteristics of the Golgi procedure (the impregnation is in most cases a partial one and usually only a few cells of the preparation are stained) the topological relationships of these two

FIG. 10. Beneath the disc basement membrane (dotted line) there are thousands of axon bundles enveloped by glial processes. Each bundle consists of eight axons (inset). Note the presence of a giant cell whose nucleus contains polytenic chromosomes. Third-instar eye-disc. *Phaenicia*. (Scale bar in μm.)

photoreceptor classes are difficult to depict. This technical limitation can be overcome by means of serial-section electron microscopy.

Studies performed in *Drosophila* and *Phaenicia* have shown that both kinds of protophotoreceptors coexist in the same ommatidial precursor (more details concerning the aggregation of photoreceptors will be provided when dealing with the origin of the retinal patterns).

The disc basement membrane is traversed by numerous groups of axons which form a conspicuous layer below the imaginal epithelium. As shown in Fig. 10 eight-axon bundles are enveloped by glial processes. Together with the glial components there are giant cells exhibiting a polytenic

organization of the nuclear material. They probably represent migrating cells derived from larval tissues.

When a comparative study is made between the electron micrographs obtained from still immature imaginal cells and those from neurally differentiated elements, important differences become evident. In the former there is little endoplasmic reticulum and the few membrane profiles found in the cytoplasm pertain to the Golgi complex. The cell is virtually packed with free ribosomes.

The neurally differentiated cells exhibit a quite different cytoplasmic aspect. As one might expect from cells involved in an active synthesis of proteins, the cisternae of rough ER are now more abundant. Free ribosomes remain as randomly scattered clusters intermixed with microvesicular bodies and lysosomes. The participation of microtubules in the development of dipteran photoreceptors has been stressed by Perry, M. (1968).

In the particular case of photoreceptors, cell differentiation implies the maturation of two functionally dissimilar poles:

(1) a phototransductive pole subserving light absorption and the initiation of the bioelectric phenomena; and
(2) a synaptic pole involved in the intercellular transmission of the light-generated signals.

In holometabolous insects differentiation of the phototransductive pole is a tardive event occurring during the quiescent pupal stage. There is now strong evidence indicating that rhabdomere differentiation is under the control of single genes.

Koenig, J. and Merriam, J. (1975), by chemical mutagenesis and appropriate inbreeding crosses, isolated a mutant in which the outer rhabdomeres are absent or poorly developed (Fig. 11). Following the accepted nomenclature the mutant was named Ora JK[84] (outer rhabdomeres absent). There is, in addition, a second mutant isolated in Benzer's laboratory in which the rhabdomere of the superior central cell (photoreceptor 7) is absent (sevenless, Sev[Ly3]). These mutants have been used to explore the spectral sensitivities and the photopigments of specific types of photoreceptors (Harris, W. et al., 1976). They are also useful to study the contribution to behavior of the different photoreceptor classes.

The onset of rhabdomere differentiation is marked by the formation of folds at the level of the photoreceptor plasma membrane facing the center of the proto-ommatidium. A few hours later these folds become more numerous and regular. It must be kept in mind that at this developmental stage the eight protophotoreceptors are not separated by a central cavity, as occurs in adult dipterans. They lie packed together forming an immature type of central rhabdom. In Phaenicia approximately 90 h after the puparium formation the separation of the different rhabdomeres begins. More or less simultaneously the rhabdomeric microvilli attain the paracrystalline organization typical of mature, fully developed photoreceptors.

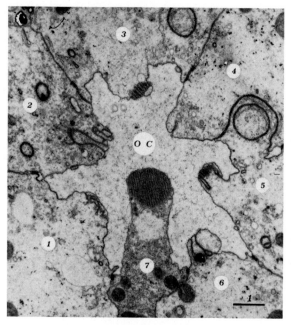

FIG. 11. In the mutant Ora the six peripheral photoreceptors (1–6) are devoid of their respective rhabdomeres, (or these are incompletely developed). Conversely, the superior central photoreceptor (7) exhibits a normally developed rhabdomere. OC, Ommatidial cavity. (Scale bar in μm.)

According to Such, J. (1975), who worked on the Phasme (Carausius morosus), the rhabdomeres derive exclusively from the apical surface of the immature photoreceptors. As a consequence of modifications of cell shape, the primitive apical surface becomes laterally oriented (facing the center of the proto-ommatidium). These movements are followed by elongation of the cellular body and the parallel development of the rhabdomeric microvilli.

As described by Sánchez y Sánchez, D. (1916–1918) in Pieris brassicae and by Bodenstein, D. (1950) in Drosophila, the size and shape of the future photoreceptors change in the course of the post-embryonic development.

Golgi studies have revealed that in Phaenicia, as in Drosophila, the protophotoreceptors appear first as elongated elements which become shorter after puparium formation. At older stages they increase in length to reach the size found in adult specimens.

Concerning the differentiation of the synaptic pole of the photoreceptor cell, it should be emphasized that the understanding of this maturative process implies some knowledge of the events occurring at the level of the optic stalks.

As mentioned, in the recently hatched larvae the eye-antenna discs and the cerebral ganglia are already anatomically linked. The connection is mediated by two thin optic stalks which contain in *Phaenicia* no more than 50 nerve fibers.

Approximately 70 h after hatching, the optic stalks have increased in diameter showing the structural organization schematically shown in Fig. 12. The electron microscope reveals that each optic stalk consists basically of a small nerve core (the larval bundle) surrounded by an extremely thick cellular sheath. Ten to twelve hours later a new population of nerve fibers is found travelling together with the larval bundle. The incoming fibers represent the axonal prolongations of the recently differentiated imaginal cells. They form the so-called "imaginal bundle" (Trujillo-Cenóz, O. and Melamed, J., 1973). At the end of larval life each optic stalk contains around 5500 nerve fibers, half of which are seen integrating conspicuous eight-axon bundles. These bundles penetrate the larval brain and terminate at the level of the lamina anlage.

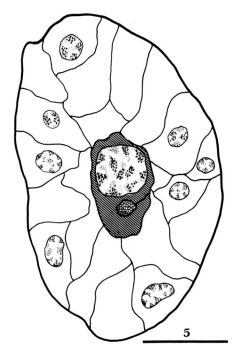

5

FIG. 12. This scheme shows the cellular organization of the optic stalk before invasion by the photoreceptor axons. The central glial cell envelops a thin axon bundle. This bundle (Bolwig's nerve) originates from larval photoreceptors. (Scale bar in μm.)

It can therefore be inferred that the primitive connections linking the larval photoreceptors to the brain lobes (Bolwig's nerves) serve as pathfinder lines for the outgrowing axons stemming from the neurally differentiated cells. A similar situation has been described in lepidopterans in which sheaths of cells derived from the stemmata nerves provide support for the centripetal growth of the photoreceptor axons.

The morphological and fine structural changes which undergo the photoreceptor terminals within the brain lobes will be described in later sections dealing with the development of the lamina.

3.5 Development of ommatidial patterns

The precise geometrical array of cells which characterizes the adult retina is already present in the third-instar eye discs of *Drosophila* and *Phaenicia*.

The systematic electron microscope exploration of sections covering large significant portions of the imaginal epithelium has allowed a better understanding of the rules governing cell distribution. The transverse groove or morphogenetic furrow divides each disc into two dissimilar cellular fields: the anterior cellular field (ACF) composed of a regular mosaic of undifferentiated cells, and the posterior cellular field (PCF) containing the ommatidia precursors. These appear as cellular groups consisting of five or eight cells.

Five-cell clusters occur close to the transverse groove and have been considered a first step during the process of aggregation of the future photoreceptors. When the imaginal discs are fixed and stained as usual for electron microscopy, these five-cell units are difficult to perceive. Identification of these precocious cell clusters is greatly facilitated by the use of ruthenium red as a plasma-membrane stain (Fig. 13).

Most of the PCF contains eight-cell clusters or proto-ommatidia. As shown in Fig. 14 in these more mature units the cells are distributed according to a fixed consistent pattern. There is a central profile which represents the long apical prolongation of cell 8 whose soma lies at a deep level (compare with Fig. 9). There are also two opposite pairs of slender cytoplasmic profiles (1–2, 5–6) which constitute the apical portions of cells whose nuclei lie at an intermediate level. The remaining three cells (7, 4, 3) have their nuclei close to the epithelium surface.

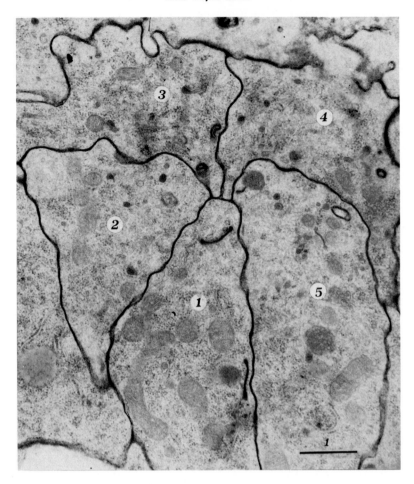

FIG. 13. Near the transverse groove or morphogenetic furrow, there are proto-ommatidia consisting of five cells (1–5). Ruthenium red stains the plasma membranes and facilitates visualization of these early aggregations of the future photoreceptor cells. Third-instar disc, *Phaenicia* sp. (Scale bar in μm.)

As noticed by Ready, D. *et al.* (1976) in *Drosophila* and by Trujillo-Cenóz, O. and Melamed, J. (1978) in *Phaenicia* the proto-ommatidia in the dorsal half of the disc exhibit cell arrangements which are the mirror images of those appearing in the ventral proto-ommatidia (Fig. 15). These two populations of enantiomorphic units meet at equator level as it occurs in the adult eye. The formation of a complete equator line seems to be a tardive event which requires fine adjustments during the pupal period.

As the result of the application of autoradiographic, mosaic-inducing techniques and transplantation experiments, the dynamics of the process leading to the ommatidium formation is now emerging.

The available evidence supports the view that the ommatidium formation is, at least in dipterans, a process mainly based on cell recruitment. In *Drosophila* the five cells composing the groups lying close to the transverse groove will become photoreceptors 2, 3, 4, 5 and 8 of the mature ommatidium. The final number is completed by recruitment of three additional cells. As stated previously, recruitment mechanism has been also proposed for explaining eye growth in *Oncopeltus*.

It is important to mention in this context the investigations made by Eley, S. and Shelton, P. (1976) trying to correlate different types of cell junctions with well-defined retina developmental stages. The studies performed in the desert locust *Schistocerca gregaria* indicate that prior to the cell cluster formation the immature cells are linked by means of punctate tight junctions, gap junctions and not fully differentiated septate desmosomes. Gap junctions

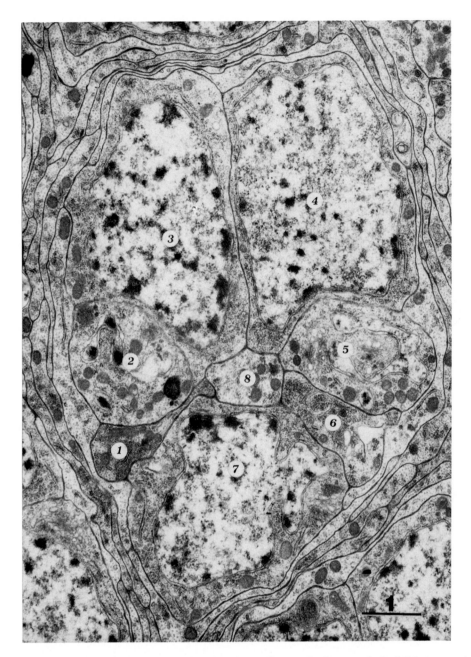

FIG. 14. Most proto-ommatidia consist of eight cells distributed according to a fixed pattern (1–8). Cell 7 always points toward the disc equator. Note that the proto-ommatidium is enveloped by several cytoplasmic layers derived from the future pigmentary cells. From Trujillo-Cenóz, O. and Melamed, J., 1978. (Scale bar in μm.)

disappear just before proto-ommatidia formation. It has been proposed that they may have some special function in cell determination. On the other hand, formation of cell clusters surely involves surface adhesive factors which may be provided by desmosomes and septate desmosomes. These two kinds of membrane specializations predominate at later developmental stages.

Table 4 and Fig. 8 provide a synthesized view of the retina development.

Table 4: *Retina development*

	Findings	Genera	References
EMBRYONIC DEVELOPMENT	The retina derives, by delamination, from the first procephalic lobe	*Mantis*	Viallanes, H. (1891)
	The retina primordia ("dorsal head discs") originate from dorso-lateral thickenings of the embryonic head epidermis	*Melophagus*	Pratt, H. (1900)
	The retina arises as a lateral thickening of the epidermis just posterior to the place where the antenna arises	*Apis*	Phillips, E. (1905)
	The neural portions of the eye and the retina develop from two different regions of the embryonic ectoderm	*Calliphora*	Schoeller, J. (1964)
	The cells which give rise to the retina-primordia ("the cephalic discs") are descendents of cells lying at the postero-dorsal edges of the presumptive ectoderm of the head	*Dacus*	Anderson, D. (1963)

	Primitive number of cells:
	  X-ray somatic crossing-over 2
	  Gynamdromorph method 20
		Drosophila
		Drosophila
			Becker, H. (1957)
			Garcia-Bellido, A. and Merriam, J. (1969)

POST-EMBRYONIC DEVELOPMENT	Historical account covering the first studies dealing with the post-embryonic development of insects. It includes information about the eye-antenna discs	—	Henneguy, L. (1904)
	In higher dipterans the compound eyes arise from the cephalic "Imaginalscheiben"	*Musca, Sarcophaga, Corethra*	Weismann, A. (1864, 1866)
	Accurate drawings of the eye-antenna discs and the optic stalks	*Volucella*	Kunckel d'Herculais, J. (1882)
	During differentiation the photoreceptors pass through a bipolar–neuron stage	*Pieris*	Sánchez y Sánchez, D. (1916, 1918)
	Transplantation experiments demonstrated that the eyes of lepidopterans are able to develop in complete independence of the central nervous system	*Lymantria*	Kopec, S. (1922)
	The various regions of the eye primordium do not differentiate at the same time, but the differentiation process spreads in a definite temporo-spatial sequence	*Ephestia*	Umbach, W. (1934)
	Transplantation experiments using young discs and older host larvae, indicated that a physiological change in the disc must occur at about 50 h after hatching	*Drosophila* (Bar)	Bodenstein, D. (1939)
	The differentiation process spreads, as a wave, from the posterior border of the eye disc	*Drosophila*	Becker, H. (1957)
	A differentiation center exists at the posterior edge of the prospective eye region	*Culex*	White, R. (1963)

Table 4: Retina development—Continued

Findings	Genera	References
In transplanted eyes, the photoreceptors become functional and give bioelectrical responses upon stimulation	*Musca*	Eichenbaum, D. and Goldsmith, T. (1968)
The rhabdomeres originate from membrane folds	*Drosophila*	Perry, M. (1968)
Approximately 70–80 h after hatching the axons stemming from the neurally differentiated imaginal cells invade the optic stalks	*Phaenicia Lucilia, Calliphora*	Trujillo-Cenóz, O. and Melamed, J. (1973) Meinertzhagen, I. (1973)
Growth and differentiation of the eye is independent of the optic lobes	*Aeschna, Anax*	Mouze, M. (1974)
The cells forming each ommatidium are derived from several, clonally unrelated cells. Determination depends not on lineage but on cellular interaction	*Oncopeltus*	Shelton, P. and Lawrence, P. (1974)
The first proto-ommatidia appear at the posterior edge of the imaginal disc, near the attachment site of the optic stalk. A transverse groove separates the population of proto-ommatidia from the still undifferentiated cells	*Phaenicia*	Melamed, J. and Trujillo-Cenóz, O. (1975)
Differentiation of photoreceptors follows a precise sequence: cell grouping, microvilli polarization, establishment of rhabdomeric patterns and formation of an equator line	*Drosophila*	Campos-Ortega, J. and Gateff, E. (1976)
Proto-ommatidia clusters in the ventral half of the disc are mirror images of clusters in the dorsal half. Pattern formation proceeds by recruitment of cells. The cells of each ommatidium are not derived from a single mother cell	*Drosophila*	Ready, D. *et al.* (1976)
Photoreceptors and large pigment cells derive from different stem cells. The cells of a single ommatidium may or may not be members of the same lineage	*Drosophila*	Campos-Ortega, J. and Hofbauer, A. (1977)
During the third larval stage, the retina anlage shows the dorso-ventral symmetry characterizing the adult organ. Pattern perturbations occur at the equator line	*Phaenicia*	Trujillo-Cenóz, O. and Melamed, J. (1978)
The compound eye of the cockroach nymph grows by the addition of new ommatidia. Transplant operations indicated that there is no recruitment of larval head-capsule epidermis into the eye	*Periplaneta, Gromphadorhina*	Nowel, M. and Shelton, P. (1980)

4 THE LAMINA

4.1 General

Few regions of the insect nervous system are organized with the degree of "immaculate precision" that is observed in the lamina. It is reasonable to conceive that the accuracy of the neural connections reflects in some way the paracrystalline array of the peripheral photoreceptors.

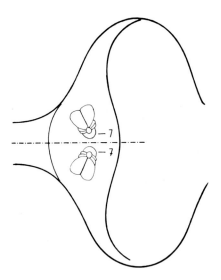

Fig. 15. As shown in this scheme the PCF contains two enantromorphic populations of proto-ommatidia. Note that cell 7 always points toward the disc equator.

Variations in thickness as well as in the relative position of the lamina with respect to the photoreceptor layer, have been evidenced by comparative studies covering different insect orders. In Diptera and Odonata, for example, the lamina is represented by a thin neural sheet lying close inside the retinal basement membrane. In other groups like Orthoptera, Dyctioptera, Coleoptera and Hemiptera, the distance between the retina and the first visual neuropile is quite extensive. In these insects the lamina integrates a macroscopically unified mass which encloses the brain and the optic centers. This feature induced Hickson, S. (1885) to believe, erroneously, that the first optic ganglion could be absent in some species (cockroaches for example).

Developmental studies have clarified the origin of the above-mentioned anatomical differences. In most species full differentiation of the lamina involves little displacement of the primordial cells. Conversely, in flies and dragonflies there is, during larval and pupal stages, a noticeable migration of the lamina-anlage toward the photoreceptor layer.

From a neurological point of view the lamina represents the first "relay station" of the insect visual pathway. At this level most of the retinal photoreceptors establish synaptic connections with the first-order visual neurons. As recognized by the first investigators interested in the arthropod visual system, these neurons are of the unipolar kind with their somata forming a distinct discontinuous layer interposed between the arriving axons and the lamina synaptic field. Along the intralaminar course each neuronal prolongation gives rise to numerous, short, dendrite-like collaterals, which mediate multiple contacts with the photoreceptor endings.

It is generally accepted that chemical transmitters mediate the passage of information from the photoreceptor axons to the lamina neurons. However, few reports have been published on the subject. In dipterans the available information points to a non-cholinergic transmitter (γ-aminobutyric acid, Campos-Ortega, J., 1974). On the other hand, histochemical methods at electron microscope level have revealed acetylcholinesterase activity in the lamina of *Apis mellifera*. The reaction product was detected both in the photoreceptor axons and in the intercellular space. The intercellular precipitation patches coincide with zones of close proximity between the photoreceptor terminals and the lateral prolongations of the lamina neurons. These data have been considered by Kral, K. and Schneider, L. (1981) to be indicative of acetylcholine as the transmitter at the first synapse of the visual pathway of *Apis mellifera*.

It is worth stressing the existence of two main classes of photoreceptor axons. The majority of them are short fibers which terminate in the lamina neuropile. There are also "long visual fibers" which, like the main neuronal prolongations, travel across the lamina and terminate in the medulla. These classes do not represent homogeneous populations and subclasses have been described in a variety of species.

It was first noticed by Viallanes, H. (1884, 1892) working on crustaceans and insects, that grouping of lamina fibers occurs in columns. He introduced the term "neuro-ommatides" to name these peculiar neural units. However, Kenyon, F. (1897) exploring the visual system of the bee was unable to confirm the findings of the French author. He stated that the neuro-ommatidia are "structures unrecognizable in the bee, and really without existence in any of the Arthropoda".

To understand Kenyon's assertion it is important to take into account that the neuro-ommatidia of Viallanes appear as well-defined anatomical entities, only in a few species (see Table 5). In the majority of insects the borders of each lamina column are difficult to perceive even under the electron microscope.

The ideas we entertain today about the synaptic nature of the lamina columns (optical cartridges using Cajal's terminology) mainly derive from the pioneer investigations of Vigier, P. (1907, 1908, 1909); Cajal, S. (1909) and Cajal, S. and Sánchez, D. (1915). They observed the transformation of the photoreceptor axons in well-characterized nerve terminals and also their aggregation around the main prolongations of the visual neurons. These findings have been confirmed with the aid of the electron microscope.

A low-power, panoramic exploration of the lamina allows recognition of the following regions:

(1) the tracheal or fenestrated region lying just below the retina basement membrane;
(2) the photoreceptor axons region;
(3) the layer of the neuron somata and
(4) the synaptic neuropile consisting of closely packed optical cartridges (Fig. 16).

This basic histological organization has been found with minor variations, in all the species so far studied.

However, the comparative studies have also revealed important differences in the nerve patterns mediating the functional coupling between the photoreceptors and the lamina neurons.

In fused-rhabdom insects (the majority of species) there are direct uncrossed connections between the ommatidia and the lamina synaptic units. In this way each ommatidium is represented at lamina level by its subjacent synaptic column. A second, very peculiar type of projection pattern has evolved in species with open rhabdoms. In these insects the ommatidial units disappear in the lamina as a consequence of a very precise rearrangement of the photoreceptor axons. The resulting optical cartridges collect information not from a single ommatidium but from several ommatidia.

It is profitable in this context to describe separately the neural organization of the lamina in these two groups of insects. This is justified not only by the differences in the retina–lamina projection patterns but also by the overwhelming body of work done on open-rhabdom species (particularly dipterans).

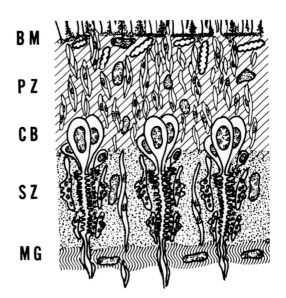

Fig. 16. Schematic drawing showing the main regions of the lamina. BM, Basement membrane; PZ, photoreceptor axons zone; CB, cell bodies layer, SZ synaptic zone; MG, marginal glia.

There is finally another aspect of the structural organization of the lamina, which is important for understanding the mechanisms involved in the generation and flow of bioelectrical signals in this visual center. It concerns the size and distribution of the extracellular space. Studies performed by Shaw, S. (1978) on *Locusta* have shown that the major extracellular compartment lies in the retina (between the ommatidia). According to the same author, below the retina basement membrane there are two main zones:

(1) the photoreceptor axons zone in which tracers fail to penetrate (the extracellular space is reduced to a minimum) and
(2) the synaptic field in which the predominance of very fine processes separated by regular interfiber clefts, determines an increase of the extracellular pool size.

The existence of a blood-eye barrier seems to occur near the photoreceptor axon zone.

4.2 The neural organization of the lamina in insects with fused-rhabdom ommatidia

As already mentioned, in fused-rhabdom species there is a direct projection of each ommatidium

Table 5: Main characteristics of the lamina in insects other than dipterans

Order	Genera	Techniques	Retina lamina projection pattern	Optic cartridges	Retinula cell axons	Types of monopolar cells	Stratification	References
Ephemeroptera	Cloeon (superposition eyes of males)	Golgi, TEM	direct	obvious	seven short, long absent	three	monolayered	Wolburg-Buchholz, K. (1977)
Odonata	Libellula and Agrion	Golgi	—	—	—	several	—	Cajal, C. and Sánchez, D. (1915)
	Aeschna (larvae)	Golgi, methylene blue	—	—	no long	two?	—	Zawarzin, A. (1913)
	Sympetrum	TEM	direct	obvious	six short two long	several	bilayered	Armett-Kibel, Ch. et al. (1977); Meinertzhagen, I. and Armett-Kibel, Ch. (1982)
	Coenagrion	Golgi	direct	—	six short two long (smooth and spiny)	several	bilayered	Strausfeld, N. (1976a)
Orthoptera	Locusta	red. silver	direct	—	—	—	—	Shaw, S. (1968)
	Schistocerca	serial semi-thin sections	direct	not obvious	—	—	—	Meinertzhagen, I. (1976)
	Locusta and Valonga	TEM	direct	not obvious	—	—	—	Shaw, S. (1978)
Dictyoptera	Periplaneta	TEM, Golgi	direct?	not obvious	seven short one long	only wide field neurons	trilayered	Ribi, W. (1977)
Hemiptera	Benacus	serial semi-thin sections	divergent	—	six short two long	—	—	Meinertzhagen, I. (1976)
	Notonecta Corixa	Golgi TEM	dispersion to different	not obvious	six short two long	four	bilayered	Wolburg-Buchholz, K. (1979)

Table 5: *Main characteristics of the lamina in insects other than dipterans* — Continued

Order	Genera	Techniques	Retina lamina projection pattern	Optic cartridges	Retinula cell axons	Types of monopolar cells	Stratification	References
	Gerris	Golgi–TEM	optical cartridges					
Coleoptera	Phausis	Golgi, Golgi–Cox. red. silver	direct	not obvious	six short two long?	five	trilayered	Ohly, K. (1975)
	Hoplia	Golgi	direct		six short two long		bilayered?	Strausfeld, N. (1976a)
Lepidoptera	Sphinx	Golgi	—		short and long	several		Cajal, S. and Sánchez, D. (1915)
	Sphinx and Pieris	Golgi–red. silver	direct	not obvious	six short three long		—	Strausfeld, N. and Blest, A. (1970)
	Pieris and Trapezites	serial semi-thin sections	direct	not obvious	six short three long			Meinertzhagen, I. (1976)
Hymenoptera	Apis (worker)	Golgi	—	not obvious	—		—	Kenyon, F. (1897)
	Apis (worker)	Golgi	—	not obvious	? short ? long	five or six	trilayered	Cajal, S. and Sánchez, D. (1915)
	Apis	TEM					—	Varela, F. (1970)
	Apis (drone)	serial semi-thin sections	direct	present	six short three long		—	Meinertzhagen, I. (1976)
	Apis (worker)	Golgi / TEM	direct	not obvious	six short three long	four	trilayered	Ribi, W. (1975b, c,– 1976, 1979c)
	Cataglypis	Golgi	direct	not obvious	six short three long	five	trilayered	Ribi, W. (1975a)

upon the subjacent optical cartridges. Well-studied examples of this kind of projection pattern are the dragonfly *Sympetrum rubicundulum* and the bee *Apis mellifera*. In both species the ommatidial units maintain their individualities in the distal regions of the lamina and the relative positions of the axons remain unchanged in the synaptic field. However, in both species there is a clear twist of each axon bundle in such a way that the primitive spatial orientation of the photoreceptors is lost at lamina level. In *Sympetrum*, clockwise and anti-clockwise rotations have been observed, but a consistent pattern has not been revealed. In *Apis*, serial section studies by Ribi, W. (1979c) have revealed mirror-image arrangements of the photoreceptor axons in different regions of the lamina. It is important to mention that according to Ribi "the direction of twist is not regularly related to the position in the eye as it is in the fly".

When the lamina of fused-rhabdom insects are explored in detail one sees that the organization of the optical cartridges is frequently complicated by the stratification of the photoreceptor endings. In the dragonfly *Sympetrum*, one pair of photoreceptor axons ends in the distal region of the lamina neuropile while the remaining two pairs terminate close to the first chiasm. A similar situation occurs in *Apis mellifera*. In this species one pair of retinular axons (those corresponding to photoreceptors 1 and 4) terminate in the medial synaptic field while the remaining two pairs end more superficially.

Menzel, R. and Blakers, M. (1976) studied the color receptors in the bee eye. Their investigations revealed that the morphology of the photoreceptor axons can be a useful guide for characterizing different photoreceptor types. The most frequently recorded "green" photoreceptors give rise to thick axons reaching the internal tangential plexus (deep short axons). They terminate by means of short irregular branches and numerous spines.

Apparently, "blue" photoreceptors are grouped into two types. Some of them have long forked terminals while others terminate by means of non-branching fibers. The aforementioned authors were also able to distinguish four types of ultraviolet receptors based on both collateral spine patterns and termination sites. In all cases the recorded and subsequently stained cells were the origin of long visual fibers.

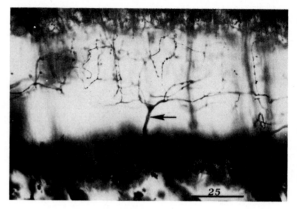

FIG. 17. Photomicrograph showing the mother trunk (arrow) giving origin to one of the systems of tangential fibers occurring in the lamina of the dragonfly *Micrathyria*. Golgi procedure. (Scale in μm.)

It is interesting to note that in the case of fused-rhabdom species the long visual fibers send out collateral processes during their intralaminar course. This feature suggests that, contrasting with dipterans, the special receptors establish synaptic connections at lamina level.

Further neural complexity results from the common occurrence of various systems of tangential fibers. These provide anatomical support for lateral interactions at different levels of the lamina. Plexuses of tangential fibers are particularly well developed in Hymenoptera and Odonata (Fig. 17). However, the synaptology of these fibers remains unexplored.

When considering the lamina neurons of fused-rhabdom species there is a general tendency to classify them into the same main classes found in dipterans.

Ribi, W. (1976), completing the classical accounts by Kenyon, F. (1897) and Cajal, S. and Sánchez, D. (1915), has furnished an accurate description of the lamina neurons of *Apis mellifera* (worker). He reported the occurrence of four neuronal classes (L1 to L4) which exhibit minor morphological variations in different eye regions. In the bee, like in most species, the main prolongations of L1 and L2 neurons occupy the center of each optical cartridge and are practically covered with very irregular, dendrite-like branches.

On closer scrutiny one sees that the branches of L1 neurons constitute a homogeneous population. They never go beyond the limits of their own cartridges. Conversely, L2 neurons have heterogeneous branches. Some of them are short, like those occurring in L1 neurons, but others are relatively long and invade neighboring cartridges.

A quite different dendritic pattern is exhibited by L3 neurons. Contrasting with the preceding neuronal types, the collateral branches are restricted to the medium synaptic stratum (stratum B). These branches are short and do not invade other cartridges.

There is finally a fourth neuronal type (L4). It is characterized by a thin radial prolongation which travels without branching through the outer and medium synaptic strata.

This thin mother trunk only ramifies at the level of the inner synaptic stratum. The branches are long and run in all directions toward distant optical cartridges. Figure 18 shows, schematically, the different neuronal types found in *Apis*.

In both bees and dragonflies the lamina appears as a geometrically organized neural sheet. This seems not to be the case for the cockroach *Periplaneta americana*. The photoreceptor axons of *Periplaneta* terminate by means of irregular varicose branches. Efforts made to disclose a regular array of the lamina fibers in this insect, have, up to now, failed. It has been speculated that the disorganized neural architecture of the lamina may be related to both the primitive phylogenetic characteristics of dyctiopterans and their nocturnal habits.

4.3 The neural organization of the lamina in insects with open-rhabdom ommatidia

Before dealing with the well-explored lamina of dipterans it is mandatory, for the sake of completeness, to furnish information concerning the other group of insects in which open-rhabdom ommatidia have evolved. As previously noted, ommatidia with anatomically independent rhabdomeres are also found in some hemipterans. Optical and electrophysiological experiments on the water-bug *Lethocerus* (Ioannides, A. and Horridge, G., 1975) have shown that in the dark-adapted state, photoreceptors lying in different ommatidia appear to be directed at the same point in space (as occurs in dipterans). However, in the light-adapted condition the fields of view of all photoreceptors

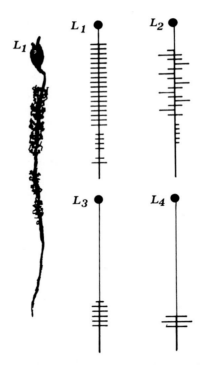

Fig. 18. Schematic representation of the four types of lamina neurons occurring in *Apis* (L1-L4) (according to Ribi, W., 1976). At the left, a camera-lucida drawing of an L1 neuron in the same species.

integrating an ommatidium coincide. This implies an optical arrangement similar to that found in fused-rhabdom species.

Light and electron microscope studies on *Benacus*, *Notonecta glauca*, *Corixa punctata* and *Gerris lacustris* (Wolburg-Buchholz, K., 1979) have revealed the existence of a divergent retina–lamina projection pattern which suggests a neural superposition mechanism of the optical signals. Nevertheless, not all short photoreceptor axons disperse as they enter the lamina synaptic field. Only the so-called "shallow short visual fibers" travel to different optical cartridges. The "deep short visual fibers" seem to project directly to their subjacent synaptic columns, as occurs in fused-rhabdom insects. It seems reasonable to suppose that the mixed neural architecture of the lamina complements the dual optical characteristics of the retina.

Since the first investigations by Vigier, P. (1907, 1908, 1909) and Cajal, S. (1909, 1910, 1915) the visual system of higher Diptera has aroused the curiosity of numerous investigators. At present the dipteran eye is one of the best known both from histological and functional points of view.

Vigier was the first to realize that the neural architecture of the lamina complements the optical organization of the retina. He observed that below the retinal basement membrane, the photoreceptor axons stemming out from a single ommatidium

"... dissociate in the most curious manner: the fascicle is twisted, it is rotated 180° along its axis; afterwards the fibers separate from each other to establish relationships with other fibers coming from neighboring ommatidia, thus forming in the periopticum such peculiar structures termed neuro-ommatides by Viallanes.... The fibers which are gathered together in the neuro-ommatides are precisely, the seven fibers conducting excitations coming from neighboring ommatidia" (Vigier, P., 1909).

That the photoreceptor axons do not follow straight pathways toward the lamina synaptic field can be easily observed in Golgi-stained material (Fig. 19).

The regular intercrossing of these axons below the retina basement membrane determines the formation of the so-called "multiple visual chiasmata region" (Cajal, S. and Sánchez, D., 1915).

By means of serial-sectioning electron microscopy and reduced silver techniques it was possible to confirm Vigier's basic histological findings (Trujillo-Cenóz, O. and Melamed, J., 1966a,b; Braitenberg, V. 1966, 1967).

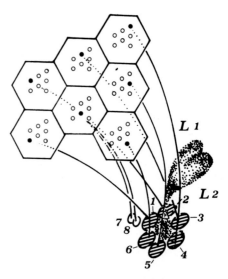

FIG. 20. The six axons which form a single cartridge originate from six photoreceptors lying in six different ommatidia. These are arranged following a trapezoidal pseudo-pupil configuration. Usually the axons originating from the superior and inferior central cells (7/8) do not establish synapses at lamina level. L1–L2, Lamina neurons.

Six short photoreceptor axons arising from a single ommatidium diverge to six different optical cartridges while the two long visual fibers (arising from the superior and inferior central cells (7–8)) bypass the lamina and project directly to the medulla. The spatial distribution of these optical cartridges is asymmetrical, forming a trapezoidal figure. On the other hand, when the six photoreceptor axons of the same single cartridge are followed back to the retina, one sees that they arise from six different ommatidia. These in turn are also arranged in a trapezoidal pseudo-pupil configuration (Fig. 20). Another important feature revealed by these studies is that the six photoreceptors projecting to the same cartridge occupy sequential angular positions in their respective ommatidia.

Kirschfeld, K. (1967), in a series of precise optical experiments performed in *Musca*, determined the ommatidial distribution of the rhabdomeres directed to the same point in space. He also inferred the neural connections required to fulfill the hypothesis that all retinular cells with identical optical alignment have their axon terminals grouped together in the same optical cartridge. Therefore there is at the cartridges a "neural superposition" of the optical

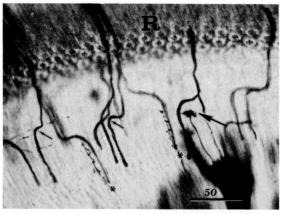

FIG. 19. In dipterans the photoreceptor axons (small arrows) do not follow rectilinear pathways toward their target neurons. The big arrow indicates a centrifugal fiber ending. The asterisks signal another type of centrifugal fibers. Golgi procedure. *Phaenicia* sp. (Scale in μm.)

signals. By this neural mechanism the light-gathering power of the dipteran eye is significantly increased.

Recent investigations indicate that "a neural superposition" mechanism may also occur in Nematocera (Bibionidae) (Zeil, J., 1979). Contrasting with higher dipterans Bibionids have ommatidia with symmetrically distributed rhabdomeres. Similarly, the histological findings suggest a symmetrical convergence of the photoreceptor axons in the lamina neuropile.

We are now in a position to explore in more detail the structural organization of the optical cartridges in Brachycera and Ciclorrapha. Cross-sections of the lamina show clearly that each cartridge consists basically of a peripheral crown formed by six photoreceptor terminals and two axial fibers (Fig. 21). These stem from the so-called giant monopolar cells (Fig. 22). In dipterans, as in Odonata, the optical cartridges appear as clearly delimited units surrounded by cytoplasmic lamellae. These arise from a peculiar type of non-neural cell that Cajal, S. (1909) termed "epithelial cell". The nuclei of the epithelial cells lie in the medial zone of the lamina between the optical cartridges. Electron microscope investigations by Boschek, C. (1971) revealed that "each cartridge is touched upon by exactly three epithelial cells". Moreover, the epithelial cells send out numerous slender prolongations which penetrate deeply within folds of the photoreceptor terminal's plasma membrane (Trujillo-Cenóz, O., 1965). Each one of these processes terminates by means of an enlarged spherical portion. Since the functional meaning of these structures is unknown, the non-committal term "capitate projections" was proposed to denominate them. Recent investigations employing freeze-fracture techniques suggest that the capitate projections may represent zones of very strong adherence between the contiguous membranes. At this level the four membrane leaflets (protoplasmic and exoplasmic faces of the receptor and glial membranes) show more particles than observed in non-specialized regions (Chi, C. and Carlson, S., 1980).

Interpreting fully the significance of the optical cartridges doubtless requires a detailed exploration of the putative synaptic loci linking the photoreceptors to the different kinds of first-order visual neurons. The most obvious synaptic specializations

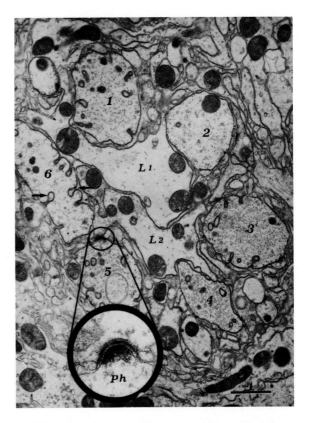

FIG. 21. Each optical cartridge consists of six photoreceptor endings (1–6) distributed around two central fibers L1–L2. These are the main prolongations of the so-called giant monopolar cells. Synaptic points are characterized by the presence of synaptic densifications (inset). The small arrows indicate capitate projections. Ph, Photoreceptor. Horsefly *Dassybasis*. (Scale in μm.)

are T-shaped presynaptic ribbons discovered in *Sarcophaga* and *Lucilia* (Trujillo-Cenóz, O., 1965). When subjected to three-dimensional studies each synaptic ribbon can be resolved in a horizontal plate and a vertical peduncle. In cross-section this latter exhibits an X-shaped profile. However, comparative studies covering different species of dipterans have revealed some variations of this basic structural plan (Trujillo-Cenóz, O., unpublished). A complete description of the fine structure of the synaptic complexes in the lamina of *Musca domestica* has been published by Burkhardt, W. and Braitenberg, V. (1976).

Quantitative studies have shown that the majority of post-synaptic elements are collateral branches of the two thick L1–L2 fibers occupying the cartridge center. By means of serial section electron

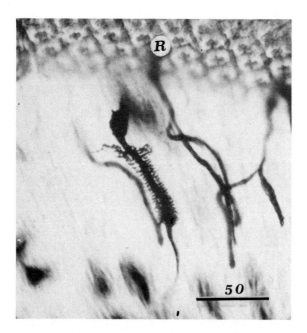

FIG. 22. One of the two axial monopolar neurons (L2). Note the numerous collateral dendrites covering the main radial prolongation. R, Retina. Golgi procedure. *Phaenicia.* (Scale in μm.)

microscopy it was determined that each of these two thick axial fibers establishes (directly or by means of branches) synaptic connections with the six photoreceptor terminals. Since the first systematic EM study of the lamina it has been evident that each cartridge constitutes a "multiple synaptic unit ensuring that information conveyed by the six phototoreceptors be funneled into each of the two second-order fibers" (Trujillo-Cenóz, O., 1965). Counts of synapse frequency in single photoreceptor terminals indicate that each receptor is presynaptic at about 200 ± 40 synaptic loci. (Nicol, D. and Meinertzhagen, I. 1982a).

As demonstrated by Golgi and electron microscopic investigations, not only the axial monopolar neurons establish connections with the photoreceptor endings. There is a third type of monopolar neuron (L3) (characterized by the polarized or asymmetric projection of its dendritic branches) which also contributes to the cartridge circuitry. Strausfeld, N. and Campos-Ortega, J. (1973a) have devoted particular attention to the synaptology of these lamina neurons. They found that the dendrites of the L3 neurons integrate, together with L1 and L2 processes, triadic or tetradic postsynaptic configurations. Through its

polarized dendritic tree the L3 neuron establishes synaptic junctions with the six photoreceptor terminals of a cartridge. However, these synaptic connections only occur in the outer one-third of the lamina synaptic field.

In higher dipterans the photoreceptor endings also behave as presynaptic elements with respect to arborizations derived from neuronal somata located outside the lamina region. There are anatomical as well as functional data suggesting that one of the varieties of long centrifugal fibers described by Cajal and Sánchez in the retina of *Musca* (*Calliphora*) *vomitoria*, the so-called nervous bags or baskets (β-fibers) (Fig. 23a) may represent a fourth centripetal pathway toward the medulla (Jarvilehto, M. and Zettler, F., 1973). It is important to emphasize that the photoreceptors do not behave exclusively as presynaptic elements.

There is some anatomical evidence indicating that the varicose processes derived from one of the tangential plexuses running in the outer zone of the lamina synaptic field (Tan 1) establish presynaptic connections with the photoreceptor terminals.

The situation remains puzzling concerning the synaptic relationship between the α-fibers (Fig. 23b) and the photoreceptor endings. At least in *Sarcophaga* and *Lucilia*, it is not rare to find typical

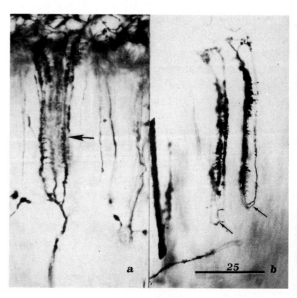

FIG. 23. **a**: Nervous bag (β-fibers); **b**: α-fibers. Note their recurrent pathways (arrows). β-fibers derive from T-cells located in the medulla. α-fibers derive from amacrine cells lying in the lamina. (Scale in μm.)

Table 6: Synaptology of short photoreceptors

Presynaptic to:	L1–L2	Vigier, P. (1908); Cajal, S. (1909); Cajal, S. and Sánchez, D. (1915); Trujillo-Cenóz, O. and Melamed, J. (1963); Trujillo-Cenóz, O. (1965); Strausfeld, N. and Blest, A. (1970)
	L3	Strausfeld, N. and Campos-Ortega, J. (1973a)
	Nervous bags (β fibers)	Campos-Ortega, J. and Strausfeld, N. (1973)
Postsynaptic to:	wide-field centrifugals (Tan 1)	Strausfeld, N. and Nässel, D. (1980)
	(α fibers) (controversial)	Trujillo-Cenóz, O. and Melamed, J. (1970)
	proximal processes of L4 neurons	Strausfeld, N. and Campos-Ortega, J. (1973b)
Symmetrical junctions	receptor to receptor	Chi, C. and Carlson, S. (1976b, 1980); Ribi, W. (1978)

presynaptic ribbons at some of the numerous zones in which the α-fibers and the photoreceptor membranes meet. However, neither Boschek, C. (1971) nor Campos-Ortega, J. and Strausfeld, N. (1973) have observed in *Musca domestica* this kind of synaptic relationship. The matter becomes still more complicated by recent findings indicating that α-like fibers may also arise from a second type of amacrine cell (Am^2) (Strausfeld, N. and Nässel, D., 1980).

Electron microscopy of Golgi-impregnated tissues has shown that some photoreceptor endings in each cartridge are also postsynaptic to the proximal collaterals of a peculiar type of monopolar lamina neuron (L4) (Strausfeld, N. and Campos-Ortega, J., 1973b).

These neurons have two sets of branches. There is a proximal set usually consisting of three dissimilar branches which synapses, as mentioned, with the photoreceptor terminals. There is also a distal set consisting of short dendrites which do not exceed the limits of the parent cartridge. The short distal processes are postsynaptic to the intrinsic system of α fibers.

The microstructure of the synaptic loci is that usually associated with the liberation of a chemical transmitter. There are, however, electron micrographs showing, in the distal portion of the cartridge, direct symmetrical junctions between photoreceptor terminals. In freeze-fracture preparations these junctions are characterized by the presence of hexagonally arrayed particles on the "P" face of the membrane leaflet. It is worth stressing that in transmission electron microscopy, a translucent space (50 Å wide) is always found between the two adjacent photoreceptor membranes. Therefore these contacts cannot be easily

catalogued together with the conventional type of gap junctions. During neural maturation of the lamina (final pupal stage) a similar type of interphotoreceptor junction occurs all along the cartridge. Table 6 summarizes the synaptology of the short axon photoreceptors (R1–6).

In the first report of Cajal, S. (1909) concerning the dipteran visual system he described optical cartridges with more than six photoreceptor axon endings. In his account, however, data referring to the topographical distribution of these atypical cartridges were lacking. More recently, by means of accurate studies based on semi-thin serial sections, Horridge, G. and Meinertzhagen, I. (1970) and Boschek, C. (1971) confirmed Cajal's findings. In addition, these modern studies revealed that "all axons that cross the equator go to cartridges with more than the usual six terminals".

The effects of this augmented presynaptic input upon the L1–L2 neurons have been the subject of a carefully morphometric analysis. In optical cartridges with 7 or 8 photoreceptor terminals, L1–L2 neurons "... produce larger individual dendrites, although maintaining the same dendrite number overall." (Nicol, D. and Meinertzhagen, I. 1982b).

Unusual cartridges with seven axon terminals also occur in a restricted area of the lamina of male flies. In this case, however, the seventh receptor terminal originates unexpectedly, from the superior central cell (cell 7). This curious projection pathway of an axon that should behave as a "long visual fiber" has been correlated with photochemical, electrophysiological and morphological data. This bulk of evidence indicates that in male flies some superior central cells closely resemble the peripheral set of six photoreceptors (Franceschini, N. *et al.*,

1981). Taking into account that at the margin of the eye less than six rhabdomeres look at the same point in space, one should expect to find in the respective lamina region optical cartridges with less than six photoreceptor terminals. This actually occurs and was described by Boschek, C. (1971) in *Musca*.

In the course of this account on the optical cartridges, four types of monopolar neurons have been introduced and partially described (the two axial monopolar L1–L2, L3 and L4). There is in addition a fifth small neuron which sends out a few short collateral processes in the distal region of the lamina (Fig. 24).

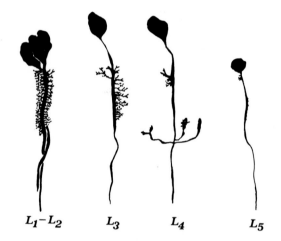

L_1–L_2 L_3 L_4 L_5

FIG. 24. These semi-schematic drawings show the different types of lamina neurons occurring in higher dipterans.

According to Strausfeld these neurons do not establish synaptic relationships with the photoreceptors but with one of the systems of tangential fibers (Tan 2) and with arborizations of the amacrine cells.

Systematic Golgi studies by Strausfeld have increased our knowledge of the origin, distribution and synaptic connections of the different classes of centrifugal fibers ramifying in the lamina.

The wide field centrifugals (Tan 1 and Tan 2) form horizontal plexuses in the distal zone of the synaptic neuropile (Fig. 25). As already mentioned, Tan 1 branches establish presynaptic contacts with the photoreceptor terminals while Tan 2 seem to establish connections with the few distal dendrites arising from the L5 neurons.

There are also two narrow field centrifugals. One type is represented by thin fibers which run parallel to the optical cartridges and give off regularly spaced branches (asterisks in Fig. 19). These reach the center of the cartridge establishing connections, as presynaptic elements, with the two axial fibers (L1–L2).

The other type of narrow field centrifugals is also represented by thin fibers which run along the optical cartridges. However, these centrifugal fibers do not send out collateral branches but terminate by means of a swollen portion in connection with the "necks" of L1–L2 and L3 neurons (arrow in Fig. 19). At this level they behave as presynaptic elements.

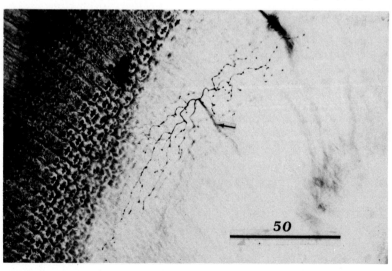

50

FIG. 25. Terminal arborization of a wide-field tangential unit. (Scale in μm.) The arrow indicates the parent trunk. Golgi procedure. *Phaenicia*.

Taking into consideration the available anatomical data, it is legitimate to speculate about possible channels of information flow in the dipteran lamina. The following are reasonable assumptions supported in some cases by experimental evidence:

(1) Visual information collected by six short-axon receptors (R1–6) located in six different ommatidia is gathered together in a single optical cartridge. Electrical coupling between the photoreceptor axons contributes to improve the signal-to-noise ratio.

(2) Lateral interactions between the lamina columns may be subserved by the L4 neurons and the amacrine cells.

(3) Centrifugal control on the photoreceptor axons may be exerted via the tangential systems, particularly Tan 1, following Strausfeld's terminology.

(4) Centrifugal control may also be exerted on the cartridge's output by means of the narrow-field centrifugal fibers.

(5) More subtle integrative phenomena may result from the multiple reciprocal connections linking the α- and the β- fiber systems.

It is important to mention, finally, that the retina of dipterans has a foveal region characterized by large facets and a denser array of the receptor cells. This foveal region "is represented by widely spaced columns in both the lamina and the medulla" (Strausfeld, N. and Nässel, D., 1980).

5   DEVELOPMENT OF THE LAMINA

The axons originating from the neurally differentiated eye-imaginal cells terminate within the larval brain close to a layer of regularly arranged neuroblasts. This layer of neuroblasts, which becomes apparent during the early third larval stage, represents the first morphological evidence of the future lamina. Further maturation of this visual ganglion operates in unison with the maturation of the retina anlage. It is important to mention in this context that the arrival of the photoreceptor axons cannot be considered a rhabdom phenomenon. Meinertzhagen, I. (1973) described a postero-anterior sequence which reflects the differentiation wave travelling along the eye disc.

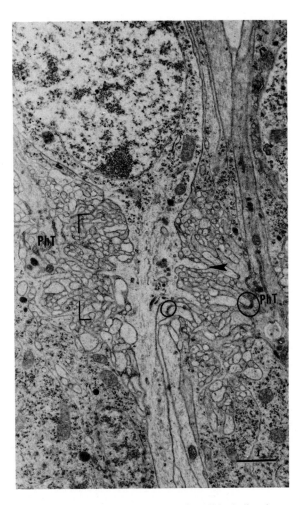

Fig. 26. Immature lamina neuron. The cell body lies close to the zone containing the photoreceptor terminals (PhT). At this developmental stage the neuropile zone (between right brackets) mainly consists of slender philopodia arising from the photoreceptor terminals. The neuron collaterals (arrow-point) are short. Note the presence of symmetric densifications at the points in which cellular processes meet (encircled). (Scale in μm.)

From the available anatomical data it can be inferred that the maturation of the dipteran lamina involves two main developmental steps. The first takes place during the larval period and conduces to both the formation of an immature type of photoreceptor ending and a transitory pattern of retina–lamina connections. The second step occurs during the late pupal stage. It leads to the formation of the optical cartridges and the parallel organization of the definitive retina–lamina projection patterns.

Light and electron microscope investigations have revealed that the immature photoreceptor terminals bear numerous filopods. During the third larval stage these processes spread out in all directions, forming a dense plexus of interwoven fibers. After puparium formation both Golgi and electron microscope images clearly show that the filopods now arise from only one side of the axon terminal mass. These thin axoplasmic projections not only intermix between them, but also establish contacts with the collateral processes stemming out from the different kinds of lamina neurons (Fig. 26). Symmetrical membrane densifications occur at the sites in which neighboring photoreceptor axons meet. Similar membrane specializations have been observed between filopods and branches of the local neurons (Trujillo-Cenóz, O. and Melamed, J., 1972, 1973). In this immature kind of neuropile the synaptic ribbons and the synaptic vesicles are absent.

At early pupal stages it is possible to recognize four well-limited zones in the lamina anlage of higher dipterans (Fig. 27). The cortical zone is composed of several layers of neuronal bodies. The latter are separated into groups by the arriving axons. (It is important to recall that at this developmental stage the lamina anlage is still remote from the retina.) The second zone consists of a single layer of regularly arranged neuronal bodies. The third zone contains the filopods of the photoreceptor terminals as well as the collateral processes stemming out from the main radial prolongations of the lamina neurons. These radial prolongations form conspicuous groups separated from each other by large cuboidal cells. These appear, when observed in low magnification electron micrographs, as forming the fourth innermost zone of the lamina anlage.

Serial section electron microscope studies (Trujillo-Cenóz, O. and Melamed, J., 1973) have revealed the occurrence of direct uncrossed connections between the proto-ommatidia of the retina primordium and the groups of photoreceptor endings in the immature lamina. In dipterans the overall picture of the developing retina–lamina complex resembles that exhibited by adult insects with fused-rhabdom ommatidia.

A very important developmental process starts 50–80 h after the puparium formation. This second main developmental step is initiated in the photoreceptor endings by the outgrowth of a thick

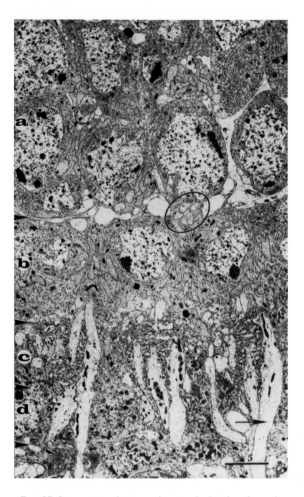

FIG. 27. Low-power electron micrograph showing the main zones of the lamina anlage during the first period of pupal life. **a**: Cortex consisting of several layers of neuronal bodies; **b**: layer of the "short neck" neurons; **c**: neuropile; **d**: layer of epithelial cells. The arrow indicates one of the groups of radial prolongations. The axon bundles (one encircled) pass through the cortex. (Scale in μm.)

lateral branch. This behaves as a big growth cone which, after reaching the set of target cells, changes its original course to run centripetally together with the radial prolongations of the lamina neurons. The remodelling of the neural connections ends with the formation of typical "optical cartridges".

The structural maturation of the synaptic loci is a more tardive event which takes place a few hours before the emergence of the imago. The development of the capitate projections is also a very tardive event.

Frohlich, A. and Meinertzhagen, I. (1982) have studied in detail the synaptogenesis in the fly's lamina (*Musca domestica*). Immature synapses have semicircular presynaptic densities, a feature which contrasts with the typical T-shaped form observed in the imago. Adult aspects of synaptic ultrastructure, such as clusters of vesicles and the complete form of the synaptic ribbon, emerge at later developmental stages.

Differentiation of the different types of lamina neurons begins in the pupa of *Phaenicia* 24–48 h after puparium formation. The first evidence of this process is the appearance in the lamina anlage of two primitive neuronal classes. One class is represented by the neurons whose somata form the cortex of the lamina anlage, the other by the "short neck" cells lying close to the immature neuropile. The next maturative step occurs 72–90 h after pupation. During this period of insect life, the first Golgi impregnations of the L4 neurons can be obtained (the presence of the set of proximal processes facilitates their early identification). L3 and L4 neurons seem to derive from the cells composing the cortex of the lamina anlage. Concerning the origin of L1–L2 neurons the situation is less clear. Apparently, they derive, like the small L5, from the layer of "short neck" cells.

In addition to the previously described cellular events the maturation of the lamina also involves important morphogenetic changes. This visual ganglion initiates development within the cephalic lobes of the larva and terminates the maturative process outside the brain mass, close to the retina basement membrane.

Experiments performed in *Phaenicia* (Trujillo-Cenóz, O. and Melamed, J., unpublished) showed that localized damage of the retina anlage prevents the migration of the corresponding lamina regions. These findings support the hypothesis that lamina displacement is not based on active movements of the lamina cells but on some sort of pulling-out mechanism. Perhaps pulling power may be generated during the dramatic shortening of the photoreceptor axons.

In species in which the larvae or the juvenile forms have compound eyes, the lamina is the site of important changes synchronized with the eye growth. In the larvae of Odonata, for example, there is at the anterior margin of the lamina a mass of undifferentiated cells (Mouze, M., 1972, 1974). This is the source of the new neurons required to meet the photoreceptor axons generated as a consequence of the increment of the ommatidia number.

On the other hand, the investigations made by Sherk, T. (1977, 1978a,b) have proved that in aeschnid larvae the foveal region of the retina exhibits important morphological changes. When the eye grows, new ommatidia are added medially and old ones are removed laterally. If the fovea has, as found in dipterans, a neural representation in the visual neuropiles it is reasonable to assume that parallel circuitry changes must occur at lamina level too.

6   THE MEDULLA AND THE FIRST OPTIC CHIASM

6.1   General

The medulla is the largest, and one of the most complex, of the visual ganglia. It corresponds to approximately 12.7% of the whole brain volume (Strausfeld, N., 1976). As revealed by the use of specific neurohistological techniques, most of the neurons contributing to the medulla neuropile have their cell bodies located within a crescent-shaped layer covering the latero-external surface of the fibrillar core. This core consists of thousands of columnar units arranged in space to conform a compact, kidney-shaped mass. Despite the existence of a modular type of organization which facilitates the histological exploration of the nerve circuits, our knowledge concerning this visual center is still far from complete.

To attempt a first approximation to the medulla circuitry each one of the medulla columns will be considered as composed of a single postsynaptic neuron connected via its lateral processes with two presynaptic axon terminals (the endings of the two giant monopolar neurons occupying the center of each optical cartridge) (Fig. 28).

Each lamina cartridge projects to a single medulla column (Trujillo-Cenóz, O., 1969; Strausfeld, N., 1971).

A better understanding of the medulla organization requires some knowledge of the spatial distribution of the incoming fibers at chiasma level.

It is a well-known fact that a crossing-over of fibers takes place in the region linking the lamina to

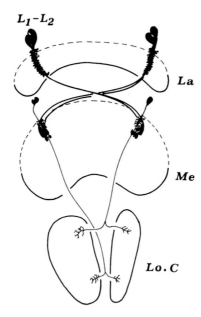

FIG. 28. Over-simplified drawing of the neural connections in the medulla. The main prolongations of two sets of lamina neurons (L1–L2) project, after crossing, to two medulla columns. These are represented as consisting of single radial medullary neurons. Their axons project via the second chiasm to the lobula complex. La, Lamina; Me, medulla; Lo.C, lobula complex.

the medulla. As revealed by classical investigations crossing of fibers occurs in such a way that the anterior part of the lamina projects to the posterior medulla, and correspondingly, the posterior lamina projects to the anterior medulla. In the vertical direction there is no crossing of fibers (Radl, E., 1912; Cajal, S. and Sánchez, D., 1915; Trujillo-Cenóz, O., 1969; Strausfeld, N., 1971).

Even cursory inspection of a single, well-stained Golgi preparation will reveal that the proposed scheme is only an oversimplified version of the actual histological organization of the medulla. The following important neurohistological features have been omitted:

(1) The input fibers which arise from the lamina constitute a heterogenous population of neural units.
(2) The medulla receives, in addition, the so-called "long visual fibers". As stated elsewhere these fibers originate from special kinds of photoreceptors.
(3) The population of medullary neurons is also heterogeneous. Taking into account the

characteristics of their dendritic arborizations several varieties of radial medullary neurons have been identified in higher dipterans.
(4) There are also local neurons which ramify exclusively within the medulla neuropile. They are known as the "amacrine cells". There are several types of amacrines which represent potential neural pathways connecting neighboring medullary columns.
(5) The presence of these amacrine cells contributes to the occurrence in each column of several synaptic levels. When sections of the medulla neuropile are examined under the light microscope, these discrete concentrations of pre- and postsynaptic fibers appear as more or less conspicuous concentric strata.
(6) The nerve fibers projecting to the medulla do not originate exclusively from cell bodies located in the lamina or the retina. There are several systems of fibers whose cell bodies lie in the contralateral medulla or in other regions of the brain.
(7) The medulla columns also contain terminal arborizations originated from T-cells which project either to the lamina or to the lobula.

If each one of the above-mentioned histological features is treated in some detail, a more natural, life-like picture of the medulla should emerge.

6.2 Input fibers from the lamina and the retina

As described in the preceding section the lamina contains various types of neurons whose main prolongations project to the medulla. In dipterans, for example, there are five types of monopolar neurons, whose axon terminals have been accurately identified in the medulla neuropile.

The so-called "giant monopolar cells" (L1–L2) terminate in the medulla by means of large bilobed masses.

As early as 1909, based on indirect evidence, Cajal proposed that these large endings may establish functional contacts with the plexus of fine dendrites located in the first synaptic stratum. A more complete investigation of the visual system of horseflies convinced him that the collateral dendrites of the radial medullary neurons are the main neural elements receiving "nervous currents" from the giant monopolar endings. On the other hand, electron microscope studies have revealed that these large endings contain presynaptic specializations (cumuli of microvesicles and typical T-shaped ribbons). Facing the presynaptic membrane there are two, three or even four very thin postsynaptic fibers. It is worth stressing that these fibers have not been accurately identified under the electron microscope. Therefore the functional inference that visual information flows from the giant monopolar to the radial medullary neurons is exclusively supported on light microscope evidence. It is interesting to mention here that Cajal, S. (1934) included this type of interneuron contact (which he termed "gearing or meshing connections") in his group of selected examples supporting the "neuron doctrine".

In the medulla of higher dipterans there are, together with the large bilobed endings, other kinds of lamina-neuron terminals. The correlation between these various kinds of endings and the different types of lamina neurons is not an easy task. However, Strausfeld, N. (1976) has been able to furnish reliable identification clues.

As already stated, the two giant monopolar cells (L1 and L2) terminate by means of gross bilobed masses within two different synaptic strata (Trujillo-Cenóz, O., 1969). L2 terminals lie superficially, whereas those representing the L1 neurons are located at a deeper level (approximately coincident with the middle zone of the medulla neuropile). As clearly demonstrated by the Golgi procedure, L1 terminals are bistratified, showing a first synaptic enlargement at the level of L2 terminals (Fig. 29). The endings of the so-called asymmetric or L3 neurons are found at an intermediate level between the synaptic masses of L2 and L1 neurons. These endings have a characteristic V-shaped appearance.

The terminal segments of the connecting neurons (L4) are relatively simple. They appear as inverted Ls practically devoid of fine lateral processes. The tangential, more superficial portion of the ending connects two neighboring medullary columns. The vertical portion establishes connections with fibers pertaining to the same parent column. This vertical, deeper segment, reaches the level of the L3 endings (Strausfeld, N. and Campos-Ortega, J., 1973b).

Concerning the intramedullary terminations of the midget or L5 neurons, the Golgi preparations

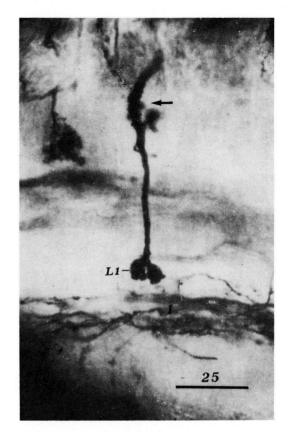

FIG. 29. Terminal segment of a L1 neuron. Note the deep bilobed terminal mass and the more superficial collateral processes (arrow). Golgi procedure. *Phaenicia.* (Scale in μm.)

reveal that they are monostratified, lying in the same synaptic stratum which contains the L2 terminals.

It was observed by Cajal, S. (1909) that "the optic lobe (medulla) of muscids receives two kind of visual fibers which differ anatomically and perhaps physiologically too: the direct fibers which connect that lobe to some rhabdoms of the retina, and the indirect fibers (colossal fibers) represented by the descending prolongation of each giant or second retinal neuron ...". In his second more complete paper dealing with the insect visual system, Cajal, S. and Sánchez, D. (1915) reported that the direct or long visual fibers arise from a peculiar type of central rhabdom occurring in the fly's ommatidium. These observations were confirmed and completed using electron microscope techniques. In higher dipterans, the two central photoreceptors (R7–R8) (from which the central rhabdom derives) give rise to two long fibers terminating directly in the medulla neuropile (Melamed, J. and Trujillo-Cenóz, O., 1968).

Exploring normal and degenerated materials, Campos-Ortega, J. and Strausfeld, N. (1972a) were able to follow under the electron microscope the pathways of the long visual fibers through different levels of the medulla. However, the synaptic connections of these endings remain unsufficiently explored.

The columnar organization of the medulla is apparent only at the peripheral region. At deeper levels the column's edges vanish rapidly. In the peripheral region each column consists of 11–12 nerve fibers (Trujillo-Cenóz, O. and Melamed, J., 1970; Strausfeld, N., 1971). However, in the deeper synaptic strata components of no less than 34 cell types have been identified in each medulla column (Campos-Ortega, J. and Strausfeld, N., 1972b).

6.3 The medullary neurons (ganglionic cells of Cajal)

Cajal, S. and Sánchez, D. (1915) described five basic types of medullary neurons (MN) in higher dipterans (blue-flies and horse-flies). More recent investigations have added numerous subclasses which have been beautifully illustrated by Strausfeld, in his atlas (Strausfeld, N., 1976a). Basically, each neuron consists of a small soma and a radial fiber traversing the whole thickness of the medulla neuropile.

The first extraneuropilic portion of the MN prolongation corresponds to the so-called "indifferent intercalar segment" of Cajal. This cell portion lacks collateral branches and follows a meandering course between the neuronal bodies. The intraneuropilic portion runs radially towards the lobula complex. The main trunk usually has one or two conspicuous groups of dendritic branches oriented perpendicularly with respect to the mother trunk (Fig. 30).

Taking into account the sites of termination of their radial prolongations two main classes of MN can be recognized in dipterans. One class is characterized by the presence of branching axons projecting to both the lobula and the lobula plate (the Y neurons of Strausfeld). Nerve cells of the other class have unbranched axons which terminate mainly in the lobula (the transmedullary neurons of Strausfeld).

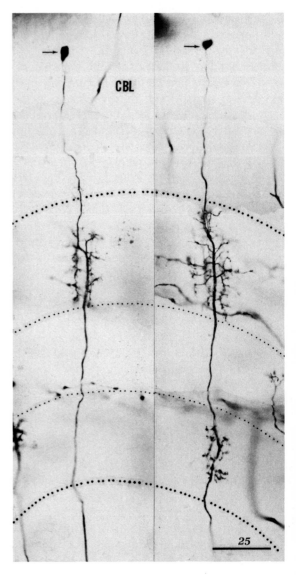

FIG. 30. Two Golgi-impregnated medullary neurons. The small somata (arrows) lie in the cell bodies layer (CBL). Thin radial prolongations project to the lobula complex. The cell at the right has two main groups of collateral dendrites ramifying within two different synaptic strata. Golgi procedure. *Eristalis* sp. (Scale in μm.)

6.4 The amacrine cells

The amacrine cells of Cajal, or local cells of Zawarzin, are particularly abundant in the medulla. Ramifications derived from different kinds of amacrine neurons are found at practically all levels of the medulla strata. The geometrical characteristics of these cells allow division into the following main types:

(1) *The horizontal unidimensional cells.* Each cell has a single prolongation which as soon as it enters the neuropile runs parallel with the medulla surface. Short irregular processes stem out from the mother trunk penetrating a few microns within the surrounding neuropile. In some Golgi-stained sections it is sometimes possible to observe the regular horizontal network formed by these cells near the surface of the medulla neuropile (Fig. 31).

(2) *The horizontal bidimensional cells.* Another superficial network of nerve fibers derived from amacrine cells is found in the medulla neuropile. In this case, however, the flat ramification of each cell extends over a large area of the medulla neuropile.

(3) *The horizontal tridimensional cells.* The preceding types of amacrine cells provide anatomical substratum for potential nerve interaction within a single synaptic stratum. The present type of amacrine cells opens the possibility of nerve interactions in a third dimension, since their recurrent terminal branches cover in depth, several microns (Fig. 32).

(4) *The radial amacrines.* Contrasting with the horizontal amacrines, this kind of anaxonal element possesses a radial prolongation traversing several medulla strata. This prolongation terminates by means of a delicate "bouquet" of extremely fine processes. Cajal, S. and Sánchez, D. (1915) described in *Calliphora* several varieties of radial amacrines.

The synaptology of the different classes of amacrine cells remains unexplored, as well as other aspects of their fine structural organization.

6.5 Input fibers from regions other than the lamina or the retina

The medulla neuropile contains conspicuous

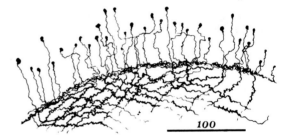

FIG. 31. Semi-schematic drawing showing the spatial arrangement of the monodimensional amacrines. Golgi procedure, mass impregnation. *Phaenicia* sp. (Scale in μm.)

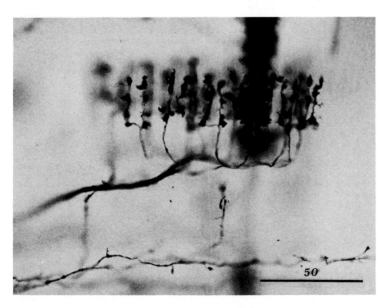

FIG. 32. Photomicrograph of the terminal recurrent arborization of a three-dimensional amacrine. Medulla of the bee-fly *Eristalis* sp. Golgi procedure. (Scale in μm.)

systems of long fibers that run parallel to the outer or inner surfaces. Cajal described in 1909 four main horizontal plexuses, and in his later account together with Sánchez y Sánchez termed them, generically, "sheets of serpentine fibers". There are two main systems of "serpentine fibers", one lies superficially at a level coincident with the apparition of the first dendritic arborizations of the medullary neurons. The other is found deeper in the neuropile, just below the layer displaying the terminal knobs of the R7 photoreceptors.

Main components of these systems are large field neurons interconnecting the two medulla neuropiles. The axonic segments of these cells project to the contralateral medulla whereas their dendritic branches expand within the homolateral synaptic field. The axons, as well as the main dendritic branches, follow tangential pathways.

According to Strausfeld, N. (1976a) the deep serpentine layer consists of two giant neurons which project their axonic prolongations to the contralateral medulla via the posterior optic tract.

6.6   The arborizations of the T-cells

Apart from these tangential systems there are other kinds of nerve fibers projecting to the medulla neuropile. Cell bodies located close to the convex external surface of the neuropile give rise to T-shaped fibers which project to both the medulla and lamina.

The most frequently Golgi-impregnated units have centripetal branches terminating within the first synaptic stratum of the medulla and a centrifugal branch, which after decussation at the level of the first chiasm ends in the lamina. The lamina endings are the so-called nervous bags (Fig. 33) (β-fibers) already described. The medullary endings appear as delicately Golgi-stained masses covered by numerous velvet-like processes. Electron microscopy of Golgi-impregnated tissues has shown that these endings establish intimate contacts with the synaptic endings of the L2 lamina neurons. Unfortunately their identification in conventional electron micrographs is a very difficult task. Consequently the synaptic relations between the terminals of the T-shaped cells and those of the L2 neurons are practically unknown.

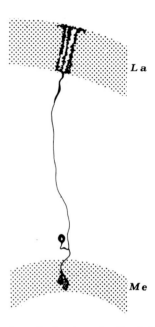

FIG. 33. Semi-schematic drawing of a T-cell linking the lamina to the medulla. The distal branch ramifies in the lamina (La) forming a "nervous bag". The proximal branch terminates in the first synaptic field of the medulla (Me).

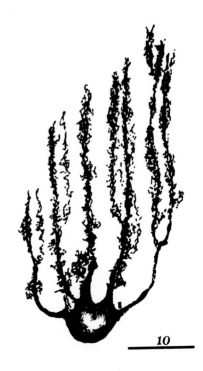

FIG. 34. Main characteristics of the most abundant type of glial cell occurring in the medulla. (Scale in μm.)

Close to the concave surface of the medulla neuropile lie other neuronal groups from which T-shaped fibers arise. The centrifugal branches terminate at different levels of the medulla neuropile. The centripetal branches end within the lobula complex. Cajal, S. and Sánchez, D. (1915) described several types of these cells both in dipterans and hymenopterans.

6.7 The glial cells

In the medulla there are various types of glial cells similar to those occurring in other regions of the brain. There is, however, a type of cell which is particularly abundant in this optic ganglion. The cell body usually lies on the neuropile surface (convex or concave surfaces). The Golgi procedure reveals that several (commonly five to six) thick branches arise from the perinuclear cytoplasm and radially penetrate the neuropile strata. Each one of these thick branches appears covered by minute leaf-like processes (Fig. 34).

7 THE LOBULA AND THE SECOND OPTIC CHIASM

7.1 General

The lobula or lobula complex is the deepest of the optic ganglia and the site of the "third synapse" in the visual pathway of insects. Coronal sections of the insect head show that the lobula has an ovoidal profile with two dissimilarly shaped poles: an external (ocular or lateral) pole lying close to the medulla and an internal (or medial) pole contacting the protocerebral ganglia. In order to obtain a more complete understanding of the three-dimensional appearance of this optic center, frontal sections, parallel to the occiput, are needed. These reveal that the lobula exhibits a kidney-like shape with its outer convex surface facing the medulla. The inner concave surface gives rise to an important nerve bundle termed by Cajal "pedúnculo del hileo". The volume of the lobula complex expressed as a percentage of the whole brain volume is (for the common house-fly) approximately 6% (Strausfeld, N., 1976a).

When this region of the insect brain is compared in different species some anatomical variations become evident. This fact led Cajal, S. and Sánchez, D. (1915) to describe three basic types of histological organization: the bee-type, in which the lobula appears as an undivided neuropile mass; the fly-type, in which the lobula is divided into two synaptic fields; and the libellula-type, in which the lobula consists of three synaptic fields. It is important to keep in mind that in modern papers dealing with species having a divided lobula, this term is reserved for the anterior neuropile mass (the "ovoid body" of Cuccati). The posterior, thinner neuropile mass (the "S-shaped body" of Cuccati or laminar ganglion of Cajal and Sanchez) receives the name of lobular plate (Table 1). The whole region is then properly known as the lobula complex.

Concerning the analogies between individual strata of the undivided lobulae and the separated neuropiles of divided lobulae see Cajal, S. and Sanchez, D. (1915) and the more recent publications by Strausfeld, N. (1976a,b).

The use of the Golgi procedure allowed Cajal to attempt the first systematic exploration of this visual center. He was able to describe the main afferent groups of nerve fibers and to disclose the major efferent tracts. He also observed the existence of intrinsic nerve circuits linking specific neuropile fields.

More recently the interesting physiological properties of visual units recorded at this level coupled with R. Pierantoni's findings (1973, 1976) of giant dendritic arborizations in the lobula plate, prompted new lines of neurohistological research.

Considering that the lobula complex contains complicated neural circuits, it seems advisable to describe separately (1) the afferent nerve fibers, (2) the efferent nerve fibers and (3) the intrinsic circuits.

7.2 The afferent fibers

Disregarding the occurrence of feedback circuits, it is generally admitted that visual information flows from the retina to the lobula through thousands of parallel channels consisting of two synaptically linked neurons. The soma of the first neuron lies in the lamina while that of the second is located in the medulla region.

Different types of radial medullary neurons project their axons to the lobula neuropiles following a main basic pattern disclosed by Cajal, S. and

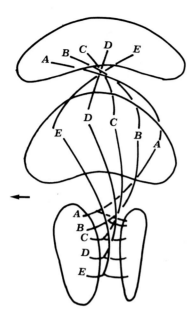

FIG. 35. Nerve fiber patterns in both the first and second chiasms. Note that the first chiasm reverses the alphabetical sequence but the second corrects it. The arrow points forward.

Sanchez, D. (1915). The radial medullary neurons lying in the most anterior region of the medulla send their axons toward the medial (oral) portions of the lobula complex. Conversely, axons originating from neurons located in the posterior regions of the medulla terminate in the lateral (ocular) portions of the lobula complex.

As a consequence of this peculiar arrangement of fibers, the crossed projection pattern of the retinal units onto the medulla is transformed to a direct uncrossed pattern at lobula level (Fig. 35). Considering that no crossing of fibers occurs within vertical planes (neither in the first nor in the second chiasm) dorsoventral projection patterns remain unchanged from the periphery to the inner optic ganglia. In insects with undivided lobulae the same basic projection pattern is maintained, but penetration of afferent fibers within the synaptic fields mainly occurs through the caudal surface of the ganglion.

A more elaborate design of the retino-topic projection at different levels of the visual pathway has derived from more recent investigations performed in *Musca, Calliphora* and *Phaenicia*. Even though this is a very active research field in which new information is continuously added, the follow-

ing can be considered a reliable conservative assembly of anatomical data.

(1) Due to the existence of shape differences between the peripheral receptor layer and the central ganglia the ommatidial lattice appears with some degree of distortion when projected onto the lobula neuropiles. In male flies distortion is increased as a consequence of the dissimilar distribution of the incoming fibers in regions representing dorsal and ventrofrontal eye regions.

(2) Corresponding to the retinal area representing the fovea (this area has a large-facet lattice) there is in the lobula complex "an enormous enhancement of spacing between columns.... This is termed the foveal expansion" (Strausfeld, N., 1979).

(3) As already described in the preceding section the projection pattern of the medulla onto the lobula complex is mediated by two main populations of radial medullary neurons.

Strausfeld, N. (1976b) has described in muscoid dipterans three medulla–lobula "projection modes" mediated by three types of radial medullary neurons with undivided axons. (He has also indicated the existence of a fourth projection mode derived from the double termination in the lobula and the lobula plate of the so-called Y cells).

Concerning the morphology of the endings of the medullary neurons little has been added since the classical account of Cajal and Sanchez. As shown in Fig. 36, there are two basic types of terminals:

(1) short simple nerve endings confined within the limits of a single synaptic stratum, and
(2) long terminals travelling through several strata; these send out longer collateral branches.

Besides the medullary afferents, the lobula complex receives nerve fibers from other nervous centers. Particularly interesting are the lobula-plate afferents described by Hausen, (1976) in Diptera. These fibers originate from heterolateral units that interconnect left and right lobula. Electrophysiological investigations have revealed that these neurons are functionally linked to the system of giant motion-detecting neurons to be described in the following sections.

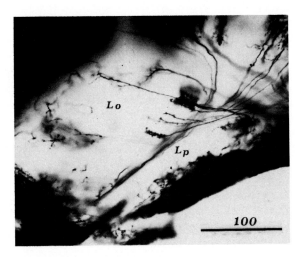

Fig. 36. Low-power photomicrograph showing the different types of medullary neuron terminals. Lo, Lobula; Lp, lobula plate. Golgi procedure. *Phaenicia* sp. (Scale in μm.)

7.3 The efferent fibers

Modern investigations have provided important new data concerning the functional anatomy of the fiber systems leaving the lobula complex. As usual in recent times, most investigations have been carried out on muscoid dipterans.

As soon as improved fixation and embedding techniques were employed to explore the histological organization of the visual system of dipterans, a group of large-diameter fibers was discovered at the lobula-plate level. Pierantoni, R. (1973, 1976) performed a careful detailed investigation of the giant fibers lying in the lobula plate of *Musca*. He was able to identify two systems of giant units whose dendritic arborizations were arranged orthogonally relative to each other.

In *Musca* there are eight or nine large fibers (10–13 μm average diameter) which after an initial horizontal course near the ganglion equator divide into two main vertical branches (Fig. 37a). The cells whose main dendritic branches run in a dorsoventral direction are known as "the vertical units". In addition to the vertical units there are three thicker fibers (average diameter 20 μm) whose dendritic ramifications run predominantly in horizontal directions. These are known as the "horizontal units" (Fig. 37b). The vertical as well as the horizontal units are accompanied by a "twin" or

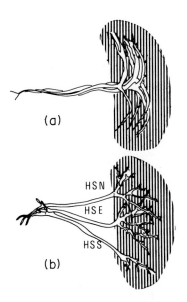

Fig. 37. **a**: The lobula plate contains 8–9 giant fibers whose main branches are vertically oriented (only four have been represented in this scheme); **b**: the lobula plate also contains three giant fibers (HSN, HSE and HSS) whose main branches are horizontally oriented. (Based on Pierantoni, R., 1973, 1976).

"mimetic system" consisting of thinner fibers. The cell bodies of these peculiar systems of fibers are located on the caudal surface of the brain between the lobula complex and the protocerebrum.

The pioneer investigations by Pierantoni also furnished information concerning the termination of these giant fibers in the brain. The fibers of the vertical system form a bundle near the equatorial level along the medial edge of the lobula plate. They follow practically straight pathways and terminate homolaterally in the periesophageal region. The axonic endings of the three horizontal units also terminate in the homolateral periesophageal region but more ventrally, near the periesophageal connective.

The electron micrographs obtained by the Italian investigator show that both the horizontal and the vertical systems are postsynaptic at the lobula plate level. Recent investigations by Hausen, K. *et al.* (1980) and Bishop, C. and Bishop, L. (1981) have confirmed that the giant vertical cells are exclusively postsynaptic in the lobula plate. On the other hand, the horizontal system seems to be pre- and postsynaptic along its course in the brain.

A detailed three-dimensional study of the giant vertical cells of *Calliphora erythrocephala* has been published by Hengstenberg, R. *et al.* (1982). In this species there are 11 vertical cells, each of which covers a particular area of the lobula plate. The gross morphological characteristics of any given cell are practically constant in different individuals. However, the fine branching pattern may show important variations.

Pierantoni's cells have been found in other species of dipterans including the fruit-fly *Drosophila*. Similar kind of cells have been described in acridids. That development of the giant fiber system is under genetic control was demonstrated by Heisenberg, M. *et al.* (1978). These investigators isolated a *Drosophila* mutant which lacks both the horizontal and the vertical cells (or in which these cells are poorly developed).

Electrophysiological experiments followed by the intracellular staining of the recorded units have proved that the giant fibers form part of a horizontal and vertical motion–detection system (Dvorak, D. *et al.*, 1975; Eckert, H. and Bishop, L., 1978).

In addition to the giant units described by Pierantoni in the lobula plate there are other systems of fibers leaving the lobula complex. Using Cajal's silver impregnation techniques one sees conspicuous argyrophilic bundles linking the lobula to ipsilateral optic foci and to the contralateral lobula. One class of these small field neurons (the so-called columnar A neurons — Col A neurons) has been the subject of recent investigations using silver-intensified cobalt impregnations (Hausen, K. and Strausfeld, N., 1980; Strausfeld, N., 1980). Each Col A neuron has distal dendritic ramifications confined within the limits of the posterior surface of the neuropile. Below the dendrite arborizations lie a plexus consisting of very thin fibers. These are collaterals arising from the first segment of the axonic prolongation.

The axons of the Col A cells seem to terminate ipsilaterally in the ventrolateral protocerebrum.

Concerning the conduction polarity of the Col A cells, electron micrographs from the distal dendritic branches indicate that they are postsynaptic to small profiles containing vesicles and T-shaped synaptic ribbons. Conversely, Col A neurons are presynaptic at the level of the axon collaterals and axon terminals. The latter establish synaptic connections with the dendritic trees of descending neurons projecting to the thoracic ganglion.

The dendritic branches derived from the Col A cells are found uniformly distributed all over the posterior synaptic field. There is, however, another neuron population whose dendritic branches are restricted within a vertical narrow neuropile band extending from the ventral margin of the lobula to several microns above the equator. These have been termed "Col B cells".

The introduction of more reliable selective staining procedures has allowed the initiation of studies directed at showing dissimilarities in the neural architecture between male and female visual centers. Observations by Strausfeld, N. (1980) in the blowfly *Calliphora* indicate the existence in the lobula of at least four classes of male-specific neurons. Some of them can be catalogued as giant cells (male lobula giants, MLG1, MLG3) while others are of the small field columnar type (Col D, Col C). It has been suggested that the male-specific neurons may form part of a neural subsystem for triggering and maintaining velocity-controlled sexual-pursuit behavior.

7.4 Intrinsic circuits

This description of the main neural components of the lobula complex has to be completed with information concerning the local or intrinsic nerve components that connect the different synaptic fields.

First described by Cajal, S. and Sanchez, D. (1915), there are in the lobula complex of dipterans small-field neurons with two groups of dendritic arborizations. One group lies within the lobula and the other within the lobula plate. The axons stemming from these cells project to the optic centers in the brain.

The lobula of dipterans also contains amacrine cells. These are less abundant than in the medulla. According to Strausfeld, N. (1976a) amacrine cells do not occur in the lobula plate.

REFERENCES

ANDERSON, D. T. (1963). The embryology of *Dacus tryoni*. 2. Development of imaginal discs in the embryo. *J. Embryol. Exp. Morph. 11*, 339–351.

ARMETT-KIBEL, C., MEINERTZHAGEN, I. A. and DOWLING, J. E. (1977). Cellular and synaptic organization in the lamina of the dragon-fly *Sympetrum rubicundulum. Proc. R. Soc. Lond. B 196*, 385–413.

AUTRUM, H. (1975). Les yeux et la vision des insectes. In *Traite de Zoologie*. Tome VII, Fasc. III. Edited by P. Grasse. Pages 742–853. Masson et Cie, Paris.

BECKER, H. J. (1957). Uber Rontgenmosaikflecken und Defektmutationen am Auge von *Drosophila* und die Entwicklungsphysiologie des Auges. *Z. Vererb. Lehre 88*, 333–373.

BERNARD, G. D. and MILLER, W. H. (1968). Interference filters in the cornea of Diptera. *Invest. Ophth. 7*, 416–434.

BERNHARD, C. G., MILLER, W. H. and MOLLER, A. R. (1965). The insect corneal nipple array. A biological, broad band impedance transformer that acts as an antireflection coating. *Acta Physiol. Scand. 63*, suppl. 243, 1–79.

BISHOP, C. and BISHOP, L. G. (1981). Vertical motion detectors and their synaptic relations in the third optic lobe of the fly. *J. Neurobiol. 12*, 281–296.

BLEST, A. D., STOWE, S., EDDEY, W. and WILLIAMS, D. S. (1982). The local deletion of a microvillar cytoskeleton from photoreceptors of tipulid flies during membrane turnover. *Proc. R. Soc. Lond. B. 215*, 469–479.

BODENSTEIN, D. (1939). Investigations on the problem of metamorphosis. V. Some factors determining the facet number in the *Drosophila* mutant Bar. *Genetics 24*, 494–508.

BODENSTEIN, D. (1950). The postembryonic development of *Drosophila*. In *Biology of Drosophila*. Edited by M. Demeree. Pages 275–367. Wiley, New York.

BODENSTEIN, D. (1953). Postembryonic development. In *Insect Physiology*. Edited by K. D. Roeder. Pages 822–865. Wiley, New York.

BOSCHEK, C. (1971). On the fine structure of the peripheral retina and lamina ganglionaris of the fly, *Musca domestica. Z. Zellforsch. 118*, 369–409.

BRAITENBERG, V. (1966). Unsymmetrische Projektion der retinula zellen auf die lamina ganglionaris bei der fliege *Musca domestica. Z. Vergl. Physiol. 52*, 212–214.

BRAITENBERG, V. (1967). Patterns of projection in the visual system of the fly. I. Retina-lamina projections. *Exp. Brain Res. 3*, 271–298.

BRAMMER, J. D. and CLARIN, B. (1976). Changes in volume of the rhabdom in the compound eye of *Aedes aegypti* L. *J. Exp. Zool. 195*, 33–40.

BULLOCK, T. H. and HORRIDGE, G. A. (1965). *Structure and Function in the Nervous System of Invertebrates*. Freeman, San Francisco.

BURKHARDT, W. and BRAITENBERG, V. (1976). Some peculiar synaptic complexes in the first visual ganglion of the fly, *Musca domestica. Cell Tiss. Res. 173*, 287–308.

BUTLER, R. and HORRIDGE, G. A. (1973). The electrophysiology of the retina of *Periplaneta americana*. L. I. Changes in receptor acuity upon light/dark adaptation. *J. Comp. Physiol. 83*, 263–278.

CAJAL, S. R. (1909). Nota sobre la estructura de la retina de la mosca. *Trab. Lab. Invest. Biol. Univ. Madr. 7*, 217–257.

CAJAL, S. R. (1910). Nota sobre la retina de los múscidos. *Bol. Soc. Esp. Hist. Nat. 10*, 92–95.

CAJAL, S. R. (1917). Contribución al conocimiento de la retina y centros ópticos de los cefalópodos. *Trab. Lab. Invest. Biol. Univ. Madr. 15*, 1–82.

CAJAL, S. R. and SÁNCHEZ, D. (1915). Contribución al conocimiento de los centros nerviosos de los insectos. *Trab. Lab. Invest. Biol. Univ. Madr. 13*, 1–164.

CAJAL, S. R. (1934). Les preuves objectives de l'unité anatomique des cellules nerveuses. *Trab. Lab. Invest. Biol. Univ. Madr. 29*, 1–38.

CAMPOS-ORTEGA, J. A. (1974). Autoradiographic localization of ³H-γ-aminobutyric acid uptake in the lamina ganglionaris of Musca and Drosophila. *Z. Zellforsch. 147*, 415–431.

CAMPOS-ORTEGA, J. A. and GATEFF, E. A. (1976). The development of ommatidial patterning in metamorphosed eye imaginal disc implants of *Drosophila melanogaster. Wilhelm Roux's Arch. 179*, 373–392.

CAMPOS-ORTEGA, J. A. and HOFBAUER, A. (1977). Cell clones and pattern formation: on the lineage of photoreceptor cells in the compound eye of *Drosophila. Wilhelm Roux's Arch. 181*, 227–245.

CAMPOS-ORTEGA, J. A. and STRAUSFELD, N. J. (1972a). The columnar organization of the second synaptic region of the fly's visual system of Musca domestica. I. Receptor terminals in the medulla. *Z. Zellforsch. 124*, 561–585.

CAMPOS-ORTEGA, J. A. and STRAUSFELD, N. J. (1972b). Columns and layers in the second synaptic region of the fly's visual system: the case for two superimposed neuronal architectures. In *Information Processing in the Visual System of Arthropods*. Edited by R. Wehner. Pages 31–36. Springer, Berlin, Heidelberg and New York.

CAMPOS-ORTEGA, J. A. and STRAUSFELD, N. J. (1973). Synaptic connections of intrinsic cells and basket arborizations in the external plexiform layer of the fly's eye. *Brain Res. 59*, 119–136.

CARLSON, S. D. and CHI, C. (1979). The functional morphology of the insect photoreceptor. *Ann. Rev. Ent. 24*, 379–416.

CHI, C. and CARLSON, S. D. (1976a). Large pigment cell of the compound eye of the housefly (*Musca domestica*), compound eye: fine structure and cytoarchitectural associations. *Cell. Tiss. Res. 170*, 77–88.

CHI, C. and CARLSON, S. D. (1976b). Close apposition of photoreceptor cell axons in the housefly. *J. Insect Physiol. 22*, 1153–1157.

CHI, C. and CARLSON, S. D. (1979). Ordered membrane particles in rhabdomeric microvilli of the housefly (*Musca domestica* L.). *J. Morph. 161*, 309–322.

CHI, C. and CARLSON, S. D. (1980). Membrane specializations in the first optic neuropile of the housefly, *Musca domestica* L. I. Junctions between neurons. *J. Neurocytol. 9*, 429–449.

CHI, C. and CARLSON, S. D. (1981). Lanthanum and freeze fracture studies on the retinular cell junction in the compound eye of the housefly. *Cell Tiss. Res. 214*, 541–552.

CLEARY, P., DEICHSEL, G. and KUNZE, P. (1977). The superposition image in the eye of *Ephestia kühniella. J. Comp. Physiol. 119*, 73–84.

CUCCATI, J. (1888). Uber die Organisation des Gehirns der Somomya erithrocefala. *Z. Wiss. Zool. 16*, 240–269.

DIETRICH, W. (1909). Die Facettenaugen der Dipteren. *Z. Wiss. Zool. 92*, 465–539.

DÖVING, K. and MILLER, W. (1969). Function of insect compound eyes containing crystalline tracts. *J. Gen. Physiol. 54*, 250–267.

DVORAK, D. R., BISHOP, L. G. and ECKERT, H. E. (1975). On the identification of movement detectors in the fly optic lobe. *J. Comp. Physiol. 100*, 5–23.

ECKERT, H. and BISHOP, L. G. (1978). Anatomical and physiological properties of the vertical cells in the third optic ganglion of *Phaenicia sericata* (Diptera, Calliphoridae). *J. Comp. Physiol. 126*, 57–86.

EICHENBAUM, D. and GOLDSMITH, T. H. (1968). Properties of intact photoreceptor cells lacking synapses. *J. Exp. Zool. 169*, 15–32.

ELEY, S. and SHELTON, P. M. J. (1976). Cell junctions in the developing compound eye of the desert locust *Schistocerca gregaria. J. Embryol. Exp. Morph. 36*, 409–423.

EXNER, S. (1891). *Die Physiologie der facettierten Augen von Krebsen und Insekten*. Deuticke, Leipzig.

FERNÁNDEZ-MORÁN, H. (1956). Fine structure of the insect retinula as revealed by electron microscopy. *Nature, 177*, 742–743.

FRANCESCHINI, N. and KIRSCHFELD, K. (1971). Etude optique *in vivo* des éléments photorécepteurs dan l'oeil composé de *Drosophila. Kybernetik, 8*, 1–13.

FRANCESCHINI, N. and KIRSCHFELD, K. (1976). Le controle automatique du flux lumineux dan l'oeil composé des Diptères. *Biol. Cybernet. 21*, 181–203.

FRANCESCHINI, N., HARDIE, R., RIBI, W. and KIRSCHFELD, K. (1981). Sexual dimorphism in a photoreceptor. *Nature, 291*, 241–244.

FROHLICH, A. and MEINERTZHAGEN, I. A. (1982). Synaptogenesis in the first optic neuropile of the fly's visual system. *J. Neurocytol. 11*, 159–80.

GARCIA-BELLIDO, A. and MERRIAM, J. R. (1969). Cell lineage of the imaginal discs in *Drosophila* gynandromorphs. *J. Exp. Zool. 170*, 61–76.

GEMNE, G. (1966). Ultrastructural ontogenesis of cornea and corneal nipples in the compound eye of insects. *Acta Physiol. Scand. 66*, 511–512.

GOLDSMITH, T. H. (1962). Fine structure of the retinulae in the compound eye of the honey-bee. *J. Cell Biol. 14*, 489–494.

GOLDSMITH, T. H. and BERNARD, G. D. (1974). The visual system of Insects. In *Physiology of Insecta*. Vol. 2. Edited by E. Rockstein. Pages 165–271. Academic Press, New York.

GOLDSMITH, T. H. and PHILPOTT, D. E. (1957). The microstructure of the compound eye of insects. *J. Biophys. Biochem. Cytol. 3*, 429–440.

GRENACHER, G. H. (1879). *Untersuchungen über das Schorgan der Arthropoden insbesondere der Spinnen, Insecten und Crustacen.* Vandenhoech u. Ruprecht. Göttingen.

GRUNDLER, O. J. (1974). Elektronenmikroskopische Untersuchungen am Auge der Honigbiene (*Apis mellifera*). I. Untersuchungen zur Morphologie und Anordnung der neun Retinulazellen in Ommatidien verschiedener Augenbereiche und zur Perzeption linear polarisierten Lichtes. *Cytobiology 9*, 203–220.

HARDIE, R. C. (1984). Properties of photoreceptors R7 and R8 in dorsal marginal ommatidia in the compound eyes of *Musca* and *Calliphora*. *J. Comp. Physiol. 154*, 157–165.

HARRIS, W. A., STARK, W. S. and WALKER, J. A. (1976). Genetic dissection of the photoreceptor system in the compound eye of *Drosophila melanogaster*. *J. Physiol. 256*, 415–439.

HAUSEN, K. (1973). Die Brechungsindices im Kristallkegel der Mehlmotte *Ephestia Kuhniella*. *J. Comp. Physiol. 82*, 365–378.

HAUSEN, K. (1976). Functional characterization and anatomical identification of motion sensitive neurons in the lobula plate of the blowfly *Calliphora erythrocephala*. *Z. Naturforsch. 31c*, 629–633.

HAUSEN, K. and STRAUSFELD, N. J. (1980). Sexually dimorphic interneuron arrangements in the fly visual system. *Proc. R. Soc. Lond. B 208*, 57–71.

HAUSEN, K., WOLBURG-BUCHHOLZ, K. and RIBI, W. A. (1980). The synaptic organization of visual interneurons in the lobula complex of flies. *Cell Tiss. Res. 208*, 371–387.

HEISENBERG, M., WONNEBERGER, R. and WOLF, R. (1978). Optomotor-blind[H 31] — a *Drosophila* mutant of the lobula plate giant neurons. *J. Comp. Physiol. 124*, 287–296.

HENGSTENBERG, R. (1982). Common visual response properties of giant vertical cells in the lobula plate of the blowfly *Calliphora*. *J. Comp. Physiol. 149*, 179–193.

HENGSTENBERG, R., HAUSEN, K. and HENGSTENBERG, B. (1982). The number and structure of giant vertical cells (VS) in the lobula plate of the blowfly *Calliphora erythrocephala*. *J. Comp. Physiol. 149*, 163–177.

HENNEGUY, L. F. (1904). *Les Insectes*. Masson et Cie, Paris.

HICKSON, S. J. (1885). The eye and optic tract of insects. *Quart. J. Mic. Sci. 25*, 25–215.

HORRIDGE, G. A. (1968). Pigment movement and the crystalline threads of the firefly eye. *Nature (Lond.) 218*, 778–779.

HORRIDGE, G. A. and BARNARD, P. B. (1965). Movement of palisade in locust retinula cells when illuminated. *Quart. J. Mic. Sci. 106*, 131–135.

HORRIDGE, G. A., DUNIEC, J. and MARCELJA, L. (1981). A 24-hour cycle in single locust and mantis photoreceptors. *J. Exp. Biol. 91*, 307–322.

HORRIDGE, G. A. and MEINERTZHAGEN, I. A. (1970). The accuracy of the patterns of connections of the first and second-order neurons of the visual system of *Calliphora*. *Proc. Roy. Soc. Lond. B 175*, 69–82.

HORRIDGE, G. A., WALCOTT, B. and IOANNIDES, A. C. (1970). The tiered retina of *Dytiscus*: a new type of compound eye. *Proc. Roy. Soc. Lond. B 175*, 83–94.

HYDE, C. A. T. (1972). Regeneration, post-embryonic induction and cellular interaction in the eye of *Periplaneta americana*. *J. Embryol. Exp. Morph. 27*, 367–379.

IOANNIDES, A. C. and HORRIDGE, G. A. (1975). The organization of visual fields in hemipteran acone eye. *Proc. R. Soc. Lond. B 190*, 373–391.

JARVILEHTO, M. and ZETTLER, F. (1973). Electrophysiological-histological studies on some functional properties of visual cells and second order neurons of an insect retina. *Z. Zellforsch. 136*, 291–306.

JOHNAS, W. (1911). Das Facettenauge der Lepidopteran. *Z. Wiss. Zool. 97*, 218–261.

KENYON, F. C. (1897). The optic lobes of the bee's brain in the light of recent neurological methods. *Amer. Nat. 31*, 365–377.

KIRSCHFELD, K. (1967). Die Projektion der optischen Unwelt auf das Raster der Rhabdomere in Komplexauge von Musca. *Exp. Brain Res. 3*, 248–270.

KIRSCHFELD, K. and FRANCESCHINI, N. (1969). Ein Mechanismus zur Steuerung des Lichtflusses in den Rhabdomeren des Komplexauges von Musca. *Kybernetik 6*, 13–22.

KIRSCHFELD, K., FRANCESCHINI, N. and MINKE, B. (1977). Evidence for a sensitizing pigment in fly photoreceptors. *Nature 269*, 386–390.

KIRSCHFELD, K. and SNYDER, W. (1975). Waveguide mode effects birefrigence and dichroism in fly photoreceptors. In *Photoreceptor Optics*. Edited by W. Snyder, and R. Menzel. Pages 56–77. Springer, Berlin.

KOENIG, J. and MERRIAM, J. R. (1975). Autosomal ERG mutants. *Drosoph. Inf. Serv.* (submitted).

KOLB, G. and AUTRUM, H. (1972). Die Feinstruktur im Auge der Biene Bei Hell-und Dunkeladaptation. *J. Comp. Physiol. 77*, 113–125.

KOPEC, S. (1922). Mutual relationship in the development of the brain and eyes of Lepidoptera. *J. Exp. Zool, 36*, 459–468.

KRAL, K. and SCHNEIDER, L. (1981). Fine structural localisation of acetylcholinesterase activity in the compound eye of the honeybee (*Apis mellifica* L.) *Cell Tiss. Res. 221*, 351–359.

KUNCKEL D'HERCULAIS, J. (1882). *Recherches sur l'organisation et le development des Volucelles, Insectes diperes de la famille des Syrphides* (an atlas) Paris. (Cited by Henneguy, J.)

KUNZE, P. (1970). Verhaltensphysiologische und optische Experimente zur Superpositionstheorie der Bildentstchung in Komplexaugen. *Deutsch. Zool. 64.* Tagung, 234–238. Fischer, Stuttgart.

LANGER, H. and THORELL, B. (1966). Microspectrophotometric assay of visual pigments in single rhabdomeres of the insect eye. In *Functional Organization of the Compound Eye*. Symp. Wenner-Gren Center (1965). Edited by C. G. Bernhard. Pages 145–149. Pergamon, London.

LASANSKY, A. and FUORTES, M. G. F. (1969). The site of origin of electrical responses in visual cells of the leech, *Hirudo medicinalis*. *J. Cell Biol. 42*, 241–252.

LAUGHLIN, S. B., MENZEL, R. and SNYDER, A. W. (1975). Membranes, dichroism and receptor sensitivity. In *Photoreceptor Optics*. Edited by A. W. Snyder and R. Menzel. Pages 237–259. Springer, Berlin.

LEYDIG, F. (1864). Vom Bau des thierischen Korpers. *Handbuch vergleichende, Anatomie*, Band 1, Tübingen.

LUDTKE, H. (1953). Retinomotorik und Adaptationsvorgange im Auge des Ruckenschwimmers (*Notonecta glauca*, L.). *Z. Vergl. Physiol. 35*, 129–152.

MEINERTZHAGEN, I. A. (1973). Development of the compound eye and optic lobe of insects. In *Developmental Neurobiology of Arthropods*. Edited by D. Young. Pages 51–104. Cambridge University Press, London.

MEINERTZHAGEN, I. A. (1976). The organization of perpendicular fibre pathways in the insect optic lobe. *Phil. Trans. Roy. Soc. B 274*, 555–596.

MEINERTZHAGEN, I. A. and ARMETT-KIBEL, CH. (1982). The lamina monopolar cells in the optic lobe of the dragonfly *Sympetrum*. *Phil. Trans. R. Soc. Lond. B. 297*, 27–49.

MEINERTZHAGEN, I. A. and FROHLICH, A. (1983). The regulation of synapse formation in the fly's visual system. *Trends Neurosci. 6*, 223–228.

MELAMED, J. and TRUJILLO-CENÓZ, O. (1968). The fine structure of the central cells in the ommatidia of Dipterans. *J. Ultrastruct. Res. 21*, 313–334.

MELAMED, J. and TRUJILLO-CENÓZ, O. (1975). The fine structure of the eye imaginal disks in muscoid flies. *J. Ultrastruct. Res. 51*, 79–93.

MENZEL, R. and BLAKERS, M. (1976). Colour receptors in the bee eye — morphology and spectral sensitivity. *J. Comp. Physiol. 108*, 11–33.

MENZER, G. and STOCKAMMER, K. (1951). Zur Polarisationsoptik der Fazettenaugen Von Insekten. *Naturwissenschaften 38*, 190–191.

MEYER-ROCHOW, V. B. (1972). The eyes of *Creophilus erythrocephalus* F. and *Sartallus signatus* Sharp (Staphylinidae: Coleoptera). *Z. Zellforsch. 133*, 59–86.

MILLER, W. H. and BERNARD, G. D. (1968). Butterfly glow. *J. Ultrastruct. Res. 24*, 286–294.

MOTE, M. and GOLDSMITH, T. (1971). Compound eyes: Localization of two color receptors in the same ommatidium. *Science 171*, 1254–1255.

MOUZE, M. (1972). Croissance et metamorphose de l'appareil visuel des Aeschnidae (Odonata). *Int. J. Insect Morph. Embryol. 1*, 181–200.

MOUZE, M. (1974). Interactions de l'oeil et du lobe optique de la croissance post-embryonnaire des Insectes odonates. *J. Embryol. Exp. Morph. 31*, 377–407.

NÄSSEL, D. R. and GEIGER, G. (1983). Neuronal organization in fly optic lobes altered by laser ablation early in development or by mutations of the eye. *J. Comp. Neurol. 217*, 86–102.

NÄSSEL, D. R. and KLEMM, N. (1983). Serotonin-like immunoreactivity in the optic lobes of three insect species. *Cell Tiss. Res. 232*, 129–140.

NÄSSEL, D. R., GEIGER, G. and SEYAN, H. S. (1983). Differentiation of fly visual interneurons after laser ablation of their central targets early in development. *J. Comp. Neurol.* 216, 421–428.

NÄSSEL, D. R., HAGBERG, M. and SEYAN, H. S. (1983). A new, possibly serotonergic, neuron in the lamina of the blowfly optic lobe: an immunocytochemical and Golgi-EM study. *Brain Res.* 280, 361–367.

NICKEL, E. and MENZEL, R. (1976). Insect UV–, and green-photoreceptor membranes studied by the freeze-fracture technique. *Cell Tiss. Res.* 175, 357–368.

NICOL, D. and MEINERTZHAGEN, I. A. (1982a). An analysis of the number and composition of the synaptic populations formed by photoreceptors of the fly. *J. Comp. Neurol.* 207, 29–44.

NICOL, D. and MEINERTZHAGEN, I. A. (1982b). Regulation in the number of fly photoreceptor synapses: the effects of alterations in the number of presynaptic cells. *J. Comp. Neurol.* 207, 45–60.

NILSSON, S. E. G. (1965). The ultrastructure of the receptor outer segments in the retina of the leopard frog *(Rana pipiens)*. *J. Ultrastruct. Res.* 12, 207–231.

NOWEL, M. and SHELTON, P. (1980). The eye margin and compound-eye development in the cockroach: evidence against recruitment. *J. Embryol. Exp. Morph.* 60, 329–343.

OHLY, K. P. (1975). The neurons of the first synaptic region of the optic neuropil of the firefly, *Phausis splendidula* L. (Coleoptera). *Cell. Tiss. Res.* 158, 89–109.

OUWENEEL, W. J. (1970). Developmental capacities of young and mature, wildtype and opht eye imaginal discs in *Drosophila melanogaster*. *Wilhelm Roux's Arch.* 166, 76–88.

PERRY, M. (1968). Further studies on the development of the eye of *Drosophila melanogaster*. I. The Ommatidia. *J. Morph.* 124, 227–248.

PHILLIPS, E. F. (1905). Structure and development of the compound eye of the honey bee. *Proc. Acad. Nat. Sci. (Phil.)* 57, 123–157.

PIERANTONI, R. (1973). Su un tratto nervoso nel cervello della Mosca. In *Atti della prima riuniore Scientifica plenaria*. (Camogli dicembre 1973) Soc. Ital. Biofis. Pura e Applicata, 231–249.

PIERANTONI, R. (1976). A look into the cock-pit of the fly. The architecture of the lobula plate. *Cell Tiss. Res.* 171, 101–122.

PRASAD, O. (1981). Optic lobe neurosecretory cells in *Poekilocerus pictus* (Orthoptera), *Periplaneta americana* (Dictyoptera), *Belostoma indicum* (Hemiptera), *Polistes hebraeus* (Hymenoptera), and *Sandrocottus de jeani* (Coleoptera). *Z. Mikrosk. Anat. Forsch.* 95, 477–483.

PRATT, H. S. (1900). The embryonic history of imaginal discs in *Melophagus ovinus* L., together with an account of the earlier stages in the development of the insect. *Proc. Boston Soc. Nat. Hist.* 29, 241–272.

RADL, E. (1912). *Neue Lehre vom Zentralen Nervensystem*. Engelman, Leipzig.

READY, D. F., HANSON, T. and BENZER, S. (1976). Development of the *Drosophila* retina, a neurocrystalline lattice. *Devel. Biol.* 53, 217–240.

RIBI, W. A. (1975a). Golgi studies of the first optic ganglion of the ant, *Cataglyphis bicolor*. *Cell Tiss. Res.* 160, 207–217.

RIBI, W. A. (1975b). The neurons of the first optic ganglion of the bee *Apis mellifera*. *Advances in Anatomy*. Vol. 50, Fasc. 4, 1–43.

RIBI, W. A. (1975c). The first optic ganglion of the bee. I. Correlation between visual cell types and their terminals in the lamina and medulla. *Cell Tiss. Res.* 165, 103–111.

RIBI, W. A. (1976). The first optic ganglion of the bee. II. Topographical relationships of the monopolar cells within and between cartridges. *Cell Tiss. Res.* 171, 359–373.

RIBI, W. A. (1977). Fine structure of the first optic ganglion (lamina) of the cockroach, *Periplaneta americana*. *Tissue Cell* 9 (1), 57–72.

RIBI, W. A. (1978). Gap junctions coupling photoreceptor axons in the first optic ganglion of the fly. *Cell Tiss. Res.* 195, 299–308.

RIBI, W. A. (1979a). Coloured screening pigments cause red eye glow hue in pierid butterflies. *J. Comp. Physiol.* 132, 1–9.

RIBI, W. A. (1979b). Do the rhabdomeric structures in bees and flies really twist? *J. Comp. Physiol.* 134, 109–112.

RIBI, W. A. (1979c). The first optic ganglion of the bee. III. Regional comparison of the morphology of photoreceptor cell axons. *Cell. Tiss. Res.* 200, 345–357.

SAINT-MARIE, R. L. and CARLSON, S. D. (1982). Synaptic vesicle activity in stimulated and unstimulated photoreceptor axons in the housefly. A freeze-fracture study. *J. Neurocytol.* 11, 747–761.

SAINT-MARIE, R. L. and CARLSON, S. D. (1983a). The fine structure of neuroglia in the lamina ganglionaris of the housefly *Musca domestica* L. *J. Neurocytol.* 12, 213–241.

SAINT-MARIE, R. L. and CARLSON, S. D. (1983b). Glial membrane specializations and the compartmentalization of the lamina ganglionaris of the housefly compound eye. *J. Neurocytol.* 12, 243–275.

SÁNCHEZ Y SÁNCHEZ, D. (1916). Datos para el conocimiento histogénico de los centros ópticos de los insectos. Evolución de algunos elementos retinianos del *Pieris brassicae* L. *Trab. Lab. Invest. Biol. Univ. Madr.* 16, 213–278.

SÁNCHEZ Y SÁNCHEZ, D. (1918). Sobre el desarrollo de los elementos nerviosos en la retina del *Pieris brassicae* L. *Trab. Lab. Invest. Biol. Univ. Madr.* 17, 1-63.

SÁNCHEZ Y SÁNCHEZ, D. (1919a). Sobre el desarrollo de los elementos nerviosos en la retina del *Pieris brassicae* L. (continuación). *Trab. Lab. Invest. Biol. Univ. Madr.* 17, 1–63.

SÁNCHEZ Y SÁNCHEZ, D. (1919b). Sobre el desarrollo de los elementos nerviosos de la retina del *Pieris brassicae* L. (continuación). *Trab. Lab. Invest. Biol. Univ. Madr.* 17, 117–180.

SCHINZ, R. H., LO, M.-V. C. and PAK, W. L. (1978). Comparison of rhabdomeric and non-rhabdomeric plasma membrane particles in freeze-fractured *Drosophila* photoreceptors. Assoc. For Res. in Vis. and Ophthalm. Spring meetings, Sarasota, Florida, p. 236 (abstract).

SCHNEIDER, L. and LANGER, H. (1969). Die Struktur des Rhabdoms im Doppelauge des Wasserläufers *Gerris lacustris*. *Z. Zellforsch.* 99, 538–559.

SCHOELLER, J. (1964). Recherches descriptives et expérimentales sur la céphalogenèse de *Calliphora erythrocephala* (Meigen) au cours des développements embryonnaire et postembryonnaire. *Arch. Zool. Exp. Gén.* 103, 1–216.

SCHOELLER-RACCAUD, J. (1977). La céphalogenèse larvaire des Diptères. In *Traité de Zoologie*. Tome VIII. Fasc V-B. Edited by P. Grassé. Pages 262–279. Masson et Cie, Paris.

SEITZ, G. (1968). Der Strahlengang im Appositionsauge von *Calliphora erythrocephala* (Meig). *Z. Vergl. Physiol.* 59, 205–231.

SEITZ, G. (1969a). Untersuchungen am dioptrischen Apparat des Leuchtkaferauges. *Z. Vergl. Physiol.* 62, 61–74.

SEITZ, G. (1969b). Polarisationsoptische Untersuchungen am Auge von *Calliphora erythrocephala*. *Z. Zellforsch.* 93, 525–529.

SHAW, S. R. (1968). Organization of the locust retina. *Symp. Zool. Soc. Lond.* 23, 135–163.

SHAW, S. R. (1978). The extracellular space and blood-eye barrier in an insect retina: an ultrastructural study. *Cell Tiss. Res.* 188, 35–61.

SHELTON, P. M. J. and LAWRENCE, P. A. (1974). Structure and development of ommatidia in *Oncopeltus fasciatus*. *J. Embryol. Exp. Morph.* 32, 337–353.

SHERK, T. E. (1977). Development of the compound eyes of dragonflies (Odonata). I. Larval compound eyes. *J. Exp. Zool.* 201, 391–416.

SHERK, T. E. (1978a). Development of the compound eyes of the dragonflies (Odonata). II. Development of the larval compound eyes. *J. Exp. Zool.* 203, 47–60.

SHERK, T. E. (1978b). Development of the compound eyes of the dragonflies (Odonata). III. Adult compound eyes. *J. Exp. Zool.* 203, 61–80.

SMOLA, U. and TSCHARNTKE, H. (1979). Twisted rhabdomeres in the Dipteran eye. I. *Comp. Physiol.* 133, 291–297.

SMOLA, U. and WUNDERER, H. (1981). Twisting of blowfly (*Calliphora erythrocephala Meigen*) (Diptera: Calliphoridae) rhabdomeres: an *in vivo* feature unaffected by preparation or fixation. *Int. J. Insect Morphol. Embryol.* 10, 331–344.

SNODGRASS, R. E. (1924). Anatomy and metamorphosis of the apple maggot, *Rhagoletis pomonella* Walsh. *J. Agric. Res.*, 28, 1–36.

SNODGRASS, R. E. (1953). The metamorphosis of a fly's head. *Smithsonian Misc. Coll.* 122, 1–25.

SNODGRASS, R. E. (1954). Insect metamorphosis. *Smithsonian Misc. Coll.* 122, 1–124.

SNYDER, A. W. and HORRIDGE, G. (1972). The optical function of changes in the medium surrounding the cockroach rhabdom. *J. Comp. Physiol.* 81, 1–8.

SNYDER, A. W., MENZEL, R. and LAUGHLIN, S. B. (1973). Structure and function of the fused rhabdom. *J. Comp. Physiol.* 87, 99–135.

STAVENGA, D. G. (1974). Refractive index of fly rhabdomeres. *J. Comp. Physiol.* 91, 417–426.

STRAUSFELD, N. J. (1971). The organization of the insect visual system. II. The projection of fibres across the first optic chiasm (light microscopy). *Z. Zellforsch. 121*, 442–454.

STRAUSFELD, N. J. (1976a). *Atlas of an Insect Brain*. Springer, Berlin.

STRAUSFELD, N. J. (1976b). Mosaic organizations, layers, and visual pathways in the insect brain. In *Neural Principles in Vision*. Edited by F. Zettler and R. Weiler. Pages 245–279. Springer, Berlin and Heidelberg.

STRAUSFELD, N. J. (1979). The representation of a receptor map within retinotopic neuropil of the fly. *Verh. Dtsch. Zool. Ges. 1979*, 167–179.

STRAUSFELD, N. J. (1980). Male and female visual neurons in dipterous insects. *Nature 283*, 381–383.

STRAUSFELD, N. J. and BLEST, A. D. (1970). Golgi studies on Insects. Part 1. The optic lobes of Lepidoptera Phil. *Trans. Roy. Soc. (Lond.) B 258*, 81–134.

STRAUSFELD, N. J. and CAMPOS-ORTEGA, J. A. (1973a). L3, the 3rd 2nd-order neuron of the 1st visual ganglion in the "neural superposition" eye of *Musca domestica*. *Z. Zellforsch. 139*, 397–403.

STRAUSFELD, N. J. and CAMPOS-ORTEGA, J. A. (1973b). The L4 monopolar neuron: a substrate for lateral interaction in the visual system of the fly *Musca domestica* (L). *Brain Res. 59*, 97–117.

STRAUSFELD, N. J. and NASSEL, D. R. (1980). Neuroarchitecture of brain regions that subserve the compound eyes of Crustacea and insects. In *Comparative Physiology and Evolution of Vision in Invertebrates*. Edited by H. Autrum. *Handbook of Sensory Physiology*. Vol. VII/6B, pages 1–133. Springer, Berlin and Heidelberg.

SUCH, J. (1969a). Observations sur l'infrastructure des cellules cristalliniennes de l'oeil du Phasme *Carausius morosus* Br. Ces cellules sont-elles de simples éléments "dioptriques"? *C.R. Acad. Sci., Paris, Série D. 268*, 356–359.

SUCH, J. (1969b). Sur la présence de structures évoquant des ébanches ciliaires abortives dans les cellules rétiniennes du jeune embryon de *Carausius morosus* Br. *C.R. Acad. Sci., Paris, Série D. 268*, 948–949.

SUCH, J. (1975). Analyse ultrastructurale de la morphogènese ommatidienne au cours du développement embryonnaire de l'oeil composè, chez le Phasme *Carausius morosus* Br. *C.R. Acad. Sci., Paris, Série D. 281*, 67–70.

TRUJILLO-CENÓZ, O. (1965). Some aspects of the structural organization of the arthropod eye. *Cold Spr. Harb. Symp. Quant. Biol. 30*, 371–382.

TRUJILLO CENÓZ, O. (1969). Some aspects of the structural organization of the medulla in muscoid flies. *J. Ultrastruct. Res. 27*, 533–553.

TRUJILLO-CENÓZ, O. and BERNARD, G. D. (1972). Some aspects of the retinal organization of *Sympycnus lineatus* Loew Diptera, Dolichopodidae). *J. Ultrastruct. Res. 38*, 149–160.

TRUJILLO–CENÓZ, O. and MELAMED, J. (1963). On the fine structure of the photoreceptor second optical neuron synapse in the insect retina. *Z. Zellforsch. 59*, 71–77.

TRUJILLO–CENÓZ, O. and MELAMED, J. (1966a). Electron microscope observations on the peripheral and intermediate retinas of Dipterans. In *The Functional Organization of the Compound Eye*. Symp. Wenner-Gren Center (1965). Edited by C. G. Bernhard. Pages 359–361. Pergamon Press, London.

TRUJILLO-CENÓZ, O. and MELAMED, J. (1966b). Compound eye of Dipterans: anatomical basis for integration — an electron microscope study. *J. Ultrastruct. Res. 16*, 395–398.

TRUJILLO-CENÓZ, O. and MELAMED, J. (1970). Light and electronmicroscope study of one of the systems of centrifugal fibers found in the lamina of muscoid flies. *Z. Zellforsch. 110*, 336–349.

TRUJILLO-CENÓZ, O. and MELAMED, J. (1972). Ontogenesis of the lamina ganglionaris in Diptera. *Rev. Mic. Elect. 1*, 144.

TRUJILLO-CENÓZ, O. and MELAMED, J. (1973). The development of the retina–lamina complex in muscoid flies. *J. Ultrastruct. Res. 42*, 554–581.

TRUJILLO-CENÓZ, O. and MELAMED, J. (1978). Development of photoreceptor patterns in the compound eyes of muscoid flies. *J. Ultrastruct Res. 64*, 46–62.

UMBACH, W. (1934). Entwicklung und Bau des Komplexauges der Mehlmotte, *Ephestia kuhniella Zeller*, nebst einigen Bemerkungen uber die Entstehung der optischen Ganglien. *Z. Morph. Okol. Tiere, 28*, 561–594.

VARELA, F. (1970). Fine structure of the visual system of the honey-bee (*Apis mellifera*). II. The lamina. *J. Ultrastruct. Res. 31*, 178–194.

VARELA, F. and PORTER, K. (1969). Fine structure of the visual system of the honeybee (*Apis mellifera*). I. The retina. *J. Ultrastruct. Res. 29*, 236–259.

VARELA, F. and WIITANEN, W. (1970). The optics of the compound eye of the honey bee (*Apis mellifera*). *J. Gen. Physiol. 55*, 336–358.

VIALLANES, H. (1884). Études histologiques et organologiques sur les centres nerveux et les organes des sens des animaux articulés. Deuxième mémoire. Le ganglion optique de la libellule (*Aeschna maculatissima*). *Ann. Sci. Nat. Zool. 18* (6), 1–34.

VIALLANES, H. (1887). Etudes histologiques et organologiques sur les centres nerveux et les organes sens des animaux articulés. 4. Le cerveau de la guêpe (*Vespa crabro* et *V. vulgaris*). *Ann. Sci. Nat. Zool. 4* (7), 5–100.

VIALLANES, H. (1891). Sur quelques points de l'histoire du développement embryonnaire de la Mante religieuse. *Ann. Sci. Nat. Zool.* 7e sér., Zool., *11*, 283–328.

VIALLANES, H. (1892). Contribution à l'histologie du système nerveux des invertébrés. La lame ganglionnaire de la langouste. *Ann. Sci. Nat. Zool.* 7e sér. Zool., *13*, 385–398.

VIGIER, P. (1907). Sur les terminations photoréceptrices dans les yeux composes des Muscides. *C.R. Acad. Sci. (Paris) 145*, 532–536.

VIGIER, P. (1908). Sur l'existence réelle et le rôle des appendices piriformes des neurones. La neurone périoptique des Diptères. *C.R. Soc. Biol. (Paris) 64*, 959–961.

VIGIER, P. (1909). Mécanisme de la synilèse des impressions lumineuses recueillies par les yeux composés des Diptères. *C.R. Acad. Sci (Paris) 148*, 1221–1223.

VRIES, H. C. DE (1956). Physical aspects of the sense organs. *Prog. Biophys. 6*, 207–264.

WADA, S. (1974). Spezielle randzonale Ommatidien von *Calliphora erythrocephala* (Diptera: Calliphoridae): Architektur der zentralen Rhabdomeren-Kolumne und Topographie im Komplexauge. *Int. J. Insect Morphol. Embryol. 3*, 397–424.

WALCOTT, B. (1969). Movements of retinula cells in insect eyes on light adaptation. *Nature, 223*, 971–972.

WATERMAN, T. H. (1982). Fine structure turnover of photoreceptor membranes. In *Visual Cells in Evolution*. Edited by J. M. Westfall. Pages 23–41. Raven Press, New York.

WEISMANN, A. (1864). Die nachembryonale Entwicklung der Musciden nach Beobachtungen an *Musca vomitoria* und *Sarcophaga carnaria*. *Z. Wiss. Zool., 14*, 187–336.

WEISMANN, A. (1866). Die Metamorphose der *Corethra plumicornis*, *Z. Wiss. Zool., 16*, 45–127.

WHITE, R. H. (1961). Analysis of the development of the compound eye in the mosquito, *Aedes aegypti*. *J. Exp. Zool. 148*, 223–240.

WHITE, R. H. (1963). Evidence for the existence of a differentiation center in the developing eye of the mosquito. *J. Exp. Zool. 152*, 139–143.

WILLIAMS, D. S. (1980). Organisation of the compound eye of the tipulid fly during the day and night. *Zoomorphologie 95*, 85–104.

WILLIAMS, D. S. (1981). Twisted rhabdomeres in the compound eye of a tipulid fly (Diptera). *Cell Tiss. Res. 217*, 625–632.

WILLIAMS D. S. (1982). Ommatidial structure in relation to turnover of photoreceptor membrane in the locust. *Cell Tissue Res. 225*, 595–617.

WOLBURG-BUCHHOLZ, K. (1977). The superposition eye of *Cloeon dipterum*: the organization of the lamina ganglionaris. *Cell Tiss. Res. 177*, 9–28.

WOLBURG-BUCHHOLZ, K. (1979). The organization of the lamina ganglionaris of the hemipteran insects *Notonecta glauca*, *Corixa punctata*, *Gerris lacustris*. *Cell Tiss. Res. 197*, 39–59.

WOLKEN, J. J., CAPENOS, J. and TURANO, A. (1957). Photoreceptor structures. III. *Drosophila melanogaster*. *J. Biophys. Biochem. Cytol. 3*, 441–448.

WUNDERER, H. and SMOLA, U. (1982a). Fine structure of ommatidia at the dorsal eye margin of *Calliphora erythrocephala* Meigen (Diptera: Calliphoridae); an eye region specialised for the detection of polarized light. *Int. J. Insect Morphol. Embryol. 11*, 25–38.

WUNDERER, H. and SMOLA, U. (1982b). Morphological differentiation of the central visual cells R7/8 in various regions of the blowfly eye. *Tissue. Cell 14*, 341–358.

ZAWARZIN, A. (1913). Histologische Studien über Insekten. IV Die optischen Ganglien der *Aeschna*-Larven. *Z. Wiss. Zool. 108*, 175–257.

ZAWARZIN, A. (1925). Einige Bemerkungen über den Bau der optischen Zentren. *Anat. Anz. 59*, 551–559.

ZEIL, J. (1979). A new kind of neural superposition eye: the compound eye of male Bibionidae. *Nature 278*, 249–250.

ADDENDUM

Since the main part of this chapter went to press, important papers dealing with different aspects of the development and neural organization of the insect compound eye have been published. This Addendum is an attempt to cover the information which has appeared in more recent articles. It is also a good opportunity to introduce relevant references omitted in the main text.

The retina

The retina of Brachycera still remains a source of interesting findings. Recent papers by Wunderer, H. and Smola, U. (1982a,b) have confirmed that the compound eye is not uniform but composed of dissimilar ommatidial populations (Wada, S., 1974). The main differences are determined by structural variations of the central rhabdomeres (those derived from the superior and inferior central cells). These peculiar photoreceptors can be distinguished by the relative length of their rhabdomeres, position of the nucleus, degree of rhabdomere twist, characteristics of the transitional zone between rhabdomere 7/8, and cross-sectional area of the photoreceptive organelles. Taking into account these characteristics, four main ommatidial classes were described in the compound eyes of *Calliphora erythrocephala* and *Musca domestica*. At the dorsal eye margin there is a conspicuous ommatidial group (termed "marg") identified by the presence of short, wide, central rhabdomeres. These are surrounded by six common photoreceptors which in this eye region bear unusually narrow rhabdomeres. Within the "marg" group the ommatidium at the eye vertex exhibits a very peculiar morphology: the central rhabdom is very short (22 μm, the cross-sectional area of rhabdomere 7 is greater than in any other marginal ommatidia, and the number of peripheral photoreceptors is reduced to two. The lateral eye region consists of an uniform ommatidial population. All ommatidia are of so-called "sl" class in which the rhabdomeres derived from the superior central cells (R 7) are short, while those derived

from the inferior central cells are long. A different situation occurs at the dorsofrontal eye region where two ommatidial classes coexist: the "sl" and the "ls". In the "ls" group the rhabdomeres derived from the superior central cells are long, while those derived from the inferior central cells are short. Finally, sex specific photoreceptors similar to those occurring in *Musca domestica* have been also found in *Calliphora erythocephala*. As in *Musca*, the superior central cells of such ommatidia show morphological features which place them in a special category between the central cells and the peripheral photoreceptors. The term "per" (from peripheral) has been proposed to name them. It is important to note that the central rhabdomeres of the dorsal marginal ommatidia, in contrast to other rhabdomeres of the fly's eye do not twist. Therefore they appear well suited for the detection of polarized light. (Concerning rhabdomere twist, see main text and also Smola, U. and Wunderer, H., 1981).

The functional properties of the central cells of the "marg" ommatidia have been recently explored by Hardie, R. (1984). Intracellular injections of Lucifer Yellow have demonstrated that these cells project to the medulla by means of "long visual fibers" as it occurs in other regions of the fly eye. The electrophysiological findings confirmed the prediction that the central cells of the dorsomarginal ommatidia are sensitive to the e-vector of polarized light. "The single feature that above all characterizes the properties of R7 and R8 cells in the marginal ommatidia is a spectacularly high polarization sensitivity" (Hardie, R., 1984).

Concerning the turnover of rhabdomere membranes, the investigations performed in tipulid flies (*Ptilogyna spectabilis* and *Leptotarsus* sp.) by Blest, A. *et al.* (1982) indicated that the process of microvilli shedding is related to the local disruption of the microvillar cytoskeleton. However, local deletion of the cytoskeleton seems to be insufficient to explain the abscission of the microvillar tips. The daily cycle of the rhabdomeres has been investigated in tipulid flies and acridids. In the tipulid *Ptilogyna*, before dawn, each rhabdomeric microvillus shows two distinct regions: a stable electron-dense portion containing an axial cytoskeletal fiber and a clear shedding region devoid of it. After dawn the shedding zones separate, vesiculate and drop off. The membrane remnants are retrieved into the cell by means of pseudopodia and phagocytotic vesicles. By dusk, a new shedding zone has appeared. Similar studies made in *Locusta migratoria* and *Valanga irregularis* have shown that there is at dusk a rapid assembly of new rhabdomeric microvilli (Williams, D., 1982). Parallely, the cross-sectional area of each rhabdom increases by 4.7 fold. The rhabdoms at night have more and longer rhabdomeric microvilli than rhabdoms during light time. Rapid shedding of photoreceptor membrane is induced by light whereas microvilli assembly is initiated by the onset of darkness. Nevertheless, both shedding and assembly of rhabdomeric microvilli are also under some kind of endogenous control.

The lamina

Advances concerning the structural organization of the insect lamina have occurred in the last years. A new neuron type (Tang 3) with very extensive tangential processes was discovered by Nassel, D. *et al.* (1983) in the lamina ganglionaris of *Calliphora erythrocephala*. Despite the fact that the somata of these neurons have not yet been accurately identified, their varicose, most distal branches have been the subject of electromicroscopical and immunocytochemical studies. The electromicrographs have shown that the varicose swellings contain electron-dense vesicles similar to those found in other invertebrate serotonergic neurons. Using antibodies to 5-HT it was found that the pattern of nerve branches that react with the antiserum is similar to that of Golgi-impregnated Tan 3 neurons. The axons seem to project to the medulla. It has been proposed that these wide field neurons may exert centrifugal control on the photoreceptor axons or on the lamina neurons. A long distance effect may be mediated by the release of 5-HT in the plexus formed by the varicose branches distal to the lamina synaptic field. 5-HT immunoreactive fibers have been also found in the lamina, medulla and lobula of other insect groups (Nässel, D. and Klemm, N., 1983).

An analysis of the number and composition of the synapses in the optical cartridges of *Musca domestica* was made by Nicol, D. and Meinertzhagen, I. (1982a). Working with recently hatched female flies, they sample (using the electron-microscope) the synaptic loci in the photoreceptor terminals. Their quantitative analyses revealed that each photoreceptor is presynaptic at about 200 ± 40 synapses. Each presynaptic point faces two pairs of post-synaptic elements. The first pair consists of branches derived from the two central monopolar cells (L1-L2). The second pair consists of fibers of dissimilar origin: α fibers, prolongations of L3 neurons or even glial processes.

The same investigators studied the frequency of synaptic contacts in optical cartridges with 7 or 8 photoreceptor terminals. These kinds of cartridges normally occur at the equator of the fly's eye. They concluded that "the synaptic population size and distribution within five depth levels of the lamina is, on average, approximately constant for all receptor terminals whether from 6R, 7R, or 8R cartridges" (Nicol, D. and Meinertzhagen, I., 1982b). However, in a more recent paper, Meinertzhagen, I. and Frohlich, A. (1983) have reported that "the total number of tetrads synapses in a cartridge innervated by eight photoreceptors is about 25% higher than in a cartridge containing the normal six receptors". The number of branches arising from the second order neurons L1 and L2 is also the same in cartridges with 6, 7, or 8 photoreceptor terminals (about 180). Nevertheless, the branching pattern shows variations conforming to the different number and distribution of the terminals within each cartridge.

The fine structure of the synaptic contacts in the cartridges has been also explored in different physiological conditions (Saint-Marie, R. and Carlson, S., 1982). In *Musca* eyes fixed during illumination, the freeze-fracture images revealed the active zones of the synapses surrounded by numerous membrane dimples. Conversely, in eyes fixed in the dark, such folds or pits were rarely found. The authors discuss these structural findings in a conceptual context which involves vesicle exocytosis and recycling as a basic mechanism for liberation of the chemical mediator.

The glial components of the fly's lamina have been explored using electromicroscopical techniques. Six morphologically distinct glial cells have been described by Saint-Marie, R. and Carlson, S. (1983a) in *Musca domestica*: 1. the fenestrated layer glia, lying just beneath the retina basement membrane; 2. the pseudocartridge glia, which invest the photoreceptor axons at the pseudo-cartridge level; 3–4. two levels of satellite glia which ensheath the somata of the different kinds of lamina neurons; 5. the epithelial cells which envelope the optical cartridges; 6. the marginal glia which invest the axons entering or leaving the lamina through its medial or optic chiasm surface. In a second paper, (Saint-Marie, R. and Carlson, S., 1983b) the same authors reported the occurrence of diverse types of membrane specializations at the points where the glial cells contact other glial elements or neuronal processes. Based on these findings they propose that the lamina is divided into two major compartments: a distal compartment containing the somata of the monopolar neurons and a distal compartment corresponding to the lamina synaptic field. Within the latter, each optical cartridge represents a separated subcompartment.

Optic lobes — the lobula plate

Since the discovery in dipterans of a dual system of giant interneurons the dendritic branches of which lie in the lobula plate, several papers have appeared dealing with its anatomical and functional characteristics. Hengstenberg, R. *et al.* (1982) have analyzed in detail the anatomy of the so-called "giant vertical cells" in the blowfly *Calliphora erythrocephala*. In this species there are 11 "vertical" neurons the main dendritic trunks of which occupy precise, fixed portions of the lobula plate neuropile. Three neurons ramify exclusively in the caudal layer of the lobula plate. In the remainder of the neurons, different portions of the dendritic tree lie at different depths of the neuropile. Functional findings by the same group (Hengstenberg, R., 1982) indicate that the "vertical" neurons perceive wide-field motions occurring when a fly performs rotatory or translatory movements in a resting environment.

Development

The use of laser microsurgery appears as a promising technical approach for studying neural development in insects. Nässel, D. *et al.* (1983) have shown that ablation of the anlagen of the lobula complex at larval stages, does not impede differentiation of the neuroblasts lying in the primordia of the lamina and medulla which are able to develop normal neuropile patterns. Results derived from more complicated experiments (Nässel, D. and Geiger, G., 1983) showed that when the primordium of the lamina is destroyed, the photoreceptors of the imago project directly to the medulla. In addition, the medulla neurons send sprouting branches towards the retina. Sprouting also occurs when the anlagen of the lobula complex is partially damaged. In these circumstances, the lobula neurons that are able to differentiate, project sprouting fibers to medulla neuropiles. Elimination of the primordia of all optic centers determines a significant reduction of the lateral midbrain. In this case the histological studies have shown that the photoreceptor axons form a disordered mass just beneath the retina.

5   The Eye: Optics

M. F. LAND

University of Sussex, Brighton, Sussex, UK

1 INTRODUCTION

This chapter is in four parts. The first is an introduction to the types of eye found in insects, based around the history of their discovery. It is intended as an overview for readers who are not particularly concerned with the detailed physics of the subject. Figure 7 contains a summary of the basic optical mechanisms in insects and is intended to supplement the text of this section. The second part deals quantitatively with the physics of simple and compound eyes, so that their performances can be compared. The limits to visual resolution and sensitivity imposed by the wave and particle (photon) nature of light are dealt with at some length, as they are fundamental in trying to understand the way eyes are designed. Sections 3 to 7 examine in more detail the structure and function of the four main eye types introduced in the first section, and section 9 is a brief summary.

1.1 The history of insect optics

The facets of the compound eyes of insects are just too small to be resolved by the naked eye, and it required the invention of the microscope before they could be properly depicted, and a start made

on the task of trying to understand the workings of eyes so different from our own. The process of elucidation, from the first drawings of insect eyes of Robert Hooke (1665) to the essentially modern account of compound eye optics by S. Exner (1891), took over 2 centuries, and although this account is still not complete, little that is conceptually new has been added in the last 90 years. The major endeavour since the turn of the century has been the study, still really in its infancy, of what the nervous system does with the images the eye presents to it.

Hooke's drawing of "The Grey Drone Fly" (probably a male tabanid) was published in his *Micrographia* only half a century after Kepler had shown how the image is formed in the human eye (Fig. 1). Jan Swammerdam made his well-known illustration of a drone honeybee, including a dissection showing its ommatidial unit structure, about 1680, but the *Biblia Naturae* in which it appears was not published until 1737, well after his death. It was Swammerdam's contemporary and friend, Antoni van Leeuwenhoek, who first looked *through* the optical array of an insect eye, and made an observation which was later to cause much controversy. In a letter to the Royal Society (published elsewhere in 1695, see Wehner, R., 1981) he wrote:

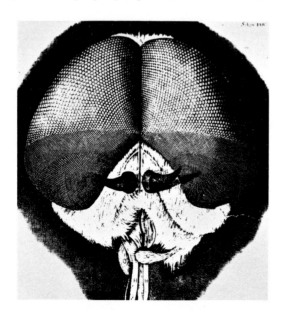
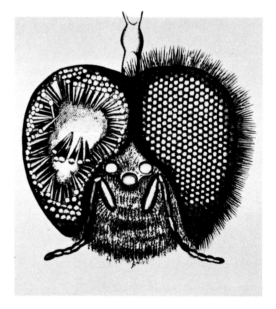

Fig. 1. *Left*: Robert Hooke's engraving of "The Grey Drone Fly" of 1665. The eye is clearly divided into upper and lower regions with facets of different sizes, as in many male dipterans (see Fig. 25). *Right*: Swammerdam's figure of the head of a male honeybee (1737), showing the ommatidial structure, and the three dorsal ocelli.

Last summer I looked at an insect's cornea through my microscope. The cornea was mounted at some larger distance from the objective of the microscope as it was usually done when observing small objects. Then I moved the burning flame of a candle up and down at such a distance from the cornea that the candle shed its light through it. What I observed by looking into the microscope were the inverted images of the burning flame: not one image, but some hundred images. As small as they were, I could see them all moving (quoted from Wehner, R., 1981).

A later letter, mentioning these findings, was published in the *Philosophical Transactions* in 1698 (see Bernhard, C., 1966).

Leeuwenhoek's demonstration was that each facet of an insect's eye produces an inverted image (Fig. 2), even though the geometry of the eye as a whole dictates that the overall image must be erect, since the axis of each ommatidium points radially from the centre of the eye. Did this mean that within each ommatidium the eight or nine receptors resolve separate regions of each tiny image? And if so how are these inverted representations incorporated into the overall image? Leeuwenhoek was in no position to answer these questions because he did not know the structure and layout of the receptors behind each facet. They were not, in fact, to be resolved until the advent of thin-section microscopy, in the mid-19th century. There is, as we shall see, no one answer.

Meanwhile, without a knowledge of Leeuwenhoek's images, Johannes Müller, (1826) explicitly ignored any optical structures the eye might have, and conceived of it as a simple set of radially arranged pigment-lined tubes, each with a "visual fibre" at the bottom (Fig. 3).

Vision of insects is based neither on dioptric (lens) nor catoptric (mirror) systems. Instead, light coming from object points and illuminating the retina from all directions is selected and restricted to single points corresponding to the object. Thus, an image is formed. In the compound eyes of insects and Crustacea, this is achieved by way of the transparent cones that are situated between cornea and visual fibres, and which are coated with opaque pigment. Each of these cones only admits as much light to the visual fibres as penetrates near the axis. All other light from the same object point meeting the cornea at oblique angles will not reach the proximal end of the cones, and therefore will not be perceived by the other visual fibres but will be absorbed by the pigment-coated sides of the cones (translated by Kunze, P., 1979).

Müller did think about the possibility of the corneal facets forming images, but, in an illuminating passage, seems to have let the theoretical clarity of his overall view deny them a function.

The convex cornea of the single facets will refract the axoparallel light towards the axis and thus concentrate it at the cone tip. The refraction of the convex surface of the cornea is, however, not sufficient to form separate small images for each facet. This would actually hamper a distinct image perception because each small image would be inverted and, therefore, in a wrong position with respect to the next small image. (Kunze, P., 1979).

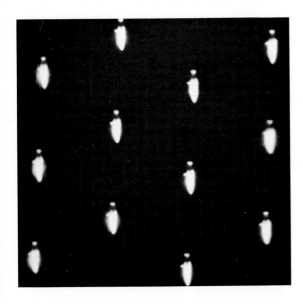

FIG. 2. An attempt to repeat Leeuwenhoek's experiment of 1695, using the cleaned cornea of a large robber-fly (Asilidae) with its inner surface in contact with a hanging drop of Ringer. The facets are 70 μm wide, and the inverted images of the candle flames about 20 μm high. In the animal the seven rhabdomeres that receive each image would scarcely occupy the width of the image of the flame (see Fig. 24).

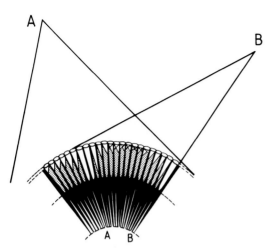

FIG. 3. Johannes Müller's concept of image formation in a compound eye (1826). The crystalline cones and associated pigment are shown as an arrangement for producing an erect image by selective shadowing. (Modified from Kunze, P., 1979.)

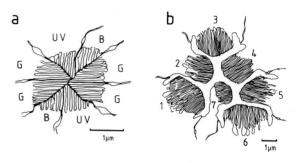

FIG. 4. **a:** Fused rhabdom of a honeybee and **b:** unfused or open rhabdom of a dipteran fly (*Drosophila*). Striations are the microvilli that make up the rhabdomeres. Both figures drawn from electron microscope cross-sections given in Eakin, R. (1972). The pigments in the bee rhabdom absorb ultraviolet (UV), blue (B) and green (G) light, and the assignment of different pigments to the rhabdomeres is based on Wehner, R. (1976). In the fly there are six peripheral rhabdomeres (1–6) and two narrower central rhabdomeres (7 and 8); 8 is continuous with 7 and lies proximal to it. See also Fig. 24.

What has become clear since Müller's time is that he was basically right that each ommatidium receives light from a restricted angle of outside space, and that the overall retinal image is upright. He was quite wrong, however, to dismiss the lenses as unimportant. His conceptual problems would certainly have been lessened if he had known what emerged from the histological studies of Schultze, M. (1868) and Grenacher, H. (1879); namely that in the eyes of most diurnal insects there is effectively only one photosensitive structure, the rhabdom, in the focal plane of each lens. Although each rhabdom contains contributions from eight or nine retinular cells, in the form of photopigment-containing microvilli, they are fused together into a single rod that behaves as a light guide (Fig. 4a), so that there is no resolution *within* the rhabdom of the image falling on its distal tip. Thus for most insects the problem of the inverted image disappears: the one rhabdom picks out a small solid angle of object space, and it obviously does not matter what the distribution of light within that solid angle happens to be. If we accept this view then the role of each lens is not so much to form an image (although it does this and thus defines the field of view of each rhabdom) as to collect a great deal more light into the rhabdom than would have been possible with one of Müller's pigment-lined tubes.

For example, a typical insect eye might have ommatidia 25 μm in diameter, with 1 μm diameter rhabdoms. For these rhabdoms to have a field of view of 1°, they would need to be at the bottom of tubes 1.4 mm long. The same field of view could be provided by a lens of focal length 57.3 μm, and this would not only reduce the size of the eye, but also increase the light reaching the rhabdom by the ratio of the areas of lens and rhabdom cross sections, i.e. $25^2 : 1^2$ or $\times 625$. This is the real function of the lens.

To complicate the issue, there are some insects, notably the Diptera, in which the rhabdomeres are not fused into a single rhabdom (Fig. 4b), and so there is here the potential for the exploitation of the image behind each lens. These "open rhabdom" eyes were noticed by Grenacher, H. (1879) and studied in detail by Dietrich, W. (1909) and Vigier, P. (1909). These early studies were largely forgotten, and the problematic structure of these eyes was rediscovered in the 1960s (Kuiper, J., 1962; Trujillo-Cenóz, O., 1965; Braitenberg, V., 1967; Kirschfeld, K., 1967). The problem, of course, is the one that Müller was anxious to avoid: how can the inverted images contribute to an overall image that is erect? The remarkable answer, first discovered by Vigier, P. (1909), is that the bundle of axons from the retinula cells of each ommatidium undergoes a 180° twist and a complicated rearrangement before entering the lamina, the first synaptic layer (Fig. 24). This means that the images project upright to the nervous system. Kirschfeld, K. (1967) completed

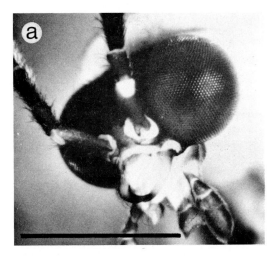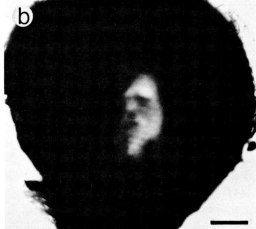

FIG. 5. Eyes of a firefly (**a**), and the image they produce (**b**). The object for the photograph in (**b**) was Julia Cameron's portrait of Charles Darwin, and it is still easily recognizable in the image. The cleaned cornea was suspended from a hanging drop, and photographed with an ordinary microscope just as with the apposition images in Fig. 2. Unlike Fig. 2, the image here is single and erect, not multiple and inverted. Scale on (**a**), 1 mm; on (**b**), 100 μm. (From Land, M., 1981.)

the story by showing that the image received in one ommatidium overlaps that in its neighbours in such a way that one of the seven receptors (actually eight, but two are stacked one above the other in any one ommatidium) has the same line of sight as one receptor from each of the adjacent six ommatidia (Fig. 7b). Furthermore, all eight receptors with a common field of view send their axons to (or in the case of the central two, through) the same laminar "cartridge". Thus, at the level of the lamina, these eyes are ordinary apposition compound eyes, with each laminar cartridge looking in a single direction in space. It would seem that the advantage of this curious arrangement is that in the lamina the contributions of six receptors are pooled, giving higher sensitivity in low light conditions without the sacrifice of resolution. Kirschfeld, K. (1967) has called these "neural superposition" eyes. They do not infringe Müller's mosaic rule, but just produce the final erect image in an unusually complicated way.

A much more radical departure from the "apposition" type of compound eye discussed so far is the "superposition" or "optical superposition" type discovered by Sigmund Exner (1891). These eyes, found in moths and some beetles, have a quite different layout from the eye of, say, a bee (see Figs 7 and 32). The receptors do not lie immediately below the surface layer of lenses, but very much deeper, typically on a hemisphere whose radius is about half

that of the eye itself. In dark-adapted eyes there is a wide zone of clear material between the optical elements and the receptors. Studying the male glow-worm, *Lampyris*, Exner made the remarkable discovery that these eyes produced a single erect image, not a series of small inverted images, and that this image lay deep in the eye, at the level of the receptors (Fig. 5). Furthermore, in contrast to apposition eyes where each rhabdom has its own "private" optical system, the image at any point on the receptor layer of *Lampyris* receives contributions from a great many corneal lenses. In some ways it is as though the whole eye surface were behaving as a single large lens as in the vertebrate type of eye, *except* that the image is erect, not inverted (Fig. 7). It was this super-imposition in the image of light from many facets that gave rise to Exner's term "superposition."

The discovery of this eye raised two related problems. Firstly, what sort of ray paths will give rise to an erect image of this kind, and secondly, how do the optical elements in the eye surface manage to produce such a ray path? The first question is reasonably straightforward: for a single image to be produced deep in the eye, each lens must bend light across its axis in such a way that the angle a ray makes with a normal to the cornea is equal and opposite to the angle the emerging ray makes with the axis of the lens (Figs 7 and 26). This is almost the same as saying that these lenses must behave as mirrors, and it is interesting that 84 years after Exner's monograph, Vogt, K. (1975) showed that crayfish eyes do in fact use mirrors to produce superposition images (see Fig. 30 and Land, M., 1980). However, in *Lampyris* and in moths the optical elements were indubitably refractile lenses, and Exner's problem was how to obtain a mirror-like ray path from a lens structure. His answer involved the postulation of a quite novel type of optical device, the "lens cylinder", which has a graded refractive index, highest along the axis and falling parabolically towards the outside. Such a structure bends light not by ordinary spherical refraction at its ends, but by continuous refraction internally, along its length. Lens-cylinder optics will be explored in detail in later sections (3 and 5) but it is worth pointing out here that Exner's conjecture, then very hard to prove, has been thoroughly confirmed in the last 15 years by interference microscopy, and that within the same period manufacturers have begun to make optically inhomogeneous structures and to exploit their properties (Land, M., 1980). Exner, it seems, was nearly a century ahead of his time.

There was a period during the 1960s when Exner's ideas, both the superposition principle and the lens cylinder, suffered a partial eclipse. Kuiper, J. (1962) failed to find evidence of inhomogeneity in the crystalline cones of crayfish eyes (he was quite right; they have mirrors instead) and suggested that the superposition principle be abandoned in favour of the idea that the various structures that cross the clear zone should be thought of as light guides, relaying light from the corneal lenses to the receptors and thus effectively converting superposition eyes back into apposition eyes. This position was enthusiastically championed by Horridge, but in a later review (Horridge, G., 1975), largely abandoned. The present consensus is that, when dark-adapted, eyes with clear zones are superposition eyes, but that during light-adaptation, when pigment migrates into the clear zone and cuts off non-axial rays, many superposition eyes do effectively convert into apposition eyes, in the sense that each rhabdom now receives light through only one facet, or a very few (Fig. 19i). This last point stresses a fundamental similarity between the two types of eye: the image on the receptor layer has the same

erect spherically symmetrical geometry in both, so that interconversion is possible (superposition eyes can become apposition eyes, though the converse is not true), and as far as the nervous system is concerned the layout of the field of view is the same.

Illuminating accounts of the evolution of our present knowledge of compound eye optics are given by Goldsmith, T. and Bernard, G. (1974); Horridge, G. (1975); Kunze, P. (1979) and Wehner, R. (1981).

In addition to the compound eyes, most insects possess small simple eyes, the ocelli. These are similar in general structure to vertebrate eyes, with a single lens and a concave receptor layer behind it. There are basically two types: the dorsal ocelli of adult insects (Figs 1 and 38) and the eyes of the larvae of some holometabolous groups (Figs 6 and 37). The former, although known to be light-sensitive since the early studies of Réaumur, M. (1740), have remained until very recently quite inscrutable as far as their function is concerned. Most flying insects have three ocelli on the top of their head directed at roughly 120° to each other (Fig. 38), with fields of view that would cover the sky rather than the terrestrial surroundings. All those who have studied the optics of the dorsal ocelli have found them to be severely out of focus, with the retina much too close to the cornea for the image to be resolved, and the question "what can an animal do with an out-of-focus upward pointing eye?" has arisen repeatedly (Goodman, L., 1981). Most past attempts to demonstrate a unique role for these organs have failed: animals with the ocelli covered seem to see perfectly well, but with the compound eyes covered and the ocelli not, they appear blind. In the last few years, however, evidence has begun to accumulate that their role is that of a fast-acting horizon detector, responsible for contributing to the maintenance of stability during level flight (Wilson, M., 1978; Stange, G., 1981). For such a detector, it can be argued, good resolution would be a hindrance rather than a help.

Larval ocelli are quite different. They are the animal's only eyes, and although small they are well-focused. They are concerned with the ordinary processes of vision: habitat selection, prey capture, predator avoidance and so on, functions which in adults are taken over by the compound eyes. Their structure varies greatly. In lepidopteran larvae there

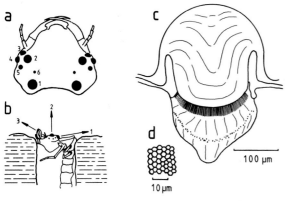

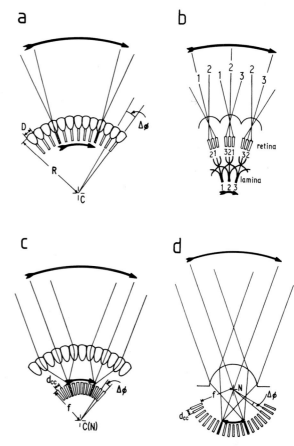

FIG. 6. Larval ocelli of tiger beetles (*Cicindela*). These are almost certainly the most impressive simple eyes in insects in terms of their resolving power, and they rival most compound eyes. **a:** Dorsal view of the head, showing the six pairs of ocelli numbered according to size. **b:** The larva lies braced in the mouth of its burrow. When a prey insect passes the larva snaps its neck back, grasps the prey, and slips down the burrow. The figure shows the axes of the three largest eyes. **c:** Median section of ocellus 1, which has 6350 receptors, and an inter-receptor angle of 1.8°. **d:** Detail of the retina in tangential section. Heavy lines are the fringes of microvilli around the receptor cytoplasm. (All figures redrawn from Friederichs, H., 1931.)

FIG. 7. Four basic types of eye in insects, indicating the features that determine the anatomical resolution ($\Delta\phi$). **a:** Apposition eye; $\Delta\phi = D/R$. **b:** Neural superposition eye in diptera. A variant of the apposition eye in which signals from separate rhabdomeres (see Fig. 4b) are recombined to form a single image in the lamina (Fig. 24). **c:** Optical superposition eye, forming a single erect deep-lying image (Fig. 5). $\Delta\phi = d_{cc}/f$. **d:** Simple eye or ocellus. $\Delta\phi = d_{cc}/f$. C is the centre of curvature, and N the nodal point. Further explanations in the text.

are usually six ocelli on each side of the head, each having a retina that is not much more than a single elaborate rhabdom (Dethier, V., 1963). At the other extreme, the largest ocelli of tiger beetle larvae have 6350 receptors in their retina (Fig. 6; Friederichs, H., 1931). One of the abiding problems of insect visual physiology, and one that shows no signs of going away, is why the larval ocellus ever came to be supplanted by the adult compound eye. For the same sensitivity and resolving power, single-lens eyes (such as those of spiders for example) can be more compact, and whilst compound eyes are limited in resolution to about $\frac{1}{2}°$ by diffraction at each tiny lens, no such restriction applies to the ocellar type of eye, with its single large lens. One cannot even argue that compound eyes have been retained through evolutionary conservatism, since at one or another time in their lives most insects possess both types. It is a mystery to which we shall return.

1.2 Basic types of insect eye

The four basic types of insect eye that were outlined in the historical introduction, and with which this chapter will be concerned, are illustrated diagrammatically in Fig. 7. The distribution of these eye types within the Insecta is given in Table 1, to the extent that details are available. Apposition eyes (Fig. 7a) are typical of diurnal insects, and are characterized optically by the fact that each rhabdom possesses its own optical system. The distal tip of each rhabdom lies in the focal plane of the corresponding lens, which is typically 50–100 μm beneath the corneal surface. There is no "clear zone" as in C. When apposition eyes are not heavily pigmented it is often possible to see a pseudopupil — the small black spot that appears to move round the eye as the observer moves round the animal (see section 3.2).

A variant of the apposition eye is the "neural super-position" eye (b). The optics are the same as in a, and the difference lies in the way the receptors are organized. Instead of the eight or nine receptors in each ommatidium contributing their microvilli to a single "fused" rhabdom, the individual rhab-domeres are physically and optically separate, giving rise to an "open" rhabdom. This receptor pattern is found in the Diptera and some Hemiptera (e.g. *Lethocerus*; Ioannides, A. and Horridge, G., 1975). In the Diptera the axons of the receptor cells rearrange themselves between the retina and the lamina in such a way that all the receptors that view the same point in space reach the same synaptic cartridge in the lamina (Kirschfeld, K., 1967). This means that from the lamina inwards these eyes are functionally the same as ordinary apposition eyes.

Optical superposition eyes (c) are usually extern-ally indistinguishable from apposition eyes, but they differ in internal structure in having a wide clear zone between the layer of optical elements and the receptor layer (Fig. 32). This zone permits light entering many facets to be focused onto single rhab-doms in the receptor layer, so that an erect image is formed across the whole retina. Some superposition eyes show eyeshine (see section 5.3, Fig. 31) a large patch of blue, green, yellow or orange glow that is visible when the eye is illuminated from the same direction that it is being viewed from. The eyeshine is due to light reflected back from a tapetum behind and around the receptors, and is analogous to the reflection from a cat's eye. During light adaptation, as pigment migrates into the clear zone, the patch of glow reduces in size or disappears. Dark-adapted superposition eyes do not have point-like pseudopupils. The optical elements in superposition eyes do not focus light, like those of apposition eyes, but redirect it across their axes (see section 5.1). Finally, the simple eye shown in d is based on the larval ocellus of a sawfly (*Perga*) (Meyer-Rochow, V., 1974). The corneal lens forms an inverted image on the receptor layer, which in this instance is in focus. In the dorsal ocelli of most adult winged in-sects the image lies well behind the receptor layer. Ocelli or "stemmata" of lepidopteran larvae have retinae that are more like a single ommatidium from an apposition eye, with only one rhabdom.

Outside the insects, compound eyes occur in the Crustacea (both apposition and superposition

types), in the Xiphosura (*Limulus*), in some advan-ced centipedes like *Scutigera* where they seem to have arisen by aggregation and multiplication of groups of ocelli, in one family of tube worms (*Sabellidae*; Kerneis, A., 1975), and in two genera of bivalve molluscs (*Arca*, *Pectunculus*; Hesse, R., 1900). Compound eyes, presumably of the app-osition type, were also present in the extinct trilobites (Levi-Setti, R., 1975) and they are thus of similar antiquity to the camera-type eyes of vertebrates and cephalopod molluscs. In spite of these parallels, there is no indication as to how com-pound eyes arose in insects. Apterygote insects, usu-ally regarded as primitive, have eyes which, al-though small, have an ommatidial structure that is recognizably the same as that of other insect orders (Paulus, H., 1975), and clearly different from that of possible insect forebears like the myriapods.

Amongst arthropods, simple eyes (d) are the only eye type found in the Arachnida, and they also occur together with compound eyes in *Limulus*. A few crustaceans, notably copepods, have simple eyes but compound eyes are more usual. *Peripatus*, often cited as a link between the annelids and arth-ropods, has a pair of simple eyes of similar general design to that shown in d (Eakin, R. and Westfall, J., 1965). Similarly, myriapods such as *Lithobius* have a small number of ocelli, each with a retina of up to 110 receptors (Bähr, R., 1974). Again, how-ever, there is no evidence to suggest that any of these is ancestral to any of the types of insect ocellus. A thorough review of the possible ancestry of com-pound and simple eyes in insects is given by Paulus, H. (1979).

2   OPTICAL PRINCIPLES

Eyes, whatever their structure, perform essentially the same task. This is to split up light according to its direction of origin in the outside world, and to collect sufficient light for a reliable signal about the luminance in each direction to be sent to the CNS. An appropriate analogy would be a series of photon counters each collecting from a small solid angle of outside space (Fig. 8). Using this analogy, we can discuss the performance of an eye, and make com-parisons between eyes, on the basis of the fineness with which the external world is sampled

Table 1: *Eye types in insects*

Order	Adults	Larvae
Apterygota		
1 Thysanura	app. or absent; ocelli absent	as adults
2 Diplura	absent	as adults
3 Protura	absent	as adults
4 Collembola	app. up to 8 ommatidia; no ocelli	as adults
Exopterygota (= Hemimetabola)		
5 Ephemeroptera	app. with 3 ocelli; male eyes enlarged	app. with dorsal ocelli
6 Odonata	app. (large); 3 ocelli	app., dorsal ocelli in late instars
7 Plecoptera	app.*; 3 ocelli	as adults
8 Grylloblattodea	app.* (reduced); no ocelli	as adults
9 Orthoptera	app.; usually 3 ocelli, absent in wingless forms	as adults
10 Phasmida	app.; ocelli in winged forms only	as adults
11 Dermaptera	app.* (sometimes absent); no ocelli	as adults
12 Embioptera	app.* (small); no ocelli	as adults
13 Dictyoptera	app. with pronounced fovea in mantids; 3 ocelli in mantids, 2 or vestigial in cockroaches	as adults
14 Isoptera	app. (often reduced in wingless forms) 2 ocelli in winged forms	depends on stage and potential caste?
15 Zoraptera	app.* reduced in apterous forms; ocelli in winged forms only.	as adults
16 Psocoptera	app.* sometimes reduced; 3 ocelli in winged forms only	as adults
17 Mallophaga	compound eyes reduced to single lens; no ocelli	as adults
18 Siphunculata	compound eyes reduced to single lens; no ocelli	as adults
19 Hemiptera	app. or possibly n.sup. Some male coccids have 6 single lens eyes replacing compound eyes; ocelli 3 (cidadas) 2 (usually) or 0	as adults
20 Thysanoptera	app. (few ommatidia); 3 ocelli in winged forms	as adults
Endopterygota (= Holometabola)		
21 Neuroptera	o.sup. in lacewings, antlions, mantispids; others? app.; ocelli 3 or absent	6 pr lateral ocelli, or blind.
22 Mecoptera	app.*; usually 3 ocelli	eyes almost compound with 20–35 ommatidia (*Panorpa*)
23 Lepidoptera	app. in most butterflies, o.sup. in skippers and moths. 2 ocelli	6 pr lateral ocelli
24 Trichoptera	probably o.sup.; ocelli 3 or absent	6 pr lateral ocelli
25 Diptera	n.sup., male eyes often enlarged; 3 dorsal ocelli	variable receptor groupings, not usually visible
26 Siphonaptera	compound eyes replaced by simple eyes; no dorsal ocelli	blind
27 Hymenoptera	app. (reduced in some ants); 3 ocelli	blind, or single pair of large ocelli (sawflies)
28 Coleoptera	app. in diurnal spp., o.sup. in nocturnal spp. (e.g. lampyrids): divided eyes common; ocelli 2, 1, or (usually) absent	6 pr lateral ocelli, or fewer or absent
29 Strepsistera	app. with abnormal ommatidia, up to 100 receptors/rhabdom; males only, females blind; no ocelli	larvae blind

Abbreviations: app. apposition compound eye
o.sup. optical superposition compound eye
n.sup. neural superposition compound eye
* implies incomplete knowledge of eye type

The taxonomic system is that of Imms, A. (1957), from which much of the information is also derived.

(resolution), and the size of the photon sample that can be obtained from each direction (sensitivity). The latter is a particularly important measure of an eye's performance at low and even moderate light intensities, because when photon numbers are small they are subject to very considerable random variation and thus any signal measuring them will be "noisy" and unreliable. The more photons that can be collected in each receptor "bin", the more reliable and useful the signal. That these considerations are important can be judged from the estimate given by Wald, G. *et al.* (1962) who conclude that at the absolute threshold of human vision each rod receives a photon about once every 40 minutes! In

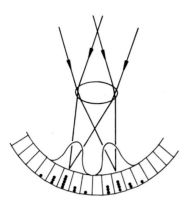

FIG. 8. Resolution and Sensitivity. Images of external objects must first be resolved by the optical system; this always results in some degree of blurring. Where the numbers of available photons are low, further statistical blurring results from random variations in the number of photons each receptor receives. The black counters in the receptors indicate the photon distribution in the final image.

the sections that follow I shall try to set out the features of an eye that establish its resolution and sensitivity, and make comparisons between the different types of insect eye. Valuable accounts which have the same general approach as this article are given by Barlow, H. (1964); Kirschfeld, K., (1974, 1976); Snyder, A. *et al.* (1977); Snyder, A. (1979) and Horridge, G., (1977).

2.1 Resolution

2.1.1 GEOMETRICAL OPTICS OF SIMPLE AND COMPOUND EYES

If an eye is to resolve fine detail the *angular* separation between the receptors must be small. If it is to resolve a grating, for example, then for each line pair in the grating there must be two corresponding receptors in the eye, one imaging the dark line and one the light (Fig. 9). More formally, the finest grating an eye can resolve will have an angular period of $2\Delta\phi$, where $\Delta\phi$ is the angle between the directions of view of two adjacent receptors. Alternatively we can say that the highest *spatial frequency* that can be resolved is $1/(2\Delta\phi)$ (line pairs per radian, or degree). We can use the highest resolvable spatial frequency (v_s) as a convenient measure of the spatial resolution of an eye: it has the virtue that it defines the limit to the fineness of vision in a way that increases as the resolution increases, unlike the minimum resolvable

angle $\Delta\phi$ which gets bigger as the eye gets "worse". Thus:

$$v_s = 1/2\Delta\phi \tag{1}$$

It is now necessary to establish from the geometry of the three types of eye (simple, apposition and superposition) what are the anatomical features that determine $\Delta\phi$.

In an eye of the *simple* type $\Delta\phi$ is given simply by the distance between the centres of two receptors d_{cc} divided by the focal length of the eye (f):

$$\Delta\phi = d_{cc}/f \text{ (radians)} \tag{2}$$

or 57.3 d_{cc}/f (degrees). The focal length in this case is the same as the posterior nodal distance, the distance between the nodal point of the eye and the image. The significance of the nodal point is that it is the point in the eye through which rays pass without angular deviation, so that an angle subtended by an object at the eye is the same as the angle subtended by its image at the nodal point. Thus angles inside and outside the eye can be simply calculated from each other by similar triangles. For a

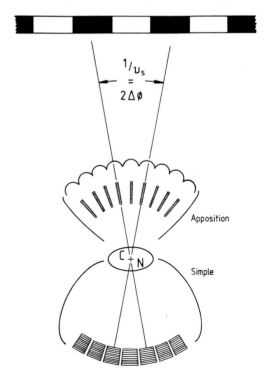

FIG. 9. Anatomical resolving power in an apposition eye (centred at C) and a simple eye (nodal point at N). v_s is the highest spatial frequency that can be resolved.

simple eye, or indeed an insect ommatidium, where the main refracting surface is a curved cornea separating air from the inside of the eye, the nodal point will be at the centre of curvature of the cornea (because radially directed rays will not be bent) and the focal length is given by:

$$f = \frac{r}{n-1} \qquad (3)$$

where r is the radius of curvature of the cornea, and n the refractive index of the inside of the eye. Thus if n is about 1.4 the focal length will be 2.5 times the corneal radius of curvature. However, where the cornea is in water, where there are important refracting surfaces in addition to the cornea, or where the optical system is based on lens cylinders, the approximation in equation 3 is inadequate, and the thick lens formula or lens cylinder formula (see section 3.1) must be used to find f. Experimentally, f can be found from the image magnification if this can be measured, provided the object is at a large distance (u) from the eye. Then:

$$f = u \times {}^I\!/_O \qquad (4)$$

where I and O are the sizes of the image and object respectively. There are few studies of insect larval ocelli that provide enough information for resolving power to be estimated accurately, although V. Meyer-Rochow's paper (1974) on the ocellus of sawfly larvae (*Perga*) is an exception. In these animals (Fig. 37) the lens has a diameter of 180–360 μm, and has two refracting surfaces in addition to the cornea, so that the simple single surface equation (3) gives too short a focal length, and ray tracing has to be used to find the focus. These tracings indicate that the nodal point to focus distance (f) is about 220 μm for an eye of corneal diameter 280 μm. Since the receptor separation in the centre of the retina is about 20 μm, $\Delta\phi$ is 20/220 radians, or about 5°. Unfortunately, sawfly larvae do not respond reliably to stripe patterns, so a behavioural measure of v_s (equation 1) cannot be obtained directly, but other behaviours suggest that the smallest objects these larvae will respond to are about 4° wide. The larval eyes of tiger beetles (*Cicindela*, Fig. 6) probably have the finest resolution of any insect simple eye, with $\Delta\phi = 1.8°$ (Friederichs, H., 1931). By comparison, human eyes, with a focal length of 16.7 mm and a minimum receptor spacing of 2.5 μm, have an anatomical resolution ($2\Delta\phi$) of

about 1′, which agrees almost exactly with the minimum resolvable grating period.

In *apposition* eyes, although each ommatidium has its own lens which behaves much as the lens of a simple eye, there is nothing that corresponds to a nodal point for the eye *as a whole*, and so the eye has no single focal length. However, provided the eye is spherically symmetrical the geometry of Fig. 7a indicates that the angle between the lines of sight of adjacent receptors ($\Delta\phi$) should be the same as the angle subtended by adjacent ommatidia at the centre of the eye. Thus, if the eye has a radius R and the facets on the surface have a diameter D,

$$\Delta\phi = D/R \qquad (5)$$

In general this only gives an approximation to the actual inter-ommatidial angle, because many insect eyes are not spherical. In some bees and grasshoppers, for example, the radius of curvature in the transverse plane may be two to three times the radius in the longitudinal plane, giving smaller values of $\Delta\phi$ measured vertically compared with horizontally (see Fig. 22). Similarly $\Delta\phi$ may vary systematically across the eye, and there may be parts of the eye where the rhabdom does not coincide with the axis of each facet. For all these reasons it is usually best to measure the inter-ommatidial angle directly, and the simplest way to do this is to map the position of the pseudopupil, when this is visible (see section 3.2). Since the pseudopupil occupies that part of the eye surface which is looking at the observer, a line joining the pseudopupil and the observer defines the line of sight of the ommatidium or ommatidia that appear dark. When the animal is rotated, the pseudopupil appears to move, and if the pseudopupil moves across five facets when the animal is turned through 15° (say) then the inter-ommatidial angle must be 3°. This method of measuring $\Delta\phi$ is reasonably free from error (Horridge, G., 1978; Stavenga, D., 1979).

Typical values for R and D in an insect are 1 mm and 25 μm. These will give a value for $\Delta\phi$ of 1.4°, and a value for v_s of 20 cycles per radian. Measured values for $\Delta\phi$ vary between 0.24° in a part of the eye of the large dragonfly *Anax junius* (Sherk, T., 1978), to 7° in the beetle *Chlorophanus* (Mazokhin-Porshnyakov, G., 1969) although some small apterygote insects will certainly have much higher values (Paulus, H., 1975). The limits to $\Delta\phi$ are set

Table 2: Definitions of symbols

f	focal length, or posterior nodal distance of a single surface, lens or lens system (equations 3 and 4)
r	radius of curvature of an image-forming surface
R	radius of a compound eye
D	diameter of a facet or lens; diameter of region admitting light to a single image point in a superposition eye
d_{cc}	centre-to-centre spacing of rhabdoms or receptors
d_r	diameter of rhabdom or receptor
l	length of rhabdom or receptor
$\Delta\phi$	angular separation of visual axes in a compound eye, or angular separation of receptors in any eye
$\Delta\rho_r$	angle subtended by tip of receptor or rhabdom at the nodal point of the lens system ($= d_r/f$ radians or 57.3 d_r/f deg.)
$\Delta\rho$	acceptance angle of receptor or rhabdom ($\sim \sqrt{(\Delta\rho_r^2 + (\lambda/D)^2)}$ in a diffraction limited system)
λ_s	spatial sampling frequency of the receptor mosaic, ($= 1/2\Delta\phi$); defines the resolution limit of any eye
v_{co}	spatial cut-off frequency of an optical system, ($= D/\lambda$, if the system is diffraction limited)
λ	wavelength of light in vacuum; typically 0.5 μm (blue-green)
S	sensitivity of an eye, the photon flux captured by a receptor viewing an extended source of unit radiance (equations 18, 22, 23)

Notes.   Other symbols defined in the text.
The symbols used here are those commonly used by most other workers on insect optics (e.g. Snyder, A., 1979) and a number of them differ from those used in an earlier review (Land, M., 1981)

by D and R as indicated in equation 5. R obviously cannot be much larger than a few mm, or the animal will not be able to support its head, and D cannot be smaller than about 10 μm because resolution then becomes severely limited by diffraction (see section 2.1.2).

In a "well-designed" apposition eye one would expect that each rhabdom would have a field of view that is contiguous with that of its neighbour, but which doesn't overlap it to any great extent. Another way of putting this is to say that the *acceptance angle* of each rhabdom ($\Delta\rho$) should be approximately equal to the inter-ommatidial angle ($\Delta\phi$). In purely geometrical terms, $\Delta\rho_r$ will be the subtense of the distal tip of the rhabdom, diameter d_r, at the nodal point of each facet lens, so that:

$$\Delta\rho_r = d_r/f \qquad (6)$$

where f is the focal length of the facet lens (Fig. 13).

In general the true acceptance angle $\Delta\rho$ will be slightly greater than equation 6 indicates, because the image on the rhabdom tip will be slightly blurred by diffraction (see section 2.1.2). In the only case where $\Delta\rho$ has been measured carefully enough by electrophysiological techniques for useful comparisons to be made — in the praying mantis *Tenodera* — its value is very close to that of $\Delta\phi$, over the whole range from 0.7° for 2.5° (Rossel, S., 1979).

In apposition eyes the highest resolvable spatial frequency can thus be thought of either as a consequence of the structure of the eye as a whole (equation 5: $1/2\Delta\phi = R/2D$) or as a result of the design of individual facets (equation 6: $1/2\Delta\phi \simeq 1/2\Delta\phi = f/2d_r$). It is comforting that the minimum stripe width that will elicit an optomotor turning reaction in a variety of insects does correspond very closely to measured values of $\Delta\phi$ (see e.g. Mazokhin-Porshnyakov, G., 1969). Theory and practice seem to agree as well for these eyes as they do for human eyes.

In *superposition* eyes the situation is different, and a little simpler because a superposition eye does have a nodal point at its centre of curvature (Fig. 7c). Since a ray along the axis of each crystalline cone is not deviated, and all such rays pass through the centre of curvature, then just as in a simple eye this can be used as the nodal point for the purpose of calculating $\Delta\phi$. The focal length of the eye, corresponding to f in equation 2, will be the distance *out* from the centre of curvature to the image, which should lie somewhere in the receptor layer. In general, the receptor layer is about half an eye radius out from the centre of curvature ($f \sim R/2$) although this need not necessarily be the case. $\Delta\phi$ is then calculable from equation 2, just as in a simple eye, with d_{cc} being the centre-to-centre spacing of the receptors. The highest potentially resolvable spatial frequency will as usual be given by $1/2\Delta\phi$ (equation 1).

To summarize, the anatomical limits to resolution in simple and superposition eyes are set by equations 1 and 2, and in apposition eyes by equations 1 and 5, or 1 and 6. As we shall see in the following section, there are limits to resolution set by the wave nature of light that must also be considered.

2.1.2   THE DIFFRACTION LIMIT

Readers familiar with other branches of optics will

recall that the resolution of an optical instrument depends on the diameter of its aperture. In the case of a telescope used to resolve stars the minimum resolvable angle is determined by the Rayleigh limit; that is, the condition when the centre of the image of one star falls on the first dark ring of the diffraction pattern caused by the other star (the Airy disc). The minimum resolvable angle (in radians) is then given by:

$$\theta = 1.22 \, \lambda/D \qquad (7)$$

where λ is the wavelength of light.

The important point is that as the diameter of the aperture of the telescope (or indeed eye) gets bigger, resolution gets better. This point is crucial for an understanding of the design of eyes, and of apposition compound eyes in particular, where the individual facets are small ($D \sim 25\mu\text{m}$) and the minimum resolvable angles correspondingly rather large. For blue-green light ($\lambda = 0.5 \, \mu\text{m}$) equation 7 gives a value for θ of one-fiftieth of a radian, or just over 1°. This is very similar to the anatomical limit (equations 5 and 6).

The diffraction limit arises from the wave nature of light. Light from a distant point reaches the aperture of an eye as a plane wave-front (Fig. 10) which is bent by the optical system into curved wave-front centred on the focus. Light from different regions of the wave-front will interfere in the region of the focus; the interference will be constructive in some places and destructive in others, and the resulting pattern for a point object is a concentric pattern with a light centre, surrounded by alternating dark and light rings of sharply decreasing brightness. Equation 7 gives the radius, in angular terms, of the first dark ring. The important point is that even an optically perfect lens is limited in its performance by the fact that point sources do not produce point images, but blur circles whose width is inversely proportional to the size of the aperture.

The diffraction limit does not only apply to point sources. The blurring effect of diffraction degrades all images, and obviously the effect will be most serious with objects that contain much fine detail. Low spatial frequencies are relatively unaffected, whereas high spatial frequencies are progressively reduced in contrast until a frequency is reached where no contrast at all is present in the image. This is the "cut-off" frequency, and is given

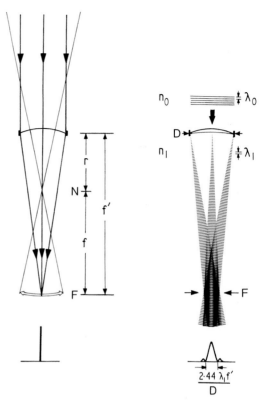

FIG. 10. Image formation according to ray optics (left) and diffraction optics (right). In ray optics a point source produces a point image, but when light is diffracted at an aperture the image produced is a blur circle (the Airy disc) whose diameter is 2.44 $\lambda_0 f/D$ (or 2.44 $\lambda_1 f'/D$, where $\lambda_1 = \lambda_0/n_1$ and $f' = n_1 f$).

in radians by:

$$v_{co} = D/\lambda \qquad (8)$$

Interestingly, the corresponding spatial period (λ/D) is almost identical to the diameter of the Airy disc at 50% of its maximum intensity (Snyder, A., 1979). If this, rather than the Rayleigh criterion, is used as a measure of the resolving power of the eye, we have:

$$\theta_{50\%} = 1/v_{co} = \lambda/D \qquad (9)$$

All these measures of resolving power (equations 7, 8 and 9) indicate that optical systems pass no information about structures in the outside world whose angular dimensions at the eye are less than λ/D radians. For larger objects, containing spatial frequencies lower than v_{co}, the quality of the image is best described by the *contrast transfer function*.

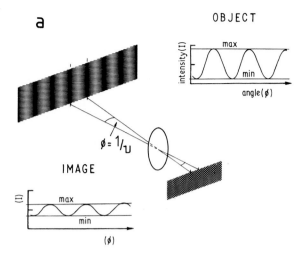

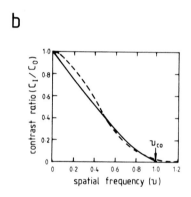

FIG. 11. Contrast transfer. **a**: Decrease in the contrast of a sinusoidally modulated grating of period ø and frequency v on passing through a lens. Contrast is defined as $(I_{max} - I_{min})/(I_{max} + I_{min})$. **b**: Contrast transfer function for a perfect diffraction-limited lens. The solid curve results when the point diffraction pattern is the Airy disc (Fig. 10), and the dashed line when the Airy disc is approximated by a Gaussian of half-width λ/D. The spatial frequency axis is scaled in relation to the cut-off frequency (v_{co}) beyond which a diffraction limited eye passes no contrast. v_{co} is equal to D/λ. (**b**: after Snyder, A., 1979.)

This is a graph which relates the ratio of image to object contrast to the spatial frequency of sinusoidally modulated gratings (Fig. 11). For a lens that is without defects other than diffraction the relationship is almost a straight line from 1 at zero spatial frequency to 0 at the cut-off frequency v_{co}. Thus from Fig. 11b it can be seen that when the spatial frequency of the object grating is half v_{co} the contrast of the image is about 40% of the contrast of the object. Where a lens has defects in addition to

diffraction (spherical and chromatic aberration, incorrect focus, etc.) the contrast transfer function is displaced towards origin and may even go negative for some spatial frequencies, so that light stripes become dark and vice-versa. In general, however, the effect of aberrations is to attenuate high frequencies further, blurring fine detail.

The importance of diffraction varies markedly between the three types of eye we are considering. In each case we need to compare the limit to resolving power set by diffraction (v_{co}; equation 8) with the limit to spatial resolution set by the sampling density of the receptor elements (v_s; equation 1).

For the simple eye of the sawfly *Perga* (Fig. 37), taking D as 180 μm and λ as 0.5 μm, v_{co} comes to 360 cycles per radian whereas v_s is only 5.5 c/rad. Clearly, the eye comes nowhere near to being limited by diffraction, and if the lens is optically good there is a great deal of "unexploited" resolution. It is generally the case with simple eyes, including our own, that diffraction is not the feature that limits their performance for the simple reason that the pupil aperture tends to be large, and hence v_s large. The human eye in bright light, with a 2 mm pupil, comes quite close to the limiting condition ($v_{co} = 4000$ c/rad, $v_s = 3340$) but here it can be argued that the match is deliberate. The pupil *can* be opened to 8 mm, giving a theoretical value for v_{co} of 16,000 c/rad, although in practice lens defects limit the performance at apertures larger than about 4 mm. Diffraction is thus not a real problem in simple eyes: the lens can always be made a little larger.

In apposition eyes the situation is quite different. Using the value for the honeybee from Table 3, v_{co} is 50 c/rad and v_s is 20.5 c/rad, so that although diffraction is not quite limiting in this case it is sufficiently close to that situation for the contrast of the image to be very substantially reduced. Referring to the contrast transfer function in Fig. 11, it can be seen that for a spatial frequency of 0.4 v_{co}, the image contrast is down to 50% of the object contrast, so diffraction is clearly impairing the eye's performance. In some insects, especially those that hover in bright sunlight, the sampling limit approaches the diffraction limit more closely (Horridge, G., 1978). In the "acute zone" of the eye of the sand wasp *Bembix palmata* the inter-ommatidial angle is as small as 0.33° ($v_s = 87$ c/rad) and since the facets in this region are about 40 μm across, the cut-off

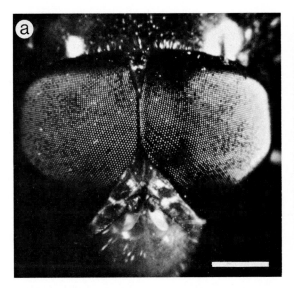
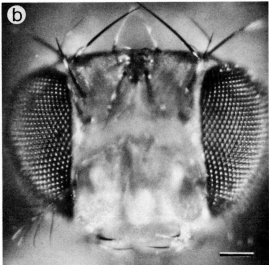

FIG. 12. Heads of a large fly (**a:** *Volucella pellucens*) and a small fly (**b:** *Drosophila melanogaster*) showing that one is not simply a scaled-up version of the other. The eye of *Volucella* has a radius about six times that of *Drosophila*, and the facet diameters are about 2.5 times larger. The scale is 1 mm on **a:** and 0.1 mm on **b:**. Barlow's finding that $D \propto \sqrt{R}$ holds for most flies.

frequency v_{co} is 80 c/rad for blue-green light — less than the sampling frequency — although in the ultraviolet ($\lambda = 350$ nm) v_{co} is higher (114 c/rad). Horridge, G. (1978) and Snyder, A. (1979) point out that the anatomical resolution of insect eyes only approaches closely to the diffraction limit when there is sufficient light for very small contrasts to be detectable (see section 2.2). Howard, J. and Snyder, A. (1983) further argue that the low contrast image at spatial frequencies close to the cut-off (v_{co}) is not usable because the "noisiness" of the transduction process makes small signals undetectable. In any event, it is clear that many apposition eyes approach the diffraction limit closely but in no case does the anatomical limit exceed it. Insect eyes do not try to resolve spatial information that is not there.

The constraints on apposition eyes imposed by the diffraction limit are best understood if one tries to work out how the resolution of such an eye might be improved. It seems intuitively right that resolution could be increased if the ommatidia were made smaller and packed more tightly together, equivalent to decreasing D in equation 5. However, decreasing D makes the resolving power of individual ommatidia worse (equation 8) so the overall resolution too will be worse, not better. If

the resolving power v_{co} is to remain greater than the sampling frequency v_s then each facet must get larger, not smaller. If they are made larger, there can be more of them because of the improved resolving power, and the eyes will get disproportinately large.

In an "ideal" apposition eye, in which the sampling frequency v_s just matches the diffraction limit v_{co}, so that the image is neither over- nor undersampled, we can calculate what its size needs to be for a given spatial resolution. From equations 1 and 5 we have:

$$v_s = \frac{R}{2D} \qquad (10)$$

Making v_{co} (equation 8) equal to v_s, and multiplying equations 8 and 10 together gives:

$$v_s^2 = \frac{R}{2\lambda} \text{ or } R = 2\lambda v_s^2 \qquad (11)$$

The radius of the eye, in other words, goes up as the square of the spatial resolution. If $\Delta\phi$ is $1°$ ($v_s = 29$ c/rad), R will be 0.82 mm, which is about right. On the other hand, if $\Delta\phi$ is $1'$, roughly as in man, R becomes 60^2 times greater, and the eye would have a radius of 3 *metres*! Clearly any animal, let alone an insect, would look very silly with such an impracticable eye. A splendid drawing illustrating this can be found in Kirschfeld, K. (1976). In *simple* eyes, where diffraction is not limiting, $\Delta\phi$ is equal to d_{cc}/f, so that from equation 1,

$$f = 2\, d_{cc}\, v_s \qquad (12)$$

The resolution term is not squared, and the eye size increases linearly with resolution. Equations 11 and 12 point out the fundamental design weakness of apposition eyes: *resolution better than about a degree cannot be attained with an eye of a size that will fit on the head*. This point was first made by A. Mallock, as early as 1894.

A related point that emerges from a discussion of the diffraction limit concerned the way that facet size and eye size are related in apposition eyes. It is clear that a large insect eye is not simply a scaled-up version of a small one (Fig. 12). A large eye has both more and larger facets than a small eye, and in fact facet size goes up as the square root of eye size. If we again make v_{co} and v_s equal (equations 8 and 10) the result is:

$$\frac{D}{\lambda} = \frac{R}{2D}, \text{ or } D = \sqrt{(R\lambda/2)} \qquad (13)$$

Barlow, H. (1952) found that D was proportional to $\sqrt{R}$ for 27 species of Hymenoptera ranging in body length from 1 to 60 mm, and R. Wehner (1981, p.309) has extended this finding to other arthropods; it turns out that even the eyes of extinct trilobites fit the same line as Barlow's data. Interestingly, however, although the square root relation holds approximately over the whole range, facet diameters (D) are about twice as large as would be predicted by equation 13. This seems to mean that insect eyes are not in general designed to operate right up to the diffraction limit (v_{co} in the contrast transfer function in Fig. 11) where there is *no* image contrast, but at spatial frequencies of around $\frac{1}{2} v_{co}$, where the image contrast is a more respectable 40%. There are exceptions, like *Bembix* whose eyes are closer to the diffraction limit, and at the other extreme the crepuscular dragonfly *Zyxomma*, where the facets are four to six times larger than equation 13 suggests (Horridge, G., 1978). In this case the limit to eye performance is certainly not diffraction, but photon scarcity in the dim environment that *Zyxomma* inhabits (see section 2.2).

One might think that *superposition* eyes would be

more like simple eyes with respect to diffraction, with the whole of the superposition pupil acting as the aperture of the eye, thereby producing very small Airy discs and potentially high resolution. This is not in fact what happens, because the condition for obtaining the sort of diffraction pattern that is illustrated in Fig. 10 is that light reaching the focus from different parts of the wave-front must all have travelled the same optical distance. If this is not the case, then constructive interference will not occur at the focus. Inspection of the ray paths contributing to a superposition image (Fig. 26) shows that they travel very different distances in reaching the focus, so that an Airy disc pattern will not be set up. The practical outcome is that in superposition eyes, as in apposition eyes, the diffraction limit is set by the diameter of *individual* facets. Resolution may be worse than this for other reasons (see section 5) but it will not be better.

A final result of diffraction is an increase in the size of the acceptance angle ($\Delta\rho$) of each receptor, or rhabdom, and this applies to all types of eye. If the image of a point source sweeps across the field of view of a receptor, simple geometrical optics would indicate that the angle $\Delta\rho_r$ over which the receptor would receive light will be d_r/f (equation 6). However, the image of a point source is not a point but a diffraction pattern of angular width $\Delta\rho_l = \lambda/D$ (equation 9), and to obtain $\Delta\rho$, the true acceptance angle of the receptor, it is necessary to take the

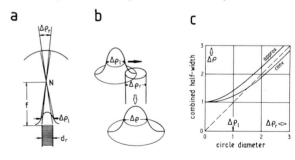

FIG. 13. Geometry of the rhabdom acceptance angle $\Delta\rho$. **a:** The field of view of a rhabdom is a combination of its angular width ($\Delta\rho_r = d_r/f$) and the point source image (half-width $\Delta\rho_l$) provided by the lens. **b:** Convolving a Gaussian function representing the image ($\Delta\rho_l$) with a circle representing the rhabdom tip (diameter $\Delta\rho_r$) gives a third function whose half-width ($\Delta\rho$) is the rhabdom's acceptance angle. **c:** Result of the convolution in **b:**. Both axes are in units of $\Delta\rho_l$ so that if $\Delta\rho_l = 2°$ and $\Delta\rho_r = 4°$, the abscissa value is $4/2 = 2$, and the corresponding ordinate is 1.9 times $\Delta\rho_l = 3.8°$. The Gaussian approximation (equation 14) gives a value of 4.5°, about 20% too high.

Table 3: Optical constants of representative eye types

	Apposition compound eye (*Apis mellifera*)	Superposition compound eye (*Ephestia kuhniella*)	Larval ocellus (*Perga affinis*)
D (μm)	25	400	180
f (μm)	60	170	220
d_r (μm)	1.5	8	10
d_{cc} (μm)	—	8	20
$\Delta\phi$ (deg)	1.4	2.7	5.2
v_s (rad^{-1})	20.5	10.6	5.5
$\Delta\rho$ (deg)	1.8	~8[1]	2.6
F_{no}	2.4	0.43	1.2
l (μm)	250	55[2]	130
S (equation 22)	0.2	114	24

[1] In theory $\Delta\rho$ could be as small as $\Delta\phi$, but the superposition image in *Ephestia* is poor, as judged from the divergence of the eye-shine. In diurnal superposition eyes (skippers, agaristids) $\Delta\rho \simeq \Delta\phi$.

[2] Because of the tapetum the effective rhabdom length is twice this (110 μm).

Most data in this table are from Wehner, R. (1981).
Definitions of symbols in Table 2.

width of the Airy pattern into account (Fig. 13). Snyder, A. (1979) makes the simplifying assumption that both the geometrical acceptance function and the Airy disc are Gaussian distributions, because this enables one to find the combination of the two (strictly, the convolution) by simple addition. Thus

$$\Delta\rho^2 = \Delta\rho_r^2 + \Delta\rho_l^2 = \left(\frac{d_r}{f}\right)^2 + \left(\frac{\lambda}{D}\right)^2 \qquad (14)$$

The amount of error introduced by making the (dr/f) term Gaussian is shown in Fig. 13c. Where receptors are narrow and the eye is close to the diffraction limit, as in most apposition eyes, the effect of the λ/D term is quite important. In the bee, for example (Table 3), the geometrical acceptance angle of a rhabdom ($\Delta\rho_r$) is 1.4°, but the true acceptance angle (i.e. the width of the acceptance function at half maximum) given by equation 14 is 1.8°.

2.1.3 RECEPTOR OPTICS

In the honeybee apposition eye the rhabdoms are 250 μm long, but the focal length of each facet lens is only 60 μm. Obviously the lens cannot form a focused image over the whole rhabdom length. What happens is that the image is formed on the distal tip of the rhabdom, and light from the image

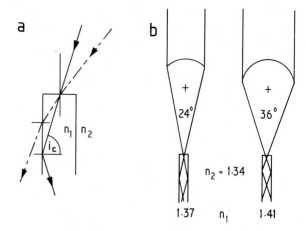

FIG. 14. Total internal reflection. **a:** Total internal reflection occurs if $i_c > \sin^{-1}(n_2/n_1)$. **b:** Cones of light trapped by receptors of refractive indices 1.37 and 1.41, the former typical of insects and the latter vertebrates.

travels down the rhabdom by total internal reflection at the interface between the relatively high refractive index rhabdom material and the surrounding cytoplasm. The rhabdoms thus behave like the light guides that are familiar from various kinds of optical instrumentation.

Whether or not total internal reflection will occur depends on the angle that the image-forming rays initially make with the wall of the rhabdom. If the angle between a ray and the normal to the rhabdom wall is greater than the critical angle (i_c), then that ray will be internally reflected. The critical angle is the incident angle for which the corresponding angle of refraction into the cytoplasm would be 90°, so that from Snell's Law:

$$\sin i_c = \frac{n_2}{n_1}, \qquad (15)$$

where n_1 is the higher refractive index of the rhabdom, and n_2 that of cytoplasm (about 1.34). The exact value of the refractive index of insect rhabdoms is not known because of formidable difficulties in making the measurements, but it is certainly in the range 1.37–1.40 (Kirschfeld, K. and Snyder, A., 1975), and the corresponding range of values for i_c is 78° to 73°. This in turn means that the cone of rays reaching the distal tip of the rhabdom must be narrower than $2(90 - i_c)$ if it is all to be held within the rhabdom. Thus the rhabdom will accept cones of light up to 24° ($n_1 = 1.37$) or 34° ($n_1 = 1.40$). The practical implication of this is that a high light-gathering power (low F-number) lens is not of much value in an insect eye that relies entirely on total internal reflection to retain light in the receptors, since the more oblique rays will escape, and indeed F-numbers around 2 (cone of light 28°) are typical for most apposition eyes. Superposition eyes often have much lower F-numbers, as low as 0.5, but there it is common for the rhabdoms to be sheathed in a reflecting layer of modified tracheoles which retain the light that enters them (see Miller, W., 1979).

A final consequence of the light guide nature of the rhabdoms is that this offers a potentially powerful, if unusual, way of producing an iris mechanism. If the external refractive index (n_2) can be increased, then i_c increases and the rhabdom's acceptance cone will become smaller. Light can also be "bled" out of the rhabdom at any point along its length (Fig. 19iii). Such a mechanism appears to exist in the "palisade" of mitochondria around the light-adapted rhabdom in locusts (Horridge, G. and Barnard, P., 1965), and in the array of small movable pigment granules around the rhabomeres in flies (Kirschfeld, K. and Franceschini, N., 1969; Franceschini, N., 1975) (see also section 2.2.4).

Besides acting as light guides, rhabdoms can also behave as *wave guides*. When a light-transmitting structure is so narrow that its diameter becomes comparable with the wavelength of light ($\sim 0.5\,\mu m$) light passes down it not so much by discrete reflection at the walls, but as a series of transverse interference patterns known as *modes*. If a rhabdom is more than a few μm wide this phenomenon can be safely ignored, but many are narrower than this, and this led Snyder and his colleagues to a long investigation of wave guide phenomena in photoreceptors during the early 1970s (review, Snyder, A., 1979). By far the most important feature of modal propagation of light, in terms of its consequences for vision, is that not all the energy in a mode travels inside the light guide. As the fibre becomes narrower an increasing amount travels along but outside it, and this has two consequences. Firstly, less light is absorbed by the photopigment, and secondly the fraction of the light energy outside the fibre becomes available for capture by *neighbouring* receptors. Since this light was not imaged on adjacent receptors, this has the effect of spoiling resolution. These considerations lead

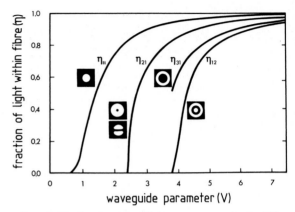

FIG. 15. The fraction (η) of light energy transmitted within the geometrical boundary of a fibre for several low order modes, as a function of the waveguide parameter V (equation 16). The insets give the appearance of the modes. The first curve (left) determines the amount of light in the fibre, so that if $V = 2$ (corresponding to a $0.9\,\mu m$ rhabdom of refractive index 1.39) 75% of the light is within the fibre and 25% outside. (Modified from Kirschfeld, K. and Snyder, A., 1975.)

to two important conclusions. *There is a minimum useful diameter, of about 1 μm, for a light-conducting photoreceptor structure, and similarly there is a limit to the packing density that a retina can have.* The wave guide properties of a rod-like structure are completely specified by the "wave guide parameter", V, which is given by:

$$V = \frac{\pi d_r}{\lambda}\,\sqrt{(n_1{}^2 - n_2{}^2)} \qquad (16)$$

where, as before, n_1 is the index inside and n_2 the index outside the fibre. Taking λ as $0.5\,\mu m$, n_1 as 1.39 and n_2 as 1.34, V is then equal to $2.32\,d_r$. Figure 15 shows how the fraction of light within the fibre varies for different modes, for different values of V, and it is evident that the important part as far as light "leak" is concerned is when only one mode is present and V is between 0 and 3. Thus when V is 1($d_r \sim 0.4\,\mu m$) only 20% of the light is inside the rhabdom, but when V is 2 ($d_r \sim 0.9\,\mu m$) 75% is inside. This finding, that rhabdoms narrower than about 1 μm are unacceptably leaky, coincides well with what is actually found. Probably the densest packing found anywhere is in the neural superposition eyes of the Diptera, where the rhabdomeres are about 1 μm in diameter, with a 1 μm gap between them.

In ordinary apposition eyes the rhabdoms are well separated, so problems of dense packing and consequent cross-talk do not really arise, and in superposition eyes the rhabdoms are usually too wide for this to be important either. However, in simple eyes such as our own it is the attainable packing density that determines the resolving power of the eye and ultimately its size, (equations 1, 2 and 12). For a given resolution an eye must have a minimum size, because there is a minimum diameter a receptor can usefully have.

2.1.4 IMAGE DEFECTS

In large optical structures like camera lenses a great deal of effort is spent in trying to minimize or eliminate the classical geometric image defects: chromatic and spherical aberration for on-axis rays, and these plus coma, astigmatism and distortion for off-axis rays. In the human eye chromatic aberration is important, and spherical aberration is reduced by the fact that the cornea is not actually spherical. However, when dealing with a structure the size of a single facet in an apposition eye these image defects are of little or no importance (McIntyre, P. and Kirschfeld, K., 1982). The reason is simply that the absolute size of the blur circle on the retina introduced by any of these defects is directly proportional to eye size, and in single ommatidia with focal lengths around $100 \, \mu m$ the blur circle is not likely to be greater than the width of a receptor, or indeed than the width of the Airy disc. I calculated elsewhere (Land, M., 1981) that an eye with a 1 cm aperture would have a chromatic blur circle $104 \, \mu m$ across—which would be serious—but with an aperture of $100 \, \mu m$, still larger than most ommatidial lenses, the blur circle would only be $1.04 \, \mu m$ wide. Spherical aberration, even if uncorrected, would be of similar magnitude. It is possible that in some of the simple larval eyes spherical aberration might just be a problem, although there is no direct evidence for this. It could in any case be eliminated, either by the presence of non-spherical surfaces, as Clarkson, E. and Levi-Setti, R. (1975) have suggested were present in Trilobite eyes, and Schwind, R. (1980) has found in *Notonecta* lenses, or by the use of varying refractive indices, as in some spiders (Blest, A. and Land, M., 1977).

It is convenient under this heading to examine the depth of focus of insect eyes, since similar considerations apply. As every photographer knows, depth of focus depends both on the relative aperture (or F-number) of a lens and on its focal length; with low F-numbers and long focal lengths the depth of focus is small. There is some arbitrariness in setting a criterion for an image to be "in focus" but if we set a fairly stringent one, that the geometrical image of a point source should not produce a blur circle larger than twice the receptor diameter $(2d_r)$, then we can at least compare the performance of different eyes.

I have shown elsewhere (Land, M., 1981) that if this condition is satisfied at infinity, then the nearest distance U at which it is also satisfied (i.e. the near point) is given by:

$$U = \frac{fD}{2d_r} \qquad (17)$$

Since f and D are likely to co-vary, this effectively says that the near point recedes as the square of the focal length. Thus for an eye with $2 \, \mu m$ diameter receptors, an F-number of 2, and a focal length of 1 cm, the near point is 12.5 m. However, for a similar eye with a 1 mm focal length it is only 12.5 cm, and for a $100 \, \mu m$ focal length eye (similar to an ommatidium in a large apposition eye) the near point is 1.25 mm. Even this minute distance is probably an overestimate, because of the fact that the Airy diffraction image of a point source is extended along the axis (Kuiper, J., 1966; see also Fig. 10) which in turn means that the range of object distances in external space that gives rise to similar light distributions in the image region is also extended.

The outcome of this discussion is that insect compound eyes, and small simple eyes, have a focus from infinity almost to the eye surface without the need for additional focusing mechanisms of the kind found in the simple eyes of vertebrates and cephalopod molluscs.

2.2 Sensitivity

In section 2 I defined an eye's sensitivity as the size of the photon sample that each receptor receives, using the analogy that an eye can be thought of as a series of photon counters, each sampling contiguous angles in the surroundings (Fig. 8). To make this a proper definition we must make it independent of the amount of light available, strictly the luminance or radiance of the field, and we must also at some stage specify the "sampling time" of a receptor — the period over which photons are counted, equivalent to the spatial sampling angle ($\Delta\phi$) considered in the last section. The sampling time, however, does not come into the definition of sensitivity, since if a receptor receives x photons per second when the surroundings are emitting y photons per square metre per steradian per second, then the sensitivity of the eye is simply x/y, and it does not matter what time period is chosen for the count. This obviously matters for the animal though: if the sampling time is short then the animal can act quickly on fresh information (a chasing fly for example can react to changes in the flight path of its target in about 20 ms, so its sampling time cannot be longer than this). On the other hand, a longer sampling time means a larger, and hence statistically more reliable, sample and one would expect to find nocturnal insects taking a longer time to make visual decisions. How the image is sampled in time is, however, essentially a problem for the animal's nervous system; it is not strictly a property of the eye itself. Our definition of sensitivity (S) is thus:

$$S = F_p/L \qquad (18)$$

where F_p is the photon flux per photoreceptor, and L the luminance or radiance (photons $m^{-2} sr^{-1} s^{-1}$) of the scene the eye is viewing. To give an idea of the numbers involved, a rhabdom in a bee's eye receives about 800 photons per second when the eye is viewing a white card at sunset, which has a luminance of $1\,cd\,m^{-2}$, or about 4×10^{15} photons $m^{-2} sr^{-1} s^{-1}$ for light in the yellow–green region of the spectrum ($\lambda = 555$ nm). The range of luminances encountered in the terrestrial environment is very large, a factor of about 10^{10}, with a white surface in bright sunlight emitting about 10^{20} photons $m^{-2}\,sr^{-1}\,s^{-1}$, about 10^{14} in moonlight, and 10^{10} in overcast starlight. The range of sensitivities, using the definition in equation 18, that is available to animals with compound eyes purely from their optical design is about 10^3 (between bee and moth for example: Kirschfeld, K., 1974), which clearly only goes part of the way towards covering the environmental range.

Sensitivity, as defined here, has the rather unusual units of m^2, because the definition of luminance includes a m^{-2} term. In fact, besides

being the multiplying factor for calculating the number of photons received by a receptor given the source luminance, the sensitivity also has the meaning of the equivalent surface area of the receptor. The sensitivity figure is the area of a flat surface outside the eye which receives the same number of photons as the receptor, when that area is receiving light from the source over a solid angle of 1 steradian. In the example of the bee, quoted above, S in m^2 would be 0.2×10^{-12}. Since compound eyes and their component receptors tend to have dimensions measured in μm rather than m, it is better to express S not in m^2 but μm^2, and dispense with the factor of 10^{12}.

In the sections that follow I will argue that the sensitivity of an eye is of crucial importance for the resolution of detail (Fig. 16), and not just at low light levels where receptors are absorbing only a few photons per second, but right up to daylight conditions. Photon starvation, in other words, is likely to affect eye design in all animals. This will be followed by an account of the features that are likely to make for high sensitivity in eyes, and based on

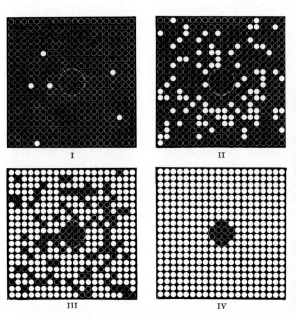

FIG. 16. Diagram representing the image of a black disc on a retina at four different illumination levels. The small circles represent receptors, and those that have received a photon in a fixed time period appear light. The situation in I approximately represents the situation in the human retina at the absolute threshold of vision. The black disc only becomes unambiguously detectable at light levels 2–3 log units higher than this, because of photon scarcity. (Pirenne, M., 1967.)

this a comparison of the sensitivities of different eye types. Under the very brightest conditions the responses of photoreceptors are liable to saturate and not give usable signals, and the final part of this section is devoted to ways in which this is prevented by the use of various kinds of iris mechanism. Other valuable discussions of the sensitivity of insect eyes can be found in Barlow, H. (1964), Kirschfeld, K. (1974) and Snyder, A. (1979). Parallel arguments, derived from consideration of the design of man-made "seeing" devices, are developed by Rose, A. (1973).

2.2.1 THE CONSEQUENCES OF PHOTON NOISE

A statistician, contemplating two sets of numbers and trying to decide whether or not the difference between them is significant, needs to know not only the means of the two sets, but also their standard deviations. Usually, if the difference in the means is greater than, say, 2 SD then the difference can be taken as real, with 95% certainty. Eyes face exactly this problem. If one receptor receives 100 photons in a specified period, and its neighbour receives 110, can this be taken as evidence that there is a real difference of brightness in the outside world between the fields of view of the two receptors? If it can, then this may mean the presence of an edge, for example. Similarly, if there is a significant difference between the counts a single receptor receives in successive time samples, this could be taken as evidence for a change, possibly a movement, in the world outside. This kind of information, though seemingly elementary, is the stuff on which all higher-order decisions of the visual system must be based. The statistical problem for the eye is how to judge the reliability of each count. What is the standard deviation of a single sample of 100 photons?

A single sample in ordinary statistics has no standard deviation, but fortunately the statistics of photon capture permit an estimate of the standard deviation to be made from the mean itself. In this type of random process, where a small sample is taken from a very large universe, Poisson statistics apply, and it is a characteristic of such processes that the standard deviation of a set of samples collected under identical circumstances is equal to the square root of their mean. In other words, a receptor that receives 100 photons can take it that

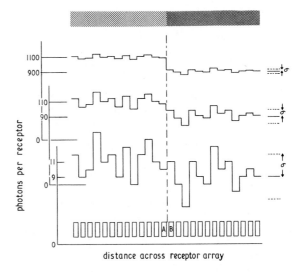

FIG. 17. The effect of light level on the detectability of small contrasts. The abscissa represent distance across the retina, and the three graphs show the magnitude of the variations in receptor photon counts when the average number of photons caught is 10 (bottom), 100, and 1000. The stripe at the top represents a luminance step of 20%. At the lowest level this is completely lost in the photon "noise". With 10 times more light the difference is detectable but only by averaging across the array, and with 10 times more light again it is clearly detectable by the adjacent receptors A and B. σ is the standard deviation of photon counts, and is equal to the square root of the average count. (Land, M., 1981.)

the standard deviation of that sample is $\sqrt{100}$, or 10. Going back to the original question, a SD of 10 is the same as the difference in counts between the two hypothetical receptors (100 and 110) and it would be an unwise statistician — or eye — that would accept this as real. If, on the other hand, two counts had been 100 and 130, there would unquestionably have been a real difference (Fig. 17).

This argument can be put in a general form which makes it possible to estimate how many photons are needed per receptor to perform different detection tasks. Suppose two receptors image adjacent parts of the visual field which differ in brightness. The contrast of the two parts, which defines their relative brightnesses irrespective of intensity, is defined as:

$$C = \frac{I_1 - I_2}{I_1 + I_2} \qquad (19)$$

Given many counts and no optical degradation, the average photon count at the receptor level will have the same ratio, i.e. if the average counts in receptors 1 and 2 are $\bar{x}_1$ and $\bar{x}_2$, then

$$C = \frac{\bar{x}_1 - \bar{x}_2}{\bar{x}_1 + \bar{x}_2} \simeq \frac{\Delta \bar{x}}{2 \bar{x}} \qquad (20)$$

With single samples, however, there will be variation due to photon noise, so that each sample x has a standard deviation of $\sqrt{x}$ and for a pair of such samples the SD of the difference is $\sqrt{[(SD\ x_1)^2 + (SD\ x_2)^2]}$, or $\sqrt{(x_1 + x_2)}$, and if x_1 and x_2 are similar this will be approximately $\sqrt{2\bar{x}}$. Now the difference between a pair of samples, $x_1 - x_2$ or Δx, can be regarded as important if this difference is larger than (say) 2 standard deviations of the difference, i.e. if

$$\Delta x > 2 \sqrt{2\bar{x}}$$

Replacing Δx with its average value, $\Delta \bar{x}$, and dividing by $2\bar{x}$ gives

$$\frac{\Delta \bar{x}}{2\bar{x}} > \frac{2\sqrt{2\bar{x}}}{2\bar{x}}$$

or, since $\Delta \bar{x} / 2\bar{x} = C$, the object contrast (equation 20), we have that C is detectable if

$$C > \frac{2\sqrt{2\bar{x}}}{2\bar{x}}, \text{ which reduces to}$$

$$\bar{x} > \frac{2}{C^2} \qquad (21)$$

In other words, a contrast of 10% (0.1) requires a minimum of about 200 photons per receptor per integration time for it to be reliably detected. A 1% contrast requires 20,000. If the integration time of the visual system is 0.1 s, then a photon count rate of 200,000 per second per receptor would be needed. The important point about this calculation is that numbers of this size are towards the upper end of the range of photon rates available to receptors, even in bright daylight conditions (recall that a bee rhabdom at sunset receives only 800 photons per second). This in turn leads to two conclusions of general importance. First, the ability of an eye to resolve contrasts in the environment is limited by the noisiness of photon samples, at *all* light levels; and second, that a major consideration in the design of *all* eyes is that their optics should provide as much light as possible for the receptors to work with.

It is important to establish that insects really do detect small contrasts, and that the numbers of photons involved in the process of seeing are similar to those predicted by equation 21. Fermi, G. and Reichardt, W. (1963) found that in houseflies the optomotor response — the tendency of animals to rotate with their environment which experimentally is usually a striped drum — is sensitive at high light levels to contrasts of about 10%. They also found that the contrast the fly could detect was approximately proportional to the square root of the luminance of the drum, which is what equation 21 would predict. More recently Dubs, A. *et al.* (1981) directly measured the photon capture rate in fly photoreceptors at the behavioural threshold of the optomotor response. They did this by measuring the photon "bumps" of a few millivolts that are produced in insect photoreceptors when they absorb single photons (Lillywhite, P., 1977). The result was that when the fly would just respond to a grating with a contrast of 1 (or 100%), individual receptors were absorbing photons at a mean rate of 1.7 per second. If the fly's integration time is 1 s, and pairs of receptors are the ultimate movement-detecting elements (see Kirschfeld, K., 1972), then equation 21 predicts that the average photon capture rate should be 2 per second. There is thus clear evidence that insects construct their visual behaviour on the basis of single-photon events, and at higher light levels on their statistical properties.

In the next section we shall consider the ways eyes can be designed to maximize the numbers of photons captured.

2.2.2 THE LIGHT-GATHERING POWER OF EYES

I am going to cheat at this point, and write down an expression for the amount of light captured by a receptor (the sensitivity S in equation 18), and justify it later.

$$S = \frac{F_p}{L} = \left(\frac{\pi}{4}\right)^2 \times \left(\frac{D}{f}\right)^2 \times d_r^2 \times (1 - e^{-kl}) \qquad (22)$$

The definitions of the terms are the same as in the section on resolution (section 2.1) and in Table 2. A full photometric derivation of the equation is given in Land, M. (1981, p.480), and here I will try to explain rather than derive the expression (see Fig. 18).

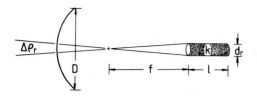

FIG. 18. Features that determine the sensitivity of an eye. See equations 22 and 23.

In photography, two lenses that have the same F-number (f/D) produce the same intensity (strictly illuminance) on the film plane, independent of their focal length. An annoying feature of F-numbers, however, is that they do not correspond directly to film plane illuminance. As this goes up by factors of 2, the F-numbers come down through a series 11, 8, 5.6, 4, 2.8, 2, ..., etc. It is easy to see that the film plane illuminance is actually proportional to (1/F-number)^2 or D^2/f^2. The same is true for eyes. The second term in equation 22 is thus the retinal illuminance. The third term, d_r^2, expresses the obvious fact that the amount of light a receptor absorbs is proportional to its cross-sectional area ($\pi d_r^2/4$). This accounts for one of the ($\pi/4$)s at the beginning, the other arises because the aperture of the eye is also circular, not square. The final term is the proportion of the light entering a receptor that is actually absorbed, and this is a function of two variables, the length of the receptor (l) and the amount of light it absorbs per unit length, taken here to be the natural extinction coefficient (k). For animals with microvillous receptors, including insects, this is about 1% per μm for the wavelength of maximal absorption. [In the lobster, whose rhabdoms are large enough for accurate photometry, Bruno, M. et al. (1977) found a value for k of 0.0067 μm^{-1}, and insect rhabdoms are likely to be similar.] By contrast, vertebrate rods are optically denser, presumably because there is less non-membrane space in their disc structure, and values for k are around 0.03 μm^{-1}. The expression is exponential rather than linear because absorption is multiplicative — if, say, 50% of light is absorbed in the first 100 μm, then 50% of what remains is absorbed in the next 100 μm, and so on. Total absorption is strictly speaking impossible. Using the figure from Bruno, M. et al. (1977) of 0.0067 μm^{-1}, we find that a 100 μm long receptor absorbs 49% of the incident light, a 200 μm receptor 74%, 500 μm 96%

and 1 mm 99.9%. Bee rhabdoms are about 250 μm long, and those of some dragonflies approach 1 mm. Clearly insect rhabdoms of these lengths will be absorbing most of the available photons.

2.2.3 RELATIVE SENSITIVITY CALCULATIONS

The full expression in equation 22 is the sensitivity S, and as mentioned in section 2.2 this is the multiplying factor for converting the luminance of the scene into the number of photons absorbed by single receptors. In the case of a bee ommatidium, $D = 25\,\mu$m, $f = 60\,\mu$m, $d_r = 1.5\,\mu$m and $l = 250\,\mu$m (Wehner, R., 1981). From equation 22, the value of S is 0.20 μm^2, or 0.2×10^{-12} m^2. Thus in bright sunlight (10^{20} photons m^{-2} sr^{-1} s^{-1}) bee rhabdoms receive 2×10^7 photons per second, which from equation 21 would be enough to detect contrasts of 0.03%, a figure far smaller than a bee would be likely to need. However, in bright moonlight ($\sim 10^{14}$ photons. m^{-2} sr^{-1} s^{-1}) this number reduces to 20 photons per second per rhabdom, which is only just above the behavioural threshold for the fly optomotor response; it is not surprising that bees do not fly in moonlight. Moths, however, do; and the reason is that equation 22 predicts for them a sensitivity about 1000 times greater than that for the bee.

For the moth *Ephestia*, with a superposition eye which gives a wider effective pupil than the apposition eye of a bee ommatidium, the corresponding values of the eye dimensions are: $D = 400\,\mu$m, $f = 170\,\mu$m, $d_r = 8\,\mu$m and $l = 55\,\mu$m (data from Table 3). Because there is a tapetum behind the rhabdoms, the effective rhabdom length can be doubled (110 μm) as light passes through it twice. Putting these values into equation 22 gives a figure for S of 114 μm^2, which is larger than that of the bee by a factor of 570 — nearly 3 log units. This should mean that a moth rhabdom will be able to operate with the same reliability as a bee rhabdom, but at light levels three orders of magnitude lower. A moth in moonlight would receive the same number of photons per receptor as a bee in room-light, for example. This result comes from purely optical considerations. If, as may indeed be the case, the visual system of a moth is slower than that of a bee, so that the integration time is longer by a factor of, say, 10, then this will extend the moth's visual working

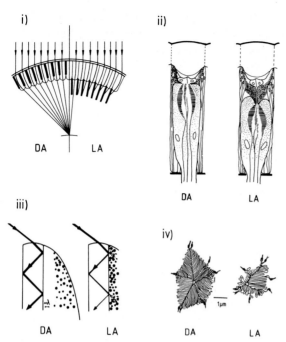

FIG. 19. Light-reducing mechanisms in insect eyes. **i**: In superposition eyes inward radial movement of pigment cuts off oblique rays. (Modified from Exner, S., 1891.) **ii**: Iris around the tip of the rhabdom in the hemipteran *Lethocerus*. (From Walcott, B., 1975.) Besides the reduction of the pigmented aperture of the rhabdom, there are gross movements of both retinular and pigment cells. **iii**: Light bleeding mechanism in Diptera. In the light small high refractive index pigment granules close around the rhabdom, preventing total internal reflexion (Fig. 14). (Kirschfeld, K. and Franceschini, N., 1969.) **iv**: Reduction in the amount of receptor membrane. Drawings from electron micrographs of rhabdoms of the locust *Valanga* (same animal, fixed in the dark at night, and the other eye in the light during the day) (Horridge, G. and Blest, D., 1980.) In all four figures DA is the dark-adapted, an LA the light-adapted state.

range downward by another log unit. One might thus expect the lower limits of intensity at which bees and moths are active to differ by about 10^4, which corresponds quite well to what one knows about the lifestyles of the two animals.

Receptors themselves have working ranges of between 3 and 5 log units (Laughlin, S., 1981) which would cover roughly the range we think of as daylight. Optical adaptations like those of moths can extend the whole range downward into the twilight and moonlight range, but this means that in daylight conditions the eye actually needs protection from saturation, which would result in a loss of usable signals from the receptors. Iris mechanisms are thus needed, both for nocturnal insects, and

even for diurnal insects — especially when their facets point directly at the sun.

2.2.4 IRIS MECHANISMS

The human eye has an iris which effectively changes the lens diameter D in equation 22, from about 8 mm in the dark to 2 mm in the light; a decrease in image brightness of 16 times. In insect eyes there are no examples of this kind of mechanism operating on the corneal lenses themselves (except in certain ocelli; Wilson, M., 1975). There are, however, several other kinds of pupil (excluding the pseudopupil, which is not a pupil at all in sense of a retinal intensity changing device, see section 3.2). These are illustrated in Fig. 19. They are:

(i) Radial pigment movements in superposition eyes, which change the effective diameter of the superposition pupil. These will be discussed in detail later (section 5.3), but they have the effect of cutting off the rays that cross the clear zone obliquely, restricting the superposition pupil to one or a few facets. This mechanism is potentially of great power. In the moth *Ephestia* the open (night-time) pupil covers about 180 facets (Kunze, P., 1979). If this area is reduced to the diameter of a single facet, retinal illuminance will drop by a factor of 180, or in practice slightly less because the outer facets contribute less than the central ones even in the open pupil. However, this type of pupil should be able to produce at least 2 log units of protection for nocturnal eyes that are exposed to daylight. The other three mechanisms operate not on the pupil, but on the receptors.

(ii) A "throttle" at the distal tip of the receptor. In many insects with apposition eyes pigment cells around the tip of the rhabdom can contract to form a narrow sphincter, which effectively reduces the diameter d_r of the rhabdom (equation 22). In the giant water-bug *Lethocerus* the pigment cells above the rhabdom tip form a narrow channel only 5 μm wide in the light, but retract in the dark to give an aperture 20 μm in diameter, almost the width of the unusually wide rhabdom (Walcott, B., 1975). This alone should give a sensitivity change of 16 times, but this could easily be greater depending on how much light is absorbed in the 30 μm long pigment channel of the light-adapted eye (see also McLean, M. and Horridge, G., 1977). This type of mechanism is essentially similar to the iris mechanism of the king-crab *Limulus* (Fahrenbach, W., 1975), and it is of most value in apposition eyes with relatively large rhabdoms. A bee's 1.5 μm rhabdom leaves little room for further constriction. An important point about an iris which is actually at the focus of the corneal lens is that it acts not only as an aperture stop, but also as a field stop. Closing it reduces the light entering the rhabdom largely because it reduces the field of view of the rhabdom. The acceptance angle $\Delta\rho_r$ of each rhabdom should decrease as the effective diameter of the rhabdom decreases (equation 6). This is indeed what happens. In *Lethocerus* Walcott, B. (1975) found a decrease in acceptance angle from 9° in the dark to 3.5° in the light. This mechanism should improve contrast transfer (though not the resolution limit which is set by ommatidial spacing: equation 1) as well as reducing sensitivity.

(iii) Bleeding light from receptors. A more subtle type of pupil, found in diurnal insects with small rhabdoms, involves the removal of light from the rhabdoms themselves. In all cases studied this is achieved by a mechanism that brings high refractive index material into contact with the rhabdom itself, weakening or destroying the refractive index step at the junction of the rhabdomere and its cell body, preventing it from operating as an effective wave guide and allowing light to "bleed" out into the surrounding pigment. In locusts the structure that accomplishes this is a palisade of mitochondria (Horridge, G. and Barnard, P., 1965), and in flies it is a system of small (0.1 μm) pigment granules. Butterflies are similar (Ribi, W., 1978). Franceschini, N. (1975) has photographed this phenomenon in flies, and one interesting feature is its speed. The pigment migration has a time constant of 2 s, and is complete after 10 s. It appears to operate at the highest light levels, where the receptors themselves are near the top of their working range.

(iv) Change of rhabdom size. A final mechanism, which is perhaps less concerned with sensitivity adjustment than with protection of photopigment, has recently been found in a number of arthropod groups. This consists of the large-scale destruction and removal of photoreceptor membrane from the rhabdom during the day, and its resynthesis and redeployment at night. The extent of this phenomenon in insects is not fully calculated yet, but in locusts, Williams, D. (1982) found a 4.7-fold increase in rhab-cross-sectional area between day and night. See also Horridge, G. and Blest, D. (1980).

This brief survey of mechanisms for reducing the sensitivity of eyes is probably not exhaustive. In particular there seems to be a wide variety of changes in the relative sizes and positions of the ommatidial components that accompany adaptation changes, but whose functions are not clear. Useful reviews are given by Walcott, B. (1975) and Stavenga, D. (1979).

2.2.5 THE CONFLICT BETWEEN RESOLUTION AND SENSITIVITY

An eye of a particular size, whether compound or simple, can have high resolution or high sensitivity, but not usually both. The reason for this is that the sensitivity equation (22) actually incorporates a term related to resolution, the acceptance angle (d_r/f, or $\Delta\rho_r$) of the receptors. It can be rewritten:

$$S = (\pi/4)^2 \times D^2 \times \Delta\rho_r{}^2 \times (1 - e^{-kl}) \qquad (23)$$

This means that S and $\Delta\rho_r{}^2$ are directly related, for an eye of a given aperture diameter D, and so there is a conflict.

This is perhaps best illustrated by trying to design eyes with doubled resolution and doubled sensitivity (Fig. 20). In the first case, resolution is doubled if $\Delta\rho$ is halved, which could be done by halving d_r ($\sim d_{cc}$) or doubling the focal length f. Ultimately d_r must reach its minimum value of about 1 μm (see section 2.1.3) which only leaves the latter option. Doubling f on its own would, however, reduce sensitivity, since D^2/f^2 would decrease, and there must be an increase in D that matches the increase in f. D and f must thus both double, and the whole eye must be scaled up isomorphically (except for d_r) in proportion to the resolution required, if

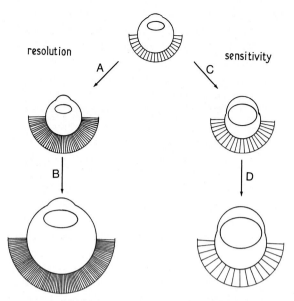

resolution

sensitivity

FIG. 20. Ways in which resolution and sensitivity can be increased. **A:** Increase in resolution by decreasing receptor diameter. Receptors increase in length to maintain sensitivity. **B:** Once receptors have reached their minimum diameter resolution can only be improved by increasing the size of the whole eye. **C:** Sensitivity increase without resolution loss can be achieved initially by increasing lens size and lengthening receptors. **D:** Thereafter the only possibility is to scale up all eye dimensions. Similar considerations apply to compound eyes, although in apposition eyes size must increase as the square of resolution (see section 2.1.2).

sensitivity is not to be sacrificed. In diffraction-limited apposition eyes (only) the situation is even worse, because D must increase for other reasons (section 2.1.2), and eye size may have to increase as the square of resolution (equation 11).

Doubling sensitivity can be achieved by doubling D^2 (i.e. increasing D by $\sqrt{2}$ or by halving $\Delta\rho_r^2$ in equation 23). If $\Delta\rho_r$ is fixed, however, only D may be increased. The problem with this is that D cannot be increased beyond a maximum of about $2f$ (an F-number of 0.5), and when that limiting condition is reached the only way to increase sensitivity without compromising $\Delta\rho_r$ is to increase D, f and d together by the same amount; in the case of doubling sensitivity they must all go up by $\sqrt{2}$.

The outcome of this discussion is that both resolution and sensitivity depend on eye size, which in a small animal can be thought of as a finite resource to be competed for. It means, for example, that if two insects have eyes of similar sizes, but one is diurnal and the other is nocturnal, the nocturnal animal is bound to have worse resolution. The change from apposition to superposition optics helps a little, by virtue of the difference in F-numbers (2 and 0.5: an increase in image brightness of 16 times), but the main difference lies in the larger receptors of nocturnal insects, resulting in high values for S but also large values for $\Delta\rho_r$.

3   APPOSITION COMPOUND EYES

An outline of the properties of an apposition eye has already been given in previous sections, and here I intend only to elaborate on some of the specializations unique to this type of eye. To recapitulate briefly: apposition eyes are the commonest type of eye in adult insects (see Table 1) and are mainly but not exclusively found in diurnal insects (the crepuscular dragonfly *Zyxomma* is an interesting exception: Horridge, G., 1978). Each ommatidium has a lens, a short clear region of crystalline cone cells behind it, and a fused rhabdom composed of the microvillous contributions of eight or nine receptor cells. The distal tip of the rhabdom lies at the focus of the lens, and light which enters a rhabdom is retained in it by virtue of its higher refractive index, and hence light-guiding properties. The receptors which make up the rhabdom may have different spectral sensitivities (up to four in some butterflies; Matić, T., 1983), and may be sensitive to light polarized in different planes (see Waterman, T., 1981; Wehner, R., 1981) but they all share the same field of view. Each ommatidium is optically isolated from its neighbours, both by the light guide properties of the rhabdom, and by the screening pigment around it.

The numbers of ommatidia vary enormously, from eight or fewer per eye in Collembola (Paulus, H., 1975) to as many as 28,000 in some dragonflies (Odonata). A worker bee has about 5000 (Mazokhin-Porshnyakov, G., 1969). Corresponding inter-ommatidial angles are 20–60° in *Podura*, and minimum of 0.24° in the foveal region of *Aeschna* (Sherk, T., 1978) and 1.4° in the bee (Wehner, R., 1981).

3.1   Lenses and lens cylinders

In the majority of apposition eyes, image formation

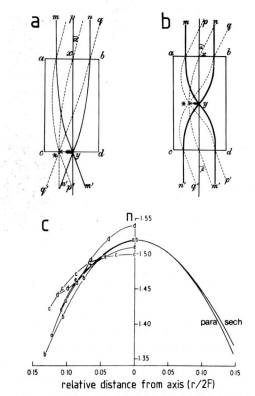

FIG. 21. Lens cylinders **a:** and **b:** Exner's diagrams of ray
paths through a focusing lens cylinder (**a**) of the type found
in the apposition eye of *Limulus*, and an "afocal" lens cylinder (**b**) found in most superposition eyes. Notice that the lens
cylinder in (**b**) is basically the same as (**a**) but twice as long.
c: Measured (left) and theoretical (right) refractive index
gradients in lens cylinders. Measured values are for a–a, a
euphausiid shrimp; b–b, a firefly; c–c, a moth; d–d, a skipper
butterfly; e–e, *Limulus*. The curves have been normalized by
taking the cone length as 2F for the superposition eyes (a–d),
and F for the apposition eye of *Limulus*. Data from various
sources; see Land, M. (1980). The theoretical curves show the
hyperbolic secant function of Fletcher, A. *et al.* (equation 24)
and the parabola of Exner (equation 25).

by each facet lens is adequately explained on the
basis of refraction at the outer surface of the cornea,
a curved interface between air and the higher refractive index body tissues (e.g. Kuiper, J., 1966). A
good approximation of the focal length can be obtained from the single spherical surface formula
(equation 3: $f = r/(n-1)$), taking a value of n
around 1.4, between that of the chitinous cornea
($\sim$ 1.5) and the fluids of the eye ($\sim$ 1.33). There are,
however, some cases where this cannot be the main
mechanism of image formation, either because the
cornea is flat, as in some mantids (Horridge, G. and
Duelli, P., 1979) or because the animal is aquatic,

and so does not have a usable refracting corneal
surface. Image formation must then be performed
by other refracting surfaces — the inner surface of
the cornea for example — or by the alternative
mechanism of the "lens cylinder", first proposed by
Exner, S. (1981) (Fig. 21). Although lens cylinder
optics have not been conclusively demonstrated in
insect apposition eyes, there are plenty of examples
from other arthropods, the most notable and well
studied being in the lateral eyes of the xiphosuran
Limulus.

   Limulus lives in water and has a flat cornea,
precluding spherical surface optics. Exner showed,
however, that each corneal facet produces a small
inverted image, close to the proximal end of each
crystalline cone, which in *Limulus* is an inward-
pointing papilla physically continuous with the cornea itself. Thus each lens is effectively a flat-ended
cylinder. Exner showed that it is possible for such a
cylinder to produce an image if there is an
appropriate refractive index gradient within it, the
index being highest along the axis and falling off
approximately parabolically towards the circumference. If one imagines a ray parallel to, but at a
distance from, the axis striking the end of such a
cylinder, it will encounter a higher refractive index
towards the axis, and so will be bent in that direction. Towards the periphery, where the gradient is
steepest, the rays will be bent the most. Ray bending
will continue until the axis is reached, and if the
gradient is the correct one, all initially parallel rays
will meet at this point, producing an image (Fig.
21a). The whole device is very similar to a simple
converging lens in its imaging properties. In one
particularly telling experiment, Exner cut parallel-
sided sections of *Limulus* crystalline cones, and
showed that they too acted as lenses, though of
longer focal length than the whole cone. Image
formation thus had to be a property of the cone's
internal optical structure, and not a result of the end
curvatures.

   The image-forming properties of lens cylinders
were studied theoretically in the 1950s by Fletcher,
A. *et al.* (1954). They found that a cylindrical structure will produce an image at a distance F from the
first face if it contains an axial gradient of the form:

$$n_r = n_0 \, \text{sech} \, (\pi r/2F) \qquad (24)$$

where n_r is the refractive index at a distance r from

the axis, and n_0 is the axial index. Exner himself believed that the refractive index gradient should be parabolic, and in fact the expression:

$$n_r = n_0 [1-\tfrac{1}{2} (\pi r/2F)^2] \qquad (25)$$

which yields a parabola, is a very close approximation to the hyperbolic secant function of Fletcher *et al.* (the latter contains further terms containing r^4, r^6, etc., of decreasing importance). For practical purposes either expression will yield an image-forming lens cylinder (Fig. 21c), and parabolic gradient glass and plastic lens cylinders are now manufactured commercially (see Land, M., 1980 and Fig. 29).

Exner was not able to determine the refractive index gradient in *Limulus* lenses directly, but the advent of interference microscopy has made this possible. The crystalline cones do have a parabolic refractive index profile (1.51 axially falling to 1.42 peripherally) and their image-forming properties correspond closely to those predicted from equation 24 (Land, M., 1979). The magnification of the image also corresponds well to that expected from the focal length of a lens cylinder (f, the distance from nodal point to image) which Fletcher, A. *et al.* (1954) give as:

$$f = 2F/\pi n_0 \qquad (26)$$

There is thus no doubt that lens cylinder optics as proposed by Exner is used to produce images in some apposition eyes. How many insects use it as an alternative to corneal refraction, however, still remains to be determined, but it is a likely candidate mechanism for aquatic forms.

The real importance of lens cylinder optics will emerge later, in connection with superposition eyes (section 5.1), where the main refracting elements are always lens cylinders of the kind shown in Fig. 21b.

3.2   The pseudopupil

In many apposition eyes which are not too deeply pigmented it is possible to see a small black spot which has the alarming property of moving round the eye as the observer moves around the insect. The term pseudopupil, coined by Leydig, F. (1855), is apt because although the black spot resembles the pupil of a vertebrate eye, it is not a pupil in the sense of a light-controlling mechanism. The phenomenon

arises because there must be some part of an apposition eye that is looking at the observer, and since this region will be absorbing light from the observer's direction, it must appear dark. Thus the pseudopupil marks the point on the eye that shares a common line of sight with the observer. This makes the phenomenon of great value in the study of apposition eyes, since the pseudopupil can be used to examine the way the outside world maps onto the ommatidial array of the eye, and to study in detail the eye's resolution in different regions (see Horridge, G., 1978). The subject of pseudopupils has been reviewed recently (Stavenga, D., 1979) and here I will only explore some of the more interesting and useful aspects of the phenomenon.

The simplest kind of pseudopupil is seen in grasshoppers, for example, and takes the form of a single black spot (Fig. 22a). Even when viewed from a distance with a narrow aperture it is considerably larger than a single facet. This does not necessarily mean that more than one facet shares each direction of view; it is because each facet lens images not only the rhabdom, but also the dark pigment around the rhabdom tip and the proximal end of the crystalline cone. The pseudopupil can be thought of as the image of the rhabdom and the dark structures around it. If a wide aperture is used to look at the pseudopupil, it is found that the best focus of the black spot is not at the cornea, but deep within the eye, at the local centre of curvature. The explanation of this is that the virtual images of the rhabdoms and associated pigment must be located along the axes of the ommatidia, and these axes coincide at the centre of curvature (Fig. 23a). Consequently, when one examines the eye with a lens whose aperture subtends an angle many times that of an ommatidium ($\Delta\phi$), what one observes are the superimposed virtual images of many rhabdom tips and encircling pigment. This can often be valuable, for example in studying the visual pigment in the rhabdoms using reflected or transmitted light, because this *deep* pseudopupil contains the pooled images of many rhabdoms, and hence provides a stronger signal for spectrophotometry (see also section 4.1). The pseudopupil is not always round. In grasshoppers, butterflies and bees it is only circular at the top and bottom of the eye; around the eye's equator it is quite clearly elongated vertically (Fig. 22a, c and d). This implies that a larger amount of

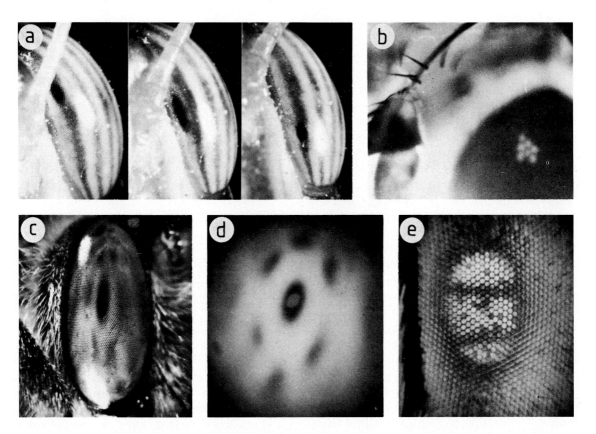

FIG. 22. Pseudopupil phenomena. **a:** Pseudopupils in the locust *Schistocerca*, photographed on the eye's equator (middle) and about 20° above and below it. The vertical elongation of the pseudopupil indicates greater vertical than horizontal resolution. **b:** Antidromic pseudopupil in *Drosophila*, produced by illuminating the eye from behind. Compare the geometry with Figs 4b and 24a. **c:** Principal and secondary pseudopupils in the eye of a bee. The eye shows a similar asymmetry to the locust in **a**. **d:** Luminous principal pseudopupil in a pierid butterfly. The central bright spot is a deep red colour, and is due to light reflected from the tapetum at the base of each rhabdom. Note the secondary pseudopupils as in **c**. **e:** Luminous pseudopupil in the nymphalid butterfly *Heteronympha*, photographed at the level of the cornea using an ophthalmoscope. The two dark stripes are lines 10° apart in the field of view, and their (apposition) image is made visible by the reflecting tapeta. The darker facets have red tapeta, the others green.

eye surface is devoted to a given vertical angle than to the same angle in the horizontal plane, i.e. that the packing density of ommatidia is greater vertically than horizontally (it might also mean that the pigment distribution around the rhabdoms is asymmetric, but I know of no examples of this). This equatorial region can be examined by rotating the animal and measuring $\Delta\phi$ using the pseudopupil (if one rotates the animal through 15° and the centre of the pseudopupil moves across five facets then $\Delta\phi$ is 3°). My own measurements on *Schistocerca* confirm those of others: the vertical inter-ommatidial angle is about 0.6° and the horizontal angle 1.8°. It is interesting that this distortion of the projection of the ommatidial array is not reflected at all in the

facet pattern, which remains a regular pattern of hexagons. Presumably the effect of this increased vertical density of sampling stations is to magnify the animal's horizon. I have used this example to illustrate the value of the pseudopupil as a tool for studying insect eyes: its shape can suggest complications in the way the eye's receptor mosaic maps on the world, and precise measurements of pseudopupil position can be used to verify the clues suggested by qualitative observations.

In addition to the dark *principal* pseudopupil, just discussed, there are a number of interesting variations in pseudopupil form. In butterflies of most families (except the Papilionidae) the centre of the pseudopupil shines brightly (Fig. 22d and e), and usually with a variety of colours, when illuminated from directly on axis (Miller, W. and Bernard, G., 1968; Ribi, W., 1979; Miller, W., 1979; Stavenga, D., 1979). This remarkable and very beautiful phenomenon is caused by the presence of a mirror-like tapetum, made from ¼-wavelength plates of chitin in modified tracheoles which lie immediately below the rhabdoms (Fig. 36a). Light passing through the rhabdoms is reflected back out, coloured by the tapetal mirrors which act as interference reflectors (see Land, M., 1972). In most butterflies the tapeta at the top of the eye reflect blue, and at the bottom red, with intermediate colours in between. This phenomenon, referred to as a *luminous* pseudopupil, has two valuable experimental attributes. Firstly, because the reflected light emerges only from the rhabdom (acting as a light guide) the luminous part of the pseudopupil is the virtual image of the rhabdom itself and can thus be easily distinguished from the dark colour of the surrounding screening pigment. Secondly, and more important, the light reflected from the basal tapetum has passed through the rhabdom twice, and any measurable changes in its intensity or spectral composition must result from changes in the way the rhabdom itself absorbs light. Changes in intensity can occur either from a "light-bleeding" pupil mechanism (a true pupil; see Fig. 19iii) or from the conversion of the rhabdom's rhodopsin to metarhodopsin. In a recent elegant study, Bernard, G. (1979) used both these effects to show that many butterflies possess red-absorbing receptors, a conclusion subsequently shown to be correct by Matić, T. (1983) by intracellular recording.

If light is shone into an insect's head from behind or below, the proximal tips of the rhabdoms are illuminated, and after travelling up the rhabdom light guides the light is re-emitted from their distal tips. Again, the pseudopupil appears as luminous but this time because of "antidromically" transmitted, rather than reflected light (Fig. 22b). These *antidromic* pseudopupils have been used extensively to study the behaviour of photopigments *in situ*, as well as the light-adjusting effect of true pupil mechanisms that remove light from rhabdoms (Franceschini, N., 1975). These studies have centred on dipteran flies, and will be considered again in section 4.1.

A final variation of the pseudopupil is the

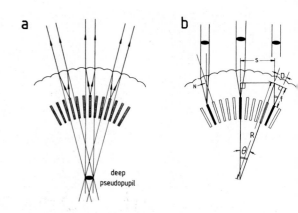

FIG. 23. Geometry of the pseudopupil. **a:** The virtual images of the rhabdom endings and associated pigment all superimpose at the local centre of curvature of the eye, which is where the principal pseudopupil is located. **b:** Secondary pseudopupils occur in some eyes when the lens of one ommatidium images the rhabdom structures of adjacent ommatidia. In the diagram $s/R = \sin\theta$, and $D/f = \tan\theta$, so that for small angles $s/R = D/f$ (equation 27); s is the distance of a secondary pseudopupil from the principal pseudopupil, and N shows the position of the nodal point of the corneal lens. Other symbols as in Table 2.

presence in many dragonfly, butterfly and bee eyes of a hexagonal arrangement of spots surrounding the principal pseudopupil but at some distance from it (Fig. 22c and d). These spots, or *accessory* (neben) pseudopupils, occur when the lens of one ommatidium images not only its own rhabdom and pigment, but that of its nearest neighbours too (Fig. 23b). These accessory pseudopupils also have interesting and useful geometrical properties.

It can be seen from Fig. 23b that the angle between the principal pseudopupil and accessory pupil, subtended at the local centre of curvature of the eye, is almost the same as the angle between rhabdoms, subtended at the nodal point of a single facet. Thus, if the measured separation of the pseudopupils is s the following approximate relation holds:

$$s/R = D/f \qquad (27)$$

The distribution of accessory pseudopupils is often asymmetric, with the hexagonal pattern elongated vertically when the eye is viewed from the side. This usually means that the radii of curvature of the eye are different in vertical and horizontal planes (smaller in the latter) and this in turn means that the interommatidial angle ($\Delta\phi = D/R$) is greater in the horizontal than the vertical plane. Thus the distribution of accessory pseudopupils can confirm

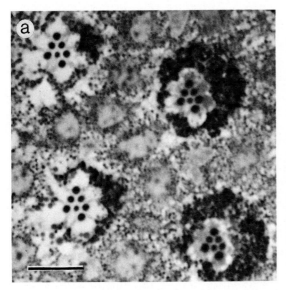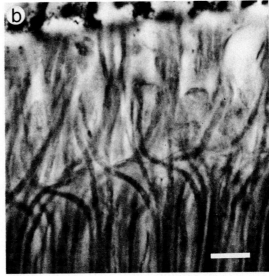

FIG. 24. Neural superposition in the dipteran retina. **a:** Characteristic pattern of rhabdomeres in four ommatidia from the eye of *Calliphora*. Light micrograph taken just at the distal tips of the rhabdomeres (compare Figs 4b and 22b). **b:** Part of the axon crossover pattern between retina and lamina, responsible for channelling signals from all receptors that image a single point onto one laminar cartridge (see Fig. 7b). (Scale bars = 10 μm.)

what can be learned from the form of the principal pseudopupil. Notice that the focal lengths of the ommatidial lenses can be found directly from equation 27 and the inter-ommatidial angle, which itself can be measured using the pseudopupil. Replacing R with $D/\Delta\phi$ in equation 27 gives:

$$f = \frac{D^2}{\Delta\phi s} \qquad (28)$$

I hope this section has made clear that far from being just an optical curiosity, pseudopupils can be of very great value in studying the way apposition eyes are designed. In particular they have shown that the majority of apposition eyes do not map uniformly onto the surroundings, but that there are considerable variations in the value of $\Delta\phi$ across the eye, in many cases justifying the use of terms like "fovea" to designate regions of increased acuity (Horridge, G., 1978; Sherk, T., 1978). I will return to the question of "foveas" in connection with fly eyes (section 4.3) which show the phenomenon to the greatest degree.

4 NEURAL SUPERPOSITION EYES

4.1 Organization of dipteran eyes

The eyes of dipteran flies are very similar in their optical construction to ordinary apposition eyes, with each corneal lens forming an inverted image at the distal tips of the receptors. The difference lies in the fact that the photoreceptive rhabdomeres of each receptor are not fused together to form a single rhabdom (as in Hymenoptera, Odonata, Orthoptera, etc.), but are separate throughout their length, and most importantly they terminate separately in the focal plane of each lens (Dietrich, W., 1909; Vigier, P., 1909; Trujillo-Cenóz, O. and Melamed, J., 1966; Kirschfeld, K., 1967). In *Musca* and other brachyceran flies there are seven receptor endings in the focal plane, each about 1 μm wide, and separated from each other by about 1 μm. There are in fact eight receptors in each ommatidium, but the rhabdoms of the central pair (R7 and R8) lie one above the other, so that only the tip of rhabdomere 7 lies at the focus. The pattern made by the receptor endings is very characteristic (Figs 4b, 22b, 24a). The outer, slightly wider rhabdomeres (R1–6) form a pattern that is hexagonal, except that one member of the hexagon is missing, and an extra rhabdomere (R3) lies outside the hexagon centred on R7 and R8.

The problem raised by this "open-rhabdom" construction is that each ommatidium is capable of resolving the inverted image produced by each lens, and the historic question was: "does it do so, and if so how is this inverted image integrated into the overall upright image?" These questions were almost answered by Vigier, P. (1909), whose studies were rediscovered more than half a century later after the problem had been solved for the second time (Kirschfeld, K., 1967). Wehner, R. (1981) has reviewed the history of the matter. The problem is resolved by two facts. First, the direction of view of each of the rhabdomeres R1–6 is the same as the direction of view of the central rhabdomere (R7/8) in an *adjacent* ommatidium (in the case of rhabdomere 3 it is the adjacent-but-one ommatidium). This means that each point in space is sampled by seven rhabdomeres in seven different ommatidia. It does not mean that there is additional resolution within the visual field of each ommatidium. The second point is that the axons of the photoreceptors leaving each ommatidium do not run to the laminar cartridge immediately beneath that ommatidium, but go instead to the cartridges of adjacent ommatidia, specifically to the cartridges of the ommatidia *whose field of view they share* (Braitenberg, V., 1967). Receptors 7 and 8 are the exceptions, being the central pair, but their axons actually run straight through to the medulla. The redirection of the axons involves a rather complicated re-wiring between the retina and lamina, as shown in Fig. 24b.

The upshot of this peculiar organization is that each laminar cartridge (the first synaptic relay) receives receptor terminals from six receptors which all have the same field of view, and consequently, as far as the lamina and subsequent optic neuropils are concerned, the arrangement of the image is exactly the same as it would be in an ordinary apposition eye. Kirschfeld, K. (1973) used the term "neural superposition" for this type of eye to indicate that the image at the laminar level is reconstructed by the neural pooling of signals from separate ommatidia; this distinguishes these eyes from optical superposition eyes (section 5) where the summation of signals is brought about optically at the receptor level. This is a different principle altogether.

Many of the optical studies that substantiate the mechanism of neural superposition eyes have been reviewed in detail by Franceschini, N. (1975). The various forms of pseudopupil have been of particular value. The pseudopupil in Diptera, being a coherent superposition of the images of the receptor tips and adjacent structures as in ordinary apposition eyes (section 3.2), has the same appearance as the layout of the receptor endings in each ommatidium (Fig. 22b) with the typical 6 + 1 arrangement. The eye can be observed either by incident light (orthodromic) or by light shone into the head (antidromic) which gives a pseudopupil, that appears to glow from within. Because the receptors of the R1–6 and R7/8 systems are separately visible in the pseudopupil they can be studied separately. For example, both systems have a (real) pupil system of the "light-bleeding" type (see Fig. 19iii), which causes the appearance of the rhabdomeres to change from deep red (in *Drosophila*) to a greenish colour as the ommachrome pigments in the receptors reach the sides of the rhabdomeres. However, the R1–6 system and the R7/8 rhabdomeres have different thresholds for activation of the pupil. It takes intensities 10–100 times greater to make the R7/8 rhabdomere go to its green (closed) state than for the R1–6 system, implying that the two systems have different working ranges — R1–6 for dimmer, and R7/8 for brighter, conditions.

4.2 Sensitivity gain

One may well ask why the Diptera have gone to such extraordinary lengths, in terms of neural organization, to secure a type of image which at the level of the lamina seems to be no different from that provided by an ordinary apposition eye. The usual and only satisfactory solution seems to be that this represents a way of increasing the size of the signal available to the laminar neurons. Since all the 1–6 receptors impinging on a cartridge have the same field of view, no resolution is lost by pooling their signals, and there is an effective gain of 6 times in the photon numbers "seen" by the lamina. This can be thought of as equivalent to increasing the facet diameters by a factor of $\sqrt{6}$, from about 25 μm to 61 μm. Another way of putting this is to say that the contrast sensitivity is improved by $\sqrt{6}$ (equation 21), or alternatively that for the same photon capture rate the integration time can be reduced by a factor of 6, making fast responses possible. Both these features are probably important. On the one hand the speed and accuracy of the chasing behaviour of male flies during sexual pursuit (with total delays of less than 30 ms; Land, M. and Collett, T., 1974) strongly suggest that a shortening of the integration time is important. On the other, the ability of some flies to detect targets considerably smaller than the inter-ommatidial angle (e.g. simuliids; Kirschfeld, K. and Wenk, P., 1976) implies that high contrast sensitivity is also at a premium. A simple increase in sensitivity, enabling the animal to extend its "working day" into dawn and dusk, is probably not the answer, since an increase in photon capture of less than a log unit represents only about $\frac{1}{4}$ h, when the ambient intensity is changing by 6 log units over 1–2 h.

In relation to the question of enhanced temporal and contrast resolution, it is of particular interest that in the anterodorsal region in the eyes of male flies, R7 also participates in the neural superposition (Franceschini, N. et al., 1981). The axons terminate in the lamina along with R1–6, unlike the situation in the rest of the eye, and throughout the eye in females, where the axons of both R7 and R8 pass straight into the medulla. In addition, the visual pigment of R7 in this region is the same as in R1–6. The implication is that in this foveal region, which is largely concerned with detecting and chasing females in flight, the number of receptors involved in tasks needing high photon numbers has been increased from six to seven, at the expense of the colour vision system normally mediated by R7 and R8.

4.3 Foveal specializations

Many diptera have eyes in which the frontal and dorsal regions have larger facets than elsewhere (Fig. 25). This is usually much more pronounced in males than in females, although in some families such as the Asilidae there is a similar development in both sexes (Dietrich, W., 1909). These regions are undoubtedly regions of higher resolution concerned with the capture of females for mating, or in the case of asilids (robber-flies) with the capture of other insects for food. The small hover-fly *Syritta pipiens* (Fig. 25a–e) can perhaps be taken as typical (Collett, T. and Land, M., 1975). Most of the male eye, and all of the female eye, has a facet lattice where the lenses are 16–20 μm in diameter, and the inter-ommatidial angles ($\Delta\phi$) are between 1.4° and 1.6°. However, the males have a frontal region only about 10° wide in which the lens diameters increase to 40 μm, and $\Delta\phi$ decreases to 0.6°. The males keep

the females within this region while they "shadow" them round prior to copulation attempts. The interesting point is that these eyes are not far from the diffraction limit. At the limit (where there is no resolution in the image) the sampling frequency v_s (equation 1) will equal the cut-off frequency v_{co} (equation 8). Combining the equations gives:

$$\frac{1}{2\Delta\phi} = \frac{D}{\lambda}, \text{ or } \Delta\phi = \frac{\lambda}{2D} \qquad (29)$$

Thus at the limit, 18 μm facets could be paired with inter-ommatidial angles as small as 0.8°, and 40 μm facets with angles of 0.36°. The corresponding values of $\Delta\phi$ in *Syritta* are actually about twice these values (1.5° and 0.6°) which is roughly what one would expect: insect eyes need some contrast to work with and never have values of $\Delta\phi$ quite as low as the "ultimate" limit (see section 2.2.1). Notice, however, that an inter-ommatidial angle of 0.6° is substantially less than could be provided by 18 μm facets. The function of the enlarged facets in the fovea of *Syritta* and other flies seems to be to provide increased resolving power without sacrifice of image contrast. Interestingly, if a male *Syritta* is shadowing a female at the limit of his ability to see her (up to 18 cm when her head subtends about 0.6° on his retina) she will not be able to see him. Presumably this enables males to capture females before they have time to evade.

Syritta has a fovea that grades smoothly into the rest of the eye. In some flies, however, there is a very distinct separation of the two parts of the eye. Some of the most dramatic examples are in the nematoceran flies, especially the simuliids (Kirschfeld, K. and Wenk, P., 1976), and the bibionids (Zeil, J., 1979). Figure 25f shows the head of a male bibionid (*Dilophus*), and whilst both male and female heads have a similar small eye (15 μm facets) the male has in addition a much larger eye overlying the smaller one, with bigger (27 μm) facets. In simuliids the situation is similar. The large (40 μm) facets of the males are capable of detecting and tracking females at distances of up to 50 cm against the dawn sky, when the females only subtend 0.2° (Kirschfeld, K. and Wenk, P., 1976). Foveas as clearly demarcated as these are relatively uncommon. They also occur in male tabanid flies (Fig. 1), and outside the diptera in the extraordinary "turbanate" eyes of male mayflies (Fig. 25e) (Wolburg-Buchholz, K., 1976;

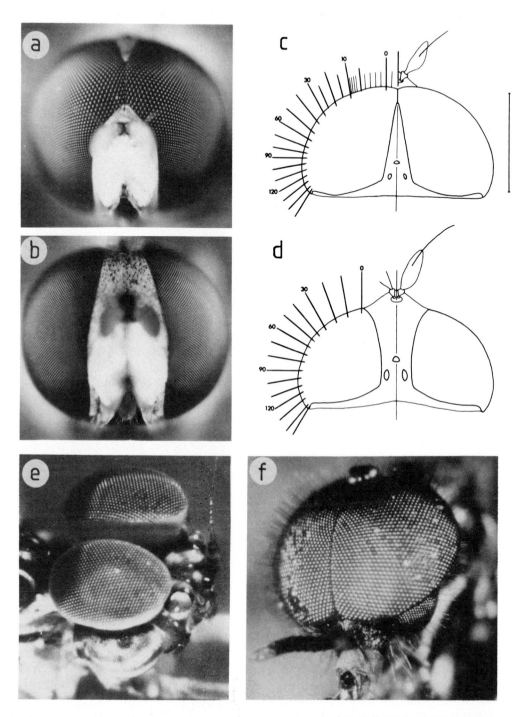

FIG. 25. Foveal specializations concerned with mate finding. **a** and **b:** Heads of male and female *Syritta pipiens* (Syrphidae) showing the anterodorsal region of enlarged facets present only in the male. **c** and **d:** Lines of sight in the horizontal plane for male and female *Syritta*, derived from pseudopupil measurements. The large facet region in the male only extends 5–10° from the midline. (**a–d** from Collett, T. and Land, M., 1975). **e:** Extreme development of the male eye in *Chloeon dipterum* (Ephemeroptera), females only have the lower part, visible beneath the "turbanate" eyes of the male. **f:** Similar divided eye in a male bibionid *Dilophus febrillis* (Diptera). See also Fig. 1.

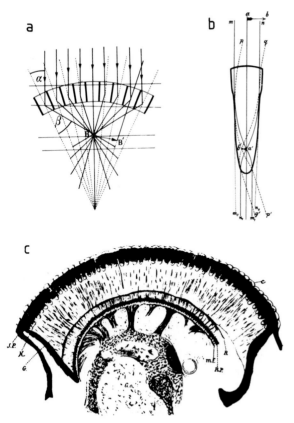

FIG. 26. Geometry of superposition eyes. **a:** Formation of the erect image at B–B' requires each optical element to bend light across its axis so that there is a constant relationship between the angle of incidence at the cornea (α) and the angle of emergence into the clear zone (β). This ratio (β/α) is the magnification (m) of each cornea–cone combination. (Modified from Exner, S., 1891.) **b:** Ray paths proposed by Exner for on-axis (m,n) and off-axis (p,q) pencils of light passing through a cornea-cone element in *Lampyris*. The structure behaves as a lens cylinder as in Fig. 21b, in which parallel light is brought to an intermediate focus (b'–a'), and then defocused again. (Exner, S., 1891.) **c:** Anatomy of the eye of *Lampyris* (from Exner, S., 1891). C, cornea; J.P., iris pigment shown in the dark-adapted state; K, tips of the crystalline cones; G, clear zone containing stained receptor cell nuclei; R, receptors; R.P., pigment associated with the receptors; m.f., basement membrane.

Horridge, G., 1976; Horridge, G. *et al.*, 1982). The more gradual type of fovea is common throughout the diptera, and there are examples from the Hymenoptera, Odonata (Sherk, T., 1978), Orthoptera and Dictyoptera, especially in the mantids where a pronounced fovea is associated with the fixation of prey before its capture (Rossel, S., 1979, 1980). For other examples the reader should consult Horridge, G. (1978) and Wehner, R. (1981).

4.4   Other neural superposition eyes

Most of our knowledge of the neural superposition mechanism comes from studies of brachyceran Diptera, especially *Musca, Drosophila* and *Calliphora*. The pattern of connections in these three species is identical (Fig. 24) but there are indications that in other Diptera there may be variations. For example, Zeil, J. (1979) found that in the dorsal eye of bibionids the pattern of rhabdomeres in each ommatidium is a true hexagon, not the odd pattern seen in the Brachycera, and that the coincidence of the rhabdomeric visual fields is not with those in neighbouring ommatidia, but with those in the next-but-one neighbours. The pattern of connections in the lamina is likewise with next-but-one neighbours. Outside the Diptera it is not yet clear whether there are any good examples of neural superposition eyes, but amongst the Hemiptera there are eyes with partially or completely "open" rhabdoms, as well as a complex pattern of projection to the lamina (Strausfeld, N. and Nässel, D., 1981; Wolburg-Buchholz, K., 1979; Ioannides, A. and Horridge, G., 1975). In *Lethocerus* (Fig. 19ii) the rhabdomeres are fused near the distal tip, but not proximally, and their axes are inclined to each other at an angle of about 9°. Since the interommatidial angle is only 3° this suggests that connections should be made with laminar neurones up to three ommatidia distant (Ioannides, A. and Horridge, G., 1975). Whether or not this is the case is not yet known, but in *Notonecta glauca* receptor axons do bend when they reach the lamina, and may run for a distance equivalent to six optic cartridges (Wolburg-Buchholz, K., 1979). The hemipterans would certainly repay further study.

5   SUPERPOSITION EYES

5.1   Geometry

In the Lepidoptera (moths and skipper butterflies), Neuroptera, Trichoptera and some Coleoptera one finds eyes with a wide space between the optical elements around the eye surface and the rhabdoms which usually lie on a hemisphere with about half the radius of curvature of the eye surface. This contrasts with the layout of apposition eyes where the distal receptor tips lie close to the corneal lenses, separated from them only by a short crystalline cone (Fig. 7). In dark-adapted animals this space, the clear zone, is devoid of pigment that might obstruct the passage of rays across it. A similar arrangement is found in some crustacea, the mysids, euphausiids and the macruran decapods — shrimp, crayfish and lobsters (Land, M., 1980). As mentioned in the Introduction, Exner, S. (1891) showed in the glow-worm *Lampyris* that eyes of this kind give a single erect image in the rhabdom region (Fig. 5), rather than a set of small inverted images directly behind the corneal lenses as in apposition eyes, and he solved the problem of how these images are formed.

If we look at Exner's figure (Fig. 26a), it becomes clear that for rays reaching different parts of the eye surface to be focused to a common image point, they must be diverted by the optics through an angle equal to roughly twice the angle of incidence at the cornea. If a ray enters the eye at an angle α to a normal to the corneal surface, and emerges into the clear zone making an angle β with the normal, then if α and β are equal, the crystalline cones act as though they were mirrors. Where α is small, this will result in all rays coming to a common focus halfway between the centre of curvature of the eye and the cornea. As with ordinary lenses, larger angles of incidence result in refracted rays that depart from the ideal, and come to a focus slightly in front of the focus for rays making small angles of incidence (a form of spherical aberration; Fig. 27b).

To produce a superposition image the optical elements do not have to bend the rays through exactly 2α (i.e. $\beta = \alpha$), if they bend it more ($\beta > \alpha$) the focus will be further out from the centre and the focal length will be longer since the centre of curvature of the eye is its nodal point (Fig. 7c). If they bend it less the focus will be closer to the centre, and the focal length shorter. The ratio β/α must remain the same, however, even if it is not 1. There is some evidence that in many eyes with superposition optics, the receptors are not always situated at half a radius of curvature from the centre of the eye, and so one might expect their crystalline cones to bend light through angles other than 2α (this is the case in Exner's own drawing of a moth cone: Fig. 26b). Another factor that must be taken into account is the refraction that occurs at the surface of the eye,

when this is an air-to-tissue surface (this surface is not important in aquatic animals). This convex surface itself behaves as a lens, bending rays towards the centre of the eye, and at the same time *reducing* the angle the rays make with the axis of each optical element (Fig. 27a). This has two interesting consequences.

(i)   The rays are bent *less*, overall, than they would be if the refractive index inside and outside the eye were the same. If the inside of the eye has a refractive index of 1.33, then the focus is at about 0.35 radii from the centre rather than 0.5 radii, when the optical elements bend the light through 2α (Fig. 27b).

(ii) Because the spherical aberration at the eye surface, and the aberration of the image produced by the superposition mechanism itself are now working in opposite directions the image quality is greatly improved for rays making large angles of incidence with the eye surface. To bring the focus back out to 0.5 ($\beta/\alpha = 1$) radii would require that the magnification factor of the elements (β/α': Fig. 27b) is increased to about 1.2.

Thus to form a superposition image, in which rays entering the eye over a large proportion of the surface are superimposed, the individual optical elements must have the property of redirecting rays across their axes. How can this be achieved, since single lenses do not have this property?

5.2   Lens cylinders in superposition eyes

A conventional optical arrangement that has the property of bending light across its axis is an inverting (astronomical) telescope. In a two-lens telescope parallel rays entering the objective at an angle α are brought to focus within the instrument, and leave the eyepiece, again parallel, at an angle $m\alpha$ with the axis, where m is the magnification (m is given by the ratio of the focal lengths of the objective and eyepiece lenses). To convert this design into one of the elements in a superposition eye, all that is required is for m to equal 1, so that the telescope redirects the entering beam, without magnifying it. This can be achieved by making the focal lengths of the two lenses equal, and placing them symmetrically about a common focus (Fig. 28).

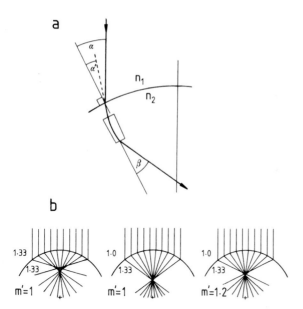

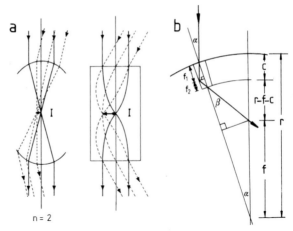

FIG. 27. Effect of the cornea acting as a lens. **a:** Refraction at the eye surface reduces the angle incident on the cone from α to α', where $n_1 \sin \alpha = n_2 \sin \alpha'$. **b:** Computer tracings of rays in eyes of different construction. *Left:* same refractive index inside and outside eye. Magnification of cone alone ($m' = \beta / \alpha' = 1$). The image shows considerable aberration and the smallest blur circle is well in front of the focus for small angles. *Middle:* with air on the outside and water inside the eye the aberrations of cornea and lens cylinders cancel each other, giving a good image lying deeper in the eye. *Right:* increasing the magnification of the optical elements to 1.2 brings the image forward with little loss of quality.

FIG. 28. **a:** Optical systems with the ray-bending properties required for superposition optics can be produced either from two surfaces arranged as a unity gain telescope (left) or using a lens cylinder (right). Each half of the lens cylinder can be regarded as a single lens. **b:** The relation of the overall eye geometry to the focal lengths (f_1, f_2) of the two halves of the lens cylinders. For small angles the cone magnification $(m) = \beta/\alpha = f_1/f_2 = f/(r-f-c)$, where f is the focal length of the whole eye, radius r, and c is the cone length. This gives:
$$f = [m(r-c)]/(1 + m).$$

Exner, S. (1891) considered the possibility that the crystalline cones of insects with superposition eyes might be simple telescopes, with the curved corneal surface acting as one lens and the proximal tip of the cone the other. He dismissed this as the principal explanation, however, on the grounds that the refractive index of the cones was not high enough, and the surface curvatures not small enough, to produce a telescope within the length of a cone. He suggested instead that the crystalline cone was constructed as a lens cylinder, similar to that proposed for the apposition eye of *Limulus* (section 3.1). The crucial difference is that in the case of *Limulus* each cone has to act as a simple lens, whereas in a moth or firefly it must act as a pair of lenses (Fig. 21). Physically this is very simply done: the lens cylinder must be made twice as long in superposition eyes compared with apposition eyes (Figs 21b and 28). One can then think of the distal half of the lens cylinder as the first lens of the

telescope and the proximal half as the second, with an internal image at the junction of the two halves. The continuously curving ray-paths in a double length cylinder of this kind are shown in Figs 26b and 28. In practice, some of the refractive power for the telescopic combination will come from the curved surfaces of the cornea and cone, so that these are really hybrid devices, using partly surface refraction and partly lens cylinder optics (see Caveney, S. and McIntyre, P., 1981). The relationship between the focal lengths of the two "lenses" in each optical element, and the focal length of the eye *as a whole*, is shown in Fig. 28b.

The construction of lens cylinders, and in particular the form of the refractive index gradient required to produce a focused image, have been discussed in section 3.1. During the past two decades graded refractive index devices (GRIN) have been manufactured in both glass and plastic, and one such device, with properties identical to those of the cones of superposition eyes, is illustrated in Fig. 29.

Exner's explanation of superposition optics, as given here, remained unchallenged until the early 1960s, when interference microscopes became available and it became possible to examine the

FIG. 29. Natural and artificial lens cylinders. **a:** Ray path, made visible with fluorescein, through a crystalline cone from a euphausiid crustacean; the magnification (β/α) is very close to 1. **b:** Glass lens cylinder of Japanese manufacture with the same ray-bending property. **c:** A property of all telescopic systems is that an object placed on the objective lens is imaged on the eyepiece. This is true of lens cylinders as well, so that here the letter "e" at one end of the lens cylinder is transferred, inverted to the other end. The explanation becomes clear if the top half of the lens cylinder drawn in Fig. 28a is removed and replaced underneath the lower half: the image at I will now be re-imaged at the bottom of the reconstituted cylinder. **d:** Correlate of (**c**) in the light-adapting eye of a sphingid moth. Pigment cells migrating around the *proximal* tips of the crystalline cones (compare Fig. 19i), are visible on the *distal* corneal surface.

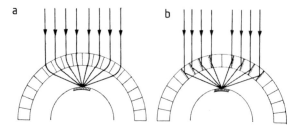

FIG. 30. Superposition images can be formed by lens cylinders **a:** or radial mirrors **b:** The latter are found in the macruran decapod crustacea. Both mechanisms are compatible with Exner's diagram (Fig. 26a).

refractive index profiles of crystalline cones directly. Unfortunately, one of the first studies was on crayfish eyes (Kuiper, J., 1962) and the crystalline cones were found to be optically homogeneous. This appeared to be a severe blow to Exner's mechanism at the time, because a refracting superposition eye *must* use lens cylinder optics. However, it turned out much later that the crayfish and its relatives do have superposition eyes (Figs 30 and 31), but they use mirrors and not telescopes to achieve the crucial ray bending (Vogt, K., 1975, 1980). Thus Kuiper's observations were good, but his conclusion, that the idea of lens cylinders should be discarded, was in retrospect an unfortunate generalization from the wrong animal. The situation was made worse by the

quite erroneous assertion by Allen (Miller, W. *et al.*, 1968) that the crystalline cones of moths were homogeneous (this was refuted convincingly by Hausen, K., 1973). At all events, by the early 1970s there was a veil of doubt surrounding the existence of both the superposition mechanism itself and the lens cylinders that made it possible, and there was much speculation as to how eyes with "clear zones" might function in the absence of these mechanisms (e.g. Horridge, G., 1971). At about this time, however, the balance of evidence began to swing the other way. Kunze, P. (1969) showed that the phenomenon of eye glow in moths was consistent with superposition theory; Seitz, G. (1969) found a refractive index gradient in firefly cones that was consistent with lens cylinder optics, and Kunze, P. and Hausen, K. (1971) showed the same thing in the moth *Ephestia*; Horridge, G. *et al.* (1972) showed the existence of a superposition image in skipper butterflies by the very direct method of cutting a hole in the cornea and looking at the retina; and finally the problem of image-formation in crayfish eyes was solved by Vogt, K. (1975), thus removing the difficulty that had started the trouble. The situation now is that most, if not all, eyes with clear zones between the crystalline cones and the retina do behave as Exner-type superposition eyes when they

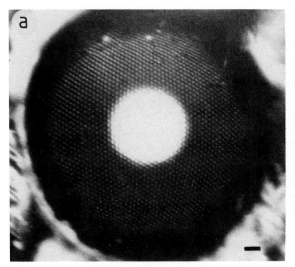 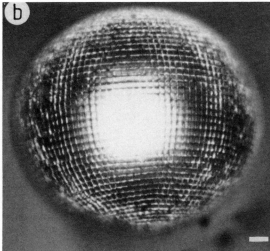

FIG. 31. Eyeshine in the refracting superposition eye of a skipper butterfly *Ochlodes venatus* **a:** and the reflecting superposition eye of a shrimp *Palaemonetes varians* **b:** The eyeshine results from reflection from a tapetum around or behind the rhabdoms, giving rise to emerging rays that are the reverse of those in Fig. 30. The glow patch is thus an indication of the size of the superposition pupil (D in Table 3). (Scale bars = 0.1 mm.; Land, M., 1981.)

are dark-adapted; and most of the doubt and speculation of a decade ago has disappeared. Unfortunately, textbooks tend to be out of date by about a decade, so that several now reflect the thinking of that period.

5.3 Eyeshine and iris mechanisms

If a dark-adapted moth is brought into the light, and viewed from the same direction as the light source, a large patch of orange "glow" or "eyeshine" is visible in the centre of the eye (Fig. 31). Over the course of a few tens of seconds the glow patch shrinks in size, becomes dimmer, and finally disappears. (Exner, S., 1891; Höglund, G., 1966). There is strong evidence that the pigment migration responsible for this is autonomous, and not the result of retinal activity (Hamdorf, K. and Höglund, G., 1981). The same phenomenon can be seen in lacewings (*Chrysopa*) except that the colour of the eyeshine is blue. Although the eye gives the impression of a circular pupil, expanding and contracting as in the human eye, the explanation of the changes in the size of the glow patch is different. Pigment migrates radially inwards from between the crystalline cones during the light, and the effect of this is to cut off, progressively, the rays that cross the clear zone at large angles (Figs 19i and 29d).

Eventually the light that reaches a single rhabdom is restricted to a single crystalline cone in the light, instead of anything up to 10^3 cones in the dark. Thus this is a true pupil mechanism, cutting down the amount of light reaching the receptors, and it is a very powerful one since it provides 2 to 3 log units of attenuation (compared with just over 1 for the human pupil). By comparison with the pupil mechanisms of flies and butterflies (Fig. 19iii), which work by moving pigment granules the short distance from the receptor cytoplasm to the rhabdom interface (section 2.2.4), the pupil mechanism in superposition eyes is slow. In the former, pupil closure is complete in about 5 s or less (Stavenga, D., 1979), whereas in the moth *Ephestia* it may take 5 min (Kunze, P., 1979). Although in neither case are the mechanisms of the pigment movements understood, there is a great difference in the distance the pigment has to travel: one or two μm in the rhabdom mechanisms, but tens or in some cases hundreds of μm in superposition pupils.

5.4 Sensitivity and image quality

Whereas apposition and neural superposition eyes tend to have high F-numbers (2 or more) and narrow rhabdoms that behave as light guides, superposition eyes have smaller F-numbers (0.5 to 1) and

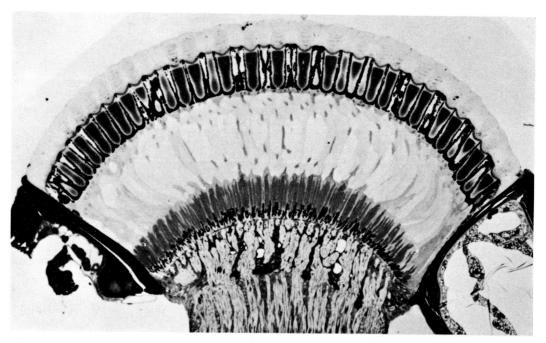

FIG. 32. Section through the eye of the nocturnal dung-beetle *Onitis westermanni*. This species has bullet-shaped crystalline cones. This beautiful section (courtesy of Dr. S. Caveney) shows all the geometrical features of superposition eyes particularly well. Compare with Exner's drawing of *Lampyris* (Fig. 26c).

hence brighter images, and relatively wide rhabdoms (8 μm in *Ephestia*, 1.5 μm in a bee). Both these features mean a greatly increased sensitivity for the moth *Ephestia* ($S = 114$; Table 3) compared with the bee ($S = 0.2$), a 570:1 ratio. The advantages of superposition optics to a nocturnal insect are obvious, but it still needs to be pointed out that the 2–3 log units of extra light provided by the optics only go part of the way towards meeting the difference of 10^6 between sunlight and moonlight. There must be neural and behavioural differences too: longer integration times and slower flight are obvious possibilities.

At what cost in resolution is this increase in sensitivity bought? This is quite difficult to estimate because it depends on the extent to which the full sensitivity of the superposition mechanism is exploited. In theory, the absolute limit to resolution imposed by diffraction is much the same in both apposition and superposition eyes, if the diameters of individual facets are similar (which they are: *Ephestia* 20 μm, *Apis* 25 μm). It is the diameter of single facets, not of the whole superposition pupil, that determines the size of the retinal diffraction pattern (see section 2.1.2).

Although the diffraction limit is much the same in the different types of eye, the wide aperture of superposition eyes imposes geometrical limits to image quality not unlike spherical aberration in large aperture simple lenses. This effect is illustrated in Fig. 27b, where it can be seen that rays making large angles of incidence with the corneal surface (α) are brought to a focus in front of the focal point for small α rays. To some extent this defect is reduced when the effect of refraction at the eye surface as a whole is taken into account, because rays which make a large angle of incidence at the eye surface (α) now make rather smaller angles (α') with the axes of the crystalline cones. This tends to offset or even cancel the aberration of the image, but unfortunately there are no direct studies of image quality in large-aperture superposition eyes. In *Ephestia kühniella*, where the pupil is about 15 facets wide, and the maximum angle of incidence at the cornea (α) is 23°, Cleary, P. *et al.* (1977) estimate by ray tracing that the width of the spot formed by the intersecting bundles in the superposition image is about 25 μm or 3 rhabdoms, in diameter. The inter-rhabdom angle ($\Delta\phi$) is 2.7°, so this implies an acceptance angle ($\Delta\rho$) of about

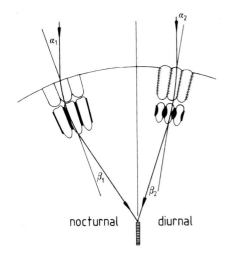

FIG. 33. Diagram showing the limiting image-forming rays in the eyes of a nocturnal dung-beetle (*Onitis*), left, and a diurnal dung-beetle (*Pachysoma*), right. The nocturnal beetles have bullet-shaped cones without a "waist", and the diurnal species hourglass-shaped cones with a pronounced waist. This waist, at the level of the intermediate image (Figs. 26b and 28a), determines the highest value of α that rays contributing to the final image can take. In *Onitis* this is about 18°, and in *Pachysoma* 10°. (Modified from Caveney, S. and McIntyre, P., 1981.)

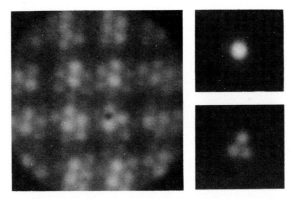

FIG. 34. Ophthalmoscopic appearance of the retina of the diurnal moth *Phalaenoides tristifica* (Agaristidae). The instrument images a plane conjugate with infinity, and hence the retinal image. The left-hand figure shows the retinal rhabdoms, imaged through the eye's own optics, together with the image of a grid of 2° lines 6° apart. Both the retinal mosaic and the grid are well resolved. *Right*: images of a point source, centred on a single rhabdom (top) and between 3 rhabdoms (below). In these eyes the half-width of the image of a point source is not measurably wider than the Airy disc due to diffraction at a single facet, even though the image is the result of the superposition of rays from about 140 facets.

three times this figure (8.1°) which is certainly poor by comparison with apposition eyes of similar dimensions.

The optical design feature which determines the maximum value that α can take when the eye is dark-adapted and the pigment retracted is the width of the crystalline cone at the location of the intermediate image (Figs 28 and 33). This image is an optical bottleneck, and the larger it is the greater the angle of incidence, and hence the diameter of the superposition pupil, can be. In a beautiful recent study, Caveney, S. and McIntyre, P. (1981) showed how the shapes of the crystalline cones of beetles are related to the effective aperture of the eyes, and to the animals' way of life. In diurnal beetles the cones are strongly "waisted" so that the intermediate images are restricted and the superposition aperture of the eye as a whole is small. In night-flying beetles the cones are more or less parallel-sided (like *Ephestia*), and the intermediate images and superposition pupils are correspondingly larger (Figs 32 and 33).

Two lepidopteran groups, the skipper butterflies and day-flying agaristid moths, have superposition eyes which are always "open", there being little or no pigment migration, and an unchanging patch of blue glow is visible in daylight (Horridge, G. *et al.*, 1972, 1977). These eyes have relatively small pupils compared with night-flying moths (α up to about 13°), and they have excellent resolution. I have examined the retinae of both groups with an opthalmoscope, using the eyes' own optics to image the retina. It is possible to see the receptors directly (Fig. 34) which means that the resolution of the image is at least as good as the retinal mosaic. The image of a point source is almost entirely contained within a single receptor, meaning that $\Delta\rho < \Delta\phi$, where $\Delta\phi$ lies between 1.5° and 2°, a value which is comparable with apposition eyes of similar size. Thus superposition optics do not necessarily provide an image that is of poor quality, provided the pupil is kept reasonably small.

6 REFLECTORS AND ANTI-REFLECTION COATINGS

6.1 Reflectors

In addition to the refracting structures which form images in insect eyes, mirrors can also be important optical structures. The best-known examples are the tapeta which produce the luminous pseudopupil in butterflies, and the glow in skippers and moths

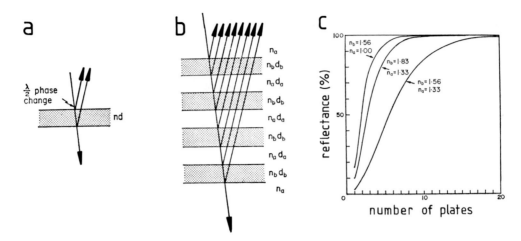

FIG. 35. Multilayer reflectors in animals. **a:** In a single film, constructive interference occurs when the optical thickness of the film (nd) is equal to $\lambda/4$. **b:** A high reflectance can be built up with alternating $\frac{1}{4}$-wavelength plates of high and low refractive index ($n_a d_a = n_b d_b = \lambda/4$).**c:** The reflectance of a stack of plates, as in (**b**), made of different combinations of materials: chitin and air ($n_b = 1.56$, $n_a = 1.00$) as in tracheal tapeta or butterfly scales; guanine and water ($n_b = 1.83$, $n_a = 1.33$) as in fish scales and many vertebrate tapeta, and chitin and water ($n_b = 1.56$, $n_a = 1.33$) a combination approximated in the reflectors producing eye colours in some diptera.

(Figs 22 and 31). In butterflies the function of the tapetum is to return light that has passed once through the rhabdom, thus effectively doubling the rhabdom length. In skippers and diurnal moths the tapetum surrounds each rhabdom, and its function is presumably to contain the light reaching the distal rhabdom tip. This is necessary because the rhabdom cannot trap the whole cone of light reaching it from the superposition pupil by total internal reflection alone. Optical isolation can be achieved either by screening pigment — which wastes light — or in this case by encompassing the rhabdoms in reflecting cases — which does not. Mirrors also occur in the corneal layers of some eyes, where they produce coloured reflections and concomitantly alter the colour of the transmitted light. Examples are the coloured eye stripes of tabanid flies and the alternately-coloured facets of dolichopodid flies (Bernard, G. and Miller, W., 1968). The functions of these reflectors are not very clear. They may serve as display or as camouflage, or they may act as colour-selective filters that enhance colour contrast. The structure and function of mirrors in insect eyes have been reviewed recently by Miller, W. (1979).

In both tapetal and corneal reflectors the reflecting structure is a set of lamellae of alternating high and low refractive index, in which each layer has an optical thickness (actual thickness times refractive index) of about $\frac{1}{4}$ wavelength (Fig. 35). This is the way mirrors are constructed throughout the animal kingdom, in fish scales, vertebrate tapeta, butterfly wings as well as in the image-forming mirrors of scallops and decapod crustacea (Land, M., 1972; Miller, W., 1979). Animals do not make metals, and the only alternative known to technology for producing a highly reflective surface is the quarter-wave stack. The principle is the same as that of the coloured reflections from oil films or soap bubbles (Fig. 35). When light is incident on a flat plate of higher refractive index than the surrounding medium, a small proportion is reflected at the upper surface, and the same proportion at the lower surface. For a single air/chitin interface the proportion of incident light reflected is 4–5%. If the plate is a quarter-wavelength thick, the light reflected from the upper and lower surfaces interferes constructively and the proportion reflected more than doubles for the ideal wavelength, and less than doubles for wavelengths longer or shorter than the ideal, so that the plate becomes a colour-selective reflector. One might think that the plate should be $\frac{1}{2}$ wavelength thick for maximal reflection, so that the extra path-length travelled by the light reflected from the lower face would be 1 wavelength. This is not so, for a reason that is not intuitively obvious: light reflected at a low-to-high refractive index interface, but not at a high-to-low one, has its phase altered by $\frac{1}{2}$ wavelength, which means that the extra optical path-length required for the ray reflected from the lower interface is only $\frac{1}{2}$ wavelength, and so the optical thickness of the plate must be $\frac{1}{4}$ wavelength. Even at the ideal wavelength such a plate only reflects about 17% (chitin in air). The way that both the reflectance and the colour selectiveness of the structure can be increased is simply to add more plates — each $\frac{1}{4}$ wavelength thick, and each separated from the next by $\frac{1}{4}$ wavelength of medium (Fig. 35b). The reflectance of such a stack increases very rapidly with the number of layers, so that at the ideal wavelength a stack made of chitin and air reflects more than 99% of the incident light with only seven chitin plates. The reflectance (R) is given by:

$$R = [(n^k_{\,b} - n^k_{\,a})/(n^k_{\,b} - n^k_{\,a})]^2 \qquad (30)$$

where n_a and n_b are the refractive indices of the low and high index components of the stack, and k is the total number of interfaces. Two extra points are important. Firstly, where $(n_b - n_a)$ is large the reflectance is high, so also is the reflected bandwidth: a chitin/air multilayer gives an almost white reflection. Reducing the refractive index difference reduces both the reflectance and the bandwidth, giving a much more saturated colour. Secondly, the bandwidth can be reduced by varying the relative optical thicknesses of the high and low index layers, so that one is less than $\frac{1}{4}$ wavelength, and the other more. Insect structural colours are in general quite saturated, so one or other of these variations is presumably being employed. Finally, unlike conventional filters, mirrors of this kind absorb no light, so what is not reflected is transmitted. They therefore act as filters that transmit the part of the spectrum that is not reflected — the complementary colour. A more complete treatment of the physics of biological reflectors is given in Land, M. (1972).

Of the two cases mentioned earlier, the tapeta in Lepidoptera are of the air/chitin type (Fig. 36a), with the plates consisting of enlarged and flattened taenidial ridges in the tracheoles that embrace the base of each rhabdom (Miller, W. and Bernard, G., 1968; Miller, W., 1979). The colours are quite saturated, implying that these are structures with layers of unequal optical thickness. The tapetal colours vary across the eye in some butterflies, and

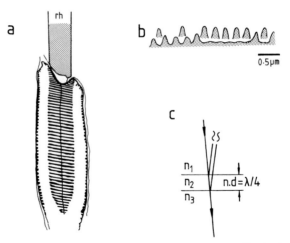

FIG. 36. Reflecting and non-reflecting structures. **a:** Tracheal tapetum in a butterfly *Junonia*. The plates are chitinous extensions of the strengthening ridges of the tracheoles. The reflected light is pastel blue, and the plates and spaces have optical thicknesses about $\frac{1}{4}$ of the wavelength of blue light. (After Miller, W., 1979.) **b:** Chitinous nipples, $\frac{1}{2}$-wavelength high, form an anti-reflection coating on the surface of the eye of a moth *Prodenia* (from a transmission electron micrograph in Miller, W. *et al.*, 1966). **c:** The normal method of blooming a lens with a $\frac{1}{4}$-wavelength film. This differs from Fig. 35a in that $n_1 < n_2 < n_3$, so that both interfaces are "low-to-high" and destructive interference results.

vary between species, but there is still some confusion as to whether there is an exact correspondence between tapetal colour, and the colour of the luminous pseudopupil (Ribi, W., 1979). In the case of the corneal colours of Diptera and probably also the orange corneal reflection of neuropterans such as *Chrysopa*, the structures responsible are chitin layers of different density within the body of the cornea itself. These colours are lost on death and dehydration, implying that the refractive index differences are the result of different degrees of hydration in the corneal layers. The layers themselves are easily visible in electron micrographs (Bernard, G. and Miller, W., 1968) although the range of refractive indices that actually occurs has not been measured, and it is difficult to see how it could be. The transmission colours of these structures are not strong (possibly implying that they are not being used as colour filters) although against the dark background of the eye pigment the reflected colours are very striking. Their function is still an enigma.

6.2 Anti-reflection coatings

The standard way of blooming a lens, to prevent

reflections and increase transmission, is to coat it with a $\frac{1}{4}$ wavelength layer of material whose refractive index is intermediate between that of air and glass. MgF_2 ($n = 1.36$) is usually used. The coating results in reflections at the air/coating and coating/glass interfaces which destroy each other by interference, and cause improved transmission instead (Fig. 36c).

The eyes of moths also have anti-reflection coatings. The eye of a fly or bee reflects about 4% of the incident light, exactly what one would expect from a surface of refractive index 1.5 in air. Moth corneas on the other hand are almost non-reflecting — 0.7 log units less than a fly cornea, or about 0.08% — and the residual reflection is slightly blue (Miller, W., 1979). The way this blooming is achieved is not quite the same as in a camera lens. Instead of a $\frac{1}{4}$ wavelength coating, moth corneas have an array of minute conical projections, or "corneal nipples", completely covering the surface (Bernhard, C. et al., 1965; Miller, W. et al., 1966). The nipples are about 200 nm high and spaced at 200 nm intervals (Fig. 36b). They do in fact constitute a $\frac{1}{4}$ wavelength plate, but unlike the simple coating, the nipple array has an effective refractive index that varies continuously from 1 to 1.5 as the base of the array is approached. The optical behaviour is very similar to that of a simple coating, except that the reflection minimum is spectrally rather broader. (The optics of nipple arrays were calculated ingeniously by Bernhard, C. et al. (1965) who used a scaled-up version of a moth cornea, made of wax, and microwave radiation to measure the transmission and reflection properties).

Miller, W. (1979) suggests three possible functions for the anti-reflection coating:

(1) Camouflage during the day. The dull surface of a moth's large eye is less likely to be a visible target.
(2) Improved transmission. Although this is only $\sim 4\%$, moths are active in conditions where all photons are valuable.
(3) The suppression of reflections emanating from the tapetum *inside* the eye which, if reflected back from the cornea, would cause a reduction in image quality.

Nipple arrays are not confined to moths; most butterflies have them too, and so do the caddis-flies (Trichoptera). Many other insect groups have some sculpturing of the corneal surface, but the protuberances are usually less than 50 nm high, and do not suppress reflections.

7   OCELLI

Insect ocelli, or simple eyes, fall into two distinct groups: the larval eyes of endopterygote insects, and the dorsal ocelli present in most winged adult insects.

7.1   Larval ocelli

In insects with a distinct larval stage the ocelli are the only eyes the larvae possess. They vary greatly in size and complexity. At one extreme, dipteran larvae have no more than a small group of light-sensitive cells in each side of the head, with no visible external structure associated with them. The larvae of Lepidoptera, Trichoptera and Neuroptera have ocelli which are more eye-like, but which nevertheless lack an extended retina (as in spider eyes) and in many respects resemble single isolated ommatidia (Fig. 37a). Dethier, V. (1942, 1943, 1963) studied the ocelli of the caterpillar *Isia isabella*, which has six on each side of the head. Each ocellus has a lens about 100 μm in diameter, which may be simple, or divided into three segments radially, each part forming a more or less separate image. Beneath the corneal lens there is a second lens, or crystalline body, and the combined effect of the two lenses is to produce an image just proximal to the inner lens; most refraction occurs at the cornea. There are seven receptors in each ocellus: three contributing to a distal rhabdom, and four to a proximal one. The distal rhabdom forms a V-shaped cup around the second lens, approximately in the plane focus, and the proximal rhabdom, beneath it, is completely fused. Although there is little possibility of image resolution within an ocellus, the fields of view of neighbouring ocelli do not overlap, and it seems that the whole array of 12 ocelli may simply provide a coarse 12-point sampling mosaic. There is now good evidence that (like the adult eyes) butterfly ocelli have receptors of at least three spectral types (Ichikawa, T. and Tateda, H., 1980, 1982). The single advantage of these ocelli would seem to be their

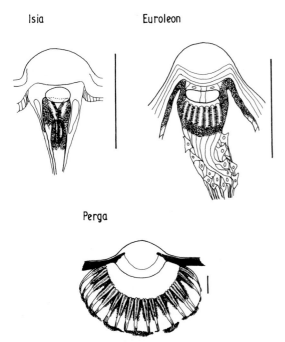

Isia   Euroleon

Perga

FIG. 37. Larval ocelli. These cover a spectrum of forms from the lepidopteran type (*Isia*) which is not much more elaborate than a single ommatidium, through the ant-lion (*Euroleon*: Neuroptera) where the rhabdoms begin to form a flat retina, to the sawfly *Perga* (Hymenoptera), which is a proper image-forming eye with an extended retina. (All scale bars = 0.1 mm.) (Compiled from Dethier, V., 1943; Jockusch, B., 1967; and Meyer-Rochow, V., 1974.)

light-gathering power. Dethier estimated the F-number of the largest ocellus as 0.5.

The ocelli of the anti-lion *Euroleon nostras* (Neuroptera) have a similar lens arrangement to those of *Isia*, but each has an extended retina of 40–50 receptors (Fig. 37b), and since the retina seems to be in focus this will provide an inter-receptor angle ($\Delta\phi$) of 5–10° (Jockusch, B., 1967), not impressive, but enough to detect a moving ant at 1 cm, which is what the eyes are required to do. Again, there are six ocelli on each side, grouped on a small turret.

The most eye-like, and best-resolving, larval insect ocelli are probably those of sawflies (*Perga*: Hymenoptera) and tiger beetles (*Cicindela*: Coleoptera). The latter have a life-style like the ant-lions, ambushing insect prey — mostly ants — as they pass their burrow (Fig. 6). There are again six ocelli on each side, but three are much larger than the others, and the largest ocellus has a lens diameter of 200 μm, and a retina containing 6350 receptors

(Friedrichs, H., 1931; see also Wehner, R., 1981). The inter-receptor angle is about 1.8°, comparable with or better than the resolution of the compound eyes of most adult insects. In *Perga* there is only one ocellus on each side of the head (Fig. 37c). It has an in-focus retina with a field of view of about 180°, and this is made up of clusters of eight receptors contributing to fused rhabdoms, very much as in a normal apposition eye. The rhabdoms are 10–15 μm in diameter, and about 20 μm apart, giving an inter-rhabdom angle ($\Delta\phi$) of 4–6° (Meyer-Rochow, V., 1974). These larvae are vegetarian, and it seems that the ocelli serve principally to direct the larvae to host plants. *Perga* larvae will, however, track moving objects with their head, and defend themselves by spitting regurgitated sap. Meyer-Rochow found objects subtending about 4° to be effective in eliciting this response, so behavioural and anatomical measures of resolution seem to match.

There is little doubt that some insect larvae, tiger beetles in particular, have simple eyes comparable in resolution to the compound eyes that later replace them. Since their sensitivity is likely to be similar too, it remains a central mystery of insect optics as to why a type of eye that is intrinsically superior — because diffraction no longer limits resolution — should be supplanted in the same animal by a compound eye that offers no clear advantages in terms of image formation, and is larger.

7.2 Adult dorsal ocelli

The dorsal ocelli of adult insects are, if anything, more problematical than the larval lateral ocelli, to which they are not related developmentally. They resemble the larval ocelli in possessing a lens, and (like the sawfly larval eyes) an extended retina (Fig. 38). Some dorsal ocelli have tapeta and some have mobile irises. Each of the ocelli — typically three, two lateral and one anteromedial — has a wide field of view of 150° or more. There can be as many as 10,000 receptors in the retina of each ocellus. So far, everything fits the description of spider eyes, or "good" larval ocelli. The crucial difference is that all authors who have tried seriously to calculate the optics of these eyes have reached the conclusion that they are profoundly out of focus (Homann, H., 1924; Wolski, A., 1931; reviews in Mazokhin-Porshnyakov, G., 1969; Wehner, R., 1981; and

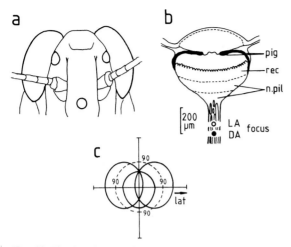

FIG. 38. The dorsal ocelli of locusts. **a:** Location of the three dorsal ocelli in *Schistocerca gregaria*. (After Goodman, L., 1981.) **b:** Structure of the median ocellus in *Austacris guttulosa*. (Wilson, M., 1978.) This ocellus has a movable iris shown here with the pigment (pig) in the light-adapted condition. The positions of the focus in the light and dark-adapted conditions are well behind both the receptors (rec) and the two layers of neuropil (n.pil). **c:** Fields of view of the three ocelli of *Locusta migratoria* seen from in front of the polar plot. Note that all three have fields bisected by the horizon. (Modified from Wilson, M., 1978.)

the compound eyes, receptors for entraining circadian rhythms, high-sensitivity detectors for use at light levels below the functional range of the compound eyes (Gould, J., 1975), and polarization detectors at low light levels (Wellington, W., 1974). These ideas are all certainly compatible with out-of-focus eyes with low F-numbers (<2). What makes the remaining idea — that the ocelli are horizon detectors (Wilson, M., 1978; Stange, G., 1981; Stange, G. and Howard, J., 1979) — more attractive than the rest is that the ocellar nerves project massively and directly into the optomotor system (Goodman, L., 1981). Presumably the reason for this is that the ocellar system provides the animal with a way of making fast corrections of pitch and tilt by reference to the horizon.

There remains, however, the question of why the compound eyes could not fulfil all these postulated tasks. They invariably cover the same fields of view as the ocelli, and the blurring of the image achieved in the ocelli simply by defocusing could equally well be performed by summation or low-pass filtering. I can think of no task that the ocelli are *uniquely* fitted to perform.

Goodman, L., 1981). In the fly *Calliphora*, for example, the receptors extend from 40 to 100 μm behind the lens, whereas the image lies at 120 μm. This conclusion, that the dorsal ocelli are severely underfocused, has been reached by too many competent people too often for it to be wrong. Dorsal ocelli are *deliberately* defocused (Fig. 38b).

We are not used to thinking about "imperfect" eyes, and so have to ask unusual questions: "what would an out-of-focus camera be useful for, for example?" Certainly not for detecting image detail, but it would be no worse than a well-focused system for determining the position of major discontinuities in its field. The most obvious of these, for an insect, will be the horizon. A defocused system will remove the clutter of leaves, houses or whatever, and only transmit information about the average position of the most important boundary: land and sky. (Another way of putting this is that high spatial frequencies will be heavily attenuated, but low spatial frequencies will not).

Many functions have been ascribed to the dorsal ocelli. They have been organs of general arousal (Wolski, A., 1933), modulators of the sensitivity of

7.3 Anomalous lateral ocelli in adults

There are a few insect orders in which the lateral compound eyes are reduced to the point where they only have a single lens. These "unicorneal" eyes occur in some lice (Mallophaga, Siphunculata and Psocoptera), in the fleas (Siphonaptera) and most remarkably in the males of scale insects (Eriococcidae: Homoptera). Little other than the anatomy is known about these eyes, but from their innervation it is clear that they are modified lateral eyes, and not dorsal ocelli (Paulus, H., 1979). In the eye of the flea *Ceratophyllus gallinae* there are about 100 receptors forming a bulky central rhabdom, surrounded by loose collections of rhabdomeres described as a lateral rhabdom (Wachmann, E., 1972). This eye does not really fit the description of either an ommatidium, or of any of the various types of ocelli. The eyes of male scale insects are even more bizarre (Duelli, P., 1978; Paulus, H., 1979). In *Eriococcus* there are three pairs of unicorneal eyes: one large dorsal pair where the compound eyes would have been, a ventral pair occupying the place of the absent mouthparts, and a small lateral pair remaining from the larval stages. Duelli, P. (1978) estimates that there are about 500 receptors in each of the larger eyes, with the inter-receptor angle ($\Delta\phi$) a quite respectable 4.7°. The other remarkable feature of these eyes is the structure of the rhabdom, which consists not of microvilli, but flattened plates resembling those of vertebrate receptors. The whole rhabdom is only 3μm long, which again must be some kind of record.

One is tempted to conclude that these eyes, so unlike anything in any other insect, either came from outer space or were put here by God to confuse scientists. More sober counsels should, however, prevail. "Since primitive Coccina have facetted eyes, these unusual lens eyes must have been derived from them. The possibility of such modifications demonstrates how easily great changes in organ structure can occur in the evolution of groups" (Paulus, H., 1979). Fortunately, evolution plays tricks like this only rarely.

8 FUTURE DEVELOPMENTS IN INSECT OPTICS

It seems certain now that insect eyes will not produce any major new optical principles comparable, say, with the superposition principle and the graded-index lens. There is no group whose eyes look genuinely inexplicable in terms of eye types that are known already. The last real revolution as far as image formation itself is concerned was that of Exner in 1891, and I cannot see that ever being supplanted. Having said that, there have been a great many smaller contributions to insect optics over the last two decades. Particular examples would be the discovery of the anti-reflection coating of lepidopteran eyes, the mechanism of action of quarter-wave reflecting stacks in tapeta, the understanding of the limits of photoreceptor performance imposed by their wave guide nature, and the direct measurement of refractive-index gradients in lens cylinders made possible by interference microscopy. Developments of this kind, resulting from improvements of technique and theory, will surely continue to take place.

The main thrust of the subject has now turned, I believe, in the direction of what might be called "ecological optics". By this I mean the study of the ways eyes are adapted to the lives animals lead in different light environments. I hope it has become clear in this chapter (expecially sections 2.1 and 2.2) that from a thorough knowledge of an eye's anatomy it is possible to make quite firm predictions about what it will be capable of resolving under different illumination conditions. Apposition eyes are actually very good subjects for such studies because their adaptations tend to be visible externally — as for example in the foveas in Fig. 25. In vertebrate eyes local adaptations for higher resolution tend only to become clear at the level of the retinal ganglion cells (Hughes, A., 1977), but such is the tyranny of the diffraction limit in compound eyes that resolution variations must be achieved by differences in the sizes of the corneal lenses themselves. Even in eyes without gross external modifications, the oval eyes of bees, butterflies and grasshoppers for example, there are interesting variations in the ommatidial sampling density in different eye regions. Inter-ommatidial angles tend to be smallest near the front and largest to the side and rear, and there are large differences around the eye's equator between vertical and horizontal inter-ommatidial angles. Walls' dictum (1942): "Everything in the vertebrate eye means something" must apply with equal force to insect eyes. Combined optical, anatomical and behavioural studies are likely to lead us to interesting conclusions not just about how insects see, but about the way visual systems must be organized in any moving animal, or for that matter in any man-made "seeing" device.

I have not mentioned electrophysiological studies, because they are outside the scope of this chapter, but there is no doubt that the coming decades are likely to produce answers to the questions of what animals do with the information that the eyes provide them. Already, the roles of a hearteningly small number of large neurones in the lobula plate of flies have been established, in relation to the animals' flight stabilization behaviour (Hausen, K., 1976). The goal of working through and making sense of an animal's visual system, from eye to behaviour, no longer seems too far out of reach.

9 SUMMARY

1 Insect eyes are of three basic types: apposition and superposition compound eyes, and simple

eyes. The "neural superposition" eyes of dipterans are a variant of the apposition type (Fig. 7). Simple eyes occur as the ocelli of holometabolous insect larvae, and as the dorsal ocelli of most winged adults.

2   Apposition eyes are typical of diurnal insects. In each ommatidium a lens forms a small inverted image on the distal tip of the rhabdom, which consists of the microvillar contributions of eight or nine receptor cells. The fact that the image is inverted is irrelevant because the rhabdom behaves as a single light guide within which resolution is not preserved. The overall image in the eye is erect.

3   Because of the small diameter of the individual lenses, the resolution of apposition eyes is limited by diffraction. A 25 μm lens cannot resolve gratings whose period at the eye subtends less than 1.1°. To improve upon this resolution, the eyes must have lenses that are larger, and to use the resolution thus gained they must have more of them. The result is that resolution much better than 1° is unattainable because the eyes become impossibly large. Predatory insects, and some males that capture females on the wing, often have "foveas": limited regions with higher resolution and larger lenses (Fig. 25).

4   In dipteran flies the individual rhabdomeres from the eight receptors are not fused, and do resolve the image presented by the lens. However, each rhabdomere images the same part of external space as a rhabdomere in a neighbouring ommatidium, and the connections between the retina and lamina are so organized that all receptors that image the same region of space connect up in the lamina. At the level of the lamina, therefore, the image is identical to that in an ordinary apposition eye. The eye's sensitivity is greater, however, because more than one receptor contributes to the signal in the lamina (Fig. 24).

5   Superposition eyes are common in nocturnal insects. Here many optical elements contribute to a single deep-lying erect image. The image is brighter than in apposition eyes (F_{no} 0.5–1, rather than around 2), and because the focal length is long the receptors can be made wider without loss of resolution. The overall sensitivity gain, compared with an apposition eye, can be as great as 1000 times.

6   The optical elements in superposition eyes are not simple lenses, but behave as though they were two-lens, unity magnification telescopes. They invariably incorporate a lens cylinder, a structure which behaves as a lens, but refracts light by virtue of its internal refractive index gradient (Figs 22 and 28).

7   Superposition eyes usually adapt to light by inward radial migration of pigment from between the optical elements (crystalline cones). This cuts off oblique image-forming rays, and effectively turns superposition eyes into apposition eyes. Some lepidoptera (skippers and agaristid moths) remain in the superposition condition in daylight. Apposition eyes have several types of pupil mechanism, including a variable aperture around the distal tip of the rhabdom, and a mechanism for moving pigment granules up to the walls of the rhabdom which breaks down its light guide property and allows light to leak out. Rhabdoms often change their diameter by loss of photo-pigment-bearing membrane between night and day (Fig. 19).

8   Apposition eyes often show a pseudopupil, a dark spot or pattern of spots which appears to rotate around the eye as the viewing direction changes. This is not a physiological pupil, but rather the virtual image of those receptors and the pigment around them that share a common line of sight with the observer. Pseudopupil phenomena are very valuable in working out the way an eye's ommatidial axes map onto the angular space around the insect (Figs 22 and 23).

9   Some eyes have reflecting tapeta behind or between the rhabdoms. These are tracheoles whose strengthening ridges have been modified into stacks of $\frac{1}{4}$-wavelength plates. In many moths and butterflies there is an anti-reflection coating on the cornea, consisting of an array of conical nipples about $\frac{1}{4}$-wavelength high (Fig. 36).

10   The larval ocelli of holmetabolous insects usually have well focused single lenses. The retina may vary from a few receptors arranged as in an ommatidium, to an extended retina with up to 6350 receptors. In the latter (tiger beetle larvae) the resolution is comparable with that of the compound eye of the adult. By contrast, the dorsal ocelli of adults are always underfocused. This is probably deliberate, and ensures that the view of

sky and horizon is not contaminated by detail. The dorsal ocelli are concerned with flight stabilization relative to the horizon.

11 There is still no satisfactory optical explanation for the adoption of compound eyes as the main organs of sight in insects. Simple eyes, which most insects possess at some stage in their lives, permit higher resolution in a smaller eye, without loss of sensitivity.

REFERENCES

BÄHR, R. R. (1974). Contribution to the morphology of chilopod eyes. *Symp. Zool. Soc. Lond. 32*, 383–404.

BARLOW, H. B. (1952). The size of ommatidia in apposition eyes. *J. Exp. Biol. 29*, 667–674.

BARLOW, H. B. (1964). The physical limits of visual discrimination. In *Photophysiology*, vol. 2. Edited by A. C. Giese. Pages 163–202. Academic Press, New York.

BERNARD, G. D. (1979). Red absorbing visual pigment of butterflies. *Science 203*, 1125–1127.

BERNARD, G. D. and MILLER, W. H. (1968). Interference filters in the corneas of Diptera. *Invest. Ophthalmol. 7*, 416–434.

BERNHARD, C. G. (1966). Opening address. In *The Functional Organization of the Compound Eye*. Edited by C. C. Bernhard. Pages 1–11. Pergamon Press, Oxford.

BERNHARD, C. G., MILLER, W. H. and MØLLER, A. R. (1965). The insect corneal nipple array. *Acta. Physiol. Scand. 63*, Suppl. 243, 1–79.

BLEST, A. D. and LAND, M. F. (1977). The physiological optics of *Dinopis subrufus* L. Koch: a fish-lens in a spider. *Proc. Roy. Soc. Lond. B. 196*, 197–222.

BRAITENBERG, V. (1967). Patterns of projection in the visual system of the fly. I. Retina-lamina projections. *Exp. Brain Res. 3*, 271–298.

BRUNO, M. S., BARNES, S. N. and GOLDSMITH, T. H. (1977). The visual pigment and visual cycle of the lobster *Homarus*. *J. Comp. Physiol. 120*, 123–142.

CAVENEY, S. and McINTYRE, P. (1981). Design of graded-index lenses in the superposition eyes of scarab beetles. *Phil. Trans. Roy. Soc. Lond. B 294*, 589–632.

CLARKSON, E. N. K. and LEVI-SETTI, R. (1975). Trilobite eyes and the optics of DesCartes and Huygens. *Nature (Lond.) 254*, 663–667.

CLEARY, P., DEICHSEL, G. and KUNZE, P. (1977). The superposition image in the eye of *Ephestia kühniella*. *J. Comp. Physiol. 119*, 73–84.

COLLETT, T. S. and LAND, M. F. (1975). Visual control of flight behaviour in the hoverfly, *Syritta pipiens* L. *J. Comp. Physiol. 99*, 1–66.

DETHIER, V. G. (1942). The dioptric apparatus of lateral ocelli. I. The corneal lens. *J. Cell. Comp. Physiol. 19*, 301–313.

DETHIER, V. G. (1943). The dioptric apparatus of lateral ocelli. II. Visual capacities of the ocellus. *J. Cell. Comp. Physiol. 22*, 115–126.

DETHIER, V. G. (1963). *The Physiology of Insect Senses*. Methuen, London.

DIETRICH, W. (1909). Die Facettenaugen der Dipteren. *Z. Wiss. Zool. 92*, 465–539.

DUBS, A., LAUGHLIN, S. B. and SRINIVASAN, M. V. (1981). Single photon signals in fly photoreceptors and first order interneurons at behavioural threshold. *J. Physiol. Lond. 317*, 317–334.

DUELLI, P. (1978). An insect retina without microvilli in the male scale insect, *Eriococcus* sp. (Eriococcidae, Homoptera). *Cell. Tiss. Res. 187*, 417–427.

EAKIN, R. M. (1972). Structure of invertebrate photoreceptors. In *Handbook of Sensory Physiology*. Vol. VII/1. Edited by H. J. A. Dartnall. Pages 625–684. Springer, Berlin, Heidelberg and New York.

EAKIN, R. M. and WESTFALL, J. A. (1965). Fine structure of the eye of *Peripatus* (Onychophora). *Z. Zellforsch. 68*, 278–300.

EXNER, S. (1891). *Die Physiologie der facettirten Augen von Kresben und Insecten*. Deuticke, Leipzig and Wein.

FAHRENBACH, W. H. (1975). The visual system of the horseshoe crab *Limulus polyphemus*. *Int. Rev. Cytol. 41*, 285–349.

FERMI, G. and REICHARDT, W. (1963). Optomotorische Reaktionen der Fliege *Musca domestica*. *Kybernetik 2*, 15–28.

FLETCHER, A., MURPHY, T. and YOUNG, A. (1954). Solutions of two optical problems. *Proc. Roy. Soc. Lond. A. 223*, 216–225.

FRANCESCHINI, N. (1975). Sampling of the visual environment by the compound eye of the fly: fundamentals and applications. In *Photoreceptor Optics*. Edited by A. W. Snyder and R. Menzel. Pages 98–125. Springer, Berlin, Heidelberg and New York.

FRANCESCHINI, N., HARDIE, R. C., RIBI, W. and KIRSCHFELD, K. (1981). Sexual dimorphism in a photoreceptor. *Nature (Lond.) 291*, 241–244.

FRIEDERICHS, H. F. (1931). Beiträge zur Morphologie und Physiologie der Sehorgane der Cicindeliden (Col.). *Z. Morphol. Okol. Tiere 21*, 1–172.

GOLDSMITH, T. H. and BERNARD, G. D. (1974). The visual system of insects. In *The Physiology of Insecta*. Vol. 2. Edited by M. Rockstein. Academic Press, New York.

GOODMAN, L. J. (1981). Organisation and physiology of the insect dorsal ocellar system. In *Handbook of Sensory Physiology*. Vol. VII/6C. Edited by H. J. Autrum. Pages 201–286. Springer, Berlin, Heidelberg and New York.

GOULD, J. L. (1975). Communication of distance information by honeybees. *J. Comp. Physiol. 104*, 161–173.

GRENACHER, H. (1879). *Untersuchungen über das Sehorgan der Arthropoden, insbesondere der Spinnen, Insecten und Crustaceen*. Vandenhoeck und Ruprecht, Göttingen.

HAMDORF, K. and HÖGLUND, G. (1981). Light induced retinal screening pigment migration independent of visual cell activity. *J. Comp. Physiol. 143*, 305–309.

HAUSEN, K. (1973). Die Brechungsindices im Kristallkegel der Mehlmotte *Ephestia kühniella*. *J. Comp. Physiol. 82*, 365–378.

HAUSEN, K. (1976). Functional characterization and anatomical identification of motion sensitive neurons in the lobula plate of the blowfly *Calliphora erythrocephala*. *Z. Naturforsch. 31c*, 629–633.

HESSE, R. (1900). Untersuchungen über die Orane der Lichtempfindung bei niederen Thieren. VI. Die Augen einiger Mollusken. *Z. Wiss. Zool. 68*, 379–477.

HÖGLUND, G. (1966). Pigment migration and retinular sensitivity. In *The Functional Organization of the Compound Eye*. Edited by C. G. Bernhard. Pages 77–101. Pergamon Press, Oxford.

HOMANN, H. (1924). Zum Problem der Ocellenfunktion bei den Insekten. *Z. Vergl. Physiol. 1*, 541–578.

HOOKE, R. (1665). *Micrographia*. J. Martyn and J. Allestry, London.

HORRIDGE, G. A. (1971). Alternatives to superposition images in clear zone compound eyes. *Proc. Roy. Soc. Lond. B. 179*, 97–124.

HORRIDGE, G. A. (1975). Optical mechanisms of clear zone eyes. In *The Compound Eye and Vision of Insects*. Edited by G. A. Horridge. Pages 255–298. Clarendon, Oxford.

HORRIDGE, G. A. (1976). The ommatidium of the dorsal eye of *Cloeon* as a specialization for photoreisomerization. *Proc. Roy. Soc. Lond. B. 193*, 17–29.

HORRIDGE, G. A. (1977). The compound eye of insects. *Sci. Amer. 237(1)*, 108–120.

HORRIDGE, G. A. (1978). The separation of visual axes in apposition compound eyes. *Phil. Trans. Roy. Soc. Lond. B 285*, 1–59.

HORRIDGE, G. A. and BERNARD, P. B. T. (1965). Movement of palisade in locust retinula cells when illuminated. *Quart. J. Mic. Sci. 106*, 131–135.

HORRIDGE, G. A. and BLEST, D. (1980). The compound eye. In *Insect Biology in the Future "VBW 80"*. Edited by M. Locke and D. S. Smith. Pages 705–733. Academic Press, New York.

HORRIDGE, G. A., GIDDINGS, C. and STANGE, G. (1972). The superposition eye of skipper butterflies. *Proc. Roy. Soc. Lond. B 182*, 457–495.

HORRIDGE, G. A., MARCELJA, L. and JAHNKE, R. (1982). Light guides in the dorsal eye of the male mayfly. *Proc. Roy. Soc. Lond B. 216*, 25–51.

HORRIDGE, G. A., McLEAN, M., STANGE, G. and LILLYWHITE, P. G. (1977). A diurnal moth superposition eye with high resolution *Phalaenoides tristifica* (Agaristidae). *Proc. Roy. Soc. Lond. B. 196*, 233–250.

HORRIDGE, G. A. and DUELLI, P. (1979). Anatomy of the regional differences in the eye of the mantis *Cuilfina*. *J. Exp. Biol. 80*, 165–190.

HOWARD, J. and SNYDER, A. (1983). Transduction as a limitation on compound eye function and design. *Proc. Roy. Soc. Lond. B. 217*, 287–307.

HUGHES, A. (1977). The topography of vision in mammals of contrasting life style: comparative optics and retinal organisation. In *Handbook of Sensory Physiology*. Vol. VII/5. Edited by F. Crescitelli. Pages 613–756. Springer, Berlin, Heidelberg and New York.

ICHIKAWA, T. and TATEDA, H. (1980). Cellular patterns and spectral sensitivity of larval ocelli in the swallowtail butterfly *Papilio*. *J. Comp. Physiol. 139*, 41–47.

ICHIKAWA, T. and TATEDA, H. (1982). Distribution of color receptors in the larval eyes of four species of lepidoptera. *J. Comp. Physiol. 149*, 317–324.

IMMS, A. D. (1957). *A General Textbook of Entomology*. Methuen, London.

IOANNIDES, A. C. and HORRIDGE, G. A. (1975). The organization of visual fields in the hemipteran acone eye. *Proc. Roy. Soc. Lond. B 190*, 373–391.

JOCKUSCH, B. (1967). Bau und Funktion eines larvalen Insektenauges. Untersuchungen am Ameisenlöwen (*Euroleon nostras* Fourcroy, Planip., Myrmel.). *Z. Vergl. Physiol. 56*, 171–198.

KERNEIS, A. (1975). Etude comparée d'organes photorécepteurs de Sabellidae (Annélides Polychetes). *J. Ultrastruct. Res. 53*, 164–179.

KIRSCHFELD, K. (1967). Die Projektion der optischen Umwelt auf das Raster der Rhabdomere im Komplexauge von *Musca*. *Exp. Brain Res. 3*, 248–270.

KIRSCHFELD, K. (1972). The visual system of *Musca*: studies on optics, structure and function. In *Information Processing in the Visual Systems of Arthropods*. Edited by R. Wehner. Pages 61–74. Springer, Berlin, Heidelberg and New York.

KIRSCHFELD, K. (1973). Das neurale Superpositionsauge. *Fortschr. Zool. 21*, 229–259.

KIRSCHFELD, K. (1974). The absolute sensitivity of lens and compound eyes. *Z. Naturforsch. 29c*, 592–596.

KIRSCHFELD, K. (1976). The resolution of lens and compound eyes. In *Neural Principles in Vision*. Edited by F. Zettler, and R. Weiler. Pages 354–370. Springer, Berlin, Heidelberg and New York.

KIRSCHFELD, K. and FRANCESCHINI, N. (1969). Ein Mechanismus zur Steurerung des Lichtflusses in den Rhabdomeren des Komplexauges von *Musca*. *Kybernetik 6*, 13–22.

KIRSCHFELD, K. and SNYDER, A. W. (1975). Waveguide mode effects, birefringence and dichroism in fly photoreceptors. In *Photoreceptor Optics*. Edited by A. W. Snyder, and R. Menzel. Pages 56–77. Springer, Berlin, Heidelberg and New York.

KIRSCHFELD, K. and WENK, P. (1976). The dorsal compound eye of simuliid flies: an eye specialized for the detection of small, rapidly moving objects. *Z. Naturforsch. 31c*, 764–765.

KUIPER, J. W. (1962). The optics of the compound eye. *Symp. Soc. Exp. Biol. 16*, 58–71.

KUIPER, J. W. (1966). On the image formation in a single ommatidium of the compound eye of Diptera. In *The Functional Organization of the Compound Eye*. Edited by C. G. Bernard. Pages 35–50. Pergamon Press, Oxford.

KUNZE, P. (1969). Eye glow in the moth and superposition theory. *Nature (Lond.). 223*, 1172–1174.

KUNZE, P. (1979). Apposition and superposition Eyes. In *Handbook of Sensory Physiology*, Vol. VII/6A. Edited by H. Autrum. Pages 441–502. Springer, Berlin, Heidelberg and New York.

KUNZE, P. and HAUSEN, K. (1971). Inhomogeneous refractive index in the crystalline cone of a moth eye. *Nature (Lond.) 231*, 392–393.

LAND, M. F. (1972). The physics and biology of animal reflectors. *Prog. Biophys. Mol. Biol. 24*, 75–106.

LAND, M. F. (1979). The optical mechanism of the eye of *Limulus*. *Nature (Lond.) 280*, 396–397.

LAND, M. F. (1980). Compound eyes: old and new optical mechanisms. *Nature (Lond.) 287*, 681–686.

LAND, M. F. (1981). Optics and vision in invertebrates. In *Handbook of Sensory Physiology*, Vol. VII/6B. Edited by H. Autrum. Pages 471–592. Springer, Berlin, Heidelberg and New York.

LAND, M. F., COLLETT, T. S. (1974). Chasing behaviour of houseflies (*Fannia canicularis*). A description and analysis. *J. Comp. Physiol. 89*, 331–357.

LAUGHLIN, S. (1981). Neural principles in the peripheral visual systems of invertebrates. In *Handbook of Sensory Physiology*, Vol. VII/6B. Edited by H. Autrum. Pages 133–280. Springer, Berlin, Heidelberg and New York.

LEVI-SETTI, R. (1975). *Trilobites. A Photographic Atlas*. University of Chicago Press, Chicago and London.

LEYDIG, F. (1855). Zum feineren Bau der Arthropoden. *Müller's Archiv. Anat. Physiol. 22*, 406–444.

LILLYWHITE, P. G. (1977). Single photon signals and transduction in an insect eye. *J. Comp. Physiol. 122*, 189–200.

MALLOCK, A. (1894). Insect sight and the defining power of composite eyes. *Proc. Roy. Soc. Lond. B. 55*, 85–90.

MATIĆ, T. (1983). Electrical inhibition in the retina of the butterfly *Papilio*: I. Four spectral types of photoreceptors. *J. Comp. Physiol.* (In press).

MAZOKHIN-PORSHNYAKOV, G. A. (1969). *Insect Vision*. Plenum Press, New York.

MCINTYRE, P. and KIRSCHFELD, K. (1982). Chromatic aberration of a dipteran corneal lens. *J. Comp. Physiol. 146*, 493–500.

MCLEAN, M. and HORRIDGE, G. A. (1977). Structural changes in light and dark-adapted compound eyes of the Australian earwig *Labidura riparia truncata* (Dermaptera). *Tissue Cell 9*, 653–666.

MEYER-ROCHOW, V. B. (1974). Structure and function of the larval eye of the sawfly, *Perga* (Hymenoptera). *J. Insect Physiol. 20*, 1565–1591.

MILLER, W. H. (1979). Ocular optical filtering. In: Handbook of Sensory Physiology. Vol. VII/6A. Edited by H. Autrum. Pages 69–143. Springer, Berlin, Heidelberg and New York.

MILLER, W. H. and BERNARD, G. C. (1968). Butterfly glow. *J. Ultrastruct. Res. 24*, 286–294.

MILLER, W. H., BERNARD, G. D. and ALLEN, J. L. (1968). The optics of insect compound eyes. *Science 162*, 760–768.

MILLER, W. H., MØLLER, A. R. and BERNHARD, C. G. (1966). The corneal nipple array. In *The Functional Organization of the Compound Eye*. Edited by C. G. Bernhard. Pages 21–33. Pergamon Press, Oxford.

MÜLLER, J. (1826). *Zur vergleichenden Physiologie des Gesichtsinnes*. C. Cnobloch, Leipzig.

PAULUS, H. F. (1975). The compound eyes of apterygote insects. In *The Compound Eye and Vision of Insects*. Edited by G. A. Horridge. Pages 3–19. Clarendon, Oxford.

PAULUS, H. F. (1979). Eye structure and the monophyly of the arthropoda. In *Arthropod Phylogeny*. Edited by A. P. Gupta. Pages 299–383. Van Nostrand Reinhold, New York.

PIRENNE, M. H. (1967). *Vision and the Eye*. Chapman & Hall, London.

RÉAUMUR, M. (1740). *Mémoires pour servir a l'Histoire des Insectes*. Tome 5ieme. L'Imprimerie Royale, Paris.

RIBI, W. (1978). Ultrastructure and migration of screening pigments in the retina of *Pieris rapae* L. (Lepidoptera, Pieridae). *Cell Tiss. Res. 191*, 57–73.

RIBI, W. (1979). Structural differences in the tracheal tapetum of diurnal butterflies. *Z. Naturforsch. 34c*, 284–287.

ROSE, A. (1973). *Vision, Human and Electronic*. Plenum Press, New York and London.

ROSSEL, S. (1979). Regional differences in photoreceptor performance in the eye of the praying mantis. *J. Comp. Physiol. 131*, 95–112.

ROSSEL, S. (1980). Foveal fixation and tracking in the praying mantis. *J. Comp. Physiol. 139*, 307–331.

SCHWIND, R. (1980). Geometrical optics of the *Notonecta* eye: adaptations to optical environment and way of life. *J. Comp. Physiol. 140*, 59–66.

SCHULTZE, M. (1868). *Untersuchungen über die zusammen gestzten Augen der Krebse und Insekten*. Cohen, Bonn.

SEITZ, G. (1969). Untersuchungen am dioptrischen Apparat des Leuchtkäferauges. *Z. Vergl. Physiol. 62*, 61–74.

SHERK, T. E. (1978). Development of the compound eyes of dragonflies (Odonata). III. Adult compound eyes. *J. Exp. Zool. 203*, 61–80.

SNYDER, A. W. (1979). Physics of vision in compound eyes. In *Handbook of Sensory Physiology*. Vol. VII/6A. Edited by H. Autrum. Pages 225–313. Springer, Berlin, Heidelberg and New York.

SNYDER, A. W., STAVENGA, D. G. and LAUGHLIN, S. B. (1977). Spatial information capacity of compound eyes. *J. Comp. Physiol. 116*, 183–207.

STANGE, G. (1981). The ocellar component of flight equilibrium control in dragonflies. *J. Comp. Physiol. 141*, 335–347.

STANGE, G. and HOWARD, J. (1979). Ocellar dorsal light response in a dragonfly. *J. Exp. Biol. 83*, 351–355.

STAVENGA, D. G. (1979). Pseudopupils of compound eyes. In *Handbook of Sensory Physiology*, Vol. VII/6A. Edited by H. Autrum. Pages 357–439. Springer, Berlin, Heidelberg and New York.

STRAUSFELD, N. J. and NÄSSEL, D. R. (1981). Neuroarchitectures serving compound eyes of crustacea and insects. In *Handbook of Sensory Physiology*, Vol. VII/6B. Edited by H. Autrum. Pages 1–132. Springer, Berlin, Heidelberg and New York.

SWAMMERDAM, J. (1737). *Biblia Naturae sive Historia Insectorum*. Edited by H. Boerhaave. Severinum & Vander, Leyden.

TRUJILLO-CENÓZ, O. (1965). Some aspects of the structural organization of the intermediate retina of dipterans. *J. Ultrastruct. Res. 13*, 1–33.

TRUJILLO-CENÓZ, O. and MELAMED, J. (1966). Electron microscope observations on the peripheral and intermediate retinas of dipterans. In *The Functional Organization of the Compound Eye*. Edited by C. G. Bernhard. Pages 339–361. Pergamon Press, Oxford.

VIGIER, P. (1909). Mécanisme de la synthése des impressions lumineuses recueillies par les yeuz composés des Dipteres. *C.R. Sci. Paris 148*, 1221–1223.

VOGT, K. (1975). Zur Optik des Flusskrebsauges. *Z. Naturforsch. 30c*, 691.

VOGT, K. (1980). Die Spiegeloptik des Flusskrebsauges. The optical system of the crayfish eye. *J. Comp. Physiol. 135*, 1–19.

WACHMANN, E. (1972). Das Auge des Hünherflohs *Ceratophyllus gallinae* (Schrank) (Insecta, Siphonaptera). *Z. Morph. Tiere, 73*, 315–324.

WALCOTT, B. (1975). Anatomical changes during light-adaptation in insect compound eyes. In *The Compound Eye and Vision of Insects*. Edited by G. A. Horridge. Pages 20–33. Clarendon, Oxford.

WALD, G., BROWN, P. K. and GIBBONS, I. R. (1962). Visual excitation: a chemo-anatomical study. *Symp. Soc. Exp. Biol. 16*, 32–57.

WALLS, G. L. (1942). *The Vertebrate Eye and its Adaptive Radiation*. Hafner, New York.

WATERMAN, T. H. (1981). Polarization sensitivity. In *Handbook of Sensory Physiology*, Vol. VII/6B. Edited by H. Autrum. Pages 281–469. Springer, Berlin, Heidelberg and New York.

WEHNER, R. (1976). Structure and function of the peripheral visual pathway in hymenopterans. In *Neural Principles in Vision*. Edited by F. Zettler, and R. Weiler. Pages 280–333. Springer, Berlin, Heidelberg and New York.

WEHNER, R. (1981). Spatial vision in arthropods. In *Handbook of Sensory Physiology*, Vol. VII/6C. Edited by H. Autrum. Pages 287–616. Springer, Berlin, Heidelberg and New York.

WELLINGTON, W. G. (1974). Bumblebee ocelli and navigation at dusk. *Science 183*, 550–551.

WILLIAMS, D. S. (1982). Ommatidial structure in relation to turnover of photoreceptor membrane in the locust. *Cell. Tiss. Res. 225*, 595–617.

WILSON, M. (1975). Autonomous pigment movement in the radial pupil of locust ocelli. *Nature (Lond.) 258*, 603–604.

WILSON, M. (1978). The functional organisation of locust ocelli. *J. Comp. Physiol. 124*, 297–316.

WOLBURG-BUCHHOLZ, K. (1976). The dorsal eye of *Clöeon dipterum* (Ephemeroptera). A light- and electronmicroscopical study. *Z. Naturf. 31c*, 335–336.

WOLBURG-BUCHHOLZ, K. (1979). The organisation of the lamina ganglionaris of the hemipteran insects *Notonecta glauca, Corixa punctata, Gerris lacustris. Cell Tiss. Res. 197*, 39–59.

WOLSKI, A. (1931). Weitere Beiträge zum Ocellenproblem — Die optischen Verhältnisse der Ocellen der Honigbiene (*Apis mellifica* L.). *Z. Vergl. Physiol. 14*, 385–389.

WOLSKI, A. (1933). Stimulationsorgane. *Biol. Rev. 8*, 370–417.

ZEIL, J. (1979). A new kind of neural superposition eye: the compound eye of male Bibionidae. *Nature (Lond.) 278*, 249–250.

6   The Eye: Electrical Activity

ROBERT D. DEVOE

The Johns Hopkins University, Baltimore, Maryland, USA
and
Indiana University, Bloomington, Indiana, USA (present address)

1 INTRODUCTION AND OVERVIEW

The scene that any eye sees is a spatial continuum, broken up into discontinuities and gradations of intensities of emitted and reflected light. Vision in insects begins first by sampling this visual field via the dioptics of the compound eye and then by focusing the sampled radiant fluxes onto discrete receptive areas of the photoreceptor cells. The specialized receiving portions of the photoreceptor cells are the rhabdoms. Phototransduction involves the absorptions of photons by the visual pigments in the rhabdoms and, after a sequence of largely unknown processes, the initiation of electrical activity. Electrical activity of the photoreceptors arises by current flows through ion-specific membrane channels, the same as in other excitable tissues. In the insect eye, the principal excitatory channels are sodium channels and their openings are mediated by light alone. When open, there is an inward sodium photocurrent, which reduces (depolarizes) the normally negative membrane potential. Insects thus have typical invertebrate photoreceptors, with depolarizing responses (Fuortes, M. and O'Bryan, P., 1972) and not the hyperpolarizing responses of vertebrate rods and cones or of a few unusual invertebrates (e.g. the distal retina of the scallop *Pectin*: Gorman, A. and McReynolds, J., 1978).

The subjects of the initial part of this chapter will be the mechanisms and processes of phototransduction: what are the relations between photons absorbed and the receptor potential? The converse is the study of sensitivities: how can the numbers of effective photons absorbed be told from the receptor potential? In either case, the electrical activity that has been measured in insects so far is the photovoltage that results from the flows of photocurrent. At the rhabdomeric sites of opened sodium channels the photovoltages result because the membrane potential moves closer to the sodium equilibrium potential. These depolarizations can, and do, initiate openings of other, opposing, electrically excitable ion channels (such as for potassium and calcium: O'Day, P. *et al.*, 1982). If the depolarizations in an insect photoreceptor are large enough there is often an initial spike due to regenerative, electrically excitable channels. Insect photoreceptors do not normally signal strengths of depolarizations by trains of impulses, however. Indeed, the only known arthropod primary photoreceptors that do are those of scorpion medial and lateral ocelli (Belmonte, C. and Stensaas, L., 1975). In these, spike frequencies are proportional to depolarizations (DeVoe, unpublished observations). Molluscan primary photoreceptors, on the other hand, are often found to spike repetitively (Hartline, H., 1938; MacNichol, E. and Love, W., 1960).

Instead of conducted, repetitive spikes, the important vehicle for information transmission in insect photoreceptors is longitudinal current. Current which enters at the light-absorbing rhabdoms, and which is not cancelled there by outward currents, must exit elsewhere in the cell. Because the rhabdom becomes depolarized with respect to the photoreceptor's axon, current will flow longitudinally between these areas of different membrane potential. A proportion of photocurrent thus flows intracellularly down the axon and exits at the receptor terminal. Since no openings of electrically excitable axonal channels are involved, this current spread is said to spread passively (electrotonically) down the axon to the terminal. Depolarization at the terminals by the electrotonic spread causes transmitter release and hence synaptic transmission onto second-order cells. Excitation of second (and higher) order cells thus results, as in any other neural network.

Photocurrents exiting photoreceptors can complete the circuit by flowing back in extracellular space, as well as by flowing back through other cells. In the latter case the cells are said to be electrically coupled. Vertebrate rods are electrically coupled in multicelled syncytia. Such coupling results in pooling of signals, so that perhaps 80% of the photovoltage recorded in a single rod is the result of activity in other rods (Fain, G., 1975). In insects there can be positive coupling of this type as well as negative coupling, resulting in hyperpolarizations of neighboring photoreceptors. Just what, if any, coupling occurs varies from species to species and depends completely on the weighted paths available for the return flows of photocurrents. In essence, however, photoreceptors can excite or inhibit each other via coupling. Second-order cells can also feed back in turn onto the receptors. In other words, the photoreceptor matrix should be looked upon as a network of spatially interacting cells: the visual field is not merely sampled punctately by neighboring photoreceptors of the compound eye, but an initial processing of intensity contrasts can also be occurring by virtue of differential photocurrent flows. A primitive neural representation of the visual image thus exists in the aggregate of all photoreceptor activity.

More sophisticated extraction from the visual image of features of interest to a particular species is carried out by cellular networks of the optic lobes, in the lamina, medulla, or lobula complex. In the lamina and distal medulla at least, the synaptic organization is repeated hundreds or thousands of times in each neural cartridge in a neurological crystal, so to speak. Each cartridge does just about what every other cartridge does for the same relatively positioned neural inputs. Absolute positional information is only weakly encoded in the distal optic lobes, compared to the proximal medulla and especially in lobular giant cells. There, positional information is apportioned among giant cells according to their dendritic spreads. The same cells also extract features of interest about the stimulus, such as its relative size, the direction it is moving, or its color contrasts. Such extracted information is then transmitted to brain centers or via the ventral nerve cord to the motor centers of flight and walking. This chapter will, however, be confined to the actual electrical activity of optic lobe cells. Their roles in visual perception will be the subject of the following chapter (chapter 7).

1.1 Historical background: photoreceptor potentials

ERGs (electroretinograms) were the earliest recordings of electrical activity in the insect eye (Hartline, H. K., 1928). An ERG is recorded by placing an active electrode on or beneath the cornea, and another in an "indifferent" location (such as in a blood space, on another eye, or in the abdomen, etc.). The corneal electrode then becomes negative to the other during illumination of the eye. Examples of ERGs from two different eyes are shown in the middle in Fig. 1; downward deflections are negative. The ERG at the left is a simple, monophasic wave, whereas the ERG at the right has additional, superimposed, positive-ON and negative-OFF effects.

ERGs result from radial flows of photocurrents in resistive extracellular space, as will be detailed below in section 2.5. When large numbers of highly aligned photoreceptors are illuminated, the massed radial current flows can develop quite large extracellular photovoltages. As a result, insect ERGs can be 10 mV or more in amplitude. Vertebrate ERGs, on the other hand, rarely exceed a few hundred μV. ON- and OFF-effects in insect ERGs result from additional radial current flows, namely those of activated laminar monopolar cells. If the currents from the monopolar cells do not flow in the same extracellular resistive paths as do the photocurrents, then there is no summation and no ON- and OFF-effects. This is what happened in the ERG at the middle left in Fig. 1.

An enormous amount of information about insect photoreceptors has been, and continues to be, obtained using recordings of ERGs. Representative areas that were pioneered with ERGs include intensity–duration reciprocity (Hartline, H. K., 1928); diurnal rhythms (Jahn, T. and Crescitelli, F., 1940); spectral sensitivities (Jahn, T., 1946; Goldsmith, T., 1960); flicker resolution (Autrum, H., 1950); polarization sensitivities (Autrum, H. and Stumpf, H., 1950); light and dark adaptation (Ruck, D., 1958), and the M-potential (Pak, W. and Lidington, K., 1974: see section 2). The ERG still continues to be the electrical response of choice under conditions where intracellular recordings are unfeasible, difficult, or too unstable for long-term experiments.

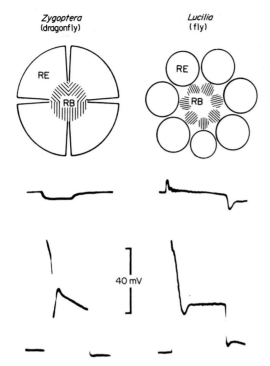

FIG. 1. Examples of simple and complex ERGs (middle row) and the corresponding intracellular recordings (lower row) from retinula cells (RE) of a dragonfly (*Agriocnemis*) and of a fly (*Lucilia*) in response to saturating flashes of light. In this and in all subsequent figures, positivity is upward and negativity is downward (depolarizations and hyperpolarizations, respectively, in intracellular recordings). The top row depicts the fused and open rhabdoms (RB) of dragonflies and flies, respectively. Reproduced from Naka, K.-I. (1961) *J. Gen. Physiol.* *44*, 571–584, by permission of the author and copyright permission of The Rockefeller University Press.

With the introduction of intracellular recordings, first in the horseshoe crab *Limulus* (Hartline, H. *et al.*, 1952) and later in insects (Kuwabara, M. and Naka, K.-I., 1959; Burkhardt, D. and Autrum, H., 1960), it could be shown that the photoreceptor response was a distal depolarization resulting from an increase in cell membrane conductance (Fuortes, M., 1959, 1963). The lower traces in Fig. 1 show the characteristic intracellular photoreceptor responses to bright lights for two species. Although the waveforms of the ERGs differ the intracellular waveforms do not. At light onset there is an abrupt, rapid depolarization, sometimes to 0 mV membrane potential (but seldom more). If the flash is long enough the depolarization declines to a lower plateau. This decline is called sensory adaptation. Sometimes the decline is oscillatory, as in Fig. 1.

Other times, there is no sustained plateau, as in green-sensitive cells of the cockroach *Periplaneta* (Mote, M. and Goldsmith, T., 1970). Finally, at light off, the receptor potential returns back towards, to, or past, baseline (the resting potential). Two kinds of afterpotentials can be recorded, prolonged depolarizing afterpotentials (PDAs) following massive conversions of rhodopsin to metarhodopsin, or shorter hyperpolarizing afterpotentials which are probably due to restorative ion pumping (see section 2.2, below). Indications of hyperpolarizing and depolarizing afterpotentials can be seen in the left and right lower traces, respectively, of Fig. 1 (see also Fig. 6).

1.2 Historical background: optic lobe responses

As for the retina, so too for the optic lobes: initially recorded, light-evoked activity was of massed extracellular slow or spike potentials (Adrian, E., 1937; Bernhard, C., 1942). Single-unit recording was first undertaken from giant afferents of the ventral nerve cord with hook electrodes (Parry, D., 1947), or with extracellular metal microelectrodes in the optic lobes (Burtt, E. and Catton, W., 1956). Rarely, however, has it been possible to know from which cells such extracellular spikes have been recorded. Some probable identifications were made from heroic tracings of the pathways of impulse activity through the optic lobes of the moth, *Sphinx* (Collett, T., 1970). Thus, the introduction in the 1970s of intracellular staining through the recording electrode made it possible to identify both spiking cells known from previous extracellular recordings (O'Shea, M. *et al.*, 1974; Hausen, K., 1976) as well as non-spiking cells in the lamina, medulla, and lobula (Järvilehto, M. and Zettler, F. 1970; DeVoe, R. and Ockleford, E., 1976; Dvorak, D. *et al.*, 1975, respectively). In particular, the structure and function of giant lobular plate cells of flies are now among the best understood of insect visual interneurons (Hausen, K., 1981).

2 PHOTORECEPTOR EXCITATION

The electrical responses of photoreceptors to light are called *receptor potentials*. Receptor potentials are graded functions of light intensities, up to saturating intensities (see Fig. 7, below). At the weakest intensities, however, a lower limit is set by the quantal nature of light (section 2.1). There are two kinds of receptor potentials: the early receptor potential (ERP) and the late receptor potential (LRP, so called because it occurs later than the ERP). ERPs are elicited by strong, brief stimuli which isomerize appreciable fractions of a cell's visual pigments. They were first recorded from monkey cones (Brown, K. and Murakami, M., 1964), and have since been found in a number of invertebrate photoreceptors (squid: Hagins, W. and McGaughy, R., 1968; the horseshoe crab *Limulus* and the barnacle *Balanus*: Hillman, P. *et al.*, 1973). ERPs are directly proportional to the numbers of photopigment molecules isomerized (Cone, R., 1965) and can persist after extremely non-physiological treatments, such as glutaraldehyde fixation. ERPs seem to result from changes in membrane charge following dipole moments of membrane-bound photopigments, due to isomerization (Hagins, W. and McGaughy, R., 1968).

ERPs due to rhodopsin isomerization are unknown in insect photoreceptors. There is, however, an ERP associated in flies with the conversion of metarhodopsin by light back to rhodopsin (see chapter 8). This has been called the M-potential (Pak, W. and Lidington, K., 1974). M-potentials can be recorded either intracellularly or with the ERG. In the fly ERG, two M-potentials can be recorded, M1 and M2. M1 is due to a depolarizing ERP of the photoreceptor, whereas M2 is a response that can also be recorded intracellularly from the second-order, laminar monopolar cells (Minke, B. and Kirschfeld, K., 1980). In the laminar monopolar cells both early and late responses are hyperpolarizing. This is depicted in Fig. 2. The second-order M-potential, M2, thus appears to result from normal transsynaptic excitation following generation of M1 in the photoreceptors. It is doubtful, however, that such ERPs play important roles in insect vision. Unphysiological adaptation states and sudden bright lights are required to elicit them, while photic transitions from metarhodopsin to rhodopsin do not otherwise excite photoreceptors (Barnes, S. and Goldsmith, T., 1977; Strong, J. and Lisman, J., 1978). On the other hand, the M-potential has been a useful tool for the study of

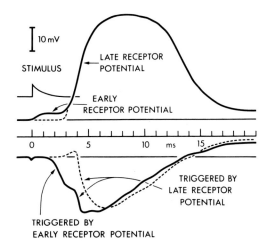

FIG. 2. Intracellular recordings of early and late potentials of peripheral photoreceptors R1–6 (top) and laminar monopolar cells (bottom) of flies (*Calliphora*). The dashed lines indicate the late potentials only. The stimulus trace shows the time course of the brief, intense flash. Reproduced from Kirschfeld, (1983), *Trends in Neurosciences, 6*, 97–101, by permission of the author and copyright permission of Elsevier Publications.

insect photopigment dynamics (Stark, W. *et al.*, 1977; Minke, B. and Kirschfeld, K., 1980).

The late receptor potential, on the other hand, occurs only under physiological conditions. In addition to being graded and showing sensory adaptation, it has a latency. This is depicted in Fig. 2 (the ERP, in contrast, has no detectable latency: Cone, R., 1967). LRPs are the identifiable results of photoreceptor transduction. They have been recorded intracellularly from eyes of a number of insect orders: Odonata, Orthoptera, Hemiptera, Neuroptera, Lepidoptera, Diptera, Coleoptera, and Hymenoptera. When tested, LRPs in insects are associated with increases in membrane conductance (Fuortes, M., 1963; Washizu, Y., 1964). Thus, the primary result of phototransduction is the opening of ionic channels, allowing inward, depolarizing photocurrents to flow. In invertebrates the locus of inward current flow appears to be the rhabdom, where the photopigments are localized. Hagins, W. (1965) showed for squid photoreceptors that inward current flowed almost entirely in the area of the rhabdom illuminated. The evidence from insect photoreceptors is consistent with initiation of inward photocurrents distally in the cells, where the rhabdoms are. First, LRPs are larger distally, where the rhabdom is, than proximally (Zettler, F., 1967;

Ioannides, A. and Walcott, B., 1971). Second, a current sink is set up in the distal photoreceptors of flies during illumination (Zimmerman, R., 1978). Thus, it is reasonable to suppose that in insect photoreceptors too, photocurrents result from openings of membrane ionic channels close to or at the loci where photons are absorbed.

Direct measurements of membrane photocurrents have been made by the voltage clamp method in *Limulus* (Millechia, R. and Mauro, A., 1969) and in *Balanus* (Brown, H. *et al.*, 1970), but not yet in any insect photoreceptor. Similarly, most studies of the biochemistry of phototransduction have been made in such giant cells as those of *Limulus*, or in vertebrate photoreceptors (cf. Fein, A. and Szuts, E., 1982). Conclusions about insect phototransduction rely heavily instead upon experimental manipulations of the LRP as a measure of underlying channel activity and photocurrents.

2.1 Quantum bumps

Light from a source is randomly emitted in quantized energy packets called photons. The energy of a single photon is inversely proportional to the wavelength of the light: shorter-wavelength photons each have more energy than long-wavelength photons. Intense light sources emit so many photons per second that there are few significant statistical fluctuations in their numbers from second to second. The quantum fluxes (photons per second) emitted from very weak light sources do vary significantly with time. Thus, a single flash of 1 s might contain 0, 1, 2 ..., etc. photons. These are the intensity levels at the absolute threshold of vision.

In 1942, Hecht, S. *et al.* tested absolute thresholds with just such weak flashes and concluded that a single photon could excite a single rod. The electrical response of a vertebrate rod to a single photon has only recently been measured unambiguously (Baylor, D. *et al.*, 1979). Earlier in an invertebrate eye, the lateral eye of *Limulus*, Yeandle, S. (1958) measured small brief electrical events that were correlated with the numbers of photons in very weak flashes (Fuortes, M. and Yeandle, S., 1964). These "quantum bumps" have since been recorded in *Locusta* (Scholes, J., 1965; Lillywhite, P., 1977), in dragonflies (Laughlin, S., 1976a), the cockroach

Periplaneta (Smola, U. and Gemperlein, R., 1973), in the flies *Musca, Calliphora* and *Drosophila* (Kirschfeld, K., 1965; Wu, C.-F. and Pak, W., 1975), in the praying mantis *Tenodera* (Rossel, S., 1979), and in the bulldog ant *Myrmecia* (Lieke, E., 1981).

Examples of intracellular recordings of quantum bumps in *Drosophila* are shown in Fig. 3. The baseline is relatively quiet in the dark, although there are spontaneous "dark" bumps. Both light and dark bumps are asymmetrical: they have faster rise than fall times. The numbers of bumps increase in proportion as the weakest intensities are increased. This has long been interpreted to mean that there is a fixed relation between the average number of photons in a flash and the average number of quantum bumps recorded (Yeandle, S., 1958). To find that relation, however, it is necessary to distinguish between light-elicited bumps and the spontaneous dark bumps. These latter are due to thermal and other excitations of the phototransduction mechanism (Fein, A. and Corson, D., 1981). Their dark frequencies can differ from one species to another, as well as with time of day (in the lateral eye of *Limulus*; Kaplan, E. and Barlow, R., 1980). A particularly quiet photoreceptor is that of *Locusta*: it has as few as 10 quantum bumps per *hour* when fully dark adapted (Lillywhite, P., 1977). The relation between bumps and incident photons is linear. From statistical analyses of such data, it appears that one bump is elicited by just one photon. The probability that a photon will produce a bump is 0.59 (Lillywhite, P., 1977). For the fly *Musca* the probability is 0.52 (Dubs, A. *et al.*, 1981). This is the quantum efficiency of bump production. The quantum efficiency of isomerization of rhodopsin is about 0.7 (Dartnall, H., 1968). Thus, in locust photoreceptors, each photon effectively absorbed by rhodopsin elicits a bump with nearly 100% probability (Fein, A. and Szuts, E., 1982).

The shape of a quantum bump is independent of the wavelength of the absorbed photon (Wu, C.-F. and Pak, W., 1975; Lillywhite, P., 1978). The spectral sensitivity of a photoreceptor depends, therefore, on the relative efficiency of quantum bump production at different wavelengths. The efficiency, in turn, depends on the absorption spectrum of the underlying photopigment (see chapter 8). The quantized energy of an absorbed

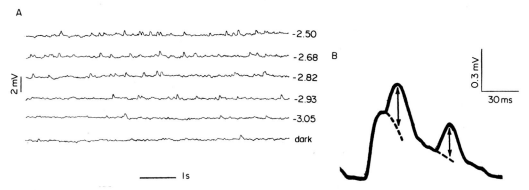

FIG. 3. Intracellularly recorded quantum bumps of photoreceptors of *Drosophila melanogaster*. Several seconds of recordings are shown in A for 450 nm continuous illumination at the relative log intensities given to the right of each trace. Summation of bumps is illustrated in B at higher resolution. The dashed lines are the extrapolated continuations of the previous bumps, showing how bump amplitudes (arrows) are measured. Reproduced from Wu, C.-F. and Pak, W. (1975), *J. Gen. Physiol. 66*, 149–168, by permission of the authors and copyright permission of The Rockefeller University Press.

photon is thus irrelevant. If absorbed, one photon produces the same physiological effect as another.

2.1.1 SUMMATIONS OF BUMPS

As light intensities and hence bump frequencies are increased, bumps begin to overlap in time and to add. This is shown in Fig. 3B, where three bumps are seen to sum. As bump frequencies and temporal overlaps increase still more, the summations result in maintained depolarizations with superimposed "noise". This can be observed in Fig. 4. The maintained depolarizations are what were called the late receptor potentials (LRP) in Fig. 2, above. For low intensities, LRPs are essentially rectangular. At medium and high intensities the initial LRP is larger than the later plateau depolarization. The plateau noise is also smaller than at low intensities. The sensory adaptation and reduced plateau noise could have a common cause. The common cause could be either production of fewer, constant-amplitude bumps with time, or production of smaller bumps. The latter appears to be the case for the wild-type *Drosophila* and for Fig. 4. Noise analyses show that the frequencies of bumps increase linearly with light intensities, up to saturation at about 3×10^5 bumps/second in *Drosophila* (Wu, C.-F. and Pak, W., 1978), and over a range of about 5 log units in the ventral eye of *Limulus* (Wong, F., 1978). Thus, the *sizes* of bumps must decrease during adaptation. This is the basis for the adapting bump model of the late receptor potential (Dodge, F. *et al.*, 1968).

In *Drosophila*, bump amplitudes are roughly constant for weak intensities, then fall by about two orders of magnitude over the range where light intensities are increasing the last four orders of

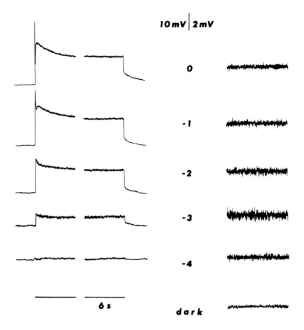

FIG. 4. Summations of quantum bumps in photoreceptors of *Drosophila melanogaster* to yield late receptor potentials (left-hand, interrupted traces) with superimposed fluctuations ("noise", shown at higher amplification in the right-hand traces). The relative log intensities of 540 nm flashes are given in the middle column. Reproduced from Wu, C.-F. and Pak, W. (1978), *J. Gen. Physiol. 71*, 249–268, by permission of the author and copyright permission of The Rockefeller University Press.

magnitude to bump-rate saturation (Wu, C.-F. and Pak, W., 1978). The constancies of amplitudes at low intensities fit with the results of Dubs, A. (1981), who found in the fly *Musca* that response amplitudes increased linearly up to 10 photons absorbed per cell. Non-linearities begin for responses of only 7–12 mV. Changes in bump amplitudes probably account for the non-linearities. In the lateral eye of *Limulus*, bump amplitudes and durations lessen when more than four quanta are absorbed per rhabdom (Borsellino, A. and Fuortes, M., 1968). However, the underlying photocurrents, during voltage clamp, increase linearly for quantal absorptions up to 100/cell in *Limulus* ventral photoreceptors (Lisman, J. and Brown, J., 1975a). Thus two processes seem to be at work, the first of which reduces the voltages but not the currents per quantum bump, and the second which reduces the currents per bump. It will be shown later (section 2.2) that voltage-sensitive potassium currents are probably responsible for voltage reductions, whereas rises in intracellular calcium ion probably reduce the photocurrents per bump (see section 2.3).

Bump shape (duration) may also change during light adaptation. In flies this occurs only over the lowest ranges of intensity (Dubs, A., 1981). For the results from *Drosophila* in Fig. 4, all light intensities were high enough that bump durations did not change at all (Wu, C.-F. and Pak, W., 1978). However, there is a reduction in both amplitudes and durations during light adaptation in *Limulus* (Dodge, F. *et al.*, 1968) and in locusts (Tsukahara, Y. and Horridge, G., 1977b).

2.1.2 LATENCIES OF BUMPS

Quantum bumps that are elicited by repeated, weak, brief flashes do not all have the same latencies. The dispersion of dark-adapted latencies varies with the species. In *Limulus*, latencies range over 0.5 to 3.5 s in the lateral eye (Fuortes, M. and Yeandle, S., 1964). In the fly *Calliphora*, the latencies vary about ±20 ms around a mean of 60–70 ms at 10° (Hamdorf, K. and Kirschfeld, K., 1980). The relative dispersion of bump latencies of locust photoreceptors, 20–100 ms, appears to be intermediate.

Some phototransduction mutants have altered bump latencies. The *norpA*[H52] mutant of *Drosophila* has a much larger range of bump latencies than does

the wild type, although the bump time courses are the same. One result is a receptor potential which can last for tens of seconds, to minutes, after a flash is turned off (Pak, W. *et al.*, 1976). Another is inability to follow flicker at frequencies above 1–2 Hz, *versus* 20 Hz for the wild type under comparable conditions (Wu, C. and Wong, F., 1977).

In addition to a dependence of durations of receptor potentials on summed bumps, the initial rates of rise of LRPs will be less if their component bumps are dispersed in time, and greater if the component bumps are more synchronous. The dispersions described are those for dark-adapted photoreceptors and weak lights. A general rule for dark-adapted photoreceptors is that latencies decrease and rates of rise of the LRP increase as light intensities increase. These results likewise seem to result from the properties of quantum bumps, namely a shortening of bump latencies.

Calliphora, with its low dispersion of dark-adapted latencies, has proven excellent for demonstrating this. Figure 5 illustrates high-gain recordings of latencies of receptor potentials elicited by short flashes. Latencies seem to decrease with each 10-fold increase in intensity. At the lowest intensity individual bumps are recorded. At a 10 times higher intensity, the limit of linear summation (Dubs, A., 1981), amplitudes are greater but the latencies in all but one trace are the same as for the weakest intensity. In that one trace, a "prebump" can be seen. Its latency is shorter, it rises more steeply, yet it has the amplitude (1 mV) of a single bump. Hamdorf, K. and Kirschfeld, K. (1980) proposed that a prebump might be due to double photon hits in the same, or neighboring, microvillus. Responses at still brighter intensities (6 in Fig. 5) all had the shorter latency of the prebump at intensity 7. It can be argued that there were many more double hits. There may also have been triple hits, indicated by a still shorter latency prebump (6 in Fig. 5). Thus, shortening of bump latencies with increased intensities may be due to increased probabilities of multiple photon hits at or near the membrane sites of bump production. Multiple photon hits in the same cell (if not the same microvillus) can also decrease *peak* times of bumps. Amplitudes add linearly for up to 10 photons absorbed per cell in *Musca*, but peak times are reduced when as few as four photons are absorbed in *Musca* (Dubs, A., 1981) and in *Limulus*

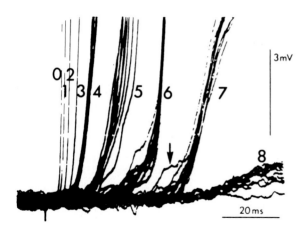

FIG. 5. Latencies of late receptor potentials recorded from *Calliphora* photoreceptors at the negative relative log intensities indicated next to each trace. Traces have been superimposed for log intensities of 10^{-3} to 10^{-8}. The times of occurrence of the brief flashes are indicated by the small, downward deflection near the beginning of the traces. The arrow indicates a "prebump". Reproduced from Hamdorf, K. and Kirschfeld, K. (1980), *Z. Naturforsch.* 35, 173–174, by permission of the authors and copyright permission of the Verlag der Zeitschreft für Naturforschung.

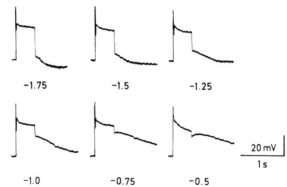

FIG. 6. Afterpotentials following saturated receptor potentials, recorded from photoreceptor cells of the dronefly *Eristalis*. The relative log intensities in each 451 nm s^{-1} flash are indicated below the traces. At log intensities -1.25 and greater, the receptor potentials are followed by prolonged depolarizing afterpotentials (PDAs). At lower intensities the afterpotentials are hyperpolarizing. Reproduced from Tsukahara, Y. *et al.* (1977), *J. Comp. Physiol.* 114, 253–266, by permission of the authors and copyright permission of Springer-Verlag, Heidelberg.

lateral eyes (Borsellino, A. and Fuortes, M., 1968). The reductions in peak time are not seen in bump currents from voltage-clamped ventral photoreceptors of *Limulus*, however. Rather, peak times remained constant up to the limits of summation of bump currents (100 photons absorbed/cell: Lisman, J. and Brown, J., 1975a). Thus, time courses as well as amplitudes of voltages during bumps may be altered by ionic currents flowing through voltage-sensitive channels.

2.1.3 BUMPS DURING PDAS

Prolonged depolarizing afterpotentials (PDAs) are recorded at the offsets of intense lights which convert appreciable fractions of rhodopsin to metarhodopsin. Figure 6 illustrates the development of the PDA in the dronefly *Eristalis* at light intensities that already saturate the LRP. The normal, abrupt repolarization of the LRP is interrupted by a lasting depolarization that slowly decays away, over tens of seconds to hours, depending on the species. The max PDA may be less than or about the same as the plateau response of the LRP. The PDA can be "killed", or knocked down, by lights which convert metarhodopsin back to rhodopsin. The relations between the thermally-stable insect visual pigments

and the PDA are presented in chapter 8 (see Hillman, P. *et al.*, 1983, for a detailed review).

In the *Drosophila* visual mutant *trp* and in UV-R7 cells of *Calliphora*, the PDA is noisy and appears to be made up of summed quantum bumps (Minke, B. *et al.*, 1975; Hardie, R. *et al.*, 1979). In *Locusta*, on the other hand, the PDA does not become resolved into individual bumps until after 15–20 min. At first, the bumps have the reduced amplitudes and durations associated with light-adapted bumps. Later, they grade into dark-adapted bumps (Tsukahara, Y. and Horridge, G., 1977b; Horridge, G. and Tsukahara, Y., 1978). During the PDA of wild-type *Drosophila* photoreceptors R1–6, sensitivities are drastically reduced, presumably because photic bump production remains strongly light-adapted (Minke, B. *et al.*, 1975).

As would be expected if both LRPs and PDAs result from summations of bumps, both are dependent on external sodium ion and both reverse at the same membrane potential in photoreceptors of the drone honey bee. Neither potassium nor chloride ions appear to be actively involved in the generation of PDAs (Baumann, F. and Hadjilazaro, B., 1972). Conductances increase during both the LRP and PDA in blowflies (Muijser, H. *et al.*, 1975) and in *Drosophila* (Wong, F. *et al.*, 1976). On the other hand, conductances decrease during the later PDA (relative to both dark and the LRP) of drone honey bees and locusts (Baumann, F. and Hadjilazaro, B., 1972; Tsukahara, Y. and Horridge, G., 1977b). These conductance decreases are unexplained.

Anoxia (replacement of air by N_2 or CO_2) abolishes the conductance changes in *Drosophila* but not the initiation or the knockdown of the PDA (Wong, F. *et al.*, 1976). This is consistent with the dependence of the PDA on photochemical, not membrane or metabolic, activity. On the other hand, metabolic activity seems essential for recovery of normal resting membrane potential and low noise after light. Azide-poisoned, locust photoreceptors, stimulated with sub-saturating lights, have after-depolarizations which normal receptors do not. Just as the PDA seems to result from sums of quantum bumps, so the after-depolarizations in poisoned receptors seem to arise from sums of $40\,\mu V$ or so potentials resulting from unitary openings of sodium channels. These unitary potentials are, however, similar to fully light-adapted quantum bumps (Payne, R., 1981). Thus the pathways from visual pigment activations to membrane activations can be affected by metabolic poisoning.

2.1.4 Mechanisms of bump production

Quantum bumps have their origins in the openings and closings of membrane ionic channels. This has been shown explicitly in voltage-clamped ventral photoreceptors of *Limulus*, where bumps are manifested as inward flows of current. The reversal potentials for bump current flow are between 0 and $+20\,mV$, the same as for the late receptor potential. The current appears to be largely carried by sodium ions, so sodium channels will be referred to here (potassium and calcium channels may also be active). The sodium channels are not electrically excitable: only light opens them (Millechia, R. and Mauro, A., 1969; Brown, J. and Mote, M., 1974). For bumps of the fully dark-adapted eye, probably 1000 or more sodium channels open asynchronously, then close independently with an exponential distribution of the lengths of time that each is open. At increasing light intensities, decreasing numbers of channels would contribute to each decreasing bump, until at saturation each bump would result from the opening of one sodium channel. Possibly

there is one sodium channel per rhodopsin molecule (Wong, F., 1978).

Quantum bumps of insect photoreceptors are generally considered to be produced by similar openings of light-activated sodium channels. Comparative quantitative data await voltage-clamping of insect photoreceptors. Sodium channel conductances of $40\,pS$ estimated for azide-poised locust photoreceptors (Payne, R., 1981) are commensurate with $18\,pS$ conductances calculated for *Limulus* (Wong, F., 1978). Transducer noise (as, for example, from current flowing in sodium channels) has been proposed in locusts to account for a greater variance in bump size than predicted from the number of quanta in a flash (Lillywhite, P. and Laughlin, S., 1981). However, the variance of bump *area*, rather than amplitude, is well predicted by quantal fluctuation noise (Cohn, T. *et al.*, 1983). This would imply that in locusts too, roughly constant numbers of sodium channels underlie each bump but have variable, unsynchronized openings and closings.

There remains the principal question of phototransduction for *any* invertebrate or vertebrate photoreceptor: how do changes in visual pigments result in openings (or closings) of sodium channels. This is an area of intense current research, without resolution at present. Approaches to the question include studies of photopigment dynamics; nucleotide (ATP, GTP) and cyclic nucleotide (cGMP) production and destruction, genetic dissection of phototransduction via visual mutants; and modeling of the dynamics of photovoltages. These approaches are complementary.

Photopigment dynamics have largely been studied in invertebrates by using electrical events (ERPs, PDAs) as indicators of photochemical states. In addition to the thermostable visual pigments rhodopsin and metarhodopsin, few hypothesized intermediates have been directly measured (Fein, A. and Cone, R., 1973). A major theoretical thrust of this approach has been "to model the LRP and PDA response phenomenologies into a single system of states and transitions of the pigment itself . . ." (Hillman, P. *et al.*, 1983).

A second approach has been to compare phototransduction to the ways that hormone signals are transduced into cyclic nucleotide syntheses.

For example, bumps can be induced in ventral photoreceptors of *Limulus* by substances which possibly activate a guanyl nucleotide binding protein (Fein, A. and Corson, D., 1981). Nucleotide roles in phototransduction are objects of intense current research (cf. Miller, W., 1981).

The genetic dissection of visual transduction can be exemplified by studies on mutants of *Drosophila*. Thus receptors of the mutant *trp* produce normal-sized quantum bumps, but the bumps do not light-adapt. Rather, the rate of bump production decreases during sustained illumination and does not remain proportional to quanta caught (Minke, B. *et al.*, 1975). The $norpA^{H52}$ mutant, discussed above (section 2.1.3), has bumps with greatly increased latencies (Pak, W. *et al.*, 1976). Clearly this is a different failure of normal bump production. Normal, light-induced changes in receptor proteins may be blocked in such mutants (Pak, W. *et al.*, 1976; Matsumoto, H. *et al.*, 1982; cf. Pak, W., 1979).

Finally, LRP photovoltage dynamics have been variously modeled to learn what the underlying transduction dynamics might be. One approach has been based on recorded frequency or pulse responses, which have been modeled in terms of *n* concatenated linear filters (DeVoe, R., 1963, 1967b; Fuortes, M. and Hodgkin, A., 1964). The concatenated filters could, for example, represent cascaded, autocalytic chemical reactions that would give photoreceptor gain (Wald, G., 1965). The problem with this approach is that the number of stages, *n*, is not necessarily constant or reasonably small. Another is that simple concatenated filters (with single poles) do not now fully account for the amplitude and phases of photoreceptor responses of the fly *Phormia* (French, A., 1980). An alternative hypothesis is that of a rising concentration of a light-elicited transmitter, with a normal distribution of thresholds of sodium channels. There results from this model a log-normal voltage waveform, which fits well the shapes of (linearly-summed) bumps of locust photoreceptors. A single parameter is all that needs to be changed to fit changes in bump durations due to light adaptation or temperature (Payne, R. and Howard, J., 1981). The hypothetical internal transmitter would help explain the reduced latencies of prebumps, in terms of synergism of transmitter released at multiple sites of photon

absorption (Hamdorf, K. and Kirschfeld, K., 1980). In the log-normal model, response sensitivity and response time scale are independent, however. This appears to be true in flies only for intensities above the range of linearity (Dubs, A., 1981). In many other photoreceptors, however, sensitivity and time scale change together (Fuortes, M. and Hodgkin, A., 1964), and this is not explained by the log-normal model.

In summary, the process of phototransduction has so far defied complete analysis.

2.2 The late receptor potential (LRP)

The thrust of the previous section was that the LRP (and PDA) are composed of summed quantum bumps, that bumps change in amplitude (and sometimes duration) with light and dark adaptation, and that these changes consist of variations in the numbers and timings of underlying openings and closing of sodium channels. Of the factors shaping quantum bumps, flows of light-initiated sodium photocurrents and the light adaptations of bumps do not arise from voltage-sensitive membrane channels, since they occur in photoreceptor cells (of *Limulus*) that are voltage clamped to the resting potential in the dark (Wong, F., 1978). When sodium photocurrents are allowed to alter membrane potentials, however, then the resulting waveforms of LRPs are affected by other, voltage-sensitive channels and ion flows in photoreceptors. It is thus necessary to view the LRP in the context of its entire ionic generation and environment.

To begin with, dark resting potentials of insect photoreceptors are most likely to be dominated by resting potassium permeabilities. In common with many other excitable cells, the resting potential of photoreceptors of the drone honey bee ($-54\,\text{mV}$) is slightly positive to the potassium equilibrium potential E_k ($-64\,\text{mV}$). Moreover, increases in extracellular potassium concentrations reduce resting potentials less than predicted by the Nernst equation. Possibly, there are significant resting permeabilities to sodium, since resting potentials increase in sodium-free environments (Fulpius, B. and Baumann, F., 1969; Coles, J. and Tsacopoulos, M., 1979). Chloride ions do seem to be passively distributed and at electrochemical equilibrium across the photoreceptor membrane. There is no

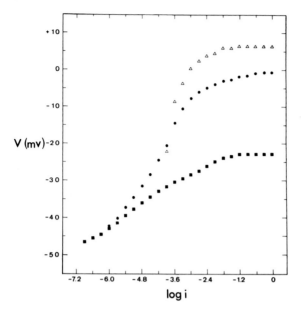

FIG. 7. $V \log I$ plots of absolute membrane potentials versus relative log intensities of long flashes. Filled squares; plateau responses: filled circles; peak responses: open triangles; amplitudes of initial spikes. Reproduced from Fulpius, B. and Baumann, F. (1969), *J. Gen. Physiol. 53*, 541–561, by permission of the authors and copyright permission of The Rockefeller University Press.

evidence for contributions of active chloride currents to receptor potentials, but passive chloride movements do reduce the photovoltage (Baumann, F. and Hadjilazaro, B., 1972).

Examples were given in Figs 1 and 4 (see also Fig. 9) of late receptor potentials elicited by illumination of the eye. As is evident in Fig. 4 for a dark-adapted cell, the waveforms of LRPs are dependent upon the light intensities. The initial responses become more phasic with increasing intensities. The potentials may also exceed the thresholds of a regenerative initial spike. A typical $V \log I$ plot of amplitudes of initial spikes, peak and plateau responses are shown versus log intensity in Fig. 7 for a photoreceptor of a drone honey bee (Fulpius, B. and Baumann, F., 1969). The initial spike is graded and not all-or-none (Benolken, R., 1965), but it does overshoot 0 mV at the brightest intensities. The peak response approaches 0 mV at the brightest intensities, and the plateau response is about 60% as great.

Changing the extracellular concentrations of sodium, potassium and calcium ions changes the amplitudes of the initial spike, the peak, and the plateau. Most of these studies have been performed

on the photoreceptors of drone honey bees. Removing extracellular sodium ion abolishes the initial spike, whereas lowering or raising calcium is without effect on the spike height (Fulpius, B. and Baumann, F., 1969). Tetrodotoxin abolishes the initial spike (Baumann, F., 1968), as it does in photoreceptors of the dragonfly median ocellus (Chappell, R. and Dowling, J., 1972). Thus, the initial spike in these two insects seems to be due to electrically excitable sodium channels. This is in contrast to the ventral photoreceptors of *Limulus*, where the initial spike is due to electrically excitable sodium plus calcium channels and is not sensitive to tetrodotoxin (O'Day, P. *et al.*, 1982).

Reducing extracellular sodium in the drone retina also reduces the peak and plateau potentials. This is to be expected if the predominant photocurrents are carried by sodium. However, the plateau response is reduced more than the peak. Conversely, increased extracellular potassium increases the absolute membrane potentials of both the peak and plateau, but the plateau more (Fulpius, B. and Baumann, F., 1969). These results can be interpreted to mean that light opens sodium channels, and the resulting photovoltages open potassium channels.

The general hypothesis that photocurrents are sodium currents in insect photoreceptors has been stated in the previous sections. Some of the evidence for this hypothesis is as follows. First, intracellular measurements with sodium-selective electrodes in drone honey bees show rises in intracellular sodium after light flashes (Coles, J. and Orkand, R., 1982). Second, at saturating light intensities, peak responses approach but rarely exceed 0 mV (Tunstall, J. and Horridge, G., 1967; Muijser, H., 1979; Mote, M. and Wehner, R., 1980; see Fig. 7). E_{Na}, the sodium equilibrium potential, is positive or close to 0 mV. In the fly *Musca*, E_{Na} is likely to be 0 to +10 mV, based on estimated intra- and extracellular sodium concentrations in brain cells (Shatoury, H., 1969). For drone honey bees, E_{Na} can be estimated to be +48 mV (data of Coles, J. and Orkand, R., 1982; Tsacopoulos, M. *et al.*, 1983). The sodium-dependent initial spike of the drone honey bee saturates at about +10 mV (Fig. 7), however, somewhat less than this value of E_{Na}. Third, at the saturated peak of the LRP, membrane resistance falls to very low values in locust, drone honeybee, and fly photoreceptors (Shaw, S., 1968, 1969; Baumann, F. and Hadjilazaro, B., 1972; Muijser, H., 1979). Subsequently, resistances increase again during the plateau response in all but fly photoreceptors (Fuortes, M., 1963; Washizu, Y., 1964; Wong, F. et al., 1976; Muijser, H., 1979). It can be calculated from M. Fuortes' (1963) data on dragonfly photoreceptors that both the peak and plateau responses had equilibrium potentials at about 0 mV. In photoreceptors of drone honey bees and the fly Calliphora, peak and plateau response also seem to reverse together for depolarizing extrinsic current. Slightly more current is needed to reverse the initial spike, consistent with a more positive saturation potential (Baumann, F. and Hadjilazaro, B., 1972; Muijser, H., 1979). Thus, saturating peak responses probably result from maximum openings of sodium channels, driving the membrane potentials toward E_{Na}. The subsequent decline to the plateau could then be due to closings of some channels, to openings of potassium channels, or both.

In voltage-clamped ventral photoreceptors of Limulus there is direct evidence for openings of fast and slow, electrically excitable potassium channels (O'Day, P. et al., 1982). The fast ("A") channels seem to be responsible for the initial oscillations after the peak of response (as in Fig. 1 at the bottom). The slow (delayed rectifier) channels open during the progression from peak to plateau response.

In photoreceptors of drone honey bees the greater sensitivities to raised extracellular potassium of plateau (compared to peak) responses (Fulpius, B. and Baumann, F., 1969) are consistent with openings of slow potassium channels. Direct measurements with potassium-selective electrodes show that potassium falls in photoreceptors during illumination, but rises extracellularly (and in the surrounding glial cells: Coles, J. and Tsacopoulos, M. 1979). Thus, there are the expected outward potassium movements.

The shape of the LRP during bright lights would thus seem to be the result of openings of sodium and potassium channels. At light onset a large number of sodium channels open, depolarizing the membrane towards the sodium equilibrium potential and perhaps initiating an initial spike. That the spike saturates at a slightly more positive potential

than the peak, suggests that some potassium channels have already opened to diminish the potential at the peak. It is also possible that some outward potassium current can flow through the sodium channels, which would likewise reduce the peak photovoltage (Brown, J. and Mote, M., 1974).

The decline of the response from peak to plateau is accompanied by both a rise in membrane resistance (except in flies), and an outward potassium current, consistent with a greater fall in sodium conductance than rise in potassium conductance. (The next section will present reasons for the fall in sodium conductance.) Blockage of the slow potassium channels with TEA (tetraethylammonium ions) in Limulus leads to a greater plateau photovoltage, but no changes in sodium photocurrents (Pepose, J. and Lisman, J., 1978). In flies, potassium conductance seems to rise as much as sodium conductance falls and may also be calcium-mediated (Muijser, H., 1979).

Finally, it is worthwhile to consider the maintenance and recovery of the ionic mechanisms of phototransduction in insect photoreceptors. Recovery but not maintenance require oxygen: resting potentials of drone honey bees are stable for at least 30 min in axonia in the dark, but resting and late receptor potentials disappear upon repeated illumination. During re-oxygenation, resting potentials recover before LRPs (Baumann, F. and Mauro, A., 1973). In both bees and flies, anoxia blocks light-induced conductance increases (Wong, F. et al., 1976). In azide-poisoned locust photoreceptors, steady membrane potentials in the dark become increasingly noisy after repeated flashes of light. Normal metabolic activity is necessary to prevent this noise, which is due to maintained openings of sodium channels in the dark (Payne, R., 1981).

After strong illumination, recovery of normal ionic distributions involves electrogenic sodium pumps. In the absence of masking by PDAs, sodium pumping can be seen to result in hyperpolarizing afterpotentials (Koike, H. et al., 1971; Tsacopoulos, M. et al., 1983). Such hyperpolarizations are evident following the responses to the weaker (but saturating) flashes in Fig. 6 (from the dronefly Eristalis). The greatest part of the oxygen consumed by the drone honey bee retina is used for the working of the sodium pump (Tsacopoulos, M. and Poitry,

S., 1982). The role of the pump in photoreceptors would be, of course, to expel all the sodium which entered as photocurrent and to take back up the potassium lost as outward current through electrically excitable channels. The time course of sodium pumping is slower than the corresponding extra oxygen consumption however, suggesting that use of ATP by the pump is not what triggers the oxygen consumption (Tsacopoulos, M. et al., 1983).

In the drone honey bee it appears that the surrounding pigment (glial) cells help to buffer the extracellular ionic concentrations. Figure 8 is a schematic illustration of potassium concentrations (activities) measured in photoreceptors, extracellular space, and glial cells during illumination. It is calculated that just about all potassium lost by photoreceptors is taken up by the glia, since the 5% of the retina that is extracellular space would be expected to show far greater maintained rises in potassium concentrations than the transient increases or even decreases found (Coles, J. and Tsacopoulos, M., 1979). Similarly small values of extracellular space are found in the locust retina (Shaw, S., 1978). The uptake of potassium by glial cells is accompanied by depolarization and hence is unlikely to be due to a pump. Rather, a sodium–potassium exchange seems most likely. The exchange in effect reduces extracellular potassium and replenishes the extracellular space with sodium (Coles, J. and Tsacopoulos, M., 1979). Such buffering of extracellular sodium may be one reason why removal of sodium from perfusion fluids does not abolish LRPs (Fulpius, B. and Baumann, F., 1969). The buffering would also work in reverse during recovery from light: sodium pumped out of photoreceptors could exchange with glial potassium, thereby providing sufficient of the obligatory extracellular potassium needed to keep the pump going. Finally, the buffering of extracellular sodium (and potassium) means that photoreceptor sensitivities would not be reduced during illumination by a deficit in extracellular sodium, as proposed by Hamdorf, K. et al. (1978).

2.3   Adaptation and calcium ions

The decline of the late receptor potential from its peak to its plateau is called sensory adaptation. Sensory adaptation has causes in common with light

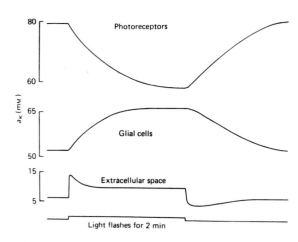

FIG. 8. Schematic illustration of changes in potassium activities a_K during illumination, in photoreceptors, glial cells and in extracellular space. The phasic alterations of extracellular a_K at light ON and OFF would result from different kinetics of release and uptake of potassium by photoreceptor and glial cells. Reproduced from Coles, J. and Tsacopoulos, M. (1979), J. Physiol. 290, 525–549, by permission of the authors and copyright permission of the Physiological Society, England.

adaptation, which is the desensitization of photoreceptors due to, e.g., continuous illumination. Some sensory adaptation results, as noted above, from delayed outward potassium currents. However, there is also sensory adaptation of photocurrents during steady illumination under voltage clamp (in Limulus), when there are no voltage-sensitive currents (Millechia, R. and Mauro, A., 1969). The major cause of the decline in the photocurrents in invertebrates is a rise in intracellular calcium ions.

In photoreceptors of drone honey bees manipulations which affect intracellular calcium concentrations have the following effects. Lowering extracellular calcium, or reducing intracellular calcium by the injection of EGTA, increases the plateau response more than the peak response and increases the amplitudes of quantum bump noise. Conversely, increasing extracellular calcium, or intracellular calcium by the injection of Ca-EGTA buffers, lowers the plateau response more than the peak (Fulpius, B. and Baumann, F., 1969; Bader, C. et al., 1976). These results are consistent with similar findings on voltage-clamped ventral photoreceptors of Limulus (Lisman, J. and Brown, J., 1975a,b).

Light adaptation occurs in parallel with sensory adaptation. An example is given in Fig. 9 for background steps of light (of the log intensities indicated at the left) and the same-intensity test flash presented at various times after the onset of the background illuminations (Baumann, F., 1975). As the background intensities were increased the amount of sensory adaptation became greater. With greater sensory adaptation came greater light adaptation: the responses to the incremental test flashes became smaller, of shorter duration (faster) and more oscillatory. Light adaptation begins some time around the peak response, since the total membrane potential is the same whether a flash is added at the time of the peak or whether an equal-intensity flash (background + incremental flash intensity) is presented to the dark-adapted eye (Bader, C. *et al.*, 1976). In voltage-clamped *Limulus* ventral photoreceptors the incremental sensitivity (in current per photon) does not begin to decline until after the peak. Then the sensitivity declines about in proportion to the decline from peak to plateau of the current elicited by the background light (Lisman, J. and Brown, J., 1975a).

The equivalent experiments with incremental flashes have not been performed for *Limulus* ventral-eye photovoltages. However, in the results from the drone honey bee in Fig. 9 the incremental voltage responses do not decrease in amplitude with time, unlike the *Limulus* photocurrents (see also Naka, K.-I. and Kishida, K., 1966; Bader, C. *et al.*, 1976). Possibly the constancy of incremental voltage response from the drone honey bee was due to partial cancellation of light-adaptation by facilitation. Facilitation of responses to a test flash after conditioning or adapting lights has been found in ERGs and in intracellular recordings from a number of invertebrate eyes (review: Ventura, D. and Puglia, N., 1977), including the locust *Schistocerca* (Giulio, L. and Lucaroni, A., 1967). These include the fly *Sarcophaga* and the milkweed bug *Oncopeltus* (Dudek, F. and Koopowitz, H., 1973), and the ant *Atta* (Ventura, D. *et al.*, 1976). Sensitivity increases are maximal for short conditioning flashes of moderate relative intensities, where these increases are then as great as 2 log units. Facilitation can last for up to 2 min. Often, facilitation is accompanied by shorter latencies and peak times (DeVoe, R., 1972). A constant intensity–duration product for the conditioning flash gives a constant decrease in latency (Dudek, F. and Koopowitz, H.,

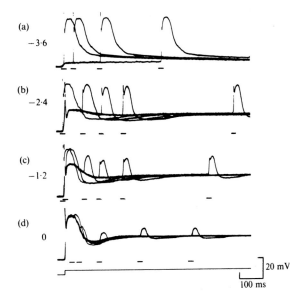

FIG. 9. Incremental responses of photoreceptors of drone honey bees, *Apis mellifera* to incremental flashes superimposed at various times upon steady background light. The onset of the background light is indicated by the bottom trace, while the times of the incremental flashes are indicated by the short lines. The relative log intensities of the background light are given at the left. Reproduced from Baumann, F. (1975), *The Compound Eye and Vision of Insects*, ed. by G. A. Horridge, pp. 53–74, by permission of the author and copyright permission of the Oxford University Press.

1973). As facilitation is not always abolished by constant illumination (Ventura, D. and Puglia, M., 1977), it could have been occurring during such responses as those in Fig. 9.

If increases in intracellular calcium concentration are responsible for light adaptation, there should be parallel changes in sensitivity and time scale during light adaptation and imposed changes of intracellular calcium. This has been shown quantitatively in *Limulus* ventral photoreceptors (Fein, A. and Charlton, J., 1977a) and qualitatively for drone honey bee receptors (Bader, C. *et al.*, 1976).

Conversely, reduced intracellular calcium may be responsible for facilitation, since facilitation can be augmented or restored with low extracellular calcium (Yamashita, S. and Tateda, H., 1976). For barnacle photoreceptors Hanani, M. and Hillman, P. (1976) suggested that facilitation is found primarily in less sensitive cells and that low sensitivities were the result of excess intracellular calcium. In this view calcium would leave during illumination. Since eyes of the ant *Atta* have circadian rhythms of sensitivity (Ventura, D. *et al.*, 1976), possibly facilitation would occur primarily in the low-sensitivity, daytime state. This was not explicitly tested, however. In any event, it is quite unclear how time scales of responses can shorten during both adaptation and facilitation, with opposite dependencies on intracellular calcium.

The adapting action of calcium is presumably on the number of sodium channels opened by absorption of a photon. This number goes down to 1 during light adaptation (Wong, F., 1978) and appears to go up in low external calcium, judging by increased photon noise (Fulpius, B. and Baumann, F., 1969) and increased sodium influx (Coles, J. and Orkand, R., 1982). Faster responses presumably occur in light adaptation because of greater synchronization of smaller bumps. Thus, in *Limulus*, both light adaptation and raised intracellular calcium reduce the variabilities of delays in response times (as well as of thresholds: Fein, A. and Charlton, J., 1977a). Light adaptation also makes incremental responses more oscillatory; in Fig. 9 (c,d) it was seen that the responses undershoot the background potential at light-off. The undershoot does not occur because light-adapted bumps become diphasic. In *Limulus* the undershoot is accompanied by a conductance decrease, implying a further transient reduction in sizes of light-adapted bumps after each incremental flash (Wong, F. and Knight, B., 1980).

It remains uncertain how light-adaptation causes rises in intracellular calcium. For *Limulus* ventral photoreceptors, calcium appears to be released from intracellular stores (see discussion in O'Day, P. *et al.*, 1982) and the release is local to the site of illumination (Fein, A. and Lisman, J., 1975). For insect photoreceptors the calcium stores are uncertain. Injections of sodium ion mimic the effects of light adaptation in drone honey bees, suggesting that normal inward sodium photocurrents themselves might trigger intracellular calcium release (Bader, C. *et al.*, 1976). However, in *Limulus*, where intracellular injections of sodium also mimic light adaptation, there is no localization of effect as there is with calcium and light (Fein, A. and Charlton, J., 1977b). Instead, perhaps a generalized sodium–calcium exchange is involved at the cell membrane, which works much more slowly than

direct injections of calcium (Fein, A. and Charlton, J., 1978). Thus, sodium-mediated calcium release seems unlikely to be significant.

2.3.1 ADAPTATION AND VISUAL PERFORMANCE

Adaptation has been shown above to have two components: opening of electrically excited ion channels (principally for potassium) to reduce the photovoltages developed by photocurrents, and rises in intracellular calcium ion to reduce the photocurrents themselves. The visual purposes served by adaptation are: (1) the compression of large stimulus ranges into smaller physiological ranges of response, (2) avoidance of sustained saturation of responses, (3) improvement in contrast sensitivities, and (4) improvements in temporal resolution. Insect photoreceptors are called upon to function over enormous variations in light inputs: from very low illuminations, where visual performance is limited by quantal catch (Fermi, G. and Reichardt, W., 1963; Dubs, A. *et al.*, 1981), up to direct sunlight. Rarely, of course, is any photoreceptor called up to perform at any time over more than a small part of this range (Laughlin, S., 1981b). Moreover, photoreceptors receive some protection from excess quantal absorptions by, e.g., movements of shielding pigments or structural changes (sensitivity losses due to low rhodopsin concentrations are described in chapter 8). Nonetheless, the major brunt of adaptation falls on the photoreceptors, and that is what will be considered here.

The sigmoidal $V \log I$ relation, typified by the results in Fig. 7, is what is meant by response compression. There, a stimulus range of 7 log units is compressed into a 50 mV range of physiological response. Moreover, within this physiological range, responses to the weakest intensities can be linear (Dubs, A., 1981) while (peak) responses to the strongest intensities saturate near the sodium equilibrium potential.

Were late receptor potentials to have little sensory adaptation, they would eventually saturate and be unable to signal still further increases in intensity. Vertebrate rods behave this way (Norman and Werblin, 1974); cones and insect photoreceptors do not. As was shown in Fig. 9, light adaptation reduces responses both to the adapting light and to subsequent incremental flashes. $V \log I$ curves from both dark- and light-adapted cells of a dragonfly *Hemicordulia* and a fly *Calliphora* are illustrated in Fig. 10 (Laughlin, S. and Hardie, R., 1978). There, the total relative receptor potential $V_t + V_a/V_{max}$ is plotted versus total flash intensity $I_t + I_a$, where V_t is the incremental response to the incremental flash I_t, V_a is the plateau response to the adapting light I_a, and V_{max} is the saturated peak response of the dark-adapted photoreceptor. Thus plotted, a sigmoidal "template" curve fitted to the dark-adapted peak responses requires only horizontal shifts along the log intensity axis in order to coincide with the light-adapted peak responses. The amounts of shift are the sensitivity changes, in log intensity units. That is, suppose a particular flash $(I_t + I_a)$ elicited a peak response which was 50% V_{max}, whereas in the dark-adapted cell an intensity 2 log units less elicited the same response. Therefore, with the adapting light, the sensitivity was 2 log units less, or one-hundredth of that in the dark. The same information could have been had by shifting the curve that fit the dark-adapted points 2 log units to the right, until it coincided with the point from the light-adapted cell. The same shift would be required for all light-adapted points, including the plateau response to the adapting light itself. Thus, the amount by which an adapting light reduces the sensitivity to any added flash is also the amount by which sensory adaptation reduces the sensitivity of the dark-adapted cell upon going from the peak to the plateau in the response to the adapting light itself (Lisman, J. and Brown, J., 1975a).

With each increase in adaptation in Fig. 10, the template curve not only moves to the right, but the operating point moves up. The operating point, of course, is the plateau response, around which there can be decremental responses (to flashes of $-\Delta I_t$) as well as the incremental responses shown. For lights where plateau responses saturate (at between 40% and 60% of V_{max}: see Figs 7 and 10), the operating point has moved to the steepest part of the template curve. The slope is a measure of contrast sensitivity, which is the amount of incremental response ΔV_t to be elicited by the same multiplicative change, $\log I_t$ in the mean illumination. Above plateau saturation, contrast sensitivity from a single template curve should be constant, obeying Weber's Law. Dawis, S. and Purple, R., (1982) have demonstrated this quantitatively, for vertebrate cones.

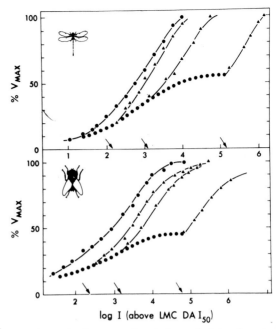

FIG. 10. V log$I$ curves for dark- and light-adapted photoreceptors of the dragonfly *Hemicordulia* and the fly *Calliphora*. Amplitudes of response are plotted as percentages of maximum deflections from the resting potentials in the dark. Intensities are plotted as the logs of test plus background intensities relative to the intensity (LMC DA I$_{50}$) which half saturates the responses of a second-order laminar monopolar cell. Arrows point to background intensities used. The lower-most curves of filled circles indicate the steady-state plateau responses to background lights, and the left-most curves show responses from dark-adapted photoreceptors. Reproduced from Laughlin, S. and Hardie, R. (1978), *J. Comp. Physiol. 128*, 319–340, by permission of the authors and copyright permission of Springer-Verlag, Heidelberg.

The same template V log I curve does not necessarily fit responses of both dark- and light-adapted arthropod eyes, as it appears to do in Fig. 10. Careful measurements of responses to both incremental and decremental flashes show that the slopes of the light-adapted curves are steeper. This was originally shown with the wolf-spider ERG (DeVoe, R., 1967a) and has now been verified with intracellular recordings from photoreceptors of locusts and flies (Matic, T. and Laughlin, S., 1981). The changes in slope of the V log I curves with light adaptation are possibly due to steady-state diminutions of the electrically excitable potassium currents (O'Day, P. *et al.*, 1982), thus increasing the photovoltages. The important point is that contrast sensitivities of light-adapted photoreceptors may be greater than can be predicted from the shape of the V log I curve of the dark-adapted photoreceptor.

Light adaptation also increases temporal resolution. This could be seen in Fig. 9 in the bottom curves, where the light-adapted responses were the more rapid and had shortened durations. One measure of the temporal resolution of a photoreceptor is its frequency response, obtained using small sinusoidal or white-noise modulations of the adapting light. This approach has been used mostly with flies (but also with beetles: Kirschfeld, K., 1961), where the "small signal" responses are linear except at the lowest frequencies (Eckert, H. and Bishop, L., 1975; Gemperlein, R. and McCann, G., 1975). The power spectrum of the late receptor potential (at 8°C) is the same as for individual bumps (French, A. and Järvilehto, M., 1978b), as would be expected from the adapting bump model. Careful averaging of responses in order to reduce noise allows detection of sensitivities to frequencies up to nearly 200 Hz in receptors R1–6 (French, A., 1980). In recordings from *Drosophila*, receptors R7–8 were not different in their frequency responses from R1–6 (Wu, C. and Wong, F., 1977). Such high-frequency responses in flies do not seem to require intense adaptation, as seen in other arthropods such as wolf spiders (DeVoe, R., 1967b). Rather, temporal changes such as shortened times-to-peak occur after the absorption of as few as four photons, and the peak times are already near their shortest for adapting lights which elicit plateau depolarizations of as little as 8–10 mV (Dubs, A., 1981). (In terms of the ordinates of Fig. 10, these depolarizations are about one-fifth of V_{max}.)

For measurements using the ERG, high gain at the receptor–laminar synapse (see section 3.1 below) makes the laminar monopolar cells the best detectors of frequency sensitivities of receptor cells. Thus receptors R1–6 of flies, which synapse in the lamina and so elicit large laminar ON- and OFF-responses in the ERG, seem to have three times higher ERG flicker fusion frequencies than do receptors R7–8, which do not synapse in the lamina (Cosens, D. and Spatz, H., 1978). The presence or absence of laminar potentials in the ERG has been one basis for distinguishing "fast" from "slow" eyes (Autrum, H., 1950). However, the lower recorded flicker fusion frequencies of slow eyes (such as of

locusts and dragonflies — see Fig. 1) may simply reflect the difficulties of detecting small amplitudes of high-frequency responses in receptors, when laminar synapses are not present to amplify and augment them (Wu, C. and Wong, F., 1977; French, A. and Järvilehto, M., 1978a).

Finally, mention must be made of the reverse of light adaptation; namely, the recoveries of sensitivities during dark adaptation. After the extinction of an adapting light, recovery may proceed first via a fast, "neural" stage and then in a slower, "photochemical" stage (chapter 8). It is the neural stage that is of interest here. In ventral photoreceptors of *Limulus*, neural adaptation far outlasts and is independent of changes in visual pigments (Fein, A. and DeVoe, R., 1973). The start of neural adaptation takes time. Immediately after offset of the adapting light the sensitivity is the same as it was before, in the plateau response (Lisman, J. and Brown, J., 1975a). Presumably this means that intracellular calcium has not yet decreased, and that neural adaptation may have to wait upon such a decrease.

Neural dark adaptation in insects is generally rapid, with most of the recovery occurring in the first minute. Full recovery may take tens of minutes, possibly because of photochemical adaptation (which could not be evaluated in early studies: Ruck, P., 1958; Goldsmith, T., 1963). Additionally, movements of shielding pigments in a number of nocturnal insects with superposition eyes provide a second, major, and much slower stage of dark adaptation. However, in these insects too, the early, neural adaptation is over within 10 min (Bernhard, C. *et al.*, 1963).

The term "neural" adaptation need not imply that sensitivity and membrane potential are related. In the dragonfly *Aeschna*, dark adaptation is not correlated with the recovery of the resting potentials and generally takes longer (Autrum, H. and Kolb, G., 1972). For locust photoreceptors, on the other hand, they are hypothesized to be related (Cosens, D., 1966; see Glantz, R., 1971).

2.4 Determinants of receptor sensitivities

The previous sections have described the transduction of photon absorptions into photovoltages. The photovoltage — the late receptor potential — can

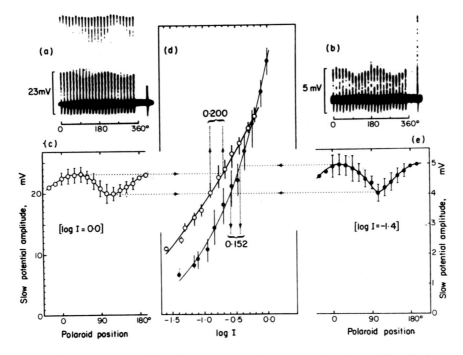

FIG. 11. Measurements of polarization sensitivities of photoreceptors of drone honey bees, *Apis mellifera*. Single runs to measure amplitudes of response *versus* angle of polarization are illustrated for strong and weak flashes in (a) and (b), respectively. Averaged responses are drawn directly below, in (c) and (e), respectively. From the *V* log *I* curves in (d), the equivalent log intensities can be found which would elicit the minimum and maximum amplitudes of response in (c) and (e) (dotted lines). The antilog of these intensities is a ratio, the polarization sensitivity. Note the different amplifications in (a), (c) compared to (b), (e). Reproduced from Shaw, S. (1969), *Vision Res. 9*, 999–1029, by permission of the author and copyright permission of Pergamon Press.

thus be thought of as a product of two terms: the numbers of photons absorbed, and the photon-to-voltage "gain". For a constant gain in, say, the dark-adapted eye it is then possible to determine from the late receptor potential the relative numbers of photons absorbed. The number of photons incident on the cornea which can reach the photopigment in a cell's rhabdom depends on the optics of each facet (chapter 5). The fraction absorbed of the photons which reach a particular photopigment depends on the wavelength and *e*-vector orientation of the photons (chapter 8). The relative number of incident photons which are transmitted to the photopigments and then absorbed is indicated by the size of the late receptor potential. However, the relation between size of potential and photons absorbed is (usually) not a linear one, so it is far less ambiguous to find the relative numbers of incident photons that elicit the *same* response. The reciprocals of these numbers are the sensitivities.

The sensitivities which are of interest here are those to spatial position (angular sensitivities),

wavelength (spectral sensitivities) and *e*-vector orientation (polarization sensitivities). These sensitivities have little significance for any individual cell, since one photon absorbed by one rhodopsin will be indistinguishable from another photon in the resulting photovoltage (Wu, C.-F. and Pak, W., 1975; Lillywhite, P., 1978). This is the principle of *univariance* (Naka, K.-I. and Rushton, W., 1966), which in essence states that single photoreceptors can only count photons absorbed. When univariance holds, there will be one *V* log *I* template curve for all photons absorbed. The probability of absorption will, however, depend on spatial position, wavelength, and polarization. Hence, relative positions of *V* log *I* curves on the log *I* axis as functions of the above variables will be measures of probabilities of absorptions: that is, of relative sensitivities.

What this means in practice is that sensitivities can be measured more simply than by the equal response method (which requires better stabilities than are sometimes possible). So long as

Table 1: *Representative angular sensitivities of insect photoreceptors*

Order	Species	Width at half-sensitivity	Comments	Reference
Odonata	*Libellula needhami, Anax junius*	1.2°–1.8° horizontal and vertical	A few multi-peaked units found (due to damage to optics?)	Horridge, G., 1969
	Hemicordulia tau	1.46° horizontal 1.31° vertical	Same angular sensitivities in soma and axon	Laughlin, S., 1974b,c
Orthoptera	*Locusta migratoria*	6.6° dark-adapted 3.4° light-adapted	Horizontal half-widths; recorded from sliced head	Turnstall, J. and Horridge, G., 1967
	Locusta migratoria	2.4° horizontal dark-adapted 2.5° vertical dark-adapted 1.5° horizontal light-adapted 1.4° vertical light-adapted	Recorded from intact eye	Wilson, M., 1975
	Valango irregularis, Locusta migratoria	1.7°, 1.9° light-adapted 2.7°, 2.8° 10–15 min in dark 4.7°, 4.9° 4 h in dark	Sensitivities, when dark-adapted, determined from bump rates	Williams, D., 1983
	Tenodera australasiae (Mantidae)	0.74° light-adapted 2.0° dark-adapted: night 1.10° dark-adapted: day 2.4° light-adapted 6° dark-adapted: night 3.2° dark-adapted: day	Foveal eye region Dorsal eye region	Rossel, S., 1979
	Periplaneta americana	6.8° dark-adapted 2.4° light-adapted		Butler, R. and Horridge, G., 1973a
Hemiptera	*Lethocerus sp. Benacus griseus*	8–9° horizontal, dark-adapted	Same angular sensitivities in soma and axon	Ioannides, A. and Walcott, B., 1971
	Lethocerus sp. Benacus griseus	9.0° dark-adapted 3.5° light-adapted		Walcott, B., 1971b
Lepidoptera	*Epargyreus claris*	2.1° dark-adapted		Doving, K. and Miller, W., 1969

	Papilio aegeus	UV cells: 2.1° dark-adapted
		            1.7° light-adapted
		Blue cells: 2.0° dark-adapted
		            1.5° light-adapted
		Green cells: 1.9° dark-adapted
		            1.7° light-adapted
				Horridge, G. *et al.*, 1983

Hymenoptera	*Apis mellifera* (drone)	2°	Central third of eye	Shaw, S. 1969
	Apis mellifera (worker)	2.5° horizontal–dark-adapted 2.7° vertical–dark-adapted	Recorded near center of eye	Laughlin, S. and Horridge, G., 1971
	Apis mellifera (worker)	2.57° frontal, ipsilateral 3.16° dorsal/dorsal rim; ipsilateral view and 10° contralateral view 5.46° dorsal rim; 10° contralateral view	Results are pooled for lateral–medial and anterior–posterior axes	Labhart, T., 1980
Diptera	*Calliphora erythrocephala*	3.1° horizontal 2.6° vertical	Wild type, white and chalky flies, central eye. In frontal eye, half-widths were about 50% smaller	Washizu, Y., *et al.*, 1964
	Calliphora erythrocephala	5.2° lateral edge of eye 3.3° frontal region	Vertical half-widths	Burkhardt, D. *et al.*, 1966
	Calliphora stygia	1.44° horizontal 1.66° vertical	Dark-adapted values, independent of wave-length of light	Horridge, G. *et al.*, 1976
	Calliphora stygia (female)	R1–6: 1.5° frontal 3° lateral R7, 8: 1.3° frontal	Dark-adapted values listed. Light adaptation reduced values for R1–6 by 20%	Hardie, R., 1979

Table 1: *Representative angular sensitivities of insect photoreceptors* — Continued

Order	Species	Width at half-sensitivity		Comments	Reference
Diptera (cont.)	*Musca domestica* (female)	2.5° horizontal		No changes found with adaptation or between soma and terminals	Scholes, J., 1969
	Musca domestica (female, white-eyed)	R1–6: 2.3° frontal R7: 1.5° frontal		Dark-adapted values	Hardie, R., 1979
	Musca domestica (Male)	R1–6: 1.7° R7r: 1.9° R8r: 1.7°		Frontal eye recordings; dark adapted R7r terminates in the lamina; R8r in the medulla like other R8s	Hardie, R. *et al.*, 1981
	Lucilia sericata	1.5°		Frontal eye regions of dark adapted eyes	Dubs, A., 1982
	Eristalis tenax	R7: 1.16° horizontal 1.10° vertical 1.24° horizontal 1.19° vertical	350 nm 450 nm	Dark-adapted values; cell identifications are tentative	Horridge, G. *et al.*, 1976
		R1–6: 1.44°		Horizontal and vertical, independent of wavelength of light	
	Boettcherisca peregrina	1.8–3.5°		Six different, non-circular patterns of spatial sensitivity profiles, which were smaller but had the same shapes with light-adaptation	Mimura, K., 1981a

univariance holds, any measured amplitude of response *V* can be converted to a relative sensitivity via a *V* log *I* curve. Figure 11 illustrates this for polarization sensitivities. Maximum and minimum responses at different angles of polarization can be converted by a static *V* log *I* curve to relative ratios of photons absorbed (0.2 log units, or a 1.58 ratio in the upper part of Fig. 11d, for example). The same would hold for amplitudes of response as functions of wavelength or spatial position.

Sometimes univariance does not hold. *V* log *I* curves may not be the same for all wavelengths of light, for example, or there may be evidence that more than one cell's responses are being recorded by an electrode that is penetrating only one. In such cases it is necessary to determine if cells are coupled together, if there is more than one rhodopsin in a single cell, etc. It will be shown below that there are often failures of univariance, and the effect can be to degrade (or enhance) each cell's selectivity. Unimportant for a single cell, selectivity is important to networks of cells that serve to compare responses of unlike single cells so as to extract position, or color, or angle of the *e*-vector. For this reason an appreciation of photoreceptors' detailed angular, spectral and polarization sensitivities is essential for subsequent evaluations of possible visual performance.

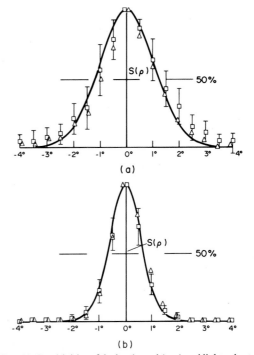

FIG. 12. Sensitivities of dark-adapted (top) and light-adapted (bottom) photoreceptors of locusts, as functions of angle from the optical axis. The solid lines are fitted Gaussian functions. The horizontal lines are drawn at one-half (50%) of the peak sensitivities. Reproduced from Wilson, M. (1975), *J. Comp. Physiol. 97*, 323–328, by permission of the author and copyright permission of Springer-Verlag, Heidelberg.

Figure 12 illustrates typical angular sensitivities of an insect photoreceptor (of a locust), when dark-adapted (top) and light-adapted (bottom). The data points in Fig. 12 are well fit by Gaussian functions (Götz, K., 1964), for both dark-adapted and light-adapted conditions. The angular sensitivity $\Delta\rho$ is defined as the angular width in degrees of such functions at half of the maximum, ON-axis sensitivity (at the horizontal lines in Fig. 12 labeled 50%). Table 1 lists representative values of angular sensitivities that have been measured electrophysiologically in insect photoreceptors. Some of the earlier measurements, giving wider angular sensitivities, may have been taken from injured preparations (Wilson, M., 1975).

Not all angular sensitivity functions are completely fit by Gaussians. Dubs, A. (1982) used drifting sine wave gratings instead of small spots to determine (via modulation transfer functions) the angular sensitivity functions of the flies *Musca*, *Calliphora*, and *Lucilia*. At wide angles (up to 10–15°) off axis, sensitivities below 10–20% were much higher than predicted by simple Gaussians. There may have been coupling within an ommatidium between receptors with different fields of view (Tsukahara, Y. and Horridge, G., 1977a; Mimura, K., 1978; Dubs, A. *et al.*, 1981). Photoreceptors in the dorsal rim of the worker honey bee, at ommatidial axes 10° contralateral to the paramedial plane, have narrow central angular sensitivities but extremely wide "brims" extending out to $\pm 30°$, at sensitivities lower than 8%. In these cells, however, there is no evidence for coupling to other cells. Rather, fine pore canals in the cornea probably scatter and/or reflect light at wide angles onto single cells (Labhart, T., 1980).

As can be seen in Table 1, horizontal and vertical angular sensitivities are sometimes but not always the same. The results, in two-dimensional contour plots of sensitivity, vary from nearly-round, iso-sensitivity contours to oval contours (Laughlin, S., 1974c; Mimura, K., 1981a). Sometimes the contour plots are wide and multi-peaked. In dragonflies the multiple peaks were ascribed to damaged optics (Horridge, G., 1969). Laughlin, S. (1974c) also found functions with small (1% of maximum sensitivity) secondary peaks in dragonflies, but he con-

cluded that they had no functional significance. On the other hand, Mimura, K. (1981a) found five types of broad, irregular, often multi-peaked angular sensitivity functions, in addition to the typical, round, narrow-field function. The sensitivities at the secondary peaks in the multiple arms of some of the star-shaped field types were less than 1% of the maximum. In long (more than 35%), extended patterns, the secondary peaks were as much as 5–25% of the maximum sensitivity. The extended patterns and secondary peaks seemed to be derived from separate, distant photoreceptor cells, since latencies and phases of polarization sensitivities (see section 2.4.2 below) altered in step fashion along the extended patterns. Despite the extended fields at low sensitivities, angular sensitivities as defined above at 50% were small for all patterns: 1.8–3.5°. Possibly some of the broad, off-axis sensitivities measured one-dimensionally by Dubs, A. (1982) might have resulted from complex, irregular fields as seen in two dimensions. The irregular fields seem to have a functional significance in phototactic behavior (Mimura, K., 1981b).

The data in Table 1 show that angular sensitivities vary from species to species. In general, the narrower angular sensitivities are found in the frontal, "foveal" parts of eyes and in species with greater visual demands for high spatial resolution. High spatial resolution comes at the cost of needing faster photoreceptor dynamics, if resolution of moving objects (or during flight) is to be retained (Srinivasan, M. and Bernard, G., 1975). However, it has been hypothesized that resolutions of moving, striped objects may be even superior to static resolution (Northrop, R., 1975).

Finally, it can be seen in Table 1 that angular sensitivities are narrower in light-adapted photoreceptors than in dark-adapted photoreceptors, during the night. Such changes in angular sensitivities are the result of retinomotor movements, structural alterations, or pigment–granule movements in the photoreceptor. For example, in giant water bugs, *Lethocerus*, movements of the rhabdom tips closer to the cornea during dark adaptation (Lüdtke, H., 1953) result in 1000 times greater absolute sensitivities and broader angular sensitivities (9.0°), compared to the light-adapted eye (3.5°: Walcott, B., 1971a,b).

Locust photoreceptors undergo two kinds of

structural changes during dark adaptation. First, packed mitochondria around the light-adapted rhabdom are displaced by a clear, vacuolous palisade. The angular sensitivity increases (from 1.9° to 2.8°: Williams, D., 1983) with the same time course as the growth of the palisade (Tunstall, J. and Horridge, G., 1967). Presumably the replacement of mitochondria by palisade increases the critical angle for internal reflection in the rhabdom and thereby its angular acceptance. Second, after dusk there is a later increase in the cross-sectional area of the rhabdom in locusts and in the amount exposed at a distal "field stop". This is correlated with a further increase in angular sensitivity to 4.9°. At dawn, excess rhabdomal membrane is shed, and the angular sensitivities narrow, even when cells are again dark-adapted (Williams, D., 1983). Thus angular sensitivities are affected by both rhabdom area and light-guide properties. Praying mantis photoreceptors also have three comparable stages of different angular sensitivities (Rossel, S., 1979: see Table 1), perhaps for the same reasons as in locusts.

In photoreceptors of many diurnal insects, pigment granules move radially up against the rhabdom during illumination, and away during dark adaptation. As in movements of mitochondria in photoreceptors of locusts, the effect is to reduce angular sensitivities by reducing total internal reflection (Kirschfeld, K. and Franceschini, N., 1969). The result is equivalent to a decrease in rhabdomal illumination (Srinivasan, M. and Bernard, G., 1980). Screening pigment movements can be optically monitored by non-invasive measurements of decreases in transmitted ("antidromic") light or increases in reflected ("orthodromic") light (Franceschini, N., 1975).

The principal pigment movements occur next to those rhabdoms which are absorbing photons. This is one anatomical method for identifying which cells in an ommatidium are sensitive to which wavelengths (Ribi, W., 1978a; see references in Lo, M. and Pak, W., 1981). The movements are also dependent upon the excitation resulting from photon absorption. In the fly Calliphora, retraction of pigment in the dark is greatly delayed after conditions which set up a PDA (strong blue illumination) but is subsequently accelerated after (red) light which knocks down the PDA (Stavenga, D. et al., 1975). Conversely, the late receptor potential elicited by a long flash is only transient in the ERG of the Drosophila visual mutant trp and so is the pigment migration. Mutants (norpA) which lack late receptor potentials also lack pigment migration (Lo, M. and Pak, W., 1981). These findings have suggested that pigment granules might move electrophoretically due to radial potential differences within the cell, set up by photocurrents. However, pigment migrations are not nearly so rapid as are the late receptor potentials, although they can have time constants as fast as 5–15 s in Hymenoptera (Stavenga, D. and Kuiper, J., 1977) or 2–7 s in Drosophila (Lo, M. and Pak, W., 1981). Moreover, in the fly Musca, pigment migrations still occur when KCl injected into the eye nearly abolishes the ERG (presumably by drastically depolarizing the photoreceptors). On the other hand, injection of EGTA into the eye, to reduce extracellular calcium, enhances the ERG (as expected: see section 2.3 above) but brings pigments into the extreme dark position, even with strong illumination (Kirschfeld, K. and Vogt, K., 1980). A delayed light-initiated rise in intracellular calcium in photoreceptors may thus be the signal for pigment migration, rather than changes in the membrane potential. It should be emphasized here that pigment migration is far less effective in reducing the sensitivity of a photoreceptor than is the calcium rise which could cause it. In butterflies and hoverflies, maximum pigment migration is equivalent only to an 0.8 log unit decrease in illumination (Srinivasan, M. and Bernard, G., 1980), whereas maximum light adaptation, due to a rise in intracellular calcium, reduces sensitivities 3 or more log units (see Fig. 13).

2.4.2 SPECTRAL AND POLARIZATION SENSITIVITIES

Spectral and polarization sensitivities are measured in much the same way as were angular sensitivities (see Fig. 11 for polarization sensitivities). The relations between spectral and polarization sensitivities, on the one hand, and the absorptions and orientations of eye pigments, on the other, are given in detail in the chapter by White in this volume. They will only be summarized here. Spectral sensitivities have been measured in a wide variety of insects, using the ERG and/or intracellular recordings

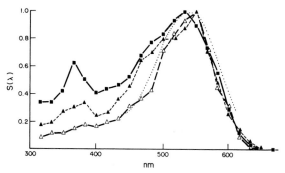

FIG. 13. Average spectral sensitivities of green-sensitive photoreceptors of worker honey bees, *Apis mellifera*, measured using plateau responses to flashes of the indicated wavelengths. The dotted line represents the relative absorption curve of a rhodopsin which absorbs maximally at 540 nm. Adapted from Menzel, R. and Blakers, M. (1976), *J. Comp. Physiol. 113*, 17–34, by permission of the author and copyright permission of Springer-Verlag, Heidelberg.

(Menzel, R., 1979). Considering all species tested, cells are found which have peak sensitivities at wavelengths in the UV, blue, green, and/or yellow-orange parts of the spectrum. For a large number of (paradigmatic) cells, the shapes of the spectral sensitivity curves can be accounted for in terms of the absorption spectra of photolabile rhodopsins (Dartnall, H., 1952) and of photostable "antennal" or screening pigments. Figure 13 illustrates spectral sensitivities of green-sensitive cells of worker honey bees. The dashed line is that of a rhodopsin absorbing maximally at 540 nm. It fits the long wavelength data and the lowermost curve at short wavelengths (Menzel, R. and Blakers, M., 1976).

There are additional, non-paradigmatic cells, however, whose spectral sensitivity curves are broader than expected from single rhodopsins, or are multi-peaked (upper curves in Fig. 13) or whose $V \log I$ curves are not parallel for all wavelengths (Wasserman, G., 1973). Broad-peaked spectral sensitivity curves of locust photoreceptors seem to be due to various mixtures of two rhodopsins in single cells (Bennett, R. *et al.*, 1967; Lillywhite, P., 1978). For dragonfly ocellar photoreceptors, $V \log I$ curves at 370 and 520 nm are not always parallel, which is also consistent with two visual pigments in each cell (Chappell, R. and DeVoe, R., 1975). These differences in $V \log I$ curves are propagated to second-order cells and to optomotor behavior (see section 6: Ocelli). They are therefore part of the visual performance of dragonfly ocelli.

The non-paradigmatic spectral sensitivities of still other photoreceptors probably have their origins in coupling between paradigmatic cells. One way to show coupling between cells is to show that they are optimally excited by different planes of polarized light. First, it is appropriate to consider polarization sensitivities.

Sensitivity as the plane of polarization is rotated was illustrated in Fig. 11 for a recording from a cell in the drone honey bee (Shaw, S., 1969). The dependence upon angle of polarization is roughly sinusoidal, which is expected since the probability of absorption is related to the square of the cosine between the *e*-vector of polarized light and the optimal absorption angle of the photopigment (Law of Malus). Polarization sensitivity is then defined as the ratio of the greatest sensitivity to the least sensitivity. Table 2 lists representative values of polarization sensitivities measured electrophysiologically in insect photoreceptors.

As for spectral sensitivities, so too for polarization sensitivities. Paradigmatically, polarization sensitivities should be related to the dichroic ratio of the ensemble of photopigment molecules in a rhabdomere. The dichroic ratio is the ratio of the maximum absorption when the *e*-vector angle of polarized light is optimal, to the minimal absorption 90° from this (Fein, A. and Szuts, E., 1982). On theoretical grounds, a dichroic ratio of 1.7 is expected in microvilli if rhodopsin is randomly oriented (Goldsmith, T. and Wehner, R., 1977). Dichroic ratios could be larger if rhodopsin was oriented non-randomly, or if there was optical coupling among the rhabdomeres of a fused rhabdom (Snyder, A. *et al.*, 1973; Snyder, A. and Laughlin, S., 1975). Many of the values of polarization sensitivity in Table 2 lie near the value of 1.7. Thus rhodopsin in their rhabdomeres is probably randomly oriented. In some cells polarization sensitivities are closer to 1 (dependent in some cases on wavelength). Possible reasons are given in the chapter by Land in this volume. These include twist in rhabdomeres (for optimal absorption of *non*-polarized light), interference of antennal pigments, and/or positive coupling between cells with preferred absorption angles which are 90° apart (Menzel, R. and Snyder, A., 1974). In still other cells, polarization sensitivities may be much larger than 1.7 because of presumed non-random

Table 2: Electrophysiologically determined polarization sensitivities

Order	Species	Polarization sensitivities	Comments	Reference
Odonata	*Anax junius,* *Libellula needhami* *Hemicordulia tau*	1.4–2 adults 4.5 juvenile 7.0 UV cells 3–6 blue cells	PS of all other cells at 2.3 peak wavelengths	Horridge, G., 1969 Laughlin, S., 1976a
Orthoptera	*Locusta* *Periplaneta americana*	2.34 5		Shaw, S., 1969 Butler, R. and Horridge, G., 1973b
Hemiptera	*Lethocerus sp*	3.5	No changes with adaptation	Walcott, B., 1971b
Lepidoptera	*Papilio aegeus*	5 UV cells 2–5 blue cells 1 or 4 green cells	PS of red cells not given	Horridge, G. *et al.,* 1983
Hymenoptera	*Apis mellifera* (drone) *Apis mellifera* (worker) *Apis mellifera* (worker) *Cataglyphis bicolor*	1.42 weak lights 1.58 25× stronger light 1.0–2.4 green cells 1.27–1.43 blue cells 1.0–1.4 UV and green cells 5 (4–9) basal UV cells dorsal rim, UV cells: 13.0 (4.5–18) no green sensitivity 5.6 10% green sensitivity 3.8 10% green sensitivity 2.0 UV cells, rest of eye 2.5 green cells 1.7 blue cells 1.5–6 UV, VIS, and UV–VIS cells 4 Ocellar cells	All cells had broad, UV-green spectral sensitivities Unchanged by light adaptation No regional eye differences PS of blue and green cells is greater in or near dorsal rim Optimum polarizations differ for UV and green light in UV–VIS cells Ocellus has UV cells only	Shaw, S., 1969 Menzel, R. and Snyder, A., 1974 Labhart, T., 1980 Mote, M. and Wehner, R., 1980
Diptera	*Calliphora erythrocephala* (male) *Calliphora erythrocephala* *Calliphora stygia* *Calliphora stygia* *Musca domestica* (female) *Musca domestica* (white-eyed) *Boettcherisca peregrina* *Eristalis tenax*	2 2.1 1.3–3.5: R1–6 2.0: R1–6 2.2 R7 1.5, 3.5: R8 R7UV: 1.5 358 nm 1.8 442 nm R7UT: 1.0 358 nm 2.3 442 nm R8: 1.8 442,541 (572) nm 1.3–2.0 2.9: R1–6 4.6 (max 6): R7 1.2–2.8 1–3	500 nm light Unchanged by light adaptation 442 nm light; no PS with UV light Maximum sensitivity was parallel to microvilli Dark-adapted values Probably R1–6	Burkhardt, D. and Wendler, L., 1960 Gemperlein, R. and Smola, U., 1973 Horridge, G. and Mimura, K., 1975 Hardie, R., 1979 Hardie, R. *et al.,* 1979 Scholes, J., 1969 Hardie, R., 1979 Mimura, K., 1978 Horridge, G. *et al.,* 1975

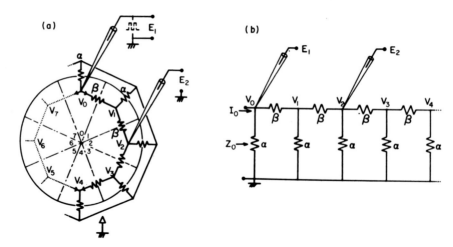

FIG. 14. Model of electrical coupling between eight photoreceptors within an ommatidium (of a honey bee or a locust). The figure illustrates how the coupling coefficients can be studied by passing current into one photoreceptor while recording the induced voltages in another. The coupling resistances α represent total membrane resistances to extracellular space (assumed to be at ground potential). Coupling resistances β represent inter-receptor resistances. From symmetry, the circular network in (a) can be simplified to that in (b). Reproduced from Shaw, S. (1969), *Vision Res. 9*, 999–1029.

orientations of rhodopsin, absence of twist (Labhart, T., 1980), or perhaps even negative coupling (Horridge, G. *et al.*, 1983). The point is that when polarization sensitivity is related to the probabilities of absorption in a rhabdomere, then each absorbed photon will give an identical quantum bump in that cell. It is the non-paradigmatic instances when the absorption in one cell contributes to the response in another that are of further interest for electrical activity.

2.4.3 COUPLING BETWEEN PHOTORECEPTORS

Coupling between insect photoreceptors has been shown directly with simultaneous intracellular recordings from two cells in locusts and in drone honey bees (Shaw, S., 1967, 1969). The ratio between current-elicited voltage in one cell and the voltage measured in the other was the coupling ratio. Strong coupling, presumably between immediate neighbors, had a ratio of 0.53 in drones and 0.15 in locusts. The simplest model for such coupling is an electrical network like that in Fig. 14, in which the dark resistance (mainly for potassium and chloride) of each cell to extracellular space is α and the coupling resistance between cells is β. Ratios between α and β of 1:0.25 and 1:2.1 would account for the above coupling ratios in drones and locust, respectively. It is not known what the coupling resistances are due to, since gap junctions

between photoreceptor somas have not been observed (Chi, C. *et al.*, 1979). However, gap junctions are found between axons in the lamina (see section 2.5).

Light induces changes in the resistances α, so that dependencies of α upon intensity must be used in calculations of coupling of light-elicited electrical activity. Assuming a dichroic ratio of 9, Shaw, S. (1969) was able to account for the low polarization sensitivities of drone and locust photoreceptors only if he also assumed that β values also changed with light (so that the ratio $\alpha:\beta$ was constant).

Different methods are needed to demonstrate coupling when recording from only one cell — the more usual situation. For example, UV bands or subpeaks in green-sensitive cells of worker honey bees (upper curves of Fig. 13) may be due to coupled UV and green cells. The bottom curve in Fig. 13 is of a green cell with low UV sensitivity. UV adaptation reduces sensitivities at all wavelengths, but those in the UV the most (Menzel, R. and Blakers, M., 1976). This is presumptive evidence for coupling of two cells, although it is not incompatible with two pigments in one cell (Chappell, R. and DeVoe, R., 1975). Stronger evidence for coupling of UV and blue-green-sensitive cells is found in UV–VIS cells of the ant *Cataglyphis*. The curve for polarization sensitivity at 340 nm is shifted 45° in phase from the curve at 517 nm (Mote, M. and Wehner, R., 1980). The angle between the microvilli of adjacent large and small cells in *Cataglyphis* is 45°, so this is evidence that when adjacent UV and green-sensitive cells are separately excited, they contribute to each other's responses (Martin, F. and Mote, M., 1980). There is similar evidence for coupling of adjacent cells in the dronefly *Eristalis* (Tsukahara, Y. and Horridge, G., 1977a) and the blowfly *Boettcherisca* (Mimura, K., 1978). In addition to phase differences in optimum polarization angle and (in *Eristalis*) spectral sensitivities, the coupled cells had different optical axes.

Dependence of polarization angle and sensitivity upon wavelength, in cells with broad spectral sensitivities, does not always indicate coupled cells, however. There may be filtering of polarized light by overlying receptors in tiered retinas, such as the dragonfly's (Laughlin, S., 1976a). UV antennal pigments in cells R1–6 of flies also have different preferred angles than do the green-absorbing rhodopsins (Horridge, G. and Mimura, K., 1975; Horridge, G., *et al.*, 1975).

Martin, F. and Mote, M. (1980) have further developed the model of electrical coupling of Shaw, S. (1969, see Fig. 14) for electrical activity elicited by light, but with constant coupling resistances β. As noted above, this model describes the coupling in UV–VIS cells of *Cataglyphis*. It also predicts that coupling will reduce polarization sensitivities in each cell and average the phases when both cells are equally excited. Green-sensitive cells of worker honey bees (Menzel, R. and Snyder, A., 1974) have very low polarization sensitivities, and this is taken to indicate coupling between two green cells with microvilli at right angles (possibly cells 2 and 3, or 6 and 7: Wehner, R. and Bernard, G., 1980). The polarization sensitivities would then be 180° out of phase and so cancel. On the other hand, strong light is predicted to uncouple cells, so it would be expected to increase polarization sensitivities. This is what is found in drones (Shaw, S., 1969; see Fig. 11), but not in worker bees (Menzel, R. and Snyder, A., 1974).

In locusts there is coupling for single quantum bumps (Shaw, S., 1967), resulting in two small (S) bumps from the two immediate neighbors for every large (L) bump generated in the impaled cell. The polarization sensitivities of S bumps are smaller than for L bumps and are shifted 30° in phase from each other (Lillywhite, P., 1978). In locusts, one

neighboring cell would have parallel microvilli and the same preferred *e*-vector orientation, the other's would be rotated 60°. The combined 30° phase shift and smaller polarization sensitivity for S bumps is consistent with the model of Martin, F. and Mote, M. (1980).

In flies too there is coupling of bumps at very low light levels, and smaller bumps are elicited by OFF-axis stimuli. Recordings of bumps from laminar monopolar cells show that they have a larger quantum catch area than at higher luminances (Dubs, A. *et al.*, 1981). This would mean that there is both summation in the lamina of the six R1–6 cells of different ommatidia but with the same angular view (Scholes, J., 1969) as well as summation within ommatidia of quantum catches by cells with different angular views. Pooling of quantum catches at low light levels in flies is thus similar to that in the vertebrate retina (Fain, G., 1975).

A word of caution about coupling must be interjected at this time. Multi-peaked or broad spectral sensitivities have been attributed to coupling between cells, as for example in worker honey bees (Menzel, R. and Blakers, M., 1976; see Fig. 13). Non-invasive measurements of spectral sensitivities of honey bee workers, using the pupillary reflex (based on screening pigment movements), have failed to find UV peaks of sensitivity in green cells (Bernard, G. and Wehner, R., 1980). Similarly, UV sensitivities in green cells recorded intracellularly in bumblebees *Bombus* (Meyer-Rochow, V., 1980) were not found using the pupillary reflex (Bernard, G. and Stavenga, D., 1978). Thus, coupling might be artifactual, resulting from damage to cells by intracellular electrodes (Wehner, R. and Bernard, G., 1980). Against this possibility is that coupling of bumps at low light levels in flies is clearly not due to electrode injury of photoreceptors, since coupling is also seen in recordings from second-order cells (Dubs, A. *et al.*, 1981). Second, the pupillary reflex probably measures only distal pigment movements in flies, and not the spatially integrated responses of photoreceptors (Vogt, K. *et al.*, 1982). Third, since screening pigment movements depend on rises of intracellular calcium (Kirschfeld, K. and Vogt, K., 1980; see section 2.4.1), they probably occur only for direct excitation of a cell by light, and not for coupled currents from a neighboring activated cell. Thus, all coupling between cells is unlikely to be artifactual, and the information received by the first synapses may well not be that from isolated photoreceptors.

Finally, there can be negative coupling between cells, as well as the positive coupling discussed above. One source of negative coupling, as seen in intracellular recordings, is the flow of photocurrents through extracellular fluids around photoreceptors. The resulting extracellular currents create in effect local ERGs. If the local ERGs resulting from excitation of one cell type by, say, long-wavelength light is in series with the potentials being measured in another, unexcited cell (a UV cell, say), then a negative "response" could be recorded. This type of negative coupling is likely to be without physiological effect, since it does not affect the photoreceptor's terminals. It could artifactually narrow the measured spectral or other sensitivity curves, however.

A more potent form of negative coupling was first hypothesized by Shaw, S. (1975) in locust photoreceptors. If there are high-resistance barriers between the retina and lamina, then photocurrent exiting at one photoreceptor's terminals in the lamina could have a lower resistance path back to the retina via other photoreceptors' terminals and axons. The return current would hyperpolarize these other cells' terminals and so reduce their release of transmitter. Thus, this type of negative coupling would have physiological effects (see section 2.5.1, below).

Negative coupling via terminals in the butterfly *Papilio* (Horridge, G. et al., 1983) has been hypothesized to underlie positive-to-negative spectral sensitivities, narrowed angular fields with hyperpolarizing surrounds that are dependent on polarization angle, and polarization sensitivities which are far more (> 30) than can be accounted for even by perfect alignment of rhodopsin in microvilli (20: Snyder, A. and Laughlin, S., 1975). Mutual negative coupling between photoreceptors with microvilli oriented 90° apart could result in such polarization sensitivities. In an analogous situation, receptors R7 and R8 of flies each inhibit turning reactions, and their microvilli are oriented at 90° to each other. The polarization sensitivity of 100 for inhibition is greater than for either cell (see Table 2), so it may likewise result from inhibitory interactions (Kirschfeld, K. and Lutz, B., 1974).

The next section will include further details of coupling at photoreceptors' terminals and in the lamina.

2.5 Conduction of excitation to axon terminals

The depolarizations elicited by light in the somas of photoreceptors are most likely conducted by passive electrotonic spread to the axon terminals (and not by spikes: see section 1). That is, depolarization of the soma by light results in potential differences between soma and terminal. Current therefore flows axially down the axon. Current flow will minimize the potential differences by depolarizing the axon and the terminal. The axial current will of course be supplied by some portion of the photocurrent. The axial current, flowing out at the terminal, then flows back through extracellular space. If there are appreciable resistance barriers in extracellular space, the return currents will set up appreciable extracellular potentials or be forced to return via other, less-depolarized axons. The release of transmitters at axon terminals depends on the transmembrane potential difference; that is, upon the difference between intracellular and extracellular potentials. Hence, transmission of excitation at the first synapse depends upon the mode of conduction to the synapse, upon interactions (coupling) around the terminals, and upon the voltage dependence of transmitter release.

Passive, electrotonic spread of excitation from soma to terminal has not been rigorously demonstrated for any insect photoreceptor, although it has for lateral and median photoreceptors of the barnacle *Balanus* (Shaw, S., 1972; Hudspeth, A. et al., 1977) and in spider optic nerves (Gallin and DeVoe, unpublished observations). Electrotonic spread would occur if there were no voltage-dependent conductances in axonal membrane and would result in amplitudes and speeds of response which decline with distance from the soma. Qualitatively, these conditions are met in insect photoreceptors (Odonata: Laughlin, S., 1974b; Hemiptera: Ioannides, A. and Walcott, B., 1971; *Musca*: Scholes, J., 1969; *Calliphora*: Zettler, F., 1967; Zettler, F. and Järvilehto, M., 1970; Smola, U. and Gemperlein, R., 1972; Rehbronn, W., 1972; Hardie, R., 1977; drone honey bee: Baumann, F., 1968). Figure 15 (left) illustrates responses from the soma and axon

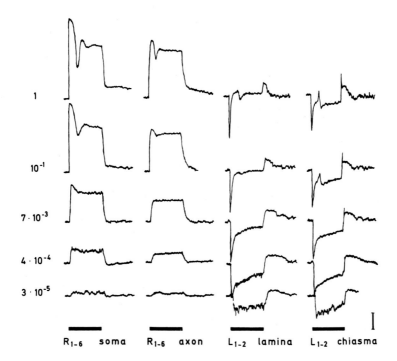

FIG. 15. Responses to flashes (indicated by solid bars at bottom) as recorded in somas and axons of peripheral photoreceptors R1–6, and in L cells penetrated within the lamina and 300 μm proximal in the outer chiasm. Relative intensities of the flashes are indicated to the left of each row. The vertical bar at lower right represents 10 mV. Reproduced from Zettler, F. and Järvilehto, M. (1973), *J. Comp. Physiol.* 85, 89–104, by permission of the authors and copyright permission of Springer-Verlag, Heidelberg.

of photoreceptors of *Calliphora*; responses in the axon are somewhat smaller, less transient, and less noisy than in the soma (Zettler, F. and Järvilehto, M., 1973). The axonal responses also have 1–2 ms longer latencies (Scholes, J., 1969; Smola, U. and Gemperlein, R., 1972).

Nothing is known of the specific membrane resistances of axonal membrane of insect photoreceptors, but in *Balanus* the axonal membrane resistances are much higher than in the soma (Shaw, S., 1972). Even if axonal and somal membrane resistances were the same in insects, the total cable resistance of a 200 μm long axon in *Locusta* is estimated to be only 1% of that of the soma (Shaw, S., 1969). Hence only a small percentage of photocurrent may actually flow axially to terminals. The rest is dissipated in the soma (Zimmerman, R., 1978).

Recordings from axon terminals in the laminae of dragonflies and waterbugs show that angular, polarization and (in dragonflies) spectral sensitivities are conserved (Laughlin, S., 1974c; Shaw, S., 1968; Ioannides, A. and Walcott, B., 1971). That

is, there is no evidence from these eyes with fused rhabdoms of presynaptic coupling between the axons, all of which originate in the same ommatidium.

In flies with unfused rhabdoms, the six axons of receptors R1–6 in a laminar cartridge come from six different ommatidia (Strausfeld, N., 1976a). In each axon, angular sensitivities of somas are conserved, but polarization sensitivities are lost (Scholes, J., 1969; Smola, U. and Gemperlein, R., 1972). Axon terminals in *Musca* can be excited by light on each of the six facets that project onto the single lamina cartridge (Scholes, J., 1969). This is evidence for presynaptic coupling between the six R1–6 photoreceptor terminals, since light on any one facet directly excites only one of the six photoreceptors. The electrical activity set up by light on one facet can also be recorded in the soma of one of the other five unilluminated cells (Shaw, S., 1981). Thus, presynaptic coupling is not due to injury in the laminar cartridge.

Presynaptic coupling in *Musca* seems to involve linear summation of the axonal responses of

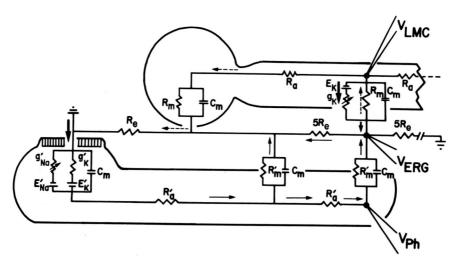

FIG. 16. Model of resistances, currents, and voltages in photoreceptors, laminar monopolar cells (LMCs) and extracellular space of flies. Light initiates depolarizing inward photocurrents in the rhabdom (heavy arrow at the left). These currents spread passively down the axon to the terminal (thin arrows) and return both in extracellular space whose resistance is Re and via the cell bodies of the LMCs (dashed arrows). Transmitter release at the terminals, a function of the intracellular potential V_{Ph} and laminar extracellular potential V_{ERG}, results in hyperpolarizing outward current in LMCs (heavy arrow at upper right). Laminar extracellular resistances are taken to be 5 times greater than extracellular resistances in the retina. Reproduced from Zimmerman, R. (1978). *J. Comp. Physiol.* **126**, 297–316, by permission of the author and copyright permission of Springer-Verlag, Heidelberg.

stimulated photoreceptors (Scholes, J., 1969, but see Shaw, S., 1981). This is another advantage of the "neural superposition" eyes of dipterans (Kirschfeld, K., 1967): the linear summation of signals of six small, weakly-excited receptors R1–6 will be greater than would be the response of one large cell with the same total quantum catch, due to the compressive function of $V \log I$ curves. (The larger cell would also have poorer angular sensitivity.) Linear summation in flies also improves signal-to-noise ratios in the terminals, compared to the somas (Gemperlein, R. and Smola, U., 1972) or compared to cockroaches *Periplaneta* which presumably lack presynaptic summation (Smola, U. and Gemperlein, R., 1973). The loss of polarization sensitivities in flies can likewise be explained by presynaptic coupling of signals from three sets of cells with microvilli oriented 120° apart. Coupling may be via gap junctions within laminar cartridges (Chi, C. and Carlson, S., 1976; Ribi, W., 1978b; Shaw, S. and Stowe, S., 1982).

There are additional potentials which can be transmitted into axon terminals. Stimulating a photoreceptor under one facet while recording in the soma of another, Shaw, S. (1981) recorded a fast, transient wave of inhibition with a 1.5–2 ms latency. He interpreted this as chemical feedback of

laminar monopolar cells onto receptor cell axons (Shaw, S., 1982). The inhibitory feedback may play a role in the cutback of the peak hyperpolarizations of laminar monopolar cells in Fig. 15 (see section 3.1, below). The feedback also emphasizes the potential complexities of the receptor–receptor–monopolar cell networks of laminar cartridges.

Spikes are sometimes also recorded in receptor somas (*Locusta*: Shaw, S., 1968) or axon terminals (dragonflies: Laughlin, S., 1974c; *Calliphora*: Rehbronn, W., 1972). It seems likely that the spikes originate in spiking cells of the lamina (see section 3.2), are coupled electrically to receptor axons, and are electronically conducted back into somas.

2.5.1 EXTRACELLULAR FIELD POTENTIALS AND THE ERG

The paths for the return flow of current from photoreceptor terminals back to the area of the rhabdom are not through a homogeneous extracellular space. Rather, the paths include high-resistance regions, so that returning current divides between extracellular space and other cells. In the model of Fig. 16, typified by flies and dragonflies, the lamina is a high-resistance region. In the model of Fig. 17, typified by locusts and perhaps butterflies, the high-resistance region lies around the photoreceptor axons. These differences between return current paths help explain the form of the ERG, especially the presence or absence of ON- and OFF-effects.

In flies, the lamina is a region of high axial resistance (Zimmerman, R., 1978). This has the following consequences. First, exiting photocurrent that flows in the high-resistance extracellular space develops high "lamina positive potentials", as great as 15 mV (Mote, M., 1970a). Lamina positive potentials of dragonflies, 12–44 mV in amplitude, have response waveforms which look like those in the soma, except they are much smoother (without bumps) and slower. They have broader angular sensitivities than axons, and almost no polarization sensitivities (Laughlin, S., 1974c). Broad angular sensitivities are also found in flies, where regions 100 μm apart contribute to the laminar positive potential (Mote, M., 1970b). All this would be expected if lamina positive potentials result from voltage drops due to summed photocurrents from the terminals of many photoreceptors.

Second, the exiting photocurrent will also flow through the lamina monopolar cells back to extracellular space, as shown in Fig. 16 (dashed arrows), if the resistance of this pathway is comparable to, or less than, the extracellular resistance in the lamina. The result would be to depolarize the distal ends of laminar monopolar cells (LMCs) with respect to the hemolymph. This effect can be seen in Fig. 15: the plateau hyperpolarizations of LMCs are reduced in laminar recordings but not in the axon (Autrum, H. et al., 1970) or in the LMC terminals in the medulla (DeVoe, R., 1980).

Third, the laminar positive potential may exert a feedback effect upon transmitter release from photoreceptor terminals (Laughlin, S., 1974c). The potential is not only due to passive current flowing outward from photoreceptor terminals, but also due to active outward current (probably potassium) from the hyperpolarizing LMCs (this will be discussed in section 3.1, below). If transmitter release is a function of *trans*membrane potential at photoreceptor terminal, the slower-rising lamina positive potential would reduce the initially large transmembrane potential due to the rapid axonal response. Together with transient inhibition of the axonal response by the LMCs (Shaw, S., 1982), there could result a transiently high initial release of

transmitter. This helps explain the rapid cutback of the peak hyperpolarizations in LMCs in Fig. 15.

Current flows as depicted in Fig. 16 also help to explain the waveforms of ERGs, such as for the fly in Fig. 1. The potential between the cornea and an indifferent electrode (the hemolymph) would have the laminar positive potential in series but electrically closer to the indifferent electrode. Thus, the photoreceptor response in the ERG would be a negative one. Because currents from the LMCs also contribute to the laminar positive potential, they are also included in the ERG. However, at ON, they flow in the opposite direction, away from the retina and towards hemolymph bathing the LMC axons. Thus, the initial hyperpolarizations of LMCs are recorded as positive ON-effects. At OFF, the LMCs transiently depolarize and currents flow from the axons towards the lamina (towards the retina). Hence OFF-effects in the ERG are negative, just as are the sustained potentials.

Consistent with this view are the depth recordings of Heisenberg, M. (1971) in *Drosophila*, in which the receptor component of the ERG was first recorded (with reference to the cornea) with an electrode proximal to a high resistance "receptor-barrier". ON- and OFF-effects were recorded still more proximally. Second, in cockroaches and flies, there are no ON- or OFF-effects in ectopic eyes that lack laminar synapses (Wolbarsht, M. et al., 1966; Eichenbaum, D. and Goldsmith, T., 1968; Sivasubramanian, P. and Stark, W., 1979). The ERGs are also smaller in ectopic eyes, as expected in the absence of current flow in a high-resistance lamina. Third, ON- and OFF-effects are missing in *Drosophila* mutants which lack functional laminar cells (Alawi, A. and Pak, W., 1971), or in *Calliphora* after blockage of synaptic transmission (Autrum, H. and Hoffman, C., 1960). It can be concluded, therefore, that "fast" eyes with ON- and OFF-effects in their ERGs (Autrum, H., 1950) have currents of photoreceptor and LMC cells flowing across a high-resistance barrier, which is in series with the pathways to the indifferent electrode.

Current flows are different in eyes with more distal high-resistance barriers around the photoreceptor axons, as in Fig. 17. In locusts a diffusion barrier seals off the retina from the blood (Shaw, S., 1977). Returning photocurrents must either pass through this barrier or, preferentially, through other, less

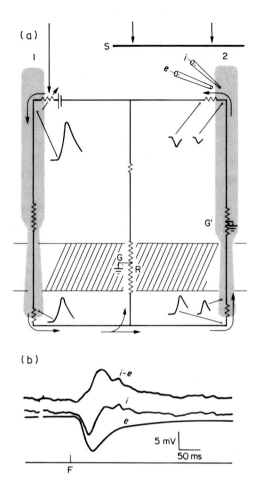

Fig. 17. Model of electrical interactions between photoreceptors in locusts. **a:** Stippled regions 1 and 2 represent
photoreceptors, and the hatched area represents a high-
resistance barrier that isolates the retinal from the laminar
parts of the cells. A grounded, indifferent electrode in
hemolymph sets up an electrical ground at G within
hemocoels in the barrier, with resistance R of the barrier
connecting G to extracellular spaces in the retina and lamina.
By analogy with a Wheatstone bridge, there is a corresponding virtual ground at G' within cell 2. As in Fig. 16, active,
inward, depolarizing photocurrent of illuminated cell 1 exits
at the terminal. The current returns via unilluminated cell 2
as well as via the barrier R. Because of the virtual ground G',
the induced intracellular potentials at the two ends of cell 2
are smaller than the immediate extracellular potentials. The
transmembrane potential, inside minus outside, is thus
depolarizing at the distal, retinal end of cell 2, as shown
in (**b**). At the proximal, laminar end, the transmembrane
potential would be hyperpolarizing. From Shaw, S. (1975),
by permission of the author. Reprinted by permission from
Nature 255, No. 5508, 480–485, copyright© 1975 Macmillan
Journals Ltd.

excited cells' axons. Currents of LMCs need not
cross the barrier, however, but can be shunted to the

indifferent electrode via hemocoels in the diffusion
barrier. The reason that ON- and OFF-effects are not
seen in corneal ERGs of slow eyes (Fig. 1, middle
left) may therefore be that there is no common high-
resistance pathway for both photo- and LMC-
currents. Local recording within the lamina itself,
where both currents are present, does yield ERGs
with ON- and OFF-effects in locusts (Cosens, D.,
1967).

Returning photocurrents will also pass into other
terminals of other cells, particularly those that are
less, or not at all, depolarized. The effects of such
currents are illustrated schematically in Fig. 17,
where photoreceptor 2 is shielded from the light
that excites receptor 1. The extracellular potential
outside the terminal of receptor 2 (the laminar
positive potential) will be greater than inside the
terminal, due to the induced voltage drop across the
membrane resistance. Thus, the transmembrane
potential, inside minus outside, will be negative: the
terminal will be hyperpolarized by the entering
passive current (at the distal end of the receptor,
exciting passive current will depolarize: Shaw, S.,
1975). The effect is clearly differential, since if two
receptors are equally excited there will be no potential differences and no current flows between their
terminals. As mentioned above (section 2.4.3), differential negative coupling between terminals may
be responsible for the types of angular, spectral, and
polarization sensitivities recorded in *Papilio* (Horridge, G. *et al.*, 1983). Conversely, if all receptors are
strongly excited, returning photocurrents must flow
through the extracellular resistances instead of
other terminals. The extracellular field potentials
could then approach the intracellular potentials,
giving not only large ERGs but also small transmembrane potentials at the terminals. This would
provide a kind of automatic gain control for synaptic output (Shaw, S., 1979).

Finally, mention must be made of standing,
radial, extracellular DC potentials in the retina and
optic lobes. In locusts, cockroaches, dragonflies,
and fleshflies, there are 20–50 mV positive potentials in the retina and 50–90 mV negative potentials
in the lamina, with respect to the hemolymph
(Burtt, E. and Catton, W., 1964; Mote, M., 1970a;
Zimmerman, R., 1978). *Drosophila* lacks the lamina
negativity (Heisenberg, M., 1971). Both positive
and negative potentials are reduced during steady

illumination, undoubtedly by flows of photocurrents. Possibly, laminar negativity arises from a synaptic current sink there in *Musca* (Zimmerman, R., 1978). Laminar negativity in locusts depends upon normal respiration and is abolished (in minutes) by anoxia (Burtt, E. and Catton, W., 1964). As it is not abolished by nicotine in locusts, it may have a different origin from that in flies.

3 THE LAMINA

The connectivity and synaptic physiology of the lamina is the best understood of any optic lobe (Strausfeld, N., 1976b; Strausfeld, N. and Campos-Ortega, J., 1977; Laughlin, S., 1981b; Shaw, S., 1981). However, the physiological properties of only four of the 12 types of laminar cells have been identified by intracellular recording and staining. These are the receptors, the largest laminar monopolar cells (LMCs) and (once) the T1 centripetal cell. The latter cells all hyperpolarize, as in Fig. 15. They serve to emphasize contrasts via high-gain synapses and to narrow angular sensitivities via lateral interactions. Another important, but unidentified, class of cells would be the laminar spiking cells. In flies, these include ON–OFF cells with large, uniform receptive fields and smaller-field sustaining cells with inhibitory flanks (Arnett, D., 1972). Certain lobular directionally selective cells (the H1 cells: see section 5.1) have cells in their input pathways with the center-flanking receptive field organization of the sustaining cells (Srinivasan, M. and Dvorak, D., 1980). Finally, still other unidentified hyperpolarizing cells (in flies) have large, complicated receptive fields (Mimura, K., 1976), possibly those of lamina amacrine cells carrying information laterally (Strausfeld, N. and Campos-Ortega, J., 1977).

3.1 Hyperpolarizing laminar monopolar cells (LMCs)

The first hyperpolarizing responses in the lamina were recorded in locusts by Shaw, S. (1968), but it was not until 1970 that such hyperpolarizing responses were identified as coming from the large LMCs (Autrum, H. *et al.*, 1970). Figure 15, right, illustrates the forms of the hyperpolarizing responses in $L_{1,2}$ for flies. Latencies are longer than for receptors, as would be expected (Shaw, S., 1968). The latency differences (the synaptic delays) are 0.5–1.0 ms in flies (French, A. and Järvilehto, M., 1978a), about 2 ms for dragonflies (Laughlin, S., 1973) and about 5 ms for locusts (Shaw, S., 1968). Following these delays, the initial responses are rapid hyperpolarizing ON-transients. The transients decline back to smaller (or no) hyperpolarizing plateaus, except at the lowest intensities. At the lowest intensities the plateaus are noisy and appear to reflect summations of quantum bumps. Only occasionally are depolarizing spike-like events seen during plateau responses, but they are variable. Sometimes frequencies of such spikes are inhibited by illumination (Shaw, S., 1968), or they are synchronized to movement-elicited oscillations (DeVoe, R. and Ockleford, E., 1976, Fig. 5g), or they are unrelated to the temporal time courses of the plateaus (DeVoe, unpublished observations). Possibly the spikes are indicative of electrically excitable channels in the LMC axons or terminals, as in second-order ocellar neurons (see section 6.3). There is no evidence that information transmission in LMC axons is by spikes. Finally, the responses at light-OFF are depolarizations above the dark potential, sometimes also with superimposed spikes. After intense flashes which set up PDAs in receptors, hyperpolarizing afterpotentials are sometimes seen in LMCs (Laughlin, S., 1974c).

In locusts, dragonflies, and fleshflies, hyperpolarizations result from resistance decreases in the LMCs. Resting resistances of dragonfly LMCs average 21 MΩ, and they are reduced an average of 5.0 and 2.7 MΩ during the ON-transient and the plateau, respectively. The ON-transient and plateau reverse at about the same potential, averaging -65 mV from the resting potential (Laughlin, S., 1974b). In locusts, about 4 MΩ resistance change occurs during the plateau, and the hyperpolarizations reverse at -25 mV from the resting potential (Shaw, S., 1968). For the fly *Phormia*, the resting potentials of LMCs average -48 mV (Zimmerman, R., 1978), about the same as the extracellular laminar negative potential or 20–30 mV positive to it in *Calliphora* (Zettler, F. and Järvilehto, M., 1971). The ON-transient and the plateau reversed at the same potential in *Phormia*, but the size of the OFF-depolarization was independent of intrinsic current (Zimmerman, R., 1978). This could happen if the OFF-depolarization was due to several conductance changes, of opposite signs (Brown, J. *et al.*, 1971). In ocellar second-order cells the OFF-depolarizations also seem to have different origins than the hyperpolarizations (Wilson, M., 1978b).

The ionic mechanisms for responses of LMCs have been studied pharmacologically primarily in *Phormia*, by Zimmerman, R. (1978). Potassium ion appears to be involved in all components. Injection into LMCs of tetraethylammonium (TEA), which blocks delayed potassium channels in *Limulus* photoreceptors (O'Day, P. *et al.*, 1981), reduces or abolishes the OFF-depolarization and the ON-transient, in that order of effectiveness. The synaptic blockers, high extracellular magnesium and cobalt, caused LMCs to hyperpolarize and their membrane resistances to increase. Complicated changes in the extracellular laminar negative potential also resulted from the application of high magnesium, but the long-term effects were decreases in the negativity, rises in apparent tissue resistance, and increases in size of the superimposed laminar positive potentials. These results suggest that there is ongoing depolarizing synaptic activity in the dark, perhaps responsible for the standing negative potential. Turning off this activity with magnesium increases the LMC membrane resistance, so that photocurrent exiting receptor terminals is less able to flow through the distal LMCs, as in Fig. 16. The result would be current flows through the higher resistance of extracellular space, resulting in greater laminar positive potentials. The effects of high magnesium are thus consistent with the return flow of photocurrent through the LMCs.

Application of bicuculline and picrotoxin, antagonists of the neuro-transmitter gamma-aminobutyric acid (GABA), could mimic the effects of high magnesium or cobalt on fly LMCs (Zimmerman, R., 1978). Only the lamina accumulates GABA (Campos-Ortega, J., 1974), which could therefore be the synaptic transmitter in flies. Acetylcholinesterase (AChE) is present on receptor terminals and LMCs of honeybees (Kral, K. and Schneider, L., 1981), so possibly in bees, as at dragonfly ocellar synapses (Klingman, A. and Chappell, R., 1978), it is acetylcholine that is the transmitter. However, α-bungarotoxin studies reveal no nicotinic receptors in the laminae of flies (Schmidt-Nielson, B. *et al.*, 1977).

The common reversal potentials of the ON-transient and the plateau indicate a cutback of transmitter release, not separate ionic mechanisms. As was suggested above (see section 2.5.1), a cutback of transmitter release would result from a cutback of transmembrane voltage at the receptor terminals. This latter could have several causes: the transient inhibition of the receptor terminals by LMCs (Shaw, S., 1982), followed by the slower rise of the lamina positive potential so as to further reduce the transmembrane potentials of the receptor terminals (Laughlin, S., 1974c). Alternatively, the synaptic input–output curve itself could change with time and voltage, as it does in the barnacle. There, the operating range for transmitter release is shifted along the terminal's voltage axis as a function of this voltage (Stuart, A. *et al.*, 1982).

There is some disagreement in the literature over whether the plateau response of LMCs is maintained or not. Recording from within the lamina of flies and dragonflies, Laughlin, S. and Hardie, R. (1978) reported that the plateau hyperpolarization decayed in 1–5 s to within $10\% V_{max}$ of the dark resting potential. This decay was stronger and faster when large-area stimuli were used. Larger stimuli elicit larger laminar positive potentials, components of which appear in the LMC's responses recorded within the lamina (Autrum, H. *et al.*, 1970). As can be seen in Fig. 15, the plateau responses are larger in chiasmal (axonal) recordings than within the lamina. This is understandable in terms of current flows in Fig. 16: photocurrents do not flow in the LMC axons. Likewise, plateaus are not reduced as much at high intensities in LMCs of locusts (Shaw, S., 1968), and this too can be explained by the different return flows of photocurrents (Fig. 17). Nonetheless, Laughlin, S. and Hardie, R. (1978) hypothesize that information about mean light intensities is not carried to the medulla by plateau responses but by conduction in restricted extracellular space of the lamina positive potential. It is not easy to evaluate their hypothesis. Others have indeed found increasing positivity in the medulla during illumination (Burtt, E. and Catton, W., 1964), but in flies this may be due to the reduction in light of a current sink in distal medullary cell bodies (Zimmerman, R., 1978). Moreover, plateau responses can be recorded from LMC terminals in the medulla, even after prolonged illumination

(DeVoe, R., 1980, and unpublished observations). Tonic medullary cells certainly do receive DC information about light intensities (Zettler, F. and Järvilehto, M., 1973; DeVoe, R. and Ockleford, E., 1976), so how this information is transmitted remains an important question.

Finally, at very low light levels, hyperpolarizing bumps can be recorded in LMCs. For flies, the LMC's bump rate is linearly proportional to intensity and is six times as great as in photoreceptors, for a point source (Dubs, A. *et al.*, 1981). With an extended source, additional small bumps made the rate 18–20 times that of receptors. The factor of 6 for point sources is what is expected from neural superposition of six receptors at each laminar cartridge. The additional small bumps with an extended source are the results of coupling between photoreceptors with different axes of view (see section 2.4.3).

The rates of laminar bumps in locusts are proportional to log intensity, not intensity itself (Shaw, S., 1968). Scatter in sizes of laminar bumps suggest inputs from distant synapses, perhaps from different ommatidia. During the PDA in photoreceptors, bumps are greatly reduced in amplitude in both the receptors and LMCs.

6.3.1 Performance of the photoreceptor–LMC synapse

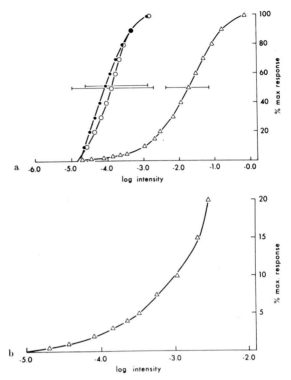

Fig. 18. Average normalized V log I curves for photoreceptors (open triangles) and LMCs (open circles: on-transients; filled circles: plateau responses) of the dragonfly *Hemicordulia tau*. Reproduced from Laughlin, S. (1973), *J. Comp. Physiol. 84*, 335–355, by permission of the author and copyright permission of Springer-Verlag, Heidelberg.

Comparisons of V log I curves from dark-adapted receptors and LMCs show that LMCs are more sensitive overall. Figure 18 illustrates this with normalized curves for the dragonfly (Laughlin, S., 1973). Observable responses of LMCs occur at 2% of the receptors' V_{max}, and responses saturate at only 20% of receptors' V_{max}. Thus, the LMCs have a narrower dynamic range of response — about 2 log units of intensity — than do the receptors, which respond over 5 log units of intensity.

The difference can be attributed to gain at the receptor–lamina synapse. For dragonflies, the gain at low (1 mV) presynaptic voltages reaches a maximum of 34 (Shaw, S., 1981). With brighter lights, gain of the on-transient is reduced to 14 or less, while the gain during the plateau is 8 or less (Laughlin, S., 1973). In flies, relative gains increase from low to high frequencies by at least 2 times (French, A. and Järvilehto, M., 1978a), or absolutely from 3 to 8 (Zettler, F. and Järvilehto, M.,

1972b). These apparent differences in gains might, however, disappear if LMC potentials were compared with transmembrane potentials of receptor terminals, rather than with potentials of somas or terminals referred to the blood (as was done). That is, the instantaneous, but not plateau, transmembrane potential difference of the terminals is likely to be the same as that recorded intracellularly, before the slower laminar positive potential can rise and back off the intracellular potential. Thus, the gain for the LMC on-transient is likely to be the "true" gain and is what is used below for consideration of gain in light-adapted eyes.

Part of the receptor–LMC gain would arise from the convergence of six photoreceptors onto each laminar cartridge, and part from multiple synapses of each receptor onto each LMC. The signal-to-noise ratio is expected to increase as the square root of the number of converging receptors (Gemperlein, R. and Smola, U., 1972) and as the square root of the number of multiple synapses (Laughlin, S., 1973). The latter would be a way of reducing synaptic noise; the former of reducing receptor noise. If there were no presynaptic summation, the signal-to-noise ratio should be better in the LMCs than presynaptically. This was not found for flies (French, A. and Järvilehto, M., 1978a), consistent with evidence for such summation in these insects, if not others (see section 2.5, above). The lower synaptic gains of flies, compared to dragonflies, could thus be limited to the gain due to multiple synapses alone.

Light adaptation increases the slopes of $V \log I$ curves of LMCs and narrows their dynamic ranges from 2 log units to about 1 log unit of intensity, in both flies and dragonflies (Laughlin, S. and Hardie, R., 1978). The increases in slopes can be accounted for by the increased slopes at the operating points on $V \log I$ curves of the receptors (see section 2.3.1, above), given constant synaptic gains of 8 to 10. For LMCs the result is increases in maximum contrast sensitivity from 50–70% when dark-adapted to 250–300% when light-adapted. Correlated with increases in slopes are horizontal range shifts of $V \log I$ curves, equal to the adapting intensities for the LMCs, although not for the receptors (see Fig. 10). These shifts occur rapidly — on the order of tenths of seconds (Laughlin, S. and Hardie, R., 1978).

The basis for range shifts of LMCs is unlikely to be multiplicative changes in sensitivity, since this predicts gain changes which are not found. Rather, range shifts could result from subtractive processes. As discussed in the previous section, reductions of the transmembrane potential of receptor terminals by subtractive action of the extracellular laminar positive potential could account for the cutback of the LMC response from ON-transient to plateau. On the grounds that the slow rise of the lamina positive potential (Laughlin, S., 1974b) is about the same as the fall of the LMC sensitivity, Laughlin, S. and Hardie, R. (1978) propose that range shifts of LMCs are also due to such subtraction. Possibly too, feedback of LMCs onto receptors (Shaw, S., 1982) and/or range shifts of transmitter release (Stuart, A. et al., 1982) play a role.

3.1.2 SENSITIVITIES OF LMCs

Convergence of many photoreceptors onto each

LMC would be expected to result in duplication of the receptors' angular sensitivities, in indications of connectivities to receptors of known spectral sensitivities, but in obliteration of polarization sensitivities. The latter is true for LMCs L1 and L2 of flies, which receive inputs from three sets of photoreceptors, with microvilli oriented 120° apart (Järvilehto, M. and Moring, J., 1974). It is also true for most, but not all, LMCs of locusts. When not absent, the polarization sensitivity is never as great as in receptors, however, implying some convergence of receptors with different microvillar orientations (Shaw, S., 1968).

Spectral sensitivities have been measured for LMCs of dragonflies, fleshflies, and worker honey bees. The spectral sensitivities of LMCs of dragonflies are like those of broad-band, "linked" UV-green cells (Laughlin, S., 1976a,b; Laughlin, S. and Hardie, R., 1978). Linked photoreceptors have polarization sensitivities of about 2 in the UV and 1 at visible wavelengths, and so too do LMCs. Thus, photoreceptors with the same microvillar orientations seem to converge on these LMCs.

Fly LMCs, for the most part, have the double-peaked, UV-green sensitivities of their R1–6 inputs (Laughlin, S. and Hardie, R., 1978). However, other LMCs (also L1 and L2) identified by staining (Moring, J., 1978) have spectral sensitivities like R7UT cells (Hardie, R. et al., 1979). Shaw, S. (1979b) found strong excitatory presynaptic coupling of R7/8 to R1–6 cells, so the LMCs with high UV sensitivities might have received inputs from such coupled receptors (Moring, J., 1978).

The LMCs of worker honey bees all seem to have inputs from blue- and green-sensitive photoreceptors, while a few had variable UV input as well (Menzel, R., 1974). The recordings may have been from L2, with input from blue- and green-sensitive cells, and L1, with additional inputs from UV-sensitive long visual fibers (Ribi, W., 1981).

From the first, angular sensitivities of LMCs have been found to be as small as, or smaller than, receptor angular sensitivities (Shaw, S., 1968; Zettler, F. and Järvilehto, M., 1972a; Laughlin, S., 1974c). This is illustrated in Fig. 19, for the fly (Zettler, F. and Weiler, R., 1976). The angular sensitivities are shown for angular widths at 5% of maximum sensitivity, rather than the usual 50%. In spite of scatter in the results, the angular sensitivities are least in LMCs, greater in photoreceptors, and greatest for depolarizations in the lamina. The latter probably included both isolated recordings from receptor terminals as well as contaminations from extracellular laminar positive potentials. Larger angular sensitivities of laminar positive potentials would be consistent with similar findings in dragonflies (Laughlin, S., 1974b).

One function of narrowing of angular sensitivities in LMCs of flies would be to match the narrower angular sensitivities of R7/8 (Zettler, F. and Järvilehto, M., 1972a; Hardie, R., 1979). There would then be the same fields of view in medullary cartridges that were excited by the R1–6 and R7/8 cells that viewed the same points in space (cf. Kirschfeld, K., 1967). The narrowed angular sensitivities of LMCs are not due to any feedback from receptors R7/8, however, since the angular sensitivity functions are independent of wavelength (Zettler, F. and Autrum, H., 1975; Mimura, K., 1976). The narrowest angular sensitivity measured in the lamina (about 1°) is that of a T1 cell, but so far only one has been identified (Järvilehto, M. and Zettler, F., 1973).

It has been hypothesized that the narrower angular sensitivities of LMCs are due to lateral inhibition (Zettler, F. and Järvilehto, M., 1972a). It is unlikely that the inhibition acts presynaptically, since angular sensitivities of receptor axons are the same as for somas (Table 1). Rehbronn, W. (1972) found no coupling between axons in different cartridges in flies but also little between what were probably LMCs. Lateral inhibition in flies has been inferred from depolarizations evoked in fly LMCs by dim annuli (Zettler, F. and Autrum, H., 1975), and from inhibitory flanks in angular sensitivities measured with drifting gratings (Dubs, A., 1982). Light on OFF-axis facets, as many as 5 away, will inhibit all but the ON-transient of an LMC in response to an ON-axis test flash (Shaw, S., 1981). The inhibition is accompanied by a transmembrane depolarization of the LMC (intra- minus extracellular potential). Dubs, A. (1982) observed similar depolarizations with OFF-axis spots and found no disinhibition. That is, depolarizations summed, which is evidence against the recurrent type of inhibition that is found in *Limulus* (Hartline, H. and Ratliff, F., 1958). The inhibitory flanks in angular sensitivities of LMCs (Dubs, A., 1982) are reduced at higher temporal frequencies of drifting gratings and at lower intensities, where angular sensitivities are broader (as in receptors). Moreover, inhibitory fields of LMCs are slightly wider and deeper in the horizontal than the vertical direction. The extracellular laminar positive potential has similar asymmetries (Dubs, A., 1982). Mimura, K. (1976) has also found asymmetric inhibition of fly LMCs. There were inhibitory fields that flanked the center horizontally, or vertically, or sometimes both.

The existing evidence for lateral inhibition of fly LMCs favors the hypothesis that it is mediated by the laminar positive potential. In this view, lateral current spread is likely within the lamina, between laminar cartridges, and results in the large angular sensitivities plotted there in Fig. 19. An OFF-axis stimulus would therefore elicit a lower response in the receptors that excite an LMC, while eliciting larger responses in off-axis receptors. The extracellular laminar positive potentials set up by current exiting the OFF-axis inhibitory receptors would still further reduce the transmembrane potentials of the excitatory terminals to the LMC. The result would be less transmitter release for OFF-axis light, thereby narrowing the LMC's angular sensitivity. Thus, the lateral inhibition need not be seen in the intracellular potentials of the receptor terminals, while the transmembrane depolarizations that are seen in LMCs (Shaw, S., 1981) would result from intercellular current flows (shown in Fig. 16). That is, the depolarizations would be the signs but not the causes of lateral inhibition.

A final point is the time course of lateral inhibition. Were inhibition to be delayed, there could be resonances at frequencies where the total phase shift of inhibition plus response is 360°. French, A. and Järvilehto, M. (1978a) failed to find any such resonances. However, Cosens, D. and LeBlanc, N. (1980) observed enhancement of every other ON-effect at 55–66 Hz in the flicker ERG of *Drosophila*. This might result if at these frequencies the slower lamina positive potential, with more phase shift than the receptors, alternately enforced and then inhibited the transmembrane potential of the receptor terminals.

3.1.3 Conduction in axons of LMCs

As described earlier, spikes play no role in conducting

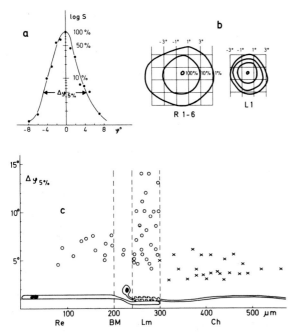

FIG. 19. Comparisons of angular sensitivities of photoreceptors R1–6 and of LMCs of the fly. A sensitivity profile for a photoreceptor is shown in (a), and receptive field plots are given in (b). Combined results from many cells and recording locations are illustrated in (c). Reproduced from Zettler, F. and Weiler, R. (1976), *Neural Principles in Vision*, ed. by F. Zettler and R. Weiler, pp. 227–236, by permission of the authors and copyright permission of Springer-Verlag, Heidelberg.

responses of LMCs to the medulla. Graded hyperpolarizations can be recorded at all points along the axons, including in the terminals. To compare points of penetration along fly LMC axons and the sizes of responses, Zettler, F. and Järvilehto, M. (1971, 1973) used an elegant method of freezing the electrode tip *in situ* before sectioning. They calculated that LMC axons should have had length constants of at least 1 mm if the small decrements in size with distance were due only to passive electrotonic spread. Their measurements of input resistances of axons were incompatible with such long length constants, so they proposed instead that there was regenerative, active conduction of the graded potentials.

In opposition to this view, Shaw, S. (1979a) argued that the low input resistances recorded could have been those of the input dendrites, not the axon (a view explicitly dismissed by Zettler, F. and Järvilehto, M., 1973). Hyperpolarizing potentials are

electrotonically conducted in second-order axons in the locust ocellus (Wilson, M., 1978b). Depolarizations of receptors R7/8 appear to be electrotonically conducted to the medulla, in axons as long as those of LMCs (Hardie, R., 1977). In sum, passive electrotonic conduction is not unfeasible in LMC axons, but definitive experiments to prove or disprove this have yet to be done.

3.2 Laminar spiking cells

A number of unidentified spiking cells have been recorded extracellularly, from the proximal edge of the lamina to the distal edge of the medulla in locusts (Horridge, G. *et al.*, 1965), grasshoppers (Northrop, R. and Guignon, E., 1970), cockroaches (Mote, M. and Rubin, L., 1981; Mote, M. *et al.*, 1981) and flies (Arnett, D., 1971, 1972; McCann, G. and Arnett, D., 1972; Mimura, K., 1974). In grasshoppers these have included multimodal neurons (sensitive to light and mechanical stimuli) as well as tonic light (L) units with occasional ON-gated mechano-responses (Northrop, R. and Guignon, E., 1970).

Spiking ON-units in the lamina–medulla projection of the cockroach *Periplaneta* receive inputs from both phasic green-sensitive (G) photoreceptors and tonic UV-sensitive receptors. At threshold the input is only from the G receptors, which partially inhibit the UV receptors' inputs (Mote, M. and Rubin, L., 1981). In half the cells tested, receptive fields were the same for violet and green stimuli, whereas in the others the centers for violet stimuli were usually more dorsal and sometimes multipeaked. Fields were large or irregular: one-third were less than 30° but most of the rest were between 30° and 90° in maximum extent (at half to one-third of maximum sensitivity). It was concluded from experiments in which part of the eye was masked that networks of G receptors inhibit each other's inputs, whereas networks of UV receptors do not (Mote, M. *et al.*, 1981). Clearly, the laminar spiking ON-units have far more extensive and complex spatial inputs than do hyperpolarizing LMCs.

Still other laminar spiking units are those in the outer chiasm and distal medulla of flies; spikes are not recorded in the lamina. First described by Arnett, D. (1971, 1972), there are both the more-sensitive sustaining units with small spikes, and the

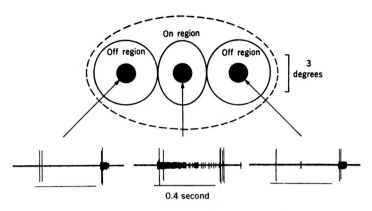

FIG. 20. Simultaneous extracellular recordings from ON–OFF units (large spikes) and sustaining units (small spikes) in the outer chiasm of the fly *Phaenicia sericata*. The dashed lines indicate the receptive field of the ON–OFF unit, the solid lines the ON and OFF regions of the sustaining fiber. Reproduced from Arnett, D. (1971), *Science, 173*, 929–931, by permission of the author and copyright permission of the AAAS; copyright 1971 by the AAAS.

ON–OFF units with large spikes. As shown from simultaneous recordings in Fig. 20, the ON–OFF units have large, elliptical receptive fields. The sustaining units have excitatory ON-regions and flanking, mediolateral OFF-regions. These regions mutually inhibit each other, as can two spots within the ON-region. More complex discharge patterns and nearly encircling inhibitory regions were also found by Mimura, K. (1974); none of the units were either motion- or directionally selective. In his experiments, an ON-region of one unit spiked at the light-ON of 440, 480 and 520 nm light but at OFF for 560 nm light. McCann, G. and Arnett, D. (1972), on the other hand, found both ON–OFF and sustaining units (ON-region) to have the spectral sensitivities of photoreceptors R1–6, with temporal congruity for any combination of wavelength or angle of polarized light. As with the ON-units of cockroaches, the sizes of receptive fields (in Fig. 20) must result from inputs of many photoreceptors.

The identity of no laminar spiking unit is known. Occasionally, intracellular recordings (without stainings) have been in the outer chiasm of flies from spiking cells (DeVoe, R. and Ockleford, E., 1976), and Hardie, R. (1978) recorded from and stained a tonic, probably small-field cell which was tentatively identified as L4. On the basis of still other arguments, Shaw, S. (1981) has proposed that sustaining units of flies may be L4, and ON–OFF units L5.

3.3   Other laminar cells

In an extensive series of measurements of receptive fields of wide-field laminar cells, Mimura, K. (1976) described one field with star-shaped spokes in horizontal and oblique directions, covering many cartridges. Responses, recorded intracellularly, were hyperpolarizing in all except two nodes dorsal and ventral to the center of the receptive field, where responses were depolarizing. The hyperpolarizing responses had the spectral sensitivity of an R7UT cell, like some stained L1,2 LMCs (Moring, 1977; see section 3.1.2 above). The penetrated cell was probably a lamina amacrine cell, which alone has processes oriented horizontally, obliquely and vertically (Strausfeld, N. and Campos-Ortega, J., 1977).

4   THE MEDULLA

The medullae are the largest optic ganglia. In flies they contain the majority of neurons: two-thirds of all visual interneurons and almost half of all neurons of the brain are in the medullae (Strausfeld, N., 1976a). It is uncertain what physiological roles necessitate so many cells. There are the same numbers of cartridges in the medulla as in the much smaller lamina. In turn, these medullary cartridges project retinotopically to cartridges in the lobula (or lobular complex in e.g., Diptera and Lepidoptera). Projection arrays of transmedullary (Tm) cells through the medulla perpendicular to its surfaces and onto lobular and other cells have been classified anatomically by Strausfeld, N. (1976b). By such anatomical criteria the medullary mosaic is both

Table 3: Recordings from presumptive medullary cells

Order/species	Recording method	Recording localization	Classifications used and properties studied	Additional recordings	Reference
Orthoptera					
Locusta migratoria	Extracellular	Depth from cornea	D, L, and ON-units: visual fields; movement sensitivities; spontaneous activity	Lobula	Burtt, E. and Catton, W., 1960
Locusta migratoria	Extracellular	Visual placement	Classes A–D (21 subclasses): visual fields; sensitivities to movement, adaptation, stimulus size, multimodal stimuli; habituation	Lamina, lobula	Horridge, G. et al., 1965
Romalea microptera	Extracellular	From lesions	Multimodal, vector edge, tonic-L, ON-L and net dimming units: visual fields; directional selectivity; selectivities to contrast, jittery movements; habituation	Lamina, lobula	Northrop, R. and Guignon, E., 1970
Gryllus campestris	Extracellular	Visual placement	Brisk or sluggish, sustained or transient, ON- of OFF-units: visual fields; directional selectivities; habituation; luminance sensitivities	Inner chiasm	Honegger, H., 1978
Gryllus campestris	Extracellular	Visual placement	Large-field ON- and OFF-units: visual fields; habituation; movement sensitivities	Inner chiasm	Honegger, H., 1980
Hemiptera					
Notonecta glauca	Extracellular	From lesions	Non-directionally selective, jittery-movement sensitive unit: visual fields; stimulus sizes; sensitivities to speeds; habituation	Inner chiasm	Schwind, R., 1978
Lepidoptera					
Bombyx mori	Intra- and extracellular	Depth from cornea	Spontaneous-OFF, silent, ON–OFF units: adaptation; luminance sensitivities		Ishikawa, S., 1962
Sphinx ligustri	Extracellular	Visual placement; spike pathway tracing	Sustained-ON (medulla–medulla projections: m:tan 1/2); ON–OFF (protocerebrum-to-medulla, giant optic lobe cells), ON–OFF (lobula-to-contralateral medulla), binocular tangential (paired protocerebrum-to-medulla): visual fields; directional selectivity; sensitivities to net luminance, movement, stimulus sizes	Lobula, protocerebrum	Collett, T., 1970
Heliconius erato adanis	Extracellular	Visual placement	Sustaining, movement-sensitive, jittery-movement, rapidly adapting units: visual fields, directional selectivities; sensitivities to extents of movements, color	Lamina, lobula, protocerebrum, ventral nerve cord	Swihart, S., 1968
Epargyreus clarus	Extracellular	Visual placement	Narrow-field ON; tonic dim-light; phasic ON–OFF, ON–OFF inhibition, ON, jittery-movement; tonic complex sustaining movement, inhibitory and excitatory sustaining units: visual fields; sensitivities to movements, and their extents, luminosities, color	Retina, other optic lobes, brain	Swihart, S., 1969
Morpho amathonte centralis	Extracelluar	Visual placement	Short-latency: luminosity; blue ON/yellow + green-OFF; medium-latency: blue-sensitive; yellow-sensitive phasic; OFF; blue-ON/green + yellow-OFF follower;		

Table 3: Recordings from presumptive medullary cells—Continued

Order/species	Recording method	Recording localization	Classifications used and properties studied	Additional recordings	Reference
			unidirectional movement; long-latency: blue-ON/green-OFF; wide-band luminosity; red–blue movement units; adaptation, directional selectivities	Lobula, protocerebrum	Swihart, S., 1972b
Agraulis, Heliconius, Danaus, Precis, Anartia, Papillio, Limnentis, Epargyrus, Phoebis and *Pieris* sps.	Extracellular	Visual placement	Visual fields: large, diffuse with holes, small, uniform; adaptation	Protocerebrum	Swihart, S. and Schümperli, R., 1974
13 species	Extracellular	Dye spots	Small, large or horizontal strip monocular fields: sensitivities to blue, green and red lights and to color contrast	Lobula, protocerebrum	Schümperli, R., 1975
Diptera					
Phaenicia sericata Musca domesticus	Extracellular	From lesions	Class I, non-directional units: visual fields; form response	Lobula, protocerebrum	Bishop, L. *et al.,* 1968
Phaenicia sericata Musca domesticus	Extracellular	From lesions	Classes Ia, Ib, Ic non-directional units: visual fields; form and intensity responses	Lobula, protocerebrum	McCann, G. and Dill, J., 1969
Boettcherisca peregrina	Extracellular	Dye spots	Non-directional, one-directional, semi-integrative; ON, ON–OFF, OFF (rare) units: visual fields; sensitivities to speeds of movements	Lobula, protocerebrum	Mimura, K., 1971
Calliphora erythrocephala	Intracellular	Visual placement, cell staining	Non-spiking; silent sustaining, ON-OFF; spontaneous sustaining, ON-OFF, other; directional selectivity; sensitivities to intensity, change-of-direction; visual fields	Outer chiasm	DeVoe, R. and Ockleford, E., 1976
Calliphora vicina	Intracellular	Visual placement	Laminar cell endings; M-cells; non-directional cells: visual fields; triangular, trapesoidal and staircase movements; sensitivities to speeds, luminance		DeVoe, R., 1980
Hymenoptera					
Apis mellifera	Extracellular	From lesions	Non-directional ON-units: visual fields	Lobula	Wiitanen, W., 1973
Apis mellifera	Extracellular, intracellular	Visual placement, cell staining	Broad-band, simple-field, sustained, inhibited or excited; broad-band complex-field: spectral sensitivities of visual fields; intensity sensitivities	Lobula	Kien, J. and Menzel, R., 1977a
Apis mellifera	Extracellular, intracellular	Visual placement, cell staining	Narrow-band/monochromatic; visual fields; intensity sensitivities	Lobula	Kien, J. and Menzel, R., 1977b
Apis mellifera	Intracellular	Cell staining	Broad and narrow-band, phasic or tonic; intensity-band cells: spatial and color antagonisms; visual fields; movement sensitivities	Lobula	Hertl, H., 1980

precise and repetitious. The large numbers of cells might be explained by repeated representations of neural networks all over the medullary mosaic, in order to compare activity in adjacent cartridges, as for example during sequential excitations by movement. In this view the medullae are large not because of complex neural processing, but because the same neural comparisons between cartridges are carried out everywhere across the medullary mosaic. The elements for carrying out such comparisons would be cells with large lateral extents, including medulla-intrinsic (amacrine) and tangential cells.

Physiologically, the medullae are the least well characterized of the optic ganglia. The reasons for this include: (1) the technical difficulties of recording from the small cells of the medullae and (2) ambiguities, in many published recordings from the "optic lobes", as to which were made from medullary cells. Table 3 lists presumptive medullary recordings from Orthoptera, Hemiptera, Lepidoptera, Diptera and Hymenoptera. Most of these recordings seem to have been made in the proximal medulla, when the recording sites could be identified at all. A few recordings have been made in the distal medulla. Tentatively, it appears from these recordings that neural processing is less complex in the distal than in the proximal medulla. That is, there seems to be a gradient of complexity from the lamina, to the distal medulla, to the proximal medulla, and eventually to the lobula (as will be illustrated below: see section 5). In only a few reports have the cells involved been identified, so it is not yet possible to judge physiologically the anatomical model of repeating neural networks all across the medullary mosaic. Instead, what will be discussed here will be a number of the response properties outlined in Table 3, including receptive field sizes, and organizations and sensitivities to wavelengths of light and movement.

4.1 Excitations of cells in the distal medulla

Inputs to the medulla form synapses from the distal-most edge of the medulla to a depth close to the serpentine layer; some cells synapse at multiple layers (see the chapter by Trujillo-Cenoz in this volume). In *Drosophila* there are two distal bands of acetylcholine receptors, as found from α-bungarotoxin binding (Schmidt-Nielson, B. *et al.*, 1977). These bands are at the approximate depths of the endings of the LMCs L1 and L2 of the lamina. The cells with the acetylcholine receptors are unidentified, but acetylcholine might be involved in medullary excitations by these L cells. There are no combined pharmacological–physiological studies yet on this question.

The predominant visual inputs to the medulla are from depolarizing receptor cells (the long visual fibers), spiking laminar cells, and the hyperpolarizing laminar cells L1–3 and T1. In flies at least, the predominant synaptic potentials recorded intracellularly, including from columnar transmedullary cells, are depolarizing (DeVoe, R. and Ockleford, E., 1976; M. Wilcox, personal communication). Hyperpolarizing responses, or OFF spike discharges, are less common although not missing (Ishikawa, S., 1962; Mimura, K., 1971; Hertl, H., 1980; see also Table 3). It might be supposed that depolarizations in the distal medulla were elicited by spiking laminar cells or by receptor cells via a sign-conserving synapse (an example of the latter is one between second- and third-order cells of the locust ocellar system: Simmons, P., 1981b). In flies, however, hyperpolarizing L cells quite likely elicit depolarizations in third-order medullary (M) cells (DeVoe, R., 1980). Figure 21 shows results of fortuitous recordings, $2 \mu m$ apart, from a hyperpolarizing L cell and an unidentified, ON–OFF depolarizing medullary cell (M1) in the distal medulla. The responses of the two cells are nearly mirror-symmetric and could have been synaptically related, L to M1. If so, this implies a second, inverting and hence chemical synapse from the retina to the medulla. A possible synaptic mechanism could be like that between hyperpolarizing rods and depolarizing bipolar cells in the vertebrate retina (Miller, R. and Dacheux, R., 1976).

In flies, spikes are recorded less often in the distal medulla than more proximally (DeVoe, R., 1980). This could imply that spike initiation zones of, e.g., M cells are at some distance from distal synapses. In grasshoppers and crickets, on the other hand, ON-units with brisk sustained spike discharges and small receptive fields (15–20°) have been recorded extracellularly in the distal medulla (Northrop, R. and Guignon, E., 1970; Honegger, H., 1978). Possibly there are species differences. It is reasonable to

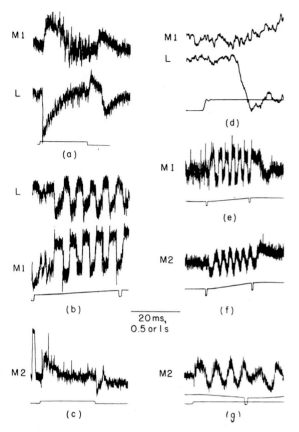

M1

L

(a)

L

M1

(b)

M2

(c)

M1
L

(d)

M1

(e)

M2

(f)

M2

(g)

20 ms,
0.5 or 1 s

FIG. 21. Comparisons of responses of a laminar cell L and of two medullary cells M1 and M2, recorded in the distal medulla of the fly *Calliphora vicina*. Cells L and M1 were penetrated 2 μm apart; M2 was from another experiment. Parts **(a)**, **(c)**, and **(d)** show responses to single flashes (10 mV calibration pulse in **(c)**). Parts **(b)**, **(e)**, **(f)**, and **(g)** show responses to a moving grating; the position of the grating is indicated by the trace immediately below the responses (pulses signify stops and starts of movement). Reproduced from DeVoe, R. (1980), *J. Comp. Physiol. 138*, 93–119, with copyright permission of Springer-Verlag; Heidelberg.

presume that M cells of flies, and the ON-units of the Orthoptera, are transmedullary cells.

4.2 Receptive fields of medullary cells

Receptive fields of medullary cells vary from small up to whole field, from uniform to complex, and ipsilateral or contralateral, monocular or binocular. Complex receptive fields include spatial and/or color opponent fields; the latter will be described in section 4.3, below. Binocular and contralateral receptive fields must involve centrifugal projections from the opposite optic lobes, such as via tangential

and other giant cells. For the most part, however, attention will be focused here upon medullary cells with columnar (perpendicular) or lateral dendritic trees.

Most columnar (perpendicular) cells are restricted to individual medullary cartridges (Strausfeld, N., 1976a). The responses of narrow-field cells recorded in *Locusta*, *Gryllus*, *Epargyreus*, *Phaenicia*, *Calliphora* and *Apis* may have come from such columnar cells, or from those spanning a number of cartridges. Narrow-field cells have ipsilateral, monocular receptive fields of from 4 to 20 degrees or larger. The lower figure, obtained in *Apis* (Wiitanen, W., 1973), is commensurate with visual fields of receptors and laminar cartridges, while the larger figures may simply reflect the resolutions of the methods used. These cells have simple ON, sustaining, or ON–OFF responses, with no antagonistic surrounds. In *Apis*, two narrow field cells were stained and were indeed columnar; one was movement-sensitive (Hertl, H., 1980). Most narrow-field cells have responded to movement, but not all (Northrop, R. and Guignon, E., 1970). It seems obvious therefore that narrow-field cells include a variety of columnar cells, within a species as well as among species.

Larger receptive fields, covering 10s of degrees, have been found in all species tested. Profiles of sensitivities have been measured in Orthoptera (see below), Diptera and Lepidoptera. In *Calliphora*, sensitivities gradually decreased in all directions from the most sensitive area (DeVoe, R., 1980). Lepidopteran receptive fields measured in high detail were non-uniform and had "holes" and "channels". Light adaptation reduced the fields from 70 degrees down to 20 degrees (Swihart, S. and Schümperli, R., 1974).

Spatially antagonistic receptive fields have been found in crickets and honey bees. Honegger, H. (1980) recorded, in *Gryllus*, large-field OFF-units with 100–120° centers, 20–30° wide ON surrounds and ON–OFF intermediate zones. The results from one such cell are shown in Fig. 22A. The area to the right of and above the dashed line is the binocular domain of the eye, where the cell was most sensitive by a factor of about 5. Whole-field illumination of a dark-adapted cell elicited an initial brief ON burst (Fig. 22B), suggesting suppression of the ON-surround by the OFF-center. Thus, these cells

A

B

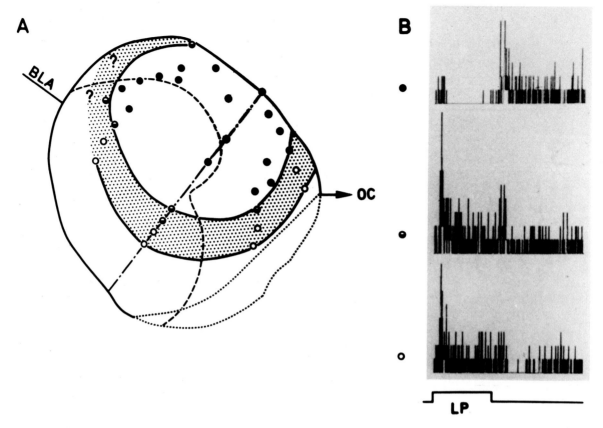

FIG. 22. Receptive field of a spatially antagonistic, large-field, OFF-center unit in the medulla of the cricket *Gryllus campestris*. The stippled region in **A** represents the ON-surround. OFF, ON–OFF, and ON responses of the PST histograms in **B** could be elicited at the points in **(a)** shown by filled, half-filled, and open circles, respectively. Reproduced from Honegger, H. (1980), *J. Comp. Physiol. 136*, 191–201, by permission of the author and copyright permission of Springer-Verlag, Heidelberg.

integrated inhibition and excitation over their entire receptive fields.

Non-concentric, flanking, antagonistic receptive fields have been recorded in *Apis*. One cell stained had two vertical-strip dendritic trees, the posterior and dorsal–anterior ones inhibitory, the ventral–anterior one excitatory (Hertl, H., 1980).

Finally, a variety of cells, with large or small receptive fields, with ipsilateral or contralateral input, have been recorded in a number of insects. Although none of these cells have been identified by intracellular staining, Collett, T. (1970) traced the axonal pathways of several of these cells to and from the medulla in *Sphinx*. Monocular, sustained ON-units without movement sensitivities projected from one medulla to the other and were probably the tangential pair *m:tan1* and *m:tan2*. Unidentified, monocular, ON–OFF, non-directional units with whole eye fields projected from one lobula to

the opposite medulla. Similar units projecting from the ipsilateral medial lobe to the lobula and medulla were probably giant optic lobe cells. It was proposed that the tangential cells signaled the net luminance on the opposite eye, while the latter two cells could have signaled total moving contrast.

4.3 Color-coded receptive fields

Spatial color coding is first encountered in the medulla (and perhaps outer chiasm). That is, different parts of receptive fields may be sensitive to different colors. Color-coded receptive fields have been investigated in Orthoptera, Lepidoptera, and Hymenoptera.

The most distal color-coded cells are the ON-type cells of *Periplaneta*, described earlier (see section 3.2). Curiously, recording locations for such cells ranged from the lamina to the lobula (Mote, M.

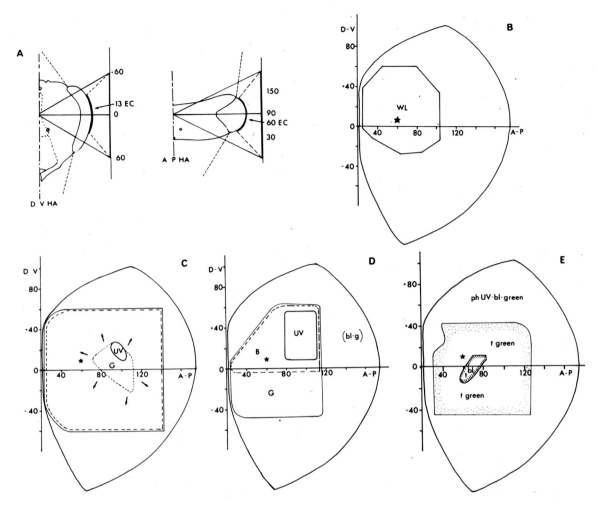

FIG. 23. Color-coded receptive fields of broad-band medullary neurons of the honey bee, *Apis mellifera*. Measurements were made on a tangent screen, shown in relation to the eye in **A** (vertical view left; horizontal view right). The eye center in **B–E** is shown by an asterisk; the outer lines outline the entire visual field. ph: phasic; t: tonic. Reproduced from Kien, J. and Menzel, R. (1977a), *J. Comp. Physiol. 113*, 17–34, by permission of the authors and copyright permission of Springer-Verlag, Heidelberg.

et al., 1981), although they were predominantly in the outer chiasm and distal medulla. Such recording locations are similar to those of multimodal neurons in *Romalea* (Northrop, R. and Guignon, E., 1970), which were not tested with color. No cells are known which span the lamina to the lobula, so at least two cells would seem to be tightly serially connected.

The receptive fields of optic lobe cells of various butterflies were tested by Schümperli, R. (1975) with random colored spots from a video color monitor. He found simple fields sensitive to blue, to green, to red, or to blue and red spots. Seven cells tested were sensitive to specific color contrasts,

sometimes with color-specific changes in field sizes. It is not clear if any medullary neurons were among those thus tested with color, however.

The medullary cells most extensively investigated for color-coded receptive fields are those of worker honey bees. Kien, J. and Menzel, R. (1977a,b) distinguished four types of color coding: broad-band; narrow-band, UV, blue or green monochromatic; polychromatic; and color opponent. The locations of most cells were not specified, and recordings were made in the lobula as well as in the medulla. However, all but polychromatic cells were found in the medulla as well as in the lobula. All medullary cells but one broad-band cell, were in the proximal

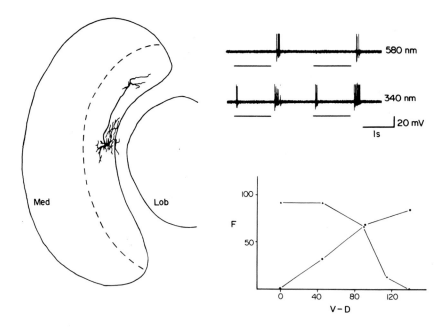

FIG. 24. Structure and response of a phasic color-opponent cell in the medulla of the honey bee, *Apis mellifera*. At the lower right, the frequencies of firing F are plotted versus dorsoventral position for 340 nm UV light (filled circles) and 580 nm orange light (filled triangles). Reproduced from Hertl, H. (1980), *J. Comp. Physiol. 137*, 215–231, by permission of the author and copyright permission of Springer-Verlag, Heidelberg.

medulla. Cells responded at ON–OFF, tonically (with excitation or inhibition), or usually with combinations of responses. Phasic and tonic components sometimes had different spectral sensitivities, and V log I curves were not always the same at all wavelengths.

Broad-band cells, with predominant UV, green, and sometimes blue peaks of sensitivity, were found in two general classes of receptive fields: simply organized, sharply bordered fields which were not color-coded, and complex fields with roughly concentric but different sized fields for UV, blue, and green wavelengths. Figure 23 illustrates both types of receptive fields. One broad-band, sustaining cell with a simple field (Fig. 23B) was a proximal medullary Y8 cell (Strausfeld, N., 1976a) that projected to the lobula. Its dendritic spread of about 20–25 medullary cartridges wide correlates well with the 60–80° diameters of simple fields (Kien, J. and Menzel, R., 1977a).

Narrow band, green-monochromatic cells were recorded in both the medulla and lobula. Like lobular UV-monochromatic cells and rare blue-monochromatic cells (of unspecified locations), spectral sensitivities were narrower than those of

the corresponding photoreceptors. This implies inhibition from the other color channels (Kien, J. and Menzel, R., 1977b).

UV^+/B^-G^- color opponent cells, both phasic and tonic, were most often recorded in the proximal medulla (Kien, J. and Menzel, R., 1977b), although Hertl, H. (1980) found only phasic medullary opponent cells. Receptive fields were uniform, or concentric for phasic and tonic responses, but they had little wavelength-dependent structure (Kien, J. and Menzel, R., 1977b). The same was true for rare G^+/UV^- cells of unstated location. Non-opponent, green-dominated polychromatic cells of the lobula had the most complex color-coded fields of all, in one case with a circular area of strong blue response and an adjacent, bar-shaped area of strong green response.

In a study of identified neurons, primarily of the medulla of the worker honey bee, Hertl, H. (1980) likewise recorded from wide-band cells and narrow-band cells, especially UV cells. These included a small-field axial fiber, a large-field cell projecting to the lobula, and two medulla amacrines. Some cells that were narrow-band, green monochromatic near threshold became broad-band with increasing intensities. Intensity-band (I-band) cells reacted to UV light and not to green light in two intensity ranges, and the opposite in two different intensity ranges. Finally, phasic color-opponent cells were found, one example of which is shown in Fig. 24. This extended, proximal amacrine cell was spatially color-coded and responded ON–OFF to UV light in its dorsal receptive field and OFF to orange light in its ventral field. It would be well suited to detect the horizon, between green foliage and UV-rich sky.

As with so much other scarce data about medullary cells, these results on small samples of color-coded cells are of necessity descriptive and cell-specific. Far more repeated recordings from identified cells will be necessary before strong generalizations can be made about the neural networks that combine the excitations from blue plus green-sensitive LMCs (Menzel, R., 1974) and UV-sensitive long visual fibers.

4.4 Movement sensitivities of medullary cells

Movement sequentially stimulates successive points in the retinal mosaic. It has been shown behaviorally that stimulation of only two such points can be sufficient to evoke optomotor responses: out-of-phase flicker on certain sets of two, adjacent, peripheral photoreceptors (R1–6) of *Musca* is all that is needed to simulate movement (Kirschfeld, K., 1972). These two photoreceptors would excite adjacent cartridges in the lamina, medulla and lobula. The central photoreceptors (R7/8, the long visual fibers) inhibit the excitations set up by the two peripheral photoreceptors, before these excitations are combined at a movement detector (Kirschfeld, K. and Lutz, B., 1974). Since the central photoreceptors in flies synapse only in the medulla, the medulla is the most distal optic ganglion where the inhibition — and hence motion detectors — could be. Biochemical and physiological studies provide evidence that there are indeed motion detectors in the medulla.

Using the autoradiographic, deoxyglucose method to mark cell activity, Buchner, E. and Buchner, S. (1980) found extra labeling in portions of *Drosophila* medullary neuropile excited by moving gratings. Comparable fields of flickering light on the other eye did not result in extra labeling in the other medullary neuropile. Receptor cells and laminar neuropile were unlabeled after either movement or flicker. Thus, the deoxyglucose method indicates that there is distinct, movement-specific activity in the medulla. The extra labeling is greatest in two distal bands, similar to the two distal bands of α-bungarotoxin binding in *Drosophila* discussed above. A more proximal band, in or near the serpentine layer, is also labeled with deoxyglucose. The labeled bands could indicate sites of intense activation (or inhibition) of motion detectors, but there is no comparable physiology as yet to test this.

Physiologically, recordings have been made from directionally selective and non-directional cells in the medulla. The most commonly found are non-directional cells, which respond to movements in any direction within a cell's receptive field. Figure 25b,c illustrates intracellular recordings from a non-directional cell in *Calliphora* to back-and-forth grating movements. Movements elicit abrupt, sustained depolarizations, which die away slowly when movements stop. Many cells throughout the medulla, with otherwise-different flash responses, have this characteristic non-directional movement response. Sometimes spikes are recorded, sometimes not. When spikes are recordable during movements, their frequencies are related to (the logarithms of) speeds over ranges of speeds of more than 100:1 (DeVoe, R., 1980 and unpublished observations).

Studies of non-directional responses in other insect medullae have been based on spikes only. Transiently moving spots, bars and edges have been used, as well as continuously moving gratings. It is always a question whether non-directional discharges elicited by transient movements are actually responses to movement (sequential stimulation), or whether the discharges simply reflect responses to transient changes in intensities (Honegger, H., 1980). However, many medullary cells do not discharge during movements, and some non-directional cells do not discharge to flashed spots (Schwind, R., 1978). Others do, but differently. Figure 25d–f shows intracellular responses to flicker of the cell that gave the non-directional movement responses (Fig. 23b,c). The flicker responses were primarily end-to-end juxtapositions (Fig. 25d) of the ON- and OFF-transients to a single flash (Fig. 23a). Clearly, this cell was able to distinguish sequential from simultaneous excitations at its inputs. That is, it detected movement.

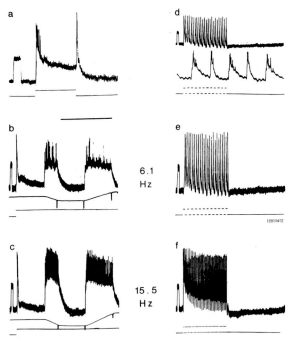

FIG. 25. Comparisons of responses to flicker (**d–f**) and movement of a grating (**b, c**) of a non-directional cell in the medulla of the fly, *Calliphora vicina*. Responses are preceded by 10 mV calibration pulses: bottom traces are from a shutter monitor. Middle traces in **b** and **c** show the grating position; pulses indicate stops and starts. Frequencies in the middle of the figure refer to spatial periods or flashes per second. The flicker response in **e** is shown at a 10 times greater sweep speed in the middle of **d**. Unpublished results of the author.

Some non-directional cells like this have large hyperpolarizations following stimulation (DeVoe, unpublished observations). It is likely that the hyperpolarizations are due to recovery processes involving ionic pumping (see section 2.2), which would also involve metabolic uptake of glucose. Thus, these cells may be among the ones stained by deoxyglucose following movements across the eye (Buchner, E. and Buchner, S., 1980).

A special case is non-directional discharge to movements of small but not large stimuli, with habituation and inhibition by large moving backgrounds (as in self-movement). Only in *Notonecta* (Schwind, R., 1978) have such cells, often called jittery-movement units, been recorded in the medulla, although they have been recorded on the surface of the medulla, in the outer chiasm, and in the lamina of the grasshopper *Romalea* (Northrop, R. and Guignon, E., 1970). Large field cells with sensitivities to small jittery movements are found in

Gryllus (Honegger, H., 1978). Jittery-movement cells in the lobula will be taken up below.

Directionally selective (DS) cells have been recorded in the medullae of Orthoptera, Lepidoptera, and Diptera, but not so far in Hymenoptera (Hertl, H., 1980). DS cells respond in certain preferred directions of movement and less in the opposite, null directions. It is reasonable to suppose that medullary DS cells would provide the inputs to DS optomotor neurons of the lobula or lobula plate. The evidence for this is primarily from behavioral demonstrations of "elementary movement detectors" which are repeated all over the visual mosaic (Buchner, E., 1976). The elementary movement detectors are presumably medullary DS cells excited by sequential stimulation of adjacent cartridges, as for example during the earlier-described stimulations of just two adjacent photoreceptors. If so, there should be at least twice as many DS medullary cells as cartridges, and most likely more (for the two vertical and the two horizontal preferred directions: Kirschfeld, K., 1972).

Physiologically, however, medullary DS cells are infrequently recorded, compared to other cells. This could simply be a technical problem, if DS cells have very fine axons. McCann, G. and Dill, J. (1969) found no DS cells, and Mimura, K. (1971) found few DS cells, in the fly medulla. Both groups found non-directional cells. With intracellular recordings, DeVoe, R. and Ockleford, E. (1976) reported a DS cell with DC shifts of potential and spiking for horizontal preferred movements, but only spiking for vertical preferred movements. This could mean that only one of two sets of synaptic potentials reached the recording site, or that there was more than one site of spike initiation.

It is of considerable interest what these movement detectors are, and from what cells they receive inputs. There is no solid evidence about their inputs, but in the fly they could include laminar (L) cells and medullary (M) cells that are not themselves movement detectors (DeVoe, R., 1980, 1983). A few stained non-directional cells in *Apis* and *Sarcophaga* have been identified as columnar or amacrine cells (Hertl, H., 1980; DeVoe, unpublished observations). Possibly, a DS cell in *Calliphora* was a T2 cell (DeVoe, R. and Ockleford, E., 1976). Finally, a centrifugal, binocular, full-field, DS tangential cell was traced physiologically from the protocerebrum into and along the distal stratum of the medulla in *Sphinx* (Collett, T., 1970). It was uncertain which tangential cell this might have been, however.

5 THE LOBULA

The lobulae, the most proximal ganglia of the optic lobes, communicate with the ipsilateral medulla, the contralateral medulla and lobula, as well as various regions of the protocerebrum, including the optic foci, optic tubercle, and the posterior slope (Strausfeld, N., 1976a). A number of extracellular recordings have been made from light-sensitive, often-multimodal neurons in the protocerebra. These include recordings from crickets (Dingle, H. and Fox, S., 1966); grasshoppers (Northrop, R. and Guignon, E., 1970), honey bees (Erber, J., 1978), moths (Blest, A. and Collett, T., 1965a,b) and butterflies (Swihart, S., 1972a; Schümperli, R. and Swihart, S., 1978). The neurons have not been identified, but quite likely they receive projections from the lobulae. Conversely, there are many lobular neurons whose structure and projections are known but whose function is not (Strausfeld, N., 1976a).

What this account will emphasize is the physiology of identified, giant, usually-unique lobular cells which subserve movement detection. These are of two types: optomotor neurons which are directionally selective for preferred horizontal or vertical movements, and jittery-movement neurons which respond to small, but not large, objects moving abruptly in any direction. The most thoroughly studied optomotor neurons are those of flies (Hausen, K., 1981) and, to a lesser extent, honey bees. The prime example of a lobular jittery-movement neuron is the giant lobular movement detector/descending contralateral movement detector (LGMD/DCMD) system of locusts.

5.1 Optomotor neurons of flies

All the giant optomotor neurons of flies are tangential cells of the lobula plate. They project to and from the posterior slope of the protocerebrum or to the opposite lobula. About 21 have been identified so far (Hausen, K., 1981). The earliest recordings from DS optomotor neurons (Bishop, L. and

Keehn, D., 1966) were made extracellularly from spiking heterolateral cells. These were classified as II-1, II-2, II-3 and II-4, responding preferentially to regressive (back-to-front), downward, progressive (front-to-back), and upward movements across the eye, respectively (McCann, G. and Dill, J., 1969). Cells were most commonly recorded contralateral to their visual fields (Bishop, L. *et al.*, 1968), in part because of greater stability and perhaps because it may be easier to record large spikes in axons that are far away from short-circuiting conductance changes of dendrites.

With intracellular recording and staining, it also became possible to identify minimally- or non-spiking DS optomotor neurons of the lobula plate (Dvorak, D. *et al.*, 1975; Hausen, K., 1976a,b), including the horizontal-system (HS) and vertical-system (VS) cells of Pierantoni, R. (1976). From cells identified by staining, the principle has emerged that horizontal movement detection is subserved by cells, or dendrites of cells, ending in the anterior layers of the lobula plate. Conversely, vertical movement detectors receive inputs in the posterior lobula plate. The terms "horizontal" and "vertical" for preferred directions should be used, however, with the following caveat: movement detection takes place along orientations set by the curving ommatidial lattice. Thus, dorsally and ventrally, "horizontal" preferred directions tend towards vertical actual directions (Hausen, K., 1981).

The inputs to the tangential cells of both the anterior and the posterior lobula plate include T and Y cells from the medulla and lobula. The retinotopic organization of cartridges in the lamina and medulla is also found in the lobula plate. To each cartridge of the lobula plate project two T4 cells from the corresponding medullary cartridge, and two T5 cells from the lobular cartridges. One pair of T4 and T5 terminate anteriorly, the other posteriorly (Strausfeld, N., 1976b; cited in Hausen, K., 1982a). On the basis of anatomical dendritic fields these T cells are presumed to be small-field columnar elements. There is no direct physiological confirmation of this, but it is consistent with the findings that the sensitivities within the receptive fields of the tangential giant cells are closely related to the cells' dendritic densities (Hausen, K., 1981). That is, if the presynaptic T cells were wide-field cells like many in the proximal medulla (see section 4 above),

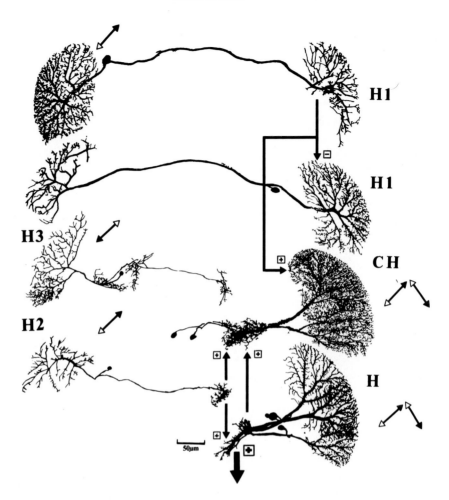

FIG. 26. Structure, directional preferences, and probable functional connections between identified giant optomotor neurons of the fly's lobula plate. Preferred and null directions of movement are shown by closed and open arrowheads, respectively. Arrows' positions, left or right, indicate the excitatory eye for each cell. Excitation between cells is denoted by a " + "; inhibition by a " − ". Adapted from Strausfeld, N. and Nässel, D. (1981), *Handbook of Sensory Physiology*, *VII/6B*, 1–132, by permission of the authors and K. Hausen, and by copyright permission of Springer, Berlin Heidelberg, New York.

then *their* receptive fields and not the postsynaptic dendritic densities should determine the receptive fields of the tangential cells. A final point is that the T cells might be directionally selective cells, with horizontally sensitive cells terminating anteriorly in the lobula plate and vertically sensitive cells terminating posteriorly. There is no physiological evidence to support (or refute) this either (Hausen, K. *et al.*, 1980).

5.1.1 THE HETEROLATERAL H1 CELL

Of the two horizontal-sensitive, spiking heterolateral cells, II-1 and II-3 (McCann, G. and Dill, J., 1969), only the II-1 cell has been intracellularly stained and identified. It has been renamed the H1 cell (Hausen, K., 1976b). The H1 cell, illustrated in Fig. 26 at the top, has extensive ipsilateral dendritic branching, an axon which courses anteriorly and dorsally to the opposite lobula, and far less extensive contralateral branching (in this account, the term "ipsilateral" will refer to the input dendritic field at a cell, usually homolateral to the soma). There are two reciprocal H1 cells, one ipsilateral to each lobula plate (Hausen, K., 1976a; Eckert, H., 1980).

Because their large spikes can be easily recorded contralaterally with extracellular electrodes, H1

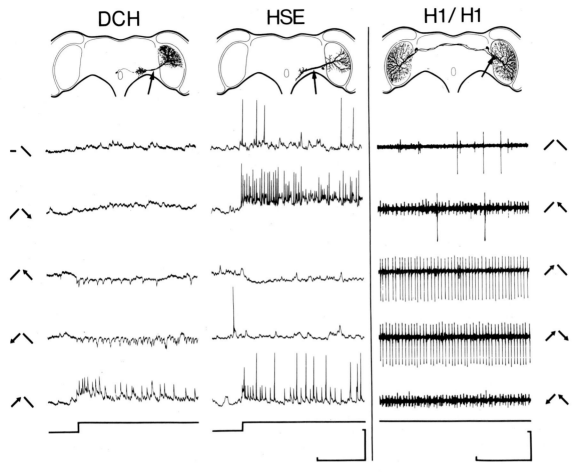

Fig. 27. Responses to moving gratings of a dorsal centrifugal horizontal cell (DCH), an equatorial horizontal cell (HSE), and from both reciprocal H1 cells. The arrows at the top show sites of recording in the fly's lobula plate. At left and right, lines without arrow-heads represent stationary gratings. Inward- and upward-pointing arrows represent regressive movement; outward and downward arrows represent progressive movement. The vertical calibration bar for DCH and HSE represents 20 mV. Reproduced from Hausen, K. (1981), *Ventr. Dtsch Zool. Ges.*, 46–70, by permission of the author and copyright permission of Gustav Fischer Verlag, Stuttgart.

cells are among the most studied of optomotor neurons. As indicated in the dual recording from both H1 cells in Fig. 27, right, they are excited and inhibited by ipsilateral regressive and progressive movements, respectively, and are inhibited by contralateral regressive movements. The latter inhibition, once thought to be direct (McCann, G. and Foster, S., 1971), now seems to involve interneurons (Hausen, K., 1981). There are no effects of contralateral progressive movements. Ipsilateral–progressive inhibition is active and not just a reduction of excitation from, e.g., a single, regressive-sensitive, DS input (Hausen, K., 1976a). This could mean that the inputs were from two DS

medullary cells of opposite preferred directions, but it could also result if there were DS synapses of non-directional T cells onto H1 cells (Torre, V. and Poggio, T., 1978).

The highest density of ipsilateral dendrites is slightly lateral to the midline and slightly ventral to the horizon (Hausen, K., 1981). It is there that H1 has its receptive field center, although H1 responds to movements in the entire ipsilateral visual field (Bishop, L. *et al.*, 1968; Eckert, H., 1980). At its receptive field center the response of H1 (and of other optomotor neurons: Hausen, K., 1981) is proportional to the cosine of the angular deviation from the preferred direction of movement (Bishop,

L. *et al.*, 1968; Srinivasan, M. and Dvorak, D., 1980).

Stepwise movements of gratings and edges in the preferred direction elicit maximum responses when the movements are about equal to the horizontal interommatidial angle (Zaagman, W. *et al.*, 1977; Mastebroek, H. *et al.*, 1980). This implies that there are sampling points for elementary movement detection (Buchner, E., 1976) that are set by the geometry of the eye. This cannot be the whole explanation, however, since the distance between sampling points increases at decreased intensities (Srinivasan, M. and Dvorak, D., 1980).

In a recent, elegant update of the behavioral experiments on sampling stations of Kirschfeld, K. (1972), Riehle, A. and Franceschini, N. (1983) recorded from H1 cells while stimulating just two photoreceptors, R1 and R6, in one ommatidium. Flashing either receptor alone or successively, or the two receptors simultaneously, elicited no response. Sequential flashes duplicating movements in the preferred or null directions elicited ON–OFF discharges by the second flash, or inhibition of the resting discharge, respectively. Apparently, excitation of the first receptor "gated" the response set up by the second receptor, but only after a delay.

Excitation of H1 cells by single receptors raises the important question of what cellular pathways lead to H1. Using counterphase gratings to determine the peripheral spatial filtering of inputs to H1 cells, Srinivasan, M. and Dvorak, D. (1980) proposed that these inputs had the angular sensitivities and the horizontal-only inhibitory flanks of sustaining laminar spiking cells (Arnett, D., 1972). The finding of phasic responses by Franceschini, N. and Riehle, A. (1983) suggests instead that the inputs to H1 cells are similar to the ON–OFF laminar spiking cells (Arnett, D., 1972), which do not, however, have inhibitory flanks (see Fig. 20). Alternatively, it could be the hyperpolarizing LMCs that are in the pathways to H1. The determinations by Mimura, K. (1976) and Dubs, A. (1982) that LMCs have inhibitory flanks is not inconsistent with the results of Srinivasan, M. and Dvorak, D. (1980). Moreover, both types of laminar spiking cells of flies have $V \log I$ curves extending over much more than 2 log units of intensity, whereas H1 cells (and LMCs, see section 3.1.1) saturate in a 1–1.5 log unit range (McCann, G. and Arnett, D., 1972).

There could of course be gain, and further narrowing of range, at subsequent synapses of lamina spiking cells. If inputs to H1 were via LMCs, the phasic ON–OFF transients could be those of the next cells, perhaps medullary M cells like the ones in Fig. 21. This would be consistent with the pathway L2 (lamina) to Tm1 to T4 (medulla) to lobula plate (Strausfeld, N., 1976b).

A final question concerns the long-held view that the H1 neuron is related to optomotor behavior. Evidence for this is, first, that the optimal contrast frequency for stimulation of H1 by moving gratings is 1.5–5 Hz (Zaagman, W. *et al.*, 1977; Eckert, H., 1980), about the same as for the optomotor torque response but less than for the landing response (Eckert, H. and Hamdorf, K., 1981). Second, the spectral sensitivity of H1 (McCann, G. and Arnett, D., 1972; Lillywhite, P. and Dvorak, D., 1981) is that of photoreceptors R1–6, as is that of the optomotor torque response (Kaiser, W., 1968; Heisenberg, M. and Buchner, E., 1977). Third, absorptions of single photons by photoreceptors are effective at absolute threshold both for optomotor behavior and for excitation of H1 (Lillywhite, P. and Dvorak, D., 1981). However, the H1 cell is not an output neuron. Moreover other output neurons (the HS cells) also have the same optimal contrast frequencies as for H1 cells and for optomotor torque (Hausen, K., 1981). The relation of H1 to ultimate optomotor behavior must therefore lie in its relations to other horizontally sensitive neurons.

5.1.2 Horizontal system (HS) cells

The largest tangential cells of the lobular plate are the HS cells of the anterior lobula plate, and the VS cells to be described later (section 5.1.4). The HS cells are shown in Fig. 26. They were first described anatomically by Pierantoni, R. (1976), who, in an inspired guess, correctly predicted that the HS cells were horizontal movement detectors. The three HS cells are termed HSN, HSE, and HSS, standing for horizontal system north, equator, and south, respectively (the terms NH, EH, and SH have also been used: Hausen, K., 1976b; Eckert, H., 1981). Their dendrites in the lobula plate are exclusively postsynaptic, whereas the axonal arborizations and terminals are mixed presynaptic and postsynaptic (Hausen, K., *et al.*, 1980; Eckert, H., 1981).

All authors agree that the three HS cells depolarize in response to ipsilateral progressive movements and hyperpolarize in response to ipsilateral regressive movements. This is shown in the middle of Fig. 27 for an HSE cell. Consistent with inputs from many columnar cells, no individual EPSPs or IPSPs can be seen (as could result if there were but a few input cells). Movements across the contralateral eye elicit only weak slow potential changes, which are of unknown origin and of unknown significance (Hausen, K., 1982b).

In some recordings spikes have also been found for contralateral stimuli (Dvorak, D. *et al.*, 1975; Eckert, H., 1981) but not for ipsilateral stimuli (Hausen, K., 1976b). With recording techniques that yield larger resting potentials, spikes have now also been observed for ipsilateral preferred (progressive) stimuli (Hausen, K., 1982a). These can be seen in Fig. 27. Low resting potentials would appear to inactivate sodium channels (Hengstenberg, R., 1977), and so, too, may strong depolarizations during ipsilateral progressive movements (Hausen, K., 1982a). It is uncertain where the spikes originate and whether they are actively propagated in the axon. They do not overshoot zero membrane potential, and their rates are very irregular compared to spikes of H1 cells (compare the middle and right columns of Fig. 27). It is possible that regenerative mechanisms in HS cells are like those in LMCs of the lamina and L cells of ocelli (see section 6.4) and have similar roles: to emphasize noisy depolarizations at axon terminals and so to augment synaptic transmission (Wilson, M., 1978b). The depolarizations themselves are conducted electrotonically to the terminals (Eckert, H., 1981; Hausen, K., 1982a).

As mentioned above, spikes are elicited in some HS cells by contralateral (regressive) movements. This is shown at the bottom middle of Fig. 27 for HSE. Small spikes are also seen, and these directly precede each large spike (Hausen, K., 1982a). The origin of these spikes is probably synaptic, from the heterolateral H2 cell (Hengstenberg, R. and Hengstenberg, B., 1980) shown at the bottom left of Fig. 26. Firing of H2 is excited by regressive movement and inhibited by progressive movement over its ipsilateral eye (the left eye in Fig. 26). The frequency of firing of H2 to regressive movement is about the same (100/s) as the frequency of EPSPs (or small spikes) in HS cells (Hausen, K., 1976a). The terminals of H2 overlap those of HS cells, and they are dye-coupled to HSN cells at least (Eckert, H., 1981). Dye-coupling between insect neurons (Strausfeld, N. and Obermayer, M., 1976) occurs across gap junctions of electrical synapses (Bassemir, U. *et al.*, 1982). Thus, the small spikes elicited in HSN and HSE cells (but not HSS cells: Hausen, K., 1982a) by contralateral regressive movements are probably electrically coupled H2 spikes. The large spikes, set up by the small spikes, could have the same electrical origins as large spikes set up by ipsilateral progressive movement.

In their activation by preferred stimuli, HS cells are much like H1 cells, except that the three HS cells divide up the area of the lobular plate. Although dendritic branching patterns vary even between HS cells of opposite lobula plates in the same fly, the areas of each HS cell's dendrites are remarkably constant. Each HS cell covers about 40% of the area of the lobula plate (Eckert, H., 1981; Hausen, K., 1982a). The centers of receptive fields occur at regions of maximum dendritic densities, just as for H1 cells (see above). Likewise, the preferred directions of movement are oriented along ommatidial *x*- and *y*-axes, so that in the dorsal and ventral parts of the eye the HSN and HSS cells, respectively, are also responding to the vertical components of movement. As for the H1 cells, amplitudes of responses vary as the cosine of the angle of movement from the preferred direction. Finally, the optimum contrast frequency for the *plateau* depolarizations and hyperpolarizations of HS cells is between 2 and 5 Hz. However, at the onsets and ends of movements there are responses to still higher contrast frequencies (Hausen, K., 1982b).

A response feature which HS cells share with cells of the vertical system (see section 5.1.4 below) is non-linear summation over the receptive field. Movements within small areas elicit somewhat smaller responses than within larger areas, but maximum responses are soon reached. These maximum responses do not appear to arise from saturation of the HS cell or from inhibition between neighboring inputs (Hausen, K., 1982b). Rather, it has been proposed that there is inhibition of presynaptic columnar inputs by integrating, large-field cells, so that the HS cells can have high gain for small movements while not saturating for movements over large areas (Poggio, T. *et al.*, 1981). The large-field, non-directional cells of the fly's proximal medulla (DeVoe, R., 1980) might possibly serve this role.

Finally, HS neurons are most likely the cells that are presynaptic to descending fibers in the ventral nerve cord (VNC). There is as yet no physiological confirmation of this from simultaneous records from HS and VNC cells. One kind of evidence is from dye coupling of HS cells to VNC cells (Eckert, H., 1981; Bassemir, U. *et al.*, 1982). A second is that VNC units have progressive-movement preferred directions, in keeping with those of HS cells (Hengstenberg, R., 1973; Hausen, K., 1976a).

5.1.3 OTHER HORIZONTALLY SENSITIVE CELLS

There are four other horizontal-movement sensitive cells to be considered: two centrifugal cells (CH), the H3 cell, and an as yet unidentified cell that is the same as the extracellularly recorded II-3 of Bishop, L. *et al.* (1968). H3 and CH cells are shown in Fig. 26.

The dorsal (DCH) and ventral (VCH) centrifugal cells are so called because their dendrites lie on the posterior slope while their telodendria divide up the area of the anterior lobula plate. The physiological responses indicate, however, that CH cells have inputs in both neuropiles, although their outputs are probably only within the lobula plate.

Spikes have not been recorded from CH cells. Rather, as seen in Fig. 27, contralateral regressive movements elicit large and small EPSPs, contralateral progressive and ipsilateral regressive movements elicit IPSPs, and ipsilateral progressive movements elicit depolarizations without distinct EPSPs. In elegant simultaneous intra/extracellular recordings, the large and small EPSPs elicited by contralateral regressive movements have been found by Hausen, K. (1976a) to be due to synaptic activation by H2 and H1 cells, respectively. The H1 cell contacts only the telodendria of the CH cells, so these telodendria are clearly both input and output regions.

The IPSPs in CH cells during contralateral progressive movements most likely result from inhibition by unit II-3 (and not H3, see below). There is no known input that would account for IPSPs in CH cells during ipsilateral regressive movements.

The depolarizations during ipsilateral progressive movements probably result from excitations by the HS cells. The latter excitation has not been directly demonstrated (Hausen, K., 1981) but seems likely in view of (1) overlapping synaptic regions on the posterior slope; (2) overlapping spatial sensitivities; (3) the same, ipsilateral progressive preferred directions; and (4) the absence of distinct EPSPs (as should be found if an ipsilateral progressive-sensitive spiking cell were exciting the CH cells). Since H2 excites both the CH cells as well as HSN and HSE, H2 probably excites the HS cells, and the HS cells excite the CH cells. The short, 0.6 ms delay for excitation of CH cells by H2 (Hausen, K., 1976a) implies that at least one of the synapses is electrical.

The H3 cell, shown in Fig. 26, is an anomaly. It is a spiking cell with an ipsilateral-progressive preferred direction. It was originally thought to be the II-3 unit (Hausen, K., 1976b), but unlike II-3 it lacks sensitivity in frontal fields of view where CH cells are sensitive (Hausen, K., 1981). Thus, it is not the contralateral H3 cell which elicits IPSPs in CH cells (Fig. 27) as originally thought, and what H3 synapses with is currently unknown.

The horizontal movement-sensitive network of the anterior lobula plate may now be summarized with reference to Fig. 26. The HS cells are directly excited by ipsilateral progressive movements and are synergistically excited via H2 by contralateral regressive movements. The HS cells of the right eye thus respond to clockwise rotations around the fly, and the HS cells of the left eye to counter-clockwise rotations.

HS cells excite descending neurons of the VNC, and they also excite CH cells. The contralateral H1 cell likewise excites the CH cells directly, whereas depolarizing current passed into a CH cell inhibits the ipsilateral H1 cell (Hausen, K., 1981). Reciprocal H1 cells are known to mutually inhibit each other (McCann, G. and Foster, S., 1971), but probably the connection is indirect, via the CH cells. As telodendria of CH cells do not directly synapse onto dendrites of H1 cells, what they may do instead is inhibit the ipsilateral-regressive input channels from the medulla (Hausen, K., 1981). CH cells would thus play a differential role in sharpening directional selectivities. For example, weak progressive movement over the right eye would partially inhibit the right H1 cell but excite the right HS and hence right CH cells. The right CH cells in turn would suppress regressive inputs from the medulla, further inhibiting the right H1 cell.

There are complex, binocular interactions possible which depend upon the bilateral H1 and H2 cells, as well as the homolateral HS and CH cells. Some of these interactions have been detailed by Hausen, K. (1982b). In view of binocular interactions involving excitatory and inhibitory feedback, it would provide stability for all interacting elements to have the same frequency responses. If they did not, then the strengths of feedback would depend on the temporal characteristics of movements. The similar optimum contrast frequencies of 1–5 Hz for all optomotor neurons (Hausen, K., 1981) as well as for optomotor torque behavior (Eckert, H. and Hamdorf, K., 1981) may have had their origins in such advantages.

Finally, in order to specify that the lobular plate neurons are optomotor neurons, it is necessary to show the connection between them and optomotor behavior. This has been done in several ways. First, direct electrical stimulation in the anterior lobula plate elicits yaw responses in freely walking and flying Calliphora. The smallest amount of current is required in the antero-lateral lobula plate, just where dendritic densities and receptive field sensitivities of H1 and HS cells are greatest. On the other hand, the direction of turning is dependent on the polarity of the stimulus: negative current (which should depolarize cells) elicits turning away from the stimulated side, whereas positive current has the opposite effect (Blondeau, J., 1981). It is therefore not obvious which cells are being stimulated (or inhibited).

Second, flies without HS (and VS) cells have impaired optomotor behavior. Yaw reactions and pattern-induced orientation are strongly reduced in the Drosophila mutant optomotor-blind[H31] (Heisenberg, M. et al., 1978). Similarly, unilateral laser ablation of larval precursors of HS and VS cells in Musca resulted in adult flies with only 50% of normal optomotor yaw responses on the treated side. They had nearly-normal, pattern-induced orientation there, however (Geiger, G. and Nässel, D., 1981), implying that cells must be absent in both lobula plates if orientation behavior is to be impaired.

Third, it appears that the positional sensitivity of HS cells to progressive movements is very similar to the position-dependent yaw responses in pattern-induced orientation. In this view the calculation of positional information is a spatial parameter in the sensitivity to motion (Wehrhahn, C. and Hausen, K., 1980). In essence, tracking of patterns could be due to the differences between responses of HS cells of the two eyes. Oscillatory self-movements (torque noise) or oscillatory movements of the pattern would excite one set of HS cells and inhibit the opposite ones, but only in the frontal regions of binocular overlap. Frontally, at the fixation point, there would be maximum changes of responses (maximum difference signals) as the pattern moved from one side to the other. Elsewhere, primarily one set of HS cells would be activated by progressive movements and would result in yaw to bring the pattern into the frontal eye fields. This hypothesis has been formulated in a formal neural network that also incorporates the non-linear summation within the receptive fields of HS cells. To a first approximation, HS cells of Calliphora seem to account for the behavioral figure–ground discrimination by relative movements as measured and predicted by the model (Reichardt, W. et al., 1983). The model does not incorporate the synergistic excitations of HS cells by the contralateral H2 cell, however.

In summary, the horizontally sensitive (H1, H2, HS, CH) cells of the lobula plate appear to provide a sufficient substratum for optomotor yaw responses. Nonetheless, there remains much to be learned about their specific inputs, the dynamics of their interactions, and their specific outputs to the ventral nerve cord.

5.1.4 VERTICALLY SENSITIVE CELLS

Directionally selective neurons in the posterior lobular plate respond primarily to vertical movements. These neurons are the vertical cells. Some vertical cells also respond to dorsal horizontally moving objects. The vertical cells of the fly's lobula plate are the giant VS (vertical system) cells, the V cells, and the small v cells.

There are 11 VS cells, VS1–VS11, in Calliphora, with homologous cells in Phaenicia and Musca (Hengstenberg, R. et al., 1982, provide comparisons

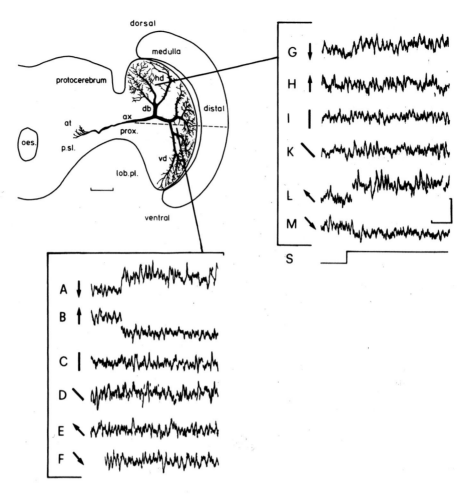

Fig. 28. Structure and function of the vertical cell VS1 in the lobula plate of the fly *Phaenicia sericata*. Recordings **A–F** were made from the more ventral caudal part of the cell; **G–M** were recorded from the more dorsal, rostral part. Diagonal arrows denote horizontal movements as in Fig. 27; vertical arrows denote vertical movements. Vertical calibration in **M** is 10 mV. Reproduced from Eckert, H. (1982), *J. Comp. Physiol. 149*, 195–205, by permission of the author and copyright permission of Springer-Verlag, Heidelberg.

among the nomenclatures used by different workers). VS1 is illustrated in Fig. 28, and a number of the VS cells and their overlaps are to be seen at the left in Fig. 29A. Just as for HS cells, there is remarkable uniformity of the size and position of a given VS cell's dendritic fields, from animal to animal, even though the fine branching patterns are variable. All VS cells span the entire dorsoventral extent of the lobula plate, but the horizontal extents differ. By plotting dendritic fields back into the retina, it is found that the equatorial horizontal fields of VS2–VS4 are between 19 and 25 ommatidia wide, or about 55–65° of visual angle. Such retinotopic calculations are consistent with receptive field

measurements of cells VS2 and VS8 (Eckert, H. and Bishop, L., 1978).

VS cells respond to increasing intensities of diffuse light on the eye with minor depolarizations, increased fluctuations of the membrane potential ("noise") and occasionally with spikes at light-ON (Hengstenberg, R., 1982). All VS cells depolarize with downward movements and hyperpolarize with upward movements. This is illustrated in the lower part of Fig. 28 for a VS1 cell. In some such recordings, as in the figure, the noise appears to decrease during hyperpolarizations, while in others it increases for hyperpolarizations as well as for depolarizations (Soohoo, S. and Bishop, L., 1980).

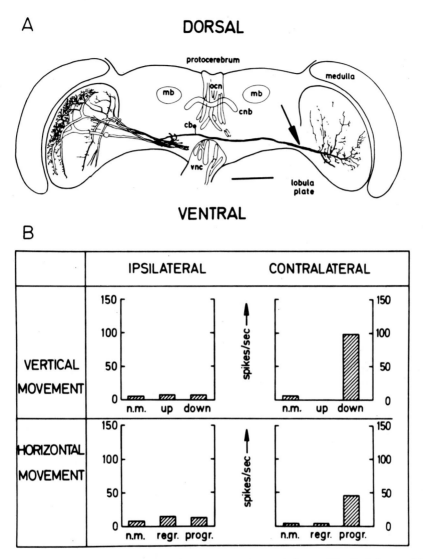

FIG. 29. The V1 cell of the fly's lobula plate (right part of **A**) and its dendrites' relations to VS cells (left part of **A**). V1 responds primarily to downward and progressive movements across the eye at its dendritic ("contralateral") end. The arrow indicates the site of penetration. Reproduced from Eckert, H. and Bishop, L. (1978), *J. Comp. Physiol. 126*, 57–86, by permission of the authors and copyright permission of Springer-Verlag, Heidelberg.

From the latter it appears that depolarizations could be due to summed EPSPs, while hyperpolarizations could be due to summed IPSPs. No conductances have been measured during activity to confirm this hypothesis. Nonetheless, it is consistent with the asymmetries of depolarizations and hyperpolarizations that are observed when resting membrane potentials change spontaneously or as a result of extrinsic currents (Hengstenberg, R., 1977, 1982).

Three of the VS cells, VS2–VS4, have their

dendrites solely in the caudal, "vertical" laminae of the lobula plate. The remaining VS cells all have dorsal dendritic arborizations in the rostral, "horizontal" laminae, the laminae occupied by dendrites of H and HS cells (Hengstenberg, R. *et al.*, 1982. It is uncertain if VS5 has such rostral dendrites). Cells VS2–S4 respond only to vertical movements, but the other VS cells respond to horizontal movements as well. An example is shown at the upper right in Fig. 28 for a VS1 cell. Its rostral dendrites receive visual excitation from the dorsolateral region of the eye. In this region regressive horizontal movements elicit depolarizations, while progressive horizontal movements elicit hyperpolarizations. Vertical movements are ineffective (Eckert, H., 1982). In cells VS5–VS10, on the other hand, depolarizations are elicited by progressive horizontal motion and hyperpolarizations by regressive motion (Eckert, H. and Bishop, L., 1978). What this means is that both forward and backward pitch could be detected (i.e. during flight), but by different VS cells (Hausen, K., 1981).

VS cells respond optimally to contrast frequencies of 1–2 Hz for movements of gratings in either vertical direction. These are frequencies close to those that are optimal in optomotor thrust behavior (Eckert, H., 1982; Hengstenberg, R., 1982). Moving gratings are of course not required to elicit responses. Moving objects which stimulate as few as 25 ommatidia are well above any threshold. For still larger objects the strengths of response increase non-linearly with the lengths of stimulus contours. That is, there is non-linear summation within the receptive field, just as for HS cells. Finally, these responses are independent of contrast polarities: both light–dark and dark–light edges must move downwards to elicit depolarizations. Hence, VS cells are truly directionally selective (Hengstenberg, R., 1982).

The second group of vertical cells are the spiking cells V1–V3 (Hausen, K., 1976a; Eckert, H. and Bishop, L., 1978). The first of these, V1, is a unique centrifugal heterolateral cell with dendrites on the posterior slope and telodendria in the opposite lobula plate. Figure 29A shows the form of V1 and the relation of its dendrites to the terminals of the VS cells. V1 responds only to stimuli at its dendritic end. As shown in Fig. 29B (where these stimuli are called contralateral), it is excited by downward movement and inhibited by upward movement. It is less strongly excited by progressive movement.

Like the H1 and HS cells of the anterior lobula plate, V1's preferred directions are oriented along the (vertical) ommatidial rows, its responses (rates of firing) are proportional to the cosine of the direction of movement relative to the preferred direction, and its optimal contrast frequency for moving gratings is about 4.5 Hz (Hausen, K., 1981). The outputs of V1 are unknown. It is likely that V1 receives its inputs from the ipsilateral VS cells, probably VS5–VS10 which also respond to progressive horizontal motion. V1 cells are the II-2 units of Bishop, L. *et al.* (1968).

V2 is a unique heterolateral cell with processes in both lobula plates, midway between the anterior and posterior laminae. It also has an arborization on the contralateral posterior slope in the region of dendrites of VNC cells. The V2 cell responds monocularly to ipsilateral upward and horizontal regressive movements (Hausen, K., 1976a, 1981). V2 is most likely the II-4 unit (Bishop, L. *et al.*, 1968; McCann, G. and Foster, S., 1971).

The third, spiking, heterolateral cell, V3, also connects both lobula plates and responds preferentially to upward movement. Its shape has not been illustrated (Hengstenberg, cited by Hausen, K., 1981).

The last group of vertical cells are the homolateral v cells, similar to but $\frac{1}{2}$–$\frac{1}{3}$ smaller than the VS cells (Bishop, C. and Bishop, L., 1981). These are spiking cells with ipsilateral downward preferred directions (Dvorak, D. *et al.*, 1975). They have also been found anatomically by Pierantoni, R. (1976).

The VS cells (and maybe V1) are the presumptive output neurons of the vertical system. Of all the vertically sensitive cells, only VS cells have been found dye-coupled to descending neurons of the VNC (Strausfeld, N. and Obermayer, M., 1976). Hengstenberg, R. *et al.* (1982) sequentially penetrated and stained in one preparation a VS9 cell and a descending neuron that seemed to be in contact. In another dual-stained preparation, a VS2 and an HSE cell had collaterals which ran towards each other but did not touch. Possibly both cells contacted still another, unknown neuron.

The involvement of the VS and V cells in behavior has been tested with direct electrical stimulation (Blondeau, J., 1981). Stimuli in the middle layer of the lobula plate, where V2 has its dendrites, elicits sideways movements in freely walking flies. Perhaps these were attempted compensations for perceived roll, which excites V2 cells (Hausen, K., 1981). Stimuli in the posterior lobula plate elicited two kinds of reactions: lift and thrust when the electrode was lateral (frontal eye fields, cells VS1–VS4), and the landing reaction when the electrode was medioventral (cell V1) (Blondeau, J., 1981).

Although these results are consistent with involvement of vertical cells in lift, Heisenberg, M. *et al.* (1978) found little degradation of lift (or landing) responses in *optomotor-blind*[H31] mutants of *Drosophila* which lacked the VS cells.

5.2 Optomotor neurons of honey bees

The best known of the neurons of honey bees is the bilateral, spiking, HR (horizontal regressive) cell, first found by Kaiser, W. and Bishop, L. (1970). With intracellular recording in the ipsilateral lobula (DeVoe, R. *et al.*, 1982), the HR cell is found to be depolarized by ipsilateral regressive movement and by contralateral progressive movement. Ipsilateral progressive movement weakly excites or inhibits, and contralateral regressive movements inhibit via hyperpolarizations, especially after a number of repetitions. These patterns of response suggest that each HR cell receives two kinds of synaptic input from the contralateral lobula: inhibition from the reciprocal HR cell, and excitation from a bilateral, horizontal progressive (HP) cell.

Reciprocal HR and HP cells have each been penetrated, but only the HR cells have been unambiguously stained (DeVoe, R. *et al.*, 1982). In form, the HR cells resemble the H1 cells of flies. Both are large-field cells whose dendrites and endings fill much or most of the lobulae, and both have axons which run between the lobulae in the anterior optic tracts and optic tubercles (Hausen, K., 1976a; Eckert, H., 1980). In function, too, HR cells resemble H1 cells: both respond best to ipsilateral regressive movements. Here the similarities stop, however. HR cells are not strongly inhibited by ipsilateral progressive movement, as are H1 cells, and H1 cells are not excited by contralateral progressive movement, as are HR cells by (presumably) HP cells. Rather, it is the H1 cells which excite the contralateral, progressive-sensitive, II-3 cells in flies (McCann, G. and Foster, S., 1971). Neither II-3 nor HP cells have been identified, thus barring any further comparisons.

HR cells have a small dynamic range of 1 log unit of intensity (Kaiser, W. and Bishop, L., 1970), somewhat smaller than the 1.5 log unit range of LMCs in honey bees (Menzel, R., 1974). These LMCs receive inputs from both blue and green photoreceptors, whereas HR cells do not. Instead,

HR cells have only the spectral sensitivities of green receptors (Kaiser, W., 1972, 1974; Menzel, R., 1973). The optomotor yaw reactions of walking and flying bees likewise receive inputs only from green receptors (Kaiser, W., 1975; Kaiser, W. and Liske, E., 1974), whereas phototactic behavior receives inputs from all three receptors (UV, blue, green), in the same eye region used for the studies on yaw (Kaiser, W. *et al.*, 1977). Thus yaw reactions are color-blind, as is the HR cell, which may thus mediate yaw reactions. Further evidence that the HR cell is an optomotor neuron is that the optimum contrast frequencies of moving gratings are about the same for the HR cell as for the yaw reaction (Kaiser, W. and Bishop, L., 1970). However, the HR cell, like the H1 cell of flies, is not an output neuron and does not project to descending neurons on the posterior slope of the protocerebrum. It is thus uncertain what are the output neurons for the horizontal system in honey bees (Guy, R. *et al.*, 1979; DeVoe, R. *et al.*, 1982). Some intracellular recordings have also been made from neurons of the vertical system of honey bees, and spiking cells similar to V1 and V2 cells of flies have been reported (Ohm, J., 1981). The outputs of the vertical system are also unknown, although one cell was stained in the lobula.

A remarkable feature of the HR neuron is its circadian cycles of sensitivity (Kaiser, W. and Steiner-Kaiser, J., 1983). In night-time hours the response to periodic, regressive movements of a (continuously illuminated) grating is greatly reduced, as is the spontaneous activity. This is shown in Fig. 30. Underlying the changes in response are shifts of $V \log I$ curves. During the same night-time hours, locomotor activity of free running bees in the dark is less than in day-time hours. The sensitivities of the photoreceptors, on the other hand, rise at night in an antiphase circadian rhythm. This is indicative that control of sensitivity of HR cells is central, not peripheral. By way of confirmation, HR cells of low sensitivity can be temporarily "aroused" and stimulated by air puffs, or by light flashes onto the other eye. The high retinal sensitivities at night during periods of low sensitivities of HR cells means that happenstance multimodal arousal of the HR cells will find the retina ready and adapted to function on demand. Excitation by the very stimuli which arouse show

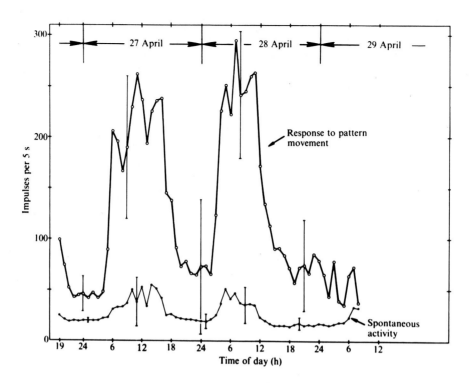

FIG. 30. Parallel circadian changes in responsivity and in spontaneous activity of the HR neuron of the lobula of the honey bee, *Apis mellifera*. From Kaiser, W. and Steiner-Kaiser, J. (1983) by permission of the authors. Reprinted by permission from *Nature 301*, No. 5902, 707–709, Copyright © 1983 Macmillan Journals Limited.

that HR cells themselves are multimodal. There is no other evidence about the mechanisms of arousal, however, or whether the circadian reductions in spontaneous activity and sensitivity occur pre- or postsynaptically. An example of postsynaptic modulation of a movement detector will be given in the next section.

5.3 The orthopteran LGMD/DCMD system

Early recordings by Parry, D. (1947) and Burtt, E. and Catton, W. (1954) from the circumesophageal connectives of locusts showed that large spikes were elicited in response to contralateral movements of small (but not large), high-contrast objects. These spikes arise in axons of the descending contralateral movement detector (DCMD). As found by intracellular staining with cobalt (O'Shea, M. *et al.*, 1974), the DCMD has its soma and integrating segment on the posterior slope of the protocerebrum, projects to an initial segment (spike initiation zone) on the opposite side, and descends in the dorsal VNC without connections until it gives off one

branch in the prothoracic ganglion and several each in the mesothoracic and metathoracic ganglia. This is illustrated in Fig. 31 at the left and the middle. The DCMD's axon is the largest in the VNC (15–17 μm diameter) and this, combined with its peripheral location in the VNC, accounts for the large (1–3 mV) spikes that are recorded extracellularly. Smaller spikes from a descending ipsilateral movement detector (DIMD) can also be recorded in the connectives of locusts; in crickets however the DIMD has the larger spikes (Rowell, C., 1971). Since the DIMD in locusts has the same conduction velocity (3 m/s) as the DCMD and so should be as large, it is possible that the very large axon seen in the center of the VNC (O'Shea, M. *et al.*, 1974) is the DIMD. Relative isolation of the DIMD in the middle of the VNC could account for the small spikes recorded extracellularly, in locusts.

At their terminations in the metathoracic ganglia, the DCMD and DIMD make synaptic connections that excite jump motoneurons (O'Shea, M. *et al.*, 1974: see Fig. 31 at the right). There is no evidence that visual or other excitation of spikes in the

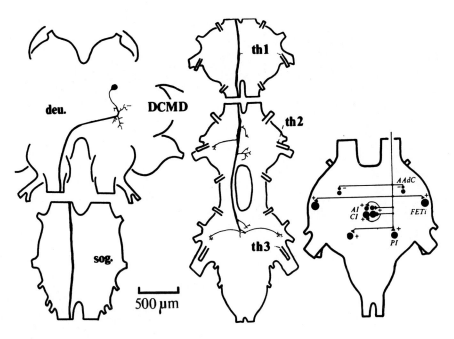

FIG. 31. Structure and output connections of the locust descending contralateral movement detector (DCMD). The metathoracic outputs of the DCMD are shown schematically at the right. Adapted from Strausfeld, N. and Nässel, D. (1981), *Handbook of Sensory Physiology, VII/6B*, 1–32, by permission of the authors and copyright permission of Springer-Verlag, Heidelberg. Figure based on O'Shea, M. *et al.* (1974).

DCMD results directly in jumping (Rowell, C., 1971). Spikes in DCMDs will sometimes elicit post-synaptic spikes in the fast extensor of the tibia (FETi) motoneuron providing, however, that the body temperature is at the preferred 30° or above (Heitler, W. *et al.*, 1977).

The DCMD and DIMD receive their primary inputs from the lobular giant movement detector (LGMD). Figure 32 illustrates the shape of the LGMD, in frontal section. The LGMD has three dendritic arborizations, labelled A, B, and C, in the distal and posterior lobula. It projects to the posterior slope, where a large proportion of its complex teleodendria are in close contact with the dendrites of the DCMD (O'Shea, M. and Williams, J., 1974; O'Shea, M. and Rowell, C., 1975a). There are chemical synapses of the LGMD onto the DIMD (O'Shea, M. and Rowell, C., 1975b), but the synapses onto the DCMD are electrical. The evidence for this includes the following: (1) The DCMD follows the LGMD spikes 1:1 up to 320 spikes/s; (2) following occurs in calcium-free solutions and at low temperatures, where synapses of other visual neurons are blocked; and (3) antidromic spikes in the DCMD result in small

(1 mV) spikes in the LGMD (O'Shea, M. and Rowell, C., 1975a). An electrical synapse thus means that the predominant pattern of spiking in the DCMD to visual stimuli results from the pattern of spiking in the LGMD.

The electrical synapse does not mean, however, that all LGMD spikes excite the DCMD. The initial segment of the DCMD, which is the spike initiation zone, appears to be some distance from the integrating segment. In the integrating segment, 60 mV presynaptic spikes of the LGMD elicit 30 mV EPSPs. These EPSPs are then electrotonically conducted to the initial segment. EPSPs that are less than 77% of their normal size are subthreshold for DCMD spikes (O'Shea, M. and Rowell, C., 1975a). Such decreases in EPSPs can be mediated by other, chemical synapses on the integrating segment (the other synapses by themselves never excite the DCMD). For example, excitation of the reciprocal DCMD results in an increase in membrane conductance to chloride ion and causes hyperpolarization, reducing EPSP amplitudes, and so blocking transmission. Tactile stimuli or stimulation of other VNC units can also block transmission (Rowell, C. and O'Shea, M., 1980). Still another suppression of

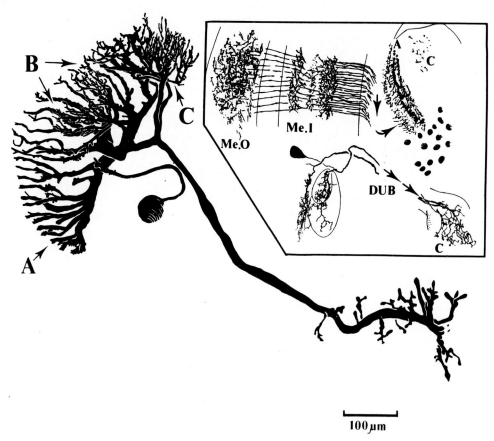

100 μm

FIG. 32. Structure of the lobula giant movement detector (LGMD) and some of its inputs. The inset depicts that small-field, habituating, medullary inputs synapse onto the A dendritic fan of the LGMD, whereas the dorsal uncrossed bundle (DUB) synapses onto the C dendritic fan. Adapted from Strausfeld, N. and Nässel, D. (1981), *Handbook of Sensory Physiology, VII/6B*, 1–132, by permission of the authors and copyright permission of Springer-Verlag, Heidelberg.

firing of the DCMD occurs during head saccades (or attempted head saccades) as a result of corollary discharges (Zaretsky, M. and Rowell, C., 1979). Although it is not known how the suppression occurs, modulation of the electrical synapse would be one possible way.

5.3.1 EXCITATION OF THE LGMD/DCMD SYSTEM

As indicated above, the DCMD responds best to movements of small, but not large, contrastive objects. The direction of movement does not matter, but the responses habituate with repeated presentations of the same stimulus at the same place in the visual field. On the other hand, large-field stimuli neither elicit responses nor habituate the DCMD (O'Shea, M. and Rowell, C., 1975b).

Since DCMD spikes originate in LGMD spikes,

the response properties just described for the DCMD are actually those of the LGMD. In practice it is far easier to record extracellular spikes from the DCMD than intracellular responses from the LGMD. The following account on excitation of the LGMD is thus based on recordings from either the LGMD or the DCMD, or both.

The small, contrastive, moving objects which excite the LGMD can be thought of as successively exciting ON- and OFF-channels in retinotopic arrays. Thus, the LGMD and DCMD should also be excited by incremental and decremental, flashed, small spots. This is indeed found, and in general responses to decreases in intensity (OFF responses) are greater than to increases (ON responses) and depend differently upon light adaptation (Rowell, C. and O'Shea, M., 1976a). The intensity ranges are as little as ± 0.5 log unit before saturation. Contrast sensitivities of the LGMD/DCMD also increase with light adaptation, just as for the receptors (Matic, T. and Laughlin, S., 1981: see section 2.3.1). Large-field or annular illumination around a test spot located anywhere reduces ON- and OFF-responses, quite possibly by lateral inhibition in the lamina (Rowell, C. and O'Shea, M., 1976b). A mechanism for large-field lateral inhibition in the lamina of locusts is that of Shaw (1975, 1979a; see section 2.5.1), in which reduction of transmembrane potentials of receptor terminals by high extracellular field potentials could provide the kind of automatic gain control observed for the LGMD/DCMD.

So far it has not been possible to record from the specific inputs to the LGMD. It seems likely that these inputs impinge on the A dendritic field of the LGMD (see Fig. 32), which does not support spikes. Rather, PSPs in the A field (the "fan") are electrotonically conducted to a spike initiating zone, which is the region of initial axonal thickening. From there spikes are conducted at 3 m/s (the same speed as in the DCMD) to the axon terminals, where there exists a second spike-initiating zone with auditory inputs (O'Shea, M. and Rowell, C., 1976).

Although no recordings have been made from the input cells to the LGMD, much has been learned about them. First of all, habituation from repeated presentations of the same stimuli (flashed or moving spots) is restricted to regions as small as 3°. This indicates inputs from small-field cells (Rowell, C., 1971). Sensitivity, after habituation, recovers over seconds or minutes, or by dishabituation, but there is no habituation at nearby but unstimulated parts of the visual field. Both ON- and OFF-responses habituate, and they habituate each other, implying that ON and OFF pathways converge before the site of habituation (O'Shea, M. and Rowell, C., 1976).

The site of habituation is probably presynaptic. There are no changes in postsynaptic threshold or conductance during habituation (nor are there measurable postsynaptic events during dishabituation). The site of habituation is between the inner chiasm and the LGMD, since local chiasmal shocks elicit habituating EPSPs in the LGMD. Latencies are consistent with a monosynaptic, chemical synapse (O'Shea, M. and Rowell, C., 1976).

A remarkable feature of the habituation is that it can be blocked by large-field moving stimuli. The blockage of habituation is dynamic, not static: stationary large-field stimuli do not block habituation to small moving spots (O'Shea, M. and Rowell, C., 1975b; Rowell, C. et al., 1977). Thus it is not the tonic, peripheral (perhaps laminar) lateral inhibition which blocks habituation.

The blockage of habituation appears to be by the same processes which block responses to large-field movements themselves. Large (20°) rotating radial gratings of sufficient spatial frequencies (0.08 to 0.4 cycles/deg) elicit few DCMD spikes but correspondingly inhibit both responses and habituation to nearby small (6°) moving targets. This dynamic inhibition is maximal after about 185 ms. At very low spatial frequencies (0.02–0.05 cycles/deg) the large gratings excite the DCMD, augment the responses to the small moving target, and induce habituation themselves (Pinter, R., 1979). Thus it is not the size alone of moving visual fields which determines the inhibition, but the contrast richness. A second point is that the effects on habituation of large fields depends on whether these fields inhibit or excite. This would only be true if inhibition (or excitation) by large fields precedes the habituating synapses.

The dynamic inhibition by large fields is presumably quite specific to afferents to the LGMD. Other visual pathways are clearly unaffected. For example, tonic optomotor neurons of locusts require wide-field stimuli (Kien, J., 1975). The same seems to be true in dragonflies, where "self-movement" optomotor units of the VNC respond tonically to large-field stimuli which do not excite DCMD-like, "object-movement" units (Olberg, R., 1981). In the locust DCMD system the location and cellular basis of the large-field inhibitory network is unknown, but it must receive both ON- and OFF-input (Rowell, C. et al., 1977).

There are still other inhibitory networks for large-field stimuli, and these feed forward directly onto the LGMD. Although movements of, e.g., large gratings, or whole-field illumination, elicit no lasting discharges in the LGMD, they can elicit brief EPSPs and sometimes spikes at the onsets of fast movements or at light ON and OFF. Such EPSPs are almost immediately terminated by large IPSPs (O'Shea, M. and Rowell, C., 1975b). The ON-IPSPs are somewhat larger than the OFF-IPSPs, reverse at a slightly more negative membrane potential, and arrive at the LGMD via a different pathway than for OFF-IPSPs. The pathway for the OFF-IPSPs is the dorsal uncrossed bundle (DUB), cells of which are shown in the inset of Fig. 32. DUB neurons synapse onto the C dendritic field of the LGMD. Cutting the DUB abolishes the OFF-IPSPs but leaves the ON-IPSPs (Rowell, C. et al., 1977). The pathway for the ON-IPSPs is uncertain but may be the median uncrossed bundle (MUB: Strausfeld, N. and Nässel, D., 1981). The function of the fast, feed-forward inhibition is to block LGMD excitation by large fields before the slower, more distal networks come into play.

The finding of two pathways, ON and OFF, for large-field, feed-forward inhibition is the best proof that there are separate ON and OFF channels distal to the combined, ON/OFF, habituating synapse. The input field of DUB neurons, in the inset of Fig. 32, lies on the proximal face of the medulla, where it picks up inputs from an ellipsoidal visual field about $8° \times 12°$. All DUB neurons together cover the entire visual field, without overlap (Rowell, C. et al., 1977). OFF channels presumably project to these same proximal medullary areas, before they combine with ON channels to excite ON/OFF units (O'Shea, M. and Rowell, C., 1976). What the ON- and OFF channels are is not known. Northrop, R. and Guignon, E. (1970) found many phasic and tonic ON (L) units in the locust medulla, but stated that OFF units were rare. They also found jittery-movement, habituating, multimodal ON/OFF units, but these are unlikely to be inputs to the LGMD/DCMD because the latter are not multimodal (Rowell, C., 1971) and because of the separation at some stage of ON and OFF channels. In cobalt stains of the LGMD, transynaptic migration of cobalt into two sets of medullary T cells (inset of Fig. 32) suggests that these latter cells might be inputs (Strausfeld, N. and Nässel, D., 1981). This will require physiological verification.

In summary, the orthopteran LGMD/DCMD system is one of the best characterized of any insect movement detecting network. This is true despite the amount that is unknown about the system: what are the ON- and OFF-input units, what are the peripheral inhibitory networks, what is the habituating synapse, how is it dishabituated and by what cells, etc.? The point is that not much is known about the specific inputs to any movement detector in insects, and for most movement detectors less is known than for the LGMD/DCMD system. After the explosion of knowledge in the last decade about unique lobular giant movement detectors, further elucidation of their complete networks will now require emphasis on the much more difficult recordings and identification of their non-unique, small-axoned, medullary inputs.

6   OCELLI

Ocelli comprise the other major photoreceptor organs in adult insects. They are principally adapted to signal changes in ambient illumination over wide fields of view. Underfocusing by ocellar optics (see chapter 5), and convergence of hundreds of photoreceptors onto a few large (L) and small (S), second-order neurons, exclude detail vision but do perhaps allow detection of gross object movements (Zenkin, G. and Pigarev, I., 1971). Ocellar second-order neurons project to the posterior slope of the protocerebrum, where they synapse onto descending fibers of the ventral nerve cord (VNC), including some that also receive inputs from lobular neurons. There appear to be ocellar connections to and from the optic lobes, as well as from thoracic efferents. There are also extensive interconnections among the various ocelli: in the ocellar cups, along the second-order axons, and on the posterior slope. The ocellar networks are thus far more complex than the simple dimming detectors that they were once imagined to be.

The pioneering electrophysiology of insect ocelli was performed with extracellular recordings by Parry, D. (1947) and Hoyle, G. (1955) on locusts, and Ruck, P. (1957, 1961) on grasshoppers, cockroaches, and dragonflies. The initial intracellular recordings were those from dragonfly median ocelli by Chappell, R. and Dowling, J. (1972). A small number of controlled behavioral experiments involving the ocelli have been performed and will be described where relevant to the electrophysiological findings. For a complete overview of ocellar structure and function the reader should refer to the excellent review by Goodman, L. (1981).

6.1   Ocellar photoreceptors

From less than a hundred to over a thousand

photoreceptor cells are found in ocelli of different insects. Their responses to light are similar to those of photoreceptors of the compound eye (see Fig. 1): (1) graded depolarizations are due to conductance increases (Labhart, T., 1977); (2) sensory adaptation occurs with bright but not dim flashes; (3) there are no tonic spike discharges, but there can be an initial spike in receptors of dragonflies, bees and moths. The initial spike in dragonfly median ocelli is abolished by tetrodotoxin (TTX), but this abolition has no effect on the postsynaptic responses of second-order, "L" cells (Chappell, R. and Dowling, J., 1972). Thus the physiological function of the initial spike is unknown. At light-off, depolarizations generally return first rapidly, then slowly back to resting potentials, except for off-oscillations in dragonfly median ocelli. The origins of these oscillations will be considered in section 6.2, below. Finally, quantum bumps have been recorded from dragonfly ocellar receptors (Stone, S. and Chappell, R., 1982; Simmons, P., 1982). Comparable photon noise in the form of discrete (hyperpolarizing) potentials has been found in L cells of locusts (Wilson, M., 1978c) and dragonflies (Simmons, P., 1982).

The photochemistry of ocellar receptors in *Drosophila* is much like those of the compound eyes. Photopigments are vitamin A-based; a stable, long-wavelength absorbing metarhodopsin is formed during illumination; and sensitivity does not fully recover during dark adaptation until the metarhodopsin is reconverted to rhodopsin (Hu, K. *et al.*, 1978). However, no PDAs have been elicited during conversions of rhodopsin to metarhodopsin (Labhart, T., 1977).

Absorption spectra of ocellar photopigments have not been measured, but spectral sensitivities give indications of what these spectra might be. Almost all ocelli studied have peak sensitivities at 340–360 nm in the UV, with equal or secondary peaks at blue (440–460 nm) and green (480–530 nm) wavelengths. The cockroach ocellus has only the green peak (Goldsmith, T. and Ruck, P., 1958), while UV + blue peaks for fly ocelli likewise seem to be due to pigments in a single type of cell (Kirschfeld, K. and Lutz, B., 1977). In other ocelli studied, two or more classes of cells are thought to be present.

Only the spectral sensitivities of dragonfly median ocelli have been determined from intracellular recordings, however. For these cells it is found that the response characteristics that are generally used to indicate two classes of photoreceptor cells can be found for single cells: waveforms of responses are wavelength-dependent (in *Libellula*); $V \log I$ curves of some cells are not parallel for all wavelengths and can even cross (Fig. 33); and spectral sensitivities of single cells can be selectively, chromatically adapted (Chappell, R. and DeVoe, R., 1975). It is uncertain if these responses are due to two photopigments in each cell or to coupling of two or more classes of cells. If the latter the cell classes are never found uncoupled. Movements of shielding pigments in dragonfly ocelli in response to light have spectral sensitivities similar to those of the cells, including selective reduction of green sensitivities by adaptation with long wavelengths (Stavenga, D. *et al.*, 1979). Thus the adaptations of cellular spectral sensitivities have physiological consequences. Likewise, the crossing of the $V \log I$ curves (as in Fig. 33) would result in greater green sensitivities at low intensities and greater UV sensitivities at high intensities. As will be shown later, such a reverse Purkinje shift is transmitted to L cells and ultimately affects ocellar-dominated visual behavior. There may also be such reverse Purkinje shifts in ocelli of the moth *Trichoplusia* (Eaton, J., 1976), but higher-order consequences, if any, are unknown.

6.2 Ocellar synapses

The synaptic activation of second-order neurons is known unambiguously only for large (L) cells. Recordings have probably been made from small (S) neurons, but the specific cells recorded from have not yet been identified. The numbers of L cells differ between medial and lateral ocellar nerves as well as among species, but in all nerves the numbers are small and range from 4 to 12. A given L cell may receive inputs from all or only a portion of photoreceptors (see Goodman, L., 1981, for details). It is not known if partial convergences result in restricted visual fields. Only the receptive fields of locust L-cells have been tested, and these included the entire 130° or more ocellar visual fields (Wilson, M., 1978a).

The postsynaptic responses of ocellar L cells

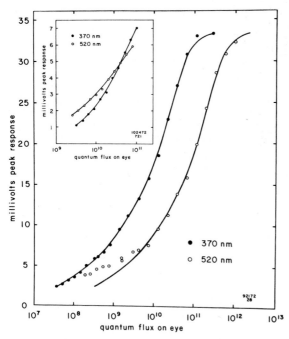

FIG. 33. $V \log I$ curves of dragonfly ocellar photoreceptors for 370 nm (UV) and 520 nm (green) flashes. The solid lines drawn through the points are parallel, but the weaker green responses converged towards the UV responses. The inset shows data from another cell in which $V \log I$ curves crossed, the cell being more green-sensitive at low intensities and more UV-sensitive at high intensities. Reproduced from Chappell, R. and DeVoe, R. (1975), *J. Gen. Physiol.* **65**, 399–419, by permission of the authors and copyright permission of The Rockefeller University Press.

from most species are quite similar to each other, and to the responses of laminar monopolar cells (see section 3.1, above). Figure 34 shows examples of these responses in dragonfly median ocelli and of the cells from which they were recorded (Mobbs, P. *et al.*, 1981). After a 3–20 ms latency there is a peak of hyperpolarization at ON; a sustained plateau (which may be hyperpolarizing, zero, or even depolarizing (Wilson, M., 1978a)), and an OFF-depolarization. The relative amplitudes of these components vary with light intensity. There may be, in addition, fast and slow transients or oscillations at ON or OFF, or quite commonly OFF-spikes (other spikes will be considered in section 6.3, below).

The sign-inverting, receptor-to-L cell synapse is undoubtedly chemical, as indicated by conductance increases in locust L cells during hyperpolarizations. Reversal potentials for the ON-peak and plateau are about −73 mV (for a resting potential of −32 mV), and the reversal potential is about

0 mV for the OFF-depolarization (during which there is also a conductance increase) (Wilson, M., 1978b). L cells of dragonfly ocelli have reversal potentials of −10 to −15 mV relative to resting potential (Simmons, P., 1982). The synapse in dragonflies seems to be cholinergic: the L cell's response is reversibly blocked by curare, and both acetylcholine (ACh) and choline acetyltransferase are found in the ocellus (Klingman, A. and Chappell, R., 1978). Acetylcholinesterase is found at ocellar synapses of the honey bee (Kral, K., 1980), which makes it possible that cholinergic transmission is a general ocellar property.

In addition to receptor-to-L cell synapses, there are reciprocal receptor–receptor, L-to-L and L cell–receptor chemical synapses in locust and dragonfly ocelli. These seem to have roles in shaping the response waveforms of receptor and L cells. Thus, curare blocks OFF-hyperpolarizations in axons of dragonfly receptor cells, implying that these hyperpolarizations arose by feedback from (now blocked) L cells. GABA antagonists potentiate OFF-hyperpolarizations of receptors during a period when OFF-depolarizations of L cells are increased and prolonged (the latter are ultimately blocked: Klingman, A. and Chappell, R., 1978; Stone, S. and Chappell, R., 1981). These findings have led to a dynamic model in which (1) ACh released from receptors inhibits other receptors as well as L cells, and (2) GABA released from L cells excites receptors and other L cells (Stone, S. and Chappell, R., 1982). In the dark, it is proposed that there is partial excitation of R cells by GABA from L cells and partial inhibition of L cells by ACh from receptors. At light-ON, there is enhanced ACh release, which results in L cell hyperpolarization, followed by receptor–receptor inhibition plus cessation of L cell-to-receptor excitation. The latter leads to the cutback in L cell hyperpolarizations. At light-OFF, ACh release stops, the L cells are no longer inhibited and depolarize, while receptors are not partially excited and hyperpolarize. Depolarized L cells release GABA and excite each other plus receptors, leading to OFF-conductance increases in L cells and oscillations in potentials as dark conditions are restored.

Using simultaneous recordings from photoreceptors and L-cells in dragonflies, Simmons, P. (1982) illustrated that hyperpolarization of an L-cell by

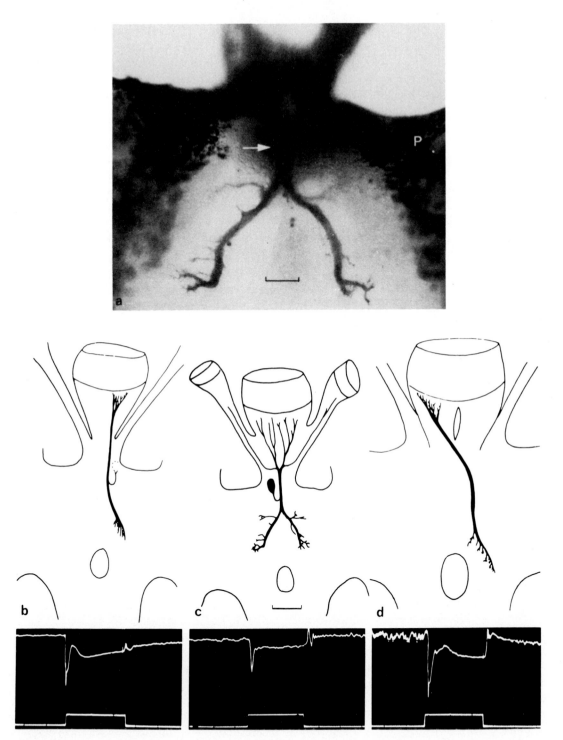

FIG. 34. Structure and responses of three L neurons of the dragonfly median ocellus. In **a** is a cobalt-filled medial-bilateral (MB) neuron with an arrow pointing to its (out-of-focus) soma; the entire cell is drawn in **c**. P:pigment sheath. The shutter traces (bottom) of 500 ms flashes also indicate 3 mV calibrations. Reproduced from Mobbs, P. *et al.* (1981), *J. Comp. Physiol. 144*, 91–97, by permission of the authors and copyright permission of Springer-Verlag, Heidelberg.

current elicited a small, slow hyperpolarization of the receptor. Although his results are consistent with a reduction of excitatory GABA output from the L-cell in the above model, he concluded that there was no feedback from L-cells to receptors. He also found no excitatory connections between L-cells (as the model supposes), only inhibitory connections or none at all. In honey bee ocelli no reciprocal synapses are found, yet there are also prominent OFF-oscillations in L-cells (Guy, R. *et al.*, 1979). Such considerations suggest that the reciprocal synapse model may have to be modified, or that additional synaptic mechanisms are at work. For example, the plateau responses of L-cells are sometimes depolarizing when intense lights are used (Chappell, R. and Dowling, J., 1972; Simmons, P., 1982), just as for contamination of plateau responses of LMCs by laminar positive potentials in laminar penetrations (see Fig. 15). Possibly, extracellular DC potentials exist at ocellar synapses too and play some role in the cutback of the L-cells' responses, as for LMCs (see section 3.1.1). This cutback occurs even if only one ocellar receptor (stimulated by current) excites an L-cell, however, (Simmons, P., 1982), and for this excitation the size of an extracellular field potential should be small. Thus there must also be other mechanisms for cutback. Wilson, M. (1978b) proposed that small (S) postsynaptic fibers might be involved in feedback at ocellar synapses of locusts. The pharmacological experiments of Stone, S. and Chappell, R. (1981) are not inconsistent with the participation in a feedback loop of cells other than receptors and L-cells. Despite electrical similarities, there could thus be profound differences between synaptic mechanisms in ocelli and in the lamina.

Finally, ocellar synapses have gains of about 4 in locusts (Patterson, J. and Goodman, L., 1974a) and as much as 9 in dragonflies (Simmons, P., 1982). As a result, L cells have 80% of their maximum amplitudes of response at intensities where receptor responses are only about 15% of maximum. This makes L cells 1 to 2 log units more sensitive (Mobbs, P. *et al.*, 1981).

6.3 Information transmission in ocellar nerves

L and S neurons carry the information traffic in ocellar nerves; receptor axons stop in the ocellar cup. Almost all identified neuronal traffic has been in L cells. Various of these project to the posterior slope of the protocerebrum, or directly to thoracic ganglia (in honey bees), or to other ocelli. Hence L cells can be afferent or efferent. In addition, L cells make axo-axonic contacts with each other within or at the roots of the ocellar nerves.

Ocellar L cells are among the largest in the insect nervous system. Transmission of hyperpolarizations from L cells' dendrites to terminals is therefore most probably by passive electrotonic spread, with only 40–60% decrement in locusts (Wilson, M., 1978b). The amount of plateau hyperpolarization of locust medial L cells can be reduced or increased by simultaneous illumination of the lateral ocelli. Reduction of plateau hyperpolarization is from summations with depolarizations produced by a conductance increase mechanism, which is blocked by picrotoxin (Taylor, C., 1981). These interactions might have occurred by the direct medial-to-lateral L cells (ML_{1-2}) or perhaps by the axo-axonic contacts. Inhibitions between ocelli would be advantageous for emphasizing differential stimulations of the separate ocelli. That is, differential stimulation of the lateral ocelli alone in dragonflies elicits behavioral roll responses of the head, whereas illumination of the median ocellus alone elicits pitch responses of the head (Stange, G., 1981).

Comparisons of L cell responses, head pitch responses, and photoreceptor responses in dragonfly median ocelli show that the reverse Purkinje shift in receptors is propagated to the behavioral level. At ranges of light intensities where intensity–response curves at 370 and 520 nm cross for photoreceptors (Fig. 33: 10.9–13.3 log quanta s^{-1} cm^{-2}), they also cross for L cells. This is shown in Fig. 35, and is true for all median ocellar L cells (Mobbs, P. *et al.*, 1981). Because of gain at the ocellar synapses, crossover occurs at 80% of the maximum response for L cells, versus about 15% for receptors. At threshold, then, such gain makes the L cells much more sensitive to green than UV light. These sensitivities in turn seem to effect the head pitch responses, which are primarily green-sensitive at threshold. As intensities are increased from threshold, however, the head pitch responses become more sensitive to UV than to green light (in "silent substitution" experiments). The crossover intensity is about the same (11–12 log quanta

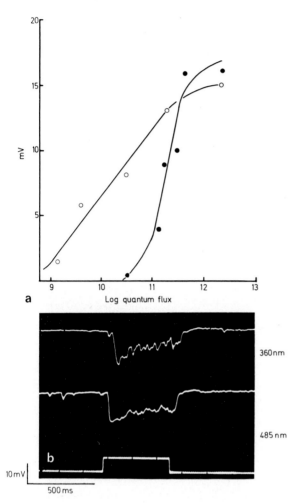

a

b

10 mV

500 ms

FIG. 35. Crossing V log I curves and responses near crossover of dragonfly ocellar L neurons for UV and green stimuli. The UV and green V log I curves cross at 11.5 log quanta s^{-1} cm^{-2}. Reproduced from Mobbs, P. *et al.* (1981), *J. Comp. Physiol.* *144*, 91–97, by permission of the authors and copyright permission of Springer-Verlag, Heidelberg.

s^{-1} cm^{-2}) as for receptors and L cells (Stange, G., 1981).

L cells also have spikes, most commonly at light-OFF but sometimes at light-ON or during steady illumination. Examples from identified L neurons of the honey bee are shown in Fig. 36 (Milde, J., 1981). Similar to neurons of the fly's lobula plate (Hausen, K., 1982a), spiking in L cells is labile (Chappell, R. and Dowling, J., 1972) and may be dependent on hyperpolarizations or on a sufficiently negative resting potential. Depolarization by current does not elicit spikes in locust L cells, but anodal break does (Wilson, M., 1978b).

Frequencies of spontaneous firing of honey bee L cells are maximum at a resting potential of -56 mV, and amplitudes of spikes overshoot zero potential at resting potentials of -60 mV or more. The result is that lights of intermediate intensities can elicit hyperpolarizations that are sufficient to maintain spike discharges, whereas stronger flashes elicit greater hyperpolarizations which abolish the spikes (Milde, J., 1981). Cockroach L cells fire 1–5 OFF-spikes, and the maximum number occurs with the intermediate intensities that elicit the largest plateau hyperpolarizations (Mizunami, M. *et al.*, 1982). This would be consistent with dependencies of spiking on hyperpolarization and anodal break.

There is no evidence where the spikes arise in honey bee L cells. They have been hypothesized to originate halfway along L cells of dragonfly medial ocelli (Rosser, B., 1974) and at L cell terminals in locusts, from which they are propagated antidromically toward the ocellus (Wilson, M., 1978b).

Additional spiking cells, mostly unidentified, have been recorded in ocellar nerves. One spiking S cell was identified in the dragonfly ocellar nerve (Mobbs, P. *et al.*, 1981). Light-excitatory (as well as inhibitory) spiking cells have been recorded, but not identified, in ocellar nerves of locusts (Wilson, M.,

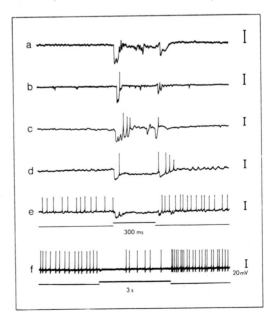

FIG. 36. Patterns of spiking in ocellar L cells of ocelli of honey bees. Reproduced from Milde, J. (1981), *J. Comp. Physiol.* *143*, 427–435, by permission of the author and copyright permission of Springer-Verlag, Heidelberg.

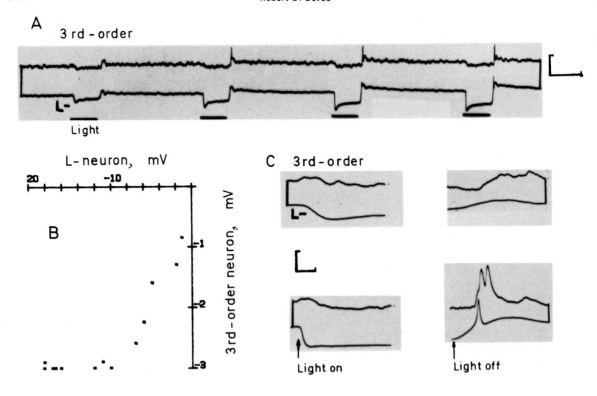

FIG. 37. Synaptic connections between an ocellar L neuron and an identified third-order, descending neuron (03) in the locust. In **A**, light intensities increase from left-to-right. **B** shows the synaptic input–output relations. The starts and ends of the first and fourth responses in **A** are shown at higher time resolution in **C**. Reproduced from Simmons, P. (1981), *J. Comp. Physiol. 145*, 265–276, by permission of the author and copyright permission of Springer-Verlag, Heidelberg.

1978a) and honey bees (Guy, R. *et al.*, 1979). Possibly they too are S cells. In dragonflies there is efferent input to lateral ocellar nerves from mechanoreceptors of the wing joints and from the compound eyes. The wing-joint efferents excite discharges in an unidentified, "large" afferent fiber (Kondo, H., 1978).

6.4 Ocellar connections in the brain and ventral nerve cord

Ocellar L cells terminate in two or more synaptic regions on each side of the posterior slope. In dragonflies, L cells from lateral ocelli terminate on either the ipsilateral or the contralateral posterior slope, while each half of the median ocellus has L cells which project either ipsilaterally or contralaterally (see Fig. 34; Mobbs, P. *et al.*, 1981). Locust lateral ocellar L cells project only ipsilaterally, and there is also one median ocellar L cell projecting to each side. Contact is made by these L cells to neurons which descend in either the ipsilateral or contralateral circumesophageal connectives (Guy, R. *et al.*, 1977). As mentioned above, five (of nine) L cells from lateral ocelli of the honey bee project directly to the thoracic ganglia (Goodman, L., 1981). Median ocelli of dragonflies, locusts, and the honey bee also have one bilateral L cell which synapses onto both sides of the posterior slope (see Fig. 34a,c).

Consistent with the above are recordings from the locust ventral nerve cord of units driven by the ipsilateral ocellus, the contralateral ocellus, or either plus the median ocellus. No units were found driven by the median ocellus alone (Guy, R. *et al.*, 1977). Tonic units inhibited by light on one lateral ocellus could be weakly excited by light on the other (Patterson, J. and Goodman, L., 1974b). Perhaps this occurred because these were depolarizations induced in median ocellar L cells by illumination of lateral ocelli (Taylor, C., 1981; see the preceding section), and these depolarizations in turn exited the unit in the ventral nerve cord.

Ocellar L cells terminate in close proximity to

terminals of giant lobular cells. In the locust,
depolarizations of L cells (at, e.g., light-OFF) elicit
EPSPs in the DCMD and can boost its responsiveness (Simmons, P., 1981a). Another identified,
descending, multimodal neuron, 03, is directly
driven by L cells via a sign-conserving, non-
adapting, conductance-increase synapse. This is
illustrated in Fig. 37. The third-order cell, 03,
follows the L cell's potential with a gain of 0.5, but
only for L cell hyperpolarizations of less than
$-10\,mV$. OFF-spikes in an L cell are effective in
depolarizing 03, but not more than are equal slow
depolarizations (Simmons, P., 1981b). Thus
regenerative OFF-spikes in L cells may serve
primarily to speed up OFF-responses in descending
neurons and so excite them sooner.

Finally, there may be ocellar interactions with the
compound eyes in addition to those with the
DCMD, or from efferent inputs. In fast walking
phototaxis at thresholds for the peripheral retinal
cells of *Drosophila*, anocellate mutants lacked
positive phototaxis (Miller, G. *et al.*, 1981). How
this might be mediated is, however, not known.

REFERENCES

ADRIAN, E. D. (1937). Synchronized reactions in the optic ganglion of *Dytiscus. J. Physiol. 91*, 66–89.

ALAWI, A. A. and PAK, W. L. (1971). On-transient of insect electroretinogram: its cellular origin. *Science 172*, 1055–1057.

ARNETT, D. W. (1971). Receptive field organization of units in the first optic ganglion of diptera. *Science 173*, 929–931.

ARNETT, D. W. (1972). Spatial and temporal integration properties of units in first optic ganglion of dipterans. *J. Neurophysiol. 35*, 429–444.

AUTRUM, H. (1950). Die Belichtungspotentiale and das Sehen der Insekten (Untersuchungen am *Calliphora* and *Dixippus*). *Z. Vergl. Physiol. 32*, 176–227.

AUTRUM, H. and HOFFMAN, C. (1960). Diphasic and monophasic responses in the compound eye of *Calliphora. J. Insect Physiol. 4*, 112–127.

AUTRUM, H. and STUMPF, H. (1950). Das Bienenauge als Analysator für polarisiertes Licht. *Z. Naturf. 56*, 116–122.

AUTRUM, H., ZETTLER, F. and JÄRVILEHTO, M. (1970). Postsynaptic potentials from a single monopolar neuron of the ganglion opticum I of the blowfly *Calliphora. Z. Vergl. Physiol. 70*, 414–424.

AUTRUM, H. and KOLB, G. (1972). The dark adaptation in single visual cells of the compound eye of *Aeschna cyanea. J. Comp. Physiol. 79*, 213–232.

BADER, C., BAUMANN, F. and BERTRAND, D. (1976). Role of intracellular calcium and sodium in light adaptation in the retina of the honey bee drone (*Apis mellifera*, L.). *J. Gen. Physiol. 67*, 475–491.

BARNES, S. W. and GOLDSMITH, T. H. (1977). Dark adaptation, sensitivity, and rhodopsin level in the eye of the lobster, *Homarus. J. Comp. Physiol. 120*, 143–159.

BASSEMIR, U., STRAUSFELD, N. J., BACON, J. P. and SINGH, R. N. (1982). Electronmicroscopy resolves special junctions in "transynaptic" cobalt-coupled nerve cells in giant fibre systems (GFS) of flies. Abstracts, Soc. Neurosci. 12th Annual Meeting, p. 530.

BAUMANN, F. (1968). Slow and spike potentials recorded from the retinula cells of the honey bee drone in response to light. *J. Gen. Physiol. 52*, 855–875.

BAUMANN, F. (1975). Electrophysiological properties of the honey bee retina. In *The Compound Eye and Vision of Insects*. Edited by G. A. Horridge. Pages 53–74. Clarendon Press, Oxford.

BAUMANN, F. and HADJILAZARO, B. (1972). A depolarizing aftereffect of intense light in the drone visual receptor. *Vision Res. 12*, 17–31.

BAUMANN, F. and MAURO, A. (1973). Effect of hypoxia on the change in membrane conductance evoked by illumination in arthropod photoreceptors. *Nature (New Biol.) 244*, 146–148.

BAYLOR, D. A., LAMB, T. D. and YAU, K. W. (1979). Responses of retinal rods to single photons. *J. Physiol. 288*, 613–634.

BELMONTE, C. and STENSAAS, L. J. (1975). Repetitive spikes in photoreceptor axons of the scorpion eye. *J. Gen. Physiol. 66*, 649–655.

BENNETT, R. R., TUNSTALL, J. and HORRIDGE, G. A. (1967). Spectral sensitivity of single retinula cells of the locust. *Z. Vergl. Physiol. 55*, 195–206.

BENOLKEN, R. M. (1965). Regenerative transducing properties of a graded visual response. *Cold Spring Harbor Symp. Quant. Biol. 30*, 445–450.

BERNARD, G. D. and STAVENGA, D. G. (1978). Spectral sensitivities of retinular cells measured in intact, living bumblebees by an optical method. *Naturwissenchaften 65*, 442–443.

BERNARD, G. D. and WEHNER, R. (1980). Intracellular optical physiology of the bee's eye. I. Spectral sensitivity. *J. Comp. Physiol. 137*, 193–203.

BERNHARD, C. G. (1942). Isolation of retinal and optic ganglion response in the eye of *Dytiscus. J. Neurophysiol. 5*, 32–48.

BERNHARD, C. G., HÖGLUND, G. and OTTOSON, D. (1963). On the relation between pigment position and light sensitivity of the compound eye in different nocturnal insects. *J. Insect Physiol. 9*, 573–586.

BISHOP, C. and BISHOP, L. G. (1981). Vertical motion detectors and their synaptic relations in the third optic lobe of the fly. *J. Neurophysiol. 12*, 281–296.

BISHOP, L. G. and KEEHN, D. G. (1966). Two types of neurons sensitive to motion in the optic lobe of the fly. *Nature 212*, 1374–1376.

BISHOP, L. G., KEEHN, D. G. and McCANN, G. D. (1968). Motion detection by interneurons of optic lobes and brain of the flies *Calliphora phaenicia* and *Musca domestica. J. Neurophysiol. 31*, 509–525.

BLEST, A. D. and COLLETT, T. S. (1965a). Micro-electrode studies of the medial protocerebrum of some lepidoptera. I. Responses to simple, binocular visual stimulation. *J. Insect Physiol. 11*, 1079–1103.

BLEST, A. D. and COLLETT, T. S. (1965b). Micro-electrode studies of the medial protocerebrum of some lepidoptera. II. Responses to visual flicker. *J. Insect Physiol. 11*, 1289–1306.

BLONDEAU, J. (1981). Electrically evoked course control in the fly *Calliphora erythrocephala. J. Exp. Biol. 92*, 143–153.

BORSELLINO, A. and FUORTES, M. G. F. (1968). Responses to single photons in visual cells of Limulus. *J. Physiol. 196*, 507–539.

BROWN, H. M., HAGIWARA, S., KOIKE, H. and MEECH, R. M. (1970). Membrane properties of a barnacle photoreceptor examined by the voltage clamp technique. *J. Physiol. 208*, 385–413.

BROWN, J. E., MULLER, K. J. and MURRAY, G. (1971). Reversal potential for an electrophysiological event generated by conductance change: Mathematical analysis. *Science 174*, 318.

BROWN, J. E. and MOTE, M. I. (1974). Ionic dependence of reversal voltage of the light response in Limulus ventral photoreceptors. *J. Gen. Physiol. 63*, 377–350.

BROWN, K. T. and MURAKAMI, M. (1964). Early receptor potential of the vertebrate retina. *Nature 204*, 736–740.

BUCHNER, E. (1976). Elementary movement detectors in an insect visual system. *Biol. Cybernetics 24*, 85–101.

BUCHNER, E. and BUCHNER, S. (1980). Mapping stimulus-induced nervous activity in small brains by H³ 2-Deoxy-D-glucose. *Cell Tiss. Res. 211*, 51–64.

BURKHARDT, D. and AUTRUM, H. (1960). Die Belichtungspotentiale einzelner Sehzellen von *Calliphora erythrocephala* meig. *Z. Naturf. 15*, 612–616.

BURKHARDT, D. and WENDLER, L. (1960). Ein direkter Beweis fur die Fähigkeit einzelner Sehzellen des Insektenauges, die Schwingungsrichtung polarisierten Lichtes zu analysieren. *Z. Vergl. Physiol. 43*, 687–692.

BURKHARDT, D., DE LA MOTTE, I. and SEITZ, G. (1966). Physiological Optics of the Compound Eye of the Blow Fly. In *The Functional Organization of the Compound Eye.* Edited by C. G. Bernhard. Pages 51–62. Pergamon Press, Oxford.

BURTT, E. T. and CATTON, W. T. (1954). Visual perception of movement in the Locust, *J. Physiol. 125*, 566–580.

BURTT, E. T. and CATTON, W. T. (1956). Electrical responses to visual stimulation in the optic lobes of the locust and certain other insects. *J. Physiol. 133*, 68–88.

BURTT, E. T. and CATTON, W. T. (1960). The properties of single-unit discharges in the optic lobe of the locust. *J. Physiol. 154*, 479–490.

BURTT, E. T. and CATTON, W. T. (1964). The potential profile of the insect compound eye and optic lobe. *J. Insect Physiol. 10*, 689–710.

BUTLER, R. and HORRIDGE, G. A. (1973a). The electrophysiology of the retina of *Periplaneta americana* L. 1. Changes in receptor acuity upon light and dark adaptation. *J. Comp. Physiol. 83*, 263–278.

BUTLER, R. and HORRIDGE, G. A. (1973b). The electrophysiology of the retina of *Periplaneta americana* L. 2. Receptor sensitivity and polarized light sensitivity. *J. Comp. Physiol. 83*, 279–288.

CAMPOS-ORTEGA, J. A. (1974). Autoradiographic localization of ^{3}H-γ-aminobutyric acid uptake in the lamina ganglionaris of *Musca* and *Drosophila. Z. Zellforsch. 147*, 415–431.

CHAPPELL, R. L. and DEVOE, R. D. (1975). Action spectra and chromatic mechanisms of cells in the median ocelli of dragonflies. *J. Gen. Physiol. 65*, 399–419.

CHAPPELL, R. L. and DOWLING, J. E. (1972). Neural organization of the median ocellus of the dragonfly. I. Intracellular electrical activity. *J. Gen. Physiol. 60*, 121–147.

CHI, C. and CARLSON, S. D. (1976). High voltage electron microscopy of the optic neuropile of the housefly, *Musca domestica. Cell. Tiss. Res. 167*, 537–545.

CHI, C., CARLSON, S. D. and ST. MARIE, R. L. (1979). Membrane specializations in the peripheral retina of the housefly *Musca domestica. Cell Tiss. Res. 198*, 501–520.

COHN, T. E., COTTERAL, R., and HARE, W. (1983). Noise in dark adapted locust retinular cell responses. *Invest. Opthal. Vis. Sci. 24*, (Suppl), 217.

COLES, J. A. and ORKAND, R. K. (1982). Sodium activity in drone photoreceptors. *J. Physiol. 332*, 16p–17p.

COLES, J. A. and TSACOPOULOS, M. (1979). Potassium activity in photoreceptors, glial cells and extracellular space in the drone retina: Changes during photostimulation. *J. Physiol. 290*, 525–549.

COLLETT, T. (1970). Centripetal and centrifugal visual cells in medulla of the insect optic lobe. *J. Neurophysiol. 33*, 239–255.

CONE, R. A. (1965). The early receptor potential of the vertebrate eye. *Cold Spring Harbor Symp. Quant. Biol. 30*, 483–491.

CONE, R. A. (1967). Early receptor potential: Photoreversible charge displacement in rhodopsin. *Science 155*, 1128–1131.

COSENS, D. J. (1966). Visual sensitivity in the light- and dark-adapted compound eye of the desert locust. *J. Insect Physiol. 12*, 871–890.

COSENS, D. J. (1967). Extracellular potentials in the locust eye and optic lobe. *J. Insect Physiol. 13*, 1373–1386.

COSENS, D. and LEBLANC, N. (1980). Frequency dependent flicker response enhancement in the lamina ganglionaris of *Drosophila. J. Comp. Physiol. 137*, 341–351.

COSENS, D. and SPATZ, H. C. (1978). Flicker fusion studies in the lamina and receptor region of the *Drosophila* eye. *J. Insect Physiol. 24*, 587–593.

DARTNALL, H. J. A. (1952). The interpretation of spectral sensitivity curves. *Brit. Med. Bull. 9*, 24–30.

DARTNALL, H. J. A. (1968). The photosensitivities of visual pigments in the presence of hydroxylamine. *Vision Res. 8*, 339–358.

DAWIS, S. and PURPLE, R. L. (1982). Adaptation in cones: a general model. *Biophys. J. 39*, 151–155.

DEVOE, R. D. (1963). Linear relations between stimulus amplitudes and amplitudes of retinal action potentials from the eye of the wolf spider. *J. Gen. Physiol. 47*, 13–32.

DEVOE, R. D. (1967a). Nonlinear transient responses from light-adapted wolf spider eyes to changes in background illumination. *J. Gen. Physiol. 50*, 1961–1991.

DEVOE, R. D. (1967b). A nonlinear model for transient responses from light-adapted wolf spider eyes. *J. Gen. Physiol. 50*, 1993–2030.

DEVOE, R. D. (1972). Dual sensitivities of cells in wolf spider eyes at ultraviolet and visible wavelengths of light. *J. Gen. Physiol. 59*, 247–269.

DEVOE, R. D. (1980). Movement sensitivities of cells in the fly's medulla. *J. Comp. Physiol. 138*, 93–119.

DEVOE, R. D. and OCKLEFORD, E. M. (1976). Intracellular responses from cells of the medulla of the fly, *Calliphora erythrocephala. Biol. Cybernet. 23*, 13–24.

DEVOE, R. D., KAISER, W., OHM, J. and STONE, L. S. (1982). Horizontal movement detectors in honey bees: Directionally-selective visual neurons in the lobula and brain. *J. Comp. Physiol. 147*, 155–170.

DINGLE, H. and FOX, S. S. (1966). Microelectrode analysis of light responses in the brain of the cricket (*Gryllus domesticus*). *J. Cell. Physiol. 68*, 45–60.

DODGE, F. A., JR., KNIGHT, B. W. and TOYODA, J. (1968). Voltage noise in *Limulus* visual cells. *Science 160*, 88–90.

DOVING, K. B. and MILLER, W. H. (1969). Function of insect compound eyes containing crystalline tracts. *J. Gen. Physiol. 54*, 250.

DUBS, A. (1981). Non-linearity and light adaptation in the fly photoreceptor. *J. Comp. Physiol. 144*, 53–59.

DUBS, A. (1982). The spatial integration of signals in the retina and lamina of the fly compound eye under different conditions of luminance. *J. Comp. Physiol. 146*, 321–343.

DUBS, A., LAUGHLIN, S. B. and SRINIVASAN, M. V. (1981). Single photon signals in fly photoreceptors and first order interneurones at behavioral threshold. *J. Physiol. 317*, 317–334.

DUDEK, F. E. and KOOPOWITZ, H. (1973). Adaptation and temporal characteristics of the insect visual response. *J. Comp. Physiol. 82*, 33–46.

DVORAK, D., BISHOP, L. G. and ECKERT, H. E. (1975). On the identification of movement detectors in the fly optic lobe. *J. Comp. Physiol. 100*, 5–23.

EATON, J. L. (1976). Spectral sensitivity of the ocelli of the adult cabbage looper moth, *Trichoplusia ni. J. Comp. Physiol. 109*, 17–24.

ECKERT, H. (1980). Functional properties of the H1 neurone in the third optic ganglion of the blowfly, *Phaenicia. J. Comp. Physiol. 135*, 29–39.

ECKERT, H. (1981). The horizontal cells in the lobula plate of the blowfly *Phaenicia sericata. J. Comp. Physiol. 143*, 511–526.

ECKERT, H. (1982). The vertical–horizontal neurone (VH) in the lobula plate of the blowfly, *Phaenicia. J. Comp. Physiol. 149*, 195–205.

ECKERT, H. and BISHOP, L. G. (1975). Nonlinear dynamic transfer characteristics of cells in the peripheral visual pathway of flies. *Biol. Cybernetics 17*, 1–6.

ECKERT, H. and BISHOP, L. G. (1978). Antomical and physiological properties of the vertical cells in the third optic ganglion of *Phaenicia sericata* (Diptera, Calliphoridae). *J. Comp. Physiol. 126*, 57–86.

ECKERT, H. and HAMDORF, K. (1981). The contrast frequency-dependence: A criterion for judging the non-participation of neurones in the control of behavioral responses. *J. Comp. Physiol. 145*, 241–247.

EICHENBAUM, D. M. and GOLDSMITH, T. H. (1968). Properties of intact photoreceptor cells lacking synapses. *J. Exp. Zool. 169*, 15.

ERBER, J. (1978). Response characteristics and after effects of multimodal neurons in the mushroom body area of the honey bee. *Physiol. Ent. 3*, 77–89.

FAIN, G. (1975). Quantum sensitivity of rods in the toad retina. *Science 187*, 838.

FEIN, A. and CONE, R. A. (1973). Limulus rhodopsin: Rapid return of transient intermediates to the thermally stable state. *Science 182*, 495.

FEIN, A. and DEVOE, R. D. (1973). Adaptation in the ventral eye of *Limulus* is functionally independent of the photochemical cycle, membrane potential, and membrane resistance. *J. Gen. Physiol. 61*, 273–289.

FEIN, A. and CHARLTON, J. S. (1977a). A quantitative comparison of the effects of intracellular calcium injection and light adaptation on the photoresponse of *Limulus* ventral photoreceptors. *J. Gen. Physiol. 70*, 591–600.

FEIN, A. and CHARLTON, J. S. (1977b). Increased intracellular sodium mimics some but not all aspects of photoreceptor adaptation in the ventral eye of *Limulus. J. Gen. Physiol. 70*, 601–620.

FEIN, A. and CHARLTON, J. S. (1978). A quantitative comparison of the time-course of sensitivity changes produced by calcium injection and light adaptation in *Limulus* ventral photoreceptors. *Biophys. J. 22*, 105–113.

FEIN, A. and LISMAN, J. (1975). Localized desensitization of *Limulus* photoreceptors produced by light or intracellular calcium ion injection. *Science 187*, 1094–1096.

FEIN, A. and CORSON, D. W. (1981). Excitation of *Limulus* photoreceptors by vanadate and by a hydrolysis resistant analog of guanosine triphosphate. *Science 212*, 555–557.

FEIN, A. and SZUTS, E. Z. (1982). *Photoreceptors: Their Role in Vision* (IUPAB biophysics series: 5). Cambridge University Press, Cambridge.

FERMI, G. and REICHARDT, W. (1963). Optomotorische Reaktionen der Fliege *Musca domestica. Kybernetik 2*, 15–28.

FRANCESCHINI, N. (1975). Sampling of the visual environment by the compound eye of the fly: fundamentals and applications. In *Photoreceptor Optics*. Edited by A. W. Snyder and R. Menzel. Pages 98–125. Springer, Berlin, Heidelberg and New York.

FRENCH, A. (1980). Phototransduction in the fly compound eye exhibits temporal resonances and a pure time delay. *Nature 283*, 200–202.

FRENCH, A. S. and JÄRVILEHTO, M. (1978a). The transmission of information by first and second order neurons in the fly visual system. *J. Comp. Physiol. 126*, 87–96.

FRENCH, A. S. and JÄRVILEHTO, M. (1978b). The dynamic behaviour of photoreceptor cells in the fly in response to random (white noise) stimulation at a range of temperatures. *J. Physiol. 274*, 311–322.

FULPIUS, B. and BAUMANN, F. (1969). Effects of sodium, potassium, and calcium ions on slow and spike potentials in single photoreceptor cells. *J. Gen. Physiol. 53*, 541–561.

FUORTES, M. G. F. (1959). Initiation of impulses in visual cells of *Limulus*. *J. Physiol. 148*, 14–28.

FUORTES, M. G. F. (1963). Visual responses in the eye of the dragon fly. *Science 142*, 69.

FUORTES, M. G. F. and HODGKIN, A. L. (1964). Changes in time scale and sensitivity in the ommatidia of *Limulus. J. Physiol. 172*, 239.

FUORTES, M. G. F. and YEANDLE, S. (1964). Probability of occurrence of discrete potential waves in the eye of *Limulus. J. Gen. Physiol. 47*, 443–463.

FUORTES, M. G. F. and O'BRYAN, P. M. (1972). Generator potentials in invertebrate photoreceptors. In *Physiology of Photoreceptors*. Edited by M. G. Fuortes. Pages 279–319. *Handbook of Sensory Physiology* II/2. Springer, Berlin, Heidelberg and New York.

GEIGER, G. and NÄSSEL, D. R. (1981). Visual orientation behavior of flies after selective laser beam ablation of interneurones. *Nature 293*, 398–399.

GEMPERLEIN, R. and SMOLA, U. (1972). Übertragungseigenschaften der Sehzelle der Schmeisfliege *Calliphora erythrocephala*. 3. Verbesserung des Signal-Störungs-Verhältnisses durch präsynaptische Summation in der Lamina ganglionaris. *J. Comp. Physiol. 79*, 393.

GEMPERLEIN, R. and SMOLA, U. (1973). Die Wirkung linear polarisierten Lichtes auf die Sehzellen von *Calliphora erythrocephala. J. Comp. Physiol. 87*, 285.

GEMPERLEIN, R. and McCANN, G. D. (1975). A study of the response properties of retinula cells of flies using nonlinear identification theory. *Biol. Cybernet. 19*, 147–158.

GIULIO, L. and LUCARONI, A. (1967). Test flash interaction in the electroretinographic response of the compound eye. *Experientia 23*, 542.

GLANTZ, R. M. (1971). Toward a general theory of visual adaptation. *Documenta Ophthalmologica 30*, 245–258.

GOLDSMITH, T. (1960). The nature of the retinal action potential, and the spectral sensitivities of ultraviolet and green receptor systems of the compound eye of the worker honeybee. *J. Gen. Physiol. 43*, 775–799.

GOLDSMITH, T. H. (1963). The course of light and dark adaptation in the compound eye of the honey bee. *Comp. Biochem. Physiol. 10*, 227–237.

GOLDSMITH, T. H. and RUCK, P. R. (1958). The spectral sensitivities of the dorsal ocelli of cockroaches and honeybees. *J. Gen. Physiol. 41*, 1171–1185.

GOLDSMITH, T. H. and WEHNER, R. (1977). Restrictions on rotational and translational diffusion of pigment in the membranes of a rhabdomeric photoreceptor. *J. Gen. Physiol. 70*, 453–490.

GOODMAN, L. J. (1981). Organization and physiology of the insect dorsal ocellar system. In *Comparative Physiology and Evolution of Vision in Invertebrates*. Edited by H. Autrum. *Handbook of Sensory Physiology*. Vol. VII/6C. Pages 201–286. Springer, Berlin, Heidelberg and New York.

GORMAN, A. L. F. and McREYNOLDS, J. S. (1978). Ionic effects on the membrane potential of hyperpolarizing photoreceptors in scallop retina. *J. Physiol. 275*, 345–355.

GÖTZ, K. G. (1964). Optomotorische Untersuchung des visuellen Systems einiger Augenmutanten der Fruchtfliege *Drosophila. Kybernetik 2*, 77–92.

GUY, R. G., GOODMAN, L. J. and MOBBS, P. G. (1977). Ocellar connections with the ventral nerve cord in the locust, *Schistocerca gregaria*: Electrical and anatomical characteristics. *J. Comp. Physiol. 115*, 337–350.

GUY, R. G., GOODMAN, L. J. and MOBBS, P. G. (1979). Visual interneurones in the bee brain: synaptic organization and transmission by graded potentials. *J. Comp. Physiol. 134*, 253–264.

HAGINS, W. A. (1965). Electrical signs of information flow in photoreceptors. *Cold Spring Harbor Symp. Quant. Biol. 30*, 403–418.

HAGINS, W. A. and McGAUGHY, R. E. (1968). Membrane origin of the fast photovoltage of squid retina. *Science 159*, 213.

HAMDORF, K. and KIRSCHFELD, K. (1980). Prebumps for double-hits at functional subunits in a rhabdomeric photoreceptor. *Z. Naturf. 35*, 173–174.

HAMDORF, K., HÖGLUND, G. and SCHLECHT, P. (1978). Ion gradient and photoreceptor sensitivity. *J. Comp. Physiol. 125*, 237–252.

HANANI, M. and HILLMAN, P. (1976). Adaptation and facilitation in the barnacle photoreceptor. *J. Comp. Physiol. 67*, 235–249.

HARDIE, R. C. (1977). Electrophysiological properties of R7 and R8 in dipteran retina. *Z. Naturf. 32c*, 887–889.

HARDIE, R. C. (1978). Peripheral visual function in the fly. Ph.D. dissertation, Australian National University, Canberra.

HARDIE, R. C. (1979). Electrophysiological analysis of fly retina. I: Comparative properties of R1–6 and R7 and 8. *J. Comp. Physiol. 129*, 19–33.

HARDIE, R. C., FRANCESCHINI, N. and McINTYRE, P. D. (1979). Electrophysiological analysis of fly retina. II. Spectral and polarization sensitivity in R7 and R8. *J. Comp. Physiol. 133*, 23–39.

HARDIE, R. C., FRANCESCHINI, N., RIBI, W. and KIRSCHFELD, K. (1981). Distribution and properties of sex-specific photoreceptors in the fly *Musca domesticus. J. Comp. Physiol. 145*, 139–152.

HARTLINE, H. K. (1928). A quantitative and descriptive study of the electric response to illumination of the anthropod eye. *Amer. J. Physiol. 83*, 466–483.

HARTLINE, H. K. (1938). The discharge of impulses in the optic nerve of Pecten in response to illumination of the eye. *J. Cell. Comp. Physiol. 11*, 465–478.

HARTLINE, H. K. and RATLIFF, F. (1958). Spatial summation of inhibitory influences in the eye of limulus, and the mutual interaction of receptor units. *J. Gen. Physiol. 41*, 1049–1066.

HARTLINE, H. K., WAGNER, H. G. and MACNICHOL, E. F., JR. (1952). The peripheral origin of nervous activity in the visual system. *Cold Spring Harbor Symp. Quant. Biol. 17*, 125–141.

HAUSEN, K. (1976a). Struktur, Function und Konnektivität bewegungsempfindlicher Interneuronen im dritten optischen Neuropil der Schmeissfliege *Calliphora erythrocephala*. Dissertation, Universität Tübingen.

HAUSEN, K. (1976b). Functional characterization and anatomical identification of motion sensitive neurons in the lobula plate of the blowfly *Calliphora erythrocephala. Z. Naturf. 31c*, 629–633.

HAUSEN, K. (1981). Monocular and binocular computation of motion in the lobula plate of the fly. *Verh. Dtsch. Zool. Ges.*, 49–70.

HAUSEN, K. (1982a). Motion sensitive interneurons in the optomotor system of the fly. I. The horizontal cells: structure and signals. *Biol. Cybernet. 45*, 143–156.

HAUSEN, K. (1982b). Motion sensitive interneurons in the optomotor system of the fly. II. The horizontal cells: Receptive field organization and response characteristics. *Biol. Cybernet. 46*, 67–79.

HAUSEN, K., WOLBURG-BUCHHOLZ, K. and RIBI, W. A. (1980). The synaptic organization of visual interneurons in the lobula complex of flies. *Cell Tiss. Res. 208*, 371–387.

HECHT, S., SCHLAER, S. and PIRENNE, M. H. (1942). Energy, quanta and vision. *J. Gen. Physiol. 45*, 819–840.

HEISENBERG, M. (1971). Separations of receptor and lamina potentials in the electroretinogram of normal and mutant *Drosophila. J. Exp. Biol. 55*, 85–100.

HEISENBERG, M. and BUCHNER, E. (1977). The role of retinula cell types in visual behavior of *Drosophila melanogaster. J. Comp. Physiol. 117*, 127–162.

HEISENBERG, M., WONNEBERGER, R. and WOLF, R. (1978). Optomotor-blind: a *Drosophila* mutant of the lobula plate giant neurons. *J. Comp. Physiol. 124*, 287–296.

HEITLER, W. J., GOODMAN, C. S. and ROWELL, C. H. F. (1977). The effects of temperature on the threshold of identified neurons in the locust. *J. Comp. Physiol. 117*, 163–182.

HENGSTENBERG, R. (1973). The effect of pattern movement on the impulse activity of the cervical connective of *Drosophila melanogaster. Z. Naturf. 28c*, 593–596.

HENGSTENBERG, R. (1977). Spike responses of "non-spiking" visual interneurone. *Nature 270*, 338–340.

HENGSTENBERG, R. (1982). Common visual response properties of giant vertical cells in the lobula plate of the blowfly *Calliphora. J. Comp. Physiol. 149*, 179–193.

HENGSTENBERG, R. and HENGSTENBERG, B. (1980). Intracellular staining of insect neurons with procion yellow. In *Neuroanatomical Techniques*. Edited by N. J. Strausfeld and T. A. Miller. Pages 307–327. Springer, Berlin, Heidelberg and New York.

HENGSTENBERG, R., HAUSEN, K. and HENGSTENBERG, B. (1982). The number and structure of giant vertical cells (VS) in the lobula plate of the blowfly *Calliphora erythrocephala. J. Comp. Physiol. 149*, 163–177.

HERTL, H. (1980). Chromatic properties of identified interneurons in the optic lobes of the bee. *J. Comp. Physiol. 137*, 215–231.

HILLMAN, P., DODGE, F. A., HOCHSTEIN, S. KNIGHT, B. W. and MINKE, B. (1973). Rapid dark recovery of the invertebrate early receptor potential. *J. Gen. Physiol. 62*, 77.

HILLMAN, P., HOCHSTEIN, S. and MINKE, B. (1983). Transduction in invertebrate photoreceptors: Role of pigment bistability. *Physiol. Rev. 63*, 668–772.

HONEGGER, H. W. (1978). Sustained and transient responding units in the medulla of the cricket *Gryllus campestris. J. Comp. Physiol. 125*, 259–266.

HONEGGER, H. W. (1980). Receptive fields of sustained medulla neurons in crickets. *J. Comp. Physiol. 136*, 191–201.

HORRIDGE, G. A. (1965). Extracellular recordings from single neurones in the optic lobe and brain of the locust. In *The Physiology of the Insect Central Nervous System*. Edited by J. E. Treherne and J.W.C. Beaumont. Pages 165–202. Academic Press, New York.

HORRIDGE, G. A. (1969). Unit studies on the retina of dragonflies. *Z. Vergl. Physiol. 62*, 1–37.

HORRIDGE, G. A. and MIMURA, K. (1975). Fly photoreceptors. I. Physical separation of two visual pigments in *Calliphora* retinula cells 1–6. *Proc. R. Soc. Lond. B. 190*, 211.

HORRIDGE, G. A. and TSUKAHARA, Y. (1978). The distribution of bumps in the tail of the locust photoreceptor afterpotential. *J. Exp. Biol. 73*, 1–14.

HORRIDGE, G. A., MIMURA, K. and HARDIE, R. C. (1976). Fly photoreceptors: Angular sensitivity as a function of wavelength and the limits of resolution. *Proc. R. Soc. Lond. B. 194*, 151–177.

HORRIDGE, G. A., MIMURA, K. and TSUKAHARA, Y. (1975). Fly photoreceptors: II. Spectral and polarized light sensitivity in the drone fly *Eristalis. Proc. R. Soc. Lond. B. 190*, 225–237.

HORRIDGE, G. A., MARCELJA, L., JAHNKE, R. and MATIC, T. (1983). Single electrode studies on the retina of the butterfly *Papilio. J. Comp. Physiol. 150*, 271–294.

HOYLE, G. (1955). Functioning of the insect ocellar nerve. *J. Exp. Biol. 32*, 397–407.

HU, K. G., REICHERT, H. and STARK, W. S. (1978). Electrophysiological characterization of *Drosophila* ocelli. *J. Comp. Physiol. 126*, 15–24.

HUDSPETH, A. J., POO, M. M. and STUART, A. E. (1977). Passive signal propagation and membrane properties in median photoreceptors of the giant barnacle. *J. Physiol. 272*, 25–43.

IOANNIDES, A. C. and WALCOTT, B. (1971). Graded illumination potentials from retinula cell axons in the bug *Lethocerus. Z. vergl. Physiol. 71*, 315–325.

ISHIKAWA, S. (1962). Visual response patterns of single ganglion cells in the optic lobe of the silkworm moth, *Bombyx Mori L. J. Insect Physiol. 8*, 485–491.

JAHN, T. L. (1946). The electroretinogram as a measure of wave-length sensitivity to light. *J. N.Y. Ent. Soc. 54*, 1–8.

JAHN, T. L. and CRESCITELLI, F. (1940). Diurnal changes in the electrical response of the compound eye. *Biol. Bull. 78*, 42–52.

JÄRVILEHTO, M. and MORING, J. (1974). Polarization sensitivity of individual retinula cells and neurons of the fly *Calliphora. J. Comp. Physiol. 91*, 387.

JÄRVILEHTO, M. and ZETTLER, F. (1970). Micro-localisation of lamina-located visual cell activities in the compound eye of the blowfly *Calliphora. Z. Vergl. Physiol. 69*, 134–138.

JÄRVILEHTO, M. and ZETTLER, F. (1973). Electrophysiological-histological studies on some functional properties of visual cells and second order neurons of an insect retina. *Z. Zellforsch. 136*, 291–306.

KAISER, W. (1968). Zur Frage des Unterscheidungsvermögens fur Spektral-farben: Eine Untersuchung der Optomotorik der königlichen Glanzfliege, *Phormia regina* meig. *Z. Vergl. Physiol. 61*, 71.

KAISER, W. (1972). A preliminary report on the analysis of the optomotor system of the honey bee — single unit recordings during stimulation with spectral lights. In *Information Processing in the Visual Systems of Arthropods*. Edited by R. Wehner. Pages 167–170. Springer, Berlin, Heidelberg and New York.

KAISER, W. (1974). The spectral sensitivity of the honeybee's optomotor walking response. *J. Comp. Physiol. 90*, 405–408.

KAISER, W. (1975). The relationship between visual movement detection and colour vision in insects. In *The Compound Eye and Vision of Insects*. Edited by G. A. Horridge. Pages 359–377. Clarendon Press, Oxford.

KAISER, W. and BISHOP, L. G. (1970). Directionally selective motion detecting units in the optic lobe of the honey bee. *Z. Vergl. Physiol. 67*, 403–413.

KAISER, W. and LISKE, E. (1974). Optomotor reactions of stationary flying bees during stimulation with spectral lights. *J. Comp. Physiol. 89*, 391–408.

KAISER, W. and STEINER-KAISER, J. (1983). Neuronal correlates of sleep, wakefulness, and arousal in a diurnal insect. *Nature 301*, 707–709.

KAISER, W., SEIDL, R. and VOLLMAR, J. (1977). The participation of all three colour receptors in the phototactic behaviour of fixed walking honey bees. *J. Comp. Physiol. 122*, 27–44.

KAPLAN, E. and BARLOW, R. B. JR. (1980). Circadian clock in *Limulus* brain increases response and decreases noise of retinal photoreceptors. *Nature 286*, 393–395.

KIEN, J. (1975). Neuronal mechanisms subserving directional selectivity in the locust optomotor system. *J. Comp. Physiol. 102*, 337–355.

KIEN, J. and MENZEL, R. (1977a). Chromatic properties of interneurons in the optic lobes of the bee. I. Broad band neurons. *J. Comp. Physiol. 113*, 17–34.

KIEN, J. and MENZEL, R. (1977b). Chromatic properties of interneurons in the optic lobes of the bee. II. Narrow band and colour opponent neurons. *J. Comp. Physiol. 113*, 35–53.

KIRSCHFELD, K. (1961). Quantitative Beziehungen zwischen Lichtreiz und monophasischem Elektroretinogramm bei Russelkäfern. *Z. Vergl. Physiol. 44*, 371–413.

KIRSCHFELD, K. (1965). Discrete and graded receptor potentials in the compound eye of the fly (Musca). In *The Functional Organization of the Compound Eye*. Edited by C. G. Bernhard. Pages 291–307. Pergamon Press, Oxford.

KIRSCHFELD, K. (1967). Die Projektion der optischen Umwelt auf das Raster der Rhabdomere im Komplexauge von *Musca. Exp. Brain Res. 3*, 248–270.

KIRSCHFELD, K. (1972). The visual system of Musca: Studies on optics, structure and function. In *Information Processing in the Visual Systems of Arthropods*. Edited by R. Wehner. Pages 61–74. Springer, Berlin, Heidelberg and New York.

KIRSCHFELD, K. (1983). Are photoreceptors optimal? *Trends Neurosci. 6*, 97–101.

KIRSCHFELD, K. and FRANCESCHINI, N. (1969). Ein Mechanismus zur Steuerung des Lichtflusses in den Rhabdomeren des Komplexauges von *Musca. Kybernetik 6*, 13–22.

KIRSCHFELD, K. and LUTZ, B. (1974). Lateral inhibition in the compound eye of the fly, *Musca. Z. Naturf. 29c*, 95–97.

KIRSCHFELD, K. and LUTZ, B. (1977). The spectral sensitivity of the ocelli of *Calliphora* (Diptera). *Z. Naturf. 32c*, 439.

KIRSCHFELD, K. and VOGT, K. (1980). Calcium ions and pigment migration in fly photoreceptors. *Naturwissenschaften 67*, 516.

KLINGMAN, A. and CHAPPELL, R. L. (1978). Feedback synaptic interaction in the dragonfly ocellar retina. *J. Gen. Physiol. 71*, 157–175.

KOIKE, H., BROWN, H. M. and HAGIWARA, S. (1971). Hyperpolarization of a barnacle photoreceptor membrane following illumination. *J. Gen. Physiol. 57*, 723–737.

KONDO, H. (1978). Efferent system of the lateral ocellus in the dragonfly: Its relationship with the ocellar afferent units, the compound eyes, and the wing sensory system. *J. Comp. Physiol. 125*, 341–349.

KRAL, K. (1980). Acetylcholinesterase in the ocellus of *Apis mellifica*. *J. Insect Physiol. 26*, 807–809.

KRAL, K. and SCHNEIDER, L. (1981). Fine structural localisation of acetylcholinesterase activity in the compound eye of the honey bee (*Apis mellifica L.*). *Cell Tiss. Res. 221*, 351–359.

KUWABARA, M. and NAKA, K.-I. (1959). Response of a single retinula cell to polarized light. *Nature 184*, (Suppl. 7), 455.

LABHART, T. (1977). Electrophysiological recordings from the lateral ocelli of *Drosophila*. *Naturwissenschaften 64*, 99–100.

LABHART, T. (1980). Specialized photoreceptors at the dorsal rim of the honeybee's compound eye: polarized and angular sensitivity. *J. Comp. Physiol. 141*, 19–30.

LAUGHLIN, S. B. (1973). Neural integration in the first optic neuropile of dragonflies. I. Signal amplification in dark-adapted second-order neurons. *J. Comp. Physiol. 84*, 335–355.

LAUGHLIN, S. B. (1974a). Resistance changes associated with the response of insect monopolar neurons. *Z. Naturf. 29c*, 449–450.

LAUGHLIN, S. B. (1974b). Neural integration in the first optic neuropile of dragonflies. II. Receptor signal interactions in the lamina. *J. Comp. Physiol. 92*, 357–375.

LAUGHLIN, S. B. (1974c). Neural integration in the first optic neuropile of dragonflies. III. The transfer of angular information. *J. Comp. Physiol. 92*, 377–396.

LAUGHLIN, S. B. (1976a). The sensitivities of dragonfly photoreceptors and the voltage gain of transduction. *J. Comp. Physiol. 111*, 221–247.

LAUGHLIN, S. B. (1976b). Neural integration in the first optic neuropile of dragonflies. IV. Interneuron spectral sensitivity and contrast coding. *J. Comp. Physiol. 112*, 199–211.

LAUGHLIN, S. B. (1981a). A simple coding procedure enhances a neuron's information capacity. *Z. Naturf. 36c*, 910–912.

LAUGHLIN, S. B. (1981b). Neural principles in the peripheral visual systems of invertebrates. In *Vision in Invertebrates, B: Invertebrate Visual Centers and Behavior I.* Edited by H. Autrum. *Handbook of Sensory Physiology.* Vol. VII/6B. Pages 133–280. Springer, Berlin, Heidelberg and New York.

LAUGHLIN, S. B. and HARDIE, R. C. (1978). Common strategies for light adaptation in the peripheral visual systems of fly and dragonfly. *J. Comp. Physiol. 128*, 319–340.

LAUGHLIN, S. B. and HORRIDGE, G. A. (1971). Angular sensitivity of the retinula cells of dark-adapted worker bees. *Z. Vergl. Physiol. 74*, 329–335.

LIEKE, E. (1981). Graded and discrete receptor potentials in the compound eye of the Australian bulldog-ant (*Myrmecia gulosa*). *Biol. Cybernet. 40*, 151–156.

LILLYWHITE, P. G. (1977). Single photon signals and transduction in an insect eye. *J. Comp. Physiol. 122*, 189–200.

LILLYWHITE, P. G. (1978). Coupling between locust photoreceptors revealed by a study of quantum bumps. *J. Comp. Physiol. 125*, 13–27.

LILLYWHITE, P. G. and DVORAK, D. R. (1981). Responses to single photons in a fly optomotor neurone. *Vision Res. 21*, 279–290.

LILLYWHITE, P. G. and LAUGHLIN, S. B. (1981). Transducer noise in a photoreceptor. *Nature 277*, 569–572.

LISMAN, J. E. and BROWN, J. E. (1975a). Light-induced changes of sensitivity in *Limulus* ventral photoreceptors. *J. Gen. Physiol. 66*, 473–488.

LISMAN, J. E. and BROWN, J. E. (1975b). Effects of intracellular injection of calcium buffers on light adaptation in *Limulus* ventral photoreceptors. *J. Gen. Physiol. 66*, 489–506.

LO, M. V. and PAK, W. (1981). Light-induced pigment granule migration in the retinular cells of *Drosophila melanogaster*. *J. Gen. Physiol. 77*, 155–175.

LÜDTKE, H. (1953). Retinomotorik und Adaptationsvorgange im Auge des Ruckenschwimmers (*Notonecta glauca L.*). *Z. Vergl. Physiol. 35*, 129–152.

MACNICHOL, E. F. JR. and LOVE, W. E. (1960). Impulse discharges from the retinal nerve and optic ganglion of the squid. In *The Visual System: Neurophysiology and Psychophysics.* Edited by R. Jung and H. Kornhuber. Pages 97–103. Springer, Berlin, Gottingen and Heidelberg.

MARTIN, F. G. and MOTE, M. I. (1980). An equivalent circuit for the quantitative description of inter-receptor couplind in the retina of the desert ant *Cataglyphis bicolor*. *J. Comp. Physiol. 139*, 277–285.

MASTEBROEK, H. A. K., ZAAGMAN, W. H. and LENTING, B. P. M. (1980). Movement detection: Performance of a wide-field element in the visual system of the blowfly. *Vision Res. 20*, 467–474.

MATIC, T. and LAUGHLIN, S. B. (1981). Changes in the intensity-response function of an insect's photoreceptors due to light adaptation. *J. Comp. Physiol. 145*, 169–177.

MATSUMOTO, H., O'TOUSA, J. E. and PAK, W. L. (1982). Light-induced modification of *Drosophila* retinal polypeptides *in vivo*. *Science 217*, 839–841.

McCANN, G. D. and ARNETT, D. W. (1972). Spectral and polarization sensitivity of the dipteran visual system. *J. Gen. Physiol. 59*, 534–558.

McCANN, G. D. and DILL, J. C. (1969). Fundamental properties of intensity, form, and motion perception in the visual nervous systems of *Calliphora phaenicia* and *Musca domestica*. *J. Gen. Physiol. 53*, 385–413.

McCANN, G. D. and FOSTER, S. (1971). Binocular interactions of motion detection fibers in the optic lobes of flies. *Kybernetik 8*, 193–203.

MENZEL, R. (1973). Spectral response of moving detecting and "sustaining" fibres in the optic lobe of the bee. *J. Comp. Physiol. 82*, 135–150.

MENZEL, R. (1974). Spectral sensitivity of monopolar cells in the bee lamina. *J. Comp. Physiol. 93*, 337–346.

MENZEL, R. (1979). Spectral sensitivity and color vision in invertebrates. In *Comparative Physiology and Evolution of Vision in Invertebrates. A: Invertebrate Photoreceptors.* Edited by H. Autrum. *Handbook of Sensory Physiology*, VII/6A. Pages 503–580. Springer, Berlin, Heidelberg and New York.

MENZEL, R. and BLAKERS, M. (1976). Colour receptors in the bee eye — Morphology and spectral sensitivity. *J. Comp. Physiol. 108*, 11–33.

MENZEL, R. and SNYDER, A. W. (1974). Polarized light detection in the bee, *Apis mellifera*. *J. Comp. Physiol. 88*, 247–270.

MEYER-ROCHOW, V. B. (1980). Electrophysiologically determined spectral efficiencies of the compound eye and median ocellus in the bumblebee *Bombus hortorum tarhakimalainen* (Hymenoptera, Insecta). *J. Comp. Physiol. 139*, 261–266.

MILDE, J. (1981). Graded potentials and action potentials in the large ocellar interneurons of the bee. *J. Comp. Physiol. 143*, 427–435.

MILLECHIA, R. and MAURO, A. (1969). The ventral photoreceptor cells of *Limulus*. III. A voltage-clamp study. *J. Gen. Physiol. 54*, 331.

MILLER, G. V., HAUSEN, K. and STARK, W. S. (1981). Phototaxis in *Drosophila*: R1–6 input and interaction among ocellar and compound eye receptors. *J. Insect Physiol. 27*, 813–819.

MILLER, R. F. and DACHEUX, R. (1976). Synaptic organization and ionic basis of On and Off channels in mudpuppy retina. I. Intracellular analysis of chloride-sensitive electrogenic properties of receptors, horizontal cells, bipolar cells and amacrine cells. *J. Gen. Physiol. 67*, 639–659.

MILLER, W. H., Editor (1981). *Current Topics in Membranes and Transport: Molecular Mechanisms of Photoreceptor Transduction.* Academic Press, New York.

MIMURA, K. (1971). Movement discrimination by the visual system of flies. *Z. Vergl. Physiol. 73*, 105–138.

MIMURA, K. (1974). Analysis of visual information in lamina neurones of the fly. *J. Comp. Physiol. 88*, 335–372.

MIMURA, K. (1976). Some spatial properties in the first optic ganglion of the fly. *J. Comp. Physiol. 105*, 65–82.

MIMURA, K. (1978). Electrophysiological evidence for interaction between retinula cells in the flesh-fly. *J. Comp. Physiol. 125*, 209–216.

MIMURA, K. (1981a). Receptive field patterns in photoreceptors of the fly. *J. Comp. Physiol. 141*, 349–362.

MIMURA, K. (1981b). Phototactic behavior of walking flies in response to some visual patterns: correlation with receptive field patterns of photoreceptors. *J. Comp. Physiol. 144*, 75–82.

MINKE, B. and KIRSCHFELD, K. (1980). Fast electrical potentials arising from activation of metarhodopsin in the fly. *J. Gen. Physiol. 75*, 381–402.

MINKE, B., WU, C. F. and PAK, W. L. (1975). Isolation of light-induced response of the central retinula cells from the electroretinogram of *Drosophila. J. Comp. Physiol. 98*, 345–355.

MIZUNAMI, M., YAMASHITA, S. and TATEDA, H. (1982). Intracellular stainings of the large ocellar second order neurons in the cockroach. *J. Comp. Physiol. 149*, 215–219.

MOBBS, P. G., GUY, R. D., GOODMAN, L. J. and CHAPPELL, R. L. (1981). Relative spectral sensitivity and reverse Purkinje shift in identified L neurons of the ocellar retina. *J. Comp. Physiol. 144*, 91–97.

MORING, J. (1978). Spectral sensitivity of monopolar neurons in the eye of *Calliphora. J. Comp. Physiol. 123*, 335–338.

MOTE, M. I. (1970a). Focal recording of responses evoked by light in the lamina ganglionaris of the fly sarcophaga bullata. *Exp. Zool. 175*, 149.

MOTE, M. I. (1970b). Electrical correlates of neural superposition in the eye of the fly *Sarcophaga bullata. J. Exp. Zool. 175*, 159.

MOTE, M. I. and GOLDSMITH, T. H. (1970). Spectral sensitivities of color receptors in the compound eye of the cockroach *Periplaneta. J. Exp. Zool. 173*, 137.

MOTE, M. I. and RUBIN, L. (1981). On type interneurones in the optic lobe of *Periplaneta americana*. I. Spectral characteristics of response. *J. Comp. Physiol. 141*, 395–401.

MOTE, M. I. and WEHNER, R. (1980). Functional characteristics of photoreceptors in the compound eye and ocellus of the desert ant, *Cataglyphis. J. Comp. Physiol. 137*, 63–71.

MOTE, M., KUMAR, V. S. N. and BLACK, K. (1981). On type interneurones in the optic lobe of *Periplanta americana*. II. Receptive fields and response latencies. *J. Comp. Physiol. 141*, 403–415.

MUIJSER, H. (1979). The receptor potential of retinal cells of the blowfly *Calliphora*: The role of sodium, potassium and calcium ions. *J. Comp. Physiol. 132*, 87–95.

MUIJSER, H., LEUTSCHER-HAZELHOFF, J. T., STAVENGA, D. B. and KUIPER, J. W. (1975). Photopigment conversions expressed in receptor potential and membrane resistance of blowfly visual sense cells. *Nature 254*, 520–522.

NAKA, K.-I. (1961). Recording of retinal action potentials from single cells in the insect compound eye. *J. Gen. Physiol. 44*, 571–584.

NAKA, K.-I. and KISHIDA, K. (1966). Retinal action potentials during dark and light adaptation. In *The Functional Organization of the Compound Eye*. Edited by C. G. Bernhard. Pages 251–266. Pergamon Press, Oxford.

NAKA, K.-I. and RUSHTON, W. A. H. (1966). An attempt to analyse colour reception by electrophysiology. *J. Physiol. 185*, 556–586.

NORMANN, R. A. and WERBLIN, F. S. (1974). Control of retinal sensitivity. I. Light and dark adaptation of vertebrate rods and cones. *J. Gen. Physiol. 63*, 37–61.

NORTHROP, R. B. (1975). A model for neural signal-to-noise ratio improvement in the insect visual system with implications for "anomalous resolution". *Biol. Cybernet. 17*, 221–235.

NORTHROP, R. B. and GUIGNON, E. F. (1970). Information processing in the optic lobes of the lubber grasshopper. *J. Insect Physiol. 16*, 691–713.

O'DAY, P. M., LISMAN, J. E. and GOLDRING, M. (1982). Functional significance of voltage-dependent conductances in *Limulus* ventral photoreceptors. *J. Gen. Physiol. 79*, 211–232.

OHM, J. (1981). Intracellular recordings from vertical movement detectors in the lobula of the bee, *Apis mellifica carnica. Verh. Dtsch. Zool. Ges.*, 172.

OLBERG, R. (1981). Object- and self-movement detectors in the ventral nerve cord of the dragonfly. *J. Comp. Physiol. 141*, 327–334.

O'SHEA, M. and ROWELL, C. H. F. (1975a). A spike-transmitting electrical synapse between visual interneurones in the locust movement detector system. *J. Comp. Physiol. 97*, 143–158.

O'SHEA, M. and ROWELL, C. H. F. (1975b). Protection from habituation of lateral inhibition. *Nature 254*, 53–55.

O'SHEA, M. and ROWELL, C. H. F. (1976). The neuronal basis of a sensory analyser, the acridid movement detector system. II. Response decrement, convergence, and the nature of the excitatory afferents to the fan-like dendrites of the LGMD. *J. Exp. Biol. 65*, 289–308.

O'SHEA, M. and WILLIAMS, J. L. D. (1974). The anatomy and output connection of a locust visual interneurone: the lobular giant movement detector (LGMD) neurone. *J. Comp. Physiol. 91*, 257–266.

O'SHEA, M., ROWELL, C. H. F. and WILLIAMS, J. L. D. (1974). The anatomy of a locust visual interneurone: The descending contralateral movement detector. *J. Exp. Biol. 60*, 1–12.

PAK, W. F. (1979). Study of photoreception function using *Drosophila* mutants. In *Neurogenetics: Genetic Approaches to the Nervous System*. Edited by X. O. Breakfield. Pages 67–99. Elsevier, North Holland and New York.

PAK, W. L. and LIDINGTON, K. J. (1974). Fast electrical potential from a long-lived, long-wavelength photoproduct of fly visual pigment. *J. Gen. Physiol. 63*, 740–756.

PAK, W. L., OSTROY, S. E., DELAND, M. C. and WU, C.-F. (1976). Photoreceptor mutant of *Drosophila*: is protein involved in intermediate steps of phototransduction? *Science 194*, 956–959.

PARRY, D. A. (1947). The function of the insect ocellus. *J. Exp. Biol. 24*, 211–219.

PATTERSON, J. A. and GOODMAN, L. J. (1974a). Intracellular responses of receptor cells and second-order cells in the ocelli of the desert locust, *Schistocerca gregaria. J. Comp. Physiol. 95*, 237–250.

PATTERSON, J. A. and GOODMAN, L. J. (1974b). Relationships between ocellar units in the ventral nerve cord and ocellar pathways in the brain of *Schistocerca gregaria. J. Comp. Physiol. 95*, 251–262.

PAYNE, R. (1981). Suppression of noise in a photoreceptor by oxidative metabolism. *J. Comp. Physiol. 142*, 181–188.

PAYNE, R. and HOWARD, J. (1981). Response of an insect photoreceptor: A simple log-normal model. *Nature 290*, 415–416.

PEPOSE, J. and LISMAN, J. E. (1978). Voltage-sensitive potassium channels in *Limulus* ventral photoreceptors. *J. Gen. Physiol. 71*, 101–120.

PIERANTONI, R. (1976). A look into the cock-pit of the fly. The architecture of the lobular plate. *Cell Tissue Res. 171*, 101–122.

PINTER, R. (1979). Inhibition and excitation in the locust DCMD receptive field: spatial frequency, temporal and spatial characteristics. *J. Exp. Biol. 80*, 191–216.

POGGIO, T., REICHARDT, W. and HAUSEN, K. (1981). A neuronal circuitry for relative movement discrimination by the visual system of the fly. *Naturwissenschaften 68*, 443–446.

REHBRONN, W. (1972). Gleichzeitige intrazellulare Doppelableitungen aus dem Komplexauge von *Calliphora erythrocephala. Z. Vergl. Physiol. 76*, 285–301.

REICHARDT, W., POGGIO, T. and HAUSEN, K. (1983). Figure-ground discrimination by relative movement in the visual system of the fly. *Biol. Cybernet. 46* (Suppl.), 1–30.

RIBI, W. A. (1978a). Ultrastructure and migration of screening pigments in the retina of *Pieris rapae L.* (Lepidoptera, Pieridae). *Cell Tiss. Res. 191*, 57–73.

RIBI, W. A. (1978b). Gap junctions coupling photoreceptor axons in the first optic ganglion of the fly. *Cell Tiss. Res. 195*, 299–308.

RIBI, W. A. (1981). The first optic ganglion of the bee. IV. Synaptic fine structure and connectivity patterns of receptor cell axons and first order interneurones. *Cell Tiss. Res. 215*, 443–464.

RIEHLE, A. and FRANCESCHINI, N. (1984). Motion detection in flies: parametric control over ON-OFF pathways. *Exp. Brain Res. 54*, 390.

ROSSEL, S. (1979). Regional differences in photoreceptor performance in the eye of the praying mantis. *J. Comp. Physiol. 131*, 95–112.

ROSSER, B. L. (1974). A study of the afferent pathways of the dragonfly lateral ocellus from extracellularly recorded spike discharges. *J. Exp. Biol. 60*, 135–160.

ROWELL, C. H. F. (1971). The orthopteran descending movement detector (DMD) neurones: A characterization and review. *Z. Vergl. Physiol. 73*, 167–194.

ROWELL, C. H. F. and O'SHEA, M. (1976a). The neuronal basis of a sensory analyser, the acridid movement detector system. I. Effects of simple incremental and decremental stimuli in light and dark adapted animals. *J. Exp. Biol. 65*, 273–288.

ROWELL, C. H. F. and O'SHEA, M. (1976b). Neuronal basis of a sensory analyser, the acridid movement detector system. III. Control of response amplitude by tonic lateral inhibition. *J. Exp. Biol. 65*, 617–625.

ROWELL, C. H. F. and O'SHEA, M. (1980). Modulation of transmission at an electrical synapse in the locust movement detector system. *J. Comp. Physiol. 137*, 233–241.

ROWELL, C. H. F., O'SHEA, M. and WILLIAMS, J. L. D. (1977). The neuronal basis of a sensory analyser, the acridid movement detector system. IV. The preference for small field stimuli. *J. Exp. Biol.* 68, 157–185.

RUCK, P. (1957). The electrical responses of dorsal ocelli in cockroaches and grasshoppers. *J. Insect Physiol.* 1, 109–123.

RUCK, P. (1958). Dark adaptation of the ocellus in *Periplaneta americana*: A study of the electrical response to illumination. *J. Insect Physiol.* 2, 189–198.

RUCK, P. (1961). Electrophysiology of the insect dorsal ocellus. I. Origin of the components of the electroretinogram. *J. Gen. Physiol.* 44, 605–627.

SCHMIDT-NIELSEN, B. K., GEPNER, J. I., TENG, N. N. H. and HALL, L. M. (1977). Characterization of an α-bungarotoxin binding component from *Drosophila melanogaster*. *J. Neurochem.* 29, 1013–1029.

SCHOLES, J. (1965). Discontinuity of the excitation process in locust visual cells. *Cold Spring Harbor Symp. Quant. Biol.* 30, 517–527.

SCHOLES, J. (1969). The electrical responses of the retinal receptors and the lamina in the visual system of the fly *Musca*. *Kybernetik* 6, 149–163.

SCHÜMPERLI, R. A. (1975). Monocular and binocular visual fields of butterfly interneurons in response to white- and coloured-light stimulation. *J. Comp. Physiol.* 103, 273–289.

SCHÜMPERLI, R. A. and SWIHART, S. (1978). Spatial properties of dark and light adapted visual fields of butterfly interneurones. *J. Insect Physiol.* 24, 777–784.

SCHWIND, R. (1978). Visual system of *Notonecta glauca*: A neuron sensitive to movement in the binocular visual field. *J. Comp. Physiol.* 123, 315–328.

SHATOURY, H. H. (1969). Intracellular concentration of electrolytes in the housefly. *Nature* 222, 82.

SHAW, S. R. (1967). Simultaneous recording from two cells in the locust retina. *Z. Vergl. Physiol.* 55, 183–194.

SHAW, S. R. (1968). Organization of the locust retina. *Symp. Zool. Soc. Lond.* 23, 135–163.

SHAW, S. R. (1969). Interreceptor coupling in ommatidia of drone honey bee and locust compound eyes. *Vision Res.* 9, 999–1029.

SHAW, S. R. (1972). Decremental conduction of the visual signal in barnacle lateral eye. *J. Physiol.* 220, 145–175.

SHAW, S. R. (1975). Retinal resistance barriers and electrical lateral inhibition. *Nature* 255, 480–485.

SHAW, S. R. (1977). Restricted diffusion and extracellular space in the insect retina. *J. Comp. Physiol.* 113, 257–282.

SHAW, S. R. (1978). The extracellular space and blood-eye barrier in an insect retina: An ultrastructural study. *Cell Tiss. Res.* 188, 35–61.

SHAW, S. R. (1979a). Signal transmission by graded slow potentials in the arthropod peripheral visual system. In *The Neurosciences: Fourth Study Program*. Edited by F. O. Schmitt and F. G. Warden. Pages 275–295. MIT Press, Cambridge, Mass.

SHAW, S. R. (1979b). Photoreceptor interaction at the lamina synapse of the fly's compound eye. *Invest. Ophthalmol. Vis. Sci.* 19 (Suppl.), 6.

SHAW, S. R. (1981). Anatomy and physiology of identified non-spiking cells in the photoreceptor-lamina complex of the compound eye of insects, especially Diptera. In *Neurones Without Impulses*. Edited by A. Roberts and B.M.H. Bush. Pages 61–116. Cambridge University Press, Cambridge, Mass.

SHAW, S. R. (1982). Synaptic gain control in insect photoreceptors. Abstracts, Soc. Neurosci. 12th Annual Meeting, page 44.

SHAW, S. R. and STOWE, S. (1982). Freeze-fracture evidence for gap junctions connecting the axon terminals of dipteran photoreceptors. *J. Cell Sci.* 53, 115–141.

SIMMONS, P. J. (1981a). Ocellar excitation of the DCMD: An identified locust interneurone. *J. Exp. Biol.* 91, 355–359.

SIMMONS, P. J. (1981b). Synaptic transmission between second- and third-order neurones of a locust ocellus. *J. Comp. Physiol.* 145, 265–276.

SIMMONS, P. J. (1982). The operation of connexions between photoreceptors and large second-order neurones in dragonfly ocelli. *J. Comp. Physiol.* 149, 389–398.

SIVASUBRAMANIAN, P. and STARK, W. S. (1979). Photoreceptor properties of an ectopic eye in the fleshfly, *Sarcophaga bullatta*. *Experientia* 36, 993–994.

SMOLA, U. and GEMPERLEIN, R. (1972). Überträgungseigenschaften der Sehzelle der Schmeissfliege *Calliphora erythrocephala*. 2. Die Abhangigkeit vom Ableitort: Retina-Lamina ganglionaris. *J. Comp. Physiol.* 79, 363–392.

SMOLA, U. and GEMPERLEIN, R. (1973). Rezeptorrauschen und Informationskapazität der Sehzellen von *Calliphora erythrocephala* und *Periplaneta Americana*. *J. Comp. Physiol.* 87, 393–404.

SNYDER, A. W. and LAUGHLIN, S. B. (1975). Dichroism and absorption by photoreceptors. *J. Comp. Physiol.* 100, 101–116.

SNYDER, A. W., MENZEL, R. and LAUGHLIN, S. B. (1973). Structure and function of the fused rhabdom. *J. Comp. Physiol.* 87, 99–135.

SOOHOO, S. and BISHOP, L. G. (1980). Intensity and motion responses of giant vertical neurones of the fly eye. *J. Neurobiol.* 11, 159–177.

SRINIVASAN, M. V. and BERNARD, G. D. (1975). The effect of motion on visual acuity of the compound eye: A theoretical analysis. *Vision Res.* 15, 515–525.

SRINIVASAN, M. V. and BERNARD, G. D. (1980). A technique for estimating the contribution of photomechanical responses to visual adaptation. *Vision Res.* 20, 511–521.

SRINIVASAN, M. V. and DVORAK, D. R. (1980). Spatial processing of visual information in the movement detecting pathway of the fly. *J. Comp. Physiol.* 140, 1–23.

STANGE, G. (1981). The ocellar component of flight equilibrium control in dragonflies. *J. Comp. Physiol.* 141, 335–347.

STARK, W. S., IVANYSHYN, A. M. and GREENBERG, R. M. (1977). Sensitivity and photopigments of R1–6, a two-peaked photoreceptor, in *Drosophila*, *Calliphora* and *Musca*. *J. Comp. Physiol.* 121, 289–305.

STAVENGA, D. G. and KUIPER, J. W. (1977). Insect pupil mechanisms. I. On the pigment migration in the retinula cells of hymenoptera (Suborder Apocrita). *J. Comp. Physiol.* 113, 55–72.

STAVENGA, D. G., FLOKSTRA, J. H. and KUIPER, J. W. (1975). Photopigment conversions expressed in pupil mechanism of blowfly visual sense cells. *Nature* 253, 740–742.

STAVENGA, D. G., BERNARD, G. D., CHAPPELL, R. L. and WILSON, M. (1979). Insect pupil mechanisms. III. On the pigment migration in dragonfly ocelli. *J. Comp. Physiol.* 129, 199–205.

STONE, S. L. and CHAPPELL, R. L. (1981). Synaptic feedback onto photoreceptors in the ocellar retina. *Brain Res.* 221, 374–381.

STONE, S. L. and CHAPPELL, R. L. (1982). Local feedback loop in neurons of the ocellar retina. Abstracts, Soc. Neurosci. 12th Annual Meeting, page 50.

STRAUSFELD, N. J. (1976a). *Atlas of an Insect Brain*. Springer, Berlin, Heidelberg and New York.

STRAUSFELD, N. J. (1976b). Mosaic organizations, layers, and visual pathways in the insect brain. In *Neural Principles in Vision*. Edited by F. Zettler and R. Weiler. Pages 245–279. Springer, Berlin, Heidelberg and New York.

STRAUSFELD, N. J. and CAMPOS-ORTEGA, J. A. (1977). Vision in insects: Pathways possibly underlying neural adaptation and lateral inhibition. *Science* 195, 894–897.

STRAUSFELD, N. J. and NÄSSEL, D. R. (1981). Neuroarchitectures serving compound eyes of crustacea and insects. In *Comparative Physiology and Evolution of Vision of Invertebrates. B: Invertebrate Visual Centers and Behavior I*. Edited by H. Autrum. *Handbook of Sensory Physiology*. Vol. VII/6B. Pages 1–132. Springer, Berlin, Heidelberg and New York.

STRAUSFELD, N. J. and OBERMAYER, M. (1976). Resolution of intraneuronal and transsynaptic migration of cobalt in the insect visual and central nervous systems. *J. Comp. Physiol.* 110, 1–12.

STRONG, J. and LISMAN, J. (1978). Initiation of light adaptation in barnacle photoreceptors. *Science* 200, 1485–1487.

STUART, A. E., HAYASHI, J. H. and MOORE, J. W. (1982). Transmission from the median photoreceptors of the giant barnacle to second-order cells. *Invest. Ophthalmol. Vis. Sci.* 22 (Suppl.), 275.

SWIHART, S. L. (1968). Single unit activity in the visual pathway of the butterfly *Heliconius erato*. *J. Insect Physiol.* 14, 1589–1601.

SWIHART, S. L. (1969). Colour vision and the physiology of the superposition eye of a butterfly (Hesperiidae). *J. Insect Physiol.* 15, 1347–1365.

SWIHART, S. L. (1972a). The neural basis of colour vision in the butterfly, *Heliconius erato*. *J. Insect Physiol.* 18, 1015–1025.

SWIHART, S. L. (1972b). Modelling the butterfly visual pathway. *J. Insect Physiol.* 18, 1915–1928.

SWIHART, S. and SCHÜMPERLI, R. (1974). Visual fields of butterfly interneurons. *J. Insect Physiol.* 20, 1529–1536.

TAYLOR, C. P. (1981). Graded interactions between identified neurons from the simple eyes of an insect. *Brain Res.* 215, 382–387.

TORRE, V. and POGGIO, T. (1978). A synaptic mechanism possible underlying directional selectivity to motion. *Proc. Roy. Soc. Lond. B. 202*, 409–416.

TSACOPOULOS, M. and POITRY, S. (1982). Kinetics of oxygen consumption after a single flash of light in photoreceptors of the drone (*Apis mellifera*). *J. Gen. Physiol. 80*, 19–55.

TSACOPOULOS, M., ORKAND, R. K., COLES, J. A., LEVY, S. and POITRY, S. (1983). Oxygen uptake occurs faster than sodium pumping in bee retina after a light flash. *Nature 301*, 604–606.

TSUKAHARA, Y. and HORRIDGE, G. A. (1977a). Interaction between two retinula cell types in the anterior eye of the dronefly *Eristalis. J. Comp. Physiol. 115*, 287–298.

TSUKAHARA, Y. and HORRIDGE, G. A. (1977b). Miniature potentials, light adaptation and afterpotentials in locust retinula cells. *J. Exp. Biol. 68*, 137–149.

TSUKAHARA, Y., HORRIDGE, G. A. and STAVENGA, D. G. (1977). Afterpotentials in dronefly retinula cells. *J. Comp. Physiol. 114*, 253–266.

TUNSTALL, J. and HORRIDGE, G. A. (1967). Electrophysiological investigation of the optics of the locust retina. *Z. Vergl. Physiol. 55*, 167.

VENTURA, D. F. and PUGLIA, N. M. (1977). Sensitivity facilitation in the insect eye. A parametric study of light adapting conditions. *J. Comp. Physiol. 114*, 35–49.

VENTURA, D. F., MARTINOYA, C., BLOCH, S. and PUGLIA, N. M. (1976). Visual sensitivity and the state of adaptation in the ant *Atta sexdens* (Hymenoptera: formicoidea). *J. Comp. Physiol. 110*, 333–342.

VOGT, K., KIRSCHFELD, K. and STAVENGA, D. G. (1982). Spectral effects of the pupil in fly photoreceptors. *J. Comp. Physiol. 146*, 145–152.

VOWLES, D. M. (1965). The receptive fields of cells in the retina of the housefly (*Musca domestica*). *Proc. Roy. Soc. Lond. B. 164*, 552–575.

WALCOTT, B. (1971a). Cell movement on light adaptation in the retina of *Lethocerus* (Belostomatidae, Hemiptera). *Z. Vergl. Physiol. 74*, 1–16.

WALCOTT, B. (1971b). Unit studies on receptor movement in the retina of *Lethocerus* (Belostomatidae, Hemiptera). *Z. Vergl. Physiol. 74*, 17–25.

WALD, G. (1965). Visual excitation and blood clotting. *Science 150*, 1028–1030.

WASHIZU, Y. (1964). Electrical activity of single retinula cells in the compound eye of the blowfly *Calliphora erythrocephala* Meig. *Comp. Biochem. Physiol. 12*, 369–387.

WASHIZU, Y., BURKHARDT, D. and STRECK, P. (1964). Visual field of single retinula cells and interommatidial inclination in the compound eye of the blowfly *Calliphora erythrocephala. Z. Vergl. Physiol. 48*, 413–428.

WASSERMAN, G. (1973). Invertebrate color vision and the tuned-receptor paradigm. *Science 180*, 268–275.

WEHNER, R. and BERNARD, G. D. (1980). Intracellular optical physiology of the bee's eye. II. Polarizational sensitivity. *J. Comp. Physiol. 147*, 205–214.

WEHRHAHN, C. and HAUSEN, K. (1980). How is tracking and fixation accomplished in the nervous system of the fly? *Biol. Cybernet. 38*, 179–186.

WIITANEN, W. (1973). Some aspects of visual physiology of the honey bee. *J. Neurophysiol. 36*, 1080–1089.

WILLIAMS, D. S. (1983). Changes of photoreceptor performance associated with the daily turnover of photoreceptor membrane in locusts. *J. Comp. Physiol. 150*, 509–519.

WILSON, M. (1975). Angular sensitivity of light and dark adapted locust retinula cells. *J. Comp. Physiol. 97*, 323–328.

WILSON, M. (1978a). The functional organization of locust ocelli. *J. Comp. Physiol. 124*, 297–316.

WILSON, M. (1978b). Generation of graded potential signals in the second order cells of locust ocellus. *J. Comp. Physiol. 124*, 317–331.

WILSON, M. (1978c). The origin and properties of discrete hyperpolarizing potentials in the second order cells of locust ocellus. *J. Comp. Physiol. 128*, 347–358.

WOLBARSHT, M. L., WAGNER, H. G. and BODENSTEIN, D. (1966). Origin of electrical responses in the eye of *Periplaneta americana*. In *The Functional Organization of the Compound Eye*. Edited by C. G. Bernhard. Pages 207–217. Pergamon Press, Oxford and New York.

WONG, F. (1978). Nature of light-induced conductance changes in ventral photoreceptors of *Limulus. Nature 275*, 76–79.

WONG, F. and KNIGHT, B. (1980). Adapting-bump model for eccentric cells of *Limulus. J. Gen. Physiol. 76*, 539–557.

WONG, F., WU, C.-F., MAURO, A. and PAK, W. A. (1976). Persistence of prolonged light-induced conductance change in arthropod photoreceptors on recovery from anoxia. *Nature 264*, 661–664.

WU, C.-F. and PAK, W. L. (1975). Quantal basis of photoreceptor spectral sensitivity of *Drosophila melanogaster. J. Gen. Physiol. 66*, 149–168.

WU, C.-F. and PAK, W. L. (1978). Light-induced voltage noise in the photoreceptor of *Drosophila melanogaster. J. Gen. Physiol. 71*, 249–268.

WU, C.-F. and WONG, F. (1977). Frequency characteristics in the visual system of *Drosophila. J. Gen. Physiol. 69*, 705–724.

YAMASHITA, S. and TATEDA, H. (1976). Hypersensitivity in the anterior median eye of a jumping spider. *J. Exp. Biol. 65*, 507–516.

YEANDLE, S. (1958). Evidence of quantized slow potentials in the eye of *Limulus. Amer. J. Ophthalmol. 46*, Part 2, 82–87.

ZAAGMAN, W. H., MASTEBROEK, H. A. K., BUYSE, T. and KUIPER, J. W. (1977). Receptive field characteristics of a directionally selective movement detector in the visual system of the blowfly. *J. Comp. Physiol. 116*, 39–50.

ZARETSKY, M. and ROWELL, C. H. F. (1979). Saccadic suppression by corollary discharge in the locust. *Nature 280*, 583–585.

ZENKIN, G. M. and PIGAREV, I. N. (1971). Optically determined activity in the cervical nerve chain of the dragonfly. *Biofizika 16*, 229–306.

ZETTLER, F. (1967). Analyse der belichtungspotentiale der sehzellen von *Calliphora erythrocephala* Meig. *Z. Vergl. Physiol. 56*, 129–141.

ZETTLER, F. and AUTRUM, H. (1975). Chromatic properties of lateral inhibition in the eye of a fly. *J. Comp. Physiol. 97*, 181–188.

ZETTLER, F. and JÄRVILEHTO, M. (1970). Histologische Lokalisation der Ableitelektrode, Belichtungspotentiale aus Retina und Lamina bei *Calliphora. Z. Vergl. Physiol. 68*, 202–210.

ZETTLER, F. and JÄRVILEHTO, M. (1971). Decrement-free conduction of graded potentials along the axon of a monopolar neuron. *Z. Vergl. Physiol. 75*, 402–421.

ZETTLER, F. and JÄRVILEHTO, M. (1972a). Lateral inhibition in an insect eye. *Z. Vergl. Physiol. 76*, 233–244.

ZETTLER, F. and JÄRVILEHTO, M. (1972b). Intraaxonal visual responses from visual cells and second-order neurons of an insect retina. In *Information Processing in the Visual Systems of Arthropods*. Edited by R. Wehner. Pages 217–222. Springer, Berlin, Heidelberg and New York.

ZETTLER, F. and JÄRVILEHTO, M. (1973). Active and passive axonal propagation of non-spike signals in the retina of *Calliphora. J. Comp. Physiol. 85*, 89–104.

ZETTLER, F. and WEILER, R. (1976). Neuronal processing in the first optic neuropile of the compound eye of the fly. In *Neural Principles in Vision*. Edited by F. Zettler and R. Weiler. Pages 227–236. Springer, Berlin, Heidelberg and New York.

ZIMMERMAN, R. P. (1978). Field potential analysis and the physiology of second-order neurons in the visual system of the fly. *J. Comp. Physiol. 126*, 297–316.

7   The Eye: Vision and Perception

MATTI JÄRVILEHTO

University of Oulu, Oulu, Finland

"Vision is not a part of the eye,
nor is perception a part of the brain:
Function always only borders the matter."

1 INTRODUCTION

Insects may be small, and this fact may limit their behavioural capabilities, but in some respect their behaviour implies a very sophisticated visual system. One should consider for example the visually controlled aerobatics of a hover-fly or the ability of ants and bees to navigate by using the natural polarized light patterns in the sky.

A particularly large portion of the head volume is taken up by the eyes in insects, since they often fly fast and/or have to use their eyes to find prey. The design of such eyes involved in the localization and identification of visual objects is rather complex (in itself).

The size of the head is one of the crucial points in this design. Therefore, instead of having a lens eye of the vertebrate type, the insect head surface is covered by tiny lenses, joined to a complex optical system, which varies among the species. For small animals compound eyes are realized, if spatial resolution is the factor according to which the design is to be stressed (Kirschfeld, K., 1976). The surprising difference with the lens eye is that this optical system does not produce any optical image.

The movable lens eye of most of the vertebrates equipped with foveal vision is able to sweep a foveal fixation point in two dimensions over the visual field and in addition to shift that fixation point along a third dimension by an accommodation process.

Usually insect compound eyes are rigid and fixed-focused. In order to examine the visual world spatially, insects must move their heads and/or bodies.

The large total visual field thus provided, sometimes even covering the full circle of the animal's surroundings, seems well suited for retrieving spatial information from the fast temporal sequences of visual events.

Apart from basic visual tasks the eyes also mediate higher psychic functions in insects, e.g. visual learning, memory and abstraction. In recent years problems of pattern recognition, either moving or stationary, have gained wide general interest. Although literature concerning visually guided behaviour has piled up, we still know surprisingly little about the neuronal mechanisms generating these functions.

The main topics of this present contribution are vision and perception. Vision as a term has to be understood as an abstraction of the total sum of the functions leading to perception; it thus goes beyond receptors and neurons. Perception is thus a "result" of the cognitive processes in the nervous system's function. The relevance of the visual cells' activity to the animal's behaviour is manifested by visual perception.

In this chapter, vision and perception are presented in two general sections where, in the first, are considered the neuronal aspects of vision and, in the second, the perceptual implications for the behaviour. First of all I shall briefly discuss what is meant by signals and codes in the sense organ.

1.1 Sense of signals and coding

The eyes are sense organs, which transduce the patterns of electromagnetic radiation energy received from the external world into neural signals. Like any other signals, these must be in some kind of code. The information processing in the animal's neuronal system gives significance to sensory signals received. This is the cognitive process of behaviour and gives sense to perception.

The coding of signals is heuristic in nature and this fact makes it difficult for the experimenter to judge the significance of recorded neuronal signals; for he may need to know the code, the relevant background stored knowledge, the situation of the organism and the tasks it is to perform, before he can make sense out of his experiments. Even though it is a component of the animal's psychophysiological entity, the role of cognition tends to be overlooked by physiologists.

Nevertheless an understanding of cognition will be necessary in order to give an adequate account of how organisms read sensory signals. Obviously cognitive concepts become more important as we move from the peripheral visual system to the central nervous system. The content of signals is read more and more by the animal's central nervous system for the purpose of elaborating predictive strategies.

Any message consists of signals, all having to carry some kind of meaning. Signals are always represented by signs and thus signals are organized together; no matter what the organization is, it is called code. The code has no logic in itself, but it has to carry a meaning. Thus a coded group of signs is called a message and can be organized in many different ways, but there is only one algorithm, which produces a message containing no ambiguous information.

By having the same group of signs, but using another algorithm, we are able to uncover new contents of information. The sign, the code, and the algorithm of the message are tied up together with the bandage of meaning. The experimenter in neurophysiology is therefore always confronted with the additional problem of the meaning of his recorded neuronal signals.

The concepts of analysis of information transfer and processing used in communication technology and neurophysiology are very similar. There is, however, one crucial difference: in the technological systems the meaning of signals is defined by the experimenter and the absolute content of information is known, whereas in the nervous system the meaning of neuronal signals is assumed by using observations of behaviour but by no means is the absolute measure of information known.

This fundamental difference creates problems which cannot be resolved in the analysis of information processing in neuronal systems. On the other hand, it does not mean that all efforts to analyse the nervous system were lost. Today we have good reason to believe that many of the methods in communication technology are useful and may be adapted to physiological research, if we keep in mind the difference mentioned above and the open-loop characteristics of biological systems.

1.2 Abbreviations and useful definitions

Throughout the literature of the present subject many of the terms are reduced to abbreviations, which only vaguely follow any rules or agreements. Therefore there is the danger of confusion. The list here is proposed to help the reader to understand the following pages, but not to provide any authority over other publications. The abbreviations are subdivided into groups of related uniformity.

1.2.1 ABBREVIATIONS

General terms

CNS	Central nervous system
d	Distance
ECS	Extracellular space
e-vector	Electrical vector of polarized light
I	Light intensity
$\Delta I/I$	Weber's fraction
MTF	Modulation transfer function
PSTH	Peristimulus time histogram
R_a	Radius of the axon
Δt	Integration time
UV-cell	Ultraviolet
λ	Wavelength of the light
λ_s	Spatial wavelength
S/N	Signal to noise ratio
V_s	Sampling frequency
VIS-cell	Wavelength sensitivity at 400–600 nm
3-D	Three-dimensional

Anatomical terms

AOT	Anterior optic tract
CH-cell	Centrifugal horizontal cell
Col A, B, C	Columnar neurons of the lobula
DIT	Dorsointermediate tract
DMT	Dorsomedial tract
EH-cell	Equatorial horizontal cell
FLG, MLG	Female (Male) lobula giant neuron
H-cell	Horizontal cell
H1, H2, H3	Horizontal cell
HS-cell	Horizontal system
L-cell	Lamina cell
L1–L5, LMC	Large monopolar neurons (L-neurons)
LOB	Lobula
MDT	Dorsomedial tract
NH-cell	North horizontal cell
R1–R6	Receptor cells, short visual fibres
R7, R8	Receptor cells, long visual fibres
SH-cell	South horizontal cell
T1	Centrifugal neuron
TCG	Tritocerebral commissure giant
V1–V9	Vertical neurons
VIT	Ventrointermediate tract
VMT	Ventromedial tract
VS-cell	Vertical system

Optical terms

D	Facet diameter
R	Eye radius
p	Eye parameter $= D \cdot \Delta\Phi = D^2/R = R \cdot (\Delta\Phi)^2$
I	Mean intensity, mean luminance
m	Modulation
V_s	Sampling frequency

Angles

α_p	Angle between arbitrary zero and object direction
α_f	Angle between arbitrary zero and flight direction
$\Delta_D\Phi$	Acceptance angle of dioptric apparatus
$\Delta_R\Phi$	Acceptance angle of fused rhabdom
$\Delta\Phi$	Divergence angle, angular spacing
$\Delta\Phi_h, \theta$	Horizontal (vertical) divergence angle
$\Delta\varphi$	Acceptance angle (functional at 50% level)
Ψ	Angle between the fly's course and its body axis, error angle (angle between object and flight direction)
ϑ	Course direction
θ	Body axis

Velocities

$\dot{\Phi}$	Angular velocity
$\dot{F}$	Forward velocity
$\dot{S}$	Sideways velocity

Electrical terms

AC	Alternating current
DC	Direct current
EPSP	Excitatory postsynaptic potential
ERG	Electroretinogram
PSP	Postsynaptic potential
RP	Receptor potential
r_i	Specific resistance of the axoplasm (ohm $\times$ cm)
r_m	Specific membrane resistance (ohm $\times$ cm^2)
r_0	Specific extracellular resistance
λ	Space constant (length unit)
V_A/V_B	Coupling ratio between two cells

Functional terms

DCMD	Descending contralateral movement detector
DIMD	Descending ipsilateral movement detector
DMD	Descending movement detector
DSMD	Directionally selective movement detector
EMD	Elementary movement detector
HAS	High-acuity system
HP	Horizontal progressive motion-sensitive cell
HR	Horizontal regressive motion-sensitive cell
HSS	High-sensitivity system
LGMD	Lobular giant movement detector
OMD	Object-movement detector
SMD	Self-movement detector
PS	Polarized light sensitivity
VS1, VS2	Vertical spiking cell
V-cell	Sensitivity to vertical stimulus movement

1.2.2 DEFINITIONS

Closed-loop mode
   The fly's torque signals are coupled to the monitor driving the drum via an electronic device which simulates the fly's free flight dynamics.

Open-loop mode
   The experimenter sets the motion of the stimulus and the fly's torque response is measured.
Fixation
   The closed-loop orientation towards an object standing in relation to a space coordinate system ($\alpha p(t) = S(t) = 0$)
Tracking
   The closed-loop orientation with constant speed and angle towards an object moving in relation to a space coordinate system ($\dot{\alpha}p(t) \neq 0; S(t) \neq 0$)
Translatory movement
   To a direction
Rotatory movement
   Turning

The direction of a horizontal pattern movement stimulating ommatidia from the anterior to the posterior regions of the eye is called *progressive*; the reverse direction of motion is called *regressive*.

The direction of motion eliciting a maximal response is called *preferred direction*; the reverse direction of motion *antipreferred*.

Contrast sensitivity: the reciprocal of the minimum contrast (m) necessary for threshold detection of a given period of grating.

The ratio of angular velocity Φ (deg s^{-1}) to spatial wavelength (λ_s) is called the *contrast frequency* Φ/λ_s (s^{-1} = Hz).

The spatial *resolution threshold* $\lambda_0 = 2\Delta\Phi$ where λ is the pattern wavelength and $\Delta\Phi$ the divergence angle between adjacent receptor units.

2 FUNCTIONAL UNITS IN THE INSECT VISUAL SYSTEM

2.1 Concept of functional units

A functional unit here is defined as a group of cells or anatomical cell types, which perform a given function, e.g. by either having low or high sensitivity combined with spectral or polarized light sensitivity characteristics.

   This idea has been adopted from the vertebrate visual system, where the retina is divided into different cell types, e.g. rods and cones, with basically different sensitivity (spectral and intensity) properties. There are crucial basic differences between the anatomy of the compound eye and the lens eye, but we can consider having groups of receptor cells forming functional units in the lens eye in the same way as it is already accepted in the compound eye (Exner, S., 1891).

The visual system, if we consider its spatial extent, must necessarily contain functional units one group within the other with different group characteristics.

The concept of functional units is arbitrary in its basic nature. To be surrounded it depends totally on the characteristic functions, which have to be defined.

Functionally, the first significant part of the visual receptor cell is called a rhabdomere (Grenacher, H., 1879), the photosensitive pigment of which is located in microvillar membrane formations. A collection of rhabdomeres inside one ommatidium is called a rhabdom. We can divide rhabdoms into three different morphological groups: open, fused continuous, and fused layered. The strategy of the subsequent neuronal connections is also obviously defined by the type of rhabdom involved. The rhabdom type determines the functional unit, e.g. for polarized light detection. The ommatidia of different compound eyes are divided according to the rhabdom length in apposition and superposition types (Exner, S., 1891). The superposition type is often complex in retinal structure with distal and proximal rhabdomeres in one ommatidium. There are some arguments whether or not the so-called superposition image exists at the level of rhabdomeres, especially in the eyes with crystalline tracts between the rhabdomeres and the optics (Horridge, G., 1968, 1969; Døving, K. and Miller, W., 1969). It seems therefore more accurate to use the term "clear zone" for all eyes where the rhabdom is discontinued in the retinal cell layer (Horridge, G., 1971). The term "superposition" would be reserved for the eyes where functional superposition image exists, as Exner, S. (1891) originally meant. (See also the chapter on Optics of the Insect Eye, by M. Land, in the present volume.)

Table 1 gives a list of different types of rhabdoms in different compound eyes and ocelli found among insects. The diversity in morphological types of rhabdoms presented here show, for example, the

Table 1: *Rhabdom types in insect eyes*

I. Compound eye
  A. Apposition eye
    1. Open rhabdom (8 retinular cells)
      Diptera
      Aquatic Hemiptera, *Notonecta, Gerris, Corixa*
      Dermaptera
      Coleoptera, Cucujiformia, Cerambycid beetle
    2. Fused rhabdom (5–11 retinular cells)
      (a) Continuous ("smooth")
        Isopoda
        Hymenoptera
        Blattoidea
        Hemiptera
        Orthoptera
      (b) Layered ("toothed")
        Thysanura
        Collembola
        Coleoptera
        Lepidoptera
  B. Clear-zone eyes
    1. Dioptric
      Lepidoptera, moth, skipper butterfly
      Firefly, *Lampyris, Phototuris*
      Coleoptera
      Neuroptera ("toothed")
    2. Catoptric
      Crustacea, crayfish, shrimp
II. Ocelli
  Opposition type, fused rhabdom with 2–10 cells contribution, interdigitation ("toothed") and large variation

correlation with the ability of the animal to perceive polarized light (Shaw, S., 1969; Snyder, A., 1973; Mote, M., 1974). Theoretical studies in that case have shown that the configuration of the whole rhabdom itself is crucial. If the rhabdomere is long and the microvillar density high, the polarization sensitivity is low and the absolute sensitivity high or vice-versa. This is valid for example for the rhabdomeres in the receptor cells R1 to R6 in the open rhabdom of the diptera.

For the fused layered rhabdom the polarization sensitivity is equal to the dichroic sensitivity. Intracellular recordings from single receptor cells in crab and crayfish eyes show that the polarization sensitivity is high indeed (Shaw, S., 1969; Waterman, T. and Fernández, H., 1970; Mote, M., 1974). These high values indicate that each microvillar layer is made up of cells of the same spectral type, or of different spectral types laid out in equal proportions.

The polarization sensitivity is difficult to determine in the case of the fused continuous rhabdom, but under special conditions a high polarization sensitivity can also be expected in fused continuous rhabdoms. The polarization sensitivity may be enhanced by optical coupling, or reduced by electrical coupling (Snyder, A. *et al.*, 1973; Menzel,R., 1975).

If a high polarization sensitivity is to be maintained despite the above-mentioned counteracting properties, different morphologies of the rhabdoms are necessary; e.g. one aberrant retinular cell can be specialized for the detection of polarized light (Menzel, R., 1975). Conversely, different morphological types of fused continuous rhabdoms already known or to be found, displaying high sensitivity to polarized light, could be explained by the above theories. One has to keep in mind that if a design of the functional unit is efficient for discriminating polarized light, it has to compete with other units having other distinguishing factors, such as intensity and wavelength.

Although we know that in fused rhabdoms the different receptor cells show functional differences, so that one ommatidium often contains cells with non-uniform spectral, polarization and intensity types and different acuities, the functional unit is not clearly defined. The electrical properties of the transduction in the different types of eyes may give rise to a division of receptors in fast and slow types (Autrum, H., 1950; Autrum, H. and Stöcker, M., 1952). This idea is based on different time constants of the responses to light stimulation. It was found that the flicker fusion frequency in the visual system of the fly *Calliphora* was about 6.6 times higher (265 Hz to 40 Hz) than that of a grasshopper, *Dixippus*. This difference is usually attributed to high absolute sensitivity and slow adaptation properties of *Dixippus*-type eye having large, fused rhabdoms.

2.2 Functional units in dipteran visual system

The open rhabdom of the fly provides a good optical isolation of individual rhabdomeres, thus allowing all possibilities for individual functional differences among cells. The fly's compound eye consists of eight retinular cells in each ommatidium, where the six peripheral rhabdomeres are optically isolated from the two situated in the middle, one on top of the other. The receptors R1 to R6 have been designated as a high-sensitivity system (HSS) by virtue of their larger rhabdomeres compared to those lying centrally (R7 and R8). The fact that each array of six cells R1 to R6 in different ommatidia respectively converges onto the same group of second-order cells in the first optical neuropile — the lamina ganglionaris — supports the concept of a system unit (Braitenberg, V., 1967). The other two cells (receptors R7 and R8) share a common tiered rhabdom (tandem rhabdomeres) and have been considered as functionally separate from the R1 to R6 group since they project their long axons to the second neuropile (the medulla externa) without having functional contacts with other cells in the lamina ganglionaris. R7 and R8 have been designated as a high-acuity system (HAS) with low absolute sensitivity, because of their narrower rhabdomeres (Kirschfeld, K., 1971, 1973).

The neuronal superposition among the R1 to R6 cells from the retinula to the lamina was first demonstrated in light microscopical silver preparations of the fly (Braitenberg, V., 1967). However the anatomical wiring diagram cannot be considered unique to the dipteran visual system, being more universal. The open rhabdom does not imply a special projection from retinular cell layer to lamina cartridges. Thus amongst the insects with those rhabdom types the waterbugs, *Notonecta glauca*, *Corixa punctata* and *Gerris lacustris*, are the only known non-dipteran insects in which the receptor axons of each ommatidium are known to diverge before they enter the lamina and terminate in different cartridges (Wolburg-Buchholz, K., 1979).

Table 2 shows the properties of those proposed two units. Recent findings on autofluorescence of the central rhabdomeres indicate that the group is not consistent (Franceschini, N. *et al.*, 1981; Hardie, R. *et al.*, 1981). There are sex-specific differences, regional differences and inter-species differences. The anatomical structure of the groups R1–6, R7 and R8 is not consistent throughout the eye (Table 3). The diameter of the rhabdomeres is significantly reduced at the dorsal eye margin of *Calliphora* (Wada, S., 1974). This eye region might be specialized for detection of polarized light, because while the R1 to R6 rhabdomeres are thinner, the R7 and R8 are several-fold larger, both in diameter and in volume (Fig. 1). Polarized light detection is supported also by the fact that these rhabdomeres do

Table 2: *The morphological and functional properties of normal photoreceptors of the fly* Musca *(Kirschfeld, K., 1971; Hardie, R., 1979)*

	R1–6	R7/8	Relative light sensitivity
Neural superposition	yes	no	6
Length of rhabdomeres (μm)	~ 200	$\sim 120/80$	1–2
Pupil mechanism = 50%	yes	no	—
	$\sim 3°$	$\sim 1.5°$	4
Polarized light sensitivity (PS)	Not all, if 1.4–2.6	yes, 1.8–?	—
Spectral sensitivity (λ_{max})	(350) 515 nm	(350) 470 nm	—

R1–6 system about 24–48 more sensitive compared with R7/8

Table 3: *The morphological and functional properties of aberrant rhabdomeres in fronto-dorsal region of the compound eye of the housefly,* Musca: *y, yellow; p, pale (Hardie, R. et al., 1981)*

	R1–6	R7r	R8r	R7/8y and p
Cell body diameter (μm)	5.1	4.2	2.3	2.4
Rhabdomere length (μm)	200	150	50	120 (7y/p), 80 (8y/p)
Microvillar length (μm)	0.61	0.53	0.44	0.36
Microvillar diameter (μm)	0.040	0.041	0.048	0.046 (7y/p), 0.057 (8y/p)
Calculated membrane surface area (μm^2)	17.200	11.000	2700	4700 (7y/p), 2400 (8y/p)
APS 50, absolute sensitivity	1.0	0.9	0.8	3.1 (R7)
(50% $V_{max/max}$)				2.1 (R8)
$\Delta\varphi°$, angular sensitivity at the 50% level	1.7°	1.9°	1.7°	1.4°
PS, polarized sensitivity measured at 490–520 nm	1.7	1.6	3.8	various
V_{max} (mV) max. peak amplitude	56 mV	59	49	53
Spectral sensitivity peaks (λnm)	360, 490	360, 490	360, 450, 520	360 (7y), 345 (7p), 540 (8y), 460 (8p)

not twist, are about half the length of normal rhabdomeres, and the microvillar axes are parallel throughout this area either in R7 and R8 (Wunderer, H. and Smola, U., 1982a,b).

Although amongst the Diptera the asymmetrical arrangement of the rhabdomere endings is widespread, the nematoceran eye shows the symmetrical organization of rhabdomeres (Zeil, J., 1979). It is not very clear how this kind of widely spaced organization of neural superposition is better adapted in the sense of evolution.

Regional and sexual specializations of insect eyes are common. Dorsal area specializations have been described, for example, in dragonfly (Laughlin, S. and McGiness, S., 1978), honeybee (Schinz, R., 1975; Labhart, T., 1980; van Praagh, J. *et al.*, 1980), the ant, *Cataglyphis bicolor*, which actually has three distinct eye regions (Herrling, P., 1976) and the male mayfly (Horridge, G. and McLean, M., 1978). Spectral properties are also found to be

regionally specific in *Ascalaphus macaronius* (Gogala, M., 1967), female *Allograpta* (Bernard, G. and Stavenga, D., 1979) and pierid butterflies (Ribi, W., 1979). Similarly high acuity of the foveal region has been described for many predatory species (Horridge, G., 1978). On the other hand no pronounced regionalization was found in spectral sensitivity distribution in the worker bee's eye when studied by using pupillary responses as a measure. All three spectral cell types of receptors are present in the regions of both dorsal and ventral poles, as well as in the frontal region of the eye (Bernard, G. and Wehner, R., 1980). Although in the dorsal retina the ninth cell is of exceptionally full length with axis oriented skyward, the peculiar structure of the cell is related to polarized light detection (Schinz, R., 1975).

Thus, as seen above, the concept of functional units is far from complete in insects, and even in the Diptera it must be considered cautiously,

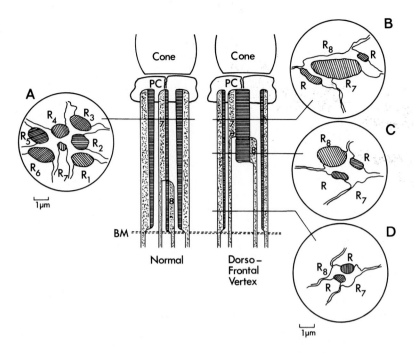

FIG. 1. Longitudinal and traversal sections of a normal and a marginal (right) ommatidium. Hatched areas show rhabdomeres with their relative proportions (after Wada, S., 1974; Wunderer, H. and Smola, U., 1982a). R, R1–R6, retinular cells; BM, basement membrane; PC, Semper cell.

as suggested by Järvilehto, M. and Moring, J. (1976).

3   RECEPTIVE FIELD ORGANIZATION AT NEURONAL LEVEL

The visual world of an insect is transferred to a neural image. How is this performed? How is the visual panorama divided into substructures? The analysis of the spatial sensitivity of single neuronal elements should help solve these problems.

First the visual analysis of environmental features requires spatially sensitive receptors and neuronal pathways. Second the patterns of receptive fields should be coupled together by the central ganglia and third interpreted by "the brain" for the purpose of visually guided behaviour. Complex mechanisms involved in this process, even in simple neuronal systems, were already indicated by the recordings from eccentric cell fibres from single ommatidia of the horseshoe crab *Limulus*, revealing both directional sensitivity and intensity dependence of the visual field (Waterman, T., 1954).

The experimental data concerning visual field patterns in different cell types of various insect species are fairly recent and we have only very few single-cell recordings from different types of eyes and their different cell types. Most probably both apposition and superposition (clear-zone, Horridge, G., 1971) types of eyes use closely related neuronal strategies in the feature analysis of the spatial visual world in spite of the crucial differences in their optical systems. By using a goniometric field of view apparatus the monoscopic and stereoscopic visual fields and blind areas can be shown, e.g. in some species of Cicindelidae (Kuster, J. and Evans, W., 1980) or a pronounced fovea in the water flea can be shown (Nilsson, D.-E. and Odselius, R., 1982).

A clear zone in the eye displays to one single rhabdom a complex structure of overlapping visual patterns from a distant stimulus source. The degree of fusion of rhabdomeres in different rhabdom types influences the coherency of the receptive fields in a single receptive cell.

A brief examination of the theoretical considerations of visual acuity will shed light on the problems of the limiting factors to resolving power, like the intensity of light, angular motion, receptor grain, lens–pupil blur, finite diameter of rhabdom and neural convergence.

3.1 Spatial resolving power

It has previously been demonstrated that receptor acuity in compound eyes is limited fundamentally by the wave (diffraction) and particle (photon noise) nature of light (Snyder, A., 1977). Because there is only a finite number of ommatidia, the image will be quantized by the receptor grain. The angular spacing ($\Delta\Phi$) and the geometry of ommatidia will define the highest spatial frequency transmitted by the receptor mosaic.

V_s = sampling frequency
   $= 1/2 \times \Delta\Phi$; square lattice of visual axes
   $= 1/\sqrt{3}\,\Delta\Phi$; hexagonal lattice of visual axes

Both lattice arrangements are found in compound eyes (Horridge, G., 1977b). Generally the physiological resolving power is lower than V_s because of other restricting conditions.

The theoretical limit is also determined by both diffraction and rhabdomere diameter (Horridge, G. et al., 1976). The angular sensitivity or acceptance function $\Delta\varphi°$ of a retinular cell (Fig. 2) characterizes the spatial uncertainty caused by the imperfect optics. The width of the function at the level of 50% sensitivity and the inclination to the axis of the ommatidium are the parameters usually measured electrophysiologically from individual receptor cells. In practice, a Gaussian function is a good fit to the measured angular sensitivity function (Götz, K., 1965; Tunstall, J. and Horridge, G., 1967). The width of the diffraction pattern (the Airy disk) is proportional to the wavelength. Recordings from receptor cells in *Calliphora* show that the acceptance angle is independent of the wavelength and also approaches the theoretical lower limit inferred from the width of the airy disk at 500 nm. The wavelength values from 300 nm to 600 nm give theoretical thickness for rhabdomeres from 1.2 μm to 3.6 μm. The rhabdomeres R1–6 in the fly, *Calliphora*, are usually about 1 μm in diameter, which gives $\Delta\varphi°$ of about 1.7° if the width of airy disk is 1.8 μm and the distance from the distal of the tip of the rhabdomere to the posterior nodal point

is 60 μm. This has also been confirmed experimentally (i.e. Horridge, G. et al., 1976).

Because neither transduction nor neural transmission is instantaneous, it is reasonable to assume a temporal uncertainty during a time sequence. If a distant point source moves at an angular velocity $\dot\Phi$, then it is displaced by an angular distance $\Delta\Phi \times \Delta t$ across the retina in the integration time Δt. The effect of an animal undergoing angular velocity is equivalent to a reduction in light intensity (I) to (I_t) by the amount $\exp[-1.78(Q_t/\Delta\Phi)^2]$ where Q_t is the amount the animal turns in one integration time. This might be the reason for *Musca* having $D \times \Delta\Phi$ about 4.5 times greater than the diffraction limit.

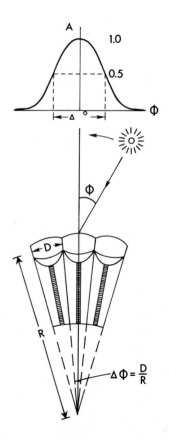

FIG. 2. The acceptance function of an apposition type rhabdom shown in a longitudinal section of ommatidia. D is the facet diameter, R the eye radius, $\Delta\Phi$ the interommatidial divergence angle and A the sensitivity in relative units. The acceptance angle ($\Delta\varphi°$) of a retinular cell is the width of the function at 50% sensitivity and the inclination of the stimulus to the axis of the ommatidium. (Reproduced from Snyder, A., 1977, with permission of Springer-Verlag, Heidelberg.)

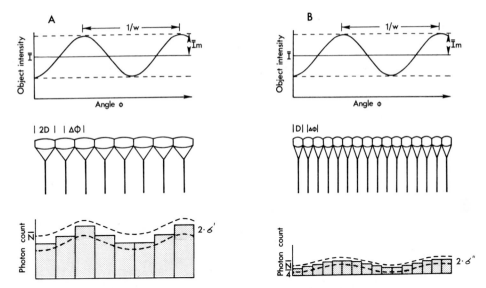

FIG. 3. The effect of photon count fluctuation on object reconstruction by the lens pupil optics and the finite angular diameter of the rhabdom. Ommatidia with a larger (A) facet diameter catch more photons than those with a smaller (B) facet diameter (if the focal lengths are equal) from the same source thus giving a greater signal to noise ratio ($\delta > \delta$). The object intensity is due to a sinusoidal grating of mean intensity $\bar{I}$, modulation $\bar{I}m$ and spatial wavelength λ_s. The amplitude of the sinusoid is given as a function ($\bar{N}_m M_l \cdot M_r$) of the mean number $\bar{N}$ of photons absorbed in one interaction time, and the modulation transfer functions of the lens pupil M_1 and finite rhabdom diameter M_r. The two dashed lines around the photon count give the limits of photon noise $\sigma_{noise} = \sqrt{\bar{N}}$ intrinsic to the photoreceptor function. (Reproduced from Snyder, A., 1977, with permission of Springer-Verlag, Heidelberg.)

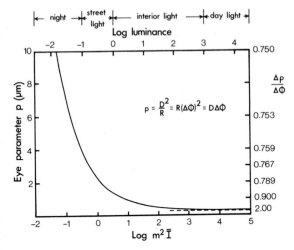

FIG. 4. The optimum eye parameter p for a given intensity parameter I and contrast or modulation m of a sinusoidal grating at the ommatidia sampling frequency $V_s = 1/2\,\Delta\Phi$. The appropriate value for the eye parameter p depends on the contrast–intensity parameter $m^2 I$. The diffraction limit is reached if $\log m^2 I = 5$ ($p = 0.29\,\mu m$ for hexagonal lattice), regardless of the arrangement of the ommatidia. Usually the value for p is higher than the diffraction limit, for the dragonfly (*Hemicordulia*) in the foveal region $p \cong 0.37\,\mu m$ and for the fly (*Musca*) *ca.* $p \cong 1.3\,\mu m$ probably due to neural summation and high angular motion velocity. D, facet diameter; $\Delta\Phi$, interommatidial angle; $\Delta\varphi^\circ$, acceptance angle of a retinular cell; $\lambda 500$ nm. (Stavenga, D., 1975; Snyder, A., 1977, with permission of Springer-Verlag, Heidelberg.)

The other fundamental limit to spatial resolving power is the photon noise. Because of the fact that photon arrival obeys a Poisson distribution in space and time, the contrast sensitivity is proportional to the intensity of light. This means that there must be an optimum relation between the contrast sensitivity and the size of ommatidia in an array. Figure 3 illustrates the noise in different spacings of receptors at a sinusoidal grating of mean intensity ($\bar{I}$) and modulation (m). The optimum eye parameter (p) can be determined for a given intensity parameter (I) and contrast or modulation parameter (m). The p-values must be different in different parts of the compound eye to perform optimally over a range of values for intensity and contrast. The regional differences in the eye of the mantis, *Ciulfina*, give different p-values to the frontal acute zone (fovea) than to lateral regions, though those values are not fully predicted by the theory (Horridge, G. and Duelli, P., 1979). Figure 4 shows a theoretical prediction consistent with the data for the dragonfly, *Zypomma*, which is most active at dusk (Horridge, G., 1976).

The neural mechanisms, such as pooling, can provide other means for increasing the angular sensitivity. Also the changes in the medium

surrounding the rhabdom during light or dark
adaptation, or even the changes in the size of the
rhabdom cross-section, will either reduce or increase the angular sensitivity. The visual field of an
ommatidium with fused rhabdom is

$$\Delta \Phi = \Delta_D \Phi \cdot \Delta_R \Phi$$

where $\Delta_D \Phi$ is the acceptance angle of the dioptric
apparatus and $\Delta_R \Phi$ the acceptance angle of the
rhabdom.

3.2 Apposition eye: the open rhabdom

Most of the data from visual fields arises from flies
(*Calliphora, Musca, Boettcherisca, Drosophila*),
which have an apposition-type eye with open rhabdom. The other types of apposition eyes with fused
rhabdomeres show different characteristics when
compared with superposition eyes of butterflies, but
those will be discussed later. The retinotopic organization of the neuropiles in the fly indicates large
sex-specific differences and pronounced foveal expansion shown already in the optics but especially
in the lobula by the neuronal organisation (Strausfeld, N., 1979).

3.2.1 RETINA–LAMINA LEVEL

The receptive field patterns in the apposition eyes are
basically determined by the geometry of the apical

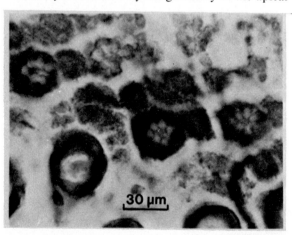

FIG. 5. A cross-section through the apical ends of the rhabdomeres R1–R7. The preparation is unifixated, freeze-dried,
embedded in Ester wax, cleared in xylol and mounted in Eukitt.
The section of the clear spots was made across the zone of the
Semper cells (cf. Fig. 1A with sections from deeper levels).

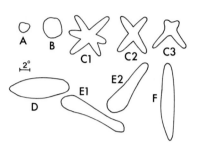

FIG. 6. Scheme of different kinds of electrophysiologically
determined receptive field patterns found in the fly's
(*Boettcherisca peregrina*) receptor cells. Types A and B
are the most commonly found in almost 70% of all
recordings. (After Mimura, K., 1981, with permission of
Springer-Verlag, Heidelberg.)

ends of single rhabdomeres and their optical limiting factors. The micrograph (Fig. 5) shows a cross-
section through the apical end of the rhabdomeres
in a dipteran (*Calliphora*) ommatidium. All the
rhabdomeres are roughly circular and 1 μm in
diameter. Thus the most probable type of receptive
field will be bell-shaped and concentric. Intracellular recordings from different types of mutants
of *Calliphora* without screening pigment show a
higher surrounding sensitivity than in normal visual
cells, but with an almost identical central receptive
field (Streck, P., 1972). By using optical methods it
was shown that the interommatidial angle in the fly
(*Musca, Calliphora*) is roughly 3° (Kirschfeld, K.,
1967). This also corresponds to other values determined electrophysiologically (Table 4) by using
intracellular recordings (Zettler, F. and Järvilehto,
M., 1970).

In the eye of the fruit-fly, *Drosophila*, the receptive field is probably larger than in the bigger flies,
Calliphora and *Musca*. The receptor spacing
estimated from the vertical stripe preference is
about 4.8° (Wehner, R. and Wehner-von Segesser,
S., 1973).

Other shapes of receptive fields have also been
reported (Mimura, K., 1981) in the photoreceptors
of the flesh-fly, *Boettcherisca peregrina*. These patterns were classified into six classes (Fig. 6), which
are surprisingly similar to those found in vertebrate
visual systems at higher levels.

Although at present Mimura's few recordings
represent only some of the different receptive field
types, it seems that they are not evenly distributed
throughout the entire eye field. Simple circular

Table 4: *Visual fields in receptors and second-order neurons in some flies, based on intracellular recordings*

Species	Specifications	Field	Ap. 50%	Author
Calliphora	(Wild-type)			
	Receptor cells R1–R6	HF	3.5	Washizu, Y. *et al.*, 1964
		HF	3.3	Burkhardt, D. *et al.*, 1965
	(dark-adapted)	HF	4.5	Zettler, F. and Järvilehto, M., 1972a,b, 1973
		VF	3.0	Washizu, Y. *et al.*, 1964
		VF	5.2	Burkhardt, D. *et al.*, 1965
	(360 nm, 495 nm)	HF, VF	2.8	Streck, P. 1972a, b
	(625 nm)	HF VF	4.0	Streck, P. 1972a, b
	(light-adapted)	NS	2-5	McCann, G. *et al.*, 1966
	Receptor cells R7, R8	HF	2.0	Järvilehto, M. and Zettler, F. 1973
	Second order neurons	L1, L2 HF	2.3	Järvilehto, M. and Zettler, F. 1972a, b; 1973
Calliphora	(White mutant)			
	Receptor cells	VF	2.9	Washizu, Y. *et al.*, 1964
		HF, VF	2.5	Streck, P. 1972a, b
Calliphora	(Chalky mutant)			
	Receptor cells	VF	3.2	Washizu, Y. *et al.*, 1964
	(360 nm, 495 nm, 625 nm)	HF, VF	2.5	Streck, P. 1972a, b
Musca				
	Receptor cells	HF	7.7	Kirschfeld, K. 1965
	(light-adapted)	HF	3.2	Vowles, D. 1966
	(light-adapted)	VF	2.5	Vowles, D. 1966
	(dark-adapted)	HF	8.5	Vowles, D. 1966
	(dark-adapted)	VF	4.5	Vowles, D. 1966
	(479 nm, dark-adapted)	HF	3.0	Scholes, J. and Reichardt, W. 1969
	(479 nm, light- and dark-adapted)	HF	2.5	Scholes, J. 1969
	(lamina ganglionaris)	NS	2.5	Scholes, J. 1969
Drosophila	(Wild-type)			
	Receptor cells			
	(optomotor response)	—	3.5	Götz, K. 1964; 1965

HF, horizontal; VF, vertical field measurements; NS, not specified

fields were found mainly in the anteromedial region, but were also scattered over the whole compound eye, whereas cells with star-like receptive fields were found only in a small area of the anteromedial region. The other types were distributed throughout the lateral regions. Their shapes are exceptional and may be due to artefactual damage of the optical system during the penetration of the retinal tissue. The recording sites were also not identified histologically, thus allowing recordings from optic neuropils other than receptor cell bodies, which could also explain the complex shapes of the receptive fields reported.

The entire receptive field has not yet been measured from R7 and R8 receptor cells, though the function for the angle of incident light at a horizontal level (Fig. 7C) indicates narrower visual fields ($\sim 1.5°$) than in receptor cells R1 to R6, $\sim 3°$ (Zettler, F. and Järvilehto, M., 1970; Järvilehto, M., 1971; Järvilehto, M. and Zettler, F., 1973), as could be predicted from cross-sections of the rhab-

domeres (Snyder, A., 1977). Hardie, R. (1979) measured values even lower than those listed above, for R7 (1.3°) and for R8 (1.2°), but recordings may originate from other regions.

The neuronal network of the fly's visual system provides good conditions for lateral interaction between cells in the neighbour cartridges. A commonly found function of incident light in the lamina is shown in Fig. 7A. It is similar to that recorded from receptors, but its shape is clearly more distended. It is not clear why the lamina fields are slightly larger than those in receptors at the level of cell bodies, but it seems to be due to signal summing and averaging in the lamina cartridges.

3.2.2 LAMINA—CHIASMA LEVEL

There are only very few identified recordings from different neuron types in the lamina ganglionaris, the first optic ganglion. Except for the large

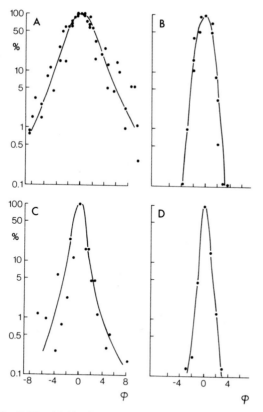

FIG. 7. Visual fields of several different cell types in the retina of the fly, *Calliphora*, recorded intracellularly. **A:** Short type receptor cell axon; **B:** second-order neuron; **C:** long type receptor cell axon (R7 or R8); **D:** second-order neuron (centrifugal cell, T1). The sensitivity of different cells is shown by relative intensity in log-% in function of the angle of the incident light. (Järvilehto, M. and Zettler, F., 1973.)

tified potentials, but they resemble ERG or potentials from some cell types recorded from the retinular cell layer. Furthermore, it is interesting to note that these cells, which possess a wide receptive field, can be UV-sensitive. The spectral sensitivity was measured by recording the hyperpolarizing potential in the middle of the receptive field, using a light spot as illumination.

Extracellularly recorded spiking responses have been reported from the lamina and the intermediate chiasma (Arnett, D., 1971, 1972; Mimura, K., 1974). Because they are rarely found, and have not been identified, their origin is still unknown. Two types of units were found: sustaining ON units and ON–OFF units, which responded by a transient discharge after the onset and cessation of a light spot; ON–OFF units are coupled together laterally in such a way that the adjacent regions antagonize the central region, for the stimulation of either one inhibites the discharge resulting from the stimulation of the central region (see Fig. 25).

Although no specifically movement-detecting units have been found in the lamina, some information processing, like lateral inhibition, could provide means for movement detection in higher centres. When an object stimulation moves over the compound eye, fast spatially produced transi₋nts in membrane potential discharges will supply a very strong visual clue to the animal's movement centres.

3.2.3 MEDULLA

In the fly's visual system each optic ganglion (lamina, medulla, lobula, lobula plate) constitutes a retinotopically ordered set of neural columns, in which each single column corresponds to one single optical axis of the visual field grain constituting the compound eye.

The synaptology is rather complicated in the lamina and is likely to be much more so in the medulla as the complexity of cell responses and the number of morphological cell types increases (Fig. 8).

The neurons show distinct branching of axons to different projective layers. The extent of branching may suggest that in a particular region the visual activity can be more pronounced than in other regions, but the branching itself does not tell anything about functional interactions in terms of what the animal sees. The recordings from the medulla of

monopolar neurons (Fig. 7B), L1 and L2, only one centrifugal cell (T1) has been recorded and identified (Fig. 7D). Its receptive field was even narrower (only about 1° at 50% sensitivity level) than that in monopolar neurons (2.7° at 50% level). The neuronal connections causing such a high acuity are still a matter of speculation (Järvilehto, M. and Zettler, F., 1973).

Some unidentified neurons in the lamina respond also to very specific stimuli, e.g. turning slits of different lengths. These properties are attributed to some narrow-field neurons (possibly other neurons than L1 or L2). Mimura, K. (1976) reported some recordings from wide-field lamina cells. These cells responded with complex potential waves having both hyperpolarizing and depolarizing components at different locations in their receptive fields. Mimura did not interpret the origin of these uniden-

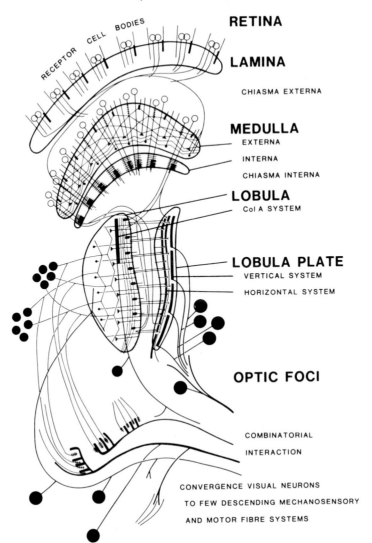

RETINA

LAMINA

CHIASMA EXTERNA

MEDULLA

EXTERNA

INTERNA

CHIASMA INTERNA

LOBULA

Col A SYSTEM

LOBULA PLATE

VERTICAL SYSTEM

HORIZONTAL SYSTEM

OPTIC FOCI

COMBINATORIAL

INTERACTION

CONVERGENCE VISUAL NEURONS

TO FEW DESCENDING MECHANOSENSORY

AND MOTOR FIBRE SYSTEMS

RECEPTOR CELL BODIES

FIG. 8. A summary diagram of principal areas in the optic lobes in the fly. Projections from receptor terminals to higher multimodal centres are shown schematically. (Based on data from Strausfeld, N., 1976b; Strausfeld, N. and Nässel, D., 1981, with permission of Springer-Verlag, Heidelberg.)

flies have shown numerous types of activity, both graded, (hyperpolarizing and depolarizing potentials) and discrete unit activities (Järvilehto, M. and Zettler, F., 1973; DeVoe, R. and Ockleford, E., 1976). Hyperpolarizing graded potentials have been recorded in the first chiasma and in the outer medulla (from monopolar axons and their terminals), whereas in the medulla relay neuron responses were characterized by spikes. These include directionally and non-directionally motion-sensitive neurons and so-called "change-of-direction" cells.

The cells recorded and dye-marked include a directionally motion-sensitive cell identified as a small-field T2 cell that terminates in the deep strata of the lobula. Another element stained represents a Y-neuron that projects from medulla to both lobula and lobula plate. Its motion sensitivity has not yet been confirmed, but it can be excited by light-on. Essentially, DeVoe, R. and Ockleford, E. (1976) found three different types of receptive fields characterized by specific responses: small-field ipsilateral monocular, large-field ipsilateral monocular, and large-field binocular. The small fields found were 20–30° in diameter without

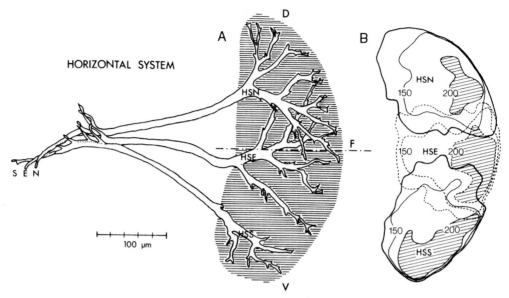

FIG. 9. The three principal types of horizontal cells in the lobula plate. **A:** Peripheral dendritic branching patterns, constructed from serial sections of cobalt-impregnated material: D, dorsal; V, ventral; F, frontal; C, caudal (Pierantoni, R., 1976). **B:** Dendritic fields and dendritic density of the north (HSN), the equatorial (HSE) and the south (HSS) horizontal cells. The dendritic density was determined from the dendritic lengths in the square arbitrary units. The arborization density of each cell is highest in its domain (hatched areas). Superimposing reveals considerable overlapping in the lower-density regions. (After Hausen, K., 1982, with permission of Springer-Verlag, Heidelberg.)

consistency in the response patterns. Among the large-field cells sustaining cells do not always respond to ipsilateral stimuli. Some cells respond to different kinds of contralateral stimuli with OFF-discharge, but basically to an ipsilateral stimulus by ON–OFF-discharge. Some degree of binocularity was assumed to exist in the medulla. The inhibitory interaction between the cells is only weak in response to spot stimulation (Mimura, K., 1972).

3.2.4 OPTIC LOBES

The third neuropil in the optic lobes of flies is divided into two separate subregions, the lobula and the lobula plate. Their large cells and manifested directional motion sensitivity have been a considerable attraction for electrophysiological investigations (Bishop, L. and Keehn, D., 1966; Dvorak, D. *et al.*, 1975; Hausen, K., 1976, 1982; Eckert, H., 1978, 1980b, 1981). The receptive fields of the cells in this neuropil show a considerable complexity depending on the type of stimulus used. The responses are usually trains of spikes with or without a graded depolarization. The responses can be elicited by using a moving grating as a stimulus or by apparent movement produced by using

sequentially flashing light spots. The direction of the stimulus movement divides the cells in the lobula plate into two distinguishable groups: the horizontal and the vertical movement system. The receptive field organization of these cell groups is complicated, because it depends on many different stimulus parameters. Movement sensitivity will be discussed in section 7, but here I will discuss some anatomical and functional aspects of receptive field organization in these neuropils.

The horizontal movement system (Fig. 9) is divided anatomically into three different regionally overlapping groups, north (NH-), equatorial (EH-) and south (SH-cell) horizontal cells, depending on their position in the lobula plate (Pierantoni, R., 1974, 1976). These cells have been subsequently described for *Musca* (Strausfeld, N., 1976a,b), for *Phaenicia* and *Sarcophaga* (Dvorak, D. *et al.*, 1975), for *Calliphora* (Hausen, K., 1976, 1982; Eckert, H., 1981) and for *Drosophila* (Heisenberg, M. *et al.*, 1978). Some ambiguity or double naming by different authors will be clarified in Table 5.

The main difference between the response characteristics of the NH-, EH-, and SH-cells seems to be found in their retinotopic organization in the ipsilateral receptive fields corresponding to the dorsal,

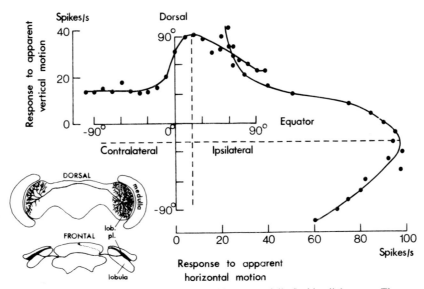

F<small>IG</small>. 10. The receptive field of a heterolateral H1-neuron determined by sequentially flashing light spots. The apparent motion sensitivity is shown by horizontal and vertical graphs. The maximum response is located at the ventro-equator level on the ipsilateral side (the crossing of broken lines). The inset figure shows the anatomical configuration of the cell with its contralateral extension. (Data from Eckert, H., 1980b., with permission of Springer-Verlag, Heidelberg.)

equatorial and ventral parts of the ipsilateral eye.

Some of these large cells can also be activated by stimulation of the contralateral visual field, but the direction of motion eliciting excitatory responses (EPSPs, action potentials, depolarizing potential shifts) will be different from that of ipsilateral stimulation. The H-cells respond to stimulation of regressive pattern motion (back-to-front movement) within the contralateral receptive field with small excitatory postsynaptic potentials and action potentials of small amplitude, whereas at the same time in the ipsilateral receptive field the stimulation consists in progressive motion (front-to-back movement) and causes an excitation. The reversal of the directions of motion stimulation induces corresponding inhibitory responses. Binocular stimulation produces about the same response as the sum of the responses induced by monocular stimulation.

Figure 10 shows the sensitivity change of the apparent movement stimulating a heterolateral H1-neuron in the lobula plate. The receptive field of this neuron extended over the whole eye, but needed to be stimulated with horizontal movement. The vertical movement caused a response which was very much smaller. The receptive field is relatively complex because the cell response extends to the contralateral eye too.

The neurons, which are movement-sensitive to vertical motion, are called V-cells in the third optic neuropil. The V-cells (Fig. 11) can be divided into 8 to 11 distinctly different types according to their retinotopics. (Pierantoni, R., 1974, 1976; Strausfeld, N., 1976; Eckert, H. and Bishop, L., 1978) and the methods of histological analysis (Hengstenberg, R. et al., 1982). The differences may be due to missing existing cells, fragmentary staining, genetic variation, physiological differences, etc.

The dendritic fields extend approx. 60° in the horizontal and 200° in the vertical direction. These fields receive information from anterior, lateral and posterior areas of the retina. There is also considerable overlapping in the receptive fields of the H-cell systems at the equatorial level.

These giant cells of the lobula plate are believed to be major output elements to descending neurons of the ventral cord and to heterolateral neural elements. Some of them (VS2) are also coupled to the horizontal system in the ventrolateral protocerebrum and to Y-shaped descending neurons providing convergence for different sensory channels (Hengstenberg, R. et al., 1982). Recent electrophysiological experiments give support to this idea (Strausfeld, N. and Obermayer, M., 1976; Eckert, H. and Bishop, L., 1978).

Beneath the above-mentioned two classes of

VERTICAL SYSTEM

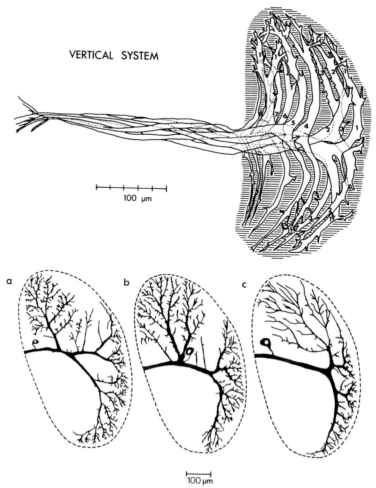

100 μm

100 μm

FIG. 11. The neuronal basis of vertical movement systems in some species of flies. The graph shows 8 vertical neurons in *Musca* analysed from semithin sections (redrawn from Pierantoni, R., 1974, 1976, with permission of Springer-Verlag, Heidelberg) and (a), (b), (c) in *Calliphora* visualized by Procion yellow injection. The brush-like dendritic arborization occupies the ventral and distal parts of the lobula and the fan-like extension is located in the dorsal region, suggesting that the structural similarity in *Musca*, *Calliphora* and *Phaenicia* (Eckert, H. and Bishop, L., 1978) of this VS1 type of neuron in different species may well have a functional homological significance. (After Hengstenberg, R. *et al.*, 1982, with permission of Springer-Verlag, Heidelberg.)

output neurons, in the main part of the lobula-complex, the lobula, the Col A cells form the third class, which is regularly spaced, columnar in structure and possibly small-field. The cells constitute a plexus of arborizations at the posterior surface of the neuropile (Fig. 8) and the axons of the cells terminate in the ventrolateral protocerebrum (Hausen, K. *et al.*, 1980). This group of neurons also show a sexual dimorphism expressed by differences in the shapes of the analogous neurons (Col A, Col B, Col C and FLG, MLG neurones) in males and females (Strausfeld, N., 1980). The differences in morphology will perhaps explain the pattern detection differences in visually guided behaviour.

3.3 Apposition eye: the fused rhabdom

3.3.1 THE RETINA LEVEL

The fused rhabdom in several insect groups (i.e. Hymenoptera, Odonata, Plecoptera, Blattoidea, Trichoptera, Orthoptera) is characterized by very small space between single rhabdomeres. Thus the cells in the same ommatidium are commonly thought to share the same visual field (Kuiper, J., 1962; Varela, F. and Wiitanen, W., 1970). In some experiments conducted for other purposes this co-incidence of overlapping was used as a criterion to ensure that the units recorded with double

Table 5: *Nomenclature of horizontal motion sensitive neurons in the lobula plate (see Fig. 9).*

Author	Fly species	Horizontal (H) cell type			Remarks
		North	Equatorial	South	
Pierantoni, R. (1974)	*Musca*	N	E	S	Homolateral
Dvorak, D. *et al.*, 1975	*Phaenicia,*	N	E	S	Homolateral
	Calliphora,	DCH		VCH	Homolateral
	Sarcophaga	H1	H2	H3	Heterolateral
Pierantoni, R. (1976)	*Musca*	HSN	HSE	HSS	Homolateral
Hausen, K. (1976)	*Calliphora*	HS1	HS2	HS3	Heterolateral
Hengstenberg, R. (1977)	*Calliphora*	HS1	HS2	HS3	Homolateral
Eckert, H. (1981)	*Phaenicia*	NH	EH1, EH2	SH	Homolateral
Hausen, K. (1982)	*Calliphora*	HSN	HSE	HSS	Homolateral

electrodes reside in the same ommatidium (Shaw, S., 1967, 1969). Furthermore, it led to the proposition that the cells in one ommatidium are likely to be electrically coupled. The functional significance of this kind of electrical arrangement is not fully understood. Some form of paired coupling of rhabdomeres would sensibly explain the double-peaked form of spectral sensitivity curves and the vanishing of polarization light sensitivity, which otherwise should exist in those cells.

Many lepidopteran larvae possess several ommatidium-like stemmata, which have a fused apposition type rhabdom consisting of several highly specialized receptor cells. The receptive fields are narrow ($\Delta\varphi° = 1.7–5°$) in proximal cells and broader ($\Delta\varphi° = 7–13°$) in distal cells. The significance of this organization is not very clear, but may be due to object versus shape and colour detection (Ichikawa, T. and Tateda, H., 1982).

The visual field of the fused rhabdomeres in complex eyes has been measured by using both optical and electrophysiological methods. The insect species which particularly use their sight for special behavioural tasks generally have narrower receptive fields in their receptor cells than the others. Light and dark adaptation will also change the size of the field. In the locust this change is about twice, i.e. from 3.4 to 6.6° measured at the half-width of the sensitivity (Tunstall, J. and Horridge, G., 1967). The visual fields of ommatidia in *Apis mellifica* and *Cataglyphis bicolor* show widths at 50% optical sensitivity of 2.6° and 8.8° respectively (Eheim, W. and Wehner, R., 1972). Dark and light adaptation produces only a minor effect on the opening angle of the ommatidia. The electrophysiological recordings from worker bee receptor cells in the dark-adapted state are only slightly different ($\Delta\varphi° = 2.5°$)

from above (Laughlin, S. and Horridge, G., 1971).

The mechanism which regulates the angle of incident light is not well known, but the actual measured data about the sensitivity of the receptor cells to the stimulation angle of rhabdomeres and incoming light path are not consistent with the anatomical facts. In flies and locusts, where some data are available, the opening angle (the size of the electrophysiologically determined visual field) is generally somewhat greater than the divergence angle. This means that the visual fields of the neighbouring ommatidia must overlap. This is only partly consistent with the mosaic theory of Müller, J. (1826), which claims that an ommatidium receives light within its morphological opening angle (the divergence angle). Since the electrophysiologically determined acceptance angle, at least in flies and locusts, is larger than the morphological angle there must be, at the neuronal level, a slightly different image than can be proposed just by using morphological data and optical methods. Thus the overlapping of the fields of single rhabdomeres must cause some fuzziness. This fact may not play any role for the animal, which is not basically interested in a sharp optical display of the surrounding world, but only in some special features within it. Thus we have to put the question to the next neuropil. Is the retinotopic organization maintained in a similar way by the neurons in the medulla or in other higher centres?

3.3.2 THE RECEPTIVE FIELDS IN HIGHER NEUROPILS

Our knowledge of the spatial processing in higher centres is based on recordings from very few different types of cells in the medulla and the lobuli. The potential responses to light stimuli are characterized by spikes and graded de- or hyper-polarizing potentials.

Responses of ON-type interneurons in the external chiasma and lamina of the cockroach *Periplaneta americana* have been recorded extracellularly. Their spectral sensitivities, using the threshold as a criterion, were similar to the class of retinal receptors, which are mainly green-sensitive (Mote, M. and Rubin, L., 1981). In these neurons the input from the retinal UV-receptors is suppressed or inhibited by activity in the green-sensitive receptors if selective adaptation is used. The discharge pattern can also be wavelength-specific.

The medullary neurons in crickets can be divided into two physiological groups. One type responds with sustained OFF-activity having an antagonistic (excitatory) surround characteristic (Honegger, H.-W., 1978). This type has a large receptive field with an OFF-centre in the medioposterior part of the eye and a heterogeneous antagonistic surround. The use of moving discs as stimuli supports the view of an antagonistic centre–surround organization. The second type of units show a sustained ON-activity. Several large and uniform receptive fields without a surround, and located indifferently in the eye, were reported (Honegger, H.-W., 1980). These cells are movement-sensitive, whereby moving targets of either positive or negative contrast suppressed the firing of the cell. In this respect, the cell's response resembled that of a certain type of ganglion cell in the vertebrate retina. These ganglion cells also lack a surround. Their activity is suppressed by contrast (Rodieck, R., 1967) and they have been termed by Levick, W. (1967) "uniformity detectors".

The transient sustained ON-units proved to be a population of neurons having different receptive fields of different size and location. Parts of these receptive fields are located in the region of binocular overlap in the frontal part of the eye. The binocular visual field covers about 46% of the total visual field. This type of units might be significant, therefore, in the detection of contrast, especially in the anterolateral visual field.

In worker bees (*Apis mellifera*) the receptive fields in the medulla were studied by intracellular recordings and the cells were identified by dye injection (Hertel, H., 1980). The responses to light stimuli show a complex correlation and therefore several different classes of potential responses were ob-

tained: (a) neurons which respond to a stimulus only with graded potentials; (b) those which respond with action potentials and graded potentials, and (c) those responding only with action potentials. Hertel, H. (1980) considers only action potential responses in his report, which he divides into several groups, according to their presence. The most common way to modulate the action potential frequency in the response is to excite the cells by ON and OFF light stimulus. The response to this kind of stimulus is usually an excitation at both ends of the stimulus. This form is also common in the vertebrate ganglion and amacrine cells. Also several recordings from ON-excitation, ON-inhibition and sustained neurons were found. Because these cells are not directly activated by a light stimulus, their responses are a combined result of the co-operation from many inputs and states of the neuronal system. Care must be taken if special receptive fields are to be attributed to certain types of cells.

The receptive fields of different neurons of the "same" anatomical type vary in size and form. Often even the receptive field of a neuron may differ, depending on the wavelength of the monochromatic stimuli. By using arbitrary criteria it is possible to distinguish two groups of field sizes: large fields ($>30°$) and small fields ($<30°$). The receptive fields of one-third of the medulla neurons are smaller than 30° but most of the lobula neurons have fields greater than 30°. Some cells show spatial antagonism. Light stimuli elicit excitation in one part of the receptive field and inhibition in the other part. These fields lack an antagonistic centre–surround characteristic, although if neurons are sensitive to stimulus movement they always possess small receptive fields in the medulla, but show a large spatial response in the lobula. The directional sensitivity can occasionally be a response mode of lobula neurons.

In the optic lobes of the cockroach *Periplaneta americana* the receptive fields have been studied by using the responses of ON-type interneurons. The receptive fields of these cells depend strongly on the criteria used for their description (Mote, M. *et al.*, 1981). Sufficiently intense light stimuli are efficient in almost any part of the visual field. The most effective receptive area is usually elliptical in shape, less than 90° in its largest extension and located in the anterior portion of the visual field. The field size can be divided arbitrarily into three groups called small (0–30°), medium (31–90°) and large (90°) fields.

Butler, R. and Horridge, G. (1973) reported a mean acceptance angle at a 50% sensitivity level for dark-adapted *Periplaneta* retinular cells of 6.7° horizontally and 6.9° vertically. At a 10% sensitivity level the acceptance angle is approx. 16°. The interommatidial angle in different eye regions is between 1° and 10°, which gives high overlapping of the receptive fields of receptor cells. The response pattern generated by on-type neurons in the dark-adapted condition can be explained by a very simple mechanism, namely a summation of excitatory inputs from the photoreceptors.

Mote, M. *et al.* (1981) also found a wavelength dependence in the form of receptive fields in the bee. The position of the maximum sensitivity in the field area shifts as a function of different wavelengths of light stimuli. The shift in the maximum response measured with short-wavelength (violet) stimuli after a selective adaptation of the green receptors clearly shows that the UV-receptors, which are the most effective in exciting the cell, in this case occupy a different relative area of the retina than occupied by the green receptors. Such a wavelength-dependence of the receptive field has also been reported in the optic lobe of the bee (Kien, J. and Menzel, R., 1977) and in crayfish, *Procambarus* (Woodcock, A. and Goldsmith, T., 1973) but seems not to occur at the lamina level in the fly (Zettler, F. and Autrum, H., 1975).

3.4 Receptive fields in superposition eyes

3.4.1 The retina

Müller, J. (1826) postulated that the function of each ommatidium is to monitor the intensity of the light from the direction that it faces. This so-called mosaic theory was extended by Exner, S. (1891) who classified the compound eyes in two groups using their anatomical differences to infer their function: the apposition and superposition eyes. The most important difference between superposition and apposition eyes is that the rhabdomeres in superposition eyes are separated from the tips of crystalline cones by a transparent layer of considerable thickness. The aperture of the rhabdom columns is shown in Fig. 12. Here a parallel beam

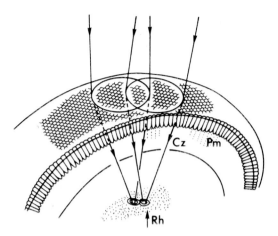

Fig. 12. The distal endings of the rhabdoms (Rh) in superposition eyes, separated by a clear-zone (Cz). Tens of facets will supply quanta from a parallel light beam to only a few rhabdoms if dark-adapted. The pigment movement (Pm) across the clear-zone will effectively reduce the aperture during light adaptation.

in the superposition eye of the skipper butterfly, *Epargyreus clarus*, converges to a single receptor via a circular patch of facets subtending about 30° at the centre of the eye (Horridge, G. *et al.*, 1972). This, though, is not always very simple. The rhabdomeres may possess very complex structures with either a simple unitary rhabdom or a divided one with proximal and distal rhabdomeres. The rhabdomeres and cones may also be connected by crystalline tracts inside one single ommatidium.

In some superposition (clear-zone) eyes this tract may act as a light guide (Horridge, G., 1968, 1976). This was also concluded from electrophysiological experiments on *Epargyrens clarus* (Døving, K. and Miller, W., 1969). The measurements of the visual fields from single dark-adapted retinal cells give a relative narrow acceptance angle (*ca.* 2°) at a 50% sensitivity level. The result, particularly with a dark-adapted eye, does not contribute to Exner's superposition theory.

On the other hand the fact that the acceptance angle in *Dytiscus* is wide (between 30 and 40°), depending on the adaptional level, may be considered as support to the superposition theory. Basically little is known about receptive fields of the receptor cells in clear-zone eyes, especially as pertains to their changes with light and dark adaptation. Some diurnal moths, such as *Phalaenoides fristifica*, with clear-zone, show an unexpectedly high resolution in the sharply focused superposition image which serves to detect small contrast differences in small objects (Horridge, G. *et al.*, 1977).

3.4.2 THE OPTIC LOBES

The higher visual centres in superposition eyes show the same general features in their anatomical structures for all insect classes. By using two independent electrodes it has been possible to study the properties of three different classes of neurons in the medulla and the other optic lobes of a moth, *Sphinx ligustri* (Collett, T., 1970). In the medulla the tangential cells cover large areas. The cells are connected to the contralateral medulla via large junctional fibres. Thus the receptive fields of these cells also cover the total binocular visual field of the eyes. Usually the units are sensitive to horizontally moving stimulation patterns with preferred direction. The firing rate of action potentials at the resting level is inhibited by movement in the opposite direction. The information carried by the tangential cells appears to be relatively simple. The monocular sustained ON-units tell the opposite medulla about the net luminance over large areas of one visual field. The ON–OFF units inform the opposite medulla, or both medullas, of the total moving contrast seen by one eye, with no special preference for movement direction. As a third function the binocular directionally selective neurons inform the medulla as to the kinds of movements the animal is making. The centrifugal cells in the medulla have been described previously in locusts (Horridge, G. *et al.*, 1965).

In butterflies (mainly Nymphalidae and Heliconiidae) monocular and binocular receptive fields have been recorded from protocerebral interneurons (Schümperli, R., 1975; Schümperli, R. and Swihart, S., 1978). Their visual fields show a large variety of shape, size and texture; but because only spike responses were analyzed their functions described here are far from complete. At least three functional groups of cells were reported; *viz.* a group of cells possessing a very large visual field (>80°), and the other group having a strip-form receptive field with horizontal or vertical extension. Some neurons are considered functionally uniform, being excited by contra- and ipsilateral stimulation, which can be considered as a binocular function.

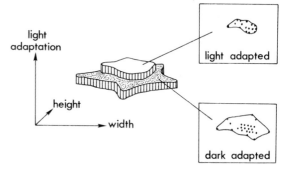

FIG. 13. The processing of visual field maps as shown by Schümperli, R. and Swihart, S. (1978). The spike responses to a particular stimulation are displayed two-dimensionally, the third dimension showing the influence of light adaptation on the receptive field.

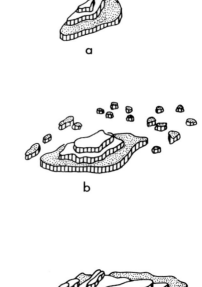

FIG. 14. Three different types of receptive fields from the butterfly visual neurons with three different adaptational levels. Lowest level dark-adapted, middle level intermediate adaptation, upper level light-adapted. The sizes of fields in the dark-adapted state are between 10° (**a**) and 50° (**b**) in diameter. (From Schümperli, R. and Swihart, S., 1978.)

Light-adaptation often reduces the field size from about 50° to 30° (Figs 13 and 14). In all these examples the neuron seems to exhibit a new adaptational strategy (Schümperli, R. and Swihart, S., 1978). Some fields become diffuse; some show no major structural changes; while some fields will consist of the main receptive area and their diffusely responding surrounding will disappear.

The frontal part of the butterflies' visual system contains small visual fields of about 6°. The corresponding neurons may be suitable for the detection and the identification of objects in the near-distance, whereas the large-field neurons (the movement sensitivity was not tested) are supposedly involved in course control and stabilization.

3.5 Receptive fields of higher order and multimodal neurons

These neurons can present a large variety of different types of potential responses, including graded potentials accompanied with spikes in several combinations. In the case of multimodal neurons the stimulus–response combinations become complex. Therefore a very important question is how to find an appropriate stimulus modality, quality and quantity combination?

Because most of the neurons converging on to large motor neurons in the thoracic ganglia respond to binocular stimulations of a complex nature, their responses are only diffusively correlated with the stimulus parameters. The visual fields are therefore structurally complicated and usually cover the whole eye field. In the cricket, *Gryllus campestris*, where four tracts connect the medulla with the lobula and medulla of the contralateral optic lobe, the receptive fields are most probably binocular and movement-sensitive, covering the eye field of both eyes (Honegger, H.-W. and Schürmann, F., 1975).

A large interneuron in the brain of the cricket, *Schistocerca gregaria*, responds to stimulation by wind blown on the head, and to light. It is bimodal with a phasitonic response to visual stimuli (Bacon, J. and Tyrer, M., 1978). The response of the neuron to both stimulus modes has been recorded at the tritocerebral commissure. Similar giant neurons have been found in some mutants of *Drosophila* (Levine, J., 1974). Complicated response patterns are initiated by flashes, which cause a triadic activity in the motor output. Descending movement detector (DMD) neurons in locusts seem to mediate similar functions: with a physiological input (vision)

and a behavioural response as output (jumping).

The descending contralateral movement detector (DCMD) neurons of locusts also respond vigorously to small targets (*ca.* 5°), but additional overlying rotating large-field stimulation can suppress the response (Pinter, R., 1979). The inhibition is decreased if the background stimulation extension is reduced below a 19° visual angle. This experiment substantiates a recently proposed lateral inhibition model for the acrididae movement-detector system (Palka, J., 1972; O'Shea, M. and Rowell, C., 1975; Pinter, R., 1979).

Another complicated group of neurons is called self-movement detectors (and will be discussed later). They are found in the ventral nerve cord of the dragonfly, *Anax junius* (Olberg, R., 1981). Those cells are multimodal and respond maximally to movement of very large patterns, or to the rotation of the animal itself. The response characteristics are unusually complicated and the input specifications are only poorly known.

All these neurons are interconnected in a complex way, and therefore their functional specialization is a matter for further investigation. The analysis offers extra difficulties because the state of the whole neuropil may change the function of a particular neuron.

4 VISUAL CELL INTERACTION

The anatomical and morphological arrangement of visual cells at different levels, in neuropils and chiasmata, provides a good basis for powerful optical, electrical and synaptic coupling between neighbour cells. The accounts of receptor and neuronal sensitivities, receptive fields and specialized organization of functional units have been made with the basic assumption that, on the other hand, each cell is electrically more or less isolated from its neighbours.

The basic claim of synaptic connections between cells is still valid, but certain other forms of interaction have recently been described. At the photoreceptor level the rhabdom structure (fused) may lead us to assume, in these ommatidium types, that there is a prominent optical coupling between receptor cells, and if they are connected by a low-resistance pathway we assume that the cells also share their electrical signals, and display a positive electrical coupling. In such a case, if the extracellular space is compartmentalized, an inhibitory coupling between cells can be proposed (Shaw, S., 1975; Zimmerman, R., 1978).

4.1 Optical coupling

Optical coupling means that several cells share the same dioptric system, through which the light has to pass before entering the cells. In certain cases optical coupling of the cells may physically process the light stimulus before it enters the receptor cell. In the layered fused rhabdoms the cells are always optically coupled (for some insect groups see Table 1). The centrally located tandem rhabdomeres of dipterans or lateral optical filter effects in the fused rhabdoms process the light stimulus in this way (Snyder, A. et al., 1973).

Optical coupling of receptor cell structures always has complex implications on the stimulus quality. The significance is found in several functions such as the photoreconversion processes, light and dark adaptations and their pigment migrations.

For further discussions of the extensive research on visual optics during the past few years see the chapter by M. Land in this volume.

4.2 Electrical coupling

Electrical coupling in the simplest form is characterized by a low-resistance pathway between neighbouring cells (electrical coupling by tight junction). Cells are not usually electrically coupled, because it would reduce the degree of freedom in the neuron's chemical network. Coupling has been found to exist in the ventral cord of different invertebrate species, e.g. crayfish giant motor nerve (Furshpan, E. and Potter, D., 1959; Watanabe, A. and Grundfest, H., 1961). If the receptor membranes in the retina have close contacts (particularly in the fused rhabdoms) some degree of electrical coupling can be expected.

Electrical coupling is usually, if it occurs, positive among the receptors, which then share their signals. If the electrical influence from the closest neighbour is inhibitory by nature, it will be related to extracellular field potentials. Electrical coupling between photoreceptors was shown to exist in the ommatidium of Limulus (Smith, T. et al., 1965;

Smith, T. and Baumann, F., 1969). The cells within each ommatidium are electrically coupled by virtue of low-resistance junctions between neighbour retinular cells and between retinular cells and the dendrite of the eccentric cell. The properties of these junctions are such that the light-evoked depolarization of the retinular cells causes a depolarization of the eccentric cell which, in turn, leads to initiation of spike potentials which are propagated along the eccentric cell axon.

By using electrical symbols we can characterize an ommatidium as a model network (Smith, T. and Baumann, F., 1969). Even if, in such a network, the electrical resistances are not completely ohmic, some dependence on the polarity and the amplitude of the voltage applied will also exist. Thus by using sufficiently large depolarization of a retinular cell the coupling resistance will increase, and as a result this retinular cell will become more and more uncoupled from its neighbours. The phenomenon may provide a mechanism for pooling of receptor cells using as a criterion the excitation (adaptation level) in receptor cells of the eye.

Another kind of coupling between the retinular cells was found in the median ocellus of Limulus, which contains two different spectral types of cells. One type is maximally sensitive to UV light, the other to green light; they are called UV- and VIS- (visible, 400–600 nm) cells, respectively (Nolte, J. and Brown, J., 1972). The retina is divided into small groups (functional units), where both UV-cells and VIS-cells can occur, even in the same group. The UV-cells in the same group are electrically coupled to each other and to a spiking arhabdomeric cell, analogous to the eccentric cell in the ommatidia of the lateral eye. The anatomical basis for a low-resistance connection was confirmed by injecting dyes of different colours into the cells.

Parallel results have been found in the crayfish retina (Kuwabara, M. and Naka, K.-J., 1959; Muller, K., 1973) and in the simple visual system of the Balanus lateral ocelli (Shaw, S., 1972). Shaw found that the voltage change in one cell led to a voltage change in the other recorded cell. The relation between the voltage changes (V_A/V_B) in the cells, the coupling ratio, varied if different pairs were compared in the same and different preparations (Fig. 15a,b).

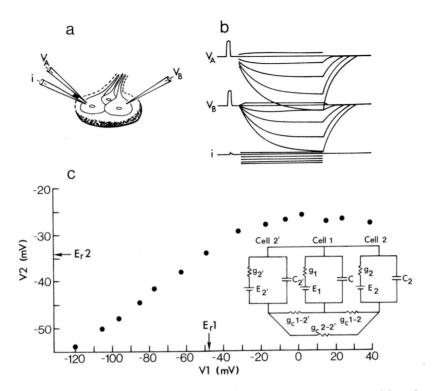

F<small>IG</small>. 15. Electrical coupling between barnacle (*Balanus*) visual receptor cells. **a** and **b:** Current (*i*) passed through an electrode will produce a voltage V_A in one cell and a slightly smaller voltage V_B in another cell. Six sweeps are superimposed (**b**) to show the coupling as a function of the passing current. Calibrations 20 mV and 0.1 s, the maximum current 12 nA (after Shaw, S., 1972, with permission of the Physiological Society). **c:** Another preparation of *Balanus* shows a relation between the membrane potential of one cell (V2) as a function of another cell (V1) in which the membrane voltage is clamped in absence of illumination, 1.5 s after step changes of membrane potential in cell 1. The cells become uncoupled, when (V1) is shifted to large positive voltages. Inset: equivalent electric network (electromotor forces, *g* conductances, *C* capacitances) of the photoreceptors in the barnacle ocellus (from Brown, H. *et al.*, 1971, with permission of the Fed. Amer. Soc. Exp. Biol.).

In *Balanus* the receptor cells uncoupled with increasing illumination (Fig. 15c) as in the *Limulus* median ocellus. The drop in the coupling ratio from 0.88 in the dark to 0.32 in steady illumination can be explained by the non-linear characteristics of the current–voltage relations of the membrane in the dark (Brown, H. *et al.*, 1971).

By simultaneous double-electrode recordings from the crayfish compound eye, combined with dye injection, it was shown that in each ommatidium the cells which are sensitive to the same orientation of the *e*-vector of polarized light are electrically coupled (Muller, K., 1973). The cells having orthogonal polarized light sensitivity were not coupled. The crayfish ommatidium contains a layered rhabdom and the electrical coupling is suggested via the tips of microvilli, because the cells with parallel microvilli lie opposite to each other. This kind of rhabdom is also found in some other

insects (Table 1), Thysanura, Coleoptera, Lepidoptera (Meyer-Rochow, V., 1971), where a similar electrical coupling mechanism can be easily accounted for.

The pairs of cells in the crayfish ommatidia showed unity coupling ratios, and the noise within them was strongly correlated. Because the pair of cells also showed both similar PS and spectral sensitivities, and in all other cases the cells were uncoupled, one can possibly also conclude that this complete coupling is an artefact caused by the double penetration of a single cell. The coupling via the tips of microvilli still seems to remain a somewhat open question in the crayfish ommatidia.

Our knowledge of electrical coupling in insect retinular cells is at present limited to only a few recordings from locust (*Locusta, Schistocerca*), dronebee (Shaw, S., 1967, 1969) and blowfly, *Calliphora* (Rehbronn, W., 1972) retinular cells.

Simultaneous recordings from two immediately adjacent receptor cells are difficult to perform. Shaw, S. (1969) has claimed electrical coupling in locust and dronebee receptor cells by using the following functional criteria for proper recordings: the cells in the same ommatidium should have identical receptive fields, and the spikes on one record should always occur or fail to occur in synchrony with those on the other recording, allowing small differences, however, in amplitude.

The electrophysiological experiments, though based only on the above criteria, do not unambiguously show that cells are electrically coupled or in the same ommatidium, and do not rule out completely the possibility of recordings originating from the same cell because the criteria logically imply coupling. Recent findings on worker bee receptors pinpoint the possibility of artefactual coupling created by the micropipette used for recording. Bernard, G. and Wehner, R. (1980) claim no electrical coupling at all between the green and the UV-receptor cells by using optical, non-invasive recording technique of pupillary responses. However, the response dynamics is very slow compared with that of the membrane potentials, and thus still leaves the question open, whether the coupling is performed by a mechanism other than that which the pupil response will reveal.

However some anatomical evidence may strongly suggest a possible coupling site in the lamina of the fly, where Ribi, W. (1978) found some gap junctions between six receptor cell axon terminals in each optic cartridge. Gap junctions are not found in the receptor cell layer in the fly, but they exist between neighbouring receptor cells in the ocellus (Eaton, J. and Pappas, L., 1977) and in the locust compound eye (Shaw, S., 1977) as pointed out before.

Recently Chi, C. and Carlson, S. (1980a,b) have demonstrated special junctions other than chemical synapses between neurons and glia cells. We have good reasons to assume that these connections are functional, and should also be considered in schemes for electrical coupling. The finding fits in well with the lamina, where the photoreceptor depolarizing potentials are summed together. The stimulation of only one single receptor cell gives a response which is only a sixth of that in the receptor cell soma if recorded from the receptor cell axon terminal (Scholes, J., 1969). The reduced level in

these potentials can be considered as further evidence for the electrical coupling of axons (Smola, U., 1975).

At the moment there is no clear picture of the possible advantages for the information processing gained by using electrical coupling. We have mentioned the regulation of the size of the receptive field during light and dark adaptation (pooling of receptor cells or their second-order cells). Furthermore Fain, G. (1975) has proposed that the single photon events in receptors will have more chance of being detected because electrical coupling spreads the signals through many synapses.

On the other hand Ribi, W. (1978); also (Laughlin, S., 1981) has suggested that some advantage may be gained by improving the signal to noise ratio before synaptic transmission. By reducing the noise level the range of voltages driving the synapse is reduced, thus minimizing the distortion introduced by non-linearities in the synaptic transmission. A likely role for photoreceptor coupling is then as follows: at low light intensities, lateral inhibition is ineffective (Barlow, H. et al., 1957), electrical coupling between photoreceptors reduces the noise in the resultant signal and perhaps also causes some loss in acuity. At moderate light intensities noise is unimportant, lateral inhibition is effective and photoreceptor coupling improves processing of signals by the neural networks of the retina (Marčelja, S., 1980). At very high light intensities the cells are uncoupled again to protect the cell membranes and synapses from overdrive. Intensity-dependent neural summation was also proposed according to the recordings from direction-sensitive motion detectors (DSMD) at different illumination levels, giving a hint of one possible adaptational strategy in the fly (Dvorak, D. and Snyder, A., 1978).

As a speculation the electrical coupling could also prevent imbalances in receptor function, caused by distortions in development and growth of synapses in lamina. We still have no clear idea how widely distributed electrical coupling will be in higher visual centres of insects.

4.3 Field potential interaction

The flow of electric current in field potentials is determined by extracellular structures. In a tightly packed array of cells the extracellular resistance in narrow channels may force the electric current to flow through neighbour cells and thus cause changes in their membrane potentials.

Large voltages, up to 40 mV (Mote, M., 1970), are found in the extracellular space (ECS) of insect eyes, especially if wide-field light stimuli are used to excite many receptor cells at the same time.

The photocurrent of crayfish (*Procambarus*) retinular cells can produce large (10 mV) negative extracellular potentials when measured near the photoreceptive membranes within ommatidia (Muller, K., 1973).

The peripheral extracellular voltage drop down to −55 mV in the fly (*Musca*) peripheral retina was located in the synaptic region of the lamina ganglionaris. An even larger, more proximally located voltage drop of −70 mV was found in the medulla (Fig. 16), although the interpretation of the recordings is not technically clear or unambiguous. In addition to direct measurements of extracellular resistances with the microelectrode, perfusion of the retina with a tracer confirms to some extent the hypothesis of a high resistance in the lamina region (Shaw, S., 1977; Zimmerman, R., 1978).

Shaw, S. (1979) reports a consistently large ECS conductivity change around the receptor cell axons in locust, occasionally as much as 90:1. The extracellular resistance is highest around the axons, down to 10- to 100-fold less near the cell bodies, and intermediate in the lamina and below.

The analysis of extracellular resistances and light-induced extracellular potentials shows that, at least in locust, the retina can be isolated from the lamina and the body cavity by a substantial resistance barrier, located just beneath the basement membrane (Shaw, S., 1975, 1977, 1978). The photocurrent entering the receptor cell may hyperpolarize the surrounding extracellular space, because it has to return either across the barrier or across the membranes of adjacent and less strongly stimulated receptor terminal membranes. A proposed equivalent diagram in Fig. 17 shows the current flow through different areas of the peripheral retina (Zimmerman, R., 1978). The voltage drops are still hypothetical and need to be determined more reliably.

Menzel, R. and Blakers, M. (1976) interpreted the hyperpolarization in their UV-sensitive cell as being caused by the inhibition of the receptors

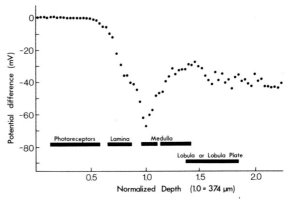

FIG. 16. The potential difference profile monitored with a microelectrode in a dark-adapted eye of *Musca* without stimulation. The markers show approximate boundaries at which named structures may be found in histological sections. In normalized depth 1.0 corresponds to approximately 374 μm. (Data after Zimmerman, R., 1978, with permission of Springer-Verlag, Heidelberg.)

sensitive to longer wavelength stimulation. However, as pointed out by Shaw, S. (1975), current passively leaving the receptor soma during this type of interaction will depolarize this part of the cell. A similar kind of interference between first- and second-order cells in the lamina was suggested by Autrum, H. *et al.* (1970).

One should analyze the complete circuit for return currents, because the potentials set up by return currents can bias recordings, which are made relative to a distant indifferent electrode in the body cavity. The quantitative application of field potential analysis is limited by certain technical problems. An accurate measurement requires the imposition of a uniform current density through the tissue, a condition not as easily produced in the fly as in the vertebrate retina (Hagins, W. *et al.*, 1970). This also means that it is not possible to relate the calculated transverse current density ($nA \cdot \mu m^{-2}$) to currents through single types of cells as performed in Fig. 17 (Zimmerman, R., 1978).

By using an extracellular broken glass pipette as an indifferent electrode placed close to the recording electrode, Shaw, S. (1975) was able to show some potential differences between different locations of the extracellular reference electrodes. It seems obvious that compartmentalization should be taken into account if careful comparisons of intra- and extracellular recordings are needed. The significance of extracellular field interaction

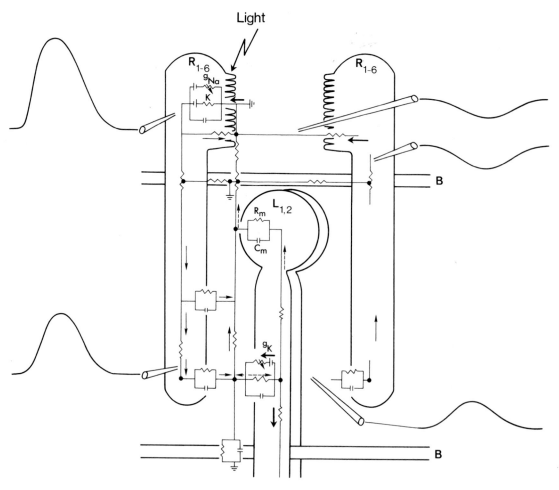

FIG. 17. A schematic presentation of photocurrents generated by a photoreceptor and second-order neurons in the lamina. The photocurrent enters an active receptor after having been initiated by light. The return current is provided by leakage from the cell body or axon. Heavy solid arrows denote currents flowing through conductance changes in the membrane of rhabdomeres and in the postsynaptic membrane of the second-order neuron. Thin solid arrows show the passive return currents flowing down the cells and through extracellular space. Thin dashed arrows show currents producing the depolarizing component in the potential response of the second-order neuron. The response polarities are shown schematically, whereas the micropipettes show their origin. The inversion of the response in an unstimulated receptor cell body is caused by high resistance barrier (B) in lamina and thus can be considered an earth (Shaw, S., 1975). For the same reason the light-induced depolarization can be recorded (bottom right) from the extracellular space in lamina. (After Zimmerman, R., 1978 and Laughlin, S., 1981, with permission of Springer-Verlag, Heidelberg.)

between different cartridges and types of cell terminals is still a matter of discussion. One explanation is the lateral suppression of potential responses via lamina electrical field to give lateral inhibition reported in the fly lamina (Zettler, F. and Järvilehto, M., 1972; Zettler, F. and Autrum, H., 1975). This idea is supported by the finding that the frequency response functions of second-order cells to point and light field stimuli can be explained by a fast interaction between the terminals in the lamina (French, A. and Järvilehto, M., 1978). Furthermore the extracellular potentials set up in the lamina must play an important role in setting the sensitivity of synapses in the lamina, though we know very little about this as yet.

4.4 Lateral inhibition

A phenomenon related to lateral inhibition was first described in sense organs by E. Mach (1866). He

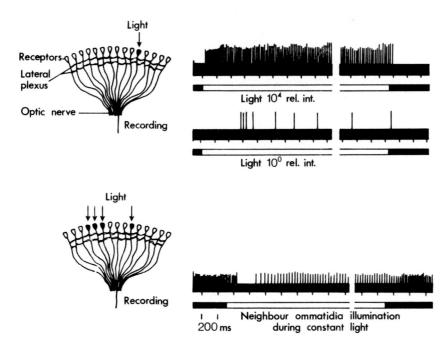

FIG. 18. A recording from a single optic nerve fibre of *Limulus*. Upper recordings: responses to single light spot stimulation with two different light intensities. Each recording interrupted for several seconds for the steady state. Lower recording: an inhibition of the activity of a steadily illuminated receptor unit produced by illumination of a number of other units near it. (Modified after Ratliff, F., 1971; recordings from Hartline, H. *et al.*, 1952, 1956.)

observed spatial transients (the Mach-bands) in the human visual system by studying perception. For nearly 80 years this phenomenon was rejected, because the effect was not accepted as a result of inhibition. This suffered from a misconception about the inhibition in the motor activity. The studies on the *Limulus* compound eye by Hartline, H. (1949) and his co-workers (Ratliff, F. and Hartline, H., 1959 and Ratliff, F. *et al.*, 1963) have stimulated considerable interest in spatial networks in the visual system.

The role of lateral inhibition is closely related to three classes of functions: image restoration, feature extraction and sensitivity adaptation. Our knowledge about the different functions is far from complete, but it is still enough to give us an idea about the significance of lateral inhibition.

4.4.1 THE CONCEPT

The visual system can be regarded as a set of spatial and temporal frequency filters. The spatial and temporal distribution of mutual inhibitory influences gives rise to powerful mechanisms of contrast enhancement. Transient responses produced by inhibition signify optimized feature detection by sacrificing the absolute sensitivity. Lateral inhibition is not only common in visual systems, but can also be found in other sensory systems.

4.4.2 THE MECHANISM

At present relatively little data are available concerning the mechanisms in insect visual systems (blowfly, *Calliphora erythrocephala*, Zettler, F. and Järvilehto, M., 1972; dragonfly, *Hemicordulia tau*, Laughlin, S., 1974). Therefore we are forced to adopt some results from one of the best-studied and closely related objects on this topic: *Limulus*. Its compound eye contains elements which are simpler than the neural elements in highly developed insect visual systems. We can generalize some of the basic findings from *Limulus* (Xiphosura) concerning lateral inhibition. The inhibitory properties of the eye are usually studied in the excised eye, but there are no qualitative differences *in situ*, though the inhibitory actions are stronger (Barlow, R. and Fraioli, A., 1978).

The fundamental experiment for studying the lateral inhibition in *Limulus* consists of singling out, arbitrarily, an ommatidium with its optic nerve fiber, illuminating it steadily, and then testing the effect on its impulse discharge of activating neighbouring receptor units. As shown in Fig. 18, a recording from single optic nerve fibres, the illumination of a small group of ommatidia neighbouring a steadily illuminated test receptor resulted in a diminishing of the discharge (Hartline, H. *et al.*, 1952, 1956). The higher the intensity of illumination on the group of neighbours, the greater was the decrease in frequency of impulse discharge by the test receptor (Hartline, H., 1959).

A similar kind of experiment shows that, within limits, the greater the number of neighbouring ommatidia illuminated, the greater is the inhibition exerted on a test receptor (spatial summation of inhibitory influence). Also, near neighbours exert a stronger inhibition on a test receptor than distant ommatidia. More exactly, somewhat stronger effects are exerted by receptors at a slight distance (*ca.* 5 ommatidia) than by the very nearest neighbours. (See the notch and hump on both sides of the edge in Fig. 19 and Fig. 21). Beyond 10 ommatidia inhibitory influences generally become imperceptible (Barlow, R., 1969; Barlow, R. and Quarles, D., 1975).

Over a wide range, the level of activity to which a test receptor is excited makes only a small difference in the decrease in frequency that is produced by a given group of neighbours, illuminated steadily with fixed intensity. Thus we may write a linear approximation

$$i \approx (e - r) \qquad (1)$$

where the magnitude of the inhibition (i) can be expressed by the differences in the frequency of discharge of the test receptor, when it is excited alone (e) and, when it is responding to the same illumination, but subject to the illumination exerted upon it (r).

The inhibition exerted on its neighbours by a receptor unit depends on its activity, but its activity is affected by the inhibition exerted on it by those very neighbours whose activity it affects. The mutually exerted inhibitory influences act recurrently — as schematically shown in Fig. 20.

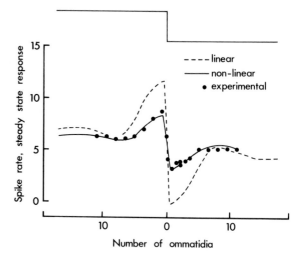

FIG. 19. Spatial transients (the Mach-bands) generated by lateral inhibition. The steady-state responses are compared with the response profiles, which are predicted from the experimentally measured inhibitory field (Barlow, R., 1969). The solid curve connects the response rates computed for the non-linear model based on the Hartline–Ratliff equation (Barlow, R. and Quarles, D., 1975). For comparison the broken line shows a computed response predicted by a piecewise linear approximation.

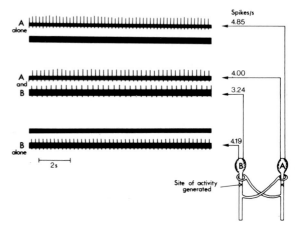

FIG. 20. A recurrently exerted mutual inhibition of two ommatidia located beside one another in the lateral eye of *Limulus*. Stimulation was performed by two adjacent small spots with fixed intensities and constant illumination. Note the decrease of the spike frequency when both ommatidia are stimulated simultaneously. (Data from Hartline, H. and Ratliff, F., 1957, reproduced from *J. Gen. Physiol., 40,* by copyright permission of the Rockefeller University Press; and from Ratliff, F. *et al.*, 1963.)

4.4.3 STEADY-STATE PROPERTIES OF LATERAL INHIBITION

As shown by Hartline, Ratliff and their co-workers, lateral inhibition generates transient activation patterns in the influenced cell response to the stimuli which are sustained in either space or time (see the review by Hartline, H. and Ratliff, F., 1972). To describe the simultaneous interaction of more than two elements, more equations (1) are required. For a group of n interacting receptor units a set of n simultaneous equations, in linear approximation, must be written, and in the equation for each unit inhibitory terms must be introduced and summed to express the inhibition on that particular unit by all of the units that act upon it.

$$R_p = E_p - \sum_{j=1}^{n} K_{p,j} (R_j - R_{p,j}^0)\, p, j \qquad (2)$$
$$= 1, 2, \ldots n, \quad p \neq j$$

In this set of equations, R_p is the response of the pth receptor, which if free of inhibition would have discharged impulses at a rate E_p, but which is subjected to the summed inhibitory influences expressed by the linear terms on the right. In each term $K_{p,j}$ is the inhibitory coefficient measuring the action of the jth receptor on the pth; $R_{p,j}^0$ is the associated threshold of the action (Hartline, H., 1969).

In the eye, receptors are deployed spatially, in a mosaic, and the strength of their interaction, as already noted, depends on their separation. In general the coefficients K decrease and the thresholds R^0 increase with increasing separation of interacting ommatidia. The spatial distribution of values of the coefficients in the inhibitory field surrounding a small group of receptors has been mapped in detail in *Limulus* (Barlow, R., 1969). It shows that detectable lateral inhibition extends over 30% of the eye's surface and is strongest 3–5 ommatidia away from the investigated unit (Fig. 21). The optical isolation of single ommatidia was experimentally carried out by removing the cornea and stimulating the individual naked ommatidial cups with a fiber optic (Johnston, D. and Wachtel, H., 1976). By incorporating into the light source an extracellular electrode, the eccentric cell activity was monitored. The inhibitory fields proved to be slightly larger at a maximum of up to 9–10 ommatidia, but showed basically the same non-linearities as mentioned before.

The receptor units in the eye are more sensitive to lateral inhibition at some levels of excitation than they are at others. The steady-state inhibition is not

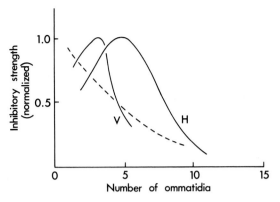

FIG. 21. The strength of inhibitory function of retinal separation measured in vertical (V) and horizontal (H) planes. Broken line shows the function without inhibition. (Data based on Barlow, R., 1969; Johnston, D. and Wachtel, H., 1976; modified after Laughlin, S., 1981, with permission of Springer-Verlag, Heidelberg.)

directly proportional to the response levels of neighbouring units. The inhibition seems to equalize the response rates of brightly and dimly illuminated ommatidia (Barlow, R. and Lange, G., 1974). Such dependence of the sensitivity to inhibition on the incident light introduces a non-linearity and we may write the equation (1) simply

$$i < (e - r) \qquad (3)$$

The linearity of the inhibition is only maintained at low intensities, when the number of activated inhibitory channels is restricted and the voltages caused will sum linearly, because of the same amount of transmitter leased for each spike. At higher spike rates a non-linearity gets more apparent. The inhibition increases with increasing spike-rate, but decreases after rates of $30–60\,\mathrm{s}^{-1}$ (Fig. 22a). The saturation of the inhibition depends on the distance between the inhibiting elements and the strength of inhibition is a non-linear function of the spike-rate of the inhibiting neuron (Fig. 22b).

The inhibition can also be facilitated if the inhibiting unit is previously stimulated. Facilitation of inhibition can be accumulated as the number of impulses in the first burst increases up to a maximum. The time constant of the decay is several seconds (Graham, N. et al., 1973). This facilitation mechanism seems to be similar to that which has been postulated for facilitation of excitatory events at the neuromuscular junction.

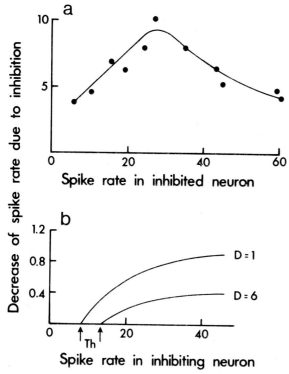

FIG. 22. The non-linearity of lateral inhibition: (a) expressed in the spike rate in inhibited neuron (after Barlow, R. and Lange, G., 1974, by copyright permission of the Rockefeller University Press) and (b) shown as an increase with the spike rate in the inhibiting cell and the distance (D) in number of ommatidia (Johnston, D. and Wachtel, H., 1976, by copyright permission of the Rockefeller University Press; Laughlin, S., 1981, with permission of Springer-Verlag, Heidelberg). Th is the threshold in different retinal separations.

Quantitative measurements of the activity of interacting receptors and groups of receptors, in various configurations, are satisfactorily accounted for by measured or postulated inhibitory fields or spatial patterns such as the Mach-bands (Reichardt, W. and MacGinitie, G., 1962; Ratliff, F. et al., 1963). The applications to other sensory systems have been discussed in great detail (von Békésy, G., 1966).

4.4.4 DYNAMICS OF LATERAL INHIBITION

Whenever, as in the natural visual world, changes occur in the patterns of light and shade on the retinal mosaic, receptor transients occur, new distributions of excitation are established, and readjustments of the inhibitory interactions are mediated over the retinal network. The interplay between excitation and inhibition is a dynamic process.

Once the dynamic properties of the excitation, self-inhibition and lateral inhibition have been expressed in terms of their individual time courses (Lange, D. et al., 1966) or, equivalently, in terms of their transfer functions (Knight, B. et al., 1970; Brodie, S. et al., 1978) the steady-state equations (2) may be extended to include the dynamics of the whole system. The latter form (transfer function) has proven to be economical and fruitful.

If I_p is the sinusoidal modulation of light intensity on ommatidium p, then the corresponding modulation of the generator potential E_p is:

$$E_p = G_p(f)I_p(f) \tag{4}$$

where $G_p(f)$ is the transfer function, the complex number, whose amplitude and phase represent the amplitude gain and phase of the generator potential E_p with respect to the modulation I_p at the frequency f.

The dynamic function above the threshold of inhibition can be written,

$$R_p = E_p - S_p T_s(f)R_p - T_l(f) \sum_{p=j} K_{p,j} \cdot R_j \tag{5}$$

where R_p and R_j are instantaneous spike rates. $T_s(f)$ is the self-inhibitory transfer function and $T_l(f)$ the lateral inhibitory transfer function, so scaled that they are unity when $f = 0$. Now, given the E_p from equation (4) properly scaled in terms of the instantaneous rate R_p, the equations (5) are in the form of a set of linear simultaneous equations, which can be solved at each modulation frequency for the amplitude and phase of the instantaneous rate R_p.

The self-inhibitory coefficient is very large ($T_s \approx 3.0$) approximately equal to the sum of all the lateral inhibitory coefficients ($\Sigma_j K_{p,j}$). The individual lateral inhibitory coefficients are very small ($K_{p,j} < 0.1$). Any particular $K_{p,j}$ varies somewhat with E_p (Hartline, H. et al., 1956). Furthermore, the $K_{p,j}$ may vary markedly with distance (Ratliff, F. and Hartline, H., 1959).

By using these approximately chosen summed lateral inhibitory coefficients within boundaries all lateral inhibition may be represented by a single transfer function (Ratliff, F. et al., 1974).

The analysis of the dynamic responses of an eccentric cell, with and without simultaneously modulated illumination of particular neighbours, indicates an effect equivalent to self-inhibition acting via a first-order low-pass filter with time constant 0.42 s, and steady-state gain about 4. The corresponding filters for lateral inhibition require time constants from 0.35 to 1 s and an effective finite delay of 50–90 ms (Biederman-Thorson, M. and Thorson, J., 1971).

Intracellular recordings from retinular and eccentric cells combined with perfusion or iontophoretically applied serotonin (5-HT) have been shown to hyperpolarize the membrane potential in both cell types (Adolph, A., 1976). Iontophoretic application of serotonin to the synaptic neuropil in *Limulus* compound eye produces an inhibition similar to lateral inhibition, and thus could be a transmitter mediating lateral inhibitory effects.

4.4.5 SELF-INHIBITION

A rapid decrease of response amplitude or in rate of discharge of impulses is observed following the onset of illumination or a step increment in illumination. Stevens, C. (1964) attributed this diminution of rate of impulses to a then hypothetical process, self-inhibition, generated by the selfsame eccentric cell. Subsequently Purple, R. (1964) and Purple, R. and Dodge, F. (1965, 1966) were able to observe this self-inhibitory (negative feedback) process directly in intracellular records of the generator potential.

A hyperpolarization of several millivolts amplitude and about 0.5 s duration follows each action potential. Although it has not been possible to isolate in detail the encoder process from self-inhibition, the assumption of a synaptic basis of self-inhibition is supported by observation of approximated linear summation of unitary hyperpolarizations associated with conductance changes (Purple, R. and Dodge, F., 1965, 1966). The effect of self-inhibition is to apply frequency-dependent negative feedback.

If the time constant of the decay is λ and if the strength of inhibition is measured by a coefficient K, a reasonably accurate approx-

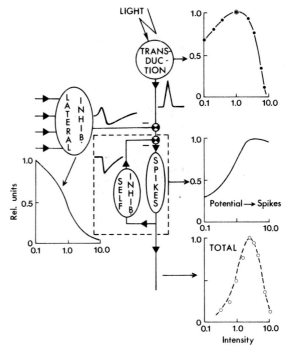

FIG. 23. A block diagram showing the frequency response functions for the processes of transduction, self-inhibition summed with spike generation process and lateral inhibition. The broken line at the bottom is a function predicted by equation 7, and follows closely the empirical data (open circles). Thick curves are impulse responses to transduction, lateral and self-inhibition. (Based on data from Knight, B. *et al.*, 1970; after Laughlin, S., 1981, with permission of Springer-Verlag, Heidelberg.)

imation of the expected frequency response in given by:

$$S(\omega) = B(\omega) \cdot (1 + K)/(1 + K/(1 + i\omega\lambda)) \qquad (6)$$

To measure the encoder transduction, a sine-wave current was injected while the receptor responded to a step of light intensity. Such data agree well with equation (6) where λ is about 0.5 s and K ranges between 2 and 3 (Knight, B. *et al.*, 1970).

The frequency response of the generator potential can be measured in isolation from spikes and self-inhibitory potentials by poisoning the spike mechanism with tetrodotoxin. The frequency response function is inhibited at low frequencies, so producing by summing pulses a large sustained hyperpolarization (Fig. 23).

If a single photoreceptor is stimulated by modulated light, its overall frequency response $R(\omega)$ should be predicted by the product of the

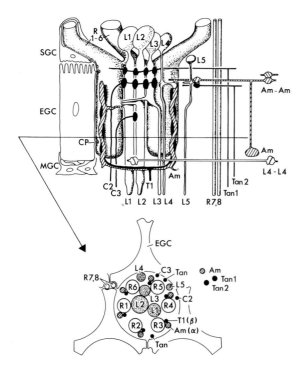

FIG. 24. The neuronal composition of a single optic cartridge in the synaptic nucleus of a dipteran eye. The synaptic interconnections are described in Table 6. The pathways in and through this neuropil can be divided into three major classes: efferent pathways, efferent feedback pathways and lateral interaction pathways. The receptive fields of different neuronal elements may be estimated by the ramifications and number of presynaptic connections to a particular cell, thus divided into roughly two groups: wide-field and narrow-field cells. The upper diagram is a section along the optical axis, and the lower one shows an approximate topography of the elements in a cross-section of the cartridge. R1–8, photoreceptor axons; L1–5, monopolar neurons; C2–3, Tan 1–2, efferent pathways; Am, lateral pathway; SGC, EGC, MGC, different glial cell types; CP, capitate projections; α, β complex fibre pair with reciprocal synapses. (After Strausfeld, N., 1976b; Shaw, S., 1981, with permission of Cambridge University Press, Cambridge.)

generator frequency response $E(\omega)$ and the self-inhibited encoder $S(\omega)$, i.e.

$$R(\omega) = E(\omega) \cdot S(\omega) \qquad (7)$$

This is shown at the foot of Fig. 23, with the measured frequency response function. The predicted response is shown as the smooth curve, and its good agreement with the measurements allows the conclusion that no important mechanism modifying the dynamic response of a photoreceptor has been neglected. The origin in self-inhibition may be

found in the feedback synapses from eccentric cell collaterals (Schwartz, E., 1971).

4.4.6 LATERAL INHIBITION IN INSECT VISUAL SYSTEMS

Lateral inhibitory connections, similar in function to those of *Limulus*, are presumed to function in the insect eye. The quantitative properties of the mechanism could markedly enhance the visual acuity in different species of insects. Indeed the lamina ganglionaris as the first synaptical coupling station (Fig. 24) contains, in higher insect orders such as Diptera and Hymenoptera, many neuronal elements which can produce many lateral connections (Table 6; Strausfeld, N. and Campos-Ortega, J., 1977; Shaw, S., 1981; see also the chapter in volume 5 by J. Campos-Ortega and V. Hartenstein).

Neural summation as it exists in the fly seems to be particularly suited for lateral interaction, but the lamina of the bee, locust or dragonfly contains a complicated lateral plexus, and many different cell types and connections. Lateral inhibitory interactions in the insect lamina have been to some extent demonstrated by means of intracellular recordings from receptor terminals and second-order neurons (Fig. 7A,B) in the first optic ganglion of the blowfly *Calliphora* (Zettler, F. and Järvilehto, M., 1972), and dragonfly (Laughlin, S., 1974). Sustaining units were recorded probably close to the lamina with central excitation flanking with inhibition on both sides (Arnett, D., 1971, 1972). This type of unit may be attributed, for example, to monopolar cell L4, with a basal interconnection to other cartridges (Fig. 25). Similar phenomena are often explained by lateral inhibitory interaction in higher centres such as the medulla, and in the lobula, where the neurons generate and transmit complex impulse patterns and graded potentials (Horridge, G. *et al.*, 1965; Mimura, K., 1972; Frantsevich, L. and Mokrushov, P., 1977; Erber, J. and Menzel, R., 1977).

The behavioural reactions of the water bug *Velia caprai* (Hemiptera, Heteroptera) indicate the existence of lateral inhibition in its visual system (Meyer, H., 1971, 1972). A series of large inhibitory actions, as shown in Fig. 26, were found at regular intervals with an exponential spatial decline in function of the retinal distance. The maximally effective stimulus was correlated with multiples of the

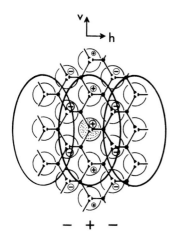

FIG. 25. The upper diagram shows a receptive field measured from a spiking neuronal element, a "sustaining unit", defined approximately by the iso-effect contours for central excitation (+) and two overlapping areas of inhibition (−) on both sides. The underlying neuronal network shows connections of monopolar neurons L4. The tripartite branching and interconnection pattern, as illustrated below, shows the supposed function of reciprocal synapses between all pairs. The recording does not exclude the influence of other cell types, such as amacrines or Tan 1–2 in producing this recorded type of lateral excitation. (Recording from Arnett, D., 1972; reciprocal synapses, Braitenberg, V. and Debbage, P., 1974; modified after Shaw, S., 1981.)

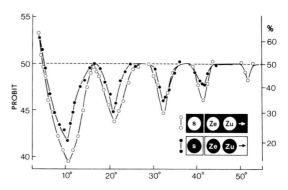

FIG. 26. Some evidence for lateral inhibitory processes in the visual system of the water strider *Velia caprai*. The relative stimulus effectiveness of black (●) and (○) circular discs, when a second disc is located at the side of the standard figure. The horizontal angular distance between the centres of the discs on the abscissa is plotted against a probit scale on the ordinate (a zero point 50% reaction). The multiples of the interommatidial angle (10.3°) show a correlation with increasing horizontal inter-stimulus distance. The curves show a comparison of optimal sized black ($\varphi4.0°$) or white ($\varphi3.4°$) central (Ze), additional (Zu) and standard (S) figures. (After Meyer, H., 1972, with permission of Springer-Verlag, Heidelberg.)

interommatidial divergence angle ($\Delta\Phi = 10.3°$). Parallel behavioural phenomena are also reported from the compound eye of *Limulus* exhibiting symmetrical bimodal inhibitory fields (Ratliff, F. *et al.*, 1969).

Lateral inhibitory mechanisms are generally supposed to exist in much wider distribution in the visual systems of insects than in the simple visual system of *Limulus*. Receptive field patterns, OFF-types of discharges, centre–surround organization, spatial and temporal frequency characteristics, and directional-sensitive units, all of which depend on complex arrangements of inhibitory connections, are found in insect and vertebrate visual systems.

Unit recording from motion-sensitive neuronal elements with preferred directions has been described for optic lobes and protocerebral regions of flies (McCann, G. and Foster, S., 1971), bees (Kaiser, W. and Bishop, L., 1970), moths (Swihart, S., 1968; Collett, T., 1971) and locust (O'Shea, M. and Rowell, C., 1975). It was also suggested that the lateral spread of inhibition could account for the spatially extended non-uniform peak response of DCMD and small visual interneurons in the locust *Schistocerca gregaria* (Catton, W., 1980). There is some evidence for inhibitory activity effectively reducing the DCMD discharge once a certain area of intensity is exceeded. The fact that the greatest rate of change in the ON/OFF ratio occurs within 2° indicates that inhibition is probably exerted over only a short range laterally, possibly no more than neighbouring ommatidia (Catton, W., 1982). In the cockroach (*Periplaneta americana*) the similar DCMDs show inhibitory interaction of the two light spots. It seems that this inhibition is non-recurrent (feed-forward) in nature, if response/intensity functions are compared in light- and dark-adapted conditions (Edwards, D., 1982a). An inhibitory effect was also found influencing visual responses in a fibre of the ventral nerve cord (Palka, J., 1967). The inhibition in higher levels is expressed more universally and will resemble habituation (Edwards, D., 1982b). Thus models of movement perception for body or object (Götz, K., 1968;

Table 6: *Dipteran lamina ganglionaris: connections to and from some classes of neurons (based on Strausfeld, N. and Campos-Ortega, J., 1977; Shaw, S., 1981, with permission of Cambridge University Press, Cambridge; see Fig. 24).*

Cell class	Lamina input from		Output to				Terminal in medulla
	Narrow	Wide	Narrow	Wide	Function	Field	
R1–R6	R1–R6 L2 C2 EGC	Tan 1	R1–R6 L1–L3 T1 (β)	Amacrine α	Photoreceptor	Narrow	—
R7/R8	—	—	—	—	Photoreceptor	Narrow	Narrow
L1	R1–R6 L2 C2/C3	L4 Tan 1	—	—	Pure output cell	Narrow	Narrow
L2	R1–R6 C2 C3	L4 Tan 1	R1–R6 L1	—	Output feed back	Narrow	Narrow
L3	R1–R6 C2	Tan 1	—	—	Output	Narrow?	Narrow
L4	other L4	Amacrines	L1, L2 L4	—	Output/lateral interaction	Narrow?	Narrow
L5	—	Amacrines Tan 2	—	—	Output	?	Narrow
T1	R1–R6	Amacrines s	EGC	Amacrines	Output?	Narrow	Narrow
Amacrine	R1–R6 T1 α⇌β	Other amacrines	T1 (α↔β)	L4 Amacrines	Intrinsic Lateral interaction	Wide	—
C2	—	—	R1–R6	—	Centrifugal	Narrow	Wider than one column
	—	—	L1–L3	—	Centrifugal		
C3	—	—	L1, L2	—	Centrifugal	Narrow	Narrow
Tan 1	—	—	R1–R6 L1–L3	—	Centrifugal	Wide	
Tan 2	—	—	L5	—	Centrifugal	Wide	

Poggio, T. and Reichardt, W., 1976) all rely on inhibition of some form.

5 ADAPTATION

Visual adaptation is a multilevel term, which in a very broad sense might comprise all those events that change the structure, form, function or behaviour of an organism. Adaptation is usually described as an optimizing process of an organism to its surroundings. The receptor properties are confronted with two opposing demands, viz. high incremental gain and large total operating range. This is accomplished by processes based on complex changes in sensitivity of the receptors which detect the stimulus. The alterations in the activity of the central nervous mechanisms underlying the response are called habituation, in distinction from adaptation (see review by Autrum, H., 1981).

The changes in sensitivity of a visual system are accomplished by several overlapping mechanisms:

(1) photomechanical adaptation (pupils, pigment migrations, morphological changes);
(2) photochemical adaptation (transformation of visual pigments, quantity of transmitters);
(3) photomembrane adaptation (membrane potentials, self-inhibition);
(4) neuronal adaptation (information processing in second- and higher-order neurons);
(5) diverse forms of adaptation (diurnal and circadian rhythms, hormonal effects, state of CNS, bilateral symmetry of stimulation).

Light- and dark-adaptation are not simply reversal processes, although they are logically coupled. Immediately after the switching ON or OFF of an adapting light the sensitivity of a photoreceptor starts to decrease or to recover. This happens even during the latency period, before any change of the membrane potential is detectable (Moring, J. *et al.*, 1979). The process usually exhibits an approximately exponential time course with a time constant that can vary from a few milliseconds to about half an hour depending on the species, cell types and stimulus conditions (Fuortes, M. and Hodgkin, A., 1964; Dodge, F. *et al.*, 1968; Autrum, H. and Kolb, G., 1972; Dörrscheidt-Käfer, M., 1972; Rosner, G., 1975).

5.1 Photomechanical adaptation

The retinomotor changes in the photoreceptor cells and accessory cells (crystalline cone, pigment cells) are called photomechanical. The movements of pigment granules are very species-specific and control the index of refraction between rhabdoms or rhabdomeres and the cytoplasm surrounding them (Stavenga, D., 1979).

The observation of so-called pseudopupils provides an optical means for studying photoreceptor processes in living and almost intact animals. Exner, S. (1891) classified and explained several different pseudopupil types: the principal pseudopupil, a single dark spot in the apposition eyes of many arthropods; the luminous pseudopupil, the eye shine visible in dark-adapted lepidopterans and crustaceans; and the accessory pseudopupils, the dark spots surrounding the principal pseudopupil in the eyes of butterflies and dragonflies.

The retinomotor phenomena can be revealed during light and dark adaptation by observing changes in pseudopupil. A large variety of structural changes in visual and accessory cells regulate the light flux to the rhabdomeres.

The following classification of insects into four groups is highly simplified (Walcott, B., 1975), but gives an idea of different strategies. We can find eyes with palisade formation and radial pigment movement (*Locusta, Periplaneta, Musca, Calliphora, Aeschna, Formica, Apis*). Some insect eyes show longitudinal pigment movements only (Bombycoidea, Saturniidae, Lampyridae, Sphingidae, Elateridae), while some others possess an iris pigment and retinular cell body movements (Megaloptera, Hydrophiloidea, Dytiscidae, Carabidae, Scarabaeoidea, Notodontidae, Tineoidea, Gelechioidea, Pyraloidea). The insects with acone eyes show extensive movements of cone cells, tracts and rhabdoms (*Notonecta, Forticula, Culex, Anoplognathus, Tenebrio,* Tipulidae, Chironomidae, Belostomatidae, Staphylinidae,

Rhodnius). The pigment movements in the perpendicular direction of the rhabdomeres in the visual cells of the fly act as a "longitudinal pupil" (Kirschfeld, K. and Franceschini, N., 1969). Here often in apposition eyes with fused rhabdom the pigment granules are separated from the rhabdom by vacuoles, forming a clear region known as the palisade (Fig. 27). The physiological effect of this change can be found as a decrease of absorption and increase of the refractive index in the surrounding rhabdom medium during dark adaptation (Snyder, A. and Horridge, G., 1972). The pupillary granules with their characteristic absorbance will change the spectral sensitivity of the eye during light adaptation, as found by Hardie, R. (1979) and explained recently (Vogt, K. *et al.*, 1982).

The mechanisms of retinomotor movements during adaptation are poorly understood (see review by Autrum, H., 1981). The microtubule and filament system has not been ruled out, and it is well known that these are present in an oriented longitudinal direction in many retinal cells (see, e.g., Meyer-Rochow, V., 1972; Miller, W. and Cawthon, D., 1974; Miller, W., 1975). If the protein tubulin is bound to colchicine the pigment moves into the light-adapted position. Thus the tubules are probably used to maintain the pigment in the dark-adapted position, but other mechanisms must also be present to account for the movement during light adaptation (Olivo, R. and Larsen, M., 1978).

Another hypothesis of retinomotor movements postulates an increase and decrease of rhabdomere volume by converting endoplasmic reticulum to microvilli in the visual cells (locust: Horridge, G. and Barnard, P., 1965; multilamellar bodies of *Lithobius forticatus*: Bähr, R., 1972).

Cell depolarization may also play a role in maintaining the pigment granule position. Prolonged depolarizing afterpotential, produced by an intense blue light stimulation on *Drosophila*, keep the pigment granules in the light-adapted state as long as the depolarization was maintained, over 20 s. Red light stimulus produces a fast-proceeding dark adaptation as well as producing pigment migration and repolarization (Lo, M.-V. and Pak, W., 1981).

One hypothesis claims that the pigment migration is based on electrophoresis driven by the membrane potential (Hagins, W. and Liebman, P., 1962; Stavenga, D. *et al.*, 1975). This probably does

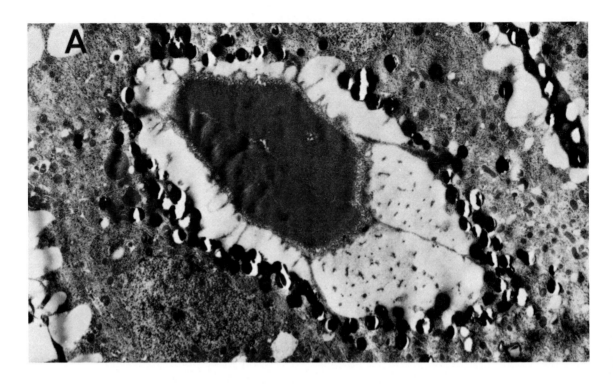

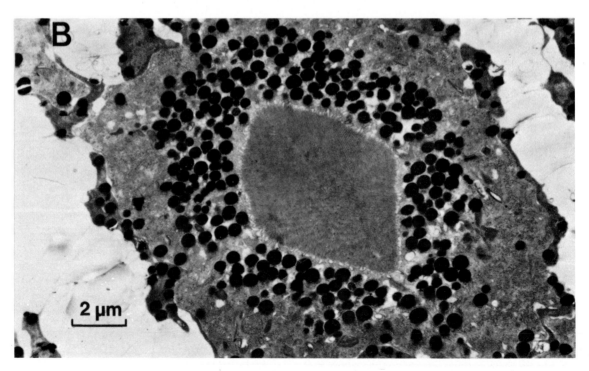

FIG. 27. Two adaptational states of an apposition eye of the stonefly (*Isoperla*) characterized by ultrastructural changes across the rhabdom. **A:** The dark-adapted rhabdom is surrounded by a clear region known as the palisade. **B:** The light-adapted state of the ommatidium with no palisade but tightly packed pigment granula surrounding the rhabdom.

not hold, because of the different time constant of membrane depolarization and because the change caused by dimethyl sulphoxide does not result in pigment migration (Miller, W., 1975; Olivo, R. and Larsen, M., 1978; Hamdorf, K. and Höglund, G., 1981). Also, stimulation with very short light pulses during the latency period of the first response shows in natural conditions that a build-up of the membrane potential is not needed for light adaptation (Moring, J. *et al.*, 1979). The mechanism for light adaptation may be explained by an inhibitory agent (Ca^{2+}?) which is released at early stages leading to the photoresponse.

The acceptance angle of the photoreceptor cells depends on the state of adaptation (Fig. 28) and the time of day in some apposition eyes, e.g. Australian mantid, *Tenodera australensis* (Rossel, S., 1979). In the foveal region the photoreceptors may increase the acceptance angle over 270% from 0.74° when light-adapted to 2° when dark-adapted at night. The night-active insects, such as *Lethocerus*, increase the acceptance angle throughout the light to dark adaptation (from 3.5 to 9°) and more than the day-active species (Fig. 29).

Many arthropods show changes in the rhabdom fine structure when illuminated in the dark-adapted state, e.g. the bee (Gribakin, F., 1969). The locust (*Valanga*) and mantis (*Orthodera*) photoreceptors also show a 24-h cycle in angular sensitivity and absolute sensitivity accompanied by the rhabdom synthesis and breakdown (Horridge, G. *et al.*, 1981). The turnover of photoreceptor membranes is maintained in equilibria after changes in light

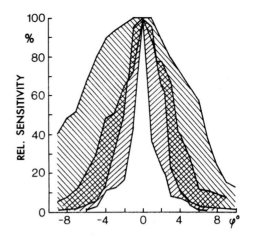

FIG. 29. The change of the angular sensitivity throughout the light-adapted (inner ruling) to dark-adapted (outer ruling) state. The overlapping area is shown by the crossed ruling. The graphs show the total run of five different cells each. The relative sensitivity is shown in the vertical axis in function of the incident light impinging the cell. (After Walcott, B., 1971, with permission of Springer-Verlag, Heidelberg.)

intensity (White, R. and Lord, E., 1975).

The retinomotor strategies in superposition eyes are characterized by extensive pigment migrations in the longitudinal direction (see review by Kunze, P., 1979). Usually the migrating pigment moves in the primary pigment cells leaving a clear zone between the crystalline cone and the distal tip of the rhabdom during the dark adaptation. When the illumination is changed, the pigment movement takes about 5–30 min, beginning only after a latency of a few minutes (Bernhard, C. *et al.*, 1963; Dreisig, H., 1981).

5.2 Photochemical adaptation

The photochemical cycle can be considered an important factor in adaptation. In insects the rhodopsin of many species is bleached by photon capture to a thermostable product, metarhodopsin. The absorption spectrum of the photoproducts is different from the membrane-active rhodopsin. The quanta, which are absorbed by the metarhodopsin, can reconvert the metarhodopsin back to light-sensitive rhodopsin. Therefore the rhodopsin concentration remains relatively high even in relatively bright light conditions (Hamdorf, K. *et al.*, 1973; Hamdorf, K. and Schwemer, J., 1975; Hamdorf, K., 1979). The rate of photoregeneration depends on the wavelength and also on the intensity of illumination.

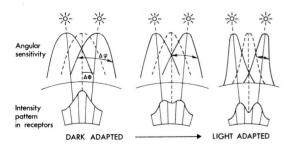

FIG. 28. The angular sensitivity shown in function of the state of adaptation, observed here in the fovea of the praying mantis (after Rossel, S., 1979). The intensity pattern distribution in receptors illustrates two-point resolution depending upon the ratio between the acceptance angle, both magnified here approximately × 10 for clarification. Note the ratio $\Delta\varphi/\Delta\Phi$, which determines the contrast transfer of two points. (Modified after Laughlin, S., 1981.)

Microspectrophotometric measurements indicate a first-order reaction (Schwemer, J. *et al.*, 1971), which has also been supported by electrophysiological experiments (Hamdorf, K. *et al.*, 1971). Usually, metarhodopsin absorption is located in the longer wavelengths while the rhodopsin absorbs in the UV, but other absorption bands are also found (*Deilephila*: Höglund, G. *et al.*, 1973).

The concentration of visual pigments (rhodopsins) is related to the sensitivity of the cell. Some of the pigment can be converted, as mentioned above, to thermostable, but also to unstable, by-products in order to reduce the sensitivity of the photoreceptor in proportion to pigment concentration (Hamdorf, K., 1971).

The photopigment might be coupled to an accessory pigment, as suggested for the fly *Musca* (Kirschfeld, K. *et al.*, 1977, 1978). This so-called sensitizing pigment might act as an antenna pigment absorbing photons and transferring energy so acquired to the photolabile pigment. The amount of the photostable antenna pigment is likely to regulate the adaptational level, at least in the fly's visual cells.

5.3 Photomembrane adaptation

The sensitivity of the visual receptors can also be interpreted in the form of an electrical potential across the membrane. Thus the adaptation will appear in the shift of the $V/\log I$ functions over the entire operating range of the cell (Järvilehto, M. and Zettler, F., 1971; Dörrscheidt-Käfer, M., 1972; Laughlin, S., 1975; Laughlin, S. and Hardie, R., 1978).

The adaptation of the photoreceptor membrane after an onset of a constant light stimulus is basically characterized by discrete potential waves, called bumps, each of which is created by the absorption of one photon (*Limulus*: Yeandle, S., 1958; fly: Kirschfeld, K., 1966; dragonfly: Lillywhite, P., 1977; see review by Fuortes, M. and O'Bryan, P., 1972). The non-linear summation of bumps in function of intensity or other stimulus parameters gives rise to the receptor potential. The size of a single bump can reach an amplitude of up to 3 mV in totally dark-adapted receptors (locust: Lillywhite, P., 1977), but light adaptation is shown in a single bump by the increase of the time course and the decrease of the

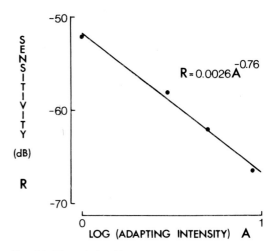

FIG. 30. The relationship between the reduction in low-frequency sensitivity and the logarithm of the mean (adapting) light level. The data fit well with a straight line with a slope of −15.13 dB/log unit. The mean changes in sensitivity produced by light adaptation corresponds to a shift of 0.76 log units along the log intensity axis for each log unit adapting light intensity. (French, A., 1979, with permission of Springer-Verlag, Heidelberg.)

amplitude, the duration and the latency. These also apply to the receptor potential (Fuortes, M. and Hodgkin, A., 1964; Dudek, F. and Koopowitz, H., 1973; Moring, J. and Järvilehto, M., 1977; Dubs, A., 1981; Howard, J., 1981). Thus the temporal resolving power is increased through adaptive changes in bumps so as to increase the visual activity in motion. On the other hand French, A. (1979) could not report any clear progression of the time constants with light adaptation in his recordings by using low-intensity white-noise stimulation.

The dynamics of light adaptation is shown in Fig. 30. If the stimulus intensity remains constant the amplitude of the response depends on the adaptational state in given time and number of absorbed quanta. The sensitivity of the membrane is reduced immediately after the onset of a light pulse (Fig. 31) and the recovery will be delayed by a few tens of milliseconds depending on the light stimulus (Moring, J. *et al.*, 1979). The time course during dark adaptation shows two phases of recovery in *Limulus*. The first phase at least seems to be strongly correlated with the decline of the intracellular calcium ion concentration measured by Arsenazo III, whereas in the second phase the correlation is not evident (Nagy, K. and Stieve, H., 1983).

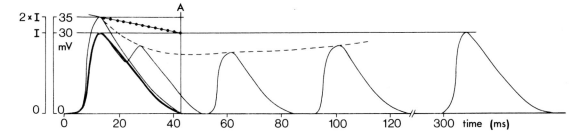

FIG. 31. Receptor cell responses to the adapting (first) and to the test (second) stimuli at different inter-stimulus intervals. The thick curve shows the response to the adapting light flash. The response shown at 0 ms interval between the flashes is a response to double intensity. The dotted line shows the theoretical course of non-linear summation of potential amplitudes without light adaptation, and the broken line shows measured peak amplitudes of the test responses. The vertical line A shows the end of non-linear summation. (Data from Moring, J. and Järvilehto, M., 1977; Moring, J. *et al.*, 1979, with permission of Springer-Verlag, Heidelberg.)

The adaptation phenomena in higher centres of optic lobes are complex and the neuronal feedback processes overlap with the membrane phenomena. Many of the changes in response configurations still remain unexplained (Burtt, E. and Catton, W., 1964; Gestri, G. *et al.*, 1980; Srinivasan, M. and Bernard, G., 1980).

The response amplitude is usually an inverse function of temperature, which means reduced sensitivity at low temperatures (French, A. and Järvilehto, M., 1978a). The light and dark adaptation at different temperatures (10–30°) applied to *Oncopeltus fasciatus* show somewhat increased relative sensitivity in dark-adapted conditions than in light-adapted ones (Dudek, F., 1975), which can be considered as a sign of non-reversal phenomena of the light and dark adaptation in the membrane or in the pigment chemistry.

5.4 Neuronal adaptation

The light intensity in the environment fluctuates over 10 or more log units. Now the usual intensity range over which the peripheral neurons and photoreceptors operate independently of adaptation is often less than 4 log units. On the other hand behavioural and physiological activities show a wide range of adaptations to all kinds of light conditions; therefore the neuronal adaptation can be understood as a link between the primary adaptational processes in photoreceptor cells and the goal-seeking operations in the central ganglia. Four different inhibitory circuits can mediate neural sensitivity at the level of the first synapse; two feedback and two feedforward circuits (Fig. 32).

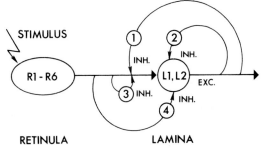

① Secondary feed-back inhibition
② Primary feed-back inhibition
③ Primary feed-forward inhibition
④ Secondary feed-forward inhibition

FIG. 32. The successive distribution of the four possible inhibitory circuits that can mediate neural sensitivity changes at the level of the first synapse in the visual system. These four are the interneuron feedback (1 and 2) and receptor feedforward (3 and 4). The roles of these pathways in accounting for neural sensitivity control in the dragonfly lamina are discussed in the text, where it is pointed out that these circuits have yet to be identified and their lateral distribution in lamina is unknown. (After Laughlin, S. and Hardie, R., 1981, with permission of Springer-Verlag, Heidelberg.)

A characteristic feature of neuronal adaptation is its probabilistic nature, and this ensures the transmission of signals, which are highly expected, like feature extraction and/or foreground–background dynamics. The signal to noise ratio (S/N), one of the limiting factors for the number of resolvable stimulus levels, is in one way expressed in Weber's fraction ($\Delta I/I = $ const.). The intensity coding follows a linear $V/\log I$ function which is shown in Fig. 33 for photoreceptors and second-order neurons in the fly, *Calliphora* (Autrum, H. *et al.*, 1970; Järvilehto, M. and Zettler, F., 1971;

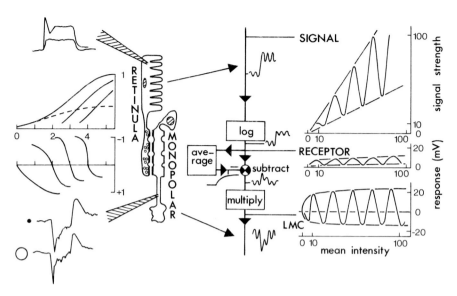

FIG. 33. Neuronal adaptation in the peripheral visual system of the fly and dragonfly described by the log transform–subtraction–amplification model as proposed by Laughlin, S. and Hardie, R. (1978, with permission of Springer-Verlag, Heidelberg). The input signal strength encompasses a constant contrast (0.5) in proportion to the mean intensity (upper graph on the right). The receptor and LMC responses undergo the log transform, averaging, subtracting and multiplication processes on an almost constant level. The signal processing at the second-order neuronal level is shown to be more dependent upon contrast than the mean luminance level. The diagram on the left shows the broad dynamic range, a relatively small range shift with light adaptation, and a standing background signal (broken line). The horizontal axis shows the relative intensity of the stimulus series and the vertical axis the relative response amplitude in arbitrary units (cf. the response on the right diagram). Comparison between the responses to a point and field stimulus show that the interneuron is subjected to lateral inhibition. (Laughlin, S., 1979.)

Laughlin, S. 1979). The waveforms of the responses and the intensity/response functions from photoreceptors and second-order neurons show only a very narrow intensity band which is linear, in common. Typical for postsynaptic responses is the high increment sensitivity (gain) during the dynamic phase of the potential (Järvilehto, M. and Zettler, F., 1971; Laughlin, S., 1976; French, A. and Järvilehto, M., 1978b), which is inversely proportional to mean intensity. This will reduce the intensity signal to a workable size and confer a degree of contrast constancy upon the visual system (Laughlin, S. and Hardie, R., 1978). In contrast efficiency the sigmoidal shape of the receptor's intensity/response function plays a major role.

Laughlin, S. and Hardie, R. (1978) proposed a "log transform–subtraction–amplification" strategy for light adaptation in the first neuropil in the fly and dragonfly (Fig. 33; on the right side in the diagram). This highly simplified scheme produces from a large-scale, low-step operating function a narrow-band, high-gain, dynamic operating function. The advantages of the three processes can be seen as follows: (1) the logarithmic product transforms the voltage responses of the neural image to function of contrast; (2) the subtraction removes the redundant background signal from monopolar neurons, improving the response range of the contrast signals; (3) the amplification protects the contrast signals from the contamination of intrinsic noise in synapses and in small-contrast signals. This simple light-adaptation scheme suggests that the receptors in different areas must operate with a similar sensitivity, and to do this the receptors should have a broad dynamic range, i.e. be capable of handling high contrasts and large areas; whereas the second-order neurons will produce a finely detailed extract from the receptor's neural image. Although many of the experimental data remain unexplained, some typical behaviour of the fly and dragonfly can find a reasonable solution, as will be shown later.

6   VISUAL INFORMATION TRANSFER

After the signal transduction process in the receptor microvillar membranes, the information of

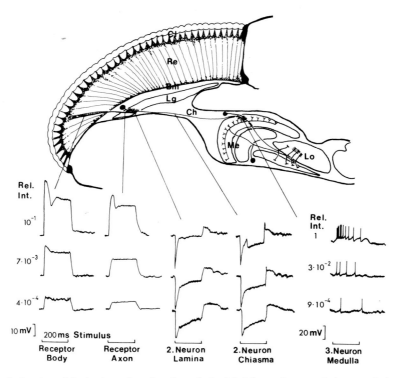

FIG. 34. A schematic drawing of the horizontal section through the fly's left eye. Some typical anatomical cell types are shown with their corresponding potential responses in function of light stimulus. Bm, basement membrane; Ch, chiasma; Cl, corneal lenses; Lg, lamina ganglionaris; Lo, optic lobe; Me, medulla; Re, retinular cell layer. (Järvilehto, M., 1978.)

excitation must be somehow transmitted to the higher optic neuropils. Usually in the insect photoreceptors the information is coded to a depolarizing potential change of the membrane potential. This potential, known as the generator potential or receptor potential (RP), is a smooth function of time and efficiency of the light stimulus. The RP amplitude increases with increasing strength of the stimulus, saturating at the level of about 50 mV.

The signal parameters and the absolute code for information transfer are not known, but the use of potential amplitude as a parameter for the analysis of information processing has been established. Other parameters of the potential wave can also be used, e.g. the rise time, the rising speed (first derivate), the integral or the power function.

In the first synaptic layer, lamina ganglionaris, the second-order neurons receive signals from receptor potentials usually in a very regular anatomical configuration. In the fly retina the receptor cells are divided into two distinct groups, one with short axons (R1–R6, 50–100 μm long)

another with long axons (R7 and R8, as much as 1000 μm long). The long axons terminate in the medulla externa without special intermediate connections.

The comparison of different potential responses recorded from receptor cell soma and axon, as well as in the second-order neuron (Fig. 34), reveals differences in their membrane properties. The soma membrane possesses properties which at least in the microvillar area are active, but the axonal membrane, which does not generate any action potentials (first in medulla), is generally accepted to be passive in terms of electrical signal propagation.

The attenuation of the passively propagated DC signal along the membrane is inversely related to the space constant of the membrane. The space constant (λ) for DC potential components depends on the membrane resistance r_m, the specific resistance of the axoplasm r_i, the extracellular specific resistance r_0 and the diameter R_a of the axon as follows:

$$\lambda = \sqrt{\frac{r_m}{r_i + r_0} \cdot \frac{R_a}{4}} \qquad (8)$$

If the extracellular resistance r_0 is low it does not influence the space constant. The inner specific resistance of the axon depends on the ionic composition of the axoplasm. The diameter of the axon can be physically measured, but is not constant in different parts of the cell. Therefore the space constant will be different in fine branches and soma or axon.

The attenuations of the high-frequency components (e.g. initial phase, spike-like component) are more pronounced (Jack, J. and Redman, S., 1971), thus showing a shorter space constant because the passive membrane properties can be depicted by a series of parallel R–C links.

The finite length of the axon with sealed ends and constant diameter may also give a different value to the space constant if calculated by using the equation, where the voltage V_0 at the beginning and V_T at the termination are taken into account

$$\frac{V_T}{V_0} = \frac{1}{\cosh(L/\lambda)} \qquad (9)$$

and where the cable length is L and the space constant λ (Weidmann, S., 1952).

Decremental conduction of the visual signals has been nicely demonstrated intra- and extracellularly by Shaw, S. (1972). The obvious advantage of passive conduction is that the graded response of the receptor is preserved. This non-linear transformation of the input signal provides accurate and high-density information content. The two factors reducing the information content are the membrane noise and the shunting of the passive membrane, thus reducing the amplitude of the potential as a function of the spreading distance. These factors are mostly avoided by adding a digital stage on the line; for instance a high-resolution voltage-to-frequency converter driven from the slow potential output of the receptor. The impulse generating mechanisms in the neurons are far from ideal, which means that quite a portion of the information may be lost if using spike coding. In all different types of developed retinae the receptor cells, and at least one or two subsequent neurons, transmit the information in the form of graded voltage fluctuations. Impulse coding is used first in the third-order neurons. This suggests information processing between the receptor cells and the third-order neurons in order to cause an information reduction by logical elements (Järvilehto, M., 1982). This can be seen in the specialization of the third- and higher-order neuron's functions.

If the depolarization simply spreads passively down the axon cable into the terminal, it must be reduced in amplitude as shown above (equations 8 and 9). The loss factor is relatively slight for short distances, a factor of two to six in the fly's retinular cell axon, for example if the attenuation is determined from the terminal (Scholes, J., 1969; Järvilehto, M. and Zettler, F., 1970).

The passive spread of the potential signal would be difficult to understand in the information channels from the lamina to the medulla. All L-neurons are thin and long, but the axons of the central retinular cells (R7 and R8) are also very thin and have about the same length as the L-neurons. The long visual fibres (R7 and R8) carry depolarizing potentials (Järvilehto, M., 1971; Järvilehto, M. and Zettler, F., 1971) to the medulla, whereas the L-neurons produce graded hyperpolarizations if stimulated by light via short visual fibres (Autrum, H. et al., 1970). That means that even if a fraction of the original potential amplitude is going to reach the synaptic terminals in the medulla a lot of information will be lost. The space constant is not known, but the membrane resistance plays a large role in that. If a common specific resistance were $10,000 \,\Omega \cdot cm^2$ ($\lambda = 1000 \,\mu m$) the assumed steady-state decrement over $500 \,\mu m$ would be around 30% of the original amplitude. The axons with a diameter less than $2 \,\mu m$ would suffer more decrement. This would be the case for the long visual fibres, which have axons less than $1 \,\mu m$ in diameter. The experimental findings (Fig. 35) do not support the concept of passive propagation of electrical signals in the long and thin axons of L-neurons (Zettler, F. and Järvilehto, M., 1971, 1973; Zettler, F., 1975). If the membrane resistance is presumed as above, or higher, this would not be consistent with the increase of the input resistance during the penetration of the cell membrane with the microelectrode. The low resistance observed cannot be explained by the low resistance shunt of the synapses in the lamina cartridges or in the medulla, because if the cell was penetrated in the chiasma between the lamina and the medulla the specific resistance of the axoplasm would cause at least $50 \,M\Omega$ increase of the input resistance and the potential measured at a distance of $300 \,\mu m$ from

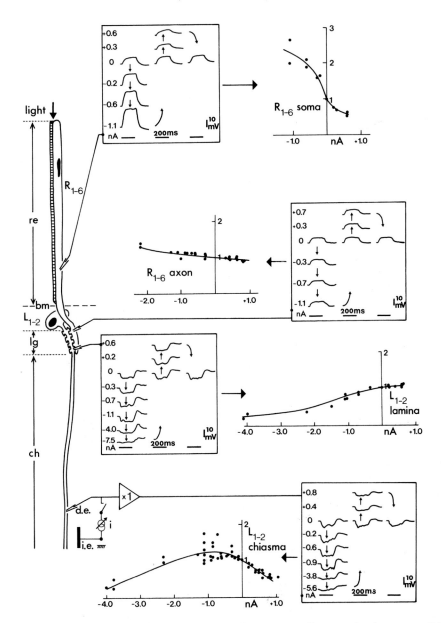

Fig. 35. Intracellular recordings from receptor cells (R1–R6) and their corresponding second-order neurons (L1–L2). The responses to a 200 ms light pulse of white light are shown in the four boxes. The dependencies of the light-induced tonic potential amplitudes on imposed membrane current (nA) are shown by arrows in the corresponding diagrams. The recording arrangement of the successive penetration of the different parts of the cells can be found on the left corner: d.e., the recording electrode; i.e., the indifferent electrode; i, the current source. The numbers in each box indicate the intensity and polarity of the current flowing through the electrode. The arrows show the temporal sequence of individual recordings: re, the retinular cell body level; bm, basement membrane; lg, the first optic ganglion (lamina ganglionaris); ch, the first optic chiasma. (Data from Zettler, F. and Järvilehto, M., 1973, with permission of Springer-Verlag, Heidelberg.)

the synapse ought only to be $1/6$ V_0. This is not supported by the experimental findings (Zettler, F. and Järvilehto, M., 1973).

The dragonfly and locust ocelli consist of cells which anatomically resemble and respond similarly to the short type receptor cells and L-neurons in the compound eye. The electrical responses of these cells in the ocellus have been characterized

(Chappell, R. and Dowling, J., 1972) and recently combined with results from dye or cobalt injections (Patterson, J. and Chappell, R., 1980). The neuronal connections in the ocellus lamina do not reveal any clearly regular columnar organization (Dowling, J. and Chappell, R., 1972) such as has been found in the compound eyes of diptera. The synaptic plexus is thus very irregular and complex in the ocellus lamina. Similarly, in the flesh-fly ocellus (*Boettcherisca peregrina*) the second-order neurons can be both pre- and postsynaptic to similar types of cells as themselves, and thus their function as either afferent or efferent elements remains open (Toh, Y. and Kuwabara, M., 1975). Chemical synapses with vesicles have been shown, but whether there are electrical synapses as well is yet unclear.

The ocellar nerve contains a few (7 or 8) large axons, and several small axons are located around the large one in small groups (Wilson, M., 1978a). The large axons are usually separated from each other by axonal membrane but occasionally they can be very thin. Those connections enable *en passant* contacts of an electrical nature between large axons. This would mean that the membrane resistance is not constant along the axon, and the axon diameter several-fold higher than supposed for the calculations of the space constant λ. This might explain the partly controversial results of the experiments in the compound eye L-neurons and the ocellar nerve giant axons. This kind of connections between the compound eye L-neurons has not yet been reported.

Interesting support for active membrane conduction of graded potentials comes from the vertebrate retina, where many of the horizontal cells possess long and thin axons. The graded potentials are generated in the soma as PSPs and conducted to the axon terminal. A comparison of the potential amplitudes in the soma and the terminal shows that this conduction is non-decremental (Weiler, R. and Zettler, F., 1979).

Whether the conduction mechanism of graded hyperpolarizations is active in some cell types as suggested by Zettler, F. and Järvilehto, M. (1971, 1973) should be tested by using similar methods as used for revealing the spiking mechanism. Unfortunately the cells, where the hyperpolarizations are found, are still very thin and long, having an un-

favourable shape for the membrane voltage clamping. Also the cell environment is complicatedly compartmentalized, so that perfusions of the cell external surface are difficult to control. To verify the hypothesis one should try to find a more suitable experimental object or improvements in the technical aids used. The nature of the information conduction along the cell pathways gives important criteria for the information processing mechanisms in the ganglia.

7 MOVEMENT DETECTION AT NEURONAL LEVEL

7.1 Peripheral retina

There are few data relating to movement detection at the retina–lamina level. For most insects with fused rhabdoms this analysis is unlikely. For the open-type rhabdomeres with neural superposition arrangements of axons in the first optic ganglion the anatomical conditions are provided. By using sophisticated techniques of single rhabdomere stimulation in the fly *Musca* Kirschfeld, K. (1972a,b) was able to find some evidence for the assumption of two sets of movement detectors at the retinal level. The detectors are orthogonally arranged in the direction 1 to 6 and 1 to 3 rhabdomeres (numbering after Dietrich, W., 1909), both influencing turning around the vertical axis. The central receptors R 7/8 were not tested and because of their UV-sensitivity they might play an essential role for other classes of movements.

No neurons which respond specifically to movement have so far been found in the lamina, only neurons responding to a brightness change.

Figure 36 shows the necessary and sufficient structure for a movement detector allowing only the very basic performance. It is based on the principle of autocorrelation (Hassenstein, B., 1958; Reichardt, W. and Varjú, D., 1959; Mastebroek, H. *et al.*, 1980). The different elements in the model cannot represent simply single neurons, but logical elements, i.e. connections between neurons.

In the next few sections the anatomical structure and the electrophysiological signals of some of the best known neurons are discussed.

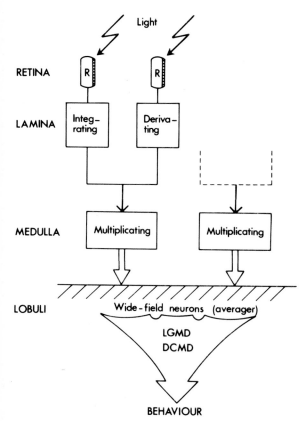

FIG. 36. Schematic representation of the basic structure for a direction-sensitive movement-detecting system. The visual inputs are followed by integrating and derivating elements in the lamina and connected together in the medulla by a multiplicator. A multiplying unit followed by the time averager are together called the correlator. A large number of units summed to wide-field neurons act as an averager, thus gating the motor neuron system and behaviour.

7.2 The medulla and lobula units

The medulla as an extremely complicated neuropil is divided into two subregions according to their morphology and developmental origin (Meinerzhagen, I., 1973), viz. the outer layer and the inner layer. The channels from the retina of the fly diverge twice: once in the lamina on to sets of monopolar cells (large monopolar neurons, LMCs) and centripetal T-neurons, and a second time to columnar relay neurons in the medulla. The endings from long visual fibres and lamina relay cells diverge to as many as 20 types of transmedullary neurons which project to the lobula complex. Currently over 200 different cell types, providing very complex connections for various

tasks, are known from recent electrophysiological studies.

Anatomical profiles of different types of neurons suggest at least a wide-field relay neuron and a columnar small-field neuron. This is confirmed by identified recordings from those types (DeVoe, R. and Ockleford, E., 1976). Hyperpolarizing slow potentials have been recorded in the chiasma and outer medulla from monopolar axons (Zettler, F. and Järvilehto, M., 1971), whereas medullary relay neurons are believed to transmit information by spikes.

Since electrophysiological techniques were improved by introducing new methods of single cell staining through the microelectrode about 15 years ago, there has been a considerable expansion of new data available from movement-detecting neurons in different species of insects. The recordings originate mainly from the lobula plate, but some also from the lobula and the medulla.

Several authors have described neuron types and activity patterns in the medulla and the lobula. The following classification is based mainly upon the spike activity of the neurons. Graded voltage changes are considered at the end of the list.

I. Spontaneously active units

(1) Spacemaker units, not excited by any light stimulus (fly: Bishop, L. and Keehn, D., 1966).

So-called "clock-spikes" have been found in the optic lobes of the blowfly *Calliphora* (Leutscher-Hazelhoff, J. and Kuiper, J., 1966; Hengstenberg, R., 1971). These neurons innervate a muscle attached to the inner frontal margin of the retina. The activity causes a regular movement of distal rhabdomere tips perpendicular to the optical axis (0.1–0.5°/10 Hz) and can be slightly influenced by light stimuli moving from front to back. The functional significance is unknown, but may play a role in the foveal vision in the binocular region.

(2) Neurons, which response to light stimuli (fly: Bishop, L. and Keehn, D., 1966; locust: Burtt, E. and Catton, W., 1960).

II. Neurons which do not respond spontaneously, but are activated by light stimuli

(1) Neurons which respond to a stationary light stimulus.

(a) "ON"-units.   Some phasic activity only at light-ON (fly: Bishop, L. and Keehn, D., 1967; locust:

Burtt, E. and Catton, W., 1960; butterfly: Swihart, S., 1969) or some tonic activity (butterfly: Swihart, S., 1969) or discharge at both light-ON and -OFF (locust: Burtt, E. and Catton, W., 1960; fly: Bishop, L. and Keehn, D., 1967; butterfly: Swihart, S., 1969).

(b) Inhibitory unit. The spontaneous activity is phasically inhibited at light-ON and -OFF, or tonically inhibited during the light stimulus (butterfly: Swihart, S., 1969).

(2) Neurons which respond to a movement of an object. Since Burtt, E. and Catton, W. (1956) described spiking neurons, which were activated by movement of an object in the visual field of the optic lobe, many other neurons of the related type have been described (locust: Burtt, E. and Catton, W., 1959, 1960; Horridge, G. *et al.*, 1965; fly: Bishop, L. *et al.*, 1968).

Movement-sensitive units have been found in the medulla, lobula, lobula plate and protocerebrum, but not yet in the lamina.

(a) Non-directional types of neurons. The movement of an object in any direction causes spike discharges in these neurons. These types of units have different sizes of receptive fields from small (<90°) to large (>90°) ones. The wide-field neurons may be activated also by contralateral stimulation. The units with a small field are described in the fly (Bishop, L. *et al.*, 1968; Mimura, K., 1971) and the butterfly (Swihart, S., 1969) as "jittery movement" fibres. The object velocity does not change the activity very much. The cells with small fields are mostly found in the medulla, and the wide-field neurons usually in the lobula and the lobula plate.

(b) Uni-directional types of neurons. Neurons in this category respond most sensitively to movement of an object in one (preferred) direction and not at all (or less) to movement in the opposite (null) direction. These cells may also have either a small or large receptive field. They are both found in the medulla and the lobula complex. The receptive field structure may be complex having components of ipsilateral, contralateral or even bilateral eye fields (Fig. 37).

Bishop, L. *et al.* (1968) found in the fly four logical types of preferred directions, two horizontal and

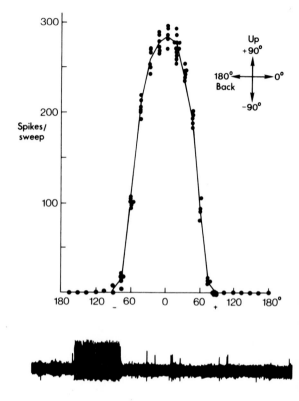

FIG. 37. A recording from a neuron of the hoverfly (*Syrphus*) responding to a wide-field ipsilateral directionally selective movement stimulus showing no spontaneous activity. The recording in the middle is an example of the response to movement stimulus (the graph below). Grating moves at 5.3° s⁻¹ with 9.5° bars in 40° circular aperture. Time mark 2 s. (From Collett, T. and King, A., 1975.)

two vertical ones. These were also confirmed by Mimura, K. (1972). Similar types were also found in the butterfly (Swihart, S., 1969). The wide-field units with contralateral interactions may form the basis for binocular vision as described by McCann, G. and Foster, S. (1971). The speed of a stimulus is coded so that the action potential frequency is approximately proportional to the logarithm of the stimulus velocity.

(c) Semi-integrative types of neurons. This type of cell is distributed all over the medulla and lobula complex. The coding of the velocity by frequency of action potentials is similar to that in the unidirectional type. The receptive fields are as complicated as if two or more receptive fields of the unidirectional type, or those responding to movements in two opposite directions, were combined in adjacent or overlapping regions of the receptive field of a higher-order neuron. Some neurons of this type discharge strongly when a spot is moved vertically in a narrow band on the midline in front of the fly (Mimura, K., 1972, 1975). The functional significance may be found in processing towards more perfect binocular vision.

(d) Integrative types of neurons. These cells respond to a pattern of different stimuli, which makes a biologically meaningful stimulus combination. This type of neuron is mainly found in the lobula complex, more precisely in the tract from lobula to protocerebrum where a multimodal input to these neurons is easily performed. The input to these neuron types comes from unidirectional or semi-integrative types and produces sensitivity increase to a complex but continuous motion within the receptive field. A continuous angular movement of an object caused by the insect turning is a common stimulus (Mimura, K., 1975).

Similar complex cells have also been found in the locust ventral cord (Burtt, E. and Catton, W., 1966, 1969). These units appear not to code the velocity of the movement, since the response magnitude does not increase gradually with the increase of the speed of movement; i.e. the response versus speed function is highly non-linear.

The units in the locust optic lobes are excited by visual, auditory and mechanical stimuli (Horridge, G. et al., 1965).

(e) Novelty type neurons. The responses of these cells habituate quickly, but respond afresh to a novel stimulus, either stationary or moving. Usually these unit types are elicited by any motion of small objects. The fast habituation serves as an alerting mechanism to new stimuli (locust: Horridge, G. et al., 1965; butterfly: Swihart, S., 1969; fly: Mimura, K., 1971, 1972, 1974).

(f) Neurons inhibited by a moving object. The spontaneous activity may be inhibited only by directional-selective stimulus or by any kind of moving-light stimulus. In one sense, all cells with directional sensitivity could be included in this class,

because the phenomena of spike discharge being suppressed in the null direction may be considered to mean that the stimulus has a positive inhibitory action rather than an indifferent one. However, some experimental results (Mimura, K., 1975) show that in some units a background activity is inhibited by involvement of a spot of light in the null direction.

(g) Self-movement type neurons. These neuron types respond maximally to movement of very large patterns or to rotation of the body in diffuse light (dragonfly: Olberg, R., 1981).

(h) The depth-motion type neurons. These neurons respond preferentially to movement towards the head. The spike frequency is a two-valued function of relative angular velocity across the eye (fly, *Phormia terraenovae*: Eriksson, E., 1980b, 1982).

III. Movement detection by non-spiking (graded) type signals

Some lobula plate neurons of the fly do not propagate action potentials for information transfer, but transmit graded signals either depolarizing or hyperpolarizing in nature (fly: Hengstenberg, R., 1977; Eckert, H. and Bishop, L., 1978; cricket: Honegger, H.-W., 1978; worker bee: Kien, J. and Menzel, R., 1977). The horizontal cells respond to stimulation of regressive motion within the ipsilateral receptive field by hyperpolarization of the cell membrane, whereas progressive motion induces a strong depolarizing membrane potential-shift with superimposed fast potential changes of a "noisy" appearance (blowfly: Eckert, H., 1981).

In the giant vertical cell in the lobula plate of the fly optic lobe, motion in the downward direction evokes a net membrane potential depolarization and upward motion results in a net hyperpolarization (SooHoo, S. and Bishop, L., 1980).

8 VISUALLY INDUCED BEHAVIOUR

Whenever a contrasted panorama is moving relative to the retina a visually induced behaviour may occur. This behaviour uses many different strategies of the visual system, as shown in previous sections of this chapter. The optomotor responses of an insect reveal the ability of an insect visual system to control and stabilize its visual world, i.e. differentiate between self-movement and object-movement. Furthermore pattern recognition is used for navigation and detection of objects. The fixation and tracking of objects are elementary functions of the movement behaviour of an insect, although stimulation may also cause anti-motion or escape reactions. Visual stimulation also mediates several other types of behaviour, e.g. feeding, learning, communication and sexual behaviour.

The optomotor responses have been studied extensively for more than 30 years. The next section considers some of the central modes of the visually induced behaviour.

8.1   Movement detection studies at behavioural level

The pioneer works in this field have shown that the movement of the optical environment caused body movements in bees (Hecht, S. and Wolf, E., 1929; Wolf, E., 1933; Hertz, M., 1933) and flies (Gaffron, M., 1934; von Gavel, L., 1939). The use of these optomotor responses for a black-box analysis of motion perception in insects was pioneered by Hassenstein, B. (1951) on the beetle *Chlorophanus*. Originally only a large-field stimulus rotation around the vertical body axis was understood as an optomotor reflex. It was measured by using a torque compensator attached to the animal (bee: Kunze, P., 1961; fly, *Musca domestica*: Fermi, G. and Reichardt, W., 1963; *Drosophila*: Götz, K., 1964).

Optomotor responses are also found in a freely flying fly which has three rotational (roll, pitch and yaw) and three translational (lift–sink, thrust–backing and shift right–left) degrees of freedom (Fig. 38). The neck (cervix), in addition, allows the head to be rotated through small angles relative to the body. Usually, though, the head is fixed on the thorax during the experiments.

The movements induced in the usual experimental set-up can be divided into five different categories if we consider the stimulus motion seen by the animal:

(1) *Object/Subject motion*. Object motion, where a neuronal correlate can be considered as a narrow-field unit. Subject motion, where the wide-field unit causes head and body movements (Fig. 39).

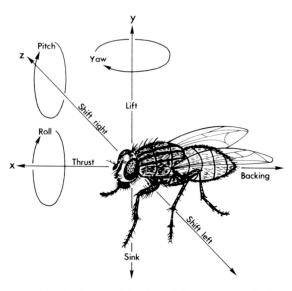

FIG. 38. The degrees of freedom of fly movements. In the three-dimensional coordinates (X, Y, Z) of movements can be described as three translational (lift–sink, thrust–backing, shifts) and three rotational (roll, pitch, yaw) movements. The movements of different parts of the body (e.g. head) are not considered here.

(2) *Apparent motion* (wobbling or flicker), where a neuronal function shows either directional sensitivity or non-directional sensitivity.

(3) *Directional motion*, where the neuronal elements can be divided into progressive or regressive sensitive units.

(4) *Binocular motion*, where the neuronal units are sensitive only to stimulation of elements in both eyes.

(5) *Depth-direction motion*, where the neuronal elements are activated by the stimulation of the expansion of the stimulus pattern.

Generally the strength of the movement-sensitive responses in insects depends on the angular velocity of a pattern or, more precisely, on the ratio of angular velocity Φ to pattern spatial wavelength λ, if the pattern is a grating (Kunze, P., 1961; Götz, K., 1964; Eckert, H., 1973).

If the optomotor turning response and the mean spike response of functionally defined horizontal lobula neurons of the fly are compared in their spatial contrast frequency, one can see that both functions have similar characteristics (Fig. 40). This is so especially if we consider the maximum response ($\Phi/\lambda = 2\,\mathrm{Hz}$). In this case it is probably obvious that

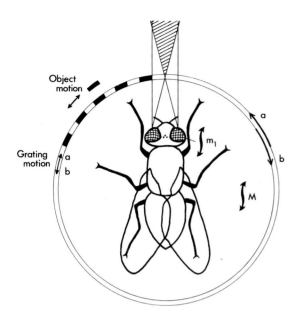

FIG. 39. An experimental set-up with stripe cylinder to test the movements of the fly to object and to grating (subject) motion. The arrow (a) shows the direction of the regressive and (b) the progressive motion. The body rotational movement is shown by (M) and head-turning movement with or without body movement is shown by (m₁). The oblique ruling characterizes the field of true binocular vision.

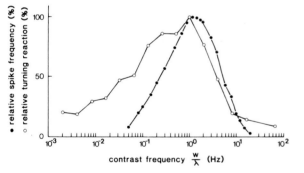

FIG. 40. The optomotor turning response (open circles) of the fly (*Musca*) compared with the relative spike response (closed circles) of a horizontal cell (H1) to stripe patterns with various contrast frequencies. (Data from Reichardt, W., 1966; Eckert, H., 1980b; after Kirschfeld, K., 1979.)

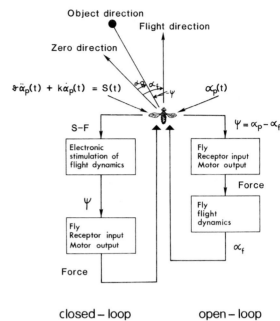

closed – loop   open – loop

FIG. 41. The angular coordinate system describing the fly's rotational degree of freedom around the vertical axis combined with signal flow diagrams of flight in closed-loop and open-loop situations. The zero direction and the flight direction are defined by the arbitrary angles $\alpha_f(t)$ and $\alpha_p(t)$ correspondingly. The location of the object on the retina of the fly is represented by $\psi(t) = \alpha_p(t) - \alpha_f(t)$, which is the "error angle" between the fly's direction of flight and the object. The open-loop signal flow diagram refers to a free flight situation. The visual-input torque-output box is followed by a box showing the properties of the flight dynamics transduced into angular displacement. The closed-loop signal flow diagram refers to the flight simulation device. The flight dynamics are simulated by analogue electronics, and the function $S(t)$ simulates the object motion. During the object movement ϑ is the measure of moment of inertia around the vertical axis of the fly and k an aerodynamic friction correction factor. Here a fly, suspended from the torque compensator, controls the velocity of a cylindrical "panorama" by its own torque signal. The transfer properties of the compensator, the motor coupling block and the servomotor approximate free flight dynamics. (After Reichardt, W. and Poggio, T., 1976.)

There are two methods to determine the optomotor response; *viz.* open-loop and closed-loop experimental configurations (Fig. 41). In the open-loop experiments the torque response evoked by the tethered fly's turning reaction is measured isometrically. Also wingbeat amplitudes during stimulation can be measured.

In closed-loop experiments the tethered fly is able to adjust the orientation of a visual pattern by exerting the appropriate torque movement on the stimulus. These experiments have revealed that the

the lobula giant neuron (H1) is a part of the optomotor turning response system at the horizontal level (Kirschfeld, K., 1979; see Fig. 10). It is also obvious that the rotational components, yaw, pitch and roll, have their own systems of giant neurons (Blondeau, J. and Heisenberg, M., 1982). The roll and yaw motions are probably mediated by similar neuronal mechanisms (Srinivasan, M., 1977).

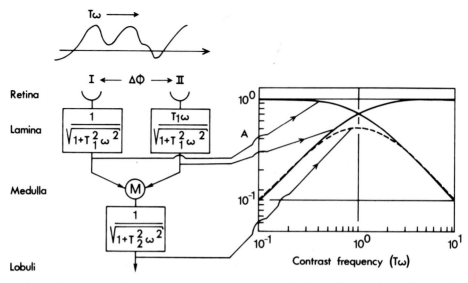

FIG. 42. Characteristics of a simple and elementary movement detector for speed and direction of a visual object. On the left are shown two input elements (I, II) with a spatial distribution of $\Delta\Phi$, one with a high-pass filter and one with a low-pass filter at the input channels. The time constant of the elements is T_1 and the time constant of the stimulus is $T\omega$. The non-linear interaction is represented by the multiplier M. The mean response is indicated at the output of a second low-pass filter with the time constant T_2. On the right the output of above-mentioned linear filters is presented as frequency response functions and indicated by arrows. (Kirschfeld, K., 1979.)

animal has a strong tendency to orient towards linear patterns such as a stripe or a linear grating, along a direction perpendicular to the equatorial plane of the eyes (Srinivasan, M., 1977).

The open-loop design serves to study the total visual field motion during the animal's self-movement, and the closed-loop design is used to study the object motion during fixation and tracking.

8.2   Fundamentals of movement detection

Movement detection can be considered as a simple but fundamental paradigm for the abstraction of an essential feature from the external visual world. The simplest network to detect directional movement consists in principle of two input elements, an interaction and an output (Fig. 42). Such a system has to be asymmetric and the interaction has to be non-linear (Reichardt, W., 1957; Reichardt, W. and Varjú, D., 1959). This so-called correlation model (see section 8.3) does not claim to represent a wiring diagram which might be directly compared to anatomical structures, thus the electrophysiologically recorded large-field movement sensitive neurons themselves cannot represent the correlation units.

The average response of the system to a uniformly moving pattern is known to be represented by a Fourier series with respect to spatial frequency (Buchner, E., 1976). If it is considered that the non-linearities are not essentially higher than second order, it is possible to calculate the strength of the coefficients that contribute to the total response by the non-linear interaction between any two input elements. The interaction element is termed "elementary movement detector, EMD".

The optomotor control requires dense networks of EMDs. In the walking fly the course control is achieved mainly by pairs of equivalent EMDs which occupy 2 o'clock and 4 o'clock positions with respect to the right eye (Buchner, E., 1976).

The EMDs show bidirectional sensitivity for the course control of turning and torque responses of legs and wings in the corresponding modes of locomotion. On the other hand the EMDs which exert the altitude control of lift and thrust lack bidirectional sensitivity. The control of lift and thrust by a pair of unidirectional EMDs obviously follows a different scheme. Götz, K. and Buchner, E. (1978) emphasized that this is valid at least in the frontolateral part of the visual field in *Drosophila* (Fig. 43). This is supported also by experiments on the mutant, optomotor-blind[H31] (*Omb*[H31]), in

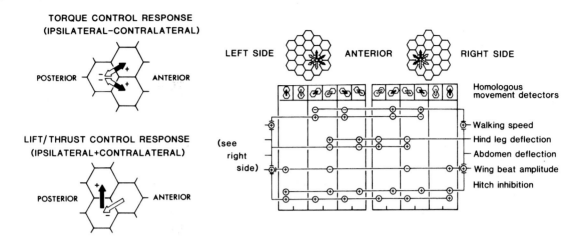

FIG. 43. The optomotor control system in *Drosophila* shown as a network with its elementary movement detectors. On the right two sets of EMDs, one for torque control response and one for linear control response. Each of the EMDs represents movement-specific non-linear interactions between two neighbouring input units in the hexagonal array of visual elements. Pattern movement in the arrow direction either increases (+) or decreases (−) the response on either side of the fly. The bidirectional movement detectors produce responses of opposite signs if movement direction is reversed. The composition of these complexes can be understood as a pair of unidirectional antagonists (Götz, K. and Buchner, E., 1978). In the wiring diagram six different groups of directionally homologous movement detectors on either side represent the input to the motor system. As presented here the control of legs, wings and body posture requires at least four different and independent signal channels on either side of the fly. (Götz, K. *et al.*, 1978.) (Reproduced with permission of Springer-Verlag, Heidelberg.)

which the giant neurons of the lobula plate are missing or severely reduced. All three optomotor torque responses (yaw, pitch, roll) are impaired, whereas other visual responses such as the optomotor lift/thrust response and the landing response (elicited by horizontal front-to-back motion) are not affected (Heisenberg, M. *et al.*, 1978).

Accepting the anatomical and electrophysiological evidence (see above), it seems likely that the activity of the EMDs is transmitted to different layers of the lobula plate according to the EMDs' orientation in the lattice of visual elements.

The regularity of the neural projections and universal location of EMDs oriented in two main axes serve as a basis for the assumption that the dendritic arborizations of HS- and VS-cells and the associated large-field tangential neurons are forming a motor link between the EMDs and the output channels (Hausen, K., 1982b).

The large-scale integration performed by the giant neurons is probably not desirable for translatory movement detection. The translatory course control deals with different angular velocities of objects in different parts of the visual field. With the extensive summation of the activity throughout the entire visual field the information of the target movement would be lost.

The small-field relay cells in the lobula of the fly (*Calliphora*) may serve for this sort of movement detection. The male flies sometimes track other flies closely. This behaviour may be mediated by special cells, the type A, B and C columnar relay cells (Col-A, B, C). The number of retinotopic pathways into the lobula from the medulla is thus enormous (Strausfeld, N. and Nässel, D., 1981).

8.3 Fundamentals of the correlation model

The concept of motion perception involves the fundamental idea that for a correct indication of speed and direction of movement a comparison is carried out at the input level between the signals from at least two visual elements. The correlation model contains the comparison of visual signals and their multiplication, and this is followed by taking time-averages of the product (Hassenstein, B. and Reichardt, W., 1953; Buchner, E., 1976; Mastebroek, H. *et al.*, 1980). The visual input spectrum (VIS) is limited by linear filters (I) and (D). The non-linear interaction of the channels is represented by the multiplier (M) (Kirschfeld, K., 1979). The processing of the signals by filters (I) and (D) and the multiplicative interaction (M) are thought to be located at the level of the retina, the lamina and the medulla (Fig. 36). Finally it is assumed that the outputs of a large number of uniformly distributed multiplying units are spatially summed by the wide-field neurons at the level of the lobula complex. In other words, the time-averaging action as present in the model is replaced by a spatial summation.

Some main properties of the model can be described (Fig. 44; Mastebroek, H. et al., 1980), if it is assumed that a spatial sinewave pattern (spatial wavelength λ_s) moves with constant velocity $\dot\Phi$ in the direction of receptor α to β. We may assume that the samplers (receptors) have an infinite narrow spatial sensitivity distribution and that their sample directions are separated by an angle $\Delta\Phi$. Then the momentary signal at input α can be described as;

$$X_\alpha(t) = \bar I + \Delta I \sin \omega_0 t \qquad (10)$$

where $\bar I$ is the mean light flux of the pattern, ΔI the modulation amplitude and ω_0 spatial frequency, which means that the absolute value of the constant velocity $\dot\Phi$ during one cycle is divided by the spatial wavelength λ_s. The signal at channel β will be similarly;

$$X_\beta(t) = \bar I + \Delta I \sin \omega_0 (t - \Delta t) \qquad (11)$$

but delayed by a fraction of time Δt, which displays the relation between angular spacing $\Delta\Phi$ and angular velocity $\dot\Phi$. The outputs $[y_\alpha(t)]$ and $[y_\beta(t - \Delta t)]$ of linear filters (I) and (D) integrating and derivating type respectively depend on the sum of the amplitude amplification multiplied by the mean intensity, the phase shift at frequency ω_0 and the modulation amplitude.

The final response $\overline{R(t)}$ now follows by taking the time average of the product $\overline{R(t)} = y_\alpha(t) \cdot y_\beta(T)$. This equation will contain a direction sensitive element

$$\omega\Delta t = 2\pi \cdot \frac{|\dot\Phi|}{\dot\Phi} \cdot \frac{\Delta\varphi}{\tau_s} \qquad (12)$$

and a direction-insensitive element if the mean luminance $\bar I$ of the pattern will be omitted. This way a so-called pattern-specific response will be extracted,

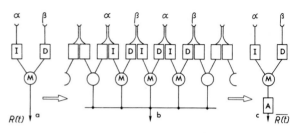

FIG. 44. A series of multiplicating units (M) will receive input from each of the two linear filters (D) and (I) of derivating and integrating type. The spatial summation can be interpreted as time-averaging output action from $R(t) \to \overline{R(t)}$. (After Mastebroek, H. et al., 1980.)

consisting of two parts: a direction-sensitive part and a direction-insensitive part (Mastebroek, H. et al., 1980). The direction-insensitive part vanishes when the (I) and (D) filters have the same time constants and they are integrating (low-pass) and respectively differentiating (high-pass). Thus a purely direction-sensitive movement detector (DSMD) results.

From the factor $\sin \omega_0 \Delta t$ it can be seen that this response changes sign for values of λ_s for which

$$\lambda_s = \frac{2\Delta\varphi}{k} \qquad (13)$$

where ($k = 1, 2, 3 \ldots$). The phenomenon of the response changing sign for smaller values of λ_s is called the reversal of the reaction, and the response which can be measured in the range $\Delta\Phi < \lambda_s < 2\Delta\Phi$ is called "reversed reaction" (Götz, K., 1964).

We assumed an infinitely sharp directional sensitivity of the input elements, but we know that these elements are retinular cells with Gaussian function-like sensitivity distribution having half-width $\Delta\varphi°$. The result, after replacing this correction in the function, is that the function must be multiplied by the factor (Götz, K., 1964):

$$g(\Delta\varphi°/\lambda_s) = \exp\{-[\pi^2/(2 \ln 2)](\Delta\varphi°/\lambda_s)^2\} \qquad (14)$$

Here the values for the half-widths of the receptors must be taken from actual measurements of the sensitivity to the angle of incident light.

8.4 Vision during flight and landing

Movement-sensitive interneurons in the visual system are converged to trigger interneurons, thus ensuring rapid stereotyped movement of the animal either to jump, fly or run in order to escape, to catch prey, or to track. The trigger interneurons for initiating a jump, in the locust for example, receive a strong excitatory input from many sensory modalities (visual, auditory, tactile and proprioceptive). This excitatory input is particularly strong from descending contralateral movement detector (DCMD) neurons, which decrease with other sensory modalities the threshold of the multimodal neurons to produce a strong isometric contraction of the extensor tibiae muscle, and thus a powerful jump in response to external stimuli (Pearson, K. *et al.*, 1980).

Whenever an animal moves it will cause its stationary surrounding to shift across its visual field. Every movement of the body causing such self-generated image motion or "reafference" (von Holst, E. and Mittelstaedt, H., 1950) is well known to play an important role in regulation of locomotion.

There are two possible sources of image motion with respect to the animal's visual system: small object motion in the environment and apparent wide field motion of the animal's surrounding. The discrimination between these two seems to be solved by two different mechanisms (Collett, T., 1980a). The first rule is concerned with the control of angular velocity and the second rule covers the control of translational velocity. Vision during flight and landing also involves other related neuronal strategies such as binocular vision, fixation, pattern detection, motion in depth.

Flight and landing are very complex behavioural tasks for the insect, thus much elementary psychophysiology is involved such as tracking, form perception, visual learning, and communication. First of all we should compare the physiological properties of object-movement detectors (OMD) and self-movement detectors (SMD). Recently it has been reported that there are two groups of descending interneurons in the dragonfly ventral cord (Olberg, R., 1981), both of which appear to perform these two different tasks. One group responds only to stimulation of movements of small objects; the other group responds principally to stimulation of movement of the visual world (Table 7).

As we have seen in the previous sections the large vertical and horizontal neurons (V- and H-group) in the lobula complex mainly serve the torque responses of flying flies. This is supported by the experimental finding that H1-neurons show a preference for a regressive pattern of stimulation direction rather than for a progressive one, and the maximal reaction strength at a contrast frequency of approx. 1.4 Hz (Eckert, H., 1980b) instead of 6–7 Hz for the landing response (Eckert, H. and Bishop, L., 1978).

The selective laser beam ablation of interneurons in the larval brain of the housefly results in various specific alterations in brain structure and behaviour of adult flies (Geiger, G. and Nässel, D., 1981). The vertical neurons in the lobula plate are considered important in thrust, lift and landing responses (Poggio, T. and Reichardt, W., 1976). However, the visually induced behaviour towards single objects found in flies lacking the large horizontal and vertical neurons on one side was not found to be significantly different from the behaviour of normal flies. This could mean that the H- and V-cells seem to be implicated in ordinary optomotor flight stabilization and the information processing of single moving objects is performed by other sets of nerve cells in the medulla and lobula (Geiger, G. and Nässel, D., 1982).

8.5 Visual fixation and tracking

A newly appeared object in the visual field of a fly causes first a fixation, which is followed by a tracking or chasing behaviour. Fixation by an animal in its natural environment is closely related to focusing of attention behaviour. The mechanisms of target and pattern fixation in insects have been investigated intensively, especially in flies.

One of the striking features in pattern fixation by insects and other animals is the perception of movements caused by retinal pattern displacements. This relation between the object pattern and experimental animal has recently given rise to several different experimental configurations.

(1) An oscillating or stationary pattern at a constant distance and in a constant mean position is presented to a fixed flying fly (Reichardt, W., 1973; Reichardt, W. and Poggio, T., 1976; Pick, B., 1976).

(2) A pattern is moved around a tethered flying fly at a constant distance (Virsik, R. and Reichardt, W., 1976).

(3) A stationary object is presented to a free-walking or free-flying fly. Under these conditions the distance between the object and the fly decreases continuously while the experimental animal approaches the pattern (Collett, T. and Land, M., 1975; Horn, E. and Wehner, R., 1975).

(4) An irregularly moving object, a fly, is chased by another fly. The movements are filmed. The actual position of the chased fly, as well as the distance from the chasing fly, are variable (Land, M. and Collett, T., 1974).

(5) In a black and white random background pattern is hidden a small identical visual object, rotating or stationary, around a semi-freely walking fly. The optomotor responses of the fly are used to control either closed-loop or open-loop experiments (Buchner, E., 1976; Bülthoff, H., 1981).

Visually induced head and body movements during fixation of stationary objects are described for the fixedly flying fly, *Calliphora erythrocephala* (Land, M., 1973) and for tethered adult and nymphal crickets, *Gryllus bimaculatus* (Tomioka, K. and Yamaguchi, T., 1980). During the fixation of a vertical stripe, rapid saccadic movements of the head occur followed by slow directional changes of the body. For the freely walking fly approaching a stationary black stripe the body movements were placed into two groups according to the frequency and amplitude of the turnings (Horn, E. and Mittag, J., 1980). Type I body movements are characterized by low frequencies, <1 Hz, with large amplitudes having the nature of visual exploration. Type II movements with high frequencies of >8.5 Hz and small amplitudes are mainly correlated with leg movements. Walking trajectories also show that the fly can fixate monocularly while approaching a pattern.

Figure 45 shows a stationary fixation to a vertically oriented black stripe (5°) as a function of the average brightness. The fly's instantaneous torque under closed-loop conditions (see Fig. 41) determines an angular displacement of the object relative to the retina according to the equation

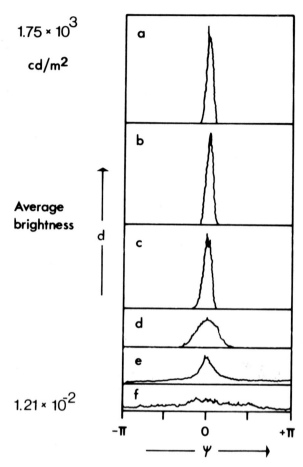

FIG. 45. The fly's orientation towards a black stripe (5° wide), placed vertically and kept stationary at any part of the eye. The visually induced orientation is called fixation if the object is standing. The histograms show the fraction of time the fly fixates any part of the visual panorama, i.e. the fraction of time is associated to any value of the error angle (ψ). Decreasing the average brightness (from a to f) of the panorama degrades fixation until the ψ-distribution is almost flat (f). The light source used here is a set of fluorescent ring bulbs (Philips TLE 40 W/34 de luxe). The sharpness of the histogram is a sign of more intense "gaze" of the fly towards the object, i.e. the error angle is approaching zero. (From Reichardt, W., 1973, with permission of Springer-Verlag, Heidelberg; Reichardt, W. and Poggio, T., 1976.)

(Reichardt, W., 1973; Reichardt, W. and Poggio, T. 1976)

$$\vartheta\ddot{\Psi}(t) + k\dot{\Psi}(t) = -F\{\Psi(t), t\} + S(t)$$

with   (15)

$$S(t) = \vartheta\ddot{\alpha}_p(t) + k\dot{\alpha}_p(t)$$

during fixation $\alpha_p(t) = 0$ when the object does not move with respect to the environment and $S(t) = \vartheta, \psi(t) = -\alpha_f$, where ϑ is the measure for the moment of inertia around the vertical axis of the fly and k an aerodynamic friction constant.

The flight behaviour controlled by the visual system is extremely complicated in the hoverflies, where both males and females are able to do many different aerobatic movements in all spatial directions. Table 8 shows some modes of flight behaviour of *Syritta* as observed by Collett, T. and Land, M. (1975).

If the fly is directing its attention towards a moving object this is called tracking. Initially this orientation of flies towards an object during flight can be considered, as mentioned before, to be a result of the difference between the flight torque response to an object motion from front to back (progressive motion) and the flight torque response to an object motion from back to front (regressive motion).

There seems to be some differences between different species and sexes, for example in *Musca* and *Calliphora* in their ability to track, but in hoverflies (*Syritta*) only the males seem to possess this behaviour. If a fly is tracking, the sex of the object seems to make no difference, i.e. males and females may track both sexes (Collett, T. and Land, M., 1975; Wehrhahn, C. and Hausen, K., 1980).

The tracking response is closely related to the torque response, which again is strongly elicited by a horizontally moving black stripe. The regressively moving stimuli are significantly less effective (Wehrhahn, C. and Hausen, K., 1980). The hypothesis was originally proposed by Reichardt, W. (1973), according to which, the torque reaction is greater for progressive motion on the retina than for regressive motion. This so-called "progressive–regressive mechanism" is also supported by findings of asymptotic oscillations in the tracking behaviour of the female flies (*Musca domestica*). The relatively long (200 ms) period of the oscillation seems to require more complex reaction dynamics than a pure single dead-time delay (Geiger, G. and Poggio, T., 1981).

The response to horizontal motion also depends on the position of the stimulus. There are at least two mechanisms for computing the optomotor response in the lower part of the fly's eye (*Musca domestica*), one performing a position-dependent velocity computation and the other depending on the position, but not on the direction of motion, of an object (Geiger, G., 1981). The regional differences in responses also show that the position-dependent component cannot be detected in the upper part of the eye. Geiger, G. and Nässel, D. (1982) concluded from laser-ablation experiments that the responses to single objects are mainly controlled by cells other than the H- and V-cells. They also suggest that there are at least two separate pathways for the processing of single-object motion and wide-field pattern motion respectively (Fig. 46).

For fixation and tracking the eye region called the "fovea" plays a special role. Many different insect species have several areas in the compound eyes, where the facets are larger and have smaller angles between adjacent visual axes. The form of the eye in many insects strongly suggests that they measure distances either by binocular overlap or by parallax (see next section). In the hoverfly, *Syritta*, the eyes of males but not females have a forward-directed region of enlarged facets where the resolution is 2 to 3 times greater than elsewhere in the eye field. The interommatidial angle in this fovea is $0.6°$ (Collett, T. and Land, M., 1975). The fixation of moving targets (tracking) within the fovea is maintained by holding the angular position of the target on the retina (ψ_p) proportional to the angular velocity of the tracking fly (Φ_p). The delay in this process is roughly 20 ms.

Outside the fovea the targets are fixated by

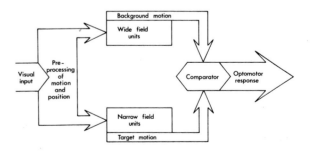

FIG. 46. Summarized view of separate pathways for visual information processing of moving single objects and wide-field patterns. The selective ablation experiments support the idea that H-cells may be responsible for wide-field processing and the lobula columnar neurons could process the narrow-field information of fast-moving targets, thus providing a scheme of structural requirements for the figure–ground discrimination. (After Geiger, G. and Nässel, D., 1982, with permission of Springer-Verlag, Heidelberg.)

Table 7: A comparison of object- and self-movement-detector properties (Olberg, R., 1981)

Object-movement detectors	Self-movement detectors
Differences:	
1. No spontaneous activity	1. Some are spontaneously active
2. Respond only to visual stimuli	2. Multimodal
3. Habituation rapid and long-lasting	3. Little habituation
4. No response to animal rotation	4. Large response to animal rotation
5. Respond only to small moving targets	5. Little or no response to small moving targets
6. No response to moving grating pattern	6. Large response to moving grating pattern
7. Important parameter for discrimination is dimension perpendicular to line of motion	7. Important parameter for discrimination is dimension parallel to line of motion
8. Receptive fields vary in size but may be small	8. Receptive fields large
9. High-velocity threshold	9. Low-velocity threshold
Similarities:	
1. Most are directionally selective	1. All are directionally selective
2. Velocity-sensitive, responding to increasing velocity with increasing spike frequency	2. Velocity-sensitive, responding to increasing velocity with increasing spike frequency
3. Some drive large wing movement	3. Some drive smaller wing movement

accurately directed, intermittent, open-loop body saccades.

Binocular fixation is often closely related to foveal fixation, either monocular or binocular. The torque response of the fly (*Musca domestica*) with a unilaterally covered visual system shows a shift of the fixation towards the unobscured eye. The fly is also tracking a moving black stripe, but if one compound eye is covered the response lags behind, by an average angle of approximately 13° off zero. The simplest hypothesis is that the monocular halves of the responses summate linearly in the binocular region where the visual fields of the two eyes overlap (Geiger, G. *et al.*, 1981). The anatomical facts agree quite well with the behavioural results (Beersma, D. *et al.*, 1977).

Careful studies on the deep pseudopupils in the binocular region of both the compound eyes of *Calliphora* have shown at least five different subareas, where the visual axes of different receptors exactly overlap (Franceschini, N. *et al.*, 1979). The functional differences of those areas are unclear, but because they are concentrated above the equator level in male flies, the function is probably related to tracking. This would correspond with their behavioural strategy.

8.6 Distance perception

The behavioural performance requires, in all

moving and preying animals, an analysis of the 3-D characteristics of the environment. We can find many examples of good behavioural capabilities for judgement of the distance in bees (von Frisch, K., 1967, 1974), in praying mantis (Maldonado, H. and Rodriguez, E., 1972; Rossel, S., 1979), in flies (Collett, T., 1980b; Poggio, T. and Reichardt, W., 1981) and in ants (Burkhalter, A., 1972).

In natural conditions (open-loop process) the tracking fly maintains a roughly constant distance (in *Syritta* in the range 5–15 cm; Collett, T. and Land, M., 1975) from the target. It is generally assumed that the two eyes are essential for distance estimation, and this is accomplished by the binocular method (Wigglesworth, V., 1953; Roeder, K., 1953).

Distance perception also requires a sophisticated neuronal apparatus. At the moment it is not clear if there is in the insect brain an equivalent one to that existing in the vertebrate brain. The problem is that we do not know what kind of neuronal circuit we are looking for. For distance perception there are other possible mechanisms available than only the binocular method (Wallace, G., 1959; Horridge, G., 1977b) such as parallax, scanning and learning through previous visual clues (Anderson, A., 1979; Eriksson, E., 1980a).

8.6.1 BINOCULAR METHOD

In the system of paired lens eye several mechanisms

Table 8: Some modes of the flight behaviour of a hoverfly, Syritta. *The entries in this table specify "inputs" which control each of the three degrees of freedom of flight during each "mode" (From Collett, T. and Land, M., 1975, with permission of Springer-Verlag, Heidelberg).*

Mode	Angular velocity (Φ)	Sideways velocity ($\dot{S}$)	Forward velocity ($\dot{F}$)
Free: Cruising	Clamped at zero by optomotor reflex. Changes of Φ occur by saccades only	Related loosely to external cues but highly variable. $\dot{S}$ and $\dot{F}$ are coupled during flight along a constant course, when $\dot{S}/\dot{F} = \tan \Phi$ (Φ is the angle between course and body axis)	
Fixation: Flight towards flowers	Saccadically or (rarely) smoothly controlled by Ψ (retinal location of flower)	Highly variable, sometimes linked to Φ as in circling	Decreases with distance from flower
Attention: Locating targets	Single saccade, specified by Ψ, the retinal error before turn	Usually not involved	Usually not involved
Tracking: Very fast	Series of saccades each controlled by Ψ before turn	Highly variable control, weakly linked to Ψ, but also predictive	Controlled by the separation (D) of the two flies, probably via size of retinal image
Moderate speed	Continuously controlled by error (Ψ less than about 8°)	ditto	ditto
Very slow	Clamped at zero	Continuously controlled by Ψ	ditto
Fixation: Circling	Coupled to $\dot{S}$ via knowledge of separation (D): $\Phi = \dot{S}/D$	Spontaneously driven, but also coupled to: $\dot{S} = D \cdot \Phi$	Usually fixed at zero
Fixation: Rape flights	Coupled to $\dot{S}$ initially (as in circling), later zero	Coupled to Φ initially, later zero	Constant acceleration. $\ddot{F} = 5\text{m s}^{-2}$.
Tracking: "Wobbling"	Controlled continuously by Ψ	Partly driven by internal oscillator and partly by Ψ	Controlled by the separation (D) of the two flies, as in tracking

contribute to the co-operation of both eyes. The obvious purpose of the coordination mechanisms such as optical aligning and convergence, the pupil mechanism, is to produce information about distance perception. Similar methods are not possible for the insect eye because of the quasi-rigid structure of the head capsule. Let us consider two compound eyes looking forward (Fig. 47). Three diverging optical axes either cut the midline at various distances from the eye, are parallel to the midline axis, or diverge without cutting the optical axes of other ommatidia. The distances where the cuts occur are noted by d and Δd. The distance of the parallel optical axis from the midline is noted by s and the minimum separable distance in the visual field is ms. In figure 47 it is obvious that the optical axes cutting the midline further and further away from the eye will cause a rapid increase of Δd, which is also the minimum interval in discrimination of distance, when the distance to the object is d.

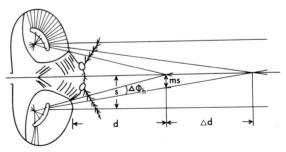

Fig. 47. A mechanism of distance perception by binocular vision. The optical axes are drawn from the adjacent ommatidia and crossed at the middle line of the visual system: $\Delta\Phi_h$, the interommatidial angle; ms, the minimum separable distance; Δd, the minimum interval in discrimination at a distance (d); s, the half-separation of the forward-looking regions of the two eyes. (After Horridge, G., 1977b.)

If we consider the interommatidial angle here in the horizontal plane $\Delta\Phi_h$ then

$$\Delta d = \frac{d^2\varphi_h}{s} \qquad (16)$$

and the frequency of sampling points in the perception of depth is

$$\frac{1}{\Delta d} = \frac{s}{d^2 \Delta \Phi_h} \qquad (17)$$

where $1/\Delta d$ is a measure of the accuracy of the distance measurement, which therefore depends directly on the distance between the two forward-looking regions of the two eyes. The field of best performance in depth perception can be approximated at a distance equal to half the span between the eyes. If the divergence angles of the ommatidia are not constant, a numerical correction should be adapted to any real insect eye (Burkhardt, D. et al., 1973).

The equation above shows that the accuracy of depth perception falls off as the inverse square of the distance. If we consider the maximum distance for depth perception in some insects by using known data, for a large insect in a favourable case we might find $s = 2$ mm and $\Delta \Phi_h = 0.5°$, so that the maximum distance for binocularity would be 20 cm by using a dragonfly as an example. In the honeybee, Apis mellifica, the binocular domain of the visual space starts about 5 mm in front of the head (Seidl, R. and Kaiser, W., 1981). In smaller insects when on the other hand $s = 1$ mm and $\Phi_h = 2°$ then $d = 25$ mm. The distances are surprisingly small, if we consider the behaviour of some of the best-performing insects. Predatory behaviour in Ranatra and in mantids indicates the importance of binocular vision if the prey is located in the range of about a few folds of the maximum reaching distance of the forelegs (up to 15 mm). Unilaterally blinded Ranatra show a clearly reduced performance (Cloarec, A., 1978). In the praying mantis the prey is centered between the foveas of both eyes before striking when the prey is within catching distance (Barrós-Pita, J. and Maldonado, H., 1970; Levín, L. and Maldonado, H., 1970). But, as was shown, the physical principles cannot give distance estimation for distances longer than 200 mm, if the insect is sitting still. The point is that, when moving, they have an alternative method available such as parallax.

8.6.2 DISTANCE PERCEPTION BY PARALLAX

Parallax means an apparent shift of position of an object due to an actual change in position of the subject. Thus the parallax method can be used only during movements of the body or the head. The head movements of the fly during flight are analyzed carefully by Land, M. (1975). While flying, but free only to rotate, the flies (Calliphora) show two kinds of head movements.

(1) Rapid saccadic movements with amplitudes of up to 20° relative to the body axis and durations of ca. 20 ms. It seems from the conclusions of Land, M. (1975) that this class of head movements, mediated by the neck muscles, are accompanied by body turns which slowly bring the body axis back into line with the head.
(2) Stabilization movements which tend to keep the axis of the head still with respect to the surroundings, in spite of fluctuations of direction of the body axis.

These movements are dependent on visual feedback. Both types of head movements are most probably made in relation to object fixation, but other mechanisms are not necessarily excluded.

8.6.3 NEURONAL RESPONSES

If distance perception is performed by parallax, then we might have to see another method such as scanning which is related to parallax. Both methods for distance perception are not simply resolved by a memory, and cannot be explained by the overlapping binocular fields of the two eyes. The structure of memory is a matter of discussion. The methods must be functionally different in scanning and in parallax, but how, is an open question. Distance judgement can also be a function of depth of motion.

The depth–motion of an object generates an expanding two-dimensional stimulus pattern at the receptor surface of the eye. Eriksson, E. (1974, 1980b, 1982) has analysed neural responses of a horizontally sensitive interneuron in the optic lobe of the blowfly (Phormia terraenovae). Usually the neurons are excited by the target motion towards the eye and inhibited by its reverse motion. This kind of depth–motion sensitivity is most appropriate to landing and approaching behaviour of the fly. The landing response does not require binocular vision since it is triggered by monocular stimulation corresponding to an approaching object (Coggshall, J., 1971; Taddei-Ferretti, C., 1973). Although at present we know very little about the mechanisms mediating distance perception, the neurons act as very precise, horizontal, velocity-sensitive, neural integrators using differential time delays of inhibitory and excitatory processes (Eriksson, E., 1982).

Another large-field interneuron in the optic lobe of *Notonecta glauca* was excited by binocular interaction of the eyes (Schwind, R., 1978). An object moving in the plane of the water surface at any distance would excite the neuron maximally if it were presented in the appropriate size, about one-quarter of the length of *Notonecta*. This indicates that the neuron is excited by the prey. The prey located at longer distances is first localized by means of the sense of vibration (Wiese, K., 1974). Visual control participates in the control of the final rush forward at the prey.

8.6.4 IMAGE EXPANSION AS DISTANCE INFORMATION

The landing response of a flying insect can be triggered by binocular visual stimulation. The flow of the continuously changing visual world contains information, which can be used to control the speed of the flight. The maximum landing response to front-to-back motion as a contrast frequency Φ/λ was observed in flies (*Musca*) at around 8 Hz. Upward and downward motion $40°$ above and below the equatorial plane were also effective (Wehrhahn, C. *et al.*, 1981). Flies reduce their flight velocity (by several kilometres per hour) before landing and control the onset of the final deceleration before the landing phase (Wagner, H., 1982). There is some evidence that the onset of deceleration is triggered when the ratio of the image expansion reaches a critical value. In a sense this gives information about the time needed to reach the target not in absolute, but in relative, units dependent on both the translation velocity and the direction of the flight with respect to a target. For houseflies (*Musca domestica*) the relative retinal expansion velocity will need a reaction of -60 ms before starting to decelerate, estimated from filmed landing trajectories (Wagner, H., 1982).

8.7 Visually mediated learning

The great success of the bee as an experimental animal over the last 70 years has proved that the bee can discriminate between different conditioned sensory signals (von Frisch, K., 1965).

There is hardly any doubt that the insect visual system, as well as brain functions, are involved in some degree of plasticity, storage and learning. Numerous training experiments of walking or flying insects in different controlled circumstances have supported this idea.

Freely flying bees can very rapidly learn to use the object patterns of their visual environment and the polarization patterns on the sky for navigation and other purposeful behaviour (von Frisch, K., 1950; von Frisch, K. and Lindauer, M., 1954; von Frisch, K. *et al.*, 1960). Visually mediated learning involves several different kinds of processes from pattern recognition, movement learning and memory-like effects, to a generalization process. It is also possible that the pattern itself does not possess any information value to the animal, but only the quality of the stimulus, such as colour (Baumgärtner, H., 1928; Hertz, M., 1937). In this context we can only touch the surface of the vast literature published about this topic.

8.7.1 PATTERN RECOGNITION

This function of the visual system is of great importance for all visual analysis of the structured optical world in order for an animal to survive. In recent years most studies on pattern recognition in insects, especially those performed in bees, have been dominated by the search for a general formalism, which would relate the chosen frequencies of an insect to certain stimulus parameters like contour density, size of figure, amount of spatial overlap between two figures and so on (Zerrahn, G., 1933; Mazochin-Porshnyakov, G. and Vishnevskaya, T., 1969; Wehner, R., 1969, 1971; von Weizsäcker, E., 1970; Anderson, A., 1972, 1977; Cruse, H., 1972). There might be, as an example, just a very simple question to ask for an animal like a stick-insect. How does it perceive shrubs, where are they usually located? Very interestingly, for them the most attractive black pattern is composed of an uninterrupted straight vertical bar, from which interrupted lines run out on both sides at an angle of less than 30° (Jander, R., 1969).

During the last 50 years there have been many different kinds of patterns used for training (Fig. 48). It might not be very surprising that, in spite of numerous attempts, there is no successful description of special pattern features generalized by bees or other insects. We might say that the insect visual system is too simple to produce any difference between the types of patterns characteristic for the vertebrate lens eye. This may be true in one sense, but most probably when we consider the structural composition of, let us suppose, 6000 ommatidia in each complex eye there are, if strongly simplified, some 10^{1800} possible combinations of light and dark points (Wehner, R., 1975). The number is increased to astronomical dimensions if the broad range of intensities is considered, which the visual system of the bee is able to discriminate. Here the perceptible difference of about 16% between two light stimuli (Kunze, P., 1961; Labhart, T., 1972) is not very exceptional and it is given by the formula

$$\%\Delta I = 2\frac{(I_1 - I_2)}{(I_1 + I_2)} \cdot 100 \qquad (18)$$

The basic problem is to find ways of defining the clusters of stimulus parameters in the animal's multidimensional space that appear to be relevant to the animal. These are possibly the sets of feature detectors that are used by the animal's pattern discriminating mechanisms. It has been so difficult to find a universal set of feature detectors that no general classification can be used at the moment. We have no idea about how many feature detectors are needed for an animal's particular performance; but the finding that some of the well-known optical illusions like Müller-Lyer figures are effective in the fly's behaviour, indicate that the feature detectors are possibly organized in a similar way to that in the human visual system (Geiger, G. and Poggio, T., 1975).

The analysis of pattern detection may be divided into different parts, where the pattern modulation is first determined by the optical properties of the complex eye. The contrast or modulation transfer functions (MTF) give the resolving power of the optical system. Here the eye parameter p gives the optimum value for the relation between the lens diameter and the interommatidial divergence angle (Fig. 2), the opening angle of one ommatidium (Snyder, A., 1977, 1979; Snyder, A. et al., 1977a,b).

8.7.2 Contrast transfer

Besides the above mentioned optical properties of the single ommatidia, pattern detection also involves the overall transfer characteristics of the subsequent stages within the nervous system. The term "visual acuity" may be introduced to refer to the neural transfer of optically-modulated patterns. The problems of the intensity characteristics of the receptors are discussed in the previous chapter and lateral inhibition elsewhere in this contribution. Some of the problems of temporal resolution of retinular cells and their subsequent neurons are discussed here.

8.7.3 Temporal frequency modulation

The dynamic transducer properties of the photoreceptors define how fast changes in light intensity can be transmitted. The high temporal resolution of retinular cells in the fly and some other insects is of functional significance. The shortening of the time constants may result in contrast enhancement produced by a hypothetical visual scanning process.

8.7.4 Associative learning

A bee is able to learn quickly the colour and the odour of flowers that yield nectar or pollen. This information is stored and used for communication.

Neural correlates for associative and non-associative learning can be found in single cells of the median protocerebrum. Neurons which display these response changes are bi- or multimodal, responding to visual, olfactory and/or gustatory stimuli (Menzel, R. and Erber, J., 1978; Erber, J. and Schildberger, K., 1980; Erber, J., 1981), while only a few non-rewarding tests will erase the conditioning.

A free-flying bee will learn to associate an odour with a sugar-water reward after a single conditioning trial, whereas three to five reward trials are necessary for colour association. An artificial shape of the food source may require approximately 20 rewards.

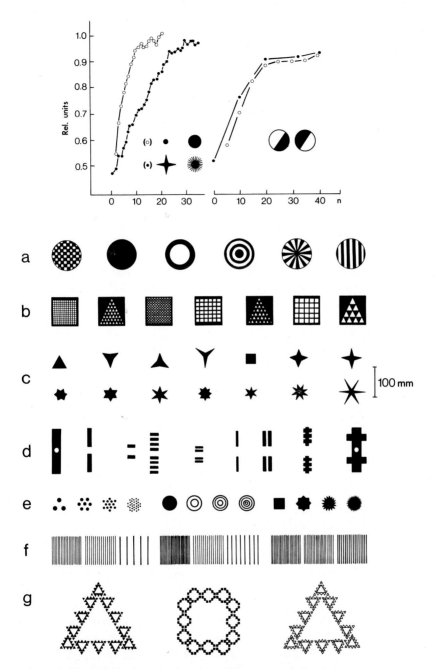

FIG. 48. The discrimination of visual patterns by honeybees, *Apis mellifera*. The diagrams on the top show learning curves of bees trained to discriminate two visual patterns. On the top left the training to a small against a large black dot (○) and to a four-pointed against a 24-pointed black star (●). On the top right two series of training experiments with bees to a circular black-and-white pattern simply rotated 180° to each other. The training and testing of ten bees has been performed either individually (○) or as a group (●). The number of reinforcements are given by (*n*) on the abscissa and the relative units are determined by the discrimination frequency $n(+)/\Sigma n$. (From Schnetter, B., 1972; Wehner, R., 1972, 1981, with permission of Springer-Verlag, Heidelberg.) The sets (a–g) of patterns are used by different authors to study pattern recognition in bees. These patterns are usually presented on a horizontal screen, except (d) and the pattern on the top right. All patterns are drawn to scale, the bar 100 mm. **a:** Zerrahn, G. (1933); **b:** Anderson, A. (1972); **c:** Cruse, A. (1972); **d:** Anderson, A. (1977a); **e:** Hertz, M. (1933); **f:** Anderson, A. (1977b); **g:** Mazochin-Porshnyakov, G. and Vishnevskaya, T. (1969).

The duration of a sugar-water reward can be as short as 100 ms and still produce a significant change in response. In many respects the organization of memory is similar to that of vertebrates. Bees have a sensory memory lasting about 2 s, a short-term memory lasting about 7 min and a long-term memory which can last for the lifetime of an adult.

The movement-sensitive neurons have axonal projections in the median protocerebrum. During the conditioning to upward movement the neurons modified their response characteristics. The response in the conditioned direction increased with the number of rewards whereas the inhibitory response for movement in the opposite direction was reduced (Erber, J., 1981). These neurons can also be multimodal in their characteristics, responding to odour, mechanical stimuli and sugar-water.

Free-flying honeybees very rapidly learn to distinguish two stripe patterns moving in opposite directions (Erber, J., 1982). They need only 5 to 10 rewards to show significant behavioural modifications. In this respect movement learning is similar to pattern learning in the bee (Wehner, R., 1971, 1972).

The training to movement alone shows clearly, up to a degree, the ability of the bee to generalize the visual stimuli. On the other hand there is no significant velocity discrimination for patterns moving in the same direction (Erber, J., 1982).

Visual learning was shown to exist in other insect groups, although their conditioning is slightly different from that of bees. Usually conditioning in *Drosophila* is produced by an aversive stimulus (Quinn, W. *et al.*, 1974; Menne, D. and Spatz, H.-Ch., 1977; Dudai, Y. and Bicker, G., 1978; Bicker, G. and Reichert, H., 1978; Folkers, E. and Spatz, H.-Ch., 1981).

Drosophila could be conditioned to blue and yellow lights, when the flies primarily respond to the colour of light (Menne, D. and Spatz, H.-Ch., 1977). A similar kind of experiment was repeated with photoreceptor degeneration mutants *rdgB* (receptor degeneration B) with functionally inactivated retinular cells R1–6. The light stimulus intensity was also altered (Bicker, G. and Reichert, H., 1978). The general visual learning ability was not impaired by the *rdgB* mutation although the conditionable intensity discrimination was clearly

higher than in the wild-type flies. For *rdgB* flies no positive evidence for the existence of colour discrimination was present.

Conditioning, at least in the wild-type *Drosophila*, is an active and flexible central process, which is shown by reversed conditioning (Folkers, E. and Spatz, H.-Ch., 1981).

Some other larger flies (*Calliphora*) have also been shown to possess memory-like effects, if recorded from the lobula elements which are stimulated with two alternating stimuli, movement and flicker (Mastebroek, H. *et al.*, 1982). This memory element has a time constant of a few minutes and thus can be considered as short-term. We do not know yet how the animal takes advantage of it.

Butterflies using different flowers for feeding show an ability to learn to choose between coloured model flowers. The conditioning experiments on *Heliconus charitonius* turn the preference to visit either green or yellow models from 2% and 9% respectively up to roughly 55% (Swihart, C. and Swihart, S., 1970). Most probably other unstudied insect groups possess better developed visual learning and memory capabilities than usually believed, though some caution must be taken if test stimuli are presented successively. If bees have to learn more than one scent signal successively, then they preferably choose the first scent signal (Koltermann, R., 1969).

9 VISUALLY MEDIATED COMMUNICATION

It is a fascinating approach to visual physiology in insects to understand "seeing" as a link to "communication". All communication is based on some kind of language. Language, on the other hand, is a name for a group of meaningful signs forming signals tied together with logic. The communication can be uni- or bidirectional, but there must always be a transmitter and a receiver. In a unidirectional system they are separated but in a bidirectional one they must be in the same system. A transmitter in a unidirectional system is usually a sign, visual or another modality; in a bidirectional system a transmitter is some kind of motor or effector output. The analysis of communication is an extremely difficult task, because of philosophical or logical problems. To understand communication, one should understand the code involved, but that is the problem of the study. All communication is therefore interpreted and translated in relation to our own communication principles. These we hope are the same as others under investigation. Whatever looks reasonable and sensible to us in insect behaviour is not necessarily that. After all this I still think that there is some sense in studying different aspects of communication among insects.

Communication in the visual domain involves the visual system with all its peripheral functions, colour vision, pattern, movement, polarized light and temporal pattern detection. This kind of communication can be found between many flowers and insects, but also often between both sexes of insects, and between prey and predator.

9.1 Unidirectional communication

9.1.1 FLOWERS AND ANIMALS

This class of communication is common between flowers and their insect visitors. Flowers transmit the information that aids pollination, and the animals receive it for another purpose. This may occur in at least three different ways. A floral pattern (1) advertises the presence of the "receptive" flower, (2) provides identification characters of the flowering species, and (3) indicates by special marking the nectar supply.

Advertising is usually communicated to a potential visitor by a contrasted reflection in the UV or another wavelength. Flowers do not look pure UV to bees, but more commonly they reflect "bee purple" (yellow + UV), yellow, "bee-violet" or blue (Daumer, K., 1958). Because the timing of the visits is very important for the flower, the floral signal is presented for only a short period. The attractiveness of the floral signal is due to both innate and learned behaviour on the part of visitor. Bees are especially attracted to figures having complex outlines.

The identification is based on the constancy and high development of the flowers. It is assumed that insects such as bees can discriminate and remember the pattern of the flower from an equally attractive but spatially differing one.

In some flowers, which probably have been identified, a "nectar-guide" marking indicates the way

to nectaries or to the entrances of floral tubes containing nectaries. Daumer, K. (1958) has demonstrated the effectiveness of UV nectar-guides in leading honeybees about on "dummy" flowers with artificial markings. Nectar-guides may contrast in "visible" and/or UV light.

9.1.2 SEXUAL ATTRACTION

Male butterflies usually respond to the visual stimuli of the female (unidirectional communication) and begin courtship, presenting visual and chemical signals to which the female responds (bidirectional communication). The sexual discrimination is made by both sexes in the white butterfly (Pieris protodice) using the sexual dimorphism in UV reflectors principally instead of other melanic markings on the wings, movement patterns or chemical signals (Rutowski, R., 1977, 1981; Silberglied, R., 1979).

The UV-reflection in the wings of female Pieris butterflies is a necessary component for the release of male courtship behaviour. There are also large interspecific differences in female behaviour. Female Colias eurytheme accept males which have strong UV reflection, but females of C. philodice do not discriminate UV-absorbing or reflecting males (Silberglied, R., 1979).

9.1.3 PREY AND PREDATOR

At least two unidirectional communication strategies are understood between prey and predator; viz. camouflage and mimicry. Some insects have a very close resemblance to their backgrounds, e.g. in UV. The distinctive UV-patterns of some lepidopterous larvae may camouflage them from visually orienting predators, such as predatoid and parasitoid wasps (Byers, J., 1975).

Some arthropod predators, on the other hand, rely upon camouflage as they wait in ambush for prey. The red marks on the female crab spider (Misumena vatia) may be warning colours aimed at vertebrate predators but invisible to insects (Hinton, H., 1976).

Mimicry in insects is widely used as signal-like warning or attraction. Visual warning signals are usually meant to frighten the predator (eye-fields in butterfly wings) or to mimic some other poisonous, unsavoury or nasty species.

9.2 Bidirectional communication

This kind of communication resembles that in many forms of higher vertebrates. The bee languages, such as the special dancing procedure, can be considered as bidirectional communication. The communication system is necessarily multimodal in nature and therefore it will be almost impossible to tell where the significance of one modality, e.g. vision, disappears and the other begins to dominate. Usually bidirectional communication is an essential part of courtship and therefore is considered here first with respect to vision.

9.2.1 COURTSHIP

Bidirectionality is a clear improvement in efficiency of communication. We can see that the perception and transmission of either excitatory or inhibitory signals will efficiently shorten the feedback between sexes of the same species, instead of using advertising as in unidirectional communication.

Vision plays two distinct roles in *Drosophila* courtship (Ewing, A., 1983). One is specific, concerning perception of inhibitory or excitatory sexual stimuli caused by display of specific postures or morphological structures. The other is non-specific, and would include aspects of behaviour such as the ability of the male to keep contact with the female and to orientate towards her, as discussed before in this contribution.

Mating is light-dependent to varying degrees in many species, depending on the dominance of the visual system. Usually *Drosophila* males show to females a characteristic circling or raising of wings in a "V". The female response is either an acceptance posture with spreading wings and vaginal plates to show willingness, or in the case of fertilized females to rock from side to side when approached by other flies. Movements are also sometimes connected with displaying body markings, which probably serve also as indicators of fertile age or readiness.

Another kind of courtship communication through the visual system is found between bioluminescent insects. The sexual signalling involves species-specific male flash patterns and coded time delays in female flash responses. The information transfer between the individuals consists of two parameters; *viz.* the emission versus spectral sensitivity and the time intervals between single events (Souček, B. and Carlson, A., 1975). The bioluminescence emission spectra in the firefly, *Photinus*, species depend on the time period of activity after sunset. This genus of fireflies contains dusk-active species which emit yellow light at 560 nm and dark-active species emitting at 558 nm or shorter in the green range. The shift in colour is suggested to contrast the emission in different environmental conditions (Lall, A. *et al.*, 1980). This is certainly an evolutionary adaptation to increase the efficiency of sexual signalling in different prevailing conditions of light environments. The ERG recordings from the receptors show corresponding sensitivities to yellow and green light stimulation.

The information produced by the male in this communication system will generate, in a receiving female, a response function composed of three intervals: initial inhibition or total blocking; acceptance; and long-term inhibition. This response function and discrimination bias generated by the short-term memory, together with background noise, determine the time window for the reception by the firefly. This time interval is used to characterize and analyse the male response function. Short-term memory is used for sensitivity adjustments during the critical time sequence (Souček, B. and Carlson, A., 1980). A precise time discrimination is needed to recognize the species-specific signal and to identify the correct partner.

9.2.2 SOCIAL ACTS

The social behaviour of insects in colonies requires communication. The types of communication or languages have been a matter of discussion. The most understood communication system is called "the dance language" (von Frisch, K., 1947). We can also find other communication systems among insects, which are based on sounds, mutual tapping or chemical exchange. Most of them are not concerned with vision.

The dance language is basically expressed by a complex series of movements combined with body vibrations. The motor acts are essentially coupled with vision. The information to be transmitted will contain the distance to the food source and the direction from the hive. The distance information is communicated, and is used by recruited bees in round dance (up to 20 m) or through transition (20–200 m) to waggle dance (> 200 m). The direction to the food is shown by the relative axis of the waggle dance in relation to the vertical direction, which represents the direction to the sun. All deviations from the vertical axis show the deviation from the direction of the sun to the direction of the food source. Here the bees may use the e-vector of polarized light and the internal clock to compensate for the time errors if the sky suddenly turns cloudy.

Food is not the only reason for dancing, which can also be performed under other circumstances. After dividing a hive, about half of the bees fly out with the old queen and cluster in a swarm. Foragers fly out and search for a suitable place to establish a new colony. When they return to the swarm they perform dances correlated with locations of new hive location (Lindauer, M., 1955).

Here an interesting question is how much vision is involved during recruitment. Although we have seen that the performance of the compound eyes in higher-level insects is very sophisticated, the role of vision through observation remains open. As shown, at least the bees "attending" the dance run after and maintain antennal contact with the dancer's body. This could also be interpreted as a sign of odour information exchange and lead to the dance-language controversy as recently discussed (Gould, J., 1976; Rosin, R., 1980).

Similar dancing-like behaviour has not been described in other insects species as thoroughly as in bees, but the ability to use polarized light for orientation may exist in numerous insect species, including a variety of bees, wasps, ants, flies, butterflies and moths (see review by Waterman, T., 1981). Furthermore, only a few species have been tested to show an ability to transpose a direction to the sun to the direction of gravity (caddis flies; Jander, R., 1963; ants: Markl, H., 1964). The transformation of the distance information by using waggling is not commonly found in other insect species except moths (Blest, A., 1960). Many basic questions of the storing of information during foraging flight still remain open but the acceptance of the bee-language as relevant to the communication process involves a silent assumption by us of a highly developed functional neural array in the insect brain.

ACKNOWLEDGEMENTS

I am very grateful to Mrs Anneli Rautio for her excellent and tenacious help throughout the preparation of this work. I also wish to express my gratitude to Miss Francine Chatelois, B.Sc. for discussing and correcting the English version.

REFERENCES

ADOLPH, A. R. (1976). Putative synaptic mechanisms of inhibition in *Limulus* lateral eye. *J. Gen. Physiol.* 67, 417–432.

ANDERSON, A. M. (1972). The ability of honey bees to generalize visual stimuli. In *Information Processing in the Visual Systems of Arthropods.* Edited by R. Wehner. Pages 207–212. Springer, Berlin and New York.

ANDERSON, A. M. (1977a). Shape perception in the honey bee. *Anim. Behav.* 25, 67–79.

ANDERSON, A. M. (1977b). Parameters determining the attractiveness of stripe patterns in the honey bee. *Anim. Behav.* 25, 80–87.

ANDERSON, A. M. (1979). Visual scanning in the honey bee. *J. Comp. Physiol.* 130, 173–182.

ARNETT, D. W. (1971). Receptive field organization of units in the first optic ganglion of *Diptera. Science* 173, 929–931.

ARNETT, D. W. (1972). Spatial and temporal integration properties of units in first optic ganglion of dipterans. *J. Neurophysiol.* 35, 429–444.

AUTRUM, H. (1950). Die Belichtungspotentiale und das Sehen der Insekten (Untersuchungen an *Calliphora* und *Dixippus*). *Z. Vergl. Physiol.* 32, 176–227.

AUTRUM, H. (1981). Light and dark adaptation in invertebrates. In *Comparative Physiology and Evolution of Vision in Invertebrates. C: Invertebrate Visual Centers and Behavior II.* Edited by H. Autrum. *Handbook of Sensory Physiology.* Vol. 7, pt 6C, pages 1–91. Springer, Berlin and New York.

AUTRUM, H. and KOLB, G. (1972). The dark adaptation in single visual cells of the compound eye of *Aeschna cyanea. J. Comp. Physiol.* 79, 213–232.

AUTRUM, H. and STÖCKER, M. (1952). Über optische Verschmelzungsfrequenzen und stroboskopisches Sehen bei Insekten. *Biol. Zentralblatt, Bd 71,* 129–152.

AUTRUM, H., ZETTLER, F. and JÄRVILEHTO, M. (1970). Postsynaptic potentials from a single monopolar neuron of the ganglion opticum I of the blowfly *Calliphora. Z. Vergl. Physiol.* 70, 414–424.

BACON, J. and TYRER, M. (1978). The tritocerebral commissure giant (TCG): a bimodal interneurone in the locust, *Schistocerca gregaria. J. Comp. Physiol.* 126, 317–325.

BÄHR, R. (1972). Licht- und dunkeladaptive Anderungen der Sehzellen von *Lithobius forficatus* L. (Chilopoda: Lithobiidae). *Cytobiologie* 6, 214–233.

BARLOW, H. B., FITZHUGH, R. and KUFFLER, S. W. (1957). Change of organization in the receptive field of the cat's retina during dark adaptation. *J. Physiol.* 137, 338–354.

BARLOW, R. B. JR. (1969). Inhibitory fields in the *Limulus* lateral eye. *J. Gen. Physiol.* 54, 383–396.

BARLOW, R. B. JR. and FRAIOLI, A. J. (1978). Inhibition in the *Limulus* lateral eye *in situ. J. Gen. Physiol.* 71, 699–720.

BARLOW, R. B. JR. and LANGE, G D. (1974). A nonlinearity in the inhibitory interactions in the lateral eye of *Limulus. J. Gen. Physiol.* 63, 579–589.

BARLOW, R. B. JR. and QUARLES, D. A. JR. (1975). Mach bands in the lateral eye of *Limulus* comparison of theory and experiment. *J. Gen. Physiol.* 65, 709–730.

BARRÓS-PITA, J. C. and MALDONADO, H. (1970). A fovea in the praying mantis eye. II. Some morphological characteristics. *Z. Vergl. Physiol.* 67, 79–92.

BAUMGÄRTNER, H. (1928). Der Formensinn und die Sehschärfe der Bienen. *Z. Vergl. Physiol. 7*, 56–143.

BEERSMA, D. G. M., STAVENGA, D. G. and KUIPER, J. W. (1977). Retinal lattice, visual field and binocularities in flies. Dependence on species and sex. *J. Comp. Physiol. 119*, 207–220.

VON BÉKÉSY, G. (1966). Mach band type lateral inhibition in different sense organs. *J. Gen. Physiol. 50*, 519–532.

BERNARD, G. D. and STAVENGA, D. G. (1979). Spectral sensitivities of retinular cells measured in intact, living flies by an optical method. *J. Comp. Physiol. 134*, 95–107.

BERNARD, G. D. and WEHNER, R. (1980). Intracellular optical physiology of the bee's eye. I. Spectral sensitivity. *J. Comp. Physiol. 137*, 193–203.

BERNHARD, C. G., HÖGLUND, G. and OTTOSON, G. (1963). On the relation between pigment position and light sensitivity of the compound eye in different nocturnal insects. *J. Insect Physiol. 9*, 573–586.

BICKER, G. and REICHERT, H. (1978). Visual learning in a photoreceptor degeneration mutant of *Drosophila melanogaster. J. Comp. Physiol. 127*, 29–38.

BIEDERMAN-THORSON, M. and THORSON, J. (1971). Dynamics of excitation and inhibition in the light-adapted *Limulus* eye *in situ. J. Gen. Physiol. 58*, 1–19.

BISHOP, L. G. and KEEHN, D. G. (1966). Two types of neurones sensitive to motion in the optic lobe of the fly. *Nature 212*, 1374–1376.

BISHOP, L. G. and KEEHN, D. G. (1967). Neural corralates of the optomotor response in the fly. *Kybernetik 3*, 288–295.

BISHOP, L. G., KEEHN, D. G. and McCANN, G. D. (1968). Motion detection by interneurons of optic lobes and brain of the flies *Calliphora phaenicia* and *Musca domestica. J. Neurophysiol. 31*, 509–525.

BLEST, A. D. (1960). The evolution, ontogeny, and quantitative control of the settling movements of some new world saturnid moths, with some comments on distance communication by honey bees. *Behaviour 16*, 188–253.

BLONDEAU, J. and HEISENBERG, M. (1982). The three-dimensional optomotor torque system of *Drosophila melanogaster*. Studies on wildtype and the mutant optomotor-blind[H31]. *J. Comp. Physiol. 145*, 321–329.

BRAITENBERG, V. (1967). Patterns of projection in the visual system of the fly. I. Retina–lamina projections. *Exp. Brain Res. 3*, 271–298.

BRAITENBERG, V. (1970). Ordnung und Orientierung der Elemente im Sehsystem der Fliege. *Kybernetik 7*, 235–242.

BRAITENBERG, V. and DEBBAGE, P. (1974). A regular net of reciprocal synapses in the visual system of the fly *Musca domestica. J. Comp. Physiol. 90*, 25–31.

BRODIE, S. E., KNIGHT, B. W. and RATLIFF, F. (1978). The spatiotemporal transfer function of the *Limulus* lateral eye. *J. Gen. Physiol. 72*, 167–202.

BROWN, H. M., HAGIWARA, S., KOIKE, H. and MEECH, R. W. (1971). Electrical characteristics of a barnacle photoreceptor. *Fed. Proc. 30*, 69–78.

BUCHNER, E. (1976). Elementary movement detectors in an insect visual system. *Biol. Cybernet. 24*, 85–101.

BÜLTHOFF, H. (1981). Figure–ground discrimination in the visual system of *Drosophila melanogaster. Biol. Cybernet. 41*, 139–145.

BURKHALTER, A. (1972). Distance measuring as influenced by terrestrial cues in *Cataglyphis bicolor* (Formicidae, Hymenoptera). In *Information Processing in the Visual Systems of Arthropods*. Edited by R. Wehner. Pages 303–308. Springer, Berlin and New York.

BURKHARDT, D., DARNHOFER-DEMAR, B. and FISCHER, K. (1973). Zum binokularen Entfernungssehen der Insekten. *J. Comp. Physiol. 87*, 165–188.

BURTT, E. T. and CATTON, W. T. (1956). Electrical responses to visual stimulation in the optic lobe of the locust and certain other insects. *J. Physiol. 133*, 68–88.

BURTT, E. T. and CATTON, W. T. (1959). Transmission of visual responses in the nervous system of the locust. *J. Physiol. 146*, 492–514.

BURTT, E. T. and CATTON, W. T. (1960). The properties of single unit discharge in the optic lobe of the locust. *J. Physiol. 154*, 479.

BURTT, E. T. and CATTON, W. T. (1964). Potential changes in the eye and optic lobe of certain insects during light- and dark-adaptation. *J. Insect Physiol. 10*, 865–886.

BURTT, E. T. and CATTON, W. T. (1966). Perception by locusts of rotated patterns. *Science 151*, 224.

BURTT, E. T. and CATTON, W. T. (1969). Resolution of the locust eye measured by the rotation of radial striped patterns. *Proc. Roy. Soc. Lond. B. 173*, 513–529.

BUTLER, R. and HORRIDGE, G. A. (1973). The electrophysiology of the retina of *Periplaneta americana* L. 1. Changes in receptor acuity upon light/dark adaptation. *J. Comp. Physiol. 83*, 263–278.

BYERS, J. R. (1975). Tyndall blue and surface white of tent caterpillars, *Malacosoma* spp. *J. Insect Physiol. 21*, 401–415.

CARTWRIGHT, B. A. and COLLETT, T. S. (1982). How honey bees use landmarks to guide their return to a food source? *Nature 295*, 560–564.

CATTON, W. T. (1980). Tonic effects of stationary luminous slits on the discharge rates of some locust visual interneurones. *J. Insect Physiol. 26*, 373–379.

CATTON, W. T. (1982). The effects of stimulus area and intensity on the on/off ratio of some locust visual interneurones. *J. Insect Physiol. 28*, 285–291.

CHAPPELL, R. L. and DOWLING, J. E. (1972). Neural organization of the median ocellus of the dragonfly. I. Intracellular electrical activity. *J. Gen. Physiol. 60*, 121–147.

CHI, C. and CARLSON, S. D. (1980a). Membrane specializations in the first optic neuropil of the housefly, *Musca domestica* L. I. Junctions between neurons. *J. Neurocytol. 9*, 429–449.

CHI, C. and CARLSON, S. D. (1980b). Membrane specializations in the first optic neuropil of the housefly, *Musca domestica* L. II. Junctions between glial cells. *J. Neurocytol. 9*, 451–469.

CLOAREC, A. (1978). Estimation of hit distance by *Ranatra. Biol. Behav. 4*, 173–191.

COGGSHALL, J. C. (1971). Sufficient stimuli for the landing response in *Oncopeltus fasciatus. Naturwissenschaften 58*, 100–101.

COLLETT, T. S. (1970). Centripetal and centrifugal visual cells in medulla of insect optic lobe. *J. Neurophysiol. 33*, 239–256.

COLLETT, T. (1971). Visual neurones for tracking moving targets. *Nature 232*, 127–130.

COLLETT, T. S. (1980a). Some operating rules for the optomotor system of a hoverfly during voluntary flight. *J. Comp. Physiol. 138*, 271–282.

COLLETT, T. S. (1980b). Angular tracking and the optomotor response. An analysis of visual reflex interaction in a hoverfly. *J. Comp. Physiol. 140*, 145–158.

COLLETT, T. and KING, A. J. (1975). Vision during flight. In *The Compound Eye and Vision of Insects*. Edited by G. A. Horridge. Pages 437–466. Clarendon Press. Oxford.

COLLETT, T. S. and LAND, M. F. (1975). Visual control of flight behaviour in the hoverfly, *Syritta pipiens* L. *J. Comp. Physiol. 99*, 1–66.

CRUSE, H. (1972). Versuch einer quantitativen Beschreibung des Formensehens der Honigbiene. *Kybernetik 11*, 185–200.

DAUMER, K. (1958). Blumenfarben, wie sie die Bienen sehen. *Z. Vergl. Physiol. 41*, 49–110.

DEVOE, R. D. and OCKLEFORD, E. M. (1976). Intracellular responses from cells of the medulla of the fly, *Calliphora erythrocephala. Biol. Cybernet. 23*, 13–24.

DIETRICH, W. (1909). Die Facettenaugen der Dipteren. *Z. Wiss. Zool. 92*, 465–539.

DODGE, F. A. JR., KNIGHT, B. W. and TOYODA, J. (1968). Voltage noise in *Limulus* visual cells. *Science 160*, 88–90.

DÖRRSCHEIDT-KÄFER, M. (1972). Die Empfindlichkeit einzelner Photorezeptoren im Komplexauge von *Calliphora erythrocephala. J. Comp. Physiol. 81*, 309–340.

DØVING, K. B. and MILLER, W. H. (1969). Function of insect compound eye containing crystalline tracts. *J. Gen. Physiol. 54*, 250–267.

DOWLING, J. E. and CHAPPELL, R. L. (1972). Neural organization of the median ocellus of the dragonfly. II. Synaptic structure. *J. Gen. Physiol. 60*, 148–165.

DREISIG, H. (1981). The dynamics of pigment migration in insect superposition eyes. *J. Comp. Physiol. 143*, 491–502.

DUBS, A. (1981). Non-linearity and light adaptation in the fly photoreceptor. *J. Comp. Physiol. 144*, 53–59.

DUDAI, Y. and BICKER, G. (1978). Comparison of visual and olfactory learning in *Drosophila. Naturwissenschaften 65*, 495–496.

DUDEK, F. E. (1975). The visual response from the compound eye of *Oncopeltus fasciatus*: effects of temperature and sensory adaptation. *J. Insect Physiol. 21*, 517–528.

DUDEK, F. E. and KOOPOWITZ, H. (1973). Adaptation and temporal characteristics of the insect visual response. *J. Comp. Physiol. 82*, 33–46.

DVORAK, D. and SNYDER, A. (1978). The relationship between visual acuity and illumination in the fly, *Lucilia sericata. Z. Naturf. 33 c*, 139–143.

DVORAK, D. R., BISHOP, L. G. and ECKERT, H. E. (1975). On the identification of movement detectors in the fly optic lobe. *J. Comp. Physiol. 100*, 5–23.

EATON, J. L. and PAPPAS, L. G. (1977). Synaptic organization of the cabbage looper moth ocellus. *Cell Tiss. Res. 183*, 291–297.

ECKERT, H. (1973). Optomotorische Untersuchungen am visuellen System der Stubenfliege *Musca domestica* L. *Kybernetik 14*, 1–23.

ECKERT, H. (1978). Response properties of dipteran giant visual interneurones involved in control of optomotor behaviour. *Nature 271*, 358–360.

ECKERT, H. (1980a). Orientation sensitivity of the visual movement detection system activating the landing response of the blowflies, *Calliphora*, and *Phaenicia*: a behavioural investigation. *Biol. Cybernet. 37*, 235–247.

ECKERT, H. (1980b). Functional properties of the H1-neurone in the third optic ganglion of the blowfly, *Phaenicia. J. Comp. Physiol. 135*, 29–39.

ECKERT, H. (1981). The horizontal cells in the lobula plate of the blowfly, *Phaenicia sericata. J. Comp. Physiol. 143*, 511–526.

ECKERT, H. and BISHOP, L. G. (1978). Anatomical and physiological properties of the vertical cells in the third optic ganglion of *Phaenicia sericata* (Diptera, Calliphoridae). *J. Comp. Physiol. 126*, 57–86.

EDWARDS, D. H. JR. (1982a). The cockroach DCMD neurone. I. Lateral inhibition and the effects of light- and dark-adaptation. *J. Exp. Biol. 99*, 61–90.

EDWARDS, D. H. JR. (1982b). The cockroach DCMD neurone. II. Dynamics of response habituation and convergence of spectral inputs. *J. Exp. Biol. 99*, 91–107.

EHEIM, W. P. and WEHNER, R. (1972). Die Sehfelder der zentralen Ommatidien in den Appositionsaugen von *Apis mellifica* und *Cataglyphis bicolor* (Apidae, Formicidae; Hymenoptera). *Kybernetik 10*, 168–179.

ELOFSSON, R. (1976). Rhabdom adaptation and its phylogenetic significance. *Zool. Scripta 5*, 97–101.

ERBER, J. (1981). Neural correlates of learning in the honeybee. *Trends Neurosci. 4*, 270–273.

ERBER, J. (1982). Movement learning of free flying honeybees. *J. Comp. Physiol. 146*, 273–282.

ERBER, J. and MENZEL, R. (1977). Visual interneurons in the median protocerebrum of the bee. *J. Comp. Physiol. 121*, 65–77.

ERBER, J. and SCHILDBERGER, K. (1980). Conditioning of an antennal reflex to visual stimuli in bees (*Apis mellifera* L.). *J. Comp. Physiol. 135*, 217–225.

ERIKSSON, E. S. (1974). A theory of veridical space perception. *Scand. J. Psychol. 15*, 225–235.

ERIKSSON, E. S. (1980a). Movement parallax and distance perception in the grasshopper (*Phaulacridium vittatum* (Sjöstedt)). *J. Exp. Biol. 86*, 337–340.

ERIKSSON, E. S. (1980b). Responses of visual interneurons in the fly optic lobe during stimulation by motion in depth. *J. Comp. Physiol. 141*, 123–130.

ERIKSSON, E. S. (1982). Neural responses to depth-motion stimulation in a horizontally sensitive interneurone in the optic lobe of the blowfly (*Phormia terraenovae*). *J. Insect Physiol. 28*, 631–639.

EWING, A. W. (1983). Functional aspects of *Drosophila* courtship. *Biol. Rev. 58*, 275–292.

EXNER, S. (1891). *Die Physiologie der facettierten Augen von Krebsen und Insekten.* Deuticke, Leipzig und Wien.

FAIN, G. L. (1975). Quantum sensitivity of rods in the toad retina. *Science 187*, 838–841.

FERMI, G. and REICHARDT, W. (1963). Optomotorische Reaktionen der Fliege *Musca domestica* (Abhängigkeit der Reaktion von der Wellenlänge, der Geschwindigkeit, dem Kontrast und der mitteren Leuchtdichte bewegter periodischer Muster). *Kybernetik 2*, 15–28.

FOLKERS, E. and SPATZ, H.-Ch. (1981). Visual learning behaviour in *Drosophila melanogaster* wildtype as. *J. Insect Physiol. 27*, 615–622.

FRANCESCHINI, N., KIRSCHFELD, K. and MINKE, B. (1981). Fluorescence of photoreceptor cells observed *in vivo. Science 213*, 1264–1267.

FRANCESCHINI, N., MÜNSTER, A. and HEURKENS, G. (1979). Aquatoriales und binokulares sehen bei der Fliege *Calliphora erythrocephala. Verh. Dtsch. Zool. Ges. 1979*, 209.

FRANTSEVICH, L. and MOKRUSHOV, P. (1977). Jittery movement fibers (JMF) in dragonfly nymphs: stimulus-surround interaction. *J. Comp. Physiol. 120*, 203–214.

FRENCH, A. S. (1979). The effect of light adaptation on the dynamic properties of phototransduction in the fly, *Phormia regina. Biol. Cybernet. 32*, 115–123.

FRENCH, A. S. and JÄRVILEHTO, M. (1978a). The dynamic behaviour of photoreceptor cells in the fly in response to random (white noise) stimulation at a range of temperatures. *J. Physiol. 274*, 311–322.

FRENCH, A. S. and JÄRVILEHTO, M. (1978b). The transmission of information by first and second order neurons in the fly visual system. *J. Comp. Physiol. 126*, 87–96.

VON FRISCH, K. (1947). The dances of the honey bee. *Brit. Bull. Anim. Behav. 5*, 1–32.

VON FRISCH, K. (1950). Die Sonne als Kompass im Leben der Bienen. *Experientia 6*, 210–221.

VON FRISCH, K. (1965). *Tanzsprache und Orientierung der Bienen.* Springer, Berlin and New York.

VON FRISCH, K. (1967). Honeybees: Do they use direction and distance information provided by their dancers? *Science 158*, 1072–1076.

VON FRISCH, K. (1974). Decoding the language of the bee. *Science 185*, 663–668.

VON FRISCH, K. and LINDAUER, M. (1954). Himmel und Erde in Konkurrenz bei der Orientierung der Bienen. *Naturwissenschaften 11*, 245–253.

VON FRISCH, K., LINDAUER, M. and DAUMER, K. (1960). Über die Wahrnehmung polarisierten Lichtes durch das Bienenauge. *Experientia 16*, 289–336.

FUORTES, M. G. F. and HODGKIN, A. L. (1964). Changes in time scale and sensitivity in the ommatidia of *Limulus. J. Physiol. 172*, 239–263.

FUORTES, M. G. F. and O'BRYAN, P. M. (1972). Responses to single photons. In *Physiology of Photoreceptor Organs.* Edited by M. G. F. Fuortes. *Handbook of Sensory Physiology.* Vol. 7, pt 2, pages 321–338. Springer, Berlin and New York.

FURSHPAN, E. J. and POTTER, D. (1959). Transmission at the giant motor synapse of the crayfish. *J. Physiol. 145*, 289–325.

GAFFRON, M. (1934). Das Bewegungssehen bei Libellenlarven, Fliegen und Fischen. *Z. Vergl. Physiol. 20*, 299–337.

VON GAVEL, L. (1939). Die "kritische Streifenbreite" als Mass der Sehschärfe bei *Drosophila melanogaster. Z. Vergl. Physiol. 27*, 89–135.

GEIGER, G. (1981). Is there a motion-independent position computation of an object in the visual system of the housefly. *Biol. Cybernet. 40*, 71–75.

GEIGER, G. and NÄSSEL, D. (1981). Visual orientation behaviour of flies after selective laser beam ablation of interneurones. *Nature 293*, 398–399.

GEIGER, G. and NÄSSEL, D. R. (1982). Visual processing of moving single objects and wide-field patterns in flies: Behavioural analysis after laser-surgical removal of interneurons. *Biol. Cybernet. 44*, 141–149.

GEIGER, G. and POGGIO, T. (1975). The Müller-lyer figure and the fly. *Science 190*, 479–480.

GEIGER, G. and POGGIO, T. (1981). Asymptotic oscillations in the tracking behaviour of the fly *Musca domestica. Biol. Cybernet. 41*, 197–201.

GEIGER, G., BOULIN, C. and BÜCHER, R. (1981). How the two eyes add together: Monocular properties of the visually guided orientation behaviour of flies. *Biol. Cybernet. 41*, 71–78.

GESTRI, G., MASTEBROEK, H. A. K. and ZAAGMAN, W. H. (1980). Stochastic constancy, variability and adaptation of spike generation: performance of a giant neuron in the visual system of the fly. *Biol. Cybernet. 38*, 31–40.

GOGALA, M. (1967). Die spektrale Empfindlichkeit der Doppelaugen von *Ascalaphus macaronius* Scop. (Neuroptera, Ascalaphidae). *Z. Vergl. Physiol. 57*, 232–243.

GOLDSMITH, T. H. and BERNARD, G. D. (1974). The visual system of insects. In *The Physiology of Insecta.* Vol. 2, 2nd edn, pages 165–272. Edited by M. Rockstein. Academic Press, New York and London.

Götz, K. G. (1964). Optomotorische Untersuchungen des visuellen Systems einiger Augenmutanten der Fruchtfliege *Drosophila*. *Kybernetik 2*, 77–92.

Götz, K. G. (1965). Die optischen Übertragungseigenschaften der Komplexaugen von *Drosophila*. *Kybernetik 2*, 215–221.

Götz, K. G. (1968). Flight control in *Drosophila* by visual perception of motion. *Kybernetik 4*, 199–208.

Götz, K. G. and Buchner, E. (1978). Evidence for one-way movement detection in the visual system of *Drosophila*. *Biol. Cybernet. 31*, 243–248.

Götz, K. G., Hengstenberg, B. and Biesinger, R. (1979). Optomotor control of wing beat and body posture in *Drosophila*. *Biol. Cybernet. 35*, 101–112.

Gould, J. L. (1976). The dance-language controversy. *Quart. Rev. Biol. 51*, 211–244.

Graham, N., Ratliff, F. and Hartline, H. K. (1973). Facilitation of inhibition in the compound lateral eye of *Limulus*. *Proc. Nat. Acad. Sci. USA 70*, 894–898.

Grenacher, H. (1879). *Untersuchungen über das Sehorgan der Arthropoder*. Vandenhoek and Ruprecht, Göttingen.

Gribakin, F. G. (1969). Cellular basis of colour vision in the honey bee. *Nature 223*, 639–641.

Hagins, W. A. and Liebman, P. A. (1962). Light-induced pigment migration in the squid retina. *Biol. Bull. 123*, 498.

Hagins, W. A., Penn, R. D. and Yoshikami, S. (1970). Dark current and photocurrent in retinal rods. *Biophys. J. 10*, 380–412.

Hamdorf, K. (1971). Die Daner der Dunkeladaptation beim Fliegenauge nach belichtung mit heterochromatischen Blitzen. *Z. Vergl. Physiol. 75*, 200–206.

Hamdorf, K. (1979). The physiology of invertebrate visual pigments. In *Comparative Physiology and Evolution of Vision in Invertebrates. A: Invertebrate Photoreceptors*. Edited by H. Autrum. *Handbook of Sensory Physiology*. Vol. 7, pt 6A, pages 145–224. Springer, Berlin and New York.

Hamdorf, K. and Höglund, G. (1981). Light induced retinal screening pigment migration independent of visual cell activity. *J. Comp. Physiol. 143*, 305–309.

Hamdorf, K. and Schwemer, J. (1975). Photoregeneration and the adaptation process in insect photoreceptors. In *Photoreceptor Optics*. Edited by A. W. Snyder and R. Menzel. Pages 263–289. Springer, Berlin and New York.

Hamdorf, K., Schwemer, J. and Gogala, M. (1971). Insect visual pigment sensitive to ultraviolet light. *Nature 231*, 458–459.

Hamdorf, K., Paulsen, R. and Schwemer, J. (1973). Photoregeneration and sensitivity control of photoreceptors of invertebrates. In *Biochemistry and Physiology of Visual Pigments*. Edited by H. Langer. Pages 155–166. Springer, Berlin and New York.

Hardie, R. C. (1979). Electrophysiological analysis of fly retina. I: Comparative properties of R1–6 and R7 and 8. *J. Comp. Physiol. 129*, 19–33.

Hardie, R. C., Franceschini, N. and McIntyre, P. D. (1979). Electrophysiological analysis of fly retina. II. Spectral and polarisation sensitivity in R7 and R8. *J. Comp. Physiol. 133*, 23–39.

Hardie, R. C., Franceschini, N., Ribi, W. and Kirschfeld, K. (1981). Distribution and properties of sex-specific photoreceptors in the fly *Musca domestica*. *J. Comp. Physiol. 145*, 139–152.

Hartline, H. K. (1949). Inhibition of activity of visual receptors by illuminating nearly retinal areas in the *Limulus* eye. *Fed. Proc. 8*, 69.

Hartline, H. K. (1959). Receptor mechanisms and the integration of sensory information in the eye. *Rev. Med. Phys. 31*, 515–523.

Hartline, H. K. (1969). Visual receptors and retinal interaction. *Science 164*, 270–278.

Hartline, H. K. and Ratliff, F. (1957). Inhibitory interaction of receptor units in the eye of *Limulus*. *J. Gen. Physiol. 40*, 357–376.

Hartline, H. K. and Ratliff, F. (1972). Inhibitory interaction in the retina of *Limulus*. In *Physiology of Photoreceptor Organs*. Edited by M. G. F. Fuortes. *Handbook of Sensory Physiology*. Vol. 7, pt 2, pages 381–447. Springer, Berlin and New York.

Hartline, H. K., Wagner, H. G. and MacNichol, E. F. Jr. (1952). The peripheral origin of nervous activity in the visual system. In *Cold Spring Harbor Symp. Quant. Biol. 17*, 125–141.

Hartline, H. K., Wagner, H. G. and Ratliff, F. (1956). Inhibition in the eye of *Limulus*. *J. Gen. Physiol. 39*, 651–673.

Hassenstein, B. (1951). Ommatidienraster und afferente Bewegungsintegration. *Z. Vergl. Physiol. 33*, 301–326.

Hassenstein, B. (1958). Über die Wahrnehmung der Bewegung von Figuren und unregelmassigen Helligkeitsmustern. *Z. Vergl. Physiol. 40*, 556–592.

Hassenstein, B. and Reichardt, W. (1953). Der Schluss von Reiz-Reaktions-Funktionen auf System-Strukturen. *Z. Naturf. 8b*, 9.

Hausen, K. (1976). Functional characterization and anatomical identification of motion sensitive neurons in the lobula plate of the blowfly *Calliphora erythrocephala*. *Z. Naturf. 31c*, 631–633.

Hausen, K. (1982a). Motion sensitive interneurons in the optomotor system of the fly. I. The horizontal cells: Structure and signals. *Biol. Cybernet. 45*, 143–156.

Hausen, K. (1982b). Motion sensitive interneurons in the optomotor system of the fly. II. The horizontal cells: receptive field organization and response characteristics. *Biol. Cybernet. 46*, 67–79.

Hausen, K., Wolburg-Buchholz, K. and Ribi, W. A. (1980). The synaptic organization of visual interneurons in the lobula complex of flies. A light and electron microscopical study using silver intensified cobalt-impregnation. *Cell Tiss. Res. 208*, 371–387.

Hecht, S. and Wolf, E. (1929). The visual acuity of the honeybee. *J. Gen. Physiol. 12*, 727–760.

Heisenberg, M., Wonneberger, R. and Wolf, R. (1978). Optomotor-blind^{H31} — a *Drosophila* mutant of the lobula giant neurons. *J. Comp. Physiol. 124*, 287–296.

Hengstenberg, R. (1971). Das Augenmuskelsystem der stubenfliege *Musca domestica*. I. Analyse der "Clock-spikes" und ihrer Quellen. *Kybernetik 9*, 56–77.

Hengstenberg, R. (1977). Spike responses of "nonspiking" visual interneurons. *Nature 270*, 338–340.

Hengstenberg, R., Hausen, K. and Hengstenberg, B. (1982). The number and structure of giant vertical cells (VS) in the lobula plate of the blowfly *Calliphora erythrocephala*. *J. Comp. Physiol. 149*, 163–177.

Herrling, P. L. (1976). Regional distribution of three ultrastructural retinula types in the retina of *Cataglyphis bicolor* Fabr. (Formicidae, Hymenoptera). *Cell Tiss. Res. 169*, 247–266.

Hertel, H. (1980). Chromatic properties of identified interneurons in the optic lobes of the bee. *J. Comp. Physiol. 137*, 215–231.

Hertz, M. (1933). Zur Physiologie des Formen- und Bewegungssehen. *Z. Vergl. Physiol. 20*, 430–449.

Hertz, M. (1937). Beitrag zum Farbensinn und Formensinn der Biene. *Z. Vergl. Physiol. 24*, 413–421.

Hinton, H. E. (1976). Possible significance of the red patches of the female crab spider *Misumena vatia*. *J. Zool. 180*, 35–39.

Höglund, G., Hamdorf, K. and Rosner, G. (1973). Trichromatic visual system in an insect and its sensitivity control by blue light. *J. Comp. Physiol. 86*, 265–279.

von Holst, E. and Mittelstaedt, H. (1950). Das Reafferenzprinzip (Wechselwirkungen zwischen Zentralnervensystem und Peripherie). *Naturwissenschaften 37*, 464–476.

Honegger, H.-W. (1978). Sustained and transient responding units in the medulla of the cricket *Gryllus campestris*. *J. Comp. Physiol. 125*, 259–266.

Honegger, H.-W. (1980). Receptive fields of sustained medulla neurons in crickets. *J. Comp. Physiol. 136*, 191–201.

Honegger, H.-W. and Schürmann, F. W. (1975). Cobalt sulphide staining of optic fibres in the brain of the cricket, *Gryllus campestris*. *Cell Tiss. Res. 159*, 213–225.

Horn, E. and Mittag, J. (1980). Body movements and retinal pattern displacements while approaching a stationary object in the walking fly, *Calliphora erythrocephala*. *Biol. Cybernet. 39*, 67–77.

Horn, E. and Wehner, R. (1975). The mechanism of visual pattern fixation in the walking fly, *Drosophila melanogaster*. *J. Comp. Physiol. 101*, 39–56.

Horridge, G. A. (1968). Pigment movement and the crystalline threads of the fire fly eye. *Nature 218*, 778–779.

Horridge, G. A. (1969). The eye of the firefly (*Photuris*). *Proc. Roy. Soc. Lond. B. 171*, 445–463.

Horridge, G. A. (1971). Alternatives to superposition images in clear-zone compound eyes. *Proc. Roy. Soc. Lond. B. 179*, 97–124.

Horridge, G. A. (1976). The ommatidium of the dorsal eye of *Cloeon* as a specialization for photoreisomerization. *Proc. Roy. Soc. Lond. B. 193*, 17–29.

HORRIDGE, G. A. (1977a). The compound eye of insects. *Scient. Amer. 237*, 108–120.

HORRIDGE, G. A. (1977b). Insects which turn and look. *Endeavour 1*, 7–17.

HORRIDGE, G. A. (1978). The separation of visual axes in apposition compound eyes. *Phil. Trans. Roy. Soc. Lond. B. 285*, 1–59.

HORRIDGE, G. A. and BARNARD, P. B. T. (1965). Movement of palisade in locust retinula cells when illuminated. *Quart. J. Sci. 106*, 131–135.

HORRIDGE, G. A. and DUELLI, P. (1979). Anatomy of the regional differences in the eye of the mantis *Ciulfina*. *J. Exp. Biol. 80*, 165–190.

HORRIDGE, G. A. and McLEAN, M. (1978). The dorsal eye of the mayfly *Atalophlebia* (Ephemeroptera). *Proc. Roy. Soc. Lond. B. 200*, 137–150.

HORRIDGE, G. A., DUNIEC, J. and MARCELJA, L. (1981). A 24-hour cycle in single locust and mantis photoreceptors. *J. Exp. Biol. 91*, 307–322.

HORRIDGE, G. A., GIDDINGS, C. and STANGE, G. (1972). The superposition eye of skipper butterflies. *Proc. Roy. Soc. Lond. B. 182*, 457–495.

HORRIDGE, G. A., MIMURA, K. and HARDIE, R. C. (1976). Fly photoreceptors. III. Angular sensitivity as a function of wavelength and the limits of resolution. *Proc. Roy. Soc. Lond. B. 194*, 151–177.

HORRIDGE, G. A., McLEAN, M., STANGE, G. and LILLYWHITE, P. G. (1977). A diurnal moth superposition eye with high resolution *Phalaenoides tristifica* (Agaristidae). *Proc. Roy. Soc. Lond. B. 196*, 233–250.

HORRIDGE, G. A., SCHOLES, J. H., SHAW, S. R. and TUNSTALL, J. (1965). Extracellular recording from single neurons in the optic lobe and brain of the locust. In *The Physiology of the Insect Central Nervous System*. Edited by J. E. Treherne and J. W. L. Beament. Pages 165–202. Academic Press, London.

HOWARD, J. (1981). Temporal resolving power of the photoreceptors of *Locusta migratoria*. *J. Comp. Physiol. 144*, 61–66.

ICHIKAWA, T. and TATEDA, H. (1982). Receptive field of the stemmata in the swallowtail butterfly *Papilio*. *J. Comp. Physiol. 146*, 191–199.

JACK, J. J. B. and REDMAN, S. J. (1971). The propagation of transient potentials in some linear cable structures. *J. Physiol. 215*, 283.

JANDER, R. (1963). Insect orientation. *Ann. Rev. Ent. 8*, 95–114.

JANDER, R. (1969). Wie erkennen Stabheuschrecken Sträucher? Attrappenversuche zum Formensehen eines Insektes. *Zool. Anz. Suppl.-Bd. 33, Verh. Zool. Ges.*, 592–595.

JÄRVILEHTO, M. (1971). Lokalisierte intrazelluläre Ableitungen aus den Axonen der 8. Sehzelle der Fliege *Calliphora erythrocephala*. Doctoral thesis, University of Munich, München, FRG.

JÄRVILEHTO, M. (1979). Receptor potentials in invertebrate visual cells. In *Comparative Physiology and Evolution of Vision in Invertebrates. A: Invertebrate Photoreceptors*. Edited by H. Autrum. *Handbook of Sensory Physiology*. Vol. 7, pt 6A, pages 315–356. Springer, Berlin and New York.

JÄRVILEHTO, M. (1982). The significance of graded potentials in the retina. *Invest. Ophthalmol. Vis. Sci. Suppl. 22*, 276.

JÄRVILEHTO, M. and MORING, J. (1976). Spectral and polarization sensitivity of identified retinal cells of the fly. In *Neural Principles in Vision*. Edited by F. Zettler and R. Weiler. Pages 214–226. Springer, Berlin and New York.

JÄRVILEHTO, M. and ZETTLER, F. (1970). Micro-localisation of lamina-located visual cell activities in the compound eye of the blowfly *Calliphora*. *Z. Vergl. Physiol. 69*, 134–138.

JÄRVILEHTO, M. and ZETTLER, F. (1971). Localized intracellular potentials from pre- and postsynaptic components in an external plexiform layer of an insect retina. *Z. Vergl. Physiol. 75*, 422–440.

JÄRVILEHTO, M. and ZETTLER, F. (1973). Electrophysiological-histological studies on some functional properties of visual cells and second order neurons of an insect retina. *Z. Zellf. 136*, 291–306.

JOHNSTON, D. and WACHTEL, H. (1976). Electrophysiological basis for the spatial dependence of the inhibitory coupling in the *Limulus* retina. *J. Gen. Physiol. 67*, 1–25.

KAISER, W. and BISHOP, L. G. (1970). Directionally selective motion detecting units in the optic lobe of the honeybee. *Z. Vergl. Physiol. 67*, 403–413.

KIEN, J. and MENZEL, R. (1977). Chromatic properties of interneurons in the optic lobes of the bee. II. Narrow band and colour opponent neurons. *J. Comp. Physiol. 113*, 35–53.

KIRSCHFELD, K. (1966). Discrete and graded receptor potentials in the compound eye of the fly (*Musca*). In *The Functional Organization of the Compound Eye*. Edited by C. G. Bernhard. Pages 291–307. Pergamon Press, Oxford.

KIRSCHFELD, K. (1967). Die Projection der optischen Umwelt auf das Raster der Rhabdomere im Komplexauge von *Musca*. *Exp. Brain Res. 3*, 248–270.

KIRSCHFELD, K. (1971). Aufname und Verarbeitung optischer Daten in Komplexaugen von Insekten. *Naturw. Rdsch. 24*, 213–214.

KIRSCHFELD, K. (1972a). Vision of polarised light. *Symp. Proc. IV Int. Biophys. Congr.*, Moscow, 7–14 Aug.

KIRSCHFELD, K. (1972b). Die notwendige Anzahl von Rezeptoren zur Bestimmung der Richtung des elektrischen Vektors linear polarisierten Lichtes. *Z. Naturf. 27*, 578–579.

KIRSCHFELD, K. (1973). Das neurale Superpositionsauge. *Fortsch. der Zool. 21*, 229–257.

KIRSCHFELD, K. (1976). The resolution of lens and compound eyes. In *Neural Principles in Vision*. Edited by F. Zettler and R. Weiler. Pages 354–370. Springer, Berlin and New York.

KIRSCHFELD, K. (1979). The visual system of the fly: physiological optics and functional anatomy as related to behavior. In *The Neurosciences: 4th Study Program*. Edited by F. O. Schmitt and F. G. Worden. Pages 297–310. MIT Press, Cambridge.

KIRSCHFELD, K. and FRANCESCHINI, N. (1968). Optische Eigenschaften der Ommatidien im Komplexauge von *Musca*. *Kybernetik 5*, 47–52.

KIRSCHFELD, K. and FRANCESCHINI, N. (1969). Ein Mechanismus zur Steuerung des Lichtflusses in den Rhabdomeren des Komplexauges von *Musca*. *Kybernetik 6*, 13–22.

KIRSCHFELD, K., FEILER, R. and FRANCESCHINI, N. (1978). A photostable pigment within the rhabdomere of fly photoreceptors No. 7. *J. Comp. Physiol. 125*, 275–284.

KIRSCHFELD, K., FRANCESCHINI, N. and MINKE, B. (1977). Evidence for a sensitising pigment in fly photoreceptors. *Nature 269*, 386–390.

KNIGHT, B. W., TOYODA, J.-I. and DODGE, F. A. JR. (1970). A quantitative description of the dynamics of excitation and inhibition in the eye of *Limulus*. *J. Gen. Physiol. 56*, 421–437.

KOLTERMANN, R. (1969). Zeitgekoppelte Lernprozesse bei der Honigbiene. *Zool. Anz. Suppl. — Bd. 33. Verh. Zool. Ges.*, 205–208.

KUIPER, J. W. (1962). The optics of the compound eye. *Symp. Soc. Exp. Biol. 16*, 58–71.

KUNZE, P. (1961). Untersuchung des Bewegungssehens fixiert fligender Bienen. *Z. Vergl. Physiol. 44*, 656–684.

KUNZE, P. (1979). Apposition and superposition eyes. In *Comparative Physiology and Evolution of Vision in Invertebrates. A: Invertebrate Photoreceptors*. Edited by H. Autrum. *Handbook of Sensory Physiology*. Vol. 7, pt 6A, pages 441–502. Springer, Berlin and New York.

KUSTER, J. E. and EVANS, W. G. (1980). Visual fields of the compound eyes of four species of *Cicindelidae* (Coleoptera). *Canad. J. Zool. 58*, 326–336.

KUWABARA, M. and NAKA, K.-J. (1959). Response of single retinula cells to polarized light. *Nature (Lond.) 184*, 455–456.

LABHART, T. (1972). The discrimination of light intensities in the honey bee. In *Information Processing in the Visual Systems of Arthropods*. Edited by R. Wehner. Pages 115–119. Springer, Berlin and New York.

LABHART, T. (1980). Specialized photoreceptors at the dorsal rim of the honeybee's compound eye: polarizational and angular sensitivity. *J. Comp. Physiol. 141*, 19–30.

LALL, A. B., SELIGER, H. H., BIGGLEY, W. H. and LLOYD, J. E. (1980). Ecology of colors of firefly bioluminescence. *Science 210*, 560–562.

LAND, M. F. (1973). Head movements of flies during visually guided flight. *Nature 243*, 299–300.

LAND, M. F. (1975). Head movements and fly vision. In *The Compound Eye and Vision of Insects*. Edited by G. A. Horridge. Pages 469–489. Oxford University Press, London.

LAND, M. F. and COLLETT, T. S. (1974). Chasing behaviour of houseflies (*Fannia canicularis*). *J. Comp. Physiol. 89*, 331–357.

LANGE, D., HARTLINE, H. K. and RATLIFF, F. (1966). The dynamics of lateral inhibition in the compound eye of *Limulus* II. In *The Functional Organization of the Compound Eye*. Edited by C. G. Bernhard. Pages 425–449. Pergamon Press, Oxford.

LAUGHLIN, S. B. (1974). Neural integration in the first optic neuropile of dragonflies. III. The transfer of angular information. *J. Comp. Physiol. 92*, 377–396.

LAUGHLIN, S. B. (1975). Receptor and interneuron light adaptation in the dragonfly visual system. *Z. Naturf. 30c*, 306–308.

LAUGHLIN, S. B. (1976). Adaptations of the dragonfly retina for contrast detection and the elucidation of neural principles in the peripheral visual system. In *Neural Principles in Vision*. Edited by F. Zettler and R. Weiler. Pages 175–193. Springer, Berlin and New York.

LAUGHLIN, S. B. (1981). Neural principles in the peripheral visual systems of invertebrates. In *Comparative Physiology and Evolution of Vision in Invertebrates. B: Invertebrate Visual Centers and Behavior I*. Edited by H. Autrum. *Handbook of Sensory Physiology*. Vol. 7, pt 6B, pages 133–280. Springer, Berlin and New York.

LAUGHLIN, S. B. and HARDIE, R. C. (1978). Common strategies for light adaptation in the peripheral visual systems of fly and dragonfly. *J. Comp. Physiol. 128*, 319–340.

LAUGHLIN, S. B. and HORRIDGE, G. A. (1971). Angular sensitivity of retinula cells of dark-adapted worker bee. *Z. Vergl. Physiol. 74*, 329–335.

LAUGHLIN, S. B. and McGINESS, S. (1978). The structure of dorsal and ventral regions of a dragonfly retina. *Cell Tiss. Res. 188*, 427–447.

LEUTSCHER-HAZELHOFF, J. T. and KUIPER, J. W. (1966). Clock-spikes in the *Calliphora* optic lobe and a hypothesis for their function in object location. In *The Functional Organization of the Compound Eye*. Edited by C. G. Bernard. Pages 483–492. Pergamon Press, Oxford.

LEVICK, W. R. (1967). Receptive fields and trigger features of ganglion cells in the visual streak of the rabbits retina. *J. Physiol. 188*, 285–307.

LEVÍN, L. and MALDONADO, H. (1970). A fovea in the praying mantis eye. III. The centring of the prey. *Z. Vergl. Physiol. 67*, 93–101.

LEVINE, J. D. (1974). Giant neuron input in mutant and wild type *Drosophila. J. Comp. Physiol. 93*, 265–285.

LILLYWHITE, P. G. (1977). Single photon signals and transduction in an insect eye. *J. Comp. Physiol. 122*, 189–200.

LINDAUER, M. (1955). Schwarmbienen auf Wohnungssuche. *Z. Vergl. Physiol. 37*, 263–324.

LO, M.-V. C. and PAK, W. L. (1981). Light-induced pigment granule migration in the retinular cells of *Drosophila melanogaster. J. Gen. Physiol. 77*, 155–175.

MACH, E. (1886). Über die physiologische Wirkung rännlich vertheilter Lichtreize. *Sitzungsber. Akad. Wiss. Wien, math.-nat. K1. 11/54*, 393–408.

MALDONADO, H. and RODRIGUEZ, E. (1972). Depth perception in the praying mantis. *Physiol. Behav. 8*, 751–759.

MARČELJA, S. (1980). Electrical coupling of photoreceptors in retinal network models. *Biol. Cybernet. 39*, 15–20.

MARKL, H. (1964). Geomenotaktische Fehlorientierung bei *Formica polyctena. Z. Vergl. Physiol. 48*, 552–586.

MASTEBROEK, H. A. K., ZAAGMAN, W. H. and LENTING, B. P. M. (1980). Movement detection: performance of a wide-field element in the visual system of the blowfly. *Vision Res. 20*, 467–474.

MASTEBROEK, H. A. K., ZAAGMAN, W. H. and LENTING, B. P. M. (1982). Memory-like effects in fly vision: spatio-temporal interactions in a wide-field neuron. *Biol. Cybernet. 43*, 147–155.

MAZOCHIN-PORSHNYAKOV, G. A. (1969). Die Fähigkeit der Bienen, visuelle Reize zu generalisieren. *Z. Vergl. Physiol. 65*, 15–28.

MAZOCHIN-PORSHNYAKOV, G. A. and VISHNEVSKAYA, T. M. (1965). Beweise der Fähigkeit der Bienen, einfache geometrische Figuren zu unterscheiden. *Zool. J. 44*, 192–197.

McCANN, G. D. and FOSTER, S. F. (1971). Binocular interactions of motion detection fibers in the optic lobes of flies. *Kybernetik 8*, 193–203.

MEINERZHAGEN, I. A. (1973). Development of the compound eye and optic lobes of insects. In *Developmental Neurobiology of Arthropods*. Edited by D. Young. Pages 51–104. University Press, Cambridge.

MENNE, D. and SPATZ, H.-C. (1977). Colour vision in *Drosophila melanogaster. J. Comp. Physiol. 114*, 301–312.

MENZEL, R. (1975). Polarization sensitivity in insect eyes with fused rhabdoms. In *Photoreceptor Optics*. Edited by A. W. Snyder and R. Menzel. Pages 372–391. Springer, Berlin and New York.

MENZEL, R. and BLAKERS, M. (1976). Colour receptors in the bee eye — morphology and spectral sensitivity. *J. Comp. Physiol. 108*, 11–33.

MENZEL, R. and ERBER, J. (1978). Learning and memory in bees. *Scient. Amer. 239*, 80–87.

MEYER, H. W. (1971). Visuelle Schlüsselreize für die Anslösung der Bentfanghandlung beim Bachwasserläufer *Velia caprai* (Hemiptera, Heteroptera). I. Untersuchung der rännlichen und zeitlichen Reizparameter mit formverschiedenen Attrappen. *Z. Vergl. Physiol. 72*, 260–297.

MEYER, H. W. (1972). Ethometrical investigations into the spatial interaction within the visual system of *Velia caprai* (Hemiptera, Heteroptera). In *Information Processing in the Visual Systems of Arthropods*. Edited by R. Wehner. Pages 223–229. Springer, Berlin and New York.

MEYER-ROCHOW, V. B. (1971). A crustacean-like organization of insect rhabdoms. *Cytobiologie 4*, 241–249.

MEYER-ROCHOW, V. B. (1972). The eyes of *Creophilus erythrocephalus* F. and *Sartallus signatus* Sharp (Staphylinidae: Coleoptera). Light-, interference-, scanning electron- and transmission electron microscope examinations. *Z. Zellf. 133*, 59–86.

MILLER, W. H. (1975). Mechanisms of photomechanical movement. In *Photoreceptor optics*. Edited by A. W. Snyder and R. Menzel. Pages 415–428. Springer, Berlin and New York.

MILLER, W. H. and CAUTHON, D. F. (1974). Pigment granule movement in *Limulus* photoreceptors. *Invest. Ophthalmol. 13*, 401–405.

MIMURA, K. (1971). Movement discrimination by the visual system of flies. *Z. Vergl. Physiol. 73*, 105–138.

MIMURA, K. (1972). Neural mechanisms, subserving directional selectivity of movement in the optic lobe of the fly. *J. Comp. Physiol. 80*, 409–438.

MIMURA, K. (1974). Analysis of visual information in lamina neurones of the fly. *J. Comp. Physiol. 88*, 335–372.

MIMURA, K. (1975). Units of the optic lobe especially movement perception units of diptera. In *The Compound Eye and Vision of Insects*. Edited by G. A. Horridge. Pages 423–436. Oxford University Press, London.

MIMURA, K. (1976). Some spatial properties in the first optic ganglion of the fly. *J. Comp. Physiol. 105*, 65–82.

MIMURA, K. (1981). Receptive field patterns in photoreceptors of the fly. *J. Comp. Physiol. 141*, 349–362.

MORING, J. and JÄRVILEHTO, M. (1977). Dark recovery of polarized light sensitive and insensitive receptor cells in the retina of the fly. *J. Comp. Physiol. 122*, 215–226.

MORING, J., JÄRVILEHTO, M. and MORING, K. (1979). The onset of light adaptation in the visual cells of the fly. *J. Comp. Physiol. 132*, 153–158.

MOTE, M. I. (1970). Focal recording of responses evoked by the light in the lamina ganglionaris of the fly *Sarcophaga bullata. J. Exp. Zool. 175*, 149–158.

MOTE, M. I. (1974). Polarization sensitivity. A phenomenon independent of stimulus intensity or state of adaptation in retinular cells of the crabs *Carcinus* and *Callinectes. J. Comp. Physiol. 90*, 389–403.

MOTE, M. I. and RUBIN, L. J. (1981). "On" type interneurons in the optic lobe of *Periplaneta americana*. I. Spectral characteristics of response. *J. Comp. Physiol. 141*, 395–401.

MOTE, M. I., KUMAR, V. S. N. and BLACK, K. R. (1981). "On" type interneurons in the optic lobe of *Periplaneta americana*. II. Receptive fields and response latencies. *J. Comp. Physiol. 141*, 403–415.

MÜLLER, J. (1826). *Zur vergleichenden Physiologie des Gesichtsinnes*. Cnobloch, Leipzig.

MULLER, K. J. (1973). Photoreceptors in the crayfish compound eye: electrical interactions between cells as related to polarized-light sensitivity. *J. Physiol. 232*, 573–595.

NAGY, K. and STIEVE, H. (1983). Changes in intracellular calcium ion concentration, in the course of dark adaptation measured by arsenazo III in the *Limulus* photoreceptor. *Biophys. Struct. Mech. 9*, 207–223.

NILSSON, D.-E. and ODSELIUS, R. (1982). A pronounced fovea in the eye of a water flea, revealed by stereographic mapping of ommatidial axes. *J. Exp. Biol. 99*, 473–476.

NOLTE, J. and BROWN, J. E. (1972). Electrophysiological properties of cells in the median ocellus of *Limulus. J. Gen. Physiol. 59*, 167–185.

OLBERG, R. M. (1981). Object- and self-movement detectors in the ventral nerve cord of the dragonfly. *J. Comp. Physiol. 141*, 327–334.

OLIVO, R. F. and LARSEN, M. E. (1978). Brief exposure to light initiates screening pigment migration in retinula cells of the crayfish, *Procambarus. J. Comp. Physiol. 125*, 91–96.

O'SHEA, M. and ROWELL, C. H. F. (1975). Protection from habituation by lateral inhibition. *Nature 254*, 53–55.

PALKA, J. (1967). An inhibitory process influencing visual responses in a fibre of the ventral nerve cord of locusts. *J. Insect Physiol. 13*, 235–248.

PALKA, J. (1972). Moving movement detectors. *Amer. Zool. 12*, 497–505.

PATTERSON, J. A. and CHAPPELL, R. L. (1980). Intracellular responses of procion filled cells and whole nerve cobalt impregnation in the dragonfly median ocellus. *J. Comp. Physiol. 139*, 25–39.

PEARSON, K. G., HEITLER, W. J. and STEEVES, J. D. (1980). Triggering of locust jump by multimodal inhibitory interneurons. *J. Neurophysiol. 43*, 257–278.

PICK, B. (1976). Visual pattern discrimination as an element of the fly's orientation behaviour. *Biol. Cybernet. 23*, 171–180.

PIERANTONI, R. (1974). An observation on the giant fiber posterior optic tract in the fly. *Biokybernetik 5*, 157–163.

PIERANTONI, R. (1976). A look into the cock-pit of the fly. The architecture of the lobula plate. *Cell Tiss. Res. 171*, 101–122.

PINTER, R. B. (1979). Inhibition and excitation in the locust DCMD receptive field: spatial frequency, temporal and spatial characteristics. *J. Exp. Biol. 80*, 191–216.

POGGIO, T. and REICHARDT, W. (1976). Visual control of orientation behaviour in the fly. Part II. Towards the underlying neural interactions. *Quart. Rev. Biophys. 9*, 377–438.

POGGIO, T. and REICHARDT, W. (1981). Visual fixation and tracking by flies: mathematical properties of simple control systems. *Biol. Cybernet. 40*, 101–112.

VAN PRAAGH, J. P., RIBI, W., WEHRHAHN, C. and WITTMAN, D. (1980). Drone bees fixate the queen with the dorsal frontal part of their compound eyes. *J. Comp. Physiol. 136*, 263–266.

PURPLE, R. L. (1964). The integration of excitatory and inhibitory influences in the eccentric cell in the eye of *Limulus*. Thesis, the Rockefeller Institute, New York.

PURPLE, R. L. and DODGE, F. A. (1965). Interaction of excitation and inhibition in the eccentric cell in the eye of *Limulus*. *Cold Spring Harbor Symp. Quant. Biol. 30*, 529–537.

PURPLE, R. L. and DODGE, F. A. (1966). Self-inhibition in the eye of *Limulus*. In *The Functional Organization of the Compound eye*. Edited by C. G. Bernard. Pages 451–464. Pergamon Press, Oxford.

QUINN, W. G., HARRIS, W. A. and BENZER, S. (1974). Conditioned behavior in *Drosophila melanogaster*. *Proc. Nat. Acad. Sci., USA 71*, 708–712.

RATLIFF, F. (1971). Contour and contrast. *Proc. Amer. Phil. Soc. 115*, 150–163.

RATLIFF, F. and HARTLINE, H. K. (1959). The response of *Limulus* optic nerve fibres to patterns of illumination on the receptor mosaic. *J. Gen. Physiol. 42*, 1241–1255.

RATLIFF, F., KNIGHT, B. W. and GRAHAM, N. (1969). On tuning and amplification by lateral inhibition. *Proc. Nat. Acad. Sci. 62*, 733–740.

RATLIFF, F., HARTLINE, H. K. and MILLER, W. H. (1963). Spatial and temporal aspects of retinal inhibitory interaction. *J. Opt. Soc. Amer. 53*, 110–120.

RATLIFF, F., KNIGHT, B. W., DODGE, F. A. and HARTLINE, H. K. (1974). Fourier analysis of dynamics of excitation and inhibition in the eye of *Limulus*: amplitude, phase and distance. *Vision Res. 14*, 1155–1168.

REHBRONN, W. (1972). Gleichzeitige intrazelluläre Doppelableitungen aus dem Komplexauge von *Calliphora erythrocephala*. *Z. Vergl. Physiol. 76*, 285–301.

REICHARDT, W. (1957). Autokorrelations — Auswertung als Funktionsprinzip des Zentralnervensystems (bei der optischen Bewegungswahrnehmung eines Insektes). *Z. Naturf. 12b*, 448–457.

REICHARDT, W. E. (1966). Detection of single quanta by the compound eye of the fly *Musca*. In *The Functional Organization of the Compound Eye*. Edited by C. G. Bernhard. Pages 267–289. Pergamon Press, Oxford.

REICHARDT, W. (1973). Musterinduzierte Flugorientierung. Verhaltensversuche an der Fliege *Musca domestica*. *Naturwissenschaften 60*, 122–138.

REICHARDT, W. and MACGINITIE, G. (1962). Zur Theorie der lateralen Inhibition. *Kybernetik 1*, 155–165.

REICHARDT, W. and POGGIO, T. (1976). Visual control of orientation behaviour in the fly. Part I. A quantitative analysis. *Quart. Rev. Biophys. 9*, 311–375.

REICHARDT, W. and VARJÚ, D. (1959). Übertragungseigenschaften im

Auswertesystem für das Bewegungssehen. *Z. Naturf. 14b*, 674–689.

RIBI, W. A. (1978). Gap junctions coupling photoreceptor axons in the first optic ganglion of the fly. *Cell Tiss. Res. 195*, 299–308.

RIBI, W. A. (1979). Coloured screening pigments cause red glow hue in pierid butterflies. *J. Comp. Physiol. 132*, 1–9.

RODIECK, R. W. (1967). Receptive fields in the cat retina: a new type. *Science 157*, 90–92.

ROEDER, K. D. (1953). *Insect Physiology*. Chapman & Hall, London.

ROSIN, R. (1980). The honey-bee "dance language" hypothesis and the foundations of biology and behavior. *J. Theor. Biol. 87*, 457–481.

ROSNER, G. (1975). Adaptation and Photoregeneration im Fliegenauge. *J. Comp. Physiol. 102*, 269–295.

ROSSEL, S. (1979). Regional differences in photoreceptor performance in the eye of the praying mantis. *J. Comp. Physiol. 131*, 95–112.

RUTOWSKI, R. L. (1977). The use of visual cues in sexual and species discrimination by males of the small sulphur butterfly *Eurema lisa* (Lepidoptera, Pieridae). *J. Comp. Physiol. 115*, 61–74.

RUTOWSKI, R. L. (1981). Sexual discrimination using visual cues in the checkered white butterfly (*Pieris protodice*). *Z. Tierpsychol. 55*, 325–334.

SCHINZ, R. H. (1975). Structural specialization in the dorsal retina of the bee *Apis mellifera*. *Cell Tiss. Res. 162*, 23–34.

SCHOLES, J. H. (1969). The electrical responses of the retinal receptors and the lamina in the visual system of the fly *Musca*. *Kybernetik 6*, 149–162.

SCHOLES, J. H. and REICHARDT, W. (1969). The quantal content of optomotor stimuli and the electrical responses of receptors in the compound eye of the fly *Musca*. *Kybernetik 6*, 74–80.

SCHNETTER, B. (1972). Experiments on pattern discrimination in honey bees. In *Information Processing in the Visual Systems of Arthropods*. Edited by R. Wehner. Pages 195–200. Springer, Berlin and New York.

SCHÜMPERLI, R. A. (1975). Monocular and binocular visual fields of butterfly interneurons in response to white- and coloured-light stimulation. *J. Comp. Physiol. 103*, 273–289.

SCHÜMPERLI, R. A. and SWIHART, S. (1978). Spatial properties of dark- and light-adapted visual fields of butterfly interneurones. *J. Insect Physiol. 24*, 777–784.

SCHWARTZ, E. A. (1971). Retinular and eccentric cell morphology in the neural plexus of *Limulus* lateral eye. *J. Neurobiol. 2*, 129–133.

SCHWEMER, J., GOGALA, M. and HAMDORF, K. (1971). Der UV-Sehfarbstoff der Insekten: Photochemie *in vitro* und *in vivo*. *Z. Vergl. Physiol. 75*, 174–188.

SCHWIND, R. (1978). Visual system of *Notonecta glanca*: a neuron sensitive to movement in the binocular visual field. *J. Comp. Physiol. 123*, 315–328.

SEIDL, R. and KAISER, W. (1981). Visual field size, binocular domain and the ommatidial array of the compound eyes in worker honey bees. *J. Comp. Physiol. 143*, 17–26.

SHAW, S. R. (1967). Simultaneous recording from two cells in the locust retina. *Z. Vergl. Physiol. 55*, 183–194.

SHAW, S. R. (1968). Organization of the locust retina. *Symp. Zool. Soc. Lond. 23*, 135–163.

SHAW, S. R. (1969). Interreceptor coupling in ommatidia of drone honey bee and locust compound eyes. *Vision Res. 9*, 999–1029.

SHAW, S. R. (1972). Decremental conduction of the visual signal in barnacle lateral eye. *J. Physiol. 220*, 145–175.

SHAW, S. R. (1975). Retinal resistance barriers and electrical lateral inhibition. *Nature 255*, 480–483.

SHAW, S. R. (1977). Restricted diffusion and extracellular space in the insect retina. *J. Comp. Physiol. 113*, 257–282.

SHAW, S. R. (1978). The extracellular space and blood-eye barrier in an insect retina: an ultrastructural study. *Cell Tiss. Res. 188*, 35–61.

SHAW, S. R. (1979). Signal transmission by graded slow potentials in the arthropod peripheral visual system. In *The Neurosciences: 4th Study Program*. Edited by F. O. Schmitt and F. G. Worden. Pages 275–295. MIT Press, Cambridge.

SHAW, S. R. (1981). Anatomy and physiology of identified nonspiking cells in photoreceptor-lamina complex of the compound eye of insects, especially Diptera. In *Neurons without Impulses*. Edited by A. Roberts and B. M. H. Bush. Pages 61–116. Cambridge University Press, Cambridge.

SILBERGLIED, R. E. (1979). Communication in the ultraviolet. *Ann. Rev. Ecol. Syst. 10*, 373–398.

SMITH, T. G. and BAUMANN, F. (1969). The functional organization within the ommatidium of the lateral eye of *Limulus*. *Brain Res. 31*, 313–349.

SMITH, T. G., BAUMANN, F. and FUORTES, M. G. F. (1965). Electrical connections between visual cells in the ommatidium of *Limulus*. *Science 147*, 1446–1448.

SMOLA, U. (1975). Übertragung von optischen Signalen durch Sehzellen. *Naturw. Rdsch. 28*, 239–250.

SNYDER, A. W. (1973). Polarization sensitivity of individual retinula cells. *J. Comp. Physiol. 83*, 331–360.

SNYDER, A. W. (1977). Acuity of compound eyes: physical limitations and design. *J. Comp. Physiol. 116*, 161–182.

SNYDER, A. W. (1979). The physics of vision in compound eyes. In *Comparative Physiology and Evolution of Vision in Invertebrates. A: Invertebrate photoreceptors*. Edited by H. Autrum. *Handbook of Sensory Physiology*. Vol. 7, pt 6A, pages 225–313. Springer, Berlin and New York.

SNYDER, A. W. and HORRIDGE, G. A. (1972). The optical function of changes in the medium surrounding the cockroach rhabdom. *J. Comp. Physiol. 81*, 1–8.

SNYDER, A. W., LAUGHLIN, S. B. and STAVENGA, D. G. (1977b). Information capacity of eyes. *Vision Res. 17*, 1163–1175.

SNYDER, A. W., MENZEL, R. and LAUGHLIN, S. B. (1973). Structure and function of the fused rhabdom. *J. Comp. Physiol. 87*, 267–286.

SNYDER, A. W., STAVENGA, D. G. and LAUGHLIN, S. B. (1977a). Spatial information capacity of compound eyes. *J. Comp. Physiol. 116*, 183–207.

SOOHOO, S. L. and BISHOP, L. G. (1980). Intensity and motion responses of giant vertical neurons of the fly eye. *J. Neurobiol. 11*, 159–177.

SOUČEK, B. and CARLSON, A. D. (1975). Flash pattern recognition in fireflies. *J. Theor. Biol. 55*, 339–352.

SRINIVASAN, M. V. (1977). A visually-evoked roll response in the housefly. Open-loop and closed-loop studies. *J. Comp. Physiol. 119*, 1–14.

SRINIVASAN, M. V. and BERNARD, G. D. (1980). A technique for estimating the contribution of photomechanical responses to visual adaptation. *Vision Res. 20*, 511–521.

STAVENGA, D. G. (1975). Optical qualities of the fly eye — An approach from the side of geometrical, physical and waveguide optics. In *Photoreceptor optics*. Edited by A. W. Snyder and R. Menzel. Pages 126–144. Springer, Berlin and New York.

STAVENGA, D. G. (1979). Pseudopupils of compound eyes. In *Comparative Physiology and Evolution of Vision in Invertebrates. A: Invertebrate Photoreceptors*. Edited by H. Autrum. *Handbook of Sensory Physiology*. Vol. 7, pt 6A, pages 357–439. Springer, Berlin and New York.

STAVENGA, D. G., FLOKSTRA, J. H. and KUIPER, J. W. (1975). Photopigment conversions expressed in pupil mechanisms of blowfly visual sense cells. *Nature 253*, 740–742.

STEVENS, C. F. (1964). A quantitative theory of neural interactions: theoretical and experimental investigations. Thesis, the Rockefeller Institute, New York.

STRAUSFELD, N. J. (1976a). Mosaic organizations, layers, and visual pathways in the insect brain. In *Neural Principles in Vision*. Edited by F. Zettler and R. Weiler. Pages 245–279. Springer, Berlin and New York.

STRAUSFELD, N. J. (1976b). *Atlas of an Insect Brain*. Springer, Berlin and New York.

STRAUSFELD, N. J. (1979). The representation of a receptor map within retinotopic neuropil of the fly. *Verh. Dtsch. Zool. Ges.*, 167–179.

STRAUSFELD, N. J. (1980). Male and female visual neurones in dipterous insects. *Nature 283*, 381–383.

STRAUSFELD, N. J. and CAMPOS-ORTEGA, J. A. (1977). Vision in insects: Pathways possibly underlying neural adaptation and lateral inhibition. *Science 195*, 894–897.

STRAUSFELD, N. J. and NÄSSEL, D. R. (1981). Neuroarchitecture of brain regions that subserve the compound eyes of crustacea and insects. In *Comparative Physiology and Evolution of Vision in Invertebrates. B: Invertebrate Visual Centers and Behavior I*. Edited by H. Autrum. *Handbook of Sensory Physiology*. Vol. 7, pt 6B, pages 1–132. Springer, Berlin and New York.

STRAUSFELD, N. J. and OBERMAYER, M. (1976). Resolution of intraneuronal and transsynaptic migration of cobalt in the insect visual and nervous systems. *J. Comp. Physiol. 110*, 1–12.

STRECK, P. (1972). Screening pigment and visual field of single retinula cells of Calliphora. In *Information Processing in the Visual Systems of Arthropods*. Edited by R. Wehner. Pages 127–131. Springer, Berlin and New York.

SWIHART, C. A. and SWIHART, S. L. (1970). Colour selection and learned feeding preferences in the butterfly, *Heliconius charitonius* Linn. *Anim. Behav. 18*, 60–64.

SWIHART, S. L. (1968). Single unit activity in the visual pathway of the butterfly *Heliconius erato*. *J. Insect Physiol. 14*, 1589.

SWIHART, S. L. (1969). Colour vision and the physiology of the superposition eye of a butterfly (*Hesperidae*). *J. Insect Physiol. 15*, 1347–1365.

TADDEI-FERRETTI, C. (1973). Landing reaction of *Musca domestica*. IV. A. Monocular and binocular vision; B. Relationships between landing and optomotor reactions. *Z. Naturf. 28c*, 579–592.

TOH, Y. and KUWABARA, M. (1975). Synaptic organization of the fleshfly ocellus. *J. Neurocytol. 4*, 271–287.

TOMIOKA, K. and YAMAGUCHI, T. (1980). Steering responses of adult and nymphal crickets to light, with special reference to the head rolling movement. *J. Insect Physiol. 26*, 47–57.

TUNSTALL, J. and HORRIDGE, G. A. (1967). Electrophysiological investigation of the optics of the locust retina. *Z. Vergl. Physiol. 55*, 167–182.

TUURALA, O., LEHTINEN, A. and NYHOLM, M. (1966). Zu den photomechanischen Erscheinungen im Auge einer Asselart *Oniscus assellus* L. *Amer. Acad. Sci. Fenn. Ser. AIV Biol. 99*, 1–8.

VARELA, F. G. and WIITANEN, W. (1970). The optics of the compound eye of the honeybee (*Apis mellifera*). *J. Gen. Physiol. 55*, 336–358.

VIRSIK, R. P. and REICHARDT, W. (1976). Detection and tracking of moving objects by the fly *Musca domestica*. *Biol. Cybernet. 23*, 83–98.

VOGT, K., KIRSCHFELD, K. and STAVENGA, D. G. (1982). Spectral effects of the pupil in fly photoreceptors. *J. Comp. Physiol. 146*, 145–152.

VOWLES, D. M. (1966). The receptive fields of cells in the retina of the housefly (*Musca domestica*). *Proc. Roy. Soc. Lond. B. Biol. Sci. 164*, 552–576.

WACHMANN, E. (1977). Vergleichende Analyse der feinstrukturellen Organization offener Rhabdomere in den Augen der *Cucuijiformia* (*Insecta, Coleoptera*), unter besonderer Berücksichtigung der *Chrysomelidae*. *Zoomorphologie 88*, 95–131.

WADA, S. (1974). Spezielle randzonale Ommatidien der Fliegen (Diptera: Brachycera): Architektur und Verteilung in den Komplexaugen. *Z. Morph. Tiere 77*, 87–125.

WAGNER, H. (1982). Flow-field variables trigger landing in flies. *Nature 297*, 147–148.

WALCOTT, B. (1971). Unit studies on receptor movement in the retina of *Lethocerus* (Belostomatidae, Hemiptera). *Z. Vergl. Physiol. 74*, 17–25.

WALCOTT, B. (1975). Anatomical changes during light adaptation in insect compound eyes. In *The Compound Eye and Vision of Insects*. Edited by G. A. Horridge. Pages 20–33. Clarendon Press, Oxford.

WALLACE, G. K. (1959). Visual scanning in the desert locust *Schistocerca gregaria*. *J. Exp. Biol. 36*, 512–525.

WASHIZU, Y. (1964). Electrical activity of single retinula cells in the compound eye of the blowfly *Calliphora erythrocephala*. Meig. *Comp. Biochem. Physiol. 12*, 369–387.

WATANABE, A. and GRUNDFEST, H. (1961). Impulse propagation at the septal and commissural junctions of crayfish lateral giant axons. *J. Gen. Physiol. 45*, 267–308.

WATERMAN, T. H. (1954). Directional sensitivity of single ommatidia in the compound eye of *Limulus*. *Proc. Nat. Acad. Sci. 40*, 252–257.

WATERMAN, T. H. (1981). Polarization sensitivity. In *Comparative Physiology and Evolution of Vision in Invertebrates. B: Invertebrate Visual Centers and Behavior. I*. Edited by H. Autrum. *Handbook of Sensory Physiology*. Vol. 7, pt 6B, pages 283–469. Springer, Berlin and New York.

WATERMAN, T. H. and FERNÁNDEZ, H. R. (1970). E-vector and wavelength discrimination by retinular cells of the crayfish *Procambarus*. *Z. Vergl. Physiol. 68*, 154–174.

WEHNER, R. (1969). Der Mechanisms der optischen Winkelmessung bei der Biene (*Apis mellifica*). *Zool. Anz. Suppl.-Bd. 33, Verh. Zool. Ges.*, 586–592.

WEHNER, R. (1971). The generalization of directional visual stimuli in the honeybee, *Apis mellifera*. *J. Insect Physiol. 17*, 1579–1591.

WEHNER, R. (1972). Dorsoventral asymmetry in the visual field of the bee, *Apis mellifica*. *J. Comp. Physiol. 77*, 256–277.

WEHNER, R. (1975). Pattern recognition. In *The Compound Eye and Vision of Insects*. Edited by G. A. Horridge. Pages 75–113. Clarendon Press, Oxford.

WEHNER, R. (1979). Mustererkennung bei Insekten: Lokalisation und Identifikation visueller Objekte. *Verh. Dtsch. Zool. Ges.*, 19–41.

WEHNER, R. (1981). Spatial vision in arthropods. In *Comparative Physiology and Evolution of Vision in Invertebrates. C: Invertebrate Visual Centers and Behavior II*. Edited by H. Autrum. *Handbook of Sensory Physiology*. Vol. 7, pt 6C, pages 287–616. Springer, Berlin and New York.

WEHNER, R. and WEHNER-VON SEGESSER, S. (1973). Calculation of visual receptor spacing in *Drosophila melanogaster* by pattern recognition experiments. *J. Comp. Physiol. 82*, 165–177.

WEHRHAHN, C. and HAUSEN, K. (1980). How is tracking and fixation accomplished in the nervous system of the fly? A behavioural analysis based on short term stimulation. *Biol. Cybernet. 38*, 179–186.

WEHRHAHN, C., HAUSEN, K. and ZANKER, J. (1981). Is the landing response of the housefly (*Musca*) driven by motion of a flow field? *Biol. Cybernet. 41*, 91–99.

WEIDMANN, S. (1952). The electrical constants of Purkinje fibres. *J. Physiol. 118*, 348–360.

WEILER, R. and ZETTLER, F. (1979). The axon-bearing horizontal cells in the teleost retina are functional as well as structural units. *Vision Res. 19*, 1261–1268.

VON WEIZSÄCKER, E. (1970). Dressurversuche zum Formensehen der Bienen, insbesondere unter wechselnden Helligkeitsbedingungen. *Z. Vergl. Physiol. 69*, 296–310.

WHITE, R. H. and LORD, E. (1975). Diminution and enlargement of the mosquito rhabdom in light and darkness. *J. Gen. Physiol. 65*, 583–598.

WIESE, K. (1974). The mechanoreceptive system of prey localization in *Notonecta*. II. The principle of prey localization. *J. Comp. Physiol. 92*, 317–325.

WIGGLESWORTH, V. B. (1953). The *Principles of Insect Physiology*, 5th edn. Page 546. Methuen, London.

WILSON, M. (1978a). The functional organisation of locust ocelli. *J. Comp. Physiol. 124*, 297–316.

WILSON, M. (1978b). Generation of graded potential signals in the second order cells of locust ocellus. *J. Comp. Physiol. 124*, 317–331.

WOLBURG-BUCHHOLZ, K. (1979). The organization of the lamina ganglionaris of the hemipteran insects, *Notonecta glauca, Corixa punctata* and *Gerris lacustris*. *Cell Tiss. Res. 197*, 39–59.

WOLF, E. (1933). Das verhalten der Bienen gegenüber flimmernden Felden und bewegten Objekten. *Z. Vergl. Physiol. 20*, 151–161.

WOODCOCK, A. E. R. and GOLDSMITH, T. H. (1973). Differential wavelength sensitivity in the receptive fields of sustaining fibres in the optic tract of the crayfish (*Procambarus*). *J. Comp. Physiol. 87*, 247–257.

WUNDERER, H. and SMOLA, U. (1982a). Fine structure of ommatidia at the dorsal eye margin of *Calliphora erythrocephala* Meigen (Diptera: Calliphoridae): An eye region specialized for the detection of polarized light. *Int. J. Insect Morph. Embryol. 11*, 25–38.

WUNDERER, H. and SMOLA, U. (1982b). Morphological differentiation of the central visual cells R7/8 in various regions of the blowfly eye. *Tiss. Cell 14*, 341–358.

YEANDLE, S. (1958). Electrophysiology of the visual system. *Amer. J. Ophthalmol. 46*, 82–87.

ZEIL, J. (1979). A new kind of neural superposition eye: the compound eye of male *Bibionidae*. *Nature 278*, 249–250.

ZERRAHN, G. (1933). Formdressur und Formunterscheidung bei der Honigbiene. *Z. Vergl. Physiol. 20*, 117–150.

ZETTLER, F. (1975). Eine neue Art der Nervenleitung: Amplitudenmodulation. *Umschau 75*, 118–120.

ZETTLER, F. and AUTRUM, H. (1975). Chromatic properties of lateral inhibition in the eye of a fly. *J. Comp. Physiol. 97*, 181–188.

ZETTLER, F. and JÄRVILEHTO, M. (1970). Histologische Lokalisation der Ableitelektrode. Belichtungspotentiale aus Retina und Lamina bei *Calliphora*. *Z. Vergl. Physiol. 68*, 202–210.

ZETTLER, F. and JÄRVILEHTO, M. (1971). Decrement-free conduction of graded potentials along the axon of a monopolar neuron. *Z. Vergl. Physiol. 75*, 402–421.

ZETTLER, F. and JÄRVILEHTO, M. (1972). Lateral inhibition in an insect eye. *Z. Vergl. Physiol. 76*, 233–244.

ZETTLER, F. and JÄRVILEHTO, M. (1973). Active and passive axonal propagation of non-spike signals in the retina of *Calliphora*. *J. Comp. Physiol. 85*, 89–104.

ZIMMERMAN, R. P. (1978). Field potential analysis and the physiology of second-order neurons in the visual system of the fly. *J. Comp. Physiol. 126*, 297–316.

8 Insect Visual Pigments and Color Vision

RICHARD H. WHITE

University of Massachusetts, Boston, Massachusetts, USA

1 INTRODUCTION AND OVERVIEW

Vision, like other photobiological processes, functions in the narrow band of light between 300 and 700 nm where photons have enough energy for photochemistry but not for the disruption of macromolecules (Wald, G. 1959). Visual pigments that initiate vision by absorbing light in this spectral region have been identified in the photoreceptors of three groups of animals: vertebrates, molluscs and arthropods.* These visual pigments are a class of membrane-associated proteins, called opsins, conjugated with a chromophore. In vertebrates, the chromophore is either retinaldehyde (retinal) or 3-dehydroretinalehyde; both are derived from vitamin A (retinol) or carotenoids such as β-carotene. Visual pigments whose chromophore is retinal are called rhodopsins, those with 3-dehydroretinal, porphyropsins. The invertebrate visual pigments that have been adequately characterized — from a few species of insects, cephalopods, crustaceans, and *Limulus* — are all rhodopsins. Retinal is the chromophore, and they resemble vertebrate rhodopsins in other respects.

1.1 Rhodopsin and retinal

Retinal is attached to opsin in a Schiff's base linkage $(-CH = N-)$ between the carbonyl group of the chromophore and an ε-amino group of lysine in the protein. There are several geometric isomers of retinal. The 11-*cis* isomer, in which the polyene chain is bent and twisted around carbon 11, is the chromophore of native rhodopsin (Fig. 1). Free retinal absorbs light in the near ultraviolet with an absorption band at 380 nm. When 11-*cis*-retinal joins opsin to form rhodopsin, the main absorption band is shifted so that the λ_{max} of different rhodopsins range from about 345 nm in the ultraviolet (Schwemer, J. *et al.*, 1971) to at least 580 nm and perhaps to 610 nm (Bernard, G. 1979) in the red. When the absorbance of the chromophore moves to longer wavelengths, it is spoken of as a bathochromatic shift, to shorter wavelengths, a hypsochromatic shift. Presumably the λ_{max} of a particular rhodopsin is determined by the particular disposition of charged opsin groups around the site of the chromophore that modify the π orbital of its polyene chain.

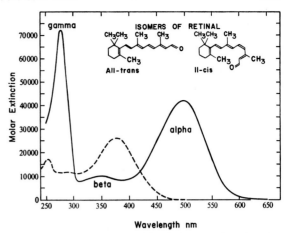

FIG. 1. Absorbance spectra of retinal $(---)$ and cattle rhodopsin (——). The inset shows the *all-trans* and *11-cis* isomers of retinal (from White, R. 1978).

The spectrum of purified cattle rhodopsin, which can be used as a standard for assessing the less well characterized insect rhodopsins, is shown in Fig. 1. The main absorption band (α-band) peaks at 499 nm in the green region of the visible spectrum. There is a β-band with lower extinction at about 350 nm. The γ-band at 280 nm results from the absorbance of the aromatic amino acids of the opsin. Plotted as a function of wavelength, as is conventional (as opposed to a frequency plot; see Wald, G. 1965), rhodopsins with absorption bands at longer wavelengths are broader than those peaking at shorter wavelengths. In order to compare different rhodopsins, nomograms based on the resonance spectrum of cattle rhodopsin have been devised for calculating theoretical α-bands at various λ_{max} (Dartnall, H. 1953; Ebrey, T. and Honig, B. 1977; Dawis, S. 1981).

Since no insect rhodopsin has been extracted to adequate purity, only their α-bands, which are delineated in difference spectra, have been well characterized. They generally match well with nomogram spectra based on cattle rhodopsin. Table 1 lists insect rhodopsins whose α-bands have been measured by spectrophotometric methods. There are also many other species, which I have not attempted to list, for which visual pigments can be

---

* A pigment similar to animal visual pigments, bacteriorhodopsin, is present in *Halobacterium halobium* bacteria, where it functions in photosynthesis as a proton pump and in a sense as a visual pigment by mediating phototaxis (Hildebrand, E. and Dencher, N. 1975).

inferred with some confidence from action spectra. Some are listed by Menzel, R. (1979) in his table 1, along with data from other arthropods. Insect rhodopsins fall into three main groups: ultraviolet sensitive ($\lambda_{max} \simeq 345$ nm), violet sensitive ($\lambda_{max} \simeq 440$ nm), and green sensitive ($\lambda_{max} \simeq 490–550$ nm).* In addition, there is evidence that several butterfly species have an exceptional red-sensitive visual pigment with $\lambda_{max} \simeq 610$ nm (Bernard, G. 1979). Insects are therefore remarkable for the spectral range of their photopigments, extending from the near ultraviolet to the red of the visible spectrum.

1.2 Photochemistry of rhodopsin

The absorption of a photon by the retinal chromophore of rhodopsin results in its isomerization from the 11-*cis* to the less strained all-*trans* configuration (Ebrey, T. and Honig, B. 1975). Vertebrate opsins then undergo a series of conformational changes, not requiring light, that culminate with the dissociation of the chromophore from the opsin. Since rhodopsin absorbs in the visible, whereas free retinal and opsin absorb in the ultraviolet, which we cannot see, this process is called bleaching. For instance, an isolated dark-adapted frog retina appears rosy-pink because of its rhodopsin. When illuminated it rapidly blanches to the orange-yellow color of retinal. After about an hour it appears colorless because the retinal has been reduced to retinol.

By slowing the process at low temperatures, or through the use of fast photometric techniques, bleaching can be resolved into a series of spectrally distinct intermediates (Fig. 2). The transition through the bleaching intermediates reflects the progressive opening up of the opsin's structure, and the loosening of the association between opsin and chromophore. For instance, reactive groups such as sulfhydryls become accessible upon bleaching, and the chromophoric site becomes vulnerable to hydrolyzing reagents such as borohydrides. For reviews of the photochemistry of vertebrate rhodopsins, see Daemen, F. (1973), Ebrey, T. and Honig, G. (1975), and Ostroy, S. (1977).

Invertebrate rhodopsins differ from vertebrate photopigments in that they generally do not bleach (Hamdorf, K. 1979). Most are transformed by light to an intermediate that is stable at physiological

Pigment	λ_{max}, nm	Half-life, s
Rhodopsin	498	stable
light ↓		
Bathorhodopsin (Prelumirhodopsin)	543	$\sim 10^{-8}$
Lumirhodopsin	497	$\sim 10^{-5}$
Metarhodopsin I	478	$\sim 10^{-4}$
Metarhodopsin II	380	~ 60
Metarhodopsin III (Pararhodopsin)	465	~ 100
H_2O ↓		
Opsin + all-*trans* retinal	389	

FIG. 2. Bleaching intermediates of vertebrate rhodopsin (Ostroy, S. E. 1977).

temperatures. That is, the isomerized chromophore remains in place. In most cases the intermediate absorbs around 480 nm, similar to the metarhodopsin I intermediate of vertebrate rhodopsin. Therefore, the stable invertebrate intermediate is called metarhodopsin.

Several milliseconds elapse between the absorption of light by a photopigment and the initiation of electrical activity in the membrane of the receptor cell. The timing of these events indicates that transduction must be initiated in vertebrates by one of the intermediate transitions prior to metarhodopsin II (Ostroy, S. 1977). Clearly, phototransduction must be triggered in invertebrates during the

---

* The relationship between wavelength, human color sensation, and color names can be best appreciated by looking at the accurately rendered spectrum published in the September 1968 issue of *Scientific American* (Feinberg, G. 1968). The wavelengths from about 600 to beyond 700 nm appear *red*; from about 580 to 600 nm, *orange*. A narrow band around 575 nm appears *yellow*. From about 550 to 570 nm, the sensation is *yellow-green*, from 490 to 550 nm, *green*. Unfortunately, there is some ambiguity in the terms applied to the shorter visible wavelengths. Sometimes the whole region from 400 to about 490 nm is spoken of as *blue*. However, that spectral region from about 400 to 480 nm is more appropriately *violet*. A distinctive narrow *blue* band from about 480 to 490 nm is sometimes called *blue-green*. The *near ultraviolet* region of the spectrum, not seen by us but important in insect vision, extends from about 300 to 400 nm. *Purple* is a non-spectral color, that is, it cannot be elicited by any monochromatic light. It is the sensation produced by a mixture of violet and red wavelengths.

transition between rhodopsin and metarhodopsin. The mechanism of transduction is currently the major unresolved problem in photoreceptor research. Invertebrate receptors with their bistable photopigments offer particularly favorable systems for investigating transduction (Hamdorf, K. 1979; Hillman, P. 1979).

1.3 Regeneration of rhodopsin

After it has been bleached or transformed by light, a rhodopsin must be regenerated. In vertebrate eyes, regeneration is accomplished by a complex biochemical pathway that enzymatically isomerizes the all-*trans* chromophore to 11-*cis*. The latter is then recombined with opsin in the photoreceptors to form rhodopsin (Bridges, C. 1976).

Another route to regeneration in vertebrates is provided by photoregeneration. As long as the all-*trans* chromophore remains attached to a bleaching intermediate (e.g. lumirhodopsin or metarhodopsin I), it can absorb another photon and reisomerize to 11-*cis* retinal. That results in the reformation of rhodopsin. Whether photoregeneration plays a significant role in vertebrate vision remains uncertain. However, photoregeneration from stable metarhodopsin is the principal means by which rhodopsin is maintained in illuminated invertebrate photoreceptors. The polychromatic light of the normal environment is absorbed by both rhodopsin and metarhodopsin. The visual pigment flips back and forth between the two stable states of the visual pigment. They coexist in a photoequilibrium determined mainly by the absorption coefficients of the two forms of the pigment, and by the spectral quality of the light (Hamdorf, K. 1979).

1.4 Photoreceptor membrane

Rhodopsins are membrane proteins. The photoreceptor membranes of vertebrates are organized as parallel flat discs isolated in appendages of the receptor cells, in their outer segments (Olive, J. 1980). Outer segment discs are either continuous with the plasma membrane of the receptor cell (in cone cells) or are derived from the plasma membrane (in rod cells). The photoreceptor membranes of arthropods, cephalopods and other invertebrates are organized as arrays of microvilli elaborated

from the plasma membrane of the receptor cell (Figs 12 and 13). The massed microvilli of a single cell are called the rhabdomere. Where the rhabdomeres of several cells are joined, as in an ommatidium, they are collectively termed the rhabdom. The insertion of photopigment molecules into cylindrical microvilli rather than flat discs has several consequences. Most notably, light absorption by rhodopsin in a microvillus is sensitive to the plane of polarized light. The high sensitivity of many arthropods to polarized light has its basis in the inherent dichroism of the rhabdomere microvillus (Waterman, T. 1975).

2 MEASUREMENT OF VISUAL PIGMENTS

The standard methods for the extracting and measuring of visual pigments, retinal and vitamin A may be found in the review by Hubbard, R. *et al.* (1971). Schwemer, J. and Langer, H. (1982) have reviewed techniques for insect photopigments.

2.1 Extraction of rhodopsin

Photoreceptor membranes from insect retinas homogenized in phosphate buffer can be recovered by flotation; they concentrate at a buffer–sucrose interface. The accessory ommochrome and pteridine pigments contained within most insect photoreceptor cells and adjacent pigment cells are a nuisance. They raise the background extinction of the extracts, and worse, may undergo absorbance changes that confuse photometric measurements. In some cases it appears that extracted accessory pigments have been mistaken for visual pigments (Bowness, J. and Wolken, J., 1959; Denys, C. 1982). Contaminating ommochrome pigments can be removed by repeated buffer washes at high pH (Schwemer, J. *et al.*, 1971; Paulsen, R. and Schwemer, J. 1972; Denys, C. 1982).

Since rhodopsins are membrane-bound hydrophobic proteins, they can only be brought into aqueous solution with the aid of detergents. Digitonin is the most satisfactory. Others that are known to remove lipids from vertebrate rhodopsins (e.g., Triton X-100; DDAO; CTAB, Ammonyx LO; octyl or nonyl glucoside) lower the thermostability of insect rhodopsins to the point where denaturation

Table 1: *Absorbance maxima (nm) of insect rhodopsins and their photoproducts*

Order/species	P	L	M	M(alk)	Ext	Reference
Blattaria						
Periplaneta americana	500	—	—	—	X	Wolken, J. and Scheer, I. (1963)
Heteroptera						
Gerris lacustris	545	—	485	—		
	460	—	520	—		
	350	—	480	—		Hamann, B. and Langer, H. (1980)
Planipennia						
Ascalaphus macaronius	345	—	475	—	X	Gogala, M. *et al.* (1970)
	345	375	475	380	X	Schwemer, J. *et al.* (1971)
Lepidoptera						
Deilephila elpenor	520	—	480	—		Hamdorf, K. *et al.* (1972)
	345	—	480	—		
	520	—	480	—	X	
	440	—	480	—	X	
	345	—	480	—	X	Schwemer, J. and Paulsen, R. (1973)
Manduca sexta	520	—	490	380	X	
	440	—	490	—	X	
	345	—	490	—	X	Schwemer, J. and Struwe, G. (from
	520	—	490	—		Schwemer, J. and Langer, H., 1982)
	440	—	490	—		
	345	—	490	—		Schwemer, J. and Brown, P. (personal communication)
Galleria mellonella	510	—	484	—		Goldman, L. *et al.* (1975)
Aglais urticae	535	—	480	—		Stavenga, D. (1975)
Spodoptera exempta	560	—	485	—		
	520	—	485	—		
	350	—	465	—		Langer, H. *et al.* (1979)
	570	—	—	—	X	
	520	—	480	—	X	
	470	—	—	—	X	
	350	—	470	—	X	Schwemer, J. and Langer, H. (1982)
Apodemia mormo	585	—	515	—		
Anartia amathea	610	—	—	—		Bernard, G. (1979; personal communication)
Hymenoptera						
Apis mellifera (drone)	445	—	505	—		Bertrand, D. *et al.* (1979)
Diptera						
Aedes aegypti	515	—	480	—		Brown, P. and White, R. (1972)

For other Diptera (Cyclorrapha) see Tables 3 and 4.

P = rhodopsin; L = lumirhodopsin; M = metarhodopsin (acid metarhodopsin); M(alk) = alkaline metarhodopsin. Those photopigments measured in extracts are so indicated; the others were measured *in situ* by MSP or reflectometry.

occurs at room temperature in darkness (Schwemer, J. and Langer, H. 1982). Generally, rhodopsins survive digitonin extraction, with their spectral characteristics unchanged. The rhodopsins of some crustaceans, however, appear to be altered by removal from the photoreceptor membrane (Bruno, M. and Goldsmith, T. 1974). The spectral sensitivity of the crayfish eye, and the absorbance of intact crayfish rhabdoms, indicate a rhodopsin with $\lambda_{max} \simeq 530$ nm. Extracts, however, contain two light-sensitive pigments absorbing at 560 and 510 nm respectively (Wald, G. 1967, 1968). Recent measurements have reinforced earlier suggestions

that these pigments are artifacts of preparation (Larrivee, D. and Goldsmith, T. 1981).

If extracted to adequate purity, a rhodopsin can be measured in a direct spectrum such as that of Fig. 1. Since vertebrate rhodopsins bleach, their α-bands also can be obtained accurately from impure extracts in difference spectra, calculated by subtracting the spectrum measured after actinic irradiation from that measured prior to bleaching. Furthermore, rhodopsins generally are stable to hydrolyzing and reducing agents such as hydroxylamine and borohydrides, that attack the Schiff's base linkage of the chromophore to the opsin, whereas intermediates are more vulnerable. These agents hasten bleaching with the production of characteristic photoproducts.

Insect visual pigments can also be measured in difference spectra. Although their metarhodopsins are usually stable in *in vivo*, they may tend to hydrolyze in extracts (Schwemer, J. and Paulsen, R. 1973). In any event the agents used to promote the bleaching of vertebrate pigments also attack insect metarhodopsins, so that difference spectra can be obtained in bleaching experiments. The application of these basic procedures for the characterization of insect visual pigments is demonstrated in section 2.4.

2.2 Microspectrophotometry

Although insect rhodopsins have been extracted, the small size of their eyes, and consequent tedium of collection and dissection, has led to the use of alternative approaches for the measurement of insect visual pigments. Microspectrophotometry (MSP) has been particularly exploited. For this technique, the beam of a spectrophotometer is passed through a microscope. Photometric measurements can then be taken within whole eyes, slices of retinas, intact cells or isolated rhabdoms. MSP was first utilized in vision research to measure the photopigments of single vertebrate cone outer segments (Brown, P. and Wald, G. 1964; Marks, W. *et al.*, 1964; Marks, W. 1965), and in the rhabdomeres of flies (Langer, H. and Thorell, B. 1966). The procedures, interpretation and limitations of MSP are covered in a review by Liebman, P. (1972). In addition to bypassing the problem of extraction, MSP allows the measurement of insect visual pigments within their natural environment of the photoreceptor membrane, and in some cases within the intact living animal.

To measure an insect visual pigment by MSP, spectra may be taken from a rhabdom and an adjacent cellular region. The subtraction of the latter from the former — a procedure analogous to the comparison of measuring and reference beams in conventional spectrophotometry — yields an approximate spectrum of the photopigment contained within the rhabdom. Alternatively, difference spectra may be calculated after transforming or bleaching the rhodopsin (see section 2.4).

Rhodopsin does not appear to be altered by the fixation of retinas with glutaraldehyde, the protein cross-linking agent used to stabilize protein crystals, and to preserve cellular structure for electron microscopy. This treatment renders tissue more stable for MSP experiments; drifting baselines can be a problem if measurements are taken in retinal slices over an extended period, or if reagents such as hydroxylamine are used.

In recent years pseudopupil MSP, a technique applicable to some apposition eyes, has been used to measure visual pigments in the eyes of living insects. Light entering the cornea and passing through the length of the rhabdom is reflected back again from a tracheal tapetum behind the retina, re-emerging from the corneal lenses of the ommatidia. Since the optical axes of the ommatidia in the curved compound eye diverge in the proximo-distal direction, the rays of this reflected light will also diverge, appearing to originate from the point located at the center of the curvature of the eye. The result is an enlarged virtual image of several superimposed adjacent retinulae. This image is called the deep pseudopupil (Franceschini, N. and Kirschfeld, K. 1971). The deep pseudopupil of apposition eyes can be observed with a microscope with reflecting optics, if the working-distance of the objective is greater than the distance from the cornea to the point behind the eye at which the ommatidial axes converge. It can be seen, for instance, in the eyes of butterflies.

Pseudopupil images can also be projected from some compound eyes by shining light through the back of the head. This approach has proved especially successful with the transparent eyes of fly mutants that lack interfering accessory pigments. The photoreceptor cells of flies bear separate rhabdomeres (Fig. 17) that function as light guides. Consequently, their distal tips stand out as bright spots against a dimmer background in antidromically illuminated eyes. Because of their very regular arrangement across the eye — a "neurocrystalline lattice" in the words of Ready, D. *et al.* (1976) — the corresponding rhabdomeres of neighboring retinulae are precisely superimposed in the deep pseudopupil image. It appears as an enlarged virtual image of the pooled separate rhabdomeres in their characteristic trapezoidal arrangement (Stavenga, D. 1979, Franceschini, N. *et al.*, 1981b).

The deep pseudopupil image can be utilized for MSP measurements, either of reflected light passing twice through the eye (Bernard, G. 1979), or of light transmitted through the back of the eye (Stavenga, D. 1976, 1979).

The characteristics of a visual pigment *in situ* may differ in various ways from those of the same pigment extracted from the rhabdomere membrane. Membrane-bound photopigments may be more stable, they show birefringence as a consequence of their orientation in the membrane (see section 6.7), and their spectra may be modified as well (see sections 8.3 and 9.1.1). The relative extinctions of rhodopsin and metarhodopsin may also be affected if the orientation of the chromophore with respect to the membrane shifts during the transformation of rhodopsin to metarhodopsin (Brown, P. and White, R. 1972). MSP has the advantage of showing the properties of a visual pigment as it functions within the receptor cells.

2.3 Action spectra

Action spectra plot the response of a cell (or a neuron, or an organism) as a function of wavelength. Although there are various ways of expressing action spectra, only a spectral sensitivity function — the reciprocal of the number of photons which elicits a constant response near threshold at different wavelengths — is appropriate for relating that response to the absorption spectra of underlying visual pigments (Burkhardt, D. 1962). This is because the response may not be linear with stimulus intensity owing to mechanisms for adjusting gain.

An action spectrum from a receptor may correspond closely with the absorption spectrum of the visual pigment it contains, but not necessarily. For instance, the spectral quality of the light reaching the photopigment may be altered by colored substances in the eye. In long rhabdomeres, self-screening, in which rhodopsin itself acts as a colored filter, may broaden the spectral sensitivity function. Or a receptor may be electrically coupled with another so that its spectral sensitivity depends on its neighbor's rhodopsin as well as its own. These effects are discussed in more detail in section 9.1.

Although extracellular mass electrical responses from whole eyes (the electroretinogram: ERG) can provide information about the spectral sensitivities of individual receptors, especially with chromatic adaptation that selectively reduces the function of one cell type or another (e.g. in the moth *Deilephila elpenor*: Höglund, G. *et al.*, 1973b), intracellular recording from individual cells is the preferred electrophysiological technique. However, the penetration of particular cells with microelectrodes is often difficult, and necessarily involves injury to the eye.

An alternative approach that has been exploited recently allows action spectra to be measured repeatedly in intact fully functioning eyes. In many insects and other arthropods, accessory pigment granules within the photoreceptor cells congregate next to the rhabdomere in light-adaptation and disperse into the cytoplasm during dark-adaptation. Kirschfeld, K. and Franceschini, N. (1969) showed in the housefly (*Musca domestica*) that the migration of pigment during light-adaptation decreases the light flux through the rhabdomere. Since the accessory pigment plays a role similar to the vertebrate pupil in regulating the amount of light transmitted to the photoreceptors, the adjustment of the pigment screen has been called the "pupillary" response. There is evidence that the pupillary response results from the light-initiated depolarization of the receptor cell (Kirschfeld, K. and Franceschini, N. 1969; Butler, R. 1971; Stavenga, D. 1979; Frixione, E. and Aréchiga, H. 1981). The pupillary response can be optically measured as either a decrease in transmittance (Franceschini, N. 1972), or change in reflectance (Bernard, G. and Stavenga, D. 1978, 1979; Bernard, G. 1982). Both approaches have been used to non-invasively measure the spectral sensitivities of receptor cells. The cellular position of retinular accessory pigment has also been used to morphologically identify receptor cells with particular spectral sensitivities in histological sections of chromatically adapted retinas (Butler, R. 1971; Menzel, R. and Knaut, R. 1973; Kolb, G. and Autrum, H. 1974).

The action spectra discussed above measure the depolarization of receptor cells and hence reflect the absorption spectra of their rhodopsins. There are other effects, discussed in section 7, that are related to the production of metarhodopsin. Arthropod receptors remain depolarized for an extended period after the cessation of a bright light that is

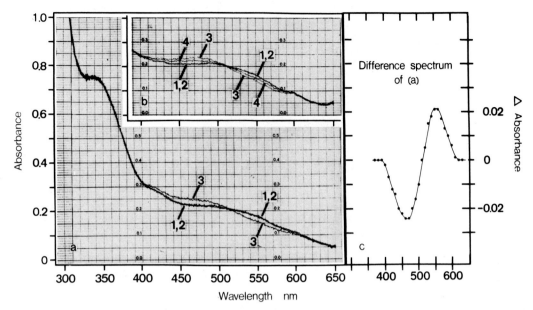

FIG. 3. MSP absorption spectra from ocelli (stemmata) of larval mosquitoes (*Aedes aegypti*). **a**: Spectra from an ocellus dissected into insect Ringer. Curves 1 and 2 were "dark" scans whose superposition shows baseline stability. Curve 3 was recorded after a bright yellow flash. The photoconversion of rhodopsin to metarhodopsin results in an absorbance change plotted as the difference spectrum (curves 1,2 − 3) in **c**. **b**: Spectra from the ocelli of a living animal. Curves 1 and 2 are dark spectra, curve 3 was recorded after a yellow flash, and curve 4 after a blue flash. Curves 3 and 4 represent different photosteady states of the visual pigment (from Brown, P. and White, R. 1972).

strongly absorbed by rhodopsin. This depolarizing afterpotential (DPA) is abolished, that is, the receptor is repolarized, by light absorbed by metarhodopsin. Therefore, the spectral sensitivity of repolarization matches the absorption spectrum of metarhodopsin (Hamdorf, K. 1979). In flies there is also an electrical response, the M potential, that is related linearly to the conversion of metarhodopsin to rhodopsin (Pak, W. and Lidington, K. 1974). Both the DPA and M potential can be utilized for measuring metarhodopsin in appropriate invertebrate photoreceptor cells.

2.4 Mosquito visual pigment

It may be helpful to demonstrate the basic procedures for identifying and characterizing an insect visual pigment by looking closely at a particular example. The larval ocelli (stemmata) of the mosquito *Aedes aegypti* are simple photoreceptors that were found to be particularly suitable for photopigment measurement by MSP (Brown, P. and White, R. 1972). As the ocelli hang free of other tissue in the hemocoel of the larval head, they can be readily dissected and mounted in a microcell faced with

quartz (ultraviolet transmitting) coverslips. The microcell, in turn, is placed on the stage of the microscope incorporated into the spectrophotometer. A white-eye mutant of *Aedes* is available that lacks the ommochrome accessory pigment which normally obscures the rhabdom. Since the head of the larval mosquito is small and transparent, measurements can also be made of intact living animals immobilized within a microcell.

For the experiments to be described, larvae were reared in darkness, and prepared for MSP under dim red light, > 650 nm, to avoid converting any rhodopsin to metarhodopsin prior to measurement. Figure 3a shows the record of a typical experiment on an isolated ocellus mounted in insect Ringer. Traces 1 and 2 are "dark" spectra; that is, they were taken before exposing the ocellus to actinic light. The congruence of these baselines indicates that the preparation was stable. The shape of the dark spectrum shows only that the overall absorbance of the receptor cells increases at shorter wavelengths.

After the dark spectra were measured, the ocellus was exposed to a flash of bright yellow light. Trace 3, taken immediately afterward, shows that the absorbance dropped at 550 nm and rose at 450 nm.

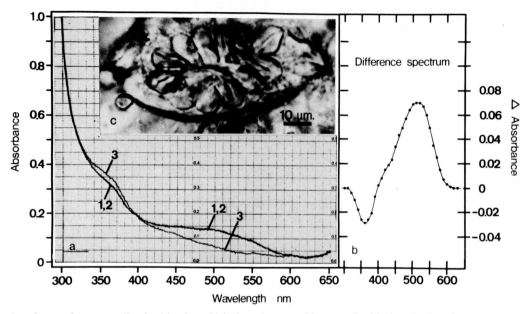

Fig. 4. **a**: Spectra from an ocellus fixed in glutaraldehyde and mounted in neutralized hydroxylamine. Curves 1 and 2 were recorded in the dark, curve 3 after the ocellus had been bleached for 10 min with yellow light. The difference spectrum plotted in **b** represents the true absorption spectrum of the rhodopsin at wavelengths greater than 450 nm. The inset (**c**) is a photomicrograph of the ocellus that was measured as it appeared in the microspectrophotometer. The lobed structure at the center of the ocellus is the rhabdom. A 30 μm central area of the rhabdom was measured (from Brown, P. and White, R. 1972).

The difference spectrum of this change, curves 1 and 2 minus curve 3, is plotted in Fig. 3c. Such a difference spectrum is characteristic of the transformation of rhodopsin to metarhodopsin where the spectra of the two forms of the visual pigment broadly overlap. Although such a rapid transformation in a visual cell is good evidence for a visual pigment, the change might also result from some other light-sensitive substance, an accessory pigment for instance.

The measurements shown in Fig. 3b provide further evidence that the spectral transformation was the response of a visual pigment. As in Fig. 3a, curves 1 and 2 are dark spectra, and trace 3 was taken after a yellow flash. Then the ocellus was irradiated with blue light, which trace 4 shows to have partially reversed the effect of the preceding yellow flash. The spectrum can be repeatedly transformed between curves 3 and 4 by alternate yellow and blue irradiation with an isosbestic point at 510 nm. Such a pattern is characteristic of a single visual pigment flipping between different rhodopsin/metarhodopsin photoequilibria.

Since the spectra of the two states of the pigment overlap, they mutually subtract in difference spectra. Although the spectra in Fig. 3 show how the photopigment responds to light *in situ*, they do not accurately characterize either state of the pigment because of that subtractive interference. In this situation, the α-band of the rhodopsin can be characterized by using reagents, such as hydroxylamine (NH_2OH), that promote the dissociation of the chromophore. Rhodopsins are generally not vulnerable to moderate concentrations of NH_2OH because the chromophore is sequestered within the protein. However, the chromophore site becomes available with the transformation to metarhodopsin. Thus, the visual pigment of the mosquito is stable to NH_2OH in darkness, but is bleached by light with the production of retinaldehyde oxime, the reaction product of NH_2OH and the chromophore (Hubbard, R. *et al.*, 1971).

Figure 4 shows a bleaching experiment. An ocellus, fixed in glutaraldehyde to increase its stability, was mounted in 0.1 M NH_2OH. Traces 1 and 2 (Fig. 4a) are dark spectra, trace 3 was taken after bleaching with yellow light. The difference spectrum (Fig. 4b) provides convincing evidence that a rhodopsin has been measured. The positive

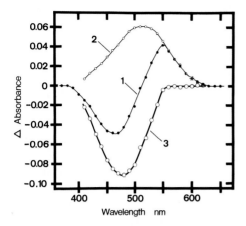

FIG. 5. Calculation of the spectrum of metarhodopsin. Curve 1 is the difference spectrum from ocelli of a living larva flashed with yellow light. Curve 2 is the calculated spectrum of the rhodopsin transformed by the flash. Curve 3 is the spectrum of the metarhodopsin; it was obtained by subtracting curve 2 from curve 1 (from Brown, P. and White, R. 1972).

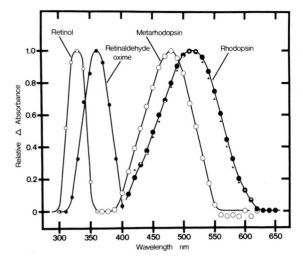

FIG. 6. Summary of difference spectra measured in larval mosquito ocelli. All spectra are arbitrarily brought to the same height. The rhodopsin spectrum is the average of seven difference spectra measured in the presence of hydroxylamine; the metarhodopsin spectrum is the average of three calculated spectra, the retinaldehyde oxime spectrum is the average of seven spectra, and the retinol spectrum is one difference spectrum. The triangles show the Dartnall nomogram of a hypothetical rhodopsin with $\lambda_{max} = 515$ nm (from Brown, P. and White, R. 1972).

component of the difference spectrum matches a rhodopsin nomogram, $\lambda_{max} = 515$ nm, the negative component matches the spectrum of retinal oxime, $\lambda_{max} = 365$ nm, proving that the chromophore is retinal. Since the oxime does not absorb above 450 nm, the difference spectrum accurately represents the absorption spectrum of the rhodopsin at wavelengths longer than 450 nm. Hydrolysis of the visual pigment with potassium borohydride (KBH$_4$) yielded a photoproduct at 330 nm (Fig. 6), either free retinol or the chromophore reduced on site to retinyl-opsin, providing further proof that the chromophore is retinal.

The photoequilibrium difference spectrum of Fig. 3c is the sum of spectral changes due both to rhodopsin and metarhodopsin. Therefore, the spectrum of the metarhodopsin, λ_{max} 480 nm, can now be calculated by subtracting the rhodopsin spectrum from the photoequilibrium spectrum as shown in Fig. 5. If it does not bleach too rapidly, the spectrum of a metarhodopsin also may be taken directly from difference spectra by subtracting successive spectra taken during the course of bleaching.

The spectral sensitivity of a photoreceptor should be related to the absorption spectrum of its rhodopsin. Figure 7 compares ERG action spectra from the mosquito ocellus with the spectrum of P515.* The action spectra are somewhat narrowed on the short wavelength side of the absorption spectrum, and

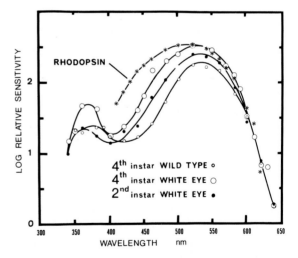

FIG. 7. Spectral sensitivity curves compared with the averaged rhodopsin difference spectrum. The curve for white-eye larvae represents the average of data from five animals, all spectral sensitivities being set equal at 520 nm. The single curve for wild type represents data from one larva. The spectral sensitivity and the rhodopsin spectrum were superimposed at long wavelengths where they coincide (from Seldin, E. *et al.*, 1972).

* Following convention, particular rhodopsins will be designated by P (for pigment) and the wavelength of their λ_{max}. Metarhodopsins will be designated by M (e.g., M480 in the mosquito).

Table 2: Molecular weights of insect rhodopsins

Species	Rhodopsin	MW (daltons)
Drosophila melanogaster	P480	37,000 (Ostroy, S., 1978)
Calliphora erythrocephala (*vicinia*)	P490	32,500 ± 1600 (Paulsen, R. and Schwemer, J., 1979)
Aedes aegypti	P520	39,000 (Stein, P. *et al.*, 1979)
Ascalaphus macaronius	P345	35,000 ± 1800 (Paulsen, R. and Schwemer, J., 1973)

peak sensitivity is shifted to 520 nm. The effect is more striking in eyes with normal accessory pigmentation. We can account for this narrowing by the selective filtering of accessory pigment and cytoplasm whose absorbance increases at shorter wavelengths. Filtering by metarhodopsin may also play a role. The secondary peak in the spectral sensitivity curve may reflect the β-band absorbance of P515, but such details cannot be inferred with confidence from action spectra.

Now that a number of arthropod rhodopsins have been characterized and their common features recognized, one is reasonably safe in identifying a visual pigment from limited data. Ideally, however, the α-band should be measured, and it should relate to spectral sensitivity. The chromophore should be identified as retinal, directly or by way of a photoproduct. In general, insect rhodopsins, at least those with λ_{max} greater than 400 nm, can be expected to match cattle rhodopsin nomograms fairly well. Thus it may be possible to infer a rhodopsin α-band from a spectral sensitivity function, or a distorted absorption spectrum, by comparing it with a nomogram.

3   MOLECULAR WEIGHT AND MOLAR EXTINCTION

The molecular weights of four insect rhodopsins have been determined from their mobilities in SDS acrylamide gel electrophoresis (Table 2). These molecular weights are similar to those of vertebrate rhodopsins (35,000–40,000; Abrahamson, E. and Fager, R. 1973), but somewhat less than the molecular weight of squid rhodopsin (49,000; Hagins, F. 1973).

The molar extinction of *Drosophila* rhodopsin was measured by Ostroy, S. (1978) using the procedure of Wald, G. and Brown, P. (1953). In that technique, the absorbance of rhodopsin is compared with that of retinal oxime (whose molar extinction is known) after bleaching a detergent extract in the presence of hydroxylamine. The extinction of P480 was found to be 35,000 mol^{-1} cm^{-1}. Using a different approach, Harris, W. *et al.* (1976) estimated the molar extinction at 33,000 mol^{-1} cm^{-1}. These values can be compared to that measured for vertebrate rhodopsin: 40,600 mol^{-1} cm^{-1} (Wald, G. and Brown, P. 1953).

It can be seen from Figs 5 and 8 that the extinction of insect metarhodopsin is characteristically about $1\frac{1}{2}$ times greater than that of rhodopsin. The same is true of vertebrate metarhodopsin I (Matthews, R. *et al.*, 1963). The molar extinction of *Drosophila* metarhodopsin (M570) has been estimated at about 56,000 mol^{-1} cm^{-1} (Ostroy, S. 1978) and 43,000 mol^{-1} cm^{-1} (Harris, W. *et al.* 1976).

4   CHROMOPHORE AND PHOTOCHEMISTRY

Several lines of evidence show that retinal is the chromophore of insect photopigments. The heads of insects contain retinal, photosensitivity is reduced when carotenoids are eliminated from insect diets, and the chromophore of a few insect rhodopsins has been directly identified as retinal.

4.1   Localization of retinal

Retinal, identified by the Carr–Price reaction with antimony trichloride (Hubbard, R. *et al.*, 1971), was found in the heads of dark-adapted honeybees, but not in their bodies (Goldsmith, T. 1958a,b). The same result has been obtained for a number of other insects including species of Orthoptera, Odonata, Lepidoptera, Coleoptera, Diptera, and Hymenoptera (Wolken, J. *et al.*, 1960; Briggs, M. 1961; Wolken, J. and Scheer, I 1963). The 11-*cis* isomer of retinal is localized in the eyes of crustaceans (Wald, G. and Burg, S. 1957; Wald, G. and Brown, P. 1957; Denys, C. 1981).

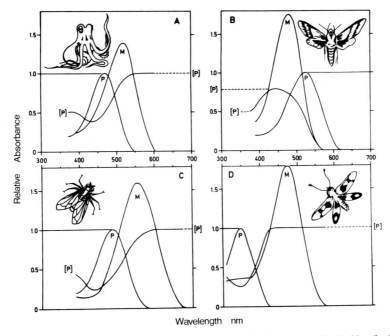

FIG. 8. Visual pigments of the cephalopod *Eledone moschata* (**A**), the moth *Deilephila elpenor* (**B**), the blowfly *Calliphora erythrocephala* (**C**), and the owlfly *Ascalaphus macaronius* (**D**). P and M are the absorption spectra of rhodopsin and metarhodopsin. The rhodopsin spectra have been normalized to an absorbance of 1.0. In all cases the extinction of metarhodopsin is greater than that of rhodopsin. The curves labelled [P] show the relative amount of rhodopsin in photoequilibria established by monochromatic adaptation assuming that the quantum efficiencies of rhodopsin and metarhodopsin are the same (from Hamdorf, K. *et al.*, 1973b).

4.2 Carotenoid deprivation

Insects, like other animals, appear to obtain their retinal only from retinol (vitamin A) and other carotenoids in their food. Since vitamin A seems to have no essential systemic function in insects, they are readily reared on carotenoid-free diets, for generations if need be. Rearing insects on artificial diets without carotenoids reduced photoreceptor sensitivity in the housefly *Musca domestica* (Goldsmith, T. *et al.*, 1964; Goldsmith, T. and Fernandez, H. 1966), *Drosophila melanogaster* (Stark, W. and Zitzmann, W. 1976), the mosquito *Aedes aegypti* (Brammer, J. and White, R. 1969), and in the moths *Bombyx mori* (Shimizu, I. *et al.*, 1981) and *Manduca sexta* (Brown, P. and Schwemer, J., personal communication; White, R. and Bennett, R., unpublished). Vision is restored in these insects when a source of retinal is added to the diet.

The photosensitivity of blowflies (*Calliphora erythrocephala*) fed as larvae on heart muscle, which lacks vitamin A, is lower in comparison with those fed vitamin A-rich liver (Razmjoo, S. and Hamdorf, K. 1976). The amount of rhodopsin in the eyes of the heart-fed animals is reduced to about the same extent as the sensitivity (Figs 15 and 16; see section 7.3). Carotenoid-deprived *Manduca* also synthesize less rhodopsin (Brown, P. and Schwemer, J., personal communication; White, R., Brown, P. and Sanentz, L., unpublished). In their experiments with *Musca*, Goldsmith, T. *et al.* (1964) took particular care to prevent microbial contamination of the carotenoid-free diet. After several generations, the visual threshold was raised by at least 4 log units. Assuming that sensitivity is proportional to the rhodopsin content of the rhabdomeres in *Musca* as it is in *Calliphora*, Hamdorf, K. (1979) calculated that only one in ten of the microvilli of these severely deficient flies contained a rhodopsin molecule.

Carotenoid deficiency also affects the ultrastructure of receptor cells. There is a reduction in the density of P face particles seen in freeze-fracture preparations of rhabdomere microvilli (*Calliphora erythrocephala*: Boschek, C. and Hamdorf, K. 1976; Schwemer, J. and Brown, P., personal communication; *Drosophila melanogaster*: Harris, W. *et al.*, 1977; Pak, W. *et al.*, 1980). Presumably the particles are rhodopsin molecules (see section 6.2). In addition, carotenoid deficiency results in the accumulation of smooth endomembrane cisternae in photoreceptor cells (White, R. and Jolie, M. 1966; Brammer, J. and White, R. 1969).

4.3 Bleaching experiments

Direct evidence that the chromophore is retinal is available for a few insect rhodopsins. Bleaching experiments of the sort described in section 2.4 have demonstrated the formation of retinaldehyde oxime from the reaction of visual pigments with hydroxylamine (*Aedes aegypti*: Brown, P. K. and White, R. H., 1972; *Galleria mellonella*: Goldman, L. J. *et al.*, 1975; *Manduca sexta*: Brown, P. K. and Schwemer, J., personal communication; *Ascalaphus macaronius*: Schwemer, J. *et al.*, 1971). Reduction with potassium borohydride releases retinol (*Aedes*: Brown, P. K. and White, R. H., 1972).

4.4 *cis-trans*-Isomerization

Only the chromophore of the ultraviolet sensitive rhodopsin (P345) of the owlfly *Ascalaphus macaronius* has been well characterized (Paulsen, R. and Schwemer, J. 1972). Retinal was measured by the Carr–Price reaction in buffer extracts of *Ascalaphus* eyes. It was bound exclusively to insoluble debris, presumably fragments of membrane. Both the all-*trans* and 11-*cis* isomers were identified by thin-layer chromatography. Other isomers were not present, nor was retinol. When P345 was denatured with Ag^{2+} in darkness — a procedure similar to one used for releasing the chromophore from vertebrate and cephalopod rhodopsins while avoiding the isomerizing effects of light (Hubbard, R. *et al.*, 1971) — only 11-*cis* retinal was recovered. The chromophore extracted in this way from *Ascalaphus* rhodopsin promoted the conversion of purified cattle opsin to rhodopsin, $\lambda_{max} = 498$ nm, a very specific assay for the 11-*cis* isomer. Therefore, the chromophore of *Ascalaphus* rhodopsin is 11-*cis* retinal, as it is in the other rhodopsins whose

chromophore has been characterized.

Irradiation of *Ascalaphus* visual pigment with ultraviolet light establishes a photoequilibrium mixture containing one-third P345 and two-thirds M475, whereas blue light drives the photopigment entirely into the rhodopsin state (Fig. 8). Denaturation after ultraviolet irradiation yielded only a third of the 11-*cis* retinal released after blue irradiation. Since only the 11-*cis* and all-*trans* isomers were present, the chromophore of the metarhodopsin must be all-*trans* retinal. Hence actinic light isomerizes the chomophore of *Ascalaphus* rhodopsin from 11-*cis* to all-*trans* as in the visual pigments of cephalopods and vertebrates.

4.5 Chromophoric site

Ascalaphus metarhodopsin is pH-sensitive. M475 is the acid form. In digitonin solution it can be converted reversibly with pK 9.2 to alkaline metarhodopsin, $\lambda_{max} = 380$ nm (Schwemer, J. *et al.*, 1971; Hamdorf, K. *et al.*, 1973b). It is evident that acid M475 will predominate at physiological pH, and this was shown by MSP measurements of *Ascalaphus* retinas. Cephalopod metarhodopsins also shift, both in extract and *in situ*, between acid (λ_{max} from 497 to 516 nm) and alkaline ($\lambda_{max} = 380$ nm) forms with pK values ranging from 6.3 to 9.1 (Brown, P. and Brown, P. 1958; Hubbard, R. and St. George, R. 1958; Hamdorf, K. *et al.*, 1968; Schwemer, J. 1969; Hara, T. and Hara, R. 1973). The spectral shifts of acid–base-sensitive intermediates result from the protonation/deprotonation of the Schiff's base linkage between retinal and opsin (Ebrey, T. and Honig, B. 1975; Ostroy, S. 1977). The different pK values presumably reflect the accessibility of the binding sites, the higher pK values, as in *Ascalaphus* metarhodopsin, indicating better-protected sites. In some other insects, the mosquito *Aedes aegypti* (Brown, P. and White, R. 1972) and the moths *Galleria mellonella* (Goldman, L. *et al.*, 1975) and *Manduca sexta* (Brown, P. and Schwemer, J., personal communication), the metarhodopsins are unaffected by pH. In this regard they are similar to those of other arthropods, e.g. the lobster *Homarus* (Wald, G. and Hubbard, R. 1957) and the spider crab *Libinia emarginata* (Hays, D. and Goldsmith, T. 1969). The metarhodopsins of some insects are also quite resistant to hydrolysis by hydroxylamine

(Goldman, L. *et al.*, 1975). For instance, we (White, R. *et al.*, 1983) have found that even with 1 M hydroxylamine, more than 2 h under bright yellow light may be required to bleach *Manduca* meta-rhodopsin in glutaraldehyde-fixed retinas. It seems that the binding site of the chromophore is generally better sequestered in insects and other arthropods than it is in cephalopods (Goldman, L. *et al.*, 1975).

Evaluation of the significance of metarhodopsin pH sensitivity in extracts must include the possibility that removing it from its native membrane alters the visual pigment so as to increase the exposure of the binding site. We know, for instance, that the metarhodopsins of the photopigments of *Deilephila elpenor* undergo thermal bleaching more readily in digitonin extracts than in the native membrane (Schwemer, J. and Paulsen, R. 1973). Early receptor potential action spectra show that the alkaline form of metarhodopsin exists in the intact retina of the squid (Hagins, W. and McGaughy, R. 1967) but there is no similar evidence that alkaline meta-rhodopsin can form in *Ascalaphus* photoreceptor membrane.

4.6 Intermediates of transformation

As outlined in section 1.2, vertebrate and squid photopigments pass through a sequence of inter-mediate states after absorbing light (Fig. 2). These intermediates have been spectrally defined in low-temperature experiments in which their rates of decay are slowed (Ebrey, T. and Honig, B. 1975). Intermediates of transformation have been iden-tified by similar experiments with *Ascalaphus* rhodopsin (Fig. 9). A photoproduct with $\lambda_{max} = 375$ nm is produced by ultraviolet irradiation of P345 at $-50°$. This intermediate was designated lumirhodopsin (L375) in accordance with the ter-minology adopted for vertebrate and cephalopod rhodopsins. When the temperature is raised to $-15°$ in darkness, L375 decays to M475 in which the chromophore is all-*trans*. As L375 → M475 is a dark reaction, the chromophore of L375 probably is also the all-*trans* isomer. Since protein conforma-tion cannot change at $-50°$, L375 can differ from P345 only in the configuration of the chromophore. The 30 nm spectral shift must reflect altered interac-tion between the chromophore and the opsin result-ing from isomerization. When L375 is warmed, the

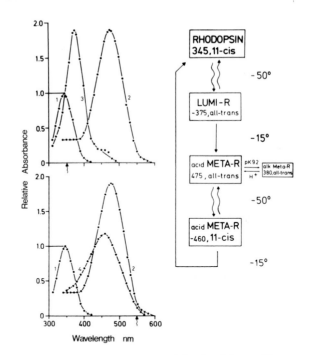

FIG. 9. Light and dark reactions of the ultraviolet-sensitive rhodopsin (P345) of *Ascalaphus macaronius*. *Left*: Absor-bance spectra of P345 (1), acid metarhodopsin (2), lumir-hodopsin (3), and 11-*cis* metarhodopsin (4). Spectra of inter-mediates 3 and 4 were calculated from difference spectra. Wavy arrows on abscissa show wavelengths used to form intermediates 3 and 4. *Right*: Scheme of photoreactions (wavy arrows) and dark reactions (straight arrows). Alkaline metarhodopsin is not present in measurable amounts at physiological pH. Its spectrum is similar to curve 3 (from Hamdorf, K. *et al.*, 1973b).

opsin relaxes into its metarhodopsin configuration.

Illumination of M475 at $-50°$ with 550 nm light produces an intermediate at about 460 nm. Warmed in darkness, this intermediate relaxes to P345. Hence the photoreaction, M475 → M460, must be the isomerization of the chromophore from all-*trans* to 11-*cis* retinal. With the 11-*cis* retinal in place, the opsin flips to its rhodopsin configuration. Similar intermediates of regeneration have been characterized in cephalopods (Suzuki, T. *et al.*, 1972, 1973; Azuma, K. *et al.*, 1975).

4.7 Isorhodopsin does not form

The irradiation of extracts of vertebrate rhodopsin at temperatures low enough to stabilize meta-rhodopsin I establishes a photoequilibrium among several forms of the visual pigment. One is iso-rhodopsin, $\lambda_{max} = 387$ nm, a form similar to rhodopsin except that it contains retinal as the 9-*cis* isomer (Hubbard, R. *et al.*, 1971). The formation of isorhodopsin seems to be inhibited *in vivo*. That this is also true for insects is indicated by the isosbestic point in photoconversions between rhodopsin and metarhodopsin (Brown, P. and White, R. 1972; Hamdorf, K. and Rosner, G. 1973; Stavenga, D. *et al.*, 1973). Apparently there is a mechanism that restricts photoisomerization to all-*trans* and 11-*cis* retinal (Goldman, L. *et al.*, 1975).

4.8 Energy transfer from protein to chromophore

Energy transfer from opsin to chromophore has been demonstrated in vertebrate visual pigments. Light absorbed by the γ-band at 280 nm initiates bleaching (Kropf, A. 1967). Goldsmith, T. and Fernandez, H. (1968) extended the measurement of the spectral sensitivity of *Musca domestica* into the ultraviolet, finding a shoulder at 280 nm. The authors suggested that this response to very short wavelengths depends on energy transfer from the opsin. Since light at wavelengths shorter than 300 nm is not present in the normal environment of insects, this has no physiological significance.

4.9 Fluorescence

Cattle rhodopsin fluoresces weakly (Guzzo, A. and Pool, G. 1968) or not at all (Busch, G. *et al.*, 1972). The same is true of fly rhodopsin, but there is evidence indicating that metarhodopsin may be fluorescent (Stark, W. *et al.*, 1979; Franceschini, N. *et al.*, 1981b). The rhabdomeres of cells R1–6 (see section 8.1) in the ommatidia of several species (*Musca, Drosophila, Calliphora, Sarcophaga* and *Eristalis*) fluoresce red ($\lambda > 620$ n) under blue excitation (Table 5). The fluorescence increases and decreases under conditions of illumination that increase and decrease the amount of metarhodopsin in the rhabdomeres. However the fluorescence develops more slowly than metarhodopsin forms. It takes about 100 s to achieve maximal emission with illumination that brings metarhodopsin to its maximum level in 0.1 s (Franceschini, N. *et al.*, 1981b). It has been suggested that the fluorescence arises from an altered form of metarhodopsin produced under the

intense ($\geq 10^{17}$ photons cm^{-2} s^{-1}) blue light used for excitation (Stavenga, D. and Franceschini, N. 1981).

While the association of rhabdomere fluorescence with metarhodopsin is somewhat uncertain in the fly, it has been firmly established in the crayfish (Cronin, T. and Goldsmith, T. 1981). Dark-adapted crayfish rhabdoms are at best only weakly fluorescent. Red fluorescence develops after they have been exposed to light, increasing in parallel with the formation of metarhodopsin. Furthermore the excitation spectrum matches the absorption spectrum of metarhodopsin, and both are shifted to shorter wavelengths at acid pH. Hence the fluorescence derives from the chromophoric site of metarhodopsin. Fluorescence emission peaks at 670 nm at neutral pH. The quantum efficiency is $1.6 \pm 0.4 \times 10^{-3}$. Although the basis of metarhodopsin fluorescence is not known, the authors note that all-*trans* retinal becomes weakly fluorescent when it is hydrogen-bonded (Takemura, T. *et al.*, 1978).

4.10 Buffer-soluble retinal–proteins

No retinol (vitamin A), only retinal, was found in the eyes of *Ascalaphus* (Paulsen, R. and Schwemer, J. 1972). Both, however, are present in the heads of honeybees, as well as an NADH-dependent enzyme that reduces retinal to retinol (Goldsmith, T. and Warner, L. 1964). In addition, Goldsmith, T. (1958a,b) extracted a retinal–protein pigment from honeybee eyes. Although the absorption spectrum of this pigment, $\lambda_{max} = 440$ nm, resembles the action spectrum of the violet receptors of honeybees, Goldsmith, T. (1970, 1972) came to the conclusion that it is not a visual pigment; for one reason, unlike all known rhodopsins, it is buffer-soluble.

It is now known that at least two buffer-soluble retinal–protein complexes with similar characteristics can be obtained from honeybee eyes (Pepe, I. and Cugnoli, C. 1980; Pepe, I. *et al.*, 1982). These pigments have been isolated by incubating crude extracts with tritiated retinol. A labeled fraction containing three proteins is then recovered by preparative gel electrophoresis. Two of the proteins — protein B, MW 27,000 ± 1000 and protein C, MW 24,000 ± 1000 — are complexed with labeled retinal. The third, protein A, either has no chromophore or it is removed during preparation. Since the retinal chromophore of proteins B and C is at least partly derived from the labeled retinol introduced during the initial incubation, the crude extract must contain a retinol dehydrogenase (Pepe, I. and Cugnoli, C. 1980).

Both proteins B and C absorb at 440 nm, and both are converted to intermediates at 370 nm by yellow light. Protein C is reconverted to its 440 nm form during a few minutes in darkness. Both proteins B and C shift to an alkaline form, $\lambda_{max} = 365$ nm, at high pH. The transformation is reversible with pH 8.4.

Protein B, about which more is known, is attacked by hydroxylamine in darkness with the production of retinal oxime. It is stable to cyanoborohydride in darkness, but the chromophore is reduced on site with irradiation. The photoproduct absorbs at 330 nm; presumably a retinyl-protein is formed.

The spectra of the acid and alkaline forms of the pigment are, respectively, characteristic of protonated and unprotonated Schiff's bases. Its reaction with hydroxylamine and borohydride also supports the conclusion that the chromophore is linked with the protein as it is in visual pigments. However, when the chromophore was removed from the 370 nm form of the protein it was found to be 11-*cis* retinal. By implication, the chromophore of P440 is the all-*trans* isomer (Pepe, I. *et al.*, 1982).

Several roles have been suggested for these proteins. They might be involved in the turnover of visual pigments, either as precursors or degradation products. They might be retinal transport proteins. However, they most closely resemble the retinochrome pigments of cephalopods (Hara, T. and Hara, R. 1973). They are similar in the stereospecificity of retinal binding, pH sensitivity, molecular weight, and most particularly, with respect to the all-*trans* to 11-*cis* photoconversion of the chromophore. They differ from the cephalopod retinochromes in their solubility; retinochromes are localized in cytoplasmic membranes within cephalopod photoreceptors, and can be extracted only with detergents.

By promoting the photoisomerization of all-*trans* to 11-*cis* retinal, retinochromes are thought to provide 11-*cis* retinal for the synthesis of rhodopsin.

Schwemer, J. (1979) has shown that 11-*cis* retinal is required for the synthesis of rhodopsin in flies, and he suggested that it is provided by an unidentified retinochrome-like pigment.

5 REGENERATION OF RHODOPSIN

As outlined in the Introduction (section 1.3), photoregeneration is the principal means by which rhodopsin is restored in illuminated insect photoreceptors. Under steady illumination a photoequilibrium will be established between rhodopsin and metarhodopsin in which the visual pigment flips back and forth between its two stable configurations. The relative amounts of rhodopsin and metarhodopsin in the photosteady state will depend on their absorption coefficients, their relative quantum efficiencies, and the spectral quality of the light. The quantum efficiencies of metarhodopsin and rhodopsin are thought to be similar; Stark, W. and Johnson, M. (1980) estimated the ratio of the former to the latter at about 0.71 in *Drosophila*. The amount of each state of the visual pigment in a particular photoequilibrium established by monochromatic light has been found in practice to depend mainly on their respective absorption coefficients at that wavelength. There is good agreement between photoequilibria calculated on that basis, and concentrations of rhodopsin and metarhodopsin, measured in extracts and by MSP (Gogala, M. *et al.*, 1970; Hamdorf, K. *et al.*, 1972, 1973a,b; Hamdorf, K. and Schwemer, J. 1975; Schwemer, J. *et al.*, 1971; Paulsen, R. and Schwemer, J. 1972; Hamdorf, K. and Gogala, M. 1973; Hamdorf, K. 1979).

Illumination of the visual pigment with monochromatic light at the wavelength of the isosbestic point establishes a photoequilibrium containing 50 per cent rhodopsin. Light at a wavelength more strongly absorbed by rhodopsin than metarhodopsin will shift the photoequilibrium to favor metarhodopsin. Rhodopsin will be favored by light that is preferentially absorbed by metarhodopsin. The molecular populations of both stable states in such a bistable pigment system approach their steady-state values under steady illumination exponentially with the same rate constant. Photoequilibria are independent of initial

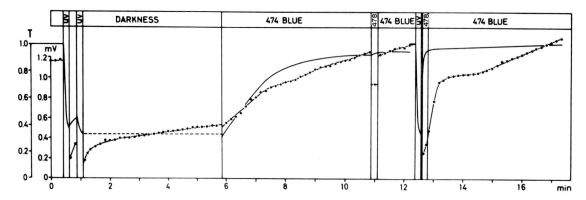

Fig. 10. Simultaneous measurement of response amplitude (dotted line) and metarhodopsin concentration (solid line) in retina of *Ascalaphus macaronius*. Transmission (T) was measured at 474 nm. As M475 increases T decreases. Ultraviolet light decreases T because P345 → M475. In the blue light of the measuring beam M475 → P345 and T increases. Hence the solid line can be taken as a measure of P345. From minutes 1–6 the measuring beam was turned off to demonstrate that during the slight increase in response the metarhodopsin content remained constant. During minutes 12–13, ultraviolet adaptation was followed immediately by bright blue irradiation (478 nm) that completely regenerated P345. The disparity between pigment concentration and response curves during minutes 1–6, and especially minutes 13–17, indicate periods of "neural" adaptation (from Hamdorf, K. and Schwemer, J. 1975).

population values and light intensity (Hochstein, S. *et al.*, 1978; Minke, B. *et al.* 1978; Hamdorf, K. 1979).

The time constant for the transition from rhodopsin to metarhodopsin measured by MSP in *Drosophila* was found to be approximately 1 ms at 5° (extrapolated to 1.25×10^{-1} ms at 25°). The rate of photoregeneration could not be determined but is considerably faster (Kirschfeld, K. *et al.*, 1978).

5.1   Photoregeneration in *Ascalaphus macaronius*

The physiological role of photoregeneration has been documented for the ultraviolet receptors of *Ascalaphus*. Since the α-bands of *Ascalaphus* rhodopsin (P345) and metarhodopsin (M475) do not overlap much, a range of photoequilibria can be established with rhodopsin contents from 100% to 30% (Fig. 8). Ultraviolet light decreases the ERG response amplitude, and there is only a slight recovery during the first minutes of subsequent dark-adaptation (Fig. 10). Simultaneous ERG and MSP measurements indicated that the photosteady state remains unchanged during that short period in darkness, so the small rise in sensitivity must result from "neural adaptation" (see section 7.3). Upon exposure to blue light at 474 nm the response returns to maximum amplitude, in concert with the photoregeneration of P345 (Fig. 10). Direct proportionality between the reciprocal of the intensity of

the blue regenerating light and the rate of recovery indicates that photoconversion is a first-order reaction (Schwemer, J. *et al.*, 1971; Hamdorf, K. and Gogala, M. 1973).

The rate of recovery of response amplitude was measured as a function of wavelength with equal irradiations from 400 nm to 590 nm. Comparing all rates to the maximal rate at 475 nm yields a curve that corresponds to the absorption spectrum of M475. This curve is the relative sensitivity curve for photoregeneration of P345 from M475. Response amplitude must therefore be directly proportional to the amount of P345 in the photoreceptors (Hamdorf, K. *et al.*, 1971; Schwemer, J. *et al.*, 1971; Hamdorf, K. and Schwemer, J. 1975; Hamdorf, K. 1979).

In order to relate these observations to the function of the eye in its normal environment, animals were exposed to mixtures of blue and ultraviolet light (Hamdorf, K. and Gogala, M. 1973). After ultraviolet alone, the response amplitude remains depressed during subsequent dark-adaptation (Fig. 11). After irradiation with blue and ultraviolet together, sensitivity rises to a level that depends on the ratio of blue to ultraviolet quantum flux. That is, sensitivity depends on the particular photosteady state established by the mix of wavelengths. The skylight to which the diurnal *Ascalaphus* is exposed has an emission maximum in the blue and a lesser peak in the ultraviolet. The high content of blue

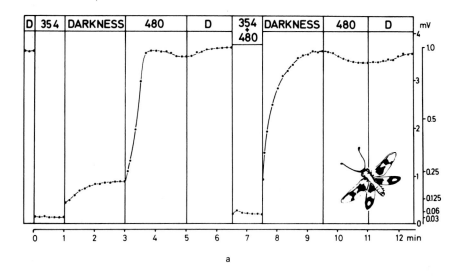

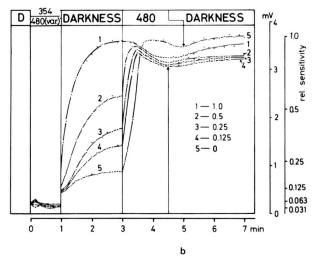

b

FIG. 11. Relationship of sensitivity to adapting wavelength in *Ascalaphus macaronius*. **a**: Amplitude of ERG responses to low-intensity ultraviolet test flashes during darkness, bright ultraviolet irradiation (354 nm), bright blue irradiation (480 nm), and irradiation with a mixture of 345 nm and 480 nm; **b**: sensitivity following irradiation with different mixtures of 354 nm and 480 nm (354 nm held constant). Curves 1–5: 480/354 = 1.0, 0.5, 0.25, 0.125 and 0 (from Hamdorf, K. and Gogala, M. 1973).

wavelengths in daylight, and the relatively high extinction of M475 maintains photoequilibria with high levels of P345. Furthermore, the primary pigment cells determining the ommatidal aperture contain pale yellow pigment granules that screens ultraviolet but transmits blue light (Langer, H. 1975; Schneider, I. *et al.*, 1978). Consequently, normal photoequilibria contain as much as 90% P345. *Ascalaphus* receptors give maximal response at that rhodopsin level (Hamdorf, K. and Gogala, M. 1973; Hamdorf, K. *et al.*, 1978).

5.2 Photoregeneration in flies

Similar results have been obtained with the higher flies (suborder Cyclorrapha), such as *Drosophila melanogaster* and *Calliphora erythrocephala*. They have rhodopsins that absorb in the blue-green, with metarhodopsins that are shifted to longer wavelengths, e.g. P480 and M570 in *Drosophila* (see section 7.8; Table 3). The characteristic red eye color of these flies appears to be an adaptation for maintaining high rhodopsin levels. The red accessory pigments absorb the shorter visible wavelengths but are transparent to longer wavelengths (Goldsmith, T. 1965; Langer, H. 1967; Stavenga, D. *et al.*, 1973). The scattered red light that bathes the photoreceptors under normal illumination is preferentially absorbed by metarhodopsin so that rhodopsin is favored.

5.3   Dark regeneration

Substantial levels of rhodopsin are maintained in illuminated insect photoreceptors by photoregeneration. Are there also mechanisms for the "dark regeneration" of rhodopsin analogous to the enzymatically mediated visual cycle of vertebrates?

No dark regeneration was found in the larval ocelli of intact *Aedes aegypti*. Previously established photoequilibria did not change during an hour in darkness (Brown, P. and White, R. 1972). However, these animals had seen no light prior to the brief illumination that established the photoequilibria. The photoreceptors of mosquito larvae reared in darkness, as these were, show morphological peculiarities, and periods of light-adaptation seem to be required to maintain certain cellular activities (White, R. and Sundeen, C. 1967; White, R. 1967, 1968). Perhaps no dark regeneration occurred because the normal mechanism had failed to develop.

Dark regeneration was monitored for several days in the moth *Galleria mellonella* by Goldman, L. *et al.* (1975). Animals were dark-adapted for 3–18 h before exposure to full sunlight for 2–4 min. The moths were then returned to darkness and the ratio of rhodopsin to the total amount of visual pigment (rhodopsin plus metarhodopsin), was periodically measured in retinal slices by MSP. The rate of recovery of rhodopsin was very slow, leveling off at about 80% of the usual dark-adapted level after 5 days. As with the mosquito measurements, however, the *Galleria* experiment involved nonphysiological conditions. The exposure of a fully dark-adapted eye to sunlight would be rare in nature. That the experiment was done in this way because sufficient conversion of rhodopsin to metarhodopsin could not be accomplished in light-adapted eyes with their protective ommochrome pigment screens extended, reinforces the sense of extreme circumstances. We know of one arthropod that normally experiences only dim light, the Nor-

way lobster (*Nephrops norvegicus*), whose photoreceptors are damaged at even moderate light levels (Loew, E. 1976, 1980).

Although dark regeneration has been reported in *Calliphora* with a half-time of about 30 min (Stavenga, D. *et al.*, 1973), other investigators have found no recovery of rhodopsin in darkness for periods of more than an hour (Rosner, G. 1975; Schwemer, J. reported in Hamdorf, K. 1979). Pak, W. and Lidington, K. (1974) measured dark regeneration in *Drosophila* with a half-time of 6 h.

Where rhodopsin is seen to increase very slowly in darkness — over a period of hours — *de novo* synthesis of visual pigment and turnover of photoreceptor membrane is likely to be involved. Morphological evidence indicates that rapid turnover of rhabdomere membrane is stimulated by light in mosquito photoreceptors (White, R. 1968; White, R. and Lord, E. 1975; Brammer, J. *et al.*, 1978) and in other arthropods (e.g. Eguchi, E. and Waterman, T. 1976). The rate at which membrane is added to the rhabdomeres of larval mosquito ocelli in darkness immediately after a period of illumination is sufficient to double the amount of membrane in 4 h. Therefore, dark regeneration based on membrane replacement could be relatively rapid in insects. Turnover of photoreceptor membrane in other arthropods occurs in diurnal cycles, suggesting reorganization at dawn and dusk (e.g. in the crab *Leptograpsus variegatus*: Stowe, S. 1980; in the spider *Dinopus subrufens*: Blest, A. 1978). Only the bulk turnover of rhabdomere membrane has been studied, not the assembly and degradation of particular constituents. Thus we can say nothing specifically about the replacement of visual pigments.

The lobster *Homarus americanus* has a means for relatively rapid dark regeneration (Bruno, M. *et al.*, 1977). Metarhodopsin was found to accumulate *in vivo* in brightly illuminated eyes at 1°. However, at 25° no metarhodopsin was measured after an hour of illumination even with red light that should favor its formation. Thus, the lobster employs a temperature-sensitive mechanism that promotes the recovery of rhodopsin independent of photoregeneration. It appears that a photosteady state is not established under physiological conditions. Regeneration was also followed in darkness after a flash that converted three-quarters of the rhodopsin to metarhodopsin. Rhodopsin was regenerated with half-times of about 25 min at 22° and 55 min at 15°. Since dark recovery was not preceded by the loss of visual pigment, either the chromophore of metarhodopsin is isomerized — perhaps enzymatically — from all-*trans* to 11-*cis* in the absence of light, or it is replaced.

Rhodopsin regenerates at a similar rate during dark-adaptation in the octopus *Eledone moschata* (Schwemer, J. 1969). The rhodopsin content rises in darkness at 20° from its photoequilibrium level by about 10% within 10 min, returning to 95% rhodopsin over a period of 2 h. Fairly rapid dark regeneration has also been measured in the butterfly *Aglais* (Stavenga, D. 1975).

Recently Bernard, G. (1981b) has found unexpected complications in the dark recovery of butterfly eyes after an orange flash that converts rhodopsin (P530) to metarhodopsin (M495). During subsequent darkness M495 decays rapidly while P530 recovers more slowly and with more complex kinetics. These results show that metarhodopsin is not a stable intermediate in some species.

From the scant evidence available, it appears that rates and modes of dark regeneration vary widely among invertebrates. In no instance, however, is the mechanism well understood.

6 PHOTORECEPTOR MEMBRANE

Several lines of evidence show that invertebrate visual pigments are components of the rhabdomeric membranes. Rhodopsin has been localized to the rhabdomeres of many photoreceptors by MSP measurements. Both the ultrastructure of the rhabdomere and its dichroic properties indicate that photopigments are associated with the microvillous membranes. Since invertebrate rhodopsins, like those of vertebrates, can be brought into solution only by detergent extraction, they must be membrane-bound.

6.1 Ultrastructure

The hexagonally packed microvilli of arthropod rhabdomeres are usually arrayed with their axes perpendicular to the optical axis of the photoreceptor unit containing them. The diameters of the microvilli range between 40 and 80 nm. Length is variable.

Rhabdomeric membrane prepared by the conventional procedures for transmission electron microscopy — glutaraldehyde fixation, osmium postfixation and lead staining — appears as a 75 Å unit membrane with a thicker or denser appearing cytoplasmic leaflet (Fig. 12). With respect to its asymmetry, thought to reflect the vectorial arrangement and distribution of protein and lipid components, rhabdomeric membrane resembles other plasma membranes. In contrast, the outer segment discs of vertebrate photoreceptors have symmetric unit membranes (Olive, J. 1980).

There is sometimes a suggestion of a globular substructure in the appearance of glutaraldehyde-fixed rhabdomeric membrane. A similar fine structure is seen in vertebrate photoreceptor membrane after osmium fixation. It is not known whether the globular appearance is related to the native molecular organization of the membrane (Olive, J. 1980). It is worth noting, however, that the apparent globular "subunits' of the membrane are about the size of rhodopsin molecules, estimated at 40–50 Å from their molecular weights (Hamdorf, K. 1979). Adjacent microvilli may be joined, as in the mosquito *Aedes* (White, R. 1967), by cross-bridges. At the core of the microvillus is a cytoskeletal scaffolding that probably plays an important role in the organization of the membrane, but about which little is known (see section 6.10).

Some aspects of the molecular morphology of rhabdomeric membrane have been inferred from freeze-fracture electron microscopy. In this technique, retinas are frozen and broken at low temperature, and platinum-shadowed carbon replicas are made of the exposed fracture surfaces. A freeze-fracture replica captures the topography of these surfaces at macromolecular dimensions. The fracture tends to follow along membranes, and several studies have supported the interpretation that it runs preferentially down the middle of the phospholipid bilayer (Pinto da Silva, P. and Branton, D. 1970). Such a cleavage exposes the "inside" of the membrane as two opposite fracture faces, generally designated the P and E faces. The P (for protoplasmic) face is the "inside surface" of the bilayer leaflet lying against the cellular protoplasm; the E (for extracellular) face is the "inside surface"

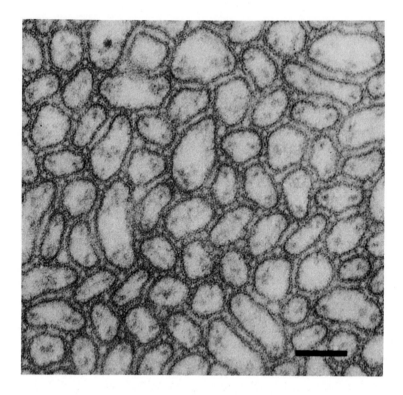

FIG. 12. Cross-sectioned rhabdomeric microvilli from a larval ocellus of *Aedes aegypti*. Glutaraldehyde-fixed, osmium- and lead-stained photoreceptor membrane has the appearance of a unit membrane. In some regions there is the suggestion of a globular substructure to the membrane. The cytoskeletons of the microvilli are not well preserved by conventional fixation, appearing here as diffuse clumps of material. The magnification bar represents 100 nm.

of the leaflet at the extracellular side of the membrane. These surfaces typically are bossed with "particles" that have been shown in several instances to be membrane protein (Bullivant, S. 1977).

The P faces of normal rhabdomeric membranes (Fig. 13) usually appear densely covered with particles (planaria: Röhlich, P. 1981; squid: Cincotta, D. and Hubbell, W. 1977; crayfish: Eguchi, E. and Waterman, T. 1976; Fernandez, H. and Nickel, E. 1976; the honeybee drone *Apis mellifera*: Perrelet, A. *et al.*, 1972; the ant *Myrmecia gulosa*: Nickel, E. and Menzel, R. 1976; *Calliphora erythrocephala*: Boschek, C. and Hamdorf, K. 1976; *Drosophila melanogaster*: Harris, W. *et al.*, 1977; *Musca domestica*: Chi, C. and Carlson, S. 1979). Sometimes the P face particles are arranged in rows pitched at an angle to the axis of the microvillus (e.g. Fernandez, H. and Nickel, E. 1976).

Rhabdomeric E faces seem more variable. Generally they present a smooth surface studded with scattered particles. However, in *Musca domestica* (Chi, C. and Carlson, S. 1979) and *Myrmecia* (Nickel, E. and Menzel, R. 1976) surfaces that have been interpreted as E faces are covered with compact ranks of particles.

6.2   P face particles and rhodopsin

There is good evidence associating the P face particles with rhodopsin molecules. Their numbers decline when animals are fed on diets deficient in carotenoids (Boschek, C. and Hamdorf, K. 1976; Harris, W. *et al.*, 1977; Pak, W. 1980; Brown, P., Schwemer, J. and Miller, K., personal communication). P faces then appear smooth with scattered particles. The same is true of *Drosophila* mutants designated *nina*, in which rhodopsin is reduced (Shinz, R. *et al.*, 1982; Larrivee, D. *et al.*, 1981). Removal of rhodopsin from crayfish rhabdomeres with digitonin also results in the loss of P face particles. Similar results have been obtained with the

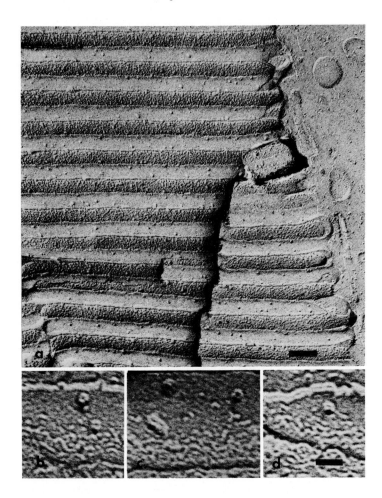

FIG. 13. Freeze-fractured glutaraldehyde-fixed rhabdomeres of *Manduca sexta*. P faces are densely particulate. E faces are smoother with scattered particles. The proximal ends of the microvilli and the cytoplasm of the photoreceptor cells are seen at the right in **a**. E face particles are shown at high magnification in **b**, **c**, and **d**. They appear as rosettes of smaller particles, often with a central depression. The magnification bar for **a** represents 100 nm; for **b**, **c**, and **d**, 20 nm.

photoreceptor membranes of vertebrates (Jan, L. and Revel, J. 1974; Mason, W. *et al.*, 1974), and particles appear in artificial membranes when they are reconstituted with rhodopsin (Darszon, A. *et al.*, 1979).

Although both P-face particles and rhodopsin are reduced in carotenoid deficiency and in the *nina* mutants of *Drosophila*, the particle density remains somewhat higher than the rhodopsin concentration. Analysis of the remaining retinal membrane protein by SDS electrophoresis indicated no excess of opsin over the measured rhodopsin (Pak, W. *et al.*, 1980). Hence the remaining particles apparently are not visual pigment molecules lacking the chromophore. Pak, W. *et al.* (1980) have estimated

that as much as 30–40% of the protein in the rhabdomere membrane is something other than photopigment. This corresponds to 10–20% nonrhodopsin protein in vertebrate outer segments (Daemen, F. 1973).

The diameters of the P face particles measured in various preparations have ranged from 70 Å to over 100 Å; most averaged around 80 Å. Particle densities measured in normal rhabdomeres have ranged from 900 to 7000 μm^{-2}. Such data are difficult to evaluate because the apparent particle size is variably enlarged by contamination of the fracture surface, and during the replication process. Furthermore, the particles may represent aggregates of protein rather than single molecules. The diameter of a rhodopsin molecule, estimated from its molecular weight (Hamdorf, K. 1979), is about half that of the P face particles. However in the crayfish (Fernandez, H. and Nickel, E. 1976), and in flies (Brown, P., Schwemer, J. and Miller, K., personal communication), the particle density in the membrane corresponds fairly well with the concentration of rhodopsin molecules in the rhabdomere calculated from MSP measurements.

6.3 E face particles

Little attention has been paid to the scattered particles usually seen on the E faces of fractured rhabdomeres. In preparations giving sufficient resolution, they appear to be groups of smaller particles about 50 Å in diameter (Fig. 13). They are reminiscent of enzyme complexes and ion channels (Henderson, R. and Unwin, P. 1975; Heuser, J. and Salpter, S. 1979; Klingenberg, M. 1981), but there is no direct evidence as to their identity or function.

6.4 Localization of rhodopsin in the membrane

Since the P face particles appear to be mainly rhodopsin, the freeze-fracture data have been interpreted as showing that the visual pigment is localized in the inner cytoplasmic leaflet of the phospholipid bilayer (Hamdorf, K. and Schwemer, J. 1975). However, this interpretation depends on the assumption that rhabdomeric membrane fractures down the center as in those membranes where the cleavage plane has been experimentally localized. Recently, Sjöstrand, F. and Kreman, M. (1978) have questioned the conventional interpretation with respect to vertebrate photoreceptor membrane, suggesting that the fracture plane might run along the surface of membranes with high protein to phospholipid ratios. As yet, no attempts have been made to determine exactly where the fracture runs in rhabdomeric microvilli.

Diverse evidence shows vertebrate rhodopsin molecules to be elongated perpendicular to the membrane. They are exposed on the cytoplasmic side of the membrane to biochemical probes, and probably are accessible from the other side also (Olive, J. 1980). There are as yet no similar data bearing on the molecular shapes of insect rhodopsins, nor on whether these molecules also span the membrane. Some current models assume that it does (Hamdorf, K. 1979, 1980). Vertebrate opsin bears two oligosaccharides near the N-terminal end. It is thought that the sugar moieties are situated at the intradiscal side of the membrane, corresponding to the extracellular side of a plasma membrane (Olive, J. 1980). Carbohydrate components of insect rhodopsins have not yet been identified.

6.5 Membrane lipids

The molar ratio of phospholipid to rhodopsin is 60:1 in the rhabdom membrane of the octopus *Eledone* (Paulsen and Zinkler, quoted in Hamdorf, K. 1979). Similar ratios are found in vertebrate photoreceptor membranes (Daemen, F. 1973). Hamdorf, K. (1979, 1980) has calculated that each rhodopsin molecule in a rhabdomeric microvillus should be surrounded by a ring of about 28 phospholipid molecules in each of the membrane leaflets. The phospholipid composition of rhabdomeric membranes is similar to that of vertebrate discs (Daemen, F. 1973). In *Calliphora erythrocephala*, the ratio of phosphatidylethanolamine to phosphatidylcholine to phosphatidylserine $\simeq$ 5:2:1 (Zinkler, D. 1975); in *Deilephila elpenor*, 4:4:1 (Zinkler, D. 1975); in *Limulus*, 2:4:1 (Benolken, R. *et al.*, 1975), and in *Eledone*, 4:4:1 (Paulsen and Zinkler, quoted in Hamdorf, K. 1979). Two to three times more cholesterol is found in rhabdomeric membranes than in disc membranes; the phospholipid to cholesterol ratio of rhabdomeres is about 2:1 (Paulsen and Zinkler, quoted in Hamdorf, K. 1979).

6.6 Orientation of vertebrate rhodopsin

The retinal molecule is a dipole that maximally absorbs polarized light whose e vector is aligned with its polyene backbone. Absorbance is minimal when the retinal molecule is illuminated end-on. Since its chromophore is dichroic, the spatial orientation of rhodopsin in photoreceptor membranes can be determined from measurements with plane polarized light. For instance, the rod outer segments of vertebrates viewed from the side are highly dichroic: polarized light whose e vector is aligned perpendicular to the rod axis measures about 10 times the absorbance seen with light polarized parallel to the rod axis. The dichroic ration of 10 indicates that rhodopsin is highly oriented with its chromophore parallel to the plane of the disc membrane (Fig. 14b). However, no dichroism is measured in rods viewed from the end, indicating that the chromophores are randomly oriented (Fig. 14a) within the plane of the membrane (Liebman, P. 1962; Brown, P. 1972).

The outer segment discs are oriented perpendicular to the axis of normal illumination entering the front of the eye. Since maximal absorbance of unpolarized light by pigment molecules is attained when their chromophores lie perpendicular to the light path but are otherwise randomly oriented, the rhodopsin molecules of the disc membrane are oriented for optimal photon capture.

6.7 Orientation of invertebrate rhodopsin

If rhodopsin is oriented in rhabdomeric membrane as it is in vertebrate rod disc membrane (in other words, if a cylindrical microvillus is essentially a rolled-up disc membrane), then the microvillus will be dichroic. A dichroic ratio of about 2 will be measured comparing the absorbance of light polarized parallel and perpendicular to the axis of the microvillus (Fig. 14c). Taking into consideration other factors that tend to reduce the dichroism due to the visual pigment, the expected ratio was calculated to be about 1.7 (Snyder, A. and Laughlin, S. 1975). Actual measurements of fly rhabdomeres range around that value (Kirschfeld, K. and Snyder, A. 1976; Aike, G. 1980). With rhabdomere microvilli arranged so their axes are perpendicular to the light path traversing the eye, as they almost invariably are, no other orientation of rhodopsin will provide more efficient absorbance of unpolarized light (Snyder, A. 1979). Light, whether polarized or not, entering a microvillus end-on,

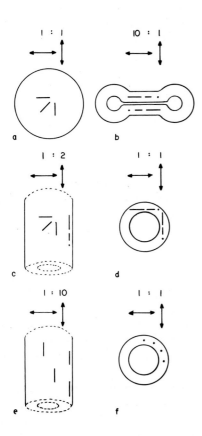

FIG. 14. Diagrammatic representation of the dichroism of photopigment molecules in photoreceptor membrane. The lines and dots represent the chromophores oriented in the membrane. The double-headed arrows represent the e vector of polarized light striking the illustrated membranes and chromophores in a beam perpendicular to the page. The likelihood that a chromophore will absorb a photon of plane polarized light is maximal when the chromophore is aligned with the e vector, and minimal when it is perpendicular to the e vector. For other chromophore orientations, absorbance is proportional to the vector component. **a:** Vertebrate disc membrane illuminated as it is normally in the retina, that is, perpendicular to the plane of the membrane. The chromophores are parallel to the membrane plane, but are otherwise randomly oriented. For simplicity, only three of the possible orientations are shown. Absorbance is the same for any plane of polarization: the dichroic ratio is 1. **b:** Vertebrate disc membrane illuminated edge-on. Since the chromophores are oriented parallel to the membrane plane, absorbance of light polarized parallel to the membranes is high, but drops off with other orientations of the e vector. Therefore the dichroic ratio is high. **c:** Rhabdomeric microvillus illuminated perpendicular to its axis, as it normally is, with chromophores oriented within the plane of the membrane but otherwise randomly disposed. Because of the different orientations of the chromophores at the "sides" of the cylindrical microvillus as compared with those on the "top" and "bottom", the absorbance of light polarized parallel to the microvillous axis is about twice that of light polarized perpendicular to it. **d:** Rhabdomeric microvillus like that in **c**, illuminated end-on. Light is absorbed equally at all orientations of the e vector, but absorbance of unpolarized light would be lower than in **c** because more of the chromophores are illuminated end-on. **e:** Rhabdomeric microvillus with all chromophores oriented parallel with the microvillous axis as well as to the membrane plane. The dichroic ratio is high. **f:** Microvillus like that in **e** illuminated end-on. Both polarized and unpolarized light in any orientation will be poorly absorbed.

assuming chromophore orientation parallel with the membrane, will be less efficiently absorbed than light striking it side-on (Fig. 14d).

Because of the inherent dichroism of their rhabdomeric photoreceptors, invertebrates are preadapted for the perception of polarized light. The neuronal circuits necessary for making use of polarization information have evolved in many insects and other arthropods (Waterman, T. 1975). The polarization sensitivities of these organisms, and the dichroic ratios of their microvilli, are often much higher than 1.7. These higher dichroic ratios can be achieved only by the alignment of the chromophore dipoles with the axes of the microvilli. Snyder, A. and Laughlin, S. (1975) showed that complete alignment should yield a dichroic ratio of about 20. Polarization sensitivities as high as 10–13 have actually been measured (Shaw, S. 1969a). Any enhanced alignment of the rhodopsin in the rhabdomere membrane leads, however, to a reduction in absolute sensitivity because light whose evector orientation is perpendicular to the axes of the microvilli will be poorly absorbed. In some cases the inherent polarization sensitivity of a receptor cell appears to be counteracted through a twisting of its rhabdomere that prevents all of the microvilli from having the same orientation (Menzel, R. 1979; Snyder, A. and McIntyre, P. 1975; Wehner, R. et al., 1975; Ribi, W. 1979; Smola, U. and Tscharntke, H. 1979; Smola, U. and Wunderer, H. 1981b; Wehner, R. and Meyer, E. 1981).

The transition from rhodopsin to metarhodopsin brings a slight change in the dichroic ratio in the frog (Tokunaga, quoted in Goldsmith, T. and Wehner, R. 1977), octopus (Täuber, U. 1975; Schlecht, P. and Täuber, U. 1975), lobster (Bruno, M. et al., 1977), and crayfish (Goldsmith, T. and Wehner, U. 1977). These measurements indicate a shift in chromophore orientation resulting from the configurational changes of the photopigment. In the octopus rhabdomere the orientation changes by about 15–20° relative to the membrane surface. It is not certain whether that tilt is toward or away from the surface of the microvillus. There are no similar data on the relative orientation of the chromophore in insect metarhodopsins. The dichroic properties of rhabdomeric membranes and photopigments is discussed quantitatively in detail by Snyder, A. (1979).

6.8 Mobility of vertebrate rhodopsin

Rhodopsin molecules show no preferential orientation in the plane of a vertebrate disc membrane because they are free to rotate and to migrate laterally. However, they cannot dip or tumble into the membrane (Brown, P. 1972; Cone, R. 1972; Poo, M. and Cone, R. 1974). Low-angle X-ray diffraction studies of outer segment membranes have also demonstrated that rhodopsin is not rigidly fixed in the disc membrane (Blaisie, J. and Worthington, C. 1969). Thus the disc membrane exemplifies the fluid-mosaic model of Singer, S. and Nicholson, G. (1972).

Rhodopsin molecules within the disc membrane can be immobilized by cross-linking with glutaraldehyde. Dichroism can then be induced in the plane of the membrane by partial bleaching with polarized light that is preferentially absorbed by those rhodopsin molecules whose chromophores are aligned with the e vector (Brown, P. 1972). Flash photolysis of unfixed outer segments with polarized light similarly induces transient dichroism. The rate of decay of the photoinduced dichroism then is a measure of the rotational mobility of the photopigment. At physiological temperature, dichroism relaxes within about 60 μs (Cone, R. 1972). Lateral diffusion has been measured by rapidly bleaching a restricted region of rod membranes and measuring the rate at which rhodopsin from surrounding areas migrates into the bleached area. Rhodopsin diffuses over a distance of about 10 μm in 10 s in the discs of amphibian photoreceptors. From these data, Poo, M. and Cone, R. (1974) calculated the viscosity of the membrane as similar to that of olive oil.

6.9 Mobility of invertebrate rhodopsin

The orientation and mobility of photopigment molecules within the confines of the rhabdomeric membrane have been studied with polarized light in crayfish (Orconectes, Procambarus) rhabdoms (Goldsmith, T. and Wehner, R. 1977). The rhabdoms are built of axial layers, each layer being a stack of parallel microvilli, an arrangement particularly suitable for measuring dichroism. Intact rhabdoms can be isolated from macerated retinas. Placed in a drop on a coverslip, they settle on their sides, with stacks of parallel microvilli well oriented for MSP measurements. Dichroic ratios of 2 were recorded in early measurements (Waterman, T. *et al.*, 1969), but improved technique yielded values greater than 3 (Goldsmith, T. 1975), indicating orientation of the rhodopsin chromophores parallel to the axes of the microvilli. Polarization sensitivities as great as 10, measured electrophysiologically in single crustacean photoreceptors (Shaw, S. 1969a; Waterman, T. and Fernandez, H. 1970; Muller, K. 1973; Mote, M. 1974) also indicate that the chromophores are highly oriented.

To prevent gross structural alterations, isolated rhabdoms must be stabilized by treatment with either glutaraldehyde or formaldehyde. These agents also promote the photobleaching of crustacean metarhodopsins at alkaline pH (Goldsmith, T. 1977). Whereas glutaraldehyde immobilizes vertebrate rhodopsin in disc membranes, formaldehyde prevents neither rotation (Brown, P. 1972) nor translational diffusion (Cone, R. 1972; Liebman, P. and Entine, G. 1974). It was reasoned that the visual pigments of rhabdoms treated with a low concentration (0.75%) of formaldehyde should likewise retain normal mobility (Goldsmith, T. and Wehner, R. 1977).

The translational diffusion of visual pigment molecules in crayfish rhabdoms was investigated by bleaching small regions at the ends of the microvilli with a narrow beam, and measuring subsequent absorbance changes in the bleached spot and further along the microvilli. Were the photopigment to diffuse laterally in the membrane, the loss of absorbance would be progressively equalized along the length of the microvilli. However, no translational diffusion of the pigment was detected over distances of about 10 μm during about 20 min in either formaldehyde- or glutaraldehyde-treated rhabdoms (Goldsmith, T. and Wehner, R. 1977).

The question of whether crayfish photopigments rotate tangentially in the microvillous membrane was approached as follows. If they are unable to rotate, then bleaching some of the visual pigment molecules with polarized light whose e vector is perpendicular to the axes of the microvilli will increase the inherent dichroic ratio by selectively removing molecules that are not oriented parallel to the axes of the microvilli. Bleaching with light polarized so that the e vector is parallel to the microvilli will decrease the dichroic ratio. On the other hand, if

rotation occurs there will be either no change in the dichroic ratio, or it will be transitory. Photoinduced dichroism was measured after partial bleaching by polarized light in both glutaraldehyde cross-linked and formaldehyde-treated rhabdoms. Therefore in neither situation was there free Brownian rotation of the visual pigment. Photoinduced dichroism was somewhat greater, however, with glutaraldehyde than formaldehyde, suggesting that the cross-linking agent had eliminated some molecular wobble that was present in the formaldehyde-treated rhabdoms.

The net orientation of the photopigment molecules can also be calculated from such experiments. To the extent that the chromophores are already aligned with the axes of the microvilli, photoinduced dichroism will be less. Experimental values fell well below theoretical functions describing the change in dichroism expected from the partial bleaching of randomly oriented chromophores by polarized light. The data indicate that the absorption vectors lie with a fan of about $\pm 50°$ with respect to the microvillous axis. That degree of orientation would provide a dichroic ratio in the unbleached rhabdom of 5–7, about twice that measured directly by MSP (Goldsmith, T. 1975), but in agreement with the average values of polarization sensitivity (Shaw, S. 1969a).

The experiments of Goldsmith, T. and Wehner, R. (1977) indicate that the visual pigment molecules in the rhabdomeric membranes of the crayfish are tightly constrained. It is crucial to the validity of this conclusion, however, that their mobility is not reduced by formaldehyde. Although the stabilization of cytoarchitecture by glutaraldeyde is well understood — it is a bifunctional reagent that links proteins (Richards, F. and Knowles, J. 1968) — the action of formaldehyde remains uncertain (Glauert, A. 1975). However, there are several reasons for believing that the immobility of crayfish visual pigment is not a consequence of formaldehyde fixation. Higher concentrations than that used on the crayfish rabdoms do not reduce the mobility of rhodopsin molecules in vertebrate outer segment discs (Brown, P. 1972), nor of membrane proteins in fibroblasts (Edidin, M. *et al*,. 1976). Some molecular motion is seen in formaldehyde-treated rhabdoms that is eliminated by glutaraldehyde fixation. Furthermore, freely rotating molecules should be fixed in random orientations, but the photo-induced dichroism measurements indicate prior axial alignment. More recently, Almagor, E. *et al.* (1979) found no translational diffusion in untreated intact barnacle (*Balanus eburneus*) photoreceptors; the visual pigment diffuses by less than 30 μm in 30 min.

Thus, the visual pigments of the crayfish and barnacle do not float freely in a fluid membrane as vertebrate rhodopsin does. While these experiments provide the most convincing evidence for the rigidity of rhabdomeric membrane, it has been inferred for other reasons also. Israelachvili, J. and Wilson, M. (1976) have argued that the tubular shape of a microvillus is itself evidence for a non-fluid membrane because of the inconstant curvature of its surface; it is neither spherical nor flat, the two stable configurations that a fluid layer might assume. The fluidity of vertebrate photoreceptor membrane may result from its high content of polyunsaturated fatty acids and low content of cholesterol (Daemen, F. 1973). The higher cholesterol content of rhabdomeres implies a more viscous membrane.

Mention should be made of a model proposed to account for the high dichroic ratios found in some rhabdomeres under the assumption of a fluid membrane and mobile visual pigment molecules. Laughlin, S. *et al.* (1975) and Snyder, A. and Laughlin, S. (1975) argued that radially symmetrical photopigment molecules will remain randomly oriented, but mobile elongate molecules will tend to align parallel to the microvillous axis. They reasoned that the ends of such a molecule would lift out of the curved microvillous membrane when oriented tangentially. Since hydrophobic proteins would encounter an energy barrier in partially emerging from the lipid membrane, they would tend to preferentially orient with the axis of the microvillus. Diffusion along the length of the microvillus will be favoured over tangential movements. However, it seems unlikely that this mechanism could provide the high dichroic values that have been measured. In any event, it could account only for photopigment orientation, not for the lack of translational diffusion.

6.10 Rhabdomeric cytoskeleton

Goldsmith, T. and Wehner, R. (1977) have suggested that rhodopsin immobility could be accounted

for if the visual pigment molecules are attached to a cytoplasmic matrix underlying the membrane. Rhabdomeric microvilli are built around a cytoskeleton consisting of an axial filament with radial fibers (White, R. 1967; Perrelet, A. 1970; Blest, A. *et al.*, 1982). Hamdorf, K. (1979) has proposed a model of the microvillus in which the regularity of its shape and the immobility of its contained rhodopsin depends on a "paracrystalline" organization of its protein and lipid components. The fact that the dimensions of the microvilli remain unchanged when the rhodopsin content is reduced by carotenoid deprivation and in *Drosophila* mutants, argues against such a rigidity of organization. Horridge, G. and Blest, A. (1980) and Stowe, S. (1980) have suggested that the cytoskeleton may function in maintaining the high curvature of the microvillous membrane, in anchoring the visual pigment molecules, and in some aspects of membrane turnover. As yet, however, we know very little about the structure or composition of the cytoskeleton of the photoreceptor microvillus.

6.11 Assembly of photoreceptor membrane

Vertebrate rhodopsin is synthesized in the rough endoplasmic reticulum of the inner segment of the photoreceptor cell, and passes through the Golgi apparatus before being assembled into the plasma membrane and discs of the outer segments. Although this general pathway is known, the details remain vague. The route from the Golgi to the plasma membrane is particularly uncertain. A popular idea is that the visual pigment is conveyed in small membrane vesicles (Olive, J. 1980; Holtzman, E. and Mercurio, A. 1980). Surprisingly little is known about the initial formation of visual pigment and photoreceptor membrane in the vertebrate embryo (Olive, J. and Recouvreur, M. 1977).

We have recently begun to look at the functional development of the retina of *Manduca sexta* (White, R. *et al.*, 1983). The rhabdomere microvilli begin to differentiate about half-way through the pupal instar. A small receptor potential can first be recorded at this stage. Eguchi, E. *et al.* (1962) also found that the receptor potential and the rhabdomere microvilli differentiate concomitantly in the developing eye of *Bombyx mori*. We found that rhodopsin, measured by MSP, appears in the

Manduca retina several days before the microvilli. This early rhodopsin may be associated with intracellular membranes that accumulate at the same time. These endomembranes later orient adjacent to the plasma membrane when the rhabdomere begins to form, as if they were contributing directly to the formation of the microvilli.

The available evidence (Bok, D. *et al.*, 1977) suggests that the chromophore is not added to vertebrate opsin until it is inserted into the plasma membrane. By contrast, *Manduca* rhodopsin must acquire its chromophore while still inside the cell. Since nascent *Manduca* rhodopsin can be measured photometrically *Manduca* photoreceptors are particularly suited for studying rhodopsin synthesis and assembly of photoreceptor membranes.

7 TRANSDUCTION AND ADAPTATION

Transduction is dealt with elsewhere in this volume (chapter 6) and only particular aspects directly connected with photopigments will be considered here.

The light-provoked transition from rhodopsin to metarhodopsin results, after a lag of several milliseconds, in the depolarization of insect and most other invertebrate receptor cells. It is this late receptor potential that transmits visual information to the central nervous system, and is customarily used to measure action spectra. The long latency indicates that time-consuming processes must intervene between the transformation of the visual pigment and membrane depolarization. Following the formation of metarhodopsin there is a period of 3–5 ms during which photoreversal prevents transduction. An additional 3–4 ms passes before the onset of the response (Hamdorf, K. and Kirschfeld, K. 1980). Thus about half the latency can be attributed to events occurring at the level of the photopigment, and the remainder to subsequent processes. It would appear that transduction is triggered by newly formed metarhodopsin if it persists for longer than 5 ms.

7.1 M potential

Since spectral sensitivity functions match rhodopsin but not metarhodopsin spectra, transduction is mediated only by the transformation of rhodopsin to metarhodopsin, not by the reverse reaction. (Barnes, S. and Goldsmith, T. 1977; Atzmon, Z. *et al.*, 1978, 1979; Lisman, J. and Strong, J. 1979; Strong, J. and Lisman, J. 1978). There is, however, an electrophysiological signal associated with photoregeneration in flies. ERGs elicited by an intense flash contain an early diphasic potential fluctuation, called the M potential, that precedes the receptor potential (Pak, W. and Lidington, K. 1974; Stephenson, R. and Pak, W. 1980). The earliest (cornea-negative) component of the M potential, M_1, is associated with the receptor cell and resembles an early receptor potential (ERP). ERP signals are thought to reflect charge displacements within the photoreceptor membrane that are closely associated with the visual pigment. They have been recorded from a number of vertebrate and invertebrate eyes (Hamdorf, K. 1979). In many cases opposite charge displacements generate opposite ERP potentials as the visual pigment flips back and forth between its configurational states. In flies, however, the M_1 potential is generated by the transition from metarhodopsin to rhodopsin, while the conversion of rhodopsin to metarhodopsin is silent. Thus the spectral sensitivity of the M_1 potential matches the spectrum of metarhodopsin (Pak, W. and Lidington, K. 1974; Minke, B. and Kirschfeld, K. 1980). A second, cornea-positive component of the M potential, M_2, is not an ERP. The M_2 component appears only in extracellular ERG records, while M_1 can be recorded intracellularly from the receptor cells. The M_2 potential appears to originate in second-order neurons that are apparently stimulated by M_1. The amplitudes of the two components of the M potential, especially of M_1, are proportional to the conversion of metarhodopsin to rhodopsin. Although its basis is not well understood, the M potential has therefore proved useful for measuring metarhodopsin in fly photoreceptors.

7.2 Prolonged depolarizing afterpotential

The prolonged depolarizing afterpotential (PDA) is another electrophysiological feature of invertebrate photoreceptors associated with metarhodopsin (for reviews see Hamdorf, K. 1979; Hillman, P. 1979; Pak, W. 1979). It is a depolarization of the photoreceptors that persists for an extended period after the cessation of a stimulus. The PDA is produced in the eyes of flies by intense blue light sufficient to convert a substantial amount of rhodopsin to metarhodopsin. The PDA lasts for several minutes in *Calliphora* (Hamdorf, K. 1979), and for over an hour in *Drosophila* (Pak, W. 1979). With a maximal PDA the receptor becomes completely insensitive even to intense stimuli. Behavioral responses to light, such as the optomotor response (Heinsenberg, M. and Buchner, E. 1977) and phototaxis (Willmund, R. and Fischbach, K. 1977; Schinz, R. and Sidorsky, L. 1978; Pak, W. 1979) are also depressed. As the PDA gradually decays in darkness, and membrane polarity is restored, the receptor becomes progressively more responsive. However, the PDA is eliminated and sensitivity restored immediately by a flash of light that regenerates rhodopsin.

The PDA has proved useful for investigating the intermediate steps that must be interposed between the transformation of the visual pigment and the change in membrane permeability. The same mechanisms of membrane conductance appear to be responsible for both the light-dependent receptor potential and the subsequent light-independent PDA. Both reflect an increase in sodium conductance (Brown, H. and Cornwall, M. 1975). As a first hypothesis one might suggest that sodium channels are opened by metarhodopsin, and remain open as long as metarhodopsin persists in the membrane. However, the gradual decay of the PDA, in darkness, does not result from the loss of metarhodopsin, which remains constant until photoregeneration. Furthermore, after subsequent photoregeneration, several minutes pass before a PDA can be elicited again by the conversion of rhodopsin to metarhodopsin even though the receptor potential is normal (Minke, B. *et al.*, 1973).

PDA defective mutations have been isolated in *Drosophila*. They fall into two classes: *ina* mutants lack PDAs but are nevertheless rendered insensitive by intense blue light, that converts most of the photopigment to metarhodopsin, just as if a PDA had been present; *nina* mutants lack PDAs, and remain responsive after blue light. Thus metarhodopsin seems to be associated with two separable phenomena, desensitization and the PDA. It is of interest that *nina* mutants have reduced amounts of visual pigment. Both rhodopsin and PDA are res-

tored when the concentration of photopigments is raised to the normal level by feeding the larvae extra retinal (Pak, W. 1979; Stephenson, R. and Pak, W. 1981). The PDA is similarly eliminated in *Calliphora* whose visual pigment is reduced by carotenoid deficiency (Razmjoo, S. and Hamdorf, K. 1976; Hamdorf, K. and Razmjoo, S. 1977).

Assessments of the various data bearing on the problem of transduction in arthropods can be found in recent reviews (Hamdorf, K. 1979; Hamdorf, K. and Razmjoo, S. 1977; Pak, W. 1979). Briefly, current models propose that the transition from rhodopsin to metarhodopsin activates "excitors" that depolarize the membrane and then are gradually eliminated, while the conversion of metarhodopsin to rhodopsin may release "inhibitors" that neutralize the excitors. The PDA would then result from an excess of excitors (Hillman, P. 1979; Hamdorf, K. 1979; Hamdorf, K. and Razmjoo, S. 1977, 1979; Minke, B. 1979), perhaps relative to the number of available sodium channels to be activated (Stark, W. and Zitzmann, W. 1976). The latter point would explain why the PDA is eliminated when the concentration of photopigment is reduced.

Like the M potential, the PDA can be used to measure metarhodopsin. Since photoconversion of metarhodopsin results in the reversal of the PDA, that response can be used to construct action spectra reflecting the absorption spectrum of metarhodopsin (Hamdorf, K. 1979).

7.3 Rhodopsin concentration and sensitivity

The way in which photoreceptor sensitivity is adjusted to the transitory conditions of illumination remains a central problem in the physiology of vision. Several factors may be involved in the adaptation of photoreceptors — changing concentrations of visual pigments, structural changes of various sorts such as pigment movements and optical adjustments that control the flux of light through the photoreceptor, alterations in the area of photoreceptor membrane (White, R. and Lord, E. 1975; Srinivasan, M. and Bernard, G. 1980), and adjustments of photoreceptor gain through modifications of the photoreceptor membrane that may be broadly categorized as "neural" (Barlow, H. 1972).

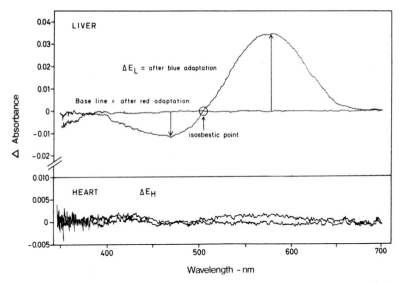

FIG. 15. Difference spectra from *Calliphora erythrocephala* fed as larvae on vitamin A rich liver or vitamin A poor heart meat. Baselines were set at 0 absorbance after red adaptation that converted all M580 to P490. Intense blue light (470 nm) was then used to maximally convert P490 to M580. The ratio of $\Delta E_L/\Delta E_H = 0.1$ measured at 580 nm shows the relative amounts of visual pigment in liver *vs.* heart flies (from Razmjoo, S. and Hamdorf, K. 1976).

Neural adaptation can be large even at low light levels. Vertebrate visual systems can show several log units of change in the stimulus intensity required to elicit a particular level of response even at light intensities that bleach only a few per cent of the rhodopsin molecules (Weinstein, G. *et al.*, 1967). Dark recovery from neural adaptation is rapid and does not depend on rhodopsin regeneration. If regeneration is prevented, the response rises to a level that reflects the amount of photopigment remaining. Thus the relationship between receptor sensitivity and rhodopsin concentration is evident after a short period of neural adaptation. One might expect, and that was the original hypothesis, that the final level of dark-adapted sensitivity would be directly proportional to photon-catching capacity, that is, to the amount of rhodopsin. In vertebrates, however, this is not the case; the relationship between rhodopsin concentration and response is logarithmic. In the rat, for instance, bleaching 17% of the rhodopsin lowers sensitivity by a factor of ten, whereas a bleach of 34% decreases the response a hundredfold (Weinstein, G. *et al.*, 1967).

The same sort of experiment relating sensitivity to rhodopsin concentration cannot be done with insects in which the visual pigment does not bleach. However, the amount of rhodopsin can be varied in insect eyes in two other ways: by varying rhodopsin concentrations in photoequilibria, and through carotenoid deficiency. Rhodopsin concentrations in photosteady states can be adjusted over a wide range in the photoreceptors of such insects as *Ascalaphus macaronius* and *Calliphora erythrocephala* whose metarhodopsins are shifted to longer wavelengths well apart from their rhodopsins. With the photoequilibrium approach, the metarhodopsin content of a photoreceptor necessarily varies at the same time in reciprocal relationship with the amount of rhodopsin. Carotenoid deficiency allows the comparison of sensitivity in photoreceptors with different absolute amounts of visual pigment, but may also introduce unknown side-effects.

The relationship between photopigment and sensitivity has been especially well documented by experiments with *Calliphora*, in which rhodopsin concentration was varied both in photoequilibria and by carotenoid deficiency, and receptor sensitivity was measured by extracellular and intracellular recording. Red adaptation results in ~ 100% P490, while blue irradiation establishes a photoequilibrium of ~ 0.2 P490: 0.8 M575 (Fig. 8). The total amount of visual pigment can be reduced by more than a factor of 10 when flies are reared on heart muscle rather than liver (Fig. 15) (see section 4.2). Fig. 16a compares liver and heart-reared

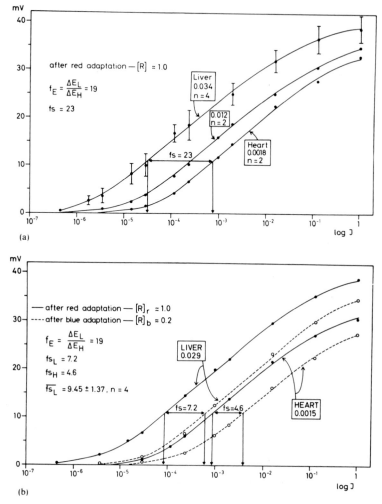

FIG. 16. Intensity–response functions from *Calliphora erythrocephala* determined by monochromatic test stimuli (498 nm) close to the isosbestic point of the visual pigment. **a**: Comparison of liver and heart flies after red (614 nm) adaptation that converted all of the visual pigment to P490. As the rhodopsin content (shown by the MSP-measured extinction values in the boxes) increases, so does the sensitivity. The sensitivity change (fs) is approximately proportional to the difference in the concentration of P490 (f). **b**: Comparison of heart and liver flies after red (614 nm) and blue (473 nm) adaptation, that established photosteady states containing 100% and 20% rhodopsin respectively. Loss of sensitivity in heart flies after blue adaptation (fs) is proportional to the change in P490 concentration (1.0/0.2 = 5) in the photoequilibria. Loss of sensitivity is somewhat higher in liver flies (from Razmjoo, S. and Hamdorf, K. 1976).

flies that had been red-adapted to eliminate meta-rhodopsin. Liver flies, with 19 times as much rhodopsin as heart flies, were 23 times as sensitive. Thus sensitivity is proportional to the rhodopsin concentration in the rhabdomeres.

With the photoequilibrium technique, sensitivity and photopigment concentrations are measured in the same cells. Blue adaptation, that reduces P490 and increases M575, decreased the sensitivity of heart flies about fivefold, directly proportional to the decrease in rhodopsin concentration. The

sensitivity of blue-adapted liver flies (after the recovery of the resting potential following the decline of the PDA) was reduced to a somewhat greater extent, by a factor of 7, although the relative change in rhodopsin concentration was the same in both liver and heart flies (Hamdorf, K. and Rosner, G. 1973; Muijser, H. *et al.*, 1975; Rosner, G. 1975; Razmjoo, S. and Hamdorf, K. 1976).

Dark-adapted sensitivity in *Calliphora* is therefore a linear function of rhodopsin concentration in contrast to the logarithmic relationship characteristic of vertebrates. This finding indicates that there is a one-to-one relationship between the number of photons absorbed by rhodopsin molecules and the excitation of the membrane. Since the slopes of the intensity-response curves are the same in different photoequilibrium mixtures of rhodopsin and metarhodopsin, quanta absorbed by the latter do not alter that relationship. However, when the concentration of visual pigment in the membrane is high, the presence of metarhodopsin seems to bring about a small additional loss of sensitivity.

With some variation, the linear relationship is generally characteristic of invertebrate photoreceptors. Carotenoid-deprived *Drosophila* are similar to heart-fed *Calliphora*. Sensitivity closely parallels rhodopsin content in photoequilibria. However, those with a full complement of visual pigment gained from a carotenoid-rich diet suffer a disproportionate loss of sensitivity after blue adaptation — several log units — in consonance with the highly developed PDA of *Drosophila* (Stark, W. and Zitzmann, W. 1976). The sensitivity of the *Ascalaphus* ultraviolet receptor actually increases a bit as the photoequilibrium concentration of rhodopsin is reduced from 100% to about 85%. Thereafter sensitivity drops somewhat more rapidly than does the rhodopsin concentration as photoequilibria are shifted further toward metarhodopsin (Hamdorf, K. *et al.*, 1978).

8 VISUAL PIGMENTS AND PHOTORECEPTORS OF DIPTERA

The compound eyes of the muscoid flies (Cyclorrapha) have been by far the most extensively studied insect visual systems. They offer particular advantages for analysis of the preliminary events of visual function. Structural and functional eye mutants are available, especially in *Drosophila* (Pak, W. 1979; Pak, W. *et al.*, 1980). The photoreceptors of flies are studied within an ample context of neurophysiological, neuroanatomical, and behavioral data on the processing of visual information.

8.1 Organization of the fly retinula

In the compound eyes of most arthropods, the

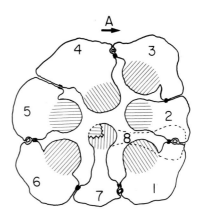

FIG. 17. Diagramatic representation of a fly retinula in distal cross-section. The receptor cells and their isolated rhabdomeres (marked by parallel lines) are numbered by convention. The retinula shown would be from the dorsal half of the eye; those in the ventral half are mirror images of the dorsal retinulae. Note the trapezoidal arrangement of R1–6. The R8 cell (dotted outline) enters the retinula proximally. Its rhabdomere lies proximal to that of R7, which has been cut away to show that the microvilli of the two rhabdomeres are oriented orthogonally (see Trujillo-Cenóz, O. and Melamed, J. 1966).

rhabdomeres of individual receptor cells are joined at the center of the retinula forming a "fused rhabdom" (Snyder, A. *et al.*, 1973). The Cyclorrapha have "open rhabdoms", in which isolated rhabdomeres project into a central fluid-filled cavity — a dilation of the extracellular space (Fig. 17). The rhabdomeres are long, slender, separate light guides differing in refractive index from the surrounding retinula cells and central cavity (Snyder, A. and Pask, C. 1973). Six receptor cells (designated R1–6) have completely isolated rhabdomeres arranged in a characteristic trapezoidal pattern around the periphery of the central cavity. The rhabdomeres of two other cells project farther into the central cavity where they are superimposed in the optical pathway. The cell bearing the distal central rhabdomere is designated R7; that contributing the underlying proximal rhabdomere is R8.

R1–6 belong to one functional visual subsystem, and R7/8 to another (Cosens, D. and Wright, R. 1975; Minke, B. *et al.*, 1975; Wehrhahn, C. 1976; Willmund, R. and Fischbach, K. 1977; Willmund, R. 1979). The former respond at low levels of illumination functioning as a high-sensitivity scotopic system and their axons terminate in the first optic neuropile, the lamina. The latter function in bright light as a high acuity photopic system, and

Table 3: Summary of data on peripheral receptors R1–R6 of the fly rhabdom

Species	Visual pigment	Spectral sensitivity
Calliphora erythrocephala (vicinia)	P490, M570–580 (i,o,p,v,aa,bb)	350, 490 (a,b,c,e,f,g,h,i,j,m,r,w)
Drosophila melanogaster	P480, M570 (k,s,u,z)	~350, 480 (c,k,t,y)
Sarcophaga bullata	P490, M575–580 (x)	
Musca domestica	P480, M580 (n,q)	350, 490 (c,i)
Chlorops sp.		350, 480 (c)
Dimocoenia spinosa		~350, 480 (c)
Eristalis arbustorum		UV?, 450 (c,d)
Eristalis tenax	P460, M550 (bb)	350, 450 (l,bb,cc)
Syrphus sp.		~350, 460 (c)
Allograpta obliqua		
ventral eye region		~350, 450 (c)
equatorial and dorsal eye regions		UV?, 495 (c)
all eye regions		~350, 480 (c)
Toxomerus marginatus		~350, 450 (c)

Visual pigments are designated by the λ_{max} of the α-bands of rhodopsin and metarhodopsin. Spectral sensitivity refers to the two response maxima in the ultraviolet and visible that characterize R1–6. Visual pigments were measured either *in situ* by MSP (references i,m,o,p,s,u,z,aa,bb) or in extracts (references k,q,s,v). Most spectral sensitivities were measured by intracellular recording except for references k (ERGs of *Drosophila* receptor mutants) and c (pupillary response). The λ_{max} of visual pigments and spectral sensitivities are in nanometers.

(a) Aike, G. (1980)
(b) Burkhardt, D. (1962)
(c) Bernard, G. and Stavenga, D. (1979)
(d) Bishop, L. (1974)
(e) Dörrscheidt-Käfer, M. (1972)
(f) Gemperlein, R. *et al.* (1980)
(g) Hamdorf, K. and Rosner, G. (1973)
(h) Hamdorf, K. *et al.* (1973b)
(i) Hamdorf, K. (1979)
(j) Hardie, R. (1979)
(k) Harris, W. *et al.* (1976)
(l) Horridge, G. *et al.* (1975)
(m) Horridge, G. and Mimura, K. (1975)
(n) Kirschfeld, K. *et al.* (1977)
(o) Langer, H. (1965, 1967, 1972)

(p) Langer, H. and Thorell, B. (1966)
(q) Marak, G. *et al.* (1970)
(r) McCann, G. and Arnett, D. (1972)
(s) Ostroy, S. *et al.* (1974)
(t) Pak, W. and Lidington, K. (1974)
(u) Pak, W. *et al.* (1976)
(v) Paulsen, R. and Schwemer, J. (1979)
(w) Rosner, G. (1975)
(x) Schwemer, J. and Brown, P. (personal communication)
(y) Stark, W. *et al.* (1979)
(z) Stark, W. and Johnson, M. (1980)
(aa) Stavenga, D. *et al.* (1973)
(bb) Stavenga, D. (1976)
(cc) Tsukahara, Y. and Horridge, G. (1977)

send their axons directly to the second optic neuropile, the medulla (Strausfeld, N. 1970). R1–6 will be considered separately from R7/8 in the following discussion.

8.2 Spectral sensitivity of R1–6

The spectral sensitivity functions of individual receptor cells were first measured in *Calliphora* by Burkhardt, D. (1962) by means of intracellular recording. The commonest type of receptor displayed two sensitivity peaks in the green and ultraviolet. These were later identified as R1–6 (McCann, G. and Arnett, D. 1972; Dorrscheidt-Käfer, M. 1972; Horridge, G. and Mimura, K. 1975; Rosner, G. 1975; Eckert, H. *et al.*, 1976; Hardie, R. 1979; Aike, G. 1980; Gemperlein, R. *et al.*, 1980). R1–6 have been shown to have double-peaked sensitivity functions from single-cell

penetrations in other species also (Table 3). Similar action spectra have been determined by other procedures: behavioral responses (Eckert, M. 1971; Schümperli, R. 1973; Jacob, K. *et al.*, 1977); mass responses (ERGs) from *Drosophila* mutants lacking various receptor types; and the pupillary response of intracellular pigment granules (Bernard, G. and Stavenga, D. 1979). The pupillary response, first utilized by Franceschini, N. (1972), has the particular advantage of being a completely non-invasive measure of photoreceptor sensitivity.

By whatever method it is determined, the normal spectral sensitivity function of R1–6 has two peaks separated by a minimum around 400 nm (Fig. 18). One peak is near 350 nm, the other ranges in different species from 450 to 495 nm (Table 3). In most species, all R1–6 cells are identical, but in Syrphidae the long wavelength maximum may vary between sexes and in different parts of the eye (Bernard, G.

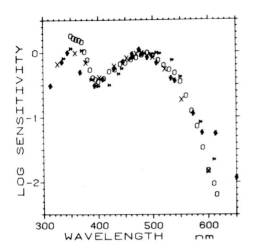

FIG. 18. Action spectra of R1–6 in *Calliphora erythrocephala*. Intracellular records: #, Burkhardt, D. (1962); ×, McCann, G. and Arnett, D. (1972); *, Dörrscheidt-Käfer, M. (1972). Pupillary responses: ○, Bernard, G. and Stavenga, D. (1979). (From Bernard, G. and Stavenga, D., 1979).

and Stavenga, D. 1979; Stavenga, D. 1979). Although sensitivity in the ultraviolet generally equals, or is greater than, blue-green sensitivity, the relative height of the ultraviolet peak may vary from cell to cell or seasonally (Horridge, G. and Mimura, K. 1975), and it is selectively reduced in carotenoid deficiency (Goldsmith, T. *et al.*, 1964; Kirschfeld, K. and Franceschini, N. 1977; Stark, W. *et al.*, 1977; Paulsen, R. and Schwemer, J. 1979; Aike, G. 1980).

8.3   Visual pigment of R1–6

The visual pigment of R1–6 in *Calliphora* was the first insect rhodopsin to be characterized (Hamdorf, K. and Langer, H. 1965; Langer, H. 1965, 1967, 1972; Langer, H. and Thorell, B. 1966). They have subsequently been measured by MSP in several species (Table 3) and in extracts from *Drosophila* (Ostroy, S. *et al.*, 1974; Pak, W. *et al.*, 1976; Ostroy, S. 1978) and *Calliphora* (Paulsen, R. and Schwemer, J. 1979). Some features of these photopigments have been discussed above: the α-bands of fly rhodopsins measured in difference spectra absorb in the blue-green, and their metarhodopsins are shifted about 100 nm into the yellow (Fig. 15). Metarhodopsins absorbing at such long wavelengths are unique to flies.

Direct MSP spectra of the rhabdomeres of R1–6 in *Musca* resemble the spectral sensitivity function

of R1–6; they exhibit two peaks of about equal height at 360 and 490 nm. The 490 nm peak is the rhodopsin α-band because it shifts to longer wavelengths with blue irradiation and is restored by red light. The ultraviolet peak, however, does not change with illumination (Kirschfeld, K. *et al.*, 1977). Ultraviolet irradiation of rhabdomeres containing 100% rhodopsin establishes a photosteady state containing about 50% metarhodopsin, with no absorbance change apparent below 400 nm. Thus the extinctions of rhodopsin and metarhodopsin are similar in the near ultraviolet, and one cannot determine the absorption spectrum of rhodopsin below 400 nm from difference spectra (Kirschfeld, K. *et al.*, 1977). Indirect approaches have been developed in attempts to calculate pigment spectra from sensitivity measurements (Tsukahara, Y. and Horridge, G. 1977; Minke, B. and Kirschfeld, K. 1979; Stark, W. and Johnson, M. 1980; Stark, W. *et al.*, 1979), using theoretical treatments of the kinetics of rhodopsin–metarhodopsin photoconversions and the relationship of sensitivity to their concentrations in photoequilibria (Hamdorf, K. and Schwemer, J. 1975; Stavenga, D. 1975, 1976; Hochstein, S. *et al.*, 1978; Hamdorf, K. 1979). The various approaches have yielded differing estimates of the short wavelength absorbance of rhodopsin, but agree that metarhodopsin and rhodopsin spectra are similar below 400 nm.

In contrast with MSP measurements of intact rhabdomeres, spectra of extracted fly rhodopsins show only the low absorbance in the near ultraviolet typical of a rhodopsin β-band (Ostroy, S. 1978; Paulsen, R. and Schwemer, J. 1979). Thus the dual sensitivity of R1–6 is a puzzle. Whereas the spectral sensitivity function with its ultraviolet peak corresponds to the absorption spectrum of the rhabdomere, extracted rhodopsin shows no unusually elevated ultraviolet absorbance.

8.4   The dual sensitivity function of R1–6

Apart from the possibility of coupling between R1–6 and some other ultraviolet-sensitive cells — ruled out by the lack of scotopic ultraviolet units and the high angular sensitivity of R1–6 (Tsukahara, Y. and Horridge, G. 1977) — four explanations of the dual spectral sensitivities of R1–6 have been proposed:

(1) they contain two rhodopsins (Burkhardt, D. 1962; Rosner, G. 1975; Horridge, G. and Mimura, K. 1975);

(2) light absorption is influenced by the waveguide properties of the narrow rhabdomeres (Snyder, A. and Miller, W. 1972; Snyder, A. and Pask, C. 1973);

(3) the rhabdomeres contain an ultraviolet-absorbing "antenna pigment" that transfers energy to the visual pigment (Ashmore, J. 1977; Kirschfeld, K. *et al.*, 1977; Kirschfeld, K. and Franceschini, N. 1977; Minke, B. and Kirschfeld, K. 1979); and

(4) rhodopsin β-band absorbance is enhanced through mutual interactions of visual pigment molecules within the rhabdomere membrane (Hamdorf, K. 1979; Paulsen, R. and Schwemer, J. 1979).

8.4.1 TWO RHODOPSINS

The strongest case for two rhodopsins in the same cell was made by Horridge, G. *et al.* (1975). They found, in *Eristalis tenax*, that the maximum response to polarized light was elicited by different orientations of the plane of polarization in the ultraviolet and blue-green. They suggested that two rhodopsins, one absorbing near 350 nm and the other at 450 nm, were located in different regions of a twisted rhabdomere. Tsukahara, Y. and Horridge, G. (1977) subsequently tested this idea in extensive adaptation experiments and concluded that only a single rhodopsin is present in R1–6. They found, for instance, that blue irradiation decreases ultraviolet sensitivity more than does ultraviolet light itself. This effect is the reverse of what would happen with separate ultraviolet and blue-sensitive rhodopsins, but is expected from a single visual pigment whose rhodopsin and metarhodopsin spectra differ at longer wavelengths but are similar in the ultraviolet; blue light transforms more rhodopsin than does ultraviolet light.

Single-cell studies in *Calliphora* showed that the two sensitivity maxima could not be shifted by selective adaptation (Burkhardt, D. 1962). Furthermore, there is an isosbestic point in rhodopsin–metarhodopsin photoconversions measured by MSP regardless of the wavelength of the actinic light (Kirschfeld, K. *et al.*, 1977). There

is abundant evidence that the dual sensitivity of R1–6 cannot be accounted for by two rhodopsins.

8.4.2 WAVEGUIDE EFFECTS

The waveguide hypothesis emerged from a theoretical treatment predicting that long-wavelength light might be preferentially excluded from rhabdomeres with diameters in the range of actinic wavelengths (Snyder, A. and Pask, C. 1973). The effect depends on the relative refractive indices of a rhabdomere and its surroundings, parameters difficult to measure with sufficient accuracy. However, measurements of rhabdomere birefringence in *Musca* provided an estimate of waveguide effects that indicated little influence on the spectral sensitivity of R1–6 (Kirschfeld, K. and Snyder, A. 1976). Thus, waveguide effects also fail to account for the elevated ultraviolet response of these cells.

8.4.3 ANTENNA PIGMENT

The novel antenna pigment hypothesis has been in vogue since its proposal by Kirschfeld, K. *et al.* (1977). They suggested that the ultraviolet absorbance of fly rhabdomeres is due to a photostable pigment capable of transferring energy to the rhodopsin whose α-band lies at longer wavelengths. Bruckler, R. and Williams, T. (1981) have also proposed that the ultraviolet sensitivity of the grasshopper *Romalea microptera* depends upon energy transfer. If there is a sensitizing pigment, it would appear to be a carotenoid derivative. Although the sensitivity of flies grown on a carotenoid-deficient medium is reduced at all wavelengths, ultraviolet sensitivity is lowered to a relatively greater extent (Goldsmith, T. *et al.*, 1964; Kirschfeld, K. and Franceschini, N. 1977; Stark, W. S. *et al.*, 1977; Aike, G. 1980).

Two modes of energy transfer have been discussed. Although dipole–dipole inductive resonance is the mechanism generally proposed (Kirschfeld, K. and Franceschini, N. 1977), fluorescence has also been considered (Stark, W. *et al.*, 1977, 1979; Franceschini, N. and Stavenga, D. 1981; Franceschini, N. *et al.*, 1981b). In fact, the current antenna pigment hypothesis was anticipated by Chance, B. (1964; see also Kay, R. 1969), who suggested that the ultraviolet sensitivity

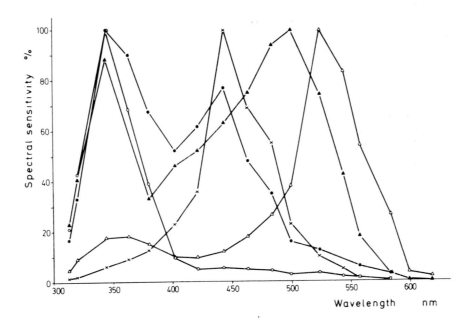

F<small>IG</small>. 19. Averaged spectral sensitivities of photoreceptors of *Calliphora erythrocephala* R1–6 (▲); R7UV (○); R7UB (●); R8B
(×); R8G (△) (from Smola, U. and Meffert, P. 1979).

of flies might be the consequence of DPNH
fluorescence. Ultraviolet excitation does elicit
fluorescence in the rhabdomeres of R1–6 (Table 5)
with a peak at about 460 nm, and it is reduced by
carotenoid deficiency. On these grounds it has been
argued that the fluorescing material may be the
antenna pigment (Stark, W. *et al.*, 1977, 1979).

The efficiency of inductive resonance depends on
the distance between the donor and recipient
molecules. There is some question as to whether
energy can be transformed with adequate efficiency
between separate light-harvesting and rhodopsin
molecules, even if they are close together and
appropriately aligned in the rhabdomere membrane
(Hagins, W. and Jennings, W. 1959; Wu, C. and
Stryer, L. 1972; Stark, W. *et al.*, 1977). As an alternative, the exciting molecule might reside on the
rhodopsin itself as a second chromophore (Ashmore, J. 1977; Kirschfeld, K. and Franceschini, N.
1977; Kirschfeld, K. 1979). Cattle rhodopsin with a
second retinal affixed artificially outside the normal
chromophoric site acquires a second high absorbance peak in the ultraviolet (Rotmans, J. *et al.*,
1974). It is not known whether this second retinal
transfers energy to the normal chromophore, but
energy can be transferred to the chromophore from

the aromatic amino acids of the opsin (Goldsmith,
T. and Fernandez, H. 1968), and from probes
attached to various sites on the opsin (Wu, C. and
Stryer, L. 1972). Preferential reduction of ultraviolet sensitivity in carotenoid deficiency would
result if the chromophoric retinal were given
precedence in rhodopsin synthesis over antennal
retinal.

8.4.4 β-B<small>AND</small> <small>ENHANCEMENT</small>

The antenna pigment hypothesis is often presented
as if the matter were settled. However, there is as yet
no conclusive evidence in support of it. The data are
also consistent with β-band enhancement that
depends upon factors present in the normal environment of the rhabdomere membrane (Hamdorf, K. 1979; Paulsen, R. and Schwemer, J. 1979).
However, the molecular basis of β-band absorbance is not well understood (Ebrey, T. and Honig,
B. 1977). According to the β-band hypothesis, the
selective reduction of short wavelength sensitivity
that accompanies carotenoid deprivation is a secondary consequence of the lowered concentration of
rhodopsin in the membrane.

It might help to decide between the antenna

Table 4: Summary of data on central receptors R7/8 of the fly rhabdom

R7 Receptor cells

Species	Visual pigments		Stable yellow pigment (P456)		Spectral sensitivity		
	R7p	R7y	R7p	R7y	R7UV	R7UB	R7UT
Calliphora erythrocephala (vicinia)			none (c,d,e)	P456 (c,d,e)	350 (a,f)	350, 440 (f)	340 + tail to 500 nm (a)
Calliphora stygia					350 (a)		
Musca domestica	PUV, M470 (c)	P430, M505 (c,e)	none (c,d,e)	P456 (c,d,e)	350 (a)		
Drosophila melanogaster	P370, M470 (b)		none (c,d,e)	P456 (c,d,e)	∼350 (only one spectral class identified from ERG) (b)		

R8 Receptor cells

	Visual Pigment	Spectral Sensitivity	
		R8B	R8G
C. erythrocephala		440 (a)	540 (a,f)
C. stygia			540 (a)
M. domestica			540 (a)
D. melanogaster	P∼490, M∼490 (b)		

(a) Hardie, R. *et al.* (1979)
(b) Harris, W. *et al.* (1976)
(c) Kirschfeld, K. (1979)
(d) Kirschfeld, K. *et al.* (1978)
(e) McIntyre, P. and Kirschfeld, K. (1981)
(f) Smola, U. and Meffert, P. (1979)

Visual pigments and the stable pigment were measured by MSP (references c,e) or in crude extracts of *Drosophila* receptor mutants (reference b). The visual pigment of R8 in *Calliphora* and *Musca* has been inferred from spectral sensitivity functions in combination with the spectrum of the stable filtering pigment, P456 (references a,f). In no instance have the visual pigments of R7/8 been well characterized. Spectral sensitivities were measured by intracellular recording, except in *Drosophila* where they were inferred from ERGs of receptor mutants (reference b). The λ_{max} of visual pigments and spectral sensitivities are in nanometers.

pigment and β-band hypotheses if rhodopsin could be reduced by some means other than carotenoid deprivation. Certain *Drosophila* visual mutants of the class designated *nina A* appear to offer that possibility. The mutant *nina A*[P228] mimics carotenoid deprivation in that the concentrations of rhodopsin and opsin are reduced, as is the particle density in the rhabdomere membrane. The mutation apparently reduces rhodopsin synthesis in R1–6, not the availability of the chromophore. It may be an allele of the structural gene for the opsin. Unlike carotenoid deprivation, *nina A*[P228] does not selectively depress the ultraviolet component of the action spectrum. Sensitivity is reduced overall, but the spectral sensitivity function retains both maxima (Larrivee, D. *et al.*, 1981). Hence, high ultraviolet sensitivity may not depend on a high density of rhodopsin in the rhabdomere.

8.5 Spectral sensitivities of R7/8

The small receptors with superimposed rhabdomeres at the center of the fly retinula have been more difficult to characterize. Initial efforts to penetrate fly receptors yielded only a few recordings that could be identified with R7/8 (Burkhardt, D. 1962; Järvilehto, M. and Zettler, F., 1973; Smola, U. and Meffert, P. 1975; Meffert, P. and Smola, U. 1976; Eckert, H. *et al.*, 1976). Alternative approaches for determining action spectra have included analysis of ERGs from *Drosophila* mutants lacking the rhabdomeres of R1–6 or R1–7 (Harris, W. *et al.*, 1976; Stark, W. 1977), isolation of the central receptors by selectively depressing the activity of the photopic R1–6 system through light adaptation (Minke, B. *et al.*, 1975; Stark, W. 1975), and behavioral studies (Eckert, M. 1971).

ERG action spectra from the combined *Drosophila* mutants *sev* and *ora* (or *sev* and *rdgB*), in which only R8 remains functional, show a single peak at 490 nm. When both R7 and R8 are present but not R1–6 (mutants *rdgB* and *ora*), there is an additional ultraviolet peak that can be selectively reduced by ultraviolet adaptation. The study of *Drosophila melanogaster* mutants therefore indicates that R7 is an ultraviolet receptor and R8 is

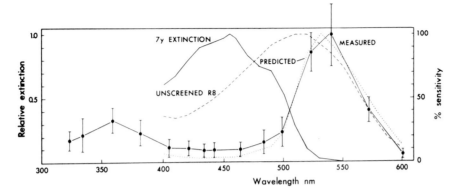

FIG. 20. Spectral sensitivity of R8G cells from *Calliphora stygia* (●) (error bars: ± 1.0 S.D.) compared with a 520 nm rhodopsin nomogram (dashed line) matched on its long wavelength side with the spectral sensitivity function. The solid line is the spectrum of the stable carotenoid, P456, measured by Kirschfeld, K. *et al.* (1978) in R7y rhabdomeres of *Musca domestica*. The dotted line is the predicted spectral sensitivity function of R8G cells if they contain P520 screened by P456 (Hardie, R. *et al.*, 1979).

a blue-green receptor (Harris, W. *et al.*, 1976; Stark, W. 1977). Similar conclusions were drawn from ERG measurements after selective adaptation in wild-type *Drosophila* with all eight receptors intact (Stark, W. 1975; Minke, B. *et al.*, 1975).

Later intracellular measurements from identified cells in *Calliphora* and *Musca* revealed at least four, and perhaps five, kinds of cells distinguished by their spectral sensitivity functions. Hardie, R. *et al.* (1979) found two types of R7 cells in *Calliphora* and *Musca*, all with a major sensitivity peak in the ultraviolet: one class (designated R7UV) with a single maximum at 340 nm, the other (designated R7UT) with a peak at 360 nm and long dropping tail of sensitivity extending to 500 nm. Smola, U. and Meffert, P. (1979) also identified two types of R7 in *Calliphora*: the strictly ultraviolet-sensitive R7UV, and a second kind (designated R7UB) with two distinct maxima at 350 nm and in the blue-green at 480 nm (Fig. 19). It is likely that R7UT and R7UB are the same units, differing for some reason in the different populations of flies used by these separate groups of investigators. Both groups found a class of R8 (designated R8G) with a sharp sensitivity peak in the green at 540–550 nm, and a broad minimum at shorter wavelengths rising again to a low peak in the ultraviolet. In addition, Smola, U. and Meffert, P. (1979) found a second type of blue-sensitive R8 (designated R8B) with a single peak at 450–460 nm. To summarize, four classes R7/R8 have been identified: R7UV, R7UB (R7UT), R8B, and R8G (Fig. 19; Table 4).

A reasonably coherent interpretation of R7 and R8 emerges when these sensitivity functions are considered in conjunction with photometric measurements. Seen in transmitted white light, two sorts of R7 cells can be distinguished in *Calliphora*, *Musca* and *Drosophila*: 65–75% have yellow rhabdomeres (these are designated R7y), the rest are pale (R7p). The two sorts of R7 seem to be randomly distributed across the retina (Kirschfeld, K. and Franceschini, N. 1977; Kirschfeld, K. *et al.*, 1978; McIntyre, P. and Kirschfeld, K. 1981). The color of the yellow rhabdomeres of R7y results from a photostable pigment (P456). It has a three-peaked spectrum very similar to that of a carotenoid such as β-carotene or xanthophyll (Fig. 20), and it is eliminated by carotenoid deficiency. Since it is dichroic, P456 is probably a component of the microvillous membranes of R7y, but unlike the characteristic birefringence of visual pigments, its absorbance is highest for light polarized perpendicular to the axes of the microvilli. It is most likely inserted parallel to the fatty acid chains of the membrane phospholipids — that is, perpendicular to the membrane surface (McIntyre, P. and Kirschfeld, K. 1981). In its various aspects, P456 resembles human macular pigment, which is xanthophyll apparently oriented within axon membranes in front of the fovea (Brown, P., personal communication).

The ratio (3:1) of R7UB:R7UV (or R7UT: R7UV) is about the same as R7y:R7p, which in turn is similar to R8G:R8B. In light of these similar ratios, it has been suggested (Hardie, R. *et al.*, 1979;

Ratio	1/4	3/4
Receptor Class	R7UV = R7p	R7UB = R7UT = R7y
Stable Pigment	None	P456
Visual Pigment	P370, M470	P430, M505
Spectral Sensitivity	350	350, 440

Receptor Class	R8B	R8G
Visual Pigment	?	?
Spectral Sensitivity	440	520 – 540

FIG. 21. Inferred relationships of the different receptor classes of the central receptors, R7/8, of fly ommatidia. The rhabdomeres of R7 lie distal to those of R8 along the optical axis. It is thought that R7UV is distal to R8B; R7UB (R7UT) distal to R8G. Spectral sensitivity maxima and inferred λ_{max} of visual pigments are given in nanometers. None of the visual pigments of these cells have been well characterized (data assembled mainly from Hardie, R. et al., 1979 and Smola, U. and Meffert, P. 1979).

Smola, U. and Meffert, P. 1979) that R7UB (or R7UT) = R7y, R7UV = R7p, and that R7UB (or R7UT) containing P456 lies distal to R8G, while R7UV overlies R8G (Fig. 21).

8.6 Visual pigments of R7/8

The visual pigments of R7 and R8 cells have not been well characterized. R7UV cells probably contain an ultraviolet-sensitive rhodopsin with a longer wavelength metarhodopsin similar to the visual pigment of *Ascalaphus* (Fig. 8). Such a photopigment was measured in retinal extracts of *Drosophila* mutants with only R7 rhabdomeres intact (Harris, W. et al., 1976), and it is suggested by MSP measurements of R7p rhabdomeres (Kirschfeld, K. 1979). Intense ultraviolet irradiation of these cells results in a prolonged depolarization (PDA) that is reversed by blue light (Hardie, R. et al., 1979). MSP difference spectra of R7y cells are consistent with a violet-absorbing rhodopsin that converts to a green-absorbing metarhodopsin (Kirschfeld, K. 1979). The high ultraviolet sensitivity of these cells (R7UB = R7y) presents the same problem as it does in R1–6.

Since no measurements of R8 rhodopsin have been made in *Calliphora* or *Musca*, they must be inferred from spectral sensitivity. Presumably the rhodopsin of R8B has a spectrum similar to the sensitivity function. The spectral sensitivity of R8G is less straightforward since the peak at 540 nm is much narrower than a typical rhodopsin α-band. However, a 520 nm rhodopsin nomogram fits the long wavelength side of the sensitivity peak. If R7y cells containing stable P456 lie in the optical path distal to R8G cells, as argued above, then blue wavelengths will be filtered out of the light reaching R8G. Subtraction of the spectrum of R7y from a 520 nm nomogram yields a good facsimile of the sensitivity function of R8G (Fig. 20). A similar sharpening of spectral sensitivity is seen in reptiles and birds in which carotenoid-containing oil droplets filter the light reaching cone outer segments (Wolbarsht, M. 1976).

The ERG measurements of *Drosophila melanogaster* mutants (Harris, W. et al., 1976; Stark, W. 1977; Stark, W. et al., 1979) indicated that R8 carries a blue-green-sensitive rhodopsin, $\lambda_{max} = 490$ nm. Since a spectrally distinct metarhodopsin could not be measured, and chromatic adaptation could not be effected, the rhodopsin and metarhodopsin probably have similar spectra. The relationship of these measurements from *Drosophila* to the classes of R8 cells identified in *Calliphora* and *Musca* by intracellular recording is uncertain.

An additional complication has recently been discovered (Franceschini, N. et al. 1981a). Most of the R7/8 cells in the frontal–dorsal region of the male *Musca* eyes have the same sensitivity as R1–6. The R7 cells in this area of the eye also bear an ultrastructural resemblance to R1–6. Whereas R7/8 in other parts of the eye extend their axons to different brain centers than do R1–6, the special R7 cells of the male synapse in the lamina along with R1–6. These receptor cell modifications may be associated with the sex-specific chasing behavior of male flies, that receives its main visual input from the dorsal–frontal region of the eye and requires tracking the small, fast-moving female. The authors suggest that adding the input of R7/8 to that of R1–6 increases the signal-to-noise ratio for this activity and possibly modifies movement detection and contrast, while sacrificing spectral information.

Table 5: *Fluorescent properties of fly rhabdomeres*

Rhabdomere	Exciting light	
	UV	Blue
R1–6	pink (metarhodopsin + antenna pigment?)	red (metarhodopsin?)
R7y	green	none
R7p	none	none
R7p (dorsal–frontal region of eye)	pink	red

From Stark, W. *et al.*, 1979; Franceschini, N. *et al.*, 1981a,b; Stavenga, D. and Franceschini, N., 1981.

8.7 Rhabdomere fluorescence

The rhabdomeres of the different classes of fly receptors have been found to possess the distinctive fluorescent properties summarized in Table 5. Rhabdomere fluorescence of *Musca* has been described in detail, but *Drosophila*, *Calliphora*, *Sarcophaga* and *Eristalis* are said to be similar. In addition to the blue-excited red fluorescence thought to arise from modified metarhodopsin (see section 4.9) ultraviolet excitation of R1–6 elicits a whitish-pink fluorescence. It has been suggested that it arises from the hypothetical antenna pigment (Franceschini, N. *et al.*, 1981a,b; Stavenga, D. and Franceschini, N. 1981). The special R7/8 cells that mimic the spectral sensitivity of R1–6 have the same fluorescence.

R7p rhabdomeres do not fluoresce under either ultraviolet or blue excitation. However, blue light excites green fluorescence in R7y rhabdomeres. The photostable carotenoid (P456) that gives R7y rhabdomeres their yellow color may be the fluorescent substance. If that is the case then P456 must be altered by its membranous environment, since carotenoids in solution hardly fluoresce at all.

8.8 Dorsal ocelli and larval photoreceptors

The dorsal ocelli of flies resemble R1–6 in having dual ultraviolet and blue-green sensitivity maxima that cannot be selectively adapted. Like R1–6 the receptor cells of the ocelli may contain only a single rhodopsin (Kirschfeld, K. and Lutz, B. 1977; Labhart, T. 1977; Hu, K. *et al.*, 1978).

Behavioral and light microscope studies indicate that *Musca* larvae have photoreceptors in the pharyngeal region of the head. The phototactic action spectrum of *Calliphora* larvae very closely matches a 504 nm rhodopsin nomogram (Strange, P. 1961). These early efforts have not yet been followed up by the application of contemporary techniques.

8.9 Nematocera

Little is known about the visual systems of the primitive flies (suborder Nematocera). The retinulae of the compound eye contain eight receptor cells arranged as they are in the higher flies except that their rhabdomeres are not separated (Brammer, J. 1970; Williams, D. 1980, 1981). A visual pigment, P520/M480, similar to that measured in the larval mosquito (see section 2.4), has been extracted from the compound eyes of *Aedes aegypti* (Stein, P. *et al.*, 1979).

9 WAVELENGTH DISCRIMINATION AND COLOR VISION

The principle of univariance states that any photon absorbed by a visual pigment has the same effect regardless of its energy, The response of a photoreceptor depends only on the number of quanta caught, and not on their energies; its response at any given wavelength can be matched at another with appropriate adjustment of intensity. Responses that are indistinguishable at the receptor level must also appear the same to the organism (Rushton, W. 1972).

Therefore if an animal is to have color vision, it must have photoreceptors with different spectral sensitivities. In general, such differences depend upon the receptors containing different visual pigments whose main absorption bands are set in different spectral regions. However, it is possible for the same rhodopsin to mediate different spectral sensitivity functions. For instance, the grasshopper *Phlaeoba* is capable of spectral discrimination, and yet its eye appears to contain only a single rhodopsin, $\lambda_{max} = 525$ nm. The eye is divided into clear and brown horizontal bands. The ommatidia of the brown bands contain an accessory pigment, absorbing maximally at about 500 nm. The brown pigment is absent from the clear bands. Whereas the sensitivity function of the receptors in the clear bands matches the spectrum of P525, in the brown bands it is shifted to about 545 nm, presumably because of the selective filtering of incident light by the accessory pigment. Behavioral tests showed that the animals can distinguish green from yellow stimuli — they are attracted to green — independent of intensity differences (Kong, K. *et al.*, 1980).

9.1 Spectral sensitivity

The various ways by which the spectral sensitivity functions of photoreceptors can be led to differ from the absorption spectra of their rhodopsins are reviewed below.

9.1.1 BROADENED AND NARROWED SPECTRAL SENSITIVITY FUNCTIONS

Self-screening by rhodopsin may broaden the spectral sensitivity of cells with long, isolated rhabdomeres, such as R1–6 in the open rhabdoms of flies. Since such rhabdomeres function as waveguides, and laterally scattered light is absorbed by screening pigments, only light entering the distal tip of the rhabdomere penetrates to the proximal end. Passing down the rhabdomere it is selectively depleted of the wavelengths most strongly absorbed by rhodopsin. Consequently the spectral sensitivity function is broadened around the absorption maximum and flattened in comparison with the rhodopsin spectrum (Snyder, A. and Pask, C. 1973; Hamdorf, K. 1979). Broadening depends on the concentration of rhodopsin in the rhabdomere and its length. With perhaps as much as 1% of the incident light absorbed per micrometre of rhabdomere (Goldstein, E. and Williams, T. 1966; Hamdorf, K. 1979; Snyder, A. 1979), the effect might be significant even in rhabdomeres of moderate length. About 90% of the incident light at the λ_{max} of rhodopsin is absorbed by the 250–300 μm rhabdomeres of *Musca* and *Calliphora* (Hamdorf, K.

1979; Hamdorf, K. and Razmjoo, S. 1979).

Although long rhabdomeres have the effect of increasing absolute sensitivity, broadened spectral sensitivities will reduce the capacity for wavelength discrimination. Rhodopsin self-screening is counteracted by several mechanisms. Stable metarhodopsin in photoequilibria also acts as a filter within the rhabdomere. If its main absorption band is shifted to longer wavelengths relative to rhodopsin it will displace the spectral sensitivity peak to a wavelength shorter than the λ_{max} of the rhodopsin (Hamdorf, K. 1979). If the metarhodopsin lies at shorter wavelengths the sensitivity function will shift to longer wavelengths. In addition the sensitivity peak will be narrowed in comparison with the rhodopsin spectrum. The sharp minimum often seen around 400 mm in ERG spectral sensitivities may at least partly result from the filtering effect of typical blue-green absorbing metarhodopsins. Screening pigment will have a similar effect if its spectrum is not exactly congruent with the rhodopsin spectrum (Fig. 7).

A twist in the rhabdomere that rotates the axes of the microvilli will also tend to counteract broadening. Since the microvilli are dichroic, absorbance by distal rhodopsin will polarize the light penetrating to deeper levels. A twist will bring the more proximal molecules into an orientation to maximally absorb that polarized light (Smola, U. and Tscharntke, H. 1979). Reports that the rhabdomeres of flies are twisted (Grundler, O. 1974; Smola, U. 1977; Smola, U. and Tscharntke, H. 1979) were later dismissed as artifacts of histological preparation (Ribi, W. 1979). The argument has stimulated efforts to decide whether in fact twist is present *in vivo*. These more recent studies indicate that the rhabdomeres of flies (Altner, I. and Burkhardt, D. 1981; Smola, U. and Wunderer, H. 1981a,b; Williams, D. 1981), and rhabdoms of other insects (Wehner, R. and Meyer, E. 1981) are indeed twisted.

Self-screening is effectively reduced when rhabdomeres with different rhodopsins are optically coupled within a fused rhabdom, such as those of Lepidoptera and Hymenoptera. This concept was introduced by Shaw, S. (1969b) and developed by Snyder, A *et al.* (1973). Light entering the tip of one rhabdomere may be absorbed by one of its neighbors in the rhabdom. As the rhabdom as a whole functions as a waveguide, light is internally reflected back and forth through the different rhabdomeres so that they function as "lateral filters" for one another. A fused rhabdom can be very long, providing high absolute sensitivity without concomitant flattening of the spectral sensitivities of the separate units. Such an arrangement is well suited for color vision at low levels of illumination.

Bernard, G. (1981b) has shown these features of the fused rhabdom particularly clearly. The total absorbance of a rhabdomere in a fused rhabdom is actually greater than it would be if the rhabdomere were alone. Depending on the absorbance maxima of other visual pigments in a rhabdom, β-band absorbance of long-wavelength rhodopsins may be reduced, thus sharpening spectral sensitivity functions.

9.1.2 SECONDARY SENSITIVITY PEAKS

In addition to shifting and sharpening spectral sensitivity peaks, accessory pigments can also produce secondary maxima (Langer, H. 1975). For instance the ERG spectral sensitivity functions of *Calliphora* and *Musca* show a peak in the red at about 620 mm (Autrum, H. and Stumpf, H. 1953). The peak is absent in mutants lacking accessory pigments, and a separate receptor with red sensitivity cannot be found among single-cell recordings. Goldsmith, T. (1965) showed that this peak results from scattered red light admitted into the eye through the red screening pigment characteristic of flies. Similar windows in the absorption spectra of screening pigment may account for anomalous secondary peaks and shoulders found in the spectral sensitivity functions of other insects.

9.1.3 ELECTRICAL COUPLING

The spectral sensitivity functions of individual receptor cells will be broadened or show multiple peaks or extended tails if the cells are electrically coupled to one another. This coupling might occur through direct contacts between neighboring receptors (proposed by Shaw, S. 1969b, for the locust and drone honeybee; see also Lillywhite, P. 1978; Mimura, K. 1978), or in the lamina by way of synaptic relationships between receptor axons (per-

haps in the dragonfly ocellus: Dowling, J. and Chappell, R. 1972; in flies: Boschek, C. 1970; Chi, C. and Carlson, S. 1976; in the bee: Varela, F. 1970). Other more indirect coupling has also been proposed: via accessory pigment cells in the honeybee drone (Bertrand, D. *et al.*, 1972; Baumann, F. 1975), or through the extracellular space in the lamina (Shaw, S. 1975). The need to invoke electrical coupling arises particularly when spectral sensitivity functions measured by intracellular penetration are broader than rhodopsin resonance spectra or show subsidiary peaks. Such coupling, however, could be a methodological artifact; the passage of the electrode through a neighboring cell might open a low-resistance channel between it and the one being measured. The suspicion of artifact presence has been reinforced by non-invasive measurements of the pupillary response (see section 2.3). Coupling seen in intracellular records from cells in the honeybee retina (Menzel, R. and Blakers, M. 1976) are not evident in pupillary action spectra (Bernard, G. and Wehner, R. 1980). On the other hand, Meyer-Rochow, V. (1980), has questioned the accuracy of the pupillary technique. In any event, positive electrical coupling, like self-screening, would reduce an insect's capacity for spectral discrimination.

9.1.4 MORE THAN ONE RHODOPSIN IN A CELL

Broadened or double-peaked sensitivity functions can also be accounted for by postulating two rhodopsins in the same cell. The hypothesis that the ommatida of cells R1–6 in flies contain a mixture of separate green- and ultraviolet-sensitive rhodopsins, and the reasons for discarding it, were discussed in section 8.4.1. The majority of receptors in the retina of *Locusta migratoria* have broad sensitivities that might result from either electrical coupling between distinct green- and blue-sensitive receptors, or two photopigments, $\lambda_{max} \simeq 450$ and 500 nm, in each cell (Bennett, R. *et al.*, 1967). Recent electrophysiological evidence supports the latter explanation (Lillywhite, P. 1978). DeVoe, R. (1972) has also suggested that cells of the median ocelli of the wolf spider contain two rhodopsins. As yet, however, no receptor has been shown by direct measurement to contain more than one species of rhodopsin.

9.1.5 Spectral transmission of the dioptric apparatus

The cornea of the dragonfly *Aeschna cyanea* is transparent across the visible into the near ultraviolet. Transmission is 95% at 540 nm and drops evenly to 88% at 330 nm. It then decreases more rapidly to 76% at 300 nm (Kolb, G. *et al.*, 1969). The corneas of moths were found to be similarly transparent (Bernhard, C. *et al.*, 1965; Carlson, S. and Philipson, B. 1972), as were those of several flies (Goldsmith, T. and Fernandez, H. 1968; Strother, G. and Superdock, D. 1972), beetles (Meyer-Rochow, V. 1975), and hymenopterans including the honeybee (Carricaburu, P. and Chardenot, P. 1967). Measurements of fly (Goldsmith, T. and Fernandez, H. 1968) and moth corneas (Bernhard, C. *et al.*, 1965) were extended to shorter wavelengths. The optical density rises sharply below 300 nm to a peak at 277 nm, probably due to absorbance by the aromatic amino acids of cuticular protein. The few measurements that have been made on the underlying crystalline cones indicate that they are as transparent as corneas at least into the near ultraviolet (Kolb, G. *et al.*, 1969; Carlson, S. and Philipson, B. 1972). The corneas of many species, especially of nocturnal insects, are covered with bumps, called corneal nipples, that provide a broad-band anti-reflection surface similar in function to commercial lens coatings. Such nipple arrays increase the overall transmission of the cornea by about 4% without altering the spectral content of the admitted light (Bernhard, C. *et al.*, 1968, 1970).

The typical transparent dioptric apparatus will have no effect on the spectral sensitivities of the underlying photoreceptors, except by reducing the response of ultraviolet receptors below 330 nm. However, a number of flies, particularly of the family Tabanidae, have colored corneas usually arranged in horizontal or vertical stripes across the eye (Bernhard, C. and Miller, W. 1968). These bright red, orange, yellow, green and blue patterns have the iridescent appearance of interference colors. Phase contrast light microscopy and electron microscopy show these colors to arise from a region of closely spaced cuticular layers of differing refractive index lying just below the corneal surface. The result is a quarter-wave interference filter whose spectral transmission and reflectance is governed by the spacing of the layers (Bernhard, C. and Miller, W. 1968; Bernard, G. 1971; Land, M. 1972; Miller, W. 1979). Bernard, G. (1971) measured the reflectance spectra from the neighboring orange and green facets of a long-legged fly (*Condylostylus*). The reflectance maxima lie at 580 and 510 nm respectively, with peak reflectance about 50% and half-band widths of about 60 nm. Although the reflectance maxima vary from species to species, they are adjusted so that in adjacent colored bands the peak of each is set at the first reflectance minimum of the other. It has been suggested that these corneal filters may function in color vision by enhancing color contrast (Bernhard, C. and Miller, W. 1968; Bernard, G. 1971). The pattern of neuronal connections in the lamina support the idea: the axons from receptors beneath different colored lenses are segregated into different cartridges (Trujillo-Cenóz, O. and Bernard, G. 1972). Although these corneal filters might serve to sharpen spectral sensitivity functions, there are few data bearing directly on their visual function.

9.1.6 Tapetal reflection

The proximal ends of the rhabdoms of Lepidoptera are invested with a bush of tracheoles that provides each ommatidium with its own individual reflecting tapetum. Light penetrating to the tapetum is reflected back as the characteristic glow of dark-adapted eyes. Tracheole tubules are strengthened with internal ridges, the taenidia. In the highly developed tapeta of butterflies, the taenidia form a multi-layered quarter-wavelength mirror. Typically, but not in all species, the color of the glow varies across a butterfly's eye from red and orange ventrally to blue dorsally. The spacing of the taenidial plates changes accordingly (Miller, W. and Bernard, G. 1968; Miller, W. 1979). Perhaps 5% of the light entering a rhabdom is reflected back again. It is evident that sensitivity will be enhanced for the wavelengths reflected, but the precise effects on individual receptors have not been determined.

9.2 Spectral classes of insect receptors

Menzel, R. (1979) has recently assembled a list of some 20 insects for which receptor sensitivities can be inferred with some confidence. The list is certainly not inclusive, considering the number of insects whose vision has been investigated by one technique or another with varying degrees of sophistication. The compilation supports the generalization that there are three principal classes of insect photoreceptors with sensitivity peaks in the ultraviolet (340–360 nm), violet to blue (420–460 nm), and green (490–550 nm) respectively. Red receptors at 610 nm in butterflies (see section 9.5.2) must now be added to these three common receptor types (Bernard, G. 1979).

There are also receptors with broad sensitivity functions. Data on such cells from half a dozen species have been assembled and discussed by Wasserman, G. (1973). Most have one peak in the ultraviolet, and a second at longer wavelengths. The only such receptors that have been studied in any detail are the peripheral R1–6 cells of the fly rhabdomere, whose broad sensitivity function is clearly a property of the individual cell (see section 8.4). In most other cases, as for the violet-green receptors of *Locusta* (Bennett, R. *et al.*, 1967), the basis of the extended response remains uncertain (Lillywhite, P. 1978). As mentioned above, there remains a lingering uncertainty as to the extent such coupling may be an artifact of intracellular penetration.

The various categories of receptor are found in different combinations in different species, and they are distributed across compound eyes in particular patterns. Only a few species have been examined with any thoroughness, so generalizations must be proposed with caution. The three principal types of receptor have been found in most of the carefully studied compound eyes (e.g. in the dragonflies *Aeschna cyanea*: Autrum, H. and Kolb, G. 1968; *Libellula*: Horridge, G. 1969; and *Hemicordulia tau*: Laughlin, S. 1975; in the hemipteran, *Notonecta glauca*: Bruckmoser, P. 1968; and in the Hymenoptera reviewed below). Some butterflies have four spectral classes of receptors, sensitive in the ultraviolet, blue-green and red (Bernard, G. 1977). The variety of receptor types present in the compound eyes of flies have been described in section 8.

Only ultraviolet and green receptors have been identified in *Ascalaphus macaronius* (Gogala, M. 1967) and *Formica polyctena* (Menzel, R. and Knaut, R. 1973). The locust is unusual in lacking ultraviolet receptors. In addition to its broad-band cells (Bennett, R. *et al.*, 1967), separate green- and blue-sensitive receptors are also present (Lillywhite, P. 1978).

A few species may have only a single receptor class. The ocelli of the larval mosquito seem to have only green receptors (Brown, P. and White, R. 1972; see section 2.4; the same may be true of the compound eyes of adult *Aedes* (Goldman, L. 1971) and the praying mantis *Tenodera sinensis* (Sontag, C. 1971).

Often ultraviolet and blue receptors are concentrated in the dorsal regions of compound eyes. The eye of *Ascalaphus* is divided into morphologically distinct dorsal and ventral parts. The dorsal half contains only ultraviolet receptors; green and ultraviolet receptors are mixed in the ventral half. In some species, violet and ultraviolet receptors may be mixed in the dorsal part of the eye, with green receptors restricted to ventral and frontal regions (e.g. in *Aeschna*: Autrum, H. and Kolb, G. 1968). The eye of the drone bee contains mainly violet receptors. The few green receptors occur only in the most ventral part (Menzel, R. 1975). The concentration of short wavelength receptors at the top of the eye may be an adaptation for increasing the contrast of small objects, such as flying prey, against skylight in which ultraviolet through blue wavelengths predominate (Hamdorf, K. and Gogala, M. 1973). Where different receptor types occur together in the same region of a compound eye they are found combined in the same rhabdoms, not segregated into different ommatidia.

9.3 Visual behavior

Since the compound eyes of most insects contain photoreceptors that are sensitive in different regions of the spectrum, it is evident that most are provided with the basis for wavelength discrimination. The extent to which that potential is realized can only be determined by behavioral tests. Some characteristic insect responses to visual stimuli are outlined below. Often the spectral sensitivity function of a particular pattern of behavior shows that it receives unequal input from the different spectral classes of photoreceptors.

9.3.1 ORIENTATION TO POLARIZED LIGHT

The ability to perceive polarized light is common in arthropods. The foundation of this capacity in the dichroism of rhodopsin molecules and rhabdomere microvilli is discussed in section 6.7. Polarized light detection in insects is generally associated with short wavelength receptors. In the few flies and hymenopterans that have been adequately studied, the system for polarized light detection does not have the capability of wavelength discrimination (Menzel, R. 1979). In the bee, for instance, responses to polarized light are mediated by special ultraviolet receptors (von Helverson, D. and Edrich, W. 1974) mainly located along the dorsal edge of the compound eye (Labhart, T. 1980; Meyer, E. and Labhart, T. 1981).

9.3.2 THE OPTOMOTOR RESPONSE

The optomotor response is a turning of the whole animal or its head in the direction opposite to a movement of its surroundings. Most often the response is elicited experimentally by a rotating striped drum. Spectral discrimination can be tested with stripes of different colors, or alternating colored and gray stripes of variable relative intensities. Color discrimination by the optomotor systems of various insects was concluded from early studies of this sort (reviewed by Kaiser, W. 1975). More recently, however, careful elimination of border-lines, and fine adjustment of intensity relationships between stripes, has indicated that the optomotor responses of the fly *Phormia regina* (Kaiser, W. 1968) and honeybee (Kaiser, W. and Liske, E. 1974) are color-blind. Although the optomotor response of the fly is insensitive to color contrasts, it receives input from both the R1–6 and R7/8 receptor systems (Kirschfeld, K. and Reichardt, W. 1970; Eckert, M. 1971; Heisenberg, M. and Buchner, E. 1977). On the other hand, only the green receptors contribute to the optomotor response of the bee. In general, the optokinetic systems of arthropods do not seem capable of wavelength discrimination (Horridge, G. 1967; Kaiser, W. 1975, Menzel, R. 1979), even in animals with well-developed color vision.

9.3.3 PHOTOTAXIS

Movement toward or away from a light source is positive and negative phototaxis respectively. Phototaxis may differ in sign or some other qualitative or quantitative aspect in different behavioral contexts. For instance, bees are positively phototactic when leaving the hive, but negatively phototactic when returning (von Frisch, K. 1965). The spectral sensitivity of phototaxis in bees is similar, however, in these opposite behaviors (Wendland, S. 1977, quoted in Menzel, R. 1979).

As attested by the success of "black-light" traps, most insects are strongly attracted to ultraviolet light (Mazokhin-Porshnyakov, G. 1969; Menzel, R. 1979). Bees, flies, and other insects that have been tested quantitatively may be much more sensitive to ultraviolet than to green light. Relatively high ultraviolet sensitivity may compensate for the relatively low intensity of the ultraviolet component of natural illumination (Laughlin, S. 1976). Mazokhin-Porshnyakov, G. (1969) has suggested that the generally strong phototactic response of insects to ultraviolet is a basic escape mechanism. Since skylight is the only natural source of ultraviolet, and most objects absorb rather than reflect at short wavelengths, positive phototaxis to ultraviolet would usually lead an insect out of enclosures.

As spontaneous phototaxis generally receives input from more than one spectral class of photoreceptor, the capacity for wavelength discrimination can be examined in the context of particular phototactic paradigms. To determine wavelength preference, insects may be given a choice of two colored lights presented simultaneously or successively.

9.3.4 CONDITIONED RESPONSES

Since wavelength discrimination is not present in all the visually driven behaviors of insects, the failure to demonstrate it in a particular response, such as the optomotor reflex or in a certain pattern of phototaxis, does not exclude the possibility of color vision. It is most likely to be found in such goal-oriented activities as food collection and mating behavior.

The most satisfactory and quantitative demonstrations of spectral discrimination in insects come from training experiments. Unfortunately, few insects have been found that are easily and appropriately conditioned to visual stimuli. So far, color vision has been extensively explored only in bees and ants, (see sections 9.6 and 9.7), insects that can be trained to make fine discriminations on the basis of wavelength (Menzel, R. 1979).

9.4 Wavelength discrimination

Color vision is defined as the ability to discriminate among stimuli on the basis of wavelength independent of intensity (Goldsmith, T. 1961; Burkhardt, D. 1964; Menzel, R. 1979). Examples have been given in the preceding sections of insect behavioral patterns that are wavelength-specific. That an animal displays different behavioral patterns with different spectral sensitivity functions is not evidence for color vision. However, the interaction of separate wavelength-specific behaviors may resemble true color vision.

The distinction between wavelength-specific behavior and wavelength discrimination can be exemplified by comparing the phototactic behavior recently examined in two insects, the whitefly *Trialeurodes vaporariorum* (Homoptera) and *Drosophila melanogaster*. In the former, phototaxis appears to be a composite of independent behavioral patterns with different spectral sensitivities. By contrast, the greater complexity of phototaxis in *Drosophila* suggests that color vision may be present.

9.4.1 Whitefly phototaxis

Above a certain threshold, whiteflies are positively phototactic to long-wavelength light (550 nm) and negatively phototactic at shorter wavelengths (400 nm). With increasing intensity the positive response to green light rises to a peak and then drops off. The negative response to violet light has the same threshold as the positive response, but increases continuously with rising intensity. The response is not univariant, more than one receptor type must be involved, and phototaxis varies in relation to both wavelength and intensity. When whiteflies are offered a choice between green and violet fields set at differing relative intensities, preference is determined at low and moderate intensities by the sum of the opposed responses. With brighter stimuli, the long-wavelength response appears to inhibit the short-wavelength response (Coombe, P. 1981). The results are interpreted as showing the independence of the long and short wavelength photoreceptor outputs, with little interaction in the central nervous system, at least at lower intensities. Whitefly phototaxis therefore appears to be a wavelength-specific behavior that does not involve color vision.

9.4.2 *Drosophila* phototaxis

Two kinds of phototaxis can be distinguished in *Drosophila melanogaster*. Undisturbed flies respond to light with slow phototaxis; fast phototaxis is the response of agitated flies. At relatively high intensities, i.e. under photopic conditions, positive slow phototaxis is shown to both ultraviolet and visible light, but given a choice, the flies find ultraviolet much more attractive. A green stimulus (550 nm) must be a thousand times as intense as an ultraviolet alternative (375 nm) if it is to be equally attractive (Schümperli, R. 1973). Furthermore, ultraviolet mixed with monochromatic light at wavelengths greater than 415 nm is less attractive to wild-type flies than ultraviolet alone, even though the mixture is more intense. Since flies prefer the brighter of two ultraviolet lights, ultraviolet mixed with longer wavelengths must be perceived as qualitatively different from pure ultraviolet. Moreover, when different colors are presented successively to the flies, ultraviolet becomes more attractive when preceded by green (Fischbach, K. 1979).

I mentioned in section 8 that receptor cells R1–6 function at low levels of illumination, while R7/8 come into play at higher intensities. Neither the strong attraction to ultraviolet light, nor the ability to perceive simultaneous and successive color contrast, is present under scotopic conditions. This is not surprising since phototaxis is then mediated only by R1–6, all of which have identical broad spectral sensitivity functions. On the other hand, differential response to wavelength is also missing under scotopic conditions in mutants lacking R1–6. Consequently, wavelength discrimination in slow phototaxis must depend on input from all eight receptors. Judging from the different responses of wild-type flies and mutants lacking various receptors, phototaxis toward ultraviolet depends on both R1–6 and R7 (Schümperli, R. 1973), while the response to longer wavelengths presumably is mediated by R8 (Willmund, R. and Fischbach, K. 1977; Willmund, R. 1979). In contrast to phototaxis in whiteflies, slow phototaxis cannot be described by a linear sum of responses initiated by different classes of photoreceptor.

"Fast" phototaxis — that is, the response of disturbed flies — shows somewhat different characteristics. Like slow phototaxis it depends on R1–6
in dim light, but under photopic conditions fast
phototaxis is mediated mainly by R7/8 (Schümperli, R. 1973; Stark, W. *et al.*, 1976; Hu, K. and
Stark, W. 1977). Training experiments under conditions that elicit fast phototaxis have shown that
Drosophila can be conditioned to discriminate between short and long wavelength stimuli. However,
it seems that the hue discrimination is not learned
entirely independently of intensity (Fukushi, T.
1976; Menne, D. and Spatz, H. 1977; Bicker, G. and
Reichert, H. 1978).

Wavelength discrimination in *Drosophila*
phototaxis is more complex than simple
wavelength-specific behavior; although its neuronal
basis is unknown, it certainly must involve the
processing of spectral information within the
central nervous system. The evidence that
Drosophila distinguishes between ultraviolet and
visible wavelengths independent of intensity, and
can be conditioned to wavelength differences, indicates at least a primitive form of color vision.
Some investigators, however, remain doubtful that
true color vision has been demonstrated in
Drosophila (Hu, K. *et al.*, 1978; Miller, G. *et al.*,
1981).

9.5   Color vision in Lepidoptera

Color vision has been inferred for various butterflies and moths from feeding behavior (Knoll, F.
1924; Ilse, D. 1928), and the responses of visual
interneurons (Swihart, S. 1969, 1970, 1972a,b).
Although color vision has not been as well
established in Lepidoptera as in Hymenoptera,
more is known about the visual pigments of the
former.

9.5.1   MOTHS: *DEILEPHILA* AND *MANDUCA*

Ultraviolet, violet, and green receptors have been
identified by ERG measurements in conjunction
with spectral adaptation (Fig. 22) in the sphingid
moths *Manduca sexta* (Höglund, G. and Struwe, G.
1970) and *Deilephila elpenor* (Höglund, G. *et al.*,
1973a,b). Three corresponding rhodopsins, P520,
P440, and P350, have been measured by MSP in

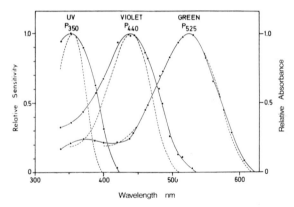

FIG. 22. ERG spectral sensitivity functions of receptor types
in *Deilephila elpenor* (continuous lines with dots). Receptor
classes were isolated by selective chromatic adaptation: UV
by violet (438 nm) + yellow (550 nm and longer
wavelengths); violet by ultraviolet (347 nm) + yellow; green
by ultraviolet + violet. Broken lines plot rhodopsin
nomograms at 350, 440, and 525 nm (from Höglund, G. *et al.*,
1973b).

Deilephila (Hamdorf, K. *et al.*, 1972, 1973a;
Schlecht, P. *et al.*, 1978) as have similar pigments in
Manduca (Schwemer, J. and Brown, P., personal
communication) and in digitonin extracts
(Schwemer, J. and Paulsen, R. 1973). From their
relative extinctions measured by MSP in retinal
slices of *Manduca* the ratio of P520 : P440 : P350 is
about 5 : 1 : 1. Schlecht, P. (1979) estimated their
relative amounts in *Deilephila* at 2 : 1 : 2 from the
size of the rhabdoms in the cells that presumably
contain them (Schlecht, P. *et al.*, 1978).

Wald, G. (1968) has pointed out, in reference to
vertebrate vision, that the maintenance of color
constancy under various conditions of adaptation
and illumination requires that the color vision pigments possess approximately parallel kinetics of
bleaching and regeneration. The same neccessity for
balance among the visual pigments of an insect
retina is met if similar photosteady states are maintained under different conditions. The three visual
pigments in the Sphingid retina have similar metarhodopsins with λ_{max} between 480 and 490 nm, near
the emission maximum of daylight (Höglund, G. *et
al.*, 1973b, Hamdorf, K. 1979). Consequently, the
three rhodopsins will be maintained at constant,
relatively high concentrations in the photoequilibria established by natural illumination.
Under skylight, the receptors contain about 90%
P350, 70% P440, and 70% P520 (Hamdorf, K. and

Schwemer, J. 1975; Hamdorf, K. 1979), and the ratio of their responses will remain constant as brightness varies, an essential condition for the maintenance of color constancy.

The sphingid retinula consists of nine cells with a fused rhabdom, that is, the rhabdomeres are contiguous down the center of the retina. Two of the rhabdomeres extend to the distal surface of the retina, six are located medially and the ninth small rhabdomere is located at the proximal end of the retinula. Structural analysis (Welsch, B. 1977) and MSP measurements of retinal slices (Schlecht, P. *et al.*, 1978; Schwemer, J. and Brown, P., personal communication) indicate that the distal rhabdomeres contain P350, while P520 is housed in the medial rhabdomeres. The cellular assignment of P440 has not been directly determined, but Welsch, B. (1977) and Schlecht, P. *et al.* (1978) have argued for the single proximal rhabdomere. However it seems to be too small to account for the amount of P440 that is measured. There have been no intracellular recordings from these cells that would serve to identify their spectral sensitivity functions.

Recently Schlecht, P. (1979) has attempted a theoretical analysis of the potential for color vision provided by the presumed arrangement of receptor cells in *Deilephila elpenor*. Although depending on certain critical assumptions, the general conclusion seems warranted: that the system is capable of good wavelength discrimination under natural conditions. A qualitative study of the feeding behavior of *Deilephila livornica* provided direct evidence for color vision in these moths (Knoll, F. 1924).

9.5.2 RED RECEPTORS IN BUTTERFLIES

Most insects, compared with vertebrates, see poorly at the red end of the spectrum. However observations of behavior (Ilse, D. 1928; Von Frisch, K. 1950), and ERG measurements (Post, C. and Goldsmith, T. 1969; Bernhard, C. *et al.* 1970; Swihart, S. and Gordon, W. 1971) have indicated exceptional red sensitivity in some butterflies. In contrast to bees, many butterflies are known to favor red flowers, and seem able to discriminate between red and orange hues (von Frisch, K. 1950). The red sensitivity of butterflies is founded upon red-absorbing rhodopsins whose λ_{max} range from 585 to 610 nm (Bernard, G. 1977, 1979).

Vertebrate visual pigments (porphyropsins) with dehydroretinal as the chromophore are shifted by 30–50 nm to longer wavelengths than the corresponding rhodopsins. For instance, the red-sensitive cones of the goldfish contain porphyropsin peaking at 620 nm (Hárosi, F. 1976). However dehydroretinal has not been found in any invertebrate photopigment. Assuming, then, that the red-absorbing visual pigments of butterflies are based on retinal, they are the most red-shifted rhodopsins known. The rhodopsin of human red-sensitive cones, for comparison, peaks at 570 nm (Brown, P. and Wald, G. 1964; Marks, W. *et al.* 1964).

Butterflies are also known that contain receptors sensitive in the ultraviolet, blue, and green (Bernard, G. 1977). Therefore it is possible that some butterflies may have color vision systems based on four receptor classes rather than three. Four visual pigments have been identified in the retina of the African armyworm moth *Spodoptera exempta* (Langer, H. *et al.*, 1979; Schwemer, J. and Langer, H. 1982). Tetrachromatic color vision is discussed in section 9.7.

9.6 Color vision in the honeybee

Color vision in the honeybee (*Apis mellifera*) was first demonstrated by the well-known experiments of von Frisch, K. (1915, 1950) showing that workers can be trained to collect food from colored cards surrounded by an array of gray cards of all shades from black to white. Later Kühn, A. (1927) repeated these experiments with spectral lights, confirming that worker bees can discriminate colors across the visible spectrum from orange to violet wavelengths, but that they are red-blind. He also discovered that bees respond to ultraviolet light. Although there have been many investigations of bee color vision since the initial experiments of von Frisch, only more recent quantitative studies will be discussed here.

9.6.1 PHOTORECEPTORS

The spectral sensitivities of single photoreceptors were first determined in bees by Autrum, H. and von Zwehl, V. (1964). Their observations were generally confirmed by the subsequent more extensive analysis of Menzel, R. and Blakers, M. (1976).

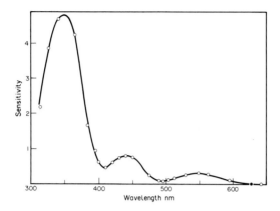

FIG. 23. Spectral sensitivity function of worker honeybees (*Apis mellifera*) measured from learning experiments (from von Helverson, O. 1972).

There are three receptor types in the worker bee retina belonging to the three principal classes of insect receptor cells: ultraviolet ($\lambda_{max} \simeq 350\,$nm), violet ($\lambda_{max} \simeq 440\,$nm), and green ($\lambda_{max} \simeq 540\,$nm) sensitive (Fig. 23). Their spectral sensitivities are well described by the corresponding rhodopsin nomograms except that most of the green- and ultraviolet-sensitive units are secondarily responsive at shorter and longer wavelengths respectively. It remains uncertain whether these secondary sensitivities arise from normal electrical coupling between the receptors or from damage that occurs during penetration (Bernard, G. and Wehner, R. 1980; Labhart, T. 1980). One would expect the capacity for wavelength discrimination to be reduced by positive electrical coupling between receptors (see section 9.1.3).

The fused rhabdom of the worker bee retinula is long (300–350 nm) and slender (about $2\,\mu m$ in diameter at the distal end). In the central part of the eye each retinula consists of eight receptor cells extending across the full depth of the retina, and a small ninth proximal cell (Varela, F. and Porter, K. 1969; Skrzipek, K. and Skrzipek, H. 1973, 1974). The mutual arrangement of the eight distal rhabdomeres is shown in Fig. 24. Each has been assigned to a sensitivity class on the basis of dye injection after intracellular recording, and — by considering the orientations of their microvilli — on the relationship between spectral and polarization sensitivity (Menzel, R. and Blakers, M. 1976; Wehner, R. and Bernard, G. 1980). Cells 1 and 5 have been identified as ultraviolet receptors, cells 3 and 7 as

green receptors, cells 4 and 8 as probably green receptors, and cells 2 and 6 as probably blue receptors. In any case, all three receptor types are represented among the long distal receptors in each ommatidium. Proximal cell 9 is probably an ultraviolet-sensitive cell that functions in polarized light detection (Menzel, R. and Snyder, A. 1974). Except along the dorsal edge of the eye, the rhabdoms are twisted (Wehner, R. and Meyer, E. 1981), thus reducing the sensitivity of the distal receptors to polarized light.

9.6.2 VISUAL PIGMENTS

None of the visual pigments from the compound eyes of worker bees have been measured directly. Digitonin extraction of worker heads yielded a pigment, $\lambda_{max} = 490\,$nm, whose spectrum matches a rhodopsin nomogram, and whose chromophore is retinal (Fernandez, H. and Bishop, L. 1973). If this is a visual pigment, it may be from the dorsal ocellus, one of whose receptor types has maximum sensitivity at 490 nm (Goldsmith, T. and Ruck, P. 1958). A rhodopsin, $\lambda_{max} = 445\,$nm, has been measured by MSP in the drone compound eye (Bertrand, D. *et al.*, 1979).

9.6.3 TRAINING EXPERIMENTS

Daumer, K. (1956) and von Helverson, O. (1972) extended the earlier measurements of color discrimination by conditioning bees to monochromatic lights and known mixtures of

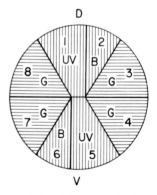

FIG. 24. Arrangement of rhabdomeres in the fused rhabdom of the worker honeybee. The orientation of the microvilli are shown, as well as the inferred sensitivities of the receptors (Menzel, R. and Blakers, M. 1976; Wehner, R. and Bernard, G. 1980).

monochromatic lights in which intensity, as well as wavelength, was controlled. The bee responds in such behavioral tests from 300 to at least 650 nm, with sensitivity dropping off at either end of the spectrum.

Spectral discrimination as a function of wavelength was measured by training bees to a particular monochromatic light, and then forcing them to choose between that wavelength and others close to it (von Helverson, O. 1972). The increment of wavelength difference required to elicit an arbitrary frequency of correct responses (in this case 70%) plotted across the spectrum is defined as the spectral or hue discrimination function (Fig. 25). There are two regions of maximal discrimination, at 390–400 nm and 480–500 nm, where bees can distinguish monochromatic lights at intervals of 4–7 nm. The region around 400 nm is of particular importance for bees. They show the fastest rate of learning from 390–410 nm (Menzel, R. 1967) where color discrimination is high, and prefer flowers with blue centers and ultraviolet edges (Daumer, K. 1958). Although vision for the bee extends beyond 600 nm, above 550 nm the capacity for wavelength discrimination is lost; that is, from 550 to 650 nm the color appears the same to the bee; only intensity differences can be recognized. The same is true below 350 nm.

A hue discrimination function should be related to the spectral sensitivity function. If input from each of the receptor types is being compared, then spectral discrimination should be highest in those spectral regions where the sensitivities of two receptor types overlap and are changing most rapidly with respect to each other. It can be seen that maximum discrimination is in fact located where the receptor sensitivity curves intersect. Discrimination is lower in the vicinity of the sensitivity maxima. Wavelengths that stimulate only one receptor, as do those greater than 550 nm, cannot be distinguished.

Saturation is another aspect of color vision that can be quantified. Monochromatic light is maximally saturated; desaturation results from the addition of white light. The more saturated a monochromatic light appears to an animal, the more white light can be added to it before it no longer can be discriminated from white light. In general, the apparent saturation of a monochromatic light depends on the ratio of

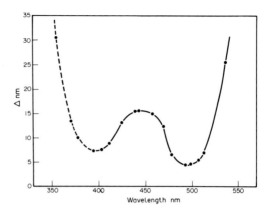

FIG. 25. Spectral discrimination function of the worker honeybee. The increment of difference between monochromatic stimuli that trained bees can distinguish (70% correct responses) is represented on the ordinate (from von Helversen, O. 1972).

chromatic to achromatic sensitivity at a particular wavelength, and is also related to the absorption spectra of the visual pigments (Jameson, D. and Hurvich, L. 1955). Saturation functions should peak, as they do in vertebrates, where the receptors are maximally sensitive (Marks, W. 1965). The data for bees are limited but consistent with this expectation. Apparent saturation is highest in the ultraviolet and violet, and lowest at about 390 nm between the ultraviolet and violet sensitivity peaks (Daumer, K. 1956).

White light for the bee must include ultraviolet as well as visible wavelengths. Daumer, K. (1956) found that three properly balanced monochromatic lights stimulating the three classes of receptors cannot be distinguished by bees from the full spectrum of a xenon lamp. Thus:

bee-white = 55% 588 nm + 30% 440 nm + 15% 360 nm.

Similarly:

bee-white = 15% 360 nm + 85% 490 nm,

defining complementary colors (Fig. 26). Mixtures of ultraviolet (360 nm) and yellow (588 nm) produce a new quality of light that bees can distinguish from other wavelengths. It is the non-spectral color "bee-purple", analogous to the unique color we see when red and blue are mixed.

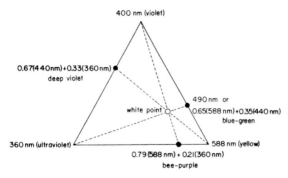

FIG. 26. Color triangle for the worker honeybee calculated by Daumer, K. (1956) from color-matching experiments. Spectral hues from 360 to 588 nm are represented along the upper sides of the triangle. The monochromatic hues at the corners are approximately those that maximally stimulate each of the three receptor classes individually. The hues along the sides of the triangle excite predominantly two of the receptor classes; the distance along a side approximating the relative stimulation of each. The hues along the bottom of the triangle are non-spectral "bee-purple" mixtures of yellow and ultraviolet. The dotted lines connect complementary colors that mixed in the correct proportions appear white, because all three receptors mechanisms receive balanced excitation.

9.6.4 TRICHROMATIC COLOR VISION

The quantitative analysis of color vision in the bee shows it to be similar to our own. It is a trichromatic system, but it is shifted about 100 nm toward shorter wavelengths. The λ_{max} of our cone cells lie at 445, 535, and 570 nm (Rushton, W. 1972), those of bee receptors at 350, 440, and 540 nm. Human spectral discrimination is highest at 575–580 nm and 480–490 nm, corresponding to the narrow spectral regions that we perceive as yellow and blue-green. In the bee those regions of high discrimination lie at 390–400 nm and 490–500 nm. As with human color vision (Rushton, W. 1972), the color vision system of the bee can be described by a color triangle (Daumer, K. 1956) (Fig. 26). Other modified versions that take into account additional data and calculated parameters have also been proposed (Goldsmith, T. 1961; Menzel, R. 1979; Neumeyer, C. 1980).

Recently, another feature of vertebrate color vision has been explored in bees: the modification of the perceived hue of a color by other colors surrounding it (Neumeyer, C. 1980). Bees were trained to one hue among others ranging from yellow to blue, presented on a gray surround, and then were tested with the same series embedded in either a blue or yellow field. Bees preferred a "bluer" hue than the one to which they were trained when the surround was blue, and a greener hue when the surround was yellow. A similar effect is seen in human subjects, for whom, for instance, the spectral locus of yellow is shifted toward longer wavelengths when surrounded by red and to shorter wavelengths when the field is green (Akita, M. et al., 1964). In the bee, the hue shift appears to result from the selective reduction of the sensitivity of the receptor mechanism most strongly stimulated by the surround. Two possible processes might be involved: either selective adaptation of the receptors in the ommatidia responding to the test field, or lateral inhibition from adjacent regions stimulated by the surround. Transverse neuronal connections within the honeybee lamina (Ribi, W. 1975) may provide lateral interactions between the color channels.

9.7 Tetrachromatic color vision in *Cataglyphis*

Behavioral experiments indicate that the ant *Cataglyphis bicolor* has tetrachromatic color vision (Kretz, R. 1979). Both spontaneous phototaxis and its modification by training were studied over a spectral range from 320 to 627 nm. The freely walking ants were offered a choice between pairs of monochromatic stimuli. The spectral sensitivity function of spontaneous phototactic choice behavior has four peaks at 342, 425, 505, and 570 nm (Fig. 27). Sensitivity at each peak can be selectively reduced by prior adaptation with light of the corresponding wavelengths, suggesting that the four maxima of the behavioral sensitivity function reflect the sensitivities of four corresponding classes of receptors.

Hue discrimination, determined from training experiments, is best at 380, 450, and 550 nm, the wavelengths corresponding to the minima of the spectral sensitivity function. Thus the wavelength discrimination function is consistent with color vision based on four receptor types rather than three. The spectral saturation function, determined in training experiments, is similar to the spectral sensitivity function except that it lacks a pronounced peak in the yellow corresponding to the long wavelength receptor.

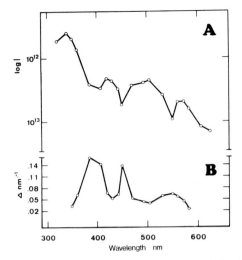

FIG. 27. Spectral sensitivity (**A**) and spectral discrimination (**B**) functions of the ant *Cataglyphis bicolor* measured by phototactic choice behavior, and in training experiments. Note that spectral discrimination is represented as the reciprocal of the wavelength difference that can be distinguished, so that the curve is inverted with respect to that plotted in Fig. 25 (from Kretz, R. 1979).

The ants can distinguish a mixture of ultraviolet and yellow from either yellow or ultraviolet alone. Therefore that mixture is "ant-purple". The ants can be trained to various hues of ant-purple; these are also distinguished from white and monochromatic lights. Appropriate mixtures of the following wavelengths are confused with white light: 342 nm + 504 nm, 434 nm + 574 nm, 382 nm + 550 nm, 449 nm + (351 nm + 574 nm = ant-purple). Thus ultraviolet/green, violet/yellow, ultraviolet-violet/yellow-green, and violet/ant-purple are complementary colors for the ant.

Although the behavioral analysis shows that *Cataglyphis* has tetrachromatic color vision, only green and ultraviolet units have been identified by intracellular recording (Mote, M. and Wehner, R. 1980). However, the selective loss of phototactic sensitivity from 550 to 650 nm after adaptation at 574 nm indicates the presence of separate long wavelength receptors. The difference spectrum of spontaneous phototactic choice after yellow-green adaptation corresponds to a receptor with $\lambda_{max} \simeq 575$ nm.

9.8 Color opponent neurons

Color sense in vertebrates depends upon neurons that compare the responses of the different classes of receptor cells. Such color opponent neurons respond differentially to each of two types of receptor, e.g., R^+G^- neurons are stimulated by red-sensitive cones and inhibited by green-sensitive cones (DeValois, R. 1972). Similar color opponent neurons have been identified in the optic centers of the bee brain (Kien, J. and Menzel, R. 1977a,b). They maintain a relatively high dark discharge frequency, responding to stimulation with either an increased or decreased discharge depending upon wavelength. The commonest unit found in the medulla (second optic ganglion) of the bee is excited by ultraviolet and inhibited by visible wavelengths (UV^+/B^-G^-). Cells that are B^+G^+/UV^- were encountered less frequently. These color opponent neurons are most sensitive to changes around 400 nm, where color discrimination is high. Unlike the chromaticity neurons of vertebrates, those of the bee do not combine spatial opponency (center–surround organization) with color opponency. Monochromatic neurons have also been found in the bee whose spectral sensitivities correspond to those of the three classes of receptors, but are narrower, presumably because of inhibitory inputs, and there are broad-band neurons with complex spectral responses, termed polychromatic neurons. Simple broad-band luminosity neurons were also found. These receive input from all three receptor types.

Color opponent neurons have also been identified in butterflies, e.g. *Heliconus papilo* and *Epargyreus* (Swihart, S. 1969, 1970, 1972a,b). These include the following types: B^+/G^-, G^+/B^-R^-, R^+/G^-O^-, B^+/G^-R^- (B = 430–480 nm, G = 490–560 nm, O = 560–620 nm, R > 620 nm). Ultraviolet wavelengths were not tested. The presence of these neurons in their optic centers is strong evidence for color vision in butterflies.

9.9 Color vision threshold

As the light dims, vertebrate vision shifts from one set of photoreceptors, the cone cells, to another, the rod cells. Cones mediate color vision and fine spatial discrimination; rods are achromatic. The switch from cones to rods in humans takes place at about the intensity of moonlight; we can just distinguish saturated colors under a bright moon. It was noted in sections 8 and 9.4.2 that the visual function of flies is divided to some extent into achromatic scotopic and wavelength-discriminating photopic systems mediated by different sets of receptors in their open rhabdoms. I have mentioned (see section 9.11) that the union of long rhabdomeres containing different rhodopsins into a single rhabdom allows them, because of optical coupling and lateral filtering, to combine high total absorbance with narrow spectral sensitivity functions. Therefore, it has been suggested that insects with such fused rhabdoms might retain color vision close to absolute threshold (Snyder, A. *et al.*, 1973). However, it has been found that honeybee vision becomes achromatic in dim light. Bees running in a T-maze were trained to a monochromatic light (533, 430, or 413 nm). They lost their ability to discriminate between the colored stimulus and white light at intensities between 10^8 and 10^9 quanta cm^{-2} s^{-1}. Absolute threshold for visual discrimination lay at about 1.5 log units lower, at intensities equivalent to around 10–50 quanta entering each ommatidium during a 30 ms test run. Since bees are more sensitive at short wavelengths, the threshold for successful discrimination at 413 and 440 nm is somewhat lower than it is at 533 nm (Menzel, R. 1981).

Bee vision has also been shown to be achromatic at low levels of natural illumination (Pflughöft, G. 1980; Rose, R. and Menzel, R. 1981). Freely flying bees, trained to color marks at a feeding station and the hive, discriminate colors best between 10^1 and 10^2 cd m^{-2}. Color vision is lost below 10^{-1} cd m^{-2}. However, they continue to fly very slowly at a luminance of 5×10^{-2} cd m^{-2}. Hence bee vision functions through an intensity range of 7–8 log units, from sunlight to moonlight, with flight orientation in very dim light depending on achromatic vision. The threshold for color vision in bees is about half a log unit higher than the photopic threshold of humans (Boynton, R. 1966).

Bees have neither specialized dim light receptors, as in vertebrates, nor achromatic receptors with broad spectral sensitivity, as in flies. However, the signals from all the receptors in an ommatidium are summed by neurons in the first optic ganglion (Menzel, R. 1974; Ribi, W. 1975). This sensitive achromatic system presumably is the basis of achromatic vision in dim light. The same receptors provide input to both the color discriminating

neuronal system, and the achromatic system. The former functions in visual orientation above the photopic threshold, the latter below it.

The honeybee's ability to discriminate color also decreases progressively at intensity levels above 10^2 cd m^{-2}. This bright light effect results from the depression of the dynamic range of the response/intensity function of visual interneurons in brighter light (Rose, R. and Menzel, R. 1981).

The capacity for wavelength discrimination at low light levels has not been studied quantitatively in other insects. However, Knoll, F. (1924) reported that the sphinx moth *Deilephila livornica* continues to feed at flowers in the late evening when he himself no longer could distinguish their colors. As something is known about the visual pigments and receptor organization of *Deiliphila* and *Manduca* (see section 9.5.1), these moths would be excellent choices for the examination of color vision in insects with superposition eyes.

9.10   Final comments

In comparing insect and vertebrate vision, we have tended to view the near-ultraviolet wavelengths as a spectral region particularly exploited by insects, but unavailable to vertebrates because human vision normally does not extend below 400 nm. Pictures taken in the ultraviolet show patterns — such as nectar guides in flowers and sexual recognition patterns in butterflies — that are important to insects but invisible to us (Eisner, T. *et al.*, 1969; Silberglied, R. and Taylor, O. 1973).

Although the visual pigments of human cones absorb in the ultraviolet, we do not see these shorter wavelengths because ultraviolet lights is filtered from the eye by yellow pigment in the lens. When the lens is removed, ultraviolet is seen as a pale bluish hue. Since chromatic aberration becomes worse at shorter wavelengths it has been suggested that their removal serves to increase the visual activity of the image-forming vertebrate eye (Wald, G. 1959). By the same reasoning, insects have been able to exploit the near-ultraviolet because chromatic aberration is not a problem for compound eyes that record a mosaic of luminous points. Particular aspects of human vision, however, are not necessarily characteristic of vertebrates in general. Certainly many other vertebrates, especially mammals, have yellowish lenses, but some are clear, and recent studies have shown that the vision of pigeons and hummingbirds, and at least some reptiles and amphibians, extends into the ultraviolet (Bowmaker, J. 1980; Goldsmith, T. 1980). Chromatic aberration in these species is presumably overcome by colored oil droplets that provide filters for the individual cone receptors (Wolbarsht, M. 1976).

Color vision is poorly developed in most mammals. Many are dichromatic, and some investigators would question whether wavelength discrimination based on two receptor types should be regarded as "true color vision" (Kretz, R. 1979). This points out our tendency to evaluate wavelength discrimination in other species in terms of our own trichromatic color vision. In recent years, however, we have become increasingly aware that human color vision is itself limited in comparison to the visual systems possessed by some other animals.

A popular explanation for the limited color vision of mammals suggests that they evolved from nocturnal ancestors whose capacity for color vision had degenerated. According to this view, the trichromatic color vision of humans and other Old World primates is a relatively recent secondary reacquisition (Bowmaker, J. 1980; Goldsmith, T. 1980). Thus it would not be surprising to find that the color vision of such highly evolved diurnal vertebrates as birds should be more complex than our own. The wavelength discrimination function recently measured in a hummingbird (Goldsmith, T. et al., 1981) indicates tetrachromatic color vision. The pigeon may have pentachromatic vision (Emmerton, J. and Delius, J. 1980).

Unlike vertebrates, most insects lack long-wavelength photoreceptors. We have seen, however, that the color space of some species extends into the yellow or red, allowing the tetrachromatic color vision. We are finding that insects as well as non-mammalian vertebrates have capacities for color vision that are richer than our own.

ACKNOWLEDGEMENTS

I am grateful to Ruth Bennett and Paul K. Brown for their help, comments, and criticism.

REFERENCES

ABRAHAMSON, E. W. and FAGER, R. S. (1973). The chemistry of vertebrate and invertebrate visual photoreceptors. Curr. Top. Bioenerg. 5, 125–200.

AIKE, G. (1980). Elektrophysiologische untersuchungen zur Spektral- und Polarisations-empfindlichkeit an den Sehzellen von Calliphora erythrocephala. In Scientia Sinica 23, 1182–1196.

AKITA, M., GRAHAM, C. H. and HSIA, Y. (1964). Maintaining an absolute hue in the presence of different background colors. Vision Res. 4, 539–556.

ALMAGOR, E., HILLMAN, P. and MINKE, B. (1979). Upper limit on translational diffusion of visual pigment in intact, unfixed barnacle photoreceptors. Biophys. Struct. Mech. 5, 243–248.

ALTNER, I. and BURKHARDT, D. (1981). Fine structure of the ommatidia and the occurrence of rhabdomeric twist in the dorsal eye of male Bibio marci (Diptera, Nematocera, Bibionidae). Cell Tiss. Res. 215, 607–623.

ASHMORE, J. (1977). The ultraviolet sense in flies. Nature 269, 373.

ATZMON, Z., HILLMAN, P. and HOCHSTEIN, S. (1978). Visual response in barnacle photoreceptors is not initiated by transitions to and from metarhodopsin. Nature 274, 74–76.

ATZMON, Z., HOCHSTEIN, S. and HILLMAN, P. (1979). Transduction in photoreceptors: Determination of the pigment transition or state coupled to excitation. Biophys. Struct. Mech. 5, 249–253.

AUTRUM, H. and KOLB, G. (1968). Spektrale Empfindlichkeit einzelner Sehzellen der Aeschniden. Z. Vergl. Physiol. 60, 450–477.

AUTRUM, H. and STUMPF, H. (1953). Elektrophysiologische Untersuchungen über das Farbensehen von Calliphora. Z. Vergl. Physiol. 35, 71–104.

AUTRUM, H. and VON ZWEHL, V. (1964). Die spektrale Empfindlichkeit einzelner Sehzellen des Bienenauges. Z. Vergl. Physiol. 48, 357–384.

AZUMA, K., AZUMA, M. and SUZUKI, T. (1975). Circular dichroism of cephalopod rhodopsin and its intermediates in the bleaching and photoregeneration process. Biochim. Biophys. Acta 393, 520–530.

BARLOW, H. B. (1972). Dark and Light Adaptation: Psychophysics. In Handbook of Sensory Physiology. Edited by D. Jameson and L. M. Hurvich. Vol. 7, pt 2, pages 1–28. Springer, Berlin and New York.

BARNES, S. N. and GOLDSMITH, T. H. (1977). Dark adaptation, sensitivity, and rhodopsin level in the eye of the lobster, Homarus. J. Comp. Physiol. 120, 143–159.

BAUMANN, F. (1975). Electrophysiological properties of the honey bee retina. In The Compound Eye and Vision of Insects. Edited by G. A. Horridge. Pages 53–74. Clarendon Press, Oxford.

BENNETT, R. R., TUNSTALL, J. and HORRIDGE, G. A. (1967). Spectral sensitivity of single retinula cells of the Locust. Z. Vergl. Physiol. 55, 195–206.

BENOLKEN, R. M., ANDERSON, R. E. and MAUDE, M. B. (1975). Lipid composition of Limulus photoreceptor membranes. Biochim. Biophys. Acta 413, 234–242.

BERNARD, G. D. (1971). Evidence for visual function of corneal interference filters. J. Insect Phsyiol. 17, 2287–2300.

BERNARD, G. D. (1977). Noninvasive microspectrophotometry of butterfly photoreceptors. J. Opt. Soc. Am. 67, 1362.

BERNARD, G. D. (1979). Red-absorbing visual pigment of butterflies. Science 203, 1125–1127.

BERNARD, G. D. (1981a). In vivo, total bleaching of invertebrate photoreceptors. Invest. Ophthal. Vis. Sci. 20 (Suppl.), 111.

BERNARD, G. D. (1981b). A comparison of vertebrate and invertebrate photoreceptors. In Vertebrate Photoreceptor Optics. Edited by J. M. Enoch and F. L. Tobey, Jr. Pages 433–463. Springer, Berlin and New York.

BERNARD, G. D. (1982). Noninvasive optical techniques for probing insect photoreceptors. In Visual Pigments and Purple Membranes. I. Edited by L. Packer. Methods in Enzymology. Vol. 81, pages 752–759. Academic Press, London, New York and San Francisco.

BERNARD, G. D. and STAVENGA, D. G. (1978). Spectral sensitivities of retinular cells measured in intact, living bumblebees by an optical method. Naturwissenschaften 65, 442–443.

BERNARD, G. D. and STAVENGA, D. G. (1979). Spectral sensitivities of retinular cells measured in intact, living flies by an optical method. J. Comp. Physiol. 134, 95–107.

BERNARD, G. D. and WEHNER, R. (1980). Intracellular optical physiology of the bee's eye. I. Spectral sensitivity. *J. Comp. Physiol.* 137, 193–203.

BERNHARD, C. G. and MILLER, W. H. (1968). Interference filters in the corneas in Diptera. *Invest. Ophthalm.* 7, 416–434.

BERNHARD, C. G., GEMNE, G. and MØLLER, A. R. (1968). Modification of specular reflexion and light transmission of biological surface structures. *Quart. Rev. Biophys.* 1, 89–105.

BERNHARD, C. G., GEMNE, G. and SÄLLSTRÖM, J. (1970). Comparative ultrastructure of corneal surface topography in insects with aspects on phylogenesis and function. *Z. Vergl. Physiol.* 67, 1–25.

BERNARD, C. G., MILLER, W. H. and MØLLER, A. R. (1965). The insect corneal nipple array — a biological broad-band impedance transformer that acts as an antireflection coating. *Acta Physiol. Scand.* 63, Suppl. 243, 1–79.

BERTRAND, D., FUORTES, G. and MURI, R. (1979). Pigment transformation and electrical responses in retinula cells of drone, *Apis mellifera* ♂. *J. Physiol.* 296, 431–441.

BERTRAND, D., PERRELET, A. and BAUMANN, F. (1972). Propriétés physiologiques des cellules pigmentaires de l'oeil du faux bourdon. *J. Physiol. (Paris)* 65, 102.

BICKER, G. and REICHERT, H. (1978). Visual learning in a photoreceptor degeneration mutant of *Drosophila melanogaster*. *J. Comp. Physiol.* 127, 29–38.

BISHOP, L. G. (1974). An ultraviolet photoreceptor in a dipteran compound eye. *J. Comp. Physiol.* 91, 267–275.

BLASIE, J. K. and WORTHINGTON, C. R. (1969). Planar liquid-like arrangement of photopigment molecules in frog retinal receptor disk membranes. *J. Mol. Biol.* 39, 417–439.

BLEST, A. D. (1978). The rapid synthesis and destruction of photoreceptor membrane by a dinopid spider: a daily cycle. *Proc. Roy. Soc. B.* 200, 463–483.

BLEST, A. D., STOWE, S. and EDDEY, W. (1982). A labile, Ca^{2+}-dependent cytoskeleton in rhabdomeral microvilli of blowflies. *Cell Tiss. Res.* 223, 553–573.

BOK, D., HALL, M. O. and O'BRIEN, P. (1977). The biosynthesis of rhodopsin as studied by membrane renewal in rod outer segments. In *International Cell Biology 1976–77*. Edited by B. R. Brinkley and K. R. Porter. Pages 608–617. Rockefeller University Press, New York.

BOSCHEK, C. B. (1970). On the structure and synaptic organization of the first optic ganglion in the fly. *Z. Naturf.* 25B, 560.

BOSCHEK, C. B. and HAMDORF, K. (1976). Rhodopsin particles in the photoreceptor membrane of an insect. *Z. Naturf.* 31c, 763.

BOWMAKER, J. K. (1980). Birds see ultraviolet light. *Nature* 284, 306.

BOWNESS, J. M. and WOLKEN, J. J. (1959). A light-sensitive yellow pigment from the housefly. *J. Gen. Physiol.* 42, 779–792.

BOYNTON, R. (1966). Vision. In *Experimental Methods and Instrumentation in Psychology*. Edited by J. B. Sidowski. Pt 3, Chap. 7. Pages 273–330. McGraw Hill, New York.

BRAMMER, J. D. (1970). The ultrastructure of the compound eye of a mosquito *Aedes aegypti* L. *J. Exp. Zool.* 175, 181–196.

BRAMMER, J. D. and WHITE, R. H. (1969). Vitamin A deficiency: effect on mosquito eye ultrastructure. *Science* 163, 821–823.

BRAMMER, J. D., STEIN, P. J. and ANDERSON, R. A. (1978). Effect of light and dark adaptation upon the rhabdom in the compound eye of the mosquito. *J. Exp. Zool.* 206, 151–156.

BRIDGES, C. D. B. (1976). Vitamin A and the role of the pigment epithelium during bleaching and regeneration of rhodopsin in the frog eye. *Exp. Eye Res.* 22, 435–455.

BRIGGS, M. H. (1961). Retinene-1 in insect tissues. *Nature* 192, 874–875.

BROWN, H. M. and CORNWALL, M. C. (1975). Ionic mechanism of a quasi-stable depolarization in barnacle photoreceptor following red light. *J. Physiol.* 248, 579–593.

BROWN, P. K. (1972). Rhodopsin rotates in the visual receptor membrane. *Nature New Biol.* 236, 35–38.

BROWN, P. K. and BROWN, P. S. (1958). Visual pigments of the octopus and cuttlefish. *Nature* 182, 1288–1290.

BROWN, P. K. and WALD, G. (1964). Visual pigments in single rods and cones of the human retina. *Science* 144, 45–52.

BROWN, P. K. and WHITE, R. H. (1972). Rhodopsin of the larval mosquito. *J. Gen. Physiol.* 59, 401–414.

BRUCKLER, R. M. and WILLIAMS, T. P. (1981). Adaptation properties of the ERG in the grasshopper, *Romalea microptera*. *Biophys. Struct. Mech.* 7, 205–208.

BRUCKMOSER, P. (1968). Die spektrale Empfindlichkeit einzelner Sehzellen des Rückenschwimmers *Notonecta glauca* L. (Heteroptera) *Z. Vergl. Physiol.* 59, 187–204.

BRUNO, M. S. and GOLDSMITH, T. H. (1974). Rhodopsin of the blue crab *Callinectes*: evidence for absorption differences *in vitro* and *in vivo*. *Vision Res.* 14, 653–658.

BRUNO, M. S., BARNES, S. N. and GOLDSMITH, T. H. (1977). The visual pigment and visual cycle of the lobster, *Homarus*. *J. Comp. Physiol.* 120, 123–142.

BULLIVANT, S. (1977). Evaluation of membrane structure: facts and artifacts produced during freeze-fracture. *J. Microsc. (Oxf.)* 111, 101–116.

BURKHARDT, D. (1962). Spectral sensitivity and other response characteristics of single visual cells in the arthropod eye. *Symp. Soc. Exp. Biol.* 16, 86–109.

BURKHARDT, D. (1964). Colour discrimination in insects. *Adv. Insect Physiol.* 2, 131–173.

BUSCH, G. E., APPLEBURY, M. L., LAMOLA, A. A. and RENTZEPIS, P. M. (1972). Formation and decay of prelumirhodopsin at room temperature. *Proc. Nat. Acad. Sci.* 69, 2802–2806.

BUTLER, R. (1971). The identification and mapping of spectral cell types in the retina of *Periplaneta americana*. *Z. Vergl. Physiol.* 71, 67–80.

CARLSON, S. D. and PHILIPSON, B. (1972). Microspectrophotometry of the dioptric apparatus and compound rhabdom of the moth *Manduca sexta* eye. *J. Insect Physiol.* 18, 1721–1731.

CARRICABURU, P. and CHARDENOT, P. (1967). Spectres d'absorption de la cornée de quelques arthropodes. *Vision Res.* 7, 43–50.

CHANCE, B. (1964). Fluorescence emission of mitochondrial DPNH as a factor in the ultraviolet sensitivity of visual receptors. *Proc. Nat. Acad. Sci.* 51, 359–361.

CHI, C. and CARLSON, S. D. (1976). Close apposition of photoreceptor cell axons in the house fly. *J. Insect Physiol.* 22, 1153–1157.

CHI, C. and CARLSON, S. D. (1979). Ordered membrane particles in rhabdomeric microvilli of the housefly *(Musca domestica L.)*. *J. Morph.* 161, 309–322.

CINCOTTA, D. E. and HUBBELL, W. L. (1977). Freeze fracture studies of squid photoreceptor membranes. *Invest. Ophthal. Vis. Sci.* (Suppl.), p. 165.

CONE, R. A. (1972). Rotational diffusion of rhodopsin in the visual receptor membrane. *Nature New Biol.* 236, 39–43.

COOMBE, P. E. (1981). Wavelength specific behavior of the whitefly *Trialeurodes vaporariorum* (Homoptera: Aleyrodidae). *J. Comp. Physiol.* 144, 83–90.

COSENS, D. and WRIGHT, R. (1975). Light elicited isolation of the complementary visual input systems in white-eye *Drosophila*. *J. Insect Physiol.* 21, 1111–1120.

CRONIN, T. W. and GOLDSMITH, T. H. (1981). Fluorescence of crayfish metarhodopsin studied in single rhabdoms. *Biophys. J.* 35, 653–664.

DAEMEN, F. J. M. (1973). Vertebrate rod outer segment membranes. *Biochim. Biophys. Acta* 300, 255–288.

DARSZON, A., VANDENBERG, C. A., ELLISMAN, M. H. and MONTAL, M. (1979). Incorporation of membrane proteins into large single bilayer vesicles. Application to rhodopsin. *J. Cell. Biol.* 81, 446–452.

DARTNALL, H. J. A. (1953). The interpretation of spectral sensitivity curves. *Brit. Med. Bull.* 9, 24–30.

DAUMER, K. (1956). Reizmetrische Untersuchung des Farbensehens der Bienen. *Z. Vergl. Physiol.* 38, 413–478.

DAUMER, K. (1958). Blumenfarben, wie sie die Bienen sehen. *Z. Vergl. Physiol.* 41, 49–110.

DAWIS, S. M. (1981). Polynomial expressions of pigment nomograms. *Vis. Res.* 21, 1427–1430.

DENYS, C. J. (1981). The visual pigment and photoreception of the Antarctic Krill, *Euphausia superba* (Crustacea, Euphausiacea). Ph.D. dissertation, De Paul University, Chicago, Il.

DENYS, C. J. (1982). Ommochrome pigments in the eyes of *Euphausia superba* (Crustacea, Euphausiacea). *Polar Biol.*, 1, 69–76.

DEVALOIS, R. L. (1972). Processing of intensity and wavelength information by the visual system. *Invest. Ophtahl.* 11, 417–427.

DEVOE, R. D. (1972). Dual sensitivities of cells in the wolf spider eyes at ultraviolet and visible wavelengths of light. *J. gen. Physiol.* 59, 247–269.

DÖRRSCHEIDT-KÄFER, M. (1972). Die Empfindlichkeit einzelner Photorezeptoren im Komplexauge von *Calliphora erythrocephala*. *J. Comp. Physiol.* 81, 309–340.

DOWLING, J. E. and CHAPPELL, R. L. (1972). Neural organization of the median ocellus of the dragonfly. II. Synaptic structure. *J. gen. Physiol.* 60, 148–165.

EBREY, T. G. and HONIG, B. (1975). Molecular aspects of photoreceptor function. *Quart. Rev. Biophys.* 8, 129–184.

EBREY, T. G. and HONIG, B. (1977). New wavelength dependent visual pigment nomograms. *Vision Res.* 17, 147–151.

ECKERT, H., BISHOP, L. G. and DVORAK, D. R. (1976). Spectral sensitivities of identified receptor cells in the blowfly *Calliphora*. *Naturwissenschaften* 63, 47–48.

ECKERT, M. (1971). Die spektrale Empfindlichkeit des Komplexauges von *Musca* (Bestimmungen aus Messungen der optomotorischen Reaktion). *Kybernetik* 9, 145–156.

EDIDIN, M., ZAGYANSKY, Y. and LARDNER, T. J. (1976). Measurement of membrane protein lateral diffusion in single cells. *Science 191*, 466–468.

EGUCHI, E., NAKA, K.-I and KUWABARA, M. (1962). The development of the rhabdom and the appearance of the electrical response in the insect eye. *J. gen. Physiol.* 46, 143–157.

EGUCHI, E. and WATERMAN, T. H. (1976). Freeze-etch and histochemical evidence for cycling in crayfish photoreceptor membranes. *Cell Tiss. Res. 169*, 419–434.

EISNER, T., SILBERGLIED, R. E., ANESHANSLEY, D., CARREL, J. E. and HOWLAND, H. C. (1969). Ultraviolet video-viewing: the television camera as an insect eye. *Science 166*, 1172–1174.

EMMERTON, J. and DELIUS, J. D. (1980). Wavelength discrimination in the "visible" and ultraviolet spectrum by pigeons. *J. Comp. Physiol. 141*, 47–52.

FEINBERG, G. (1968). Light. *Scient. Amer. 219(3)*, 50–59.

FERNANDEZ, H. R. and BISHOP, L. G. (1973). Photosensitive pigment from the worker honeybee, *Apis mellifera*. *Vision Res. 13*, 1379–1381.

FERNANDEZ, H. R. and NICKEL, E. (1976). Ultrastructural and molecular characteristics of crayfish photoreceptor membranes. *J. Cell Biol. 69*, 721–732.

FISCHBACH, K. F. (1979). Simultaneous and successive colour contrast expressed in "slow" phototactic behaviour of walking *Drosophila melanogaster*. *J. Comp. Physiol. 130*, 161–171.

FRANCESCHINI, N. (1972). Pupil and pseudopupil in the compound eye of *Drosophila*. In *Information Processing in the Visual Systems of Arthropods*. Edited by R. Wehner. Pages 75–82. Springer, Berlin and New York.

FRANCESCHINI, N. and KIRSCHFELD, K. (1971). Etude optique *in vivo* des éléments photorécepteurs dans l'oeil composé de *Drosophila*. *Kybernetik 8*, 1–13.

FRANCESCHINI, N. and STAVENGA, D. G. (1981). The ultra-violet sensitizing pigment of flies studied by *in vivo* microspectrofluorimetry. *Invest. Ophthal. Vis. Sci. 20*, (Suppl.), 111.

FRANCESCHINI, N., KIRSCHFELD, K. and MINKE, B. (1981b). Fluorescence of photoreceptor cells observed *in vivo*. *Science 213*, 1264–1267.

FRANCESCHINI, N., HARDIE, R., RIBI, W. and KIRSCHFELD, K. (1981a). Sexual dimorphism in a photoreceptor. *Nature 291*, 241–244.

FRISCH, K. VON (1915). Der Farbensinn und Formensinn der Biene. *Zool. Jahrb. Abt. Allgem. Zool. Physiol. 35*, 1–182.

FRISCH, K. VON (1950). *Bees, their Vision, Chemical Senses, and Language*. Cornell University Press, New York.

FRISCH, K. VON (1965). *Tanzsprache und Orientierung der Bienen*. Springer, Berlin and New York.

FRIXIONE, E. and ARÉCHIGA, H. (1981). Ionic dependence of screening pigment migrations in crayfish retinal photoreceptors. *J. Comp. Physiol. 144*, 35–43.

FUKUSHI, T. (1976). Classical conditioning to visual stimuli in the housefly, *Musca domestica*. *J. Insect Physiol. 22*, 361–364.

GEMPERLEIN, R., PAUL, R., LINDAUER, E. and STEINER, A. (1980). UV fine structure of the spectral sensitivity of flies visual cells. *Naturwissenschaften 67*, 565–566.

GLAUERT, A. M. (1975). Fixation, dehydration and embedding of biological specimens. In *Practical Methods in Electron Microscopy*. Edited by A. M. Glauert. Vol. 3, pt I, pages 37–38. North Holland, Amsterdam.

GOGALA, M. (1967). Die spektrale Emfindlichkeit der Doppelaugen von *Ascalaphus macaronius* Scop. (Neuroptera, Ascalaphidae). *Z. Vergl. Physiol. 57*, 232–243.

GOGALA, M., HAMDORF, K. and SCHWEMER, J. (1970). UV-Sehfarbstoff bei Insekten. *Z. Vergl. Physiol. 70*, 410–413.

GOLDMAN, L. J. (1971). The electroretinogram and spectral sensitivity of the compound eye of *Aedes aegypti*. University of Florida Ph.D. thesis (University Microfilms).

GOLDMAN, L. J., BARNES, S. N. and GOLDSMITH, T. H. (1975). Microspectrophotometry of rhodopsin and metarhodopsin in the moth *Galleria*. *J. gen. Physiol. 66*, 383–404.

GOLDSMITH, T. H. (1958a). The visual system of the honeybee. *Proc. Nat. Acad. Sci. 44*, 123–126.

GOLDSMITH, T. H. (1958b). On the visual system of the bee *(Apis mellifera)*. *Ann. N.Y. Acad. Sci. 74*, 223–229.

GOLDSMITH, T. H. (1961). The color vision of insects. In *Light and Life*. Edited by W. D. McElroy and B. Glass. Pages 771–794. John Hopkins Press, Baltimore.

GOLDSMITH, T. H. (1965). Do flies have a red receptor? *J. gen. Physiol. 49*, 265–287.

GOLDSMITH, T. H. (1970). The retinaldehyde of the honeybees. *Ophthalm. Res. 1*, 292–301.

GOLDSMITH, T. H. (1972). The natural history of invertebrate visual pigments. In *Photochemistry of Vision*. Edited by H. J. A. Dartnall. *Handbook of Sensory Physiology*. Vol. 7, pt 1, pages 685–719. Springer. Berlin and New York.

GOLDSMITH, T. H. (1975). The polarization sensitivity–dichroic absorption paradox in arthropod photoreceptors. In *Photoreceptor Optics*. Edited by A. W. Snyder and R. Menzel. Pages 392–409. Springer, Berlin and New York.

GOLDSMITH, T. H. (1978). The spectral absorption of crayfish rhabdoms: pigment, photoproduct, and pH sensitivity. *Vis. Res. 18*, 463–473.

GOLDSMITH, T. H. (1980). Hummingbirds see near ultraviolet light. *Science 207*, 786–788.

GOLDSMITH, T. H. and FERNANDEZ, H. R. (1966). Some photochemical and physiological aspects of visual excitation in compound eyes. In *The Functional Organization of the Compound Eye*. Edited by C. G. Bernhard. Pages 125–143. Pergamon Press, New York.

GOLDSMITH, T. H. and FERNANDEZ, H. R. (1968). The sensitivity of housefly photoreceptors in the mid-ultraviolet and the limits of the visible spectrum. *J. Exp. Biol. 49*, 669–677.

GOLDSMITH, T. H. and RUCK, P. R. (1958). The spectral sensitivities of the dorsal ocelli of cockroaches and honey bees. An electrophysiological study. *J. gen Physiol. 41*, 1171–1185.

GOLDSMITH, T. H. and WARNER, L. T. (1964). Vitamin A in the vision of insects. *J. gen. Physiol. 47*, 433–441.

GOLDSMITH, T. H. and WEHNER, R. (1977). Restrictions on rotational and translational diffusion of pigment in the membranes of a rhabdomeric photoreceptor. *J. gen. Physiol. 70*, 453–490.

GOLDSMITH, T. H., BARKER, R. J. and COHEN, C. F. (1964). Sensitivity of visual receptors of carotenoid-depleted flies: a vitamin A deficiency in an invertebrate. *Science 146*, 65–67.

GOLDSMITH, T. H., COLLINS, J. S. and PERLMAN, D. L. (1981). A wavelength discrimination function for the hummingbird *Archilochus alexandri*. *J. Comp. Physiol. 143*, 103–110.

GOLDSTEIN, E. B. and WILLIAMS, T. P. (1966). Calculated effects of "screening pigments". *Vision Res. 6*, 39–50.

GRUNDLER, O. J. (1974). Elektronenmikroskopische Untersuchungen am Auge der Honigbiene *(Apis mellifica)*. I. Untersuchungen zur Morphologie und Anordnung der neun Retinulazellen in Ommatidien verschiedener Augenbereiche und zur Perzeption linear polarisierten Lichtes. *Cytobiologie 9*, 203–220.

GUZZO, A. V. and POOL, G. L. (1968). Visual pigment fluorescence. *Science 159*, 312–314.

HAGINS, F. M. (1973). Purification and partial characterization of the protein component of squid rhodopsin. *J. Biol. Chem. 248*, 3298–3304.

HAGINS, W. A. and JENNINGS, W. H. (1959). Radiationless migration of electronic excitation in retinal rods. *Discuss. Faraday Soc. 27*, 180–190.

HAGINS, W. A. and McGAUGHY, R. E. (1967). Molecular and thermal origins of fast photoelectric effects in the squid retina. *Science 157*, 813–816.

HAMANN, B. and LANGER, H. (1980). Sehfarbstoffe im Auge des Wasserlaufers *Gerris lacustris* L. *Verh. Dtsch. Zool. Ges. 73*, 337.

HAMDORF, K. (1979). The physiology of invertebrate visual pigments. In *Comparative Physiology and Evolution of Vision in Invertebrates. A: Invertebrate Photoreceptors.* Edited by H. Autrum. *Handbook of Sensory Physiology.* Vol. 7, pt 6A, pages 145–224. Springer, Berlin and New York.

HAMDORF, K. (1980). Wie "sehen" Moleküle? *Umschau 80*, 562–565.

HAMDORF, K. and GOGALA, M. (1973). Photoregeneration und Bereicheinstellung der Empfindlichkeit beim UV-Rezeptor. *J. Comp. Physiol. 86*, 231–245.

HAMDORF, K. and KIRSCHFELD, K. (1980). Reversible events in the transduction process of photoreceptors. *Nature 283*, 859–860.

HAMDORF, K. and LANGER, H. (1965). Veränderung der Lichtabsorption im Facetenauge bei Belichtung. *Z. Vergl. Physiol. 51*, 172–184.

HAMDORF, K. and RAZMJOO, S. (1977). The prolonged depolarizing afterpotential and its contribution to the understanding of photoreceptor function. *Biophys. Struct. Mechanism 3*, 163–170.

HAMDORF, K. and RAZMJOO, S. (1979). Photoconvertible pigment states and excitation in *Calliphora*; the induction and properties of the prolonged depolarising afterpotential. *Biophys. Struct. Mechanism 5*, 137–161.

HAMDORF, K. and ROSNER, G. (1973). Adaptation und Photoregeneration im Fliegenauge. *J. Comp. Physiol. 86*, 281–292.

HAMDORF, K. and SCHWEMER, J. (1975). Photoregeneration and the adaptation process in insect photoreceptors. In *Photoreceptor Optics.* Edited by A. W. Snyder and R. Menzel. Pages 263–289. Springer, Berlin and New York.

HAMDORF, K., GOGALA, M. and SCHWEMER, J. (1971). Beschleunigung der Dunkeladaptation eines UV-Rezeptors durch sichtbare Strahlung. *Z. Vergl. Physiol. 75*, 189–199.

HAMDORF, K., GOGALA, M. and STUSEK, P. (1978). Methods of simultaneous photometry and electrophysiology of insect eyes. *Biol. Vestn. 26*, 107–130.

HAMDORF, K., HÖGLUND, G. and LANGER, H. (1972). Mikrophotometrische Untersuchungen an der Retinula des Nachtschmetterlings *Dielephila elpenor. Verh. Dtsch. Zool. Ges. 65*, 276–281.

HAMDORF, K., HÖGLUND, G. and LANGER, H. (1973a). Photoregeneration of visual pigments in a moth. A microphotometric study. *J. Comp. Physiol. 86*, 247–263.

HAMDORF, K., PAULSEN, R. and SCHWEMER, J. (1973b). Photoregeneration and sensitivity control of photoreceptors of invertebrates. In *Biochemistry and Physiology of Visual Pigments.* Edited by H. Langer. Pages 155–166. Springer, Berlin and New York.

HAMDORF, K., SCHWEMER, J. and TÄUBER, U. (1968). Der Schfarbstoff, die Absorption der Rezeptoren und die spektrale Empfindlichkeit der Retina von *Eledone moschata. Z. Vergl. Physiol. 60*, 375–415.

HARA, T. and HARA, R. (1973). Biochemical properties of retinochrome. In *Biochemistry and Physiology of Visual Pigments.* Edited by H. Langer. Pages 181–191. Springer, Berlin and New York.

HARDIE, R. C. (1979). Electrophysiological analysis of fly retina. I: Comparative properties of R1–6 and R7 and 8. *J. Comp. Physiol. 129*, 19–33.

HARDIE, R. C., FRANCESCHINI, N. and MCINTYRE, P. D. (1979). Electrophysiological analysis of fly retina. II. Spectral and polarization sensitivity in R7 and R8. *J. Comp. Physiol. 133*, 23–39.

HÁROSI, F. I. (1976). Spectral relations of cone pigments in goldfish. *J. Gen. Physiol. 68*, 65–80.

HARRIS, W. A., READY, D. F., LIPSON, E. D., HUDSPETH, A. J. and STARK, W. S. (1977). Vitamin A deprivation and *Drosophila* photopigments. *Nature 266*, 648–650.

HARRIS, W. A., STARK, W. S. and WALKER, J. A. (1976). Genetic dissection of the photoreceptor system in the compound eye of *Drosophila melanogaster. J. Physiol. 256*, 415–439.

HAYS, D. and GOLDSMITH, T. H. (1969). Microspectrophotometry of the visual pigment of the spider crab *Libinia emarginata. Z. Vergl. Physiol. 65*, 218–232.

HEISENBERG, M. and BUCHNER, E. (1977). The role of retinula cell types in visual behavior of *Drosophila melanogaster. J. Comp. Physiol. 117*, 127–162.

HELVERSON, O. VON (1972). Zur spektralen Unterschiedsempfindlichkeit der Honigbiene. *J. Comp. Physiol. 80*, 439–472.

HELVERSON, O. VON and EDRICH, W. (1974). Der Polarisationsempfänger im Bienenauge: ein Ultraviolettrezeptor. *J. Comp. Physiol. 94*, 33–47.

HENDERSON, R. and UNWIN, P. N. T. (1975). Three-dimensional model of purple membrane obtained by electron microscopy. *Nature 257*, 28–32.

HEUSER, J. E. and SALPETER, S. R. (1979). Organization of acetylcholine receptors in quick-frozen, deep-etched, and rotary-replicated *Torpedo* postsynaptic membrane. *J. Cell Biol. 82*, 150–173.

HILDEBRAND, E. and DENCHER, N. (1975). Two photosystems controlling behavioural responses of *Halobacterium halobium. Nature 257*, 46–48.

HILLMAN, P. (1979). Introduction to the symposium on bistable and sensitizing pigments in vision. *Biophys. Struct. Mechanism 5*, 113–116.

HOCHSTEIN, S., MINKE, B., HILLMAN, P. and KNIGHT, B. W. (1978). The kinetics of visual pigment systems. I. Mathematical analysis. *Biol. Cybernetics 30*, 23–32.

HÖGLUND, G. and STRUWE, G. (1970). Pigment migration and spectral sensitivity in the compound eye of moths. *Z. Vergl. Physiol. 67*, 229–237.

HÖGLUND, G., HAMDORF, K., LANGER, H., PAULSEN, R. and SCHWEMER, J. (1973a). The photopigments in an insect retina. In *Biochemistry and Physiology of Visual Pigments.* Edited by H. Langer. Pages 167–174. Springer, Berlin and New York.

HÖGLUND, G., HAMDORF, K. and ROSNER, G. (1973b). Trichromatic visual system in an insect and its sensitivity control by blue light. *J. Comp. Physiol. 86*, 265–279.

HOLTZMAN, E. and MERCURIO, A. M. (1980). Membrane circulation in neurons and photoreceptors: Some unresolved issues. *Int. Rev. Cytol. 67*, 1–67.

HORRIDGE, G. A. (1967). Perception of polarization plane, colour and movement in two dimensions by the crab *Carcinus. Z. Vergl. Physiol. 55*, 207–224.

HORRIDGE, G. A. (1969). Unit studies on the retina of dragonflies. *Z. Vergl. Physiol. 62*, 1–37.

HORRIDGE, G. A. and BLEST, A. D. (1980). The compound eye. In *Insect Biology in the Future.* Edited by M. Locke and D. S. Smith. Pages 705–733. Academic Press, New York.

HORRIDGE, G. A. and MIMURA, K. (1975). Fly photoreceptors. I. Physical separation of two visual pigments in *Calliphora* retinula cells 1–6. *Proc. Roy. Soc. B. 190*, 211–224.

HORRIDGE, G. A., MIMURA, K. and TSUKAHARA, Y. (1975). Fly photoreceptors. II. Spectral and polarized light sensitivity in the drone fly *Eristalis tenax. Proc. Roy. Soc. B. 190*, 225–237.

HU, K. G. and STARK, W. S. (1977). Specific receptor input into spectral preference in *Drosophila. J. Comp. Physiol. 121*, 241–252.

HU, K. G., REICHERT, H. and STARK, W. S. (1978). Electrophysiological characterization of *Drosophila* ocelli. *J. Comp. Physiol. 126*, 15–24.

HUBBARD, R. and ST. GEORGE, R. C. C. (1958). The rhodopsin system of the squid. *J. gen. Physiol. 41*, 501–528.

HUBBARD, R., BROWN, P. K. and BOWNDS, D. (1971). Methodology of Vitamin A and visual pigments. In *Methods in Enzymology.* Edited by D. B. McCormick and L. D. Wright. Vol. 18c, pages 615–653. Academic Press. New York, London.

ILSE, D. (1928). Über den Farbensinn der Tagfalter. *Z. Vergl. Physiol. 8*, 658–692.

ISRAELACHVILI, J. N. and WILSON, M. (1976). Absorption characteristics of oriented photopigments in microvilli. *Biol. Cybernetics 21*, 9–15.

JACOB, K. G., WILLMUND, R., FOLKERS, E., FISCHBACH, K. F. and SPATZ, H.-C. (1977). T-Maze phototaxis of *Drosophila melanogaster* and several mutants in the visual systems. *J. Comp. Physiol. 116*, 209–225.

JAMESON, D. and HURVICH, L. M. (1955). Some quantitative aspects of an opponent-colors theory I. Chromatic responses and spectral saturation. *J. Opt. Soc. Amer. 45*, 546–552.

JAN, L. Y. and REVEL, J. P. (1974). Ultrastructural localization of rhodopsin in the vertebrate retina. *J. Cell Biol. 62*, 257–273.

JÄRVILEHTO, M. and ZETTLER, F. (1973). Electrophysiological-histological studies on some functional properties of visual cells and second order neurons of an insect retina. *Z. Zellforsch. 136*, 291–306.

KAISER, W. (1968). Zur Frage des Unterscheidungsvermögens für Spektralfarben: Eine Untersuchung der Optomotorik der königlichen Glanzfliege *Phormia regina* Meig. *Z. Vergl. Physiol. 61*, 71–102.

KAISER, W. (1975). The relationship between visual movement detection and colour vision in insects. In *The Compound Eye and Vision of Insects.* Edited by G. A. Horridge. Pages 359–377. Clarendon Press, Oxford.

KAISER, W. and LISKE, E. (1974). Die optomotorischen Reactionen von fixiert fliegenden Bienen bei Reizung mit Spektrallichtern. *J. Comp. Physiol. 89*, 391–408.

KAY, R. E. (1969). Fluorescent material in insect eyes and their possible relationship to ultra-violet sensitivity. *J. Insect Physiol. 15*, 2021–2038.

KIEN, J. and MENZEL, R. (1977a). Chromatic properties of interneurons in the optic lobes of the bee. I. Broad band neurons. *J. Comp. Physiol. 113*, 17–34.

KIEN, J. and MENZEL, R. (1977b). Chromatic properties of interneurons in the optic lobes of the bee. II. Narrow band and colour opponent neurons. *J. Comp. Physiol. 113*, 35–53.

KIRSCHFELD, K. (1979). The function of photostable pigments in fly photoreceptors. *Biophys. Struct. Mechanism. 5*, 117–128.

KIRSCHFELD, K. and FRANCESCHINI, N. (1969). Ein Mechanismus zur Steuerung des Lichtflusses in den Rhabdomeren des Komplexauges von *Musca. Kybernetik 6*, 13–22.

KIRSCHFELD, K. and FRANCESCHINI, N. (1977). Photostable pigments within the membrane of photoreceptors and their possible role. *Biophys. Struct. Mech. 3*, 191–194.

KIRSCHFELD, K. and LUTZ, B. (1977). The spectral sensitivity of the ocelli of *Calliphora* (Diptera). *Z. Naturforsch. 32c*, 439–441.

KIRSCHFELD, K. and REICHARDT, W. (1970). Optomotorische Versuche an *Musca* mit linear polarisiertem Licht. *Z. Naturforsch. 25B*, 228.

KIRSCHFELD, K. and SNYDER, A. W. (1976). Measurement of a photoreceptor's characteristic waveguide parameter. *Vision Res. 16*, 775–778.

KIRSCHFELD, K., FEILER, R. and FRANCESCHINI, N. (1978). A photostable pigment within the rhabdomere of fly photoreceptors no. 7. *J. Comp. Physiol. 125*, 275–284.

KIRSCHFELD, K., FRANCESCHINI, N. and MINKE, B. (1977). Evidence for a sensitizing pigment in fly photoreceptors. *Nature 269*, 386–390.

KLINGENBERG, M. (1981). Membrane protein oligomeric structure and transport function. *Nature 290*, 449–454.

KNOLL, F. (1924). Lichtsinn und Blütenbesuch des Falters von *Deilephila livornica. Z. Vergl. Physiol. 2*, 329–380.

KOLB, G. and AUTRUM, H. (1974). Selektive Adaptation und Pigmentwanderung in den Sehzellen des Bienenauges. *J. Comp. Physiol. 94*, 1–6.

KOLB, G., AUTRUM, H. and EGUCHI, E. (1969). Die spektrale Transmission des dioptrischen Apparates von *Aeschna cyanea* Müll. *Z. Vergl. Physiol. 63*, 434–439.

KONG, K.-L., FUNG, Y. M. and WASSERMAN, G. S. (1980). Filter-mediated color vision with one visual pigment. *Science 207*, 783–786.

KRETZ, R. (1979). A behavioural analysis of colour vision in the ant *Cataglyphis bicolor* (Formicidae, Heymenoptera). *J. Comp. Physiol. 131*, 217–233.

KROPF, A. (1967). Intramolecular energy transfer in rhodopsin. *Vision Res. 7*, 811–818.

KÜHN, A. (1927). Über den Farbensinn der Bienen. *Z. Vergl. Physiol. 5*, 762–800.

LABHART, T. (1974). Behavioral analysis of light intensity discrimination and spectral sensitivity in the honey bee, *Apis mellifera. J. Comp. Physiol. 95*, 203–216.

LABHART, T. (1977). Electrophysiological recordings from the lateral ocelli of *Drosophila. Naturwissenschaften 64*, 99–100.

LABHART, T. (1980). Specialized photoreceptors at the dorsal rim of the honeybee's compound eye: polarizational and angular sensitivity. *J. Comp. Physiol. 141*, 19–30.

LAND, M. F. (1972). The physics and biology of animal reflectors. *Progr. Biophys. 24*, 77–105.

LANGER, H. (1965). Nachweis dichroitischer Absorption des Sehfarbstoffes in den Rhabdomeren den Insektenauges. *Z. Vergl. Physiol. 51*, 258–263.

LANGER, H. (1967). Über die Pigmentgranula im Facettenauge von *Calliphora erythrocephala. Z. Vergl. Physiol. 55*, 354–377.

LANGER, H. (1972). Metarhodopsin in Single Rhabdomeres of the Fly, *Calliphora erythrocephala.* In *Information Processing in the Visual Systems of Arthropods.* Edited by R. Wehner. Pages 109–113. Springer, Berlin and New York.

LANGER, H. (1975). Properties and functions of screening pigments in insect eyes. In *Photoreceptor Optics.* Edited by A. W. Snyder and R. Menzel. Pages 429–455. Springer, Berlin and New York.

LANGER, H. and THORELL, B. (1966). Microspectrophotometry of single rhabdomeres in the insect eye. *Exp. Cell Res. 41*, 673–677.

LANGER, H., HAMANN, B. and MEINECKE, C. C. (1979). Tetrachromatic visual system in the moth *Spodoptera exempta* (Insecta, Noctuidae). J. Comp. Physiol. 129, 235–239.

LARRIVEE, D. C. and GOLDSMITH, T. H. (1981). Properties of crayfish rhodopsin. *Invest. Ophthal. Vis. Sci. 20* (Suppl.), 215.

LARRIVEE, D. C., CONRAD, S. K., STEPHENSON, R. S. and PAK, W. L. (1981). Mutation that selectively affects rhodopsin concentration in the peripheral photoreceptors of *Drosophila melanogaster. J. Gen. Physiol. 78*, 521–545.

LAUGHLIN, S. B. (1975). Receptor function in the apposition eye: An electrophysiological approach. In *Photoreceptor Optics.* Edited by A. W. Snyder and R. Menzel. Pages 479–498. Springer, Berlin and New York.

LAUGHLIN, S. B. (1976). The sensitivities of dragonfly photoreceptors and the voltage gain of transduction. *J. Comp. Physiol. 111*, 221–247.

LAUGHLIN, S. B., MENZEL, R. and SNYDER, A. W. (1975). Membranes, dichroism and receptor sensitivity. In *Photoreceptor Optics.* Edited by A. W. Snyder and R. Menzel. Pages 237–259. Springer, Berlin and New York.

LIEBMAN, P. A. (1962). *In situ* microspectrophotometric studies on the pigments of single retinal rods. *Biophys. J. 2*, 161–178.

LIEBMAN, P. A. (1972). Microspectrophotometry of photoreceptors. In *Photochemistry of Vision.* Edited by H. J. A. Dartnall. *Handbook of Sensory Physiology.* Vol. 7, pt 1, pages 481–528. Springer, Berlin and New York.

LIEBMAN, P. A. and ENTINE, G. (1974). Lateral diffusion of visual pigment in photoreceptor disk membranes. *Science 185*, 457–459.

LILLYWHITE, P. G. (1978). Coupling between locust photoreceptors revealed by a study of quantum bumps. *J. Comp. Physiol. 125*, 13–27.

LISMAN, J. E. and STRONG, J. (1979). The initiation of excitation and light-adaptation in *Limulus* ventral photoreceptors. *J. gen. Physiol. 73*, 219–243.

LOEW, E. R. (1976). Light, and photoreceptor degeneration in the Norway lobster, *Nephrops norvegicus* (L.). *Proc. Roy. Soc. B. 193*, 31–44.

LOEW, E. R. (1980). Visual pigment regeneration rate and susceptibility to photic damage. In *The Effects of Constant Light on Visual Processes.* Edited by T. P. Williams and B. N. Baker. Pages 297–306. Plenum, New York and London.

MARAK, G. E., GALLIK, G. J. and CORNESKY, R. A. (1970). Light-sensitive pigments in insect heads. *J. Ophthal. Res. 1*, 65–71.

MARKS, W. B. (1965). Visual pigments of single goldfish cones. *J. Physiol. 178*, 14–32.

MARKS, W. B., DOBELLE, W. H. and MACNICHOL, E. F. (1964). Visual pigments of single primate cones. *Science 143*, 1181–1183.

MASON, W. T., FAGER, R. S. and ABRAHAMSON, E. W. (1974). Structural response of vertebrate photoreceptor membranes to light. *Nature 247*, 188–191.

MATTHEWS, R. G., HUBBARD, R., BROWN, P. K. and WALD, G. (1963). Tautomeric forms of metarhodopsin. *J. gen. Physiol. 47*, 215–240.

MAZOKHIN-PORSHNYAKOV, G. A. (1969). *Insect Vision.* Plenum Press, New York.

MCCANN, G. D. and ARNETT, D. W. (1972). Spectral and polarization sensitivity of the dipteran visual system. *J. gen. Physiol. 59*, 534–558.

MCINTYRE, P. and KIRSCHFELD, K. (1981). Absorption properties of a photostable pigment (P456) in rhabdomere 7 of the fly. *J. Comp. Physiol. 143*, 3–15.

MEFFERT, P. and SMOLA, U. (1976). Electrophysiological measurements of spectral sensitivity of central visual cells in eye of blowfly. *Nature 260*, 342–344.

MENNE, D. and SPATZ, H. C. (1977). Colour vision in *Drosophila melanogaster. J. Comp. Physiol. 114*, 301–312.

MENZEL, R. (1967). Untersuchungen zum Erlernen von Spektralfarben durch die Honigbiene, *Apis mellifica. Z. Vergl. Physiol. 56*, 22–62.

MENZEL, R. (1974). Spectral sensitivity of monopolar cells in the bee lamina. *J. Comp. Physiol. 93*, 337–346.

MENZEL, R. (1975). Colour receptors in insects. In *The Compound Eye and Vision of Insects.* Edited by G. A. Horridge. Pages 121–153. Clarendon Press, Oxford.

MENZEL, R. (1979). Spectral Sensitivity and Color Vision in Invertebrates. In *Comparative Physiology and Evolution of Vision in Invertebrates A. Invertebrate Photoreceptors*. Edited by H. Autrum, *Handbook of Sensory Physiology*. Vol. 7, pt 6A, 503–580. Springer, Berlin and New York.

MENZEL, R. (1981). Achromatic vision in the honeybee at low light intensities. *J. Comp. Physiol. 141*, 389–393.

MENZEL, R. and BLAKERS, M. (1976). Colour receptors in the bee eye-morphology and spectral sensitivity. *J. Comp. Physiol. 108*, 11–33.

MENZEL, R. and KNAUT, R. (1973). Pigment movement during light and chromatic adaptation in the retinula cells of *Formica polyctena* (Hymenoptera, Formicidae). *J. Comp. Physiol. 86*, 125–138.

MENZEL, R. and SYNDER, A. W. (1974). Polarized light detection in the bee, *Apis mellifera*. *J. Comp. Physiol. 88*, 247–270.

MEYER, E. P. and LABHART, T. (1981). Pore canals in the cornea of a functionally specialized area of the honey bee's compound eye. *Cell Tiss. Res. 216*, 491–501.

MEYER-ROCHOW, V. B. (1975). Axonal wiring and polarization sensitivity in eye of the rock lobster. *Nature 254*, 522–523.

MEYER-ROCHOW, V. B. (1980). Electrophysiologically determined spectral efficiencies of the compound eye and median ocellus in the bumblebee *Bombus hortorum tarhakimalainen* (Hymenoptera, Insecta). *J. Comp. Physiol. 139*, 261–266.

MILLER, G. V., HANSEN, K. N. and STARK, W. S. (1981). Phototaxis in *Drosophila*: R1–6 input and interaction among ocellar and compound eye receptors. *J. Insect Physiol. 27*, 813–819.

MILLER, W. H. (1979). Ocular optical filtering. In *Comparative Physiology and Evolution of Vision in Invertebrates. A. Invertebrate Photoreceptors*. Edited by H. Autrum. *Handbook of Sensory Physiology*. Vol. 7, pt 6A, pages 69–143. Springer, Berlin and New York.

MILLER, W. H. and BERNARD, G. D. (1968). Butterfly glow. *J. Ultrastruct. Res. 24*, 286–294.

MIMURA, K. (1978). Electrophysiological evidence for interaction between retinula cells in the flesh-fly. *J. Comp. Physiol. 125*, 209–216.

MINKE, B. (1979). Transduction in photoreceptors with bistable pigments: intermediate processes. *Biophys. Struct. Mech. 5*, 163–174.

MINKE, B. and KIRSCHFELD, K. (1979). The contribution of a sensitizing pigment to the photosensitivity spectra of fly rhodopsin and metarhodopsin. *J. gen. Physiol. 73*, 517–540.

MINKE, B. and KIRSCHFELD, K. (1980). Fast electrical potentials arising from activation of metarhodopsin in the fly. *J. gen. Physiol. 75*, 381–402.

MINKE, B., HOCHSTEIN, S. and HILLMAN, P. (1973). Early receptor potential evidence for the existence of two thermally stable states in the barnacle visual pigment. *J. Gen. Physiol. 62*, 87–104.

MINKE, B., HOCHSTEIN, S. and HILLMAN, P. (1978). The kinetics of visual pigment systems. II. Application to measurements on a bistable pigment system. *Biol. Cybernetics 30*, 33–43.

MINKE, B., WU, C.-F. and PAK, W. L. (1975). Isolation of light-induced response of the central retinula cells from electroretinogram of *Drosophila*. *J. Comp. Physiol. 98*, 345–355.

MOTE, M. I. (1974). Polarization sensitivity: a phenemenon independent of stimulus intensity or state of adaptation in retinular cells of the crabs *Carcinus* and *Callinectes*. *J. Comp. Physiol. 90*, 389–403.

MOTE, M. I. and WEHNER, R. (1980). Functional characteristics of photoreceptors in the compound eye and ocellus of the desert ant, *Cataglyphis bicolor*. *J. Comp. Physiol. 137*, 63–71.

MUIJSER, H., LEUTSCHER-HAZELHOFF, J. T., STAVENGA, D. G. and KUIPER, J. W. (1975). Photopigment conversions expressed in receptor potential and membrane resistance of blowfly visual sense cells. *Nature 254*, 520–522.

MULLER, K. J. (1973). Photoreceptors in the crayfish compound eye; electrical interactions between cells as related to polarized light sensitivity. *J. Physiol. 232*, 573–595.

NEUMEYER, C. (1980). Simultaneous color contrast in the honeybee. *J. Comp. Physiol. 139*, 165–176.

NICKEL, E. and MENZEL, R. (1976). Insect UV- and green-photoreceptor membranes studied by the freeze-fracture technique. *Cell Tiss. Res. 175*, 357–368.

OLIVE, J. (1980). The structural organization of mammalian retinal disc membrane. In *International Review of Cytology*. Edited by G. H. Bourne and J. F. Danielli. Vol. 64, pages 107–169. Academic Press, New York and San Francisco.

OLIVE, J. and RECOUVREUR, M. (1977). Differentiation of retinal rod disc membranes in mice. *Exp. Eye Res. 25*, 63–74.

OSTROY, S. E. (1977). Rhodopsin and the visual process. *Biochem. Biophys. Acta 463*, 91–125.

OSTROY, S. E. (1978). Characteristics of *Drosophila* rhodopsin in wild-type and *norpA* vision transduction mutants. *J. gen. Physiol. 72*, 717–732.

OSTROY, S. E., WILSON, M. and PAK, W. L. (1974). *Drosophila* rhodopsin: photochemistry, extraction and differences in the *norp* A^{P12} phototransduction mutant. *Biochem. Biophys. Res. Comm. 59*, 960–966.

PAK, W. L. (1979). Study of photoreceptor function using *Drosophila* mutants. In *Neurogenetics: Genetic Approaches to the Nervous System*. Edited by X. O. Breakefield. Pages 67–99. Elsevier, New York and Oxford.

PAK, W. L. and LIDINGTON, K. J. (1974). Fast electrical potential from a long-lived, long-wavelength photoproduct of fly visual pigment. *J. gen. Physiol. 63*, 740–756.

PAK, W. L., CONRAD, S. K., KREMER, N. E., LARRIVEE, D. C., SCHINZ, R. H. and WONG, F. (1980). Photoreceptor function. In *Development and Neurobiology of Drosophila*. Edited by O. Siddiqi, P. Babu, L. M. Hall, and J. C. Hall. Pages 331–346. Plenum, New York and London.

PAK, W. L., OSTROY, S. E., DELAND, M. C. and WU, C.-F. (1976). Photoreceptor mutant of *Drosophila*: is protein involved in intermediate steps of phototransduction? *Science 194*, 956–959.

PAULSEN, R. and SCHWEMER, J. (1972). Studies on the insect visual pigment sensitive to ultraviolet light: retinal as the chromophoric group. *Biochim. Biophys. Acta. 283*, 520–529.

PAULSEN, R. and SCHWEMER, J. (1973). Proteins of invertebrate photoreceptor membranes. Characterization of visual-pigment preparation by gel electrophoresis. *Eur. J. Biochem. 40*, 577–583.

PAULSEN, R. and SCHWEMER, J. (1979). Vitamin A deficiency reduces the concentration of visual pigment protein within blowfly photoreceptor membranes. *Biochim. Biophys. Acta 557*, 385–390.

PEPE, I. M. and CUGNOLI, C. (1980). Isolation and characterization of a water-soluble photopigment from honeybee compound eye. *Vision Res. 20*, 97–102.

PEPE, I. M., SCHWEMER, J. and PAULSEN, R. (1982). Characteristics of retinal-binding proteins from the honeybee retina. *Vision Res. 22*, 775–781.

PERRELET, A. (1970). The fine structure of the retina of the honey bee drone. An electron microscope study. *Z. Zellforsch. 108*, 530–562.

PERRELET, A., BAUER, H. and FRYDER, V. (1972). Fracture faces of an insect rhabdome, *J. Microscop. 13*, 97–106.

PFLUGHÖFT, G. (1980). Zum Farbunterschiedungsvermögen der Honigbiene bei niedrigen Umgebungshelligkeiten. *Z. Naturf. 35c*, 1114–1116.

PINTO DA SILVA, P. and BRANTON, D. (1970). Membrane splitting in freeze-etching: covalently bound ferritin as a membrane marker. *J. Cell Biol. 45*, 598–605.

POO, M.-M. and CONE, R. A. (1974). Lateral diffusion of rhodopsin in the photoreceptor membrane. *Nature 247*, 438–441.

POST, C. T. JR and GOLDSMITH, T. H. (1969). Physiological evidence for color receptors in the eye of a butterfly. *Ann. Ent. Soc. Amer. 62*, 1497–1498.

RAZMJOO, S. and HAMDORF, K. (1976). Visual sensitivity and variation of total photopigment content in the blowfly photoreceptor membrane. *J. Comp. Physiol. 105*, 279–286.

RAZMJOO, S. and HAMDORF, K. (1980). In support of the "photopigment model" of vision in invertebrates. *J. Comp. Physiol. 135*, 209–215.

READY, D. F., HANSON, T. E. and BENZER, S. (1976). Development of the *Drosophila* retina, a neurocrystalline lattice. *Devel. Biol. 53*, 217–240.

RIBI, W. A. (1975). The first optic ganglion of the bee. I. Correlation between visual cell types and their terminals in the lamina and medulla. *Cell Tiss. Res. 165*, 103–111.

RIBI, W. A. (1979). Do the rhabdomeric structures in bees and flies really twist? *J. Comp. Physiol. 134*, 109–112.

RICHARDS, F. M. and KNOWLES, J. R. (1968). Glutaraldehyde as a protein cross-linking reagent. *J. Mol. Biol. 37*, 231–233.

RÖHLICH, P. (1981). Structure of retinal photoreceptor membranes as seen by freeze-fracturing. *Acta Histochem.* (Suppl.) *23*, 123–136.

ROSE, R. and MENZEL, R. (1981). Luminance dependence of pigment color discrimination in bees. *J. Comp. Physiol. 141*, 379–388.

ROSNER, G. (1975). Adaptation und Photoregeneration im Fliegenauge. *J. Comp. Physiol. 102*, 269–295.

ROTMANS, J. P., DAEMEN, F. J. M. and BONTING, S. L. (1974). Biochemical aspects of the visual process. XXVI. Binding site and migration of retinaldehyde during rhodopsin photolysis. *Biochim. Biophys. Acta 357*, 151–158.

RUSHTON, W. A. H. (1972). Pigments and signals in colour vision. *J. Physiol. 720*, 1–31P.

SCHINZ, R. H. and SIDORSKY, L. R. P. (1978). Color-dependent modulation of phototactic behavior of *Drosophila melanogaster*. *Invest. Ophthalmol. Vis. Sci.* (Suppl.), p. 195.

SCHLECHT, P. (1979). Colour discrimination in dim light: an analysis of the photoreceptor arrangement in the moth *Deilephila*. *J. Comp. Physiol. 129*, 257–267.

SCHLECHT, P. and TÄUBER, U. (1975). The photochemical equilibrium in rhabdomeres of *Eledone* and its effect on dichroic absorption. In *Photoreceptor Optics*. Edited by A. W. Snyder and R. Menzel. Pages 316–335. Springer, Berlin and New York.

SCHLECHT, P., HAMDORF, K. and LANGER, H. (1978). The arrangement of colour receptors in a fused rhabdom of an insect. A microspectrophotometric study on the Moth *Deilephila*. *J. Comp. Physiol. 123*, 239–243.

SCHNEIDER, I., DRAŠLAR, K., LANGER, H., GOGALA, M. and SCHLECHT, P. (1978). Feinstruktur und Schirmpigment-Eigenschaften der Ommatidien des Doppelauges von *Ascalaphus* (Insecta, Neuroptera). *Cytobiologie 16*, 274–307.

SCHÜMPERLI, R. A. (1973). Evidence for colour vision in *Drosophila melanogaster* through spontaneous phototactic choice behaviour. *J. Comp. Physiol. 86*, 77–94.

SCHWEMER, J. (1969). Der Sehfarbstoff von *Eledone moschata* und seine Unsetzung in der lebenden Netzhaut. *Z. Vergl. Physiol. 62*, 121–152.

SCHWEMER, J. (1979). Molekulare Grundlagen der Photorezeption bei der Schmeissfliege *Calliphora erythrocephala* Meig. Habilitationsschrift Abt. Biologie, Ruhr-Universität, Bochum (W. Germany).

SCHWEMER, J. and LANGER, H. (1982). Insect visual pigments. In *Visual Pigments and Purple Membranes*. I. Edited by L. Packer. *Methods in Enzymology*. Vol. 81, pages 182–190. Academic Press, London, New York and San Francisco.

SCHWEMER, J. and PAULSEN, R. (1973). Three visual pigments in *Deilephila elpenor* (Lepidoptera, Sphigidae). *J. Comp. Physiol. 86*, 215–229.

SCHWEMER, J., GOGALA, M. and HAMDORF, K. (1971). Der UV-Sehfarbstoff der Insekten: Photochemie *in vitro* und *in vivo*. *Z. Vergl. Physiol. 75*, 174–188.

SELDIN, E. B., WHITE, R. H. and BROWN, P. K. (1972). Spectral sensitivity of larval mosquito ocelli. *J. gen. Physiol. 59*, 415–420.

SHAW, S. R. (1969a). Sense-cell structure and interspecies comparisons of polarized light absorption in arthropod compound eyes. *Vision Res. 9*, 1031–1040.

SHAW, S. R. (1969b). Interreceptor coupling in ommatidia of drone honeybee and locust compound eyes. *Vision Res. 9*, 999–1029.

SHAW, S. R. (1975). Retinal resistance barriers and electrical lateral inhibition. *Nature 255*, 480–483.

SHIMIZU, I., KITABATAKE, S. and KATO, M. (1981). Effect of carotenoid deficiency on photosensitivities in the silkworm, *Bombyx mori*. *J. Insect Physiol. 27*, 593–599.

SHINZ, R. H., LO, M.-V. C., LARRIVEE, D. C. and PAK, W. L. (1982). Freeze-fracture study of the *Drosophila* photoreceptor membrane: mutations affecting membrane particle density. *J. Cell Biol. 93*, 961–969.

SILBERGLIED, R. E. and TAYLOR, O. R. (1973). Ultraviolet differences between the sulphur butterflies, *Colias eurytheme* and *C. philodice*, and a possible isolating mechanism. *Nature 241*, 406–408.

SINGER, S. J. and NICHOLSON, G. L. (1972). The fluid mosaic model of the structure of cell membranes. *Science 175*, 720–731.

SJÖSTRAND, F. S. and KREMAN, M. (1978). Molecular structure of outer segment disks in photoreceptor cells. *J. Ultrastruct. Res. 65*, 195–226.

SKRZIPEK, K.-H. and SKRZIPEK, H. (1973). Die Anordnung der Ommatidien in der Retina der Biene. *Z. Zellforsch. 139*, 567–582.

SKRZIPEK, K.-H. and SKRZIPEK, H. (1974). The ninth retinula cell in the ommatidium of the worker honey bee (*Apis mellifica* L.). *Z. Zellforsch. 147*, 589–593.

SMOLA, U. (1977). Das "Twisten" der Rhabdomere der Sehzellen im Auge von *Calliphora erythrocephala*. *Verh. Dtsch. Zool. Ges. 1977*, 234.

SMOLA, U. and MEFFERT, P. (1975). A single-peaked UV receptor in the eye of *Calliphora erythrocephala*. *J. Comp. Physiol. 103*, 353–357.

SMOLA, U. and MEFFERT, P. (1979). The spectral sensitivity of the visual cells R7 and R8 in the eye of the blowfly *Calliphora erythrocephala*. *J. Comp. Physiol. 133*, 41–52.

SMOLA, U. and TSCHARNTKE, H. (1979). Twisted rhabdomeres in the dipteran eye. *J. Comp. Physiol. 133*, 291–297.

SMOLA, U. and WUNDERER, H. (1981a). Fly rhabdomeres twist *in vivo*. *J. Comp. Physiol. 142*, 43–49.

SMOLA, U. and WUNDERER, H. (1981b). Twisting of blowfly (*Calliphora erythrocephala* Meigen) (Diptera, Calliphoridae) rhabdomeres: an *in vivo* feature unaffected by preparation or fixation. *Int. J. Insect Morphol. Embryol. 10*, 331–344.

SNYDER, A. W. (1979). Physics of vision in compound eyes. In *Comparative Physiology and Evolution of Vision in Invertebrates. A: Invertebrate Photoreceptors*. Edited by H. Autrum. *Handbook of Sensory Physiology*. Vol. 7, pt 6A, pages 225–313. Springer, Berlin and New York.

SNYDER, A. W. and LAUGHLIN, S. B. (1975). Dichroism and absorption by photoreceptors. *J. Comp. Physiol. 100*, 101–116.

SNYDER, A. W. and MCINTYRE, P. D. (1975). Polarization sensitivity of twisted fused rhabdoms. In *Photoreceptor Optics*. Edited by A. W. Snyder and R. Menzel. Pages 388–391. Springer, Berlin and New York.

SNYDER, A. W. and MILLER, W. H. (1972). Fly colour vision. *Vision Res. 12*, 1389–1396.

SNYDER, A. W. and PASK, C. (1973). Spectral sensitivity of Dipteran retinula cells. *J. Comp. Physiol. 84*, 59–76.

SNYDER, A. W., MENZEL, R. and LAUGHLIN, S. B. (1973). Structure and function of the fused rhabdom. *J. Comp. Physiol. 87*, 99–135.

SONTAG, C. (1971). Spectral sensitivity studies on the visual system of the praying mantis *Tenodera sinensis*. *J. gen. Physiol. 57*, 93–112.

SRINIVASAN, M. and BERNARD, G. D. (1980). A technique for estimating the contribution of photomechanical responses to visual adaptation. *Vision Res. 20*, 511–521.

STARK, W. S. (1975). Spectral sensitivity of visual response alterations mediated by interconversions of native and intermediate photopigments in *Drosophila*. *J. Comp. Physiol. 96*, 343–356.

STARK, W. S. (1977). Sensitivity and adaptation in R7, an ultraviolet photoreceptor, in the *Drosophila* retina. *J. Comp. Physiol. 115*, 47–59.

STARK, W. S. and JOHNSON, M. A. (1980). Microspectrophotometry of *Drosophila* visual pigments: determinations of conversion efficiency in R1–6 receptors. *J. Comp. Physiol. 140*, 275–286.

STARK, W. S. and ZITZMANN, W. G. (1976). Isolation of adaptation mechanisms and photopigment spectra by vitamin A deprivation in *Drosophila*. *J. Comp. Physiol. 105*, 15–27.

STARK, W. S., FRAYER, K. L. and JOHNSON, M. A. (1979). Photopigment and receptor properties in *Drosophila* compound eye and ocellar receptors. *Biophys. Struct. Mech. 5*, 197–209.

STARK, W. S., IVANYSHYN, A. M. and GREENBERG, R. M. (1977). Sensitivity and photopigments of R1–6, a two-peaked photoreceptor, in *Drosophila*, *Calliphora*, and *Musca*. *J. Comp. Physiol. 121*, 289–305.

STARK, W. S., IVANYSHYN, A. M. and HU, K. G. (1976). Spectral sensitivities and photopigments in adaptation of fly visual receptors. *Naturwissenschaften 63*, 513–518.

STARK, W. S., STAVENGA, D. G. and KRUIZINGA, B. (1979). Fly photoreceptor fluorescence is related to UV sensitivity. *Nature 280*, 581–583.

STAVENGA, D. G. (1975). Dark regeneration of invertebrate visual pigments. In *Photoreceptor Optics*. Edited by A. W. Snyder and R. Menzel. Pages 290–295. Springer, Berlin and New York.

STAVENGA, D. G. (1976). Fly visual pigments. Difference in visual pigments of blowfly and dronefly peripheral retinula cells. *J. Comp. Physiol. 111*, 137–152.

STAVENGA, D. G. (1979). Pseudopupils of compound eyes. In *Comparative Physiology and Evolution of Vision in Invertebrates. A. Invertebrate Photoreceptors*. Edited by H. Autrum. *Handbook of Sensory Physiology*. Vol. 7, pt 6A, pages 357–439. Springer, Berlin and New York.

STAVENGA, D. G. and FRANCESCHINI, N. (1981). Fly visual pigment states, rhodopsin R490, metarhodopsins M and M', studied by transmission and fluorescence microspectrophotometry *in vivo*. *Invest. Ophthal. Vis. Sci. 20* (Suppl.), 111.

STAVENGA, D. G., ZANTEMA, A. and KUIPER, J. W. (1973). Rhodopsin processes and the function of the pupil mechanism in flies. In *Biochemistry and Physiology of Visual Pigments*. Edited by H. Langer. Pages 175–180. Springer, Berlin and New York.

STEIN, P. J., BRAMMER, J. D. and OSTROY, S. E. (1979). Renewal of opsin in the photoreceptor cells of the mosquito. *J. gen. Physiol.* 74, 565–582.

STEPHENSON, R. S. and PAK, W. L. (1980). Heterogenic components of a fast electrical potential in *Drosophila* compound eye and their relation to visual pigment conversion. *J. gen. Physiol.* 75, 353–379.

STEPHENSON, R. S. and PAK, W. L. (1981). *Drosophila* visual mutants defective in Vitamin A utilization. *Invest. Ophthalmol. Vis. Sci.* 20 (Suppl.), 112.

STOWE, S. (1980). Rapid synthesis of photoreceptor membrane and assembly of new microvilli in a crab at dusk. *Cell Tiss. Res. 211*, 419–440.

STRANGE, P. H. (1961). The spectral sensitivity of *Calliphora* maggots. *J. Exp. Biol.* 38, 237–248.

STRAUSFELD, N. J. (1970). Golgi studies on insects. II. The optic lobes of Diptera. *Phil. Trans. Roy. Soc. B. 258*, 135–223.

STRONG, J. and LISMAN, J. E. (1978). Initiation of light adaptation in barnacle photoreceptors. *Science 200*, 1485–1487.

STROTHER, G. K. and SUPERDOCK, D. A. (1972). *In situ* absorption of *Drosophila melanogaster* visual screening pigments. *Vision Res. 12*, 1545–1547.

SUZUKI, T., SUGAHARA, M., AZUMA, K., AZUMA, M., SAIMI, Y. and KITO, Y. (1973). Studies on cephalopd rhodopsin conformational changes in chromophore and protein during the photoregeneration process. *Biochim. Biophys. Acta 333*, 149–160.

SUZUKI, T., SUGAHARA, M. and KITO, Y. (1972). An intermediate in the photoregeneration of squid rhodopsin. *Biochim. Biophys. Acta. 275*, 260–270.

SWIHART, S. L. (1969). Color vision and the physiology of the super-position eye of a butterfly (Hesperiidae). *J. Insect Physiol. 15*, 1347–1365.

SWIHART, S. L. (1970). The neural basis of colour vision in the butterfly *Papilio troilus. J. Insect Physiol. 16*, 1623–1636.

SWIHART, S. L. (1972a). Modelling the butterfly visual pathway. *J. Insect Physiol. 18*, 1915–1928.

SWIHART, S. L. (1972b). The neural basis of colour vision in the butterfly, *Heliconius erato. J. Insect Physiol. 18*, 1015–1025.

SWIHART, S. L. and GORDON, W. C. (1971). Red photoreceptor in butterflies. *Nature, 231*, 126–127.

TAKEMURA, T., DAS, P. K., HUG, G. and BECKER, R. S. (1978). Visual pigments. 8. Hydrogen bonding effects on fluorescence properties of retinals. *J. Amer. Chem. Soc. 100*, 2626–2630.

TÄUBER, U. (1975). Photokinetics and dichroism of visual pigments in the photoreceptors of *Eledone* (Ozoena) *moschata*. In *Photoreceptor Optics*. Edited by A. W. Snyder and R. Menzel. Pages 296–316. Springer, Berlin and New York.

TRUJILLO-CENÓZ, O. AND BERNARD, G. D. (1972). Some aspects of the retinal organization of *Sympycnus lineatus* Loew (Diptera, Dolichopodidae). *J. Ultrastruct. Res. 38*, 149–160.

TRUJILLO-CENÓZ, O. and MELAMED, J. (1966). Electron microscope observations on the peripheral and intermediate retinas of Dipterans. In *The Functional Organization of the Compound Eye*. Edited by C. G. Bernhard. Pages 339–361. Pergamon Press, Oxford.

TSUKAHARA, Y. and HORRIDGE, G. A. (1977). Visual pigment spectra from sensitivity measurements after chromatic adaptation of single drone-fly retinula cells. *J. Comp. Physiol. 114*, 233–251.

VARELA, F. G. (1970). Fine structure of the visual system of the honeybee *Apis mellifera*. II. The lamina. *J. Ultrastruct. Res. 31*, 178–194.

VARELA, F. G. and PORTER, K. R. (1969). Fine structure of the visual system of the honeybee *(Apis mellifera)*. I. The retina. *J. Ultrastruct. Res. 29*, 236–259.

WALD, G. (1959). Life and light. *Scient. Amer. 201, no. 4*, 92–108.

WALD, G. (1965). Frequency or wavelength? *Science 150*, 1239–1240.

WALD, G. (1967). Visual pigments of crayfish. *Nature 215*, 1131–1133.

WALD, G. (1968). Single and multiple visual systems in arthropods. *J. gen. Physiol. 51*, 125–156.

WALD, G. and BROWN, P. K. (1953). The molar extinction of rhodopsin. *J. gen. Physiol. 37*, 189–200.

WALD, G. and BROWN, P. K. (1957). The vitamin A of a Euphausiid Crustacean. *J. gen. Physiol. 40*, 627–634.

WALD, G. and BURG, S. P. (1957). The vitamin A of the lobster. *J. gen. Physiol. 40*, 609–625.

WASSERMAN, G. S. (1973). Invertebrate colour vision and the tuned-receptor paradigm. *Science 180*, 268–275.

WATERMAN, T. H. (1975). The optics of polarization sensitivity. In *Photoreceptor Optics*. Edited by A. W. Snyder and R. Menzel. Pages 339–371. Springer, Berlin and New York.

WATERMAN, T. H. and FERNANDEZ, H. R. (1970). E-vector and wavelength discrimination by retinular cells of the crayfish *Procambarus. Z. Vergl. Physiol. 68*, 154–174.

WATERMAN, T. H., FERNANDEZ, H. R. and GOLDSMITH, T. H. (1969). Dichroism of photosensitive pigment in rhabdoms of the crayfish *Orconectes. J. gen. Physiol. 54*, 415–432.

WEHNER, R. and BERNARD, G. D. (1980). Intracellular optical physiology of the bee's eye. II. Polarizational sensitivity. *J. Comp. Physiol. 137*, 205–214.

WEHNER, R. and MEYER, E. (1981). Rhabdomeric twist in bees — artefact or *in vivo* structure? *J. Comp. Physiol. 142*, 1–17.

WEHNER, R., BERNARD, G. D. and GEIGER, E. (1975). Twisted and non-twisted rhabdoms and their significance for polarization detection in the bee. *J. Comp. Physiol. 104*, 225–245.

WEHRHAHN, C. (1976). Evidence for the role of retinal receptors R7/8 in the orientation behaviour of the fly. *Biol. Cybernetics 21*, 213–220.

WEINSTEIN, G. W., HOBSON, R. R. and DOWLING, J. E. (1967). Light and dark adaptation in the isolated rat retina. *Nature, 215*, 134–138.

WELSCH, B. (1977). Ultrastruktur und funktionelle Morphologie der Augen des Nachfalters *Deilephila elpenor* (Lepidoptera, Sphingidae). *Cytobiologie 14*, 378–400.

WHITE, R. H. (1967). The effect of light and light deprivation upon the ultrastructure of the larval mosquito eye. II. The rhabdom. *J. Exp. Zool. 166*, 405–426.

WHITE, R. H. (1968). The effect of light and light deprivation upon the ultrastructure of the larval mosquito eye. III. Multivesicular bodies and protein uptake. *J. Exp. Zool. 169*, 261–278.

WHITE, R. H. (1978). Insect visual pigments. In *Advances in Insect Physiology*. Edited by J. E. Treherne, M. J. Berridge and V. B. Wigglesworth. Vol. 13, pages 35–67. Academic Press, London, New York and San Francisco.

WHITE, R. H. and JOLIE, M. A. (1966). The effects of light and beta-carotene upon the endoplasmic reticulum of the mosquito photoreceptor cell. *J. Cell. Biol. 31*, 122A.

WHITE, R. H. and LORD, E. (1975). Diminution and enlargement of the mosquito rhabdom in light and darkness. *J. gen. Physiol. 65*, 583–598.

WHITE, R. H. and SUNDEEN, C. D. (1967). The effect of light and light deprivation upon the ultrastructure of the larval mosquito eye. I. Polyribosomes and endoplasmic reticulum. *J. Exp. Zool. 164*, 461–478.

WHITE, R. H., BROWN, P. K., HURLEY, A. K. and BENNETT, R. R. (1983). Rhodopsins, retinula cell ultrastructure, and receptor potentials in the developing pupal eye of the moth *Manduca sexta. J. Comp. Physiol. 150*, 153–163.

WILLIAMS, D. S. (1980). Organization of the compound eye of a tipulid fly during the day and night. *Zoomorphologie 95*, 85–104.

WILLIAMS, D. S. (1981). Twisted rhabdomeres in the compound eye of a tipulid fly (Diptera). *Cell Tiss. Res. 217*, 625–632.

WILLMUND, R. (1979). Light induced modification of phototactic behavior of *Drosophila melanogaster*. II. Physiological aspects. *J. Comp. Physiol. 129*, 35–41.

WILLMUND, R. and FISCHBACH, K. F. (1977). Light induced modification of phototactic behaviour of *Drosophila melanogaster* wildtype and some mutants in the visual system. *J. Comp. Physiol. 118*, 261–271.

WOLBARSHT, M. L. (1976). The function of intraocular color filters. *Fed. Proc. 35*, 44–50.

WOLKEN, J. J. and SCHEER, I. J. (1963). An eye pigment of the cockroach. *Exp. Eye Res.* 2, 182–188.

WOLKEN, J. J., BOWNESS, J. M. and SCHEER, I. J. (1960). The visual complex of the insect: retinene in the housefly. *Biochim. Biophys. Acta.* 43, 531–537.

WU, C.-W. and STRYER, L. (1972). Proximity relationships in rhodopsin. *Proc. Nat. Acad. Sci.* 69, 1104–1108.

ZINKLER, D. (1975). Zum Lipidmuster der Photorezeptoren von Insekten. *Verh. Dtsch. Zool. Ges. 1974*, 28–32.

9 Hearing and Sound

AXEL MICHELSEN and OLE NÆSBYE LARSEN

Odense University, Odense, Denmark

1   INTRODUCTION

A sense of hearing has evolved independently at least a dozen times in different groups of insects, and these ears have further differentiated into a large number of functional types. Similarly, mechanisms for generating sound waves have evolved independently in many groups of insects. Until recently, the differences and similarities within sound generators and within hearing organs were thought to generally reflect the evolutionary prehistory of the animals and thus to follow their systematic positions. However, we are now beginning to understand that some features of both the sound generators and the hearing organs are simple consequences of the physical nature of sound waves. Furthermore, some insect ears are adapted for the reception of sounds from predators (e.g. the ultrasonic cries of hunting bats) and other ears for receiving conspecific songs. The former group of ears tend to be more simple than the latter, and this difference probably reflects the complexity of the tasks performed. The properties of the hearing organs have to match those of the sound-generating mechanisms, if conspecific communication is to be efficient. There is some evidence that the kind of sound signal used by an animal may depend upon the acoustic properties of its environment. There may, then, be three reasons for the rich diversity of mechanisms used for sound communication in insects: the size of the insect relative to the wavelength of sound plays a role (see below), as does its evolutionary history; then the acoustic properties of the environment may be a third element determining the properties of sound signals and thus of the sound generators and hearing organs.

Several reviews have been published on sound communication in general and on limited areas such as sound production, hearing, or the acoustic properties of the environment (e.g. Busnel, R. -G., 1963; Elsner, N. and Popov, A., 1978; Michelsen, A., 1978; Michelsen, A. and Nocke, H., 1974; Sebeok, T., 1977; Wiley, R. and Richards, D., 1978; Zhantiev, R., 1981).

The definition of acoustic communication is arbitrary, since the boundaries between phonoreception and other forms of mechanoreception are not sharp. We have not attempted to make strict definitions, and the selection of material is therefore a matter of taste. The reception of air-borne sound in the extreme near field of the sound source is included here, but the communication through solid substrates is not.

2   SOUND WAVES

Sound waves are longitudinal, i.e. the particles of the medium oscillate in the direction of propagation of the wave. Sound waves may be described and measured as fluctuations in pressure propagating away from the sound source with a certain velocity. The presence of sound may therefore be detected by devices sensitive to pressure (pressure receivers and pressure difference receivers, see below). It is important to remember that sound pressure is a scalar quantity, although sound waves have directional properties. The movements of the particles of the medium are directional, however, and sound receivers sensitive to this component of sound are therefore inherently directional.

The frequency range used by insects for communication or orientation extends from some hundred Hz to more than 100 kHz, i.e. about 3 orders of magnitude. In air, the wavelengths (λ) range from about 1 m to about 3 mm ($\lambda = c/f$, where c is the velocity of propagation and f the frequency). The behaviour of both sound-emitting and sound-receiving systems depends very much upon the relationship between the wavelength of sound and the size of the sound emitter or receiver. The physical mechanisms operating in one part of this large frequency range may therefore be rather different from those at other frequencies.

The physics of sound waves is described in many textbooks (e.g. Beranek, L., 1954; Morse, P., 1948), but most of the literature on acoustics is not easily accessible to biologists or even to physicists without proper training in the subject. Many books and papers are filled with equations, and the newcomer may be forgiven the impression that all acoustical problems can be solved by means of calculations. However, it is very important to realize that most equations in acoustics are approximations which are valid only under the conditions specified when the equations were derived. Such equations may be used in simple situations, e.g. when the obstacles are very small compared with the wavelength of sound, or when they are very large. Unfortunately, many problems of biological interest are somewhere between these extremes. Also, in many cases, it is not possible to find the "correct" equation, since the mechanical properties of many biological materials are not known. But while calculations may lead to very misleading results, experiments without a proper theoretical basis are not likely to lead to any real understanding. The ideal approach, then, is a combination of experiments and calculations.

2.1 Sound fields

Sound waves behave in a complicated manner close to the sound source. From textbooks of acoustics one may get the impression that it is possible to calculate the behaviour of such *near fields*, but in practice this is rarely possible. Biological sound sources or loudspeakers are normally far from the ideal sound sources assumed in such calculations (e.g. small pulsating spheres), and the near fields of such real sound sources may be more complicated than those of the ideal ones. In the near field the sound pressure varies rapidly with the distance to the source, and other components of the sound wave (the pressure gradient and the vibration velocity of the particles of the medium) are even more dependent on the distance to, and geometry of, the sound source (see e.g. Skudrzyk, E., 1971). It may be virtually impossible to guess the magnitude of the forces acting upon ears sensitive to these components of sound if the ear is in the near field of the sound source (e.g. when two grasshoppers are courting each other at short distance).

The sound may behave in a much simpler way at greater distances from the sound source (in the *far field*), especially if no redirected sound components are present (a situation called a *free field*). The boundary between the near field and the far field is not sharp, although one may often get this impression from the literature. It is possible to calculate the distance range of the near field, if one is dealing with ideal sound sources (which vibrate in a simple way and are small compared with the wavelengths of interest). In practise, and provided that the sound source is "small" (the term "small" and "large" indicate the size relative to the wavelength of sound), one has to be at least one wavelength away from the sound source, in order to be in the far field.

2.2 Some measures of sound

In a free sound field the front of the propagating wave acting on a sound receiver can be considered almost plane. One now finds a simple relationship between the vibration velocity (v) of the particles of the medium and the sound pressure (p):

$$v = p/c\rho \qquad (1)$$

where c is the propagation velocity of the sound wave, and ρ is the density of the medium. In *water*, $c =$ about $1430\,\mathrm{m\,s^{-1}}$ and $\rho = 1000\,\mathrm{kg\,m^{-3}}$, whereas in *air* $c =$ about $340\,\mathrm{m\,s^{-1}}$ and $\rho = 1.2\,\mathrm{kg\,m^{-3}}$. $c\rho$ is thus about 3500 times larger in water than in air. In other words, at a certain sound pressure (p), the vibration velocity (v) of the particles is about 3500 times smaller in water than in air. This has important consequences for the reception of sound, both in animals living in water and in terrestrial animals equipped with water-filled ears.

The relationship expressed in equation (1) reflects the physical fact that water is much more difficult to compress than air. A local change in pressure will therefore cause only small changes in volume, and the water particles will move very little. It is important to realize, however, that equation (1) describes the relationship between v and p, when these components are only determined by the properties of the medium itself. Equation (1) is therefore true, for example, for a volume of water surrounded by large amounts of water (e.g. in the middle of an ocean), but not for water close to the surface or close to an air bubble. In such cases the presence of the readily compressible air will allow much larger movements of the water particles than expected from equation (1).

The power carried by a propagating sound wave is given by $p \cdot v$ (this is analogous to the power carried by alternating current, which is the voltage times the current). *Sound intensity* (I) is the sound energy passing a unit area perpendicular to the direction of propagation of the sound wave per unit time. In the free sound field the intensity of the sound wave may be found simply by measuring p:

$$I = p \cdot v = p^2/c\rho \qquad (2)$$

It should be noted that the power cannot be estimated in this simple way when the relationship between v and p is not determined by $c\rho$ alone. One also has to remember that equations (1) and (2) are only true when there are no redirected sound waves present. This is rarely the case in practice when one is studying the sound communication of insects. In the laboratory, a free sound field of air-borne sound may be created by covering the walls, floor and ceiling of the room with wedges made of mineral wool.

Sound levels are normally indicated in dB (decibel). One often meets dB values in the literature without a clearly defined reference. 56 dB without a reference is just as meaningless as 56% without reference. A *sound pressure level* (SPL) is indicated as a number of decibels relative to a reference sound pressure. The reference pressure normally used is $2 \times 10^{-5} \, \text{N m}^{-2}$ ($= 2 \, 10^{-4} \text{dyn cm}^{-2}$, about the human threshold at 1 kHz), and this reference pressure is also used here when not otherwise stated. It should be noted, however, that other reference levels may occur in the literature (e.g. 1 dyn cm^{-2} = 1 μbar, 1 Pascal = 1 N m^{-2}). One can easily calculate the SPL value of a sound pressure (p) by means of the equation

$$\text{SPL} = 20 \log_{10} \frac{p}{p_r} \quad \text{dB} \qquad (3)$$

where p_r is the reference pressure. A 10 times change in pressure thus corresponds to 20 dB, 100 times to 40 dB, 2 times to 6 dB, and so on.

The dB value is also used for the energy emitted from a sound source or the energy received by a microphone or an ear. The energy is proportional to p^2 (see equation 2), and the definition of dB for sound intensities (sound powers) takes this into consideration. For example, the *intensity level* (IL) of a sound is defined by

$$\text{IL} = 10 \log_{10} \frac{I}{I_r} \quad \text{dB} \qquad (4)$$

where I is the intensity in question and I_r is the reference intensity. I_r is often taken to be 10^{-12} watt m^{-2}. Under standard atmospheric conditions and for a free sound field this reference corresponds approximately to the commonly used reference sound pressure ($2 \times 10^{-5} \, \text{N m}^{-2}$). This means that SPL and IL values may refer to the same sound level, provided that the references for the dB values correspond to each other. Note that a 10 times change in intensity (power) corresponds to only 10 dB.

3 SOUND EMISSION

Sound waves are normally generated by the action of a vibrating structure on the surrounding medium. Muscles vary greatly in their properties, and even the asynchronous ("myogenic") muscles of insects are not able to contract faster than about 1000 times per second. The contraction curves are rather smooth and do not contain high-frequency components. The frequency range of insect songs would therefore be limited to frequencies below about 1 kHz if the sound-emitting structures were vibrated directly by muscular contractions. On the other hand, for the sound emission to be efficient, the diameter of the sound source should be about the same order of magnitude as the wavelength of sound (λ) — or larger (Fig. 1). The sound-emitting structures of insects are often so small compared with the wavelength of the communication sounds that they radiate sound with less than the maximum efficiency. For example, the harp of crickets has a radius of about 3 mm, and such small structures do not emit air-borne sounds with maximum efficiency below 30 kHz. In general, even large insects can only obtain a reasonably efficient sound emission in the kHz range, and small insects (except at very close range) must use ultrasonic frequencies if they use air-borne sound waves for communication.

The sound production of vibrating bodies depends on the way in which they vibrate, as well as on their surroundings. Sound sources behaving like pulsating and oscillating spheres are called, respectively, *monopoles* and *dipoles* (or sound sources of order zero and order one). Their behaviour can be calculated quantitatively, but the calculations are far from easy. In textbooks (e.g. Beranek, L., 1954) one can find diagrams of the so-called radiation impedances for various kinds of sound sources. The *radiation impedance* represents the load to the vibrating structure caused by the reaction of the medium, and it consists of a reactive part and a resistive part. At low frequencies one may imagine the reactive part as representing the load of a thin layer of air which is moved with the vibrating structure. The resistive part (R_r) represents the load of the radiation of sound energy from the sound emitter. When consulting the literature one should note that different types of impedances may be used (mechanical, acoustic, specific acoustic).

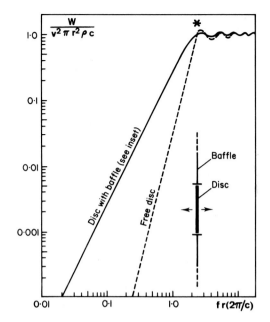

FIG. 1. The emitted sound power (W) of a free disc and of a disc surrounded by an "infinite" baffle. The symbols are explained in the text.

In a pulsating sphere all parts of the surface "agree" on either producing a compression or a rarefaction. If the sphere oscillates to and fro, one side of the sphere will produce a compression, while the opposite side produces a rarefaction — and viceversa when the sphere moves in the opposite direction a little later. The compression and rarefaction are separated by a small distance, if the sphere is small relative to the wavelength (which is the distance that the disturbance can spread during one oscillation period), and they may therefore cancel. This phenomenon is called "acoustic short-circuiting", and it causes the sound output from a small oscillating sphere to be much smaller than that from a pulsating sphere of the same size and vibration frequency. We are avoiding this phenomenon when we place a loudspeaker in a cabinet, thus separating the front and back surfaces.

The physical effect of acoustic short-circuiting is illustrated in Fig. 1 for a free disk and a disk with a baffle (a disk in an "infinite" wall). The diagrams of the resistive part of the radiation impedances from Beranek, L. (1954) have been redrawn to show the radiated sound power (measured in watts) as a function of the frequency and radius of the vibrating disk. The sound power (W) is related to the resistive part of the (mechanical) radiation impedance (R_{rm}) by the equation

$$W = R_{rm} \, v^2 \qquad (5)$$

where v is the velocity of motion of the sound emitter. In Fig. 1 W is divided by v^2 times πr^2 (the area of the disk) times $c\rho$ (see equation 1). From the figure it is apparent that there may be a considerable difference in the sound power emitted from a disk with and without acoustic short-circuiting.

The problem of acoustic short-circuiting has been solved completely only by the cicadas, which emit sound from two cuticular membranes in the first abdominal segment, the tymbals (Fig. 2). This is similar to a loudspeaker backed by a closed cabinet, but in some cicadas the walls of the "cabinet" (abdomen) may be caused to vibrate, so the abdomen operates like the pulsating sphere discussed above. The tymbal is buckled inwards by the twitch contraction of a large muscle and springs back to the resting position after the muscle relaxes (Pringle, J., 1954). Other insects suffer some acoustic short-circuiting. In the European field cricket (*Gryllus campestris*), for example, the sound-emitting structure is the harp, a part of the wing which is about 7 mm from the distal border. The calling song is at 4–5 kHz, i.e. λ is about 8 cm. From theory, therefore, one would expect some acoustic short-circuiting between the upper and lower surfaces of the harp (Nocke, H., 1971). This is in fact the case and it has been further confirmed by studies of some South African tree crickets (*Oecanthus burmeisteri*), which cut a hole in the middle of large leaves and place themselves in the holes in such a way that they avoid short-circuiting. The sound pressure produced from this position is about 10 dB higher than from the same cricket without this self-made baffle (Prozesky-Schulze, L. et al., 1975).

Monopoles and dipoles differ not only in terms of power output; they also have widely different directivity patterns and near fields. Monopoles radiate sound waves evenly in all directions, and their directivity pattern (drawn on a plane intersecting the centre of the sound source) is circular. The directivity pattern of a dipole is a figure of "8", i.e. the sound pressure is a maximum in the axial direction of the vibrating dipole. It is easy to imagine that the compression and rarefaction components cancel in a plane perpendicular to the axial direction (e.g. in

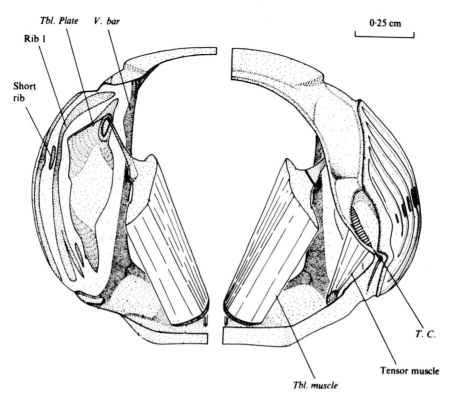

FIG. 2. Anterior view of the sound-producing structures in a bladder cicada. *Left*: a cut through the tymbal; *right*: a cut through the anterior boundary of the first abdominal segment. Tbl.: tymbal; V: vertical; T.C.: tymbal chordotonal organ. (From Simmons, P. and Young, D., 1978.)

the plane of a vibrating free disk), and that the sound pressure must be zero in this direction. This does not mean, however, that other components of the sound wave are also zero in this direction. The particles of the medium move in parallel to the axial direction, and there is also a pressure gradient. This is an interesting situation from a biological point of view. Many insect ears are sensitive to either the movement or pressure gradient components of sound, and such ears may be subjected to intense stimulation close to a dipole (a singing insect) although a microphone sensitive to the pressure components may indicate a very small sound pressure. Space does not permit a more detailed discussion of the near field of dipoles and more complicated sound sources. In passing on, we may note that virtually nothing is known about the properties of the sound fields generated by singing non-monopole insects.

3.1 Frequency multiplication

As we have discussed, all sound-producing insects

are so small that their sound emission would be very inefficient if they were to drive the sound-radiating structures directly with muscle contractions. The solution to this problem is to let the muscles act on a frequency multiplier. This is a device which converts each muscle twitch into a vibration of higher frequency. Several types of frequency multipliers exist. First, we shall examine the stridulatory organs, which occur in many insects.

Numerous examples of stridulation apparatuses have been described (see e.g. Dumortier, B., 1963). Despite their different evolutionary origin and anatomical position on the animals, they are all built in a similar way (Fig. 3). A "scraper" located on one part of the body is moved over a "file" on another part of the body. The scraper may have the shape of a ridge, and the file may consist of a row of "teeth". Typically, the teeth are a few μm high. During each muscular contraction the scraper may hit many teeth, thus producing a higher-frequency vibration. One may easily imitate this process on a larger scale by moving the blade of a knife over the teeth in a comb.

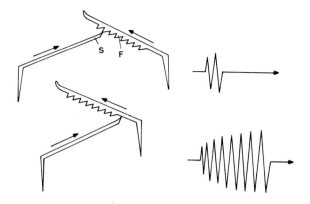

FIG. 3. Schematic diagram of the stridulatory organ and the production of the primary vibration in a cricket. F, file; S, scraper. (From Huber, F., 1970).

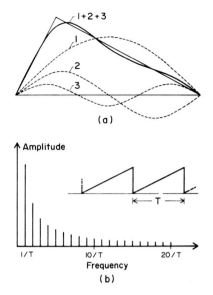

FIG. 4. A sawtooth vibration and its Fourier spectrum. **a**: The shape of a sawtooth is approached by the addition of three harmonic sine waves; **b**; the first 20 components of the line spectrum. Note the relationship between the period time (T) and the frequency of the first line in the spectrum.

The vibrations of the stridulatory apparatus do not produce much air-borne sound, because the stridulatory apparatus is too small (cf. Fig. 1 and the discussion above). The vibrations must spread to a larger structure, which can act as sound radiator. Both the pathway to the sound radiator and the mechanics of the sound radiator may affect the vibrations, and we therefore have to distinguish between the primary vibration of the stridulatory apparatus and the secondary vibration of the sound radiator. Finally, because of the physical laws governing the emission of sound, the characteristics of the air-borne sound waves (especially the frequency spectrum) may differ from those of the secondary vibration. We may expect to find rather little change in these characteristics if the sound radiator is heavily damped (i.e. without large resonances) and behaving as a linear system, whereas considerable changes may be found if one of these conditions is not fulfilled.

Although some attempts have recently been made to measure the primary vibration pattern in the stridulatory apparatus we do not know the exact shape of the vibration. One may speculate that it might, for example, be a sawtooth series (Fig. 4; Markl, H., 1968). Such a complicated vibration pattern, which consists of a long series of identical vibrations, has a frequency spectrum consisting of a number of lines (Fig. 4). The lowest (fundamental) frequency component corresponds to the time period between each tooth impact; the other components are harmonics of this fundamental frequency. Such line spectra are to be expected for

any kind of repetitive non-sinusoidal vibration, so for this argument it does not matter whether the primary vibration has the shape of a sawtooth or some other shape. Differently shaped vibrations with the same fundamental frequency differ in their frequency spectra only by the relative magnitude of the harmonic frequency components.

In some insects, e.g. short-horned grasshoppers, the sound-radiating structures do not appear to cause much distortion of the primary vibration, and here the individual tooth impacts may be observed in the sound recorded from one-legged animals (which only have one "instrument"). Here, the individual impulses (caused by a single tooth impact) may last about $200\,\mu s$ (Fig. 5D). The frequency spectrum of such an impulse is broad and continuous (Fig. 5E). When repeated with a constant repetition rate the impulses produce a line spectrum (Figs 5F and 5G). Note that the amplitude of each of the lines is determined by the spectrum of the individual impulses, and that there are many fewer lines at high repetition frequencies than at low ones because of the harmonic nature of the series.

These spectra are, of course, theoretical, since they pre-suppose that the scraper only hits each tooth once during each movement, that the tooth

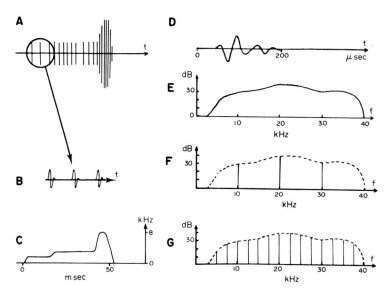

Fig. 5. Analysis of the song of the short-horned grasshopper *Omocestus viridulus*. **A**: Schematic diagram of one chirp in the song; **B**: Individual impulses in the chirp. **C**: The impulse rate during a chirp (schematically). (The time scale applies to **A** and **C**.) **D**: An impulse shown in more detail. **E**: One impulse has a continuous frequency spectrum. **F**, and **G**: The frequency spectrum of a series of impulses with constant repetition rate consists of a number of discrete frequencies (lines). The intensity of each of these frequencies is determined by the spectrum of the individual impulse. The frequency of the lines depends upon the impulse repetition frequency (10 kHz in **F**, 2.5 kHz in **G**). (From Skovmand, O. and Pedersen, S., 1978; changed.)

impact rate remains constant, and that the duration of the series of impacts is "infinite" (i.e. longer than the "memory" of the frequency analyser). In some insects both the file and the scraper consist of many teeth or ridges, so here the spectrum is likely to be much more complicated. In other animals the tooth impact rate varies with time, and the "lines" in the spectra are therefore not lines, but broad-band maxima, the position of which move during the song. But although the actual spectra may not be quite the same as the theoretically expected ones, recent evidence suggests that the line-structure in actual songs may still be so prominent that it will cause an uneven stimulation of different groups of receptor cells in a frequency-analysing ear (Skovmand, O. and Pedersen, S., 1978). In other words, although the average spectrum of a communication song may correspond rather nicely to the frequency sensitivity of the ears, the spectrum of minor parts of the song (having different tooth impact rates) may not do so. How far the insects are able to use the information encoded in these fine details of the song depends upon their capability for auditory time resolution, about which we known very little. Much research is needed on these problems.

The process of stridulation is the most commonly used mechanism of frequency multiplication in insects. Other mechanisms include, for example, click-producing and aerodynamic mechanisms. High-frequency clicks may be produced by means of a bi-stable mechanical structure, which is deflected by muscular action and jumps between two stable positions. Such a clicker is located on the wings of some butterflies and used for scaring away bats in the winter when the bats and insects hibernate together (Møhl, B. and Miller, L., 1976). The mechanism of sound production in cicadas has been much debated, but most of the disagreements between the authors appear to be due to genuine differences between the mechanisms used by different groups of cicadas. In some groups, each inward movement of the tymbal is associated with two clicks, whereas in other groups 3 to 5 or even 10 clicks occur during each inward movement (references in Simmons, P. and Young, D., 1978). The clicks are caused by the sudden buckling of ribs in the tymbal membrane, but the number of click sounds during each inward movement is not equal to the number of ribs, since some ribs may buckle simultaneously. Each inward movement is caused by a twitch of the tymbal muscle. Some cicadas have ordinary synchronous tymbal muscles, i.e. each twitch is initiated by a depolarization of the muscle membrane, which again is preceded by neural activity. In contrast, other cicadas have asynchronous ("myogenic") tymbal muscles, in which there is no such direct correspondence between the electrical and mechanical activity. When attached to a resonant load these muscles can have much higher contraction frequencies than the frequency of nerve impulses in the motor nerve (Pringle, J., 1954; Josephson, R. and Young, D., 1981).

Aerodynamic mechanisms are also used by some insects. For example, some cockroaches hiss by expelling air from a pair of specialized abdominal spiracles (Nelson, M., 1979). In most of these examples the frequency range of the vibrations and sound extends far above the upper frequency limit of the muscular contractions involved.

3.2 Sound radiation

Many investigators have tried to locate the sound radiator by removing parts of the insect body and observing the change in sound pressure after the operation. Such experiments are interesting to perform, but the results may be very difficult to interpret. The intensity and other characteristics of the emitted sound depend not only on the properties of the sound radiator itself, but also on the amount of acoustic short-circuiting and on the size and shape of the structures surrounding the sound radiator. If one observes, for example, a decrease in sound intensity when a part of a bush cricket wing is removed, then one may either have removed a part of the sound radiator, or one may have affected its output in other ways (or both). Similarly, the decrease in sound intensity caused by the removal of most of the abdomen in a bladder cicada may be due either to the destruction of a resonance chamber in the abdomen or to a decrease in the surface of the sound radiator — or both. Instead of trying to locate the sound radiators by means of such operations, the investigators should measure the vibrations of various parts of the body. Several methods are available for this purpose (see e.g. Michelsen, A., 1982).

By the use of some sort of frequency multiplier many insects are able to compensate for their size and overcome the physical obstacle to sound emission illustrated in Fig. 1. Some insects are so small, however, that they would have to communicate by means of *very* high frequencies, if they "insisted" on using air as the medium for their sound signals. There is one alternative open to such animals: they may use another medium. In the ordinate of Fig. 1 the radiated power is indicated relative to the magnitude of $c\rho$ of the medium (cf. equation 1). The emitted power may therefore become much larger, if the sound (vibration) waves are transmitted through a heavier medium. Soil, water and wood all have densities (ρ) about 1000 times larger than air, whereas the propagation velocities for elastic waves (c) are roughly of the same order of magnitude as in air. Many small insects communicate through their host plants by means of songs in the kHz range, which can hardly be radiated to the surrounding air (Michelsen, A. *et al.*, 1982). Other insects can emit their calls into both air and solids, but they use different frequency ranges for these media. For example, buried ants may emit distress calls with a maximum intensity at 1–3 kHz through the soil, whereas the same animals emit most of their air-borne sound between 20 and 60 kHz (Markl, H., 1968).

At a certain body size and vibration frequency the radiated sound power is determined by the amplitude of vibration. In lightly damped structures the amplitude will be at a maximum at the resonance frequency(ies). Many insects make use of this possibility for obtaining a more efficient sound emission. The European field cricket, *Gryllus campestris*, for example, radiates its calling songs by means of a lightly damped part of the wing (the harp), which is tuned to about 4.5 kHz with a Q_{3dB} (explained in Fig. 6) of about 28 (Nocke, H., 1971). In order for the sound radiation to be efficient with such a highly tuned sound resonator, the cricket has to drive the harp at the resonance frequency by adjusting the tooth impact rate during the movements of the stridulatory organ.

There is an upper limit to the amplitude which can be obtained by resonance, since the sound radiation represents a frictional load to the vibrating sound emitter. There is also a price to be paid for boosting up the sound output by means of resonances. First, the insects may be limited to the resonance frequency(ies) of their sound radiator, and they may not be able to signal their message by

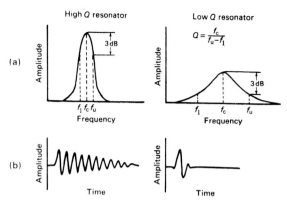

FIG. 6. The behaviour of a lightly (left) and a heavily (right) damped mechanical oscillator. The upper figures show the frequency response of the two systems. The lower figures show the impulse response (the waveform produced when the oscillator is excited by a sudden shock of broad band energy). The degree of damping is often indicated by the Q-value (a high Q means a small amount of damping). (From Sales, G. and Pye, D., 1974.)

means of frequency sweeps. Secondly, highly tuned resonators are mechanically "slow" systems (Fig. 6). Such systems are therefore less suited for transmitting rapid modulations of the sound. This shortcoming in resonance systems appears to be an integral part of their behaviour, and the insects cannot do much to affect the time resolution. Insects may therefore have to "choose" between either having a fast radiating system or being able to produce high sound intensities.

It is possible to build a sound "booster", which is also "fast". The solution to this problem is to have a horn-shaped structure in front of the sound radiator. Horns may have resonances in the same way as tubes, and in this case they will also be "slow". Ideal horns may transmit sounds over a wide frequency range without resonance. Ideal horns work as sound boosters by acting as acoustical transformers. The effective area radiating sound is equal to the area of the horn mouth — not to the area of the sound radiator itself (cf. Fig. 1). Horns are used by singing mole crickets, which place themselves in the "throat" of their horn-shaped burrows (Bennet-Clark, H., 1970). The mole cricket's song is similar to that of the cricket in being a rather pure frequency, and the mole cricket does not seem to make use of the theoretical possibility of boosting up fast sound signals. Horn-shaped structures are also found in many ears (see below).

The acoustical and mechanical properties of the sound radiator(s) may not only affect the sound level produced during singing, but also cause the frequency spectrum and fine structure of the emitted sound to differ from those of the primary vibrations produced by the frequency multiplying mechanism. In theory, there are several possibilities: the sound radiator may be rather heavily damped; it may have one or a few prominent resonances; or it may have a number of resonances. As illustrated in Fig. 6, a heavily damped structure will not cause much distortion of the fine structure of the signals, whereas a short impact on a lightly damped structure will cause a long vibration (which may or may not be close to a sinusoidal vibration of gradually decreasing amplitude, see Schiolten, P. et al., 1981). Short impulses corresponding to the individual tooth impacts (Fig. 5D) are emitted as sound in "non-resonant" singers (acridid grasshoppers, stridulating ants, some bush crickets), so the sound radiator is probably rather heavily damped in these animals. In contrast, many "resonant singers" emit rather pure tones, and here the tooth impact rate almost perfectly corresponds to the natural frequency of the lightly damped sound radiator (many crickets, mole crickets, some bush crickets). Numerous studies illustrate these different strategies of sound emission (for references see Elsner, N. and Popov, A., 1978).

The frequency spectrum of a single tooth impact is rather broad in a "non-resonant" singer, but the spectrum of a series of tooth impacts may be a more or less perfect line spectrum (see above and Fig. 5). In contrast, in "resonant singers" there is not much difference between the spectrum of single tooth impacts and those of series of impacts, and the spectra normally contain one or a few maxima. These findings correspond very well with the behaviour expected for mechanically linear systems. The sound-emitting system is not necessarily linear, however. A highly non-linear sound emitter has recently been found in the tree cricket *Oecanthus nigricornis* F., and here the relationship between the tooth impact rate and the spectrum of the emitted sound is much more complex than in the linear systems (Sismondo, E., 1979). As in *Gryllus*, the normal carrier frequency of the song is equal to the tooth impact rate, but the frequency (and rate) varies with temperature from about 3 kHz at 15° to about 4.6 kHz at 32°. Within this frequency range a large number of resonances have been found, corresponding to various vibrational modes of the wing. Exceptional individuals may stridulate with higher carrier frequencies, but in this case the carrier does not equal the tooth impact rate or one of its regular harmonics (cf. Fig. 4). In contrast, the carrier is excited as an ultra-subharmonic of the tooth-impact rate (e.g. the carrier may be 8/5 of the tooth impact rate). This is possible in non-linear systems, but there are several possibilities for the identity of the non-linear component. This problem deserves further study.

Stridulation is an expensive way to attract a mate. Measurements of the oxygen consumption of resting and trilling crickets show that the cost of singing varies from 6 to 16 times resting values in three species of crickets (Prestwich, K. and Walker, T., 1981). One of the species studied spends about one-fourth of its daily energy budget on singing for only about 45 min at sundown. The cost of singing increases with the number of wing strokes per second and with the number of teeth struck per wing stroke. These crickets are poikilothermic, but similar energy costs have been found in thermoregulating bush crickets (Stevens, E. and Josephson, R., 1977). Most of the energy is spent on moving the wings and on overcoming the resistance offered by the teeth when struck by the scraper. Only a small fraction of the energy is radiated as sound waves. In the bladder cicada, *Cystosoma saundersii* less than 1% of the energy is transformed into sound (MacNally, R. and Young, D., 1981), but these animals are rather small compared with the wavelength of the emitted sound. Loud insects singing at higher frequencies may be more efficient in converting metabolic energy into sound, but the efficiency is probably below 10% even in the most efficient animals.

4 THE ENVIRONMENTAL FILTER

The environment may play a dominant role in determining the kinds of signals which may be used for communication. In section 2 we noted the very different acoustical properties of air and water, and it has long been known that both the hearing organs and sound-emitting structures have to be adapted to the properties of the medium. Air and water are homogeneous media, and the problems concerning the effect of their properties on the generation and reception of sound waves are, in principle, rather simple. The problems confronting students of the transmission of sound waves in terrestrial environments are much greater, since these environments contain many components of different material properties and geometrical shapes.

Relatively little is known about the acoustic conditions of relevance to animals in terrestrial environments (reviews: Michelsen, A., 1978; Wiley, R. and Richards, D. 1978). The best-understood area is the propagation of sound over the ground in environments with little vegetation (see e.g. Donato, R., 1976), whereas the propagation of sound in forests is hardly understood at all. In this context it is important to distinguish between collections of reliable data and real understanding. Several studies exist, which have aimed at analysing the effect of the vegetation on the propagation of sound waves. Some of these studies probably give a fair description of the acoustical properties of the particular environments studied, but one cannot use the data for other environments because one does not know, for example, how much of the observed effects were due to the leaves and stems of the vegetation itself, how much to the ground, and how much to meteorological factors. Furthermore, very little is known about the physical processes important in, for example, the interaction between a plant and a sound wave (see e.g. Martens, M. and Michelsen, A., 1981).

The attenuation of sound energy due to *geometric spreading*, i.e. the decrease of sound pressure with distance from the sound source, is a simple consequence of the geometry of the air space occupied by the sound energy. Monopoles radiate sound energy evenly in all directions, and the attenuation due to geometric spreading is here 6 dB (2 times) decrease in sound pressure per doubling of the distance (abbreviated: 6 dB/dd). A similar decrease is found far away from most small sound sources. In the near field, however, more complicated sound sources may have other magnitudes of attenuation due to geometric spreading. Close to a dipole (see section 3) the attenuation is 12 dB/dd. In some other cases, e.g. cylindrical spreading, the attenuation may be smaller than the "normal" 6 dB/dd.

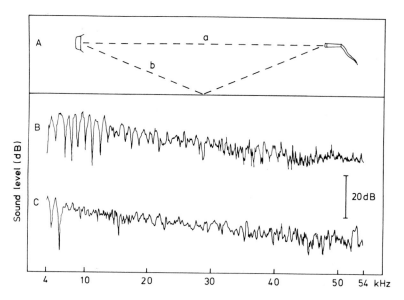

Fig. 7. Reflection from flat ground. **A**: The loudspeaker and microphone are 1 m above ground and 8 m apart. The broken lines indicate the direct (a) and reflected (b) sound waves, which may arrive in phase or out of phase at the microphone. **B**: Reflection from gravel on a road causes a large frequency filtering up to about 15 kHz and less filtering up to about 25 kHz. **C**: Much less frequency filtering is found over short grass. The decrease in sound level towards the high frequencies in both **A** and **B** is caused by the dissipation of sound energy in the air.

The simple geometric laws of *reflection*, similar to those describing reflections from mirrors in optics, are valid when the surface is several wavelengths long and broad, and when the dimensions of the surface roughness are well below a wavelength. It is important to realize that waves are reflected whenever they meet a change in the acoustic impedance (*cρ*) of the medium. Reflections may therefore occur not only at air–soil or air–water interfaces, but also at the interfaces of volumes of air with different temperatures or humidities. Such partially restricted air volumes with different meteorological parameters often occur in nature, e.g. on a warm summer day when masses of warm air leave the surface of the ground and move upwards. It should also be noted that reflections occur, both when the impedance becomes larger and smaller. For example, the vibrational songs of small insects transmitted through the host plant are reflected both at the top of the plant and at the root (Michelsen, A. *et al.*, 1982).

The propagation of sound close to the ground is a rather complicated physical process, where the propagation occurs by an interaction between the air above ground and the air in the pores in the ground. This sound propagation is known as a *ground wave*. It is fairly efficient at low frequencies, but not above about 1 kHz. This means that, for example, crickets on the ground may only be able to hear each other at distances of a few metres, whereas they might be able to hear one another about 50 m away, if they were at some distance from the ground. In fact, some crickets behave as if they "know" this. By climbing the vegetation before singing, they considerably increase their range of communication (Paul, R. and Walker, T., 1979). However, there is a price to be paid for this longer range if the animal is using sounds below about 10 kHz: reflections from the ground will arrive at the receiver and add to the direct wave. The result will be a large sound pressure if the direct and reflected waves arrive in phase — and a small pressure if they are out of phase, since they will now tend to cancel. The phase relationship, of course, depends on the extra distance travelled by the reflected wave — how many wavelengths — and thus on the frequency of sound. The reflection of sound at the surface of the ground therefore causes a *frequency filtering* of the sound reaching the receiver (Fig. 7). This filtering can be quite dramatic, and it depends on the position of the sender and receiver relative to the ground. It may therefore not be a good strategy for animals to use pure tone signals below about 10 kHz, if they live 1 or a few metres above ground. However, this may not be so at higher frequencies. It was said above that the simple laws of reflection are only valid when the surface roughness is well below a wavelength. At higher frequencies, and for grounds covered with some vegetation, the surface roughness becomes so large relative to the wavelengths of the sounds that the reflection at the ground will be rather diffuse. Ultrasonic signals may therefore propagate parallel to the ground almost without being affected by the presence of the ground (Michelsen, A. and Larsen, O., 1983).

The *absorption* of sound includes the dissipation of sound energy to heat and the transmission of sound into other media. It is, however, often difficult to distinguish between absorption and other causes of attenuation. Good absorbers of sound energy in air are porous materials such as mineral wool, which offer a large surface to frictional interactions with the oscillating air particles. The air itself can also dissipate sound energy, especially at high frequencies. Ultrasonic sound signals are therefore not suitable for long-distance communication in air. The effect of absorption is an attenuation of a certain number of dB per metre.

The deflection of sound waves known as *refraction* is caused by an abrupt or gradual change in the propagation velocity of the sound wave. The velocity of sound depends upon the temperature and the density of the medium. The velocity of sound relative to an observer also depends upon the velocity of the medium, the wind. In air the velocity increases about $0.6 \, \text{m s}^{-1}$ for each increase of $1°$, i.e. about 0.2%. Close to the ground or to the surface of objects one often finds gradients of temperature and wind. On sunny days the temperature decreases with distance from the ground, as does the velocity of sound. The sound waves will therefore bend upwards, leaving the shadow zone beyond a certain distance from the source. The situation at other times of the day may be quite different. On a clear night the temperature often increases with distance from the ground, and the sound waves may then be "caught" in a channel over the ground. Sounds may then be heard at much larger distances than during the day.

The term "reflection" was used above for describing the redirection of a sound wave impinging upon a large surface. If the object is much smaller than the wavelength of sound, the sound will hardly affect the object (cf. the interaction between light and a virus particle in a light microscope). If the size of the object is of the same order of magnitude as the wavelength, the sound (or light) will be redirected in all directions in a very complicated way. This redirection is called *diffraction* or *scattering* (the terminology is not quite fixed). The change of sound pressure occurring at the surface of an object is illustrated in Fig. 21. Very complicated interactions between the direct and the redirected sound waves also occur at some distance from the object, where the two sound waves are sometimes in phase and sometimes out of phase. Scattering may be a major contributor to the attenuation of the sound when sound passes obstacles such as vegetation. The additional attenuation caused by scattering may be up to 6 dB/dd, i.e. the same as the attenuation due to geometric spreading in the case of spherical radiation.

The effects of vegetation are poorly understood and will not be considered here. The propagation of most animal calls is further complicated by temporary fluctuations in the acoustical conditions, which are caused by atmospheric turbulence and by the wind moving the vegetation. These effects may limit the use of amplitude modulations as a parameter carrying information in animal calls (see e.g. Wiley, R. and Richards, D., 1978).

5 THE FUNCTIONAL ANATOMY OF INSECT EARS

A sense of hearing has evolved at least a dozen times in various insects, and the original ears again evolved into a large number of functional types, making use of a wide variety of physical mechanisms. The number of receptor cells in the insect ears varies from just one cell in some moths to more than a thousand in mosquitoes and cicadas. Space does not allow a full description of all the known types of ears, and the aim of this section is only to give the reader the impression of the main properties.

The scolopophorous receptor units (Fig. 8) in insect ears take their name from the scolopale (also called auditory spike or Hörstift), a concentric arrangement of rods, which at the distal end are fused into a tube. The scolopale is an intracellular structure in the scolopale cell, one of the four satellite cells in each sensory unit (sensillum). The dendrite of the receptor cell is contained in an extracellular space within the scolopale, and its tip projects into a channel in the scolopale cap, an extracellular structure which fits like a lid on the scolopale tube. Other satellite cells are the glial cell, the fibrous sheath cell, and the cap cell (see Fig. 8). In some ears a fifth satellite cell, the accessory cell, may be interconnected between the cap cell and the hypodermis cell of the cuticle. The receptor cells are bipolar neurons, and the distal part of the dendrite contains a cilium-like structure. The receptor units are thus surprisingly complicated, but nothing is known about the function of the individual parts. In most insect ears the dendrites point towards the tympanum or some other vibrating structure. However, in some other ears the dendrites point away from the vibrating structure but these ears appear to be just as sensitive as the first type. From the mechanical point of view it is also surprising that at least two, presumably "soft", cells are interconnected between the tip of the dendrite and the cuticle.

5.1 The lepidopteran ears

A sense of hearing has evolved independently within several groups of moths and butterflies (Kennel, J. and Eggers, F., 1933). The position on the body varies (metathorax, base of the wings, first two abdominal segments, seventh abdominal segment), and so does the number of receptor cells (1, 2 or 4). In most ears the dendrites point towards the tympanum, but in some moths the dendrites point away from the tympanum. Apart from these differences the anatomy of the ears is surprisingly uniform. The ears normally consist of a tympanum (about 1 μm thick) backed by a tracheal air sac. The receptor organ attaches to the centre of the tympanum and is suspended between the tympanum and the wall of the air sac. The ear of noctuid moths is shown in Fig. 9. Most lepidopteran ears, including some very curious ones on the palps of hawkmoths (see below), are used for detecting the cries of bats and are therefore mainly sensitive to ultrasound, but some ears are used for detecting social sound signals in the kHz range (Roeder, K., 1967a; Swihart, S., 1967).

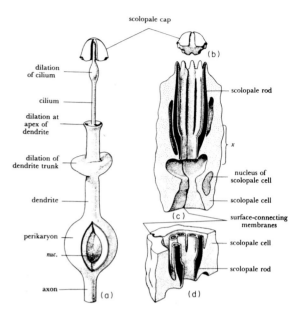

FIG. 8. The scolopoid region of the receptor cell (locust); (**d**) is the region marked x in (**c**). (From Gray, E., 1960; courtesy of The Royal Society.)

5.2 Fluid-filled ears

Most ears detect the variation in sound pressure in the air by means of a thin tympanal membrane backed by an air-filled tracheal sac. The air inside the traceal sac offers little resistance to the vibrational movements of the tympanum, and the ear is therefore well suited for detecting the small pressure variations associated with sound of moderate intensity. From physics textbooks one may easily get the impression that the sound waves would suffer almost total reflection if the tympanum was backed by fluid. This is not true, however, if the dimensions of the ear are small compared with the wavelength of sound, and some insects do in fact have fluid-filled ears.

Lacewings (Chrysopidae, Neuroptera). These have their ears in the radial vein of the forewings. A swelling of the vein (Fig. 10) is bounded laterally and dorsally by thick cuticle and ventrally by thin cuticle (Miller, L., 1970). Most of the swelling is filled with fluid, except for a small tracheal tube running through the dorsal part of the swelling. The physics of this ear has not been investigated, but calculations show that the membrane and the fluid may move as a whole and work against the elasticity of the air inside the

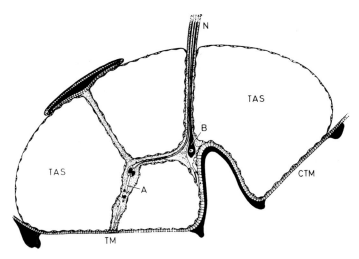

FIG. 9. The ear of noctuid moths. A, A-cells (the two auditory receptor cells); B, B-cell; CTM, countertympanic membrane; N, nerve; TAS, tympanic air sac; TM, tympanic membrane. (From Roeder, K., 1967a.).

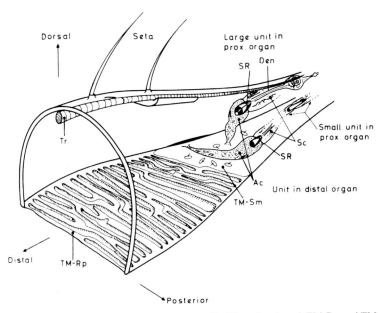

FIG. 10. The lacewing ear. Ac, attachment cell; Den, dendrite; Sc, scolopale cells; SR, scolopale rod; TM-Rp and TM-Sm, rippled and smooth portion of the tympanic membrane; Tr, trachea. (From Miller, L., 1970.)

tracheal tube. This model gives a resonance frequency around the best frequency of the ear (40–50 kHz). The sensitivity of the ear is rather low (threshold around 60 dB), but this is sufficient for the avoidance of hunting bats (Miller, L., 1971). The membrane vibrations are detected by about 25 receptor units, which attach in two groups to the tympanum.

Water bugs (Hydrocorisae, Heteroptera). These have three pairs of scolopale organs on the meso- and metathorax and the first abdominal segment, respectively. The scolopale organs consist of a membrane and two receptor units situated at the caudal border of the membrane or even further away. In some water bugs (the water boatmen, Corixidae) an air sac is in direct contact with the membrane in the

A.

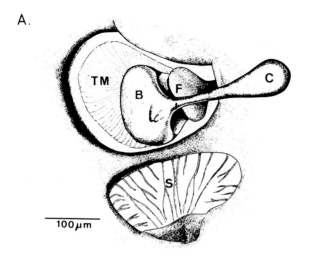

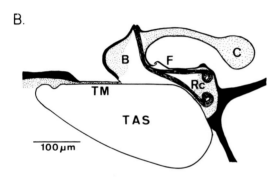

B.

Bush crickets (Tettigoniidae), crickets (Gryllidae) and mole crickets (Gryllotalpidae) belong to the suborder Ensifera of the order Orthoptera (which also includes the acridid grasshoppers). The ensiferan insects have their ears in the front legs, just below the "knee". Most of the anatomy was clarified in a classical paper by Schwabe, J. (1906), but additional information about the fine structure and development has appeared in recent years (Young, D. and Ball, E., 1974; Michel, K., 1974; Schumacher, R., 1975). The anatomy varies between the groups, and the mechanics, physiological properties and fine anatomy may differ between the species within the groups. The problem of hearing is further complicated by the presence of several sense organs in the region of the ear, which may or may not have different functions.

Sound waves can act on the outer surface of one or two tympanal membranes. The membranes are backed on the inside by the wall of a trachea (the main leg trachea and a branch from this trachea in the region of the ear). The trachea runs through the leg and opens on the lateral surface of the thorax. Sound waves can enter the trachea and travel to reach the internal surface of the tympanum, which is therefore driven by the difference in sound pressure across its two surfaces. In bush crickets (Fig. 12) the trachea is permanently open, whereas crickets (Fig. 13) can close the trachea by closing a spiracle. A tracheal tube, interrupted at the midline by a septum, may connect the trachea from one ear with that from the contralateral ear, thus providing a sound guide for acoustical interactions between the two ears. This system of acoustic inputs and tubes inside the animal is important for directional hearing.

The mole cricket ear differs from the other ensiferan ears in that the receptor cells attach, through some attachment cells, to the tympanum (Friedrich, H., 1930). In crickets and bush crickets the receptor cells are normally situated on the wall of the anterior trachea, some distance away from the tympanal membrane(s). In crickets, 60–70 scolopophorous receptor units form the tracheal organ, which may be divided into two or three parts. The scolopidia of the receptor units are attached, through cap cells, to a sheath of accessory cells. The

FIG. 11. The ear of the water boatman. **A**: seen from the outside. **B**: The same in cross-section; B, base; C, club; F, bottle-shaped structure; Rc, receptor cells; S, spiracle; TAS, tracheal air space; TM, tympanum. (From Prager, J., 1976.)

mesothoracic ear. In *Nepa cinerea*, however, the nearest air sac is about 200–400 μm away from the tympanum (Arntz, B., 1975).

The attachment part of the tympanum may be very complex in these animals. In *Nepa* it is a simple cuticular body, but in the water boatmen the receptor units attach to the base of a large, club-shaped structure (Fig. 11), which is attached to the membrane at its base and protrudes into the air outside (Prager, J., 1976). Although the water bugs are often submerged they always carry with them a layer of air, so the outer surface of the ears is never exposed to water. The club vibrates in a very complicated way when the ear is activated by sound. This ear is remarkable also in another way: the right and left ears have different frequency characteristics and absolute sensitivities.

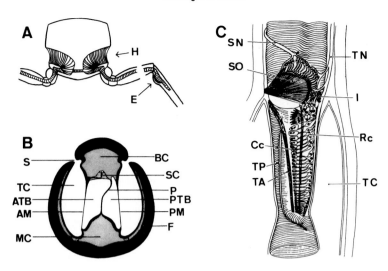

Fig. 12. Schematic diagrams of the hearing organ of a bush cricket. **A**: Section through the front legs and thorax; H, opening of hearing trumpet; E, ear. **B**: Section through the ear in the region of the leg just below the "knee"; AM, anterior tympanal membrane; ATB, anterior tracheal branch; BC, blood channel; F, flap; MC, muscle channel; P, partition; PM, posterior tympanal membrane; PTB, posterior tracheal branch; S, slit; SC, sense cells; TC, tympanal cavity. **C**: Lengthwise section through the ear. Cc, cap cell; I, intermediate organ; RC, receptor cell; SN, subgenual nerve; SO, subgenual organ; TA, anterior trachea; TC, tympanic cavity; TN, tympanic nerve; TP, posterior trachea. (From Michelsen, A. and Larsen, O., 1978 and Schwabe, J., 1906.)

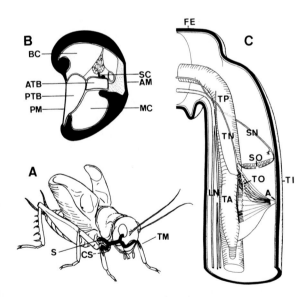

Fig. 13. Schematic diagram of the hearing organ of a cricket. **A**: The H-shaped tracheal system; CS, cross-section shown in **B**; S, spiracle; TM, tympanal membrane. **B**: Section through the ear; AM, anterior tympanal membrane; ATB, anterior tracheal branch; BC, blood channel; MC, muscle channel; PM, posterior tympanal membrane; PTB, posterior tracheal branch; SC, sense cells. **C**: Lengthwise section through the ear. A, attachment to leg wall; FE, femur; LN, leg nerve; SN, subgenual nerve; SO, subgenual organ; TA, anterior trachea; TI, tibia; TN, tympanic nerve; TO, tympanic (tracheal) organ; TP, posterior trachea. (From Michel, K., 1974, and Larsen, O. and Michelsen, A., 1978.)

sheath fastens to the wall of the leg at a small attachment area (Fig. 13). Another sense organ, the subgenual organ, traditionally thought to be a vibration receptor, is situated proximally to the tracheal organ. The afferent fibres from one half of the subgenual organ join the afferents from the tracheal organ so close to the ear that it is very difficult for the experimentalists to record the nervous response from only one of the organs.

The cricket ear generally has a large posterior tympanum and a small anterior tympanum. In bush crickets the tympana are normally of approximately equal size. The receptor units are here situated in a row called the crista acustica. The receptor units do not attach to the wall of the leg, but are covered by a tent-like tectorial membrane. They vary gradually in size from about 16 μm at the distal to 21 μm at the proximal end. In addition to a subgenual organ, a third receptor organ, the intermediate organ, is present.

In both crickets and bush crickets all investigators so far have had great problems in trying to ascribe the functional types of nervous response recorded from the nerve to a particular peripheral origin. Unfortunately, the dyes used for staining neurons during recording do not spread to the cell bodies and dendrites in the periphery. Another

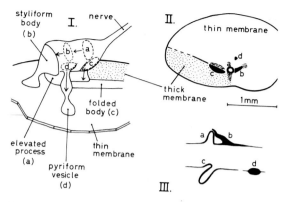

FIG. 14. The ear of a locust. **I**: The left ear of a locust with
groups of receptor cells (a–d) and attachment bodies. **II**: The
right ear seen from outside the animal; attachment bodies in
black. **III**: Schematic section through the attachment bodies.
Most of the elevated process is hidden by the receptor cells
and satellite cells in **I**. The scale of 1 mm applies only to **II**.
(After Schwabe, J., 1906, and Michelsen, A., 1971.)

reason for the present confusion is that the units
recorded from the nerve may respond to either
sound, substrate vibration or both. It is far from
easy to distinguish between these sensory
modalities, since sound waves may cause quite
vigorous vibrations in the legs (Larsen, O. and
Michelsen, A., 1978). The animals may also have a
problem in distinguishing between these sensory
modalities.

5.4 The acridid ear

Acridid grasshoppers have their ears on the first
abdominal segment. Each tympanal organ (ear)
consists of a sclerotized ring encircling the tympanum (Fig. 14). The tympanum is backed by an air
sac, and a series of air sacs connect the two ears,
thus allowing sound waves to pass from one ear to
the other and providing the ear with directional
properties at low frequencies (see section 6.4). The
tympanum is somewhat thicker in the ventral–
anterior corner than elsewhere (Schwabe, J., 1906;
Gray, E., 1960). A total of 60–80 receptor units are
attached in four groups to four modified parts of the
tympanum at the junction of the thin and thick
parts of the tympanum (Fig. 14). The attachment
parts of the tympanum are of different and complicated shapes, and some of these cuticular bodies
extend for more than 100 μm from the membrane.
The four groups of receptor units are called the a,

b, c, and d cells. The a, c, and d cells are oriented in
three almost perpendicular planes, with the b cells
in almost the same plane as the a cells. This complicated anatomy of the tympanum, receptor cells,
and attachment bodies causes the vibration pattern
of the ear to be very frequency-dependent, and the
ear thus acts as a frequency analyser (see section
6.3).

5.5 The cicada ear

Cicadas have their ears in the second abdominal
segment. More than 1000 receptor units attach to a
cuticular body in the tympanum, the tympanal
apodeme (Fig. 15; Michel, K., 1975; Young, D. and
Hill, K., 1977; Doolan, J. and Young, D., 1981).
Another cuticular body in the tympanum, the tympanal ridge, is a thick structure extending one-third
to one-half of the way across the tympanum. The
two bodies are separated by thin cuticle, which apparently acts as a hinge. The mode of vibration of
these structures is not known. In the cicada ear the
receptor units are pointing away from the tympanum and have their caps close to an attachment
area on the distant wall of the auditory capsule
which encloses the receptor organ.

The ears of males and females differ somewhat,
because the ear of the male is in close association
with the sound-producing structures. Females do
not produce sound (this was already known to the
Greek writer Xenarch, who remarked that the
cicadas are happy, because their wives are dumb!).
Rather little is known about the hearing of cicadas,
because they are difficult to keep in the laboratory.

5.6 Ears without tympanal membranes

Ears without membranes may be quite efficient in
picking up sound energy if they are reasonably large
compared with the wavelength of sound. The
curious ears of hawkmoths (Choerocampinae,
Sphingidae) are an example of this (Fig. 16). These
animals use their balloon-shaped palps for detecting the ultrasonic cries of hunting bats (Roeder, K.,
1972). The first and second segment are almost
entirely filled with an air sac, and the palps are lightly
poised and buoyant. The receptor organs, the
pilifers, nest in a depression of the palpal wall. The
balloons are so large compared with the wavelength

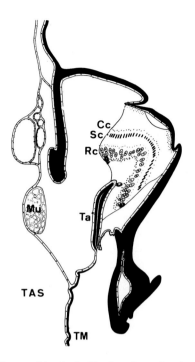

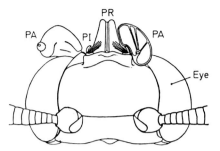

FIG. 16. Ear of a hawkmoth. Left labial palp (PA) deflected laterally to expose the pilifer (PI). The right palp transected. PR, proboscis. (From Roeder, K. *et al.*, 1970; Copyright 1970 by the American Association for the Advancement of Science.)

FIG. 15. The ear of the cicada *Cicada orni* contains about 1300 receptor cells attached to the tympanal apodeme (Ta) and to the distal wall of the cuticular (in solid black) auditory capsule. Cc, cap cells; Mu, muscle; Rc, receptor cells; Sc, scolopals; TAS, tracheal air space; TM, tympanum (the tympanal ridge is further away from the receptor organ). (From Michel, K., 1975.)

of the sounds activating them that the force caused by the difference in phase of the sound acting on the opposite outer surfaces is not much less than that acting on a pressure receiver of similar size. The threshold of these ears (about 45 dB) is about the same as in the ears of noctuid moths.

The force acting on the *sensory hairs* of insects is much smaller, however, and these hairs are not very sensitive sound detectors. In caterpillars the threshold is at a particle displacement in the air of about 2 μm at 150 Hz, corresponding to a sound pressure in a far field of about 92 dB. However, in the near field of sound sources the particle displacements are so large that the caterpillars may detect flying wasps at distances up to 70 cm (Tautz, J., 1979). The antennae of fruit-flies (*Drosophila*) are more sensitive, but here the particle movement caused by the male wing strokes at 5 mm distance only corresponds to that in a far field of 75 dB sound pressure level (Bennet-Clark, H., 1971). Male mosquitoes can detect the flight noise of the female

by means of the antennae. Thresholds around 0 dB have been reported (Tischner, H., 1953), but they were measured as a sound pressure in the extreme near field. The neural threshold is at a vibration amplitude of 0.1–0.2 μm at 300 Hz, when the antenna is subjected to mechanical vibration, and male mosquitoes are probably not able to detect the females at a greater range than about 1 m (Markl, H., 1973).

The forces acting upon these hairs and antennae are almost entirely caused by the frictional (viscous) interaction between the surface of the receiver and the vibrating medium, and these ears are therefore movement receivers. The difference between a movement receiver and a pressure difference receiver may be a question of the size relative to the wavelength of the sounds of interest. In a thin insect hair responding at some hundred Hz the force caused by the difference in phase of the sound pressures acting at opposite surfaces is negligible in comparison with the viscous force (Fletcher, N., 1978). In an insect leg caused to vibrate by sound at some kHz, the opposite is true.

The sensory hairs are simple mechanoreceptors with one receptor cell at their base. The receptor organ of the insect antenna (Johnston's organ) is surprisingly complicated. In mosquitoes thousands of scolopophorous sensilla are attached in two layers to the wall of the second segment (pedicellus) and terminate at the spinous processes of a plate formed by the base of the third antennal segment (Risler, H. and Schmidt, K., 1967). The antenna distal to the pedicellus forms a flagellum covered with numerous hairs. In flies the third segment carries a feathery arista pointing sideways from an apical joint. The function of the arista is debated. Some investigators think that the arista is the auditory receptor (Bennet-Clark, H. and Ewing, A., 1970), whereas others consider it to be an olfactory organ (Averhoff, W. *et al.*, 1979).

6 PERIPHERAL INFORMATION PROCESSING

The information of interest to a listening animal can be summarized as "who? what? where?", i.e. who is the sound source, what does it say, and where is it? The who and what parts of the sound signal are coded in the frequency, time and intensity parameters of the sound, whereas the direction of the sound wave can tell the listening animal where the sound source is located.

6.1 Intensity

The mechanical parts of most insect ears behave as rather linear systems at moderate sound levels, i.e. the amplitude of vibration at a certain frequency is linearly related to the driving force, which in most ears is linearly related to the sound pressure. The relationship between the sound pressure level (in dB) and the number of nerve impulses per unit time is normally sigmoid. The intensity–response curve does not fit exactly to a logarithmic nor to a power function, although both functions fit reasonably well to limited parts of the curve (Adams, W., 1971). Various values for the intensity range covered by the individual receptor cells (from threshold to saturation) can be found in the literature, because the definitions of threshold and saturation vary. The intensity–response curves obtained in different insect ears are alike, however, and generally cover 20–30 dB (noctuid moths: Roeder, K., 1967a; locusts: Michelsen, A., 1971; Römer, H., 1976; crickets: Esch, H. *et al.*, 1980; bush crickets: Kalmring, K. *et al.*, 1978b). In contrast, the dynamic range of single cells (hair cells, primary afferents) in vertebrate ears is 40–50 dB.

Some range fractionation with respect to intensity seems to exist in most insect ears, i.e. receptor cells with similar frequency sensitivities may have different thresholds. In the ears of noctuid moths the extent of this range fractionation is obvious

from simple recordings from the auditory nerve: there are only two receptor cells with identical frequency sensitivity, and one cell is about 20 dB more sensitive than the other. In other insect ears one may easily get the impression of a considerable range fractionation if one is pooling threshold data for single cells from different animals. This may be very misleading, however, because the sensitivity of the entire ear may be affected by various factors. In the locust ear, for example, the threshold of single cells determined in different animals vary by 15 and 26 dB in low- and high-frequency receptor cells, respectively (Römer, H., 1976). Within an individual ear, however, the thresholds of low-frequency cells vary less than 5 dB, whereas the high-frequency receptor cells show a variation of 18 dB. A range fractionation thus exists for the high-frequency cells, but hardly for the low-frequency cells. The different sensitivities to low frequencies observed in different locusts are correlated with the amount of fat and other tissue (e.g. ovaries) between the ears, but this factor hardly affects the sensitivity to high frequencies (Michelsen, A., 1971).

A variation of up to 30 dB has been found for the thresholds for single receptor cells tuned to the same frequency and observed in the same ear of bush crickets (Kalmring, K. *et al.*, 1978b). Less substantial evidence exists for a range fractionation in crickets. Most insect ears contain relatively few receptor cells, and the "price" for a range fractionation with respect to intensity may therefore be a smaller frequency resolution. One may speculate that the animals could avoid this reduction in frequency resolution, if the overlap between threshold curves of units with different frequency sensitivity could be exploited for coding of intensity.

The thresholds reported in the literature for insect ears vary a great deal, but part of the variation is due to differences in the definition of threshold. For example, the A1 cell in the ear of noctuid moths may be said to have a threshold of 25 or 40–46 dB, depending on the definition of threshold (Adams, W., 1971; Roeder, K., 1967a). Similarly, the thresholds determined by various authors by recording the activity of the entire auditory nerve in locusts vary by no less than 40 dB (Michelsen, A., 1971). Such mass recordings are also biased in favour of large axons or large groups of receptor cells.

The data for the vibration amplitude necessary for activating the receptor cells are not very precise, because both the definition of threshold and the design of the experiments vary. Furthermore, it is far from clear whether the receptor cells are sensitive to the displacement, velocity or acceleration amplitude of the vibration, and it is even possible that they may (also?) respond to the force acting on them. In noctuid moths the threshold appears to be determined mainly by the velocity amplitude, whereas in the locust the threshold is at a certain acceleration amplitude at low frequencies and at a certain displacement amplitude at high frequencies (Fig. 19B). When measured as a displacement amplitude a vibration amplitude of about 1 Å is sufficient in the most sensitive frequency range of the ear for activating the receptor cells of noctuid moths, locusts, bush crickets and crickets (Adams, W., 1971; Michelsen, A., 1973; Michelsen, A. and Larsen, O., 1978; Larsen, O. and Michelsen, A., 1978).

As already mentioned, the sensitivity of the locust ear is affected by the amount of tissue between the ears. The physical reason for this damping effect is not known, but the damping is probably caused by friction in the tissues behind the ear. Some insects may be able to control the sensitivity of their ears by activating some muscles. A muscle may attach to the rim of the tympanum, and from the anatomy such muscles have been called a tensor tympani (acridid grasshoppers) or detensor tympani (cicadas). In theory, crickets should be able to vary their auditory thresholds by opening or closing the spiracles of the auditory tracheae. We do not know, however, whether the insects make use of these possibilities for efferent control.

6.2 Time

Not much is known about the time resolution of hearing organs, either in insects nor in vertebrates. Several investigators have measured the ability of receptor cells to respond to individual "clicks" when stimulated with a series of click sounds. For example, the receptor cells in the locust ear (except group c) are able to give a synchronous response to clicks at repetition rates up to $250 \, s^{-1}$ (Michelsen, A., 1966). Up to 50 clicks s^{-1} all the cells respond to each click, whereas at higher repetition rates the

cells alternate. In contrast, the c-cells respond more slowly and are unable to respond to the individual clicks at repetition rates above $20 \, s^{-1}$. The receptor cells in the ears of cicadas are specialized to follow amplitude modulations of at least 300–500 Hz, and most cells respond to each click at repetition rates below 100 Hz (Huber, F. et al., 1980; Popov, A., 1969).

The time resolution measured with such clicks is not easy to interpret. A click is not a well-defined stimulus, unless its frequency spectrum is known. Furthermore, the frequency spectra of pairs or series of clicks differ from those of single clicks. Many animal calls are much more complicated than the rather simple sounds normally used in studies of hearing. In theory, such complicated sound signals may carry a great deal of information, but we are vastly ignorant about the extent to which the hearing organs and auditory pathways can extract the specific information from such signals.

The mechanical time resolution for some rather simple ears has recently been studied with laser vibrometry (a technique for recording the vibration of a membrane by means of a laser beam which does not present any load to the delicate structures in the ears). In some insect ears the oscillatory decay of the tympanum was observed when the ear was activated by very short sound pulses. The response of the tympanum to a very short pulse (called Dirac's Delta function and ideally having a duration of less than 10% of the time constant of the system under study) is the *impulse response*. The time constant for the tympanum of noctuid moths was found to be about 60 μs (Schiolten, P. et al., 1981). The attachment area of the receptor cells in these ears can thus separate short sound impulses arriving with time intervals larger than 150–200 μs (Fig. 17). The tympanum of acridid grasshoppers is also very fast when responding to sounds above 10 kHz (time constant around 50–100 μs), but these ears react more slowly to sounds containing lower frequencies (time constants around 1 ms).

It is rather easy to measure these mechanical time constants and to understand their effect on the transmission of the sound stimulus to the receptor cells. The "time constants" of the receptor cells and higher-order auditory neurons, however, are more difficult to define, measure, and interpret. In noctuid moths the shape of the post-stimulus–time

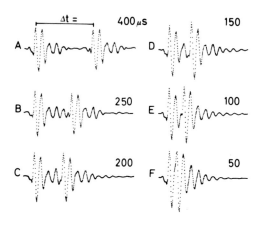

FIG. 17. The response of the tympanal membrane of a noctuid moth to double-impulse sounds. The different responses have been obtained by varying the time interval, Δt, between the impulse sounds as indicated in **A**. Each impulse sound has a duration of about 16 μs. (From Schiolten, P. *et al.*, 1981.)

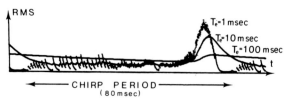

FIG. 18. The sound level of a chirp in the song of a grasshopper measured with integration time 1, 10 and 100 ms and expressed as the root mean square (RMS) value. (From Skovmand, O. and Pedersen, S., 1978.)

histogram of the occurrence of action potentials in response to short sounds of low intensity has been considered a measure of the time constant of the cell membrane (Adams, W., 1971). This time constant is about 1 ms, i.e. about 16 times longer than the mechanical time constant. It is far from certain, however, that this time constant determines the time resolution of the receptor cells when they are activated so as to respond with several spikes.

We probably have to look for several "time constants" in auditory systems. Some will be small and may account for the "flicker fusion frequency" and similar phenomena which may be studied with short sound pulses of the type discussed above. Other time constants are likely to be much larger. For example, the "integration time" for human loudness perception is around 200 ms (Zwislocki, J., 1960). This is, loosely speaking, the time interval up to which the loudness of sound of constant intensity is proportional to the duration of the sound.

The magnitude of these time constants is not a trivial matter, since they may have a dramatic effect on both the perceived amplitude (loudness) and the amount of detail heard in a complex sound signal. Figure 18 illustrates this for a detector, which integrates sound energy from a piece of grasshopper song with different time constants.

6.3 Frequency

In theory there is no reason to distinguish between the time and frequency representations of events. One can be transformed into the other, and viceversa, provided that only ideal systems are considered. This is not so in biological systems. Although the timing of sound signals affects their frequency spectra, as already discussed, time is also handled as a separate parameter in hearing organs and by the nervous system.

Two different principles of frequency analysis have been exploited by auditory receptors. The receptor cells may differ as to characteristic (best) frequency (i.e. the frequency of maximum sensitivity), thus providing the nervous system with information about frequency. This is called the place principle and is commonly used by both insects and vertebrates. The other mechanism (the telephone or volley principle) is based upon the tendency for phase locking at low frequencies: the nerve impulses are mainly triggered at a certain phase of the sound wave, and the train of impulses travelling in the auditory nerve therefore carries information about frequency. The central nervous system of vertebrates can extract this information up to some kHz. This ability has not been proven in insects, but behavioural observations suggest that it may be used. The flight sounds of mosquitoes are around 400 Hz, and the neural response in the auditory (antennal) nerve is phase locked to the sound wave (Tischner, H., 1953). Male mosquitoes respond to a new frequency when they have become adapted to one frequency (Roth, L., 1948). The flight sounds of mosquitoes are of much lower frequency than most insect sound signals, and the volley principle would therefore be less useful in most insects. Furthermore, although the nerve impulses are phase locked to the sound wave at low frequencies in insect ears activated by, for example, mechanical vibration, the phase locking disappears at frequencies above 200–400 Hz (Michelsen, A., 1973).

Insects appear to use at least three different mechanisms for analysing frequencies by means of the place principle. The tympanum may be used both for receiving the sound and for analysing its frequency content (locusts, mole crickets). In most insects, however, the tympanum is only used for picking up the sound energy, and the frequency analysis takes place in an "inner ear" (crickets, bush crickets). A third mechanism may be based upon frequency-dependent, oscillatory rocking movements of heavy structures in the ear (water boatmen, locusts).

Locusts and other acridid grasshoppers have ears, in which four groups of receptor cells (called a, b, c, and d) attach to four cuticular bodies on the tympanum (see section 5.4 and Fig. 14). Microelectrode recordings both from the periphery (Michelsen, A., 1971) and from primary auditory neurons entering the nervous system (Römer, H., 1976) show that the four groups of receptor cells have different frequency sensitivities (Fig. 19A). Frequency analysis in the locust ear is a purely mechanical phenomenon. Most of the frequencies at which one finds a sensitivity maximum correspond to the expected and observed resonance frequencies of the tympanal membrane. The tympanal membrane acts as a band-pass filter: The ear does not respond to sounds below 1 kHz or above 40 kHz. In contrast, the receptor cells are able to respond over a wide frequency range when the isolated receptor organ is subjected to controlled mechanical vibration (Fig. 19B). The receptor cells now respond down to 50 Hz and probably also to lower frequencies. For technical reasons the sensitivity has not been determined above 16 kHz, but responses to 100 kHz have been observed (Michelsen, A., 1973).

The tympanum vibrates both in its basic mode and at higher circularly symmetrical modes. A large number of other modes are possible, however, since the membrane is of irregular shape and inhomogenous. There are two sets of dominant modes of vibration, corresponding to the thin part of the tympanum (see Fig. 14) and to the entire tympanum, respectively. These sets are rather localized and occupy different "territories" on the tympanum (Michelsen, A., 1971). The receptor cells attach to cuticular bodies, which are situated between these "territories". The relative influence of the two sets of vibrations depends upon frequency, and the

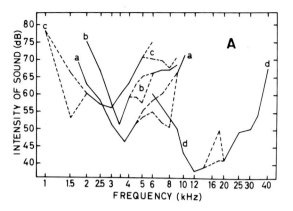

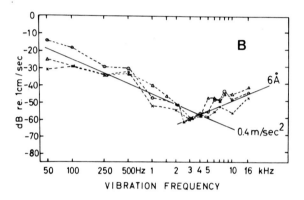

Fig. 19. **A**: The threshold curves for the four groups of receptor cells in isolated locust ears stimulated by sound. Broken lines indicate variations in threshold curves for different cells within each group. **B**: The threshold curves from three experiments, where the isolated receptor organ was stimulated by mechanical vibration. (From Michelsen, A., 1971, 1973.)

receptor cells may avoid responding to a vibrational mode by being at a nodal circle. The mechanism of frequency analysis is, however, complicated by the fact that the receptor cells are not only responding to the vibrations of the tympanum. The cuticular bodies and the receptor cells are also able to perform complicated vibrations of their own (Michelsen, A., 1973; Stephen, R. and Bennet-Clark, H., 1982). The interaction between these vibrations and those of the tympanum has not been studied in detail at physiological sound pressure levels.

Mole crickets use their front legs for digging, and the tympanum borders a small cavity, which is connected to the outside through a narrow slit. The receptor cells attach, through various cells, to the tympanum, which in *Gryllotalpa* is about 0.5 × 1.4 mm across at its widest. The threshold curve of the intact ear has a maximum sensitivity at around 1.5–2 kHz and again at 15–18 kHz, and some animals also have a maximum around 6 kHz (Zhantiev, R. and Korsunovskaya, O., 1973). Different receptor cells appear to be specialized for responding to these frequencies.

The velocities of the tympanal vibrations have been measured in preparations where the cavity in front of the tympanum had been opened so as to allow a laser beam to reach the tympanum (Michelsen, A., 1979). At the proximal end of the tympanum the velocity is at a maximum around 3 kHz, whereas the distal end has a maximum at around 20 kHz, and the membrane is in between, at around 6 kHz. The parts of the tympanum having these maxima include parts of the attachment area of the receptor cells. The receptor cells thus appear to derive their different frequency sensitivities from their position on the tympanum, in which localized vibrations occur. This system should be studied in more detail.

Crickets have a tracheal organ with 60–70 receptor cells and a subgenual organ in the fore legs between the two tympanal membranes (the small anterior tympanum is probably not important for hearing), see Fig. 13. The response properties of the afferent fibres in the auditory nerve have been studied by several investigators, but only in a few species of crickets. Most investigators have found that two (or three?) groups of receptor cells are tuned to different frequencies: that of the calling song around 5 kHz and that of the courtship song around 15 kHz (references in Esch, H. *et al.*, 1980). The response from a third group with an optimum around 1–2 kHz has also been described (Nocke, H., 1972), but this may be the response of the subgenual organ to the sound-induced vibrations in the leg (Larsen, O. and Michelsen, A., 1978). In contrast, a recent investigation on another species shows that the characteristic frequencies of most units are distributed over much of the frequency range studied (0.5–42 kHz), and some units even appear to have their characteristic frequency above 42 kHz (Hutchings, M. and Lewis, B., 1981). This very different finding may be due either to the fact that most investigators have concentrated on the frequencies characteristic of the songs or to a real difference between the species.

Three different theories have been advanced to explain the frequency selectivity of the cricket ear: the tuning of the tympanal membranes, acoustic resonances in the leg trachea, and the properties of an unknown frequency analyser inside the leg. The first-mentioned theory was supported by observations of tympanal vibrations by means of Mössbauer spectroscopy (Johnstone, B. *et al.*, 1970), but later studies with laser beams (which do not provide any mechanical loading of the membrane) have not supported this idea. The large posterior tympanum vibrates in its basic mode (as a piston) in the frequency range 1–30 kHz, and the tuning of the tympanum is much broader than that of the receptor cells (Dragsten, P. *et al.*, 1974; Larsen, O. and Michelsen, A., 1978). The frequency analyser must therefore be located elsewhere. The theory of tracheal resonance is supported by experiments where the frequency sensitivities of higher-order auditory neurons were shifted when the leg trachea was shortened or filled with helium (Paton, J. *et al.*, 1977). These procedures affect several properties of the hearing organ, however, and other observations do not support the idea of tracheal resonances (Ball, E. and Hill, K., 1978; Hill, K. and Boyan, G., 1977).

The nature of the frequency analyser in the cricket ear is not known, but by selective dissection of auditory nerve branches the low- and high-frequency receptor units have been shown to be located in the proximal and distal parts of the tracheal organ, respectively (Zhantiev, R. and Tshukanov, V., 1972a). We are totally ignorant about the modes of vibration of the structures inside the leg, and we also need simultaneous single-cell recordings and dye markings so close to the ear that the location of the different receptors can be definitely established.

Bush crickets have no less than three receptor organs in the region of the tympanal membrane: the crista acustica, the intermediate organ, and the subgenual organ (Fig. 12). Selective cutting of auditory nerve branches suggested that the intermediate organ responds to sounds between 1 and 5 kHz, whereas the crista acustica responds between 1 and 100 kHz (Zhantiev, R., 1971). Recent single-cell recordings suggest that the proximal units in the crista are most sensitive around 6 kHz, the distal units around 40–50 kHz, and the units in the middle of the crista at frequencies between these extremes (Fig. 20; Zhantiev, R. and Korsunovskaya, O.,

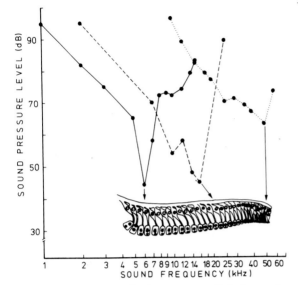

FIG. 20. Threshold curves for three receptor cells in the ear of a bush cricket. (Redrawn from Zhantiev, R. and Korsunovskaya, O., 1978.)

1978). The threshold increases with frequency from about 40 dB at the proximal end to about 60 dB at the distal end of the crista. The intermediate organ is situated close to the proximal end of the crista, and it also appears to be most sensitive around 5–7 kHz, but its threshold is higher (60–80 dB).

Other investigators find the situation to be more complex (Rheinlaender, J., 1975; Kalmring, K. et al., 1978b). Recordings from the auditory nerve at its entry to the prothoracic ganglion demonstrate three main groups of units: pure sound receptors, mixed sound and vibration receptors, and pure vibration receptors. The mixed sound and vibration receptors have their best frequencies at around 1–5 kHz for both modalities, but the best frequency for a particular unit is not always the same for sound and vibration. Unfortunately, the sound-induced vibrations of the leg were not measured in these experiments. The tuning of the auditory units varied, and it was not possible to divide the units into groups on the basis of threshold. Although the afferents can be stained during the recording at the level of the ganglion, the stain does not spread to the periphery. On the other hand, with the recordings close to the ear one cannot be sure that the presence of the microelectrode has not affected the reception of sound by the ear. It has not been possible so far to record the activity of units at a "safe" distance

from the ear and at the same time to stain their cell bodies in the periphery.

The tympanal membranes of bush crickets are rather broadly tuned and do not appear to contribute to the fine frequency analysis (Michelsen, A. and Larsen, O., 1978). In most species the leg trachea also seems to transmit the sound over a broad frequency range, but some species may be specialized for obtaining a fine tuning of the ear by means of the transmission properties of the leg trachea (Nocke, H., 1975).

Water boatmen are very interesting because the two ears have different frequency sensitivities (Prager, J., 1976). In *Corixa punctata* the best frequency of the A1 cell in the left mesothoracic ear is around 2.3 kHz, whereas the A1 cell in the right ear is most sensitive at 1.7 kHz. The threshold at the best frequency also differs: the left A1 cell is more sensitive than the right A1 cell. The A2 cells are less sensitive than the A1 cells, have flatter threshold curves, and a less pronounced right–left difference in frequency response.

Anatomically, the ears of water boatmen are curious because they have a large club attached to the tympanum (Fig. 11). Microscopic observations at high sound intensities, as well as recent laser measurements at moderate sound levels, indicate that the club vibrates in two planes (around its long axis and normal to this). Furthermore, the clubs of the left and right ears differ with respect to the frequencies of maximum vibration, which are in reasonable agreement with the best frequencies of the A1 cells (Prager, J. and Larsen, O., 1981).

6.4 Direction

Three different mechanisms are available for obtaining information about the direction of the sound: the diffraction of sound around the body, the difference in time of arrival of the sound at the ears, and the use of a receiver which is inherently sensitive to the direction (a movement receiver or a pressure difference receiver). However, not all animals can make use of each of these mechanisms. Insects are small, and the sound arrives at the ears almost instantaneously. Sound travels in air with a velocity of about 340 m s^{-1}, so the maximum difference in the *time of arrival* at ears separated by, for example, 1 cm is about 30 μs. Higher vertebrates can detect differences in the time of arrival of less than 10 μs, but the ears of these animals contain a great number of receptor cells and nerve fibres, and the information about time of arrival can be carried by hundreds of nerve fibres. A single nerve fibre does not signal the time of arrival very accurately, and this method probably requires a good deal of statistical analysis in the brain. Most insects seem to have too few receptor cells in their ears to exploit this mechanism. The situation is more favourable in, for example, cicadas, but we do not know whether the central nervous system in cicadas can handle time differences of less than 50 μs. Although the time of arrival mechanism is probably not used by most insects a neuronal time cue is available to the nervous system if the forces acting on the two ears differ. This time cue may be many times larger than the difference in the time of arrival of the sound at the two ears. In the locust, for example, the latency of the neural response may change by about 6 ms with the direction of the sound (22 ms tone bursts, 25 dB above threshold) — as compared with the change of about 30 μs expected from the time of arrival of the sound (Mörchen, A. *et al.*, 1978).

The *diffraction mechanism* is also difficult to use for small animals such as insects, unless the animal is listening to sound of a high frequency. Diffraction is the reflection of sound from objects, which are of about the same size as the wavelength of sound (see section 4). Objects much smaller than the wavelength do not "disturb" the wave (cf. a virus particle placed under a light microscope), but a surplus pressure starts to build up at the surface facing the sound source when the object is about ten times smaller than the wavelength. At higher frequencies (smaller wavelength) a complicated pattern of surplus pressures and "shadows" is formed at the surface of the object. Figure 21 illustrates this for a spherical object. More complicated patterns are found around objects of more complicated shapes. The exact pattern of diffraction around insect bodies is only known in the case of the cricket (Kleindienst, H., 1980), but the approximate interaural difference in sound pressure is known as a function of the direction of sound for some insects (see below). Although some insects (e.g. moths listening to the cries of hunting bats) are able to determine the direction of sound very well by means of the diffraction mechanism alone, most insects are so

small compared to the wavelengths of the sounds of interest that the interaural difference in sound pressure caused by diffraction is negligible. At 3–4 kHz, for example, the wavelength is about 10 cm, and the difference in sound pressure at ears situated on opposite surfaces of the cylindrical body of a locust is below 1 dB (Michelsen, A., 1971). Slightly larger differences are found for animals resting on the ground or flying, since the presence of the ground/wings affects the diffraction (Kleindienst, H., 1980), but still the difference in pressure is only a few dB. Much larger differences are found at high frequencies (small wavelengths), however. In a locust, the difference is about 7 dB at 15 kHz (Miller, L., 1977). In large species of noctuid moths (wingspan 6–8 cm) listening to 60 kHz, the difference may be no less than 40 dB, when the wings are extended and the sound is coming from a direction where the "shadow" caused by the wings is a maximum (Payne, R. *et al.*, 1966). Diffraction also plays an important role in the directional hearing of bush crickets listening to 10–30 kHz (see below).

Close to sound sources the amplitude of oscillation of the medium may be rather large (see section 2),

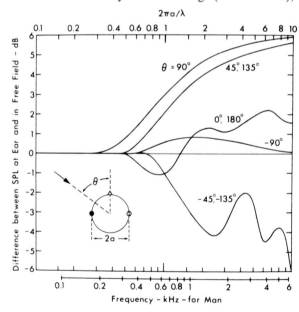

FIG. 21. Calculated transformation of sound pressure level from a free field to a simple ear (point receiver) on the surface of a hard spherical head of radius a as a function of $2\pi a/\lambda$ (where λ = wavelength of sound) for various values of azimuth θ of the incident plane waves. The frequency scale at the bottom is for a sphere of radius $a = 8.75$ cm. (From Shaw, E., 1974.)

and the insects may therefore use inherently directional *movement receivers* for detecting the presence of, and direction to, the sound source. Many insects have mechanoreceptive hairs, which are caused to vibrate by the oscillatory movements of the air particles. The directionality of these hairs is caused first, by the hairs being insensitive to sound propagating in the direction of the hair; second, the base of the hair may be suspended in such a way as to allow movements in only one plane — or the directivity may be caused by the shape of the hair. For example, in some caterpillars, curved hairs are directional, whereas straight hairs on the same animal are non-directional (Tautz, J., 1977). The directionality disappears if the curved hairs are cut so as to leave a straight stump. Finally, the sensory cells may be attached to the hair in such a way that movements in one direction may be more effective in causing the cell to discharge nerve impulses than movements in other directions (Tautz, J., 1978).

In the far field of the sound source the movements of the medium are generally too small to be detected by means of movement receivers. Furthermore, in most insects the body is so small compared with the wavelength of the sounds of interest that these animals cannot use the diffraction mechanism for directional hearing. The solution to this problem was first realized in 1940 by H. Autrum, who pointed out that the observed directionality could be accounted for if the ears were *pressure difference receivers*, i.e. if the sound waves were able to reach both the outer and inner surface of a tympanal membrane (Fig. 22). This is possible in many animals, since the ears are often connected by an air-filled passage. Alternatively, the sound waves may enter the body and reach the inner surface of the tympanal membrane through some other route (e.g. a tracheal tube: bush crickets, crickets). We now know that the directional hearing of the majority of small animals, both insects and vertebrates, is based upon this kind of sound receiver mechanism. Although Autrum's theory is now over 40 years old, the textbooks are still dominated by the "displacement receptor" theory, which was suggested by R. Pumphrey, also in 1940. This theory postulated that the ears of insects are movement receivers, i.e. following the movements of the particles of the medium and thus inherently directional. Although correct for sensory hairs in insects, Pumphrey's theory is not correct for ears with tympanal membranes (Michelsen, A. and Nocke, H., 1974).

The directionality of a pressure difference (pressure gradient) receiver may be understood by considering a small, free membrane (or a piece of paper) placed in a free acoustic field (as defined above). The driving force acting to move the membrane (paper) is at a maximum when the membrane is perpendicular to the direction of the sound wave. In this position the amplitudes of the sound pressure are often the same on the front and back surfaces of the membrane, but they are somewhat out of phase, because the sound wave reaches the back surface a little later than it reaches the front surface. The difference between the two pressures will therefore have a certain magnitude. The force is zero when the free membrane is parallel to the direction of the sound wave, since the sound pressure now has the same amplitude and phase on both sides of the membrane. If the distance from the centre of a directional diagram indicates the magnitude of the driving force acting on the receiver, then the directional diagram of a pressure receiver is circular, while for a pressure difference receiver it may be a figure-of-8 or kidney-shaped (cardioid).

The advantage of the pressure difference receiver mechanism in comparison with the pressure receiver is not only that directional hearing is possible even when the body of the animal is very small, but also that the directivity at medium frequency can be much larger than in a pressure receiver. The expected change in pressure (driving force) is below 10 dB (= 3 times), when a pressure receiver is rotated at moderate frequencies (Fig. 21). In pressure difference receivers the change may, in theory, be infinite, and it is commonly around

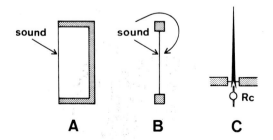

FIG. 22. A pressure receiver (**A**), a pressure difference receiver (**B**), and a movement receiver (**C**) Rc. receptor cell.

20 dB (10 times). In pressure receivers, changes of this magnitude are only possible at high frequencies.

In order to demonstrate the existence of a directional pressure difference receiver it is not enough to find an air-filled passage from one tympanum to the other (or from another sound input to the back surface of the tympanum). The problems are how much sound gets through the passage, and how the phase of the sound inside the animal is affected by the direction from which the sound wave reaches the outer surface of the animal. These problems are not simple, and there are a number of pitfalls in both theory and experiments. The passage may attenuate the sound waves to such an extent that — for practical purposes — the ear is behaving like a pressure receiver. The ears of noctuid moths provide an example of this: although connected by a series of air sacs, the ears behave as pressure receivers (Schiolten, P. et al., 1981).

Some insects need their ears to work as pressure difference receivers in the lower end of their frequency range (where there is very little diffraction), whereas the ears may be pressure receivers at high frequencies (where the diffraction is providing sufficient directionality). This may be achieved in two different ways. In the locust, low frequency sound (1–8 kHz) is transmitted with only little attenuation through the series of air sacs connecting the two ears, and each ear is therefore a pressure difference receiver (Michelsen, A., 1971; Miller, L., 1977). Above 10 kHz, however, the air sacs appear to act as an acoustic low-pass filter in much the same way as a silencer on a car. Very little sound therefore gets through, and the ears are almost pressure receivers. A similar situation has recently been found in birds (Hill, K. et al., 1980). An alternative way of obtaining a similar situation is to increase (rather than decrease) the transmission of sound to the back surface of the tympanum at high frequencies. This appears to be the case in some bush crickets. As already discussed, a horn-shaped trachea connects the back surface of the two tympanal membranes in the bush cricket ear with the air at the surface of the thorax. At low frequencies the trachea has a gain of one, and the ear is therefore a pressure difference receiver (Michelsen, A. and Larsen, O., 1978). At high frequencies, however, the sound pressure acting on the back surface of the tympana becomes much larger than that acting on

their front surfaces, and the ear now behaves almost as a pressure receiver, being driven from the inside (Hill, K. and Oldfield, B., 1981; Larsen, O., 1981). In all the cases studied so far the transition from pressure difference receiver to pressure receiver characteristics appears to happen at a frequency where diffraction begins to play a considerable role.

The physics of directional hearing is very complicated in bush crickets and crickets, and the literature reflects a considerable disagreement between the investigators. A part of this disagreement is due to the complicated nature of the physical problems studied, but recently it has become evident that there are some differences in the acoustical properties of different species of crickets (and bush crickets?), although the anatomy of their auditory systems are apparently rather similar. For example, the gain of the ipsilateral tracheal tubes in crickets to high-frequency sound impulses is about 1.5 in two species, but about 3.5 in two other species (Larsen, O., 1981). It therefore does not make sense to try to fit all observations into a common model for "the cricket ear".

Space does not allow us to go into details on these very interesting systems, but a brief comparison of the two most-studied species of crickets should give an impression of the different strategies used by these animals for locating the sound sources.

The directionality of the ear in the cricket Teleogryllus commodus is a maximum (about 14–15 dB difference in sensitivity to ipsi- and contralateral sound) in a narrow frequency band around the species' calling song at 3.7 kHz (Hill, K. and Boyan, G., 1977). The measured directionality could be accounted for by assuming that sound waves from the ear ipsilateral to the sound source can travel through the tracheal system (Fig. 13) with very little attenuation (this theory was first proposed by Zhantiev, R. et al. (1975), who observed that the directionality disappeared in the ear of the cricket Gryllus bimaculatus when the contralateral sound inputs were blocked). The directionality of the Teleogryllus ear persists when the thoracic spiracles are occluded, but it is abolished when only the tympanum of the opposite ear is covered with Vaseline. Hill and Boyan therefore concluded that transmission of sound from the tympanum of one ear to that of the other ear is responsible for directional hearing. They further assumed that a favourable transmission of sound through the tracheal tubes is restricted to the narrow frequency band around the calling song frequency, because they found the directionality to be much less outside this range. These physiological results (obtained mainly by recording the neural activity of units in the cervical connectives) have been confirmed by behavioural studies of the orientation of female crickets in the closely related species *Teleogryllus oceanicus* (Oldfield, B., 1980). It has also been possible to fit the experimental results into the framework of a theoretical model (Fletcher, N. and Thwaites, S., 1979).

A transmission with only little attenuation from one tympanum to the back surface of the other is not present, however, in the cricket *Gryllus campestris*, in which the magnitude of this transmission has been measured with very different techniques in two different laboratories. In one study the vibration velocity of the tympanal membrane was measured with laser vibrometry. A simple vector analysis of the vibrations measured when the tympanum of the other ear was intact or blocked indicated that the sound entering the tracheal system through the contralateral tympanum is reduced in amplitude to about 16% of the sound pressure outside the animal's body (Larsen, O. and Michelsen, A., 1978). A value of about 20% has been found by enclosing the front legs (and ears) in two sound cavities (see Fig. 27A), in which the amplitude and phase angle of the sound can be controlled independently for each ear (Kleindienst, H. *et al.*, 1981). In these experiments the motion of the tympanum was monitored by observing the excitation of a central auditory neuron under conditions where it only receives input from one ear, and assuming that the central neuron becomes silent when the sound pressures acting on the two surfaces of the tympanum are exactly equal in amplitude and phase. A transmission of only 20% corresponds to a directionality of only a few dB.

Recent behavioural studies by two research groups, both using a Kramer sphere for obtaining quantitative measurements of the orientation behaviour of unrestrained crickets listening to sound from loudspeakers, have shown that *Gryllus campestris* females are, in fact, able to locate a sound source when both their auditory spiracles are closed (Thorson, J. *et al.*, 1982; Schmitz, B. *et al.*,

1982). The accuracy of orientation is much improved, however, when the spiracles are intact.

A vector analysis of the tympanum in crickets with intact and closed contralateral spiracles showed that the sound waves entering through this acoustic input are reduced in amplitude to about 35% when they reach the back surface of the tympanum (Larsen, O. and Michelsen, A., 1978). The directionality expected from this input is about twice that expected from the sound wave originating at the contralateral tympanum. An even larger directionality may be derived from the sound wave entering the ipsilateral spiracle, but the exact amplitude and phase of this contribution to the sound pressures acting on the tympanum are not known. Furthermore, the different sound waves contributing to the total sound pressure at the back surface of the tympanum cannot be added in a simple way, because the state of each of the inputs may affect the sound waves from the other inputs (Kleindienst, H. *et al.*, 1981).

In summary, there may be some differences in acoustic properties between the peripheral auditory systems in different species of crickets. In some cases a qualitative agreement has been found between the results of behavioural experiments and those of biophysical and physiological studies. In other cases such an agreement is not in sight. A similar situation exists for different species of bush crickets. We are still far from having a quantitative understanding of the directional hearing in insects with pressure difference receivers.

7 CENTRAL PROCESSING OF SOUND INFORMATION

The number of investigations on central processing of sound in insects has increased so rapidly within the last 10 years that space does not allow for a comprehensive reviewing of the considerable literature. Although insect ears vary in structure according to their evolutionary origin, the organization of the central auditory pathways and the central processing of sound information seem to be rather similar in the different insect groups studied so far. Therefore we shall consider the general principles of anatomy and integration properties in detail in one of the most extensively studied groups, the crickets (Gryllidae). The other groups (e.g.

Tettigoniidae and Acrididae) are only considered to an extent that will show differences from crickets or accentuate mechanisms, which have not (yet) been found here.

For other aspects of central processing of sound information in insects readers are referred to Schwartzkopff, J. (1974); Huber, F. (1974, 1977, 1978); and to Elsner, N. and Popov, A. (1978).

7.1 Gryllidae

7.1.1 ANATOMY

In crickets the overall structure of the auditory pathways on each side of the ventral nerve cord is that of a T. The auditory information is carried along receptor fibres to a general switchboard, the prothoracic auditory neuropile. From here the information flow is diverted in two directions via bilaterally interacting interneurons of first and probably higher order to reach the brain and the thoracic motor centres, respectively.

The anatomy of the first level of information transfer is clear: the receptor fibres originate in the tympanal and associated organs, and they project into a well-defined auditory neuropile. Here, two main problems have been investigated. First, the genuine auditory receptor fibres must be distinguished from those also originating in the tibial sensory complex but carrying information from other sensory modalities. Second, the origin and projections of all different, physiologically characterized types of auditory receptors must be mapped. The first problem has been solved to some extent but we are still far from solving the second problem.

The anatomy of the second level of information transfer is more problematic, since even in the most recent experiments the order of a given interneuron is determined only indirectly. It seems reasonable to assume, however, that there are four classes of first-order auditory interneurons: the segmental ones, the ascending interneurons, the descending ones, and the T-interneurons (Elsner, N. and Popov, A., 1978). The difficulties arise when considering the variety of interneurons within these classes, since different neurons in different cricket species have been characterized physiologically and/or anatomically to some degree and named differently by several authors. Taking into consideration, how-

ever, the conservative nature of the ventral cord morphology and the low estimates of the number of auditory interneurons (about 12 — Popov, A. and Shuvalov, V., 1974) it seems reasonable to assume that all the different, more-or-less fragmentary, reports may eventually fit to a few types of anatomically homologous interneurons found in all cricket species. In the following sections we shall try to make such generalizations in characterizing each neuronal type by shape, position, and input–output connections.

The anatomy of higher levels of auditory information transfer has received only little attention so far. The anatomical and physiological information about the anteriorly directed information flow is based almost entirely on data from unimodal brain neurons (Boyan, G., 1980, 1981). The target cells of the posteriorly directed information flow are not known.

(a) *Central projections of auditory receptor fibres* The auditory receptors of the tympanal organ project into the prothoracic ganglion to form a crescent-shaped, strictly ipsilateral intermediate neuropile, as shown in Fig. 23A (Rehbein, H., 1973; Esch, H. *et al.*, 1980). The 50–70 auditory receptor cells may be divided into several types according to their characteristic frequencies, CFs (see below). In order to decide whether the auditory neuropile is tonotopically organized, some attempts have been made to follow the course of physiologically known single fibres. Eibl, E. and Huber, F. (1979) were the first to show that auditory fibres tuned to the calling song frequency in *Gryllus bimaculatus* and *Gryllus campestris* exclusively project into the crescent-shaped neuropile with most of the projections and the more massive arborizations in the anterior–medial part. On the assumption that the auditory receptors are divided into only three groups, i.e. those which are tuned to the frequency ranges 4–5 kHz, below 2 kHz (LF fibres), and above 5 kHz (HF fibres), respectively, Esch, H. *et al.* (1980) recorded extracellularly from single fibres and cobalt-stained fibres close to the electrode tip. Their results supported those of Eibl, E. and Huber, F. (1979) as regards the 4–5 kHz fibres. Their results furthermore suggested that the LF fibres have three end branches. One projects into the crescent, like the 4–5 kHz receptors; another projects into the dorsal motor neuropile; while the third one projects

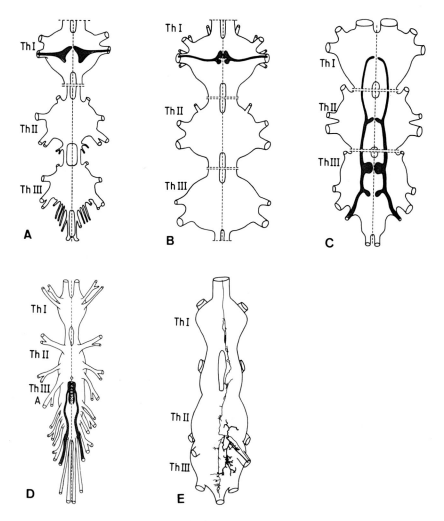

FIG. 23. Central projections of tympanal receptors of five major insect groups. **A**: Crickets (*Gryllus campestris* and *Gryllus bimaculatus*; redrawn from Eibl, E. and Huber, F., 1979). **B**: Bush crickets (*Decticus verrucivorus*; redrawn from Kalmring, K. *et al.*, 1978b). **C**: Locusts (*Locusta migratoria*; redrawn from Rehbein, H., 1976). **D**: Cicadas (*Magicicada septemdecim*; redrawn from Huber, F. *et al.*, 1980). **E**: Moths (*Agrotis segetum*; redrawn from Surlykke, A. and Miller, L., 1982). Thin line, A_1; thick line, A_2; ThI-III, the pro-, meso-, and metathoracic ganglion.

into the lateral part of the crescent. The projections of the HF fibres were not firmly established, but they seem to project into almost the same areas as the 4–5 kHz receptors. According to Hutchings, M. and Lewis, B. (1981), however, it may be an oversimplification to divide auditory receptors into three groups only, since (in *Teleogryllus oceanicus* at least) their CFs range from 0.5 kHz to 42 kHz and probably higher. Furthermore, some receptors are broad-band with no "preferred" frequencies.

The question of a possible tonotopical organiz-

ation of the auditory neuropile should therefore be reinvestigated over a wider frequency range, preferably using intracellular recording and staining techniques. The peripheral origin of these fibres should also be determined.

(b) *Ventral cord interneurons*

Segmental auditory interneurons. Since the auditory neuropiles are strictly ipsilateral, some "bridging" interneurons are needed to compare and jointly process information from the tympanal organs of the two forelegs, especially as regards directional information. Among such neurons are the so-called omega neurons (ONs) described in the species *Teleogryllus oceanicus, Gryllus bimaculatus,* and *Gryllus campestris* (Casaday, G. and Hoy, R., 1977; Popov, A. *et al.*, 1978; Wiese, K., 1978; Wohlers, D. and Huber, F. 1978, 1982). The ONs belong to the class of spiking local (segmental) interneurons. In the prothoracic ganglion there are believed to be no more than two such types, the ON1s and ON2s (Wohlers, D. and Huber, F., 1982), each type represented by only two neurons which are mirror images of each other. Typical representatives of each neuronal type are shown in Fig. 24. For clarity the mirror image cells have been omitted.

The most easily accessible cell is the ON1, the dendritic fields of which penetrate both the right and the left acoustic neuropiles and have no branches outside these regions. The ON1 receives information from all parts of the soma-ipsilateral neuropile and delivers it to all parts of the soma-contralateral neuropile (Popov, A. *et al.*, 1978). The ON1 is believed to be a first-order acoustic inter-neuron, since there is a close proximity between the receptor projections and the ON1's dendritic fibres,

because latency measurements do not rule out the possibility of monosynaptic connections, and finally because the time course of the summed EPSP in the ON1 indicates a synchronized input from the similar presynaptic fibres (Wohlers, D. and Huber, F., 1982). There are four candidates for the neurons receiving information from a given ON1: the contralateral ON1, the ON2s, the intersegmental auditory neurons and the contralateral receptor projections. Synaptic input to the contralateral ON1 is well established (Wohlers, D. and Huber, F., 1978; Kleindienst, H. *et al.*, 1981), whereas nothing is known about the other possible connections. It is, however, most likely that the ON1 delivers information to intersegmental interneurons. In summary then: the ON1 receives information from the soma-ipsilateral receptor projections and from the contralateral ON1, and it delivers information to the soma-contralateral ON1 and to soma-contralateral dendritic fields of intersegmental auditory interneurons.

The ON2 has now been identified in three cricket species, *Gryllus campestris, Teleogryllus oceanicus,* and *Acheta domesticus* (Wohlers, D. and Huber, F., 1982) but has not yet been studied as intensively as the ON1. Its soma and dendritic fields are adjacent to those of ON1 (Fig. 24). It is distinguished from the ON1 by its smaller diameter and by a large process crossing the midline in the lower part of the Ω-shape. Nothing is known about the input or output connections, but unlike the ON1 it receives information from both ears, most probably through branches originating from the soma-ipsilateral dendritic fields (Wohlers, D. and Huber, F., 1982).

Ascending auditory interneurons. The ascending auditory interneurons comprise a class of inter-neurons carrying auditory information from the prothoracic auditory neuropiles to the supraoesophageal ganglion, the brain. Apart from the HF1 neuron (Popov, A. in Elsner, N. and Popov, A., 1978), none of these interneurons have been stained completely so far, but partial stainings and physiological evidence suggest that the number of types is very low. Two such types are the AN1 and the AN2, each type comprising only one pair of mirror image neurons in *G. campestris* (Wohlers, D. and Huber, F., 1982), see Fig. 24.

The AN1 type has probably been described in several species by different authors: the pulse-coder

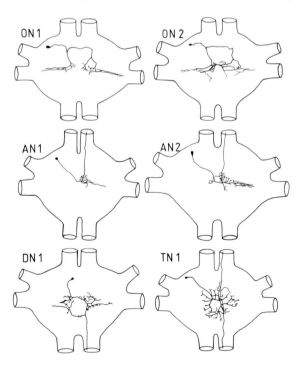

FIG. 24. Six types of auditory interneurons located in the prothoracic ganglion of *Gryllus campestris*. (Redrawn from Wohlers, D. and Huber, F., 1982.) Fine side branches in the neuropile have been omitted. For further explanation see text.

in *G. campestris* (Stout, J. and Huber, F., 1972); the pulse rate predetector (Popov, A. and Shuvalov, V., 1974); the LF1 neuron in *G. bimaculatus* (Rheinlaender, J. *et al.*, 1976), and the type 1 neuron in *G. campestris* (Stout, J. and Huber, F., 1981).

Similarly, the AN2 type may be identical with or homologous to: the chirp-coder in *G. campestris* (Stout, J. and Huber, F., 1972); the chirp duration predetector neuron (Popov, A. and Shuvalov, V., 1974); the HF1 neuron in *G. bimaculatus* (Rheinlaender, J. *et al.*, 1976); the Int-1 in *T. oceanicus* (Casaday, G. and Hoy, R., 1977); the AIAA in *G. campestris* (Wohlers, D. and Huber, F., 1978); and the type 3 neuron in *G. campestris* (Stout, J. and Huber, F., 1981).

Other auditory interneurons with ascending branches have only been characterized physiologically such as the L (LAU, LHU) and S (STU) neurons in *T. oceanicus* (Hill, K., 1974; Ball, E. and Hill, K., 1978; Boyan, G., 1979a) and the type 2 neuron in *G. campestris* (Stout, J. and Huber, F., 1981). It is not known whether some of these names refer to other types of neurons, or whether some of them might be homologous with either type AN1 or type AN2.

The proximal part of AN1 is rather simple (Fig. 24). Its soma is located close to those of the ONs. The neurite crosses the midline to form a condensed dendritic field located entirely within the median part of the soma-contralateral auditory neuropile, and the axon courses anteriorly in the medioventral part of the cervical connective. The projections into the sub- and supraoesophageal ganglia are not known at present (Wohlers, D. and Huber, F., 1982). According to the same authors the AN1 is likely to be a first-order interneuron receiving information exclusively from the receptor fibre populations of the soma-contralateral ear.

The structure of the AN2 is more complicated as regards its dendritic field, whereas the positions of the soma, the neurite, and the axon do not deviate appreciably from those of the AN1. Some dendritic branches are found on the soma-ipsilateral side, but the more extensive arborizations are found on the soma-contralateral side, consisting of a smaller diffuse field and a large elongated area stretching towards the leg nerve. The AN2 may also be a first-order interneuron receiving auditory information from the soma-contralateral ear, but additionally it

receives inputs from the ipsilateral ear, probably via other first-order interneurons (Wohlers, D. and Huber, F., 1982). If it proves to be homologous with the HF1 (Rheinlaender, J. *et al.*, 1976), its projection areas in the supraesophageal ganglion may be similar to those of the latter with most of the terminals in the anterior ventral protocerebrum.

Descending auditory interneurons. All evidence suggests that the interneuron types ON1, ON2, AN1, and AN2 are single pairs of bilaterally symmetrical, large-diameter auditory interneurons. In contrast, the desending auditory interneurons from the prothoracic ganglion comprise a class of many types of descenders with smaller or larger axon diameters (Elephandt, A. and Popov, A., 1979; Boyan, G., 1979b). One of the largest found in *Gryllus campestris* is shown in Fig. 24 (Wohlers, D. and Huber, F., 1982). This so-called DN1, which is also represented as a single mirror-image pair, has major arborizations in the medial and lateral parts of the soma-contralateral auditory neuropile. It has large dendritic branches on the soma-ipsilateral side, but these are ventral to the auditory neuropile. The DN1 may be a first-order interneuron receiving its main auditory input via receptor fibres of the soma-contralateral ear (Wohlers, D. and Huber, F., 1982). Nothing is known about its target cells in the other thoracic ganglia.

Other descending interneurons have only been characterized physiologically: the D neuron in *Teleogryllus oceanicus* (Boyan, G., 1978, 1979b), and the five different types of fibres reported by Elephandt, A. and Popov, A. (1979). The descending fibres studied by Zhantiev, R. and Tshukanov, V. (1972b) are similar to the brain descenders studied by Zhantiev, R. and Kalinkina, I. (1977) and may, according to Boyan, G. (1979b), be identical with the latter ones.

Auditory T interneurons. T interneurons, i.e. interneurons passing information along an ascending and descending axon, have been known for years in other orthopterans (Suga, N. and Katsuki, V., 1961; Suga, N., 1963; Rheinlaender, J., 1975; Rheinlaender, J. and Kalmring, K., 1973, 1975). In *G. campestris*, however, such a neuron, TN, has been identified only recently (Wohlers, D. and Huber, F., 1982), see Fig. 24. Also, this neuron type is believed to be represented within the prothoracic ganglion as only one set of mirror image partners. The soma of the TN is located contralaterally to the ascending and descending axon, which leaves a large, diffuse dendritic field, the most lateral branches of which may receive synaptic input directly from the ipsilateral and contralateral receptor fibres. Since many of the dendritic fibres of the TN project dorsally as well as ventrally, relative to the auditory neuropile, it is suggested that the TN is multimodal (Wohlers, D. and Huber, F., 1982). Nothing is known about its anterior and posterior projections.

According to B. D. Lewis (personal communication) at least four types of T-shaped interneurons are found in *T. oceanicus*, the central acoustic pathways of which may differ from those of *G. campestris* in this respect.

(c) *Auditory interneurons in the brain* Auditory interneurons located entirely within the brain (the supraoesophagael ganglion) have recently been described by Boyan, G. (1980, 1981), who found two types of unimodal neurons in *Gryllus bimaculatus*: the unisegmental auditory brain neuron (UABN) and the plurisegmental auditory brain neurons (PABN1 and PABN2).

All processes of the UABN remain within the protocerebrum (Fig. 25). It has no clearly defined axon, but a dense dendritic field close to the calyx. Nothing is known about the input and output connections, but it shares projection regions and physiological properties with the HF1 neuron (Rheinlaender, J. *et al.*, 1976), which may be homologous with the AN2 described above (Boyan, G., 1981).

One of two plurisegmental interneurons in the brain, the PABN2 (Fig. 25) is the most complex, with its three main branches penetrating large parts of the proto- and deuterocerebrum. Its arborizations in the anterior ventral protocerebrum may be the input locations, since ascending auditory interneurons terminate here, while the dorsal deuterocerebral arborizations may be output locations, since they overlap the dendritic fields of the descending interneurons described below (Boyan, G., 1981).

(d) *Interneurons descending from the brain* So far only two types of sound responsive neurons descending from the brain have been described anatomically as well as physiologically. The ipsilateral descending brain neuron, IDBN, and the contralateral descending brain neuron, CDBN (Boyan, G. and Williams, J.,

1981 — see Fig. 26) found in *G. bimaculatus* are named after the position of their respective cell bodies relative to their axons. They have dense arborizations occupying the same space in the deuterocerebrum as the PABNs mentioned above. The IDBN and the CDBN may therefore receive auditory and several other inputs from the plurisegmental interneurons. Nothing is known about the projections of the DBNs, but it seems reasonable to suggest that they reach all three thoracic ganglia.

(e) *Information flow in the auditory system* From the description above it appears that a number of auditory interneurons are reasonably well known as regards their position and shape, but their probable input and output relations can only be guessed at. This is certainly one of the most important problems for the future, since the organization of the interneurons appears to be more network-like than

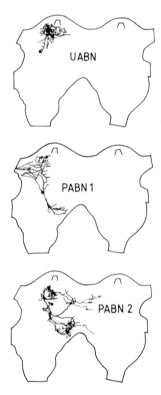

FIG. 25. Three types of auditory neurons in the supraoesophageal ganglion (the brain) of the cricket *Gryllus bimaculatus*. (Redrawn after Boyan, G., 1980, and Boyan, G. and Williams, J., 1981). The brain is viewed from a ventral aspect and the three pairs of cut nerves are, from top to bottom: the optic stalks, the antennal nerves, and the circumoesophageal connectives. For further explanation see text.

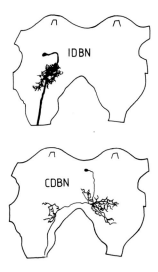

FIG. 26. Two types of interneurons descending from the brain of the cricket *Gryllus bimaculatus*. (Redrawn from Boyan, G. and Williams, J., 1981.) Aspect and shape of the brain as in Fig. 25.

serial. According to the serial concept, auditory information must flow along interneurons of increasingly higher order to reach the brain, where the auditory information is integrated along with that from other sensory modalities. Instructions resulting from this integration then flow along descending interneurons to initiate or modulate motor activity in the thorax.

Such an information flow does exist but for the following anatomical reasons it is not exclusively serial. Auditory information may be transferred directly to motor neuropiles in the prothorax by receptor fibres (e.g., the LF fibres — Eibl, E. and Huber, F., 1979). Possible first-order interneurons descend to the posterior thoracic ganglia (e.g. the DN1 and TN). The overlapping dendritic fields of the possible first-order interneurons in the prothoracic ganglion suggest network properties, and so does the structure of the brain neurons identified so far. Furthermore, it is hard to believe that all auditory interneurons have been identified. The ones described above all have rather large-diameter axons. There might well be other small-diameter interneurons carrying auditory information. Only spiking segmental interneurons have been identified (the ONs and the UABN). In other systems non-spiking segmental interneurons prevail (e.g. Pearson, K. and Fuortner, C., 1975; Burrows, M. and

Siegler, M., 1976, 1978) and might well occur here. Input–output connections are, however, hard to establish firmly and so are the network properties of neurons. Totally new techniques similar to the elegant one of Selverston, A. and Miller, J. (1980) used in motor systems may need to be developed for use on auditory pathways.

7.1.2 Central processing

Recordings from auditory neurons have been used in two main contexts: for investigating the physiology of the ear, and in a search for internal "recognizers" of relevant sound information. In the first kind of experiment the position of the sound source has been changed and/or parts of the ears have been manipulated. In such experiments auditory neurons have been used mainly as monitors of the peripheral manipulations. In the second kind of experiment the properties of the neuron in question have been examined by changing the sound parameters: investigators cheered if they were able to report some tuning to the conspecific song, or more often, regrettedly stated that no pronounced tuning to the main parameters of the conspecific song was found.

These methods of attack are of course perfectly legitimate and have probably been necessary during the first decades of research in these matters. It should be borne in mind, however, that in an experiment where the response of an interneuron is recorded while the sound parameters are changed, the information processing system studied *is not only* the interneuron: it is *a cascade of information-processing elements* which includes the forces created on the ears by the sound waves, the mechanical systems of the ears, the different populations of receptor cells, a network of interneurons, and only finally the recorded interneuron. Such an experiment may tell us how sound stimuli are *represented* in the activity of that particular neuron or type of neuron, but not how the stimuli came to be represented exactly that way. It does not tell us which (if any) of the normal mechanisms of neural integration was used, i.e. temporal and spatial summation of excitatory and inhibitory presynaptic input, facilitation, fatigue, habituation, and "hard-wiring" of the connections. If the cascade of elements is serial one may be able to determine the properties of an nth-order neuron by recording from the $n-1$st-order neuron and the nth-order neuron under the same stimulus conditions and then decide what kind of information processing has taken place from level $n-1$ to n. However, as stated above, the cascade is not entirely serial, but is more network-like, and it becomes increasingly more complicated, the "higher" up in the nervous system the neuron in question is located. In summary, if a change in behaviour of an nth-order neuron is observed upon a change in sound parameters, this change may be brought about by the integrating properties of this neuron, but it may equally well have taken place at a lower level, the interneuron only passively relaying this processing.

We shall not arrive at a thorough understanding of the central processing until two requirements have been fulfilled. First, that all types of auditory interneurons are known, especially as regards their input and output connections. Second, that intracellular recordings at all levels of the auditory pathway are carried out, paying special attention to the normal mechanisms of integration, e.g. by simultaneous recordings of pre- and postsynaptic activity. These requirements have not been met so far, and the following description of central processing is therefore rather tentative.

In the following section we shall try to outline our present knowledge about central processing of the sound parameters of frequency, intensity, time (i.e. amplitude modulation, AM), and direction, the first three of which have been studied in connection with one of the main themes of auditory physiology in insects, "recognition of conspecific songs". The processing of directional information is the object of the other main theme, "phonotaxis". (If not otherwise stated the frequency responses and tuning properties described refer to ipsilateral stimulation, i.e. perpendicular to the body axis.)

It should be pointed out here that the reaction of an auditory neuron, i.e. the graded and/or spiking response, is seldom dependent on just one sound parameter. This means, for instance, that it is nonsense to talk about, e.g., the frequency sensitivity of an auditory neuron, unless the stimulus intensity, the sound direction relative to the body, and the AM of the sound are constant and specified in the given frequency range. The occurrence of a given number of spikes with a certain repetition frequency in an auditory neuron may be caused by several combinations of the four parameters mentioned. Thus, an ipsilaterally presented sound with a certain AM, a high frequency and a high intensity may cause the same spike response as an ipsilaterally presented sound with the same AM, but a low intensity and a middle range frequency, or a contralaterally presented sound with the same AM, but a high intensity and middle range frequency.

Reduction of such ambiguity is what central processing in the auditory pathways is all about.

(a) *Representation of sound parameters on the receptor level* The filtering and processing of sound by the mechanical parts of the ear is described in section 6. The information transferred from stimulus sound to vibrations of the mechanical parts of the ear is transferred to the receptor level without much loss. At the receptor level we need only consider the coding of frequency, intensity, and time, since variation in sound direction causes a smaller or larger vibration amplitude of the mechanical parts of the ear, and thus reduces to the coding of intensity.

In crickets the main theme of conspecific song recognition has received so much attention, that only the frequency range 2–20 kHz is reasonably covered at all levels. Some recent investigations, however, indicate that this frequency range is too limited, if we are to appreciate the total power of frequency processing and eventually to understand the total acoustic behaviour (Moiseff, A. *et al.*, 1978; Hutchings, M. and Lewis, B., 1981; Kämper, G. and Dambach, M., 1981). The largest number of receptor fibres (between one-third and two-thirds of the total — Hutchings, M. and Lewis, B. (1981) and Esch, H. *et al.* (1980), respectively) are tuned to the carrier frequency of the calling song which in crickets is typically in the range 4–5 kHz (Popov, A., 1969; Nocke, H., 1972; Zhantiev, R. and Tshukanov, V., 1972a; Hill, K. and Boyan, G., 1977; Eibl, E. and Huber, F., 1979; Zaretsky, M. and Eibl, E., 1978; Esch, H. *et al.*, 1980; Hutchings, M. and Lewis, B., 1981). This means that if all parameters except frequency are kept constant, the number of nerve impulses per presentation will have a maximum at this frequency. The tuning is often rather broad, typical values of Q_{10dB} being about 2 (Q_{10dB} equals the frequency of maximum response (= characteristic frequency, CF) divided by the bandwidth 10 dB below this maximum — Kiang, N.,

1966). In the most recent investigations very sharp tunings of the 4–5 kHz receptors near threshold are reported (Q_{10dB} about 15 — as judged from the data of Hutchings, M. and Lewis, B., 1981). A second and smaller population of receptors is tuned to a frequency in the range 12–16 kHz, which is close to the third harmonic of the calling song frequency and to the main frequency component of the courtship song (Zhantiev, R. and Tshukanov, V., 1972b; Nocke, H., 1972; Esch, H. *et al.*, 1980) with a Q_{10dB} of 1–2. Finally, there is a small population of receptors with a low tuning whose CF is below 2 kHz (Popov, A., 1969; Nocke, H., 1972; Esch, H. *et al.*, 1980). The observations of Hutchings, M. and Lewis, B. (1981) on *Teleogryllus oceanicus*, however, suggest that the number of receptor populations characterized by their frequency responses is much larger than three, since they found receptor fibres with CFs distributed over the whole range of frequencies investigated (0.5–42 kHz). Moreover, they found receptor fibres with two "CFs" and receptors with no CF at all, but responding equally well over a broad band of frequencies. As regards the range of frequency sensitivity then, the tympanal organ of the cricket may resemble that of bush crickets (Kalmring, K. *et al.*, 1979) more than previously anticipated. The tunings mentioned above are for ipsilateral stimulation. They are not absolute quantities, but are largest near threshold and disappear almost totally at high intensities (over 100 dB SPL — Esch, H. *et al.*, 1980).

The frequency sensitivity of a first-order interneuron is therefore determined first by the number and type of receptor fibres synapsing with it, and second by the sound intensity and direction.

At the receptor level a change in sound intensity is coded as a change in spike frequency and a simultaneous change in latency, i.e. the time elapsed from the onset of a stimulus sound to the occurrence of the first spike. For the 4–5 kHz receptors the threshold at CF is typically 40 dB SPL (cf. section 6.1) and the dynamic range is typically about 30 dB, i.e. the sound pressure must be increased by 30 dB above threshold sound pressure level to reach the saturation level of the receptor fibre (Nocke, H., 1972; Eibl, E. and Huber, F., 1979; Esch, H. *et al.*, 1980). But other groups of receptors may differ in this respect (Hutchings, M. and Lewis, B., 1981) by having dynamic ranges of 3–20 dB, or by covering

the same frequency range, but with a threshold differences of 20–30 dB, thereby expanding the dynamic range of an interneuron receiving input from two such receptor populations by a factor of two or more. This phenomenon of "range fractionation" is also found in bush crickets (Rheinlaender, J., 1975) and moths (Roeder, K. and Treat, A., 1957), and may be a rather general principle in insects. While the spike frequency increases with intensity, the latency decreases, typically following a power function and varying between 25 ms and 5 ms at the entrance to the prothoracic ganglion (Nocke, H., 1972; Hutchings, M. and Lewis, B., 1981; Esch, H. *et al.*, 1980). According to the latter authors, latency at the receptor level is not only dependent on the intensity of sound, but also on the number of presentations, since it increases when the receptor is presented with just four sound pulses.

Most of the receptor fibres are phasic–tonic or pure tonic, faithfully reproducing the duration of the stimulus sound (at least in the range 10–350 ms in *Gryllus campestris* — Nocke, H., 1972) with only little adaptation at moderate sound levels, but larger at higher intensities. Marked adaptations are found in receptors tuned to 12–16 kHz (Nocke, H., 1972). The repetition rate of the pulses is also reproduced by the receptors, which can follow repetition frequencies in at least the range 15–40 Hz (Esch, H. *et al.*, 1980). Again, the extensive investigations of Hutchings, M. and Lewis, B. (1981) draw a more varied picture. They report highly phasic receptor fibres responding with only one spike per sound pulse in *T. oceanicus*.

In summary, the sound information propagating in one channel outside the animal is, at the receptor level, carried by 50–70 different channels from each of the two ears, each channel or channel population accentuating certain aspects of the information.

A first-order interneuron receiving information from both ears may have a dynamic range 2–3 times larger than that of a typical receptor cell, according to how many different range fractionating fibres synapse with it. Directional information, which typically gives rise to differences in vibration amplitude of the two ears, will not reach the central neuron simultaneously, since the latency is larger in fibres from the ear with the lowest vibration amplitude. In addition, the spike frequency is lower in these fibres. The interneurons may be sensitive to a smaller or larger frequency range depending on the fibre populations impinging on it. However, the rather large number of 4–5 kHz receptors makes it probable that the interneuron has its largest sensitivity in this frequency range. Since it may be sensitive over a wide frequency and intensity range, it is likely to be sharply tuned to certain frequencies at low intensities and less tuned, or not tuned at all, but responding to a wide band of frequencies, at high sound levels. Finally, the accuracy of AM coding in the receptors preserves the information on sound pulse duration and repetition over a wide band of durations and frequencies.

(b)   *Representation of sound parameters in first order auditory interneurons*
The omega interneurons.   In the ON1s spikes are generated in the soma-ipsilateral dendritic fields and propagate without any filtering due to cell morphology along the axonic arch to the contralateral dendritic field (Wohlers, D. and Huber, F., 1978). The number of spikes per sound presentation may, however, be changed due to presynaptic inputs other than those of tympanal fibres. In a series of elegant experiments (Wohlers, D. and Huber, F., 1978; Kleindienst, H. *et al.*, 1981) it was shown, first, that the number of spikes propagating to the contralateral side is determined by two factors, the momentary sound intensity and the prevailing resting potential of the neuron, and second, that the

ipsilateral ON1 normally tends to hyperpolarize the contralateral one through inhibitory presynaptic input (Fig. 27). The result of this mutual inhibition is an enhancement of the contrast between ipsilateral and contralateral sound information, i.e. the difference between the number of spikes per sound presentation in the ipsilateral ON1 and that of the contralateral ON1 is larger than it would have been, had there been no mutual inhibition. For middle-range sound intensities this difference is increased by a factor of about 1.6, if the sound levels at the two ears are different, which is almost always the case (Kleindienst, H. *et al.*, 1981). Such a mechanism is, of course, very useful as regards sound localization, since even the slightest deviation from equal sound pressure at the two ears will cause a difference in the number of spikes propagating along the axons of the two neurons. The mechanism would allow a network of motor neurons to keep a course towards or away from a sound source. However, neither the target cells nor the actual directional sensitivity of the ON1s are known; yet, the ON1s might act as the cross-inhibiting interneurons accentuating the difference in spike number of the ascending interneurons, as proposed by Boyan, G. (1979a).

The number of spikes may also be reduced, if low-frequency (0.1–2 kHz) substrate vibration is presented together with a 4.5 kHz sound stimulus

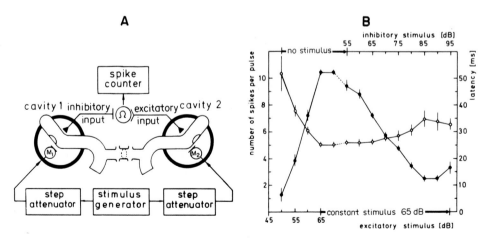

FIG. 27. Analysis of contralateral inhibition in the omega cell of *Gryllus campestris* in response to 5 kHz sound signals. **A**: Closed sound field arrangement for external and internal isolation of excitatory and inhibitory inputs to the omega cell. The H-shaped central figure symbolizes the acoustic prothoracic tracheal apparatus of a cricket lying ventral side up with the ears enclosed in two cavities in which condenser microphones (M1 and M2) act as loudspeakers. The sound stimuli are transmitted via the tracheal system from one ear to the other and the resulting inhibitory and excitatory stimuli are monitored in the activity of the omega cell. **B**: Intensity response curves (●–●) and latency (○–○) of an omega cell for various excitatory and inhibitory 5 kHz, 50 ms duration stimuli. (From Kleindienst, H. *et al.*, 1981.)

(Wiese, K., 1981), or if the ipsilateral ear is stimulated with sounds in the frequency range below 2 kHz (Kleindienst, H. *et al.*, 1981). The inhibition is a maximum, if the substrate vibration precedes the sound stimulus by 15–20 ms, and the number of spikes may be reduced between 10% and 60%, according to the relative stimulus strengths and temporal patterns (Wiese, K., 1981). The experiments further suggest that the inhibition is mediated by the bimodal receptor fibres (LF fibres — Nocke, H., 1972) originating in the part of the tibial sensory complex termed the subgenual organ (Dambach, M., 1972). Similar mechanisms are found in bush crickets (section 7.2.2).

The ON1s are sensitive to sounds in the frequency range 1–40 kHz (at least) with a relatively sharp tuning at threshold and low sound intensities in the frequency range 4–5 kHz (Popov, A. *et al.*, 1978). Judging from the threshold curves, the Q_{10dB} values vary between 1–2 in *Gryllus campestris* (Wohlers, D. and Huber, F., 1978) and 4–5 in *Gryllus bimaculatus* (Popov, A. *et al.*, 1978). On the assumption that relatively high Q_{10dB} values of 4–5 kHz receptors are not confined to *Teleogryllus oceanicus* (Hutchings, M. and Lewis, B., 1981), it is reasonable to assume that no frequency sharpening takes place from the receptor cells to the ON1. The frequency response of the ON1 is therefore most likely to be a simple consequence of the convergence of excitatory inputs from different populations of ipsilateral receptor fibres.

The hypothesis that the ON1 integrates the inputs from many different receptor fibre populations is supported by the fact that its dynamic range is very large at 4–5 kHz. Here, the intensity–response curve increases monotonically, with a dynamic range of 55–60 dB above threshold. This corresponds to the integration of inputs of several "range fractionating" receptor fibre populations (Popov, A. *et al.*, 1978; Wohlers, D. and Huber, F., 1978). As a consequence of this integration, intensity coding is dependent upon frequency, sound direction, and temporal parameters. The dynamic range decreases with frequency (to 20–25 dB at 10 kHz, and to 10–15 dB at 16 kHz) and changes with direction (the dynamic range at 5 kHz is about 30 dB larger for ipsilateral than for contralateral stimulation) (Wohlers, E. and Huber, F., 1978). The dynamic range is increased for sound pulse repetition rates in the range 26–36 Hz by 6–7 dB relative to that of repetition rates between 11 and 22 Hz (Popov, A. *et al.*, 1978). The latency, which at the receptor level decreases linearly with intensity, also decreases at the ON1 level, but as a function of both intensity and frequency, thus reflecting the properties of the presynaptic receptor fibre populations and the mutual inhibition between the ON1s. At 5 kHz the latency is about 14 ms at high intensities, increasing to about 25 ms near threshold, whereas it is 1–2 ms higher at higher frequencies (Popov, A. *et al.*, 1978).

The slowly-adapting ON1 (50% spike frequency reduction over 1 s — Popov, A. *et al.*, 1978) codes time parameters as spike trains over a wide range of durations. Since the summed EPSP of the ON1 has a certain time constant, a pause between two sounds can only be coded as a pause between spike trains if the summed EPSP is allowed to fall to a certain level after the cessation of one sound pulse, before it rises again as a result of the next sound pulse. The critical pause-length of the ON1 is about 15 ms (Wohlers, D. and Huber, F., 1982). This defines the upper limit for repetition frequencies of sound pulses (Fig. 28). Two sound pulses of e.g. 2 ms duration will not be coded as two spike bursts unless the repetition period is more than 17 ms, corresponding to an upper rate limit of 59 Hz. For longer pulse duration the upper repetition rate is correspondingly lower. Pulse durations of up to at least 140 ms are coded by the duration of the spike trains (Wohlers, D. and Huber, F., 1982). In Fig. 29 the dependence of syllable coding on critical pause length and sound incidence is shown for six types of interneurons.

The integration properties of the ON2s are less well-known. Their frequency response is broadband (at least 2–20 kHz) with no pronounced tuning and relatively high threshold intensities. Since their critical pause length is about 40 ms their upper repetition rate limit is lower than that of the ON1, and they do not regularly code the syllables of the natural calling song, which in *G. campestris* has a repetition rate of about 25 Hz (Fig. 29).

Ascending auditory interneurons. The information flow along the segmental, the descending, and the T neurons of the first order is rather well-established, since their properties are described in relatively few, predominantly concordant, investigations. This is not so as regards the ascending first-order interneurons. There may be only two (Wohlers, D. and

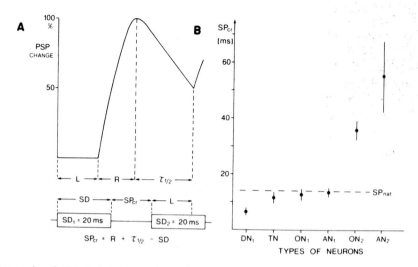

FIG. 28. Critical pause length (SPcr), definition and magnitude in different interneurons of *Gryllus campestris*. **A**: Schematic diagram showing the relationship between postsynaptic potential change (PSP change) in percentages, with respect to SPcr, which is the sum of the rise time (R) and the 50% decay time ($\tau_{\frac{1}{2}}$) minus pulse duration (SD). L indicates latency. **B**: The magnitude in six types of auditory interneurons. (From Wohlers, D. and Huber, F., 1982.)

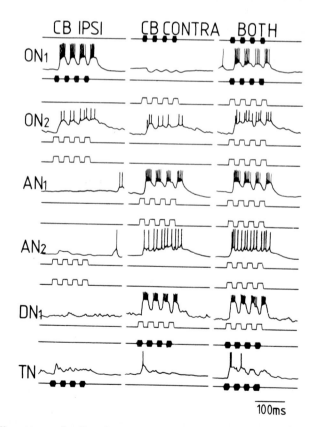

FIG. 29. Responses of six different types of auditory interneurons in *Gryllus campestris* to 4.5 kHz, 80 dB sound pulses presented in a closed sound field (cf. Fig. 27A). CB IPSI and CONTRA indicate that identical sound pulse trains are presented ipsilaterally and contralaterally to the cell body side, respectively. Note by comparing with Fig. 28 how the pulse coding depends on SPcr. (From Wohlers, D. and Huber, F., 1982.)

Huber, F., 1982) or three types of ascending inter-neurons (Boyan, G., 1979a; Stout, J. and Huber, F., 1981), or there may be as many as the bewildering variety of names suggests (see above). In the following we shall concentrate on the integration properties of the two types AN1 (LF1) and AN2 (HF2) (Wohlers, D. and Huber, F., 1982), and as regards directional information on the types S and L neurons (Boyan, G., 1979a).

The AN1 type, which receives input exclusively from the soma-contralateral side, is relatively sharply tuned to a frequency in the range 4–5 kHz and it has, judged from threshold curves, a Q_{10dB} of 1–3. Furthermore it has a threshold of about 40 dB SPL and a monotonically increasing intensity curve with a dynamic range of about 50 dB at 5 kHz (Stout, J. and Huber, F., 1972, 1981; Rheinlaender, J. et al., 1976; Wohlers, D. and Huber, F., 1982). These properties might be a simple consequence of a convergence of populations of "range-fractioning" 4–5 kHz receptors on the AN1. Another kind of information processing which occurs in the AN1, is the coding of time. The "critical pause-length" of the AN1 is equal to that of ON1 (Fig. 28B), about 15 ms (Wohlers, D. and Huber, F., 1982), which means that two sound pulses of duration d ms will not be coded as two separate spike trains, unless $(d + 15)$ ms has elapsed from the onset of the first pulse to the onset of the next. This means that syllable lengths of, e.g., 25 ms can only be coded as separate bursts if the repetition frequency is 25 Hz or lower. This may be the explanation why G. bimaculatus only codes the syllables of its conspecific calling song and one of another species with low repetition frequency, but not those of other sympatric cricket species with similar power spectra but higher repetition rates (Rheinlaender, J. et al., 1976).

The AN2-type interneuron covers a much wider frequency range than the AN1, having no sharp tuning, but at lower intensities accentuating frequencies near 15 kHz. The HF1, which may be homologous with the AN2, responds to stimulus frequencies up to above 50 kHz (Rheinlaender, J. et al., 1976), and the same may be the case for the AN2. Wohlers, D. and Huber, F. (1982), using intracellular recordings, have found that the response to frequencies below 10 kHz is highly variable from animal to animal, often with inhibition or

suppression of the response, and possibly being dependent on the recording site. Such inhibition at low frequencies is also reported for the HF1, the Int-1, and the AIAA (Rheinlaender, J. et al., 1976; Casaday, G. and Hoy, R., 1977; Wohlers, D. and Huber, F., 1978), but the source of inhibition remains unknown. The response of the AN2 (in HF1 spike repetition frequencies may reach 700 Hz (Rheinlaender, J. et al., 1976)) typically outlasts the stimulus pulse, and the (probably) homologous or identical neurons have been termed "chirp-coders" (Stout, J. and Huber, F., 1972; Popov, A., 1973). The mechanism again most probably resides within the interneuron itself, since the time constant of the cell membrane is very long (see Fig. 28B), yielding a "critical pause length" of 40–80 ms (Wohlers, D. and Huber, F., 1982). This means that if the sound stimulus is the conspecific song consisting of four pulses of 25 ms duration and 25 Hz repetition rate, the EPSP will not have had time to decrease between sound pulses. The neuron will therefore remain excited and consequently generate spikes during the whole chirp (Fig. 29). The courtship song, however, which contains maximum energy in the range 14–16 kHz and which consists of "ticks" repeated at 2–3 Hz, will be coded very accurately, partly owing to the low repetition rate and partly because the highly phasic response ensures a very large dynamic range.

The possession of two symmetrically situated, directionally sensitive ears enables an animal to locate a sound source by assessing the amount of asymmetry in the auditory information, reaching the CNS through the auditory receptors. The activity of the bilateral pairs of ascending first-order auditory interneurons therefore also reflects the information asymmetry. In order to assess the information processing as regards directionality at this level, it is necessary to monitor the information flow bilaterally. This has been done by Boyan, G. (1979a) recording the response of the ascending S and L neurons in Teleogryllus commodus. Here the difference in number of spikes per sound presentation is dependent on the angle of sound incidence. Frontal sound incidence (0°) could be expected to generate an equal number of spikes in both left and right S and L neuron, i.e., the left–right difference (L–R) could be expected to be zero. However, the animals showed an inherent bias to either the left or the right, ranging from $-17°$ to $+38°$. This asymmetry is not likely to be caused by nervous integration, since the same trend was found in both S and L neurons, but it may be due to asymmetries in the mechanical system of the ears. The intensity curve (i.e. the number of spikes per stimulus as a function of sound intensity) of the right S neuron is identical with that of the left S neuron, and the same applies to the L neurons (Fig. 30A). If the sound intensity and frequency is kept constant, but the sound source moved by an angle of, e.g., $90°$, the vibration amplitude of the mechanical parts of the directionally sensitive ears will increase or decrease according to which direction the sound source is moved. Consequently, the intensity curves of the S and L neurons appear to shift to the left or the right (cf. Fig. 30B and C), which for a given sound intensity produces a difference in the number of spikes between the left and right interneurons (Fig. 30D). The lower figure also shows that the (L–R) of the S neurons is a maximum about 20 dB lower than that of the L neurons but, taken together, the directional response covers an intensity range of more than 50 dB. Thus the principle of "range fractionation" may also apply here. If such measurements are made for a number of angles between $0°$ and $360°$, curves similar to those of Fig. 31 result, showing the potential directional information available to higher order neurons from the S neurons. Most of this information is probably present already at the receptor level, which means that only little integration of directional information takes place at the level of the ascending interneurons. The necessary information processing, i.e. the comparison between the right and left information channels, must

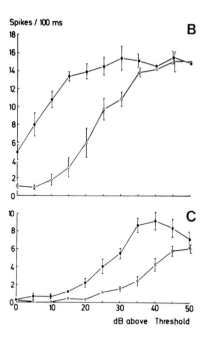

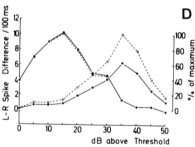

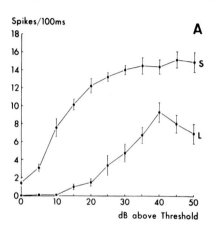

FIG. 30. Intensity–response curves of ascending auditory interneurons (S and L neurons) in *Teleogryllus commodus*. **A**: Frontal (symmetrical) stimulation. The curves from bilateral pairs of interneurons are almost identical. **B**: Intensity–response curves simultaneously recorded from left (○) and right (●) S neuron. Sound stimulus ipsilateral to the right ear. **C**: Intensity–response curves simultaneously recorded from left (○) and right (●) L neuron. Sound stimulus ipsilateral to the right ear. **D**: Difference in spike number between left and right S neuron (●) and L neuron (■), respectively. □----□ represents spike differences for the L neurons, normalized as a percentage of each maximum response. (From Boyan, G., 1979a.)

take place in neurons or neural networks within the brain.

In summary, the ascending auditory interneurones represent highly evolved information-processing devices, which act as band-pass filters for modulation frequency, and as more or less passive relay stations of directional information.

Descending auditory interneurons. In crickets,

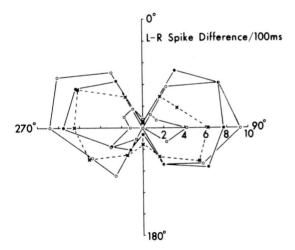

FIG. 31. Difference in spike number between simultaneously recorded left and right S interneurons to sounds from different angles of incidence, plotted on polar co-ordinates. ●, ○, ×, and □ indicate responses at 10, 15, 20, and 30 dB above threshold, respectively. (From Boyan, G., 1979a.)

detailed auditory information flows caudally along a whole class of what are probably first-order interneurons, possibly to make synaptic contacts with thoracic motor neurons. The representation of auditory information in such interneurons has received some attention, but in different species and most probably in non-homologous neurons (*Teleogryllus commodus* — the D neuron, *Gryllus campestris* — the DN1 interneuron, and *Gryllus bimaculatus* — five types, studied by Boyan, G. (1978, 1979b); Wohlers, D. and Huber, F. (1982); and Elephandt, A. and Popov, A. (1979), respectively).

The identified neuron, DN1, does not respond to soma-ipsilateral sound, but responds energetically to soma-contralateral or bilateral sound (Fig. 29) with a sensitivity maximum at or below 2 kHz, a dynamic range of about 30 dB, and a latency of 12–15 ms at 80 dB SPL. Furthermore, it faithfully copies the temporal structure of the calling song, being able to code pause lengths of above 8–10 ms (Fig. 28). Such properties are also found among the receptor fibres (see section 7.1.2a), and the properties of the DN1 described so far could be brought about simply by a certain combination of axon-ipsilateral, presynaptic receptor fibres. Tentatively, therefore, the main function of the DN1 may be to relay information about the temporal structure of low-frequency sounds to motor neuropiles of posterior thoracic ganglia.

The extracellularly recorded, but only physiologically identified, D neuron in *T. commodus* is so far the most extensively studied descender. It has properties in common with the DN1 as regards coding of time and frequency sensitivity. The duration of a sound pulse is coded by the duration of the spike train at least up to 150 ms (the tonic component), and by the total number of spikes up to 80 ms duration (the phasic component?) — see Fig. 32. The bursting activity of this non-habituating neuron also codes the pulse repetition rate of the conspecific calling song (about 15 Hz). The D neuron is broadly tuned in the frequency range 0.5–12 kHz, with rather sharp sensitivity peaks at 0.9 kHz and 4 kHz (Q_{10dB} about 1 and 2, respectively). This tuning (expressed as the number of spikes per stimulus as a function of frequency) is a maximum at an intensity about 10 dB above the threshold at 4 kHz and gradually disappears at higher intensities. Information on sound direction is also well represented in this neuron. It is not clear, however, how much integration takes place here, since the representation of directional information in the different populations of receptor fibres is not known. Directional information might be enhanced by integration of ipsi- and contralateral inputs. Severing of the contralateral leg nerve causes a small increase (about 2 dB) in the threshold intensities, and a decrease in latency at threshold to about 30 ms (cf. Fig. 32B and C). At higher intensities the latency stays constant at about 20 ms, whereas the number of spikes reduces by 25–40% — all for ipsilateral presentation. At contralateral presentation the same changes occur at threshold intensities, but at higher intensities the latency decreases, whereas the number of spikes per stimulus shows some increase. It is obvious from such an experiment that the D neuron does integrate information from both ears, but it is difficult to say exactly what kind of integration takes place. Intracellular recordings are badly needed. Most of the directionality must be derived from the ipsilateral ear. The directional sensitivity is dependent on the frequency and intensity of the stimulus sound, being at a maximum at 4 kHz and at an intensity 10 dB above threshold (Fig. 33).

So, a motor neuron or a network of motor neurons receiving and comparing inputs from the right and left D neuron obtain precise information on the temporal structure of the sound stimulus. In

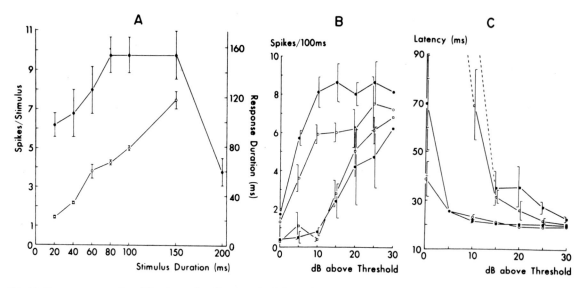

FIG. 32. Response properties of the descending D neuron in *Teleogryllus commodus*. **A**: Duration of the spike train (○) and number of spikes per response (●) in response to stimuli of increasing duration and constant intensity (10 dB above threshold at 4 kHz). **B**: Intensity–response curves for ipsi- (●) and contralateral (■) stimulation in intact preparations. The same after cutting the contralateral leg nerve (○ and □ are ipsi- and contralateral presentation, respectively). **C**: Latency as a function of intensity. Symbols as in **B**. (From Boyan, G., 1979b.)

addition the sound direction can be calculated from the number of spikes relative to the spike train duration and the relative latency. In principle this can be illustrated by considering Fig. 32. A 4 kHz sound pulse of 100 ms duration, and presented ipsilaterally to the right ear at 10 dB above threshold, will generate a spike train of about 100 ms duration and consist of about eight spikes in the right D inter-

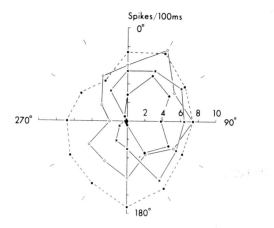

FIG. 33. Response of the descending D neuron in *Teleogryllus commodus* to 4 kHz sound presented at different angles and at different intensities above threshold: (●), 5 dB above; (×), 10 dB above; (○), 15 dB above; and (■), 25 dB above threshold. (From Boyan, G., 1979b.)

neuron (Fig. 32A). In the left D neuron only one spike will be generated (Fig. 32B) about 90 ms later than the first spike of the right D interneuron (Fig. 32C). It is questionable, however, whether any such combination of spikes in the right and left D interneuron represents a single set of stimulus parameters, or whether there is some ambiguity, since the data presented by Boyan, G. (1979b) are not comprehensive enough in this respect.

The interneurons recorded in the mesothoracic ganglion of *G. bimaculatus* by Elephandt, A. and Popov, A. (1979) may be true descending interneurons from the prothorax, but the possibility that they, or some of them, are descending branches of T interneurons or descending neurons from the brain cannot be excluded. The type 1 neurons share many properties with the D neuron and are probably first-order interneurons with latencies in the range 12–25 ms. Special units encountered were the pulse-markers, coding neither intensity nor stimulus duration but sound pulse onset, and the syllable-coding, broad-band units covering the frequency range 1.5–25 kHz. This type of unit had a very complicated response and adaptation pattern, depending on the frequency of the stimulus sound, probably resulting from the convergence of many types of presynaptic fibre populations.

In summary, the class of descending first-order interneurons from the prothorax, of which two types are reasonably well described, carries remarkably detailed information about temporal structure, location, and intensity of stimulus sounds to the posterior ganglia. The largest neurons carry information mainly on low and middle-range sound frequencies, while smaller neurons may cover a wider frequency range. The time parameters are not processed in the large descenders, but to some degree in the smaller ones. Information on direction and intensity are received from both ears, but mainly from the ipsilateral one. Some integration of these parameters may take place in the descenders, but it may predominantly have the character expected of the convergence of several populations of presynaptic receptor fibres on the single descender.

The T-neurons. Only little is known about information processing in the TNs recently described by Wohlers, D. and Huber, F. (1982) in *Gryllus campestris*. With dendritic branches in both prothoracic acoustic neuropiles, the TN may receive direct input from the auditory receptors. The threshold of the TN is rather high (70–80 dB) and it typically responds with subthreshold excitations in the frequency range 4–10 kHz, but may respond at high intensities to sounds in the range 2–15 kHz. The small response to unilateral stimulation is enhanced, often with spike production, if the animal is stimulated bilaterally (Fig. 29). Sound duration is coded in the duration of the EPSP, and single spikes may code the temperal occurrence of sound pulses of low repetition frequencies and low duty cycle. When presented with chirps of the conspecific calling song the TN adapts very quickly, recovering only over a period of several seconds.

Some of the properties of this neuron may reflect the receptor populations synapsing with it, whereas others, such as the fast adaptation and the enhancement with bilateral stimulation, may be due to integrating properties of the TN. Since no target cells are known, and since only some of its integrating properties have been investigated, it is hard to ascribe to the TN a special role in the processing of auditory information.

In some T interneurons (and one ascending neuron) of *Teleogryllus oceanicus* a special mechanism enhancing the syllable-coding seems to exist (Lewis, personal communication). At high sound intensities simulated song (18 kHz carrier frequency) will generate an almost continuous spike-train in the T interneurons. If, however, the 18 kHz carrier is mixed with a 4.5 kHz sound, the syllable coding dramatically improves. This is found in crickets with the contralateral auditory nerve severed. So, receptors from one ear may converge on the same interneuron, HF receptors exciting while some LF neurons inhibit the interneuron. Similar mechanisms enhancing temporal or spectral components are also found at higher levels of the cricket auditory pathways (see section 7.1.2b) and in other insects (see section 7.2.2).

(c) *Representation of sound parameters in higher-order interneurons*
At the receptor level a simple description of the representation of the four sound parameters in different receptor populations will suffice, since it indicates what kind of information is available to the central nervous system. At the level of first-order neurons, such a description may still be useful, since — as indicated above — it allows us to infer part of the integration which takes place here, by comparison with the representation at the receptor level. Still, such a comparison will elucidate only parts of the information processing at this level. We also need to know the subsequent integration of the information carried by the first-order interneurons. But at the level of higher-order interneurons a mere description of the representation of sound parameters is probably too simple-minded. It does describe part of reality, but essential aspects of the information processing may escape attention, unless more sophisticated experimental procedures are developed. It may prove useful, for instance, to consider the techniques developed for the study of vertebrate hearing, as has been done by Boyan, G. (1981) in two-tone inhibition experiments on brain neurons, and more recently in determining critical bands in crickets (Ehret, G. et al., 1982).

Among the unimodal interneurons of the auditory pathway described so far, probably only the segmental and plurisegmental brain neurons (Boyan, G., 1980, 1981) can with certainty be regarded as higher-order auditory neurons. All three types of neurons have dendritic arborizations overlapping with the projections of the HF1 described above. Furthermore, they share many properties with the latter, which may therefore be one source of input. All three types of interneurons have thresholds of about 40 dB SPL at CF and are broadly tuned in the range 8–18 kHz with a CF at about 16 kHz. The EPSP typically outlasts the sound pulse, the relative duration increasing with intensity. This maintained discharge is not caused by a long time constant of the brain neurons, but must be due to presynaptic inputs, such as an AN2/HF1 with a long time constant or inputs from units with different thresholds and consequently different latencies. Boyan, G. (1980) speculates that this retention of auditory information within brain circuits via maintained discharges may raise the central excitatory state and thus increase the probability of the same or other circuits responding to novel stimuli. In the plurisegmental brain neurons the intensity coding follows the normal pattern both as regards spike number and latency with dynamic ranges of 30–35 dB and latencies of 14–37 ms. In the unisegmental brain neuron, however, an active presynaptic inhibition suppresses the response to frequencies lower than 16 kHz, thus reducing the dynamic range at all other frequencies (Fig. 34), while retaining a large dynamic range at frequencies near those of the courtship song. The source of this active inhibition is unknown. Another mechanism of enhancing spectral components at the PABN2 is described by Boyan, G. (1981). In an elegant series of experiments on two-tone inhibition he showed that the response of PABN2 is selectively suppressed when a lower frequency sound is added, and that the frequencies near that of the conspecific calling song are most effective in suppressing the response (Fig. 35). The suppression also operates on the second harmonic of the calling song, 10 kHz, and the hypothesis is proposed that this mechanism of selective suppression more clearly defines the information content of the calling and courtship songs, since ambiguous spectral information may be present (intense calling song with harmonics of 10 and 15 kHz versus courtship song of 16 kHz) and the broad-band brain neuron may, without such a mechanism, respond equally well to both sounds. The origin of the suppression is not clear, but interneurons such as the AN1 might inhibit the normal input such as the AN2 to the PABN2. This would cause an inhibition when the stimulus was the calling song, but not the courtship song. Such LF-suppression of HF components is also found at lower levels of the auditory pathways (cf. section 7.1.2b).

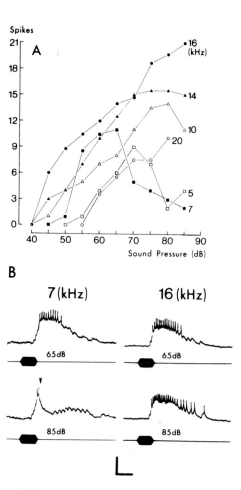

Fig. 34. Response properties of the UABN (unisegmental auditory brain neuron) in *Gryllus bimaculatus*. **A**: Intensity–response curves at six different frequencies. Note the depression at frequencies different from 16 kHz. **B**: Intracellular recorded responses to 7 kHz and 16 kHz at two intensities. Arrow indicates the cut-off after the second spike due to inhibition; compare with the summed response at 16 kHz. Calibration bars: 10 mV v. 20 ms. (From Boyan, G., 1981.)

(d) *Descending brain interneurons: the role in auditory processing* The anteriorly directed flow of auditory information is integrated in the brain with that of other sensory modalities (Huber, F., 1974, 1978). Some of the resulting information is then directed via descending interneurons to the thoracic ganglia, where it may initiate or modulate motor activity (Otto, D., 1971; Bentley, D., 1977; Pollack, G. and Hoy, R., 1981). So, although a stimulus sound may trigger some activity in descending interneurons, they are not auditory interneurons in the proper sense of the word.

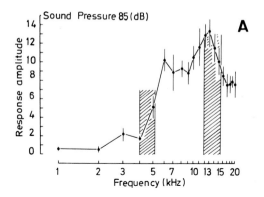

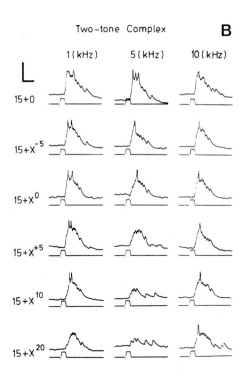

FIG. 35. Responses of a PABN2 (plurisegmental auditory brain neuron) to sounds of different frequencies. **A**: Frequency sensitivity expressed as the area of the compound PSP in response to sounds of constant intensity as indicated. Vertical bands represent the frequency range of the calling (5 kHz) and courtship (15 kHz) songs. **B**: Responses to two-tone complexes. Topmost response in each column (15 + 0) is to the control tone of 15 kHz, 85 dB SPL. Other responses are two-tone complexes in which control tones are mixed with test tones (×) or 1, 5, and 10 kHz. The "exponents" indicate the sound intensities of the test tones in dB relative to 85 dB SPL. Calibration bars: 15 mV *vs.* 40 ms. (From Boyan, G., 1981.)

The interneurons described by Boyan, G. and Williams, J. (1981), do respond to at least two modalities — sound and light. Both types are sensitive to sounds in the frequency range 13–16 kHz, but have very high thresholds (about 80 dB SPL). While the latency of the response of the IDBM is independent of sound intensity, the spike latency of the CDBN may be as large as 120 ms at 90 dB SPL. Both types may, according to their frequency range, be involved in courtship behaviour, while other types of descending interneurons only physiologically described (Zhantiev, R. and Korsunovskaya, O., 1977) may be involved in the behaviour related to the calling song.

7.2 Tettigoniidae

7.2.1 ANATOMY

The central projection of the tibial mechanoreceptor complex in bush crickets is very similar to that of crickets (Rehbein, H., 1973; Kalmring, K. *et al.*, 1978b) forming a crescent-shaped neuropile and being strictly ipsilateral in the prothoracic ganglion (Fig. 23B).

At the level of first- or second-order interneurons, the bifurcating information flow is more pronounced in bush crickets than in crickets, since a number of large T-shaped auditory interneurons carry identical information to the brain and via the thoracic ganglia also to the first abdominal ganglion. The largest-diameter neuron of the "acoustic bundle" readily found in the ventral cord of bush crickets is the S1 interneuron present as one mirror image pair (Kalmring, K. *et al.*, 1979; Kühne, R. *et al.*, 1980) (see Fig. 36). The S1 may be homologous or identical to those neurons whose different physiological aspects have been investigated by Suga, N. and Katsuki, Y. (1961); Suga, N. (1963); McKay, J. (1969, 1970); Rheinlaender, J. and Kalmring, K. (1973 — the I2 type); and Rheinlaender, J. and Römer, H. (1980). The postsynaptic region (input) of this strictly ipsilateral interneuron is located in the ipsilateral prothoracic acoustic neuropile, where different populations of receptor cells converge on it (directly or via interneurons). The fine branches in the brain and in the ventromedial neuropile of the suboesophageal and thoracic ganglia are probably presynaptic (output)

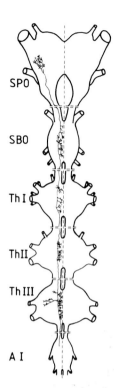

FIG. 36. The S1 interneuron in *Decticus verrucivorus*. In this semidiagrammatic drawing the fine branches are not drawn to scale and the right interneuron has been omitted for clarity. SPO, supraoesophageal ganglion (brain); SBO, suboesophageal ganglion; ThI-III, pro-, meso-, and metathoracic ganglion; AI, first abdominal ganglion. (Redrawn from Kalmring, K. *et al.*, 1979.)

terminals. Physiological evidence suggests the existence of a number of types of T-shaped interneurons as well as ascending and descending interneurons (Rheinlaender, J. and Kalmring, K., 1973). However, none of these have been described morphologically, and the same applies to interneurons in the brain (Rheinlaender, J., 1975) and interneurons descending from the brain. Omega interneurons like those in crickets are found in bush crickets (Rheinlaender, personal communication), but they have not yet been described either anatomically or physiologically.

7.2.2 CENTRAL PROCESSING

Space does not allow a comprehensive review of all the literature on central processing of auditory information in bush crickets, and readers are referred to the work of Rheinlaender, J. (1975) for a general description of the representation of sound parameters at the level of receptor cells, prothoracic interneurons, and brain neurons, respectively.

An important aspect in which bush crickets may differ from crickets as regards central processing of auditory information has emerged from a recent series of investigations by Kalmring and co-workers (Kalmring, K. *et al.*, 1978b, 1979; Kühne, R. *et al.*, 1980; Silver, S. *et al.*, 1980; Kalmring, K. and Kühne, R., 1980). It appears that the ascending and T-shaped interneurons of the auditory pathway that have been encountered at the level of the prothoracic ganglion all integrate information from the whole of the tibial mechanoreceptor complex, i.e. information in response to both air-borne sound and vibration stimuli. One group of interneurons, the VS interneurons, respond with spiking activity to vibration as well as to sound. A second group, the S interneurons, respond with spiking activity to airborne sound alone, but the activity is modified by excitatory or inhibitory subthreshold input from the vibration-sensitive receptors. The opposite is true for the third group, the V interneurons. It should be noted that since earlier investigations did not cover the frequency range up to 100 kHz nor take into account vibration stimuli, the more recent data are generally rather difficult to compare with the earlier classifications such as that of Rheinlaender, J. *et al.* (1972) and Rheinlaender, J. and Kalmring, K. (1973). The exceptions are the largest interneurons in the auditory pathway, the S1 interneurons, which are probably identical to the I2 interneurons. Furthermore, the investigations up to now have focused on a classification of the interneurons from the concept of serial information processing. Network-like interactions between the interneurons have not been considered, except for some ideas on contralateral influences, or on the complex properties of some higher-order interneurons.

Some of the response characteristics of the VS, V, and S interneurons are listed in Table 1. Latency measurements and the complex response patterns indicate that the S3 and the V1 are higher-order, or at least second-order, interneurons, while the other types may be first-order or more probably second-order interneurons. From the response range and pattern it is therefore not possible to make definitive

Table 1: *Response characteristics of V, S and VS neurons*

Type		Air-borne sound stimuli		Vibration stimuli		Habituation
Morpho-logical	Functional	Response pattern	Response range (kHz)	Response pattern	Response range (Hz)	
T	VS1	phasic	7–100	phasic/phase-locked	20–100	strong
T	VS2	phasic	4–100	tonic to phasic	2–3000	some
A (?)	VS3	tonic	7–100	tonic	20–5000	none
T	VS4	on-bust	4–100	on-bust	20–5000	strong
?	VS5	only response to V+S	—	only response to V+S	—	
T	S1	phasic	4–100	—	—	strong
T	S2	phasic	4–100	—	—	strong
A (?)	S3	phasic (long latency)	7–30	—	—	strong
A (?)	S4	tonic	7–100	—	—	none
A (?)	S5	on-bust	4–100	—	—	some
A (?)	V1	—		tonic	20–500	none
A (?)	V2	—		tonic	20–5000	none
T	V3	unspecific reaction		on-bust to tonic	20–5000	some
T	V4	—		phase-locked	up to 200	none
?	V5	—		phase-locked	up to 200	none

(Adapted from Kalmring, K. and Kühne, R., 1980, and Silver, S. *et al.*, 1980.)

statements about receptor populations converging on the different interneurons. Kühne, R. *et al.* (1980) nevertheless suggest that the VS1 receives input from the mid- and/or high-frequency sound receptors and from some campaniform sensillae of the ipsilateral fore leg, the contralateral inputs only contributing with minor inhibition. The other VS interneurons also receive mainly ipsilateral information from midrange sound, mixed sound and vibration, and from pure vibration receptors. The S1 interneuron integrates the activity of mid- and high-frequency sound receptors like the VS1, but it is inhibited by mixed sound and vibration receptors. The S4 and S5 also receive their main inputs from the ipsilateral ear, probably from mid-range receptors, while direct receptor convergence on the S3 is highly unlikely, owing to its long latency and labile response pattern. The same holds true for the V1 interneurons, while the V2 and V3 neurons may receive input from mixed sound and vibration receptors. The V4 type seems to receive input exclusively from campaniform receptors. Nothing is known about the target cells of these interneurons.

The typically large dynamic range, the very low thresholds, and the broad-band characteristics with no pronounced best frequencies (cf. Fig. 37A and B) make these neurons well adapted for carrying information on the typically broad-band bush cricket songs. The phasic response pattern of most of the neurons allow for some syllable coding of the songs, though a thorough examination of the time-coding capabilities has not been undertaken.

The S and VS interneurons respond tonically to the first chirps of a song, provided that the stimulation has been preceded by a longer pause. The strong habituation characteristic for most of these neurons, however, makes them respond only phasically and unspecifically to each chirp when the chirp repetition rate is increased as is the case for the typical bush-cricket song. This unspecific reaction of the habituated interneurons is dramatically changed, enhancing the response and the syllable coding, when vibration stimuli, amplitude-modulated in the pattern of the conspecific song, are presented simultaneously with the air-borne sound stimuli. Unfortunately, nothing is known about the central processing generating this dishabituation.

So the process of enhancing syllable-coding by adding stimuli of another frequency range described above for crickets (section 7.1.2c) is also found in bush crickets. The fact that low-frequency vibration stimuli produce the enhancement in bush crickets, whereas the interaction of low- and high-frequency sound components are effective in crickets, may reflect the different ecological niches of the two groups. Thus, vibration of the plant stem

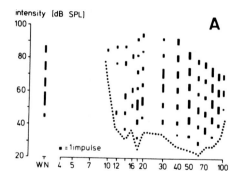

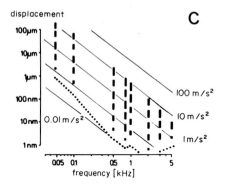

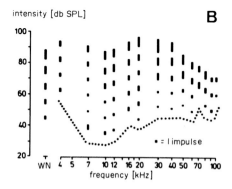

FIG. 37. Threshold curve and response magnitude of auditory interneurons in *Tettigonia cantans*. **A**: Of an S2 interneuron. **B**: A VS4 response to air-borne stimuli over the whole of its frequency range. **C**: The response of the same VS4 to vibration stimuli. (From Silver, S. *et al.*, 1980.)

may indicate the presence of a singing bush cricket on the same plant as the receiver, and intense high-frequency components of the cricket song may indicate that a singing and perhaps visually obscured male is sitting on the ground, close to the listening cricket.

Precise information on sound direction is en-

coded as an asymmetry in the neural response of the bilaterally symmetrical interneurons ascending to centres in the brain and/or descending to the thoracic ganglia as exemplified by the recordings of Rheinlaender, J. and Römer, H. (1980) from the two (presumed) S1 interneurons in *Tettigonia viridissima* (Fig. 38). However, neither the target cells (or network of cells) nor the precise bilateral interactions and enhancing processes are known.

7.3   Acrididae

7.3.1   ANATOMY

The central projections of the locust ear differ from those of crickets in two respects: first, they enter the posterior part of the metathoracic ganglion; second, they are intersegmental (Fig. 23C). The receptor fibres form four acoustic neuropiles (Rehbein, H., 1973; Rehbein, H. *et al.*, 1974): the caudal metathoracic neuropile (about 10% of the fibres branching here — mainly a-receptors, Römer, H., 1976), the large frontal metathoracic neuropile (about 70% of the fibres branching here), the frontal mesothoracic neuropile, and the prothoracic neuropile (about 5% of the fibres terminating here). The receptor fibres occur in at least two different forms, the LF type with branches both in the caudal and in the frontal metathoracic neuropile, and the HF type with metathoracic branches only in the frontal acoustic neuropile (Rehbein, H. *et al.*, 1974).

The auditory ventral-cord interneurons of locusts may be divided into three morphological classes: the intrathoracic interneurons, the ascending interneurones, and the T-shaped interneurons (Rehbein, H., 1976); see Fig. 39. A class of descending interneurons is not found, obviously, because the location of the ear, and consequently the receptor projections, make such a structure superfluous. The presence of omega-shaped interneurons has not been reported either. Except for the intrathoracic interneurons, all the interneurons described below are probably second- or third-order interneurons (Kalmring, K., 1975). On the other hand recent recordings by Römer, H. *et al.* (1981) suggest that at least some of the ascending interneurons receive inputs directly from receptor projections as well as from intrathoracic interneurons.

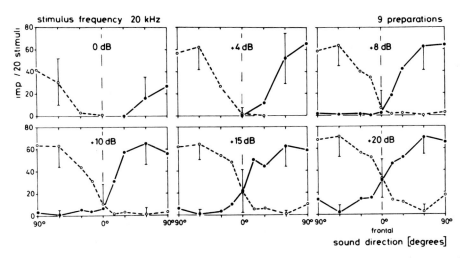

FIG. 38. Averaged directional characteristics of nine pairs of S1 interneurons at different sound intensities above threshold as indicated (stimulus frequency 20 kHz). The number of nerve impulses per stimulus of the left interneuron equals that of the right interneuron, if the animal is facing the sound source (0°), but differs if the sound source is positioned elsewhere relative to the body axis. ○---○, left IN; ●——●, right IN. (From Rheinlaender, J. and Römer, H., 1980).

The auditory ventral-cord interneurons have been classified physiologically in the types A-K (Popov, A., 1967; Kalmring, K., 1971; Kalmring, K. et al., 1972a,b) and VS/S/V (Silver, S. et al., 1980). Only in a few cases, however, has it been possible to unambiguously identify the interneurons. This applies for instance to the B_1 (= S1 — Silver, S. et al., 1980), the B_2, and the G_1, all of which are present bilaterally as one mirror image pair. Other early classifications have since been abolished, e.g. the I type, which turned out to be nothing but the descending part of the T-shaped G_1 interneuron. This example shows that circumspection is necessary when classifying interneurons on physiological data only, even if this is based on laborious "shotgun" approaches like those of Kalmring, K. et al. (1972a,b). This is especially true when new sets of characterization stimuli are introduced (Silver, S. et al., 1980). However, groups of small homologous interneurons may escape unambiguous identification, and for such interneurons their response to a battery of different stimuli combined with the general morphology and input–output relations may be the only realistic classification.

The intrathoracic interneurons are probably of the first order, receiving inputs directly from ipsilateral receptor projections and delivering inhibitory outputs to ascending and T-shaped auditory interneurons on both sides of the ventral cord (Rehbein, H. et al., 1974; Rehbein, H., 1976; Römer, H. et al., 1981). There are probably several types of intrathoracic interneurons (Kalmring, K., 1975), but only one type, "the thoracic low-frequency interneuron", is present as a mirror image pair (shown diagrammatically in Fig. 39), and has been thoroughly studied. This interneuron, with its soma close to the entrance of the tympanic nerve, is thought to receive synaptic inputs from the a-receptor projections in the caudal metathoracic acoustic neuropile. The output sections, probably delivering inhibitory input to bilateral ascending interneurons are located in the other ipsilateral and contralateral thoracic acoustic neuropiles, thus providing a connection between the neuropiles on the two sides of each ganglion.

Ascending auditory interneurons such as the large-diameter B interneurons receive synaptic input in the frontal acoustic neuropile of the metathoracic ganglion (Fig. 39), probably from first-order interneurons. They carry the information to the auditory neuropiles of the supra-oesophageal ganglion, giving off smaller, strictly ipsilateral and medially directed branches in the meso-, pro-, and suboesophageal ganglia (Rehbein, H., 1976).

The T-shaped auditory interneurons exemplified by the large-diameter G_1-interneuron are also shown diagrammatically in Fig. 39. The soma and two branches projecting into the ipsilateral and

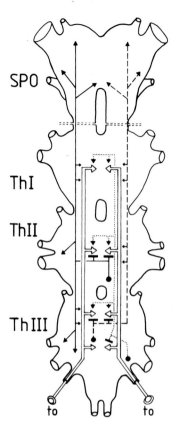

SPO

ThI

ThII

ThIII

to   to

FIG. 39. Diagrammatic representation of the three classes of auditory interneurones relative to the central projections of receptor cells in the ventral cord of *Locusta migratoria*. Each class of interneuron is represented bilaterally in the animal, but for clarity it is drawn unilaterally here. ····, intrathoracic interneurons; ----, ascending interneurons; ——, T-shaped, interneurons; ⊣, input area; →, output area; ⇨, central projections; to, tympanal organ. (Redrawn from Rehbein, H., 1976.)

The acoustic information carried by the three populations of LF and one population of HF receptors flows into the ventral cord, is distributed along the central projections to be integrated in the large ascending and T-shaped auditory interneurons, eventually reaching the brain and the motor neuropiles. The response properties of the receptors and of the ascending and T-shaped interneurons are rather well known from several works using *extracellular* recordings (Römer, H., 1976; Kalmring, K., 1975). However, the response pattern of the ascending and T-shaped interneurons is much more complicated than can be accounted for by simple convergence on the interneurons of different receptor populations either directly or via interneurons of the "repeater type". The first *intracellular* recordings from thoracic auditory interneurons recently made by Römer, H. *et al.* (1981) have helped bridge the gap in our understanding between the receptor properties and the properties of the large interneurons; postsynaptic potentials (PSPs) reveal the activity of intercalated interneurons such as the "intrathoracic low-frequency interneurons". Unfortunately, no stainings were made in this work, so it is difficult to ascribe the findings to any of the morphological types of interneurons mentioned above.

The general principle of central processing of acoustic information emerging from the work of Römer, H. *et al.* (1981) is that the large auditory interneurons receive excitatory input directly from receptor fibres or via interneurons of the "repeater" type (i.e. without much integration), giving rise to a depolarization of the large interneurons which typically outlasts the stimulus duration. In more than 50% of the interneurons, however, this depolarization is superimposed on inhibitory potentials which occur with different durations and time delays relative to the depolarization. The inhibitory postsynaptic potentials are produced by intrathoracic interneurons, such as the "intrathoracic low-frequency interneuron", which relays the activity of the type 1 (a-type) receptors in the caudal metathoracic acoustic neuropile to the large interneurons.

An illustrative example of this mechanism is shown in Fig. 40, where threshold curves of EPSPs

contralateral acoustic neuropiles are located in the mesothoracic ganglion, which appears to be the input section of this giant neuron. From the mesothorax identical information is passed along the soma-contralateral main nerve trunk, caudally to the metathorax and rostrally to the lateral neuropile region of the protocerebrum. In addition to small medially directed branches in the neuropile regions, larger lateral branches are found in the dorsolateral motor neuropiles of the meso- and metathoracic ganglia, probably making synaptic contact with motor neurons (Rehbein, H., 1976).

and IPSPs of a large auditory interneuron strongly suggest the integration of excitatory information directly from type 3 and 4 (c- and d-type) receptors (response latency 10–20 ms) and inhibitory information from the "intrathoracic low-frequency interneuron". From Fig. 40 it appears that the interaction and strength of the excitatory and inhibitory components are determined both by frequency and by intensity, creating three different response areas in the frequency–intensity field: above the curve to the left of the hatched area only inhibitory PSPs occur, in the hatched area both types of PSPs are found, and finally between the curve of open circles and the hatched area only excitatory PSPs occur.

The relative occurrence of EPSPs and IPSPs is not only dependent on frequency and intensity, but also on sound direction as exemplified in Fig. 41A and B. The interneuron in Fig. 41A always receives the IPSP after the onset of the EPSP, i.e. only the number of spikes varies with sound direction. In the

interneuron in Fig. 41B the PSPs occur in the reverse order and the sound direction is reflected both in spike number and in response latency. Clearly, future investigations into the central processing of auditory information in acridids should include intracellular recordings and modern staining techniques, as shown by the strengths and weaknesses of this first experiment.

The role of response latency in directional hearing of locusts has received attention in a series of experiments by Mörchen (Mörchen, A. *et al.*, 1978; Rheinlaender, J. and Mörchen, A., 1979; Mörchen, A., 1980). At the receptor level the response latency decreases linearly with the number of spikes per presentation. The number of spikes in turn is dependent on the vibration amplitude of the tympanal

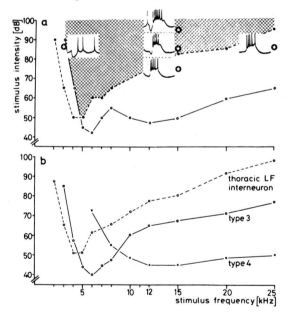

Fig. 40. **a**: Threshold curves of EPSP (solid line) and IPSP (dashed line) of an auditory interneuron in *Locusta migratoria*. The hatched area represents the compound action of EPSP and IPSP of this neuron. The asterisks indicate the frequency– intensity values at which the inserted displays were established. **b**: Threshold curves of type 3 and type 4 receptor fibres and the intrathoracic low-frequency interneurons. Note the congruence between the threshold curves of type 3 and 4 receptors with the EPSP response in **a** and between the tuning of the intrathoracic LF interneuron with the IPSP response. (From Römer, H. *et al.*, 1981.)

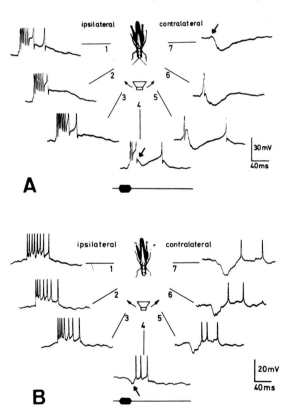

Fig. 41. Intracellular responses of directionally sensitive interneurons at different sound directions, changed in 30° steps in the frontal field of the preparation (sound frequency: 20 kHz; sound intensity: 20 dB above threshold). The stimulus starts at the onset of the trace as indicated in position 4. **A**: An interneuron in which the IPSP appears after the onset of the EPSP; arrow points to first appearance. **B**: An interneuron in which the IPSP starts the EPSP; arrow points to first appearance. (From Römer, H. *et al.*, 1981.)

membrane which is directionally sensitive and therefore — apart from symmetrical stimulation — is different at the two ears. So, although the outer distance between the two ears gives a maximum interaural time difference of the order of 30 μs, the receptor responses of the two ears will arrive at the metathoracic ganglion with a time difference of up to 10 ms, depending on the difference in sound intensity and consequently in vibration amplitude of the two ears. In this way the physical interaural time difference is enhanced by a factor of 300.

In accordance with the findings of Römer, H. et al. (1981) there seem to be three modes of coding sound direction at the interneuronal level (Mörchen, A., 1980). One population of directionally sensitive interneurons codes direction in the same way as the receptors, i.e. by changing both the response latency and the number of spikes, but more drastically than at the receptor level, while the two other populations of directionally sensitive interneurons employ either latency or spike number for coding directionality.

In the large auditory ascending and T-shaped interneurons information from the vibration-sensitive subgenual organs of all six legs and from the two tympanal organs are integrated in much the same way as described for tettigoniids (Čokl, A. et al., 1977; Silver, S. et al., 1980).

The giant interneurons carrying bimodal information in one or two directions over relatively long distances might change the spike sequence along the main axon, thus further integrating the responses. However, simultaneous extracellular recordings at several levels of a given interneuron show that the spike pattern passes unchanged to the brain (Kalmring, K. et al., 1978a). Only at points where the axon diameter is reduced from that of the large axon trunk to that of small side branches can various time-dependent filtering processes occur, e.g. creating reflection potentials (cf. Spira, M. et al., 1976 and Parnas, I. et al., 1976) and changing the activity of the side branch relative to that of the main axon trunk.

7.4 Cicadidae

7.4.1 Anatomy

The ventral nerve cord of the cicadas is condensed into a prothoracic ganglion, and a mesothoracic

ganglion visually almost inseparable from the metathoracic–abdominal ganglionic complex with which it is fused. The auditory receptors project almost exclusively ipsilaterally into this complex, reflecting its metameric structure as shown in Fig. 23D (Wohlers, D. et al., 1979; Doolan, J. and Young, D., 1981; Popov, A., 1981). Here they form an intermediate neuropile extending the length of the complex.

So far, only Huber, F. et al. (1980) have reported successful staining of auditory interneurons in the cicada Okanagana rimosa (Fig. 42). In this ascending type of auditory interneuron a presumably dendritic zone of arborization overlaps the anterior part of the intermediate neuropile in the metathoracic–abdominal ganglionic complex mentioned above. It is not known whether the neuron receives direct input from the receptor fibres. The axon of the interneuron courses through the meso- and prothoracic ganglion with some branching in the mesothoracic ganglion, and may even reach the brain.

7.4.2 Central processing

Since the very first reports on recordings from the tympanal nerve (Pringle, J., 1954; Katsuki, Y. and Suga, N., 1958, 1960), only little progress has been made in understanding nervous processing of sound in the cicadas. This is due partly to difficulties in keeping the animals, and partly to their anatomy.

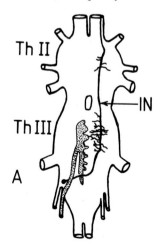

Fig. 42. An ascending auditory interneuron in the ventral cord of the cicada Okanagana rimosa. (Redrawn after Huber, F. et al., 1980.)

Thus, the large number of axons in the tympanal nerve (1300–2000) and their small diameter (less than 1 μm) have impeded reliable recordings from single receptor fibres (Huber, F. *et al.*, 1980; Doolan, J. and Young, D., 1981).

At the receptor level, whole nerve recordings show that the tympanal organ is rather sharply tuned to a frequency within or close to the frequency interval of maximum energy in the continuous, conspecific song (Katsuki, Y. and Suga, N., 1960; Enger, P. *et al.*, 1969; Simmons, J. *et al.*, 1971; Young, D. and Hill, K., 1977). In most instances the threshold curves have been measured at frequencies close to the dominant frequency of the conspecific song, and probably never above 16 kHz. The maximum sensitivity of the receptors (whole nerve recording) varies from about 40 dB SPL in the bladder cicada *Cystosoma saundersii* (Young, D. and Hill, K., 1977) to about 95 dB SPL in the 17-year cicada *Magicicada cassini* (Simmons, J. *et al.*, 1971). In the latter case the high threshold may be an adaptation to the very high sound pressure levels experienced by *M. cassini* males in the centre of an aggregation (about 135 dB — calculated from the data of Simmons, J. *et al.*, 1971).

A strong correlation between the energy spectrum of the tymbal-produced spontaneous song and the threshold curve at the receptor level does not exist in all species. Popov, A. (1981) found that the energy spectrum of the spontaneous song of *Cicadetta sinuatipennis* had its maximum in the frequency range 13–20 kHz. The threshold curve of the tympanic nerve, however, had a peak of sensitivity at about 6 kHz. Consequently, the tympanal nerve did not respond to tymbal sounds below about 70 dB SPL, and it was suggested that such sounds can only be used for intraspecific communication at very close range. However, *C. sinuatipennis* possesses an accessory sound-producing apparatus localized in the forewings. The short, intense clicks produced by rapid movements of the forewings have maximum energy in the frequency range 5–10 kHz, and are therefore well suited for long-range communication. The sharp transients further facilitate the auditory localization of the singer, whereas the rather continuous tymbal sound seems to be much more difficult to localize and it may primarily act as a bird-repellent, as suggested by Simmons, J. *et al.* (1971). It is not known if this

communication system has any significance in other species, but Huber, F. *et al.* (1980) report strong activity in both the auditory nerve and interneurones in response to wing-flipping.

Coding of direction at the receptor level has only been reported by Young, D. and Hill, K. (1977), who measured a pronounced directionality from the tympanal nerve of female *C. saundersii* (ipsilateral/contralateral threshold differences of up to 15 dB), but very little directionality in the ears of the sedentary males. The directionality is a function of frequency, being at a maximum close to the carrier frequency of the continuous, conspecific song.

Coding of time has been investigated by stimulating the ear with clicks or short pulses of the conspecific song, presented with different repetition rates. The rapidly adapting compound action potential of the tympanal nerve can follow repetition rates up to about 100 Hz (Pringle, J., 1954) without changing its shape. At higher frequencies (up to 500 Hz), however, the amplitude of the compound action potential rapidly diminishes after the initial "on"-response (Enger, P. *et al.*, 1969), but there are indications that single receptor cells or groups of receptor cells may follow the sound pulses resulting from the individual tymbal rib movements at such high frequencies (Huber, F. *et al.*, 1980).

Apart from the notion of Simmons, J. *et al.* (1971) that a "sharpening" effect possibly exists at higher levels of the nervous system, the only information on central acoustic processing in cicadas is given by Huber, F. *et al.* (1980). The ascending, non-habituating auditory interneurons recorded in *Magicicada cassini* and *Magicicada septemdecim* respond to sounds in the same frequency range as the receptors, and code the temporal structure of the songs. The interneurons respond vigorously to sharp amplitude modulations of the songs. The interneuron encountered in *M. septemdecim* shows a high degree of species specificity as regards natural songs when presented with the conspecific "calling" and "courtship" songs *versus* those of *M. cassini*. It is, however, not clear how this species specificity is brought about.

We are still ignorant about basic questions such as: why should cicadas have 1300–2000 receptor cells in each ear, when other insects can do with 1–100 auditory receptors? Do all receptors respond in the same way, or are there classes with special responses, which are masked in the whole nerve recordings? Is the cicada ear confined to frequencies in the lower sonic range? Single-fibre recordings from the tympanal nerve might provide some answers.

We still lack the mapping of the central auditory pathways, and an answer to the question: how do the interneurons integrate the information from all the (different?) receptor cells?

In the Cicadidae the physics and physiology of sound production is reasonably well described, whereas quantitative information is still lacking about other kinds of acoustic behaviour such as aggregation, chorusing, mate-attraction, and prey–predator relationships. Since such information provides the ultimate key to understanding the function of the auditory system, quantitative studies on acoustic behaviour of cicadas are urgently needed. Last, but not least, studies at all the different levels should concentrate on a single species to avoid comparing incommensurate properties. The studies on *C. saudersii* (Young, D. and Hill, K., 1977; Young, D., 1980; Doolan, J. and Young, D., 1981; Doolan, J., 1981; Doolan, J. and MacNally, R., 1981) may be a step in this direction.

7.5   Noctuidae

The auditory system of moths and green lacewings (see below) are, in contrast to those of orthopterans and cicadas, adapted exclusively to interspecific communication, i.e. to the detection and avoidance of ultrasound-emitting insectivorous bats. A hunting bat approaching a noctuid moth may be detected by the most sensitive tympanal receptor cells, the A_1s, at a distance of 30–40 m, causing the moth to turn and fly directly away from the bat. Once the bat is so close that the less sensitive receptor cells, the A_2s, are excited, the moth switches from avoidance to evasive flight patterns by diving to the ground in an unpredictable manner. These observations and many important aspects of the underlying central nervous processing have been reported by K. D. Roeder and co-workers in a long series of studies (for reviews see, e.g., Roeder, K., 1975a,b). Treat, A. (1955) demonstrated that even decapitated moths are able to change their flight patterns

in response to pulsed ultrasound. This and other observations brought about the hypothesis that the sound information is processed in two loops: a local one and a more complex one. The local loop connects the tympanal receptors with the motorneurons to the wing muscles, while the more complex loop involves ascending interneurons to the brain, brain neurons, and descending interneurons to modulate the information flow in the local loop. The experiments suggest that two such loops do exist, but the investigations were carried out before modern staining techniques were available, so the number of neurons, their morphology, and connections can only be guessed, and should be reinvestigated with modern techniques.

Below we shall concentrate on noctuid moths, since central processing has not been investigated in other families of moths with tympanal organs sensitive to ultrasound, such as geometrids and species of sphingids (Eggers, F., 1919; Roeder, K., 1974; Roeder, K. *et al.*, 1970).

7.5.1   ANATOMY

In noctuid moths (Noctuidae) the two auditory receptor cells, the A_1 and the A_2, of each ear project into the pterothoracic ganglion dividing into anterior and posterior branches (Paul, D., 1973; Surlykke, A. and Miller, L., 1982) Fig. 23E. The thin A_1 axon, occupying a medioventral position, sends a posterior branch into the abdominal connective, giving off fine branches that appear to cross the midline in the metathoracic ganglion. The anterior branch extends at least as far as the cervical connective. The projections of the A_2 receptor are strictly ipsilateral, terminating in the central portion of the mesothoracic ganglion and the posterior region of the metathoracic ganglion, respectively. In another family of moths (Notodontidae) having only one auditory receptor cell, the projections resemble those of the A_2 in Noctuidae (Surlykke, personal communication). No auditory interneurons have been described anatomically so far. Physiological evidence, however (Roeder, K., 1966, 1969a,b, 1975a,b; Paul, D., 1974) suggests that there are more than four types of thoracic interneurons, some of which ascend to the brain, and at least two types of interneurons descending from the brain.

7.5.2 Central processing

The threshold curves of the two receptor cells are congruent, but shifted about 20 dB relative to one another (Roeder, K. and Treat, A., 1957). Frequency information in the kHz range is therefore lost, coding by phase-locking occurring to only the much slower sound pulse repetition rate and the pulse duration. Sound direction can be determined from the difference in spike information of the two sets of receptors correlated with the wing beat cycle (Payne, R. *et al.*, 1966), while the coding of intensity is expressed by the classical sigmoid intensity curve with a dynamic range of 20–25 dB and by the linearly decreasing latency dependence on sound intensity (see, e.g., Adams, W., 1971).

In the thoracic ganglia four or five types of auditory interneurons have been described physiologically (Roeder, K., 1966; Paul, D., 1974). The evidence suggests that the posterior ventromedial mesothoracic ganglion the A_1 projection make synaptic contacts with some first-order interneurons, the *repeater neurons*, which appear to be relatively large-diameter neurons coursing side by side with the A_1 projections to the cervical connectives and on to the brain. Repeater neurons also cross the midline carrying ipsilateral information to the contralateral side of the thoracic ganglia, but they probably do not receive information from the contralateral ear. The term, repeater neuron, refers to the fact that integration of the A_1 spike-train is minimal, though the repetition is not exactly 1:1, since the phase-locking is not exactly 100%, and its adaptation to long sound pulses is larger than that of the A_1. The conduction velocity of the repeater neurons is about twice that of the A_1, and their probable main function is therefore to secure a quick distribution of A_1 information to the brain and to wide areas of the thoracic ganglia. This will accentuate the leading "edge" of spike trains resulting from long sound pulses or rapid successions of short pulses. The repeater neurons can be classified as stable or labile followers according to their ability to follow the sound pulse repetition rate. The stable followers have the shortest latency and will follow pulse repetition rates of above 10 Hz, whereas the labile followers are unable to respond to sound pulses repeated above 5 Hz. The labile followers may be involved in the process of distinguishing

between ultrasonic friends and foes, since 5 Hz corresponds to the lower limit of ultrasonic pulse repetition rates emitted by cruising bats (Paul, D., 1974).

The second type of interneuron is the *pulse-marker* localized in the ipsilateral part of the mesothoracic ganglion. In the pulse-marker interneurons some integration of the A_1 information takes place. A train of spikes in the A_1 projections generates only one spike in the pulse-marker, and only on condition that the interspike interval is less than 2 ms and the A_1 train has been preceded by a "quiet" period without spikes. However, the pulse-marker may be brought to fire repetitively if the ear is stimulated with a train of short pulses with pulse repetition rates of up to 40 Hz. In the pulse-markers therefore, not only frequency but also intensity and pulse-length information is lost, leaving only information on sound pulse repetition rate and pulse train length. Also, the pulse-marker neurons may be divided into groups of labile and stable ones using the same criteria as for the repeaters. The labile pulse-markers are likely to be second-order neurons, since their latency is more than 2 ms larger than that of the stable ones.

The pulse-markers may be involved in the local loop controlling the activity of motorneurons supplying the direct flight muscles concerned with steering during flight (Roeder, K., 1966, 1967b; Paul, D., 1974), while the repeater neurons may constitute the ascending part of the more complex loop.

Roeder, K. (1969a,b) followed repeater neurons ascending to the protocerebral ganglia, noting that unilateral inputs were carried via bilateral interneurons through the suboesophageal ganglion to the brain. One type of integration was found in a small tonic unit which encodes intensity over a range of 40 dB by increasing the number of spikes per sound stimulus almost linearly with intensity. In contrast, a large unit only increased the number of spikes over an intensity range of 20 dB. At higher intensities the number of spikes decreased and the response was entirely suppressed at 40 dB above threshold. The experiments suggest that inputs from A_1 and A_2 receptors converge as excitatory PSPs on the small unit, thus extending its dynamic range, whereas in the large unit only the A_1-receptor input is excitatory while the A_2-receptor input is inhibitory. A second indication of integration in the ascending interneurons was the observation of "lapses" in the response, i.e. the sensitivity abruptly reduced by 10–20 dB or more, the "lapses" occurring with unpredictable frequency and duration. The "lapses" could be induced by visual stimulation, but their physiological background and significance are unknown. A third type of integration at this level was the combination of inputs from both tympanic organs, showing a directional sensitivity with the ipsilateral side 3–4 dB more sensitive than the contralateral one.

The modulating effect of the more complex loop on the local one could be expected to consist of a sustained influence on the activity initiated by the local loop, making it continue some time after the cessation of the initial threat stimulus (Roeder, K., 1975b). Sustained activity with some correlation to the sound stimuli was actually found in descending interneurons from the brain (Roeder, K., 1975a,b). One type of unit fired a prolonged afterdischarge, lasting for more than 100 ms in response to repeater spike-trains produced by a 10 ms sound pulse. A second type of descending interneuron showed the inverse pattern, having a free-running spike activity, which was suppressed for 200–600 ms after the repeater spike-train.

It should be emphasized that many of the considerations on different loops are highly speculative, and must await confirmation from future experiments applying the modern staining and recording techniques.

7.6 Chrysopidae

The auditory system of green lacewings (*Chrysopa* sp.) is adapted to the detection of insectivorous bats, since in the presence of hunting bats or of pulsed ultrasound alone, they will dive to the ground in an unpredictable manner (Roeder, K., 1962; Miller, L., 1975; Miller, L. and Olesen, J., 1979; Olesen, J. and Miller, L., 1979).

The central projections of auditory receptor fibres are unknown, and so is the morphology of auditory interneurons.

At the receptor level pulsed sound in the frequency range 13–120 kHz is represented by spike-trains that can code the intensity of pulses as short as 1 ms, and with repetition rates of up to 150 Hz (Miller, L., 1971).

Among presumedly first-order auditory interneurons, *repeater neurons* have been described physiologically (Miller, L. and Olesen, J., 1978), having tunings and sensitivities similar to the receptor fibres and producing two to five spikes per 10 ms sound pulse. It is not known whether these interneurons alone convey auditory information to uncouple the motor neurons involved in wing folding and "last-time maneuvers" (Miller, L. and Olesen, J., 1979).

7.7 Other insects

Virtually nothing is known about the central processing of sound information in the large number of other insect families possessing ears and displaying a distinct acoustical behaviour. This applies to major insect orders such as Hemiptera (bugs, water boatmen, and pond skaters), Coleoptera, and Diptera (mosquitoes and fruit-flies).

ACKNOWLEDGEMENTS

The authors are grateful to Dr B. Lewis for critically reading this manuscript, and to C. Naesbye Larsen for redrawing some of the figures.

REFERENCES

Adams, W. B. (1971). Intensity characteristics of the noctuid acoustic receptor. *J. Gen. Physiol. 58*, 562–579.
Arntz, B. (1975). Das Hörvermögen von *Nepa cinerea* L. Zur Funktionsweise der thorakalen Scolopalorgane. *J. Comp. Physiol. 96*, 53–72.
Autrum, H. (1940). Über Lautäusserungen und Schallwahrnehmung bei Arthropoden. II. Das Richtungshören von *Locusta* und Versuch einer Hörtheorie für Tympanalorgane vom Locustidentyp. *Z. Vergl. Physiol. 28*, 326–352.
Averhoff, W. W., Ehrman, L., Leonard, J. E. and Richardson, R. H. (1979). Antennal signal receptors in *Drosophila* mating. *Biol. Zbl. 98*, 1–12.
Ball, E. E. and Hill, K. G. (1978). Functional development of the auditory system of the cricket, *Teleogryllus commodus. J. Comp. Physiol. 127*, 131–138.
Bennet-Clark, H. C. (1970). The mechanism and efficiency of sound production in mole crickets. *J. Exp. Biol. 52*, 619–652.
Bennet-Clark, H. C. (1971). Acoustics of insect song. *Nature* (Lond.). *234*, 255–259.
Bennet-Clark, H. C. and Ewing, A. W. (1970). The love song of the fruit fly. *Sci. Amer. 223*, 84–92.
Bentley, D. R. (1977). Control of cricket song patterns by descending interneurons. *J. Comp. Physiol. 116*, 19–38.
Beranek, L. L. (1954). *Acoustics.* McGraw-Hill, New York.
Boyan, G. S. (1978). Coding of directional information by a descending interneuron in the auditory system of the cricket. *Naturwissenschaften 65*, 212–213.

BOYAN, G. S. (1979a). Directional responses to sound in the central nervous system of the cricket *Teleogryllus commodus* (Orthoptera: Gryllidae). I. Ascending interneurons. *J. Comp. Physiol.* 130, 137–150.

BOYAN, G. S. (1979b). Directional responses to sound in the central nervous system of the cricket *Teleogryllus commodus* (Orthoptera: Gryllidae). II. A descending interneuron. *J. Comp. Physiol.* 130, 151–159.

BOYAN, G. S. (1980). Auditory neurons in the brain of the cricket *Gryllus bimaculatus* (De Geer). *J. Comp. Physiol.* 140, 81–93.

BOYAN, G. S. (1981). Two-tone suppression of an identified auditory neuron in the brain of the cricket *Gryllus bimaculatus* (De Geer) *J. Comp. Physiol.* 144, 117–125.

BOYAN, G. S. and WILLIAMS, J. L. D. (1981). Descending interneurons in the brain of the cricket. *Naturwissenschaften* 67, 486.

BURROWS, M. and SIEGLER, M. V. S. (1976). Transmission without spikes between locust interneurons and motoneurons. *Nature* 262, 222–224.

BURROWS, M. and SIEGLER, M. V. S. (1978). Graded synaptic transmission between local interneurons and motor neurons in the metathoracic ganglion of the locust. *J. Physiol.* 285, 231–255.

BUSNEL, R. -G. (Ed., 1963). *Acoustic Behaviour of Animals.* Elsevier, Amsterdam.

CASADAY, G. B. and HOY, R. R. (1977). Auditory interneurons in the cricket *Teleogryllus oceanicus*: Physiological and anatomical properties. *J. Comp. Physiol.* 121, 1–13.

ČOKL, A., KALMRING, K. and WITTIG, H. (1977). The responses of auditory ventral-cord neurons of *Locusta migratoria* to vibration stimuli. *J. Comp. Physiol.* 120, 161–172.

DAMBACH, M. (1972). Vibrationssinn der Grillen. I. Schwellenmessungen an Beinen freibeweglicher Tiere. II. Antworten von Neuronen im Bauchmark. *J. Comp. Physiol.* 79, 281–324.

DONATO, R. J. (1976). Propagation of a spherical wave near a plane boundary with complex impedance. *J. Acoust. Soc. Amer.* 60, 34–39.

DOOLAN, J. M. (1981). Male spacing and the influence of female courtship behaviour in the bladder cicada, *Cystosoma saundersii*. *Behav. Ecol. Sociobiol.* 9, 269–276.

DOOLAN, J. M. and MACNALLY, R. C. (1981). Spatial dynamics and breeding ecology in the cicada, *Cystosoma saundersii*: the interaction between distribution of resources and intraspecific behaviour. *J. Anim. Ecol.* 50, 925–940.

DOOLAN, J. M. and YOUNG, D. (1981). The organization of the auditory organ of the bladder cicada, *Cystosoma saundersii*. *Phil. Trans. Roy. Soc. B* 291, 525–540.

DRAGSTEN, P. R., WEBB, W. W., PATON, J. A. and CAPRANICA, R. R. (1974). Auditory membrane vibrations: measurements at sub-Angstrom levels by optical heterodyne spectroscopy. *Science* 185, 55–57.

DUMORTIER, B. (1963). Ethological and physiological study of sound emissions in Arthropoda. In *Acoustic Behaviour of Animals* Edited by R.-G. Busnel. Pages 583–654. Elsevier, Amsterdam.

EGGERS, F. (1919). Das thoracale bitympanale Organ einer Gruppe der Lepidoptera Heterocera. *Zool. Jb.* 41, 273–276.

EHRET, G., MOFFAT, A. J. M. and TAUTZ, J. (1982). Behavioural Determination of Frequency Resolution in the Ear of the cricket. *Teleogryllus oceanicus*. *J. Comp. Physiol.* 148, 237–244.

EIBL, E. and HUBER, F. (1979). Central projections of tibial sensory fibres within the three thoracic ganglia of crickets (*Gryllus campestris* L., *Gryllus bimaculatus* DeGeer). *Zoomorphologie* 92, 1–17.

ELEPHANDT, A. and POPOV, A. V. (1979). Auditory interneurons in the mesothoracic ganglion of crickets. *J. Insect Physiol.* 25, 429–441.

ELSNER, N. and POPOV, A. V. (1978). Neuroethology of acoustic communication. *Adv. Insect Physiol.* 13, 229–355.

ENGER, P. S., AIDLEY, D. J. and SZABO, T. (1969). Sound reception in the Brazilian cicada *Fidicina rana* Walk. *J. Exp. Biol.* 51, 339–345.

ESCH, H., HUBER, F. and WOHLERS, D. W. (1980). Primary auditory neurons in crickets: physiology and central projections. *J. Comp. Physiol.* 137, 27–38.

FLETCHER, N. H. (1978). Acoustic response of hair receptors in insects. *J. Comp. Physiol.* 127, 185–189.

FLETCHER, N. H. and THWAITES, S. (1979). Acoustical analysis of the auditory system of the cricket *Teleogryllus commodus* (Walker). *J. Acoust. Soc. Amer.* 62, 350–357.

FRIEDRICH, H. (1930). Weitere vergleichende Untersuchungen über die tibialen Scolopalorgane bei Orthopteren. *Z. Wiss. Zool.* 137, 30–54.

GRAY, E. G. (1960). The fine structure of the insect ear. *Phil. Trans. Roy. Soc. B.* 243, 75–94.

HILL, K. G. (1974). Carrier frequency as a factor in phonotactic behaviour of female crickets (*Teleogryllus commodus*). *J. Comp. Physiol.* 93, 7–18.

HILL, K. G. and BOYAN, G. S. (1977). Sensitivity to frequency and direction of sound in the auditory system of crickets. *J. Comp. Physiol.* 121, 79–97.

HILL, K. G., LEWIS, D. B., HUTCHINGS, M. E. and COLES, R. B. (1980). Directional hearing in the Japanese quail (*Coturnix coturnix japonica*). I. Acoustic properties of the auditory system. *J. Exp. Biol.* 86, 135–151.

HILL, K. G. and OLDFIELD, B. P. (1981). Auditory function in Tettigoniidae (*Orthoptera: Ensifera*) *J. Comp. Physiol.* 142, 169–180.

HUBER, F. (1970). Nervöse Grundlagen der akustischen Kommunikation bei Insekten *Rheinisch-Westfäl. Akad. Wiss.*, Vorträge 205. Pages 41–91. Westdeutscher Verlag, Opladen.

HUBER, F. (1974). Neural integration (central nervous system). In *The Physiology of Insecta.* Edited by M. Rockstein. Vol. IV, pages 3–100. Academic Press, New York and London.

HUBER, F. (1977). Lautäusserungen und Lauterkennen bei Insekten (Grillen). *Rheinisch-Westfäl. Akad. Wiss.*, Vorträge 265. Pages 15–66. Westdeutscher Verlag, Leverkusen.

HUBER, F. (1978). The insect nervous system and insect behaviour. *Anim. Behav.* 26, 969–981.

HUBER, F., WOHLERS, D. W. and MOORE, T. E. (1980). Auditory nerve and interneurone responses to natural sounds in several species of cicadas. *Physiol. Ent.* 5, 25–45.

HUTCHINGS, M. and LEWIS, B. (1981). Response properties of primary auditory fibers in the cricket *Teleogryllus oceanicus* (Le Guillou). *J. Comp. Physiol.* 143, 129–134.

JOHNSTONE, B. M., SAUNDERS, J. C. and JOHNSTONE, J. R. (1970). Tympanic membrane response in the cricket. *Nature* 227, 625–626.

JOSEPHSON, R. K. and YOUNG, D. (1981). Synchronous and asynchronous muscles in cicadas. *J. Exp. Biol.* 91, 219–237.

KALMRING, K. (1971). Akustische Neuronen im Unterschlundganglion der Wanderheuschrecke *Locusta migratoria*. *Z. Vergl. Physiol.* 72, 95–110.

KALMRING, K. (1975). The afferent auditory pathway in the ventral cord of *Locusta migratoria* (Acrididae). I. Synaptic connectivity and information processing among auditory neurons of the ventral cord. *J. Comp. Physiol.* 104, 103–141.

KALMRING, K. and KÜHNE, R. (1980). The coding of airborne sound and vibration signals in bimodal ventral-cord neurons of the grasshopper *Tettigonia cantans*. *J. Comp. Physiol.* 139, 267–275.

KALMRING, K., RHEINLAENDER, J. and REHBEIN, H. (1972a). Akustische Neuronen im Bauchmark der Wanderheuschrecke *Locusta migratoria*. *Z. Vergl. Physiol.* 76, 314–332.

KALMRING, K., RHEINLAENDER, J. and RÖMER, H. (1972b). Akustische Neuronen im Bauchmark von *Locusta migratoria*. Der Einfluss der Schallrichtung auf die Antwortmuster. *J. Comp. Physiol.* 80, 325–352.

KALMRING, K., KÜHNE, R. and MOYSICH, F. (1978a). The auditory pathway in the ventral cord of the migratory locust (*Locusta migratoria*). Response transmission in axons. *J. Comp. Physiol.* 126, 25–33.

KALMRING, K., LEWIS, B. and EICHENDORF, A. (1978b). The physiological characteristic of the primary sensory neurons of the complex tibial organ of *Decticus verrucivorus* L. (Orthoptera, Tettigoniidae). *J. Comp. Physiol.* 127, 109–121.

KALMRING, K., REHBEIN, H. and KÜHNE, R. (1979). An auditory giant neuron in the ventral cord of *Decticus verrucivorus* (Tettigoniidae). *J. Comp. Physiol.* 132, 225–234.

KÄMPER, G. and DAMBACH, M. (1981). Responses of the cercus-to-giant interneuron system in crickets to species-specific song. *J. Comp. Physiol.* 141, 311–317.

KATSUKI, Y. and SUGA, N. (1958). Electrophysiological studies on hearing in common insects in Japan. *P. Japan. Acad.* 34, 633–638.

KATSUKI, Y. and SUGA, N. (1960). Neural mechanisms of hearing in insects. *J. Exp. Biol.* 37, 279–290.

KENNEL, J. V. and EGGERS, F. (1933). Die abdominalen Tympanalorgane der Lepidopteren. *Zool. Jahrb, Abt. Anat.* 57, 1–104.

KIANG, N. (1966). *Discharge Patterns of Single Fibres in the Cat's Auditory Nerve*. MIT Press, Massachusetts.

KLEINDIENST, H. U. (1980). *Biophysikalische Untersuchungen am Gehörsystem von Feldgrillen*. Dissertation, Bonn University.

KLEINDIENST, H. U., KOCH, U. T. and WOHLERS, D. W. (1981). Analysis of the cricket auditory system by acoustic stimulation using a closed sound field. *J. Comp. Physiol.* 141, 283–296.

KÜHNE, R., LEWIS, B. and KALMRING, K. (1980). The responses of ventral cord neurons of *Decticus verrucivorus* (L) to sound and vibration stimuli. *Behav. Proc.* 5, 55–74.

LARSEN, O. N. (1981). Mechanical time resolution in some insect ears. II. Impulse sound transmission in acoustic tracheal tubes. *J. Comp. Physiol.* 143, 297–304.

LARSEN, O. N. and MICHELSEN, A. (1978). Biophysics of the ensiferan ear. III. The cricket ear as a four-input system. *J. Comp. Physiol.* 123, 217–227.

MACNALLY, R. and YOUNG, D. (1981). Song energetics of the bladder cicada, *Cystosoma saundersii. J. Exp. Biol.* 90, 185–196.

MARKL, H. (1968). Die Verständigung durch Stridulationssignale bei Blattschneider-ameisen. II. Erzeugung und Eigenschaften der Signale. *Z. Vergl. Physiol.* 60, 103–150.

MARKL, H. (1973). Leistungen des Vibrationssinnes bei Wirbellosen Tieren. In *Fortschritte der Zoologie*. Vol. 21, pages 100–120. Gustav Fischer Verlag, Stuttgart.

MARTENS, M. J. M. and MICHELSEN, A. (1981). Absorption of acoustic energy by plant leaves. *J. Acoust. Soc. Amer.* 69, 303–306.

MCKAY, J. M. (1969). The auditory system of *Homorocoryphus* (Tettigonioidea, Orthoptera). *J. Exp. Biol.* 51, 787–802.

MCKAY, J. M. (1970). Central control of an insect sensory interneuron. *J. Exp. Biol.* 53, 137–145.

MICHEL, K. (1974). Das Tympanalorgan von *Gryllus bimaculatus* Degeer (Saltatoria, Gryllidae). *Z. Morph. Tiere* 77, 285–315.

MICHEL, K. (1975). Das Tympanalorgan von *Cicada orni* L. (Cicadina, Homoptera). Eine Licht- und elektronmikroskopische Untersuchung. *Zoomorphologie* 82, 63–78.

MICHELSEN, A. (1966). Pitch discrimination in the locust ear: observations on single sense cells. *J. Insect Physiol.* 12, 1119–1131.

MICHELSEN, A. (1971). The physiology of the locust ear. I. Frequency sensitivity of single cells in the isolated ear. II. Frequency discrimination based upon resonances in the tympanum. III. Acoustical properties of the intact ear. *Z. Vergl. Physiol.* 71, 49–128.

MICHELSEN, A. (1973). The mechanics of the locust ear: an invertebrate frequency analyzer. In *Mechanisms in Hearing*. Edited by Å. Møller. Pages 911–934. Academic Press, London.

MICHELSEN, A. (1978). Sound reception in different environments. In *Sensory ecology, review and perspectives*. Edited by M. A. Ali. Pages 345–373. Plenum Press, New York.

MICHELSEN, A. (1979). Insect ears as mechanical systems. *Amer. Sci.* 67, 696–706.

MICHELSEN, A. (1982). Laser techniques in studies of hearing. In *Biomedical Applications of Laser-light Scattering*. Edited by D. B. Satelle, W. J. Lee, and B. R. Ware. Pages 357–370. Elsevier, Amsterdam.

MICHELSEN, A. and LARSEN, O. N. (1978). Biophysics of the ensiferan ear. I. Tympanal vibrations in bushcrickets (Tettigoniidae) studied with laser vibrometry. *J. Comp. Physiol.* 123, 193–203.

MICHELSEN, A. and LARSEN, O. N. (1983). Strategies for acoustic communication in complex environments. In *Neuroethology and behavioral physiology*. Edited by F. Huber and H. Markl. Pages 321–331. Springer, Heidelberg.

MICHELSEN, A. and NOCKE, H. (1974). Biophysical aspects of sound communication in insects. *Adv. Insect Physiol.* 10, 247–296.

MICHELSEN, A., FINK, F., GOGALA, M. and TRAUE, D. (1982). Plants as transmission channels for insect vibrational songs. *Behav. Ecol. Sociobiol.* 11, 269–281.

MILLER, L. A. (1970). Structure of the green lacewing tympanal organ (*Chrysopa carnea*, Neuroptera). *J. Morphol.* 131, 359–382.

MILLER, L. A. (1971). Physiological responses of green lacewings (*Chrysopa*, Neuroptera) to ultrasound. *J. Insect Physiol.* 17, 491–506.

MILLER, L. A. (1975). The behaviour of flying green lacewings, *Chrysopa carnea*, in the presence of ultrasound. *J. Insect Physiol.* 21, 205–219.

MILLER, L. A. (1977). Directional hearing in the locust *Schistocerca gregaria* Forskål (Acrididae, Orthoptera). *J. Comp. Physiol.* 119, 85–98.

MILLER, L. A. and OLESEN, J. (1978). Neuronal activity and avoidance behaviour in green lacewings. *Soc. Neurosci. Abst.* 4, 363.

MILLER, L. A. and OLESEN, J. (1979). Avoidance behaviour in green lacewings. I. Behaviour of free flying green lacewings to hunting bats and ultrasound. *J. Comp. Physiol.* 131, 113–120.

MOISEFF, A., POLLACK, G. S. and HOY, R. R. (1978). Steering responses of flying crickets to sound and ultrasound: Mate attraction and predator avoidance. *Proc. Natl. Acad. Sci. USA* 75, 4052–4056.

MORSE, P. M. (1948). *Vibration and Sound*. McGraw-Hill, New York.

MØHL, B. and MILLER, L. A. (1976). Ultrasonic clicks produced by the peacock butterfly: a possible bat-repellent mechanism. *J. Exp. Biol.* 64, 639–644.

MÖRCHEN, A. (1980). Spike count and response latency. Two basic parameters encoding sound direction in the CNS of insects. *Naturwissenschaften* 67, 469–470.

MÖRCHEN, A., RHEINLAENDER, J. and SCHWARTZKOPFF, J. (1978). Latency shift in insect auditory nerve fibers. A neuronal time cue of sound direction. *Naturwissenschaften* 65, 656.

NELSON, M. C. (1979). Sound production in the cockroach, *Gromphadorhina portentosa*: the sound-producing apparatus. *J. Comp. Physiol.* 132, 27–38.

NOCKE, H. (1971). Biophysik der Schallerzeugung durch die Vorderflügel der Grillen. *Z. Vergl. Physiol.* 74, 272–314.

NOCKE, H. (1972). Physiological aspects of sound communication in crickets (*Gryllus campestris* L.). *J. Comp. Physiol.* 80, 141–162.

NOCKE, H. (1975). Physical and physiological properties of the tettigoniid ("grasshopper") ear. *J. Comp. Physiol.* 100, 25–57.

OLDFIELD, B. P. (1980). Accuracy of orientation in female crickets, *Teleogryllus oceanicus* (Gryllidae): Dependence on song spectrum. *J. Comp. Physiol.* 141, 93–99.

OLESEN, J. and MILLER, L. A. (1979). Avoidance behaviour in green lacewings. II. Flight muscle activity. *J. Comp. Physiol.* 131, 121–128.

OTTO, D. (1971). Untersuchungen zur zentralnervösen Kontrolle der Lauterzeugung von Grillen. *Z. Vergl. Physiol.* 74, 227–271.

PARNAS, I., HOCHSTEIN, S. and PARNAS, H. (1976). Theoretical analysis of parameters leading to frequency modulation along an inhomogeneous axon. *J. Neurophysiol.* 39, 909–923.

PATON, J. A., CAPRANICA, R. R., DRAGSTEN, P. R. and WEBB, W. W. (1977). Physical basis for auditory frequency analysis in field crickets (Gryllidae). *J. Comp. Physiol.* 119, 221–240.

PAUL, D. H. (1973). Central projections of the tympanic fibres in noctuid moths. *J. Insect Physiol.* 19, 1785–1792.

PAUL, D. H. (1974). Responses to acoustic stimulation of thoracic interneurons in noctuid moths. *J. Insect Physiol.* 20, 2205–2218.

PAUL, R. C. and WALKER, T. J. (1979). Arboreal singing in a burrowing cricket, *Anurogryllus arboreus. J. Comp. Physiol.* 132, 217–223.

PAYNE, R. S., ROEDER, K. D. and WALLMAN, J. (1966). Directional sensitivity of the ears of noctuid moths. *J. Exp. Biol.* 44, 17–31.

PEARSON, K. G. and FOURTNER, C. R. (1975). Nonspiking interneurons in the walking system of the cockroach. *J. Neurophysiol.* 38, 33–52.

POLLACK, G. S. and HOY, R. R. (1981). Phonotaxis in flying crickets: Neural correlates. *J. Insect Physiol.* 27, 41–45.

POPOV, A. V. (1967). Synaptic transmission at the level of the first synapses of the auditory system in *Locusta migratoria*. In *Evolutionary Neurophysiology and Neurochemistry*. (In Russian.) Edited by E. M. Kreps. Pages 55–67. Nauka, Leningrad.

POPOV, A. V. (1969). Comparative analysis of sound signals and some principles of auditory system organization in cicadas and Orthoptera. In *Modern Problems of Structure and Function of the Nervous System of Insects*. (In Russian.) Pages 182–221. Nauka, Leningrad.

POPOV, A. V. (1973). Frequency selectivity of the reaction of auditory neurons in the 1st thoracic ganglion of the cricket *Gryllus bimaculatus* DeGeer. (In Russian.) *Zh. Evol. Biokim. Fiziol.* 9, 265–277.

POPOV, A. V. (1981). Sound production and hearing in the cicada, *Cicadetta sinuatipennis* Osh. (Homoptera, Cicadidae). *J. Comp. Physiol.* 142, 271–280.

POPOV, A. V. and SHUVALOV, V. F. (1974). Time-characteristics of communication sounds and their analysis in the auditory system of crickets. *Acustica* 31, 315–319.

POPOV, A. V., MARKOVICH, A. M. and ANDJAN, A. S. (1978). Auditory neurons in the prothoracic ganglion of the cricket, *Gryllus bimaculatus* DeGeer. I. The large segmental auditory neuron (LSAN). *J. Comp. Physiol.* 126, 183–192.

PRAGER, J. (1976). Das mesothorakale Tympanalorgan von *Corixa punctata* Ill. (Heteroptera, Corixidae). *J. Comp. Physiol.* 110, 33–50.

PRAGER, J. and LARSEN, O. N. (1981). Asymmetrical hearing in the water bug *Corixa punctata* observed with laser vibrometry. *Naturwissenschaften 68*, 579.

PRESTWICH, K. N. and WALKER, T. J. (1981). Energetics of singing in crickets: Effects of temperature in three trilling species (*Orthoptera: Gryllidae*). *J. Comp. Physiol.* 143, 199–212.

PRINGLE, J. W. S. (1954). A physiological analysis of cicada song. *J. Exp. Biol. 31*, 525–560.

PROZESKY-SCHULZE, L., PROZESKY, O. P. M., ANDERSON, F. and VAN DER MERWE, G. J. J. (1975). Use of a selfmade sound baffle by a tree cricket. *Nature 255*, 142–143.

PUMPHREY, R. J. (1940). Hearing in insects. *Biol. Rev. 15*, 107–132.

REHBEIN, H. (1973). Experimentell-anatomische Untersuchungen über den Verlauf der Tympanalnervenfasern im Bauchmark von Feldheuschrecken, Laubheuschrecken und Grillen. *Verh. Dtsch. Zool. Ges. 66*, 184–189.

REHBEIN, H. (1976). Auditory neurons in the ventral cord of the locust: Morphological and functional properties. *J. Comp. Physiol.* 110, 233–250.

REHBEIN, H., KALMRING, K. and RÖMER, H. (1974). Structure and function of acoustic neurons in the thoracic ventral nerve cord of *Locusta migratoria* (Acrididae). *J. Comp. Physiol. 95*, 263–280.

RHEINLAENDER, J. (1975). Transmission of acoustic information at three neuronal levels in the auditory system of *Decticus verrucivorus* (*Tettigoniidae, Orthoptera*). *J. Comp. Physiol. 97*, 1–53.

RHEINLAENDER, J. and KALMRING, K. (1973). Die afferente Hörbahn im Bereich des Zentralnervensystems von *Decticus verrucivorus* (Tettigoniidae). *J. Comp. Physiol. 85*, 361–410.

RHEINLAENDER, J., KALMRING, K. POPOV, A. V. and REHBEIN, H. (1976). Brain projections and information processing of biologically significant sounds by two large ventral-cord neurons of *Gryllus bimaculatus* DeGeer (Orthoptera, Gryllidae). *J. Comp. Physiol. 110*, 251–269.

RHEINLAENDER, J., KALMRING, K. and RÖMER, H. (1972). Akustische Neuronen mit T-Struktur im Bauchmark von Tettigoniiden. *J. Comp. Physiol. 77*, 208–224.

RHEINLAENDER, J. and MÖRCHEN, A. (1979). "Time-intensity trading" in locust auditory interneurones. *Nature 281*, 672–674.

RHEINLAENDER, J. and RÖMER, H. (1980). Bilateral coding of sound direction in the CNS of the bushcricket *Tettigonia viridissima* L. *J. Comp. Physiol. 140*, 101–111.

RISLER, H. and SCHMIDT, K. (1967). Der Feinbau der Scolopidien im Johnstonschen Organ von *Aëdes aegypti* L. *Z. Naturf. 22b*, 759–762.

ROEDER, K. D. (1962). The behaviour of free flying moths in the presence of artificial ultrasonic pulses. *Anim. Behav. 10*, 300–304.

ROEDER, K. D. (1966). Interneurons of the thoracic nerve cord activated by tympanic nerve fibres in noctuid moths. *J. Insect Physiol. 12*, 1227–1244.

ROEDER, K. D. (1967a). *Nerve Cells and Insect Behaviour.* 2nd edn. Harvard University Press.

ROEDER, K. D. (1967b). Turning tendency of moths exposed to ultrasound while in stationary flight. *J. Insect Physiol. 13*, 873–888.

ROEDER, K. D. (1969a). Acoustic interneurons in the brain of noctuid moths. *J. Insect Physiol. 15*, 825–838.

ROEDER, K. D. (1969b). Brain interneurons in noctuid moths: Differential suppression by high sound intensities. *J. Insect Physiol. 15*, 1713–1718.

ROEDER, K. D. (1972). Acoustic and mechanical sensitivity of the distal lobe of the pilifer in choerocampine hawkmoths. *J. Insect. Physiol. 18*, 1249–1264.

ROEDER, K. D. (1974). Responses of less sensitive acoustic sense cells in tympanic organs of some noctuid and geometrid moths. *J. Insect Physiol. 20*, 55–66.

ROEDER, K. D. (1975a). Neural transactions during acoustic stimulation of noctuid moths. In *Sensory Physiology and Behaviour.* Edited by R. Galun, P. Hillman, I. Parnas and R. Werman. Pages 99–115. Plenum, New York.

ROEDER, K. D. (1975b). Neural factors and evitability in insect behaviour. *J. Exp. Zool. 194*, 75–88.

ROEDER, K. D. and TREAT, A. E. (1957). Ultrasonic reception by the tympanic organ of noctuid moths. *J. Exp. Zool. 134*, 127–157.

ROEDER, K. D., TREAT, A. E. and BERGE, J. S. V. (1970). Distal lobe of the pilifer: an ultrasonic receptor in choerocampine hawkmoths. *Science 170*, 1098–1099.

ROTH, L. M. (1948). A study of mosquito behaviour. An experimental laboratory study of the sexual behaviour of *Aëdes aegypti* (Linnaeus). *Amer. Midland Naturalist 40*, 265–352.

RÖMER, H. (1976). Die Informationsverarbeitung tympanaler Rezeptorelemente von *Locusta migratoria* (Acrididae, Orthoptera). *J. Comp. Physiol. 109*, 101–122.

RÖMER, H., RHEINLAENDER, J. and DRONSE, R. (1981). Intracellular studies on auditory processing in the metathoracic ganglion of the locust. *J. Comp. Physiol. 144*, 305–312.

SALES, G. and PYE, D. (1974). *Ultrasonic Communication by Animals.* Chapman & Hall, London.

SCHIOLTEN, P., LARSEN, O. N. and MICHELSEN, A. (1981). Mechanical time resolution in some insect ears. I. Impulse responses and time constants. *J. Comp. Physiol. 143*, 289–295.

SCHMITZ, B., SCHARSTEIN, H. and WENDLER, G. (1982). Phonotaxis in *Gryllus campestris* L. (Orthoptera, Gryllidae). I. Mechanism of acoustic orientation in intact female crickets. *J. Comp. Physiol. 148*, 431–444.

SCHUMACHER, R. (1975). Scanning-electron-microscope description of the tibial tympanal organ of the *Tettigonioidea* (*Orthoptera, Ensifera*). *Z. Morph. Tiere 81*, 209–219.

SCHWABE, J. (1906). Beiträge zur Morphologie und Histologie der tympanalen Sinnesapparate der Orthopteren. *Zoologica* (Stuttg.) 50, 1–154.

SCHWARTZKOPFF, J. (1974). Principles of signal detection by the auditory pathways of invertebrates and vertebrates. In *Symposium Mechanoreception. Rheinisch-Westfäl. Akad. Wiss. Vorträge 53.* Pages 331–346. Westdeutscher-Verlag.

SEBEOK, T. A. (Ed.), (1977). *How Animals Communicate.* Indiana University Press, Bloomington and London.

SELVERSTON, A. I. and MILLER, J. P. (1980). Mechanisms underlying pattern generation in lobster stomatogastric ganglion as determined by selective inactivation of identified neurons. I. Pyloric system. *J. Neurophysiol. 44*, 1102–1121.

SHAW, E. A. G. (1974). The external ear. *In Auditory System, Anatomy, Physiology (Ear). Handbook of Sensory Physiology.* Vol. V/1, pages 455–490. Edited by W. D. Keidel and W. D. Neff. Springer, Berlin and New York.

SILVER, S., KALMRING, K. and KÜHNE, R. (1980). The responses of central acoustic and vibratory interneurons in bushcrickets and locusts to ultrasonic stimulation. *Physiol. Ent. 5*, 427–435.

SIMMONS, J. A., WEVER, E. G. and PYLKA, J. M. (1971). Periodical cicada: Sound production and hearing. *Science 171*, 212–213.

SIMMONS, P. and YOUNG, D. (1978). The tymbal mechanism and song patterns of the bladder cicada, *Cystosoma saundersii. J. Exp. Biol. 76*, 27–45.

SISMONDO, E. (1979). Stridulation and tegminal resonance in the tree cricket *Oecanthus nigricornis* (Orthoptera: Gryllidae: Oecanthinae). *J. Comp. Physiol. 129*, 269–279.

SKOVMAND, O. and PEDERSEN, S. B. (1978). Tooth impact rate in the song of a shorthorned grasshopper: a parameter carrying specific behavioural information. *J. Comp. Physiol. 124*, 27–36.

SKUDRZYK. E. (1971). *The Foundations of Acoustics.* Springer, Berlin.

SPIRA, M. E. YAROM, Y. and PARNAS, I. (1976). Modulation of spike frequency by regions of special axonal geometry and by synaptic inputs. *J. Neurophysiol. 39*, 882–899.

STEPHEN, R. O. and BENNET-CLARK, H.C. (1982). The anatomical and mechanical basis of stimulation and frequency analysis in the locust ear. *J. Exp. Biol. 99*, 279–314.

STEVENS, E. D. and JOSEPHSON, R. K. (1977). Metabolic rate and body temperature in singing katydids. *Physiol. Zool. 50*, 31–42.

STOUT, J. F. and HUBER, F. (1972). Response of central auditory neurons of female crickets (*Gryllus campestris* L.) to the calling song of the male. *Z. Vergl. Physiol. 76*, 302–313.

STOUT, J. F. and HUBER, F. (1981). Responses to features of the calling song by ascending auditory interneurons in the cricket *Gryllus campestris. Physiol. Ent. 6*, 199–212.

SUGA, N. (1963). Central mechanism of hearing and sound location in insects. *J. Insect Physiol. 9*, 867–873.

SUGA, N. and KATSUKI, Y. (1961). Central mechanism of hearing in insects. *J. Exp. Physiol. 38*, 545–558.

SURLYKKE, A. and MILLER, L. A. (1982). Central branchings of three sensory axons from a moth ear (*Agrotis segetum*), Noctuidae. *J. Insect Physiol. 28*, 357–364.

SWIHART, S. L. (1967). Hearing in butterflies (*Nymphalidae: Heliconius, Ageronia*). *J. Insect Physiol. 13*, 469–476.

TAUTZ, J. (1977). Reception of medium vibration by thoracal hairs of caterpillars of *Barathra brassicae* L. (Lepidoptera, Noctuidae). I. Mechanical properties of the receptor hairs. *J. Comp. Physiol. 118*, 13–31.

TAUTZ, J. (1978). Reception of medium vibration by thoracal hairs of caterpillars of *Barathra brassicae* L. (Lepidoptera, Noctuidae). II. Response characteristics of the sensory cell. *J. Comp. Physiol. 125*, 67–77.

TAUTZ, J. (1979). Reception of particle oscillation in a medium — an unorthodox sensory capacity. *Naturwissenschaften 66*, 452–461.

THORSON, J., WEBER, T. and HUBER, F. (1982). Auditory behavior of the cricket. II. Simplicity of calling-song recognition in *Gryllus* and anomalous phonotaxis at abnormal carrier frequencies. *J. Comp. Physiol. 146*, 361–378.

TISCHNER, H. (1953). Über den Gehörsinn von Stechmücken. *Acustica 3*, 335–343.

TREAT, A. E. (1955). The response to sound of certain Lepidoptera. *Ann. Ent. Soc. Amer. 48*, 272–284.

WIESE, K. (1978). Negative Rückkopplung in der akustischen Bahn von *Gryllus bimaculatus* als Grundlage temporalen Filterns. *Verh. Dtsch. Zool. Ges. 168*.

WIESE, K. (1981). Influence of vibration on cricket hearing: Interaction of low frequency vibration and acoustic stimuli in the omega neuron. *J. Comp. Physiol. 143*, 135–142.

WILEY, R. H. and RICHARDS, D. G. (1978). Physical constraints on acoustic communication in the atmosphere: Implications for the evolution of animal vocalizations. *Behav. Ecol. Sociobiol. 3*, 69–94.

WOHLERS, D. W. and HUBER, F. (1978). Intracellular recording and staining of cricket auditory interneurons (*Gryllus campestris* L., *Gryllus bimaculatus* DeGeer). *J. Comp. Physiol. 127*, 11–28.

WOHLERS, D. W. and HUBER, F. (1982). Processing of sound signals by six types of neurons in the prothoracic ganglion of the cricket, *Gryllus campestris* L. *J. Comp. Physiol. 146*, 161–173.

WOHLERS, D. W., WILLIAMS, J. L. D., HUBER, F. and MOORE, T. E. (1979). Central projections of fibres in the auditory and tensor nerves of cicadas (Homoptera: Cicadidae). *Cell Tissue Res. 203*, 35–51.

YOUNG, D. (1980). The calling song of the bladder cicada, *Cystosoma saundersii*: A computer analysis. *J. Exp. Biol. 88*, 407–411.

YOUNG, D. and BALL, E. (1974). Structure and development of the auditory system in the prothoracic leg of the cricket *Teleogryllus commodus* (Walker). I. Adult structure *Z. Zellforsch. 147*, 293–312.

YOUNG, D. and HILL, K. G. (1977). Structure and function of the auditory system of the cicada *Cystosoma saundersii*. *J. Comp. Physiol. 117*, 23–45.

ZARETSKY, M. D. and EIBL, E. (1978). Carrier frequency-selective primary auditory neurons in crickets and their anatomical projections to the central nervous system. *J. Insect Physiol. 24*, 87–95.

ZHANTIEV, R. D. (1971). Frequency characteristic of tympanal organs in grasshoppers (*Orthoptera, Tettigoniidae*). (In Russian). *Zool. J. 50*, 507–514.

ZHANTIEV, R. D. (1981). *Bioacoustics of Insects.* (In Russian.) Moscow University Press.

ZHANTIEV, R. D. and KALINKINA, I. N. (1977). Sound reaction of descending neurons in abdominal part of the central nervous system of Orthoptera. (In Russian.) *Nauch. Dokl. Vyss. Shkoly. Biol. Nauk. 8*, 66–71.

ZHANTIEV, R. D., KALINKINA, I. N. and TSHUKANOV, V. S. (1975). The characteristics of the directional sensitivity of tympanal organs in *Gryllus bimaculatus* Deg. (*Orthoptera, Gryllidae*). (In Russian). *Rev. Ent. USSR. 54*, 249–257.

ZHANTIEV, R. D. and KORSUNOVSKAYA, O. S. (1973). Sound communication and some characteristics of the auditory system in mole crickets (*Orthoptera, Gryllotalpidae*). (In Russian.) *Zool. J. 52*, 1789–1801.

ZHANTIEV, R. D. and KORSUNOVSKAYA, O. S. (1977). Reaction to sound of descending neurons in cervical connections of the cricket *Gryllus bimaculatus* DeGeer (Orthoptera, Gryllidae). (In Russian.) *Ent. Obozr. 54*, 248–257.

ZHANTIEV, R. D. and KORSUNOVSKAYA, O. S. (1978). Morphological organization of tympanal organs in *Tettigonia cantans* (*Orthoptera, Tettigoniidae*). (In Russian.) *Zool. J. 57*, 1012–1016.

ZHANTIEV, R. D. and TSHUKANOV, V. S. (1972a). Frequency characteristics of tympanal organs of the cricket *Gryllus bimaculatus* Deg. (*Orthoptera, Gryllidae*). (In Russian.) *Vestn. Moscow Univ. 2*, 3–8.

ZHANTIEV, R. D. and TSHUKANOV, V. S. (1972b). Reaction of the auditory system of *Gryllus bimaculatus* (Orthoptera, Gryllidae) to intraspecific sound signals. *Zool. Zh 51*, 983–993.

ZWISLOCKI, J. (1960). Theory of temporal auditory summation. *J. Acoust. Soc. Amer. 32*, 1046.

10 Gravity

EBERHARD HORN

Universität Karlsruhe, Karlsruhe, FRG

1 INTRODUCTION

Gravity is a natural stimulus which has been present throughout animal evolution. It has therefore become an important reference system for orientation, the control of body posture and equilibrium during walking, swimming, flying or digging, and in communication, etc. In all animal groups, gravity receptive systems work on the principle of the displacement of a heavy body towards the earth's centre of gravity or a light body away from it. Land-living animals therefore rely on the downward displacement of a heavy body, while in water-living animals, equivalent information is drawn from the upward displacement of an air bubble.

Although insects have successfully conquered many environments, e.g. water, land and air, for a long time it was difficult to understand the part played by sensory responses to gravity in the control of well-developed behavioural acts such as stable flight of many insects, the digging of some beetles or the waggle dance of honeybees. The reason was that, in contrast to many other animals (for example, vertebrates, molluscs, crustaceans, coelenterates and tunicates), statocyst-like sense organs composed of a cavity with a sensory epithelium and a movable heavy body were rarely found in insects, being reported only in some water-living larvae.

Since the end of the 1950s, however, it has become clear that most insects use proprioceptors or specialized cuticular sensillae for gravity reception. In land-living species and in the water-living larvae of dragonflies and stone-flies, these receptors are stimulated by displacements, either of parts of the body or of the whole body, and therefore form a proprioceptive gravity receptor system (PGR system). In crickets and cockroaches, specialized bulb-like cuticular sensillae, which are clearly non-proprioceptive, function as gravity receptors. Some water-living species, on the other hand, detect the direction of gravity by using the buoyancy of a gas bubble (buoyancy organs).

2   LOCATION OF RECEPTORS

2.1   Water-living insects

2.1.1   BUOYANCY ORGANS

Buoyancy organs seem to be common in Hemiptera. In the Corixidae, Naucocoridae and Notonectidae it has been suggested that the buoyancy of an air bubble between head and prothorax deflects the antennae according to the animal's spatial position, stimulating antennal receptors. Rabe, W. (1953) proved that in *Notonecta glauca* the antennal receptors are necessary only when the animals are swimming, during which period their sensory input is supposed to improve manoeuvrability. Animals at rest will always lie ventral side upwards, because the centre of gravity is dorsal.

Nepidae, such as *Nepa rupra* or *Ranatra lineatus*, possess buoyancy organs which are connected with the animal's air supply. Their importance for gravity orientation has been proved convincingly by Baunacke, W. (1912) who investigated it using an underwater seesaw in complete darkness. On this seesaw, animals which are deprived of their air supply normally orientate upwards and return to the water surface. Tilting the swing during upward walking of the animals elicits a turning reaction so that the bug comes back to the correct direction. After destruction of the buoyancy organs, however, *Nepa* arrive at the water surface only by chance.

The buoyancy organs of the larvae and the adults of the Nepidae have different structures (Baunacke, W. 1912). The larvae possess two grooves on the ventral side which are filled with air for respiration. These grooves are formed by the sternites and the paratergites. In *Nepa*, large bristles which insert on both sides of the grooves prevent the air from leaking away. The bristles of one type look like paper-cornets, while the others are very long with normally shaped cuticular bristles. In the third to sixth abdominal segments the paratergites have a small

inlet where the long bristles are absent. Here, fan-shaped bristles are directed into the air groove. Behind them is a water-filled cavity. At the contact surface between the respiratory air of the ventral groove and the water of this cavity, there are some sensory bristles (Fig. 1). According to the spatial position of the animal the rostral or caudal bristle rows are stimulated differently. The pattern of this excitation could be used as a measure of the spatial position of the animal with respect to gravity via strong connections with the motor system. The buoyancy organs of other Nepidae are similar in principle, though they may be simpler or more complex in design.

Adults possess no ventral groove. Their buoyancy organs are located in the fourth to sixth abdominal segments at the spiracle openings. Each organ is composed of two types of receptors: small sensory papillae with unknown function and umbrella-shaped cuticular sensillae which have been identified as water pressure detectors (Bonke, D. 1975, Thorpe, W. and Crisp, D. 1947). In *Nepa*, each buoyancy organ contains nearly 100 umbrella-shaped hairs which overlap the sensory papillae. From the structure of these hairs it can be concluded that they receive information about the displacements of the air within the tracheae. The anatomical structures of the sensory part of this sensilla resemble that of the typical cuticular hair sensilla of insects (see Fig. 4C) except that the orientation of the tubular body is perpendicular to the longitudinal axis of the outer segment of the receptor cell (Figs. 3 and 4).

2.1.2   STATOCYST-LIKE ORGANS

There are some descriptions of statocyst-like organs in water-living insects (Weber, H. 1933), but in no case has their participation in gravity reception been proved. All these organs occur in larvae with the exception of the Palmén-organ of the Ephemera which also occurs in adults. The organ of Palmén is composed of a chitinous mass which is connected with four blind-ending tracheal branches. It may be that the displacements of the chitinous mass stimulate receptor endings inserted on to the tracheae.

Larvae of the Limnobiidae possess cavities containing grains of sand or chitin which are sometimes

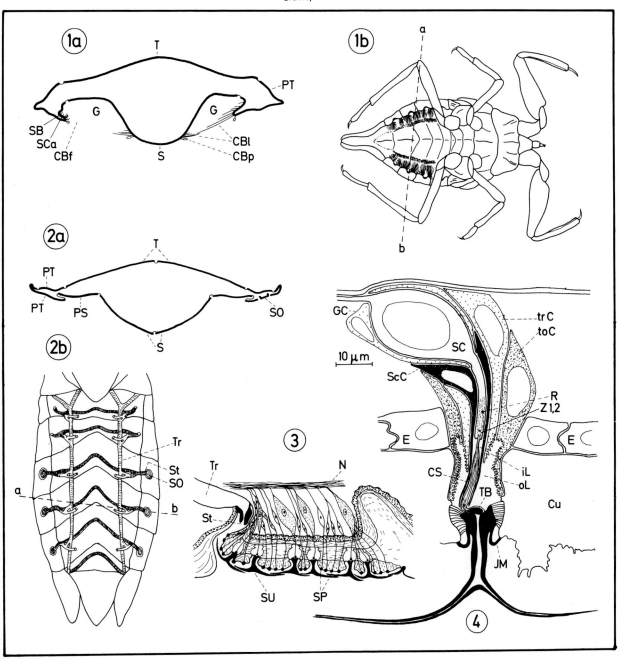

FIG. 1. Location and structure of the buoyancy organ of larval (1) and adult (2–4) *Nepa*. 1b and 2b: arrangement of the sense organs on the abdomen; 1a and 2a: slightly oblique transverse sections through the abdomen. The plane of this section is indicated by the dotted line (a, b) in 1b and 2b. 3 and 4: arrangement and fine structure of the umbrella-shaped sensilla in the buoyancy organ. CBf, filter bristles; CBl, large cover bristles; CBp, paper-bag-like cover bristles; CS, cuticular sheath; Cu, cuticle; E, epidermal cells; G, ventral groove; GC, glial cell; iL and oL, inner and outer liquor cavity; N, nerve; PS, parasternite; PT, paratergite; R, root fibres; S, sternite; SB, sensory bristles; SC, sensory cell; SCa, sensory cavity; SO, sense organ; SP, sense papillae; St, stigma; SU, umbrella-shaped hair sensilla; T, tergite; TB, tubular body; toC, tormogene cell; trC, trichogene cell; Tr, trachea; Z1 and Z2, centriol-like structures. (Combined and modified from Bonke, D. 1975 (4); Baunacke, W. 1912 (1, 2); Thorpe, W. and Crisp, D. 1947 (3).)

deeply invaginated into the body. These structures are known as pelotactic organs. The opening of every cavity is covered with long bristles and the inner walls carry sensory sensillae. The contraction of some very small muscles attached to the outside of the cavity wall whirls up the inclusion bodies which, according to their direction of fall, stimulate certain of the sensory hairs. This informs the animals about the direction of gravity. Elimination of the sense organs, however, does not abolish the geotactic behaviour of these larvae.

2.2 Land-living insects

2.2.1 Leg PGR systems

Nearly all insects possess a PGR system in their legs (Horn, E. 1975b). It is stimulated by displacements of the whole body. They have been demonstrated even in the water-living larvae of dragonflies or stone-flies.

First evidence for the importance of leg proprioceptors in gravity orientation came from experiments by Crozier, W. and Stier, T. (1929) for the beetle *Tetraopes*, although these authors did not accept the possibility of a gravity sense in insects. They suggested that the animal's orientation on an inclined plane is simply the result of a turning response which continues until the sensory input coming from both sides of the body is equal. Later, it was suggested that the whole body acts like the statolith, thus stimulating the proprioceptors of the legs (Vowles, D., 1954; Lindauer, M. and Nedel, J., 1959). The existence of gravity receptors in the legs, however, was first demonstrated by Markl, H. (1962) in honeybees, wasps and ants, and later by Bässler, U. (1965) in the stick insect *Carausius morosus*, by Horn, E. (1970) in the mealworm beetle, *Tenebrio molitor*, and Jander, R. *et al.* (1970) in a number of diverse insect species. Horn, E. (1973, 1975a) presented further evidence for the existence of a PGR system in the legs of bees. All of these experiments were based on the analysis of geotactic behaviour, for example the accuracy of negative geotactic orientation, the frequency of turning movements on a seesaw, or the orientation during competing light and gravity stimulation. The importance of receptors on these reactions was analysed by elimination of certain proprioceptors in

different joints of the legs, either by shaving single cuticular bristle fields or by loading the body of the animal with weights, and thus simulating an increased gravitational stimulus to the leg proprioceptors. With the exception of the honeybees *Apis mellifera* and *Apis cerana* (Horn, E. 1975a) elimination of the receptors generally impairs orientation in the negative geotactic direction, while loading the animals with weights improves the orientation in the direction of gravity.

However, it is only in the stick insect that single receptors important in gravity reception have been positively identified in the legs (Bässler, U., 1965). The leg PGR system of *Carausius morosus* is composed of bristle fields on the subcoxa, and a chordotonal organ in the femur (Fig. 2). The subcoxal bristle fields are stimulated by body displacements along the longitudinal axis of the animal, while the femoral chordotonal organ is sensitive to displacements along the transverse axis.

In contrast to this experimental procedure of measuring geotactic behaviour, Horn, E. and Lang, H.-G. (1978) introduced a method in which compensatory head reflexes are used to determine the function of PGR systems in the legs. They showed that flies (*Calliphora erythrocephala*) and bees (*Apis mellifera*) perform compensatory roll and/or pitch movements of the head when walking in a direction which does not coincide with that of gravity. In tethered walking flies and bees, the reaction is elicited only when the animal holds a walking ball. The ball simulates the displacements of the body (free-ball situation) (Fig. 3). The reflexes do not occur if these stimuli are lacking, for example when the ball is mounted on an axle (fixed-ball situation). The amount of the head reflexes depends sinusoidally on the tilt and roll angles. There are many functional similarities between the leg PGR system of the fly *Calliphora erythrocephala* and the statocyst system of vertebrates, crustaceans and molluscs (Table 1). Both receptor systems are composed of (a) a tonic and (b) a modulatory component. The tonic influence is shown by the fact that elimination of one, two or three legs on only one side of the body induces an asymmetrical displacement of the head from its normal position and therefore a shift of the reflex function along the ordinate. It is mediated by receptors of all legs. In contrast, the modulatory influence determines the amount of the reflexes.

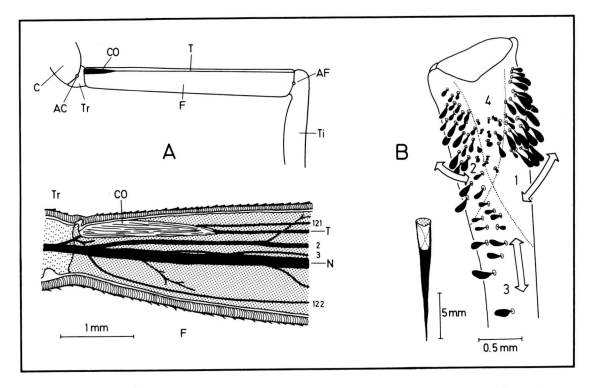

FIG. 2. Identified gravity receptor systems in insects. **A:** Structure of the femoral chordotonal organ of the stick insect, *Carausius morosus*. AC, AF, axis of movement in the joints between coxa/trochanter and femur/tibia; C, coxa; CO, chordotonal organ; F, femur; N, main leg nerve; T, tendon of the chordotonal organ; Ta, tarsus; Ti, tibia. **B:** Distribution of the club-shaped sensilla of the cercus of the cricket, *Gryllus bimaculatus*. Area 1 is ventral, area 2 dorsal; Arrows indicate main plane of hair movement. (Modified and combined from Bässler, U., 1965 (**A**); Bischof, H.-J., 1974 (**B**).)

This is mediated only by proprioceptors of the forelegs and middle legs (Horn, E. 1982a) (Fig. 3). The most important difference between the two types of gravity sense organs, however, is that the leg PGR systems of the flies and bees perceive the direction of gravity only during walking, while the statocyst systems act even in the resting animal (Horn, E. and Lang, H.-G., 1978).

2.2.2 OTHER PGR SYSTEMS

Much interest has been focused on proprioceptors controlling the position of extremely mobile parts of the body such as the head, the antennae and, in Hymenoptera, the abdomen. Elimination of such receptors, by either cutting their nerves, shaving their bristles or immobilizing the joints, have been used to identify those receptors involved in gravity reception. In the bee, *Apis mellifera*, Lindauer, M. and Nedel, J. (1959) proved that the bristle field of

the episternal process (neck organ) is responsible for gravity reception. This sense organ is innervated by a branch (nervus cervicalis) of the second prothoracic nerve (Fig. 4). Cutting this nerve causes disorientation. Markl, H. (1962) and Horn, E. (1973) came to similar general conclusions although their results differed considerably from those of Lindauer and Nedel. Markl found only a weak impairment in the discrimination between upward and downward orientation after immobilization of the head, while Horn found smaller angular deviations of the spontaneous geomenotactic walking directions from the upward direction (negative geotactic basic direction) after cutting the nervus cervicalis. Bumblebees, *Bombus terrestris* and *Bombus lapidarius*, use the same sense organ in the perception of the gravitational stimulus, but, in contrast to the honeybee, elimination of this input by cutting the nerve causes a great deterioration in the orientation in the basic negative geotactic direction

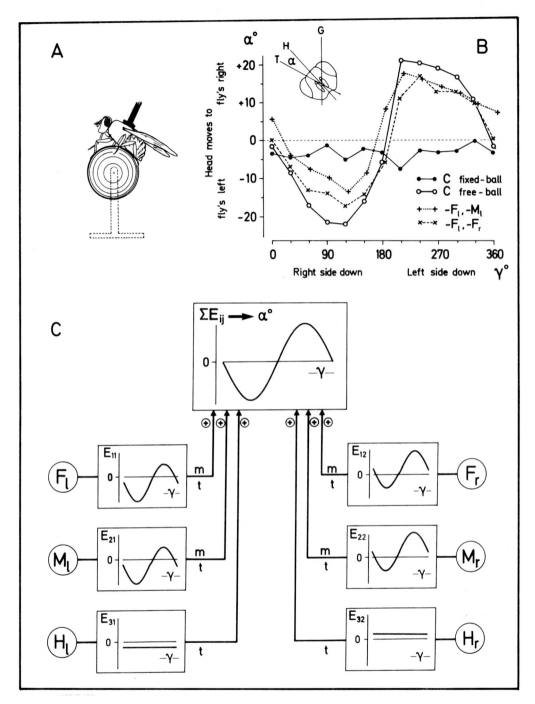

FIG. 3. The leg gravity receptor system in the fly, *Calliphora erythrocephala*. **A:** Experimental situation; dotted line characterizes the fixed-ball situation. **B:** Strength $\alpha°$ of the compensatory head reflexes in intact flies (C) for the free-ball and the fixed-ball situations, for flies without forelegs ($-F_l$, $-F_r$), and for flies without forelegs and middle legs of one body side ($-F_l$, $-M_l$). **C:** Hypothetical scheme of the interaction between the leg afferences. E_{ij} afferent excitation, the indices i and j correspond to the forelegs, middle legs and hind legs (i = 1, 2, 3), and to right and left (j = 1, 2); F, foreleg; H, hind leg; M, middle leg; l, left; m, modulatory influences; r, right; t, tonic influences; α, angle between the dorsoventral axis of the head and the thorax; γ, tilt angle. (Modified from Horn, E. and Lang, H.-G., 1978; Horn, E., 1982a.)

Table 1: *Similarities between the statocyst systems of vertebrates, molluscs and crustaceans, and the leg PGR system of flies (taken from Horn, E., 1982a)*

	Molluscs		Crustaceans	Vertebrates	Flies*
	Cephalopods	Non-Cephalopods			
Response characteristics of behavioural reactions	sinusoidal	sinusoidal	sinusoidal	sinusoidal	sinusoidal
Type of interaction between different gravity sense organs	additive		additive	additive	additive
Behavioural defects caused by unilateral elimination of the gravity sense organ	head position	rotation (not in Aplysia)	1. rotation 2. eye position	1. rotation 2. eye position 3. limb posture	head position
Defects are directed to	intact side	intact side	defect side	defect side	intact side
Defects can be compensated		yes	yes	yes	yes

*Only valid for walking flies.

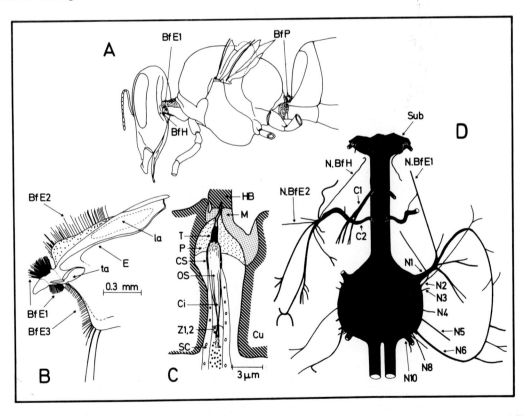

FIG. 4. Position (**A, B**), structure (**C**) and innervation (**D**) of the gravity receptors in the neck of the bee, *Apis mellifera*. BfE1 – BfE3, bristle fields of the episternum, with BfE1 located on the episternal process (neck organ, compare Lindauer, M. and Nedel, J., 1959); BfH, bristle field located ventral to the occipital foramen; BfP, petiole organ with its dorsal and ventrolateral bristle fields; C1, C2, nerves of the connective between prothorax and suboesophageal ganglion; Ci, ciliary structure; CS, cuticular sheath; Cu, cuticle; E, episternum; HB, hair base; la, longitudinal apodem; M, joint membrane; N1 – N10, prothoracic nerves; N.BfE1, N.BfE2, N.BfH, nerves innervating the corresponding bristle fields; OS, outer segment of the sensory cilium; P, cap; SC, Schwann cell; Sub, suboesophageal ganglion; T, tubular body; ta, transverse apodem; Z1, Z2, centriole-like structures. (Modified and combined from Lindauer, M. and Nedel, J., 1959 (**A**); Markl, H., 1962 (**B**); Markl, H., 1966 (**D**); Thurm, U., 1965 (**C**).)

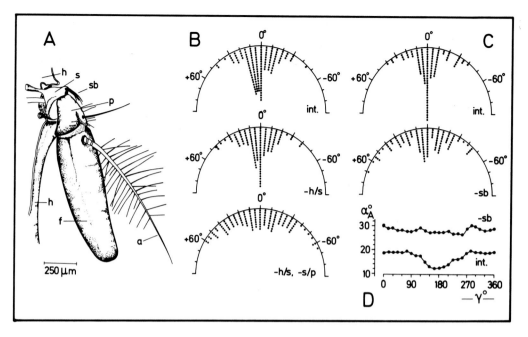

FIG. 5. The antennal PGR system of the fly, *Calliphora erythrocephala*. **A:** Morphology of the antennae. a, arista; f, funicle; h, head; p, pedicelle; s, scapus; sb, scapus bristles. **B:** Frequency distributions of the walking directions of flies on a vertical surface before (int.), after immobilization of the head-scapus-joint (−h/s), and after additional immobilization of the joint between scapus and pedicelle (−h/s, −s/p). **C:** Frequency distributions of the walking direction of the flies before (int.) and after shaving the scapus bristles (−sb). 0° negative geotaxis. **D:** The angular position of the antennae during rotation of the flies around their transverse axis before (int.) and after shaving the scapus bristles (−sb). α_A angle between the dorsoventral axis of the head and the longitudinal axis of the antennae; $\gamma = 0°$ the normal position, $\gamma = 90°$ the head-down position and $\gamma = 270°$ the head-up position. (Combined and modified from Horn, E., 1975 (**A, B, C**); Horn, E. and Kessler, W., 1975 (**D**).)

(Horn, E., 1973) although the animals are still able to detect the direction of gravity. Proprioceptors of the neck are also responsible for gravity perception in the ant, *Formica polyctena*, but it is still unknown which of the three different bristle fields of the episternum (similar to those in the honeybee, Fig. 4B) are involved in gravity perception, because Markl, H. (1962) investigated discrimination between negative and positive geotaxis only after immobilization of the whole neck.

Other PGR systems of the honeybees, bumblebees, ants and wasps are located in the antennae, the petiole joint and the gaster joint. Elimination of these receptors by shaving the bristle fields (Horn, E., 1973, 1975b), or by immobilizing the joints (Markl, H., 1962) aways impairs the accuracy of negative geotactic orientation or the ability to distinguish between upward and downward orientation on a seesaw. In the wasp, *Vespa vulgaris*, however, antennal receptors seem to be unimportant.

Antennal receptors are included in the gravity receptive system of the stick insect, *Carausius*

morosus (Bässler, U., 1971; Wendler, G., 1965) in the blowfly, *Calliphora erythrocephala* (Horn, E. 1969) and in the cricket, *Gryllus bimaculatus* (Horn, E. and Bischof, H.-J., 1983), but only in the blowfly were both the receptors and the mechanisms by which they operate in gravity reception, identified (Horn, E., 1975b; Horn, E. and Kessler, W. 1975). Immobilization or amputation of complete antennae in the fly impairs recognition of the gravity direction. The receptors responsible for this are bristles inserting dorsally onto the scapus. Immobilization of the joint between scapus and pedicellus, as well as elimination of the bristles only, impaired the ability of orienting negatively geotactically, while immobilization of the joint between head and scapus had no effect on the precision of the negative geotactic response. Horn, E. and Kessler, W. (1975) made it clear that control of the angular position of the antennae is necessary for gravity reception because (a) the scapus bristles can only be stimulated by an upward movement of the antennae, and (b) elimination of these bristles

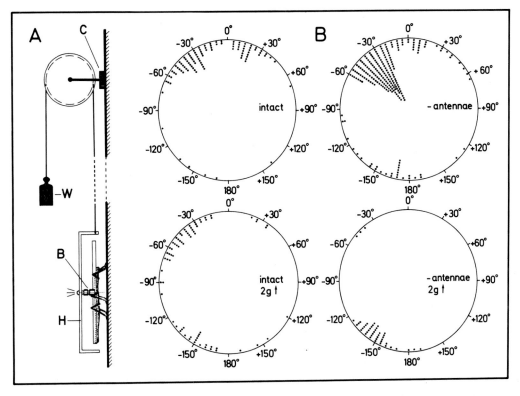

FIG. 6. The antennal PGR system of the stick insect, *Carausius morosus*. **A:** Experimental procedure: B, ball-bearing; C, slide rail; H, holder; W, counterweight. **B:** Frequency distributions of the walking directions of the stick insect before (intact) and after amputation of the antennae (− antennae). **2g↑** indicates that the stick insect is counterbalanced by a load of twice its own weight. (Modified and combined from Bässler, U., 1971 (**B**); Wendler, G. 1969 (**A**).)

results in an overshoot of the antennal position for all pitch angles of the fly (Fig. 5).

The existence of the antennal PGR system in the stick insect was first shown by a very elegant method devised by Wendler, G. (1965). He changed the direction of the natural downward-orientated force on the animal to an upward-orientated one by using the experimental equipment demonstrated in Fig. 6. While no difference in the directional orientation of the intact and antennaeless stick insect could be found when they walked in darkness on a vertical plane, the change of the direction of the effective "body weight", and therefore of the stimulus direction to the pedal PGR system, caused antennaeless stick insects to walk only downwards. He concluded that there are antennal PGR systems and this conclusion was confirmed by Bässler, U. (1971), using the same experimental trick but investigating the orientation of the stick insect in competing light and gravity fields (Fig. 6). Additionally, his experiments proved that the antennal

receptors affect only the sign of the geotactic response, but not its magnitude.

The antennal PGR system of crickets, however, elicits compensatory head movements if the animal are rolled around their long axis (Horn, E. and Bischof, H.-J. 1983). It exerts no tonic influence on the neck muscle system, because crickets with only one antenna amputated still keep their head in a body symmetric position (Horn, E. and Föller, W. 1984).

2.2.3 CLUB-SHAPED SENSILLAE

At the base of the cerci of crickets and cockroaches there are specialized cuticular sensillae. In the cricket, *Gryllus bimaculatus*, they are club-shaped. The whole area of insertion can be divided into four parts, each containing hairs which can be deflected along the same plane of movement (Fig. 7C; Bischof, H.-J., 1974). In the cockroach, *Arenivaga* sp., each sensillum consists of a spherical mass

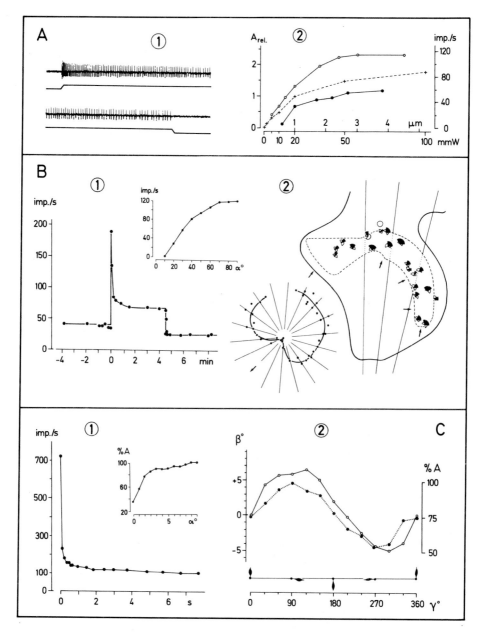

FIG. 7. Physiology of single receptors of the PGR systems. **A:** Umbrella-shaped sensillae of the buoyancy organ of *Nepa cinerea*. **1:** Extracellular recorded action potentials during an inward movement of 1 μm. **2:** Intensity characteristic of the tonic component of two receptors (○, ●) (upper abscissa, right ordinate) and for the integrated activity of the whole organ (lower abscissa, left ordinate). **B:** Hair sensillum of the neck organ (BfE1, see Fig. 4) of the honeybee, *Apis mellifera*. **1:** Extracellularly recorded activity of a single receptor during a deflection of the cuticular hair. (Inset: intensity characteristic for the tonic component of the response.) **2:** Directions of maximal sensitivity for the sensillae of the episternal process. Solid segments: angular area with ≥ 90% maximal sensitivity; adjacent white sectors with ≥ 50%. (Inset: one example of the directional sensitivity of a single hair sensillum. Arrows indicate the direction of hair bending, α° angular degree of deflection.) **C:** Club-shaped sensillae of the cerci of the cricket, *Gryllus bimaculatus*. **1:** Extracellularly recorded activity during a deflection of single hairs of 3.8°. (Inset: intensity characteristic for the tonic component of the response.) **2:** ●···●: Activity of the receptor during a complete rotation of the cricket (mean of two measurements); the activity was standardized to the maximal activity (= 100% A). ○—○: Deflection β° of the hairs during a complete rotation of the cricket. γ, tilt angle. (Modified and combined from Bischof, H.-J., 1974, 1975 (**C**); Bonke, D., 1975 (**A**); Thurm, U., 1963 (**B**).)

positioned at the distal end of the slender shaft. They are arranged in two rows of seven or eight sensillae (Hartman, H. *et al.*, 1979). Because of their mechanical properties, these sensilla seem to be gravity receptors as the angle of deflection of the hairs from their resting position depends sinusoidally on the spatial position of the cricket (Fig. 7C; Bischof, H.-J., 1974).

This suggestion was confirmed by Horn, E. and Bischof, H.-J. (1983) who showed that these sensillae elicit compensatory head movements when crickets are rolled around their longitudinal axis. The amplitude of the head response depends sinusoidally on the roll angle. The club-shaped sensillae, however, exert no tonic influence on the neck muscle system because crickets with only one cercus removed keep their head in a body symmetric position (Horn, E. and Föller, W. 1984).

3 PHYSIOLOGY

3.1 Physiology of receptors

The receptors of the buoyancy organ in *Nepa* are not spontaneously active, but are of the phasic–tonic type. The tonic component increases with increasing water pressure and the intensity response curve is linear within the physiological range of stimuli (Fig. 7A; Bonke, D. 1975).

There is not much information available on the physiology of the gravity receptive cells of land-living insects. Thurm, U. (1963) made recordings from single bristles of the neck organ of the honey-bee, *Apis mellifera*; Bischof, H.-J. (1974) investigated response characteristics of the sensory cells of the club-shaped hairs on the cerci of the cricket, *Gryllus bimaculatus*. These receptors are spontaneous-active; deflection of the cuticular hair in one direction increases the impulse activity, while deflection in the opposite direction decreases the impulse frequency. A ramp-like deflection induces a phasic–tonic response which is maximal for the mechanical preferred plane of movement. The amount of the tonic component depends on the strength of the deflection but, while in the bee's hairs the tonic plateau increases up to deflection angles of 70°, in the cricket's sensillae there is only an increase up to an angle of 4°. These values corres-

pond to the range of extent of deflection angles of these hairs. Rotation of the cercus around an axis lying perpendicular to the plane of movement of the investigated sensillum induces a sinusoidal modulation of impulse activity from the club-shaped hairs (Fig. 7B and 7C).

3.2 Physiology of central neurones

The influence of the spatial position of insects on the activity of central neurones has been investigated only in the cockroach, *Arenivaga*. Giant neurones which receive input from cercal afferents produce action potentials which depend on the direction of roll and pitch movements (Hartman, H. *et al.*, 1979; Walthall, W. and Hartman, H., 1981). A roll to the left elicits activity in the right giant neurone; the same neurone is inactivated after changing the direction of the roll movement. As the roll angle approaches zero, i.e. the animal is horizontal, interneurones in the left connective become active. During pitch movements corresponding interneurones in both connectives are active simultaneously. The frequency of firing increases always with increasing displacement angle and when the cockroach is maintained in a certain position the neurones fire tonically. These interneurones are known to be driven by club-shaped hairs of the cerci, because the regular changes of impulse activity elicited by pitch-and-roll movements disappear after they are removed.

3.3 The adequate stimulus for PGR systems

In all physiological experiments it becomes clear that in land-living insects deflection of the cuticular hairs is the effective (adequate) stimulus for the single receptor cells. However, while the angular deflection of the club-shaped hairs of crickets seems to be a direct measure of the angular position of the animal, this information can be obtained by two different mechanisms in insects which use PGR systems. Lindauer, M. and Nedel, J. (1959) suppose that parts of the body are dislocated in the field of gravity in an analogous way to statoliths in the statocysts, and that the sensory bristles of these joints are deflected according to the relative positions of the parts of the body (position hypothesis). In contrast, Wendler, G. (1972)

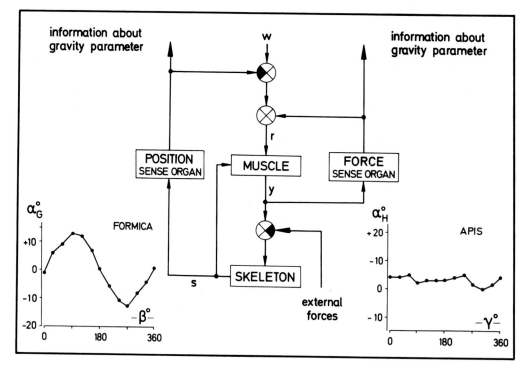

FIG. 8. Control pattern of gravity reception in insects, including the position hypothesis and the strength hypothesis (left side and right side of the flow diagram, respectively). Inset figures show the characteristics of the angular position of the gaster α_G of the ant *Formica* during orientation on a vertical inclined surface in relation to the angle of orientation ($= \beta°$) and the angular position of the head α_H of the honeybee *Apis* in relation to the tilt angle during rotation around its longitudinal axis. (Modified from Horn, E., 1982b.)

suggested that the receptors only control the position of these parts of the body, e.g. the head or the antennae, while other receptors receive the strength which is necessary to adjust this position. This hypothesis is a new formulation of an old suggestion by Crozier, W. and Stier, T. (1929) who wrote that "the conception of geotropic orientation is limited by the tensions applied to the musculature of the body or of appendages when the body is supported upon an inclined surface". This strength of responses depends on the spatial orientation of the animal (strength hypothesis). Campaniform sensillae are suitable receptors for this because they are adapted to measure strains in the body cuticle which occur during muscle contraction.

The position theory and the strength theory may both be valid, depending on the animal involved. The strength hypothesis could, for example, explain why the bee's head remains in a constant position during a complete roll (Horn, E., 1982a; Fig. 8, right side) even though the neck has receptors which are important for gravity reception as is demonstrated

by the change in geotactic behaviour after elimination of the episternal bristle field (Lindauer, M. and Nedel, J., 1959; Horn, E., 1973). On the other hand, the position theory explains why the abdomen of freely walking ants is deflected by gravity, this deflection being sinusoidally dependent on the walking direction on the vertical plane (Markl, H. 1962). Supposing that ants are able to distinguish upwards and downwards, then the movement of the gaster and the corresponding stimulation of the petiole organ give direct information about the direction of the gravitational stimulus in relation to the ant's walking direction (Fig. 8, left side).

4   BEHAVIOUR

4.1   Geotaxis

Geotaxis is defined as a behavioural response in relation to the field of gravity. Some insects spontaneously respond in only a small number of prescribed directions which are defined as the geotactic basic directions (Jander, R., 1963 a, b). Under these circumstances the insects are in a state of physiological symmetry, which is maintained even under strong disturbances. Positive and negative geotaxis are common in most walking insects. The transverse geotaxis has only been shown in a few species, for example in some *Hydrocorisae* or some flies (e.g. see Jander, R., 1963a). After complete amputation of the abdomen, the bee, *Apis mellifera*, shows transverse geotaxis. After additional cutting of the nervus cervicalis (Fig. 2), the preferential transverse directions disappear (Horn, E., 1973). According to Kühn's (1919) classification system of mechanisms of orientation, negative and positive geotaxis may be a tropotactic mechanism.

Jander, R. and Daumer, K. (1974) described a geoclinotactic behaviour in termites which walk on the edges of leaves, sticks, branches, etc., to find their forage. These termites perform right and left turns; the direction of the turns changes when the slope becomes too steep. The threshold inclination which induces the opposite turning movement is 12–13° relative to the horizontal.

The precision of the geotactic response depends in some insects on the inclination of the walking plane. Jander, R. (1963b) defined these reactions as progeotactic if the mean walking direction changed with changing inclination of the walking plane, and metageotactic if the direction of orientation was independent of the slope. Because there are insects such as the bee, *Apis mellifera* (Horn, E., 1973), which react progeotactically under simultaneous stimulation by light and gravity, but metageotactically in darkness, this classification is not very valuable without knowledge of the underlying mechanisms of central information processing.

The sign of the geotactic orientation is influenced by many factors; by gravity itself and other external stimuli like light, humidity or temperature. Lack of these stimuli changes the ratio between upward and downward orientation, a fact that can lead to some misinterpretation in extirpation experiments. One example is the controversy over the importance of antennal receptors in the gravity reception of the mealworm beetle, *Tenebrio molitor* (Bässler, U., 1961; Horn, E., 1970). Bässler investigated the discrimination between upward and downward orientation on a seesaw, and came to the conclusion

that there are antennal PGR systems. In contrast, Horn determined the strength of the geotactic behaviour under competing light and gravity stimulation, and found that only the pedal PGR system is responsible for gravity orientation. His conclusion was that Bässler eliminated the afference caused by humidity or temperature which determine only the sign of gravity. It is interesting that many temperature and humidity receptors are located on the antennae. But the answer to this controversy remains open because there is enough evidence that certain gravity receptors are adapted to specific gravity-induced behaviour.

Furthermore, the sign of geotaxis depends on internal factors including hormonal, genetic or biorhythmic ones (Jander, R., 1963b). Periodic changes of sign have been described for many social insects like ants or bees. *Apis mellifera* is more negatively geotactic in the morning, while positive geotaxis is dominant in the evening. In ants and bees starting their collection trip, a negative geotaxis predominates. However, this changes to a positive geotaxis at the end of their collecting trip (see Jander, R., 1963b).

Interest in genetic factors has been focused by the investigations of Erlenmeyer-Kimling, L. and Hirsch, J. (1961); Hirsch, J. and Erlenmeyer-Kimling, L. (1961); and Dobzhansky, Th. and Spassky, B. (1967) on *Drosophila*. These authors showed that it is possible to separate a positive geotactic line and a negative geotactic line, which means that the sign of the spontaneous geotaxis in the wild-type *Drosophila* is composed of a large genetic component. The major changes which occur with selection for geotaxis can be seen in the three large chromosomes X, II and III (Erlenmeyer-Kimling, L. and Hirsch, J., 1961). This leads to the conclusion that the genes affecting the sign of geotaxis are located on each of the three major chromosomes, and that the smaller fourth one has no importance in the determination of the sign. One cannot assume, however, that it has no influence in geotaxis at all. In cross-breeding experiments to produce special stocks carrying several marker genes and chromosomal inversions, Hirsch, J. and Erlenmeyer-Kimling, L. (1962) confirmed the hypothesis for the polygenic determination of the sign of geotaxis. They found that in the unselected wild-type population of *Drosophila melanogaster*,

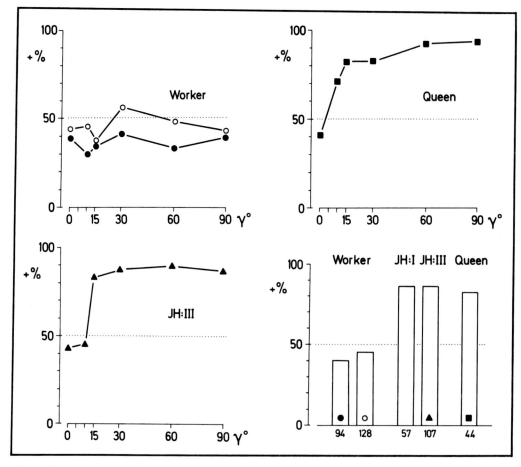

FIG. 9. Effect of juvenile hormone (JH) on the geotactic orientation of honeybee worker larvae compared to the geotactic response of queen larvae. Ordinate: relative number of positive geotactic responses; abscissa: inclination of the artificial cell relative to the horizontal plane. ○, untreated, and ●, sham-treated worker larvae; ▲, JH: III-treated worker larvae; ■, queen larvae. The histogram on the lower right presents the results of a continuous rotation of the larvae during the whole time of development. Ordinate: relative number of positive geotactic responses; the numbers under the columns indicate the number of test larvae. (Modified from Ebert, R., 1980.)

chromosomes X and II contribute to positive geotaxis, while chromosome III contributes to negative geotaxis. In the strain which had been selected for positive geotaxis all three major chromosomes contributed to the determination of the sign of geotaxis, while in the strain selected for negative geotaxis, the positive effect of X and II was strongly reduced although still present, while the negative effect of III was considerably increased.

Hormonal factors can be important during development. This has been convincingly proved by Ebert, R. (1980) who investigated the influence of juvenile hormone (JH) on the behaviour of the female honeybee larvae (*Apis mellifera*) during the final orientation at the outset of metamorphosis.

The ultimate orientation of the queen larvae which develop in hanging cells with the opening directed downwards is controlled by gravity; they display a positive geotaxis. In worker bees, however, which develop in slightly inclined cells (inclination angle in relation to the horizontal is 13° with the opening directed upwards), the final orientation of the larvae, and therefore the pupae, is not affected by gravity although all worker bee larvae finally settle down with the head facing the capping of their cells. This orientation, which has been interpreted for a long time as a negative geotaxis, is only caused by the shape of the cell endings, i.e. by tactile cues. In artificial cells with equal cappings, therefore, no preferential orientation can be observed. This geotactic indifference, however, is replaced by a positive geotaxis in worker larvae treated with JH:I or JH:III during the third day of larval development. Ebert concludes that the JH titre not only controls the determination of morphological and anatomical caste characteristics but also affects the orientation behaviour of the spinning larvae in relation to gravity (Fig. 9).

Despite this fixed behavioural pattern of positive, negative or transverse geotaxis, most insects can change their walking direction on an inclined walking plane and maintain the new direction for a long period. This occurs either spontaneously or as a result of learning. This behavioural response is called a geomenotaxis. Social insects such as ants or bees learn a geomenotactic direction very well, and reproduce the learned direction with small errors (Markl, H. 1964, 1966a). In honeybees, geomenotactic behaviour is included in a highly developed communication system known as the waggle dance, the importance of which was first discovered by von Frisch, K. (1946). A bee coming home from a collecting flight performs this waggle dance, from which other bees can get information about the direction to an attractive food source. The angle between the waggle direction and the direction upwards is identical to the angle between the direction of the food source and the sun. The dancing direction, which describes the location of a certain food source, must change during the day, and even bees which dance spontaneously during the night closely maintain the correct direction although they cannot see the sun (Lindauer, M., 1957). There are, however, small systematically changing errors in the dancing direction throughout the day (misdirection = "Missweisung").

Several attempts have been made to discover the underlying mechanisms for this complex and fascinating behaviour (Markl, H., 1966b). Up till now we have found no answer to the question of how the PGR system acts in controlling this behaviour. The only precise fact known is that the error in the dance is not caused by the PGR system as was suggested by Markl, H. (1962, 1964, 1966b) and Jander, R. (1963a). Lindauer, M. and Martin, H. (1968) and Martin, H. and Lindauer, M. (1977) proved convincingly that the systematic error in the waggle dance is caused by the magnetic field of the earth. This error depends on the inclination and orientation of the honeycomb, but for the same experimental conditions it is always predictable from the characteristics of the earth's magnetic field. Therefore, there is a difference between the misdirection of bees dancing on the honeycomb and those hanging under the honeycomb. Only when the comb lies in the inclination plane of the earth's magnetic field are the daily error functions identical on both sides. Martin, H. and Lindauer, M. (1977) suggest that the magnetic field stabilizes the dancing behaviour of the bee against other disturbances in the hive, e.g. the light.

4.2 Evolutionary aspects of gravity reception and orientation

Jander, R. et al. (1970) suggest that the pedal PGR systems of the Pterygota are homologous, and that the loss of importance of a pedal PGR system is a plesiomorphic character. But this is an oversimplification as several factors must be taken into account:

(1) leg PGR systems can differ in structure and function;
(2) the central information processing based on the afferents of these systems can be different (Horn, E. 1975a), and
(3) the PGR system of the legs can be important only for certain inclination angles of the walking plane as was demonstrated for the bumblebees, Bombus terrestris and Bombus lapidarius (Horn, E. 1973).

Therefore, intensive comparative investigations of geotactic reactions and the identification of the receptors, as well as the type of information processing they employ, are necessary to reveal evolutionary trends in geotaxis of insects and its underlying mechanisms. One example is given for the higher bees (Apoidea). In honeybees, except Apis florea, the angle of orientation is independent of the inclination of the walking plane (metageotaxis), while in most bees (including A. florea), the angle of orientation decreases with increasing slope of the substrate (progeotaxis) (Jander, R. and Jander, U. 1970). This is due to a difference in the information processing of the afferent input from the leg PGR systems (Horn, E. 1973, 1975a). In the honeybees, Apis cerana and Apis mellifera, as well as in the

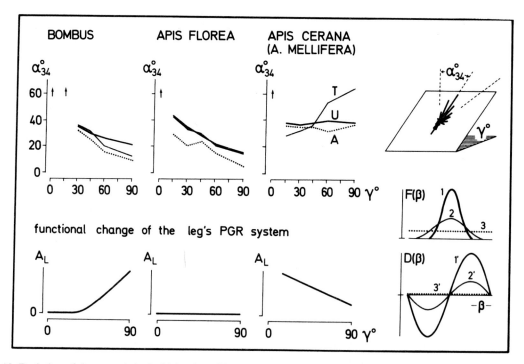

FIG. 10. Evolution of the geotaxis in the higher bees. Upper row: Relationship between the precision of geotactic orientation (defined by the angular sector lying symmetrical to the negative geotactic basic direction and including $2 \times 34\%$ of the orientation directions) and the inclination of the surface ($= \gamma$). Arrows indicate disorientation. A, T, bees with a load on the abdomen and thorax, respectively; U, unloaded bees. Lower row: the functional change of the leg's PGR system can be explained by the value A_L. A_L describes the amount of the proprioceptors of the legs on the geotactic turning tendency D. Large A_L causes a turning tendency function $D(\beta)$ with a large differential quotient for the stable zero position (index 1') and therefore an orientation with high precision (compare hypothetical frequency distribution $F(\beta)$ of the walking directions, index 1). The smaller is A_L, the smaller this differential quotient, and therefore the worse the recognition of the direction of gravity (indexes 2' and 2). If A_L equals zero then no preferential orientation direction occurs. (Modified from Horn, E., 1975, 1982b.)

bumblebees, *Bombus lapidarius* and *B. terrestris*, this orientation is controlled by the leg PGR system and the petiole organ. While in the bumblebees, an increase of the stimulus strength on each PGR system decreases the angle of orientation relative to the negative geotactic direction, in the honeybees this holds only for the petiole organ. The influence of the leg PGR system is antagonistic, which means that the stronger the stimulation on the receptors of the legs, either by an increase of the slope of the surface or by loading the thorax of these honeybees, the larger the angular deviation of the walking direction from the negative geotactic basic direction. In *A. florea,* however, the pedal PGR system does not contribute to this type of geotactic behaviour, while the petiole organ has the same influence as in all the other bees (Fig. 10). It must be stressed that *A. florea* is the only honeybee which does not dance on a vertical honeycomb. Its dancing floor is horizontal and the reference stimulus

field is, as during flight, the sun. In none of the bees does loading the abdomen change the type of geotactic orientation, because this manipulation increases the stimulus strength to both the leg PGR system and the petiole organ.

4.3 Equilibrium during flight

The posture of insects during flight is mainly controlled and influenced by visual as well as mechanical stimuli. Among the visually induced reactions, the dorsal light reaction (see refs in Tomioka, K. and Yamaguchi, T. 1980) and the optomotor response (see refs in Götz, K. *et al.,* 1979) are dominant. Among the mechanical ones, stimuli which cause linear and angular accelerations are of importance.

The question of whether the gravitational stimulus is important during flight can only be answered by investigations with freely flying insects.

Under these conditions gravity can act on receptors of the wings as they are stimulated by the weight of the body hanging from the wings. A first step in such an analysis was made by May, M. *et al.* (1980), who observed changes of flight behaviour in the moth, *Manduca sexta*, in an aircraft carrying out parabolic dives. Under these conditions gravitational fields varying from 0 to 2*g* are experienced. The most important behavioural change from flights at 1*g* to flights at 0*g* is that while moths never flew upside-down or head-down at 1*g*, these attitudes were as likely as any other at 0*g*. This method has the disadvantage, however, that the general activity of the animals is greatly reduced under 0*g* conditions.

A common principle for flight stabilization is to bring the insertion points of the wings above the centre of gravity of the insect body. Under these conditions it is only necessary to measure and compensate for yaw movements. Fraser, P. (1977) described compensatory wing reflexes in tethered flying cockroaches, *Periplaneta americana*, which may depend on gravitational stimulation. During stationary tilts of the animal, its wings perform asymmetrical wing abduction. The same reflexes can be elicited after removal of one cercus, but not if both cerci are removed. Fraser, however, pointed out that the detection of velocity changes of wind stimuli caused by the spontaneous angular displacements of the cockroach during flight may be responsible for this reflex.

As in vertebrates, flight equilibrium is mainly controlled by dynamic sense organs which measure angular accelerations. Vertebrates use the afferents from the crista organs of the semicircular canals, which are long and of small diameter, especially so in birds which are good fliers. Insects, however, have a different set of specialized dynamic sense organs.

In dragonflies, mechanoreceptors of the neck measure the displacements of the movable head during body tilt (Mittelstaedt, H. 1950), but only during acceleration-induced movements. Mittelstaedt proved that the structure of the neck is adapted for the reception of angular accelerations. He summarized the mechanical properties of the head insertion in the dragonfly, *Anax imperator*:

(1) the shape of the head is nearly circular and the masses of the heaviest parts of the head such as mandibles, eyes, etc. are concentrated on the periphery;

(2) the friction of the joint is low because of the type of head insertion and the soft cuticle of the membranes of the joint; and

(3) the torque diameter of the rotator muscles of the head is small and these muscles are very small compared with the other muscles in the head and the prothorax.

This means that the inertia of the head is very large compared with the joint friction and the elastic properties of the rotator muscles. Therefore, the head must tend to remain in its initial spatial position at the first moment of a rapid tilt movement of the thorax. Indeed, Mittelstaedt observed compensatory wing movements under these circumstances. During other behavioural reactions, e.g. feeding, the head can be held fixed by specialized muscles so that this very sensitive sensory apparatus is not disturbed (Fig. 11B).

The halteres, the modified hind wings, are the most important apparatus for flight equilibrium in flies (Fig. 11C). The underlying mechanism can be described by the following reflex arc: angular velocity experienced during a turn results in a torque being applied by the vibrating halteres to the campaniform sensillae at their bases. The afferents of these sensillae then elicit efferent signals and therefore bring about corrective movements of the wings (gyroscope principle, Pringle, J. 1948). Additional deflections of the halteres out of their main vibrating plane due to passive movements of the body induce head reflexes (Sandeman, D. 1980); on the other hand, artificial deflections of the head elicit wing reflexes which tend to bring the body in line with the head (Liske, E. 1977).

However, Schneider, G. (1953) proved that even the mechanical characteristics of the haltere system during the up-and-down vibratory movement are sufficient to stabilize the fly's flight. After the destruction of the sensillae at the haltere base by heat radiation without changing the mechanical properties of the vibrating haltere, there was no significant change in their influence on the maintenance of equilibrium during flight.

One interesting system, present in the beetle *Sisyphus schaefferi*, may also be important in the control of equilibrium during flight. Immediately

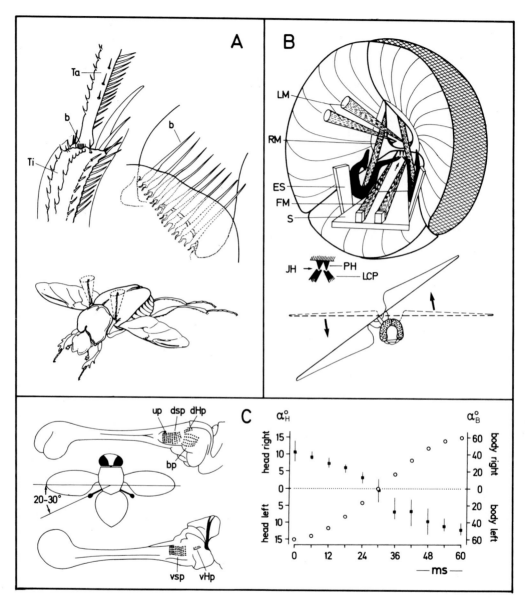

FIG. 11. Dynamic sense organs in insects. **A**: A proposed equilibrium organ (b) in the beetle *Sisyphus schaefferi* at the joint between tibia (Ti) and tarsus (Ta) of the middle leg. Note strange posture of the middle legs at the moment of flight initiation. **B**: View onto the occiput of the dragonfly *Anax imperator* which demonstrates the movable suspension of the head (JH = head joint), including the prothoracic sclerites LCP (latero-cervicale portator) and a pair of cuticular processes of the head (PH). Note that, after a short push of the body, the head maintains its initial position and that wing reflexes occur. ES, episternum; FM, flexor muscle; LM, levator muscle; RM, rotator muscle; S, sternite. **C**: Structure and function of the halteres in flies. Left: sensory areas on the halteres of *Lucilia sericata*. Inset demonstrates the orientation of the vibration plane of the halteres in *Calliphora erythrocephala*. Note that in the fly *Lucilia sericata* oscillatory movements of the body ($\bigcirc$, α_B°) induce reflexes of the head ($\blacksquare$, α_H°) around its dorsoventral axis. These reflexes disappear after elimination of the halteres. bp, basal plate; dHp (vHp), dorsal (ventral) Hicks papillae; dsp (dvp), dorsal (ventral) scalpal plate; up, undifferentiated papilla. (Combined and modified from Wigglesworth, V., 1946 (**A**); Mittelstaedt, H., 1950 (**B**); Pringle, J., 1948; Sandeman, D., 1980 (**C**); Schneider, G., 1953 (**C**).)

before flight, this beetle applies the tibiae of the middle legs to a groove in the side of the thorax and then sets the tarsae into rapid vibrations (Lengerken, H. 1934). In contrast to the related beetle *Geotrupes sylvaticus*, which does not perform this behavioural response, *Sisyphus* possesses a straight row of (sensory?) hairs, extending from the tibia over the base of the tarsus on its outer aspect (Wigglesworth, V. 1946). Because the middle leg is held in this position during flight, it would be interesting to find out whether it is used to detect angular displacements during flight using a similar mechanism to that of the halteres of flies. This question has not been investigated in detail.

ACKNOWLEDGEMENT

I thank Dr A. D. Watson (Cambridge, UK) for correcting the English text.

REFERENCES

BÄSSLER, U. (1961). Zum Schweresinn von Mehlkäfern (*Tenebrio molitor*) und Stechmücken (*Aedes aegypti*). *Z. Naturf.* 16b, 264–267.

BÄSSLER, U. (1965). Propriorezeptoren am Subcoxal- und Femur-Tibia-Gelenk der Stabheuschrecke *Carausius morosus* und ihre Rolle bei der Wahrnehmung der Schwerkraftrichtung. *Kybernetik* 2, 168–193.

BÄSSLER, U. (1971). Zur Bedeutung der Antennen für die Wahrnehmung der Schwerkraftrichtung bei der Stabheuschrecke *Carausius morosus*. *Kybernetik* 9, 31–34.

BAUNACKE, W. (1912). Statische Sinnesorgane bei Nepiden. *Zool. Jb. Abt. Anat.* 34, 179–345.

BISCHOF, H.-J. (1974). Verteilung und Bewegungsweise der keulenförmigen Sensillen von *Gryllus bimaculatus* DEG. *Biol. Zbl.* 93, 449–457.

BISCHOF, H.-J. (1975). Die keulenförmigen Sensillen auf den Cerci der Grille *Gryllus bimaculatus* als Schwererezeptoren. *J. Comp. Physiol.* 98, 277–288.

BONKE, D. (1975). Der Bau und die Antwortcharakteristika des Schirmrezeptors aus dem Statoorgansystem von *Nepa cinerea* L. (Hemiptera, Rhynchota). *Verh. Dtsch. Zool. Ges.* 1974, 42–45.

BÜCKMANN, D. (1954). Die Leistungen der Schwereorientierung bei dem im Meeressand grabenden Käfer *Bledius bicorius* GRM (Staphilinidae). *Z. Vergl. Physiol.* 36, 488–507.

CROZIER, W. J. and STIER, T. J. (1929). Geotropic orientation in arthropods. II. *Tetraopes*. *J. Gen. Physiol.* 12, 675–693.

DOBZHANSKY, TH. and SPASSKY, B. (1967). Effects of selection and migration on geotactic and phototactic behavior of *Drosophila*. *Proc. Roy. Soc. B.* 168, 27–47.

EBERT, R. (1980). Influence of juvenile hormone on gravity orientation in the female honeybee larva (*Apis mellifera* L.). *J. Comp. Physiol.* 137, 7–16.

ERLENMEYER-KIMLING, L. and HIRSCH, J. (1961). Measurements of the relations between chromosomes and behavior. *Science* 134, 1068–1069.

FRASER, P. J. (1977). Cercal ablation modifies tethered flight behavior of cockroach. *Nature* 268, 523–524.

FRISCH, K. VON. (1948). Gelöste und ungelöste Rätsel der Bienensprache. *Naturwissenschaften* 35, 12–23, 38–43.

GÖTZ, K. G., HENGSTENBERG, B. and BIESINGER, R. (1979). Optomotor control of wing beat and body posture in *Drosophila*. *Biol. Cybernet.* 35, 101–112.

HARTMAN, H. B., WALTHALL, W. W., BENNETT, L. B. and STEWART, R. R. (1979). Giant interneurons mediating equilibrium reception in an insect. *Science* 205, 503–505.

HIRSCH, J. and ERLENMEYER-KIMLING, L. (1961). Sign of geotaxis as a property of the genotype. *Science* 134, 835–836.

HIRSCH, J. and ERLENMEYER-KIMLING, L. (1962). Studies in experimental behavior genetics: IV. Chromosomes analyses for geotaxis. *J. Comp. Physiol. Psychol.* 55, 732–739.

HORN, E. (1969). Die Bedeutung von Körpergelenken für die geotaktische Orientierung von *Calliphora*. *Zool. Anz., Suppl.-Bd.* 33, 570–574.

HORN, E. (1970). Die Schwerkraftrezeption bei der Geotaxis des laufenden Mehlkäfers (*Tenebrio molitor*). *Z. Vergl. Physiol.* 66, 343–354.

HORN, E. (1973). Die Verarbeitung des Schwereizes bei der Geotaxis der höheren Bienen (Apidae). *J. Comp. Physiol.* 82, 379–406.

HORN, E. (1975a). Mechanisms of gravity processing by leg and abdominal gravity receptors in bees. *J. Insect Physiol.* 21, 673–679.

HORN, E. (1975b). The contribution of different receptors to gravity orientation in insects. *Fortschr. Zool.* 23, (1), 1–20.

HORN, E. (1982a). Gravity reception in the walking fly, *Calliphora erythrocephala*: tonic and modulatory influences of leg afference on the positional head reflexes. *J. Insect Physiol.* 28, 713–721.

HORN, E. (1982b). *Vergleichende Sinnesphysiologie*. Gustav Fischer Verlag, Stuttgart-New York.

HORN, E. and BISCHOF, H.-J. (1983). Gravity reception in crickets: the influence of cercal and antennal afferences on the head position. *J. Comp. Physiol.* 150, 93–98.

HORN, E. and FÖLLER, W. (1984). Tonic and modulatory subsystems of the complex gravity receptor system of crickets, *Gryllus bimaculatus*. (in prep.)

HORN, E. and KESSLER, W. (1975). The control of antennae lift movements and its importance on the gravity reception in the walking blowfly, *Calliphora erythrocephala*. *J. Comp. Physiol.* 97, 189–203.

HORN, E. and LANG, H.-G. (1978). Positional head reflexes and the role of the prosternal organ in the walking fly, *Calliphora erythrocephala*. *J. Comp. Physiol.* 126, 137–146.

JANDER, R. (1963a). Grundleistung der Licht- und Schwereorientierung von Insekten. *Z. Vergl. Physiol.* 47, 381–430.

JANDER, R. (1963b). Insect orientation. *Ann. Rev. Entomol.* 8, 95–114.

JANDER, R. and DAUMER, K. (1974). Guide-line and gravity orientation of blind termites foraging in the open (Termitidae: *Macrotermes*, *Hospitalitermes*). *Insectes Soc.* 21, 45–69.

JANDER, R. and JANDER, U. (1970). Über die Phylogenie der Geotaxis innerhalb der Bienen (Apoidea). *Z. Vergl. Physiol.* 66, 355–368.

JANDER, R., HORN, E. and HOFFMANN, M. (1970). Die Bedeutung des Körpergewichtes für die Geotaxis der höheren Insekten (Pterygota). *Z. Vergl. Physiol.* 66, 326–342.

KÜHN, A. (1919). *Die Orientierung der Tiere im Raum*. Gustav Fischer Verlag, Jena.

LENGERKEN, H. V. (1934). Beine als Schwirrorgane. *Biol. Zbl.* 54, 646–650.

LINDAUER, M. (1957). Sonnenorientierung der Bienen unter der Äquatorsonne und zur Nachtzeit. *Naturwiss.* 44, 1–6.

LINDAUER, M. and MARTIN, H. (1968). Die Schwereorientierung der Bienen unter dem Einfluss des Erdmagnetfeldes. *Z. Vergl. Physiol.* 60, 219–243.

LINDAUER, M. and NEDEL, J. O. (1959). Ein Schweresinnesorgan bei der Honigbiene. *Z. Vergl. Physiol.* 42, 334–364.

LISKE, E. (1977). The influence of head position on the flight behaviour of the fly, *Calliphora erythrocephala*. *J. Insect Physiol.* 23, 375–379.

MARKL, H. (1962). Borstenfelder an den Gelenken als Schweresinnesorgane bei Ameisen und anderen Hymenopteren. *Z. Vergl. Physiol.* 45, 475–569.

MARKL, H. (1964). Geomenotaktische Fehlorientierung bei *Formica polyctena* Förster. *Z. Vergl. Physiol.* 48, 552–586.

MARKL, H. (1966a). Peripheres Nervensystem und Muskulatur im Thorax der Arbeiterin von *Apis mellifica* L., *Formica polyctena* Förster und *Vespa vulgaris* L. und der Grundbauplan der Innervierung des Insektenthorax. *Zool. Jb. Anat. Bd.* 83, 107–184.

MARKL, H. (1966b). Schwerkraftdressuren an Honigbienen. II. Die Rolle der schwererezeptorischen Borstenfelder verschiedener Gelenke für der Schwerekompassorientierung. *Z. Vergl. Physiol.* 53, 353–371.

MARTIN, H. and LINDAUER, M. (1966). Sinnesphysiologische Leistungen beim Wabenbau der Honigbiene. *Z. Vergl. Physiol. 53*, 372–404.

MARTIN, H. and LINDAUER, M. (1977). Der Einfluss des Erdmagnetfeldes auf die Schwereorientierung der Honigbiene (*Apis mellifica*). *J. Comp. Physiol. 122*, 145–187.

MAY, M. L., WILKIN, P. J., HEATH, J. E. and WILLIAMS, B. A. (1980). Flight performance of the moth, *Manduca sexta*, at variable gravity. *J. Insect Physiol. 26*, 257–265.

MITTELSTAEDT, H. (1950). Physiologie des Gleichgewichtssinnes bei fliegenden Libellen. *Z. Vergl. Physiol. 32*, 422–463.

PRINGLE, J. W. S. (1948). The gyroscopic mechanism of the halteres of Diptera. *Phil. Trans. Roy. Soc. B. 233*, 347–384.

RABE, W. (1953). Beiträge zum Orientierungsproblem der Wasserwanzen. *Z. Vergl. Physiol. 35*, 300–325.

SANDEMAN, D. C. (1980). Angular acceleration, compensatory head movements and the halteres of flies (*Lucilla serricata*). *J. Comp. Physiol. 136*, 361–367.

SCHNEIDER, G. (1953). Die Halteren der Schmeissfliege (*Calliphora*) als Sinnesorgane und als mechanische Flugstabilisatoren. *Z. Vergl. Physiol. 35*, 416–458.

THORPE, W. H. and CRISP, D. J. (1947). Studies on plastron respiration. III. The orientation response of *Aphelocheirus* (Hemiptera, Aphelocheiridae (Naucoridae)) in relation to plastron respiration; together with an account of specialised pressure receptors in aquatic insects. *J. Exp. Biol. 24*, 310–328.

THURM, U. (1963). Die Beziehungen zwischen mechanischen Reizgrössen und stationären Erregungszuständen bei Borstenfeld-Sensillen von Bienen. *Z. Vergl. Physiol. 46*, 351–382.

THURM, U. (1965). An insect mechanoreceptor. I: Fine structure and adequate stimulus. *Cold Spring Harbor Symp. Quant. Biol. 30*, 75–82.

TOMIOKA, K. and YAMAGUCHI, T. (1980). Steering responses of adult and nymphal crickets to light, with special reference to the head rolling movement. *J. Insect Physiol. 26*, 47–57.

VOWLES, D. M. (1954). The orientation of ants. II. Orientation to light, gravity and polarized light. *J. Exp. Biol. 31*, 356–375.

WALTHALL, W. W. and HARTMAN, H. B. (1981). Receptors and giant interneurons signaling gravity orientation information in the cockroach *Arenivaga*. *J. Comp. Physiol. 142*, 359–369.

WEBER, H. (1933). *Lehrbuch der Entomologie*. Gustav Fischer Verlag, Stuttgart.

WENDLER, G. (1965). Über den Anteil der Antennen an der Schwererezeption der Stabheuschrecke *Carausius morosus* BR. *Z. Vergl. Physiol. 51*, 60–66.

WENDLER, G. (1969). Was messen die Beine von *Carausius morosus* beim Ermitteln der Schwerkraftrichtung? *Verh. Dtsch. Zool. Ges., Innsbruck, 1968*, 439–444.

WENDLER, G. (1972). Körperhaltung bei der Stabheuschrecke: ihre Beziehung zur Schwereorientierung und Mechanismen ihrer Regelung, *Verh. Dtsch. Zool. Ges. 1971*, 214–219.

WIGGLESWORTH, V. B. (1946). Organs of equilibrium in flying insects. *Nature 157*, 655.

11 Clocks and Circadian Rhythms

TERRY L. PAGE

Vanderbilt University, Nashville, Tennessee, USA

1 INTRODUCTION: CLOCKS AND CIRCADIAN ORGANIZATION

As a consequence of the motion of the earth relative to the sun and moon the immediate environment of most organisms exhibits dramatic daily and seasonal fluctuations. The response of natural selection has been the evolution of regulatory systems whose primary function is to match these cyclic environmental challenges with appropriate, periodic alterations in physiology, biochemistry, and behavior. Such systems are found in most, if not all, eukaryotic organisms and are a pervasive feature of biological organization.

The ability of these regulatory systems to generate an adaptive "temporal program" (Pittendrigh, C., 1981a,b) for the organism depends first on a reasonably accurate registration of the passage of time, and second on some mechanism to synchronize the program to local time. In consequence, the systems function as clocks, providing the organism with both a measure of the lapse of time and a recognition of the local time of day.

These "biological clocks" are prominent and particularly well studied in insects. They underlie the daily rhythmicity characteristic of many aspects of insect physiology and behavior, are involved in the control of seasonal cycles of development and reproduction, and are used in time-compensated sun orientation. The biological clocks of insects share many common features with clocks in other organisms. With a few important exceptions they are based on endogenously generated, self-sustaining oscillations whose periods approximate those of naturally occurring environmental cycles. These oscillations will persist in the laboratory in the absence of any identifiable periodic external cues, but can be synchronized (or entrained) by appropriate periodic stimuli. The significance of entrainment is the establishment of a specific, stable phase relationship between the oscillation and the environmental cycle — i.e., entrainment sets the clocks to local time.

The most common and most thoroughly studied oscillations are those with period lengths that are *circadian* — near the 24 h of the solar day. However, oscillations with periods that have evolved as approximations of other environmental cycles (circa-annual, circa-lunar) as well as non-oscillatory (hourglass) clocks have been discovered in insects, and in a few cases have been the subject of detailed analysis.

The focus of this review will be on circadian clocks and the overt rhythms they control. Three general questions have been the subject of extensive investigation over the past 30 years:

(1) What sorts of processes do circadian clocks regulate and what is the adaptive significance of this control?

(2) What is the mechanism by which the oscillations are phased (entrained) to local time?

(3) What is the anatomical and physiological substrate for the generation of the oscillation?

Each of these questions are dealt with in some detail below but a few general comments seem appropriate.

The diversity of processes in insects that are under the control of circadian oscillators is extensive, ranging from the activities of specific enzymes through the regulation of general metabolic rate to the initiation of specific, complex sequences of behavior that involve both the nervous and endocrine systems. The adaptive significance of this pervasive regulatory control, that must either primarily or secondarily impact virtually every aspect of the individual's biology, surely lies, in part, with restricting those processes that are best undertaken at a particular phase of the environmental cycle to a particular time of day. The provision for entrainment of the oscillator (by daily cycles of light or temperature) ensures that its phase is appropriate to local conditions. There is also a growing conviction that a second major function of circadian oscillations may be to provide a temporal framework for an internal organization of various activities relative to each other (Pittendrigh, C., 1981a,b; Moore-Ede, M. and Sulzman, F., 1981). In this view internal *circadian organization* is an important adjunct to temporal order with respect to the external environment.

In the past 20 years there has been a substantial, and largely successful, effort to identify the anatomical correlates and physiological mechanisms of circadian clocks and circadian organization. Two general experimental strategies have been used. The first essentially treats the circadian system as a black box and its properties are inferred from the behavior of overtly expressed rhythms in response to external stimuli, usually light. The results of this approach have led to a sophisticated understanding of the entrainment process and its central role (in at least some insect species) in photoperiodic time measurement (Pittendrigh, C., 1967; Saunders, D., 1981a,b).

The second strategy has been to employ the traditional experimental approaches of neurophysiology and endocrinology — primarily lesioning and transplantation techniques — in an effort to "dissect" the circadian system. The results have confirmed the expectation that the machinery for generating circadian oscillations and regulating circadian organization resides primarily, though not exclusively, with the nervous and neuroendocrine systems.

Finally, the results of both approaches have led to the general proposition that circadian organization is derived from a population of anatomically discrete circadian oscillators whose temporal order is maintained by their submission to entrainment — either *via* direct access to temporal cues from the environment or by other oscillators within the population (Pittendrigh, C., 1981b; Page, T. 1981b,c). This emerging view, developed largely by C. Pittendrigh (1960, 1974, 1981a,b), suggests one or a few driving oscillators or "pacemakers" impose temporal order on the diverse processes they regulate through entrainment of a population of secondary or "slave" oscillators that in turn are responsible for the timing of overt rhythmicity.

The purpose of this review is to summarize the experimental foundations of our current understanding of biological clocks and their role in temporal organization in insects. It will be apparent that much work remains to be done before this understanding is complete. Nevertheless, substantial progress has been made in defining the general principles, sharpening the salient experimental questions and identifying preparations especially suitable for analysis. A solid and extensive foundation on which to build a comprehensive understanding of temporal organization in insects has been laid.

2 CIRCADIAN RHYTHMICITY IN INSECTS

Endogenously generated daily rhythms have been observed in a wide variety of behavioral, physiological, and metabolic processes of a large number of insect species. The examples of insect circadian rhythms presented in Table 1 illustrate both the range of species investigated and the types of rhythms that have been reported. This list is by no means exhaustive and more extensive tables can be found in Saunders, D. (1976a) and Beck, S. (1980). The following pages are a summary of circadian

Table 1: Examples of processes for which regulation by the circadian system has been demonstrated

	Process	Species	Reference
A.	Locomotor activity	*Acheta domesticus*	Nowosielski, J. and Patton, R. (1963)
		Apis mellifera	Spangler, H. (1972)
		Culex pipiens	Jones, M. (1976)
		Drosophila melanogaster	Konopka, R. and Benzer, S. (1971)
		Glossina morsitans	Brady, J. (1972)
		Leucophaea maderae	Roberts, S. (1960)
B.	Stridulation	*Ephippiger ephippiger*	Dumortier, B. (1972)
		Teleogryllus commodus	Loher, W. (1972)
C.	Oviposition	*Aëdes aegypti*	Haddow, A. and Gillett, J. (1957)
		Chorthippus curtipennis	Loher, W. and Chandrashekaran, M (1970)
		Ostrinia nubilalis	Skopik, S. and Takeda, M. (1980)
		Pectinophora gossypiella	Minis, D. (1965)
D.	Spermatophore formation	*Teleogryllus commodus*	Loher, W. (1974)
E.	Sperm release from testes	*Anagasta kuhniella*	Riemann, J. *et al.* (1974)
F.	Egg hatch	*Masonia titillans*	Nayar, J. *et al.* (1973)
		Pectinophora gossypiella	Minis, D. and Pittendrigh, C. (1968)
G.	Pupation	*Aëdes taeniorhynchus*	Nayar, J. (1967)
		Drosophila victoria	Rensing, L. and Hardeland, R. (1967)
H.	Pupal eclosion	*Antheraea pernyi*	Truman, J. and Riddiford, L. (1970)
		Drosophila pseudo-obscura	Pittendrigh, C. (1954)
		Pectinophora gossypiella	Pittendrigh, C. and Minis, D. (1971)
		Sarcophaga argyrostoma	Saunders, D. (1976)
I.	PTTH secretion	*Manduca sexta*	Truman, J. (1972)
		Samia cynthia	Fujishita, M. and Ishizaki, H. (1981)
J.	Daily cuticle growth	*Leucophaea maderae*	Lukat, R. (1978)
		Oncopeltus fasciatus	Dingle, H. *et al.* (1969)
		Schistocerca gregaria	Neville, A. (1965)
K.	Pheromone release	*Megoura viciae*	Marsh, D. (1972)
		Trichoplusia ni	Sower, L. *et al.* (1970)
		Trogoderma glabrum	Hammack, L. and Burkholder, W. (1976)
L.	Oxygen consumption	*Periplaneta americana*	Richards, A. (1969)
M.	Insecticide sensitivity	*Musca domesticus*	Shipp, E. and Otton, J. (1976)
N.	Electroretinogram amplitude	*Dysticus fasciventris*	Jahn, T. and Crescitelli, J. (1940)
		Blaps gigas	Koehler, W. and Fleissner, G. (1978)

rhythms in insects. The discussion of activity rhythms, in particular, will provide a background on which to describe some general properties of circadian oscillations (e.g. stability of period, effects of light, temperature, etc.). In a few cases the study of particular rhythms has been truly extensive, comprising two or more decades of intense investigation such as the rhythms of eclosion behavior in fruit-flies and locomotor activity in cockroaches. As a result data from these organisms are heavily relied upon to illustrate general properties.

2.1 Locomotor activity rhythms and general properties of circadian oscillations

Locomotor activity has long been a favorite behavior to study for investigations on circadian rhythmicity, both because of the ease with which it can be recorded for long periods of time and because it can be measured with minimal disturbance to the organism under study. When individual insects are maintained in natural or artificial cycles of light and/or temperature, activity is primarily restricted to a particular time period during the day. In cockroaches, for example, the majority of activity occurs at night and typically is only limited and scattered during the daylight hours (e.g. Roberts, S., 1962) (Fig. 1). In contrast, the diurnal flight activity of the bee *Apis mellifera* occurs throughout the day and there is little activity at night (Spangler, H., 1972, 1973). Frequently (especially in diurnal forms) the activity is bimodal

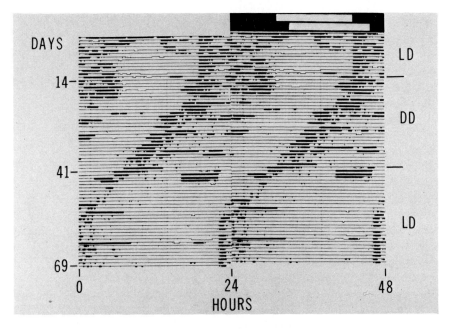

FIG. 1. Event recording of running wheel activity of a cockroach *Leucophaea maderae*. Data for successive days are placed one below the other in chronological order, and the record has been duplicated to provide a 48 h time base to aid in visual inspection of the data. The bars at the top of the record represent the light cycles to which the animal was exposed, with the shaded portion of the bar corresponding to the dark phase of the *LD 12* : 12 cycle. The record begins with the animal entrained to the LD cycle illustrated by the top bar (temperature was constant at 24.5°). On day 14 the LD cycle was discontinued and the animal began to freerun in DD with a period of about 23.5 h. On day 41 the animal was exposed to the second LD cycle. After several transient cycles the rhythm becomes phase locked to the LD cycle (Page, unpublished.)

with peaks occurring near dawn and dusk. This is particularly clear, for example, in the flight behavior of the tsetse fly *Glossina morsitans* (Brady, J., 1972).

These cycles of activity and rest, as well as the specific details of activity patterns of particular individuals, are influenced by a number of factors, and often appear to be the result of both endogenously and exogenously driven processes. The endogenous nature of periodic activity/rest cycles is most easily revealed by placing the organism in constant environmental conditions, generally constant light and temperature. In most cases a rhythm of activity persists (Fig. 1) and in constant conditions the rhythm is described as "free-running". Free-running rhythms behave formally as self-sustaining oscillations (Pittendrigh, C. and Bruce, V., 1957), and the free-running period (τ) of the rhythm (e.g. the time from beginning of one cycle of activity and rest to the next) is taken to be a reflection of the "natural" period of an underlying endogenous oscillator or pacemaker that drives the periodic behavior. Typically the periods of free-running rhythms are close to, but not exactly, a day, and

may be either slightly longer or slightly shorter than 24 h.

Free-running periods of individuals within a species are, in general, restricted to a small range of values, but may vary within this range from one individual to the next. For example, in adult male *Leucophaea maderae* that have been raised in a 24 h light cycle consisting of 12 h of light alternated with 12 h of darkness (*LD 12* : 12) the range of free-running periods measured in constant darkness (DD) at 24.5° is about 23.5–24.0 h (Page, T. *et al.*, 1977; Page, T. and Block, G., 1980). The range of τ values for a species is at least partly under genetic control, but it may also be affected by external factors such as environmental conditions during postembryonic development (see section 7.3).

Within the individual, τ is also subject to variation. In cockroaches both spontaneous (unexplained) and "induced" changes in individual τ values have been observed (Roberts, S., 1960; Page, T. and Block, G., 1980). In the latter case it was also shown that exposure of adults for several weeks to 22 h (*LD 11* : 11) or 26 h (*LD 13* : 13) light cycles caused a slight shortening or lengthening, respectively, of τ measured in DD (Page, T. and Block, G., 1980). This "history-dependence" of τ on prior lighting conditions appears to be a common feature of circadian rhythms in metazoans. These τ changes are usually referred to as "after-effects" (Pittendrigh, C., 1960; Pittendrigh, C. and Daan, S., 1976).

An important and apparently universal feature of circadian rhythmicity is that the free-running period is at most only slightly affected by changes in the constant level of ambient temperature (not to be confused by cycles in ambient temperature). The significance of the "temperature compensation" of τ was first explicitly recognized by Pittendrigh, C. (1954), who demonstrated the near independence of τ from temperature for the rhythm in eclosion behavior in *Drosophila pseudo-obscura*. He pointed out that if the circadian oscillator was to function effectively as a "clock" then the period must necessarily be relatively independent of temperature. This proposition was recently expanded by Pittendrigh, C. and Caldarola, P. (1973) who noted that it is a functional necessity that the frequency of circadian oscillations be "homeostatically" conserved in the face of all possible changes in the oscillator's environment.

The activity record in Fig. 2 illustrates the phenomenon of temperature compensation in *Leucophaea maderae*, first demonstrated by Roberts, S. (1960) and subsequently investigated in an extensive series of experiments by Caldarola, P. (1974). Interestingly, the slight dependence of τ on $T°$ in *Leucophaea* is a non-monotonic function of the temperature (Fig. 3). For lower temperatures (17–22°) the Q_{10} is slightly less than 1, and for higher temperatures (23–35°) the Q_{10} is slightly greater than 1 (Caldarola, P., 1974). Similar results have also been obtained with another cockroach *Byrsotria fumigata* (Caldarola, P., 1974). Temperature compensation has been demonstrated in several other insect species including activity rhythms of *Periplaneta americana* (Bünning, E., 1958) and *Tenebrio molitor* (Lohmann, M., 1964). It should be emphasized that the stability of τ with respect to a constant level of ambient temperature should not be construed to reflect a complete independence of the oscillator from temperature effects. Temperature cycles are fully effective in entraining

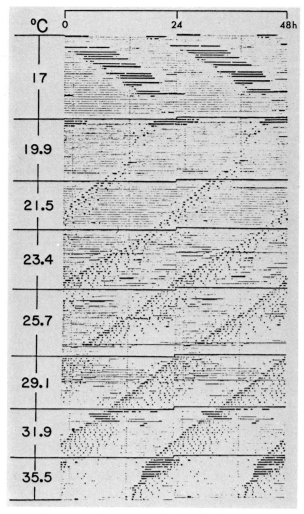

FIG. 2. Temperature compensation of the free-running period of the activity rhythm in *Leucophaea maderae*. The animal was maintained in constant darkness through the record. Ambient temperatures are given to the left. (Adapted from Caldarola, P., 1974.)

circadian rhythms in insects, and temperature pulses or steps can cause shifts in the phase of the oscillation (see section 3.6).

In several insect species free-running activity rhythms also persist in constant light (LL). Typically the period of the rhythm in LL is different from the period in DD (Fig. 4). In many cases the change in τ follows "Aschoff's Rule" (Pittendrigh, C., 1960) which is an empirically derived principle that states that for diurnal animals τ decreases with increasing light intensity (or, more simply, $\tau_{LL} < \tau_{DD}$), and for nocturnal animals τ increases with increasing light intensity ($\tau_{LL} > \tau_{DD}$). There are

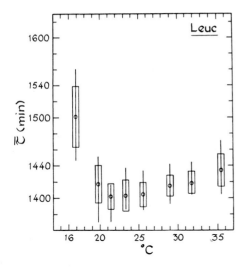

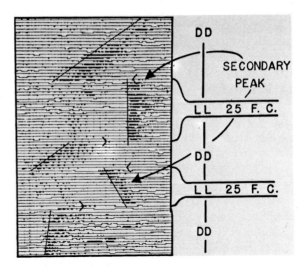

FIG. 3. Free-running period of the activity rhythm of *Leucophaea maderae* as a function of temperature. Shown are mean τ values (circle), standard deviations (rectangle); and the range (vertical line). (From Caldarola, P., 1974.)

FIG. 4. Activity record of *Leucophaea maderae* in constant darkness (DD) and constant light (LL, 25 foot candles). The record illustrates the lengthening of the period and the emergence of a secondary peak of activity in LL. (From Roberts, S., 1960.)

however, several exceptions to the rule including the nocturnal pit-building activity rhythm of the ant-lion *Myrmeleon obscura* (Youthed, G. and Moran, V., 1969a), and the diurnal flight activity rhythm of the mosquito *Aëdes aegypti* (Taylor, B. and Jones, M., 1969). The results of several studies on the effects of light intensity on τ have been summarized by Aschoff, J. (1979).

The detailed pattern of activity of free-running rhythms can also be influenced by several factors including light intensity, temperature, and reproductive or developmental state. In the cockroach *Leucophaea maderae* the activity rhythm is occasionally bimodal in DD with a second minor peak of activity occurring a few hours after the major peak. In LL this secondary peak is invariably expressed as the major peak of activity (Fig. 4) (Roberts, S., 1960). In the mosquito *Culex pipiens* LL has a different effect on the pattern of activity. The flight activity rhythm in DD is clearly bimodal, whereas in LL the pattern becomes unimodal (Jones, M., 1976). In other cases exposure to constant light may affect the amount of activity as in the mating behavior of *Dacus tryoni* (Tychsen, P. and Fletcher, B., 1971) or may cause damping in the amplitude of the rhythm as in the flight activity of the mosquito *Anopheles gambiae* (Jones, M., 1973) and the tsetse fly *Glossina morsitans* (Brady, J., 1972).

Patterning of activity can also be reproducibly affected by ambient temperature. In *G. morsitans* both high (35°) and low (19°) constant temperature reduce spontaneous activity by about 80% and disproportionately depress the evening peak (Brady, J. and Crump, A., 1978). In *Leucophaea maderae* low temperatures (17–20°) typically lead to the expression of the activity peak as a single band with a sharp onset and offset. At higher temperatures (20–30°) the activity usually occurs as several short bursts within a peak that has a regular onset but variable offset (Fig. 2) (Caldarola, P., 1974).

The entrainment of the activity rhythm by light or temperature (discussed in more detail in the next section) can also affect the pattern of activity by "reflexively" triggering behavior exclusive of any effect on the oscillatory system that drives the free-running rhythm. In extreme cases this exogenous influence may be primarily responsible for rhythmic behavior. For example, locomotor activity and oviposition of the stick insect *Carausius morosus* are clearly periodic in LD 12:12, but the rhythms are frequently lost upon transfer of individuals to constant darkness (Godden, D., 1973). Similar behavior has been observed in the locust *Schistocerca gregaria* (Odhiambo, T., 1966). In other cases only a component of the rhythmic activity in the presence of an environmental cycle may be exogenously driven. In a light cycle the mosquito *Anopheles gambiae* exhibits a strong peak of activity at the dark to light transition that does not persist in constant conditions, although the primary diurnal peak free-runs in DD (Jones, M. *et al.*, 1972). This "lights-on" response also disappears if the change in light intensity is gradual rather than abrupt. Exogenously driven peaks are also evident in some, though not all, individuals of several species of cockroaches maintained in LD cycles at constant temperature (Roberts, S., 1962; Nishiitsutsuji-Uwo, J. *et al.*, 1967; Ball, H., 1972) or maintained in a temperature cycle in DD (Page, unpublished), and in the cricket *Teleogryllus commodus* the light to dark transition inhibits stridulation (Loher, W., 1972; Sokolove, P., 1975a) and triggers locomotor activity (Sokolove, P., 1975a).

Patterns of activity or even the expression of rhythmic behavior can also be affected by the developmental (see section 7) or reproductive state of the organism. In female *A. gambiae* both insemination and oviposition cause changes in the amount and pattern of activity in LD 12:12 and DD (Jones, M. and Gubbins, S., 1977, 1978). Similarly, the expression of rhythmicity in female cockroaches *Leucophaea maderae* (Leuthold, R., 1966) and *Periplaneta americana* (Lipton, G. and Sutherland, D., 1970) and in the ant *Camponotus clarithorax* (McCluskey, E. and Carter, C., 1969) appears to be affected by the reproductive state. In the latter case it appears that rhythmicity in the female is lost permanently as a result of mating.

It is clear (and not surprising) from the results discussed that the control of the temporal distribution of activity in insects is a complex process under the influence of factors in both the external and internal environment of the organism. The primary concern here is with the endogenous oscillation that imposes a circadian or diurnal periodicity on the activity. While the pattern and amount of activity are significantly affected by this oscillation it is equally certain that processes other than those directly related to the generation of sustained rhythmicity will also have a major impact. For this reason one must be very cautious in interpretations that attempt to relate either pattern or amplitude to the state of the "biological clock", and as has been explicitly pointed out (e.g. Pittendrigh, C., 1976),

the only parameters of the overt rhythm that can reasonably safely be assumed to be taken as measures of the state of the circadian oscillator are the period of the rhythm and its phase *in steady-state*.

2.2 Rhythms in reproductive behavior

There are now several examples of various processes associated with reproductive activity that are under the control of the circadian system. These include sex pheromone-release, calling-song behavior, mating, oviposition, and spermatophore production.

Of these sex-pheromone release has probably been the most extensively documented, and persistent rhythmicity in constant conditions has been demonstrated in several species of moth including *Trichoplusia ni* (Sower, L. *et al.*, 1970), *Anagasta kuhniella* (Traynier, R., 1970), *Adoxophyes fasciata* (Nagata, K. *et al.*, 1972) and *Dioryctria abietella* (Fatzinger, C., 1973). Similar rhythms have been reported in the dermestid beetle *Trogoderma glabrun* (Hammack, L. and Burkholder, W. 1976) and in the aphids *Megoura viciae* (Marsh, D., 1972) and *Schizaphis gramium* (Eisenbach, J. and Mittler, T., 1980). In general the release of sex pheromone occurs during the active phase of the potential mate's locomotor rhythm (Bartell, R. and Shorey, H., 1969; Hammack, J. and Burkholder, W., 1976) and may correspond to a peak of rhythmic sensitivity of the opposite sex to the pheromone (Shorey, H. and Gaston, L., 1965; Traynier, R., 1970). Both factors would likely improve reproductive success.

Circadian rhythms of calling song activity have been described in a few species of Orthoptera. Stridulation in *Teleogryllus commodus* occurs primarily at night in animals maintained in LD 12:12 (Loher, W., 1972; Sokolove, P., 1975a). The rhythm free-runs both in constant darkness and constant light. Similar rhythms have been reported in several species of *Ephippiger* (Dumortier, B., 1972).

Oviposition behavior also appears to be "gated" (see section 2.3) by the circadian system in several insect species so that eggs are deposited only at particular times of day. This behavior has been studied in some detail in the mosquito *Aëdes aegypti* (Haddow, A. and Gillett, J., 1957) and in the pink bollworm *Pectinophora gossypiella* (Pittendrigh, C.

and Minis, D., 1964, 1971; Minis, D., 1965). Oviposition in *A. aegypti* populations maintained in an artificial light cycle or observed under natural conditions in the field occurs as a well-defined peak near dusk. In conditions of constant darkness a clear circadian rhythm persists for several cycles (Haddow, A. *et al.*, 1961). In *P. gossypiella* oviposition occurs almost exclusively in the dark in animals maintained in light cycles (Fig. 5). In constant darkness the rhythm persists with a period of about 22.7 h (Pittendrigh, C. and Minis, D., 1964; Minis, D., 1965). Free-running rhythms in oviposition behavior have also been reported for the moth *Crambus topiarius* (Crawford, C., 1967), the grasshopper *Chorthippus curtipennis* (Loher, W. and Chandrashekaran, M., 1970), and the European cornborer *Ostrinia nubilalis* (Skopik, S. and Takeda, M., 1980). An interesting example of rhythmicity in oviposition that apparently is not under circadian control has been reported for the milkweed bug *Oncopeltus fasciatus* (Rankin, M. *et al.*, 1972). In this case the periodicity persists in LL, but the period is strongly temperature-dependent. It appears that the periodic oviposition under these conditions is a consequence of the temperature-dependent developmental cycle of the egg.

A final example of a reproductively related rhythm is that of spermatophore formation in the cricket *Teleogryllus commodus*. Loher, W. (1974) found that in *T. commodus* one spermatophore is formed per day. It is first present 1–5 h prior to the onset of stridulation, is retained during the stridulatory phase, and is disposed of a variable period of time after the end of stridulation. The periodicity and its phase relationship to stridulation persists in constant conditions (LL or DD). A similar rhythm — that of the movement of sperm bundles into the seminal vesicles and ejaculatory duct — has been described in the Mediterranean flour-moth *Anagasta kuhniella* (Riemann, J. *et al.*, 1974; Thorson, B. and Riemann, J., 1977). The rhythm is abolished by constant light but persists in DD.

2.3 "Gated" rhythms in development

There are a number of developmental events that occur only once in the life of the insect. These

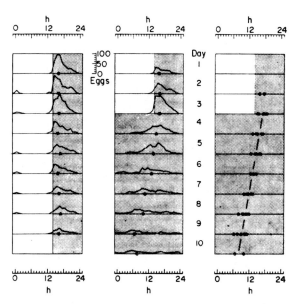

FIG. 5. Oviposition rhythm of the moth *Pectinophora gossypiella*. The left panel shows the entrained rhythm (*LD 14 : 10*). The middle panel shows the rhythm free-running in DD (days 4–10). The right-hand panel shows medians of seven free-running rhythms. τ is about 22.6 h. (From Minis, D., 1965.)

include processes such as egg hatch, a particular larval ecdysis, pupation, and eclosion. Frequently the timing of these events is under the control of a circadian clock and has been the subject of some of the most extensive investigations of circadian rhythmicity. Because the individual performs the act (e.g. adult emergence) only once in its life, rhythmicity is only observed in populations of insects that are developmentally asynchronous, but whose circadian clocks are in relatively close synchrony, An example of such a rhythm is shown in Fig. 6, which shows the rhythm of adult emergence in 12 replicate populations of pupae of *Drosophila pseudo-obscura* of mixed developmental age (Pittendrigh, C., 1967). In *LD 12 : 12* the flies' emergence is restricted to the hours shortly after the dark to light transition. When transferred to DD the periodicity of eclosion activity persists with a period very close to 24 h, and emergence is still restricted to a sharply defined peak of activity each day.

Often rhythmicities of unitary events in populations of organisms are referred to as "gated" rhythms. In descriptive terms the event — for example emergence — is directly controlled by a "gate" that, in turn, is controlled by a circadian oscillation.

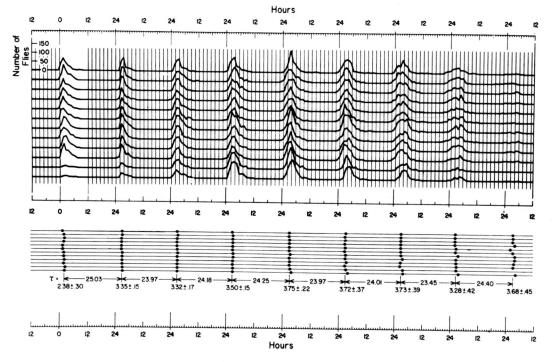

FIG. 6. Eclosion rhythm of *Drosophila pseudo-obscura*. Upper panel plots the number of emergences as a function of time for 12 replicate populations entrained to *LD 12*: 12 and released into DD. Lower panel shows medians for each eclosion peak of the 12 populations. (From Pittendrigh, C., 1966.)

The gate is opened for a specified period of time once each day. During the time the gate is open, flies that have completed development are able to emerge. Those flies that are not developmentally competent to eclose in that gate must wait for the next (or later) open gates. In developmentally asynchronous populations if the circadian oscillators of individuals are synchronous, gates will open synchronously, and within the population open gates will recur at τ intervals in constant conditions.

Gated rhythms of pupal eclosion have been described in several species of *Drosophila* (Brett, W., 1955; Pittendrigh, C. and Skopik, S., 1970) as well as in a number of other Diptera, Lepidoptera, and Coleoptera. These include the giant silkmoths (Truman, J., 1971b), the pink bollworm *Pectinophora gossypiella* (Minis, D., 1965), several species of marine midges (Neumann, D., 1976a), and the flesh-fly *Sarcophaga argyrostoma* (Saunders, D., 1976b).

Other examples of gated developmental events include pupation — for example, in the mosquitos *Aëdes taeniorhynchus* (Nayar, J., 1967a,b) and *Anopheles gambiae* (Jones, M. and Reiter, P., 1975),

and prepupae formation in *Drosophila victoria* (Rensing, L. and Hardeland, R., 1967). In *P. gossypiella* both the time of eclosion and the time of egg hatching are gated by the circadian system. Both occur early in the day, although the phase of the eclosion peak is a few hours later than that of egg hatch (Pittendrigh, C. and Minis, D., 1971).

There are also several instances worth noting where particular developmental processes are not under control of the circadian system. For example, in both *D. melanogaster* and *D. psuedo-obscura* there is clear evidence that pupariation is independent of the time of day, as is adult emergence of the mosquito *Aëdes aegypti* (Haddow, A. *et al.*, 1959). In some cases the temporal control of one developmental event will lead to a "fortuitous" periodicity of a subsequent event that occurs a fixed time after the first. This has been demonstrated for eclosion in the mosquito *Aëdes taeniorhynchus*. Pupation is timed by a circadian oscillator, and adult emergence occurs at a fixed (but temperature-dependent) time after pupation (Nayar, J., 1967b).

A particularly interesting example of what could be called fortuitous gating (Beck, S., 1980) has been described in the tobacco hornworm (Truman, J., 1972a). When the time of ecdysis between various larval instars was examined it was found that each instar ecdysed at a characteristic time of day, although as the animal developed the time of each ecdysis occurred later in the day than the previous one, and the distribution of ecdysis in the population was broadened. The explanation of these facts appears to be that ecdysis itself is not gated but that the release of prothoracicotropic hormone (PTTH), used to initiate each molt, is. The synchrony in larval ecdysis simply reflected the fact that molt initiation was occurring at the same time of day 1.5–2 days before when PTTH was released. Because larger larvae require more time to form the larger cuticle, the completion of development required more time; and therefore ecdysis for older animals was delayed to later times of day. Similar results on the rhythmic release of PTTH and the timing of larval ecdysis have recently been described for the saturnid *Samia cynthia* (Fujishita, M. and Ishizaki, H., 1981).

The idea that rhythmicity in a population may not be a function of a circadian oscillation in the individual, but simply occurs as an emergent property of the time it takes to complete all the prior developmental steps, has in some cases been misapplied. Harker, J. (1964, 1965a,b), for example, suggested that the eclosion rhythm observed in *Drosophila melanogaster* was due to the facts that (a) the time interval between successive developmental steps is a function of the phase of the LD cycle at which the first event occurs, and (b) the fly would simply emerge when development is complete. Thus in Harker's view the synchrony in emergence in the population resulted not from gating by a circadian oscillator but from the exogenous effect of the LD cycle in controlling developmental rate in a way that makes developmentally asynchronous pupae developmentally synchronous.

Pittendrigh, C. and Skopik, S. (1970) and Skopik, S. and Pittendrigh, C. (1967) performed a thorough test of this proposition in three species of *Drosophila* in which they failed to substantiate Harker's view. Their approach was to create a developmentally synchronous population of flies by collecting newly formed puparia in a 1h collection window. The developmental progress of these flies was then followed by scoring the times of three subsequent

events in development that could be readily visualized through the pupal case. Besides the time of pupariation (t_0), they determined the time of head eversion (t_h), bristle pigmentation (t_b), yellow eye pigmentation (t_y) and the time of eclosion (t_e). In the first series of experiments groups of flies were maintained in constant light which damps the emergence rhythm. They found that there was considerable interindividual variation in developmental rate — for example the 1 h range for pupariation (t_0) was amplified to 14 h for bristle pigmentation (t_b) and to a range of 28 h for emergence events (t_e). As expected, flies emerged at all times of day. The rate of development was also found to be temperature-dependent.

When developmentally synchronous flies were transferred to constant darkness or to an LD cycle, the results showed a striking difference. Figure 7 summarizes data from 27 populations of developmentally synchronous flies (raised in LL) that were transferred to DD at different times relative to t_0. The top panel plots medians for t_y, t_b, and t_e. It is clear that while t_y and t_b occur at a fixed interval after t_0, t_e does not. The time of eclosion only occurs within gates (~ 6 h wide) that recur at ~ 24 h intervals after the LL/DD transition $+ 15$ h. These results suggest the LL/DD transition initiates or resets the oscillation to a fixed phase. The oscillation then persists with a period of about 24 h. The lower three panels of Fig. 7 show the distribution of developmental events (histograms) for populations 14–27 from above compared with predicted distributions (smooth curves) assuming that each developmental event occurs with probability (from Fig. 7) based on time from t_0. The observed distributions of t_y and t_b are a close fit to prediction, there is no evidence that these events are gated. In contrast the time of eclosion t_e is clearly different from the distribution predicted on basis of no gating. Similar results were obtained for flies in *LD* 18:6 (Pittendrigh, C. and Skopik, S., 1970). The only plausible interpretation of these results is that t_y and t_b are unaffected by either an exogenous light cycle or an endogenous oscillation. Emergence, on the other hand, is clearly "gated". Animals in which development is not complete at the opening of one gate must wait a full 24 h for the next gate. Furthermore, the phases of the gates can be set by the light/dark transition or light pulses (Skopik, S. and

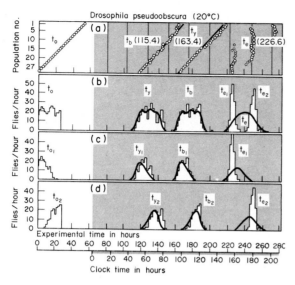

FIG. 7. Illustrates developmental events that are gated by a circadian oscillator (eclosion) and ungated events (eye and bristle pigmentation) in *Drosophila pseduo-obscura* males after a step from constant light to constant dark at 20°. In (a) the medians of times of emergence (t_e), yellow eye pigmentation (t_y) and bristle pigmentation (t_b) are plotted for 27 populations of developmentally synchronous pupae; t_o represents the time of a 1 h window during which prepupae of each population were collected. In panel (b) populations 14–27 from (a) are treated as a single (synthetic) population. In (c) and (d) the flies that made up the two peaks of emergence shown in (b) are treated separately. In the lower three panels (b,c,d) histograms show the actual distribution of each event, the smoothed curves are the predicted distributions assuming the events are not gated (see text for more detail). (From Pittendrigh, C. and Skopik, S., 1970.)

Pittendrigh, C., 1967), early in development. The lack of evidence for gating of bristle or eye pigmentation therefore is not because the oscillation is not present at these earlier stages in development (in fact the oscillation is in motion in the first instar larvae (see section 7.1), but simply reflects the fact that these events are not coupled to the circadian system.

A similar story has been described for the gating of oviposition in the mosquito *A. aegypti* (Gillett, J., 1962, 1972). Once eggs have completed development, the female waits for the next available egg-laying gate to deposit the eggs. If all the eggs have matured by the time the gate opens all are deposited in a single period. If, however, only some of the eggs are mature, the female will only deposit these eggs and wait for the next gate to deposit the remainder. Thus eggs that mature in a single frequency distribution may be deposited in two discrete batches

whose timing is determined by an ongoing circadian oscillation. Furthermore, since the rate of development is temperature-dependent, females could be forced to utilize later gates for oviposition by a lower temperature.

2.4 Physiological and metabolic rhythms

While persistent behavioral rhymicity has received by far the most attention in the study of circadian periodicity in insects, a number of other rhythmic processes have been described. These include rhythms in oxygen consumption, luminescence, chromatophore dispersion, cuticular growth, and insecticide resistance. In many studies experiments were not carried out in constant conditions and the question remains in these cases whether or not the periodicities are under the control of a circadian clock. Nevertheless there are examples of persistent rhythmicity (in constant conditions) in several cases.

Oxygen consumption rates have been examined in several insect species. Rhythmicity in LD cycles has been observed in the grasshopper *Romalea microptera* (Fingerman, M. *et al.*, 1958), the cockroaches *Periplaneta americana* (Janda, V. and Mrciak, M., 1957) and *Blattella germanica* (Beck, S., 1963), the mealworm *Tenebrio molitor* (Campbell, B., 1964), and in *Drosophila melanogaster* larvae, pupae, and adults (Rensing, L., 1966; Belcher, K. and Brett, W., 1973). In at least one case, periodicity of O_2 consumption has been shown to persist in constant conditions — in *P. americana* the rhythm continues for several days in constant light (Richards, A., 1969).

In most cases these rhythms appear to be coincident with rhythms of activity and are likely to be simply a response to the increased metabolic load, although in some instances the relationship between O_2 consumption and activity is uncertain. For example, Beck, S. (1963) reported that although peak O_2 consumption in *B. germanica* occurred in the early portion of the dark period of an LD cycle during the time of expected peak activity, O_2 consumption increased steadily throughout the day when little activity was expected. There are also two examples of rhythmic O_2 consumption in diapausing (and therefore quiescent) lepidoptera larvae. Both the cornborer *Ostrinia nubilalis* (Beck, S.,

1964) and the coddling moth *Laspeyresia pomonella* (Hayes, D. *et al.*, 1972) exhibit complex trimodal rhythms when held in LD *12* : 12.

In a few cases hemolymph sugar concentrations have also been shown to exhibit daily rhythmicity. Trehalose levels in the hemolymph of *Acheta domesticus* (Nowosielski, J. and Patton, R., 1964), *P. americana* (Hilliard, S. and Butz, A., 1969), and fourth-instar larvae of the mosquito *Aëdes taeniorhynchus* (Nayar, J., 1969) all vary periodically in LD. In the mosquito *Culex tarsalis* glycogen levels were also reported to exhibit a rhythm in LD (Takahashi, S. and Harwood, R., 1964).

One interesting and widespread example of a physiological process that is under the control of the circadian system is the growth of the endocuticle early in adult life. The deposition of the cuticle occurs as alternately lamellate and non-lamellate layers which appear as alternating light and dark bands when examined under crossed polaroids. The periodic nature of this growth has been extensively studied by Neville (see Neville, A., 1975 for review). In *Schistocerca gregaria* growth of the adult cuticle occurs primarily after ecdysis, continuing for 2–3 weeks. In animals maintained in a light and temperature cycle during this period a lamellate layer is laid down each night and a non-lamellate layer is laid down each day (Neville, A., 1965). This daily periodicity persists in constant darkness for at least 2 weeks as a circadian rhythm with a period of about 23 h. Furthermore the period of the rhythm is temperature-compensated with a Q_{10} of 1.04. In *S. gregaria* the rhythm is damped in constant light, but it persists in LL in some other species such as *P. americana* (Neville, A., 1965) or in the milkweed bug *Oncopeltus fasciatus* (Dingle, H. *et al.*, 1969). The rhythm in *O. fasciatus* has also been shown to be temperature compensated (Dingle, H. *et al.*, 1969).

In all, rhythmic cuticular growth has been described in several species from nine insect orders (Neville, A., 1970). Only in Coleoptera does it appear that the rhythm is independent of circadian control (Zelazny, B. and Neville, A., 1972).

A number of interesting studies, described in more detail below, have been carried out on localization of the source of the oscillation controlling these rhythms (section 10.1). Although the data

are not conclusive, the results suggest the epidermal cells that secrete the cuticular material may be autonomously rhythmic and contain photoreceptors for entrainment. It also appears that the control of periodicity in cuticular growth is independent of the circadian system responsible for rhythmic locomotor activity (Lukat, R., 1978).

Another epidermal rhythm of interest is that of color change in the stick insect, *Carausius morosus*. In the light cycle *Carausius* is normally a light color during the day and darker at night. This rhythm will persist for several weeks in constant darkness, but is absent from insects maintained in DD from the time of hatching (Schliep, W., 1910, 1915; references from Saunders, D., 1976a).

2.5   Rhythmic sensitivity to insecticides and X-rays

A potentially important case of insect periodicity is the sensitivity to various insecticides. While this topic has probably not received the attention it deserves, there are several reports that susceptibility to a number of poisons varies significantly throughout the day in insects maintained in light cycles. The earliest studies were carried out by Beck, S. (1963) on the cockroach *B. germanica*. Animals held in LD *12* : 12 were injected with a variety of substances at different times of day, then transferred to LL to assess mortality. While susceptibility to DDT, dinitrophenol, sodium azide, and sodium fluoride showed no clear rhythmicity, the effects of Dimetilan, Dichlorvos and potassium cyanide were dependent on the time of injection. Sensitivity to Dichlorvos (an organophosphate) showed a bimodal rhythm with peaks before "dawn" and before "dusk". Dimetilan (a carbamate) and cyanide produced unimodal rhythms in mortality with sensitivity to Dimetilan peaking in the middle of the night and sensitivity to cyanide being highest shortly after "dusk".

Since this early study several other cases of diurnal periodicity in sensitivity to insecticides have been reported, including the susceptibility of the spider mite *Tetranychus urticae* to DDVP (Polcik, B. *et al.*, 1964); houseflies and cockroaches to pyrethrum (Sullivan, W. *et al.*, 1970); and *Tenebrio molitor* (Fondacaro, J. and Butz, A., 1970), *Pectinophora gossypiella* (Ware, G. and McComb, M., 1970), and *Drosophila* (Rothert, H., 1970) to organophosphates. In none of these cases has an effort been made to determine whether or not these rhythms in sensitivity persist in constant conditions. It has been reported, however, that a rhythm of susceptibility of the housefly *Musca domesticus* to DDT and dieldrin does continue in DD (Shipp, E. and Otton, J., 1976). Circadian rhythmicity in sensitivity to pharmacological agents has also been demonstrated in vertebrates (e.g. Halberg, F., 1960) and it seems likely that the rhythmic susceptibility of insects to insecticides is generally dependent on circadian organization.

A similar periodicity in LD in susceptibility has been reported in the response to X-irradiation of *D. melanogaster* (Rensing, L., 1969) and *Pectinophora gossypiella* (Haverty, M. and Ware, G., 1970). In *D. melanogaster* mortality was highest at "dawn" and "dusk", while in *P. gossypiella* a minimum occurred at "dawn" in *LD 14*:10.

3 ENTRAINMENT

A major adaptive feature of circadian oscillations is that they provide the organism with an innate "temporal program" (Pittendrigh, C., 1981a,b) that allows the organism to exploit the predictability of fluctuations in the environment that they must deal with on a day-to-day basis. To successfully perform this function the oscillation must be appropriately phased to the geophysical cycles of the immediate environment — the clocks must be set to local time. Oscillators in general are subject to synchronization by external forces. Circadian oscillations are no exception, and their "access" to external time cues from the environment results in their synchronization (or entrainment) which correctly phases the oscillations to local conditions.

Environmental cues that are effective in entrainment have been referred to as "zeitgebers". The most important zeitgeber is the daily cycle of light and dark which, as the most reliable (noise-free) environmental periodicity, has been universally chosen by natural selection as the dominant entraining stimulus.

In poikilotherms in general, and in insects in particular, temperature is also very effective in entrainment of circadian rhythms. The possibility that other stimuli such as social interaction, food availability, or sound may also be effective as zeitgebers has hardly been investigated in insects; but it is interesting to note that there are examples from vertebrates of entrainment of circadian rhythmicity by each of these stimuli.

Two important aspects of the effect of entrainment on circadian rhythms are illustrated in the data shown in Fig. 1. The first is that in entrainment the period of rhythm (τ) assumes the period of the zeitgeber (T). Thus when $T = 24$ h, τ is changed to $\tau^* = T$. The second is that the rhythm establishes a particular phase relationship (denoted as ψ_{RL}) to the zeitgeber cycle. Typically the effect on τ is slight, and the major functional significance of entrainment is derived from the resulting control of phase.

There are a number of empirical observations that allow us to formulate some generalizations about the response of circadian rhythms to entrainment. Some of the general properties are of functional significance to the organism. Others are only evident in light cycles the organism would never experience in nature, but nevertheless must be explained by any general model of the entrainment process.

One of the most important features of entrainment to 24 h light/dark cycles is that the steady-state phase angle difference between the rhythm and the zeitgeber (ψ_{RL}) is a function of the duration of the light portion of the cycle (photoperiod). This aspect of entrainment has apparently been exploited by natural selection in the use of the circadian system by some insects to measure photoperiod duration as a cue to the time of year. Phase angle difference is also dependent on the period of the zeitgeber (or, more generally, on the ratio of τ/T, Pittendrigh, C., 1981c), Typically ψ_{RL} becomes more positive as T increases.

It is also a general property of self-sustaining oscillators that they can be entrained only to periods that are "near" the natural period of the oscillation. Beyond these "limits of entrainment" when T is either too short to too long the rhythm "free-runs" with a period different from T, but the frequency is modulated (relative coordination) as it changes its phase relation to the zeitgeber in the course of its free-run. Limits of entrainment vary widely from one species to the next. In the cockroach *Leucophaea maderae*, for example, moderate-intensity light cycles will entrain from about

$T = 22\,h$ up to $T = 26\,h$ (Page, unpublished), whereas in *Drosophila pseudo-obscura* the limits are nearer 19 h and 29 h (Pittendrigh, C., 1967). In very short or very long light cycles entrainment may actually occur *via* "frequency demultiplication" or "frequency multiplication" in which the period of the oscillation assumes a period that is equal to T either divided or multiplied by some integral. Thus when *D. pseudo-obscura* is subjected to a 12 h light cycle a precise 24 h rhythm is expressed (see section 3.4).

3.1 Phase response curves

The effect of a light cycle in achieving entrainment is to alter the period of the oscillator by an amount $\tau - T$ (Pittendrigh, C. and Minis, D., 1964). In general there are two broadly defined mechanisms by which this process can be accomplished (Bruce, V., 1960). The zeitgeber may modulate the angular velocity of the oscillator over a large part or the whole of its cycle. This class of entrainment mechanisms has generally been referred to as "parametric" or, more recently, "continuous" entrainment. Alternatively, specific aspects of the zeitgeber wave form (e.g. the light to dark transition in a square-wave light cycle) may effect a discrete phase shift in the oscillator, rapidly sending it from one phase of its cycle to a new, earlier or later phase. This process is referred to as "non-parametric", "discontinuous", or "discrete" entrainment. There is no doubt that circadian oscillators are affected by both continuous and discrete mechanisms, and both undoubtedly function in entrainment.

A major step in the analysis of entrainment by light was the discovery that a single light pulse of short duration could shift the phase of circadian rhythms free-running in constant darkness, and furthermore, that the magnitude and direction (advance or delay) of the phase shift ($\Delta\varphi$) depended on the phase of the rhythm at which the stimulus was given. The curve depicting the dependence of $\Delta\varphi$ on φ is referred to as a "phase response curve" (PRC). PRCs to light pulses have been measured in a large number of animals and all share several common features. Figure 8 shows the PRCs of *Drosophila pseudo-obscura* and *Drosophila melanogaster* to 15 min pulses of light. The pacemakers are relatively insensitive to light during the subjective day[1],

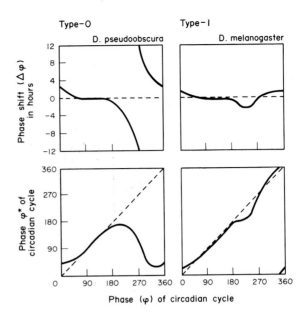

FIG. 8. Phase response curves (upper panels) and phase transition curves (lower panels) for the *Drosophila pseudo-obscura* and *Drosophila melanogaster* eclosion rhythms to 15 min light pulses. The phase response curve plots the phase shift characteristic of each phase at which the pulse occurs; the phase transition curve plots the new phase (after the pulse) as a function of the phase of the pulse. (From Pittendrigh, C., 1981c.)

respond with phase delays in the early subjective night, and with phase advances in the late subjective night. These features are typical for *every* phase response curve to light that has been measured, including those of unicellular organisms, plants, vertebrates, and invertebrates. The *precise* form of the PRC does vary from one species to another. PRCs in general can be classified according to one of two types, based on the average slope of a curve that plots the new phase (after the pulse) *vs.* the phase at the time of the pulse (Winfree, A., 1970) (Fig. 8). For *D. pseudo-obscura* this curve, called a phase transition curve, has an average slope of 0, and its PRC is a type 0. For *D. melanogaster* the

---

[1] In discussing phase response curves and circadian oscillations it is often useful to describe that part of the circadian cycle that would normally occur in darkness *if* the organism were entrained to *LD 12 : 12* as the "subjective night". Similarly "subjective day" refers to the portion of the cycle that would occur in the light. For the purpose of identifying particular phases in the oscillation one full cycle is divided into 24 "circadian hours" where each circadian hour is equal to $\tau/24\,h$ of real time. By convention circadian time (ct) 12 is designated as the beginning of the subjective night. The middle of the subjective night then is ct 18, or 6 circadian hours after ct 12, and so on.

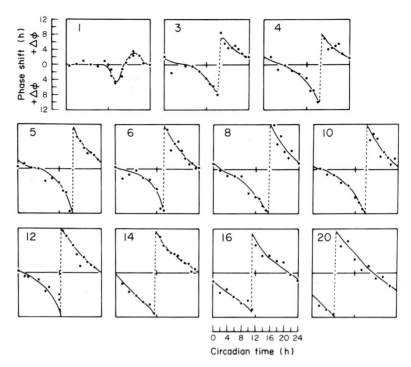

FIG. 9. Phase response curves (PRCs) for the eclosion rhythm of *Sarcophaga argyrostoma* to pulses of light 1–20 h in duration. For 1 h pulses the PRC is a type 1. Longer pulses (e.g. 4 h) lead to type 0 resetting. (From Saunders, D., 1978a.)

average slope is 1 and its PRC is a type 1. The two curves do not necessarily reflect qualitative differences in pacemaker structure. In response to a weaker stimulus the *D. pseudo-obscura* PRC becomes a type 1 (e.g. see Pittendrigh, C., 1960). The transition from "weak" (type 1) to "strong" (type 0) resetting can be seen in the family of PRCs Saunders, D. (1978a) has measured for different-duration light pulses in *Sarcophaga* (Fig. 9).

3.2 Rhythm transients

The phase shift response of circadian rhythms to light pulses is often not immediate and the full magnitude of the phase shift may not be apparent for several days during which time the rhythm may go through several "transient" cycles (Pittendrigh, C., 1960). This is illustrated in Fig. 10 which shows data from an experiment measuring the PRC of the *D. pseudo-obscura* eclosion rhythm. In this experiment 25 populations of pupae were released from *LD* 12 : 12 into DD. One population served as a control and remained in DD throughout the experiment. The medians for successive eclosion peaks of the control are shown above and below the box

enclosing data from experimental populations. The other 24 populations were exposed to a single light pulse during the first 24 h in DD, with each population receiving the pulse at a different hour. For populations pulsed during the subjective day the time of eclosion was not substantially altered from that of the controls. Pulses in the early subjective night delayed the eclosion peaks and the complete shift is expressed on the first day after the pulse. Pulses given later in the night led to a phase advance which is not complete until the third or fourth day after the pulse (Fig. 10). The meaning of the transient cycles and their bearing on current views of the organization of the circadian system are discussed in some detail below (section 10.2). At this point it is sufficient to note that the evidence from *D. pseudo-obscura* clearly indicates that the transient cycles do not reflect transients in the driving oscillator, which is reset to a new phase almost immediately, but represent transient motion in a driven process.

3.3 PRC as a measure of pacemaker time course

The fact that the magnitude of the phase shift in response to a standard light pulse is dependent on

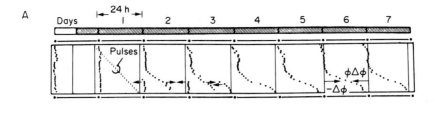

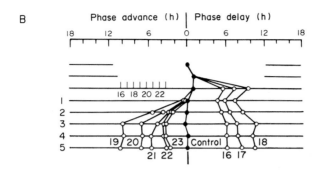

FIG. 10. Transient behavior in the *Drosophila pseudo-obscura* eclosion rhythm following exposure to single 15 min pulses of light (data from Pittendrigh, C., 1960). **A**: Midpoints of eclosion peaks plotted for 24 experimental populations and one control. Each population received a light pulse (shown as the short bars) on the first day of a DD free-run (see text). (From Pittendrigh, C., 1981c.) **B**: Transient cycles following a single light pulse in populations pulsed at ct 16, 17 ... 23. The pulses at ct 16, 17, and 18 caused delaying transients which end rapidly. Pulses between ct 19 and 23 caused phase shifts with advancing transients. With advances there are several days of transients before the phase shift is complete. Filled circles are data from a control population. (From Pittendrigh, C. and Minis, D., 1964.).

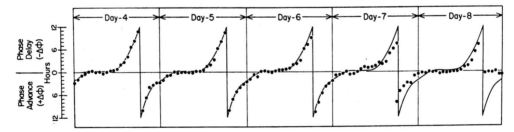

FIG. 11. Phase response of the eclosion rhythm of *Drosophila pseudo-obscura* during that last 5 days of pupal life showing a free-running rhythm in the sensitivity of the oscillation to light. (From Pittendrigh, C., 1966.)

the phase of the pacemaking oscillation at the time of the pulse suggests the PRC can be used to directly assay the time course of the pacemaker — the PRC is a measure of the succession of phases through which the pacemaker passes in the course of its cycle (Pittendrigh, C., 1976). Figure 11 shows the results from an experiment in which each of 60 replicate cultures of *D. pseudo-obscura* free-running in DD were exposed to a light pulse from 0 to 60 h after entry into constant conditions. The plot of $\Delta\varphi$ as a function of the time of the pulse of each culture traces a "free-running" PRC.

3.4 Discrete entrainment model

In the mid-1960s Pittendrigh, C. (1965, 1967; Pittendrigh, C. and Minis, D., 1964) developed a formal model of the entrainment process in *D. pseudo-obscura* for light cycles involving only brief (discrete) pulses of light. The success of this model in explaining the general features of entrainment is now well documented not only in insects but also in several vertebrates. The model has also found widespread utility in investigations of insect photoperiodic clocks, discussed below, and forms the foundation for our current understanding of the

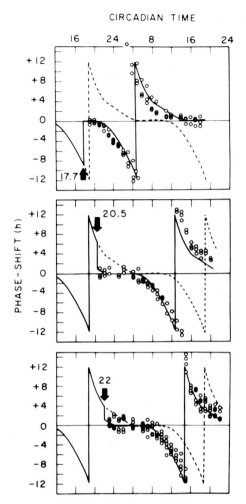

FIG. 12. Plot of the time course of the PRC of *Drosophila pseudo-obscura* in three "2-pulse" experiments. Circles show the data; solid lines are the predicted PRC time courses assuming the pacemaker is instantaneously reset by the first light pulse; dashed lines show the predicted time course of the unperturbed pacemaker. For both delay phase shifts (upper panel) and advance phase shifts (middle and lower panels) the pacemaker is immediately reset by the first pulse. (Pittendrigh, C., 1976.)

significance of both "discrete" and "continuous" mechanisms in entrainment.

The model proposes that in light cycles involving periodic 15 min pulses of light the effect of each pulse is to phase shift the pacemaker by an amount $\Delta\varphi$. The value of $\Delta\varphi$ is a function of the phase at which the pulse hits the oscillator and is predicted by the PRC to single pulses of light. In the entrained steady state the light pulse must fall at a particular phase such that $\Delta\varphi = \tau - T$.

One important assumption of the model is that the motion of the oscillator from its old phase to a new phase in response to a light pulse is very rapid — for present purposes instantaneous — and that it resumes its normal motion immediately after the pulse. A rigorous test of this assumption was carried out by Pittendrigh, C. (1976) in what have been referred to as "two-pulse" experiments (Fig. 12). Replicate populations of *Drosophila pseudo-obscura* pupae all receive a first, identical pulse of light at the same time. Second pulses are then given (one to each population) at hourly intervals after the first. The magnitude of the subsequent steady-state phase shift generated by the second pulse is a measure of pacemaker phase after the first pulse, and the complete PRC determined from the several populations is a measure of the pacemaker phase and time course in the period immediately after the first pulse. The predictions and results of three such experiments are shown in Fig. 12 and clearly verify the proposition that the *Drosophila* pacemaker is immediately reset to its new phase following a 15 min light pulse — there are no transients in the pacemaker's behavior that reflect those of the observed eclosion rhythm.

One clear implication of the model is that in light cycles in which T is close to but not equal to τ the phase angle difference between the eclosion peak and the light pulse should be predictable. For $T < \tau$ the pulse should fall in the late subjective night effecting a phase advance ($+\Delta\varphi$), and for $T > \tau$ the pulse should fall in the early subjective night generating a phase delay ($-\Delta\varphi$). The exact time of the pulse will be dependent on the magnitude of the difference between T and τ and on the phase of the PRC where the pulse must fall to generate a $\Delta\varphi = \tau - T$. The predicted curve and the observed points for this experiment are shown in Fig. 13. Results and expectation match very well except at the extreme ends of the curve.

It should also be possible to predict the limits of entrainment on the basis of this model, where:

$$T_{max} = \tau - (\text{maximum phase delay})$$
$$T_{min} = \tau + (\text{maximum phase advance})$$

Experimental data were again found to match predictions but with the caveat, derived from a perturbation analysis of the entrainment model, that stable entrainment is only possible where the

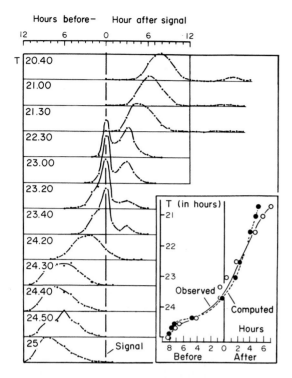

FIG. 13. The phase of the *Drosophila pseudo-obscura* eclosion rhythm entrained to 15 min pulses of light as a function of the period of the LD cycle (*T*). The inset shows the medians of the eclosion peaks (open circles) compared to the values predicted from the entrainment model (filled circles). (From Pittendrigh, C. and Minis, D., 1964.)

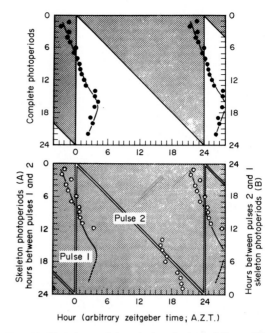

FIG. 14. The phase of the eclosion rhythm of *Drosophila pseudo-obscura* in complete (upper) and skeleton (lower) photoperiods. The circles in the two panels plot the medians for emergence peaks in steady-state entrainment. In the lower panel the solid curve plots the phase of the medians from the upper curve for comparison. Simulation of complete photoperiods by the skeleton is satisfactory up to photoperiods of about 14 h. (From Pittendrigh, C. and Minis, D., 1964.)

PRC slope lies between 0 and −2 (Otteson, E., 1965).

An important discovery in the investigation of entrainment by brief pulses of light was that two light pulses 1–14 h apart could simulate entrainment to complete photoperiods of the same duration (Pittendrigh, C. and Minis, D., 1964; Pittendrigh, C., 1965, 1967). The light pulses formed what were termed "skeleton photoperiods" (PP$_s$) in which one pulse was taken to simulate "dawn" and the other "dusk" of a complete photoperiod (PP$_c$). It was found that the dependence of ψ_{RL} on "photoperiod" in these skeletons could be predicted from the entrainment model. In this case the two pulses fall in the same cycle so that $\tau - T = \Delta\varphi_1 + \Delta\varphi_2$ where $\Delta\varphi_1$ and $\Delta\varphi_2$ are the successive phase shifts caused by the pulses defining the skeleton. For skeletons beyond 14 h the model predicts, and data confirm, that the simulation of complete photoperiods fails (Fig. 14). The pacemaker is forced into a "ψ-jump" in which the "dusk" pulse is

now taken as "dawn". Every skeleton photoperiod is open to two "interpretations" in which the shorter interval can either be taken as "day" or as "night". The position of the "ψ-jump" defined the "minimum tolerable night" of the skeleton regime. In complete photoperiods the continuous action of light prevents the ψ-jump and greatly increases the "minimum tolerable night".

For a narrow range of skeleton photoperiods centred around $\tau/2$ there are two possible steady states. Either of the light pulses can be interpreted as dawn for skeleton photoperiods ranging from about 10 to 14 h. The steady-state phase relationship ultimately assumed by the pacemaker depends on two "initial conditions": (a) the phase of the first pulse, and (b) the duration of the first dark interval. Figure 15 shows predictions of the entrainment model and results of an experiment in which the time of the first pulse and the duration of the first interval were varied for the skeleton of *LD 11*:*13* (PP$_s$ 11:13). The "bistability" of the entrainment

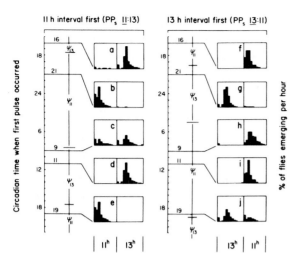

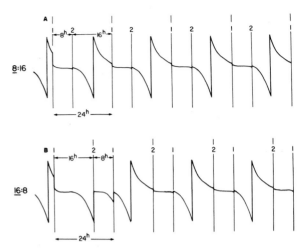

FIG. 15. The bistability of entrainment of the eclosion rhythm of *Drosophila pseudo-obscura* to skeleton photoperiods (PP$_s$) of *11*:13 and *13*:11. There are two possible steady states of each skeleton depending on the phase of the first light pulse in the entraining cycle. In one steady state (ψ_{11}), the eclosion peaks falls in the 11 h interval; in the other (ψ_{13}) it falls in the 13 h interval. Predictions of which steady state will be realized based on the entrainment model are given at the left of each data panel. (From Pittendrigh, C., 1981c.)

FIG. 16. Entrainment of *Drosophila pseudo-obscura* by the skeleton photoperiods of LD *8*:16 and LD *16*:8. In both examples the first pulse hits at ct 21 causing a large phase advance. In **A** subsequent phase shifts alternate between delays and advances, but in **B** there are two successive delays, which forces the pacemaker into the LD *8*:16 "interpretation" of the skeleton. (From Pittendrigh, C., 1981c.)

response is clear and the match with prediction is remarkably good.

The consequences for entrainment of shortening the skeleton photoperiod and the reason for the ψ-jump are illustrated in Fig. 16 which traces the PRC through several cycles during exposure to PP$_s$ 8:16 (= PP$_s$ 16:8). When the shorter interval occurs in the subjective night as though the photoperiod were 16 h, the duration between the two pulses is short enough that the pacemaker is subjected to two successive phase delays. The sum of the two $\Delta\varphi$s is so large that the pacemaker is forced into the alternative steady state.

Entrainment is also possible for "asymmetric" skeletons in which one pulse is much longer than the other. As with the "symmetric" skeletons the pacemaker can be forced into a ψ-jump as the simulated photoperiod duration increases past some maximum tolerable value.

These various phenomena of entrainment involving brief pulses of light (including skeleton photoperiods, both symmetric and asymmetric, ψ-jumps, bistability, and the dependence of ψ_{RL} on T) have been of great importance in the design and interpretation of experiments on the mechanism of photoperiodic time measurement discussed in some detail below.

The data also suggest that entrainment to natural photoperiods is a complex process involving both discrete responses to the phasic signals at dawn and dusk and the continuous action of the tonic effect of the complete photoperiod during the day which prevents the ψ-jump as day length increases beyond the maximum tolerable range defined by "dawn" and "dusk".

3.5 Effects of long photoperiods

Some clues as to the mechanism of action of long photoperiods presumably can be obtained from investigating the effects of constant light. In general, the period of free-running rhythms is a function of the light intensity, and in many species, as light intensity is increased, rhythmicity is lost completely. In some cases the intensity at which arrhythmicity ensues may be quite low. In *D. pseudo-obscura*, for example, the eclosion behavior becomes rapidly aperiodic at constant light intensities above 0.05 lux (Pittendrigh, C., 1981c). In other cases (e.g. the cockroach, *Leucophaea maderae* — Roberts, S., 1960) rhythmicity persists at relatively high light intensities. In *D. pseudo-obscura* (Pittendrigh, C., 1960, 1966), *Sarcophaga argyrostoma* (Saunders, D., 1978a) and *Anopheles gambiae* (Jones, M., 1973) where constant light arrhythmicity has been studied in some detail it has been found that when the system is transferred to darkness, rhythmicity is rapidly resumed. Furthermore, the restored rhythm begins at LL/DD transition from (or close to) the beginning of the subjective night (ct 12) regardless of the length of the LL treatment (assuming it was at least 12–14 h). In general this result has been taken to mean that the oscillation had stopped at that phase and simply resumed its motion on entry into DD. This interpretation may well be incorrect (Chandrashekaran, M. and Loher, W., 1969; Pittendrigh, C., 1981c). Recent results from *Drosophila pseudo-obscura*, for example, show that even in very low-intensity constant light where the rhythm persists, transfer into constant darkness resets the rhythm to ct 12 (Pittendrigh, C., 1981c). Data have also been obtained from the *S. argyrostoma* eclosion rhythm that show (1) although cultures made behaviorally arrhythmic by exposure to LL always begin *near* ct 12 upon transfer to DD, the precise phase fluctuates around ct 12 with a period near 24 h; and (2) the phase coherence of the population in DD varies as a periodic function of the duration of the LL treatment (Saunders, D., 1976b; Peterson, E. and Saunders, G., 1980). The results clearly suggest a continuing oscillation in LL. Thus neither the overt aperiodicity or the return to a specific phase upon transfer to DD necessarily mean the oscillation has stopped its motion.

All effects of constant light are most easily explained by assuming the pacemaker behaves as a limit cycle oscillator, and many of the responses of circadian rhythms to light pulses and constant light are readily understood on the basis of the topology of limit cycle behavior (e.g. Pavlidis, T., 1968; Winfree, A., 1970; Peterson, E., 1980; Pittendrigh, C., 1981c). A detailed discussion of limit cycles is beyond the scope of this review. Interested readers are directed to a particularly lucid discussion of E. Peterson's (1980) for an introduction to the limit cycle approach to circadian oscillations.

Another approach to understanding the effects of long light pulses was suggested by DeCoursey, P. (1959) and explored in some detail by Daan, S. and Pittendrigh, C. (1976). They suggested that the effect of constant light is to modulate the angular velocity of the circadian oscillator in a phase-dependent fashion, with the change in velocity at any particular phase being proportional to the amplitude of the PRC at that phase. Thus the effect of LL is to speed up the oscillator during the advance portion of the PRC and slow it down in the delay region. For any individual the overall effect on τ would depend on the precise shape of its PRC and, in particular, on the relative amplitude of the advance and delay phases of the response curve.

In summary entrainment is seen as a mixture of the discrete effects of light pulses at the transitions between day and night and the continuous effects on angular velocity of the light pulse in between these transitions. For those animals that are exposed to very long photoperiods in nature the functional significance of the continuous effects of light is clear (Pittendrigh, C., 1976). It prevents the ψ-jump in long photoperiods that would be inevitable were the entrainment mechanism exclusively via discrete phase shifts at the light/dark transitions. In *D. pseudo-obscura* or *S. argyrostoma*, for example, the phase of the oscillation is reset to the beginning of the subjective night at the end of any light pulse exceeding about 12 h. Thus in nights as short as 6 h the discrete phase shift caused by the morning light is always an advance; never a delay that would force the oscillation into a ψ-jump.

3.6 Entrainment by temperature cycles

The free-running period of circadian oscillators are relatively insensitive to the absolute value of a constant ambient temperature within the physiological range. However, transients in temperature can effect major phase shifts in the pacemaker, and temperature cycles have been shown to be fully effective in entrainment of circadian rhythms in several insect species.

Figure 17 shows the effect of a 24 h sinusoidal temperature cycle on the activity rhythm of the cockroach *Leucophaea maderae* (Roberts, S., 1962). For the first 17 days of the record the animal was in DD and the temperature was a constant 25°. The imposition of a temperature cycle between days 17 and 44 entrained the rhythm to a 24 h period with the onset of activity occurring near the high point of the temperature. The rhythm began to free-run following return to constant conditions on day 44. Similar results have been obtained by Rence, B. and Loher, W. (1975) working with the rhythm of

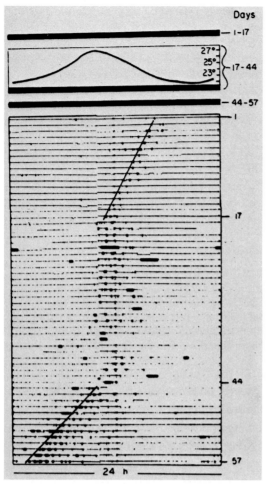

FIG. 17. Activity record of a cockroach *Leucophaea maderae*, illustrating entrainment to a 24 h temperature cycle (shown above). The animal was in constant darkness throughout the experiment and was exposed to the temperature cycle between days 17 and 44. (From Roberts, S., 1962.)

Entrainment of the eclosion rhythm of *D. pseudo-obscura* by square wave temperature cycles has been investigated in some detail by Zimmerman, W. *et al.* (1968) who studied the phase shifting effects of temperature steps (up and down) as well as pulses. They found that (a) temperature steps-up generated only phase advances, (b) steps-down generated only phase delays and, (c) temperature pulses generated both delays and advances. For steps and pulses the magnitude of $\Delta\varphi$ depended on the phase of the pulse. Interestingly, the $\Delta\varphi$ caused by a temperature pulse was found to be predictable from the sum of the $\Delta\varphi$s generated by the two steps that comprised the pulse. The results suggested that the $\Delta\varphi$ generated by a temperature step occurs relatively rapidly and entrainment by temperature pulses is analogous to entrainment by skeleton photoperiods where in steady state the two temperature steps (up and down) occurring in each cycle elicit phase shifts such that $\tau - T = \Delta\varphi_{up} + \Delta\varphi_{down}$. The fact that the dependence of ψ on pulse duration could be predicted from the phase response curves for the temperature steps (Zimmerman, W. *et al.*, 1968) supports this view.

A few experiments have also been done in which organisms were exposed simultaneously to light and temperature cycles presented in various phase relationships (Pittendrigh, C., 1960). Both the locomotor activity rhythm of the cockroach *Leucophaea maderae* and the eclosion rhythm of *D. pseudo-obscura* behave similarly. In both cases as the low point of a sinusoidal temperature cycle was steadily delayed relative to "dawn" of a fixed light cycle (*LD 12*: 12), the rhythm maintained a fixed phase relationship to the temperature cycle up to the point at which the low point of the temperature cycle coincided with the onset of darkness. As the low point of the temperature cycle moved into the dark phase, the rhythms were forced into a phase-jump of about 180°. Similar behavior has also been observed in *Pectinophora gossypiella* (Pittendrigh, C. and Minis, D., 1971). The results, which suggest there is a zone of 180° of forbidden phase relations between the light cycle and pacemaker, are consistent with the behavior of a self-sustained oscillator subjected to two conflicting entraining stimuli (Pittendrigh, C., 1960).

The results of studies on entrainment by light and temperature cycles suggest that entrainment of

stridulatory activity in *Teleogryllus commodus*. They found a nearly square wave cycle of temperature of 10° amplitude was effective in entrainment with activity beginning near the stepdown. Temperature cycles have also been found to be effective zeitgebers for the eclosion rhythms of *D. pseudo-obscura* (Zimmerman, W. *et al.*, 1968), *Dacus tryoni* (Bateman, M., 1955) and *Anagasta kuhniella* (Moriarty, F., 1959).

While much less is known about the mechanism of entrainment by temperature cycles, work with *D. pseudo-obscura* suggests that the formalism developed to explain entrainment by light pulses is applicable to entrainment by temperature pulses.

insect rhythms in nature must indeed be a complex process. Much more data on the effects of temperature and its relation to entrainment by light are needed before we can begin to assess the ecological significance and relative importance of light and temperature in assuring an appropriate phasing of the circadian system.

4 TIME-MEMORY AND TIME-COMPENSATED ORIENTATION

Some of the most remarkable experiments on the biological clocks of insects are the reports from von Frisch's laboratory on the "time-sense" of bees. Indeed, it has been suggested (Pittendrigh, C., 1974) that these studies, along with the classic experiments by Kramer, E. (1952) on time-compensated sun orientation in starlings, were in a large part responsible for the surge of interest in circadian rhythms after 1950.

4.1 Zeitgedächtnis

Prompted by an observation by Forel, A. (1910) that bees could "remember" to appear at a food source at a particular time of day, von Frisch successfully attempted to train honeybees (*Apis mellifera*) to a feeding time (see von Frisch, K., 1967 and Renner, M., 1960 for accounts of the early studies of bees' "time-memory"). This observation was followed up by Beling, I. (1929) who trained bees to appear at a feeding station by offering a sugar solution for a brief period (e.g. 2 h) each day at the same time for several days. Bees that came to feed during this period were individually marked. After this training period an empty dish was set out for a day and the time of arrival of the marked bees was noted. Bees were found to approach the empty food dish only during the 2 h period when food was expected (Fig. 18). Beling, I. (1929) further demonstrated that bees could be trained to appear at an artificial food source at any time of day; and it was found that bees could be trained to appear at a feeding place three, four and occasionally five separate times during the course of a day. The adaptive significance of this behavior was explained with the discovery that the amount of nectar produced by flowers varies in the course of the day and the

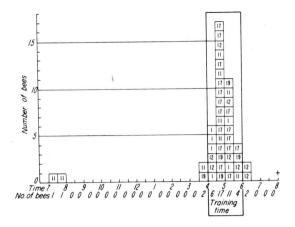

FIG. 18. Illustrates "zeitgedächtnis" in honeybees. Individually marked bees were trained to come to a food source between 4 and 6 p.m. Histogram shows the time of appearance of bees at the empty food dish the day after training. (From von Frisch, K., 1967; after Beling, I., 1929).

peak time of nectar secretion varies among plant species (Kleber, E., 1935).

Two other important observations on the bees' ability to learn the time of day a food source will be available were subsequently made. First, Beling, I. (1929) found bees could be trained in a closed room maintained at constant temperature and constant light. It appeared, therefore, that bees were not using cues from the external environment to determine the time of day. This conclusion was confirmed by Wahl, O. (1932) who successfully trained bees in a colony that had been transported to the bottom of a salt mine to eliminate all daily environmental fluctuations. The second important observation was that bees could only be trained to return to a food source at 24 h intervals (Beling, I., 1929; Wahl, O., 1932). Attempts to train bees to appear at a food source, say every 19 h, failed completely; and when trained to a 48 h schedule bees arrived, at the appropriate time of day, every 24 h.

These results suggested that the bee was relying on an internal clock whose motion continued in constant conditions and whose period approximated 24 h. Renner, M. (1955, 1957) provided strong support for this view in a "translocation" experiment. Bees were trained in a closed room in LL and constant temperature to appear at a food station at a particular time of day in Paris. The bees were then flown overnight to New York, and tested the following day under conditions similar to those in Paris. Although the translocation amounted to a 5 h change in *local time* the bees returned to the food source 24 h after the last feeding period in Paris; the clocks of the bees were still "set" to Paris time and were apparently unaffected by the movement to New York. Renner, M. (1959) also showed in a subsequent translocation experiment between New York and California that if tested outdoors, the bees' clocks gradually reset to the new local time. Finally, it was shown that in constant conditions (LL) the interval between return trips to a feeding station after training was 23.4–23.8 h. The rhythm free-ran with a circadian period (Bennet, M. and Renner, M., 1963). The results are all consistent with the view that bees associate the presence of food at a particular time and location with a particular phase of an endogenous circadian oscillator, and that they are capable of using that information to return to a food source at the appropriate time of day.

Other investigations of the bee's *zeitgedächtnis* have shown the clock can be entrained by light (Beier, W. and Lindauer, M., 1970), and can be phase delayed by low temperature pulses (4–5°) (Renner, M., 1957). CO_2 narcosis has also been shown to effect the behavior but in a more complicated way. Following exposure to CO_2 bees were found to return to the food source at the original training time and again at some later time. The time of the second peak was determined by the duration of narcosis as well as the CO_2 concentration used (Medugorac, I. and Lindauer, M., 1967). It was suggested that these results indicated two clocks were involved (Medugorac, I. and Lindauer, M., 1967).

A very interesting recent study suggests that it is possible to "transplant" the learned feeding time from one bee to another (Martin, U. *et al.*, 1978). Following transplantation of the corpora pedunculata from trained bees to "naive" bees the recipient animals, after 1–2 days recovery, exhibited a preference to approach a food source at the time food had been presented to the donor animals. It was also reported that the tranplantation of learned feeding time was successful even if the tissue has been frozen (Martin, U. *et al.*, 1978). The results suggest the storage of phase information on feeding time may not require an ongoing oscillation.

4.2 Time-compensated sun orientation

The original discovery that animals were able to directionally orient using celestial cues dates back to the experiments of Santschi, F. (1911) who showed that ants were able to use the sun's azimuth as a "compass" to find their direction back to the nest. Subsequently, sun orientation was discovered in several organisms including vertebrates as well as other invertebrates. The sun, of course, moves across the sky, but because its movement is very slow ($15°$ arc h^{-1}) this creates no problem for rapid trips to and from the nest. However, it is clear that long-term orientation in a fixed compass direction is only possible if the angle of orientation to the sun is "time-compensated". Early work on ants (Brun, R., 1914) and bees (Wolf, E., 1927) suggested that such compensation was not occurring, and the possibility was discounted until 1950 with the simultaneous and independent demonstration by von Frisch and Kramer that honeybees and starlings (respectively) were capable of time-compensated sun-orientation.

In bees two types of experiments have confirmed the suggestion that the motion of the sun is compensated for. The first is illustrated in Fig. 19 (Lindauer, M., 1960). Bees were trained in the *afternoon* to come to a food source 180 m distant from the hive in a particular compass direction (NW). That night the hive was moved to a new location (to eliminate landmarks as a guide to the food) and four feeding stations were placed 180 m from the hive at its new site in the NW, NE, SE, SW directions. When the bees were tested the next *morning* they predominantly appeared at the NW station even though the sun at this time of day was at a completely different location in the sky (von Frisch, K. and Lindauer, M., 1954).

A second approach was used by Meder, E. (1958). Following a period of training to a food source in a fixed compass heading from the hive, Meder captured the bees and maintained them in darkness for one or more hours. When the bees were subsequently released, the flight to the hive was in the appropriate direction, suggesting that the bees had compensated for the movement of the sun during their imprisonment in the dark.

Foraging honeybees, having found a distant food source, are able to return to the hive and communicate

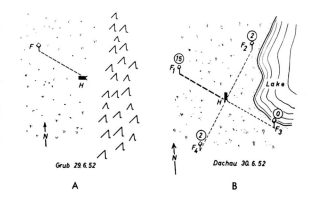

FIG. 19. Sun-compass navigation in the honeybee. **A**: Bees were trained to a feeding station (F) 180 m NW of the hive (H); **B**: during the night the hive was moved to a new location and bees had to choose between four feeding stations located 180 m NE, NW, SE, and SW. The majority of the bees appeared at the NW station (numbers of bees given in circles). (From Lindauer, M., 1960).

both the direction and distance of the food to other workers via a "dance-language". The ability to communicate by dance language was originally postulated by von Frisch, K. (1946) and following a period of controversy was subsequently confirmed (Gould, J., 1975). For distant food sources the dance takes the general form of a figure eight. The bee runs clockwise in a semi-circle, then connects the two ends "waggling" its abdomen along the way, then runs in counterclockwise semi-circle, completing the figure eight. The information on the direction of the food source with respect to the sun's azimuth is given in the dance by the angle of the "waggle runs" (on the verticle hive) with respect to gravity. While these dances typically last only a short period of time bees occasionally have been observed to dance for hours at a time (Lindauer, M., 1954, 1955), signalling direction without further exploratory flights. Even without any further view of the sky these "marathon dancers" continuously compensated for the movement of the sun, changing the angle of the waggle run slowly, counterclockwise at about 15° per hour. These dancers often even continued to perform at night when they still indicated the correct azimuth even though they could not see the sun's position (Lindauer, M., 1960). From these results it is clear that bees are able to use an internal clock to compensate for the movement of the sun across the sky. This ability allows them to use the sun's azimuth as a compass to main-

tain a constant directional heading even in the absence of any fixed landmark cues.

Studies from other insects have suggested that sun compass orientation may be relatively widespread. Besides bees (*Apis mellifera*), the ants *Formica rufa* and *Lasius niger* (Jander, R., 1957), the pond skater *Velia currens* (Birukow, G., 1960) and the beetle *Geotrupes sylvaticus* (Birukow, G., 1960) have been reported to exhibit time-compensated orientation with the sun.

In starlings it has been clearly shown that the properties of the clock utilized in the sun-compass behavior parallels those of the circadian clock that controls the daily activity rhythm (Hoffmann, K., 1960). Both free-run with similar periods in constant light, and both can be entrained by light. These results suggest the clock controlling both behaviors may be the same. Comparable data are not yet available in insects although bees, for example, do exhibit a free-running activity rhythm in constant conditions (e.g. Spangler, H., 1972).

The results with time-compensated sun orientation and *zeitgedächtnis* provide the most vivid examples that a circadian oscillation can function as a "clock" that can be "continuously consulted" for time-of-day information. In the case of bees, the ability to measure lapse of time has been shown to be impressive. Data from "marathon dancers" suggest bees are able "read" intervals as small as 5–10 min (von Frisch, K., 1967).

More recently, Gould has published two additional interesting observations on the bee's ability to navigate by the sun. The first concerns the bees' ability to compensate for the fact that although the rate of movement of the sun is constant, the rate of movement of the sun's azimuth varies with the position of the sun in the sky. Bees compensate for this variation by extrapolation, using the rate of movement of the azimuth when last observed in "computing" the rate of movement during a 2 h period when they are trapped in the dark (Gould, J., 1980). Thus it appears bees can measure the velocity of the sun's azimuth and use the measurement to compute the expected new position. The second observation of interest is that bees are able to navigate on cloudy days based on a memory of the diurnal course of the sun's position with respect to local landmarks (Dyer, F. and Gould, J., 1981). In this situation bees utilize a local landmark as a reference and orient their dances as though the sun were moving past it with normal time course. These results, like those from marathon dancers, support the view that the clock used to keep track of the sun's position provides bees with information on where the sun *should* be even if it may not be visible for long periods of time.

5 CLOCKS RELATED TO THE LUNAR CYCLE

Several important environmental periodicities are associated with the revolution of the moon about the earth and with the position of the moon with respect to the sun and earth. These cycles include (1) the intensity of moonlight, which varies with a period of 29.5 days; (2) tidal cycles corresponding to the lunar day of 24.8 h; and (3) the "semi-lunar" (14.8-day) rhythm of the amplitude of the tides which varies between the neap and spring tides. Endogenous rhythms corresponding to each of these periodicities have been described in various insect species. The most extensively studied of these rhythms are the semi-lunar rhythms of emergence in various species of the intertidal midges of the genus *Clunio*. Little work beyond description of rhythms related to the lunar cycles has been done on other species.

There is only one reported example of an endogenous tidal rhythm in an insect (although tidal rhythmicity is common in intertidal crustaceans — Neumann, D., 1981). Evans, W. (1976) found that the intertidal carabid beetle *Thalassotrechus barbarae* forages and mates at night only during periods of low tide. Both circadian and circa-tidal components persist for a few days in constant conditions. The circa-tidal rhythm functions to inhibit nocturnal activity during periods of high tide.

Several field studies have described rhythms of activity or adult emergence related to the lunar month (see Saunders, D., 1976a; Neumann, D., 1981). In two cases the rhythms have been found to persist in constant conditions. Hartland-Rowe, R. (1955, 1958) found that mayflies *Povilla adusta* on Lake Victoria emerged regularly on the second night after the full moon. The timing of emergence was maintained in larvae that had been kept in constant darkness for 10 days and therefore appears to be under the control of an endogenous oscillation.

In the second example Youthed, G. and Moran, V. (1969b) reported that populations of the ant-lion *Myrmeleon obscrurus* exhibit a monthly rhythm in pit-building activity (using mean daily pit volume as an index of activity) with peak activity occurring at the full moon. The rhythm persisted for at least two cycles ($\sim$ 2 months) in constant darkness, but damped out in constant light.

This lunar rhythm of activity was superimposed on a circadian rhythm of pit-building activity in *Myrmeleon obscura* that had been described in an earlier paper (Youthed, G. and Moran, V., 1969a; see section 2.1), and further analysis of the daily activity records of *M. obscura* larvae revealed the presence of a lunar-day component to the rhythm (Youthed, G. and Moran, V., 1969b). It was found that activity tended to be higher about 4 h after moonrise. It was suggested that the monthly occurrence of peak activity resulted from the coincidence, that would occur once each month, of the lunar-day and solar-day (circadian) peaks of activity. This hypothesis was tested by placing larvae in a "reversed" artificial light cycle in which dusk occurred in the morning and dawn in the evening. If the hypothesis were true, the monthly peak of activity should occur near the time of the new moon (when moonrise occurs in the morning). This prediction was confirmed by the results of the experiment, suggesting that the lunar month rhythm emerges from an interaction between the solar- and lunar-day components. The adaptive significance of the rhythmic components of the pit-building activity is unclear (Youthed, G. and Moran, V., 1969b).

An extensive analysis of lunar and circadian periodicities has been carried out by Neumann (reviewed in Neumann, D., 1976a,b, 1981) on the marine midge *Clunio marinus*. Populations of this European species inhabit a variety of rocky shore habitats, most situated in the lower midlittoral zone. Reproduction of the imagoes which are short-lived (about 2 h) involves the attachment of egg masses to exposed substrates of the larval habitat. Typically emergence and reproduction of the adults is synchronized with the extremely low afternoon waters which occur during the spring tides shortly after either the new or full moon. The expression of this semilunar rhythm in reproduction is under the control of two endogenous systems — one that has a circa-semilunar period ($\sim$ 15 days) and sets the

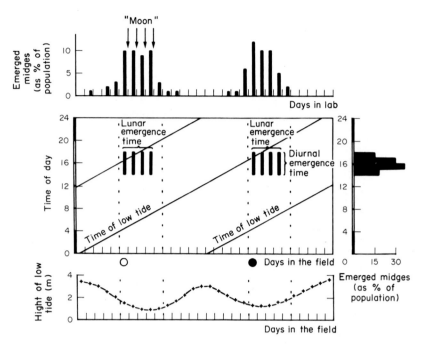

FIG. 20. Laboratory and field observations on emergence distribution of *Clunio marinus*. *Top*: Semilunar emergence rhythm under LD *16* : 8 and artificial moonlight. *Middle*, time of emergence in the field with respect to the lunar cycle and time of low tide. *Lower*, amplitude of low tide over one lunar cycle. *Right*, daily emergence time. (From Neumann, D., 1967.)

time of pupation of the larvae so that adult emergence coincides with the spring tide, and a second with a circadian period that controls the time of day of adult emergence (Neumann, D., 1966). Figure 20 shows data on emergence times of populations of midges from the Normandy coast where low tide occurs twice each day at 12.4 h intervals. The amplitudes of the tides reach their highest point (spring tides) twice a month (every 14.77 days). Low tide at these times occurs either in the early morning or early evening. In the field the midges emerge over a period of 4–5 days, in synchrony with the spring tides, in gates 24 h apart which are "open" during the time of low tide in the morning (Fig. 20, middle panel). Laboratory populations in an artificial 24 h cycle (LD *16* : 8) exposed to artificial moonlight (0.4 lux for 4 nights every 30 days) exhibit similar behavior (Fig. 20 top and far right).

The adaptive significance of the precise timing of emergence is clear. Synchronization at low, spring tide guarantees that there will be a concentration of the short-lived adults for mating, and that the larval habitat will be exposed for deposition of the eggs by the female adult (Neumann, D., 1976b). Neumann, D. (1976b, 1978; Neumann, D. and Heimbach, F.,

1979) has carried out an extensive series of investigations on the mechanism of entrainment of the semilunar rhythm in emergence of *Clunio marinus*. Several artificial times cues were used, including daily light cycles, artificial moonlight, and water turbulence (to simulate tidal stimuli). In studying several populations representing local geographical races, Neumann discovered interesting and ecologically relevant differences between the strains. Results from experiments with the Normandy populations are shown in Fig. 21. When larvae were bred in LD *12* : 12 without other time cues, emergence was aperiodic (Fig. 21, top). The semilunar rhythm could be initiated by exposure to artificial moonlight every 30 days (Fig. 21, middle) or to a cycle of water turbulence every 12.4 h (Fig. 21, bottom) superimposed on the LD *12* : 12 light cycle (Neumann, D., 1976b). While some southern European populations (below 49°N latitude) responded similarly to both artificial moonlight or tides, others responded only to moonlight. In contrast, in northern populations results indicated that moonlight was ineffective, possibly because the low height of the full moon and the shortened nights of midsummer in northern latitudes make moonlight

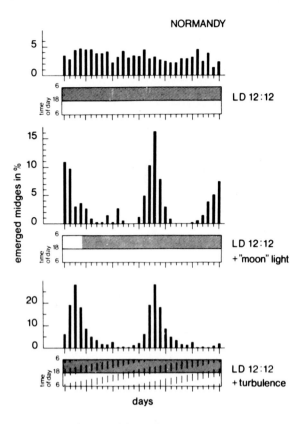

FIG. 21. Entrainment of the semilunar emergence rhythm in *Clunio marinus*. The three panels plot the distribution of emergence as a function of time. *Top*, Control; *middle*, 12 h of 0.4 lux dim light every 30 days on consecutive nights; *lower*, 12.4 h cycles of mechanical disturbance. Zeitgeber conditions are shown schematically below each record. (From Neumann, D., 1976a.)

a less reliable time cue. Entrainment of the northern population appears to depend exclusively on the combined tidal and daily light cycles. The results suggested that the rhythmic recurrence of identical phase angle differences between the two stimuli, which cycles with a period of about 15 days, was the effective cue for entrainment of the semilunar rhythm (Neumann, D., 1978; Neumann, D. and Heimbach, F., 1979).

Another interesting difference between geographical races of *Clunio marinus* is evident in the time of day of adult emergence. Although the days of the spring tides are the same in different regions, the phase of low tide relative to local time can vary considerably. This variation is reflected in the phases of the emergence rhythms of different populations (Neumann, D., 1976a). These dif-

ferences in emergence phase have been shown to be under genetic control (see section 8.2).

The results of studies on periodicities related to the lunar cycle reinforce the view that "biological clocks" have evolved to exploit predictable periodicities in the environment. The internally generated oscillations provide a temporal program that organizes behavior in concert with the temporal fluctuations in environmental challenge and opportunity, with provision for the synchronization of the clocks to local conditions by appropriate environmental stimuli.

6   PHOTOPERIODIC TIME MEASUREMENT AND SEASONAL CYCLES

The changing seasons which recur with an annual periodicity represent a major environmental challenge for most organisms, and no doubt exert a strong selective pressure for the ability to anticipate and cope with the challenges of extreme changes in ambient temperature, rainfall and food availability. The most reliable environmental cue to the time of year is the length of the daily photoperiod (or dark period) and a wide diversity of plants and animals have evolved systems which can measure photoperiod duration with remarkable precision. The information derived from the daily light cycle can then be used to affect a "switch" from one metabolic pathway to another appropriate to the prevailing, or in most cases anticipated, changes in the environment.

Since the initial discovery that photoperiod plays a major role in seasonal adjustments in plants was made by Garner, W. and Allard, H. in 1920 a massive literature has emerged on photoperiodic effects. The problem has been intensively investigated in insects and has been the focus of several books and reviews (e.g., Danilevskii, A., 1965; Saunders, D., 1976a, 1981a,c; Becks, S., 1980).

In the insects a wide variety of processes are known to be at least partially under control of photoperiod. By far the most commonly investigated *photoperiodic response* is the induction of diapause which has been described in well over 200 species representing about 12 orders (Saunders, D., 1981a). Other events which have been shown to be responsive to photoperiod in some insects include termination of diapause, the appearance of seasonal morphs, growth rates, migration, coloration, sexual behavior, sex ratio, fecundity, insecticide sensitivity, and recovery from heat stress. Extensive tables with references can be found in Saunders, D. (1976a) or Beck, S. (1980).

The primary focus of the following discussion revolves around the specific question of how the length of the day (or night) is measured. Two facts will become apparent. First, the problem of exactly how photoperiodic time is measured in insects has been investigated in detail in only a handful of species, most notably *Megoura viciae* (Lees, A., 1959, 1963, 1973), *Pectinophora gossypiella* (Adkisson, P., 1964; Pittendrigh, C. and Minis, D., 1971), *Sarcophaga argyrostoma* (Saunders, D., 1971, 1973b, 1975b, 1976b, 1978a, 1979), and *Nasonia vitripennis* (Saunders, D., 1965a,b; 1966a,b, 1967, 1968, 1970, 1974). Secondly, the results of these efforts to devise a model to explain photoperiodic time measurement (PTM) are very complex; and it seems safe to conclude that a completely satisfactory explanation has not yet emerged. Nevertheless, several important features of PTM have been elucidated from studies on insect photoperiodism that do provide some insight into the mechanism of insect photoperiodic clocks.

6.1 Photoperiodic response curves

Typically the description of a photoperiodic response utilizes a photoperiodic response curve which plots the frequency of a response, for example, induction of diapause, as a function of the duration of the photoperiod in a 24 h day. Figure 22 shows a series of photoperiodic response curves for "long day" insects, i.e. those that develop, are active, and/or reproduce when the photoperiod *exceeds* a certain critical daylength. This type of response curve is particularly common in multi-voltine species which show a facultative winter diapause. Less common are short-day insects in which diapause is averted by exposure to photoperiods shorter than some critical daylength. There are also a few species, for example, *Mamestra brassicae* (Masaki, S., 1956, 1968) that undergo both a summer (aestival) and a winter (hibernal) diapause which are triggered by long and short days respectively. The utility of the photoperiodic response curve is that it serves to

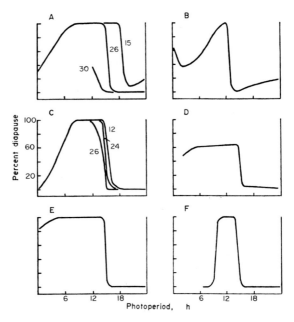

FIG. 22. Photoperiodic response curves for a variety of long-day species. **A**: *Acromycta rumicus* (after Danilevskii, A., 1965); **B**: *Pectinophora gossypiella* (after Pittendrigh, C. and Minis, D., 1971); **C**: *Pieris brassicae* (after Danilevskii, A., 1965); **D**: *Nasonia vitripennis* (at 15°) (after Saunders, D., 1966a); **E**: *Megoura viciae* (at 15°) (after Lees, A., 1965); **F**: *Ostrinia nubilalis* (after Beck, S., 1962). In **A** and **C** curves are given for several temperatures which are given in °C. (From Saunders, D., 1976a.)

define for each species the functional expression of the time-measuring system. An explanation or model of the photoperiodic clock for a particular species is in essence directed toward an explanation of the response curve.

Several points regarding the ecological significance of these response curves are worth comment. First, as noted by Danilevskii, A. (1965) and reiterated by Saunders, D. (1976a), only a small portion of the curve is ecologically relevant. At the extreme ends of the curve the photoperiods are longer or shorter than those typically experienced in nature. On the other hand, the entire curve is representative of the physiological processes underlying the PTM, and any effort to explain the physiology of photoperiodism must contend with the response over the entire range of both natural and experimental photoperiods.

Second, response curves are typically derived by exposing organisms to a variety of fixed or "stationary" photoperiods, whereas natural photoperiods are either constantly lengthening or shortening depending on the time of year. The direction of change may in fact be an important clue to the time of year for some insects (e.g. Norris, M., 1965), although in many species, particularly those that are rapidly developing, it is of no consequence whether the photoperiods are increasing, decreasing or stationary (Lees, A., 1970).

Third, the details of response curves may or may not vary significantly with the mean temperature or with exposure to temperature cycles. This fact is not only of ecological importance but also bears directly on the physiology of the photoperiodic clock. In general, high constant temperatures act synergistically with long days in promoting a long day response, and antagonistically with short days. Conversely, low temperature is synergistic with short days and antagonistic with long days. In *Sarcophaga argyrostoma*, for example, the "short" day, *LD 10* : 14 induces diapause in nearly 100% of the pupae at 15°. At 18° the percentage drops, and at 25° and above *LD 10* : 14 (or in fact even shorter days) is completely ineffective in inducing diapause (Saunders, D., 1971). In other insects such as the Indian meal moth (Masaki, S. and Kikukawa, S., 1981), the effectiveness of short days in promoting diapause may be reduced, but not eliminated by high temperatures. Saunders, D. (1966a, 1971) has suggested an explanation of the loss of diapause induction on the basis of the interaction between developmental rate and temperature and the "photoperiodic counter" (see below).

In some cases changes in ambient temperature can also significantly affect the value of the critical day length. For example in *Acromycta rumicus* critical day length is shortened by $1\frac{1}{2}$ h for each 5° increase in temperature (Fig. 22A) (Danilevskii, A., 1965). In many species, however, the effect of temperature on critical day length is minor or non-existent (e.g. Fig. 22C). These include *S. argystroma* (Saunders, D.. 1971) and *Plodia interpunctella* (Masaki, S. and Kikukawa, S., 1981) as well as *Pieris brassicae* (Bünning, E. and Joerrens, G., 1960), *Ostrinia nubilalis* (Beck, S. and Hanec, W., 1960), and *Megoura viciae* (Lees, A., 1963). This "temperature-compensation" of critical day length and of the precision of the time measurement (reflected in the steepness of the photoperiodic response curve at critical day length) are remarkable features of the time-measuring system and are

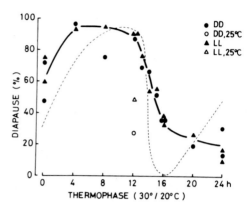

FIG. 23. Diapause response of the Indian meal moth *Plodia interpunctella* in response to 24 h temperature cycles. Percentage diapause is plotted as a function of the duration of the warm portion (thermophase) of the temperature. Data for both DD and LL lighting conditions are shown. The dashed line shows the photoperiodic response curve for comparison. (From Masaki, S. and Kikukawa, S., 1981.)

reminiscent of both the precision and temperature-compensation of time measurement observed for circadian oscillations.

Finally, there are now several examples that show that 24 h cycles of temperature are fully as effective as 24 h light cycles in their ability to regulate the response. In *Nasonia vitripennis* (Saunders, D., 1966a, 1973a), *Diatraea grandiosella* (Chippendale, G. and Reddy, A., 1973; Chippendale, G. *et al.*, 1976), *Pieris brassicae* (Dumortier, B. and Brunnarius, J., 1977, 1981), *Plodia interpunctella* (Masaki, S. and Kikukawa, S., 1981) and *Ostrinia nubilasis* (Beck, S., 1982) "long-day" temperature cycles in which the warm phase of the cycle exceeds a critical day length prevents diapause, and "short day" cycles promote diapause (Fig. 23).

Clearly, in a natural setting the "photoperiodic response" in insects will be a complex function of the daily light and temperature cycles as well as the mean ambient temperatures. Only in the last few years has the effort to understand the interactions between these variables really begun.

6.2 Sensitive period

Typically insects are sensitive to the photoperiod at a stage in the life cycle prior to the stage in which the photoperiodic response is expressed. Furthermore, the period of sensitivity is usually (but not always) restricted to a portion of the life cycle. For example, in *S. argyrostoma* photoperiods experienced during either the embryonic (Denlinger, D., 1972) or larval stages (Saunders, D., 1971) determine whether or not the pupae will enter diapause, while the pupal stage itself is insensitive to photoperiod. In other cases sensitivity may extend through the diapause stage (e.g., *Antheraea pernyi*; Williams, C. and Adkisson, P., 1964), and termination as well as induction of diapause may be photoperiodically determined. There are also several examples in which the sensitive and response stages occur in successive generations. In the aphid *Megoura viciae* the photoperiod to which a virginoparous female is exposed determines whether her offspring will be virginoparous or oviparous. In short days the effect is even carried over to the following generation and determines the diapause condition of the eggs deposited by the oviparous daughter (Lees, A., 1959).

An important feature of the process of photoperiodic time measurement is that a specific photoperiod usually must be experienced for a minimum number of days to have a complete effect. In cases where the period during which the insect is sensitive to day length is relatively short, this required day number (RDN) (Saunders, D., 1976a) may explain the loss of photoperiod control at higher temperatures. In *S. argyrostoma* the RDN is 13–14 days and relatively independent of temperature ($Q_{10} = 1.4$). The rate of development, on the other hand, is highly temperature-dependent ($Q_{10} = 2.7$). At high temperatures development proceeds rapidly enough (9 days at 26°) so that the pupal stage is reached before the larvae can be subjected to a sufficient number of short days to affect diapause induction (Saunders, D., 1971, 1976a). Similar explanations for the loss of photoperiodic control at warmer temperatures have been forwarded for *Nasonia vitripennis* (Saunders, D., 1966a), *Acronycta rumicis* (Goryshin, N. and Tyshchenko, V., 1970) and *Aëdes atropalpus* (Beach, R., 1978).

6.3 Models of the photoperiodic clock

Ever since the discovery that both plants and animals are capable of using a precise measurement of day length for redirecting metabolic pathways in anticipation of seasonally recurring changes in the environment a substantial effort has been made to determine exactly how the time measurement is effected. Because of our basic ignorance of the anatomical and physiological basis of the photoperiodic clock, experimentalists have used the "black-box" approach to understanding the mechanisms of measuring day length or night length. The experiments involve subjecting organisms to a series of exotic and often complicated light cycles and determining whether the organism exhibits its typical long day or its short day response. Based on the results an effort is then made to construct a model of the time-measuring process. Currently there are nearly as many models as there are insects which have been subjected to this approach, and one general conclusion that has emerged is that it seems likely that mechanisms for measuring day length have evolved separately several times, and that it is unlikely the mechanism will be the same in all insects. As Pittendrigh, C. (1976) and others have pointed out, functional unity is no sure guide that underlying mechanisms are similar.

In 1936 Erwin Bünning put forth the idea that circadian oscillations, as clocks, were somehow involved in photoperiodic time measurement. There is no doubt now about the validity of this proposition for many plants and animals, including some insects. Whether or not the circadian system functions in PTM in all organisms, however, is still an open question, and in at least one case, *Megoura viciae*, extensive research has failed to produce any evidence that a circadian clock is involved.

Apart from the general proposition that the circadian clock is involved in PTM, Bünning also proposed a specific hypothesis as to how it might be involved. He suggested that there is an endogenous (circadian) rhythm of cellular activity which is divided into two half-cycles of 12 h each, one of them occurring in the light termed the photophil and the second occurring (primarily) in the dark called the scotophil (Fig. 24). It was assumed the photophil always begins at dawn and that on long days the "long day" response is affected when daylight extends into the beginning of the scotophil. On short days the scotophil is not illuminated and consequently the organism exhibits the short day response. This version of Bünnings' general hypothesis that circadian clocks were involved in PTM has come to be known as the "external coincidence" model since the response is determined

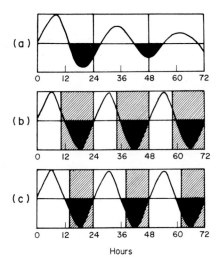

FIG. 24. Schematic representation of the Bünning hypothesis of photoperiodic time measurement by a circadian oscillation. **A**: An oscillation in constant conditions of alternating photophil (light) and scotophil (dark) phase; **B**: in short days the scotophil occurs entirely in the dark; **C**: in long days a portion of the scotophil is illuminated. (From Bünning, E., 1960.)

by the coincidence of light with a particular portion of endogenous cycle. An important aspect of the model which deserves emphasis, since it forms the basis for interpretation of many experiments designed to test it, is that in the absence of light (at constant temperature) the periodic alternation of photophil and scotophil persists as an endogenous oscillation with a period near 24 h.

As formulated by Bünning, the external coincidence model ignores complexities in the entrainment process which were subsequently recognized by Pittendrigh, C. and Minis, D. (1964) who proposed a more explicit version. They pointed out that it is necessary to recognize that light has two separate functions in the external coincidence model:

(1) light entrains and consequently phases the oscillation of scotophil and photophil, and
(2) light is directly involved in affecting the response by its coincidence, or lack of it, with the scotophil.

With one or two exceptions (e.g. Bünsow, R., 1960) the implications of (1) seem not to have been recognized in early efforts to test the model. In particular the important fact that the phases of circadian oscillations are not necessarily locked to dawn or

dusk but can show a strong dependence on the duration of the photoperiod was apparently ignored.[2] In their more explicit formulation of the external coincidence model Pittendrigh, C. and Minis, D. (1964) also recognized that while the "scotophil" of Bünning encompassed a complete 12 h, photoperiodic induction is contingent on illumination of a much smaller portion of the cycle. They suggested the concept of a more restricted "photoperiodically inducible phase" (φ_i) to denote the portion of the cycle where light is effective in induction (Pittendrigh, C. and Minis, D., 1964; Pittendrigh, C., 1966). The term "subjective night" essentially replaced scotophil and the term "subjective day" refers to Bünning's photophil. Thus in its most recent formulation (Pittendrigh, C., 1966) the external coincidence model suggests there is a circadian oscillation in φ_i. During steady-state entrainment of this oscillation, φ_i is illuminated in some (long) photoperiods but not in other (short) photoperiods.

Another major contribution of Pittendrigh, C. and Minis, D. (1964; see also Minis, D., 1965) was their emphasis on the importance of the development of a theory of entrainment of circadian oscillations in general, and more specifically on the importance of the analysis of entrainment and the photoperiodic response in a single organism. The successes of this approach (e.g., Saunders, D., 1981b) and some of its potential pitfalls (e.g., Pittendrigh, C. and Minis, D., 1971) are discussed below.

A second model to account for photoperiodic time measurement which has received much attention in the literature suggests that a transition from light to dark (or, less frequently, dark to light) sets some light-sensitive process in motion. On short days (long nights) the process runs to completion, but on long days dawn "arrives early" and illuminates a light-sensitive phase interrupting the course of the process before it can be completed. The most explicit formulation of this hypothesis and its greatest experimental support comes from work by Lees, D. (1968, 1973) on the aphid *Megoura viciae* (see below). This model, referred to

---

[2] On the other hand it is important to remember for subsequent discussion that in some organisms (e.g. *Drosophila pseudoobscura*, *Sarcophaga argyrostoma*) long photoperiods invariably do reset the circadian pacemaker to ct 12 or near ct 12. Thus, for photoperiods > 12 h the oscillation *is* phase locked to dusk.

as the "hourglass" hypothesis, is strictly analogous to the external coincidence model with the exception that the hourglass requires an external signal, e.g. a minimum period of illumination, to restart the photosensitive process. In contrast, in the external coincidence model the photosensitive stage behaves as a self-sustaining oscillator with no requirement for an external signal to restart it. Tests to distinguish between these two models focus on this difference, and in general are designed to determine whether or not the photosensitive phase repeats with a period of about 24 h in extended nights. An important constraint on the "hourglass" model that is worth noting is that the photosensitive phase maintains a constant phase relationship with the L to D transition (or more generally with at least some phase of the light regime). This is not a necessary constraint on the external coincidence model although, as mentioned above, a circadian oscillation in φ_i could be phase locked to the L/D transition in long photoperiods and still be consistent with known properties of circadian rhythms.

A third hypothesis to explain photoperiodic time measurement, also based on Bünning's suggestion that a circadian oscillation is causally involved, was independently formulated by Tyshchenko, V. (1966) and Pittendrigh, C. (1972), and has been recently embellished (Pittendrigh, C., 1981a). The original formulation of the model was based on the idea that the phase relationship between circadian oscillations and the entraining light cycle (ψ_{OL}) can vary with photoperiod, and that the precise dependence of ψ_{OL} on photoperiod will vary with specific properties of the oscillator. It was suggested that in multicellular organisms, in which there may be several oscillators with slightly different properties, a change in the photoperiod of the entraining light cycle would lead to a change in the "internal" phase relationships of oscillators within the organism. In certain light regimes therefore some phases of the oscillations would be coincident; in others the same phases would occur at different times. The presence or absence of "internal coincidence" of phases could then provide a basis for distinction between long and short photoperiods. Two hypotheses have been proposed. In one there are two oscillators, one which maintains a relatively stable phase relationship with dawn, the other with dusk. At very long or very short photoperiods the two oscillations are forced "in-phase", whereas at intermediate photoperiods the oscillations are held out of phase (Danilevskii, A. et al., 1970; Tyshchenko, V., 1966).

A somewhat different version of "internal coincidence" has recently been proposed by Pittendrigh, C. (1981a) and is based on his detailed analysis of the eclosion rhythm in D. pseudo-obscura. Stated in its simplest form the model suggests there is a single pacemaking oscillation which is coupled to the light cycle and in turn entrains two or more "secondary" or "slave" oscillators within the organism, that themselves are not directly affected by light. Pittendrigh showed in computer simulations of such a model that, in different photoperiods, slave oscillators with slightly different properties could exhibit rather dramatic changes in their phase relationships. Therefore, coincidence (or lack of it) between specific phases of the slaves' cycles could provide a basis for photoperiodic time measurement.

A final variation on "internal coincidence" has been proposed by Truman, J. (1971d) and elaborated by Beck, S. (1974a,b, 1975, 1977, 1980). It proposed that it is the phase relationship between a circadian oscillation that is entrained by light and a second hourglass-like process timed from dawn that determines the photoperiodic response.

Before proceeding with a detailed discussion of the efforts of experimentalists to distinguish between the various models of the photoperiodic clock some general comments on the utility of specific hypotheses, and indeed on this general approach to understanding the mechanism of photoperiodic time measurement, are in order. First, it should be pointed out (and has been emphasized by Pittendrigh, C., 1966, 1981a) that the major contribution the "modeling approach" has made toward our understanding of photoperiodic time measurement is derived from the remarkably insightful suggestion of Bünning nearly 50 years ago that circadian rhythmicity was causally involved in photoperiodic time measurement, whatever the details of the mechanism (internal coincidence, external coincidence, or some other). However, efforts to focus on the specifics of how the circadian system is involved have been less convincing. The major difficulty with external coincidence is that in its simplest form it cannot account for *all* the details of the photoperiodic response even in the one organism,

Sarcophaga argyrostoma, for which the most experimental evidence in support of this model has been obtained (Saunders, D., 1981b). Further, difficult-to-test assumptions are needed, in particular, to explain the low incidence of diapause in DD or very short photoperiods. The various internal coincidence models, on the other hand, suffer from a general lack of specificity, and as a consequence face a different problem: these models can be manipulated to explain almost any experimental result. The real difficulty here is the lack of any identified and measurable physiological or biochemical correllate of the various elements of these models. Some progress in vertebrates, and particularly mammals (e.g. Follett, B. and Follett, D., 1981) has been made toward this end but the effort has barely begun in insects (although see Lees, A. and Hardie, J., 1981). Future gains will depend heavily on progress in describing the anatomy, physiology, and metabolism of the photoperiodic time-measuring system. Until at least some of these parameters of the mechanism are described it is likely that it will be very difficult to put the kinds of constraints on these models that will allow us to devise specific experimental tests of the various hypotheses.

At present the black box approach is adequate only to determine whether there is a circadian oscillation involved or if the system behaves purely as an hourglass. Currently we have examples of each type of clock. Lees, A. (1963, 1966, 1970) has presented very good evidence (described below) that photoperiodic time measurement for seasonal morph determination in the aphid is strictly an hourglass or interval timer that measures night length. Extensive efforts to find an underlying circadian oscillation have been negative (Lees, A., 1970; Hillman, W., 1973). In contrast there are now several cases where the involvement of a circadian rhythm is firmly established — most extensively in *Nasonia vitripennis* and *Sarcophaga argyrostoma* (Saunders, D., 1973b, 1974). In other cases the data may not fit clearly into one model or another. In the pages that follow, work on the mechanism of PTM in three species of insects is reviewed in some detail, both to illustrate the experimental approach and to underscore the apparent diversity in the basis of photoperiodic time measurement.

6.4 The "hourglass" of *Megoura*

In the aphid, *Megoura viciae*, the development of female offspring into sexual, egg-laying morphs (oviparae) *vs.* asexual parthenogenetic morphs (virginoparae) is under photoperiodic control. Long summer days give rise to successive generations of virginoparae, but as the days begin to shorten and night length increases beyond about 10 h, oviparous females that lay diapausing eggs are produced (Lees, A., 1970). Lees has done an extensive series of experiments involving a variety of experimental light cycles to investigate the mechanism for PTM. The results have failed to provide any evidence that a circadian oscillation is involved, and suggest the response depends on an "hourglass" measurement of night length.

Two early experiments pointed to the importance of night length for PTM in the aphid. In the first, light periods of various durations were coupled to a period of darkness that was greater than the critical night length. Short day responses were consistently observed, even when the photoperiod was extended to 40 h, far in excess of a normal long day. In a second experiment aphids were exposed to light cycles in which the photoperiod was fixed at 8 h and the dark period was varied from 4 to 72 h. As shown in Fig. 25, once the night length was extended beyond the critical 10 h the short day response was dominant. The results suggested the photoperiodic response depended on a measurement of night length, and that once it exceeded the critical period, the short day response was fixed. Lees further investigated the properties of night length measurement by exploring the effect of brief light pulses during the dark period. One such experiment is illustrated in Fig. 26, which shows the effect of 1.0 h light pulses during the dark period of light cycles that consisted of 13.5 h of light and 10.5 h of darkness. As noted above an uninterrupted night of this length leads to the production of oviparae. It was found that this response was reversed when animals were exposed to the light pulses at either one of two times during the dark. The first of these light-sensitive points occurred between 0 and 3 h after the L/D transition. This was followed by a period of insensitivity to light (3–4 h) and a second light-sensitive period that occurred 6–10.5 h after light offset. The timing of these light-sensitive periods

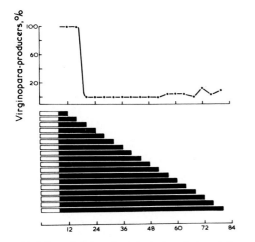

FIG. 25. Effect of the length of the dark period on viginopara production in *Megoura viciae*. The light period was constant at 8 h and the dark period was varied between 4 and 72 h. Once the dark period exceeds the critical night length (9.5 h) virginopara production remains at a low level. (After Lees, A., 1973, from Saunders, D., 1981a.)

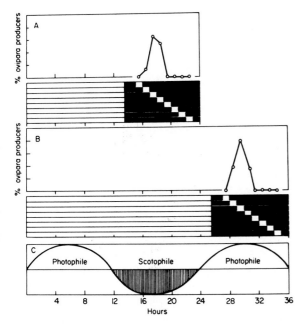

FIG. 26. Night interruption experiments in *Megoura viciae* with photoperiods of 13.5 h (**A**) and 25.5 h (**B**) and a dark period of 10.5 h. In cycles with an uninterrupted 10.5 h dark period all aphids are ovipara producers (not shown). The position of the light-sensitive points are phase-locked to the L/D transition and are independent of photoperiod duration. **C**: The predicted photophil and scotophil phases of the Bünning hypothesis. (From Lees, A., 1968.)

with respect to the L/D transition was unaffected when the light period was extended to 25.5 h or

reduced to 8 h (Lees, A., 1966). In the "external coincidence" terminology, φ_i was phase-locked to the L/D transition.

If a circadian oscillation was involved in the photoperiodic time measurement the period of photosensitivity would be expected to recur at approximately 24 h intervals in extended periods of darkness. Two separate experiments provided evidence that this is not the case in *Megoura*. In the first (alluded to above) aphids were exposed to light cycles in which the photoperiod was fixed at 8 h and the dark period varied from 4 to 72 h. As shown in Fig. 25, once the night length was extended beyond the critical duration, the short day response was dominant and there was no indication of a recurring period of light sensitivity. This protocol, typically referred to as a "resonance" experiment, has proved one of the most powerful for detecting a circadian rhythm of "photoperiodic photosensitivity" (see below). In the second experiment, a photoperiod of 8 h was combined with a dark period of 64 h and the dark period was "scanned" at 4 h intervals with 1 h pulses of light (Fig. 27). The expected peak of sensitivity to lights was evident 8 h after the onset of darkness, but there was no evidence that it recurred at subsequent 24 h intervals (i.e. at 32 and 56 h after L/D). Thus these "night interruption" experiments also provided no support for the involvement of the circadian system in photoperiodic time measurement. Another interesting aspect of these data is that the light pulse that falls 8 h after the onset of darkness reverses the response even though the subsequent dark period is substantially longer than the critical night length. In contrast the effect of light falling at the early photosensitive phase can be reversed if it is followed by a period of darkness greater than the critical night length (Lees, A., 1970).

Taken together the results are persuasive evidence that night length is measured by an "hourglass"-type process. The function of the main photoperiod is to "prime" the mechanism, or by analogy "turn the hourglass over". Light pulses early in the night reset the timer and the timing process is resumed from the "zero" position. The presence or absence of light during the late photosensitive period determines whether or not the cycle is "read" as a short or long night. In natural light cycles this photosensitive stage would be

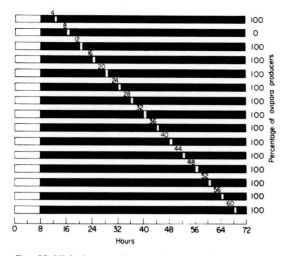

FIG. 27. Night interruption experiments in *Megoura viciae*. Following an 8 h photoperiod a 72 h night was interrupted by a 1 h pulse. The percentage of short day responses is given to the right of each light regime. Only the pulse 8 h after the onset produced the long day response. Note that this photosensitive phase does not recur approximately 24 h later. (From Lees, A., 1968.)

encroached upon by dawn in short nights and result in virginoparae production. Further experiments have also shown that the main photoperiod must be 4 or more hours long to be effective in "priming" the timing mechanism. This provides an explanation for the observation that the 1 h light break late in the night does not reset the timing process to its zero position. In a final test of the possible involvement of circadian organization in the aphid photoperiodic response Hillman, W. (1973) made use of several skeleton photoperiods which, depending on initial conditions, were expected to entrain the circadian system in one of two steady states (see discussion of "bistability" in section 3.4). Depending on which dark interval is taken as night, the skeletons could be "interpreted" by the circadian system as either long or short photoperiods (see Pittendrigh, C., 1966). In the short photoperiod interpretation, light pulses would fall at the extreme ends of the subjective night or in subjective day. In the long photoperiod interpretation, light pulses would fall near the middle of the subjective night presumably illuminating φ_i. A positive result in this experiment is a powerful indication of the involvement of circadian organization since the inductive effect of the light cycle depends entirely on the phase of the circadian system. Since the same light cycle can be

inductive or non-inductive the response cannot be simply the duration of the light or dark periods.

Hillman, W. (1973) found no evidence from his experiments with skeleton photoperiods to support the hypothesis that the circadian system was in any way involved in the photoperiodic response of *Megoura*. Furthermore, all the results were consistent with the hourglass model of the photoperiodic clock. In summary, the data from *Megoura* argue strongly against any model of photoperiodic time measurement that depends on a circadian oscillation and against any version of Bünning's hypothesis.

6.5 The photoperiodic clock of the pink bollworm — a test of the external coincidence model

In their development of an explicit version of the external coincidence model, Pittendrigh, C. and Minis, D. (1964) emphasized the importance of recognizing the dual role of light as both an entraining agent and an inductive stimulus. They found that, on the basis of the external coincidence hypothesis coupled with a theory of entrainment derived from studies on the eclosion rhythm of *Drosophila pseudo-obscura*, they could readily explain W. Hillman's (1964) published data on photoperiodic induction of flowering in the duckweed *Lemna* in skeleton photoperiods as well as the results of night interruption experiments in several species of insects which show two peaks of light sensitivity (Pittendrigh, C. and Minis, D., 1964; Pittendrigh, C., 1966). These initial successes of the external coincidence model were promising, but in view of the known differences in the details of entrainment in general, and phase response curves in particular, among species, it was recognized that the potential of the experimental approach could only be realized by investigating the relationship between entrainment and photoperiodic induction in a single species (Pittendrigh, C. and Minis, D., 1971). This was attempted using the pink bollworm moth *Pectinophora gossypiella*.

Since the circadian oscillation that was to underlie the rhythm in φ_i for induction of diapause postulated by external coincidence could not be measured directly, Pittendrigh, C. and Minis, D. (1971) were forced to make the fundamental assumption that the position of φ_i bore a fixed phase relationship to another, overt, rhythm that could be directly assayed. Three such rhythms were studied: oviposition, egg hatch, and eclosion. In practice φ_i was typically defined with respect to the eclosion rhythm. The definition of the position of φ_i was obtained by repeating P. Adkisson's (1964) "night interruption" (or asymmetric skeleton photoperiod) experiments using a main photoperiod of 8 h with 1 h light breaks during the night while monitoring the effects of the light cycles on the entrainment of pupal eclosion rhythm and on the induction of diapause. The data (Fig. 28) reconfirmed the earlier report of two peaks of photosensitivity during the night, but also showed that because of the pattern of entrainment to the asymmetric skeleton it was unneccessary to assume there were two photo-inducible phases. As the "night interruption" moved across the early subjective night (hours 10–16), the eclosion peak moved across the subjective day. On the assumption that φ_i occurred approximately 5 h before the eclosion peak, this increase in the negative phase angle would drive φ_i out into the main photoperiod, which would cause the first peak of photoperiodic reversal of diapause. As the light break moved later into the night, the eclosion peak was forced back toward the beginning of the main photoperiod. Initially φ_i would be back in the dark, but eventually the light break would coincide with φ_i (at about hour 21) and the second peak of diapause prevention would occur.

Having defined the position of φ_i on the basis of these experiments, several tests of the external coincidence were devised. In the first, concurrent cycles of light and temperature were used to manipulate the phase of the eclosion rhythm. As seen in Fig. 28, as the low point of the temperature cycle moved across the subjective day, the eclosion peak was also forced across the subjective day, eventually pulling φ_i into the main photoperiod and subsequently, into the early subjective night. As predicted there was a clear decrease in diapause in those cycles in which φ_i was illuminated.

While these experiments provided some encouragement for the external coincidence model, two aspects of the data were not readily accounted for as Pittendrigh, C. and Minis, D. (1971) noted. First, in the night interruption experiments, there was an unexpected discrepancy in the amplitudes of the two peaks of diapause suppression. Second, in

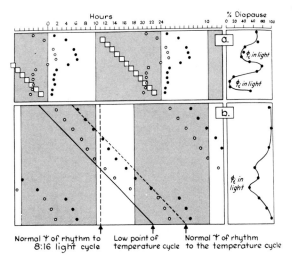

FIG. 28. Entrainment of the circadian rhythm in eclosion and its relation to photoperiodic induction in *Pectinophora gossypiella*. **a**: Asymmetric photoperiods; open circles, postulated photoinducible phase (φ_i); filled circles (left), median of eclosion peak. Populations were maintained in LD 8:16 with a light break of 1 h (squares) during the night. The panel at the right plots the corresponding data for diapause induction. **b**: Concurrent cycles of light (LD 8:16) and temperature (20–29°, sinusoidal) in various phase relationships. Symbols same as in (**a**). Panel at right plots incidence of diapause (see text). (From Pittendrigh, C. and Minis, D., 1971.)

the experiments with conflicting light and temperature cycles, suppression of diapause was generally lower than expected. Subsequent experiments provided data that conflicted even more dramatically with predictions from the external coincidence model.

One of these involved entrainment to light cycles of 20 to 27 h (all with 8 h photoperiods). On the basis of entrainment theory, in T cycles longer than τ, steady-state entrainment would occur when light pulses fall in the early subjective night where they would affect a phase delay. In cases where $T < \tau$, however, light pulses would fall in the late subjective night of the entrained pacemaker, causing a phase advance. In this latter case φ_i, which was postulated to occur in the late subjective night, would be illuminated, leading to the prediction of a reduced incidence of diapause. In the former case ($T > \tau$), φ would escape illumination and the rate of diapause induction should remain high. The behavior of the eclosion rhythm was exactly as predicted in the experiments: however, none of the cycles substantially reduced diapause. This result provided strong evidence against the external coincidence model.

A final series of experiments provided further evidence against external coincidence and, in fact, seemed to preclude any form of participation of the circadian system in PTM in this species. This involved subjecting populations of animals to "resonance" light cycles, in which fixed photoperiods were coupled with various periods of darkness. No evidence of positive resonance effect was obtained in either the initiation of diapause (Adkisson, P., 1964, 1966; Pittendrigh, C. and Minis, D., 1971) or in the termination of diapause (which is also under photoperiodic control in *Pectinophora* (Pittendrigh, C. and Minis, D., 1971).

A further difficulty arose when it was discovered that although the circadian system could not be entrained by red light (600 nm) the photoperiodic system could readily discriminate between 14 h of red light per day (17% diapause) and 12 h of red light per day (98%). These results called into question the original assumption that φ_i bears any fixed phase relationship to the overt circadian rhythms of the organism.

Another interesting approach used to test external coincidence involved creating strains of *Pectinophora* by artificial selection in which the phase of the eclosion rhythm was either earlier or later than normal (see section 8.2). If φ_i were tightly coupled to the phase of eclosion it was predicted that the critical day length would be different for the various strains. In the "early" strain φ_i would occur earlier in the late subjective night and its illumination would require longer-than-normal days. By similar argument critical day length in "late" would occur at shorter photoperiods. The results showed that the critical daylength of "early" was not measurably different from "stock". The results for "late", on the other hand, were more complex — the entire photoperiodic response curve was depressed. Maximum diapause still occurred in a 12 h photoperiod but was only 60% for the late strain *vs.* 100% for the early and stock strains. The results are not readily explained by external coincidence, but it is intriguing, and as yet not understood, that selection on the circadian system dramatically, if not predictably, affected the photoperiodic response.

In summary, the results of this extensive series of experiments provide no support for "external coincidence" in *Pectinophora*. On the other hand neither are the data easily explained by the "hourglass"

approach. It is clear that the facts, and their explanation, are more complex than can be accounted for by simple versions of any current models of the photoperiodic clock. In this context these experiments are especially important in two other respects. First they clearly point to some of the limitations of the "black box" approach to understanding photoperiodic time measurement. Second they illustrate the utility of a comprehensive understanding of the role of light in entrainment in investigations of the causal relationship between circadian organization and photoperiodic time measurement.

6.6   External coincidence as an explanation for PTM in *Sarcophaga argyrostoma*

The approach suggested by Pittendrigh, C. and Minis, D. (1971) has been embraced and extended in an analysis of the properties of the photoperiodic clock in the flesh-fly *Sarcophaga argyrostoma*. A complete account of the extensive series of experiments is beyond the scope of this review and is more completely covered in Saunders, D. (1981b).

These experiments involved a comparison of the analytical properties of entrainment of the circadian rhythm of eclosion with the photoperiodic induction of diapause. Interpretation of the results was based on the assumption that φ_i bears a fixed (or nearly fixed) phase relationship to the eclosion peak. The definition of the position of φ_i was based on the observation that in photoperiods longer than 12 h the circadian oscillation that controls eclosion is reset to ct 12 and in darkness free-runs from that phase point. Since the critical night length in this species is 9.5 h, φ_i was postulated to occur at a phase 9.5 h after ct 12 or at ct 21.5. Thus, in an LD cyle with a 14.5 h photoperiod and a 9.5 h dark phase, dawn would occur at φ_i.

Justification for this assumption and evidence for a circadian oscillation in photoperiodic photosensitivity were provided by the data illustrated in Fig. 29. Pupae were exposed to photoperiods of 4–20 h duration coupled with variable lengths of darkness. Figure 29 shows the effects of these light cycle on diapause induction; there is a clear "positive resonance" effect with peaks of induction occurring at approximately 24 h intervals. It can also be seen that as the photoperiod extends beyond 12 h the

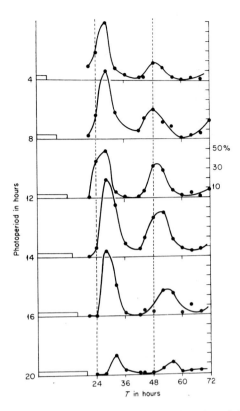

FIG. 29. Induction of pupal diapause as a function of the period (T) of the LD cycles with photoperiods from 4 to 20 h. Induction shows peaks approximately 24 h apart. Temperature, 17° (from Saunders, D., 1973b.)

peaks are delayed. This is more clearly seen in Fig. 30 which plots the resonance peaks (open circles) as a function of photoperiod duration. For comparison the phases of the peaks of the pupal eclosion rhythm (φ_R) (closed circles) in the same LD cycles are shown. In photoperiods longer than 12 h both the circadian rhythm of eclosion and the peak of photoperiodic photosensitivity (φ_i) exhibit a constant phase relationship to the light to dark transition.

Saunders, D. (1978a) also measured phase response curves of the eclosion rhythm to light pulses of 1–20 h. In a subsequent series of experiments involving a variety of light treatments, the position of φ_i and φ_R could be computed (using the Pittendrigh entrainment model) and compared with measured values for incidence of diapause and the phase of the eclosion peak. The experiments involved (a) 24 h light cycles with complete photoperiods, (b) 24 h "skeleton" light cycles in which the beginning and end of the photoperiod were defined by 1 h light pulses, and (c) "asymmetrical skeletons" (night interruption experiments) in which dark periods that followed a longer (8–10 h) photoperiod were interrupted at various times by 1 h light pulses. With few exceptions data on both the eclosion rhythm and diapause induction closely matched the response predicted by entrainment

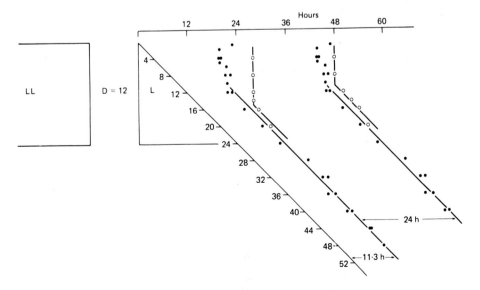

FIG. 30. Peaks of pupal eclosion (filled circles) and the photoperiodic response (open circles) of *Sarcophaga argyrostoma* following exposure to various photoperiods. In photoperiods greater than 12 h, eclosion and photoperiodic response peaks occur at a fixed duration after the end of the photoperiod. (From Saunders, D., 1976b.)

theory and the external coincidence model of photoperiodic time measurement.

Examples of these experiments are shown in Figs. 31 and 32. In the first the incidence of diapause in complete photoperiods of 0–24 h is compared with predicted phases of the circadian rhythm of φ_i in the same LD cycles. The postulated photosensitive phase (φ_i) remains in the dark in photoperiods up to about 14 h; thereafter it occurs in the early portion of the photoperiod. The correlation of the illumination or non-illumination of the putative φ_i with the incidence of diapause and the position of critical day length is clear, and except for very short photoperiods where diapause induction declines the data match expectations from the external coincidence model quite well.

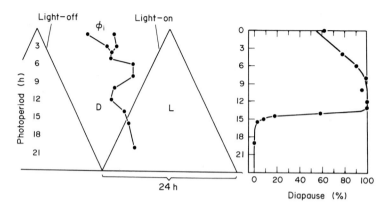

FIG. 31. *Sarcophaga argyrostoma*. *Left*: computed phase relationship between the photoinducible phase (φ_i) and the light cycle in complete photoperiods. In the 1 h and 3 h photoperiods steady state was not reached in the 18 cycles of the computation and the two plotted points for these photoperiods show the range of phase relationships. *Right*: photoperiodic response curve. The critical day length coincides closely with the movement of φ_i into the light. (From Saunders, D., 1981b.)

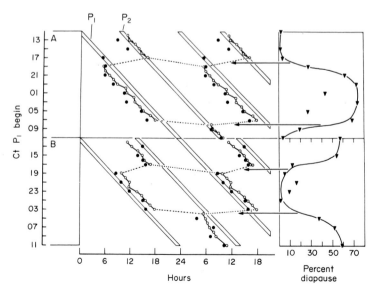

FIG. 32. "Bistability" and photoperiodic induction in *Sarcophaga argyrostoma*. Open circles are the computed phases of the eclosion rhythm; filled circles are experimentally determined phases of eclosion. **A**: PPs 11 (*LD 1:9 1:13*) with the first light pulse (P_1) occurring at all circadian times. **B**: PPs 15 (*LD 1:13:1:9*). Horizontal arrows indicate the positions of the phase-jumps. Panels at right show the incidence of diapause corresponding to these light treatments. Induction of diapause correlates with the phase of the circadian oscillation. (From Saunders, D., 1978a.)

In a second experiment Saunders, D. (1975b, 1981b) investigated the effects of skeleton photoperiods that were in the zone of "bistability" for *Sarcophaga* (see section 3.4). The results are shown in Fig. 32 for two skeleton light cycles *LD* $1:9:1:13$ and *LD* $1:13:1:9$. The final steady-state phase relationship between eclosion varies as a function of the time of the first pulse. In some cases the shorter dark interval is "interpreted" as day; in other cases it is the longer dark interval. When φ_R occurs prior to a light pulse, incidence of diapause is high, but when the predicted φ_R occurs shortly after or during a light pulse, diapause is averted. The transitions from the diapause to non-diapausing state are clearly correlated with the "phase-jump" from one "interpretation" of the skeletons to the other. This experiment provides powerful evidence for the involvement of the circadian system in photoperiodic time measurement in this species and is consistent with the "external coincidence" model.

A final experiment of interest utilized light cycles whose period was near 24 h (Saunders, D., 1979). As noted above, this experiment failed to produce diapause reversal in *Pectinophora* and in that species provided good evidence against external coincidence. In contrast, this approach provided a remarkably good match between the expected and predicted incidence of diapause in *Sarcophaga*. The data are shown in Fig. 33. In cycles in which T is short ($T < 24$ h), the 1 h light pulse is predicted to occur in the late subjective night, illuminating φ_i and averting diapause. In longer T cycles ($T > 24$ h)

the light pulse hits in the early subjective night, φ_i occurs in darkness, and the incidence of diapause is high.

These data and others (Saunders, D., 1981b) show good agreement between the predictions of "external coincidence" and the control of the incidence of diapause by light. As Saunders, D. (1981b) has noted, it seems to be a good working hypothesis to explain photoperiodic time measurement in *Sarcophaga*, but a strict application of the model fails to explain a number of properties of the photoperiodic response, the most troublesome being the decline in diapause in very short days and in continuous darkness.

6.7 Oscillatory and non-oscillatory photoperiodic clocks

Besides *S. argyrostoma* there are several species of insects in which "resonance" or "night interruption" experiments have provided evidence for the involvement of a circadian oscillation in photoperiodic time measurement (Table 2). It is not clear in any of these species, however, exactly how the circadian system is involved. The data from *S. argyrostoma*, for example, while generally consistent with the "external coincidence" model, can be accounted for by internal coincidence. At this point it is not possible to unambiguously distinguish between the two hypotheses experimentally. However, details of the photoperiodic response to "resonance" light cycles are suggestive of external coincidence in some species (e.g. *S. argyrostoma*) and internal coincidence in other (e.g. *Nasonia vitripennis*) (Saunders, D., 1978b).

Non-oscillatory, "hourglass" clocks appear to be functional in PTM in several other species (Table 3) based on negative results from "night interruption" and "resonance" experiments. It should be noted, however, that the detection of positive resonance may depend on ambient temperature. There are two cases, *S. argyrostoma* (Saunders, D., 1973b) and *Drosophila aurariae* (Page and Pittendrigh, unpublished) in which positive results are obtained at one temperature and negative results at another. Furthermore, the lack of a circadian rhythm in "photoperiodic photosensitivity", suggested by a negative result in a resonance experiment, does not necessarily completely rule out the involvement of

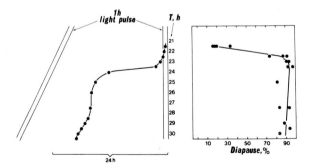

FIG. 33. *Left*: Computed phase of φ_i during entrainment to 1 h light pulses in cycles ranging from periods of 21.5 to 30.5 h. *Right*: Incidence of diapause in the same T cycles. The percentage diapause is low when φ_i is illuminated and high when φ_i is in the dark portion of the LD cycles. (From Saunders, D., 1981b.)

circadian organization in photoperiodic time measurement (see Pittendrigh, C., 1966).

In summary it appears there is significant diversity in the photoperiodic clocks of insects. Further progress in discovering the details of this diversity will likely depend heavily on elucidation of the physiological and anatomical basis of the time measuring system.

6.8 Circannual clocks

While it is clear that photoperiod is the dominant cue for seasonal cycles in physiology and behavior in many organisms, another approach to the timing of seasonally appropriate events has been the evolution of circannual clocks, self-sustaining oscillators with free-running periods of about a year. The existence of circannual clocks in a variety of vertebrate species is well documented (see Gwinner, E., 1981 for review). However, one of the earliest demonstrations of an endogenous annual clock comes from work on the carpet beetle *Anthrenus verbasci* (Blake, G., 1958, 1959, 1963).

The larval stage in *A. verbasci* lasts 2 (or in some cases 3 or 4) years with winters being spent in larval diapause. At the end of the second diapause the larvae pupate and the adults emerge the following spring. Pupation is a gated event, and in the laboratory larval populations maintained in conditions of constant temperature and constant darkness ex-

hibit a free-running rhythm of pupation with a period of 41–44 weeks (Fig. 34). Furthermore, the free-running period is temperature-compensated and also appears to be unaffected by the duration of the stationary photoperiod (Blake, G., 1960, 1963). Temperature, however, does influence developmental rate and affects the proportion of larvae that pupate during a particular gate (Blake, G., 1959). In Fig. 34 it can be seen that at high temperature development proceeds rapidly and all larvae are able to use the first gate. As the temperature is

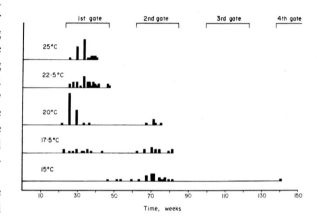

Fig. 34. The circannual rhythm of pupation in *Anthrenus verbasci*. Data from animals maintained in constant darkness and at constant temperature and humidity. As the temperature is lowered animals are forced into later gates. The free-running period of the rhythm is about 42 weeks. (From Blake, G., 1959.)

Table 2: *Species in which photoperiodic time measurement appears to involve a circadian oscillator (from Saunders, D., 1981)*

Species	Function controlled	References
Nasonia vitripennis	larval diapause	Saunders, D. (1970)
Sarcophaga argyrostoma	pupal diapause	Saunders, D. (1973b)
Aëdes atropalpus	embryonic diapause	Beach, R. and Craig, G. (1977)
Pterostichus nigrita	reproductive diapause	Thiele, H.-U. (1977)
Pieris brassicae	pupal diapause	Claret, J. *et al.* (1981); Dumortier, B. and Brunnarius, J. (1981)
Drosophila aurariae	reproductive diapause	Page and Pittendrigh (unpublished)
Dendroides canadensis	synthesis of "antifreeze" proteins	Horwath, K. and Duman, J. (1982)

Table 3: *Species in which photoperiodic time measurement appears to involve an "hourglass" clock (from Saunders, D., 1981)*

Species	Function controlled	References
Megoura viciae	virginoparae production	Lees, A. (1973)
Pectinophora gossypiella	larval diapause	Pittendrigh, C. and Minis, D. (1971)
Ostrinia nubilalis	larval diapause	Bowen, M. and Skopik, S. (1976)
Plodia interpunctella	larval diapause	Takeda, M. and Masaki, S. (1976)
Diatraea grandiosella	larval diapause	Takeda, M. (1978)

lowered development is slowed and more and more larvae delay pupation to the second gate. As Saunders, D. (1976a) has pointed out, the results of these remarkable experiments on the circannual rhythm of pupation in *A. verbasci* exactly parallel the data from the gated circadian rhythm of eclosion in *Drosophila pseudo-obscura* (Pittendrigh, C., 1966; Skopik, S. and Pittendrigh, C., 1967).

Entrainment of circannual rhythms in vertebrates to an exact 52-week period appears to depend on the seasonal change in daylength (Gwinner, E., 1981). In *A. verbasci* the dominant zeitgeber appears to be the seasonal change in temperature (or perhaps thermoperiod); however, photoperiod may also play some role (Blake, G., 1963).

7 ONTOGENY

As discussed in some detail above, there have been a large number of studies on the control of developmental events in insects by the circadian system. However, little attention has been paid to the ontogeny of the circadian system itself. There are several problems of interest, including the time of pacemaker differentiation, changes in the circadian system during development, and environmental effects on the ontogeny of circadian organization.

7.1 Time of differentiation

There is now evidence from several holometabolous insects that suggests differentiation of a circadian pacemaker occurs relatively early in development, In cases where overt rhythmicity is absent or not easily measured, evidence for the presence of an oscillator has been obtained principally by showing that the phase of some overt rhythm observed later in development (e.g. eclosion time) can be set by light or temperature in earlier stages. For example, in *Pectinophora gossypiella* egg hatching is gated by a circadian oscillation whose phase can be set by light or temperature steps or pulses approximately midway through embryogenesis (Minis, D. and Pittendrigh, C., 1968). Stimuli on the 6th (or later) day after oviposition can synchronize the population rhythm. In *Drosophila* (Brett, W., 1955; Zimmerman, W. and Ives, D., 1971) and in *Sarcophaga*

(Saunders, D., 1978a), the phase of the adult eclosion rhythm can be set by light in the first larval instar, and in the mosquito *Anopheles gambiae* the adult flight rhythm can be phased by light at least as early as the second larval instar (Jones, M. and Reiter, P., 1975). In *Sarcophaga argyrostoma* complete phase response curves were measured through the first 5 days of larval development, directly demonstrating the presence of a circadian oscillation in the larval stages. In other insects the demonstration that egg hatch (Nayar, J. *et al.*, 1973) or the endocrine events of larval ecdysis (Truman, J., 1972a; Fujishita, M. and Ishizaki, H., 1981) are gated by a circadian oscillator also provides direct evidence that differentiation of a pacemaker occurs relatively early in development.

In hemimetabolous insects data on the presence of a larval circadian pacemaker are more limited. In an early study Eidmann, H. (1956) reported a rhythm in the peak activity of *Carausius morosus* in its first instar, although activity occurred throughout the day. In the house cricket, Nowosielski, J. and Patton, R. (1963) recorded activity from the last instar of a large number of individuals and found only a few which exhibited any evidence of rhythmicity, and most animals exhibited no, or in some cases little and scattered, activity. The clearest evidence for the expression of rhythmicity in early larval stages comes from a recent study on locomotor activity in cockroach nymphs (*Leucophaea maderae*). Approximately one-third of first-instar nymphs were found to show a clear rhythm in activity recorded within the first few days of larval life (Page, T. and Block, G., 1980) (Fig. 35). Activity of most other animals became clearly periodic within 4–8 weeks (second to third instar). The nymph activity rhythms were similar to those of the adults with the clear exception that nymphs typically alternated several days of activity with several days of inactivity (Fig. 35). However, the variation in activity levels, which may have been related to the molting cycle, had no discernible effect on the phase of the pacemaker underlying the rhythm which apparently continued in motion even though overt rhythmicity was no longer evident (Page, T. and Block, G., 1980). Thus in *Leucophaea* the pacemaker that controls activity is present at the time of hatching, but no information is available on the time of differentiation prior to egg hatch.

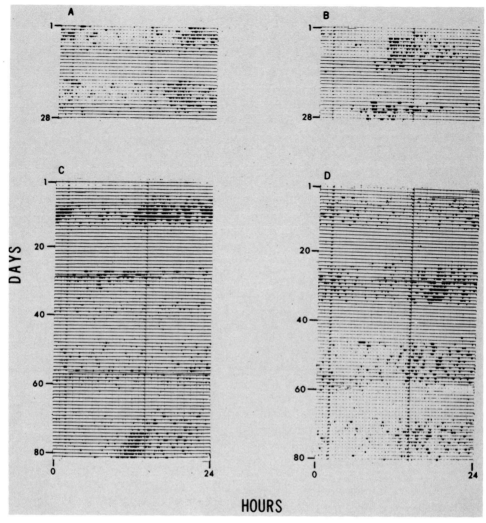

FIG. 35. Activity records of four cockroach nymphs recorded in constant darkness. Each record begins with a few days of hatching from the egg. The record shown in **A** was obtained from a nymph hatched in constant darkness (the mother had been maintained in DD for over 6 months). The other three records are from nymphs hatched from females which had been maintained in constant light for over 6 months. (From Page, T. and Block, G., 1980.)

7.2   Effects of post-embryonic development

The fact that the data in insects suggest a circadian pacemaker is differentiated early in development raises the general question of what effect, if any, subsequent, post-embryonic development may have on the system. Of particular interest is whether the overt rhythms expressed at various stages of development are controlled by the same, different, or developmentally modified circadian oscillators.

In *Pectinophora gossypiella* egg hatch, pupal ecdysis, and oviposition are rhythmic processes that occur at quite different stages of development (Pit-

tendrigh, C. and Minis, D., 1971). Each of the rhythms shows some unique characteristics: (a) each rhythm has a unique phase relationship to the light cycle in the entrained steady state (Fig. 36); (b) the period of the oscillation gating egg hatch, which is about 24 h, is much longer than the periods of the eclosion or oviposition rhythms which are about 22.5 h; (c) constant red light (600 nm) shortens τ of the eclosion rhythm but fails to affect the period of oviposition; (d) in response to artificial selection on the time of eclosion, two strains were developed in which emergence time in *LD 12*:12 differed by about 5 h. In these strains the time of egg hatch was

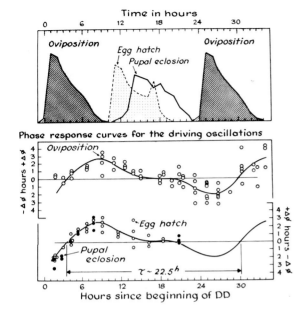

FIG. 36. Rhythms and phase response curves of *Pectinophora gossypiella*. *Upper panel*: daily distribution of oviposition, egg hatch, and pupal eclosion in LD *14*:10. *Lower panel*: PRCs for each of the rhythms to 15 minute pulses of light. (From Pittendrigh, C. and Minis, D., 1971.)

(Konopka, R. and Benzer, S., 1971). All mutations have essentially identical effects on τ of both the eclosion and adult activity rhythms.

Caution in interpreting similarities or differences in free-running periods or PRCs at different developmental stages is warranted. As in *Pectinophora* the PRCs for activity and eclosion rhythms are similar in the wild-type *D. melanogaster* (Konopka, R., 1981), but in contrast, the free-running periods and PRC of the eclosion and activity rhythms in *Drosophila pseudo-obscura* are quite different (Engelmann, W. and Mack, J., 1978). These differences do not necessarily suggest separate pacemakers. This fact is illustrated by results from *L. maderae* and *S. argyrostoma* that show both τ and the PRC of a circadian oscillation can change during development. Activity records from *Leucophaea* nymphs suggest the period of the pacemaker that drives the locomotor activity rhythm increases significantly 2–3 weeks after egg hatch (Page, T. and Block, G., 1980). In *S. argyrostoma* the PRC of the circadian oscillator that controls eclosion systematically changes from type 0 to type 1 in response to 12 h light pulses given through the first 5 days of larval life (Saunders, D., 1978a). Ultimately, an unambigous resolution to the question of whether different oscillators subserve different rhythms in the course of development will probably depend on anatomical localization of the pacemaking systems. The only directly relevant data are J. Truman's (1972b, 1974b) which suggest that the silkmoth midbrain controls both eclosion and locomotor activity; however, more precise localization of either oscillator has not yet been reported.

7.3 Environmental effects

There are only limited data on the potential influence of the environment on the development of the circadian system. One fact is relatively secure: insects do not need to be exposed to a 24 h fluctuation in the environment for the expression of a free-running rhythm. For example, fruit-flies (Brett, W., 1955; Pittendrigh, C., 1966) and cockroaches (Page, T. and Block, G., 1980) which have never been subjected to a periodic environment still exhibit clear overt rhythms (Fig. 35). Less is known about more subtle effects of the environment on

similarly affected, but selection had no effect on the oviposition rhythm.

The only features which appeared to be similar among the three rhythms were the phase response curves to light (Fig. 36). Clearly there are some differences in the circadian control of the three rhythms, but whether these differences reflect the fact that the pacemaker driving each rhythm is anatomically and physiologically unique is uncertain. It is plausible that the pacemaking system is identical for each rhythm, and the differences relate to different or developmentally altered output pathways, or the developmental modification of input pathways to the pacemaker. Of the properties of the rhythms investigated, the phase response curves (PRCs) are most nearly a direct measure of pacemaker properties — their similarity suggests that even if the oscillators are separate entities, the fundamental structure may be similar. Support for this view comes from work on genetic mutants in *Drosophila* (see next section). In *Drosophila melanogaster*, three mutations at a single locus on the X-chromosome have been isolated that either cause a significant change in τ or cause aperiodicity system development. Two studies, however, suggest that the environment during post-embryonic development may significantly influence circadian organization in the adult. Truman, J. (1973b) found that when male *Antheraea pernyi* were maintained in 25° constant temperature (*LD 16*:8) during development and as adults, flight activity of the adult consisted of a burst of intense activity that began 5 h after onset of darkness and terminated at the onset of light. In contrast, individuals exposed to 12° during development then transferred to 25° as adults exhibit a more dispersed activity pattern with activity occurring throughout most of the night. The locus of the temperature effect has not been identified. Temperature may effect the circadian pacemaker or its input or output pathways.

In another study it was found that exposure of *Leucophaea* to non-24 h light cycles or to constant darkness during postembryonic development has a major impact on the free-running period of the adult activity rhythm (Page, T. and Block, G., 1980). The average τ for animals raised in *LD 11*:11 is about 1.5 h less than animals raised in *LD 13*:13; animals exposed to *LD 12*:12 or DD are intermediate (Fig. 37). In contrast exposure of animals as adults to these same light cycles has little effect on τ. The difference in τ between nymphs exposed to 22 h and 26 h light cycles is remarkably stable, appears to persist with only a slight dimunition throughout adult life (Page, T. and Block, G., 1980; Page, unpublished), and persists through transplantation of the driving oscillator from one animal to another (see below; Page, T., 1982c). Here the environment most likely has a fundamental effect on the pacemaker mechanism. The possibility, suggested by these studies on moths and roaches, that certain parameters of the circadian system may be adaptively "fine-tuned" by the environment during development is intriguing, but more data are needed to determine the generality and adaptive significance of these responses.

8 GENETICS

Some of the work of the genetic analysis of circadian organization has been alluded to in previous sections. Studies on clock genetics have dealt with both the effects of single-gene mutations and with polygenic inheritance. In the former case chemical mutagens have been used profitably to isolate mutations at a single locus that affect some aspect (e.g. free-running period) of overt rhythmicity. In the latter case studies have involved analysis of strain differences in natural populations or the creation of new strains by artificial selection.

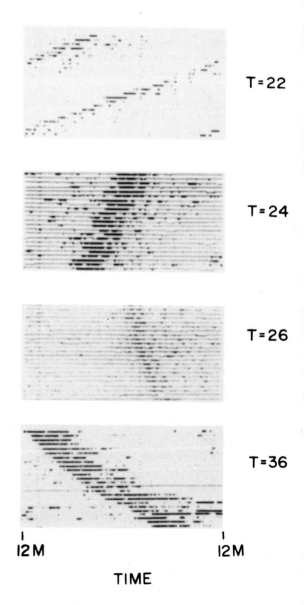

FIG. 37. Free-running activity rhythms (in DD) of *Leucophaea maderae* raised in light cycles with periods (*T*) of 22, 24, 26, and 36 h. (Data from Page, T. and Block, G., 1980 and Page, unpublished.)

8.1 Single-gene mutations

Four X-chromosome-linked mutant strains of *D. melanogaster* have been isolated. Three of these strains are allelic at the *per* locus (Konopka, R. and Benzer, S., 1971). Flies of the strain designated *per⁰* exhibited aperiodicity of eclosion and adult activity in constant conditions. The other two strains were rhythmic but had free-running periods substantially different from the wild-type strain in which τ of both the eclosion and activity rhythms is about 24.5 h (Fig. 38). The period of *per^s* rhythms was 19–20 h while *per^l* individuals exhited a period of 28–29 h. In both *per^s* and *per^l* activity rhythms are temperature-compensated; however, the slight dependence of τ on T^0 is opposite in the two strains. With increasing temperature the period of *per^s* decreases while the period of *per^l* increases (Konopka, R., 1979).

Complementation tests were carried out by examining activity rhythms of heterozygous individuals. The results suggested that *per⁰* and *per^l* were recessive to wild-type; rhythms of *per⁰/+* and *per^l/+* had essentially normal periods. Rhythms of the *per^s/+* heterozygote had an intermediate period to the mutant and wild-type strains, thus *per^s* is partially dominant. Finally, *per^s/per⁰* and *per^l/per⁰* exhibited periods nearly identical to *per^s* and *per^l* homozygotes respectively. The results suggest the *per⁰* allele is inactive.

Another effect of the *per^s* mutation is that it modifies the resetting behavior of the pacemaker in response to a light pulse (Winfree, A. and Gordon, H., 1977; Konopka, R., 1972, 1979). Wild-type

Drosophila melanogaster show type 1 resetting, whereas *per^s* exhibits type 0. Furthermore, the period of light-insensitivity during the subjective day is reduced from about 12 h in wild-type to about 7 h in *per^s*, while the duration of the light-sensitive period remains about the same in both strains. Interestingly, the reduced period of light-insensitivity in *per^s* is correlated with a reduction in the duration of activity (which also occurs in the subjective day) in each cycle.

A final recent discovery on behavioral effects of the *Drosophila* mutants comes from a study of short-term fluctuations in the male courtship song (Kyriacou, C. and Hall, J., 1980). It was found that the courtship song, which is produced by wing vibration, shows a rhythmic fluctuation with a period of about 1 min. Each of the *per* alleles affects this rhythm in the same manner it affects the circadian rhythms of activity and eclosion — *per^s* shortened the period of the song rhythm, *per^l* lengthened it, and *per⁰* abolished it. While the results of these experiments are clear, their implications for the role of the *per* allele in the circadian system are less obvious.

Five chemically induced, X-chromosome-linked mutations have been isolated in *Drosophila pseudoobscura* (Pittendrigh, C., 1974). Eclosion behavior of all five strains is aperiodic in constant conditions, although two strains show a clear rhythm in a light cycle. Heterozygotes of some pairs of strains are also aperiodic, while others show incomplete complementation. The eclosion patterns of these latter females are rhythmic in constant conditions, but the periods are longer than wild-type, and the phases of the eclosion peaks are later.

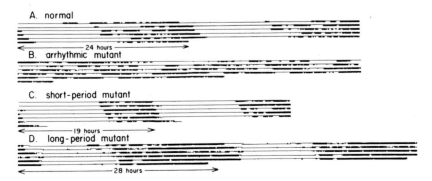

A. normal

← 24 hours →

B. arrhythmic mutant

C. short-period mutant

← 19 hours →

D. long-period mutant

← 28 hours →

Fig. 38. Activity records of *Drosophila melanogaster* normal and mutant strains in infrared light at 25°. The records are "double plotted". The time base is 24 h for **A** and **B**, 19 h for **C**, and 28 h for **D**. (From Konopka, R. and Benzer, S., 1971.)

8.2 Multi-gene analysis

Artificial selection has been used both in *Drosophila* and *Pectinophora* to create genetic strains that differ in the phase relationship of circadian rhythms of eclosion to an entraining LD cycle (Pittendrigh, C., 1967, Pittendrigh, C. and Minis, D., 1971; Clayton, D. and Paietta, J., 1972). Many of the details of results of experiments with these strains are discussed elsewhere in this chapter; they will be covered only briefly here.

In *D. pseudo-obscura* two strains differing in time of eclosion were created by selection over 50 generations (Pittendrigh, C., 1967). At 20° the "early" strain exhibited an eclosion peak that occurred about 4 h prior to the peak of the "late" strain following release of flies from *LD 12*:12 to DD (Fig. 39). Wild-type is intermediate between the two strains. Interestingly, selection had little effect on the driving oscillator — the PRCs of the three strains (early, late, stock) were essentially identical (Fig. 39). There were slight differences in τ of the three strains, but these were not in the direction that would account for the differences in eclosion time. The results suggested, and subsequent work (Pittendrigh, C., 1981a) has confirmed, that selection had operated primarily on a system secondary to the driving oscillator.

Selection on the *Pectinophora* eclosion rhythm likewise resulted in the isolation of two strains that differed in eclosion time from the wild-type moths (Pittendrigh, C. and Minis, D., 1971). Two other rhythms, egg hatch and oviposition, were also examined in these strains to determine whether selection on one rhythm had similar effects on all rhythms. It was discovered that selection on eclosion had affected the egg hatch rhythm; phases of the rhythm in the strains were different by about the same amount as the eclosion rhythms. In contrast, selection had had no effect on the rhythm of oviposition; phases were essentially identical in early, late, and wild-type. Similar selection experiments have been carried out in *D. melanogaster*, both on a laboratory stock and a wild strain (Clayton, D. and Paietta, J., 1972). Selection for early and late in the wild strain produced a smaller variation than in the laboratory strain. Possibly this reflects a greater genotypic variability in the laboratory animals due to a relaxation of selection pressures in

the course of laboratory breeding (Clayton, D. and Paietta, J., 1972).

Another approach to multigene analysis has been the investigation of the effects of crossing geographical races. Crosses between two strains of the midge *Clunio marinus*, whose eclosion times are adapted to local tidal conditions (see section 5) and differ by about 4.5 h resulted in an F₁ generation with a peak of emergence intermediate between the two parental stocks. Backcrossing the F₁ hybrids with one parental strain likewise resulted in an intermediate eclosion time (Neumann, D., 1967).

It is clear from these results that the timing system controlling rhythmicity in insects is under genetic control. On the other hand, it is equally certain

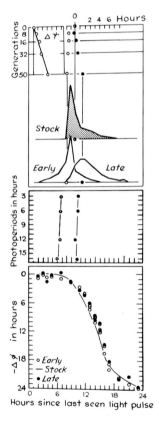

FIG. 39. Selection on the phase of eclosion with respect to the light cycle (*LD 12*:12) in *Drosophila pseudo-obscura. Upper panel*: Change in ψ_{RL} over 50 generations of selection that produces a strain that emerged "early" and a strain that emerged "late" with respect to the stock strain. *Middle panel*: Plots ψ_{RL} for the "early" and "late" strains as a function of photoperiod. *Lower panel*: Phase response curves for "early" (open circles) and "late" (filled circles) are not measurably different from "stock" (solid line). (From Pittendrigh, C. and Minis, D., 1971.)

(particularly from results on *D. pseudo-obscura*) that effects on manipulation of the overt rhythm do not necessarily reflect alterations in the driving oscillator underlying those rhythms, and one must be cautious in interpreting results of genetic manipulation. Nevertheless, the genetic analysis of circadian oscillations in insects has proved its utility in the analysis of circadian organization and future efforts promise to provide new insights into the properties and perhaps molecular basis of circadian rhythmicity.

9   METABOLIC ASPECTS

While some progress has been made in studies on single-cell organisms (e.g. *Gonyaulax*) or *in vitro* culture systems (*Aplysia* eye) on the molecular basis of circadian oscillations there is very little information from insects. A few treatments (low temperature, anoxia, D_2O) that should have very general affects on metabolism have been shown to affect the phase or period of circadian rhythms in insects.

Pulses of very low temperature (~ 5–$7°$) cause major phase shifts in free-running rhythms of *Leucophaea* (Roberts, S., 1960; Wiedenmann, G., 1977a; Page, T., 1981c). The phase shifts are always delays and the magnitude of the shift is phase-dependent. Following pulses of greater than 12 h duration the pacemaker always begins its motion at the beginning of the subjective night (ct 12). The results are consistent with the view that low temperature greatly slows or stops the oscillation, but this interpretation should be accepted only with caution in view of results with constant light (see section 3.5).

Some data are available on effects of anoxia on the *D. pseudo-obscura* eclosion rhythm (Pittendrigh, C., 1974). Replacing the air surrounding a pupal population with nitrogen causes phase-dependent phase shifts (both advances and delays). The results have been interpreted to suggest that the oscillator's cycle consists of an energy-dependent "charge" phase and an energy dissipating "discharge" phase (Pittendrigh, C., 1974).

Deuterium oxide has been found to consistently act to slow circadian oscillations (increase τ) in several vertebrates and invertebrates. In *D. pseudo-*

obscura the effect of D_2O on τ is very small. A greater effect was found on the phase angle of the entrained rhythm with the light cycle (Pittendrigh, C. *et al.*, 1973). In *Leucophaea* the effect of D_2O on τ has been investigated extensively (Caldarola, P., 1974). D_2O consistently lengthens τ at all temperatures, but the magnitude of the effect is temperature-dependent.

The effects of these various treatments provide a general, and unsurprising, confirmation that cellular metabolism is involved in circadian oscillations and suggest that the consequences of disrupted metabolism are phase-dependent. Because each of these treatments is expected to affect virtually every aspect of cellular function, more specific conclusions about the molecular mechanism of the clock system are not possible.

10   NEURAL AND ENDOCRINE CONTROL MECHANISMS

In the past 15 years there has been a substantial effort made to discover the neural and endocrine basis of circadian organization in insects. Invariably, the investigations begin with the model, initially suggested by Pittendrigh, C. and Bruce, V. (1957), illustrated in Fig. 40. There are four functionally defined components of the circadian system that control overt rhythmicity: an oscillator mechanism, a photoreceptor for entrainment, and two coupling pathways — one that mediates the flow of information from the photoreceptor to the oscillator and a second between the oscillator and the process it controls.

There are two general problems in understanding the physiology of the circadian system described by this model. The first involves identifying the anatomical correlates of the functionally defined components. Second is the elucidation of the mechanisms by which these components interact to form an organized regulatory system. Substantial progress has been made on both fronts, and two points of general interest have emerged.

Although in no instance have circadian oscillators or photoreceptors been linked to particular, identified cells, it is clear that restricted regions of the nervous and neuroendocrine systems can fulfill the functions of various components in Fig. 40. Oscillators, photoreceptors, and steps in the

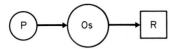

FIG. 40. Oscillator model of the circadian system. Os, oscillator; P, photoreceptor; R, overt rhythm. Coupling pathways are designated by arrows.

coupling pathways have been localized in various insects. On the other hand, there is much evidence that the model is incomplete as a representation of the circadian system. In any one individual there may be several circadian oscillators, in some cases independent in function, in other cases interactive. The emerging picture of circadian organization in insects is similar to that for other metazoan groups — temporal order within the individual is derived from a *population* of anatomically discrete oscillators and depends not only on the properties of the individual pacemakers but also on the coupling relationships between them (Page, T., 1981b,c).

In this section the current state of progress in the efforts to understand the neural and endocrine basis of circadian organization in insects is summarized. Four general problems are discussed. The first is the location, numbers, and interactions of circadian oscillators in several species. The second focuses on the coupling pathways between circadian pacemakers and the processes they control. The localization of photoperiodic clocks is then discussed. Finally, the location and physiology of photoreceptors utilized in entrainment pathways and in photoperiodic time measurement is reviewed.

10.1 Localization of circadian pacemakers in the nervous system

Most of the research directed toward localizing circadian pacemakers has involved efforts to identify the oscillator that controls the timing of some particular, overtly expressed rhythm. These investigations have often employed lesions as a technique, at least in early stages. The procedure involves the destruction of a portion of the nervous system followed by an assay for the presence or absence of rhythmicity in a specified behavioral or physiological function. These experiments suffer from a major interpretive difficulty which has been frequently

noted in the literature. Since the rhythmic expression of a behavior may involve pathways and processes quite distinct from the pacemaker, loss of rhythmicity does not necessarily reflect pacemaker destruction. As Pittendrigh, C. (1976) pointed out, the only parameters of an overt behavior which can be considered to reflect the state of the pacemaker underlying that behavior are the phase and free-running period of the system in steady state. Thus results of experiments utilizing surgical or electrolytic lesions to the nervous system are rarely unambiguous, and interpretations must be made cautiously. One experimental technique which has been used to overcome these interpretive obstacles is that of tissue transplantation, where an effort is made to demonstrate that transplantation of the putative pacemaker tissue from one animal to another, in which the tissue has been removed, both restores a free-running rhythm and confers either the phase or the period of the donor's rhythm on the rhythm of the host. There are now four cases in insects in which data from lesioning and transplantation experiments have provided good evidence for localization of circadian pacemakers that control behavioral rhythmicity to the supraesophageal ganglion.

10.1.1 SILKMOTHS

In a series of now classic experiments on silkmoths (Truman, J. and Riddiford, L., 1970; also Truman, J., 1972b) convincingly demonstrated that the circadian pacemaker which controls the time of eclosion is located in the brain. They worked with two species of silkmoth, *Hyalophora cecropia* and *Antheraea pernyi*. In both species the emergence of the pharate adult from the pupal case is "gated" by a circadian oscillator; and when placed into constant darkness populations exhibit a daily peak of eclosion (Truman, J., 1971b). The time of day that each species emerges is different, however. When raised on a 24 h light cycles consisting of 17 h of light followed by 7 h of darkness *H. cecropia* emerge during a gate approximately 8 h in duration which opens 1 h after lights-on. *A. pernyi*, on the other hand, emerges in the last 5.5 h before lights-off (Fig. 41). While removal of the brains of these animals did not prevent eclosion, "brainless" moths emerged at random times throughout the day (Fig. 41).

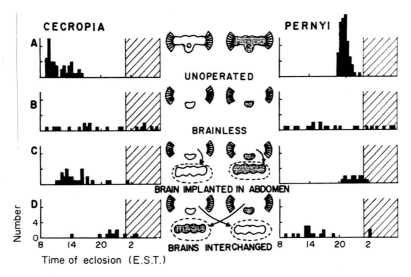

FIG. 41. The time of eclosion of *Hyalophora cecropia* and *Antheraea pernyi* moths in a 17 h light : 7 h dark regimen showing the effects of brain removal, transplantation of the brain to the abdomen, and interchange of brains between two species. After brain exchange, the host emerges at the eclosion time characteristic of the donor species. (From Truman, J., 1971a.)

If the brain was removed from the head but reimplanted in the abdomen, normal rhythmicity was restored — populations exhibited a free-running eclosion rhythm in constant conditions (Truman, J., 1972b) and emerged at the appropriate time of day in a light cycle (Fig. 41) (Truman, J. and Riddiford, L., 1970; Truman, J., 1972b). These results strongly implicated the brain as the site of a clock which controlled the time of emergence via release of an eclosion hormone. The most convincing evidence, however, was provided by experiments in which brains were removed from the head of one species and transplanted into the abdomen of the other. Moths which had received these "switched brain" transplants exhibited the normal stereotypic eclosion behavior for the host species. However, the phase of the eclosion rhythm in LD *17* : *7* was characteristic of the donor and not the host; *A. pernyi* which had received *H. cecropia* brains emerged just after dawn, while *H. cecropia* receiving *A. pernyi* brains emerged near dusk (Fig. 41). The demonstration that the transplanted brains not only restored rhythmicity but also determined the phase of the rhythm in the host moths leaves little doubt that the circadian pacemaker that controls the time of emergence is located in the brain. In an attempt to determine which part of the brain is responsible for timing the eclosion behavior, the brain was subdivided prior to transplantation (Truman, J.,

1972b, 1974a). The results suggested that the optic lobes are unnecessary since intact cerebral lobes adequately gated eclosion. When the cerebral lobes were subdivided, however, by cuts made lateral to the medial neurosecretory cell groups which appear to be the site of production of the hormone which triggers eclosion behavior (Truman, J., 1973a), the isolated median piece did not gate the emergence. This suggested the possibility that the lateral portion of the cerebral lobes contains the pacemaker. Truman, J. (1974b) has also investigated the flight activity rhythm of adult silkmoths. Removal of the brain or severance of its neural connections with the thorax abolished the rhythm, while ablation of the optic lobes did not disrupt rhythmicity. The results suggest the pacemaker for the adult activity rhythm also resides in the cerebral lobes.

10.1.2 FRUIT-FLIES

The transplantation technique has also been used in adult *Drosophila melanogaster*. The experiments made use of two X-chromosome-linked mutant strains. One of the strains (per^0) invariably exhibits aperiodic locomotor activity. The other strain (per^s) exhibits an activity rhythm with a period in constant darkness of about 19–20 h — markedly shorter than the wild-type period of about 24 h

(Konopka, R. and Benzer, S., 1971). Brains of short-period mutants were transplanted to the abdomens of aperiodic mutants (Handler, A. and Konopka, R., 1979). Four of the 55 animals which survived the surgery exhibited three or more cycles of periodic activity in constant conditions, and the periods of these rhythms were between 16 and 20 h. Although the success rate was low, the fact that some per^0 hosts expressed a rhythm with a period similar to that of the per^s donors indicates that the brain is the site of the pacemaker that controls activity, and that its effects can be mediated *via* a humoral pathway. In this context it is interesting to note that the morphology of an identifiable group of neurosecretory cells is frequently abnormal in the aperiodic mutant of *Drosophila melanogaster* and in aperiodic mutants of *Drosophila pseudo-obscura* (Konopka, R. and Wells, S., 1980). This raises the possibility these cells may be involved in circadian rhythmicity in *Drosophila*.

10.1.3 COCKROACHES

Transplantation experiments attempted in the cockroach in the mid-1950s yielded controversial results. The experiments were based on evidence for hormonal involvement in the control of the locomotor activity rhythm of the cockroach *Periplaneta americana* that originated with a short report by Harker, J. (1954). She found that when an arrhythmic roach was parabiotically linked (back to back) with a normally rhythmic, but immobilized, roach rhythmicity was restored for about a week. The design and interpretation of this experiment has come under a great deal of criticism for several reasons including: (a) there were no control pairs with unconnected hemocoels, and (b) the host animal was made arrhythmic by exposing it for several days to constant light, although it is now clear that cockroaches may remain rhythmic indefinitely in LL (Roberts, S., 1965b; Cymborowski, B. and Brady, J., 1972). Nevertheless, the result led Harker to search for a hormonal clock. Harker had discovered that beheaded roaches no longer exhibited an activity rhythm, and she attempted to restore rhythmicity by implanting the subesophageal ganglion (SEG) of one donor animal into another, headless host. It was reported that the procedure did restore rhythmicity for at least a few

days (Harker, J., 1956), and that the phase of the newly restored host rhythm was the same as that of the donor prior to transplantation (Harker, J., 1960).

Unfortunately, attempts to repeat these very important transplantation experiments have failed, and a large body of other evidence (reviewed by Brady, J., 1969, 1971, 1974; Page, T., 1981b,c) argues convincingly against a hormonal clock mechanism in the SEG. Roberts, S. (1966) was unable to restore rhythmicity in 19 attempts at SEG transplants in headless roaches. Brady, J. (1967b) obtained records from 19 recipient animals and found clear evidence for rhythmicity for 2–3 days in only two cases — a result not significantly different from random expectation. Furthermore, it has been shown that destruction of the neurosecretory cells *in situ* in the SEG by microcautery (Brady, J., 1967b) or by surgical removal (Nishiitsutsuji-Uwo, J. and Pittendrigh, C., 1968b) does not affect the rhythm in otherwise intact animals.

While these results are difficult to reconcile with Harker's hypothesis, it is interesting to note that the original parabiosis experiments have been successfully repeated. Cymborowski, B. and Brady, J. (1972), cognizant of several objections to the design of Harker's original experiments, have convincingly demonstrated in both *P. americana* and the cricket *Acheta domesticus* that rhythms could be driven in headless animals by a blood-borne factor released by a parabiotically coupled donor. They were, however, cautious in their interpretation and suggested several alternatives to Harker's conclusion that the rhythm is normally hormonally controlled. In Brady's view (Brady, J., 1974) the most likely of these is that the donor animal responds to being pinioned upside-down by secreting a neuroactive stress factor (Cook, B. *et al.*, 1969; Brady, J., 1967a) in response to aborted CNS commands at its normal activity time.

The apparent failure of Harker's hypothesis that the SEG contained the circadian pacemaker that controlled the activity rhythm in the cockroach led the search elsewhere in the CNS. Nishiitsutsuji-Uwo, J. and Pittendrigh, C. (1968b) focused attention on the optic lobes of the brain. They reported that complete ablation of these structures in *Leucophaea maderae* caused persistent arrhythmicity without significantly affecting the level of activity. Similar results have been obtained with two other species of cockroach, *Periplaneta americana* (Roberts, S., 1974) and *Blaberus fuscus* (Lukat, K. and Weber, F., 1979). These results suggested the possibility that the optic lobes were the source of the driving oscillation for the rhythm. This suggestion has been confirmed by results from several subsequent studies.

In the cockroach unilateral optic lobe ablation does not abolish rhythmicity (Nishiitsutsuji-Uwo, J. and Pittendrigh, C., 1968b; Page, T. *et al.*, 1977). This fact has been exploited to further localize the cells crucial to maintaining the activity rhythm in the cockroach. Roberts, S. (1974), working with both *Leucophaea maderae* and *Periplaneta americana*, removed all of one optic lobe and with surgical lesions removed various amounts of the contralateral optic lobe. He found the most distal neuropile area, the lamina, to be dispensable for the maintenance of the free-running rhythm. However, more proximal cuts which removed the medulla invariably caused arrhythmicity. Animals which had part of the medulla removed gave an intermediate result; some animals remained rhythmic while others were aperiodic. Sokolove, P. (1975b) published a similar study on *Leucophaea* in which he utilized small electrolytic rather than surgical lesions. His results were similar to Roberts' in that he found no evidence for involvement of the lamina and aperiodicity in only 3 of 31 animals with lesions in the medulla. However, lesions in the ventral half of the lobe in the region of the second optic chiasm and lobula were frequently effective in abolishing the rhythm. Sokolove hypothesized that it was the somata of cells in this area that were crucial, and suggested that Roberts' surgical lesions through the medulla were also likely to have damaged these cells. Page, T. (1978) has obtained results essentially identical to those of Sokolove, and has also concluded that the cells in the ventral half of the lobe near the lobula are critical to maintenance of rhythmicity, and that the medulla is probably not involved.

The finer localization of the portion of the optic lobe involved in sustaining rhythmicity did not remove the uncertainty that these cells may be necessary only for the expression of the rhythm and do not necessarily constitute the pacemaker which times it. Two recent studies, however, have provided evidence which strongly supports the notion that the optic lobes are involved in pacemaker activity.

Although bilateral sectioning of the optic tracts invariably abolishes the free-running rhythm of locomotor activity in *L. maderae*, it was recently discovered that if the optic lobes were left *in situ* the rhythm consistently reappeared in 3–5 weeks, with a free-running period near that of the rhythm prior to surgery (Page, T., 1982c). The return of rhythmicity appeared to depend on the regeneration of neural connections between the optic lobe and the midbrain (Page, T., 1982b,c). These observations prompted an effort to transplant the optic lobes from one animal into another whose own optic lobes had been removed. The experiment involved two groups of animals that had been raised in *LD* *11*:11 (*T* = 22) or *LD* *13*:13 (*T* = 26) and therefore had substantially different free-running periods (see section 7.3); Page, T. and Block, G., 1980). Optic lobes were exchanged both between individuals of the two groups (Page, T., 1982c) and between individuals within a group (Page, unpublished). Activity was disrupted and aperiodic for several weeks after transplantation; however, in several animals a clear circadian rhythm of activity reappeared within 4–8 weeks (Fig. 42). In every case in which a rhythm was re-established the free-running period was near the period of the donor animal's rhythm prior to transplantation of the optic lobes (Fig. 43). Histological examination of the brains of these animals showed that structural connections had been established between the transplanted optic lobe and the midbrain (Page, T., 1982c).

These results demonstrated that the transplanted optic lobe was capable of supporting rhythmicity and, critically, also determined the free-running period of the host rhythm. This strongly supported the idea that the circadian pacemaker was located in the optic lobes.

Evidence has also been obtained that shows the optic lobes control the phase of the activity rhythm (Page, T., 1981a). The experiments were based on the observation that the phase of the rhythm can be reset by pulses of low temperature (< 12°) (Roberts, S., 1962; Wiedenmann, G., 1977a). There was reason to believe these pulses act directly on the pacemaker and not on the light entrainment pathway, since it had been shown that the phase

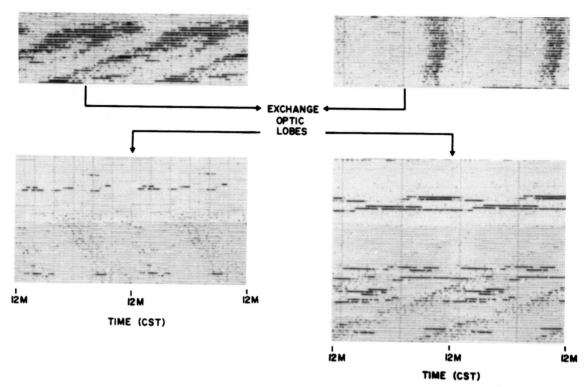

FIG. 42. Activity records of *Leucophaea maderae* showing effects of optic lobe transplantation on the free-running rhythm. The animals whose activity is shown to the left was raised in *LD 11* : 11 and the animal shown on the right was raised in *LD 13* : 13. Animals were allowed to free-run in DD for several weeks then the optic lobes were exchanged between the pair. For 3–7 weeks after transplantation the activity was sparse and apparently aperiodic. A clear rhythm of activity subsequently reappeared in both animals with the period of the rhythms near the period of the donor rhythm before surgery. (Data from Page, T., 1982c.)

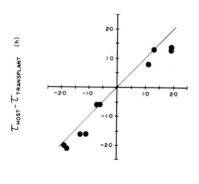

FIG. 43. *Leucophaea maderae*. Change in the period of host animals ($\tau_{Host} - \tau_{Transplant}$) after optic lobe transplantation, plotted as a function of the difference in τ between the host and donor prior to surgery. The diagonal line is the predicted relationship if τ of the regenerated rhythm was equal to τ of the donor's rhythm. (From Page, T., 1982c.)

response curve for short-duration (6 or 8 h) low $T°$ pulses is composed of all delaying phase shifts (Wiedenmann, G., 1977a) while the PRC for light includes both delays and advances (Roberts, S., 1962; Wiedenmann, G., 1977b). Furthermore, low-temperature pulses generate large phase shifts at phases where the pacemaker is relatively insensitive to light. These observations suggested that if the pacemaker were in fact located in the optic lobe localized cooling of this structure should cause a phase delay in the activity rhythm.

Localized cooling was accomplished by positioning an insect pin attached to a peltier block just ventral to the optic lobe. The temperature of the contralateral optic lobe was controlled by another insect pin attached to a resistor. By separately adjusting the current through the peltier and the resistor, one optic lobe could be cooled to about 7.5° while the other was maintained near 25°. Animals free-running in constant darkness were removed from activity monitors, treated with low-temperature pulses (for 6 h beginning at activity onset), and returned to the monitors.

In animals in which one optic tract had been cut, cooling the intact optic lobe consistently caused a phase delay of several hours (Fig. 44A). In contrast, cooling either the neurally isolated optic lobe (Fig. 44B) or the midbrain (Page, T., 1981a) had little or no effect on the phase of the rhythm. The demonstration that the optic lobes were involved in control of phase, as well as period, left virtually no doubt that they were involved in generating the driving oscillation for the activity rhythm. Furthermore, the fact that each optic lobe is sufficient to sustain rhythmicity raised the possibility that the pacemaking system is composed of a bilaterally distributed pair of circadian oscillators. This suggestion, and the evidence for mutual coupling between these two oscillators, is discussed in more detail below.

10.1.4 CRICKETS

Substantial work has been done on the localization of components of the circadian system in crickets. In general, results of lesion experiments parallel those obtained with cockroaches. Loher, W. (1972) found that bilateral ablation of the optic lobes in *Teleogryllus commodus* completely disrupted the rhythm in stridulation, and it was later shown that optic lobe ablation also abolished both the circadian rhythms of spermatophore formation (Loher, W., 1974) and locomotion (Sokolove, P. and Loher, W., 1975). As in cockroaches either optic lobe is sufficient to sustain rhythmicity (Loher, W., 1972). These results are consistent with the idea that the circadian pacemaker also consists of two oscillators, one located in each optic lobe.

Further support that the circadian pacemaker is in the brain comes from work on the locomotor activity rhythm in the house cricket *Acheta domesticus*. These experiments involved transplantation of brains from another cricket into animals made arrhythmic by destruction of the *pars intercerebralis* (see below). Following transplantation of the brain to the abdomen, rhythmicity was restored for two or three cycles with a free-running period similar to that previously exhibited by the donor (Cymborowski, B., 1981). The results provide evidence for a circadian pacemaker in the brain, and further suggest the brain can release a humoral factor that can drive the activity rhythm.

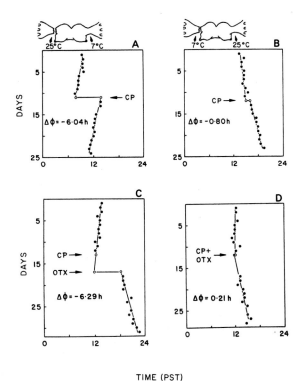

TIME (PST)

FIG. 44. Examples of data from animals that were treated with localized low-temperature pulses. ●, Time of activity onset for each day; ○, projected phases of the rhythms before and after the pulse; lines are linear regressions. Pulses were 6 h in duration and began at activity onset. In A the intact optic lobe of an animal in which one optic tract was cut was cooled (CP) while the neurally isolated optic lobe was maintained at 25°; in B the neurally isolated optic lobe was cooled while the intact lobe was maintained at 25°. Cooling the intact lobe caused a large phase delay ($\Delta\varphi$) while cooling the neurally isolated lobe had little effect. C and D illustrate the effects of a low-temperature pulse to one lobe on the rhythm driven by the contralateral lobe. In C the optic tract of the treated lobe was cut (OTX) 4 days after the pulse (CP). The subsequent rhythm, driven by the untreated lobe, is phase-delayed by several hours. In D the optic tract of the treated lobe was sectioned 0.5 h after the pulse. Optic tract section shortly after the pulse prevents the phase shift in the rhythm. (Data from Page, T., 1981b.)

In summary, efforts to localize circadian pacemakers that control various behavioral rhythms in several insect species uniformly suggest that the driving oscillation originates in the supraesophageal ganglion. In some cases (e.g. cockroach, cricket) the locus of pacemaker activity has been further localized to the optic lobes, while in other instances (e.g. silkmoth) the optic lobes are unnecessary. It is interesting to note that in those cases where the optic lobes are required for rhythmicity the photoreceptors for entrainment are located in the compound eyes, while extraretinal photoreceptors are utilized in those insects where the pacemaker appears to be in the main cerebral lobes (see section 10.6).

10.2 Multiple oscillators

The data presented in the previous section clearly indicate that circadian oscillators can be localized in insects. However, the results also suggest the circadian system in the individual may be composed of more than one oscillator. There is good evidence that the driving oscillators are bilaterally paired in cockroaches and crickets where either optic lobe is independently capable of sustaining rhythmicity. Evidence has also been obtained from the beetle *Blaps gigas* for bilateral distribution of pacemakers that control the rhythms in amplitude of the electroretinogram in the two eyes (Koehler, W. and Fleissner, G., 1978; see below). Finally, there are now several reports that provide evidence for "secondary" oscillators that may be interposed between the pacemaking oscillations and a rhythmic behavior, or for non-neural oscillators that may be independent of neural pacemakers.

In both cockroaches and crickets there is some evidence that a strongly damped oscillator survives ablation of the optic lobes. Rence, B. and Loher, W. (1975) reported that following optic lobe ablation, *T. commodus* in a 24 h temperature cycle restricted their stridulatory activity to the first few hours after the high-temperature (W) to low-temperature (C) transition. The conclusion that the rhythmicity was not simply forced by the change in temperature was prompted by the observations that: (a) animals did not exhibit a rhythm when subjected to a temperature cycle with a period that was outside the circadian range (W:C, 15:15) and (b) there was evidence of one or two transient cycles of rhythmicity following a single 12 h low-temperature pulse. Following optic lobe ablation the cockroaches *Blaberus fuscus* (Lukat, K. and Weber, F., 1979) and *Leucophaea maderae* (Page, T., 1981b) will also exhibit a rhythm of locomotor activity when placed in a 24 h temperature cycle although, unlike crickets, the rhythm is also evident in a 30 h temperature cycle (Lukat, K. and Weber, F., 1979). Interestingly, in *Leucophaea* the phase-angle be-

tween the onset of activity and the W to C transition varies as a function of the period of the temperature cycle (Page, T., 1981b). These observations on crickets and roaches are difficult to explain as a simple reflexive response to a change in ambient temperature, and are at least consistent with the idea that the temperature cycle is acting through a strongly damped oscillator.

The best evidence for a circadian oscillator outside the optic lobes in cockroaches comes from a recent study on the rhythm of cuticular deposition in cockroaches. In many insects the formation of the endocuticle early in adult life involves the daily deposition of alternately lamellate and non-lamellate chitin (reviewed by Neville, A., 1975). This results in the formation of growth layers which are visible under polarized light. It was found in the cockroach, *Blaberus fuscus*, that after imaginal ecdysis the circadian rhythm in the formation of the endocuticle was still evident in animals whose optic lobes had been removed in the last larval instar or following decapitation of the adult shortly after imaginal ecdysis (Lukat, R., 1978). The results demonstrate the existence of an extra-cephalic pacemaker, and raise the possibility that the cells of the epidermis are autonomously rhythmic. Evidence has also been presented that suggests a circadian oscillator is present in the abdomen of the flour moth, *Anagasta kueniella*. In this moth there is a daily rhythm of release of sperm from the testes that persists in DD (Riemann, J. *et al.*, 1974). Thorson, B. and Riemann, J. (1977) found that following isolation of the abdomen from the thorax there was a significant reduction in the amount of sperm released, but the normal periodicity of release continued in both LD and DD. Furthermore, in LD the rhythm could be phase shifted by a shift in the phase of the LD cycle.

Finally, Weitzel, G. and Rensing, L. (1981) have recently presented evidence for persistent rhythmicity in isolated salivary glands of *Drosophila melanogaster* larvae. In these experiments, variations in the fluorescence intensity in glands labeled with a cyanine dye were measured for both wild-type and *per[0]* mutant strains. In both strains there was evidence for rhymicity in DD.

Thus the results from several insect species suggest that the circadian organization involves a multi-oscillator system. This fact prompts the question: what are the mechanisms for maintaining temporal order among these oscillators? Are appropriate phase relationships among oscillators maintained by coupling between them, or does temporal order depend on direct synchronization of individual oscillators by external cues such as the light/dark cycle? There are two cases in which the bilateral distribution of circadian pacemakers in insects has made possible investigations of the coupling relationships between oscillators.

In *Leucophaea maderae* the suggestion that each optic lobe could function as a driving oscillator for the activity rhythm prompted a systematic investigation of the effects of unilateral optic lobe ablation in which two questions were raised (Page, T. *et al.*, 1977). First, if there are two pacemakers are they functionally equivalent? Second, does ablation of one optic lobe have any effect on the rhythm that would suggest an interaction between optic lobe pacemakers? In these experiments one optic lobe was removed (or the optic tract cut) in 39 animals that were free-running in constant darkness (23 right lobe, 16 left lobe). The results indicated that, at least as measured by (a) ability to maintain rhythmicity, and (b) average free-running period, the optic lobes were functionally redundant. However, it was found that ablation of either the right or left optic lobe led to a small, but consistent and significant, increase in τ. In the 39 animals studied, $\bar{\tau}$ before surgery was about 23.7 h (23.73 ± 0.26) whereas rhythms driven by either the right or left lobe alone had average periods of about 24.0 hours ($\bar{\tau}_{\text{left}} = 23.95 \pm 0.28$; $\bar{\tau}_{\text{right}} = 23.96 \pm 0.24$). The surgical control of optic nerve section had no effect on τ. These results prompted the suggestion that the bilaterally redundant oscillators in the optic lobes were mutually coupled to form a compound pacemaker whose period was shorter than that of either constituent oscillator. At least two general mathematical models of coupled oscillators exhibit similar behavior (Daan, S. and Berde, C., 1978; Kawato, M. and Suzuki, R., 1980).

The coupling hypothesis was reinforced by the finding that either one of the compound eyes was sufficient to entrain both oscillators. The experiments involved animals in which one optic lobe was isolated from input from its own compound eye by optic nerve section. After maintaining the animals for several days in a light cycle which imposed a major phase shift on the activity rhythm, the unoperated optic lobe was removed. The animals began to free-run in the light cycle driven by the optic lobe which had been isolated from its own eye with a phase determined by the light cycle (Page, T. *et al.*, 1977). These data suggested the model illustrated schematically in Fig. 45. Each oscillator (Os), which independently has a free-running period of 24 h, receives input from photoreceptors (P) of the ipsilateral compound eye. In the intact animal the oscillators are mutually entrained with τ of the coupled system being about 23.7 h. Mutual coupling would also provide a pathway by which either compound eye could entrain both oscillators.

The hypothesis of mutually coupled optic lobe oscillators has been recently tested in a series of experiments utilizing low-temperature pulses (Page, T., 1981a). In these experiments low-temperature pulses were given to one optic lobe of intact animals. When the treated lobe was removed 4 days after the pulse, the subsequent rhythm (driven by the untreated contralateral lobe) was phase delayed by several hours (Fig. 44C); but, if the treated lobe was removed only 0.5 h after the pulse, the phase shift in the contralateral optic lobe pacemaker was prevented (Fig. 44D). The small phase delay was essentially the same as that caused by optic tract section alone (Page, T., 1981a). The results indicated that the low-temperature pulse

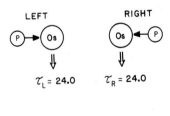

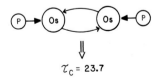

FIG. 45. Schematic representation of the hypothesis of mutually coupled optic lobe oscillators in *Leucophaea maderae*. Os, Oscillator; P, photoreceptors for entrainment in the compound eye. Each oscillator receives photoreceptive input from its own compound eye and has a free-running period of about 24 h. The coupled system has a period of 23.7 h and allows either compound eye to entrain both oscillators.

caused a phase shift in the pacemaker of the treated optic lobe (without *directly* affecting the phase of the contralateral pacemaker) that was subsequently transmitted to the contralateral pacemaker. Mutual coupling also appeared to reduce the amplitude of the phase shift generated by localized low-temperature pulses. Cooling one optic lobe of an intact animal resulted in an average steady-state phase shift that was nearly 2 h less than the phase shift obtained when the optic lobe contralateral to the pulse was neurally isolated. This observation suggests that following desynchronization of the two optic lobe pacemakers, the motion of the coupled system to steady state involves a phase advance in the treated pacemaker as well as the phase delay in the untreated pacemaker.

The validity of the model shown in Fig. 45 also rested heavily on the assumption that the same optic lobe cells that were involved in sustaining rhythmicity were also responsible for maintaining the normal period and for mediating entrainment of the contralateral pacemaker. That this was the case was shown in a study utilizing small electrolytic lesions. The sufficiency of one compound eye to entrain the pacemaker of the contralateral lobe and the maintenance of the normal free-running period depended on the integrity of cells in the ventral portion of the lobe near the lobula (Page, T., 1978), the same specific region identified as being crucial to the maintenance of rhythmicity (Sokolove, P., 1975b; Page, T., 1978).

In summary, the evidence to support the hypothesis that bilaterally paired pacemakers are mutually coupled in *Leucophaea maderae* is extensive. As yet little is known about the anatomy of the coupling pathway beyond the fact that the known monosynaptic connections between the optic lobes (Roth, R. and Sokolove, P., 1975) are not involved (Page, T., 1978).

In contrast to the cockroach, results from a recent study in the beetle *Blaps gigas* suggest the ERG amplitude rhythms in the two compound eyes are controlled by independent circadian pacemakers. The rhythms from the two eyes of an individual, monitored in constant darkness, have been shown to free-run with different periods (Koehler, W. and Fleissner, G., 1978). Furthermore, the phase and period of the rhythm in one eye can be controlled by local illumination of that eye (either LD or LL)

without any consistent effect on the rhythm in the contralateral eye (Fig. 46). The results indicate the ERG rhythm in each eye is controlled by its own pacemaker, and that the two pacemakers are either independent, or only very weakly coupled.

In *Blaps* it appears that the circadian system depends on daily cues from the environment to maintain synchrony between the bilaterally paired oscillators. In the cockroach, however, there is no evidence to suggest that the optic lobe pacemakers either spontaneously desynchronize, or that they can be driven out of phase with unilateral illumination (Page, T. *et al.*, 1977). In this case maintenance of synchrony is accomplished by mutual entrainment between the oscillators in the optic lobes.

The role of multiple (and in this case "non-redundant") oscillators has also been investigated in the regulation of eclosion behavior in *Drosophila pseudo-obscura* (Pittendrigh, C., 1981a,b). As early as 1960 Pittendrigh (Pittendrigh, C. *et al.*, 1958; Pittendrigh, C., 1960) had suggested that eclosion in *Drosophila* was controlled by a light-sensitive pacemaker (or A-oscillator) which entrained a second slave (or B-oscillator). The B-oscillator, in turn, was believed to "gate" eclosion. This

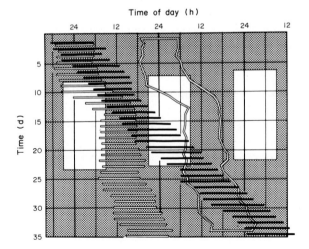

FIG. 46. The independence of the circadian ERG amplitude rhythms in the beetle *Blaps gigas*. The horizontal bars mark the time during which the ERG of the left eye (open bars) and right eye (closed bars) were in the night phase (high amplitude). Desynchronization begins almost immediately in DD (days 1–5). On days 9–23 the left eye was exposed and entrained by an LD cycle. The right eye, which remained in DD throughout the experiment, was unaffected. The envelope of night phases of the left eye is repeated once. (From Koehler, W. and Fleissner, G., 1978.)

hypothesis was originally based, to a large extent, on the observation of transient cycles in the rhythm, following a single phase shifting stimulus (see section 3.2), and was supported by subsequent experiments that showed the pacemaker itself was "instantaneously" reset by light pulses (Pittendrigh, C., 1976), suggesting the transients in the rhythm must have been due to a driven process.

This general model of the *Drosophila* gating system has been explored in detail utilizing computer simulations of an explicit mathematical version of the model developed by Pavlidis (Pittendrigh, C., 1981a). Central to the success of these investigations was the recognition that the phase relationship between the pacemaker and the eclosion rhythm (ψ_{EP}) exhibited a substantial lability. Besides the transient behavior following a single light shock Pittendrigh, C. (1981b) identified several other examples, both genetic and phenotypic, of the lability in ψ_{EP}. These included sex differences (at 20° ψ_{EP} is more negative in males than females) and strain differences in ψ_{EP}, as well as photoperiod and temperature-dependent variation. The model was tested for its ability to explain these variations. The essential propositions of the model were: (1) that the eclosion event is triggered by a particular phase in the B-oscillator's cycle; (2) that the B-oscillator is entrained by the A-oscillator (or pacemaker); and (3) that the observed variation in ψ_{EP} is explained by variations in the phase relationship between the A- and B-oscillators (ψ_{BA}).

The computer simulations were remarkably successful in explaining and predicting several detailed features of eclosion behavior. For example, the model could account for the dependence of the temporal distribution of eclosion on temperature given certain assumptions about the temperature dependence of specific parameters (e.g., the strength of coupling between the A- and B-oscillators or the period of the B-oscillators). These assumptions led to the wholly unexpected prediction that at low temperatures the phase relationship between the Early and Late strains of *D. pseudo-obscura* should be inverted with flies of the Late strain emerging earlier in *LD 12:12* than flies of the Early strain. The results of the subsequent successful test of this prediction are shown in Fig 47. The success of the model in predicting these, and other, data, as well as explaining previously observed features of eclosion

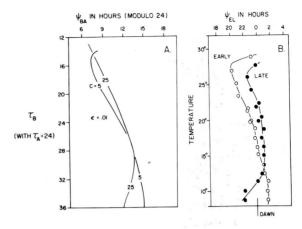

FIG. 47. Reversal in phase relationships between eclosion peaks in *Drosophila pseudo-obscura*. **A**: Plots ψ_{BA} as a function of τ_B for two B oscillators in computer simulations. The two oscillators differ in their strength of coupling (C) to the A oscillator, but have the same "damping" coefficient. **B**: The phases of the eclosion rhythms of Early and Late strains (entrained to a 24 h LD cycle) as a function of temperature. At low temperatures there is an inversion in the phase relationship between Early and Late. Other data suggested a temperature dependence of τ_B that led to the prediction of this phase reversal. (From Pittendrigh, C., 1981a.)

behavior, leave little doubt about its general validity. The implications of these results for our understanding of circadian organization are discussed more fully below.

10.3 Circadian rhythms in neural and neuroendocrine systems

In the previous section the problem of localization of the driving oscillators for the circadian system was considered in detail. Results from work on behavioral rhythmicity in several insects indicate that in the individual a localized region of the nervous system can function as a circadian pacemaker that exerts control over the timing of overt rhythmicity.

In this section another question central to understanding circadian organization is examined. How do these circadian oscillators impose periodicity on various physiological and behavioral processes? With few exceptions this question appears to have been asked only with the context of identifying the location of the oscillator, without addressing the more general and equally intriguing question of how the circadian system is able to impose rhythmicity on a large number of diverse processes and behaviors simultaneously. Is it by a change in "central excitatory state" as has been suggested by Brady, J. (1975, 1981), at least for behavioral rhythms, by peripheral modulation of the sensory input which elicits various behaviors, or is it by direct control of motor or immediately premotor elements in the nervous system? Or, as seems likely, is it through a combination of these mechanisms?

Many studies have focused on the neuroendocrine systems of insects, and rhythmicity within neurosecretory and endocrine cells has been described for several species. Rhythms of nuclear size, histological appearance, RNA synthesis, spectrophotometric absorbance, and ultrastructural changes have been reported (Table 4). In many instances these results have prompted various hypotheses concerning the hormonal control of circadian rhythmicity. However, the relationship between these rhythms and the overt behavior and physiology of the organisms is obscure. As Brady, J. (1974) has emphasized, any two processes which exhibit a daily rhythm are inevitably temporally correlated, but are not necessarily causally related. Furthermore, most of these studies have revealed nothing about the origin of the oscillations described; nor do they directly address the crucial question of whether or not these structural and histochemical rhythms are reflected in the amount of active circulating hormone. Finally, in most cases no evidence is offered that the rhythms persist in constant conditions. This is all the more critical because of the demonstration that the electrical activity of the medial neurosecretory cells in the cockroach *Periplaneta americana* can be affected "reflexively" by ambient light (Cook, D. and Milligan, J., 1972).

Although the study of rhythmicity in neurosecretory cells has, as yet, revealed little about their role in circadian organization, there are several cases in which there is almost certainly a neuroendocrine link in the pathway by which the circadian oscillator normally imposes rhythmicity on some process. In the case of the control of eclosion behavior in the silkmoth this link is well documented.

The importance of the silkmoth brain in gating eclosion, as well as the hormonal nature of the trigger, was demonstrated in the transplantation experiments described in the previous section (Truman, J. and Riddiford, L., 1970; Truman, J., 1972b). The eclosion hormone is probably produced in neurosecretory cells of the pars intercerebralis and released *via* the corpora cardiaca (Truman, J., 1973a). During the larval and pupal stages extracts of the brains and corpora cardiaca exhibit almost no hormonal activity, but as adult development takes place the activity levels in the brain and corpora cardiaca gradually increase (Truman, J., 1973a). The onset of the eclosion gate is accompanied by both a rapid decrease in the brain

Table 4: *Examples of endocrine cells for which there is evidence of daily rhythmicity*

Cell type (location)*	Organism	Assay	Lighting conditions	Reference
CA	*Carabus nemoralis*	nuclear volume	LD	Klug, H. (1958)
Medial NSC (brain)	*Carabus nemoralis*	histological appearance	LD	Klug, H. (1958)
NSC (brain, SEG, CC)	*Drosophila melanogaster*	nuclear size	LD	Rensing, L. (1964, 1966); Rensing, L. *et al.* (1965)
CA, prothoracic gland	*Drosophila melanogaster*	nuclear size	LD	Rensing, L. (1966, 1969)
Salivary gland	*Drosophila melanogaster*	nuclear size, fluorescence of cyanin dye treated cells	LD, LL, DD	Rensing, L. (1969); Weitzel, G. and Rensing, L. (1981)
NSC (SEG)	*Periplaneta americana*	nuclear size	LD	Brady, J. (1967c)
NSC (brain, SEG)	*Periplaneta americana*	histological appearance	LD	Cymborowski, B. and Flisinika-Bojanowska, A. (1970)
"A"-cells (ventral nerve cord)	*Leucophaea maderae*	histological appearance	LD	De Bess, N. (1965)
NSC (brain, SEG)	*Acheta domesticus*	RNA synthesis	LD	Cymborowski, B. and Dutkowski, A. (1969)
NSC (brain)	*Acheta domesticus*	Ultrastructural appearance	LD	Dutkowski, A. *et al.* (1971)

*CA, corpora allata; NSC, neurosecretory cells; SEG, subesophageal ganglion; CC, corpora cardica

and corpora cardiaca levels of eclosion hormone and an appearance of activity in the blood.

The release of the eclosion hormone triggers a stereotyped sequence of behavior, primarily involving abdominal movements, which ultimately results in emergence of the adult (Truman, J., 1971c). There is evidence to suggest that the motor score is pre-programmed in the abdominal nerve cord. The isolated abdomen when injected with crude hormone extract performs the pre-eclosion behavior (Truman, J., 1971c), and the motor neuron activity responsible can be recorded with suction electrodes on the motor roots of a completely deafferented nerve cord (Truman, J. and Sokolove, P., 1972). It appears then that the circadian pacemaker triggers the release of the eclosion hormone at the appropriate time of day. This hormone in turn initiates a motor program, present in the neural circuitry of the abdomen, that subsequently leads to emergence of the adult.

Investigations into the mechanisms of circadian regulation in other insects have also provided evidence that humoral pathways may be important. The results of transplantation experiments on both *Drosophila melanogaster* (Handler, A. and Konopka, R., 1979) and the house cricket *Acheta domesticus* (Cymborowski, B., 1981) indicate that the brains of these organisms release a diffusible factor that restores rhythmicity in locomotor activity in arrhythmic animals. A humoral influence on activity of *A. domesticus* and the cockroach *P. americana* was also indicated by the parabiosis experiments of Cymborowski, B. and Brady, J. (1972) (see section 10.1). In general there has been caution in interpreting these results, and it is not certain that in the intact animal the humoral pathway is the only or even "primary" coupling mechanism between the pacemaker system and motor activity (Cymborowski, B. and Brady, J., 1972; Handler, A. and Konopka, R., 1979).

In crickets and cockroaches much of the work directed toward the anatomical identification of the coupling pathway between the pacemaker and locomotor or stridulatory activity has focused on the pars intercerebralis (PI) which contains a large complement of neurosecretory cells.

Roberts, S. (1966) first called attention to the PI with a report that surgical lesions in the area "generally evoke arrhythmicity" in cockroaches. Shortly

afterward in a systematic study on the effects of surgical ablation of the PI in *Leucophaea maderae* it was found that lesions often resulted in loss of rhythmicity (Nishiitsutsuji-Uwo, J. *et al.*, 1967). Furthermore, there was a correlation between the presence or absence of neurosecretory cells in the PI region of lesioned animals and the postoperative regeneration of rhythmicity. In a subsequent study Nishiitsutsuji-Uwo, J. and Pittendrigh, C. (1968b) reported that bilateral section of the circumesophageal connectives (CEC) isolating the brain from the rest of the central nervous system did not affect the activity rhythm. On the basis of these results it was suggested that a hormonal pathway involving neurosecretory cells in the PI coupled the thoracic locomotor circuits to a circadian pacemaker located in the optic lobes (Nishiitsutsuji-Uwo, J. and Pittendrigh, C., 1968b). The strength of this hypothesis rested heavily on the observation that rhythmicity persisted following CEC section. Roberts, S. *et al.* (1971), however, in attempting to repeat this crucial experiment found that complete severance of the CEC invariably led to arrhythmicity in *Leuocphaea*. That CEC section does cause arrhythmicity has been verified (Brady, J., 1967b; Page, unpublished observation, and see Pittendrigh's comment in discussion in Roberts, S. *et al.*, 1971), and it is now generally believed that the link between the pacemaker (optic lobes) and thoracic motor centers includes a neuronal pathway in the CEC.

Other work has also failed to support the proposition that neurosecretion from PI cells is necessary for rhythmicity. It has been reported that removal of the corpora cardiaca, the major neurohemal organs of brain neurosecretory cells, does not disrupt the activity rhythm (Roberts, S., 1966; Brady, J., 1967a) and in a series of microlesion experiments Brady, J. (1967a) found that nearly all of the medial neurosecretory cells of the PI could be destroyed without abolishing rhythmicity in *P. americana*. As yet, however, there has been no clear demonstration that the lateral neurosecretory cells of the PI are not involved in the circadian system. Before discussing the meaning of these results it will be useful to review the work on the crickets *Teleogryllus commodus* and *A. domesticus*, since many of the observations parallel those in cockroaches.

The involvement of the pars intercerebralis in the control of circadian rhythmicity in the cricket has been the subject of numerous investigations. Lesions in the PI frequently abolish the circadian rhythms of stridulation (Loher, W., 1974; Sokolove, P. and Loher, W., 1975), locomotion (Cymborowski, B., 1970a, 1973; Sokolove, P. and Loher, W., 1975) and spermatophore production (Loher, W., 1974). In the case of locomotor activity in *Acheta domesticus* this result led to the proposal that the rhythm is humorally controlled (Cymborowski, B., 1970a,b, 1973). However, the arguments for a humoral control mechanism involving the pars intercerebralis were based mainly on the effects of lesions and on the observed rhythm (in LD) of neurosecretory cell function (see above), and were not compelling. Sokolove, P. and Loher, W. (1975), on the other hand, proposed an intermediary role for the PI in the control of rhythmicity which emphasized the presence of non-neurosecretory neurons in this region. They suggested that the PI was a major coupling pathway between a circadian pacemaker located elsewhere in the brain (probably the optic lobes) and several overt rhythms. They also suggest that different cells in the PI subserve different periodicities, some of which may be hormonally driven (e.g. spermatophore production) while others were driven by purely neuronal (electrical) signals (e.g. locomotion and stridulation). Support for this view came from the observation of differential effects of PI lesions on the locomotor and stridulatory rhythms (Sokolove, P. and Loher, W., 1975). Some lesions were found to abolish stridulation without disturbing the locomotor rhythm ($N = 1$), abolish locomotion but leave the stridulatory rhythm intact ($N = 2$), or cause arrhythmicity in locomotor activity with no effect on stridulation ($N = 1$). Although the numbers of observations are small, the data do support the suggestion that different cells in the PI control different behaviors.

In summary, efforts to identify the coupling mechanism between the circadian system and overt rhythms of activity in cockroach (*Periplaneta americana* and *L. maderae*) and crickets (*T. commodus* and *A. domesticus*) have not yet provided an unambiguous answer. The pars intercerebralis is likely to be involved, but whether or not the neurosecretory cells in the region are important is

unclear. In cockroaches, at least (the appropriate experiment needs to be done in crickets), the axonal connections of the CEC are crucial, but in what capacity has not been determined. This latter result, coupled with the inability to identify specific neurosecretory cell groups or neurohemal organs that affect rhythmicity, has generally led to the view that hormonal factors are not involved (as a major pathway) in coupling the pacemaker to activity (Brady, J., 1971, 1974; Sokolove, P. and Loher, W., 1975; Page, T., 1981b). However, this does not account for the positive results from the parabiosis experiments on *P. americana* and *A. domesticus* (Cymborowski, B. and Brady, J., 1972) or the more recent demonstration that transplanted brains can, at least for two or three cycles, drive via a humoral pathway the activity rhythm in *A. domesticus* made arrhythmic by PI lesions (Cymborowski, B., 1981). These results certainly suggest that the brain, at least under certain conditions, periodically releases a diffusible substance that can modulate activity. Further work is needed to determine the significance of this fact in the normal regulation of activity levels in these insects.

10.4 Circadian modulation of sensory input

Besides the considerable degree of control exerted by the circadian system over the effector pathways in the nervous and neuroendocrine systems of insects there is also some evidence to suggest that there is a circadian modulation of the excitability of sensory systems, particularly with respect to visual input. Probably the earliest demonstration of a circadian rhythm in a sensory response to a standard stimulus was the demonstration in 1940 of a rhythm in the amplitude of the electroretinogram (ERG) of several carabid beetles (e.g., *Dysticus fasciventris*) (Jahn, T. and Crescitelli, J., 1940; Jahn, T. and Wulff, V., 1943). More recently rhythmicity in ERG amplitude has been observed in the tenebrionid beetles *Blaps gigas* and *Blaps requienii* (Fig. 46) (Koehler, W. and Fleissner, G., 1978).

Typically, the rhythm of ERG response reflects a low sensitivity during the daytime and an increased sensitivity at night. In *B. gigas*, for example, the amplitude of the ERG evoked by a standard light pulse is 10–100 times higher during the subjective night than during the subjective day (Koehler, W.

and Fleissner, G., 1978). The rhythms in sensitivity likely reflect, in part, daily changes in the position of retinal shielding pigments which modulate the amount of light that reaches the photoreceptors. Rhythmic movement of eye pigments has been reported, for example, in *Tenebrio molitor* (Wada, S. and Schneider, G., 1968). However, the ERG rhythms may also depend on a change in the sensitivity of the primary photoreceptors (Jahn, T. and Wulff, V., 1943).

A circadian rhythm of spontaneous neural activity from the eye of the housefly *Musca domestica* has also been reported (Shipp, E. and Gunning, R., 1975; Gunning, R. and Shipp, E., 1976). Neural impulses were recorded with external electrodes placed on the compound eye. In *LD 12:12* the frequency of impulses was several-fold higher during the day than at night. This rhythmicity persisted for several days in both LL and DD. More recently, Kaiser, W. (1979) has found a circadian rhythm in the sensitivity of single visual interneurons in the optic lobe of the bee *Apis mellifica*.

Circadian rhythms in visual sensitivity and spontaneous activity have also been described in crustaceans, scorpions, and the horseshoe crab (see Page, T., 1981b for review), and therefore may be a common feature of visual system physiology in arthropods.

Some effort has also been made to determine whether or not sensory systems responsive to stimuli other than light exhibit rhythms in sensitivity. It appears that in moths there is no rhythm in the primary sensory response of males to female sex pheromone (measured by electroantennograms), although there is a rhythm in the behavioral response (Payne, T. *et al.*, 1970; Riddiford, L., 1974). Similarly, while there is a rhythm in the proboscis extension reflex of blowflies in response to stimulation of the tarsi with sugar, there is no evidence of a rhythm sensitivity in the primary afferents (Hall, M., 1980). Finally, the behavioral response of tsetse flies to visual stimuli is rhythmic, but other psychophysical evidence suggests it is not due to a change in sensitivity in the eyes (Brady, J., 1975). In each of these cases, then, no evidence was obtained for periodicity in the primary sensory system's sensitivity although the behavioral response to stimulation was clearly rhythmic.

10.5 Localization of photoperiodic clocks

There are two cases in which evidence suggests the clocks responsible for photoperiodic time measurement in insects are located in the brain. In the oak silkmoth *Antheraea pernyi* lesion and transplantation studies suggest that the mechanism for photoperiod control of diapause resides in the cerebral lobes of the brain (Williams, C. and Adkisson, P., 1964; Williams, C., 1969). The results of an extensive series of surgical lesions on *A. pernyi* are presented in Table 5 (from Williams, C., 1969). Only those procedures that involved substantial damage to the cerebral lobes of the brain had any disruptive effect on the photoperiodic control of diapause termination; and the response was completely abolished only if the lateral regions of the cerebral lobes were completely removed.

Further evidence for the importance of the brain came from the demonstration that sensitivity to photoperiod depended upon illumination of the brain (Williams, C. and Adkisson, P., 1964). Diapausing pupae were placed in holes in an opaque partition which separated two chambers that could be put on different light regimes. Pupae could then be subjected to one LD cycle at their anterior end (e.g. a long day, *LD 16:8*) and another at the posterior end (a short day, *LD 8:16*). It was found that the pupae responded to the LD cycle that illuminated the anterior end. The importance of the brain in this response was demonstrated by transplanting the brains to the abdomens of a group of diapausing moths and exposing either the anterior or posterior ends of these animals to a short day (*LD 8:16*) and the opposite end to DD (which causes a long day response in intact moths). 100% of those "loose-brain" animals whose abdomens were exposed to *LD 8:16* exhibited the short day response (maintained diapause) while 71% whose abdomens were exposed to DD broke diapause. It was found that transplantation of the brain to the abdomen transplanted the site of photosensitivity. The results demonstrated the photoreceptor for the photoperiodic response was located in the brain, and supported the hypothesis that the brain contained the clock mechanism for photoperiodic time measurement.

Recent data from the aphid *Megoura viciae* also point to the importance of the brain in photoper-

Table 5: *Surgical procedures on the central nervous system of diapausing pupae* of* Antheraea pernyi *and effects on the subsequent response to "long-day" and "short-day" light cycles (from Williams, C., 1969)*

Procedure	No. of viable pupae	Percentage initiating development in 2 months	
		SD†	LD‡
Cut circumesophageal connectives	40	0	100
Cut tracheal connections to brain	40	0	100
Cut nerves to antennae and eyes	40	0	100
Cut nerves to corpora cardiaca	40	0	100
Excise pigmented tips of brain	40	0	100
Bisect brain at midline	40	0	100
Dissociate optic lobes from cerebral lobes	40	45	80
Excise entire brain; reimplant cerebral lobes	40	45	85
Excise entire brain; reimplant dorsal half of cerebral lobes	40	30	75
Excise entire brain; reimplant pars intercerebralis	40	25	25

* First brood pupae stored at 2–3° for 8–11 weeks.
† Short day = LD *12:12*
‡ Long day = LD *17:7*

iodic time measurement. Steel, C. and Lees, A. (1977) have shown that lesions that destroy the Group I neurosecretory cells of the protocerebrum abolished the photoperiodic control of production of sexual and parthenogenetic morphs. The response was unaffected by damage to other neurosecretory cell groups or to the compound eyes and optic lobes. They also found that photoperiod control could be abolished by lesions of cells just lateral to the Group I neurosecretory cells. They proposed that these cells function as the photoperiodic clock and that they regulate the response by controlling the release of neurosecretory material from the Group I cells.

10.6 Photoreception in circadian rhythmicity and photoperiodic time measurement

In the past two decades a large number of investigations have been devoted to the identification of the photoreceptive pathways that are involved in entrainment of circadian oscillations and in photoperiodic responses in insects (for reviews see Truman, J., 1976; Page, T., 1982a). While information has been obtained on both the general location of photoreceptors involved, and on their spectral sensitivity (see next section) there are a number of important, but unanswered, questions that concern both the properties of the photoreceptors as well as the transduction and

transmission of the temporal information provided by the daily light cycle. Future progress will depend heavily on the precise identification of the photoreceptors and pathways involved.

10.6.1 LOCALIZATION OF PHOTORECEPTORS FOR ENTRAINMENT

Although "organized" photoreceptors are frequently bypassed in entrainment in invertebrates there are at least two cases, the cockroach and the cricket, in which the compound eyes appear to be the exclusive photoreceptors for entrainment. It was suggested in early publications in which opaquing techniques were used that the ocelli were either necessary (Cloudsley-Thompson, J., 1953) or sufficient (Harker, J., 1955, 1956) photoreceptors for entrainment of the cockroach activity rhythm. More recent work, however, has demonstrated convincingly that the compound eyes are the sole photoreceptive pathway for entrainment. Roberts, S. (1965a) provided the original evidence. He showed that when the eyes of intact roaches that were entrained to an LD cycle were painted over with black lacquer the animals began to free-run, even though the single pair of ocelli remained intact and exposed to the LD cycle (Fig. 48). In contrast, surgical ablation of the ocelli had no effect on entrainment. Roberts' conclusion that the eyes were the necessary photoreceptive pathway for entrainment was subsequently verified by Nishiitsu-tsuji-Uwo, J. and Pittendrigh, C. (1968a) who were able to repeat Roberts' observations on the effect of painting the compound eyes, and furthermore showed that surgical ablation of the eyes or section of the optic nerve also abolishes entrainment by light. They confirmed that the ocelli were not involved. Finally, Driskill, R. (1974) has shown that, even at very high light intensities of 22,700 lux, optic nerve section effectively abolishes entrainment, thus ruling out the possibility that the much lower inten-

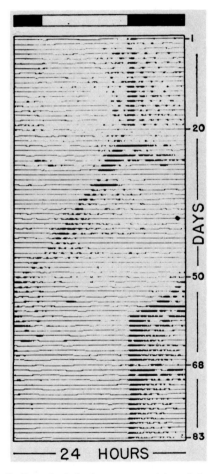

FIG. 48. Record of the locomotor activity of the roach, *Leucophaea maderae*, illustrating the necessity of the compound eyes for entrainment. The animals is maintained on a light–dark cycle indicated by the bar at the top of the record. On day 20 the compound eyes were painted with black lacquer and the animal began to free-run. On day 50 the paint was peeled off and the roach became entrained. Surgical ablation of the ocelli on day 68 had no effect on entrainment. (From Roberts, S., 1965a. Copyright 1965, by the American Association for the Advancement of Science. Reprinted with permission.)

sities used by Roberts, S. (1965a) and Nishiitsutsuji-Uwo, J. and Pittendrigh, C. (1968a) (40–275 lux) were merely insufficient for entrainment via either the ocelli or other, extraretinal photoreceptive pathways.

Similar results have been obtained with the cricket *Teleogryllus commodus*. Severance of the optic nerves abolishes entrainment of the stridulatory rhythm causing the animal to free-run in the presence of a light cycle; destruction of the ocelli has no effect (Loher, W., 1972). There is also limited evidence that entrainment of the activity rhythm involves the compound eyes (Sokolove, P. and Loher, W., 1975).

There are several cases where extraretinal photoreceptors have been implicated as the primary pathway for entrainment in insects. In the silkmoths *Antheraea pernyi* and *Hyalophora cecropia* the circadian oscillator that controls the time of adult emergence is located in the brain and its effects are mediated via a hormone produced in the neurosecretory cells of the brain (section 10.1). In animals in which the brain is neurally isolated or transplanted from the head to the abdomen, the circadian rhythm in eclosion persists and can be entrained by light (Truman, J. and Riddiford, L., 1970; Truman, J., 1972b). Truman, J. (1972b) demonstrated that the photoreceptor for entrainment was located in the brain. The experiment is illustrated in Fig. 49. Brains were removed from a population of pupae and were either replaced in the head region or transplanted to the abdomen. The pupae were then placed in holes in a partition which separated two chambers in which there were two light cycles 180° out of phase. Animals that had brains replaced in the head region entrained to the light cycle to which the anterior end of the pupae was exposed. Those in which brains had been transplanted to the abdomen were entrained by the light cycle presented to the posterior end. The results provided a clear demonstration that the photoreceptor resides in the brain. The optic lobes are not required for entrainment, and thus the photoreceptor is probably located in the main cerebral lobes (Truman, J., 1972b). Entrainment of flight activity rhythms in adult silkmoths has also been shown to involve extraretinal photoreceptors that are probably located in the brain (Truman, J., 1974b).

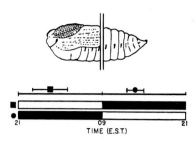

FIG. 49. Time of eclosion (mean and SD) of two groups of *Hyalophora cecropia* which differed in the site of brain implantation. The anterior end of each pupa was exposed to light from 21 : 00 h to 09 : 00 h (top bar) and the posterior half was exposed from 09 : 00 to 21 : 00 h (lower bar). ■, Brain implanted into head; ●, brain implanted into abdomen. Moths entrained to the light cycle to which the brain was exposed. (From Truman, J., 1972b.)

In *Drosophila melanogaster* the eclosion rhythm of the eyeless (and ocelliless) mutant, *sine oculis*, can be readily entrained by light, clearly implicating extraretinal photoreception (Engelmann, W. and Honneger, H., 1966; Zimmerman, W. and Ives, D., 1971). Evidence from experiments with local illumination suggested the photoreceptor is at the anterior end of the pupae (Kalmus, H., 1938; Zimmerman, W. and Ives, D., 1971), and it seems likely that, as in the silkmoth, the receptor is located in the brain.

Other experiments with both the fruit-fly and silkmoths further suggest the eyes are not only unnecessary, but that they are not sufficient for entrainment. In intact silkmoth adults, covering the entire head with wax *except* for the compound eyes frequently abolished entrainment of the flight activity rhythm (Truman, J., 1974b). In the white-eyed mutant of *Drosophila pseudo-obscura* the photoreceptors of the compound eyes are 2 log units more sensitive to light because of a loss of screening pigment, but there was no increase in the light sensitivity of the circadian rhythm (Zimmerman, W. and Ives, D., 1971). Furthermore, while *D. melanogaster* raised on a carotenoid-free medium, exhibited a 1000-fold drop in ERG amplitude, the sensitivity to a phase shifting stimulus was unaffected (Zimmerman, W. and Goldsmith, T., 1971).

Evidence for extraretinally mediated entrainment has also been obtained from surgical ablation experiments in several other insects. Rhythms of oviposition in *Chorthippus curtippenis* (Loher, W. and Chandrashekaran, M., 1970) and *Carausius*

morosus (Godden, D., 1973) and of stridulation in *Ephippiger ephippiger* and *Ephippiger bitterrensis* (Dumortier, B., 1972) could still be driven by light following surgical removal of the compound eyes and/or ocelli.

Another interesting example of the involvement of extraretinal photoreception in insect circadian rhythms has been described by Neville, A. (1967). In many insects the formation of the endocuticle occurs as a daily deposition of alternately lamellate and non-lamellate chitin. The rhythm of deposition persists in constant darkness but is abolished by constant light. In locusts (*Schistocerca gregaria* and *Locusta migratoria*) the effect of constant light persisted after ablation of the compound eyes and ocelli, and the results of opaquing experiments indicated the site of photoreception was within the epidermal cells (Neville, A., 1967).

10.6.2 LOCALIZATION OF PHOTORECEPTORS FOR PHOTOPERIODIC CLOCKS

In insects there are several cases where photoperiodic responses are mediated by extraretinal photoreceptors. In the two most thoroughly investigated species, the aphid *Megoura viciae* and the silkmoth *Antheraea pernyi*, the site of photoreception has been localized to the brain.

In *Megoura* the development of female offspring into sexual, egg-laying morphs *vs.* asexual, parthenogenetic morphs is determined by the length of the photoperiod to which the mother is exposed. Lees, A. (1964) found that cauterization of the compound eyes did not affect the photoperiodic response indicating extraretinal photoreceptors were involved. Further localization was carried out with fine light guides which were used to extend a short day exposure of the whole aphid into a long day with localized supplemental illumination (Fig. 50). The results suggested the photoreceptor was in the head region and probably in the protocerebrum. A subsequent study (Steel, C. and Lees, A., 1977) involving small brain lesions provided evidence consistent with this view (see section 10.5).

A brain-centered photoreceptor has also been implicated in the photo-periodically controlled termination of diapause in *A. pernyi*. When the anterior and posterior ends of intact pupae were exposed to different photoperiods, the maintenance or termination of diapause depended on the

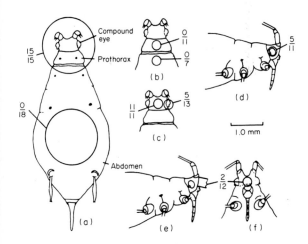

FIG. 50. Results obtained from selectively illuminating various regions of *Megoura viciae*. Circles shows the area of illumination. The denominator in each fraction indicates the number of aphids tested and the numerator gives the number of positive responses (see text). (From Lees, A., 1964.)

photoperiod to which the anterior end was exposed. When the brain was transplanted to the abdomen the ability to discriminate between long and short days was retained but the site of photosensitivity had been transferred to the abdomen (Williams, C. and Adkisson, P., 1964). The results are similar to those obtained in studies on entrainment of the eclosion rhythm (see section 10.6) and indicate the locus of the photoreceptor is in the brain. Similar transplantation experiments have implicated the brain as the photosensitive pathway for photoperiodically controlled initiation of diapause in the cabbage butterfly *Pieris brassicae* (Claret, J., 1966; Seuge, J. and Veith, K., 1976).

Studies on two species of beetles have produced contrasting results. In the Colorado potato-beetle *Leptinotarsa decemlineata* photoperiodically controlled reproductive diapause in females did not require the compound eyes (deWilde, J. *et al.*, 1959). In contrast, reproductive diapause in male *Pterostichus nigrita* appears to involve retinal photoreception. Animals in which the compound eyes were ablated responded as if they were in constant darkness regardless of photoperiod (Ferenz, H., 1975). This latter example appears to be the only case described thus far in which the photoreception for photoperiodic time measurement in insects requires the compound eyes.

The central role of extraretinal photoreception in the temporal organization of insects is a curious fact whose functional significance is not well understood. This reliance on extraretinal pathways is shared by other invertebrate groups (Page, T., 1982a) as well as most vertebrates (Menaker, M., 1976) and thus seems to be a general feature of circadian and photoperiodic systems. As yet no extraretinal photoreceptor involved in these processes has been identified. In insects (and other organisms) the evidence from localization experiments suggests the photoreceptors are in the central nervous system; and it seems reasonable to conclude that photoreceptive neurons are involved. But many of the detailed questions about the physiological properties of the photoreceptors, as well as the mechanisms utilized in the transduction, processing, and transmission of the temporal information provided by the daily light cycle, await a more precise identification of the cells responsible.

10.7 Action spectra and sensitivity

The spectral sensitivity of the photoperiodic and entrainment responses of insects has been examined in several species, although relatively complete action spectra have only been obtained in a few cases.

In *Drosophila pseudo-obscura*, in which protoreception for entrainment is extraretinal the effectiveness of 15 min pulses of various wavelengths of light from 375–600 nm in evoking either a delay or advance phase shift has been investigated. Both action spectra showed a broad region of sensitivity from 375 to 500 nm with the most effective wavelengths being between 420 and 480 nm. Above 500 nm sensitivity fell off sharply (Frank, K. and Zimmerman, W., 1969). A more detailed spectrum for delay phase shifts was subsequently obtained (Klemm, E. and Ninnemann, H., 1976). The data revealed a peak sensitivity at 457 nm with lower maxima at 473, 435, 407, and 375 nm. These results suggested that the photoreceptor pigment was a flavin or flavoprotein (Klemm, E. and Ninnemann, H., 1976), a view consistent with the observation that *Drosophila* raised on a carotenoid-free medium exhibited no change in sensitivity to a phase shifting stimulus (Zimmerman, W. and Goldsmith, T., 1971, and see above.).

Mote, M. and Black, K. (1981) have recently determined the action spectrum for entrainment of the locomotor activity rhythm in the cockroach *Periplaneta americana*. In this case the photoreceptors involved are located in the compound eyes. The spectral characteristics of the eyes have been determined from ERG measurements (Walther, J., 1958) and intracellular recording (Mote, M. and Goldsmith, T., 1970). The results indicated there are two receptor types, one maximally sensitive to UV and the other responding best to green (Fig. 51). The entrainment action spectrum suggested the entrainment mechanism predominantly involves the green-sensitive receptors with peak sensitivity occurring near 495 nm (Fig. 51).

In *Pectinophora gossypiella* the spectral sensitivity for initiation of the egg hatching rhythm has been investigated (Bruce, V. and Minis, D., 1969). The most effective wavelengths were between 400 nm and 480 nm. Wavelengths above 500 nm were completely ineffective. Similar results were obtained for entrainment of the eclosion and oviposition rhythms, which were found to entrain readily to 480 nm light, but were insensitive to 600 nm.

A relatively complete action spectrum for the photoperiodic response has been obtained for the aphid *Megoura viciae* (Lees, A., 1971, 1981). Aphids maintained in a light cycle consisting of 13.5 h of light and 10.5 h of dark exhibit a "short day" response. This response can be reversed by exposing the animals to a brief (0.5–1.0 h) pulse of light at one or two times during the dark period. The first of these light-sensitive points, which occurs between 0 and 3 h after the onset of darkness, exhibited a peak sensitivity at 450–470 nm and falls off rapidly above 475 nm and below 450 nm. The second light-sensitive period, which occurs several hours later (6–10.5 h after onset of darkness), was also found to be maximally sensitive at 450–470 nm, but its sensitivity extended above 550 nm and down to 400 nm. These results are consistent with a view that the photoreception involves a carotenoprotein, and suggest either that there are two pigment systems or that there is a single pigment system whose properties are modified during the course of the night (Lees, A., 1971, 1981).

In *Antheraea pernyi* the action spectrum for diapause termination was measured by extending a 10 h (short) day with an additional 6 h of monochromatic light (Norris, K. *et al.*, 1969). The most effective wavelengths were between 400 and 500 nm with sensitivity declining steadily between 500 and 600 nm.

The general trend of peak sensitivity of the photoperiodic response to blue-green light has been observed in several species (see Truman, J., 1976) including, for example, *Pieris brassicae* (Seuge, J. and Veith, K., 1976) and the coddling moth *Laspeyresia pomonella* (Norris, K. *et al.*, 1969; Hayes, D., 1971). However, the photoperiodic responses of several other species such as *Pectinophora gossypiella* (Pittendrigh, C. *et al.*, 1970) and *Nasonia vitripennis* (Saunders, D., 1975a) are sensitive to longer wavelengths. It is also interesting to note that while the photoperiodic response in *Pectinophora gossypiella* is sensitive to red light (Pittendrigh, C. *et al.*, 1970), the photoreceptors for entrainment of the circadian system are unaffected by red light (see above).

The data on action spectra in insects have also been used to obtain a quantitative estimate of the sensitivity of extraretinal photoreceptors (Truman, J., 1976). The results indicated the photoreceptors are quite sensitive, and compare favorably with, for example, the phytochrome system in plants. Standard phase shifting or photoperiodic responses are obtained with incident energies on the order of 10^{-1} to 1 Joule-meter^{-2}. Mote, M. and Black, K. (1981) have reported an extraordinary sensitivity of the

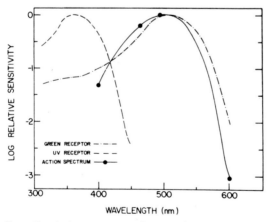

FIG. 51. Action spectrum for entrainment (–●–) in *Periplaneta americana* compared to the spectral sensitivies of the two classes of photoreceptors in the compound eye. ––––––, UV photoreceptor; –·–·–·–, green photoreceptor. (From Mote, M. and Black, K., 1981.)

entrainment in *P. americana*. They have estimated that a mean flux of about 5 photons eye^{-1} s^{-1} is sufficient for entrainment in *LD 12*:12.

11   CONCLUDING REMARKS: TEMPORAL ORGANIZATION IN INSECTS

The information summarized in the preceding pages illustrates the pervasive importance of temporal regulation in the lives of insects. In large part this regulation appears to be a function of physiological systems, based on self-sustaining oscillations, that have, because of their unique properties, been regarded as endogenous clocks. These clocks serve to (1) organize, in time, those processes that are best restricted to a particular time of day, phase of the lunar cycle, or season of the year, and (2) provide a temporal program within the individual to ensure that events are appropriately timed with respect to each other.

Our current understanding of the physiological and anatomical substrate of the regulatory systems responsible for maintaining temporal organization is far from complete. Nevertheless, substantial progress has been made on this problem in the last 20 years, and a general view of the physiological basis of temporal regulation is emerging, particularly with respect to circadian systems.

A major theme of this view is that temporal (circadian) organization is derived from a population of anatomically distributed oscillators and is dependent not only on the properties of the individual oscillators but also on their interactions with each other and with the environment through entrainment pathways. Recent work has left no doubt about the validity of this general proposition (see section 10.2), and has suggested that there are a variety of "configurations" of the oscillating components within the system. Oscillators may be mutually coupled, unidirectionally coupled, or be completely independent. A specific variation of the multiple-oscillator hypothesis may also explain how a single pacemaker controls a variety of overt periodicities. There are many examples in the Metazoa where a single pacemaking system appears to drive several overt rhythms that have quite different phase relationships with environmental zeitgebers. In the cricket *Teleogryllus commodus*, for example, locomotor activity is primarily diurnal, stridulation is nocturnal, and spermatophore formation occurs in late afternoon (Lohor, W., 1972, 1974; Sokolove, P., 1975). All three rhythms appear to be controlled by a circadian pacemaker in the optic lobes of the brain. How does a pacemaker temporally organize several rhythms?

One attractive hypothesis is that a "central pacemaker" controls several overt rhythms via entrainment of a population of slave oscillators, that in turn each control a specific process or processes. In this view differing phase relationships among various rhythms derive from differences in the properties of the slaves (e.g., free-running period) or in the coupling between the slaves and the pacemaker that leads various steady-state phase angle differences (Pittendrigh, C., 1981a,b). Experimental support for a pacemaker–slave organization is currently strong (in insects) only in the case of the *Drosophila* eclosion rhythm (see section 10.2), but is consistent with experimental observations in other organisms. For example, the hypothesis prompts the speculation that the residual rhythmicity in activity following optic lobe (pacemaker) ablation in cockroaches and crickets reflects the presence (perhaps in the pars interecerebralis?) of a temperature-sensitive slave oscillator.

It is important to note that other coupling relationships between circadian oscillators within the individual are not precluded. Oscillators may exhibit strong mutual coupling as in the pacemaking system in the cockroach optic lobes; or there may be several independent pacemaking systems, as appears to be the case in the control of the ERG rhythms in the two eyes of the beetle *Blaps gigas* and in the control of locomotor activity and cuticular deposition rhythms in the cockroach. In these latter two examples, maintenance of internal temporal organization must depend on the access of pacemakers to common temporal cues from the external environment.

It is clear that understanding the physiological mechanisms of entrainment by external and internal periodicities is of central importance in understanding the mechanisms of temporal organization. Substantial progress has been made in describing in quantitative detail the formal, analytical properties of the entrainment process (Pittendrigh, C., 1981b) and, in the case of light entrainment, localizing the

primary receptors responsible for transducing the
zeitgeber stimulus. However, virtually nothing is
known about the anatomical and physiological
pathways by which this information is transmitted
to the oscillating system. Even less information is
available on pathways and mechanisms involved in
internal coupling among constituent oscillators of
the circadian system.

Certainly much work remains before we can
claim to understand how circadian (and other) temporal programs are generated. Progress made thus
far in unraveling the intricacies of the biological
clocks of insects suggests future efforts will be
rewarding.

ACKNOWLEDGEMENTS

Supported by PHS-NIH Grant GM 30039 to T.L.P.
and BRSG Grant PHS-NIH 2-507-RR07201-02 to
Vanderbilt University.

REFERENCES

ADKISSON, P. L. (1964). Action of the photoperiod in controlling insect
    diapause. *Amer. Nat.* 98, 357–374.
ADKISSON, P. L. (1966). Internal clocks and insect diapause. *Science 154*,
    234–241.
ASCHOFF, J. (1979). Circadian rhythms: influences of internal and external
    factors on the period measured in constant conditions. *Z. Tierp-
    sychol.* 49, 225–249.
BALL, H. J. (1972). Photic entrainment of circadian activity rhythms by
    direct brain illumination in the cockroach *Blaberus craniifer. J. Insect
    Physiol.* 18, 2449–2455.
BARTELL, R. J. and SHOREY, H. H. (1969). A quantitative bioassay for the
    sex pheromone of *Epiphyas postvittana* (Lepidoptera) and factors
    limiting male responsiveness. *J. Insect Physiol.* 15, 33–40.
BATEMAN, M. A. (1955). The effect of light and temperature on the rhythm
    of pupal ecdysis in the Queensland fruit-fly, *Dacus (Strumeta) tryoni*
    (Frogg.) *Aust. J. Zool.* 3, 22–33.
BEACH, R. F. (1978). The required day number and timely induction of
    diapause in geographic strains of the mosquito *Aëdes atropalpus. J.
    Insect Physiol.* 24, 449–455.
BEACH, R. F. and CRAIG, G. B. (1977). Night length measurements by the
    circadian clock controlling diapause induction in the mosquito *Aëdes
    atropalpus. J. Insect Physiol.* 23, 865–870.
BECK, S. D. (1962). Photoperiodic induction of diapause in an insect. *Biol.
    Bull. Mar. Lab., Woods Hole, 122*, 1–12.
BECK, S. D. (1963). Physiology and ecology of photoperiodism. *Bull. Ent.
    Soc. Amer.* 9, 8–16.
BECK, S. D. (1964). Time-measurement in insect photoperiodism. *Amer.
    Nat.* 98, 329–346.
BECK, S. D. (1974a). Photoperiodic determination of insect development
    and diapause. I. Oscillators, hourglasses, and a determination model.
    J. Comp. Physiol. 90, 275–295.
BECK, S. D. (1974b). Photoperiodic determination of insect development
    and diapause. II. The determination gate in a theoretical model. *J.
    Comp. Physiol.* 90, 297–310.
BECK, S. D. (1975). Photoperiodic determination of insect development
    and diapause. III. Effects of nondiel photoperiods. *J. Comp. Physiol.*
    103, 227–245.

BECK, S. D. (1977). Dual system theory of the biological clock: effects of
    photoperiod, temperature, and thermoperiod on the determination
    of diapause. *J. Insect Physiol.* 23, 1363–1372.
BECK, S. D. (1980). *Insect Photoperiodism.* 2nd edition. Academic Press,
    New York.
BECK, S. D. (1982). Thermoperiodic induction of larval diapause in
    the European corn borer, *Ostrinia nubilalis. J. Insect Physiol.* 28,
    273–277.
BECK, S. D. and HANEC, W. (1960). Diapause in the European corn borer,
    Pyrausta nubilalis (Hübn.). *J. Insect Physiol.* 4, 304–318.
BEIER, W. and LINDAUER, M. (1970). Der Sonnenstand als Zeitgeber für
    die Biene. *Apidologie, 1*, 5–28.
BELCHER, K. and BRETT, W. J. (1973). Relationship between a metabolic
    rhythm and emergence rhythm in *Drosophila melanogaster. J. Insect
    Physiol.* 19, 277–286.
BELING, I. (1929). Über das Zeitgedächtnis der Bienen. *Z. Vergl. Physiol.*
    9, 259–338.
BENNETT, M. F. and RENNER, M. (1963). The collecting performances of
    honey bees under laboratory conditions. *Biol. Bull. 125*, 416–430.
BIRUKOW, G. (1960). Innate types of chronometry in insect orientation.
    Cold Spring Harb. Symp. Quant. Biol. 25, 403–412.
BLAKE, G. M. (1958). Diapause and regulation of development in *Anth-
    renus verbasci* (L.) (Col., Dermestidae). *Bull. Ent. Res.* 49, 751–775.
BLAKE, G. M. (1959). Control of diapause by an "internal clock" in *Anth-
    renus verbasci* (L.) (Col., Dermestidae). *Nature 183*, 126–127.
BLAKE, G. M. (1960). Decreasing photoperiod inhibiting metamorphosis
    in an insect. *Nature 188*, 168–169.
BLAKE, G. M. (1963). Shortening of a diapause-controlled life cycle by
    means of increasing photoperiod. *Nature 198*, 462–463.
BOWEN, M. F. and SKOPIK, S. D. (1976). Insect photoperiodism: the "T
    Experiment" as evidence for an hourglass mechanism. *Science 192*,
    59–60.
BRADY, J. (1967a). Control of the circadian rhythm of activity in the
    cockroach. I. The role of the corpora cardiaca, brain and stress. *J.
    Exp. Biol.* 47, 153–163.
BRADY, J. (1967b). Control of the circadian rhythm of activity in the
    cockroach. II. The role of the sub-oesophageal ganglion and ventral
    nerve cord. *J. Exp. Biol.* 47, 165–178.
BRADY, J. (1967c). Histological observations on circadian changes in the
    neurosecretory cells of cockroach suboesophageal ganglia. *J. Insect
    Physiol.* 13, 201–213.
BRADY, J. (1969). How are insect circadian rhythms controlled? *Nature
    223*, 781–784.
BRADY, J. (1971). The search for the insect clock. In *Biochronometry.*
    Edited by M. Menaker. Pages 517–524. National Academy of
    Sciences, Washington, DC.
BRADY, J. (1972). Spontaneous, circadian components of tsetse fly ac-
    tivity. *J. Insect Physiol.* 18, 471–484.
BRADY, J. (1974). The physiology of insect circadian rhythms. *Adv. Insect
    Physiol.* 10, 1–115.
BRADY, J. (1975). Circadian changes in central excitability — the origin of
    behavioral rhythms in tsetse flies and other animals? *J. Ent. 50*,
    79–95.
BRADY, J. (1981). Behavioral rhythms in invertebrates. In *Biological
    Rhythms.* Edited by J. Aschoff. *Handbook of Behavioral Neuro-
    biology.* Vol. 4, pages 125–144. Plenum Press, New York.
BRADY, J. and CRUMP, A. J. (1978). The control of circadian activity
    rhythms in tsetse flies: environment or physiological clock? *Physiol.
    Ent.* 3, 177–190.
BRETT, W. J. (1955). Persistent diurnal rhythmicity in *Drosophila* emer-
    gence. *Ann. Ent. Soc. Amer.* 48, 119–131.
BRUCE, V. G. (1960). Environmental entrainment of circadian rhythms.
    Cold Spring Harb. Symp. Quant. Biol. 25, 29–48.
BRUCE, V. G. and MINIS, D. H. (1969). Circadian clock action spectrum
    in a photoperiodic moth. *Science 163*, 583–585.
BRUN, R. (1914). *Die Raumorientierung der Ameisen und das Orientierungs-
    problem im Allgemeinen.* Gustav Fischer, Jena.
BÜNNING, E. (1936). Die endogene Tagesrhythmik als Grundlage der
    Photoperiodischen Reaktion. *Berichte der Deutschen Botanischen
    Gesellschaft 54*, 590–607.
BÜNNING, E. (1958). Über den Temperatureinfluss auf die endogene
    Tagesrhythmik besonders bei *Periplaneta americana. Biol. Zbl. 77*,
    141–152.

BÜNNING, E. (1960). Circadian rhythms and the time measurement in photoperiodism. *Cold Spring Harb. Symp. Quant. Biol.* 25, 249–256.

BÜNNING, E. and JOERRENS, G. (1960). Tagesperiodische antagonistische Schwankungen der Blau-violett und Gelbrot-Empfindlichkeit als Grundlage der photoperiodischen Diapause-Induktion bei *Pieris brassicae*. *Z. Naturf.* 15, 205–213.

BÜNSOW, R. C. (1960). The circadian rhythm of photoperiodic responsiveness in *Kalanchoë*. *Cold Spring Harb. Symp. Quant. Biol.* 25, 257–260.

CALDAROLA, P. C. (1974). The interaction of temperature and heavy water on circadian pacemakers. Ph.D thesis, Stanford University.

CAMPBELL, B. O. (1964). Solar and lunar periodicities in oxygen consumption by the mealworm, *Tenebrio molitor*. Ph.D. thesis, Northwestern University. (Quoted from Beck, S., 1980.)

CHANDRASHEKARAN, M. K. and LOHER, W. (1969). The effect of light intensity on the circadian rhythm of eclosion in *Drosophila pseudoobscura*. *Z. Vergl. Physiol.* 62, 337–347.

CHIPPENDALE, G. M. and REDDY, A. S. (1973). Temperature and photoperiodic regulation of diapause of the southwestern corn borer, *Diatraea grandiosella*. *J. Insect Physiol.* 19, 1397–1408.

CHIPPENDALE, G. M., REDDY, A. S. and CATT, C. L. (1976). Photoperiodic and thermoperiodic interaction in the regulation of the larval diapause of *Diatraea grandiosella*. *J. Insect Physiol.* 22, 823–828.

CLARET, J. (1966). Mise en evidence du rôle photorécepteur du cerveau dans l'induction de la diapause chez *Pieris brassicae* (Lepido.). *Ann. Endocr.* 27, 311–320.

CLARET, J., DUMORTIER, B. and BRUNNARIUS, J. (1981). Mise en evidence d'une composante circadienne dans l'horloge biologique de *Pieris brassicae* (Lepidoptera), lors de l'induction photopériodique de la diapause. *C.R. Acad. Sci. Paris Série III* 292, 427–430.

CLAYTON, D. L. and PAIETTA, J. V. (1972). Selection for circadian eclosion time in *Drosophila melanogaster*. *Science* 178, 994–995.

CLOUDSLEY-THOMPSON, J. L. (1953). Studies on diurnal rhythms. III. Photoperiodism in the cockroach *Periplaneta americana* (L.). *Ann. Mag. Nat. Hist.* 6, 705–712.

COOK, B. J., DE LA CUESTA, M. and POMONIS, J. G. (1969). The distribution of factor S in the cockroach, *Periplaneta americana*, and its role in stress paralysis. *J. Insect Physiol.* 15, 963–975.

COOK, D. J. and MILLIGAN, J. V. (1972). Electrophysiology and histology of the medial neurosecretory cells in adult male cockroaches, *Periplaneta americana*. *J. Insect Physiol.* 18, 1197–1214.

CRAWFORD, C. S. (1967). Oviposition rhythm studies in *Crambus topiarius* (Lepidoptera: Pyralidae: Crambinae). *Ann. Ent. Soc. Amer.* 60, 1014–1018.

CYMBOROWSKI, B. (1970a). Investigations of the neurohormonal factors controlling circadian rhythm of locomotor activity in the house cricket (*Acheta domesticus* L.). I. The role of the brain and the subeosophageal ganglion. *Zool. Polon.* 20, 103–126.

CYMBOROWSKI, B. (1970b). Investigations on the neurohormonal factors controlling circadian rhythm of locomotor activity in the house cricket *Acheta domesticus* L. II. Daily histochemical changes in the neurosecretory cells of the pars intercerebralis and subseophageal ganglion. *Zool. Polon.* 20, 127–153.

CYMBOROWSKI, B. (1973). Control of the circadian rhythm of locomotor activity in the house cricket. *J. Insect Physiol.* 19, 1423–1440.

CYMBOROWSKI, B. (1981). Transplantation of circadian pacemaker in the house cricket, *Acheta domesticus* L. *J. Interdisciplinary Cycle Res.* 12, 133–140.

CYMBOROWSKI, B. and BRADY, J. (1972). Insect circadian rhythms transmitted by parabiosis — A re-examination. *Nature* 236, 221–222.

CYMBOROWSKI, B. and DUTKOWSKI, A. (1969). Circadian changes in RNA synthesis in the neurosecretory cells of the brain and suboesophageal ganglion of the house cricket. *J. Insect Physiol.* 15, 1187–1198.

CYMBOROWSKI, B. and FLISINSKA-BOJANOWSKA, A. (1970). The effect of light on the locomotor activity and structure of neurosecretory cells of the brain and subeosophageal ganglion of *Periplaneta americana* L. *Zool. Poloniae* 20, 387–399.

DAAN, S. and BERDE, C. (1978). Two coupled oscillators: simulations of the circadian pacemaker in mammalian activity rhythms. *J. Theor. Biol.* 70, 297–313.

DAAN, S. and PITTENDRIGH, C. S. (1976). A funcional analysis of circadian pacemakers in nocturnal rodents. III. Heavy water and constant light: homeostasis of frequency? *J. Comp. Physiol.* 106, 267–290.

DANILEVSKII, A. S. (1965). *Photoperiodism and Seasonal Development of Insects*. 1st English edition. Oliver & Boyd, Edinburgh and London.

DANILEVSKII, A. S., GORYSHIN, N. I. and TYSHCHENKO, V. P. (1970). Biological rhythms in terrestrial arthropods. *Ann. Rev. Ent.* 15, 201–244.

DE BESSÉ, N. (1965). Recherches histophysiologiques sur la neurosecretion dans la chaîne nerveuse ventral d'une blatte, *Leucophaea maderae* (F.). *C.R. Hebd. Séanc. Acad. Sci., Paris* 260, 7014–7017.

DECOURSEY, P. J. (1959). Daily activity rhythms in the flying squirrel *Glaucomys volans*. Ph.D. thesis. University of Wisconsin.

DENLINGER, D. L. (1972). Induction and termination of pupal diapause in *Sarcophaga* (Diptera: Sarcophagidae). *Biol. Bull.* 142, 11–24.

DINGLE, H., CALDWELL, R. L. and HASKELL, J. B. (1969). Temperature and circadian control of cuticle growth in the bug *Oncopeltus fasciatus*. *J. Insect Physiol.* 15, 373–378.

DRISKILL, R. J. (1974). The circadian locomotor rhythm of the cockroach: An examination of the photoreceptive system operative in entrainment. Master's thesis, University of Delaware.

DUMORTIER, B. (1972). Photoreception in the circadian rhythm of stridulatory activity in *Ephippiger* (Ins., Orthoptera); Likely existence of two photoreceptive systems. *J. Comp. Physiol.* 77, 80–112.

DUMORTIER, B. and BRUNNARIUS, J. (1977). L'Existence d'une composante circadienne dans l'induction thermopériodique. *C. R. Acad. Sci. Paris, D.* 285, 361–364.

DUMORTIER, B. and BRUNNARIUS, J. (1981). Involvement of the circadian system in photoperiodism and thermoperiodism in *Pieris brassicae* (Lepidoptera). In *Biological Clocks in Seasonal Reproduction Cycles*. Edited by B. K. Follett and D. E. Follett. Pages 83–99. Wright, Bristol.

DUTKOWSKI, A. B., CYMBOROWSKI, B. and PRZELECKA, A. (1971). Circadian changes in the ultrastructure of the neurosecretory cells of the pars intercerebralis of the house cricket. *J. Insect Physiol.* 17, 1763–1772.

DYER, F. C. and GOULD, J. L. (1981). Honey bee orientation: a backup system for cloudy days. *Science* 214, 1041–1042.

EIDMANN, H. (1956). Über rhythmische Erscheinungen bei der Stabheuschrecke *Carausius morosus* Br. *Z. Vergl. Physiol.* 28, 370–390.

EISENBACH, J. and MITTLER, T. E. (1980). An aphid circadian rhythm: factors affecting the release of sex pheromone by oviparae of the greenbug, *Schigaphis graminum*. *J. Insect Physiol.* 26, 511–515.

ENGELMANN, W. and HONEGGER, H. W. (1966). Tagesperiodische Schlüpfryhthmik einer augenlosen *Drosophila melanogaster*-Mutante. *Naturwissenschaften* 53, 588.

ENGELMANN, W. and MACK, J. (1978). Different oscillators control the circadian rhythm of eclosion and activity in *Drosophila*. *J. Comp. Physiol.* 127, 229–237.

EVANS, W. G. (1976). Circadian and circatidal locomotory rhythms in the intertidal beetle *Thalassostrechus barbarae* (Horn): Carabidae. *J. Exp. Mar. Biol. Ecol.* 22, 79–90.

FATZINGER, C. W. (1973). Circadian rhythmicity of sex pheromone release by *Dioryctria abietella* (lepidoptera: Pyralidae (Phycitinae)) and the effect of a diel light cycle on its precopulatory behavior. *Ann. Ent. Soc. Amer.* 66, 1147–1154.

FERENZ, H. J. (1975). Photoperiodic and hormonal control of reproduction in male beetles, *Pterostichus nigrita*. *J. Insect Physiol.* 21, 331–341.

FINGERMAN, M., LAGO, A. D. and LOWE, M. E. (1958). Rhythm of locomotor activity and oxygen consumption of the grasshopper *Romalea microptera*. *Amer. Midl. Nat.* 59, 58–66.

FOLLETT, B. K. and FOLLETT, D. E., Editors (1981). *Biological Clocks in Seasonal Reproductive Cycles*. Wright, Bristol.

FONDACARO, J. D. and BUTZ, A. (1970). Circadian rhythm of locomotor activity and susceptibility to methyl parathion of adult *Tenebrio molitor* (Coleoptera: Tenebrionidae). *Ann. Ent. Soc. Amer.* 63, 952–955.

FOREL, A. (1910). *Das Sinnesleben der Insekten*. Reinhardt. Münich.

FRANK, K. D. and ZIMMERMAN, W. F. (1969). Action spectra for phase shifts of a circadian rhythm in *Drosophila*. *Science* 163, 688–689.

VON FRISCH, K. (1946). Die Tänze der Bienen. *Österr. Zool. Z.* 1, 1–48.

VON FRISCH, K. (1967). *The Dance Language and Orientation of Bees*. English edition. Belknap Press of Harvard University Press, Cambridge, Mass.; Oxford University Press, London.

VON FRISCH, K. and LINDAUER, M. (1954). Himmel und Erde in Konkurrenz bei der Orientierung der Bienen. *Naturwissenschaften 41*, 245–253.

FUJISHITA, M. and ISHIZAKI, H. (1981). Circadian clock and prothoracicotropic hormone secretion in relation to the larval-larval ecdysis rhythm of the saturnid *Samia cynthia ricini*. *J. Insect Physiol. 27*, 122–128.

GARNER, W. W. and ALLARD, H. A. (1920). Effect of the relative length of day and night and other factors of the environment on growth and reproduction in plants. *J. Agric. Res. 18*, 553–606.

GILLETT, J. D. (1962). Contributions to the oviposition cycle by the individual mosquitoes in a population. *J. Insect Physiol. 8*, 665–681.

GILLETT, J. D. (1972). *The Mosquito. Its Life, Activities, and Impact on Human Affairs.* Doubleday & Co. Inc., Garden City, New York.

GODDEN, D. H. (1973). A re-examination of circadian rhythmicity in *Carausius morosus*. *J. Insect Physiol. 19*, 1377–1386.

GORYSHIN, N. I. and TYSHCHENKO, V. P. (1970). Thermostability of the process of perception of photoperiodic information in the moth *Acronycta rumicis* (Lepidoptera, Noctuidae). *Dokl. Akad. Nauk. SSSR 193*, 458–461.

GOULD, J. L. (1975). Honey bee recruitment: the dance-language controversy. *Science 189*, 685–693.

GOULD, J. L. (1980). Sun compensation by bees. *Science 207*, 545–547.

GUNNING, R. AND SHIPP, E. (1976). Circadian rhythm in endogenous nerve activity in the eye of *Musca domestica* L. *Physiol. Ent. 1*, 241–248.

GWINNER, E. (1981). Circannual systems. In *Biological Rhythms.* Edited by J. Aschoff. *Handbook of Behavioral Neurobiology.* Vol. 4, pages 391–410. Plenum Press, New York.

HADDOW, A. J. and GILLETT, J. D. (1957). Observations on the oviposition-cycle of *Aëdes (Stegomyia) aegypti* (Linnaeus). *Ann. Trop. Med. Parasit. 51*, 159–169.

HADDOW, A. J., GILLETT, J. D. and CORBET, P. S. (1959). Laboratory observations on pupation and emergence in the mosquito *Aëdes (Stegomyia) aegypti* (Linnaeus). *Ann. Trop. Med. Parasit. 53*, 123–131.

HADDOW, A. J., GILLETT, J. D. and CORBET, P. S. (1961). Observations on the oviposition-cycle of *Aëdes (Stegomyia) aegypti* (Linnaeus) V. *Ann. Trop. Med. Parasit. 55*, 343–356.

HALBERG, F. (1960). Temporal coordination of physiologic function. *Cold Spring Harb. Symp. Quant. Biol. 25*, 289–310.

HALL, M. J. (1980). Circadian rhythm of proboscis extension responsiveness in the blowfly: central control of threshold changes. *Physiol. Ent. 5*, 223–233.

HAMMACK, L. and BURKHOLDER, W. E. (1976). Circadian rhythm of sex pheromone-releasing behavior in females of the dermestid beetle, *Trogoderma glabrum*: regulation by photoperiod. *J. Insect Physiol. 22*, 385–388.

HANDLER, A. M. and KONOPKA, R. J. (1979). Transplantation of a circadian pacemaker in *Drosophila*. *Nature 279*, 236–238.

HARKER, J. E. (1954). Diurnal rhythms in *Periplaneta americana* L. *Nature 173*, 689–690.

HARKER, J. C. (1955). Control of diurnal rhythms of activity in *Periplaneta americana* L. *Nature 175*, 733.

HARKER, J. E. (1956). Factors controlling the diurnal rhythm of activity in *Periplaneta americana*. *J. Exp. Biol. 33*, 224–234.

HARKER, J. E. (1960). Internal factors controlling the subesophageal ganglion neurosecretory cycle in *Periplaneta americana*. *J. Exp. Biol. 37*, 164–170.

HARKER, J. E. (1964). *The Physiology of Diurnal Rhythms.* Cambridge University Press, London and New York.

HARKER, J. E. (1965a). The effect of a biological clock on the development rate of *Drosophila* pupae. *J. Exp. Biol. 42*, 323–337.

HARKER, J. E. (1965b). The effect of photoperiod on the development rate of *Drosophila* pupae. *J. Exp. Biol. 43*, 411–423.

HARTLAND-ROWE, R. (1955). Lunar rhythm in the emergence of an Ephemeropteran. *Nature 176*, 657.

HARTLAND-ROWE, R. (1958). The biology of a tropical mayfly *Povilla adusta* Navas (Ephemeroptera, Polymitarcidae) with special reference to the lunar rhythm of emergence. *Revue Zool. Bot. Afr. 58*, 185–202.

HAVERTY, M. I. and WARE, G. W. (1970). Circadian sensitivity and dosage-rate response to x-irradiation in the pink bollworm. *J. Econ. Ent. 63*, 1296–1300.

HAYES, D. K. (1971). Action spectra for breaking diapause and absorption spectra of insect brain tissue. In *Biochronometry*. Edited by M. Menaker. Pages 392–402. National Academy of Sciences, Washington, DC.

HAYES, D. K., HORTON, J., SCHECHTER, M. S. and HALBERG, F. (1972). Rhythm of oxygen uptake in diapausing larvae of the codling moth at several temperatures. *Ann. Ent. Soc. Amer. 65*, 93–96.

HILLIARD, S. D. and BUTZ, A. (1969). Daily fluctuations in the concentrations of total sugar and uric acid in the hemolymph of *Periplaneta americana*. *Ann. Ent. Soc. Amer. 62*, 71–74.

HILLMAN, W. S. (1964). Endogenous circadian rhythms and the response of *Lemna perpusilla* to skeleton photoperiods. *Amer. Nat. 98*, 323–328.

HILLMAN, W. S. (1973). Non-circadian photoperiodic timing in the aphid *Megoura*. *Nature 242*, 128–129.

HOFFMANN, K. (1960). Experimental manipulation of the orientational clock in birds. *Cold Spring Harb. Symp. Quant. Biol. 25*, 379–387.

HORWATH, K. L. and DUMAN, J. G. (1982). Involvement of the circadian system in photoperiodic regulation of insect antifreeze proteins. *J. Exp. Zool. 219*, 267–270.

JAHN, T. L. and CRESCITELLI, J. (1940). Diurnal changes in the electrical responses of the compound eye. *Biol. Bull. 78*, 45–52.

JAHN, T. L. and WULFF, V. J. (1943). Electrical aspects of a diurnal rhythm in the eye of *Dytiscus fasciventris*. *Physiol. Zool. 16*, 101–109.

JANDA, V. and MRCIAK, M. (1957). Gesamtsoffwechsel der Insekten. VI. Die Bewegungsaktivität der Schabe *Periplaneta americana* L. während des Tages und ihre Beziehung zum Sauerstoffverbrauch. *Acta. Soc. Zool. Bohem. 21*, 244–255.

JANDER, R. (1957). Die optische Richtungsorientierung der Roten Waldameise (*Formica rufa* L.). *Z. Vergl. Physiol. 40*, 162–238.

JONES, M. D. R. (1973). Delayed effect of light on the mosquito "clock". *Nature 245*, 384–385.

JONES, M. D. R. (1976). Persistence in continuous light of a circadian rhythm in the mosquito *Culex pipiens fatigans* Wied. *Nature 261*, 491–492.

JONES, M. D. R. and GUBBINS, S. J. (1977). Modification of circadian flight activity in the mosquito *Anopheles gambiae* after insemination. *Nature 268*, 731–732.

JONES, M. D. R. and GUBBINS, S. J. (1978). Changes in the circadian flight activity of the mosquito *Anopheles gambiae* in relation to insemination, feeding and oviposition. *Physiol. Ent. 3*, 213–220.

JONES, M. D. R. and REITER, P. (1975). Entrainment of the pupation and adult activity rhythms during development in the mosquito *Anopheles gambiae*. *Nature 254*, 242–244.

JONES, M. D. R., CUBBIN, C. M. and MARSH, D. (1972). The circadian rhythm of flight activity of *Anopheles gambiae*: the light response rhythm. *J. Exp. Biol. 57*, 337–346.

KAISER, W. (1979). Circadiane Empfindlichkeitsänderungen einzelner visueller Interneurone der Biene *Apis mellifica carnica*. *Verh. Dtsch. Zool. Ges. 1979*, 211.

KALMUS, H. (1938). Die Lage des Aufnahmeorgans für die Schlupfperiodik von *Drosophila*. *Z. Vergl. Physiol. 26*, 362–365.

KAWATO, M. and SUZUKI, R. (1980). Two coupled neural oscillators as a model of the circadian pacemaker. *J. Theor. Biol. 86*, 547–575.

KLEBER, E. (1935). Hat das Zeitgedächtnis der Bienen eine biologische Bedeutung? *Z. Vergl. Physiol. 22*, 221–262.

KLEMM, E. and NINNEMANN, H. (1976). Detailed action spectrum for the delay shift in pupae emergence of *Drosophila pseudoobscura*. *Photochem. Photobiol. 24*, 369–371.

KLUG, H. (1958). Histo-physiologische Untersuchungen über die Aktivitätsperiodik bei Carabiden. *Wiss. Z. Humboldt. Univ. Berlin, Math.-Naturw. Reihe, 8*, 405–434.

KOEHLER, W. K. and FLEISSNER, G. (1978). Internal desynchronization of bilaterally organized circadian oscillators in the visual system of insects. *Nature 274*, 708–710.

KONOPKA, R. J. (1972). Circadian clock mutants of *Drosophila melanogaster*. Ph.D. thesis, California Institute of Technology, Pasadena.

KONOPKA, R. J. (1979). Genetic dissection of the *Drosophila* circadian system. *Fed. Proc. 38*, 2602–2605.

KONOPKA, R. J. (1981). Genetics and development of circadian rhythms in invertebrates. In *Biological Rhythms*. Edited by J. Aschoff. *Handbook of Behavioral Neurobiology.* Vol. 4, pages 173–181. Plenum Press, New York.

KONOPKA, R. and BENZER, S. (1971). Clock mutants of *Drosophila melanogaster*. *Proc. Nat. Acad. Sci. 68*, 2112–2116.

KONOPKA, R. J. and WELLS, S. (1980). *Drosophila* clock mutations affect the morphology of a brain neurosecretory cell group. *J. Neurobiol. 11*, 411–415.

KRAMER, G. (1952). Experiments on bird orientation. *Ibis 94*, 265–285.

KYRIACOU, C. P. and HALL, J. C. (1980). Circadian rhythm mutation in *Drosophila melanogaster* affect short-term fluctuations in the male's courtship song. *Proc. Nat. Acad. Sci. 77*, 6729–6733.

LEES, A. D. (1959). The role of photoperiod and temperature in the determination of parthenogenetic and sexual forms in the aphid *Megoura viciae* Buckton — I. The influence of these factors on apterous virginoparae and their progeny. *J. Insect Physiol. 3*, 92–117.

LEES, A. D. (1963). The role of photoperiod and temperature in the determination of parthenogenetic and sexual forms in the aphid *Megoura viciae* Buckton — III. Further properties of the maternal switching mechanism in apterous aphids. *J. Insect Physiol. 9*, 153–164.

LEES, A. D. (1964). The location of the photoperiodic receptors in the aphid *Megoura viciae* Buckton. *J. Exp. Biol. 41*, 119–133.

LEES, A. D. (1966). The control of polymorphism in aphids. *Adv. Insect Physiol. 3*, 207–277.

LEES, A. D. (1968). Photoperiodism in insects. In *Photophysiology*. Vol. IV. Edited by A. C. Giese. Pages 47–137. Academic Press, New York.

LEES, A. D. (1970). Insect clocks and timers. Inaugural Lecture, Imperial College of Science and Technology, 1 December 1970.

LEES, A. D. (1971). The relevance of action spectra in the study of insect photoperiodism. In *Biochronometry*. Edited by M. Menaker. Pages 372–380. National Academy of Sciences, Washington, DC.

LEES, A. D. (1973). Photoperiodic time measurement in the aphid *Megoura viciae*. *J. Insect Physiol. 19*, 2279–2316.

LEES, A. D. (1981). Action spectra for the photoperiodic control of polymorphism in the aphid *Megoura viciae*. *J. Insect Physiol. 27*, 761–771.

LEES, A. D. and HARDIE, J. (1981). The photoperiodic control of polymorphism in aphids: neuroendocrine and endocrine components. In *Biological Clocks in Seasonal Reproductive Cycles*. Edited by B. K. Follett and D. E. Follett. Pages 125–135. Wright, Bristol.

LEUTHOLD, R. (1966). Die Bewegungsaktivität der weiblichen Schabe *Leucophaea maderae* (F.) im Laufe des Fortpflanzungszyklus und ihre experimentelle Beeinflussung. *J. Insect. Physiol. 12*, 1303–1331.

LINDAUER, M. (1954). Dauertänze im Bienenstock und ihre Beziehung zur Sonnenbahn. *Naturwissenschaften 41*, 506–507.

LINDAUER, M. (1955). Schwarmbienen auf Wohnungsuche. *Z. Vergl. Physiol. 37*, 263–324.

LINDAUER, M. (1960). Time-compensated sun orientation in bees. *Cold Spring Harb. Symp. Quant. Biol. 25*, 371–377.

LIPTON, G. R. and SUTHERLAND, D. J. (1970). Activity rhythms in the American cockroach, *Periplanata americana*. *J. Insect Physiol. 16*, 1555–1566.

LOHER, W. (1972). Circadian control of stridulation in the cricket, *Teleogryllus commodus* Walker. *J. Comp. Physiol. 79*, 173–190.

LOHER, W. (1974). Circadian control of spermatophore formation in the cricket *Teleogryllus commodus* Walker. *J. Insect Physiol. 20*, 1155–1172.

LOHER, W. and CHANDRASHEKARAN, M. K. (1970). Circadian rhythmicity in the oviposition of the grasshopper *Chorthippus curtipennis*. *J. Insect Physiol. 16*, 1677–1688.

LOHMANN, M. (1964). Der einfluss von Beleuchtungsstärke und Temparatur auf die Tagesperiodische Laufaktivität des Mehlkäfers, *Tenebrio molitor* L. *Z. Vergl. Physiol. 49*, 341–389.

LUKAT, R. (1978). Circadian growth layers in the cuticle of behaviorally arrhythmic cockroaches (*Blaberus fuscus*, Ins., Blattoidea). *Experientia 34*, 477.

LUKAT, K. and WEBER, F. (1979). The structure of locomotor activity in bilobectomized cockroaches (*Blaberus fuscus*). *Experientia 35*, 38–39.

MARSH, D. (1972). Sex pheromone in the aphid *Megoura viciae*. *Nature 238*, 31–32.

MARTIN, U., MARTIN, H. and LINDAUER, M. (1978). Transplantation of a time-signal in honeybees. *J. Comp. Physiol. 124*, 193–201.

MASAKI, S. (1956). The local variation in the diapause pattern of the cabbage moth, *Barathra brassicae* Linné, with particular reference to the aestival diapause (Lepidopetera: Noctuidae). *Bull. Fac. Agr. Mie Univ. 13*, 29–46.

MASAKI, S. (1968). Geographic adaptation in the seasonal life cycle of *Mamestra brassicae* (Linné) (Lepidoptera: Noctuidae). *Bull. Fac. Agr. Hirosaki Univ. 14*, 16–26.

MASAKI, S. and KIKUKAWA, S. (1981). The diapause clock in a moth: response to temperature signals. In *Biological Clocks in Seasonal Reproductive Cycles*. Edited by B. K. Follett, and D. E. Follett. Pages 101–112. Wright, Bristol.

MCCLUSKEY, E. S. and CARTER, C. E. (1969). Loss of rhythmic activity in female ants caused by mating. *Comp. Biochem. Physiol. 31*, 217–226.

MEDER, E. (1958). Über die Einberechnung der Sonnenwanderung bei der Orientierung der Honigbiene. *Z. Vergl. Physiol. 40*, 610–641.

MEDUGORAC, I. and LINDAUER, M. (1967). Das Zeitgedächtnis der Bienen unter dem Einfluss von Narkose und von sozialen Zeitgebern. *Z. Vergl. Physiol. 55*, 450–474.

MENAKER, M. ed. (1976). Symposium on extraretinal photoreception in circadian rhythms and related phenomena. Introduction. *Photochem. Photobiol. 23*, 213–306.

MINIS, D. H. (1965). Parallel peculiarities in the entrainment of a circadian rhythm and photoperiodic induction in the pink bollworm (*Pectinophora gossypiella*). In *Circadian Clocks*. Edited by J. Aschoff. Pages 333–343. North-Holland, Amsterdam.

MINIS, D. H. and PITTENDRIGH, C. S. (1968). Circadian oscillation controlling hatching: its ontogeny during embryogenesis of a moth. *Science 159*, 534–536.

MOORE-EDE, M. C. and SULZMAN, F. M. (1981). Internal temporal order. In *Biological Rhythms*. Edited by J. Aschoff. *Handbook of Behavioral Neurobiology*. Vol. 4, pages 215–241. Plenum Press, New York.

MORIARTY, F. (1959). The 24-hr rhythm of emergence of *Ephestia kuhniella* Zell. from the pupa. *J. Insect Physiol. 3*, 357–366.

MOTE, M. I. and BLACK, K. R. (1981). Action spectrum and threshold sensitivity of entrainment of circadian running activity in the cockroach *Periplaneta americana*. *Photochem. Photobiol. 34*, 257–265.

MOTE, M. I. and GOLDSMITH, T. H. (1970). Spectral sensitivities of color receptors in the compound eye of *Periplaneta americana*. *J. Exp. Zool. 173*, 137–146.

NAGATA, K., TAMAKI, Y., NOGUCHI, H. and YUSHIMA, T. (1972). Changes in sex pheromone activity in adult females of the smaller tea tortrix moth *Adoxophyes fasciata*. *J. Insect Physiol. 18*, 339–346.

NAYAR, J. K. (1967a). Endogenous diurnal rhythm of pupation in a mosquito population. *Nature 214*, 828–829.

NAYAR, J. K. (1967b). The pupation rhythm in *Aëdes taeniorhynchus* (Diptera: Culicidae). II. Ontogenetic timing, rate of development, and endogenous diurnal rhythm of pupation. *Ann. Ent. Soc. Amer. 60*, 946–971.

NAYAR, J. K. (1969). The pupation rhythm in *Aëdes taeniorhynchus*. V. Physiology of growth and endogenous diurnal rhythm of pupation. *Ann. Ent. Soc. Amer. 62*, 1079–1087.

NAYAR, J. K., SAMARAWICKREMA, W. A. and SAUERMAN, D. M. (1973). Photoperiodic control of egg hatching in the mosquito *Mansonia titillans*. *Ann. Ent. Soc. Amer. 66*, 831–835.

NEUMANN, D. (1966). Die lunare und tägliche Schlüpfperiodik der Mücke *Clunio*: Steuerung und Abstimmung auf die Gezeitenperiodik. *Z. Vergl. Physiol. 53*, 1–61.

NEUMANN, D. (1967). Genetic adaptation in emergence time of *Clunio* populations to different tidal conditions. *Helgoländer Wiss. Meeresunters. 15*, 163–171.

NEUMANN, D. (1176a). Entrainment of a semilunar rhythm. In *Biological Rhythms in the Marine Environment*. Edited by P. J. DeCoursey. Pages 115–127. University of South Carolina Press, Columbia.

NEUMANN, D. (1976b). Adaptations of chironomids to intertidal environments. *Ann. Rev. Ent. 21*, 387–414.

NEUMANN, D. (1978). Entrainment of a semilunar rhythm by simulated tidal cycles of mechanical disturbance. *J. Exp. Mar. Biol. Ecol. 35*, 73–85.

NEUMANN, D. (1981). Tidal and lunar rhythms. In *Biological Rhythms*. Edited by J. Aschoff. *Handbook of Behavioral Neurobiology*. Vol. 4, pages 351–380. Plenum Press, New York.

NEUMANN, D. and HEIMBACH, F. (1979). Time cues for semilunar reproduction rhythms in European populations of *Clunio*. I. The influence of tidal cycles of mechanical disturbance. In *Cyclical Phenomena in Marine Plants and Animals*. Edited by E. Naylor. Pages 423–433. Pergamon Press, Oxford.

NEVILLE, A. C. (1965). Circadian organization of chitin in some insect skeletons. *Quart. J. Mic. Sci. 106*, 315–325.

NEVILLE, A. C. (1967). A dermal light sense influencing skeletal structure in locusts. *J. Insect Physiol. 13*, 933–939.

NEVILLE, A. C. (1970). Cuticle ultrastructure in relation to the whole insect. *Symp. R. Ent. Soc. Lond. 5*, 17–39.

NEVILLE, A. C. (1975). *Biology of the Arthropod Cuticle*. Springer, Berlin.

NISHIITSUTSUJI-UWO, J. and PITTENDRIGH, C. S. (1968a). Central nervous system control of circadian rhythmicity in the cockroach. II. The pathway of light signals that entrain the rhythms. *Z. Vergl. Physiol. 58*, 1–13.

NISHIITSUTSUJI-UWO, J. and PITTENDRIGH, C. S. (1968b). Central nervous system control of circadian rhythmicity in the cockroach. III. The optic lobes, locus of the driving oscillation? *Z. Vergl. Physiol. 58*, 14–46.

NISHIITSUTSUJI-UWO, J., PETROPULOS, S. F. and PITTENDRIGH, C. S. (1967). Central nervous system control of circadian rhythmicity in the cockroach. I. Role of the pars intercerebralis. *Biol. Bull. 133*, 679–696.

NORRIS, K. H., HOWELL, F., HAYES, D. K., ADLER, V. E., SULLIVAN, W. N. and SCHECHTER, M. S. (1969). The action spectrum for breaking diapause in the codling moth, *Laspeyresia pomonella* (L.) and the oak silkworm *Antheraea pernyi* Guer. *Proc. Nat. Acad. Sci. 63*, 1120–1127.

NORRIS, M. J. (1965). The influence of constant and changing photoperiods on imaginal diapause in the red locust (*Nomadacris septemfasciata* Serv.). *J. Insect Physiol. 11*, 1105–1119.

NOWOSIELSKI, J. W. and PATTON, R. L. (1963). Studies on circadian rhythms of the house cricket, *Gryllus domesticus* L. *J. Insect Physiol. 9*, 401–410.

NOWOSIELSKI, J. W. and PATTON, R. L. (1964). Daily fluctuation in the blood sugar concentration of the house cricket, *Gryllus domesticus* L. *Science 144*, 180–181.

ODHIAMBO, T. R. (1966). The metabolic effects of the corpus allatum hormone in the male desert locust. II. Spontaneous locomotor activity. *J. Exp. Biol. 45*, 51–63.

OTTESON, E. O. (1965). Analytical studies on a model for the entrainment of circadian systems. Bachelor's thesis, Princeton University.

PAGE, T. L. (1978). Interactions between bilaterally paired components of the cockroach circadian system. *J. Comp. Physiol. 124*, 225–236.

PAGE, T. L. (1981a). Effects of localized low-temperature pulses on the cockroach circadian pacemaker. *Amer. J. Physiol. 240*, R144–R150.

PAGE, T. L. (1981b). Localization of circadian pacemakers in insects. In *Biological Clocks in Seasonal Reproductive Cycles*. Edited by B. K. Follett and D. E. Follett. Pages 113–124. Wright, Bristol.

PAGE, T. L. (1981c). Neural and endocrine control of circadian rhythmicity in invertebrates. In *Biological Rhythms*. Edited by J. Aschoff. *Handbook of Behavioral Neurobiology*. Vol. 4, pages 145–172. Plenum Press, New York.

PAGE, T. L. (1982a). Extraretinal photoreception in entrainment and photoperiodism in invertebrates. *Experientia. 38*, 1007–1013.

PAGE, T. L. (1982b). Regeneration of the optic tracts and circadian pacemaker activity in the cockroach. *Neurosci. Abs.* (In press.)

PAGE, T. L. (1982c). Transplantation of the cockroach circadian pacemaker. *Science 216*, 73–75.

PAGE, T. L. and BLOCK, G. D. (1980). Circadian rhythmicity in cockroaches: effects of early post-embryonic development and aging. *Physiol. Ent. 5*, 271–281.

PAGE, T. L., CALDAROLA, P. C. and PITTENDRIGH, C. S. (1977). Mutual entrainment of bilaterally distributed circadian pacemakers. *Proc. Nat. Acad. Sci. 74*, 1277–1281.

PAVLIDIS, T. (1968). Studies of biological clocks: a model for the circadian rhythms in nocturnal animals. In *Lectures on Mathematics in the Life Sciences*. Vol. 1, pages 88–112. American Mathematical Society, Providence, RL.

PAYNE, T. L., SHOREY, H. H. and GASTON, L. K. (1970). Sex pheromones of noctuid moths: factors influencing antennal responsiveness in males of *Trichoplusia ni*. *J. Insect Physiol. 16*, 1043–1055.

PETERSON, E. L. (1980). A limit cycle interpretation of a mosquito circadian oscillator. *J. Theor. Biol. 84*, 281–310.

PETERSON, E. L. and SAUNDERS, D. S. (1980). The circadian eclosion rhythm in *Sarcophaga argyrostoma*: a limit cycle representation of the pacemaker. *J. Theor. Biol. 86*, 265–277.

PITTENDRIGH, C. S. (1954). On temperature independence in the clock system controlling emergence time in *Drosophila*. *Proc. Nat. Acad. Sci. 40*, 1018–1029.

PITTENDRIGH, C. S. (1960). Circadian rhythms and the circadian organization of living systems. *Cold Spring Harb. Symp. Quant. Biol. 25*, 159–184.

PITTENDRIGH, C. S. (1965). On the mechanism of entrainment of a circadian rhythm by light cycles. In *Circadian Clocks*. Edited by J. Aschoff. Pages 277–297. North-Holland, Amsterdam.

PITTENDRIGH, C. S. (1966). The circadian oscillation in *Drosophila pseudoobscura* pupae: a model for the photoperiodic clock. *Z. Pflanzenphysiol. 54*, 275–307.

PITTENDRIGH, C. S. (1967). Circadian systems I. The driving oscillation and its assay in *Drosophila pseudoobscura*. *Proc. Nat. Acad. Sci. 58*, 1762–1767.

PITTENDRIGH, C. S. (1972). Circadian surfaces and the diversity of possible roles of circadian organization in photoperiodic induction. *Proc. Nat. Acad. Sci. 69*, 2734–2737.

PITTENDRIGH, C. S. (1974). Circadian oscillations in cells and the circadian organization of multicellular systems. In *The Neurosciences: Third Study Program*. Edited by F. O. Schmitt and F. G. Worden. Pages 437–458. MIT Press, Cambridge, Mass.

PITTENDRIGH, C. S. (1976). Circadian clocks: what are they? In *Molecular Basis of Circadian Rhythms*. Edited by J. W. Hastings and H. Schweiger. Pages 11–48. Dahlem Konferenzen, Berlin.

PITTENDRIGH, C. S. (1981a). Circadian organization and the photoperiodic phenomena. In *Biological Clocks in Seasonal Reproductive Cycles*. Edited by B. K. Follett and D. E. Follett. Pages 1–35. Wright, Bristol.

PITTENDRIGH, C. S. (1981b). Circadian rhythms: general perspective. In *Biological Rhythms*. Edited by J. Aschoff. *Handbook of Behavioral Neurobiology*. Vol. 4, pages 57–80. Plenum Press, New York.

PITTENDRIGH, C. S. (1981c). Circadian systems: entrainment. In *Biological Rhythms*. Edited by J. Aschoff. *Handbook of Behavioral Neurobiology*. Vol. 4, pages 95–124. Plenum Press, New York.

PITTENDRIGH, C. S. and BRUCE, V. G. (1957). An oscillator model for biological clocks. In *Rhythmic and Synthetic Processes in Growth*. Edited by D. Rudnik. Pages 75–109. Princeton University Press, Princeton, N.J.

PITTENDRIGH, C. S. and CALDAROLA, P. C. (1973). General homeostasis of the frequency of circadian oscillations. *Proc. Nat. Acad. Sci. 70*, 2697–2701.

PITTENDRIGH, C. S., CALDAROLA, P. C. and COSBEY, E. S. (1973). A differential effect of heavy water on temperature-dependent and temperature-compensated aspects of the circadian system of *Drosophila pseudoobscura*. *Proc. Nat. Acad. Sci. 70*, 2037–2041.

PITTENDRIGH, C. S. and DAAN, S. (1976). A functional analysis of circadian pacemakers in nocturnal rodents. I. The stability and lability of spontaneous frequency. *J. Comp. Physiol. 106*, 223–252.

PITTENDRIGH, C. S., EICHHORN, J. H., MINIS, D. H. and BRUCE, V. G. (1970). Circadian systems VI. Photoperiodic time measurement in *Pectinophora gossypiella*. *Proc. Nat. Acad. Sci. 66*, 758–764.

PITTENDRIGH, C. S. and MINIS, D. H. (1964). The entrainment of circadian oscillations by light and their role as photoperiodic clocks. *Amer. Nat. 98*, 261–294.

PITTENDRIGH, C. S. and MINIS, D. H. (1971). The photoperiodic time measurement in *Pectinophora gossypiella* and its relation to the circadian system in that species. In *Biochronometry*. Edited by M. Menaker. Pages 212–250. National Academy of Sciences, Washington, DC.

PITTENDRIGH, C. S. and MINIS. D. H. (1972). Circadian systems: longevity as a function of circadian resonance in *Drosophila melanogaster*. *Proc. Nat. Acad. Sci. 69*, 1537–1539.

PITTENDRIGH, C. S. and SKOPIK, S. D. (1970). Circadian systems. V. The driving oscillation and the temporal sequence of development. *Proc. Nat. Acad. Sci. 65*, 500–507.

PITTENDRIGH, C. S., BRUCE, V. G. and KAUS, P. (1958). On the significance of transients in daily rhythms. *Proc. Nat. Acad. Sci. 44*, 965–973.

POLCIK, B., NOWOSIELSKI, J. W. and NAEGELE, J. A. (1964). Daily sensitivity rhythm of the two-spotted mite, *Tetranychus urticae*, to DDVP. *Science 145*, 405.

RANKIN, M. A., CALDWELL, R. L. and DINGLE, H. (1972). An analysis of a circadian rhythm of oviposition in *Oncopeltus fasciatus*. *J. Exp. Biol. 56*, 353–359.

RENCE, B. and LOHER, W. (1975). Arrhythmically singing crickets: Thermoperiodic reentrainment after bilobectomy. *Science 190*, 385–387.

RENNER, M. (1955). Ein Transozeanversuch zum Zeitsinn der Honigbiene. *Naturwissenschaften 42*, 540–541.

RENNER, M. (1957). Neue Versuche über den Zeitsinn der Honigbiene. *Z. Vergl. Physiol. 40*, 85–118.

RENNER, M. (1959). Über ein weiteres Versetzungs-experiment zur Analyse des Zeitsinnes und der Sonnenorientierung der Honigbiene. *Z. Vergl. Physiol. 42*, 449–483.

RENNER, M. (1960). The contribution of the honey bee to the study of time-sense and astronomical orientation. *Cold Spring Harb. Symp. Quant. Biol. 25*, 361–367.

RENSING, L. (1964). Daily rhythmicity of corpus allatum and neurosecretory cells in *Drosophila melanogaster* (Meig.). *Science 144*, 1586–1587.

RENSING, L. (1966). Zur circadianen Rhythmik des Sauerstoffverbrauches von *Drosophila*. *Z. Vergl. Physiol. 53*, 62–83.

RENSING, L. (1969). Ein circadianer Rhythmus der Empfindlichkeit gegen Röntgenstrahlen bei *Drosophila*. *Z. Vergl. Physiol. 62*, 214–220.

RENSING, L. and HARDELAND, R. (1967). Zur Wirkung der circadianen Rhythmik auf die Entwicklung von *Drosophila*. *J. Insect Physiol. 13*, 1547–1568.

RENSING, L., THACH, B. T. and BRUCE, V. G. (1965). Daily rhythms in the endocrine glands of *Drosophila*. *Experientia 21*, 103–104.

RICHARDS, A. G. (1969). Oxygen consumption of the American cockroach under complete inanition prolonged to death. *Ann. Ent. Soc. Amer. 62*, 1313–1316.

RIDDIFORD, L. M. (1974). The role of hormones in the reproductive behavior of female wild silkmoths. In *Experimental Analysis of Insect Behavior*. Edited by L. Barton Brown. Pages 278–285. Springer-Verlag, Berlin.

RIEMANN, J. G., THORSON, B. J. and RUDD, R. L. (1974). Daily cycle of release of sperm from the testes of the Mediterranean flour moth. *J. Insect Physiol. 20*, 195–207.

ROBERTS, S. K. (1960). Circadian activity in cockroaches. I. The freerunning rhythm in steady-state. *J. Cell. Comp. Physiol. 55*, 99–110.

ROBERTS, S. K. (1962). Circadian activity in cockroaches. II. Entrainment and phase-shifting. *J. Cell Comp. Physiol. 59*, 175–186.

ROBERTS, S. K. (1965a). Photoreception and entrainment of cockroach activity rhythms. *Science 148*, 958–959.

ROBERTS, S. K. (1965b). Significance of endocrines and central nervous system in circadian rhythms. In *Circadian Clocks*. Edited by J. Aschoff. Pages 198–213. North-Holland, Amsterdam.

ROBERTS, S. K. (1966). Circadian activity rhythms in cockroaches. III. The role of endocrine and neural factors. *J. Cell Comp. Physiol. 67*, 473–486.

ROBERTS, S. K. (1974). Circadian rhythms in cockroaches: effects of optic lobe lesions. *J. Comp. Physiol. 88*, 21–30.

ROBERTS, S. K., SKOPIK, S. D. and DRISKILL, R. J. (1971). Circadian rhythms in cockroaches: does brain hormone mediate the locomotor cycle? In *Biochronometry*. Edited by M. Menaker. Pages 505–515. National Academy of Sciences, Washington DC.

ROTH, R. L. and SOKOLOVE, P. G. (1975). Histological evidence for direct connections between the optic lobes of the cockroach, *Leucophaea maderae*. *Brain Res. 87*, 23–29.

ROTHERT, H. (1970). Tagesperiodische Schwankungen der Empfindlichkeit von *Drosophila melanogaster* gegenüber Parathion. *Z. Angew. Ent. 65*, 403–409.

SANTSCHI, F. (1911). Observations et remarques critiques sur le mecanisme de l'orientation chez les fourmis. *Revue Suisse Zool. 19*, 303–338.

SAUNDERS, D. S. (1965a). Larval diapause induced by a maternally-operating photoperiod. *Nature 206*, 739–740.

SAUNDERS, D. S. (1965b). Larval diapause of maternal origin: Induction of diapause in *Nasonia vitripennis* (Walk.) (Hymenoptera: Pteromalidae). *J. Exp. Biol. 42*, 495–508.

SAUNDERS, D. S. (1966a). Larval diapause of maternal origin — II. The effect of photoperiod and temperature on *Nasonia vitripennis*. *J. Insect Physiol. 12*, 569–581.

SAUNDERS, D. S. (1966b). Larval diapause of maternal origin — III. The effect of host shortage of *Nasonia vitripennis*. *J. Insect Physiol. 12*, 899–908.

SAUNDERS, D. S. (1967). Time measurement in insect photoperiodism: reversal of a photoperiodic effect by chilling. *Science 156*, 1126–1127.

SAUNDERS, D. S. (1968). Photoperiodism and time measurement in the parasitic wasp, *Nasonia vitripennis*. *J. Insect Physiol. 14*, 433–450.

SAUNDERS, D. S. (1970). Circadian clock in insect photoperiodism. *Science 168*, 601–603.

SAUNDERS, D. S. (1971). The temperature-compensated photoperiodic clock "programming" development and pupal diapause in the flesh-fly, *Sarcophaga argyrostoma*. *J. Insect Physiol. 17*, 801–812.

SAUNDERS, D. S. (1973a). Thermoperiodic control of diapause in an insect: theory of internal coincidence. *Science 181*, 358–360.

SAUNDERS, D. S. (1973b). The photoperiodic clock in the flesh-fly, *Sarcophaga argyrostoma*. *J. Insect Physiol. 19*, 1941–1954.

SAUNDERS, D. S. (1974). Evidence for "dawn" and "dusk" oscillators in the *Nasonia* photoperiodic clock. *J. Insect Physiol. 20*, 77–88.

SAUNDERS, D. S. (1975a). Spectral sensitivity and intensity thresholds in *Nasonia* photoperiodic clock. *Nature 253*, 732–734.

SAUNDERS, D. S. (1975b). "Skeleton" photoperiods and the control of diapause and development in the flesh-fly, *Sarcophaga argyrostoma*. *J. Comp. Physiol. 97*, 97–112.

SAUNDERS, D. S. (1976a). *Insect Clocks*. Pergamon Press, Oxford and New York.

SAUNDERS, D. S. (1976b). The circadian eclosion rhythm in *Sarcophaga argyrostoma*: some comparisons with the "photoperiodic clock". *J. Comp. Physiol. 110*, 111–133.

SAUNDERS, D. S. (1978a). An experimental and theoretical analysis of photoperiodic induction in the flesh-fly, *Sarcophaga argyrostoma*. *J. Comp. Physiol. 124*, 75–95.

SAUNDERS, D. S. (1978b). Internal and external coincidence and the apparent diversity of photoperiodic clocks in the insects. *J. Comp. Physiol. 127*, 197–207.

SAUNDERS, D. S. (1979). External coincidence and the photoinducible phase in the *Sarcophaga* photoperiodic clock. *J. Comp. Physiol. 132*, 179–189.

SAUNDERS, D. S. (1981a). Insect photoperiodism. In *Biological Rhythms*. Edited by J. Aschoff. *Handbook of Behavioral Neurobiology*. Vol. 4, pages 411–447. Plenum Press, New York.

SAUNDERS, D. S. (1981b). Insect photoperiodism: entrainment within the circadian system as a basis for time measurement. In *Biological Clocks in Seasonal Reproductive Cycles*. Edited by B. K. Follett and D. D. Follett. Pages 67–81. Wright, Bristol.

SAUNDERS, D. S. (1981c). Insect photoperiodism — the clock and the counter: a review. *Physiol. Ent. 6*, 99–116.

SCHLIEP, W. (1910). Der Farbenwechsel von *Dixippus morosus* (Phasmidae). *Zool Jb. (Physiol.) 30*, 45–132.

SCHLIEP, W. (1915). Über die Frage nach der Beteiligung des Nervensystems beim Farbenwechsel von *Dixipuss*. *Zool. Jb. (Physiol.) 35*, 225–232.

SEUGE, J. and VEITH, K. (1976). Diapause de *Pieris brassicae*: role des photorecepteurs cephaliques, etude des cartonenoids cerebraux. *J. Insect Physiol. 22*, 1229–1235.

SHIPP, E. and GUNNING, R. V. (1975). Endogenous rhythm of nerve activity in the housefly eye. *Nature 258*, 520–521.

SHIPP, E. and OTTON, J. (1976). Circadian rhythms of sensitivity to insecticides in *Musca domestica* (Diptera: Muscidae). *Ent. Exp. App. 19*, 163–171.

SHOREY, H. H. and GASTON, L. K. (1965). Sex pheromones of Noctuid moths. V. Circadian rhythm of pheromone-responsiveness in males of *Autographa californica, Heliothis virescens, Spodoptera exigua*, and *Trichoplusia ni* (Lepidoptera: Noctuidae). *Ann. Ent. Soc. Amer. 58*, 597–600.

SKOPIK, S. D. and PITTENDRIGH, C. S. (1967). Circadian systems, II. The oscillation in the individual *Drosophila* pupae; its independence of developmental stage. *Proc. Nat. Acad. Sci. 58*, 1862–1869.

SKOPIK, S. D. and TAKEDA, M. (1980). Circadian control of oviposition activity in *Ostrinia nubilalis*. *Amer. J. Physiol. 239*, R259–R264.

SOKOLOVE, P. G. (1975a). Locomotory and stridulatory circadian rhythms in the cricket, *Teleogryllus commodus*. *J. Insect Physiol. 21*, 537–558.

SOKOLOVE, P. G. (1975b). Localization of the cockroach optic lobe circadian pacemaker with microlesions. *Brain Res. 87*, 13–21.

SOKOLOVE, P. G. and LOHER, W. (1975). Role of eyes, optic lobes, and pars intercerebralis in locomotory and stridulatory circadian rhythms of *Teleogryllus commodus*. *J. Insect Physiol. 21*, 785–799.

SOWER, L. L., SHOREY, H. H. and GASTON, L. K. (1970). Sex pheromones of noctuid moths, XXI. Light-dark cycle regulations and light inhibition of the sex pheromone release by females of *Trichoplusia ni*. *Ann. Ent. Soc. Amer. 63*, 1090–1092.

SPANGLER, H. G. (1972). Daily activity rhythms of individual worker and drone honey bees. *Ann. Ent. Soc. Amer. 65*, 1073–1076.

SPANGLER, H. G. (1973). Role of light in altering the circadian oscillations of the honey bee. *Ann. Ent. Soc. Amer. 66*, 449–451.

STEEL, C. G. and LEES, A. D. (1977). The role of neurosecretion in the photoperiodic control of polymorphism in the aphid *Megoura viciae*. *J. Exp. Biol. 67*, 117–135.

SULLIVAN, W. N., CAWLEY, B., HAYES, D. K., ROSENTHAL, J. and HALBERG, F. (1970). Circadian rhythm in susceptibility of house flies and Madeira cockroaches to pyrethrum. *J. Econ. Ent. 63*, 159–163.

TAKAHASHI, S. and HARWOOD, R. F. (1964). Glycogen levels of adult *Culex tarsalis* in response to photoperiod. *Ann. Ent. Soc. Amer. 57*, 621–623.

TAKEDA, M. (1978). Photoperiodic time measurement and seasonal adaptation of the southwestern corn borer, *Diatraea grandiosella* Dyar (Lepidoptera: Pyralidae). Ph.D. dissertation, University of Missouri–Columbia.

TAKEDA, M. and MASAKI, S. (1976). Photoperiodic control of larval development in *Plodia interpunctella*. *Proceedings of the Joint United States/Japanese Seminar on Stored Product Insects, Manhattan, Kansas, 1976*. Pages 186–201. Kansas State University.

TAYLOR, B. and JONES, M. D. R. (1969). The circadian rhythm of flight activity in the mosquito *Aëdes aegypti* (L.): the phase-setting effects of light-on and light-off. *J. Exp. Biol. 51*, 59–70.

THIELE, H.-U. (1977). Measurement of day length as a basis for photoperiodism and annual periodicity in the carabid beetle *Pterostichus nigrita* F. *Oecologia 30*, 331–348.

THORSON, B. J. and RIEMANN, J. G. (1977). Abdominally entrained periodicities of testis and vas deferens activity in the mediterranean flour moth. *J. Insect Physiol. 23*, 1189–1197.

TRAYNIER, R. M. M. (1970). Sexual behaviour of the mediterranean flour moth, *Anagasta kuhniella*: some influences of age, photoperiod, and light intensity. *Canad. Ent. 102*, 534–540.

TRUMAN, J. W. (1971a). Circadian rhythms and physiology with special reference to neuroendocrine processes in insects. In *Proceedings of the International Symposium of Circadian Rhythmicity*. Pages 111–135. Pudoc Press, Wageningen, Netherlands.

TRUMAN, J. W. (1971b). Hour glass behavior of the circadian clock controlling eclosion of the silkmoth *Antherea pernyi*. *Proc. Nat. Acad. Sci. 68*, 595–599.

TRUMAN, J. W. (1971c). Physiology of insect ecdysis. I. The eclosion behavior of saturniid moths and its hormonal control. *J. Exp. Biol. 54*, 805–814.

TRUMAN, J. W. (1971d). The role of the brain in the ecdysis rhythm of silkmoths: comparison with the photoperiodic termination of diapause. In *Biochronometry*. Edited by M. Menaker. Pages 483–504. National Academy of Sciences, Washington, DC.

TRUMAN, J. W. (1972a). Physiology of insect rhythms. I. Circadian organization of the endocrine events underlying the moulting cycle of larval tobacco hornworms. *J. Exp. Biol. 57*, 805–820.

TRUMAN, J. W. (1972b). Physiology of insect rhythms. II. The silk moth brain as the location of the biological clock controlling eclosion. *J. Comp. Physiol. 81*, 99–114.

TRUMAN, J. W. (1973a). Physiology of insect ecdysis. II. The assay and occurrence of the eclosion hormone in the chinese oak silkmoth, *Antheraea pernyi*. *Biol. Bull. 114*, 200–211.

TRUMAN, J. W. (1973b). Temperature sensitive programming of the silkmoth flight clock: a mechanism for adapting to the seasons. *Science 182*, 727–729.

TRUMAN, J. W. (1974a). Circadian release of a prepatterned neural program in silkmoths. In *The Neurosciences: Third Study Program*. Edited by F. O. Schmitt and F. G. Worden. Pages 525–529. MIT Press, Cambridge, Mass.

TRUMAN, J. W. (1974b). Physiology of insect rhythms. IV. Role of the brain in the regulation of the flight rhythm of the giant silkmoths. *J. Comp. Physiol. 95*, 281–296.

TRUMAN, J. W. (1976). Extraretinal photoreception in insects. *Photochem. Photobiol. 23*, 215–225.

TRUMAN, J. W. and RIDDIFORD, L. M. (1970). Neuroendocrine control of ecdysis in silkmoths. *Science 167*, 1624–1626.

TRUMAN, J. W. and SOKOLOVE, P. G. (1972). Silkmoth eclosion: Hormonal triggering of a centrally programmed pattern of behavior. *Science 175*, 1491–1493.

TYCHSEN, P. H. and FLETCHER, B. S. (1971). Studies on the rhythm of mating in the Queensland fruit fly, *Dacus tryoni*. *J. Insect Physiol. 17*, 2139–2156.

TYSHCHENKO, V. P. (1966). Two-oscillatory model of the physiological mechanism of insect photoperiodic reaction. *Zhur. obshch. Biol. 27*, 209–222.

WADA, S. and SCHNEIDER, G. (1968). Circadianer rhythmus der pupillenweite in ommatidium von *Tenebrio molitor*. *Z. Vergl. Physiol. 58*, 395–397.

WAHL, O. (1932). Neue Untersuchungen über das Zeitgedächtnis der Bienen. *Z. Vergl. Physiol. 16*, 529–589.

WALTHER, J. B. (1958). Changes induced in spectral sensitivity and form of retinal action potential of the cockroach eye by selective adaptation. *J. Insect Physiol. 2*, 142–151.

WARE, G. W. and McCOMB, M. (1970). Circadian susceptibility of pink bollworm moths to azinphosmethyl. *J. Econ. Ent. 63*, 1941–1942.

WEITZEL, G. and RENSING, L. (1981). Evidence for cellular circadian rhythms in isolated fluorescent dye-labelled salivary glands of wild type and arrhythmic mutant of *Drosophila melanogaster*. *J. Comp. Physiol. 143*, 229–235.

WIEDENMANN, G. (1977a). Two activity peaks in the circadian rhythm of the cockroach *Leucophaea maderae*. *J. Interdiscipl. Cycle Res. 8*, 378–383.

WIEDENMANN, G. (1977b). Weak and strong phase shifting in the activity rhythm of *Leucophaea maderae* (Blaberidae) after light pulses of high intensity. *Z. Naturf. 32c*, 464–465.

DE WILDE, J., DUINTJER, C. S. and MOOK, L. (1959). Physiology of diapause in the adult Colorado beetle (*Leptinotarsa decemlineata* Say) — I. The photoperiod as a controlling factor. *J. Insect Physiol. 3*, 75–85.

WILLIAMS, C. M. (1969). Photoperiodism and the endocrine aspects of insect diapause. *Symp. Soc. Exp. Biol. 23*, 285–300.

WILLIAMS, C. M. and ADKISSON, P. L. (1964). Physiology of insect diapause, XIV. An endocrine mechanism for the photoperiodic control of pupal diapause in the oak silkworm, *Antheraea pernyi*. *Biol. Bull. 127*, 511–525.

WINFREE, A. T. (1970). The temporal morphology of a biological clock. In *Lectures on Mathematics in the Life Sciences*. Edited by M. Gerstenhaber. Pages 111–150. American Mathematical Society, Providence, RI.

WINFREE, A. T. and GORDON, H. (1977). The photosensitivity of a mutant circadian clock. *J. Comp. Physiol. 122*, 87–109.

WOLF, E. (1927). Über das Heimkehrvermögen der Bienen. *Z. Vergl. Physiol. 6*, 221–254.

YOUTHED, G. J. and MORAN, V. C. (1969a). The solar-day activity rhythm of Myrmeleontid larvae. *J. Insect Physiol. 15*, 1103–1116.

YOUTHED, G. J. and MORAN, V. C. (1969b). The lunar-day activity rhythm of Myrmeleontid larvae. *J. Insect Physiol. 15*, 1259–1271.

ZELAZNY, B. and NEVILLE, A. C. (1972). Endocuticle layer formation controlled by non-circadian clocks in beetles. *J. Insect Physiol. 18*, 1967–1979.

ZIMMERMAN, W. F. and GOLDSMITH, T. H. (1971). Photosensitivity of the circadian rhythm and of visual receptors in carotenoid depleted *Drosophila*. *Science 171*, 1167–1168.

ZIMMERMAN, W. F. and IVES, D. (1971). Some photophysical aspects of circadian rhythmicity in *Drosophila*. In *Biochronometry*. Edited by M. Menaker. Pages 381–391. National Academy of Sciences, Washington, DC.

ZIMMERMAN, W. F., PITTENDRIGH, C. S. and PAVLIDIS, T. (1968). Temperature compensation of the circadian oscillation in *Drosophila pseudoobscura* and its entrainment by temperature cycles. *J. Insect Physiol. 14*, 669–684.

12 Multimodal Convergences

EBERHARD HORN

Universität Karlsruhe, Karlsruhe, FRG

1 INTRODUCTION

Modality can be defined in two ways. In human psychology a group of similar sensations which reach "conscious awareness" is called a modality. This definition implies some classical modalities such as vision, hearing, feeling, taste, or olfaction which correspond to specific types of sense organs. It includes, however, difficulties. Pain — one component of the feeling-modality — can be elicited by different stimuli such as electric shock, strong mechanical stimuli, chemicals, etc. (Hensel, H., 1966). On the other hand, vestibular and especially otolith sensations do not reach "conscious awareness" (Guedry, F., 1974), and would therefore be excluded from being a modality. Furthermore, we do not know anything about consciousness in non-human beings so that this definition of modality is useless for the comparative standpoint.

The other possibility of defining modality is given by the physical description of the effective stimulus. Light, for example, is a modality which induces vision; high-frequency mechanical waves are another one, which induce hearing, etc. In this respect gravity is also a modality, which is connected with the reflexive control of equilibrium. Another advantage of this view is that even internal chemical or mechanical stimuli which arise during nutrition or stimuli produced by movements of the animals (proprioceptive stimuli) can be regarded as a special modality. On the other hand, pain is in this respect caused by a multimodal input in which each modality alone is effective.

In defining what is multimodal convergence, we therefore have to consider:

(1) that there are many modalities acting simultaneously on the animal at any point in its life;
(2) that there are specialized organs activated by single modalities;
(3) that stimuli also activate specific sensory pathways and elicit behavioural responses; and
(4) that there are anatomical and physiological connections between different sensory pathways located in specialized neurones, which are therefore activated by different modalities.

The pairing of pathways originating from exteroceptors with those from proprioceptors is probably the most distributed type of connection, whereas the pairing of pathways originating from different exteroceptors is specialized for the control or elicitation of certain behavioural reactions.

The basic character of multimodal convergence, therefore, is the uptake of information by functionally different sense organs and the connection between their neuronal activities in certain parts of the sensory pathway, the so-called multimodal nuclei which contain multimodal neurones. The anatomical and physiological interaction can occur in any part of the sensory pathway, i.e. at the receptor level, in ascending or descending neurones, or even in motoneurones (Fig. 1). The probability of finding a multimodal neuron is higher for neural centres of higher than lower order. In *Apis mellifera*, for example, multimodal inputs play a greater role in the lobular neurones than in the medullar ones (Hertel, H., 1980).

There is also the possibility, however, that one receptor is specialized in responding to more than one modality without any participation of synaptic transmission. Such receptors are called bimodal if two modalities only are effective, and multimodal if many modalities cause activity changes for biologically relevant stimulus strengths.

2   MULTIMODAL CONVERGENCE AND BEHAVIOUR

Multimodal convergence becomes obvious in many types of behaviour, for example:

(1) the determination of the sign of orientation;
(2) compromise behaviour in competing fields of stimuli or the stabilization of responses in bimodal fields when the reference directions of both stimuli agree;
(3) those responses which are necessarily initiated by one modality but controlled by another one (trigger effect);
(4) those responses which are only successfully performed if two different modalities act simultaneously on the animal — each modality alone, however, is uneffective (coupling effect); and
(5) the proprioceptive control of movements elicited by external stimuli.

Two modalities, however, can also affect the behaviour at different times, but with strong correlations between the responses in both unimodal fields. This situation holds for intersensory transfer and transposition.

2.1   Sign of orientation

There are many external and internal stimuli which affect the sign of orientation. Pertunnen, V. (1963) pointed out: "One of these modifying factors causing reversals and changing in the intensity of the reaction is temperature, and another important factor is desiccation. Fine adjustments must exist between the nature of the light reaction, on the one hand, and the environmental temperature, humidity, light intensity, state of water balance, state of nutrition, and perhaps some other factors, on the other". This makes clear that temperature and humidity are important external stimuli which control whether an insect exhibits positive or negative phototaxis. The beetle *Pissodes strobi* and the horsefly *Haematopota insidiatrix* reverse the sign of phototaxis from positive to negative with increasing temperature (Barrass, R., 1960; Sullivan, C., 1959) while in the beetle *Sitona cylindricollis*, Hans, H. and Thornsteinson, A. (1961) observed the reverse. In *Eristalis tenax* reversal of the photopositive reaction can be produced either by an increase or decrease of temperature. Only between 10° and 30° does the hoverfly respond highly photopositively (Dolley, W. and Golden, L., 1947); the exact reversal temperature for the photonegative response depends on light intensity (Dolley, W. and White, J., 1951). In some animals the reversal temperature is influenced by previous temperature conditioning (Sullivan, C., 1959); other examples are given in Table 1. Similar influences of temperature and humidity also exist for the sign of geotaxis and anemotaxis (Jander, R., 1963a; Markl, H., 1974).

The influence of internal factors on the determination of the sign of phototaxis and geotaxis was

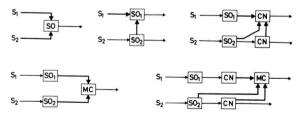

FIG. 1. Types of interactions between different sensory pathways. $S_{1,2}$, modality 1 and 2, resp.; $SO_{1,2}$, the corresponding sense organs; CN, central neurones including ascending and descending ones; MC motoneurones. *Upper row left*: bimodal receptor; *upper row middle*: convergence at the receptor level (not realized in insects). (Horn, E., original.)

Table 1: *The influence of external and internal stimuli on the sign of spontaneous phototactic orientation.*

Species	Sign of phototaxis	Effective or suggested stimulus	References
Hymenoptera			
Apis mellifera, Formica rufa, Bombus hortorum	positive when leaving the nest negative when homing	mechanical?: stretch of the crop?	Jacobs-Jessen, U. (1959); Jander, R. (1957); Rau, G. (1970)
Apis mellifera, Bombus hortorum, Formica rufa, Myrmica laevinodis	positive when swarming	chemical?	Wilson, E. (1971)
Apis mellifera	positive negative	high temperature low temperature	Müller, E. (1931)
Diptera			
Haematopota insidiatrix	positive: choice of sunny areas negative: choice of shaded areas	low temperatures high temperatures	Barrass, R. (1960)
Eristalis tenax	positive negative	medium temperature low, high temperature	Dolley, W. and Golden, L. (1947); Dolley, W. and White, J. (1951)
Calliphora erythrocephala	positive negative	high temperature low temperature	Jander, R. (1963a)
Aedes aegypti, Culex molestus, Anopheles maculipennis (larvae)	induction of negative positive or indifferent, when returning to the water surface	mechanical shock	Folger, H. (1948); Mellanby, K. (1958)
Coleoptera			
Trypodendron lineatum	positive if flight-inexperienced negative after flight experience	mechanical: stretch of the intestine	Graham, K. (1961)
Ips curvidens	negative during nutrition positive during flight	mechanical?, chemical?	Hierholzer, O. (1950)
Pissodes strobi	positive negative	low temperature high temperature	Sullivan, C. (1959)
Sitona cylindricollis	positive negative	high temperature low temperature	Hans, H. and Thornsteinson, A. (1961)
Tenebrio molitor (adults)	indifferent negative	high temperature medium, low temperature	Pertunnen, V. and Paloheimo, L. (1963)
Blastophagus piniperda	positive negative	medium temperature low, high temperature	Pertunnen, V. (1959)
Tenebrio molitor (adults)	strongly negative weakly negative	medium, low humidity high humidity	Pertunnen, V. and Lahermaa, M. (1963)

? indicates suggested stimuli.

proved by Pertunnen, V. (1963) for fruit-flies and beetles, and by Jacobs-Jessen, U. (1959) and Jander, R. (1957) in social Hymenoptera. Pertunnen investigated the effect of desiccation on the light reactions in *Drosophila melanogaster, Tenebrio molitor, Calandra granaria, Calandra oryzae, Rhizopertha dominica*, and *Acanthoscelides obsoletus*. From his experiments it was not clear how the internal conditions of the beetles and the fruit-fly act on the neuronal mechanisms responsible for orientation. In contrast, the photonegative and geopositive response of homing bees and ants, and the photopositive and geonegative responses of these Hymenoptera when leaving the nest (Jacobs-Jessen,

U., 1959; Jander, R., 1957) are probably caused by the excitation of internal mechanoreceptors measuring the stretch of the filled abdomen (Rau, G., 1970) (Fig. 2).

The change of the sign of orientation can be caused by convergence of neuronal activity originating in the eyes or the gravity receptors on the one hand, and in antennal temperature and humidity receptors or even internal mechanoreceptors of the abdomen on the other. Temperature and humidity may also have a direct influence on the stimulus-conducting systems of sense organs, thus changing their sensitivity by swelling or other deformations or by changing the degree of optic

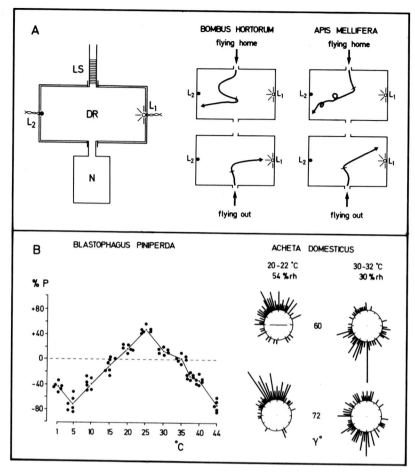

FIG. 2. The influence of internal and external factors on the sign of phototaxis and geotaxis. **A**: *Bombus hortorum, Apis mellifera*. The phototactic response of bees leaving or returning to their nest. *Left side*: experimental set-up with $L_{1,2}$ light sources (L_1 is switched on); LS, light trapped entrance to the darkroom (DR); N, nest. *Right side*: walking trails of the bees after they had entered the dark room. The point at which light was switched on is marked by the bar. **B**: *Left*: the influence of temperature on the sign of phototaxis in the beetle *Blastophagus piniperda*. %P, percentage of positive ($+$) or negative ($-$) responses. *Right*: the influence of temperature and humidity (%rh) on the sign of the geotaxis in the cricket *Acheta domesticus* while walking on an inclined surface (inclination angle γ). (Modified and combined from Jacobs-Jessen, U., 1959 (**A**); Pertunnen, V., 1959 (**B**, left); Horn, E., 1983 (**B**, right).)

transmission. This mechanism resembles the principles of stimulus transduction in temperature or humidity receptors (Altner, H. *et al.*, 1978).

2.2 Compromise behaviour

Compromise responses are mostly elicited under artificial experimental conditions as, for example, in the competition between the phototaxis and geotaxis (Jander, R., 1963b). While under unimodal stimulation by light or gravity, insects mostly prefer the direction towards or away from the light source or the centre of gravity (basic directions); the direction of orientation in the bimodal situation deviates from these basic directions. This experimental procedure was often used to investigate the properties of the gravity receptor systems (Bässler, U., 1962, 1965; Horn, E., 1970, 1973; Jander, R. *et al.*, 1970). Recently, a compromise behaviour between geotaxis and optomotor response was found in the walking fly, *Calliphora erythrocephala* (Fig. 3 left). In honeybees, *Apis mellifera*, compromise behaviour occurs for the spontaneous orientation (Horn, E., 1973) as well as for learnt behaviour (Edrich, W., 1977). In the waggle dance, however, the compromise behaviour disappears if the angle

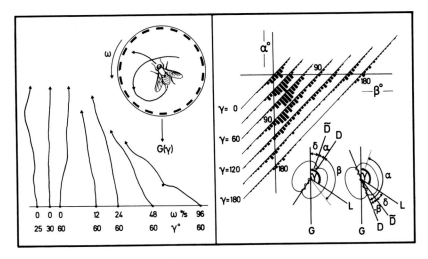

FIG. 3. Compromise behaviour in bimodal fields of stimulus. *Left*: Compromise between optomotor response and geotaxis in the fly, *Calliphora erythrocephala*. The fly walked in the centre of a striped rotating cylinder on a Kramer-sphere. This apparatus compensates each forward movement of the fly by a counter-rotation of the sphere so that the animal walks stationary. The graphs are walking trajectories of a fly. γ°, inclination of the sphere apparatus, i.e. angle between the horizontal planes of the space and the animal; ω, angular velocity of the pattern (pattern wavelength 20°). Upwards direction corresponds to the negative geotaxis; note that with increasing pattern velocity the fly's walking direction deviates from the upwards direction (Horn, E., Scharstein, H. and Wendler, G., unpublished). *Right*: interaction of light and gravity in the orientation of the waggle dance in the honeybee, *Apis mellifera*. α, β are the angular deviations from the waggle dance direction (D) with respect to the gravity (G) and light (L) direction, respectively. γ, angle between light and gravity direction with $\alpha + \beta = \gamma$; δ, direction of the waggle dance with respect to the light and gravity direction for the unimodal condition. (Modified from Edrich, W., 1977.)

between the reference directions of light and gravity exceeds 150°. Under these conditions the waggle dance direction is connected either with the light or with the gravity direction (Fig. 3 right).

If two modalities influence the same behavioural reaction, and if additionally their reference directions agree, a stabilization or improvement of this response can be observed. The underlying convergence became a useful tool in the control of equilibrium by visual and gravitational stimuli. Other insects (*Geotrupes sylvaticus* and some hawkmoths) improve their mate-finding using odour and wind simultaneously, although each of these stimuli alone can lead the animal to its partner by means of a tropotactic mechanism (Markl, H., 1974). Grasshoppers improve their mate-finding by using airborne sound and vibration simultaneously (Kalmring, K., 1983; Kalmring, K. and Kühne, R., 1980).

2.3 Trigger effect

Mate-, host- and food-finding is generally a multimedia approach. Mostly, wind and odour are used, but also gustatory stimuli, temperature, humidity, tactile and even optic stimuli are involved. In some animals the different modalities act alternatively: one modality, the trigger stimulus, activates the insect and at the same time switches on the control mechanism of orientation with respect to the other modality. After losing the triggering stimulus the animal performs searching movements until the stimulus is detected again; then the directional response continues. The combination of scent and wind is used by *Geotrupes sylvaticus*, *Drosophila melanogaster* and *Schistocerca gregaria* to find a food source (Kennedy, J. and Moorhouse, J., 1969; Steiner, G. 1953, 1954) while male moths (*Bombyx mori*, *Anagasta kühniella*) are activated by the scent of a female and then use anemotactic orientation while approaching the partner (Schwinck, I. 1954; Traynier, R., 1968).

2.4 Coupling effect

The coupling effect describes a mechanism of mate- or host-finding in which two fields of stimuli must be necessarily present simultaneously if the response is to be successful. One example is the host-finding of the myrmecophil beetle *Atemeles pubicollis*, which is attracted by the specific odour of

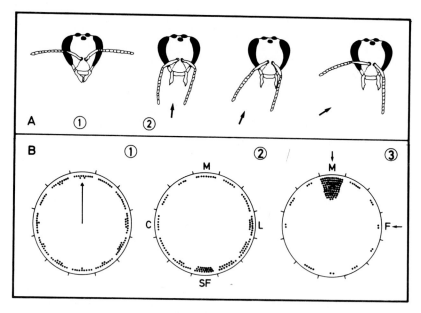

FIG. 4. Coupling effect. **A**: *Apis mellifera.* Movements of the antennae induced by odour and wind. (1) Constant odourless airflow (0.8 m s⁻¹). (2)–(4) Wind and odour from different directions, indicated by the arrows. **B**: *Atemeles pubicollis.* Frequency distributions of beetles in an arena under various conditions. (1) A trail scented by *Myrmica* was presented; no airflow. (2) At the periphery of the arena nests of four species of ants are presented; no airflow. C, *Camponotus ligniperda*; L, *Lasius niger*; M, *Myrmica laevinodis*; SF, *Serviformica fusca.* (3) At the periphery of the arena, a nest of *Myrmica laevinodis* (M) and *Formica polyctena* (F) was presented over which air flows at 0.5 m s⁻¹ (arrows). (Combined and modified from Suzuki, H., 1975 (**A**); Hölldobler, B., 1970 (**B**).)

its host *Myrmica* only in the presence of wind. The chemically marked trail of the ants is ineffective. Additionally, neither wind nor scent alone causes the directional response towards a *Myrmica* nest (Hölldobler, B., 1970) (Fig. 4B). As with the beetles, the orientational movements of the antennae of honeybees (*Apis mellifera*) are only elicited if wind is paired with odour (Suzuki, H., 1975) (Fig. 4A).

2.5 Proprioceptive control of behaviour

Directional responses of parts of the body or the whole animal elicited by visual, olfactory, gravitational or other external stimuli influence the activity of proprioceptors. The afferent input to the CNS can converge with the input originating in the exteroceptors. By this convergence the extent of the orientational response can be controlled according to intensity and direction of the external stimulus. Three types of responses belong to this group:

(1) *Goal-directed movements*, such as the rapid movement of the head before the stroke of the forelegs in the preying mantis (Fig. 5A), or the directional response of the honeybee's antennae during wind/scent stimulation (Mittelstaedt, H., 1957, 1962; Suzuki, H., 1975);

(2) *Pursuit movements* following the initial goal-directed saccadic jump of the head and/or the body of hunting insects such as robber-flies or dragonflies, or which are performed by the antennae towards a visual stimulus (Fig. 5B) (Honegger, H.-W., 1981; Kirmse, W. and Lässig, P., 1972; Mittelstaedt, H., 1957, 1962);

(3) *Compensatory movements* of parts of the body, such as the compensatory head movements of flies *Calliphora erythrocephala* in the field of gravity (Horn, E. and Lang, H.-G., 1978) or following haltere stimulation in the fly *Lucilia sericata* (Sandeman, D., 1980), or such as the light-induced head movements of *Calliphora* (Horn, E., unpublished).

In the preying mantis, Mittelstaedt, H. (1957, 1962) investigated this interaction between visual and proprioceptive input during the goal-directed head movement which is elicited when a target comes into sight. He proved that both inputs are connected at

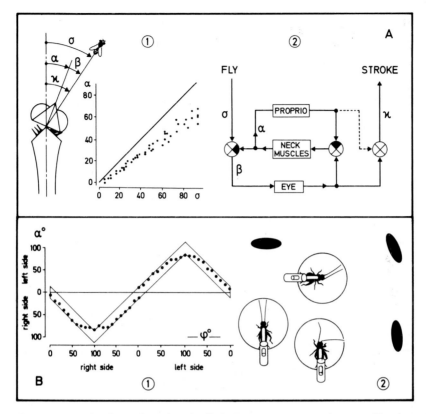

Fig. 5. Interaction between external and proprioceptive stimuli. **A**: *Parastagmatoptera unipunctata, Tenodera aridifolia sinensis*. Prey capture. (1) Head angles α produced by small-amplitude tracking saccades as a function of the stimulus direction σ. Note the error β in the direction of head fixation. The right side presents the function diagram of the mechanism underlying localization and stroke in the prey-catching behaviour of the mantids. κ stroke direction. **B**: *Gryllus campestris*. Antennal tracking of moving targets (black disc, diameter 26°). α, angular position of the antennae; φ, angular position of the disk with φ = 0° = target in front of the cricket. (Combined and modified from Lea, J. and Mueller, C., 1977 (**A**); Mittelstaedt, H., 1957 (**A**); Honegger, H.-W., 1981 (**B**).)

two points in the CNS whose locations are still unknown, but whose importance is clear (Fig. 5 A2). The first connection causes a rivalry between the visual induced command to move the head exactly towards the prey, and the counteracting proprioceptive system of the neck which tends to bring the head into its normal position. This rivalry is seen in the error of the fixation movement of the head which was observed by Mittelstaedt, and later confirmed by Lea, J. and Mueller, C. (1977) (Fig. 5 A1). The second interaction improves the stroke precision because the visual, as well as the proprioceptive, system can transmit information about the target direction. Both items of information superpose and, despite the large error of the head movement, render possible a stroke success of 90–95% (Mittelstaedt, H., 1957).

2.6 Intersensory transfer, transposition

Intersensory transfer is defined as the ability of animals to use information received via a certain sensory pathway for pattern recognition, communication or orientation with respect to another sensory system and modality. In the case of orientation, it is called transposition.

Transposition is common in insects. It was discovered in the communication system of the honeybee, *Apis mellifera*, the waggle dance by K. von Frisch (1946; cited in von Frisch, K., 1965), and later described for the beetle *Geotrupes silvaticus* by Birukow, G. (1954), for ants by Vowles, D. (1954), for Trichoptera by Jander, R. (1960), and for some other insects by Linsenmair-Ziegler, C. (1970). Intersensory transfer in relation to pattern

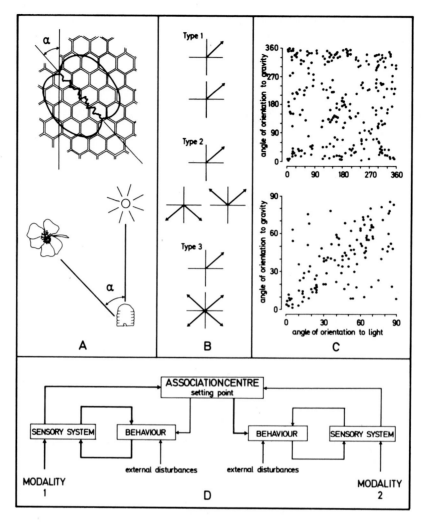

FIG. 6. Transposition. **A**: *Apis mellifera*. The transposition of the flight direction α with respect to the sun to the waggle dance direction α with respect to gravity. **B**: Types of transposition. Type 1: unequivocal type; each direction corresponds to only one direction in the second field of stimulus. Type 2: equivocal type; each direction corresponds to two directions in the second field of stimulus. Type 3: biequivocal type; each direction corresponds to four directions in the second field of stimulus. **C**: *Myrmica laevinodis*. Correlation between the angular orientation in the light and gravity field. In the lower graph the angles of orientation are measured as the smallest angle between the track and the line of action of the stimulus, irrespective of the actual direction. **D**: Functional diagram explaining the intersensory transfer. A behavioural response is controlled by a sensory system sensitive to modality 1, according to the setting point. This setting point, however, can also be used as a reference of a similar behavioural response controlled by another sensory system, sensitive to modality 2. (Modified and combined from von Frisch, K. 1965 (**A**); Horn, E., original (**B, D**); Vowles, D., 1954 (**C**).)

recognition and identification, however, is unknown in insects; but it would be worthwhile to know if insects possess similar capacities as are known from vertebrates. Perhaps the visual estimation of distances by using retinal pattern displacements and the subsequent jump in locusts (*Schistocerca gregaria*) (Wallace, G., 1959), which cannot be visually guided, may be based on an intersensory transfer from the visual to the proprioceptive neural network.

The transposition from phototaxis to geotaxis, and vice versa, is the best-investigated one. It is characterized by a high correlation between the direction of runs performed successively in the light and gravity field (Fig. 6 A,C). Three types of transposition have been described: the unequivocal, the equivocal and the biequivocal (Fig. 6 B). The relation between the interacting (converging) pathways is shown in Fig. 6 D. This scheme is based first on

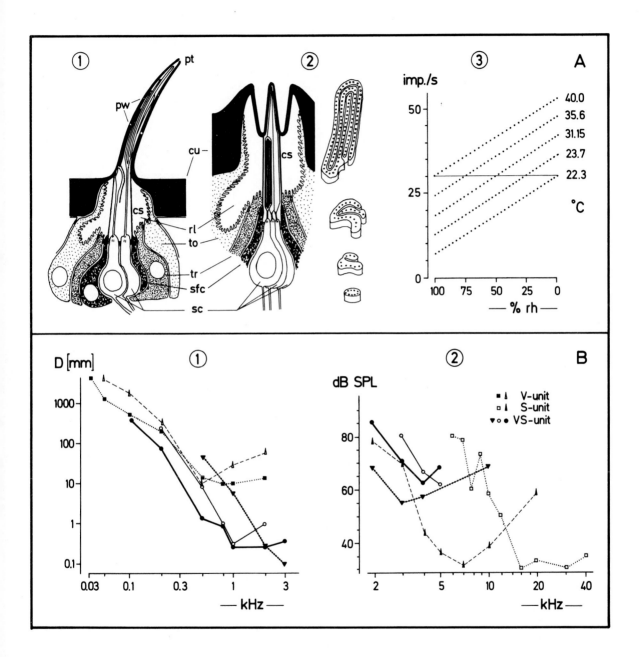

FIG. 7. Multimodal sense organs and bimodal receptors. **A**: Structural and physiological characteristics of multimodal hair sensillae. (1) Structural components of hair sensillae, containing chemoreceptors and one mechanoreceptor. (2) Structural components of a triad-type sensillum containing thermosensitive and hygrosensitive receptor cells. Note the lamellated structure of the outer segment of the receptor cell extending only to the base of the sensillum. (3) *Periplaneta americana*. The relation between the activity of the receptor cell and the relative humidity (% rh) at different temperatures of a bimodal receptor cell on the antenna of the cockroach. **B**: *Decticus verrucivorus*. Threshold curves of unimodal and bimodal receptor units of the complex tibial organ. S-units are sensitive only to airborne sounds while V-units respond only to vibration. VS-units are bimodal. (1) Vibration; (2) airborne sound. — cs, cuticular sheath; pt, terminal pore; pw, wall pore; rl, receptor lymph space; sc, sensory cells; sfc, sheath-forming cell; to, tormogene cell; tr, trichogene cell. (Combined and modified from Altner, H., 1977 (**A**); Loftus, R., 1976 (**A**); Kalmring, K. *et al.*, 1978 (**B**).)

Table 2: *Structural characteristics and modality specifity in cuticular sensilla of insects.*

Structural characteristics	Modality		Sensillum-type
No-pore sensillum			
socket flexible			
one cell, dendrite with tubular body	hair:	mechanical	S. chaeticum
	cap:	mechanical	S. campaniformicum
socket inflexible			
3–4 cells, one cell with infoldings	hair:	thermosensitive + hygrosensitive	S. coeloconicum
Terminal-pore sensillum			
socket flexible			
10 cells, one cell with tubular body	hair:	gustatory + mechanical	S. chaeticum, trichodeum, basiconicum, styloconicum
socket inflexible			
9 cells, no tubular body	hair:	gustatory	
	cap:	gustatory	
Wall pore sensillum			
single-walled			
pore tubules, 1–40 cells, dendrites branched and unbranched	hair:	olfactory (+ mechanical in Collembola)	S. basiconicum, trichodeum, coeloconicum
	plate:	olfactory	S. placodeum
double-walled			
2–4 cells, unbranched dendrites	hair:	olfactory (+ thermosensitive in *Periplaneta*)	S. basiconicum, coeloconicum

Data from Altner, H., 1977; Altner, H. *et al.*, 1983.

the fact that transfer is bidirectional, and second on the assumptions from Vowles, D. (1954) who distinguished four steps in the transfer process:

(1) the animal starts running;
(2) the constitution of the sensory field determines the initial "setting" of the orientation centre;
(3) the animal continues to orient to the first stimulus in a way determined by the "setting"; and
(4) the insect orients to the second stimulus in a way similarly determined by the initial "setting" of the orientation centre.

The terms "setting" or "setting-point" correspond to the term "central turning command" used later by Jander, R. (1957).

3 MULTIMODAL RECEPTORS AND SENSE ORGANS

Convergence at the receptor level is defined by a direct synaptic contact between receptor cells which are sensitive for different modalities. This type of convergence, however, is unknown in insects. The only animal in which this type of convergence is realized is a small mollusc (*Hermissenda crassicornis*) in which the activity of a photoreceptor cell influences monosynaptically the activity of hair cells in the statocysts, and vice versa (Alkon, D., 1973).

In insects, multimodal sense organs, however, are widely distributed. Many cuticular sensilla contain receptor cells which are sensitive to different modalities and which are characterized by modality-specific structures (Altner, H., 1977; Altner, H. *et al.*, 1983). Mechanosensitivity, chemosensitivity and sensitivities for temperature and humidity are common for these sensilla. Mechanoreceptors are often combined with gustatory ones, thermoreceptors with humidity receptors; sometimes one sensillum contains receptor cells for olfaction, temperature and humidity; in other sensilla there is a triad of one cold receptor, one dry and one moist receptor (Table 2).

Modality-specific structures are the tubular body of mechanoreceptors and the pores of the cuticular structures of chemosensitive sensilla. One pore mostly indicates a taste sensillum, many pores olfactory ones. Thermo- and hygrosensitivity are normally neither coupled with a special type of sensillum nor with a special anatomical structure of

Table 3: Bimodal and multimodal neurones in the insect's CNS

	Species	Visual	Acoustic	Vibratory	Wind	Touch	Temperature	Olfaction	Taste	Degree of modality	References
Optic lobes	locust	×	×	×		×				multi (M)	Horridge, G. (1964)
	Boettcherisca peregrina	×			×					bi	Mimura, K. (1974)
	Schistocerca vaga	×	×							bi (LGMD)	O'Shea, M. (1975)
	Apis mellifera	×			×			×	×	multi (M,L)	Hertel, H. (1980)
	Apis mellifera	×			×					bi	Kaiser, W. *et al.*, (1981)
Protocerebrum	locust	×	×	×		×				multi	Horridge, G. (1964)
	Periplaneta americana	×				×				bi	Dingle, H. and Caldwell, R. (1967)
	Boettcherisca peregrina	××			×			×		bi	Mimura, K. *et al.* (1969, 1970)
	Apis mellifera	××			×			×		bi	Suzuki, H. *et al.* (1976)
	Apis mellifera	×			×	×		×	×	multi	Erber, J. (1978, 1981)
	Acheta domesticus	×	×		×	×		×		multi	Schildberger, K. (1981)
Deutocerebrum	*Boettcherisca peregrina*	××			×			×		bi	Mimura, K. *et al.* (1969, 1970)
	Periplaneta americana					×	×	×		multi	Waldow, U. (1975); Boeckh, J. *et al.* (1976)
Tritocerebrum	*Boettcherisca peregrina*	××			×			×		bi	Mimura, K. *et al.* (1969, 1970)
	Schistocerca gregaria	×			×					bi	Bacon, J. and Tyrer, M. (1978)
Suboesophageal ganglion	*Boettcherisca peregrina*	××			×			×		bi	Mimura, K. *et al.* (1969, 1970)
Ventral nerve cord	*Schistocerca gregaria*	×	×		×	×				bi, tri	Catton, W. and Chakraborty, A. (1969)
	Locusta migratoria	×	×		×	×				bi, tri	Catton, W. and Chakraborty, A. (1969)
	Periplaneta americana	××			×	×				bi	Cooter, R. (1973)
	Schistocerca vaga	×	×							bi (DCMD)	O'Shea, M. (1975)
	Tettigonia cantans		×	×						bi	Kalmring, K. and Kühne, R. (1980); Kalmring, K. (1983)
	Gryllus bimaculatus		×	×						bi*	Wiese, K. (1981)

Neurones in the CNS area cited respond to combinations of the modalities marked by ×. ×× indicates that this stimulus is paired with a second one. M, medullar neurones; L, lobular neurones. *, the bimodal omega-neurone of the cricket *Gryllus bimaculatus* also receives efferent input.

the sensillum or the receptor cell. They are common for sensilla with and without pores. In some receptor cells the outer segment of the cilium is lamellated (Altner, H. *et al.*, 1978). This lamellation is especially seen in sensilla of the triad type; but it still remains open whether this lamellation is a modality-specific structure (Fig. 7 A).

Bimodal receptor cells were identified by electrophysiological methods. Loftus, R. (1976) described sensory cells in the antennae of the cockroach *Periplaneta americana* which were activated by temperature and humidity. These receptor cells increase their activity with increasing temperature and decreasing humidity. At first glance this relation seems to be trivial and not related to a special bimodal property, because of the physical relation between temperature and relative humidity: the warmer the air the drier the air, supposing the same absolute water content. But the activity changes elicited by increasing temperatures at different levels of relative humidity do not coincide with that expected from physical calculations. There is a residual dependence on temperature which cannot be explained by a dependence on the relative

664   Eberhard Horn

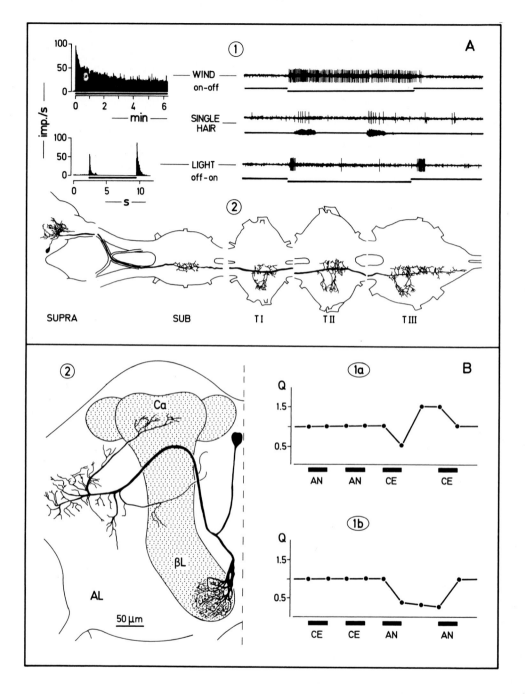

FIG. 8. Anatomy and physiology of multimodal neurones. **A**: A bimodal neurone of the desert locust, *Schistocerca gregaria*. (1) The neurone is phasic-tonic if stimulated by wind on the head but phasic with an OFF–ON activity if stimulated with light. The response of a single hair of the head stimulated by displacing it with a pipette is presented in the middle. (2) The anatomy of this neurone. SUPRA, supra-oesophageal ganglion; SUB, suboesophageal ganglion; TI–TIII, thoracic ganglia. **B**: *Acheta domesticus*. A multimodal neurone of the mushroom-body. (1) Reaction spectra of this neurone if the ipsilateral antenna (AN) and cercus (CE) are stimulated mechanically. Q, quotient of the actual action potential frequency and mean spontaneous frequency. The frequencies at the onset and offset of the stimulus are shown. Stimulus duration is 1 s. (2) The anatomy of the neurone with AL, antennal lobe; βL, β-lobe; Ca, calyx. (Combined and modified from Bacon, J. and Tyrer, M., 1978 (A); Schildberger, K., 1981 (B).)

humidity. These receptors, therefore, are bimodal (Fig. 7 A3).

Other bimodal primary units are known from the grasshopper (*Decticus verrucivorus*). Kalmring, K. *et al.* (1978) found 34 bimodal units when they recorded from primary fibres in the ventral cord. The best frequencies for the vibratory stimuli are near 1000 Hz (1000–3000 Hz), and near 4500 Hz (3000–5000 Hz) for the acoustic stimuli. One unit has the same optimum frequency, at 3000 Hz, for both stimuli. The functional meaning and the identity of these units in the receptor complex near the joint between femur and tibia, however, remain still unclear (Fig. 7B).

4 MULTIMODAL NEURONES

Multimodal neurones exist in all parts of the insect's CNS. Most investigators focused their interest on the protocerebrum; but it has become clear that there are strong multimodal convergences on single neurones even at low neuronal levels. The antennal afferences containing olfactory, mechanical, temperature and humidity information converge in the deutocerebrum (Waldow, U., 1975; Boeckh, J. *et al.*, 1976) while vibratory and airborne sound information converges on neurones of the ventral cord (Kalmring, K., 1983; Kalmring, K. and Kühne, R., 1980; Wiese, K., 1981).

Older investigations only show that there are such convergences. Visual stimuli, such as different light intensities presented to the complex eyes or the ocelli, are the most-used standard stimuli which were paired with other modalities (Table 3), but recently it has become clear that the reaction of a multimodal neurone to each modality can differ, for example, in the time-course of excitation. One modality can induce a tonic, the other a phasic, response (Fig. 8A). Additionally, the relation of the multimodal character of neurones to behavioural responses becomes more and more important, so that precisely defined multimodal fields are applied to the animals, such as, for example, the combination of airborne sound and vibration or wind and scent (Table 3). Multimodal units of the protocerebrum become attractive in the analysis of learning. Erber, J. (1978) described multimodal neurones of the mushroom-body area which

showed increasing sensitivity for a scent after a number of olfactory conditionings. The increase of sensitivity in these neurones is correlated with an increase of the spontaneous activity.

On the other hand it has to be taken into consideration that multimodal neurones are involved in a non-specific system which influences the general state of activity in the CNS. This possibility holds especially for those neurones with widespread arborizations (Mimura, K. *et al.*, 1969, 1970; Horridge, G., 1964), or for neurones which are excited by many modalities. Such neurones can be found in lower neuronal areas but they are common in higher centres, as for example in the protocerebrum (Table 3).

A remarkable characteristic was described for protocerebral neurones of the cricket *Acheta domesticus*. These neurones are not activated by any stimulus (mechanical, acoustical, optical, olfactory) given alone; but if a cercal stimulus follows an antennal one, an inhibitory response is exhibited with long-lasting excitatory after-effects. The reverse situation causes an inhibition which will be terminated after a second antennal stimulus (Fig. 8 B1). Such neurones have arborizations in the β-lobes and in the calyx of the mushroom-body and, additionally, near the entrance of the optic lobes (Fig. 8 B2) (Schildberger, K., 1981). Neurones with similar properties depending on wind and scent may be involved in the orientational responses which are determined by the trigger and coupling effect (see sections 2.3 and 2.4 of this chapter).

The convergence of two modalities on one central neurone may cause a complete loss of unimodal information transmission by subsequent neurones of the sensory pathway. Horridge, G. (1964) supposed that "because [the neurones] are multimodal they cannot be much use for specific responses". This suggestion, however, is questionable after O'Shea succeeded in the proof of modality-specific initiation places of action potentials (O'Shea, M., 1975) in the LGMD neurone of the lobula in locusts (*Schistocerca gregaria*). This neurone is bimodal; light changes elicit a sustained activity while a sound causes only a strong phasic response at the onset and end of stimulation. The LGMD is connected with the DCMD neurone (descending contralateral movement detector giant) by means of an electric synapse. The DCMD is also bimodal; its

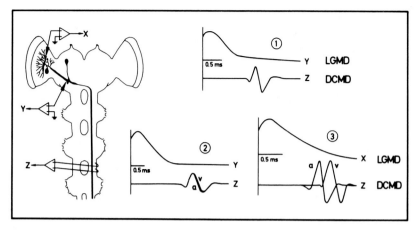

FIG. 9. *Schistocerca vaga*. Modality-specific sites of axonal spike initiation in a bimodal giant neuron of the lobula (LGMD). *Left side*: a diagrammatic representation of the anatomy of the LGMD and the synaptically coupled DCMD (descending contralateral movement detector). X, Y, Z, positions of the recording electrodes. *Right side*: different stimulations of the neurone and simultaneous recordings of electric activity with the electrodes X, Y, Z. (1) upper trace is an intracellular recording of the LGMD at Y and the lower trace an extracellular recording of the DCMD at Z. Action potentials in the LGMD were initiated by the injection of depolarizing current through the microelectrode. (2) As in (1), but with visual (v) and auditory (a) stimulation. (3) Upper trace is an intracellular recording of the LGMD at X and the lower trace an extracellular recording of the DCMD at Z. Note that there are two distinct classes of latency. The latency in response to the auditory stimulus is shorter by 0.38 ms than that to visual stimulation. In all recordings the oscilloscope was triggered by the LGMD spikes. (Modified from O'Shea, M., 1975.)

soma is located in the protocerebrum, its axon descends contralaterally to the third thoracic ganglion.

Electric stimulation of the LGMD near the electric synapse elicits action potentials in the DCMD after a short latency. Also light and acoustical stimuli activate the DCMD with no latency differences with respect to the LGMD response if the recording electrode for the LGMD activity is located near the presynaptic region. If, however, this electrode is positioned near the origin of the LGMD's dendritic arborizations the latency between the onset of the LGMD and DCMD activity is smaller for sound than for light stimulation. O'Shea explained this difference by different initiation places for action potentials. Supposing a conducting velocity of action potentials of 3 m s^{-1}, the initiation places for acoustically induced action potentials is nearer to the electric synapse, 576 μm away from that for light stimuli (Fig. 9).

This property has consequences for the signal transmission from the LGMD to the DCMD. The DCMD is especially activated by visual stimuli penetrating rapidly into its receptive field. On the other hand it is also activated by acoustic stimuli via the LGMD. It is, therefore, possible that certain visual inputs to the LGMD cause a postsynaptic inhibition without changing the acoustically induced synaptic transmission of excitation. This means that the DCMD behaves like a unimodal neurone despite a bimodal stimulus input to the animal.

Modality-specific information transmission is also maintained in the DCMD by the time-course of excitation which is phasic for acoustic stimuli but phasic–tonic for the visual ones. This characteristic holds also for a large interneurone of *Schistocerca gregaria*, whose cell body lies in the brain and whose axon descends to the suboesophageal and thoracic ganglia via the tritocerebral commissure (Bacon, J. and Tyrer, M., 1978). This neurone responds phasically to light stimuli but phasic–tonically to wind stimulation (Fig. 8 A).

Based on anatomical investigations in flies, *Musca domestica* and *Calliphora erythrocephala* Strausfeld, N. and Bacon, J. (1983) described the multimodal convergence on descending neurones (DN) which originate in the brain and terminate in the thoracic ganglia and which are arranged in heteromorphic (deuto-cerebral DNs) or isomorphic clusters (suboesophageal DNs). These clusters are places of multimodal inputs with each type of ending having a precise location in the cluster which reiterated from one individual to the next:

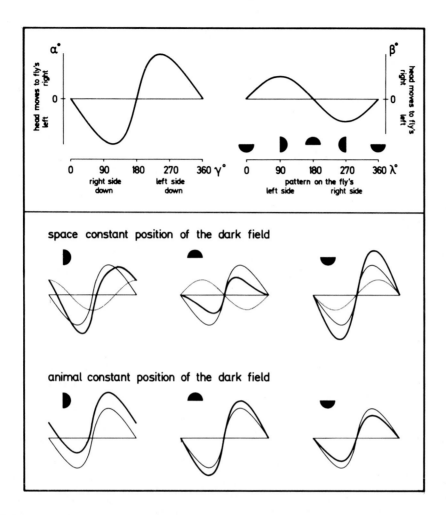

FIG. 10. *Calliphora erythrocephala*. Head movements of the flies, elicited under unimodal and bimodal stimulation by light and/or gravity. *Upper part*: the responses of the flies under unimodal stimulation. The gravitational stimulus is simulated by a ball (see chapter 10 of this volume). The flies are moved around the longitudinal axis (left graph). The visual stimulus is a dark area which covers 180° of the visual field of the flies. If this area is moved around the longitudinal axis of the flies, the head moves to the light side (right graph). *Lower part*: The responses of the flies under bimodal stimulation. Now the dark area is presented either in a space-constant position with its contrast edge either vertical or horizontal, or it is presented in a fly-constant position irrespective of the spatial position of the flies; i.e. the dark area and the flies are rotated simultaneously. For the space-constancy experiment the black half-sector indicates the position of the dark black area with respect to gravity, while in the animal-constancy experiment this inset indicates the position of the dark area with respect to the fly. Bold line is bimodal response, light line is unimodal response to gravity. (From Horn, E., 1983.)

(1) relay neurones from the olfactory lobes converge at the junction between the main posterior branch and the lateral trunk of the giant DNs;

(2) ascending terminals from the thoracic ganglia end at dendrites derived from the lateral trunk of the giant DNs;

(3) visual inputs terminate onto a second group of specialized giant DN dendrites;

(4) the contralateral male lobula giant neurone terminates on the initial segment of some axons in the cluster; and

(5) fibres from Johnston's organ invest the main posterior branch of the giant DNs.

These differences in the location of cellular contact may also contribute to the preservation of modality-specific information transmission in the insect's CNS.

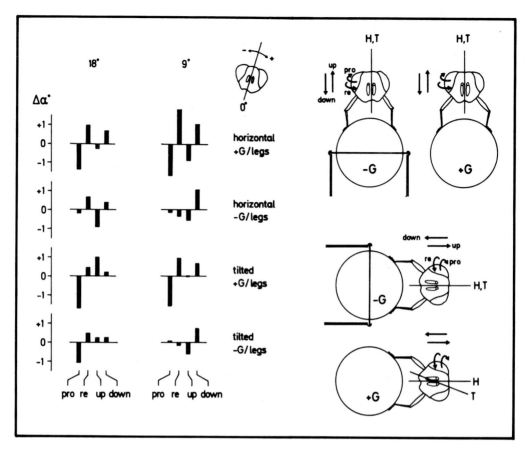

FIG. 11. *Calliphora erythrocephala*. Convergence between visual- and gravity-induced excitations in the control of head movements in the walking fly. Gravity (+ G) induces compensatory head movements via the proprioceptive gravity receptor system of the legs (see chapter 10 of this volume). If the gravity-dependent leg afferences are eliminated (− G; the walking ball is suspended by a spindle) no head movements occur. Moving visual patterns induce head movements whose amplitudes depend on the direction of pattern movement. Pattern movement from back to front (regressive = re) elicits stronger movements of the head than the pattern movement from front to back (progressive = pro). This asymmetry, however, disappears (pattern wavelength 9°) or is strongly reduced (pattern wavelength 18°) if no gravitational force influences the leg gravity receptors. 0° = mean of the head position for all four directions of pattern movement (re, pro, up, down). The mean absolute deviation of the head from its normal position is about 5–7° for the stimulus combinations horizontal/− G, horizontal/+ G, and tilted/− G, but 18° for tilted/+ G. Δα, angular deviation of the head from the 0° − position; Δα is positive (negative) if the head deviation from the normal position increases (decreases). (From Horn, E. 1983.)

5   PERSPECTIVES

The analysis of multimodal convergence is multi-disciplinary. It is complicated by the complexity of the CNS with its convergent and divergent network. One sense organ can influence many output systems, but one output system is mostly linked with many input systems. Therefore there must be an enormous number of parallel computing systems in the CNS to obtain unambiguous information transmission from sense organs to the muscular system.

After it has become clear that there are multimodal neurones in most parts of the CNS, the neurophysiological approach of multimodality must be restricted to the search for neurones involved in biologically relevant situations. It is easy to understand the meaning of neurones which respond to wind and scent, but what is the importance of a responsiveness to acoustical and visual stimuli, or wind and visual stimuli?

From the examples presented in the description of behaviour (see section 2 of this chapter) some attractive questions arise. What are the physiological characteristics of neurones involved in mate- or host-finding? Are there neurones which are only activated under a bimodal situation? The existence of obligatory bimodal units is out of question (compare coupling effect, section 2.4), but is there any possibility of finding such neurones? Schildberger's investigation (1981) is a first step in this direction (Fig. 8).

Other questions are, how unambiguous information is available from ambiguous information, or where the interaction is between external and proprioceptive stimuli during goal-directed behaviour or compensatory movements? Mittelstaedt's investigations (1957) point to the latter case, but they give no answer.

Another attractive problem is which are the converging parameters of different modalities? Are there receptors of a sense organ which project to multimodal neurones while others do not? Or, in other words, are there stimuli whose efficacy is invariant or non-invariant with respect to stimulus transformations, for example, spatial displacements? Edrich, W. (1979) showed that in the waggle dance of the honeybee, *Apis mellifera*, the occurrence of compromise behaviour between visual and gravitational stimulation (see Fig. 3) also depends on the wavelength of the light source. The action spectrum of this behaviour has peaks at 450 nm (blue) and 550 nm (green-yellow). There is no obvious contribution from the UV-receptors except when the light is polarized.

In the fly, *Calliphora erythrocephala*, Horn, E. (1983) described different types of interaction between gravitational and visual stimuli. Both modalities elicit head movements whose amplitude depends on many parameters such as direction, size, movements, colour, etc. (Horn, E., unpublished). If large dark objects are competing with gravitational input to the proprioceptive gravity receptor system of the legs (see chapter 10, of this volume), the responses in the bimodal field can be quantitatively calculated by addition of the responses in the unimodal fields, indicating that the convergence take place at the motor level (Fig. 10). In contrast, the gravitational input has only a modulatory effect on the stimulus efficiency of moving patterns, which cannot be expected from the properties of the visual interneurones investigated under unimodal stimula-

tion. In the fly, *Phaenicia sericata*, lobular neurones are depolarized by progressive (from front to back) and hyperpolarized by regressive (from back to front) movements of striped patterns (Eckert, H. and Bishop, L., 1978). A similar asymmetry is pronounced for the amplitude of head movements of *Calliphora erythrocephala*, if the flies are simultaneously stimulated by moving patterns and by gravity acting on the leg PGR-system. Failure of this mechanical input, however, decreases or even abolishes this asymmetry of stimulus efficiency between progressive and regressive pattern movement (Fig. 11) (Horn, E., 1983). The conclusion of these experiments, therefore, is that the stimulus efficacy of a dark hemisphere is independent of (= invariant with respect to) the flies' spatial position while the efficacy of moving patterns is non-invariant with respect to spatial transformations of the flies.

Edrich's and Horn's experiments therefore stressed that there must be different types of convergences between two modalities, even if the same behavioural response is influenced. Finally, we cannot exclude that, by the interaction between two modalities, a new modality arises which is outside our own experience.

REFERENCES

ALKON, D. L. (1973). Intersensory interactions in *Hermissenda*. *J. Gen. Physiol.* 62, 185–202.

ALTNER, H. (1977). Insektensensillen: Bau und Funktionsprinzipien. *Verh. Dtsch. Zool. Ges. 1977*, 139–153.

ALTNER, H., LOFTUS, R., SCHALLER-SELZER, L. and TICHY, H. (1983). Modality-specificity in insect sensilla and multimodal input from body appendages. In *Multimodal Convergences in Sensory Systems*. Edited by E. Horn. *Fortschr. Zool.* 28, pages 17–31. Gustav Fischer Verlag, Stuttgart and New York.

ALTNER, H., TICHY, H. and ALTNER, I. (1978). Lamellated outer dendritic segments of a sensory cell within a poreless thermo- and hygroreceptive sensillum of the insect *Carausius morosus*. *Cell Tiss. Res.* 191, 287–304.

BACON, J. and TYRER, M. (1978). The tritocerebral commissure giant (TCG): a bimodal interneurone in the locust, *Schistocerca gregaria*. *J. Comp. Physiol.* 126, 317–325.

BARRASS, R. (1960). The settling of female *Haematopota insidiatrix* Austen (Diptera, Tabanidae) on cloth screens. *Ent. Exp. Appl.* 3, 257–266.

BÄSSLER, U. (1962). Zum Einfluß von Schwerkraft und Licht auf die Ruhestellung der Stabheuschrecke *(Carausius morosus)*. *Z. Naturf.* 17b, 477–480.

BÄSSLER, U. (1965). Proprioceptoren am Subcoxal- und Femur-Tibia-Gelenk der Stabheuschrecke *Carausius morosus* und ihre Rolle bei der Wahrnehmung der Schwerkraftrichtung. *Kybernetik* 2, 168–193.

BIRUKOW, G. (1954). Photo-Geomenotaxis bei *Geotrupes silvaticus* Panz. und ihre zentral-nervöse Koordination. *Z. Vergl. Physiol.* 36, 176–211.

BOECKH, J., ERNST, K.-D., SASS, H. and WALDOW, U. (1976). Zur nervösen Organisation antennaler Eingänge bei Insekten unter besonderer Berücksichtigung der Riechbahn. *Verh. Dtsch. Zool. Ges. 1976*, 123–139.

CATTON, W. T. and CHAKRABORTY, A. (1969). Single neurone responses to visual and mechanical stimuli in the thoracic nerve cord of the locust. *J. Insect Physiol. 15*, 245–258.

COOTER, R. J. (1973). Visual and multimodal interneurones in the ventral nerve cord of the cockroach, *Periplaneta americana. J. Exp. Biol. 59*, 675–696.

DINGLE, H. and CALDWELL, R. L. (1967). Multimodal interneurones in cockroach cerebrum. *Nature 215*, 63–64.

DOLLEY, JR, W. D. and GOLDEN, L. H. (1947). The effect of sex and age on the temperature at which reversal in reaction to light in *Eristalis tenax* occurs. *Biol. Bull. 92*, 178–186.

DOLLEY, W. D. and WHITE, J. D. (1951). The effect of illuminance on the reversal temperature in the drone fly, *Eristalis tenax. Biol. Bull. 100*, 84–89.

ECKERT, H. and BISHOP, L. G. (1978). Anatomical and physiological properties of the vertical cells in the third optic ganglion of *Phaenicia sericata* (Diptera, Calliphoridae). *J. Comp. Physiol. 126*, 57–86.

EDRICH, W. (1977). Interaction of light and gravity in the orientation of the waggle dance of the honeybees. *Anim. Behav. 25*, 342–363.

EDRICH, W. (1979). Honeybees: photoreceptors participation in orientation behaviour to light and gravity. *J. Comp. Physiol. 133*, 111–116.

ERBER, J. (1978). Response characteristics and after effects of multimodal neurones in the mushroom body area of the honeybee. *Physiol. Ent. 3*, 77–89.

ERBER, J. (1981). Neural correlates of learning in the honeybee. *Trends Neurosci. 4*, 270–273.

FOLGER, H. T. (1948). The reactions of *Culex* larvae and pupae to gravity, light, and mechanical shock. *Physiol. Zool. 19*, 190–202.

FRISCH, K. v. (1965). *Tanzsprache und Orientierung der Bienen.* Springer, Berlin, Heidelberg and New York.

GRAHAM, K. (1961). Air-swallowing: a mechanism in photic reversal of the beetle *Trypodendron. Nature 191*, 519–520.

GUEDRY, F. E. (1974). Psychophysics of vestibular sensation. In *Vestibular System, Part 2: Psychophysics, Applied Aspects and General Interpretations.* Edited by H. H. Kornhuber. *Handbook of Sensory Physiology.* Vol. 6, pt. 2, pages 3–154. Springer, Berlin and New York.

HANS, H. and THORNSTEINSON, A. J. (1961). The influence of physical factors and host plant odour on the induction and termination of dispersal flights in *Sitona cylindricollis* Fahr. *Ent. Exp. Appl. 4*, 165–177.

HENSEL, H. (1966). *Allgemeine Sinnesphysiologie, Hautsinne, Geschmack, Geruch.* Springer, Berlin, Heidelberg and New York.

HERTEL, H. (1980). Chromatic properties of identified interneurones in the optic lobes of the bee. *J. Comp. Physiol. 137*, 215–231.

HIERHOLZER, O. (1950). Ein Beitrag zur Frage der Orientierung von *Ips curvidens* Germ. *Z. Tierpsychol. 7*, 588–620.

HÖLLDOBLER, B. (1970). Zur Physiologie der Gast-Wirt-Beziehungen (Myrmecophilie) bei Ameisen. II. Das Gastverhältnis des imaginalen *Atemeles pubicollis* Bris. (Col. Staphylinidae) zu *Myrmica* und *Formica* (Hym. Formicidae). *Z. Vergl. Physiol. 66*, 215–250.

HONEGGER, H.-W. (1981). A preliminary note on a new optomotor response in crickets: antennal tracking of moving targets. *J. Comp. Physiol. 142*, 419–421.

HORN, E. (1970). Die Schwerkraftreception bei der Geotaxis des laufenden Mehlkäfers *(Tenebrio molitor). Z. Vergl. Physiol. 66*, 343–354.

HORN, E. (1973). Die Verrechnung des Schwerereizes bei der Geotaxis der höheren Bienen (Apidae). *J. Comp. Physiol. 82*, 379–406.

HORN, E. (1983). Behavioural reactions of insects in bimodal fields of stimuli. In *Multimodal Convergences in Sensory Systems.* Edited by E. Horn. *Fortschr. Zool. 28*, pages 179–196. Gustav Fischer Verlag, Stuttgart and New York.

HORN, E. and LANG, H.-G. (1978). Positional head reflexes and the role of the prosternal organ in the walking fly, *Calliphora erythrocephala. J. Comp. Physiol. 126*, 137–146.

HORRIDGE, G. A. (1964). Multimodal interneurones of locust optic lobes. *Nature 204*, 499–500.

JACOBS-JESSEN, U. (1959). Zur Orientierung der Hummeln und einiger anderer Hymenopteren. *Z. Vergl. Physiol. 41*, 597–641.

JANDER, R. (1957). Die optische Richtungsorientierung der roten Waldameise *(Formica rufa* L.). *Z. Vergl. Physiol. 40*, 162–238.

JANDER, R. (1960). Menotaxis und Winkeltransponieren bei Köcherfliegen (Trichoptera). *Z. Vergl. Physiol. 43*, 680–686.

JANDER, R. (1963a). Insect orientation. *Ann. Rev. Ent. 8*, 95–114.

JANDER, R. (1963b). Grundleistungen der Licht- und Schwereorientierung von Insekten. *Z. Vergl. Physiol. 47*, 381–430.

JANDER, R., HORN, E. and HOFFMANN, M. (1970). Die Bedeutung von Gelenkrezeptoren in den Beinen für die Geotaxis der höheren Insekten (Pterygota). *Z. Vergl. Physiol. 66*, 326–342.

KAISER, W., DEVOE, R. and OHM, J. (1981). The influence of non-optic stimuli on the sensitivity of identified visual interneurones in the optic lobe of the honeybee. *Verh. Dtsch. Zool. Ges. 1981*, 173.

KALMRING, K. (1983). Convergence of auditory and vibratory senses at the neuronal level of the ventral cord in grasshoppers; its probable importance for behaviour in the habitat. In *Multimodal Convergences in Sensory Systems.* Edited by E. Horn. *Fortschr. Zool. 28*, pages 129–141. Gustav Fischer Verlag, Stuttgart and New York.

KALMRING, K. and KÜHNE, R. (1980). The coding of airborne-sound and vibration signals in bimodal ventral-cord neurones of the grasshopper *Tettigonia cantans. J. Comp. Physiol. 139*, 267–275.

KALMRING, K., LEWIS, B. and EICHENDORF, A. (1978). The physiological characteristics of the primary sensory neurones of the complex tibial organ of *Decticus verrucivorus* L. (Orthoptera, Tettigonioidae). *J. Comp. Physiol. 127*, 109–121.

KENNEDY, J. S. and MOORHOUSE, J. E. (1969). Laboratory observations on locust responses to wind-borne grass odour. *Ent. Exp. Appl. 12*, 487–503.

KIRMSE, W. and LÄSSIG, P. (1971). Strukturanalogie zwischen dem System der horizontalen Blickbewegungen der Augen beim Menschen und dem System der Blickbewegungen des Kopfes bei Insekten mit Fixierreaktionen. *Biol. Zbl. 90, 175–193.*

LEA, J. Y. and MUELLER, C. G. (1977). Saccadic head movements in mantids. *J. Comp. Physiol. 114*, 115–128.

LINSENMAIR-ZIEGLER, C. (1970). Vergleichende Untersuchungen zum photo-geotaktischen Winkeltransponieren pterygoter Insekten. *Z. Vergl. Physiol. 68*, 229–262.

LOFTUS, R. (1976). Temperature-dependent dry receptor on antenna of *Periplaneta*. Tonic response. *J. Comp. Physiol. 111*, 153–170.

MARKL, H. (1974). Insect behaviour: functions and mechanisms. In *The Physiology of Insecta.* Edited by M. Rockstein. Vol. III, pages 3–148. Academic Press, New York and London.

MELLANBY, K. (1958). The alarm reaction of mosquito larvae. *Ent. Exp. Appl. 1*, 153–160.

MIMURA, K. (1974). Units of the optic lobe, especially movement perception units of diptera. In *The Compound Eye and Vision of Insects.* Edited by G. A. Horridge. Pages 423–436. Claredon Press, Oxford.

MIMURA, K., TATEDA, H., MORITA, H. and KUWABARA, M. (1969). Regulation of insect brain excitability by ocellus. *Z. Vergl. Physiol. 62*, 382–394.

MIMURA, K., TATEDA, H., MORITA, H. and KUWABARA, M. (1970). Convergence of antennal and ocellar inputs in the insect brain. *Z. Vergl. Physiol. 68*, 301–310.

MITTELSTAEDT, H. (1957). Prey capture in mantids. In *Recent Advances in Invertebrate Physiology.* Edited by B. T. Scheer, Pages 51–71. University of Oregon Publications, Oregon.

MITTELSTAEDT, H. (1962). Control systems of orientation in insects. *Ann. Rev. Ent. 7*, 177–198.

MÜLLER, E. (1931). Experimentelle Untersuchungen an Bienen und Ameisen über die Funktionsweise der Stirnocellen. *Z. Vergl. Physiol. 14*, 348–384.

O'SHEA, M. (1975). Two sites of axonal spike initiation in a bimodal interneurone. *Brain Res. 96*, 93–98.

PERTUNNEN, V. (1959). Effect of temperature on the light reactions of *Blastophagus piniperda* L. (Col., Scolytidae). *Ann. Ent. Fenn. 25*, 65–71.

PERTUNNEN, V. (1963). Effect of desiccation on the light reactions of some terrestrial arthropods. *Ergebn. Biol. 26*, 90–97.

PERTUNNEN, V. and LAHERMAA, M. (1963). The light reactions of the larvae and adults of *Tenebrio molitor* L. (Col., Tenebrionidae) and their interference with the humidity reactions. *Ann. Ent. Fenn. 29*, 83–106.

PERTUNNEN, V. and PALOHEIMO, L. (1963). Effect of temperature and light intensity on the light reactions of the larvae and adults of *Tenebrio molitor* L. (Col., Tenebrionidae). *Ann. Ent. Fenn. 29*, 171–184.

RAU, G. (1970). Zur Steuerung der Honigmagenfüllung sammelnder Bienen an einer künstlichen Futterquelle. *Z. Vergl. Physiol. 66*, 1–21.

SANDEMAN, D. C. (1980). Angular acceleration, compensatory head movements and the halteres of flies *(Lucilia serricata)*. *J. Comp. Physiol. 136*, 361–367.

SCHILDBERGER, K. (1981). Some physiological features of mushroom-body linked fibres in the house cricket brain. *Naturwissenschaften 68*, 623–624.

SCHWINCK, I. (1954). Experimentelle Untersuchungen über Geruchssinn und Strömungswahrnehmung in der Orientierung bei Nachtschmetterlingen. *Z. Vergl. Physiol. 37*, 19–56.

STEINER, G. (1953). Zur Duftorientierung fliegender Insekten. *Naturwissenschaften 40*, 514–515.

STEINER, G. (1954). Über die Geruchs-Fernorientierung von *Drosophila melanogaster* in ruhender Luft. *Naturwissenschaften 41*, 287.

STRAUSFELD, N. J. and BACON, J. P. (1983). Multimodal convergence in the central nervous system of dipterous insects. In *Multimodal Convergences in Sensory Systems*. Edited by E. Horn. *Fortschr. Zool. 28*, pages 47–76. Gustav Fischer Verlag, Stuttgart and New York.

SULLIVAN, C. R. (1959). The effect of light and temperature on the behaviour of adults of the white pine weevil, *Pissodes strobi* Peck. *Canad. Ent. 91*, 213–232.

SUZUKI, H. (1975). Antennal movements induced by odour and central projection of the antennal neurones in the honeybee. *J. Insect Physiol. 21*, 831–847.

SUZUKI, H., TATEDA, H. and KUWABARA, M. (1976). Activities of antennal and ocellar interneurones in the protocerebrum of the honeybee. *J. Exp. Biol. 64*, 405–418.

TRAYNIER, R. M. M. (1968). Sex attraction in the mediterranean flour moth, *Anagasta kühniella*: location of the female by the male. *Canad. Ent. 100*, 5–10.

VOWLES, D. M. (1954). The orientation of ants. I. The substitution of stimuli. *J. Exp. Biol. 31*, 341–355.

WALDOW, U. (1975). Multimodale Neurone im Deutocerebrum von *Periplaneta americana*. *J. Comp. Physiol. 101*, 329–341.

WALLACE, G. K. (1959). Visual scanning in the desert locust *Schistocerca gregaria* Forskal. *J. Exp. Biol. 36*, 512–525.

WIESE, K. (1981). Akustische, vibratorische und efferente Eingänge am Omega-Neuron der Grillenhörbahn. *Verh. Dtsch. Zool. Ges. 1981*, 168.

WILSON, E. O. (1971). *The Insect Societies*. Harvard University Press, Cambridge, Massachusetts.

13   Visual Guidance of Flies During Flight

CHRISTIAN WEHRHAHN

Max-Planck-Institut für Biologische Kybernetik, Tübingen, FRG

This chapter is dedicated to Professor Reichardt on the occasion of his 60th birthday.

1   INTRODUCTION

Flies are able to fixate or track objects. This task requires high spatial and temporal resolution in information processing of the visual system and aerobatic capabilities of the flight motor system. This review briefly outlines: (1) which basic perceptual principles are used by flies during tracking; and (2) how these principles might actually be realized in the 'hardware' of the brain.

1.1   Description of the fly visual system

1.1.1   STRUCTURE

Flies have two compound eyes, the visual fields of which roughly cover the left and right hemisphere respectively. They nevertheless have a good spatial resolving power because each compound eye is built up from several thousand facets (e.g. 3600 in houseflies) or ommatidia, see Fig. 1. Thus the whole surround is sampled visually in about 7200 'points'. Flies also have three lens eyes (or 'ocelli'), the visual fields of which are directed more or less upwards and whose resolving power is very poor. They probably have nothing or very little to do with the behaviour considered here.

Each of the 7200 points to which the optical axes of the ommatidia are directed is monitored by eight photoreceptors (R1–R8) in the retina belonging to two major classes (indicated by the small and the large block in Fig. 1) which differ in their spectral sensitivity and also in their pattern of neural projection onto the optic ganglia (review: Franceschini, N., 1983). The first optic ganglion (lamina) contains the same number of subunits (cartridges) of identical neural structure as there are ommatidia. Each cartridge is served by the six terminals of the first class of photoreceptors (large block in Fig. 1) all

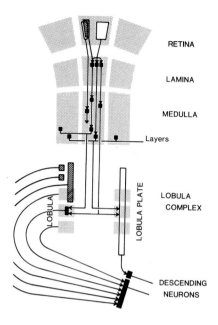

RETINA

LAMINA

MEDULLA

Layers

LOBULA

LOBULA
COMPLEX

LOBULA PLATE

DESCENDING
NEURONS

FIG. 1. Schematic diagram of the basic connections found in the visual system of the fly. Blocks and lines represent classes of cells and connections respectively. Arrows indicate functional connections (synapses). Retina, lamina and medulla represent the functional subunit corresponding to one ommatidium. Shaded blocks represent sex-specific cell classes found only in male flies. For detailed explanation see text.

sampling the same point in space and adding their signals onto first-order interneurons. Thus the ratio between signal and noise in a first-order interneuron is enhanced by a factor of $\sqrt{6}$ compared to that of a single photoreceptor axon but without loss in spatial resolution (review: Shaw, S., 1981). The axons of the remaining two photoreceptors of one ommatidium R7 and R8 (small block in Fig. 1) pass the lamina and first make synaptic contacts in a subunit (column) of the second optic ganglion (medulla) of which there are the same number as there are ommatidia and cartridges. The interneurons of the first optic ganglion project into the corresponding subunits of the second optic ganglion (Strausfeld, N. and Nässel, D., 1981). In contrast to the lamina, where the organization indicates that the signal flow is mainly along cartridges, the structure of the medulla exhibits several prominent horizontal layers containing amacrine and tangential cells, thus indicating a high degree of lateral interaction among the signals of different columns. The output cells of the medullary columns project onto corresponding subunits into

the third optic ganglion (lobula complex). This ganglion is subdivided into two parts: the lobula and the lobula plate.

The remarkable features of the lobula plate are 20–30 classes of large tangential cells which receive input from larger fractions or all ipsilateral columns of the medulla (review: Hausen, K., 1983). There are also some columnar elements in the lobula plate. These converge at their output site onto the descending neurons. The other part of the third optic ganglion (lobula) is dominated by columnar small field elements receiving input from one or a few medullary columns. Again these columnar elements converge at their synapse to the next stage (descending neurons, Fig. 1) or serve as inputs to the contralateral lobula complex. Also in the lobula some large tangential cells exist, some of which will be considered below. Hence spatial integration takes place at the level of the lobula complex or at its output. Therefore we must assume that all essential steps of spatiotemporal information processing, for which local information is needed, have been carried out at this level. A simplified scheme showing the essential features of neural connectivity in the optic lobes of the flies is shown in Fig. 1.

An extra structure in male flies. Recently several parts of what seems to be an extra structure have been discovered in the visual system of male flies. The characteristic feature of almost all these structures is their location in the dorsal frontal part of the field of view. At the photoreceptor level in this region the second subclass (R7/8) have the same visual pigment (indicated by hatched block in Fig. 1) as the other class (R1–6), which is different from all photoreceptors R7/8 everywhere else in male flies and everywhere in female flies (Franceschini, N., 1983). The photoreceptor R7 does not project to the medulla but terminates in the lamina. Male-specific fibres in lamina and medulla have not yet been described. In the lobula, however, spectacular giant large field interneurons have been found (dashed block in Fig. 1) which are accompanied by an extra set of small field columnar neurons (Hausen, K. and Strausfeld, N., 1980). The extra system also terminates at the descending neurons. Thus, in this case, parallel pathways of spatially integrated information and local information are also formed in the third optic ganglion.

Interestingly in male *Bibio* flies the sex-specific

structure has evolved into an extra eye. The optic lobes belonging to this extra eye are completely separated from the 'normal' eye (Zeil, J., 1983b).

The retina obviously transduces the light fluxes selected by the array of the ommatidial lenses into electrical signals which can then be handled by the nervous system. The functional role of the first optic ganglion seems to be that the steady background intensity level acting at the photoreceptors is subtracted to concentrate on purely transient local light flux changes within the receptive field of an ommatidium (Laughlin, S. and Hardie, R., 1978). The phase relation between the light fluxes, coming from neighbouring points, is essential for the processing of motion information (Reichardt, W., 1969), thus the signal present at the output cells of the lamina is ideally suited for that purpose. Indeed the few recordings made from cells of the medulla indicate that (among other tasks) small field motion detection is carried out at this level (Mimura, K., 1971; DeVoe, R., 1980; Buchner, E. and Buchner, S., 1983) also in accordance with the structural features of the medulla. It should be kept in mind, however, that via R7/8 information about the background intensity level is still preserved in the medulla.

The tangential cells of the lobula plate are sensitive to motion selective for different directions depending on the specific cell. The functional properties and interrelations of many of these cells are well known and the block drawn in Fig. 1 is in reality a complicated network in itself with feedback loops and orderly layered sheets of large cells according to their function (review: Hausen, K., 1983).

Nothing is known about the physiology of the lobula columnar neurons and very little about tangentials. It seems, however, safe to say that at the output of the lobula complex local (via the columnar element) and global (via the tangential elements) information are both present and perhaps converge only at the subsequent synapse.

In summary the compound eye and the optic lobes of the fly can be understood as a parallel network diverging and converging at all three levels of processing with respect to the information coming from a single ommatidium.

Nothing much is known about the physiology of the sex-specific system except the spectral sensitivity of the R7/8 and their synaptic connections in the lamina (Hardie, R., 1983).

2 BEHAVIOUR

2.1 Free flight analysis of female flies

Houseflies kept in a cage will fly around as they do in houses. Both female and male flies will track other flies differing greatly, however, in their success. Their flights can be filmed in the cage simultaneously from two sides and subsequently the flight trajectories reconstructed in three dimensions from the films through frame-by-frame analysis (Wehrhahn, C., 1979; Bülthoff, H. et al., 1980; Wehrhahn, C. et al., 1982). As is known from everyday experience a fly (female or male) may cruise through a contrasted environment keeping its course without being visibly influenced by the flow of visual contrast generated by the flight motion. In some cases, however, a fly will locate a resting object and land on it or follow a moving object and perhaps even try to catch up with it. About the processes causing a transition from the cruising state into other states nothing can be said within this analysis. Apparently some properties of the moving objects are necessary but not sufficient to elicit specific responses as described below.

Figure 2 shows a flight episode from two female houseflies (Musca). The first fly is apparently flying without specific goal; the second, however, is clearly trying to keep track with the first one. More specifically the second fly tries to align itself with respect to the first fly such that its body axis points towards it. A quantitative evaluation of the turning response $\dot{\alpha}_v$ as a function of the position ψ_A of the target fly in the horizontal plane shows that a 'best fit' is achieved when the turning response is shifted by 20–30 ($\pm$ 10) ms with respect to the target position. Thus the time required by the fly to compute a turning response from the target position and to generate a response is very short. If we assume a delay of about 8 ms for the photoreceptors (Hardie, unpublished) and see from the diagram in Fig. 1 that the signal has to cross at least four synaptic

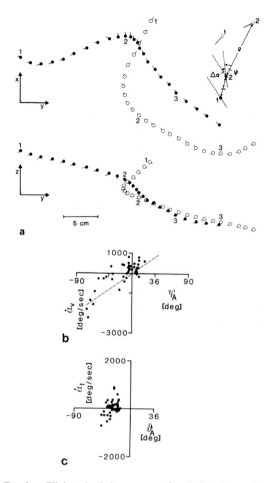

a

b

c

FIG. 2. **a**: Flight episode between two female flies, *Musca*, in the horizontal (xy)-plane (upper graph) and vertical (yz)-plane (lower graph). The leading fly is represented by the open circles; the tracking one by solid circles. Lines represent body axis. Numbers beside dots indicate corresponding time instants. Interval between two successive points is 20 ms. (From Wagner, unpublished). For definition of co-ordinates used see inset. **b**: Turning response around the vertical axis α_v of the tracking fly plotted as a function of the error angle ψ_A in the horizontal plane under which the leading fly is seen by the tracking fly shifted by 20 ms in time. The turning response depends on the position of the target. If we approximate the relation between α_v and ψ_A by $\alpha_v = -k\psi_A$, then $k \sim 20\,\text{s}^{-1}$. **c**: Turning response around the horizontal axis perpendicular to the direction of flight (pitch) α_t plotted as a function of the vertical error angle ϑ_A. It can be seen that the target is fixated in the lower frontal part of the field of view ($\vartheta_A < 0$).

junctions (of about 1 ms each) before leaving the head, we are left with a lower limit of about 12 ms for synapses in the thoracic ganglia and the generation of the motor response. On the other hand the velocity of a fly is about 1 m s^{-1} thus in 20 ms the fly

moves 2 cm. For a reasonable tracking performance in free flight such a fast response is very important. The turning response apparently depends on the position ψ_A of the target, as can be concluded from Fig. 2b. Finally Fig. 2c shows that during the whole pursuit the target is held by the tracking fly in the lower frontal part of its field of view.

2.2 Tethered flight. Open-loop experiments with female flies

The turning response of a fly can also be determined in tethered flight with a highly sensitive torque meter (Fermi, G. and Reichardt, W., 1963). Flies which are attached to such a device (and are thus fixed in space) will fly for up to several hours and their flight torque to visual stimuli, whose parameters under these circumstances are well defined, can be measured. Under these circumstances the fly is completely decoupled from its environment and hence its responses do not change its visual input: the 'motor loop' of the animal is 'opened'. Such a situation is often called an 'open-loop' experiment (Hassenstein, B., 1951; Mittelstaedt, H., 1950; Von Holst, E. and Mittelstaedt, H., 1950). Figure 3 shows averaged open-loop flight torque responses of 20 houseflies (*Musca domestica* L.) to a narrow vertical black stripe on a bright cylinder which was rotated clockwise (upper trace) and counter-clockwise (lower trace) around the test animals. The responses shown in Fig. 3 clearly depend on the position of the stripe. Many experiments have shown that this is in part due to a direction-insensitive response component (Pick, B., 1974; Reichardt, W., 1979; Wehrhahn, C., 1981; Strebel, J., 1982) and a direction-sensitive response component (Reichardt, W., 1973; Wehrhahn, C. and Hausen, K., 1980; Wehrhahn, C., 1981); that is both response components depend on the position of the respective stimulus in the receptive field. The contribution of the respective response components to the fixation and tracking process is at present not completely clear. Theoretical considerations have shown that either component would be suitable for a fixation and tracking mechanism of the kind used by free-flying houseflies (Reichardt, W. and Poggio, T., 1976; Wehrhahn, C. and Hausen, K., 1980; Poggio, T. and Reichardt, W., 1981).

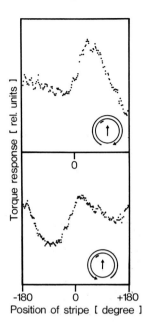

FIG. 3. Average flight torque of female flies, *Musca*, recorded as a function of the position of a dark vertical stripe on a bright background rotated clockwise (upper trace) and counter-clockwise (lower trace) around the fly. The response strongly depends on the position of the target. (From Wehrhahn, C. and Hausen, K., 1980.) The average luminance of the panorama is 60 cd m^{-2}. The stripe was 3° wide and 45° high. The response amplitude at $\psi = 45°$ is about 10^{-7} Nm. Average of 20 flies.

The orientation behaviour of tethered flying male and female houseflies has been compared recently by Strebel, J., (1982). Both sexes behave very similarly in tethered flight as well as in free flight during cruising and landing (Wagner, H., 1980) but not when tracking other flies (see section 6).

3 QUANTITATIVE COMPARISON OF BEHAVIOUR AND ELECTROPHYSIOLOGY

Electrophysiological recordings with extracellular electrodes revealed the existence of motion-sensitive interneurons in the lobula plate which were hypothesized to take part in the neural network responsible for the generation of the optomotor response (Bishop, L. and Keehn, D., 1966). The significance of the lobula plate for the control of flight behaviour has been tested by electrical stimulation and simultaneous recordings of flight responses in *Calliphora* (Blondeau, J., 1981). Two sets of giant interneurons were described in the lobula plate

(Braitenberg, V., 1972; Pierantoni, R., 1976): the horizontal cells and the vertical cells. The horizontal cells, which are of particular interest here, are a class of three giant output cells whose axons terminate on descending interneurons, which in turn project to the motor centres (Fig. 1). The inputs to the horizontal cells are columnar elements from the medulla, and the lobula with a receptive field of six ommatidia (Strausfeld, N., 1983). They have not yet been characterized functionally, thus we do not know whether motion is computed at the level of the medulla or lobula plate. Two of the three horizontal cells, the ipsilateral receptive fields of which extend from below the equator to the upper limit of the field of view, receive additional contralateral input near their terminal. Many intracellular recordings and subsequent identifications have been achieved from all three horizontal cells (review by Hausen, K., 1983). As expected from the structure, they receive input from the ipsilateral eye. Their preferred direction is oriented horizontally from front to back. The intracellular signals of these horizontal cells, as well as the flight torque responses of tethered flying female flies (*Calliphora erythrocephala*), have been analysed quantitatively with identical visual stimuli consisting of a moving periodic pattern. The main difference between the signal from a horizontal cell and a torque signal (Fig. 4a) seems to be that the latter is much more noisy and apparently subject to influences other than motion. The amplitude of the two signals depends similarly upon the angular orientation of the moving stimulus (Fig. 4b), the position of the moving stimulus relative to the fly (Fig. 4c), the angular velocity (Fig. 4d), and the average luminance of the pattern (Fig. 4e). The signals from which the points in Fig. 4e are plotted are action potential frequencies recorded from the H1 cell, a large-field motion-sensitive neuron in the lobula plate of flies (Hausen, K., 1983). In particular the spatial sensitivity distribution (Fig. 4c) reveals that the intracellular signals, as well as the flight torque response, are heavily dependent on the position of the stimulus quite similar to the function depicted in Fig. 3 for the flight torque response to a single stripe. The horizontal cells receiving input from the contralateral eye are sensitive to horizontal back-to-front motion from that eye. The three horizontal cells in the right optic lobe form a system which is sensitive to clockwise motion and is

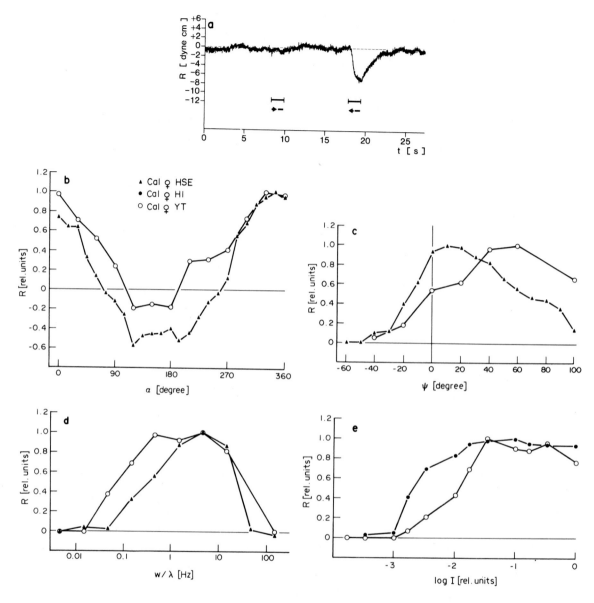

FIG. 4. Yaw torque responses of tethered flying female blowflies, *Calliphora* (open circles), intracellular recordings of the equatorial horizontal cell (triangles), and extracellular recordings of the H1-cell (solid circles) in female animals of the same species to the motion of a periodic grating (angular diameter 40°, spatial wavelength of grating 13.3°, pattern contrast 75%, average luminance 60 cd m^{-2}). **a**: An example for a torque response averaged from 12 animals over 200 sweeps to horizontal back to front and front to back motion of the grating positioned at − 40° to the left of the direction of flight in the equatorial plane. There is only a response to motion from front to back but none to motion from back to front. The data in the plots of Fig. 4b to e are the response amplitudes (i.e. maximal responses). A measure for the error inherent in the experiments may be taken from the fluctuations of the response when no stimulus is given. It is at most 2 scale units. This holds for all behavioural data of Fig. 4. The error inherent to the electrophysiological data is smaller. **b**: Dependence of yaw torque and horizontal cell response upon the angular orientation of the grating positioned at 40° to the right from the direction of flight. Maximal response is elicited for motion from front to back. **c**: Dependence of yaw torque and horizontal cell response upon different positions of the grating (moving front to back) along the equatorial plane. Both responses strongly depend on the position of the grating. The maximum in both cases is found in the frontolateral part of the visual field. **d**: Dependence of yaw torque and horizontal cell responses upon the contrast frequency w/λ. Both responses have their maximum at 4–5 Hz. **e**: Dependence of yaw torque and extracellular signal of H1 cell upon the absolute average luminance of the grating moving front to back at 40° lateral to the direction of flight in the equatorial plane. Threshold and saturation coincide.

strongly dependent upon the position of the stimulus in the ipsilateral eye, very similar to the response of the upper trace in Fig. 3. Thus the horizontal cells of the right optic lobe may represent one input to the flight torque motor system for clockwise motion, and those of the left optic lobe may represent the input for counterclockwise motion (Wehrhahn, C. and Hausen, K., 1980).

Recently the figure-ground discrimination of flies was analysed. A set of experiments compared intracellular recordings from horizontal cells in *Calliphora* to flight torque responses of tethered flying flies to different figure-ground stimuli. It was found that functional properties of the neuronal circuitry proposed for the figure-ground discrimination are in part consistent with properties of the horizontal cells (Reichardt, W. *et al.*, 1983).

4   LESION EXPERIMENTS

A direct test for the hypothesis outlined at the end of the preceding paragraph would be to selectively block the horizontal cells of, say, the right optic lobe in a fly. If the hypothesis were correct such an animal should show no response in the flight torque to clockwise motion but nevertheless respond to counterclockwise motion. Such an experiment was carried out recently with female blowflies (Hausen, K. and Wehrhahn, C., 1983). The stimulus consisted of two projectors with circular screens which were placed on either side of the animals (Fig. 5). On the screens a periodic pattern could be moved from front to back or from back to front. The stimulus sequence was binocular motion from front to back, binocular motion from back to front, clockwise and counterclockwise motion. The flight torque responses of the fixed flying animals were recorded with the highly sensitive torque transducer mentioned above. The result of an experiment with normal flies is shown in Fig 5a. A slight oscillatory response to binocular motion from front to back is seen first. This response should be zero because the stimulus is in principle symmetric. Since both eyes are very sensitive to front-to-back motion (see Fig. 3) the oscillations in the torque signal are probably the result of a slight misalignment of the animals in the experiments. The response to binocular motion from back to front is negligible, as expected.

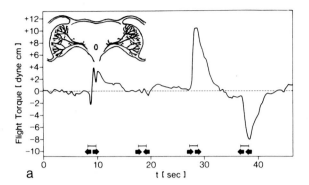

a

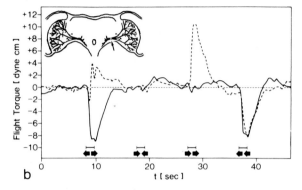

b

FIG. 5. Average torque response of female blowflies, *Calliphora erythrocephala*, tethered to a torque meter recorded as a function of binocular horizontal motion of two periodic gratings positioned at the two sides in the equatorial plane of the animal. The stimulus is indicated below. **a**: Normal animals: flies respond to binocular clockwise and counterclockwise motion. **b**: Continuous line: flies lesioned in the right optic lobe (see text). Flies respond to binocular front-to-back motion and to counterclockwise motion. The response to clockwise motion present in normal animals has disappeared. The horizontal cells abolished by the lesion take part in the optomotor control of flight. (From Hausen, K. and Wehrhahn, C., 1983.)

Binocular clockwise and counterclockwise motion elicit strong responses following the direction of motion.

After having contributed to this experiment the flies were subject to an operation the details of which are described elsewhere (Hausen, K. and Wehrhahn, C., 1983) and which interrupted the axonal connection between horizontal cells and descending neurons (cf. Fig. 1). The experiment described above was repeated when the flies had recovered from the operation. The response of these animals differs considerably from normal ones (Fig. 5b, continuous line). Binocular motion from front to back is seen by these animals as monocular motion from front to back at the left side. Binocular

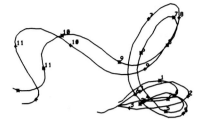 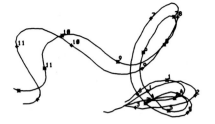

FIG. 6. Three-dimensional reconstruction of a chase between two male flies seen from above. The numbers correspond to 100 ms intervals. The plot should be observed with standard stereo glasses. (From Wehrhahn, C. *et al.*, 1982.)

motion from back to front and counterclockwise motion elicit responses similar to the normal case. The response to clockwise motion which is very strong in normal animals has completely disappeared in the operated animals.

This is to be expected on the basis of the earlier hypothesis that the horizontal cells of the right optic lobe respond to clockwise motion and that they transmit this information via the descending neurons to the motor centres controlling yaw torque.

Monocular stimulation experiments with lesioned animals revealed the existence of an additional system contributing to the flight torque response. It is responsive to motion in any direction and is inhibited by contralateral motion. This second system could possibly be responsible for the direction-insensitive component found earlier in the flight torque signal using flickered stripes (Pick, B., 1974; Reichardt, W., 1979; Wehrhahn, C., 1981) or moving edges behind very small slits (Strebel, J., 1982).

Some recent findings on mutants of the fruit-fly *Drosophila* are of interest here. The mutant *lobula plate less*[N684], which still shows rudiments of the horizontal cells, also displays rudimentary flight torque responses to binocular horizontal motion (Paschma, R., 1983). The mutant *omb*[H31] (optomotor blind), which has virtually no horizontal cell, shows a strongly reduced flight torque response to binocular horizontal motion but an almost normal response to a monocular stimulus (Heisenberg, in preparation). Finally the mutant *sol* (small optic lobes), which has horizontal cells reduced in size, shows almost normal flight torque responses to binocular horizontal motion (Heisenberg, in preparation). Laser-induced ablation of horizontal cell precursors in *Musca* larvae leads to a reduction

of the flight torque response to monocular horizontal motion in the adult flies as well as a reduction in amplitude in an experiment very similar to that shown in Fig. 3 (Geiger, G. and Nässel, D., 1981, 1982). Unfortunately no binocular experiment was performed with these animals. So far the results are not in conflict with those described here.

5 CONCLUSIONS FROM EXPERIMENTS WITH FEMALE FLIES

1. Female flies fixate and track objects in free flight.
2. The object is held preferably in the lower frontal part of their field of view.
3. The turning response around the vertical axis in free flight depends on the position of the tracked object.
4. Open-loop experiments with tethered flying houseflies show that their torque signal depends on the position of the stimulus.
5. Electrophysiological recordings from horizontal cells in *Calliphora* show qualitatively a functional similarity of the signals to flight torque recordings in *Calliphora*.
6. Experiments with operated flies, *Calliphora*, show that the directionally sensitive part of the torque response is controlled by a set of well-described directionally sensitive interneurons situated in the third optic ganglion.
7. Tracking may be considered to be one of several functions of the horizontal cells in visually guided flight.

6 MALE CHASING

Much higher performance is achieved by the male

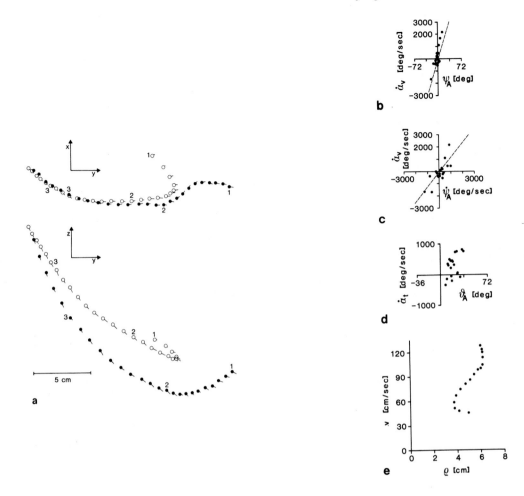

FIG. 7. **a**: Reconstruction of a shorter chase between two male flies, *Musca*, in the *xy*-plane (upper graph) and in the *zy*-plane (lower graph). Open circles = leading fly, solid circles = chasing fly. Interval between two successive points is 12.5 ms. **b**: Turning response around the vertical axis α_v plotted as a function of the error angle ψ_A; α_v is delayed by 12.5 ms with respect to ψ_A. The turning response strongly depends on the target position. If we approximate the relation between α_v and ψ_A with $\alpha_v = -k\psi_A$ then $k \sim 80 \, \mathrm{s}^{-1}$. **c**: Turning response around the vertical axis α_v plotted as a function of the error rate $\dot\psi_A$; α_v is delayed by 12.5 ms with respect to $\dot\psi_A$. The turning response also depends on the error rate. **d**: Pitch response (see legend to Fig. 2b) plotted as a function of the vertical error angle ϑ_A. The target is fixated in the upper part of the field of view ($\vartheta_A > 0$). **e**: Relation between forward velocity v and distance ρ for a shift of 72.5 ms.

chasing system than by the female tracking system. As can be seen from Fig. 6 the flight path of the leading fly is very tortuous, including three more or less sharp bends at the beginning and a spectacular looping. All manoeuvres are followed easily by the chasing fly. A performance like that would rarely be seen when observing female flies.

A closer look at some properties of the chasing system is possible by examining a reconstruction of a shorter chase which now includes the body axis of the flies. It is easy to recognize that body axis of the chasing fly and flight trajectory coincide only in the

minority of the observed instances (Fig. 7a). Quantitative evaluation of this chase shows that the turning response around the vertical axis of the chasing fly is strongly dependent upon the horizontal position ψ_A of the target fly (Fig. 7b). In this case the "best fit" is achieved when the turning response is shifted by 12.5 ms with respect to the target position. Thus for the male system the total dead time may be even shorter than for the female system.

In addition the slope of the regression line connecting the points is much steeper than that of Fig. 2b. Thus the gain in the relation error angle/turning response is much higher in males than in females. This greatly improves tracking performance. The delay contained in this feedback system represents a potential problem of instability. If k is the value relating $\dot{\alpha}v$ and $\dot{\psi}_A$ and ε the dead time involved in the system, the "aperiodic limit", that is when no oscillations occur, is given by $k = e^{-1}/\varepsilon$ (Land, M. and Collett, T., 1974). For $\varepsilon = 12.5$ ms this gives $k \sim 30\,\text{s}^{-1}$. The values of k in the chases evaluated until now are between 40 and $80\,\text{s}^{-1}$, which means that slight oscillations or overshoots are to be expected in the flight response of the chasing fly, and this is indeed found (Land, M. and Collett, T., 1974; Wehrhahn, C. *et al.*, 1982). However, for small amplitudes this error becomes very small and, in addition, the turning response also depends on the error rate (Fig. 7c). The small amount of instability is damped out probably by this effect. Thus the operation of the chasing system seems to be adapted to turning as fast as possible without becoming unstable. In contrast the female system seems to operate at longer dead times and, correspondingly, lower values of k (Fig. 2b).

Fig. 7d shows the turning response around the horizontal axis (pitch) plotted as a function of the vertical error angle ϑ_A. It can be seen from these data that the target is held by the chasing male in the upper frontal part of the field of view. Also the visual field of the sex-specific photoreceptors found in the retina and the sex-specific neurons found in the lobula are located in the upper frontal part of their field of view. Thus it is tempting to speculate that the system found in male flies is the physiological counterpart for the behaviour of male flies.

Finally, another property of the male chasing system should be mentioned. Plotting the three-dimensional forward velocity v against the distance ρ to the target reveals a linear relation $v = \chi\rho$ (Fig. 7e). v was delayed with respect to ρ by 72.5 ms in this plot. This is long compared to the 12.5 ms used by the turning system. Thus male flies, at least in some cases, may regulate their forward velocity as a function of the distance to their target. This represents a further optimization of the system (Wehrhahn, C. *et al.*, 1982). In female flies a kind of regulation of the forward velocity is also observed (Wagner, unpublished).

The fact that distance is used by the chasing fly to regulate its forward velocity raises the question of how this might be achieved. The distance on the compound eyes between corresponding ommatidia is about 1 mm. The angular resolution of about $1.5°$ (i.e. divergence angle between two neighbouring ommatidia) would give very large errors if the target distance were computed from triangular measurements. Collett, T. and Land, M. (1975) have proposed for syrphid flies a mechanism which comprises objects of specific size as targets for the chasing system, an assumption implying that only objects of the 'right' size (= species) are tracked properly.

In males of the march-fly *Bibio* during some periods of chasing, a linear relation between distance to target and forward velocity is found. The exact nature of this control is uncertain (Zeil, J., 1983a).

A problem connected to the obvious specificity of the male system for small objects is the sensitivity to large field stimuli and possible interactions with the 'female' system described above which is present also in males. This can be shown by comparing the closed-loop behaviour in fixed flight of male and female flies, which is exactly the same (Strebel, J., 1982). In particular stimulation with a periodic grating of the upper frontal part of the field of view elicits virtually no yaw torque responses of tethered flying female and male flies, irrespective of the direction of motion (Wehrhahn, unpublished).

Obviously the male chasing system is triggered by small dark objects (Collett, T. and Land, M., 1975). It is known, from syrphid flies, that the optomotor system is in full operation during chasing (Collett, T., 1980). Thus male flies may also have to solve the problem of separating a figure (the target fly) from a textured background; but this will occur only rarely because the chasing fly fixates the target fly in the dorsofrontal part of its field of view. Thus in most cases the sky will serve as background, giving the chasing male optimal contrast and luminance conditions. The diameters of ommatidial lenses are especially large in this region, and the photoreceptors are tuned to the UV and blue-green, thus further optimizing the chasing performance of male flies.

7   CONCLUSIONS FROM OBSERVATIONS ON MALE FLIES

1. Male flies chase other flies.

2. The target is held during chases in the upper frontal part of the field of view.

3. The turning response around the vertical axis depends on the position of the tracked target.

4. The forward velocity of the chasing fly depends on the distance to the target.

5. The male chasing system is physiologically different from the female tracking system. It is optimally tuned in many respects for chasing.

ACKNOWLEDGMENT

I am grateful to H. Wanger for the permission to use the unpublished data of Fig. 2 and, together with E. Buchner, K. Hausen and W. Reichardt, for comments on the manuscript; also to H. Hadam for drawing the figures and I. Geiss for typing the manuscript.

REFERENCES

BISHOP, L. G. and KEEHN, D. G. (1966). Two types of neurons sensitive to motion in the optic lobe of the fly. *Nature 212*, 1374–1376.

BLONDEAU, J. (1981). Electrically evoked course control in the fly *Calliphora erythrocephala*. *J. Exp. Biol. 92*, 143–153.

BRAITENBERG, V. (1972). Periodic structures and structural gradients in the visual ganglia of the fly. In *Information Processing in the Visual System of Arthropods*. Edited by R. Wehner. Pages 1–15. Springer, Berlin and New York.

BUCHNER, E. and BUCHNER, S. (1983). Neuroanatomical mapping of visually induced nervous activity in insects by 3H-Deoxyglucose. In *Photoreception and Vision in Arthropods*. Pages 623–634. Edited by M. A. Ali. Plenum Press, New York and London.

BÜLTHOFF, H. and WEHRHAHN, C. (1983). Computation of movement and position in the visual system of the fly (*Musca*): experiments with uniform stimulation. In *Localization and Orientation in Technics and Biology*. Edited by D. Varjú and H. U. Schnitzer. Springer, Heidelberg and New York.

BÜLTHOFF, H., POGGIO, T. and WEHRHAHN, C. (1980). 3-D Analysis of flight trajectories of flies (*Drosophila melanogaster*). *Z. Naturf. 35c*, 811–815.

COLLETT, T. S. and LAND, M. F. (1975). Visual control of flight behaviour in the hoverfly, *Syritta pipiens* L. *J. Comp. Physiol. 99*, 1–66.

COLLETT, T. S. (1980). Angular tracking and the optomotor response. An analysis of visual reflex interaction in a hoverfly. *J. Comp. Physiol. 140*, 145–158.

DEVOE, R. D. (1980). Movement sensitivities of cells in the fly's medulla. *J. Comp. Physiol. 138*, 93–119.

FERMI, G. and REICHARDT, W. (1963). Optomotorische Reaktionen der Fliege *Musca domestica*. *Kybernetik 2*, 15–28.

FRANCESCHINI, N. (1983). The retinal lattice of the fly compound eye. In *Photoreception and Vision in Arthropods*. Edited by M. A. Ali. Pages 439–455. Plenum Press, New York and London.

GEIGER, G. and NÄSSEL, D. R. (1981). Visual orientation behaviour of flies after selective laser beam ablation of interneurons. *Nature 293*, 398–399.

GEIGER, G. and NÄSSEL, D. R. (1982). Visual processing of moving single objects and wide-field patterns in flies: Behavioural analysis after laser-surgical removal of interneurons. *Biol. Cybernet. 44*, 141–149.

HARDIE, R. C. (1983). Morphology of sex-specific photoreceptors in the compound eye of the male housefly (*Musca domestica*). *Cell Tissue Res. 233*, 1–21.

HASSENSTEIN, B. (1951). Ommatidienraster und afferente Bewegungsintegration. *Z. Vergl. Physiol. 33*, 301–326.

HAUSEN, K. (1983). The lobula-complex of the fly: structure, function and significance in visual behaviour. In *Photoreception and Vision in Arthropods*. Edited by M. A. Ali. Pages 523–560. Plenum Press, New York and London.

HAUSEN, K. and STRAUSFELD, N. J. (1980). Sexually dimorphic interneuron arrangements in the fly visual system. *Proc. Roy. Soc. Lond. B 208*, 57–71.

HAUSEN, K. and WEHRHAHN, C. (1983). Microsurgical lesion of horizontal cells changes optomotor yaw responses in the blowfly *Calliphora erythrocephala*. *Proc. Roy. Soc. Lond. B. 219*, 211–216.

HEISENBERG, M. (in prepration). Vision in *Drosophila*. In *Studies of Brain Function*. Springer, Heidelberg and New York.

HOLST, E. v. and MITTELSTAEDT, H. (1950). Das Reafferenzprinzip. (Wechselwirkungen zwischen Zentralnervensystem und Peripherie). *Naturwissenschaften 37*, 464–476.

LAND, M. F. and COLLETT, T. S. (1974). Chasing behaviour of houseflies (*Fannia canicularis*). A description and analysis. *J. Comp. Physiol. 89*, 331–357.

LAUGHLIN, S. B. and HARDIE, R. C. (1978). Common strategies for light-adaptation in the peripheral visual systems of fly and dragonfly. *J. Comp. Physiol. 128*, 319–340.

MIMURA, K. (1971). Integration and analysis of movement information by the visual system of flies. *Nature, Lond. B 226*, 964–966.

MITTELSTAEDT, H. (1950). Physiologie des Gleichgewichtssinnes bei fliegenden Libellen. *Z. Vergl. Physiol. 32*, 422–463.

PASCHMA, R. (1982). Strukturelle und funktionelle Defekte der *Drosophila*-Mutante *lobula plate less*[N684]. Diplomarbeit, Universität Würzburg.

PICK, B. (1974). Visual flicker induces orientation behaviour in the fly *Musca*. *Z. Naturf. 29c*, 310–312.

PIERANTONI, R. (1976). A look into the cockpit of the fly. The architecture of the lobula plate. *Cell Tiss. Res. 171*, 101–122.

POGGIO, T. and REICHARDT, W. (1981). Visual fixation and tracking by flies: Mathematical properties of simple control systems. *Biol. Cybernet. 40*, 101–112.

REICHARDT, W. (1969). Movement perception in insects. In *Processing of Optical Data by Organisms and by Machines*. Rendiconti S.I.F. XLIII. Edited by W. Reichardt. Pages 465–493. Academic Press, London and New York.

REICHARDT, W. (1973). Musterinduzierte Flugorientierung. *Naturwissenschaften 60*, 122–138.

REICHARDT, W. (1979). Functional characterization of neural interactions through an analysis of behaviour. In *The Neurosciences, Fourth Study Program*. Edited by F. O. Schmitt and F. G. Worden. Pages 81–103. The MIT Press. Cambridge, Mass. and London, England.

REICHARDT, W. and POGGIO, T. (1976). Visual control of orientation behaviour in the fly. Part I. A quantitative analysis. *Quart. Rev. Biophys. 9*, 311–375.

REICHARDT, W., POGGIO, T. and HAUSEN, K. (1983). Figure-ground discrimination by relative movement in the visual system of the fly. Part II: Towards the neural circuitry. *Biol. Cybernet. Suppl. 46*, 1–30.

SHAW, S. R. (1981). Anatomy and physiology of identified non-spiking cells in the photoreceptor-lamina complex of the compound eye of insects, especially Diptera. In *Neurons without Impulses*. Edited by A. Roberts and B. M. H. Bush. Pages 61–116. Cambridge University Press, Cambridge.

STRAUSFELD, N. (1983). Functional neuroanatomy of the blowfly's visual system. In *Photoreception and Vision in Arthropods*. Pages 483–522. Edited by M. A. Ali. Plenum Press, New York and London.

STRAUSFELD, N. J. and NÄSSEL, D. R. (1981). Neuroarchitectures serving compound eyes of Crustacea and insects. *Handbook of Sensory Physiology*, Vol. VII/6BI, 1–132. Springer, Heidelberg and New York.

STREBEL, J. (1982). Eigenschaften der visuell induzierten Drehmomenten-Reaktion von fixiert fliegenden Stubenfliegen *Musca domestica* L. und *Fannia canicularis* L. Dissertation, Eberhard-Karls-Universität, Tübingen.

WAGNER, H. (1980). Messung und Beschreibung von Landetrajektorien der Stubenfliege *Musca d.* Diplomarbeit, Eberhard-Karls-Universität, Tübingen.

WEHRHAHN, C. (1979). Sex-specific differences in the chasing behaviour of houseflies (*Musca*). *Biol. Cybernet. 32*, 239–241.

WEHRHAHN, C. (1981). Fast and slow flight torque responses in flies and their possible role in visual orientation behaviour. *Biol. Cybernet. 40*, 213–221.

WEHRHAHN, C. and HAUSEN, K. (1980). How is fixation and tracking accomplished in the nervous system of the fly? A behavioural analysis based on short time stimulation. *Biol. Cybernet. 38*, 179–186.

WEHRHAHN, C., POGGIO, T. and BÜLTHOFF, H. (1982). Tracking and chasing in houseflies. An analysis of 3-D flight trajectories. *Biol. Cybernet. 45*, 123–130.

ZAAGMAN, W. H., MASTEBROEK, H. A. K., BAYSE, I. and KUIPER, J. W. (1977). Receptive field characteristics of a directionally selective movement detector in the visual system of the fly. *J. Comp. Physiol. 116*, 39–50.

ZAAGMAN, W. H., MASTEBROEK, H. A. K. and KUIPER, J. W. (1978). On the correlation model. Performance of a movement detecting neural element in the fly visual system. *Biol. Cybernet. 31*, 163–168.

ZEIL, J. (1983a). Sexual dimorphism in the visual system of flies: the free flight behaviour of male Bibionidae (Diptera). *J. Comp. Physiol. 150*, 395–412.

ZEIL, J. (1983b). Sexual dimorphism in the visual system of flies: the divided brain of male Bibionidae (Diptera). *Cell Tiss. Res. 229*, 591–610.

Species Index

685

Author Index

Subject Index